NANO
and
MOLECULAR
ELECTRONICS
Handbook

Nano- and Microscience, Engineering, Technology, and Medicine Series

Series Editor
Sergey Edward Lyshevski

Titles in the Series

NANO
and
MOLECULAR
ELECTRONICS
Handbook

Edited by
Sergey Edward Lyshevski

CRC Press
Taylor & Francis Group
Boca Raton   London   New York

CRC Press is an imprint of the
Taylor & Francis Group, an informa business

CRC Press
Taylor & Francis Group
6000 Broken Sound Parkway NW, Suite 300
Boca Raton, FL 33487-2742

© 2007 by Taylor & Francis Group, LLC
CRC Press is an imprint of Taylor & Francis Group, an Informa business

No claim to original U.S. Government works
Printed in the United States of America on acid-free paper
10 9 8 7 6 5 4 3 2 1

International Standard Book Number-10: 0-8493-8528-8 (Hardcover)
International Standard Book Number-13: 978-0-8493-8528-5 (Hardcover)

Library of Congress Cataloging-in-Publication Data

Nano and molecular electronics handbook / editor, Sergey E. Lyshevski.
    p. cm. -- (Nano- and microscience, engineering, technology, and
medicine series)
    Includes bibliographical references and index.
    ISBN-13: 978-0-8493-8528-5 (alk. paper)
    ISBN-10: 0-8493-8528-8 (alk. paper)
    1. Molecular electronics--Handbooks, manuals, etc. I. Lyshevski, Sergey Edward. II. Title. III. Series.

TK7874.8.N358 2007
621.381--dc22
                          2006101011

Visit the Taylor & Francis Web site at
http://www.taylorandfrancis.com

and the CRC Press Web site at
http://www.crcpress.com

The Editor

Sergey Edward Lyshevski was born in Kiev, Ukraine. He received his M.S. (1980) and Ph.D. (1987) degrees from Kiev Polytechnic Institute, both in electrical engineering. From 1980 to 1993, Dr. Lyshevski held faculty positions at the Department of Electrical Engineering at Kiev Polytechnic Institute and the Academy of Sciences of Ukraine. From 1989 to 1993, he was the Microelectronic and Electromechanical Systems Division Head at the Academy of Sciences of Ukraine. From 1993 to 2002, he was with Purdue School of Engineering as an associate professor of electrical and computer engineering. In 2002, Dr. Lyshevski joined Rochester Institute of Technology as a professor of electrical engineering. Dr. Lyshevski serves as a Full Professor Faculty Fellow at the U.S. Air Force Research Laboratories and Naval Warfare Centers. He is the author of ten books (including *Logic Design of NanoICs*, coauthored with S. Yanushkevich and V. Shmerko, CRC Press, 2005; *Nano- and Microelectromechanical Systems: Fundamentals of Micro- and Nanoengineering*, CRC Press, 2004; *MEMS and NEMS: Systems, Devices, and Structures*, CRC Press, 2002) and is the author or coauthor of more than 300 journal articles, handbook chapters, and regular conference papers. His current research activities are focused on molecular electronics, molecular processing platforms, nanoengineering, cognitive systems, novel organizations/architectures, new nanoelectronic devices, reconfigurable super-high-performance computing, and systems informatics. Dr. Lyshevski has made significant contributions in the synthesis, design, application, verification, and implementation of advanced aerospace, electronic, electromechanical, and naval systems. He has made more than 30 invited presentations (nationally and internationally) and serves as an editor of the Taylor & Francis book series *Nano- and Microscience, Engineering, Technology, and Medicine.*

Contributors

Rajeev Ahuja
Condensed Matter Theory
    Group
Department of Physics
Uppsala University
Uppsala, Sweden

Richard Akis
Center for Solid State
    Engineering Research
Arizona State University
Tempe, Arizona, USA

Andrea Alessandrini
CNR-INFM-S3
NanoStructures and
    BioSystems at Surfaces
Modena, Italy

Supriyo Bandyopadhyay
Department of Electrical and
    Computer Engineering
Virginia Commonwealth
    University
Richmond, Virginia, USA

Valeriu Beiu
United Arab Emirates
    University
Al-Ain, United Arab Emirates

Robert R. Birge
Department of Chemistry
University of Connecticut
Storrs, Connecticut, USA

A.M. Bratkovsky
Hewlett-Packard Laboratories
Palo Alto, California, USA

J.A. Brown
Department of Physics
University of Alberta
Edmonton, Canada

K. Burke
Department of Chemistry
University of California
Irvine, California, USA

Horacio F. Cantiello
Massachusetts General Hospital
and
Harvard Medical School
Charlestown, Massachusetts,
    USA

Aldo Di Carlo
Università di Roma
    Tor Vergata
Roma, Italy

G.F. Cerofolini
STMicroelectronics
Post-Silicon Technology
Milan, Italy

J. Cuevas
Grupo de Física No Lineal
Departamento de Física
    Aplicada I
ETSI Inform Universidad
    de Sevilla
Sevilla, Spain

Shamik Das
Nanosystems Group
The MITRE Corporation
McLean, Virginia, USA

John M. Dixon
Massachusetts General Hospital
and
Harvard Medical School
Charlestown, Massachusetts,
    USA

J. Dorignac
College of Engineering
Boston University
Boston, Massachusetts, USA

Rodney Douglas
Institute of Neuroinformatics
Zurich, Switzerland

J.C. Eilbeck
Department of Mathematics
Heriot-Watt University
Riccarton, Edinburgh, UK

James C. Ellenbogen
Nanosystems Group
The MITRE Corporation
McLean, Virginia, USA

Christoph Erlen
Technische Universität
    München
München, Germany

F. Evers
Institut für Theorie der
    Kondensierten Materie
Universität Karlsruhe
Karlsruhe, Germany

Paolo Facci
CNR-INFM-S3
NanoStructures and
    BioSystems at Surfaces
Modena, Italy

David K. Ferry
Center for Solid State
    Engineering Research
Arizona State University
Tempe, Arizona, USA

Danko D. Georgiev
Laboratory of Molecular
    Pharmacology
Faculty of Pharmaceutical
    Sciences
Kanazawa University Graduate
    School of Natural Science
    and Technology
Kakuma-machi Kanazawa
Ishikawa, Japan

James F. Glazebrook
Department of Mathematics
    and Computer Science
Eastern Illinois University
Charleston, Illinois, USA

Anton Grigoriev
Condensed Matter Theory
    Group
Department of Physics
Uppsala University
Uppsala, Sweden

Rikizo Hatakeyama
Department of Electronic
    Engineering
Tohoku University
Sendai/Japan

Thorsten Hansen
Department of Chemistry and
    International Institute for
    Nanotechnology
Northwestern University
Argonne, Evanston,
    Illinois, USA

Jason R. Hillebrecht
Department of Molecular and
    Cell Biology
University of Connecticut
Storrs, Connecticut, USA

Walid Ibrahim
United Arab Emirates
    University
Al-Ain, United Arab Emirates

Giacomo Indiveri
Institute of Neuroinformatics
Zurich, Switzerland

Dustin K. James
Department of Chemistry
Rice University
Houston, Texas, USA

Bhargava Kanchibotla
Department of Electrical and
    Computer Engineering
Virginia Commonwealth
    University
Richmond, Virginia, USA

Jeremy F. Koscielecki
Department of Chemistry
University of Connecticut
Storrs, Connecticut, USA

Mark P. Krebs
Department of Ophthalmology
College of Medicine
University of Florida
Gainesville, Florida, USA

Craig S. Lent
Department of Electrical
    Engineering
University of Notre Dame
Notre Dame, Indiana, USA

Takhee Lee
Department of Materials
    Science and Engineering
Gwangju Institute of Science
    and Technology
Gwangju, Korea

Paolo Lugli
Technische Universität München
München, Germany

Sergey Edward Lyshevski
Department of Electrical
    Engineering
Rochester Institute of
    Technology
Rochester, New York, USA

Lyuba Malysheva
Bogolyubov Institute for
    Theoretical Physics
Kiev, Ukraine

Thomas Marsh
University of St. Thomas
St. Paul, Minnesota, USA

Duane L. Marcy
Department of Electrical
    Engineering and Computer
    Science
Syracuse University
Syracuse, New York, USA

Robert M. Metzger
Laboratory for Molecular
    Electronics
Department of Chemistry
University of Alabama
Tuscaloosa, Alabama, USA

M. Meyyappan
Center for Nanotechnology
NASA Ames Research Center
Moffett Field, California, USA

Lev G. Mourokh
Physics Department
Queens College of the City
    University of New York
Flushing, New York, USA

Vladimiro Mujica
Department of Chemistry and
    International Institute for
    Nanotechnology
Northwestern University
Evanston, Illinois, USA
and
Argonne National Laboratory
    Center for Nanoscale
    Materials
Argonne, Illinois, USA

Alexander Onipko
IFM
Linkping University
Linkping, Sweden

Alexei O. Orlov
Department of Electrical
    Engineering
University of Notre Dame
Notre Dame, Indiana, USA

F. Palmero
Grupo de Física No Lineal
Departamento de Física
ETSI Inform Universidad
    de Sevilla
Sevilla, Spain

Alessandro Pecchia
Università di Roma
    Tor Vergata
Roma, Italy

Carl A. Picconatto
Nanosystems Group
The MITRE Corporation
McLean, Virginia, USA

Sandipan Pramanik
Department of Electrical and
    Computer Engineering
Virginia Commonwealth
    University
Richmond, Virginia, USA

Avner Priel
Department of Physics
University of Alberta
Edmonton, Alberta, Canada

Mark A. Ratner
Department of Chemistry and
    International Institute for
    Nanotechnology
Northwestern University
Evanston, Illinois, USA

Mark A. Reed
Departments of Electrical
    Engineering, Applied
    Physics, and Physics
Yale University
New Haven, Connecticut, USA

R.A. Römer
Department of Physics and
    Centre for Scientific
    Computing
University of Warwick
Coventry, UK

F.R. Romero
Grupo de Física No Lineal
Departamento de FAMN
Facultad de Física
Universidad de Sevilla
Sevilla, Spain

Garrett S. Rose
Department of Electrical
    and Computer Engineering
Polytechnic University
Brooklyn, New York, USA

Anatoly Yu. Smirnov
Quantum Cat Analytics Inc.
Brooklyn, New York, USA

Gregory L. Snider
Department of Electrical
    Engineering
University of Notre Dame
Notre Dame, Indiana, USA

Gil Speyer
Center for Solid State
    Engineering Research
Arizona State University
Tempe, Arizona, USA

Jeffrey A. Stuart
Department of Chemistry
University of Connecticut
Storrs, Connecticut, USA

William Tetley
Department of Electrical
    Engineering and Computer
    Science
Syracuse University
Syracuse, New York, USA

James M. Tour
Department of Chemistry
Rice University
Houston, Texas, USA

Jack A. Tuszynski
Department of Physics
University of Alberta
Edmonton, Alberta, Canada

James Vesenka
University of New England
Biddeford, Maine, USA

Wenyong Wang
Semiconductor Electronics
    Division
National Institute of Standards
    and Technology
Gaithersburg, Maryland, USA

Bangwei Xi
Department of Chemistry
Syracuse University
Syracuse, New York, USA

Bin Yu
Center for Nanotechnology
NASA Ames Research Center
Moffett Field, California, USA

Matthew M. Ziegler
IBM T. J. Watson Research
    Center
Yorktown Heights, New York,
    USA

Contents

Section III   Biomolecular Electronics and Processing

Section IV   Molecular and Nano Electronics: Device-Level Modeling and Simulation

Preface

It was a great pleasure to edit this handbook, which consists of outstanding chapters written by acclaimed experts in their field. The overall objective was to provide coherent coverage of a broad spectrum of issues in molecular and nanoelectronics (e.g., covering fundamentals, reporting recent innovations, devising novel solutions, reporting possible technologies, foreseeing far-reaching developments, envisioning new paradigms, etc.). Molecular and nanoelectronics is a revolutionary theory- and technology-in-progress paradigm. The handbook's chapters document sound fundamentals and feasible technologies, ensuring a balanced coverage and practicality. There should be no end to molecular electronics and molecular processing platforms (^{M}PPs), which ensure superior overall performance and functionality that cannot be achieved by any envisioned microelectronics innovations.

Due to inadequate commitments to high-risk/extremely-high-pay-off developments, limited knowledge, and the abrupt nature of fundamental discoveries and enabling technologies, it is difficult to accurately predict when various discoveries will mature in the commercial product arena. For more than six decades, large-scale focused efforts have concentrated on solid-state microelectronics. A matured $150-billion microelectronics industry has profoundly contributed to technological progress and societal welfare. However, further progress and envisioned microelectronics evolutions encounter significant fundamental and technological challenges and limits. Those limits may not be overcome. In attempts to find new solutions and define novel inroads, innovative paradigms and technologies have been devised and examined. Molecular and nanoelectronics have emerged as one of the most promising solutions.

The difference between molecular- (nano) and micro-electronics is not the size (dimensionality), but the profoundly different device- and system-level solutions, the device physics, and the phenomena, fabrication, and topologies/organizations/architectures. For example, a field-effect transistor with an insulator thickness less than 1 nm and a channel length less than 20 nm cannot be declared a nanoelectronic device even though it has the subnanometer insulator thickness and may utilize a carbon nanotube (with a diameter under 1 nm) to form a channel. Three-dimensional topology molecular and nanoelectronic devices, engineered from atomic aggregates and synthesized utilizing *bottom-up* fabrication, exhibit quantum phenomena and electrochemomechanical effects that should be uniquely utilized. The topology, organization, and architecture of three-dimensional molecular integrated circuits (^{M}ICs) and ^{M}PPs are entirely different compared with conventional two-dimensional ICs.

Questions regarding the feasibility of molecular electronics and ^{M}PPs arise. No conclusive evidence exists of the overall feasibility of *solid* ^{M}ICs and there was no analog for solid-state microelectronics and ICs existed in the past. In contrast, an enormous variety of biomolecular processing platforms are visible in nature. These platforms provide one with undeniable evidence of feasibility, soundness, and unprecedented supremacy of a molecular paradigm. Though there have been attempts to utilize and prototype biocentered electronics, processing, and memories, these efforts have faced—and still face—enormous fundamental, experimental, and technological challenges. Superior organizations and architectures of ^{M}ICs and ^{M}PPs can be devised utilizing biomimetics, thus examining and prototyping brain and central nervous system functions. Today, many unsolved problems plague biosystems—from the baseline functionality of neurons to the capabilities of neuronal aggregates, from information processing to information measures, from the phenomena utilized to the cellular mechanisms exhibited, and so on. Even though significant challenges still exist, rapid progress and new discoveries have been made in recent years on both fundamental and technological forefronts. This progress and some of its major findings are covered in this handbook. The handbook consists of four sections, providing coherence in its subject matter. The six chapters of Section I: *Molecular and Nano Electronics: Device- and System-Level* are as follows:

- Electrical Characterization of Self-Assembled Monolayers
- Molecular Electronic Computing Architectures
- Unimolecular Electronics: Results and Prospects
- Carbon Derivatives
- System-Level Design and Simulation of Nanomemories and Nanoprocessors
- Three-Dimensional Molecular Electronics and Integrated Circuits for Signal and Information Processing Platforms

These chapters report the device physics of molecular devices (^{M}devices), the synthesis of those ^{M}devices, the design of ^{M}ICs, and devising ^{M}PPs. Meaningful results on device- and system-level fundamentals are offered, and envisioned technologies and engineering practices are documented.

Section II: *Nanoscaled Electronics* consists of the following six chapters:

- Inorganic Nanowires in Electronics
- Quantum Dots in Nanoelectronic Devices
- Self Assembly of Nanostructures Using Nanoporous Alumina Templates
- Neuromorphic Networks of Spiking Neurons
- Allowing Electronics to Face the TSI Era—Molecular Electronics and Beyond
- On Computing Nano-Architectures using unreliable Nanodevices or on Yield-Energy-Delay Logic Designs

These chapters focus on nano- and nanoscaled electronics. Various practical solutions are reported.

Section III: *Biomolecular Electronics and Processing* covers recent innovative results in biomolecular electronics and memories. The six chapters included are

- Properties of "G-Wire" DNA
- Metalloprotein Electronics
- Localization and Transport of Charge by Nonlinearity and Spatial Discreteness in Biomolecules and Semiconductor Nanorings. Aharonov–Bohm Effect for Neutral Excitons
- Protein-Based Optical Memories
- Subneuronal Processing of Information by Solitary Waves and Stochastic Processes
- Electronic and Ionic Conductivities of Microtubules and Actin Filaments, Their Consequences for Cell Signaling and Applications to Bioelectronics

Each chapter is of practical importance regarding the envisioned biomolecular platforms, and will help in comprehending significant phenomena in biosystems.

The eight chapters of Section IV: *Molecular and Nano Electronics: Device-Level Modeling and Simulation* focus on various aspects of high-fidelity modeling, heterogeneous simulations, and data-intensive analysis. The chapters included consist of the following:

- Simulation Tools in Molecular Electronics
- Theory of Current Rectification, Switching, and the Role of Defects in Molecular Electronic Devices
- Complexities of the Molecular Conductance Problem
- Nanoelectromechanical Oscillator as an Open Quantum System
- Coherent Electron Transport in Molecular Contacts: A Case of Tractable Modeling
- Pride, Prejudice, and Penury of *ab initio* Transport Calculations for Single Molecules
- Molecular Electronics Devices
- An Electric Cotunneling Model of STM-Induced Unimolecular Surface Reactions

These chapters provide the reader with valuable results that can be utilized in various applications, with a major emphasis on the device-level fundamentals.

The handbook's chapters report the individual authors' results. Therefore, in reading different chapters, the reader may observe some variations and inconsistencies in style, definitions, formulations, findings, and vision. This, in my opinion, is not a weakness but rather a strength. In fact, the reader should be aware of the differences in opinions, the distinct methods applied, the alternative technologies pursued, and the various concepts emphasized. I truly enjoyed collaborating with all the authors and appreciate their valuable contribution. It should be evident that the views, findings, recommendations, and conclusions documented in the handbook's chapters are those of the authors', and do not necessarily reflect the editor's opinion. However, all the chapters in the book emphasize the need for further research and development in molecular and nanoelectronics, which is today's engineering, science, and technology frontier.

It should be emphasized that no matter how many times the material has been reviewed, and effort spent to guarantee the highest quality, there is no guarantee this handbook is free from minor errors, and shortcomings. If you find something you feel needs correcting, adjustment, clarification, and/or modification, please notify me. Your help and assistance are greatly appreciated and deeply acknowledged.

Acknowledgments

Many people contributed to this book. First, I would like to express my sincere thanks and gratitude to all the book's contributors. It is with great pleasure that I acknowledge the help I received from many people in preparing this handbook. The outstanding Taylor & Francis team, especially Nora Konopka (Acquisitions Editor, Electrical Engineering), Jessica Vakili, and Amy Rodriguez (Project Editor), helped tremendously, and assisted me by offering much valuable and deeply treasured feedback. Many thanks to all of you.

Sergey Edward Lyshevski
Department of Electrical Engineering
Rochester Institute of Technology
Rochester, NY, 14623-5603, USA
E-mail: Sergey.Lyshevski@rit.edu
Web cite: www.rit.edu/~seleee

I

Molecular and Nano Electronics: Device- and System-Level

1

Electrical Characterization of Self-Assembled Monolayers

Wenyong Wang

Takhee Lee

Mark A. Reed

Abstract

Electrical characterization of alkanethiol self-assembled monolayers (SAMs) has been performed using a nanometer-scale device structure. Temperature-variable current-voltage measurement is carried out to distinguish between different conduction mechanisms and temperature-independent transport characteristics are observed, revealing that tunneling is the dominant conduction mechanism of alkanethiols. Electronic transport through alkanethiol SAMs is further investigated with the technique of inelastic electron tunneling spectroscopy (IETS). The obtained IETS spectra exhibit characteristic vibrational signatures of the alkane molecules used, presenting direct evidence of the presence of molecular species in the device structure. Further investigation on the modulation broadening and thermal broadening of the spectral peaks yields intrinsic linewidths of different vibrational modes, which may give insight into molecular conformation and prove to be a powerful tool in future molecular transport characterization.

1.1 Introduction

The research field of nanoscale science and technology has made tremendous progress in the past decades, ranging from the experimental manipulations of single atoms and single molecules to the synthesis and possible applications of carbon nanotubes and semiconductor nanowires [1–3]. As the enormous literature has shown, nanometer scale device structures provide suitable testbeds for the investigations of novel physics in a new regime, especially at the quantum level, such as single electron tunneling or quantum confinement effect [4,5]. On the other hand, as the semiconductor device feature size keeps decreasing, the traditional top-down microfabrications will soon enter the nanometer range, and further continuous downscaling will become scientifically and economically challenging [6]. This will motivate researchers around the world to find alternative ways to meet future increasing computing demands.

With a goal of examining individual molecules as self-contained functioning electronic components, molecular transport characterization is an active part of the research field of nanotechnology [2,3]. In 1974, a theoretical model of a unimolecular rectifier was proposed, according to which a single molecule consisting of an electron donor region and an electron acceptor region separated by a σ bridge would behave as a unimolecular p-n junction [7]. However, an experimental realization of such a unimolecular device was hampered by the difficulties of both the chemical synthesis of this type of molecule and the microfabrication of reliable solid-state test structures. A publication in 1997 reported an observation of such a unimolecular rectification in a device containing Langmuir–Blodgett (L-B) films; however, it is not clear if the observed rectifying behavior had the same mechanism since it was just shown in a single current-voltage [I(V)] measurement [8]. In the meantime, instead of using L-B films, others proposed to exploit self-assembled conjugated oligomers as the active electronic components [9,10] and started the electrical characterization of monolayers formed by the molecular self-assembly technique [2].

Molecular self-assembly is an experimental approach to spontaneously forming highly ordered monolayers on various substrate surfaces [11,12]. Earlier research in this area includes the pioneering study of alkyl disulfide monolayers formed on gold surfaces [13]. This research field has grown enormously in the past two decades and self-assembled monolayers (SAMs) have found their modern-day applications in various areas, such as nanoelectronics, surface engineering, biosensing, etc. [11].

Various test structures have been developed in order to carry out characterizations of self-assembled molecules, and numerous reports have been published in the past several years on the transport characteristics [2,3,14,15]. Nevertheless, many of them have drawn conclusions on transport mechanisms without performing detailed temperature-dependent studies [14,15], and some of the molecular effects were shown to be due to filamentary conduction in further investigations [16–21], highlighting the need to institute reliable controls and methods to validate true molecular transport [22]. A related problem is the characterization of molecules in the active device structure, including their configuration, bonding, and even their very presence.

In this research work, we conduct electrical characterization of molecular assemblies that exhibit understood classical transport behavior and can be used as a control for eliminating or understanding fabrication variables. A molecular system whose structure and configuration are well-characterized such that it can serve as a standard is the extensively studied alkanethiol [$CH_3(CH_2)_{n-1}SH$] self-assembled monolayer [11,22–25]. This system forms a single van der Waals crystal on the Au(111) surface [26] and presents a simple classical metal–insulator–metal (MIM) tunnel junction when fabricated between metallic contacts because of the large HOMO–LUMO gap (HOMO: highest occupied molecular orbital; LUMO: lowest unoccupied molecular orbital) of approximately 8 eV [27]. Utilizing a nanometer scale device structure that incorporates alkanethiol SAMs, we demonstrate devices that allow temperature-dependent I(V) [I(V,T)] and structure-dependent measurements [24]. The obtained characteristics are further compared with calculations from accepted theoretical models of MIM tunneling, and important transport parameters are derived [24,28].

Electronic transport through alkanethiol SAM is further investigated with the technique of inelastic electron tunneling spectroscopy (IETS) [25,29]. IETS was developed in the 1960s as a powerful spectroscopic tool to study the vibrational spectra of organic molecules confined inside metal–oxide–metal tunnel junctions [29–31]. In our study, IETS is utilized for the purpose of molecule identification, and the investigation of chemical bonding and the conduction mechanism of the "control" SAM. The exclusive presence of well-known characteristic vibrational modes of the alkane molecules used is direct evidence of the molecules in the device structure, which is the first unambiguous proof of such an occurrence. The spectral lines also yield intrinsic linewidths that may give insight into molecular conformation, and may prove to be a powerful tool in future molecular device characterization [22,25].

1.2 Theoretical Background of Tunneling

1.2.1 Electron Tunneling

Tunneling is a purely quantum mechanical behavior [32,33]. During the tunneling process, a particle can penetrate through a barrier—a classically forbidden region corresponding to negative kinetic energy—and transfer from one classically allowed region to another. This happens because the particle also has wave characteristics. Since the development of quantum mechanics, tunneling phenomena have been studied by both theorists and experimentalists on many different systems [34,35].

One of the extensively studied tunneling structures is the metal–insulator–metal tunnel junction. If two metal electrodes are separated by an insulating film, and the film is sufficiently thin, current can flow between the two electrodes by means of tunneling [34,35]. The purpose of this insulating film is to introduce a potential barrier between the metal electrodes. The tunneling current density for a rectangular barrier can be expressed as [34–36]:

$$J = \left(\frac{e}{4\pi^2 \hbar\, d^2}\right)\left\{\left(\Phi_B - \frac{eV}{2}\right)\exp\left[-\frac{2(2m)^{1/2}}{\hbar}\left(\Phi_B - \frac{eV}{2}\right)^{1/2} d\right]\right.$$
$$\left. -\left(\Phi_B + \frac{eV}{2}\right)\exp\left[-\frac{2(2m)^{1/2}}{\hbar}\left(\Phi_B + \frac{eV}{2}\right)^{1/2} d\right]\right\} \tag{1.1}$$

where m is electron mass, d is barrier width, Φ_B is barrier height, $h(= 2\pi\hbar)$ is Planck's constant, and V is applied bias. In the low bias range, Equation (1.1) can be approximated as [36]:

$$J \approx \left(\frac{e^2(2m\Phi_B)^{1/2}}{h^2 d}\right) V \exp\left[-\frac{2}{\hbar}(2m\Phi_B)^{1/2} d\right] \tag{1.2}$$

which indicates that the tunneling current increases linearly with the applied bias. It also shows that the current depends on the barrier width exponentially as $J \propto \exp(-\beta_0 d)$. The decay coefficient β_0 can be expressed as:

$$\beta_0 = \frac{2(2m)^{1/2}}{\hbar}(\Phi_B)^{1/2} \tag{1.3}$$

An empirical model related to the complex band theory is the so-called Franz two-band model proposed for an MIM junction in the 1950s [37–41]. Unlike the Simmons model, the Franz model considered the contributions from both the conduction band and valence band of the insulating film by taking into account the energy bandgap of E_g [37]. Instead of giving a tunneling current expression, it empirically predicted a non-parabolic energy-momentum dispersion relationship inside the bandgap [37]:

$$k^2 = \frac{2m*}{\hbar^2}E\left(1 + \frac{E}{E_g}\right) \tag{1.4}$$

where m^* is the electron's effective mass, and E is referenced relative to the conduction band.

The Franz model is useful for finding the effective mass of the tunneling electron inside the band gap [38–41]. From the non-parabolic E(k) relationship of Equation (1.4), the effective mass can be deduced by knowing the barrier height of the MIM tunnel junction [41]. But when the Fermi level of the metal electrodes is aligned close to one energy band, the effect of the other distant band on the tunneling transport is negligible, and the Simmons model is a good approximation of the Franz model, as shown in the previous analysis [37,42].

1.2.2 Inelastic Electron Tunneling

Inelastic electron tunneling due to localized molecular vibrational modes was discovered by Jaklevic and Lambe in 1966 when they studied the tunneling effect of metal–oxide–metal junctions [29]. Instead of finding band structure effects due to metal electrodes as they initially hoped, they observed structures in the d^2I/dV^2 characteristics which were related to vibrational excitations of molecular impurities contained in the insulator [29,43]. IETS has since been developed into a powerful spectroscopic tool for various applications such as chemical identification, bonding investigation, trace substance detection, and so on [30,31].

Figure 1.1 shows the energy band diagrams of a tunnel junction and the corresponding I(V) plot. When a negative bias is applied to the left metal electrode, the left Fermi level is lifted. An electron from an occupied state on the left side tunnels into an empty state on the right side, and its energy is conserved (process a). This is the elastic process discussed in Section 1.2.1. During this process, the current increases linearly with the applied small bias [Figure 1.1(b)]. However, if there is a vibrational mode with a frequency of ν localized inside this barrier, then when the applied bias is large enough such that $eV \geq h\nu$, the electron can lose a quantum of energy of $h\nu$ to excite the vibration mode and tunnel into another empty state (process b) [44,45]. This opens an inelastic tunneling channel for the electron and its overall tunneling probability is increased. Thus, the total tunneling current has a kink as a function of the applied bias [Figure 1.1(b)]. This kink becomes a step in the differential conductance (dI/dV) plot, and turns into a peak in the d^2I/dV^2 plot. However, since only a small fraction of electrons tunnel inelastically, the conductance step is too small to be conveniently detected. In practice, people use a phase-sensitive detector ("lock-in") second harmonic detection technique to directly measure the peaks of the second derivative of I(V) [44].

After an IETS spectrum is obtained, the positions, widths, and intensities of the spectral peaks need to be comprehended. The peak position and width can be predicted on very general grounds, independent

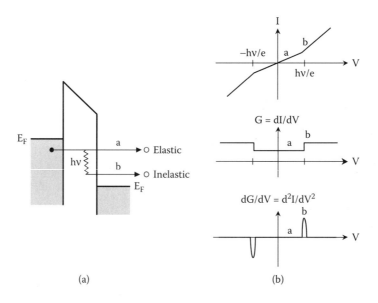

(a)   (b)

FIGURE 1.1   (a) Energy band diagram of a tunnel junction with a vibrational mode of frequency ν localized inside. a is the elastic tunneling process, while b is the inelastic tunneling process. (b) Corresponding I(V), dI/dV, and d^2I/dV^2 characteristics.

of the electron–molecule interaction details. However, the peak intensity is more difficult to be calculated since it depends on the detailed aspects of the electron-molecule couplings [44].

1.2.2.1 Peak Identification

As discussed earlier, an inelastic process can only start to occur when the applied bias reaches $V_i = h\nu_i/e$ [29]. Therefore, a peak at a position of bias V_i corresponds directly to a molecular vibrational mode of energy $h\nu_i$. This conclusion is based on energy conservation and is independent of the mechanism for the electron–molecule coupling. By referring to the huge amount of assigned spectra obtained by other techniques such as infrared (IR), Raman, and high-resolution electron energy loss spectroscopy (HREELS), the IETS peaks can be identified individually [43–45].

1.2.2.2 Peak Width

According to IETS theoretical studies, the width of a spectral peak includes a natural intrinsic linewidth and two width broadening effects: thermal broadening that is due to the Fermi level smearing effect, and modulation broadening that is due to the dynamic detection technique used to obtain the second harmonic signals [44].

The thermal broadening effect was first studied by Lambe and Jaklevic [43,46]. Assuming that the voltage dependence of the tunneling current is only contained in the Fermi functions of the metal electrodes, and the energy dependence of the effective tunneling density of states is negligible, the predicted thermal linewidth broadening at half maximum is $5.4\ kT/e$ [43,46]. This broadening prediction has been confirmed by experimental studies [47].

The broadening effect due to the finite modulation technique was first discussed by Klein *et al.* [46]. Assuming a modulation voltage of V_ω at a frequency of ω is applied to the tunnel junction, the full width at half maximum for the modulation broadening is $1.2\ V_\omega$, or $1.7\ V_{rms}$, the rms value of the modulation voltage, which is usually measured directly [44,46].

Of these two broadening contributions, the modulation broadening is more dominant [45]. By lowering the measurement temperature, the thermal broadening effect can be reduced—for example, at liquid helium temperature it gives a resolution of 2 meV. In order to make the modulation broadening comparable to the thermal effect, the modulation voltage should be less than 1.18 mV. However, since the second harmonic signal is proportional to the square of the modulation voltage and the signal-to-noise improvements varies with the square root of the averaging time, at such a small modulation the measurement time would be impractically extended. Therefore, little is gained by further lowering measurement temperature since the modulation broadening is more dominant [45].

The experimentally obtained spectral peak linewidth, W_{exp}, consists of three parts: the natural intrinsic linewidth, $W_{intrinsic}$; the thermal broadening $W_{thermal}$ that is proportional to 5.4 kT; and the modulation broadening $W_{modulation}$ that is proportional to 1.7 V_{rms}. These three contributions add as squares [43,48]:

$$W_{\text{exp}} = \sqrt{W_{intrinsic}^2 + W_{thermal}^2 + W_{modulation}^2} \qquad (1.5)$$

1.2.2.3 Peak Intensity

After the experimental discovery of inelastic electron tunneling due to molecular vibrations, several theoretical models were proposed for the purpose of quantitative analysis of the IETS spectra. The first theory was developed by Scalapino and Marcus in order to understand the interaction mechanism [49]. They treated the electron–molecule coupling as a Coulomb interaction between the electron and the molecular dipole moment and considered the case where the molecule of dipole moment is located very close to one of the electrodes so that the image dipole must be included. The interaction potential was treated as a perturbation on the barrier potential that was assumed to be rectangular. Using the WKB approximation, they could estimate the ratio of the inelastic conductance to the elastic one and predict that the intensities in a tunneling spectrum should be the same as in an infrared spectrum. However, it is found experimentally that although large peaks in IR spectra usually correspond to large peaks in tunneling spectra, the proportionality is not exact. Furthermore, peaks that are completely absent in IR spectra also appear in tunneling spectra [44].

Lambe and Jaklevic studied other mechanisms for electron–molecule interactions and generalized the preceding treatment to include the Raman type of interaction, where the electron induces a dipole moment in the molecule and interacts with this induced dipole [43]. Their calculation showed that the Raman-type interaction produces inelastic conductance changes of nearly the same order of magnitude as the IR-type electron–dipole interaction.

The preceding dipole approximations provided clear physical pictures of the interaction mechanisms of the tunneling electron and the localized molecular vibration; however, the calculations were oversimplified. Using the transfer Hamiltonian formalism [50,51], Kirtley *et al.* developed another theory for the intensity of vibrational spectra in IETS [44,52,53]. Rather than making the dipole approximation, they assumed that the charge distribution within the molecule can be broken up into partial charges, with each partial charge localized on a particular atom. These partial charges arise from an uneven sharing of the electrons involved in the bonding. The interaction potential between the tunneling electron and the vibrating molecule is thus a sum of Coulomb potentials with each element in the sum corresponding to a partial charge. This partial charge treatment allows one to describe the interaction at distances comparable to interatomic length. The inelastic tunneling matrix element, which corresponds to the tunneling transmission coefficient, can be calculated considering the WKB wave functions and the partial charge interaction potential [52]. The calculation results show that molecular vibrations with net dipole moments normal to the junction interface have larger inelastic cross sections than vibrations with net dipole moments parallel to the interface for dipoles close to one electrode. This is because when this close to a metal surface the image dipole adds to the potential of a dipole normal to the interface but tends to cancel out the potential of a dipole parallel to the interface. However, the case is different for vibrational modes localized deep inside the tunnel junction, where dipoles oriented parallel to the junction interface are favored, although at a lower scattering amplitude [44,52,53].

1.3 Experimental Methods

1.3.1 Self-Assembled Monolayers of Alkanethiols

Molecular self-assembly is a chemical technique to form highly ordered, closely packed monolayers on various substrates via a spontaneous chemisorption process at the interface [11]. Alkanethiol is a thiol-terminated *n*-alkyl chain molecular system [$CH_3(CH_2)_{n-1}SH$] [11]. As an example, Figure 1.2(a) shows the chemical structure of octanethiol, one of the alkanethiol molecules. It is well known that when self-assembled on Au(111), surface alkanethiol forms a densely packed, crystalline-like structure with the alkyl chain in an all-trans conformation [13]. The SAM deposition process is shown in Figure 1.2(b), where a clean gold substrate is immersed into an alkanethiol solution and, after time, a monolayer is formed spontaneously on the gold surface via the following chemical reaction [11,12,54]:

$$RS - H + Au \rightarrow RS - Au + 0.5H_2$$

where R is the backbone of the molecule. This chemisorption process has been observed to undergo two steps: a rapid process that takes minutes (depending on the thiol concentration) and gives $\sim 90\%$ of the film thickness, followed by a second, much slower process that lasts hours and reaches the final thickness and contact angles [11,12]. Research has shown that the second process is governed by a transition from a SAM lying-down phase into an ordered standing-up phase, and it is also accompanied or followed by a crystallization of the alkyl chains associated with molecular reorganization [11,55–57]. Three forces likely determine this SAM formation process and the final monolayer structure: the interaction between the thiol head group and gold lattice, the dispersion force between alkyl chains (the van der Waals force, etc.), and the interaction between the end groups [11,12].

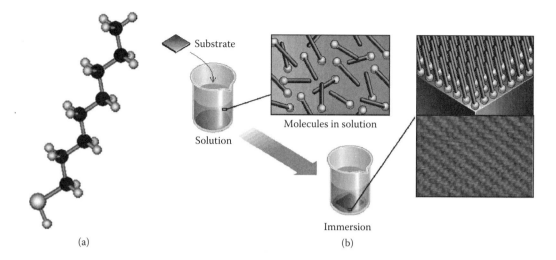

FIGURE 1.2 (a) Chemical structure of an octanethiol molecule. (b) Schematic of the SAM deposition process, after Ref. 54 and 61. It also shows an STM image of the SAM (see Figure 1.3).

Various surface analytical tools have been utilized to investigate the surface and bulk properties of alkanethiol SAMs, such as infrared (IR) and Fourier transform infrared (FTIR) spectroscopy [13,58], x-ray photoelectron spectroscopy (XPS) [59], Raman spectroscopy [60], scanning tunneling microscopy (STM) [23,61], etc. As an example, Figure 1.3(a) shows a constant current STM image of a dodecanethiol SAM formed on an Au(111) surface (adapted from Reference 61). Figure 1.3(b) is the schematic of the commensurate crystalline structure that alkanethiol SAM adopts, which is characterized by a c(4 × 2) superlattice of the ($\sqrt{3} \times \sqrt{3}$)R30° lattice [23,62]. In Figure 1.3(b), the large circular symbols represent the alkanethiol molecules and the small circular symbols represent the underlying gold atoms, and a and b are lattice vectors of the molecular rectangular unit cell with dimensions of 0.8 and 1.0 nm, respectively [61]. Investigations have also shown that the standing-up alkyl chains of alkanethiol SAMs on the Au(111)

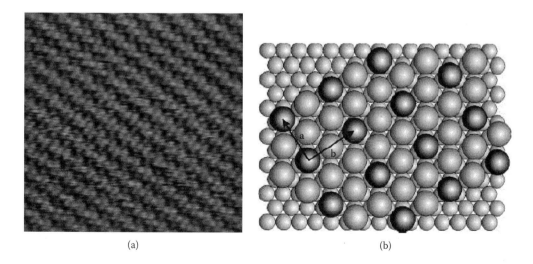

FIGURE 1.3 (a) STM image of a dodecanethiol SAM formed on Au(111) surface. The image size is 13 × 13 nm². (b) Schematic of the alkanethiol SAM commensurate crystalline structure. Large circular symbols represent the alkanethiol molecules, and small circular symbols represent the underlying gold atoms. a and b are lattice vectors of a rectangular unit cell with dimensions of 0.8 and 1.0 nm, respectively. After Ref. 61.

surface are tilted $\sim 30°$ from the surface normal [62] and the bonding energy between the thiolate head group and the gold lattice is ~ 40 kcal/mol (~ 1.7 eV) [11].

Studies have revealed that defects, such as pinholes or grain boundaries, exist in the self-assembled monolayers, and the domain size of an alkanethiol SAM usually is on the order of several hundred Ångstroms [11,23]. In addition to the irregularities introduced during the self-assembly process, another source of the defects is the roughness of the substrate surface. For example, although frequently called "flat" gold, grain boundaries exist on the Au surface layer, which introduce defects into the assembled monolayer [23]. However, surface migration of thiolate–Au molecules, the so-called SAM annealing process, is found to be helpful for healing some of the defects [11,23].

Alkanethiols are large HOMO–LUMO gap (~ 8 eV [27,63–66]) molecules with short molecular lengths ($\sim$ several nanometers), therefore the electronic transport mechanism is expected to be tunneling. Electrical characterizations have been performed on alkanethiol SAMs and will be discussed in the next section.

1.3.2   Methods of Molecular Transport Characterization

A correct understanding of the electronic transport properties through self-assembled molecules requires fabrication methods that can separate the effects of contacts from the intrinsic properties of the molecular layer. However, such transport measurements are experimentally challenging due to the difficulties of making repeatable and reliable electrical contacts to a nanometer-scale layer. A number of experimental characterization methods have been developed to achieve this goal, and in the following we briefly review some of the major techniques.

Various scanning probe–related techniques have been utilized for the study of molecular electronic structures, which include STM and atomic force microscopy (C-AFM). STM has been used widely at the early stage of molecular characterization due to its capability to image, probe, and manipulate single atoms or molecules [67–69]. Transport measurement on a single molecule contacted by STM has also been reported [70–73]. However, for such a measurement, the close proximity between the probe tip and the sample surface could modify what is being measured by tip-induced modification of the local surface electronic structure. The presence of a vacuum gap between the tip and the molecule also complicates the analysis [74]. Besides, contamination could occur if the measurement is taken in ambient conditions; therefore, inert gas (nitrogen or argon) filled or vacuum STM chamber is preferred [75,76].

The C-AFM technique also has been employed recently for the purpose of electrical characterizations of SAMs [77–80]. For example, Wold *et al.* reported C-AFM measurements on alkanethiol molecules [77]; Cui *et al.* bound gold nanoparticles to alkanedithiol in a monothiol matrix and measured its conductance [78]. However, in this technique the C-AFM tip might penetrate and/or deform the molecular layer as well as create a force-dependent contact junction area. Adhesion force analysis (to rule out deformation or penetration) and a complimentary temperature-dependent characterization need to be performed to make C-AFM measurements a broadly applicable method for determining molecular conductivity [28].

Another important characterization method is the mechanically controllable break junction technique [81–84]. It can create a configuration of a SAM sandwiched between two stable metallic contacts, and two-terminal I(V) characterizations can be performed on the scale of single molecules [81]. In the fabrication process, a metallic wire with a notch is mounted onto an elastic bending beam and a piezo electric element is used to bend the beam and thus break the wire. The wire breaking is carried out in the molecular solution and after the breaking the solvent is allowed to evaporate, then the two electrodes are brought back together to form the desired molecular junction [84]. A lithographically fabricated version of the break junction uses e-beam lithography and the lift-off process to write a gold wire on top of an insulating layer of polyimide on a metallic substrate. The polyimide is then partially etched away and a free standing gold bridge is left on the substrate. The suspended gold bridge is then bent and broken mechanically using a similar technique to form a nanometer scale junction [83]. Using the break junction method, Reed *et al.* measured the charge transport through a benzene-1,4-dithiol molecule at room temperature [81]. Using a similar technique, Kergueris *et al.* [82] and Reichert *et al.* [83] performed conductance measurements on SAMs and concluded that I(V) characterizations of a few or individual molecules were achieved.

Recently, another type of break junction that utilizes the electromigration properties of metal atoms has been developed [85–87]. For this testbed, a thin gold wire with a width of several hundred nanometers is created via e-beam lithography and angle evaporation [85]. Bias is then applied and a large current passing through this nanoscale wire causes the gold atoms to migrate, thus creating a small gap a few nanometers wide. Molecules are deposited on the wire at room temperature before electro-breaking at cryogenic temperatures [86]. The advantage of this technique is that a third gating electrode can be introduced; therefore, three-terminal characterizations can be achieved. Using this electromigration break junction technique, Park *et al.* measured two types of molecules at cryogenic temperatures and observed Coulomb blockade behavior and the Kondo effect [86]. Similar Kondo resonances in a single molecular transistor were also observed by Liang *et al.* using the same test structure on a different molecular system [87]. However, in these measurements the molecules just serve as impurity sites [87], and the intrinsic molecular properties have yet to be characterized.

The cross-wire tunnel junction is a test structure reported in 1990 in an attempt to create an oxide-free tunnel junction for IETS studies [88]. It is formed by mounting two wires in such a manner that the wires are in a crossed geometry with one wire perpendicular to the applied magnetic field. The junction separation is then controlled by deflecting this wire with the Lorentz force generated from a direct current [88]. Using this method, Kushmerick *et al.* recently studied various molecules and observed conductance differences due to molecular conjugation and molecular length differences [89,90]. The drawback of this method is that it is very difficult to control the junction gap distance: the top wire might not touch the other end of the molecules or it might penetrate into the monolayer. Furthermore, temperature-variable measurement has not been reported using this test structure.

Other experimental techniques utilized in molecular transport studies include the mercury-drop junction [91,92] and the nanorod [93], among many others. For example, the mercury-drop junction consists of a drop of liquid Hg, supporting an alkanethiol SAM, in contact with the surface of another SAM supported by a second Hg drop [91,92]. This junction has been used to study the transport through alkanethiol SAMs, but the measurement can only be performed at room temperature [91].

For the research conducted in this work, we mainly use the so-called nanopore technique [24,94,95]. Using the nanopore method, we can directly characterize a small number of self-assembled molecules ($\sim$ several thousand) sandwiched between two metallic contacts. The contact area is around 30 to 50 nm in diameter, which is close to the domain size of the SAM [11]. Thus, the adsorbed monolayer is highly ordered and mostly defect free. This technique guarantees good control over the device area and intrinsic contact stability and can produce a large number of devices with acceptable yield so that statistically significant results can be achieved. Fabricated devices can be easily loaded into cryogenic or magnetic environments; therefore, critical tests of transport mechanisms can be carried out.

1.3.3   Device Fabrication

Figure 1.4 shows the process flow diagram of the nanopore fabrication. The fabrication starts with double-side polished 3-inch (100) silicon wafers with a high resistivity ($\rho > 10$ $\Omega \cdot$ cm). The thickness of the substrate is 250 μm. Using the low pressure chemical vapor deposition (LPCVD) method, a low stress Si_3N_4 film of 50 nm thick is deposited on both sides of the wafer. A low stress film is required in order to make the subsequent membrane less sensitive to mechanical shocks. Next, a 400 μm $\times$ 400 μm window is opened on the backside of the substrate via standard photolithography processing and reactive ion etching (RIE). Before the photolithography step, the topside of the substrate is coated with FSC (front side coating) to protect the nitride film. This FSC is removed after RIE by first soaking in acetone and then isopropanol alcohol. The exposed silicon is then etched through by anisotropic wet etching with the bottom nitride as the etch mask. The etchant is an 85% KOH solution heated to 85 to 90°C, and during the etching a magnetic stirrer is used to help the gas byproducts escape. At the end of the KOH etching, an optically transparent membrane of 40 μm $\times$ 40 μm is left suspended on the topside of the wafer. Figure 1.5(a) is an optical image of the suspended transparent membrane.

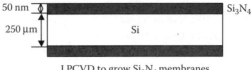

LPCVD to grow Si$_3$N$_4$ membranes

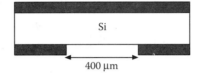

Photolithography & RIE to open the backside window

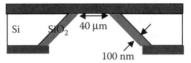

KOH to etch through the silicon and wet
oxidation to grow SiO$_2$ on the sidewalls

E-beam lithography & RIE to open the pore on the membrane

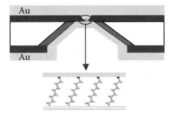

Final metal-SAM-metal junction

FIGURE 1.4   Schematics of the nanopore fabrication process.

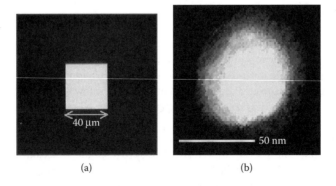

(a)   (b)

FIGURE 1.5   (a) Optical image of the membrane (topside view). (b) TEM image of an etched-through nanopore.

The wafer is carefully rinsed in water and then immersed in an isotropic silicon etchant (HNO_3:H_2O:HF = 300:150:2) for 5 minutes to remove any remaining silicon nodules on the membrane and to round out the sharp edges. The wafer is subsequently cleaned with the standard RCA cleaning process to remove any organic and metallic contaminations and then loaded into a wet oxidation furnace to oxidize the exposed silicon sidewalls for the purpose of preventing future electrical leakage current through the substrate. In order to reduce the thermal stress to the membrane caused by this high-temperature process, the wafer is loaded very slowly in and out of the furnace. A wet oxidation processing at $850°C$ for 60 minutes grows $\sim 1000\,\text{Å}$ SiO_2 on the sidewalls, which is enough to provide a good electrical insulation.

The last, and most critical, steps are the electron beam (e-beam) lithography and subsequent RIE etching to open a nanometer scale pore on the membrane. For the e-beam patterning, the PMMA thickness is 200 nm (4% 495K in anisole spun at 3500 rpm) and the e-beam dosage is between 40 to 300 mC/cm^2. After the exposure, the wafer is developed in MIBK:IPA of 1:1 for 60 seconds and then loaded into an RIE chamber to transfer the developed patterns. A CHF_3/O_2 plasma is used to etch the hole in the membrane and the etching time is varied from 2 to 6 minutes for a 50-nm thick nitride film. The RIE chamber has to be cleaned thoroughly by an O_2 plasma before the etching and every 2 minutes during the etching to remove the hydrocarbon residues deposited in the chamber. The etching is severely impeded deep in the pore due to the redeposition of hydrocarbon on the sidewalls; therefore, the opening at the far side is much smaller than that actually patterned, rendering a bowl-shaped cross section. After the etching is completed, the PMMA residue is striped off in the O_2 plasma.

SEM and TEM (Transmission Electron Microscope) examination and metallization have been used to determine if a pore is etched through. If not, further etching is performed until the hole is completely open. As an example, a TEM picture of an etched nanopore is shown in Figure 1.5(b). The size of the hole is roughly 50 nm in diameter, small enough to be within the domain size of both the evaporated gold film and the SAM layer. However, SEM and TEM examination is very time consuming and a more practical way to verify whether the pore is etched open is to deposit metal contacts on both sides of the membrane and measure the junction resistance. For a completely etched pore, I(V) measurement on a regular probe station usually shows a good ohmic short with a resistance of several ohms. For a non-etched-through device, I(V) measurement shows an open-circuit characteristic with a current level of $\sim$ pA at 1.0 Volt.

After the nanofabrication, 150 nm of gold is thermally evaporated onto the topside of the membrane to fill the pore and form one of the metallic contacts. The device is then transferred into a molecular solution to deposit the SAM. This deposition is done for 24 hours inside a nitrogen filled glove box with an oxygen level of less than 100 ppm. The sample is then rinsed with the deposition solvent and quickly loaded in ambient conditions into an evaporator with a cooling stage to deposit the opposing Au contact. A challenging step in fabricating molecular junctions is to make the top electrical contact. During the fabrication of metal–SAM–metal junctions, metallic materials deposited on the top of molecules often either penetrate through the thin molecular layer or contact directly with the substrate via defect sites (such as grain boundaries) in the monolayer, causing shorted circuit problems. Examination showed that $\sim 90\%$ of the devices were shorted with ambient temperature evaporation [74]; therefore, a low-temperature deposition technique is adopted [24,95]. During the thermal evaporation under the pressure of $\sim 10^{-8}$ Torr, liquid nitrogen is kept flowing through the cooling stage to minimize the thermal damage to the molecular layer. This technique reduces the kinetic energy of evaporated Au atoms at the surface of the monolayer, thus preventing Au atoms from punching through the SAM. For the same reason, the evaporation rate is kept very low. For the first 10 nm of gold evaporated, the rate is less than 0.1 Å/s. Then the rate is increased slowly to 0.5 Å/s for the remainder of the evaporation, and a total of 200 nm of gold is deposited to form the contact.

Preliminary I(V) measurements are carried out on a probe station at room temperature to screen out the functioning devices from those exhibiting either short circuit (top and bottom electrodes are shorted together) or open circuit (the nitride membrane is not etched through). The wafer is then diced into individual chips and the working devices are bonded onto a 16-pin packaging socket for further electrical characterizations.

1.3.4 Lock-in Measurement for IETS Characterizations

The IETS signal, which is proportional to the second derivative of I(V), is usually measured by an AC modulation method, the so-called lock-in technique [29,31]. Theoretically, the signal can also be determined by the mathematical differential approach that computes the numerical derivatives of the directly measured I(V) characteristics [96]. But this is generally not feasible in practice. On the contrary, the lock-in second harmonic detection technique measures a quantity directly proportional to d^2I/dV^2 [30,31]. During the lock-in measurement, a small sinusoidal signal is applied to modulate the voltage across the device, and the response of the current through the device to this modulation is studied. The detection of the first (ω) and second (2ω) harmonic signals give the scaled values of the first and second derivatives of I(V), respectively.

Experimentally, this modulation detection is realized by a lock-in amplifier. In our experiment, a typical DC source is used as the DC voltage provider, and a synthesized function generator is used as the AC modulation source, as well as to provide the reference signal to the lock-in amplifier. The DC bias and AC modulation are attenuated and mixed together by a custom-built voltage adder and then applied to the device under test (DUT). If a higher bias range is desired, a voltage shifter is included in the measurement setup before the DUT to increase the DC base voltage. An I-V converter is used if the voltage input of the lock-in amplifier is chosen for the measurement. The output of the lock-in amplifier is read by a digital multimeter.

1.4 Electronic Conduction Mechanisms in Self-Assembled Alkanethiol Monolayers

1.4.1 Conduction Mechanisms of Metal-SAM-Metal Junctions

In a metal-SAM-metal system, just as in a metal-semiconductor-metal junction, the Fermi level alignment is critical in determining the charge transport mechanism [97]. Created by the overlap of the atomic orbitals of a molecule's constituents, two molecular orbitals, lowest unoccupied molecular orbital (LUMO) and highest occupied molecular orbital (HOMO), play similar roles as conduction band and valence band in a semiconductor, respectively. The energy difference between them, the HOMO–LUMO gap, is typically of the order of several electron volts [2,3]. In general, the Fermi level of the metallic contacts does not energetically align with either the HOMO or the LUMO of the molecule, but instead lies close to the center of the gap [98]. This energy level mismatch gives rise to a contact barrier, and depending on the height and thickness of this barrier and the presence of defects, charge transport in such a metal-SAM-metal system exhibits a variety of behaviors. Table 1.1 gives a summary of possible conduction mechanisms with their characteristic behavior, temperature dependence, and voltage dependence [22,24,99–101].

Based on whether thermal activation is involved, the conduction mechanisms fall into two distinct categories: (1) thermionic or hopping conduction which has temperature-dependent I(V) behavior, and (2) direct tunneling or Fowler–Nordheim tunneling which does not have temperature-dependent I(V) behavior. Thermionic emission is a process in which carriers overcome the metal-dielectric barrier by thermal agitation, and the current has a strong dependence on temperature. The extra voltage term in

TABLE 1.1   Possible Conduction Mechanisms

Conduction Mechanism	Characteristics Behavior	Temperature Dependence	Voltage Dependence
Direct Tunneling	$J \sim V \exp(-\frac{2d}{\hbar}\sqrt{2m\Phi})$	None	$J \sim V$
Fowler-Nordheim Tunneling	$J \sim V^2 \exp(-\frac{4d\sqrt{2m}\Phi^{3/2}}{3q\hbar V})$	None	$\ln(\frac{J}{V^2}) \sim \frac{1}{V}$
Thermionic Emission	$J \sim T^2 \exp(-\frac{\Phi - q\sqrt{qV/4\pi\varepsilon d}}{kT})$	$\ln(\frac{J}{T^2}) \sim \frac{1}{T}$	$\ln(J) \sim V^{\frac{1}{2}}$
Hopping Conduction	$J \sim V \exp(-\frac{\Phi}{kT})$	$\ln(\frac{J}{V}) \sim \frac{1}{T}$	$J \sim V$

J is the Current Density, *d* is the Barrier Width, *T* is the Temperature, *V* is the Applied Bias, and Φ is the Barrier Height.
After Reference 99.

the exponential is due to image-force correction and it lowers the barrier height at the metal–insulator interface. Hopping conduction usually is defect-mediated, and in a hopping process the thermally activated electrons hop from one isolated state to another, and the conductance also depends strongly on temperature. However, unlike thermionic emission, there is no barrier-lowering effect in hopping transport. Tunneling processes (both direct and Fowler–Nordheim tunnelings) do not depend on temperature (to first order), but strongly depend on film thickness and voltage [99–101]. After a bias is applied, the barrier shape of a rectangular barrier is changed to a trapezoidal form. Tunneling through a trapezoidal barrier is called direct tunneling because the charge carriers are injected directly into the electrode. However, if the applied bias becomes larger than the initial barrier height, the barrier shape is further changed from trapezoidal to a triangular barrier. Tunneling through a triangular barrier, where the carriers tunnel into the conduction band of the dielectric, is called Fowler–Nordheim tunneling or field emission [99,100].

For a given metal-insulator-metal system, certain conduction mechanism(s) may dominate in certain voltage and temperature regimes. For example, thermionic emission usually plays an important role for high temperatures and low barrier heights. Hopping conduction is more likely to happen at low applied bias and high temperature if the insulator has a low density of thermally generated free carriers in the conduction band. Tunneling transport will occur if the barrier height is large and the barrier width is thin.

Temperature-variable I(V) characterization is an important experimental technique to elucidate the dominant transport mechanism and to obtain key conduction parameters such as effective barrier height. This is especially crucial in molecular transport measurements where defect-mediated conduction often complicates the analysis. For example, previous work on self-assembled thiol-terminated oligomers illustrated that one can deduce the basic transport mechanisms by measuring the I(V,T) characteristics [95]. It has been found that the physisorbed aryl-Ti interface gave a thermionic emission barrier of approximately 0.25 eV [95]. Another study on Au-isocyanide SAM-Au junctions showed both thermionic and hopping conductions with barriers of 0.38 and 0.30 eV, respectively [74].

In this research work, we investigate the charge transport mechanism of self-assembled alkanethiol monolayers. I(V,T) characterizations are performed on certain alkanethiols to distinguish between different conduction mechanisms. Electrical measurements are also carried out on alkanethiols with different molecular length to further examine length-dependent transport behavior.

1.4.2   Previous Research on Alkanethiol SAMs

Alkanethiol SAM [$CH_3(CH_2)_{n-1}SH$] is a molecular system whose structure and configuration are sufficiently well-characterized such that it can serve as a test standard [11]. This system is useful as a control since properly prepared alkanethiol SAM forms a single van der Waals crystal [11,23]. This system also presents a simple classical MIM tunnel junction when fabricated between two metallic contacts due to its large HOMO–LUMO gap of $\sim$ 8 eV [27,63–66].

Electronic transport through alkanethiol SAMs have been characterized by STM [70,73], conducting atomic force microscopy [77–80], mercury-drop junctions [91,92,102,103], cross-wire junctions [89], and electrochemical methods [104–106]. However, due to the physical configurations of these test structures it is very hard, if not impossible, to perform temperature-variable measurements on the assembled molecular layers; therefore, these investigations were done exclusively at ambient temperature, which is insufficient for an unambiguous claim that the transport mechanism is tunneling (which is expected, assuming that the Fermi level of the contacts lies within the large HOMO–LUMO gap). In the absence of I(V,T) characteristics, other transport mechanisms such as thermionic, hopping, or filamentary conduction can contribute and complicate the analysis. Previous I(V) measurements performed at room and liquid nitrogen temperatures on Langmuir–Blodgett alkane monolayers exhibited a large impurity-dominated transport component [107,108], further emphasizing the need and significance of I(V,T) measurement in SAM characterizations.

Using the nanopore test structure that contains alkanethiol SAMs, we demonstrate devices that allow I(V,T) and length-dependent measurements [24,25], and show that the experimental results can be compared with theoretical calculations from accepted models of MIM tunneling.

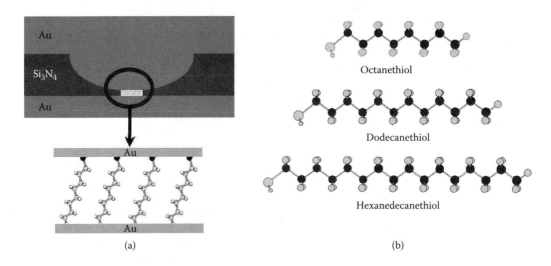

FIGURE 1.6 (a) Schematic of the nanopore structure used in this study. (b) Chemical structures of octanethiol, dodecanethiol, and hexadecanethiol.

1.4.3 Sample Preparation

Electronic transport measurement on alkanethiol SAMs is performed using the nanopore testbed [Figure 1.6(a)]. After 150 nm of gold is thermally evaporated onto the topside of the wafer, the sample is transferred into a molecular solution to deposit the SAM layer. For our experiments, a $\sim$ 5 mM alkanethiol solution is prepared by adding $\sim$ 10 μL of alkanethiols into 10 mL of ethanol. The deposition is done in solution for 24 hours inside a nitrogen-filled glove box with an oxygen level of less than 100 ppm. Three alkanemonothiol molecules of different molecular lengths—octanethiol [$CH_3(CH_2)_7SH$; denoted as C8, for the number of alkyl units]; dodecanethiol [$CH_3(CH_2)_{11}SH$, denoted as C12]; and hexadecanethiol [$CH_3(CH_2)_{15}SH$, denoted as C16]—were used to form the active molecular components. The chemical structures of these molecules are shown in Figure 1.6(b). The sample is then rinsed with ethanol and transferred to the evaporator for the deposition of 200 nm of gold onto the bottom side. Next, it is packaged and loaded into a low-temperature cryostat for electrical characterizations.

In order to statistically determine the pore size, test patterns (arrays of pores) were created under the same fabrication conditions (e-beam dose and etching time) as the real devices. Figure 1.7 shows a scanning electron microscope image of one such test pattern array. This indirect method for the measurement of device size is adopted because SEM examination of the actual device can cause hydrocarbon contamination of the device and subsequent contamination of the monolayer. Using SEM, the diameters have been examined for 112 pores fabricated with an e-beam dose of 100 mC/cm^2 and an etching time of 4.5 minutes, 106 pores fabricated with an e-beam dose of 100 mC/cm^2 and an etching time of 7 minutes, 130 pores fabricated with an e-beam dose of 200 mC/cm^2 and etching time of 6 minutes, and 248 pores fabricated with an e-beam dose of 300 mC/cm^2 and an etching time of 6 minutes. These acquired diameters were used as the raw data input file for the statistics software Minitab. Using Minitab, a regression analysis has been conducted on the device size as a function of e-beam dose and etching time, and a general size relation is obtained:

$$\text{Size(in nm)} = 35.0 + 0.027 \times \text{dose(in mC/cm}^2) + 1.63 \times \text{time(in min)}$$

Using the same software, a device size under particular fabrication conditions can be predicted via entering the fabrication dose and etching time. The error rage of the size is determined by specifying a certain confidence interval. For example, the fabrication conditions for the C8, C12, and C16 devices that are used in the length dependence study are an e-beam dose of 100 mC/cm^2 and an etching time of

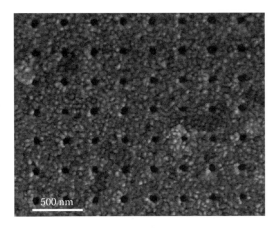

FIGURE 1.7 A representative scanning electron microscope image of an array of pores used to calibrate device size. The scale bar is 500 nm.

7 minutes, 88 mC/cm^2 and 4.5 minutes, and 85 mC/cm^2 and 5 minutes, respectively. From the regression analysis, the device sizes of the C8, C12, and C16 samples are predicted as 50 ± 8, 45 ± 2, and 45 ± 2 nm in diameters with a 99% confidence interval, respectively. We will use these device sizes as the effective contact areas. Although one could postulate that the actual area of metal that contacts the molecules may be different, there is little reason to propose it would be different as a function of length over the range of alkanethiols used, and at most it would be a constant systematic error.

1.4.4 Tunneling Characteristics of Alkanethiol SAMs

1.4.4.1 I(V,T) Characterization of Alkane SAMs

In order to determine the conduction mechanism of self-assembled alkanethiol molecular systems, I(V,T) measurements in a sufficiently wide temperature range (300 to 80 K) and resolution (10 K) on dodecanethiol (C12) were performed. Figure 1.8 shows representative I(V,T) characteristics measured with the device structure shown in Figure 1.6(a). Positive bias in this measurement corresponds to electrons injected from the physisorbed Au contact [the bottom contact in Figure 1.6(a)] into the molecules. By using the contact area of 45 ± 2 nm in diameter determined from the SEM study, a current density of

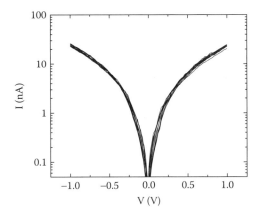

FIGURE 1.8 Temperature-dependent I(V) characteristics of dodecanethiol. I(V) data at temperatures from 300 to 80 K with 20 K steps are plotted on a log scale.

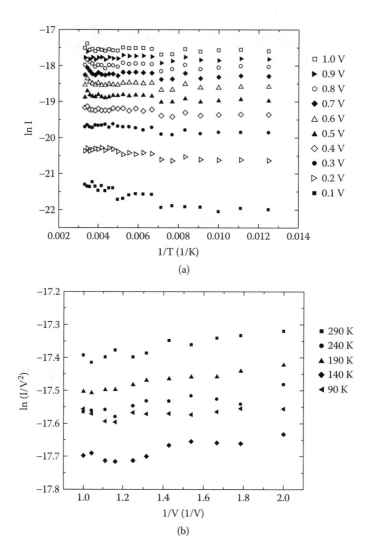

FIGURE 1.9 (a) Arrhenius plot generated from the I(V,T) data in Figure 1.8 at voltages from 0.1 to 1.0 Volt with 0.1 Volt steps. (b) Plot of $\ln(I/V^2)$ versus $1/V$ at selected temperatures to examine the Fowler–Nordheim tunneling.

1500 ± 200 A/cm^2 at 1.0 Volt is obtained. No significant temperature dependence of the characteristics from $V = 0$ to 1.0 Volt is observed over the temperature range from 300 to 80 K. An Arrhenius plot (ln I versus $1/T$) is shown in Figure 1.9(a), exhibiting little temperature dependence in the slopes of ln I versus $1/T$ at different biases, and thus indicating the absence of thermal activation. Therefore, we conclude that the conduction mechanism through alkanethiol is tunneling contingent on demonstrating correct molecular length dependence.

Based on the applied bias as compared with the barrier height (Φ_B), tunneling through a SAM layer can be categorized into either direct ($V < \Phi_B/e$) or Fowler–Nordheim ($V > \Phi_B/e$) tunneling. These two tunneling mechanisms can be distinguished by their distinct voltage dependencies (see Table 1.1). Analysis of $\ln(I/V^2)$ versus $1/V$ [in Figure 1.9(b)] of the C12 I(V,T) data shows no significant voltage dependence, indicating no obvious Fowler–Nordheim transport behavior in the bias range of 0 to 1.0 Volt and thus determining that the barrier height is larger than the applied bias, i.e., $\Phi_B > 1.0$ eV. This study is restricted to applied biases ≤ 1.0 Volt and the transition from direct to Fowler–Nordheim tunneling requires higher bias.

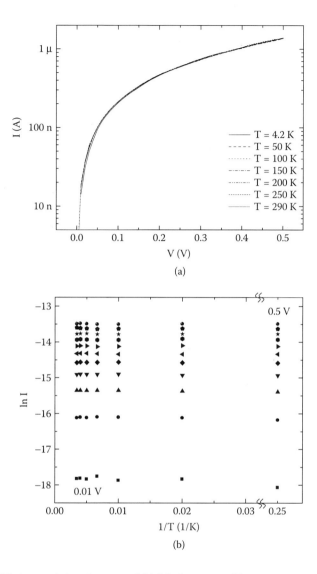

FIGURE 1.10 (a) I(V,T) characteristics of an octanedithiol device measured from room temperature to 4.2 K (plotted on a log scale). (b) Arrhenius plot generated from the I(V,T) data in (a) at voltages from 0.1 to 0.5 Volt with 0.05 Volt steps.

I(V,T) characterizations have also been done on other alkane molecules. As an example, Figure 1.10(a) shows the I(V,T) measurement of an octanedithiol device from 290 to 4.2 K. As the corresponding Arrhenius plot [Figure 1.10(b)] exhibits, there is no thermal activation involved, confirming that the conduction through alkane SAMs is tunneling.

As discussed in the previous section, temperature-variable I(V) measurement is a very important experimental method in molecular transport characterizations. This importance is demonstrated by Figure 1.11. Figure 1.11(a) shows a room-temperature I(V) characteristic of a device containing C8 molecules. The shape of this I(V) looks very similar to that of a direct tunneling device. Indeed, it can be fit using the Simmons model (see the next subsection), which gives a barrier height of 1.27 eV and an α of 0.96 (though a larger value; see the next section). However, further I(V,T) measurements display an obvious temperature dependence [Figure 1.11(b)], which can be fit well to a hopping conduction model (Table 1.1) with a well-defined activation energy of 190 meV, as illustrated by Figure 1.11(c). Another example is shown in

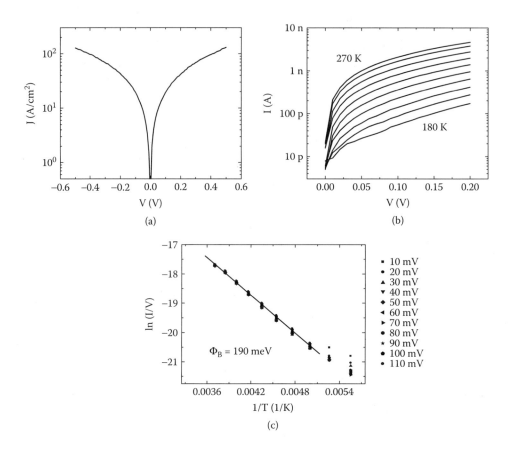

FIGURE 1.11 (a) I(V) characteristic of a C8 device at 270 K. (b) Temperature dependence of the same device from 270 to 180 K (in 10-K increments). (c) Plot of ln(I/V) versus 1/T at various voltages. The line is the linear fitting, and a hopping barrier of 190 meV is determined from this fitting.

Figure 1.12: Figure 1.12(a) shows the I(V) for a C12 device measured at 4.2 K, while Figure 1.12(b) is the corresponding numerical differential conductance. Instead of displaying a direct tunneling conduction, this device exhibits a Coulomb blockade behavior with an energy gap of $\sim$ 60 meV, which corresponds to a device capacitance of 3×10^{-18}F. These impurity-mediated transport phenomena are indicative of the unintentional incorporation of a trap or defect level in the devices and I(V,T), and subsequent IETS characterizations are needed to discover the correct conduction mechanism.

1.4.4.2 I(V) Fitting Using the Simmons Model

Having established tunneling as the main conduction mechanism of alkanethiols, we can now obtain the transport parameters, such as the effective barrier height, by comparing our experimental I(V) data with theoretical calculations from a tunneling model.

The current density (J) expression in the direct tunneling regime ($V < \Phi_B/e$) from the Simmons model is expressed as [Equation (1.1); 36,91]:

$$J = \left(\frac{e}{4\pi^2\hbar\, d^2}\right)\left\{\left(\Phi_B - \frac{eV}{2}\right)\exp\left[-\frac{2(2m)^{1/2}}{\hbar}\alpha\left(\Phi_B - \frac{eV}{2}\right)^{1/2}d\right]\right.$$
$$\left. -\left(\Phi_B + \frac{eV}{2}\right)\exp\left[-\frac{2(2m)^{1/2}}{\hbar}\alpha\left(\Phi_B + \frac{eV}{2}\right)^{1/2}d\right]\right\} \tag{1.6}$$

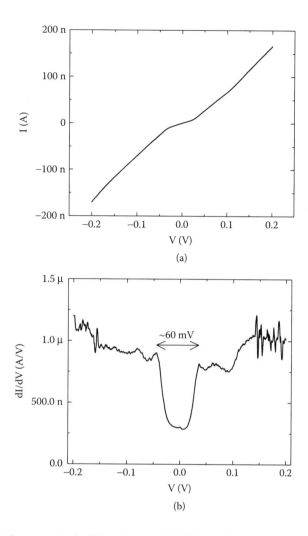

FIGURE 1.12 (a) I(V) characteristic of a C12 device at 4.2 K. (b) Numerical derivative of the I(V) in (a) exhibits a gap due to the Coulomb blockade effect.

For molecular systems, the Simmons model has been modified with a unitless adjustable parameter α [36,78,91]. α is introduced to account for the effective mass ($m^\star$) of the tunneling electrons through a molecular wire. $\alpha = 1$ corresponds to the case of a bare electron, which previously has been shown not to fit I(V) data well for some alkanethiol measurements at fixed temperature (300 K) [91]. By fitting individual I(V) data using Equation (1.6), Φ_B and α can be found.

Equation (1.6) can be approximated in two limits: low bias and high bias, as compared with the barrier height Φ_B. In the low bias range, Equation (1.6) can be approximated as:

$$J \approx \left(\frac{(2m\Phi_B)^{1/2} e^2 \alpha}{h^2 d} \right) V \exp \left[-\frac{2(2m)^{1/2}}{\hbar} \alpha \, (\Phi_B)^{1/2} \, d \right]. \qquad (1.7)$$

To determine the high bias limit, we compare the relative magnitudes of the first and second exponential terms in Equation (1.6). At high bias, the first term is dominant and thus the current density can be approximated as

$$J \approx \left(\frac{e}{4\pi^2\hbar \, d^2} \right) \left(\Phi_B - \frac{eV}{2} \right) \exp\left[-\frac{2(2m)^{1/2}}{\hbar}\alpha \left(\Phi_B - \frac{eV}{2} \right)^{1/2} d \right] \qquad (1.8)$$

The tunneling currents in both bias regimes have exponential dependence on the barrier width d. In the low bias regime, the tunneling current density is $J \propto \frac{1}{d} \exp(-\beta_0 d)$, where β_0 is a bias-independent decay coefficient:

$$\beta_0 = \frac{2(2m)^{1/2}}{\hbar}\alpha \, (\Phi_B)^{1/2} \qquad (1.9)$$

While in the high bias regime $J \propto \frac{1}{d^2} \exp(-\beta_V d)$, where β_V is a bias-dependent decay coefficient:

$$\beta_V = \frac{2(2m)^{1/2}}{\hbar}\alpha \left(\Phi_B - \frac{eV}{2} \right)^{1/2} = \beta_0 \left(1 - \frac{eV}{2\Phi_B} \right)^{1/2} \qquad (1.10)$$

At high bias, β_V decreases as bias increases, which results from a barrier lowering effect due to the applied bias.

The preceding distinction between low and high bias in the direct tunneling regime may seem unnecessary at first. However, it is needed to clarify the confusion and misleading conclusions present in current literature and to deduce the decay coefficient expressions from a solid tunneling model. For example, in previous publications [78,80], the expression of the decay coefficient β_V [Equation (1.10)] has been postulated and applied in the entire bias range from 0 to 1 [78] and 3 Volts [80], which are incorrect according to Equation (1.6) since over these bias ranges there is no simple exponential dependence of $J \propto \exp(-\beta_V d)$. In another published report [92], the correct Simmons equation [Equation (1.6)] has been utilized to fit the measured I(V) data, but again the β_V expression is used for the whole bias range. Some groups [27,77–80,91] used the general quantum mechanical exponential law

$$G = G_0 \exp(-\beta \, d) \qquad (1.11)$$

to analyze the length dependence behavior of the tunneling current, but this equation is incapable of explaining the observed bias dependence of the decay coefficient β. On the contrary, in our study the Simmons equation (1.6) is used to fit the I(V) data in the direct tunneling regime and is reduced to Equation (1.7) in the low bias range to yield a similar bias-independent decay coefficient as Equation (1.11). While in the high bias range, the exponential term of $e^{-C\sqrt{\Phi_B - \frac{eV}{2}}}$ in Equation (1.6) dominants, and thus Equation (1.6) is approximated by Equation (1.8), giving a bias-dependent coefficient β_V. This distinction between the low and high biases will be seen to explain the experimental data very well in a later subsection.

Using the modified Simmons Equation (1.6), by adjusting two parameters Φ_B and α a nonlinear least squares fitting has been performed on the measured C12 I(V) data. The tunneling gap distance is the length of the adsorbed alkanethiol molecule, which is determined by adding an Au-thiol bonding length of 2.3 Å to the length of the free molecule [77]. For C12, the length (therefore the gap distance) is calculated as 18.2 Å. By using a device size of 45 nm in diameter, the best fitting parameters (minimizing χ^2) for the room temperature C12 I(V) data were found to be $\Phi_B = 1.42 \pm 0.04$ eV and $\alpha = 0.65 \pm 0.01$, where the error ranges of Φ_B and α are dominated by potential device size fluctuations of 2 nm. Figure 1.13(a) shows this best-fitting result (solid curve) as well as the original I(V) data (circular symbol) on a linear scale. A calculated I(V) for α 1 and $\Phi_B = 0.65$ eV (which gives the best fit at low bias range) is shown as the dashed curve in the same figure, illustrating that with α1 only a limited region of the I(V) curve can be fit (specifically here, for $|V| < 0.3$ Volt). The same plots are shown on a log scale in Figure 1.13(b). The value of the fitting parameter αs obtained earlier, corresponds to an effective mass m* ($= \alpha^2$ m) of 0.42 m.

Likewise, I(V) measurements have also been performed on octanethiol (C8) and hexadecanethiol (C16) SAMs. The Simmons fitting on C8 with an adsorbed molecular length of 13.3 Å (tunneling gap distance)

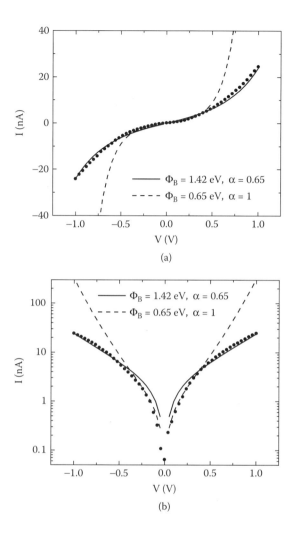

FIGURE 1.13 Measured C12 I(V) data (circular symbol) is compared with calculation (solid curve) using the optimum fitting parameters of $\Phi_B = 1.42$ eV and $\alpha = 0.65$. The calculated I(V) from a simple rectangular model ($\alpha = 1$) with $\Phi_B = 0.65$ eV is also shown as the dashed curve. Current is plotted on (a) linear scale and (b) log scale.

and a device diameter of 50 ± 8 nm yields values of $\{\Phi_B = 1.83 \pm 0.10$ eV and $\alpha = 0.61 \pm 0.01\}$. Same fitting on C16 with a length of 23.2 Å and a device diameter of 45 ± 2 nm gives a data set of $\{\Phi_B = 1.40 \pm 0.03$ eV, $\alpha = 0.68 \pm 0.01\}$. The I(V) data and fitting results are shown in Figure 1.14(a) and (b) for C8 and C16, respectively.

Nonlinear least square fittings on C12 I(V) data at different measurement temperatures allow us to determine $\{\Phi_B, \alpha\}$ over the entire temperature range (300 to 80 K) and the fitting results show that Φ_B and α values are temperature-independent. From these fittings, average values of $\Phi_B = 1.45 \pm 0.02$ eV and $\alpha = 0.64 \pm 0.01$ are obtained $[1\sigma_M$ (standard error)].

In order to investigate the dependence of the Simmons model fitting on Φ_B and α, a fitting minimization analysis is undertaken on the individual Φ_B and α values, as well as their product form of $\alpha\Phi_B^{1/2}$ in Equation (1.9). $\Delta(\Phi_B, \alpha) = (\Sigma |I_{exp,V} - I_{cal,V}|^2)^{1/2}$ is calculated and plotted, where $I_{exp,V}$ is the experimental current value and $I_{cal,V}$ is the calculated one from Equation (1.6). A total of 7500 different $\{\Phi_B, \alpha\}$ pairs are used in the analysis with Φ_B, ranging from 1.0 to 2.5 eV (0.01 eV increment) and α from 0.5 to 1.0 (0.01 increment). Figure 1.15(a) is a representative contour plot of $\Delta(\Phi_B, \alpha)$ versus Φ_B and α generated for the C12 I(V)

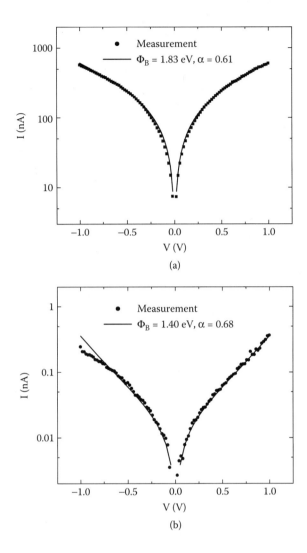

FIGURE 1.14 (a) Measured C8 I(V) data (symbol) are compared with calculations (the solid curve) using the optimum fitting parameters of $\Phi_B = 1.83$ eV and $\alpha = 0.61$. (b) Measured C16 I(V) data (symbol) are compared with calculations (the solid curve) using the optimum fitting parameters of $\Phi_B = 1.40$ eV and $\alpha = 0.68$.

data where darker regions represent smaller $\Delta(\Phi_B, \alpha)$, and various shades correspond to $\Delta(\Phi_B, \alpha)$ with half order-of-magnitude steps. The darker regions also represent better fits of Equation (1.6) to the measured I(V) data. In the inset in Figure 1.15(a), one can see there is a range of possible Φ_B and α values yielding good fittings. Although the tunneling parameters determined from the previous Simmons fitting $\{\Phi_B = 1.42$ eV and $\alpha = 0.65\}$ lie within this region, there is also a distribution of other possible values.

A plot of $\Delta(\Phi_B, \alpha)$ versus $\alpha\Phi_B^{1/2}$ is shown in Figure 1.15(b). As it exhibits, except for the minimum point of $\Delta(\Phi_B, \alpha)$, different Φ_B and α pairs could give the same $\Delta(\Phi_B, \alpha)$ value. For this plot, the $\Delta(\Phi_B, \alpha)$ is minimized at $\alpha\Phi_B^{1/2}$ of 0.77 (eV)$^{1/2}$, which yields a β_0 value of 0.79 Å^{-1} from Equation (1.9). The C8 and C16 devices show similar results, confirming that the Simmons fitting has a strong $\alpha\Phi_B^{1/2}$ dependence. For the C8 device, although Φ_B obtained from the fitting is a little larger, combined α and Φ_B give a similar β_0 value within the error range as the C12 and C16 devices. The values of Φ_B and α for C8, C12, and C16 devices are summarized in Table 1.2, as well as the β_0 values calculated from Equation (1.9).

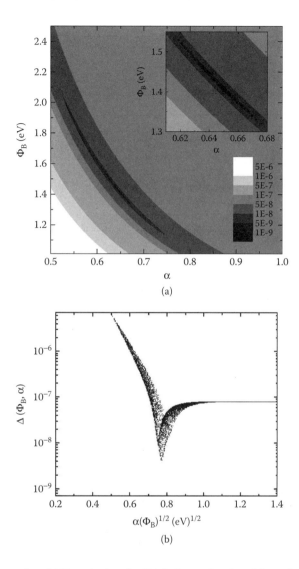

FIGURE 1.15 (a) Contour plot of $\Delta(\Phi_B, \alpha)$ values for C12 device as a function of Φ_B and α, where the darker region corresponds to a better fitting. Inset shows detailed minimization fitting regions. (b) Plot of $\Delta(\Phi_B, \alpha)$ as a function of $\alpha\Phi_B^{1/2}$.

TABLE 1.2 Summary of Alkanethiol Tunneling Parameters Obtained Using the Simmons Model

Molecules	J at 1 V (A/cm^2)	Φ_B (eV)	α	m* (m)	β_0 (Å^{-1})
C8	$31{,}000 \pm 10{,}000$	1.83 ± 0.10	0.61 ± 0.01	0.37	0.85 ± 0.04
C12	$1{,}500 \pm 200$	1.42 ± 0.04	0.65 ± 0.01	0.42	0.79 ± 0.02
C16	23 ± 2	1.40 ± 0.03	0.68 ± 0.01	0.46	0.82 ± 0.02

1.4.4.3 Length Dependence of the Tunneling Current through Alkanethiols

As discussed in Subsection 1.4.4.2 [Equations (1.7) and (1.8)], the tunneling currents in the low and high bias ranges have an exponential dependence on the molecular length as $J \propto \frac{1}{d} \exp(-\beta_0 d)$ and $J \propto \frac{1}{d^2} \exp(-\beta_V d)$, respectively, where β_0 and β_V are the decay coefficients. In order to study this length-dependent tunneling behavior, I(V) characterizations are performed on three alkanethiols of different

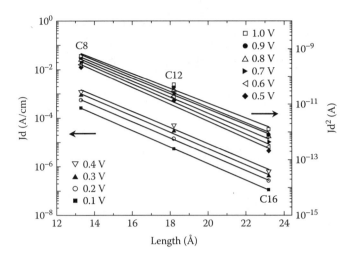

FIGURE 1.16   Log plot of tunneling current densities (symbols) multiplied by molecular length d at low bias and by d^2 at high bias versus molecular length. The lines through the data points are linear fittings.

molecular length: C8, C12, and C16 [Figure 1.6(b)]. The adsorbed molecular length of C8, C12, and C16 are 13.3, 18.2, and 23.2 Å, respectively, as used in the Simmons fitting. To define the boundary of the high and low bias ranges, the relative magnitudes of the first and second exponential terms in Equation (1.6) are evaluated. Using $\Phi_B = 1.42$ eV and $\alpha = 0.65$ obtained from nonlinear least squares fitting of the C12 I(V) data, the second term becomes less than $\sim 10\%$ of the first term at 0.5 Volt, which is chosen as the bias boundary.

Figure 1.16 is a semi-log plot of the tunneling current density multiplied by molecular length—Jd at low bias and Jd^2 at high bias—as a function of the molecular length for these alkanethiols. As seen in this figure, the tunneling currents (symbols) show exponential dependence on molecular lengths. The decay coefficient β can be determined from the slopes of the linear fittings (lines in Figure 1.16) on the measured data. The obtained β values at each bias are plotted in Figure 1.17(a) and the error bar of an individual β value in this plot is determined by considering both the device size uncertainties and the linear fitting errors.

As Figure 1.17(a) shows, in the low bias range (V < 0.5 V) the β values are almost independent of bias, while in the high bias range (V > 0.5 V)β has bias dependence: β decreases as bias increases due to the barrier lowering effect. From Figure 1.17(a), an average β_w of $0.77 \pm 0.06\,\text{Å}^{-1}$ can be calculated in the low bias region. According to Equation (1.10), β_V^2 depends on bias V linearly in the high bias range. Figure 1.17(b) is a plot of β_V^2 versus V in this range (0.5 to 1.0 Volt) and a linear fitting of the data. $\Phi_B = 1.35 \pm .20$ eV and $\alpha = 0.66 \pm 0.04$ are obtained from the intercept and slope of this fitting, respectively, which are consistent with the values acquired from the nonlinear least squares fitting on the I(V) data in the previous subsection.

Table 1.3 is a summary of previously reported alkanethiol transport parameters obtained by different techniques [28]. The current densities (J) listed in Table 1.3 are for C12 monothiol or dithiol devices at 1 V, which are extrapolated for some techniques from published results of other length alkane molecules using the exponential law of Equation (1.11). The large variation of J among reports can be attributed to the uncertainties in device contact geometry and junction area, as well as complicating inelastic or defect contributions. The β_o value ($0.77 \pm 0.06\,\text{Å}^{-1} \approx 0.96 \pm 0.08$ per methylene) for alkanethiols obtained in our study using the Simmons model is comparable to previously reported values as summarized in Table 1.3.

Length-dependent analysis using the exponential equation (1.11) in the entire applied bias range (0 to 1.0 V) has also been performed in order to compare with these reported β values. This gives

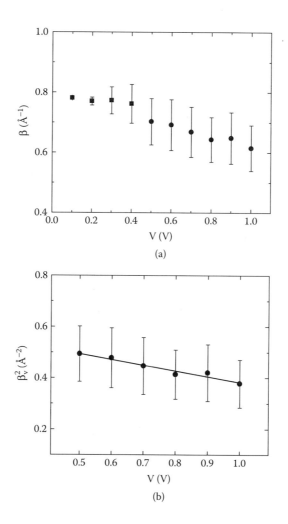

FIGURE 1.17 (a) Plot of β versus bias in the low bias range (square symbol) and high bias range (circular symbol). (b) β_V^2 versus bias plot (symbol) with a linear fitting (solid curve).

β values from 0.84 to 0.73 Å^{-1} in the bias range from 0.1 to 1.0 Volt, which are similar to the reported values. For example, Holmlin *et al.* reported a β value of 0.87 Å^{-1} by mercury-drop experiments [91], Wold *et al.* have reported β of 0.94 Å^{-1}, and Cui *et al.* reported β of 0.64 Å^{-1} for various alkanethiols by using a conducting atomic force microscope technique [79,80]. However, these reported β were treated as bias-independent quantities, contrary to the results from our study and those observed in a slightly different alkane system (ligand-encapsulated nanoparticle/alkane-dithiol molecules) [78]. Since all of these experiments were performed at room temperature, the reported parameters have not been checked with a temperature-dependent analysis and non-tunneling components can dramatically affect the derived values.

1.4.4.4 I(V) Fitting Using the Franz Model

We can also analyze our experimental data using the Franz two-band model [37–42]. By considering the contributions from both the conduction band and valence band, the Franz model empirically predicted a non-parabolic E(k) relationship inside the bandgap, as expressed by Equation (1.4). By using this equation, the effective mass of the tunneling electron can be deduced by knowing the barrier height of the tunnel junction [41,109]. However, since there is no reliable experimental data on the Fermi level alignment in the

TABLE 1.3 Summary of Alkanethiol Tunneling Parameters Obtained by Different Test Structures

Junction	β (Å^{-1})	J (A/cm^2) at 1 V	Φ_B (eV)	Technique	Ref.
(bilayer) monothiol	0.87 ± 0.1	25–200[a)]	2.1[e)]	Hg-junction	91
(bilayer) monothiol	0.71 ± 0.08	0.7–3.5[a)]		Hg-junction	102
monothiol	0.79 ± 0.01	1500 ± 200[b)]	1.4[e)]	Solid M-I-M	24
monothiol	1.2			STM	70
dithiol	0.8 ± 0.08	$3.5–5 \times 10^{5}$[c)]	5 ± 2[f)]	STM	73
monothiol	0.73–0.95	1100–1900[d)]	2.2[e)]	CAFM	77
monothiol	0.64–0.8	10–50[d)]	2.3[e)]	CAFM	78
dithiol	0.46 ± 0.02	$3–6 \times 10^{5}$[c)]	1.3–1.5[e)]	CAFM	80
monothiol	1.37 ± 0.03		1.8[f)]	Tuning fork AFM	141
monothiol	0.97 ± 0.04			Electrochemical	104
monothiol	0.85			Electrochemical	105
monothiol	0.91 ± 0.08			Electrochemical	106
monothiol	0.76	2×10^4 (at 0.1 V)	1.3–3.4[g)]	Theory	122
monothiol	0.76			Theory	142
monothiol	0.79			Theory	143

Note:
(1) Some decay coefficient β are converted into the unit of Å^{-1} from the unit of per methylene.
(2) The junction areas are estimated by an optical microscope[a)], SEM[b)], assuming a single molecule[c)], and Hertzian contact theory[d)].
(3) Current densities (J) for C12 monothiol or dithiol at 1 V are extrapolated from published results of other length molecules using the exponential law of G $\propto$ exp($-\beta$b).
(4) Barrier height values are obtained from Simmons equation[e)], bias-dependence of β[f)], and theoretical calculation[g)].

Au-alkanethiol SAM-Au system, Φ_B is unknown and is thus treated as an adjustable parameter together with m* in our analysis. The imaginary k value is related to the decay coefficient β [$k^2 = -(\beta/2)^2$] obtained from the length-dependent study. Using an alkanethiol HOMO–LUMO gap of 8 eV, a least squares fitting has been performed on the experimental data, and Figure 1.18 shows the resultant E(k) relationship and the corresponding energy band diagrams. The zero of energy in this plot is chosen as the LUMO energy. The best fitting parameters obtained by minimizing χ^2 are $\Phi_B = 1.49 \pm 0.51$ eV and m* = 0.43 ± 0.15 m,

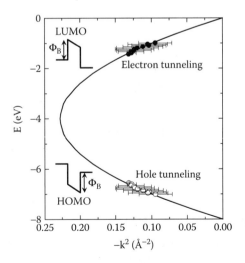

FIGURE 1.18 E(k) relationship generated from the length-dependent measurement data of alkanethiols. Solid and open symbols correspond to electron and hole conductions, respectively. The solid curve is the Franz two-band E(k) plot for m* = 0.43 m and E_g = 8 eV. The insets show the corresponding energy band diagrams.

where the error ranges of Φ_B and m* are dominated by the error fluctuations of β. Both electron tunneling near the LUMO, and hole tunneling near the HOMO can be described by these parameters. $\Phi_B = 1.49$ eV indicates that the Fermi level is aligned close to one energy level in either case. The Φ_B and m* values obtained here are in reasonable agreement with previous results deduced from the Simmons model.

1.5 Inelastic Electron Tunneling Spectroscopy of Alkanethiol SAMs

1.5.1 A Brief Review of IETS

As discussed previously, IETS was discovered by Jaklevic and Lambe in 1966 when they studied tunnel junctions containing organic molecules and the vibrational modes of the molecules were detected by electrons that tunneled inelastically through the barrier [29,43]. In the earlier stage of IETS, the tunnel barrier was usually made of a metal oxide, therefore the choice of the metallic material was crucial since it must be capable of forming a coherent and stable oxide layer with a thickness of several nanometers [30,31,44]. For this purpose, aluminum was often utilized because of its good oxide quality. The molecular species were then introduced by either vapor phase exposure or liquid solution deposition on the surface of the barrier. Care also needed to be taken for top electrode deposition since high temperature evaporation may destroy the adsorbed molecular layer [44]. IETS has been mostly used in the spectra range of 0 to 500 meV (0 to 4000 cm^{-1}), which covers almost all molecular vibrational modes [30,31,44].

In the 1990s, another type of tunneling barrier was reported for IETS measurements [88]. This so-called cross-wire structure replaces the metal oxide barrier with an inert gas film. In order to form the tunnel junction, molecular species are mixed with the inert gas at a predetermined composition, and then are introduced into the vacuum chamber, where they then condense on the wire surfaces [88,110]. Recently, this test structure has been used again for the investigation of vibronic contributions to charge transport across molecular junctions [111]. However, due to the difficulties in controlling the exact position of the top wire, the top wire might not touch the other end of the molecules to form a perfect metal–SAM–metal junction, or it might penetrate into the monolayer. Besides, no temperature-dependent measurement has been reported using this structure.

Another important advance in this field is the realization of single molecular vibrational spectroscopy by STM-IETS [112]. The possibility of performing IETS studies utilizing STM was discussed soon after its invention [67]. However, due to the difficulties in achieving the extreme mechanical stability that is necessary to observe small changes in tunneling conductance, this technique has only been realized recently [112]. In the STM implementation of IETS, the metal–oxide–metal tunnel junction is replaced by an STM junction consisting of a sharp metallic tip, a vacuum gap, and a surface with the adsorbed molecules. Using STM-IETS, imaging and probing can be performed at the same time, and vibrational spectroscopy studies on a single molecule can be achieved [112].

The advantage of inelastic tunneling spectroscopy over conventional optical vibrational spectroscopy such as IR and Raman is its sensitivity [30,31]. IR spectroscopy is a well-developed technology and has been used widely for studies of adsorbed species. It does not require cryogenic temperature measurement and can be applied to a variety of substrates [30]. Raman spectroscopy is used when IR is difficult or impossible to perform, such as for seeing vibrations of molecules in solvents that are infrared opaque or for vibrations that are not infrared active [30]. Both IR and Raman have lower sensitivities compared with IETS: they require 10^3 or more molecules to provide a spectrum. Since the interaction of electrons with molecular vibrations is much stronger than that of photons, as small as one monolayer of molecules is enough to produce good IETS spectra [44,45]. Additionally, IETS is not subject to the selection rules of infrared or Raman spectroscopy. It has an orientational preference, as discussed earlier, but there are no rigorous selection rules. Both IR and Raman active vibrational modes appear in IETS spectra with comparable magnitudes [44,45].

After its discovery, IETS found many applications in different areas such as surface chemistry, radiation damage, and trace substance detection, among many others [30,31]. It is a powerful spectroscopic tool for chemical identification purposes: the vibrational spectra can be used as fingerprints to identify the molecular species confined inside a tunnel junction. It can also be used for chemical bonding investigations—in a solid state junction, the breaking of various bonds can be monitored by the decrease in intensities of the corresponding vibrational peaks, and the formation of new bonds can be monitored by the growth of new vibrational peaks [30,31]. The application of IETS has branched out to the modern silicon industry as well, where it is utilized to study phonons in silicon, the nature of the SiO_2 tunneling barrier, interface states in metal-oxide-semiconductor (MOS) systems, and high-k dielectrics [113,114].

In our study, IETS is utilized to identify the molecular species confined inside a solid state junction [25]. The measurement is performed using the nanopore test structure. Unlike earlier tunnel junctions, the nanopore uses the self-assembled molecules themselves as the tunnel barrier; thus, it creates oxide-free junctions, and intrinsic molecular properties can be investigated. Because the tunneling current depends exponentially on the barrier width, in the cross-wire and STM tunnel junctions a small change in the tunneling gap distance caused by vibration of the top electrode can produce a large change in the junction conductance, which can mask the conductance change associated with the inelastic channels. Compared to these systems, the nanopore structure has direct metal-molecule contacts and a fixed top electrode, and both ensures intrinsic contact stability and eliminates the preceding problems. The molecular species used are the "control" molecules—alkane SAMs—which have been shown to form good insulating layers and present well-defined tunnel barriers in previous studies.

1.5.2 Alkanethiol Vibrational Modes

Various spectroscopic techniques have been developed to help chemists investigate the chemical structures of molecules and to study their interactions. These include mass spectrometry, nuclear magnetic resonance (NMR) spectroscopy, infrared (IR) spectroscopy, ultraviolet (UV) spectroscopy, Raman spectroscopy, and high-resolution electron energy loss spectroscopy (HREELS) [115–119]. The majority of these spectroscopic tools analyze molecules based on differences in how they absorb electromagnetic radiation [116]. A very important concept in molecular spectroscopies is the so-called group frequency. A molecule usually consists of many atoms, and even though these atoms will move during a normal mode of vibration, most of the motion can be localized within a certain molecular fragment that vibrates with a characteristic frequency. Thus, the existence of a functional group can be inferred by the appearance of an absorption band in a particular frequency range. In other words, we can detect the presence of a specific functional group in a molecule by identifying its characteristic frequency [116,117]. By identifying individual functional groups in a molecule, we can determine the molecule's chemical composition.

As for the case of alkanethiol molecules, the important vibrational modes include the stretching modes of C-C and C-S groups and various vibrations of the CH_2 group. Figure 1.19 illustrates the available CH_2 group vibrational modes, which include the symmetric and antisymmetric stretching modes, in-plane scissoring and rocking modes, and out-of-plane wagging and twisting modes [25,116]. Each of the different vibrational modes gives rise to a characteristic frequency in a spectroscopic spectrum.

Vibrational structures of self-assembled alkanethiols on Au(111) surface have been investigated by spectroscopic tools such as IR, Raman, and HREELS, and a large literature exists on the subject. References 118–121 are representative publications in this field. For example, IR measurement was conducted at the earlier stage to characterize the packing and orientation of the alkanethiol SAMs formed on the Au(111) surface. The results suggest they are densely packed in a crystalline arrangement [26, 62]. It has also been used by Castiglioni *et al.* to study the CH_2 rocking and wagging vibrations and to obtain related characteristic group frequencies [120]. Using Raman spectroscopy, Bryant *et al.* have investigated the C-C stretching bands of alkanethiols on Au surfaces since these bands are weak in the IR spectra. They have also characterized other vibrational features such as the C-S, S-H, and C-H stretching modes [121]. Duwez *et al.* and Kato *et al.* utilized HREELS to study various vibrational structures of alkanethiol SAMs and the Au-S bonding [118,119].

Table 1.4 is a summary of the alkanethiol vibrational modes obtained using the aforementioned spectroscopic methods [25,118–121]. In this table, the symbols of $\delta_{s,r}$ and $\gamma_{w,t}$ denote in-plane scissoring (s)

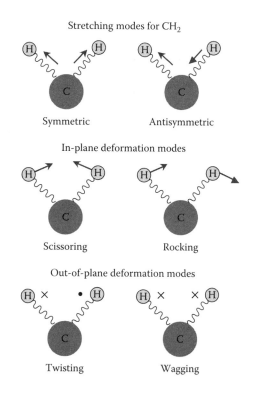

Stretching modes for CH$_2$

Symmetric   Antisymmetric

In-plane deformation modes

Scissoring   Rocking

Out-of-plane deformation modes

Twisting   Wagging

FIGURE 1.19   CH$_2$ vibrational modes. After Ref. 115.

TABLE 1.4   Summary of Alkanethiol Vibrational Modes Obtained from IR, Raman, and HREELS

Modes	Methods	Wavenumber (cm^{-1})	(meV)
ν(Au-S)	HREELS	225	28
ν(C-S)	Raman	641	79
	Raman	706	88
δ_r(CH$_2$)	HREELS	715	89
	IR	720	89
	IR	766	95
	IR	925	115
ν(C-C)	HREELS	1050	130
	Raman	1064	132
	Raman	1120	139
$\gamma_{w,t}$(CH$_2$)	IR	1230	152
	HREELS	1265	157
	IR	1283	159
	IR	1330	165
δ_s(CH$_2$)	HREELS	1455	180
ν(S-H)	Raman	2575	319
ν_s(CH$_2$)	Raman	2854	354
	HREELS	2860	355
ν_{as}(CH$_2$)	Raman	2880	357
	Raman	2907	360
	HREELS	2925	363

After Refs. 118–121.

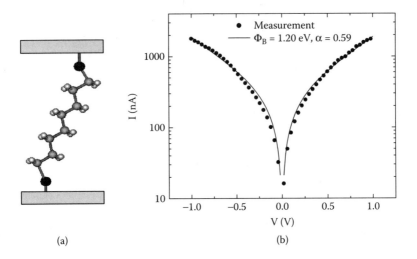

(a)   (b)

FIGURE 1.20   (a) Schematic of an octanedithiol device. (b) I(V) measurement data at room temperature (circular symbol) and the fitting from Simmons equation (solid curve).

and rocking (r) and out-of-plane wagging (w) and twisting (t) modes, respectively. ν and $\nu_{s,as}$ denote tretching and CH$_2$ group symmetrical (s) and antisymmetrical (as) stretching modes, respectively. These characteristic group frequencies will be compared to the signal peaks in our acquired IETS spectra to identify the molecular species confined in the device junction.

1.5.3   IETS of Octanedithiol SAM

Electrical measurements on octanedithiol SAM are performed with the nanopore structure discussed earlier. The molecular solution is prepared by adding $\sim$ 10 μL octanedithiol to 10 mL ethanol. SAM formation is carried out for 24 hours inside a nitrogen-filled glove box with an oxygen level of less than 5 ppm. Figure 1.20(a) shows the schematic of the device configuration. I(V,T) measurement from 4.2 to 290 K shows a tunneling transport behavior (see Figure 1.10). Figure 1.20(b) is the room temperature I(V) measurement result with the fitting from the Simmons equation. Using a junction area of 51 $\pm$ 5 nm in diameter obtained from statistical studies of the nanopore size with SEM, a current density of (9.3 ± 1.8) $\times$ 10^{4}A/cm^2 at 1.0 Volt is calculated. Using the modified Simmons model [Equation (1.6)], the transport parameters of $\Phi_B = 1.20 \pm 0.03$ eV and $\alpha = 0.59 \pm 0.01$ (m* = 0.34 m) are obtained for this C8 dithiol device. As a comparison, the C8 monothiol device used in the length-dependent study has a current density of $(3.1 \pm 1.0) \times 10^4$A/cm^2 at 1.0 Volt, a barrier height of 1.83$\pm$0.10 eV, and an α of 0.61$\pm$0.01 (m* = 0.37 m). That the observed current density of the C8 dithiol device is approximately three times larger than that of monothiol is consistent with previously published theoretical calculations and experimental data [122]. For example, Kaun *et al.* performed first-principle calculations on alkane molecules in a metal–SAM–metal configuration using nonequilibrium Green's functions combined with density functional theory [122]. They found that in an Au-alkanedithiol-Au device, although both Au leads are contacted by a sulfur atom, the transport behavior is essentially the same as that of an alkanemonothiol device where only one Au lead is contacted by sulfur. However, the current through alkanedithiols is found to be approximately ten times larger than that through alkanemonothiols, which, they suggest, indicates that the extra sulfur atom provides a better coupling between the molecule and the lead [122]. Experimental measurement on alkanedithiol molecules has also been performed by Cui *et al.* using the conducting AFM technique, and the result shows that alkanedithiol has $\sim$ 100 times larger current than alkanemonothiol has [78,80].

IETS measurements are performed on the molecular devices using the lock-in technique. The second harmonic signal (proportional to d^{2}I/dV^2) is directly measured with a lock-in amplifier, which has also been

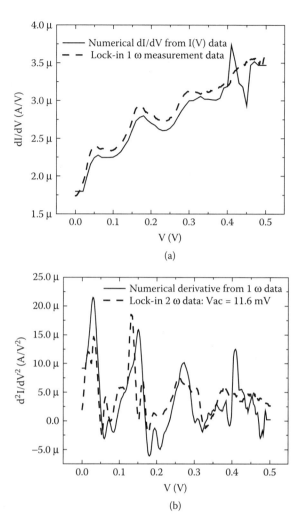

FIGURE 1.21 (a) Lock-in 1 ω data and the numerical dI/dV obtained from I(V) measurement data. (b) Lock-in 2 ω data and the numerical derivative of the lock-in 1 ω data in (a). All measurement data are taken at 4.2 K.

checked to be consistent with the numerical derivative of the first harmonic signal. As an example, Figure 1.21(a) shows the lock-in first harmonic measurement data compared with the numerical derivative of the I(V) of the C8 dithiol device, while Figure 1.21(b) is the second harmonic measurement result checked with the numerical derivative of the first harmonic signal (all of the data are taken at 4.2 K). As Figure 1.21(b) demonstrates, the IETS spectrum calculated from the numerical differential method is compatible with that obtained from the lock-in second harmonic measurement; however, the lock-in measurement yields a much more resolved spectrum.

Figure 1.22 shows the inelastic electron tunneling spectrum of the same C8 dithiol SAM device obtained at T = 4.2 K. An AC modulation of 8.7 mV (rms value) at a frequency of 503 Hz is applied to the sample to acquire the second harmonic signals. The spectra are stable and repeatable upon successive bias sweeps. The spectrum at 4.2 K is characterized by three pronounced peaks in the 0 to 200 mV region at 33, 133, and 158 mV. From comparison with previously reported IR, Raman, and HREEL spectra of alkanethiol SAMs on Au(111) surfaces (Table 1.4) [118–121], these three peaks are assigned to Au-S stretching, C-C stretching, and CH$_2$ wagging modes of a surface bound alkanethiolate. The absence of a strong S-H stretching signal at $\sim$ 329 mV suggests that most of the thiol groups have reacted with the gold bottom

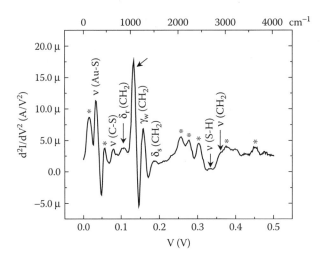

FIGURE 1.22 Inelastic electron tunneling spectrum of a C8 dithiol SAM obtained from lock-in second harmonic measurement with an AC modulation of 8.7 mV (rms value) at a frequency of 503 Hz (T = 4.2 K). Peaks labeled * are most probably background due to the encasing Si_3N_4.

and top contacts. Peaks are also reproducibly observed at 80, 107, and 186 mV. They correspond to C-S stretching, CH_2 rocking, and CH_2 scissoring modes. The stretching mode of the CH_2 groups appears as a shoulder at 357 meV. The peak at 15 mV is due to vibrations from either Si, Au, or δ(C-C-C) since all three materials have characteristic frequencies in this energy range [123,125]. We note that all alkanethiol peaks without exception or omission occur in the spectra. Peaks at 58, 257, 277, and 302, as well as above 375 mV are likely to originate from Si-H and N-H vibrations related to the silicon nitride membrane [123,126,127], which forms the SAM encasement. Measurement of the background spectrum from Si_3N_4 of an "empty" nanopore device with only gold contacts is hampered by either too low (open circuit) or too high (short circuit) currents in such a device.

According to the IETS theory [128], molecular vibrations with net dipole moments perpendicular to the tunneling junction interface have stronger peak intensities than vibrations with net dipole moments parallel to the interface. In our device configuration [Figure 1.20(a)], the vibrational modes of Au-S, C-S, and C-C stretching and CH_2 wagging are perpendicular to the junction interface, while the vibrations of the CH_2 group rocking, scissoring, and stretching modes are parallel to the interface. In the obtained IETS spectrum (Figure 1.22), the vibrations perpendicular to the junction interface produce peaks of stronger intensities, while those vibrations parallel to the interface generate less dominant peaks. This experimental observation of the relative IETS peak intensities is in good agreement with the theory.

1.5.4 Spectra Linewidth Study

In order to verify that the obtained spectra are indeed valid IETS data, the peak width broadening effect is examined as a function of temperature and applied modulation voltage. IETS measurements have been performed with different AC modulations at a fixed temperature, and at different temperatures with a fixed AC modulation. Figure 1.23 shows the modulation dependence of the IETS spectra obtained at T = 4.2 K, and the modulation voltages used are 11.6, 10.2, 8.7, 7.3, 5.8, 2.9, and 1.2 mV (rms values). According to theoretical analysis, AC modulation will bring in a linewidth broadening of 1.7 V_{rms} for the full width at half maximum (FWHM) [46]. Besides, the Fermi level smearing effect at finite temperature will also produce a thermal broadening of 5.4 kT [43], and these two broadening effects add as squares [43,48]. In order to determine the experimental FWHMs, a Gaussian distribution function is utilized to fit the spectra peaks [48,129] and an individual peak is defined by its left and right minima. Figure 1.24 shows the modulation broadening analysis of the C-C stretching mode at T = 4.2 K. The circular symbols are FWHMs of the

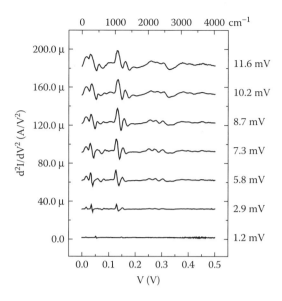

FIGURE 1.23 Modulation dependence of IETS spectra obtained at 4.2 K.

experimental peaks obtained from the Gaussian fitting, and the square symbols are calculated values. The error range of the experimental data is also determined by the Gaussian fitting. As shown in Figure 1.24, the agreement is excellent over most of the modulation range; however, the saturation of the experimental linewidth at low modulation bias indicates the existence of a non-negligible intrinsic linewidth.

Taking into account the known thermal and modulation broadenings, and including the intrinsic linewidth (W_I), the measured experimental peak width (W_{exp}) is given by Equation (1.5):

$$W_{exp} = \sqrt{W_I^2 + W_{thermal}^2 + W_{modulation}^2}$$

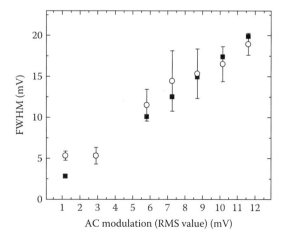

FIGURE 1.24 Line broadening of the C-C stretching mode as a function of AC modulation. The circular symbols are experimental FWHMs and the square symbols are theoretical calculations including both thermal and modulation broadenings.

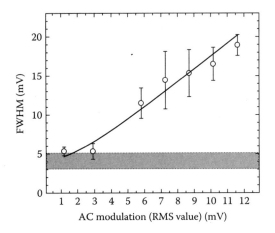

FIGURE 1.25 Nonlinear least squares fitting (solid line) on the modulation broadening data (circular symbol) to determine the intrinsic linewidth of the C-C stretching mode. The shaded bar indicates the expected saturation due to this intrinsic linewidth and the thermal contribution at 4.2 K.

By treating W_I as a fitting parameter, a nonlinear least squares fitting using Equation (1.5) on the AC modulation data can be performed. Figure 1.25 shows the fitting result, and from this fitting an intrinsic linewidth of 3.73 ± 0.98 meV can be obtained for the C-C stretching mode (the error range is determined by the NLS fitting). The shaded bar in Figure 1.25 denotes the expected saturation due to this derived intrinsic linewidth (including a 5.4 kT thermal contribution).

The broadening of the linewidth due to thermal effect can also be independently checked at a fixed modulation voltage. Figure 1.26 shows the temperature dependence of the IETS spectra obtained with an AC modulation of 8.7 mV (rms value) at temperatures of 4.2, 20, 35, 50, 65, and 80 K. Figure 1.27 shows the thermal broadening analysis of the same C-C stretching mode. The circular symbols (and corresponding error bars) are experimental FWHM values determined by the Gaussian fitting (and error of the fitting) to

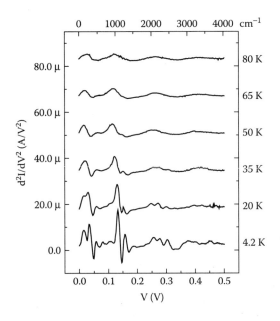

FIGURE 1.26 Temperature dependence of the IETS spectra obtained at a fixed modulation of 8.7 mV.

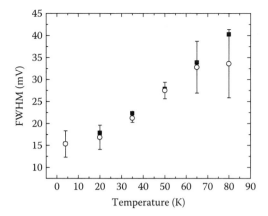

FIGURE 1.27 Line broadening of the C-C stretching mode as a function of temperature. The circular symbols are experimental FWHMs and the square symbols are calculations including thermal and modulation broadenings and the intrinsic linewidth.

the experimental lineshapes. The square symbols are calculations included from thermal broadening, modulation broadening, and the intrinsic linewidth of 3.73 meV determined from the modulation broadening analysis. The error ranges of the calculation (due to the intrinsic linewidth error) are approximately the size of the data points. The agreement between theory and experiment is very good, spanning a temperature range from below ($\times 0.5$) to above ($\times 10$) the thermally broadened intrinsic linewidth.

Similar linewidth investigation has also been carried out on other vibrational modes. For example, Figure 1.28 shows the modulation broadening analysis on the Au-S stretching mode at 33 meV and the CH_2 wagging mode at 158 meV. For the Au-S stretching mode, the deviation of experimental data from calculated values is little, indicating that its intrinsic linewidth is small. A linewidth upper limit of 1.69 meV is determined for this vibrational mode. For the CH_2 wagging mode, a nonlinear least squares fitting using Equation (1.5) [the solid curve in Figure 1.28(b)] gives an intrinsic linewidth of 13.5 ± 2.4 meV. For other vibrational modes (because of the weak spectral peaks), the obtained FWHMs from the lineshape fitting have large error ranges; thus, the intrinsic linewidths cannot be well resolved.

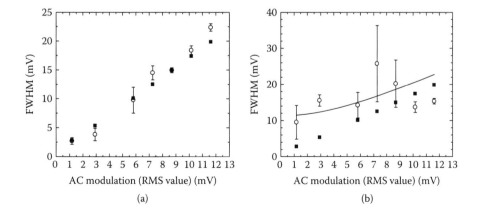

FIGURE 1.28 Line broadenings as a function of AC modulation obtained at 4.2 K for (a) the Au-S stretching mode and (b) the CH_2 wagging mode. The circular symbols are experimental FWHMs and the square symbols are calculations including both modulation and thermal contributions. A nonlinear least squares fitting using Equation (1.5) to determine the intrinsic linewidth is shown as the solid curve in (b).

The inspection of intrinsic linewidth was not generally considered in the classical IETS literatures [30,31]. Linewidth broadening effects due to thermal and modulation contributions have been explored; however, the intrinsic linewidth was usually treated as negligible [44]. A recent report of the IETS intrinsic linewidth comes from the STM-IETS study on the C-H stretching peak of a single HCCH molecule adsorbed on the Cu(001) surface [48]. The reported value of 4 ± 2 meV is found to be consistent with the value of 6 ± 2 meV estimated for the hindered rotation of CO on Cu(001) from similar STM-IETS studies of the same research group [48,129]. Nevertheless, by comparing to the intrinsic linewidth value of ~ 6 cm^{-1} (~ 0.75 meV) obtained from an IR study on the same type of molecules [130], the authors argue that the natural linewidth is negligible and that this intrinsic linewidth may be dominated by instrumental broadening originating from control electronics or the environment [48].

The preceding STM-IETS study reported only one intrinsic linewidth from the obtained spectrum [48]. Our nanopore-based IETS characterization produces a spectrum with multiple peaks originated from different vibrational modes. The obtained intrinsic linewidths are different for different peaks; therefore, they cannot be attributed to one systematic broadening effect, but rather are due to intrinsic molecular device properties. Furthermore, analysis on Raman or IR spectra of alkanethiols on gold shows that the spectral linewidths could be much larger than 1 meV, and different spectral peaks could have similar linewidths. For example, linewidth fittings using both Lorentzian and Gaussian distribution functions on a Raman spectrum [131] containing both Au-S stretching and CH$_2$ wagging peaks yield linewidths of ~ 6 meV and ~ 5 meV, respectively. Therefore, such comparison provides little help in the understanding of the origin of the intrinsic linewidths in our case.

A recent theoretical study by Galperin et al. on the linewidths of vibrational features in inelastic electron tunneling spectroscopy proposes that the intrinsic IETS linewidths are actually dominated by the couplings of molecular vibrations to electron-hole pair excitations in the metallic electrodes [132]. Using a nonequilibrium Green's function (NEGF) approach, the authors have investigated a junction consisting of two electrical leads bridged by a single molecule. After self-consistently solving the related Green's functions and self-energies, important junction characteristics such as the total tunneling current and intrinsic linewidth of the vibrational feature can be estimated. It is found that the interaction of the bridge phonon and the thermal environment contributes little (less than 0.1 meV) to the linewidth, and the dominant part of the intrinsic linewidth comes from the coupling between the bridge phonon and the electronic states of the electrodes [132]. Calculations show that the dominant part of the intrinsic linewidth has a dependence on the bridge-electrode electrical couplings. For coupling parameters corresponding to the nanopore structure, the calculated linewidth value exceeds 1 meV, which has the same order of magnitude as that obtained from the experiment [132].

One might assume that an inhomogeneous contribution would be a dominant part of the measured intrinsic linewidths because the nanopore junction contains several thousand molecules; however, it is very unlikely such a contribution based on the number of molecules would give different linewidths for different vibrational modes. Furthermore, the characterized linewidths from the nanopore method have a similar order of magnitude to the STM-IETS measurement results, where only a single molecule is examined [48,129].

The asymmetric line shapes and negative values of our IETS spectra such as those at 33 mV (Au-S stretching) and 133 mV (C-C stretching) can also be explained by the same theoretical model [132–137]. Asymmetric features in IETS spectra have been observed in several cases in an aluminum oxide tunnel junction and STM-IETS studies [138–140]. Theoretical investigations based on the same molecule-induced resonance model found that the inelastic channel always gives positive contribution to the tunneling conductance, while depending on the junction energetic parameters the contribution from the elastic channel could be negative and, furthermore, could possibly overweight the positive contribution from the inelastic channel and result in a negative peak in the IETS spectra [132,134,135]. The source of the negative contribution of the elastic channel, which only happens at the threshold voltage of $V = \hbar\omega/e$, is the interference between the purely elastic current amplitude that does not involve electron-phonon interaction and the elastic amplitude associated with the excitation and reabsorption of virtual molecular vibrations. By setting certain values of the couplings of the bridging molecular state with the electrodes in the previously discussed model, numerical calculations have been performed to examine the change of the IETS spectrum as a function of the molecular energy level [132,135]. However, an analytical expression is needed from theoretical studies in order to fit the experimental data reported here and better understand such features.

The experimental study presented in this work has also stimulated theoretical investigations, especially first-principle simulations, of the alkanethiol IETS spectra in order to, for instance, understand the effect of the molecule-metal contact geometry change on the tunneling spectra or to provide further details in the peak assignments [144,145]. For example, Solomon *et al.* have used the Green's function density-functional tight-binding ("gDFTB") method to exam an octanedithiol molecule sandwiched between two gold electrodes [145]. As Figure 1.29 shows, the reported calculation result showed good agreement with the experimental data [25, 145]. Based on the theoretical calculations, the authors proposed that some experimental spectra peaks in the low bias region, which has previously been attributed to the Si_3N_4 matrix [25], could actually have molecular origins such as from the C-C-C scissoring vibrations [145]. By comparing the calculated IETS spectra of different molecular binding configurations, the authors have also studied the effect of the contact geometry on the intensities of the peaks and showed that the IETS spectra should have considerable variations with subtle changes of the binding sites [145].

In summary, our observed intrinsic linewidths of spectral peaks of different vibrational modes are dominated by intrinsic molecular properties. Theoretical inspections using nonequilibrium Green's function formalism on a simplified metal-single bridge molecule-metal model suggests that the coupling of the molecular vibrational modes to the electronic continua of the electrodes makes a substantial contribution to the spectral line shape and linewidth. The observed intrinsic linewidth differences can be qualitatively explained by the linewidth dependence on the threshold voltage. By choosing appropriate junction parameters, a quantitative comparison between theory and experiment is expected.

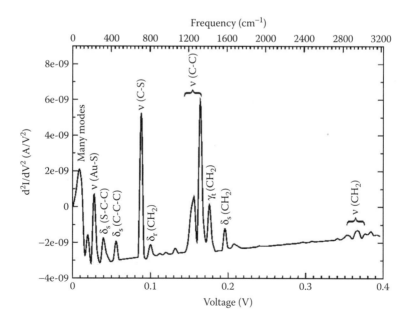

FIGURE 1.29   The calculated IETS spectrum for octanedithiol between two gold electrodes, which suggests that some experimental peaks in the low bias region that have previously been attributed to the Si_3N_4 matrix could actually be from molecular vibrations such as the C-C-C scissoring vibrations. After Ref. 145.

1.6 Conclusion

Using a nanometer-scale device structure, we have performed temperature-dependent I(V) characterization for the first time on alkanethiol SAMs, and demonstrated unambiguously that tunneling is the dominant conduction mechanism. Comparing to a standard model of metal–insulator–metal tunneling, important transport parameters such as the barrier height have been derived, which qualitatively described the tunneling process. In addition, the inelastic electron tunneling spectroscopy technique has been applied to the study of molecular transport. This technique is used to fingerprint the chemical species inside the molecular junction. The obtained spectra exhibit characteristic vibrational signatures of the confined molecular species, presenting direct evidence of the presence of molecules in a molecular transport device for the first time.

The field of "molecular electronics" is rich in proposals and promises of plentiful device concepts, but unfortunately has a dearth of reliable data and characterization techniques upon which to test these ideas. As our results have shown, a well-prepared self-assembled alkanethiol monolayer behaves as a good, thin insulating film and shows understood "canonical" tunneling transport behavior. This molecular system should be used as a standard control structure for future molecular transport characterizations. The IETS technique has been proven to be a dependable tool for the identification of chemical species. It has especially indispensable applications in solid state molecular devices, where other spectroscopic tools such as IR or Raman are hard, if not impossible, to employ. The spectroscopic study conducted in this research verified the characterization of intrinsic molecular properties; therefore, it should be generally utilized for any future molecular transport investigations. Understanding the fundamental charge transport processes in self-assembled monolayers is a challenging task. However, the model control system and the reliable characterization methods presented in this research work should assist in guiding future research work toward more interesting and novel molecular transport systems.

References

[1] Timp, G., Ed., *Nanotechnology*, Springer-Verlag, Berlin, Germany, 1999.
[2] Aviram, A. and Ratner, M.A., Eds., *Molecular Electronics: Science and Technology*, The Annals of the New York Academy of Sciences, 852, The New York Academy of Sciences, New York, 1998.
[3] Reed, M.A. and Lee, T., Eds., *Molecular Nanoelectronics*, American Scientific Publishers, 2003.
[4] Grabert, H. and Devoret, M.H., Eds., *Single Charge Tunneling*, Plenum, New York, 1991.
[5] Reed, M.A. and Seabaugh, A.C., Prospect for semiconductor quantum devices, in *Molecular and Biomolecular Electronics*, Birge, R.R., Ed., American Chemical Society, Washington, 1994.
[6] International Technology Roadmap for Semiconductors; http://public.itrs.net/.
[7] Aviram, A. and Ratner, M.A., *Chem. Phys. Lett.*, 29, 277, 1974.
[8] Metzger, R.M. et al., *J. Am. Chem. Soc.*, 119, 10455, 1997.
[9] Reed, M.A., U.S. Patent No. 5,475,341, 1995.
[10] ———, U.S. Patent No. 5,589,629, 1996.
[11] Ulman, A., *An Introduction to Ultrathin Organic Films from Langmuir–Blodgett to Self-Assembly*, Academic Press, Boston, 1991.
[12] ———, *Chem. Rev.*, 96, 1533, 1996.
[13] Nuzzo, R.G., and Allara, D. L., *J. Am. Chem. Soc.*, 105, 4481, 1983.
[14] Aviram, A. et al., Eds., *Molecular Electronics II*, The Annals of the New York Academy of Sciences, 960, The New York Academy of Sciences, New York, 2002.
[15] Reimers, J. et. al., Eds., *Molecular Electronics III*, The Annals of the New York Academy of Sciences, 1006, The New York Academy of Sciences, New York, 2003.
[16] Chen, Y. et al., *Nanotechnology*, 14, 462, 2003.
[17] Beyond Silicon: Breakthroughs in Molecular Electronics, http://www.hpl.hp.com/ research/qsr/ (Hewlett-Packard Quantum Science Research).
[18] Luo, Y. et al., *Chem. Phys. Chem.*, 3, 519, 2002.

[19] Collier, C.P. et al., *Science*, 289, 1172, 2000.

[20] Stewart, D.R. et al., *Nano Lett.*, 4, 133, 2004.

[21] Lau, C.N. et al., *Nano Lett.*, 4, 569, 2004.

[22] Wang, W., Lee, T., and Reed, M.A., *Rep. Prog. Phys.*, 68, 523, 2005.

[23] Poirier, G.E., *Chem. Rev.*, 97, 1117, 1997.

[24] Wang, W., Lee, T., and Reed, M.A., *Phys. Rev. B.*, 68, 035416, 2003.

[25] Wang, W. et al., *Nano Lett.*, 4, 643, 2004.

[26] Nuzzo, R.G., Dubois, L.H., and Allara, D.L., *J. Am. Chem. Soc.*, 112, 558, 1990.

[27] Ratner, M.A. et al., in *Molecular Electronics: Science and Technology*, The Annals of the New York Academy of Sciences, 852, Aviram, A. and Ratner, M., Eds., The New York Academy of Sciences, New York, 1998.

[28] Lee, T. et al., *J. Phys. Chem. B.*, 108, 8742, 2004.

[29] Jaklevic, R.C. and Lambe, J., *Phys. Rev. Lett.*, 17, 1139, 1966.

[30] Wolfram, T., Ed., *Inelastic Electron Tunneling Spectroscopy*, Springer, New York, 1978.

[31] Hansma, P.K., Ed., *Tunneling Spectroscopy: Capabilities, Applications, and New Techniques*, Plenum, New York, 1982.

[32] Bohm, D., *Quantum Theory*, Prentice-Hall, New York, 1951.

[33] Landau, L.D. and Lifshitz, E.M., *Quantum Mechanics (Non-Relativistic Theory)*, 3rd ed., Pergamon Press, New York, 1977.

[34] Burstein, E. and Lundqvist, S., Eds., *Tunneling Phenomena in Solids*, Plenum Press, New York, 1969.

[35] Duke, C.B., *Tunneling in Solids*, Academic Press, New York, 1969.

[36] Simmons, J.G., *J. Appl. Phys.*, 34, 1793, 1963.

[37] Franz, W., in *Handbuch der Physik*, 17, 155, Flugge, S., Ed., Springer-Verlag, Berlin, Germany, 1956.

[38] Lewicki, G. and Mead, C.A., *Phys. Rev. Lett.*, 16, 939, 1966.

[39] Stratton, R. et al., *J. Phys. Chem. Solids.*, 27, 1599, 1966.

[40] Parker, G.H. and Mead, C.A., *Phys. Rev. Lett.*, 21, 605, 1968.

[41] Brar, B. et al., *Appl. Phys. Lett.*, 69, 2728, 1996.

[42] Maserjian, J. and Petersson, G.P., *Appl. Phys. Lett.*, 25, 50, 1974.

[43] Lambe, J. and Jaklevic, R.C., *Phys. Rev.*, 165, 821, 1968.

[44] Hansma, P.K., *Phys. Reports* (Sec. C of *Phys. Lett.*), 30, 145, 1977.

[45] Adkins, C.J. and Phillips, W.A., *J. Phys. C.*, 18, 1313, 1985.

[46] Klein, J. et al., *Phys. Rev. B.*, 7, 2336, 1973.

[47] Jennings, R.J. and Merrifi, J.R., *J. Phys. Chem. Solids*, 33, 1261, 1972.

[48] Lauhon, L.J. and Ho, W., *Rev. Sci. Instrum.*, 72, 216, 2001.

[49] Scalapino D.J. and Marcus, S.M., *Phys. Rev. Lett.*, 18, 459, 1967.

[50] Bardeen, J., *Phys. Rev. Lett.*, 6, 57, 1961.

[51] Harrison, W.A., *Phys. Rev.*, 123, 85, 1961.

[52] Kirtley, J. et al., *Phys. Rev. B.*, 14, 3177, 1976.

[53] Kirtley, J. and Hall, J.T., *Phys. Rev. B.*, 22, 848, 1980.

[54] Reed, M. A. and Tour, J.M., *Scientific American*, 282, 86, 2000.

[55] Bain, C.D. et al., *J. Am. Chem. Soc.*, 111, 321, 1989.

[56] Grunze, M., *Phys. Scr.*, T49B, 711, 1993.

[57] Noh, J. and Hara, M., *Langmuir*, 18, 1953, 2002.

[58] Nuzzo, R.G. et al., *J. Am. Chem. Soc.*, 109, 733, 1987.

[59] Walczak, M.W. et al., *J. Am. Chem. Soc.*, 113, 2370, 1991.

[60] Widrig, C.A. et al., *J. Electroanal. Chem.*, 310, 335, 1991.

[61] Schäfer, A.H. et al., *Adv. Mater.*, 10, 839, 1998.

[62] Porter, M.D. et al., *J. Am. Chem. Soc.*, 109, 3559, 1987.

[63] Although the HOMO–LUMO gap of alkyl chain type molecules has been reported, there is no experimental data on the HOMO–LUMO gap for the Au/alkanethiol SAM/Au system. 8 eV is commonly used as the HOMO–LUMO gap of alkanethiol.

[64] Boulas, C. et al., *Phys. Rev. Lett.*, 76, 4797, 1996.
[65] Fujihira, M. and Inokuchi, H., *Chem. Phys. Lett.*, 17, 554, 1972.
[66] Lias, S.G. et al., Gas-phase ion and neutral thermochemistry, *J. Phys. Chem.*, Ref. Data, 17, 1, 1988.
[67] Binnig, G. et al., *Phys. Rev. Lett.*, 49, 57, 1982.
[68] Eigler, D.M. and Schweizer, E. K., *Nature*, 344, 524, 1990.
[69] Lee, H.J. and Ho, W., *Science*, 286, 1719, 1999.
[70] Bumm, L.A. et al, *Science*, 271, 1705, 1996.
[71] Donhauser, Z.J. et al., *Science*, 292, 2303, 2001.
[72] Dorogi, M. et al., *Phys. Rev. B.*, 52, 9071, 1995.
[73] Xu, B. and Tao, N., *Science*, 301, 1221, 2003.
[74] Chen, J., Molecular wires, switches, and memories, Ph.D. thesis, Yale University, 2000.
[75] Onipko, A.I. et al., *Phys. Rev. B.*, 61, 11118, 2000.
[76] Cygan, M. T. et al., *J. Am. Chem. Soc.*, 120, 2721, 1998.
[77] Wold, D.J. et al., *J. Phys. Chem. B.*, 106, 2813, 2002.
[78] Cui, X.D. et al., *Science*, 294, 571, 2001.
[79] Wold, D.J. and Frisbie, C.D., *J. Am. Chem. Soc.*, 122, 2970, 2000.
[80] Cui, X.D. et al., *Nanotechnology*, 13, 5, 2002.
[81] Reed, M.A. et al., *Science*, 278, 252, 1997.
[82] Kergueris, C. et al., *Phys. Rev. B.*, 59, 12505, 1999.
[83] Reichert, J. et al., *Phys. Rev. Lett.*, 88, 176804, 2002.
[84] Zhou, C., Atomic and molecular wires, Ph.D. thesis, Yale University, 1999.
[85] Park, H. et al., *Appl. Phys. Lett.*, 75, 301, 1999.
[86] Park, J. et al., *Nature*, 417, 722, 2002.
[87] Liang, M.P. et al., *Nature*, 417, 725, 2002.
[88] Gregory, S., *Phys. Rev. Lett.*, 64, 689, 1990.
[89] Kushmerick, J.G. et al., *J. Am. Chem. Soc.*, 124, 10654, 2002.
[90] Kushmerick, J.G. et al., *Phys. Rev. Lett.*, 89, 086802, 2002.
[91] Holmlin, R. et al., *J. Am. Chem. Soc.*, 123, 5075, 2001.
[92] Rampi, M. A. and Whitesides, G. M., *Chem. Phys.*, 281, 373, 2002.
[93] Mbindyo, J.K.N. et al., *J. Am. Chem. Soc.*, 124, 4020, 2002.
[94] Ralls, K.S. et al., *Appl. Phys. Lett.*, 55, 2459, 1989.
[95] Zhou, C. et al., *Appl. Phys. Lett.*, 71, 611, 1997.
[96] Horiuchi, T. et al., *Rev. Sci. Instrum.*, 60, 994, 1989.
[97] Datta, S. et al., *Phys. Rev. Lett.*, 79, 2530, 1997.
[98] Xue, Y. et al., *J. Chem. Phys.*, 115, 9, 4292, 2001.
[99] Sze, S.M., *Physics of Semiconductor Devices*, Wiley, New York, 1981.
[100] Lamb, D.R., *Electrical Conduction Mechanisms in Thin Insulating Films*, Methue, London, 1967.
[101] Rhoderick, E.H. and Williams, R.H., *Metal-Semiconductor Contact*, Clarendon Press, Oxford, UK, 1988.
[102] Slowinski, K. et al., *J. Am. Chem. Soc.*, 121, 7257, 1999.
[103] York, R.L. et al., *J. Am. Chem. Soc.*, 125, 5948, 2003.
[104] Smalley, J.F. et al., *J. Phys. Chem.*, 99, 13141, 1995.
[105] Weber, K. et al., *J. Phys. Chem. B.*, 101, 8286, 1997.
[106] Slowinski, K. et al., *J. Am. Chem. Soc.*, 119, 11910, 1997.
[107] Mann, B. and Kuhn, H., *J. Appl. Phys.*, 42, 4398, 1971.
[108] Polymeropoulos, E. E. and Sagiv, J., *J. Chem. Phys.*, 69, 1836, 1978.
[109] Joachim, C. and Magoga, M., *Chem. Phys.*, 281, 347, 2002.
[110] Zimmerman, D.T. et al., *Appl. Phys. Lett.*, 75, 2500, 1999.
[111] Kushmerick, J.G. et al., *Nano Lett.*, 4, 639, 2004.
[112] Stipe, B.C. et al., *Science*, 280, 1732, 1998.

[113] Bencuya, I., Electron tunneling in metal/tunnel-oxide/degenerate silicon junctions, Ph.D. thesis, Yale University, 1984.

[114] Lye, W–K., Inelastic electron tunneling spectroscopy of the silicon metal-oxide-semiconductor system, Ph.D. thesis, Yale University, 1998.

[115] Vollhardt, K.P. and Schore, N.E., *Organic Chemistry: Structure and Function*, 3rd ed., W.H. Freeman and Company, New York, 1999.

[116] Kemp, W., *Organic Spectroscopy*, John Wiley and Sons, New York, 1975.

[117] Cooper, J.W., *Spectroscopic Techniques for Organic Chemists*, John Wiley and Sons, New York, 1980.

[118] Duwez, A.–S. et al., *Langmuir*, 16, 6569, 2000.

[119] Kato, H. S. et al., *J. Phys. Chem. B.*, 106, 9655, 2002.

[120] For sample IR data, see Castiglioni, C. et al., *Chem. Phys.*, 95, 7144, 1991.

[121] For sample Raman data, see Bryant, M.A. and Pemberton, J.E., *J. Am. Chem. Soc.*, 113, 8284, 1991.

[122] Kaun, C–C. and Guo, H., *Nano Lett.*, 3, 1521, 2003.

[123] Molinary, M. et al., *Mat. Sci. Eng. B.*, 101, 186, 2003.

[124] Bogdanoff, P.D. et al., *Phys. Rev. B.*, 60, 3976, 1999.

[125] Mazur, U. and Hipps, K.W., *J. Phys. Chem.*, 86, 2854, 1982.

[126] ———, *J. Phys. Chem.*, 85, 2244, 1981.

[127] Kurata, H. et al., *Jap. J. Appl. Phys.*, 20, L811, 1981.

[128] Kirtley, J., *The Interaction of Tunneling Electrons with Molecular Vibrations in Tunneling Spectroscopy*, P.K. Hansma, Ed., Plenum, New York, 1982.

[129] Lauhon, I.J. and Ho, W., *Phys. Rev. B.*, 60, R8525, 1999.

[130] Hirschmugl, C.J. et al., *Phys. Rev. Lett.*, 65, 480, 1990.

[131] Joo, S.W. et al., *J. Phys. Chem. B.*, 104, 6218, 2000.

[132] Galperin, M. et al., *Nano Lett.*, 4 1605, 2004.

[133] Persson, B.N.J. and Baratoff, A., *Phys. Rev. Lett.*, 59, 339, 1987.

[134] Persson, B.N.J., *Phys. Scr.*, 38, 282, 1987.

[135] Mii, T. et al., *Phys. Rev. B.*, 68, 205406, 2003.

[136] ———, *Surface Science*, 502–503, 26, 2002.

[137] Tikhodeev, S.G. et al., *Surface Science*, 493, 63, 2001.

[138] Bayman, A. et al., *Phys. Rev. B.*, 24, 2449, 1981.

[139] Hahn, J.R. et al., *Phys. Rev. Lett.*, 85, 1914, 2000.

[140] Pascual, J.I. et al., *Phys. Rev. Lett.*, 86, 1050, 2001.

[141] Fan, F.F. et al., *J. Am. Chem. Soc.*, 124, 5550, 2002.

[142] Piccinin, S. et al., *Chem. Phys.*, 119, 6729, 2003.

[143] Tomfohr, J.K. et al., *Phys. Rev. B.*, 65, 245105, 2002.

[144] Jiang, J. et al., *Nano Lett.*, 5 1551, 2005.

[145] Solomon, G.C. et al., *J. Chem. Phys.*, 124, 094704, 2006.

2

Molecular Electronic Computing Architectures

James M. Tour

Dustin K. James

2.1 Present Microelectronic Technology

Technology development and industrial competition have been driving the semiconductor industry to produce smaller, faster, and more powerful logic devices. The concept that the number of transistors per integrated circuit will double every 18–24 months due to advancements in technology is commonly referred to as Moore's Law, after Intel founder Gordon Moore, who made the prediction in a 1965 paper with the prophetic title "Cramming More Components onto Integrated Circuits" [1]. At the time, he thought his prediction would hold until at least 1975; however, the exponentially increasing rate of circuit densification has continued into the present (see Graph 2.1).

In 2000, Intel introduced the Pentium 4, containing 42 million transistors, an amazing engineering achievement. The increases in packing density of the circuitry are achieved by shrinking the linewidths of the metal interconnects, by decreasing the size of other features, and by producing thinner layers in the multilevel device structures. These changes are only brought about by the development of new fabrication techniques and materials of construction. As an example, commercial metal interconnect linewidths have decreased to 0.13 mm. The resistivity of Al at 0.13-mm linewidth, combined with its tendency for electromigration (among other problems), necessitated the substitution of Cu for Al as the preferred interconnect metal in order to achieve the 0.13-mm linewidth goal. Cu brings along its own troubles, including its softness, a tendency to migrate into silicon dioxide (thus requiring a barrier coating of Ti/TiN), and an inability to deposit Cu layers via the vapor phase. New tools for depositing copper using

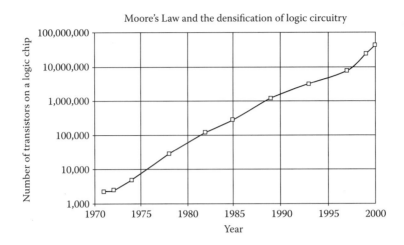

GRAPH 2.1 The number of transistors on a logic chip has increased exponentially since 1972. (Courtesy of Intel Data.)

electroless electroplating, and new technologies for removing the metal overcoats—because copper does not etch well—had to be developed to meet these and other challenges. To integrate Cu in the fabrication line, innovations had to be made all the way from the front end to the back end of the process. These changes did not come without cost, time, and herculean efforts.

2.2   Fundamental Physical Limitations of Present Technology

This top-down method of producing faster and more powerful computer circuitry by shrinking features cannot continue due to fundamental physical limitations related to the material of construction of the solid-state-based devices, which cannot be overcome by engineering. For instance, charge leakage becomes a problem when the insulating silicon oxide layers are thinned to about three silicon atoms deep, which will have been reached commercially by 2003–2004. Moreover, silicon loses its original band structure when it is restricted to very small sizes. The lithography techniques used to create the circuitry on the wafers has also neared its technological limits, although derivative technologies such as e-beam lithography, extreme ultraviolet lithography (EUV) [2], and x-ray lithography are being developed for commercial applications. A tool capable of x-ray lithography in the sub-100 nm range has been patented [3].

Financial roadblocks to continued increases in circuit density exist. Intel's Fab 22, which opened in Chandler, Arizona, in October 2001, cost $2 billion to construct and equip; and it is slated to produce logic chips using copper-based 0.13-mm technology on 200-mm wafers. The cost of building a Fab is projected to reach $15 to 30 billion by 2010 [4] and could be as much as $200 billion by 2015 [5]. The staggering increase in cost is due to the extremely sophisticated tools needed to form the increasingly small features of the devices. It is possible manufacturers may be able to take advantage of infrastructures already in place to reduce the projected cost of introducing these new technologies, but much is uncertain because the methods for achieving further increases in circuit density are unknown or unproven.

As devices increase in complexity, defect, and contamination, control becomes even more important since defect tolerance is very low (nearly every device must work perfectly). For instance, cationic metallic impurities in the wet chemicals, such as sulfuric acid, used in the fabrication process are measured in the part-per-billion (ppb) range. With decreases in linewidth and feature size, the presence of a few ppb of metal contamination could lead to low chip yields. Therefore, the industry has been driving suppliers to produce chemicals with part-per-trillion (ppt) contamination levels, raising the cost of the chemicals used.

Depending on the complexity of the device, the number of individual processing steps used to make them can be in the thousands [6]. It can take 30 to 40 days for a single wafer to complete the manufacturing process. Many of the steps are cleaning steps, requiring some fabs to use thousands of gallons of ultra-pure water per minute [7]. The reclaim of waste water is gaining importance in semiconductor fab operations [8]. The huge consumption of water and its subsequent disposal can lead to problems where aquifers are low and waste emission standards require expensive treatment technology.

A new technology that addresses only one of the potential problems we have discussed would be of interest to the semiconductor industry. Indeed, it would be revolutionary if it produced faster and smaller logic and memory chips, reduced complexity, saved days to weeks of manufacturing time, and reduced the consumption of natural resources.

2.3 Molecular Electronics

How do we overcome the limitations of the present solid-state electronic technology? Molecular electronics is a fairly new and fascinating research area that is firing the imagination of scientists as few research topics have [9]. For instance, *Science* magazine labeled the hook-up of molecules into functional circuits the breakthrough of the year for 2001 [10]. Molecular electronics involves the search for single molecules or small groups of molecules that can be used as the fundamental units for computing (i.e., wires, switches, memory, and gain elements) [11]. The goal is to use these molecules, designed from the bottom up to have specific properties and behaviors, instead of present solid-state electronic devices constructed using lithographic technologies from the top down.

The top-down approach is currently used in the silicon industry, wherein small features such as transistors are etched into silicon using resists and light, but the ever-increasing demand for densification is stressing the industry. The bottom-up approach, on the other hand, implies the construction of functionality into small features, such as molecules, with the opportunity to have the molecules further self-assemble into higher-ordered structural units such as transistors. Bottom-up methodologies are quite natural in that all systems in nature are constructed bottom-up. For example, molecules with specific features assemble to form higher-order structures such as lipid bilayers. Further self-assembly, albeit incomprehensibly complex, causes assembly into cells, and finally into higher-life forms. Hence, utilization of a diversity of self-assembly processes could lead to enormous advances in future manufacturing processes once scientists learn to further control specific molecular-level interactions.

Ultimately, given technological advancements, molecular electronics proponents believe purposeful bottom-up design will be more efficient than the top-down method, and that the incredible structural diversity available to the chemist will lead to more effective molecules, thus approaching optional functionality for each application. A single mole of molecular switches, weighing about 450 g and synthesized in small reactors (a 22-L flask might suffice for most steps of the synthesis), contains 6×1023 molecules—a number greater than all the transistors ever made. While we do not expect to build a circuit in which each single molecule is both addressable and connected to a power supply (at least not in the first few generations), the extremely large numbers of switches available in a small mass illustrate one reason molecular electronics can be a powerful tool for future computing development.

The term molecular electronics covers a broad range of topics. Petty, Bryce, and Bloor recently explored molecular electronics [12], and using their terminology, here we will focus on molecular-scale electronics instead of molecular materials for electronics. Molecular materials for electronics deal with films or crystals (i.e., thin-film transistors or light-emitting diodes) that contain many trillions of molecules per functional unit, the properties of which are measured on the macroscopic scale. Molecular-scale electronics, on the other hand, deal with one to a few thousand molecules per device.

2.4 Computer Architectures Based on Molecular Electronics

In this section, we will initially discuss three general architectural approaches that researchers are considering to build computers based on molecular-scale electronics and the advances made in these three areas in the years 1998 to 2001. In addition, we will touch upon the progress made in measuring the electrical characteristics of molecular switches and in designing logic devices using molecular electronics components.

The first approach to molecular computing, based on quantum cellular automata (QCA), was briefly discussed in our prior review [11]. This method relies on electrostatic field repulsions to transport information throughout the circuitry. One major benefit of the QCA approach is that heat dissipation is less of an issue because only one to fractions of an electron are used rather than the 16,000 to 18,000 electrons needed for each bit of information in classical solid-state devices.

The second approach is based on the massively parallel solid-state Teramac computer developed at Hewlett-Packard (HP) [4] and involves building a similar massively parallel computing device using molecular electronics–based crossbar technologies proposed to be very defect-tolerant [13]. When applied to molecular systems, this approach is proposed to use single-walled carbon nanotubes (SWNT) [14–18] or synthetic nanowires [14,19–22] for crossbars. As we will see, logic functions are performed either by sets of crossed and specially doped nanowires, or by molecular switches placed at each crossbar junction.

The third approach uses molecular-scale switches as part of a nanocell, a new concept that is a hybrid between present silicon-based technology and technology based purely on molecular switches and molecular wires (in reality, the other two approaches will also be hybrid systems in their first few generations) [23]. The nanocell relies on the use of arrays of molecular switches to perform logic functions but does not require that each switching molecule be individually addressed or powered. Furthermore, it utilizes the principles of chemical self-assembly in construction of the logic circuitry, thereby reducing complexity. However, programming issues increase dramatically in the nanocell approach.

While solution phase–based computing, including DNA computing [24], can be classified as molecular-scale electronics, it is a slow process due to the necessity of lining up many bonds, and it is wedded to the solution phase. It may prove to be good for diagnostic testing, but we do not see it as a commercially viable molecular electronics platform; therefore, we will not cover it in this review.

Quantum computing is a fascinating area of theoretical and laboratory study [25–28], with several articles in the popular press concerning the technology [29,30]. However, because quantum computing is based on interacting quantum objects called qubits, and not molecular electronics, it will not be covered in this review. Other interesting approaches to computing such as "spintronics" [31] and the use of light to activate switching [32] will also be excluded from this review.

2.4.1   Quantum Cellular Automata (QCA)

Quantum dots have been called artificial atoms or boxes for electrons [33] because they have discrete charge states, energy-level structures similar to atomic systems, and can contain from a few thousand electrons to only one. They are typically small electrically conducting regions, 1 mm or less in size, with a variety of geometries and dimensions. Because of the small volume, the electron energies are quantized. No shell structure exists; instead, the generic energy spectrum has universal statistical properties associated with quantum chaos [34]. Several groups have studied the production of quantum dots [35]. For example, Leifeld and co-workers studied the growth of Ge quantum dots on silicon surfaces that had been precovered with a 0.05 to 0.11 monolayer of carbon [36], (i.e., carbon atoms replaced about five to ten of every 100 silicon atoms at the surface of the wafer). It was found that the Ge dots grew directly over the areas of the silicon surface where the carbon atoms had been inserted.

Heath discovered that hexane solutions of Ag nanoparticles, passivated with octanethiol, formed spontaneous patterns on the surface of water when the hexane was evaporated [37]; and he prepared superlattices of quantum dots [38,39]. Lieber investigated the energy gaps in "metallic" single-walled carbon nanotubes [16] and used an atomic-force microscope to mechanically bend SWNT in order to create quantum dots less than 100 nm in length [18]. He found that most metallic SWNT are not true metals, and that by bending the SWNT, a defect was produced that had a resistance of 10 to 100 kW. Placing two defects less than 100 nm apart produced the quantum dots.

One proposed molecular computing structural paradigm that utilizes quantum dots is termed a quantum cellular automata (QCA) wherein four quantum dots in a square array are placed in a cell such that electrons are able to tunnel between the dots but are unable to leave the cell [40]. As shown in Figure 2.1, when two excess electrons are placed in the cell, Coulomb repulsion forces the electrons to occupy dots on opposite corners. The two ground-state polarizations are energetically equivalent and can be labeled logic "0" or "1." Flipping the logic state of one cell (for instance, by applying a negative potential to a lead near the quantum dot occupied by an electron) results in the next-door cell flipping ground states in order to reduce Coulomb repulsion. In this way, a line of QCA cells can be used to do computations.

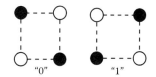

FIGURE 2.1 The two possible ground-state polarizations, denoted "0" and "1," of a four-dot QCA cell. Note that the electrons are forced to opposite corners of the cells by Coulomb repulsion.

A simple example is shown in Figure 2.2, the structure of which could be called a binary wire, where a "1" input gives a "1" output. All of the electrons occupy positions as far away from their neighbors as possible, and they are all in a ground-state polarization. Flipping the ground state of the cell on the left end results in a domino effect, where each neighboring cell flips ground states until the end of the wire is reached.

An inverter built from QCA cells is shown in Figure 2.3—the output is "0" when the input is "1." A QCA topology that can produce AND and OR gates is called a majority gate [41] and is shown in Figure 2.4, where the three input cells "vote" on the polarization of the central cell. The polarization of the central cell is then propagated as the output. One of the inputs can be designated a programming input and determines whether the majority gate produces an AND or an OR. If the programming gate is a logic 0, then the result shown in Figure 2.4 is OR while a programming gate equal to logic 1 produces a result of AND.

A QCA fan-out structure is shown in Figure 2.5. Note that when the ground state of the input cell is flipped, the energy put into the system may not be enough to flip all the cells of both branches of the structure, producing long-lived metastable states and erroneous calculations. Switching the cells using a quasi-adiabatic approach prevents the production of these metastable states [42].

Amlani and co-workers have demonstrated experimental switching of 6-dot QCA cells [43–45]. The polarization switching was accomplished by applying biases to the gates of the input double-dot of a cell fabricated on an oxidized Si surface using standard Al tunnel junction technology, with Al islands and leads patterned by e-beam lithography, followed by a shadow evaporation process and an in situ oxidation step. The switching was experimentally verified in a dilution refrigerator using the electrometers capacitively coupled to the output double-dot.

A functioning majority gate was also demonstrated by Amlani and co-workers [46], with logic AND and OR operations verified using electrometer outputs after applying inputs to the gates of the cell. The experimental setup for the majority gate is shown in Figure 2.6, where the three input tiles—A, B, and C—were supplanted by leads with biases equivalent to the polarization states of the input cells. The negative or positive bias on a gate mimicked the presence or absence of an electron in the input dots of the tiles A, B, and C that were replaced. The truth table for all possible input combinations and majority gate output is shown in Figure 2.7. The experimental results are shown in Figure 2.8. A QCA binary wire has been experimentally demonstrated by Orlov and co-workers [47], and Amlani *et al.* have demonstrated a leadless QCA cell [48], Bernstein *et al.* demonstrated a latch in clocked QCA devices [49].

While the use of quantum dots in the demonstration of QCA is a good first step in reduction to practice, the ultimate goal is to use individual molecules to hold the electrons and pass electrostatic potentials

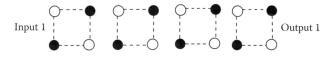

FIGURE 2.2 A simple QCA cell logic line where a logic input of 1 gives a logic output of 1.

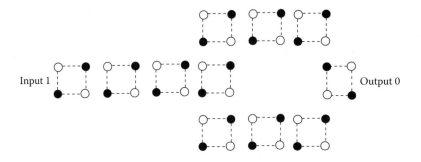

FIGURE 2.3   An inverter built using QCA cells such that a logic input of 1 yields a logic output of 0.

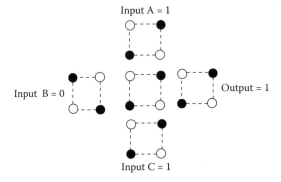

FIGURE 2.4   A QCA majority cell in which the three input cells A, B, and C determine the ground state of the center cell, which then determines the logic of the output. A logic input of 0 gives a logic output of 1.

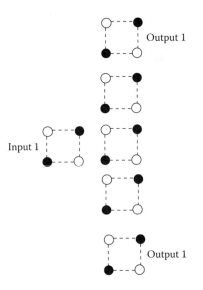

FIGURE 2.5   A fan-out constructed of QCA cells. A logic input of 1 produces a logic output of 1 at both ends of the structure.

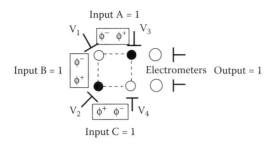

FIGURE 2.6 A QCA majority cell as set up experimentally in a nonmolecular system.

A	B	C	Output
0	0	0	0
0	0	1	0
0	1	1	1
0	1	0	0
1	1	0	1
1	1	1	1
1	0	1	1
1	0	0	0

FIGURE 2.7 The logic table for the QCA majority cell.

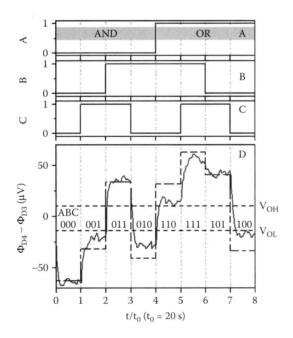

FIGURE 2.8 Demonstration of majority gate operation, where A to C are inputs in Gray code. The first four and last four inputs illustrate AND and OR operations, respectively. (D) An output characteristic of majority gate where $t0 = 20$ s is the input switching period. The dashed stair step–like line shows the theory for 70 mK; the solid line represents the measured data. Output high (V_{OH}) and output low (V_{OL}) are marked by dashed horizontal lines. (Reprinted from Amlani, I., Orlov, A.O., Toth, G., Bernstein, G.H., Lent, C.S., and Snider, G.L. *Science*, 284, 289, 1999. © 1999 American Association for the Advancement of Science. With permission.)

down QCA wires. We have synthesized molecules shown by ab initio computational methods to have the capability of transferring information from one molecule to another through electrostatic potential [50]. Synthesized molecules included three-terminal molecular junctions, switches, and molecular logic gates.

The QCA method faces several problems that need to be resolved before QCA-based molecular computing can become reality. While relatively large quantum-dot arrays can be fabricated using existing methods, a major problem is that placement of molecules in precisely aligned arrays at the nanoscopic level is very difficult to achieve with accuracy and precision. Another problem is that degradation of only one molecule in the array can cause failure of the entire circuit. There has also been some debate about the unidirectionality (or lack thereof) of QCA designs [47,51–52]. Hence, even small examples of two-dots have yet to be demonstrated using molecules, but hopes remain high and researchers continue their efforts.

A	B	C	Output
0	0	0	0
0	0	1	0
0	1	1	1
0	1	0	0
1	1	0	1
1	1	1	1
1	0	1	1
1	0	0	0

2.4.2   Crossbar Arrays

Heath, Kuekes, Snider, and Williams recently reported on a massively parallel experimental computer that contained 220,000 hardware defects yet operated 100 times faster than a high-end single processor workstation for some configurations [4]. The solid-state-based (not molecular electronic) Teramac computer built at HP relied on its fat-tree architecture for its logical configuration. The minimum communication bandwidth necessary to the fat-tree architecture was determined by utilizing Rent's rule, which states that the number of wires coming out of a region of a circuit should scale with the power of the number of devices (n) in that region, ranging from n1/2 in two dimensions to n2/3 in three dimensions. The HP workers built in excess bandwidth, putting in many more wires than needed. The reason for the large number of wires can be understood by considering the simple but illustrative city map depicted in Figure 2.9. To get from point A to point B, one can take local streets, main thoroughfares, freeways, interstate highways, or any combination thereof. If there is a house fire at point C, and the local streets are blocked, then by using the map it is easy to see how to go around that area to get to point B. In the Teramac computer, street blockages are stored in a defect database. When one device needs to communicate with another device, it uses the database and the map to determine how to get there. The Teramac design can therefore tolerate a large number of defects.

In the Teramac computer (or a molecular computer based on the Teramac design), the wires that make up the address lines controlling the settings of the configuration switches and the data lines that link the logic devices are the most important and plentiful part of the computer. It is logical that a large amount of research has been done to develop nanowires (NWs) that could be used in the massively parallel molecular computer. Recall that nanoscale wires are needed if we are to take advantage of the smallness in size of molecules.

Lieber has reviewed the work done in his laboratory to synthesize and determine the properties of NWs and nanotubes [14]. Lieber used Au or Fe catalyst nanoclusters to serve as the nuclei for NWs of Si and GeAs with 10-nm diameters and lengths of hundreds of nanometer. By choosing specific conditions, Lieber was able to control both the length and the diameter of the single crystal semiconductor NW [20]. Silicon

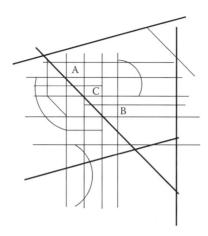

FIGURE 2.9   A simple illustration of the defect tolerance of the Teramac computer. In a typical city, many routes are available to get from point A to point B. One who dislikes traffic might take only city streets (thin lines), while others who want to arrive faster may take a combination of city streets and highways (thick lines). If there was a house fire at point C, a traveler intent on driving only on city streets could look at the map and determine many alternate routes from A to B.

NW doped with B or P were used as building blocks by Lieber to assemble semiconductor nanodevices [21]. Active bipolar transistors were fabricated by crossing n-doped NW with p-type wire base. The doped wires were also used to assemble complementary inverter-like structures.

Heath reported the synthesis of silicon NW by chemical vapor deposition using SiH4 as the Si source and Au or Zn nanoparticles as the catalytic seeds at 440°C [22,53]. The wires produced varied in diameter from 14 to 35 nm and were grown on the surface of silicon wafers. After growth, isolated NWs were mechanically transferred to wafers; and Al contact electrodes were put down by standard e-beam lithography and e-beam evaporation such that each end of a wire was connected to a metallic contact. In some cases, a gate electrode was positioned at the middle of the wire (Figure 2.10). Tapping AFM indicated the wire in this case was 15 nm in diameter.

Heath found that annealing the Zi–Si wires at 550°C produced increased conductance attributed to better electrode/nanowire contacts (Figure 2.11). Annealing Au–Si wires at 750°C for 30 minutes increased current about 104, as shown in Figure 2.12—an effect attributed to doping of the Si with Au, and lower contact resistance between the wire and the Ti/Au electrodes.

Much research has been done to determine the value of SWNT as NW in molecular computers. One problem with SWNT is their lack of solubility in common organic solvents. In their synthesized state, individual SWNT form ropes [54] from which it is difficult to isolate individual tubes. In our laboratory, some solubility of the tubes was seen in 1,2-dichlorobenzene [55]. An obvious route to better solubilization is to functionalize SWNT by attachment of soluble groups through covalent bonding. Margrave and Smalley found that fluorinated SWNT were soluble in alcohols [56], while Haddon and Smalley were able to dissolve SWNT by ionic functionalization of the carboxylic acid groups present in purified tubes [57].

We have found that SWNT can be functionalized by electrochemical reduction of aryl diazonium salts in their presence [58]. Using this method, about 1 in 20 carbon atoms of the nanotube framework are reacted. We have also found that the SWNT can be functionalized by direct treatment with aryl diazonium tetrafluoroborate salts in solution or by in situ generation of the diazonium moiety using an alkyl nitrite reagent [59]. These functional groups give us handles with which we can direct further, more selective derivatization.

Unfortunately, fluorination and other sidewall functionalization methods can perturb the electronic nature of the SWNT. An approach by Smalley [54,60] and Stoddart and Heath [17] to increase the solubility without disturbing the electronic nature of the SWNT was to wrap polymers around the SWNT to break

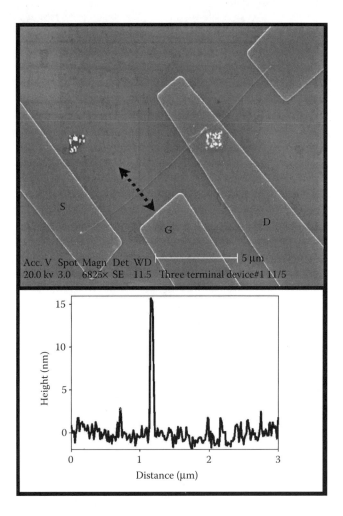

FIGURE 2.10 (Top) SEM image of a three-terminal device, with the source (S), gate (G), and drain (D) labeled. (Bottom) Tapping mode AFM trace of a portion of the silicon nanowire (indicated with the dashed arrow in the SEM image), revealing the diameter of the wire to be about 15 nm. (Reprinted from Chung, S.–W., Yu, J.–Y, and Heath, J.R., *Appl. Phys. Lett.*, 76, 2068, 2000. © 2000 American Institute of Physics. With permission.)

up and solubilize the ropes but leave individual tube's electronic properties unaffected. Stoddart and Heath found the SWNT ropes were not separated into individually wrapped tubes; the entire rope was wrapped. Smalley discovered that individual tubes were wrapped with polymer; the wrapped tubes did not exhibit the roping behavior. While Smalley was able to demonstrate removal of the polymer from the tubes, it is not clear how easily the SWNT can be manipulated and subsequently used in electronic circuits. In any case, the placement of SWNT into controlled configurations has been by a top-down methodology for the most part. Significant advances will be needed to take advantage of controlled placement at dimensions that exploit a molecule's small size.

Lieber proposed an SWNT-based nonvolatile random access memory device comprising a series of crossed nanotubes, wherein one parallel layer of nanotubes is placed on a substrate, and another layer of parallel nanotubes, perpendicular to the first set, is suspended above the lower nanotubes by placing them on a periodic array of supports [15]. The elasticity of the suspended nanotubes provides one energy minima, wherein the contact resistance between the two layers is zero and the switches (the contacts between the two sets of perpendicular NWs) are OFF. When the tubes are transiently charged to produce attractive electrostatic forces, the suspended tubes flex to meet the tubes directly below them and a contact is made,

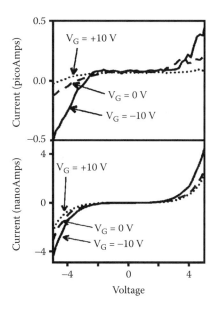

FIGURE 2.11 Three-terminal transport measurements of an as-prepared 15-nm Si nanowire device contacted with Al electrodes (top) and the same device after annealing at 550°C (bottom). In both cases, the gating effect indicates p-type doping. (Reprinted from Chung, S.–W., Yu, J.–Y, and Heath, J.R., *Appl. Phys. Lett.*, 76, 2068, 2000. American Institute of Physics. With permission.)

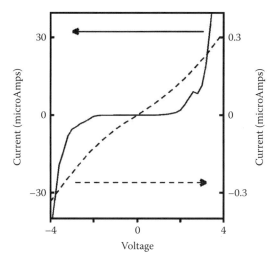

FIGURE 2.12 I(V) characteristics of Au-nucleated Si nanowires contacted with Ti/Au electrodes, before (solid line, current axis on left) and after (dashed line, current axis on right) thermal treatment (750°C, 1 h). After annealing, the wire exhibits metallic-like conductance, indicating it has been heavily doped. (Reprinted from Chung, S.–W., Yu, J.–Y, and Heath, J.R., *Appl. Phys. Lett.*, 76, 2068, 2000. American Institute of Physics. With permission.)

representing the ON state. The ON/OFF state can be read by measuring the resistance at each junction and can be switched by applying voltage pulses at the correct electrodes. This theory was tested by mechanically placing two sets of nanotube bundles in a crossed mode and measuring the I(V) characteristics when the switch was OFF or ON (Figure 2.13). Although they used nanotube bundles with random distributions

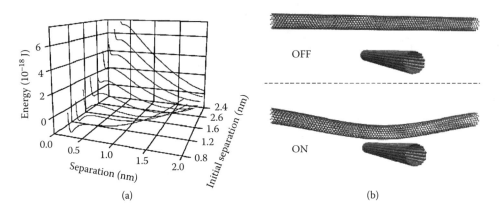

FIGURE 2.13   Bistable nanotubes device potential. (a) Plots of energy, Et = EvdW + Eelas, for a single 20-nm device as a function of separation at the cross point. The series of curves corresponds to initial separations of 0.8, 1.0, 1.2, 1.4, 1.6, 1.8, 2.0, 2.2, and 2.4 nm, with two well-defined minima observed for initial separations of 1.0 to 2.0 nm. These minima correspond to the crossing nanotubes being separated and in cdW contact. (b) Calculated structures of the 20-nm (10, 10) SWNT device element in the OFF (top) and ON (bottom) states. The initial separation for this calculation was 2.0 nm; the silicon support structures (elastic modulus of 168 Gpa) are not shown for clarity. (Reprinted from Rueckes, T., Kim, K., Joselevich, E., Tseng, G.Y., Cheung, C.–L., and Lieber, C.M., *Science*, 289, 94, 2000. © 2000 American Association for the Advancement of Science. With permission.)

of metallic and semiconductor properties, the difference in resistance between the two modes was a factor of 10, enough to provide support for their theory.

In another study, Lieber used scanning tunneling microscopy (STM) to determine the atomic structure and electronic properties of intramolecular junctions in SWNT samples [16]. Metal–semiconductor junctions were found to exhibit an electronically sharp interface without localized junction states while metal–metal junctions had a more diffuse interface and low-energy states.

One problem with using SWNT or NW as wires is how to guide them in formation of the device structures—i.e., how to put them where you want them to go. Lieber has studied the directed assembly of NWs using fluid flow devices in conjunction with surface patterning techniques and found that it was possible to deposit layers of NW with different flow directions for sequential steps [19]. For surface patterning, Lieber used NH2-terminated surface strips to attract the NW; in between the NH2-terminated strips were either methyl-terminated regions or bare regions, to which the NW had less attraction. Flow control was achieved by placing a poly(dimethylsiloxane) (PDMS) mold, in which channel structures had been cut into the mating surface, on top of the flat substrate. Suspensions of the NW (GaP, InP, or Si) were then passed through the channels. The linear flow rate was about 6.40 mm/s. In some cases, the regularity extended over mm-length scales, as determined by scanning electron microscopy (SEM). Figure 2.14 shows typical SEM images of their layer-by-layer construction of crossed NW arrays.

While Lieber has shown it is possible to use the crossed NWs as switches, Stoddart and Heath have synthesized molecular devices that would bridge the gap between the crossed NWs and act as switches in memory and logic devices [61]. The UCLA researchers have synthesized catenanes (Figure 2.15) and rotaxanes (Figure 2.16) that can be switched OFF and ON using redox chemistry. For instance, Langmuir–Blodgett films were formed from the catenane in Figure 2.15, and the monolayers were deposited on polysilicon NW etched onto a silicon wafer photolithographically. A second set of perpendicular titanium NW was deposited through a shadow mask, and the I(V) curve was determined. The data, when compared to controls, indicated that the molecules were acting as solid-state molecular switches. As yet, however, there have been no demonstrations of combining the Stoddart switches with NW.

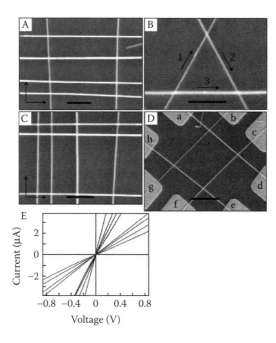

FIGURE 2.14 Layer-by-layer assembly and transport measurements of crossed NW arrays. (A and B) Typical SEM images of crossed arrays of InP NW obtained in a two-step assembly process with orthogonal flow directions for the sequential steps. Flow directions are highlighted by arrows in the images. (C) An equilateral triangle of GaP NW obtained in a three-step assembly process, with 60° angles between flow directions, which are indicated by numbered arrows. The scale bars correspond to 500 nm in (A), (B), and (C). (D) SEM image of a typical 2-by-2 cross array made by sequential assembly of n-type InP NW with orthogonal flows. Ni/In/Au contact electrodes, which were deposited by thermal evaporation, were patterned by e-beam lithography. The NW were briefly (3 to 5 sec) etched in 6% HF solution to remove the amorphous oxide outer layer before electrode deposition. The scale bar corresponds to 2 mm. (E) Representative I(V) curves from two terminal measurements on a 2-by-2 crossed array. The solid lines represent the I(V) of four individual NW (ad, by, cf, eh), and the dashed lines represent I(V) across the four n–n crossed junctions (ab, cd, ef, gh). (Reprinted from Huang, Y., Duan, X., Wei, Q., and Lieber, C.M., *Science*, 291, 630, 2001. © 2001 American Association for the Advancement of Science. With permission.)

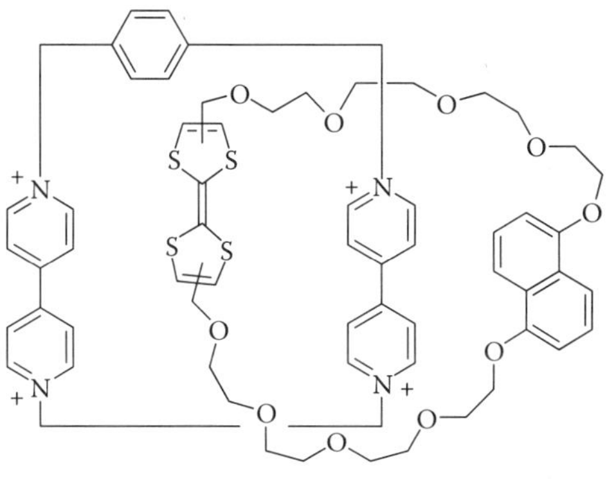

FIGURE 2.15 A catenane. Note the two ring structures are intertwined.

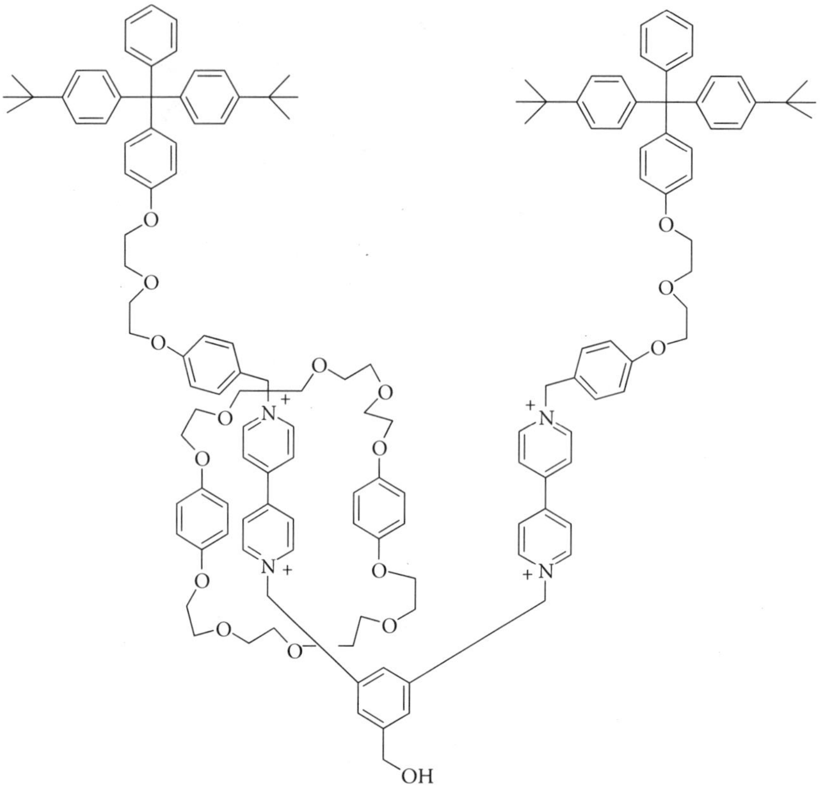

FIGURE 2.16   A [2] rotaxane. The two large end groups do not allow the ring structure to slip off either end.

Carbon nanotubes are known to exhibit either metallic or semiconductor properties. Avouris and co-workers at IBM have developed a method of engineering both multiwalled nanotubes (MWNT) and SWNT using electrical breakdown methods [62]. Shells in MWNT can vary between the metallic or semiconductor character. Using electrical current in air to rapidly oxidize the outer shell of MWNT, each shell can be removed in turn because the outer shell is in contact with the electrodes and the inner shells carry little or no current. Shells are removed until arrival at a shell with the desired properties.

With ropes of SWNT, Avouris used an electrostatically coupled gate electrode to deplete the semi-conductor SWNT of their carriers. Once depleted, the metallic SWNT can be oxidized while leaving the semiconductor SWNT untouched. The resulting SWNT, enriched in semiconductors, can be used to form nanotubes-based field-effect transistors (FETs) (Figure 2.17).

The defect-tolerant approach to molecular computing using crossbar technology faces several hurdles before it can be implemented. As we have discussed, many very small wires are used in order to obtain the defect tolerance. How is each of these wires going to be accessed by the outside world? Multiplexing, the combination of two or more information channels into a common transmission medium, will have to be a major component of the solution to this dilemma. The directed assembly of the NW and attachment to the multiplexers will be quite complicated. Another hurdle is signal strength degradation as it travels along the NW. Gain is typically introduced into circuits through the use of transistors. However, placing a transistor at each NW junction is an untenable solution. Likewise, in the absence of a transistor at each cross point in the crossbar array, molecules with very large ON:OFF ratios will be needed. For instance, if a switch with a 10:1 ON:OFF ratio were used, then ten switches in the OFF state would appear as an ON switch. Hence, isolation of the signal via a transistor is essential. Presently, however, the only solution for

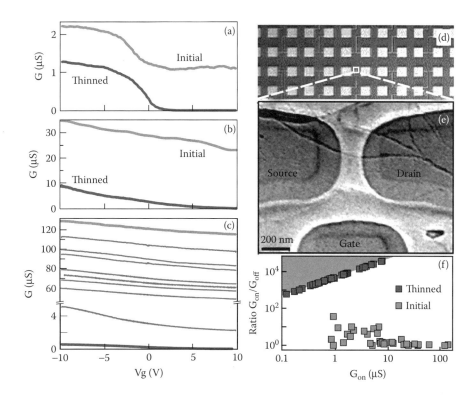

FIGURE 2.17 (a and b) Stressing a mixture of s- and m-SWNT while simultaneously gating the bundle to deplete the semiconductors of carriers resulted in the selective breakdown of the m-SWNT. The G(Vg) curve rigidly shifted downward as the m-SWNT were destroyed. The remaining current modulation is wholly due to the remaining s-SWNTs. (c) In very thick ropes, some s-SWNT must also be sacrificed to remove the innermost m-SWNT. By combining this technique with standard lithography, arrays of three-terminal, nanotubes-based FETs were created (d and e) out of disordered bundles containing both m- and s-SWNT. Although these bundles initially show little or no switching because of their metallic constituents, final devices with good FET characteristics were reliably achieved (f). (Reprinted from Collins, P.G., Arnold, M.S., and Avouris, P., *Science*, 292, 706, 2001. © 2001 American Association for the Advancement of Science. With permission.)

the transistor's introduction would be for a large solid-state gate below each cross point, again defeating the purpose for the small molecules.

Additionally, if SWNT are to be used as the crossbars, connection of molecular switches via covalent bonds introduces sp3 linkages at each junction, disturbing the electronic nature of the SWNT and possibly obviating the very reason to use the SWNT in the first place. Noncovalent bonding will not provide the conductance necessary for the circuit to operate. Therefore, continued work is being done to devise and construct crossbar architectures that address these challenges.

2.4.3 The Nanocell Approach to a Molecular Computer: Synthesis

We have been involved in the synthesis and testing of molecules for molecular electronics applications for some time.[11] One of the synthesized molecules, the nitro aniline oligo(phenylene ethynylene) derivative (Figure 2.18), exhibited large ON:OFF ratios and negative differential resistance (NDR) when placed in a nanopore testing device (Figure 2.19) [63]. The peak-to-valley ratio (PVR) was 1030:1 at 60 K.

The same nanopore testing device was used to study the ability of the molecules to hold their ON states for extended periods of time. The performance of molecules 1 thru 4 in Figure 2.20 as molecular memory devices was tested, and in this study only the two nitro-containing molecules 1 and 2 were found to exhibit storage characteristics. The write, read, and erase cycles are shown in Figure 2.21. The I(V) characteristics of the Au-(1)-Au device are shown in Figure 2.22. The characteristics are repeatable to high accuracy with no degradation of the device noted even after 1 billion cycles over a one-year period.

The I(V) characteristics of the Au-(2)-Au were also measured (Figure 2.23, a and b). The measure logic diagram of the molecular random access memory is shown in Figure 2.24.

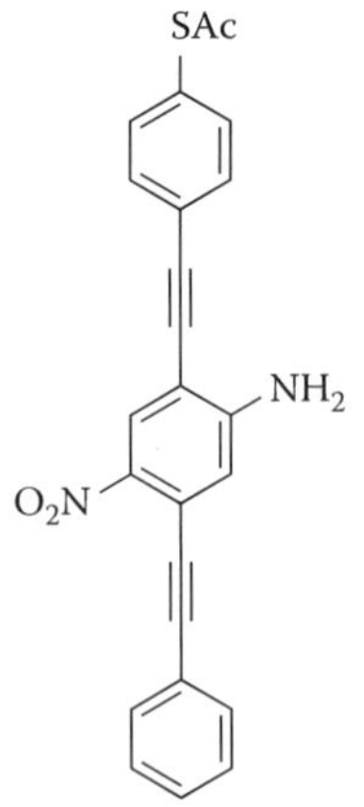

Seminario has developed a theoretical treatment of the electron transport through single molecules attached to metal surfaces [64] and has subsequently done an analysis of the electrical behavior of the four molecules in Figure 2.20 using quantum density functional theory (DFT) techniques at the B3PW91/6–31G* and B3PW91/LAML2DZ levels of theory [65]. The lowest unoccupied molecular orbit (LUMO) of nitro-amino functionalized molecule 1 was the closest orbital to the Fermi level of the Au. The LUMO of neutral 1 was found to be localized (nonconducting). The LUMO became delocalized

FIGURE 2.18 The protected form of the molecule tested in Reed and Tour's nanopore device.

(conducting) in the −1 charged state. Thus, ejection of an electron from the Au into the molecule to form a radical anion leads to conduction through the molecule. A slight torsional twist of the molecule allowed the orbitals to line up for conductance and facilitated the switching.

Many new molecules have recently been synthesized in our laboratories, and some have been tested in molecular electronics applications [66–69]. Since the discovery of the NDR behavior of the nitro aniline derivative, we have concentrated on the synthesis of oligo(phenylene ethynylene) derivatives. Scheme 2.1 shows the synthesis of a dinitro derivative. Quinones, found in nature as electron acceptors, can be easily reduced and oxidized, thus making them good candidates for study as molecular switches. The synthesis of one such candidate is shown in Scheme 2.2.

The acetyl thiol group is called a protected alligator clip. During the formation of a self-assembled monolayer (SAM) on a gold surface, for instance, the thiol group is deprotected in situ, and the thiol forms a strong bond ($\sim$ 2 eV, 45 kcal/mole) with the gold.

Seminario and Tour have done a theoretical analysis of the metal–molecule contact [70] using the B3PW91/LANL2DZ level of theory as implemented in Gaussian-98 in conjunction with the Green function

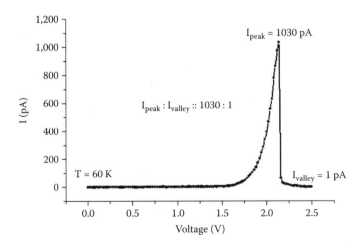

FIGURE 2.19 I(V) characteristics of an Au-(2′-amino-4-ethynylphyenyl-4′-ethynylphenyl-5′-nitro-1-benzene-thiolate)-Au device at 60 K. The peak current density is $\sim$ 50 A/cm^2, the NDR is $\sim$ −400 mohm$\sum$cm^2, and the PVR is 1030:1.

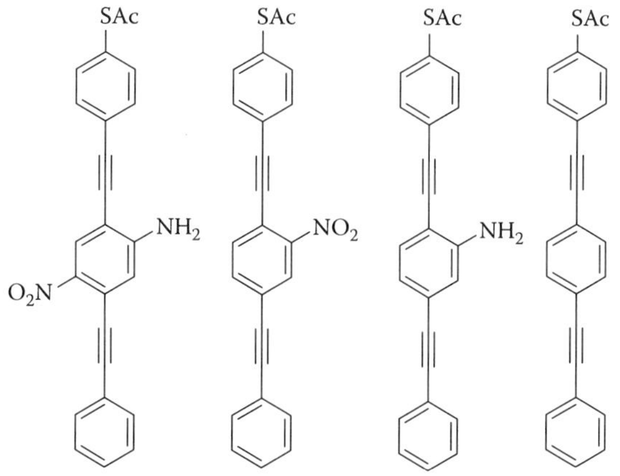

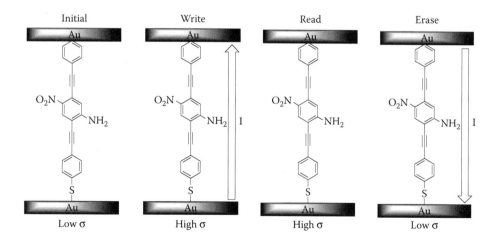

FIGURE 2.20 Molecules 1 through 4 were tested in the nanopore device for storage of high- or low-conductivity states. Only the two nitro-containing molecules 1 and 2 showed activity.

approach that considers the "infinite" nature of the contacts. They found that Pd was the best metal contact, followed by Ni and Pt; Cu was intermediate, while the worst metals were Au and Ag. The best alligator clip was the thiol clip, but they found it was not much better than the isonitrile clip.

We have investigated other alligator clips such as pyridine end groups [68], diazonium salts [67], isonitrile, Se, Te, and carboxylic acid end groups [66]. Synthesis of an oligo(phenylene ethynylene) molecule with an isonitrile end group is shown in Scheme 2.3.

We have previously discussed the use of diazonium salts in the functionalization of SWNT. With modifications of this process, it might be possible to build the massively parallel computer architecture using SWNT as the crosswires and oligo(phenylene ethynylene) molecules as the switches at the junctions of the crosswires, instead of the cantenane and rotaxane switches under research at UCLA

FIGURE 2.21 The memory device operates by the storage of a high- or low-conductivity state. An initially low-conductivity state (low σ) is changed into a high-conductivity state (high σ) upon application of a voltage. The direction of current that flows during the write and erase pulses is diagrammed by the arrows. The high σ state persists as a stored bit. (Reprinted from Reed, M.A., Chen, J., Rawlett, A.M., Price, D.W., and Tour, J.M., *Appl. Phys. Lett.*, 78, 3735, 2001. © 2001 American Institute of Physics. With permission.)

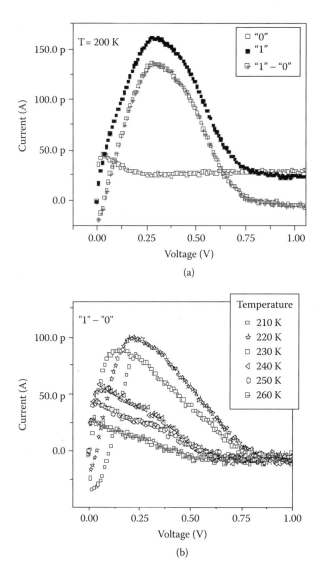

FIGURE 2.22 (a) The I(V) characteristics of a Au-(1)-Au device at 200 K. The number 0 denotes the initial state, 1 the stored written state, and 1–0 is the difference between the two states. Positive bias corresponds to hole injection from the chemisorbed thiol-Au contact. (b) Difference curves (1–0) as a function of temperature. (Reprinted from Reed, M.A., Chen, J., Rawlett, A.M., Price, D.W., and Tour, J.M., *Appl. Phys. Lett.*, 78, 3735, 2001. American Institute of Physics. With permission.)

(see Figure 2.25). However, the challenges of the crossbar method would remain as described earlier. The synthesis of one diazonium switch is shown in Scheme 2.4. The short synthesis of an oligo(phenylene ethynylene) derivative with a pyridine alligator clip is shown in Scheme 2.5.

2.4.4   The Nanocell Approach to a Molecular Computer: The Functional Block

In our conceptual approach to a molecular computer based on the nanocell, a small 1 mm^2 feature is etched into the surface of a silicon wafer. Using standard lithography techniques, 10 to 20 Au electrodes are formed around the edges of the nanocell. The Au leads are exposed only as they protrude into the nanocell's core;

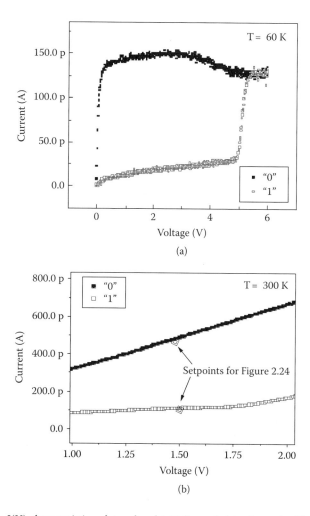

FIGURE 2.23 (a) The I(V) characteristics of stored and initial/erased states in an Au-(2)-Au device at 60 K and (b) ambient temperatures (300 K). The setpoints indicated are the operating point for the circuit of Figure 2.24. (Reprinted from Reed, M.A., Chen, J., Rawlett, A.M., Price, D.W., and Tour, J.M., *Appl. Phys. Lett.*, 78, 3735, 2001. American Institute of Physics. With permission.)

all other gold surfaces are nitride-coated. The silicon surface at the center of the nanocell (the molehole—the location of "moleware" assembly) is functionalized with HS(CH2)3SiOx. A two-dimensional array of Au nanoparticles, about 30–60 nm in diameter, is deposited onto the thiol groups in the molehole. The Au leads (initially protected by alkanethiols) are then deprotected using UV/O$_3$; and the molecular switches are deposited from the solution into the molehole, where they insert themselves between the Au nanoparticles and link the Au nanoparticles around the perimeter with the Au electrodes. The assembly of nanoparticles combined with molecular switches in the molehole will form hundreds to thousands of complete circuits from one electrode to another (see Figure 2.26 for a simple illustration). By applying voltage pulses to selected nanocell electrodes, we expect to be able to turn interior switches ON or OFF, especially with the high ON:OFF ratios we have achieved with the oligo(phenylene ethynylene)s. In this way, we hope to train the nanocell to perform standard logic operations such as AND, NAND, and OR. The idea is that we construct the nanocell first, with no control over the location of the nanoparticles or the bridging switches, and train it to perform certain tasks afterwards. Training a nanocell in a reasonable amount of time will be critical. Eventually, trained nanocells will be used to teach other nanocells. Nanocells

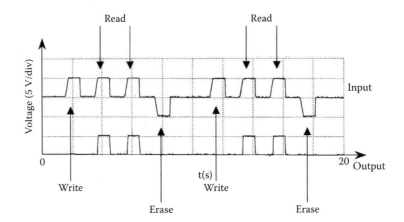

FIGURE 2.24 Measured logic diagram of the molecular random access memory. (Reprinted from Reed, M.A., Chen, J., Rawlett, A.M., Price, D.W., and Tour, J.M., *Appl. Phys. Lett.*, 78, 3735, 2001. American Institute of Physics. With permission.)

will be tiled together on traditional silicon wafers to produce the desired circuitry. We expect to be able to make future nanocells 0.1 mm^2 or smaller if the input/output leads are limited in number—i.e., one on each side of a square.

While we are still in the research and development phase of the construction of an actual nanocell, we have begun a program to simulate the nanocell using standard electrical engineering circuit simulation programs such as SPICE and HSPICE, coupled with genetic algorithm techniques in three stages [23]:

1. With complete omnipotent programming, wherein we know everything about the interior of the constructed nanocell such as the location of the nanoparticles, how many switches bridge each nanoparticle pair, and the state of the conductance of the switches, and that we have control over turning specific switches ON or OFF to achieve the desired outcome without using voltage pulses from the outside electrodes;

2. With omniscient programming, where we know what the interior of the nanocell looks like and know the conductance state of the switches, but we have to use voltage pulses from the surrounding electrodes to turn switches ON and OFF in order to achieve the desired outcome; and

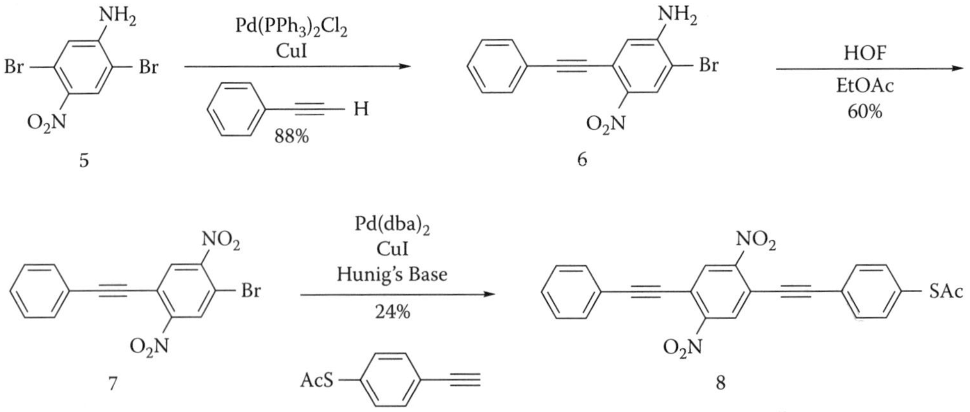

SCHEME 2.1 The synthesis of a dinitro-containing derivative. (Reprinted from Dirk, S.M., Price, D.W. Jr., Chanteau, S., Kosynkin, D.V., and Tour, J.M., *Tetrahedron*, 57, 5109, 2001. © 2001 Elsevier Science. With permission.)

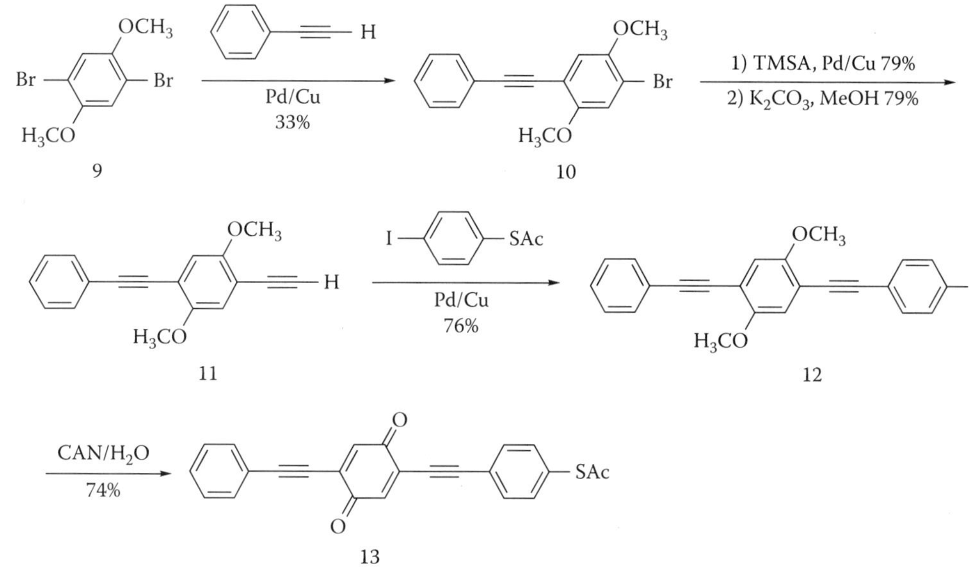

SCHEME 2.2 The synthesis of a quinone molecular electronics candidate. (Reprinted from Dirk, S.M., Price, D.W. Jr., Chanteau, S., Kosynkin, D.V., and Tour, J.M., *Tetrahedron*, 57, 5109, 2001. Elsevier Science. With permission.)

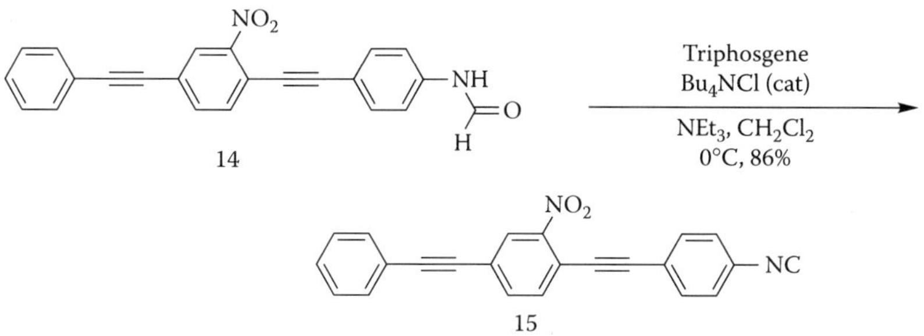

SCHEME 2.3 The formation of an isonitrile alligator clip from a formamide precursor.

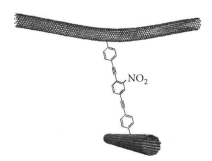

FIGURE 2.25 Reaction of a bis-diazonium-derived nitro phenylene ethynylene molecule with two SWNT could lead to functional switches at cross junctions of SWNT arrays.

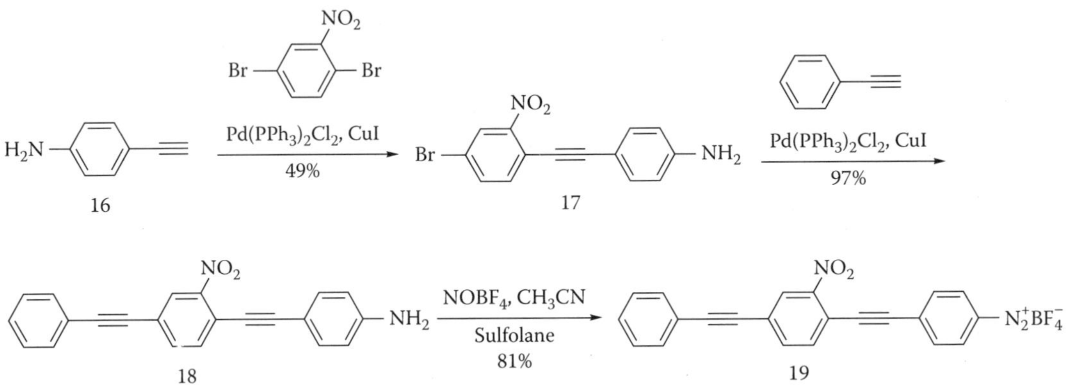

SCHEME 2.4 The synthesis of a diazonium containing a molecular electronics candidate. (Reprinted from Dirk, S.M., Price, D.W. Jr., Chanteau, S., Kosynkin, D.V., and Tour, J.M., *Tetrahedron*, 57, 5109, 2001. Elsevier Science. With permission.)

3. With mortal programming, where we know nothing about the interior of the nanocell and have to guess where to apply the voltage pulses. We are just beginning to simulate mortal programming; however, it is the most critical type since we will be restricted to this method in the actual physical testing of the nanocell.

Our preliminary results with omnipotent programming show we can simulate simple logic functions such as AND, OR, and half-adders.

The nanocell approach has weaknesses and unanswered questions just as the other approaches do. Programming the nanocell is going to be our most difficult task. While we have shown that in certain circumstances our molecular switches can hold their states for extended periods of time, we do not know if that will be true for the nanocell circuits. Will we be able to apply voltage pulses from the edges that will bring about changes in the conductance of switches on the interior of the nanocell through extended distances of molecular arrays? Deposition of the SAMs and packaging the completed nanocells will be monumental development tasks. However, even with these challenges, the prospects for a rapid assembly of molecular systems with few restrictions to fabrication make the nanocell approach enormously promising.

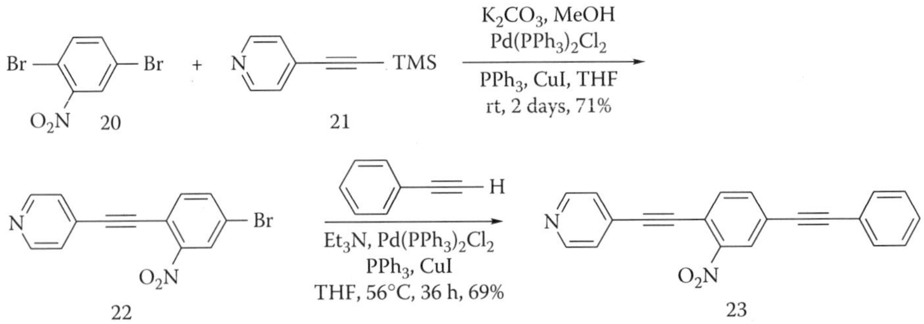

SCHEME 2.5 The synthesis of a derivative with a pyridine alligator clip. (Reprinted from Chanteau, S. and Tour, J.M., Synthesis of potential molecular electronic devices containing pyridine units, *Tet. Lett.*, 42, 3057, 2001. Elsevier Science. With permission.)

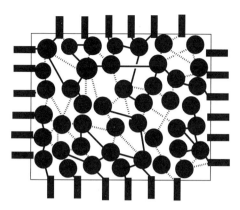

FIGURE 2.26 The proposed nanocell, with electrodes (black rectangles) protruding into the square molehole. Our simulations involve fewer electrodes. The metallic nanoparticles, shown here as black circles with very similar sizes, are deposited into the molehole along with organic molecular switches, not all of which are necessarily the same length or contain the same functionality. The molecular switches, with alligator clips on both ends, bridge the nanoparticles. Switches in the ON state are shown as solid lines while switches in the OFF state are shown as dashed lines. Because there would be no control of the nanoparticle or switch deposition, the actual circuits would be unknown. However, thousands to millions of potential circuits would be formed, depending on the number of electrodes, the size of the molehole, the size of the nanoparticles, and the concentration and identity of the molecular switches. The nanocell would be queried by a programming module after assembly in order to set the particular logic gate or function desired in each assembly. Voltage pulses from the electrodes would be used to turn switches ON and OFF until the desired logic gate or function was achieved.

2.5 Characterization of Switches and Complex Molecular Devices

Now that we have outlined the major classes of molecular computing architectures under consideration, we will touch upon some of the basic component tests that have been done. The testing of molecular electronics components has been recently reviewed [11,71]. Seminario and Tour developed a density functional theory calculation for determination of the I(V) characteristics of molecules, the calculations from which corroborated well with laboratory results [72].

Stoddart and Heath have formed solid-state, electronically addressable switching devices using bistable [2] catenane-based molecules sandwiched between an n-type polycrystalline Si bottom electrode and a metallic top electrode [73]. A mechanochemical mechanism, consistent with the temperature-dependent measurements of the device, was invoked for the action of the switch. Solid-state devices based on [2] or [3] rotaxanes were also constructed and analyzed by Stoddart and Heath [74,75].

In collaboration with Bard, we have shown it is possible to use tuning fork–based scanning probe microscope (SPM) techniques to make stable electrical and mechanical contact to SAMs [76]. This is a promising technique for quick screening of molecular electronics candidates. Frisbie has used an Au-coated atomic-force microscope (AFM) tip to form metal–molecule–metal junctions with Au-supported SAMs. He has measured the I(V) characteristics of the junctions, which are approximately 15 nm,[2] containing about 75 molecules [77]. The I(V) behavior was probed as a function of the SAM thickness and the load applied to the microcontact. This may also prove to be a good method for quick screening of molecular electronics candidates.

In collaboration with Allara and Weiss, we have examined conductance switching in molecules 1, 2, and 4 (from Figure 2.20) by scanning tunneling microscopy (STM) [78]. Molecules 1 and 2 have shown NDR effects under certain conditions, while molecule 4 did not [63]. SAMs made using dodecanethiol are known to be well packed and to have a variety of characteristic defect sites such as substrate step edges, film-domain boundaries, and substrate vacancy islands where other molecules can be inserted. When 1, 2, and 4 were separately inserted into the dodecanethiol SAMs, they protruded from the surrounding molecules due to their height differences. All three molecules had at least two states that differed in height by about 3 Å when observed by STM over time. Because topographic STM images represent a combination of the electronic and topographic structure of the surface, the height changes observed in the STM images could be due to a change in the physical height of the molecules, a change in the conductance of the molecules, or both. The more conductive state was referred to as ON, and the less conductive state was referred to as OFF. SAM formation conditions can be varied to produce SAMs with lower packing density. It was found that all three molecules switched ON and OFF more often in less ordered SAMs than in more tightly packed SAMs. Because a tightly packed SAM would be assumed to hinder conformational changes such as rotational twists, it was concluded that conformational changes controlled the conductance switching of all three molecules.

McCreery has used diazonium chemistry to form tightly packed monolayers on pyrolyzed photoresist film (PPF), a form of disordered graphitic material similar to glassy carbon [79]. Electrochemical reduction of stilbene diazonium salt in acetonitrile solvent in the presence of PPF forms a strong C–C bond between the stilbene molecule and carbons contained in the PPF. The I(V) characteristics of the stilbene junction was measured using Hg-drop electrode methods.

Lieber and co-workers constructed logic gates using crossed NW, which demonstrated substantial gain and were used to implement basic computations [80]. Avouris used SWNT that had been treated to prepare both p- and n-type nanotubes transistors to build voltage inverters, the first demonstration of nanotube-based logic gates [81]. They used spatially resolved doping to build the logic function on a single bundle of SWNT. Dekker and co-workers also built logic circuits with doped SWNT [82]. The SWNT were deposited from a dichloroethane suspension, and those tubes having a diameter of about 1 nm and situated atop preformed Al gate wires were selected by AFM. Schön and co-workers demonstrated gain for electron transport perpendicular to a SAM by using a third gate electrode [83]. The field-effect transistors based on SAMs demonstrate five orders of magnitude of conductance modulation, and gain as high as six. In addition, using two-component SAMs, composed of both insulating and conducting molecules, three orders of magnitude changes in conductance can be achieved [84].

2.6   Conclusion

It is clear that giant leaps remain to be made before computing devices based on molecular electronics are commercialized. The QCA area of research, which has seen demonstrations of logic gates and devices earlier than other approaches, probably has the highest hurdle due to the need to develop nanoscopic quantum dot manipulation and placement. Molecular-scale quantum dots are in active phases of research but have not been demonstrated. The crossbar-array approach faces similar hurdles since the advances to date have only been achieved by mechanical manipulation of individual NWs; thus, it is still very much a research-based phenomenon and nowhere near the scale needed for commercialization. Pieces of the puzzle, such as flow control placement of small arrays, are attractive approaches but need continued development. To this point, the self-assembly of the crossbar arrays, which would simplify the process considerably, has not been a tool in development. The realization of mortal programming and development of the overall nanocell assembly process are major obstacles facing those working in the commercialization of the nanocell approach to molecular electronics. As anyone knows who has had a computer program crash for no apparent reason, programming is a task in which one must take into account every conceivable perturbation, while at the same time not knowing what every possible perturbation is—a difficult task, to say the least. Many cycles of testing and feedback analysis must occur with a working nanocell before we know the programming of the nanocell is successful.

Molecular electronics as a field of research is rapidly expanding with almost weekly announcements of new discoveries and breakthroughs. Those practicing in the field have pointed to Moore's Law and the inherent physical limitations of the present top-down process as reasons to make these discoveries and breakthroughs. They are aiming at a moving target, as evidenced by Intel's recent announcements of the terahertz transistor and an enhanced 0.13-mm process [85–87]. One cannot expect that companies with "iron in the ground" will stand still and let new technologies put them out of business. While some may be kept off the playing field by this realization, for others it only makes the area more exciting. Even as we outlined computing architectures here, the first insertion points for molecular electronics will likely not be for computation. Simpler structures such as memory arrays will probably be the initial areas for commercial molecular-electronics devices. Once simpler structures are refined, more precise methods for computing architecture will be realized. Finally, by the time this review is published, we expect our knowledge will have greatly expanded, and our expectations as to where the technology is headed will have undergone some shifts compared with where we were when we wrote these words. Hence, the field is in a state of rapid evolution, which makes it all the more exciting.

Acknowledgments

The authors thank DARPA administered by the Office of Naval Research (ONR); the Army Research Office (ARO); the U.S. Department of Commerce, National Institute of Standards and Testing (NIST); National Aeronautics and Space Administration (NASA); Rice University; and the Molecular Electronics Corporation for financial support of the research done in our group. We also thank our many colleagues for their hard work and dedication. Dustin K. James thanks David Nackashi for providing some references on semiconductor manufacturing. Dr. I. Chester of FAR Laboratories provided the trimethylsilylacetylene used in the synthesis shown in Scheme 2.2.

References

[1] Moore, G.E., Cramming more components onto integrated circuits, *Electronics*, 38, 1965.

[2] Hand, A., EUV lithography makes serious progress, *Semiconductor Intl.*, 24(6), 15, 2001.

[3] Selzer, R.A. et al., Method of improving x-ray lithography in the sub-100 nm range to create high-quality semiconductor devices, U.S. patent 6, 295, 332, 25 September 2001.

[4] Heath, J.R. et al., A defect-tolerant computer architecture: opportunities for nanotechnology, *Science*, 280, 1716, 1998.

[5] Reed, M.A. and Tour, J.M., Computing with molecules, *Sci. Am.*, 292, 86, 2000.

[6] Whitney, D.E., Why mechanical design cannot be like VLSI design, *Res. Eng. Des.*, 8, 125, 1996.

[7] Hand, A., Wafer cleaning confronts increasing demands, *Semiconductor Intl.*, 24 (August), 62, 2001.

[8] Golshan, M. and Schmitt, S., Semiconductors: water reuse and reclaim operations at Hyundai Semiconductor America, *Ultrapure Water*, 18 (July/August), 34, 2001.

[9] Overton, R., Molecular electronics will change everything, *Wired*, 8(7), 242, 2000.

[10] Service, R.F., Molecules get wired, *Science*, 294, 2442, 2001.

[11] Tour, J.M., Molecular electronics, synthesis and testing of components, *Acc. Chem. Res.*, 33, 791, 2000.

[12] Petty, M.C. et al., *Introduction to Molecular Electronics*, Oxford University Press, New York, 1995.

[13] Heath, J.R., Wires, switches, and wiring: a route toward a chemically assembled electronic nanocomputer, *Pure Appl. Chem.*, 72, 11, 2000.

[14] Hu, J. et al., Chemistry and physics in one dimension: synthesis and properties of nanowires and nanotubes, *Acc. Chem. Res.*, 32, 435, 1999.

[15] Rueckes, T. et al., Carbon nanotubes-based nonvolatile random access memory for molecular computing, *Science*, 289, 94, 2000.

[16] Ouyang, M. et al., Atomically resolved single-walled carbon nanotubes intramolecular junctions, *Science*, 291, 97, 2001.

[17] Star, A. et al., Preparation and properties of polymer-wrapped single-walled carbon nanotubes, *Angew. Chem. Intl. Ed.*, 40, 1721, 2001.

[18] Bozovic, D. et al., Electronic properties of mechanically induced kinks in single-walled carbon nanotubes, *App. Phys. Lett.*, 78, 3693, 2001.

[19] Huang, Y. et al., Directed assembly of one-dimensional nanostructures into functional networks, *Science*, 291, 630, 2001.

[20] Gudiksen, M.S. et al., Synthetic control of the diameter and length of single crystal semiconductor nanowires, *J. Phys. Chem. B*, 105, 4062, 2001.

[21] Cui, Y. and Lieber, C.M., Functional nanoscale electronic devices assembled using silicon nanowire building blocks, *Science*, 291, 851, 2001.

[22] Chung, S.–W. et al., Silicon nanowire devices, *App. Phys. Lett.*, 76, 2068, 2000.

[23] Tour, J.M. et al., A method to compute with molecules: simulating the nanocell, submitted for publication, 2002.

[24] Adleman, L.M., Computing with DNA, *Sci. Am.*, 279, 54, 1998.

[25] Preskill, J., Reliable quantum computing, *Proc.R. Soc. Lond. A*, 454, 385, 1998.

[26] ———, Quantum computing: pro and con, *Proc.R. Soc. Lond. A*, 454, 469, 1998.

[27] Platzman, P.M. and Dykman, M.I., Quantum computing with electrons floating on liquid helium, *Science*, 284, 1967, 1999.

[28] Kane, B., A silicon-based nuclear spin quantum computer, *Nature*, 393, 133, 1998.

[29] Anderson, M.K., Dawn of the QCAD age, *Wired*, 9(9), 157, 2001.

[30] ———, M.K., Liquid logic, *Wired*, 9(9), 152, 2001.

[31] Wolf, S.A. et al., Spintronics: a spin-based electronics vision for the future, *Science*, 294, 1488, 2001.

[32] Raymo, F.M. and Giordani, S., Digital communications through intermolecular fluorescence modulation, *Org. Lett.*, 3, 1833, 2001.

[33] McEuen, P.L., Artificial atoms: new boxes for electrons, *Science*, 278, 1729, 1997.

[34] Stewart, D.R. et al., Correlations between ground state and excited state spectra of a quantum dot, *Science*, 278, 1784, 1997.

[35] Rajeshwar, K. et al., Semiconductor-based composite materials: preparation, properties, and performance, *Chem. Mater.*, 13, 2765, 2001.

[36] Leifeld, O. et al., Self-organized growth of Ge quantum dots on Si(001) substrates induced by sub-monolayer C coverages, *Nanotechnology*, 19, 122, 1999.

[37] Sear, R.P. et al., Spontaneous patterning of quantum dots at the air–water interface, *Phys. Rev. E*, 59, 6255, 1999.

[38] Markovich, G. et al., Architectonic quantum dot solids, *Acc. Chem. Res.*, 32, 415, 1999.

[39] Weitz, I.S. et al., Josephson coupled quantum dot artificial solids, *J. Phys. Chem. B*, 104, 4288, 2000.

[40] Snider, G.L. et al., Quantum-dot cellular automata: review and recent experiments (invited), *J. Appl. Phys.*, 85, 4283, 1999.

[41] Snider, G.L. et al., Quantum-dot cellular automata: line and majority logic gate, *Jpn. J. Appl. Phys. Part I*, 38, 7227, 1999.

[42] Toth, G. and Lent, C.S., Quasiadiabatic switching for metal-island quantum-dot cellular automata, *J. Appl. Phys.*, 85, 2977, 1999.

[43] Amlani, I. et al., Demonstration of a six-dot quantum cellular automata system, *Appl. Phys. Lett.*, 72, 2179, 1998.

[44] Amlani, I. et al., Experimental demonstration of electron switching in a quantum-dot cellular automata (QCA) cell, *Superlattices Microstruct.*, 25, 273, 1999.

[45] Bernstein, G.H. et al., Observation of switching in a quantum-dot cellular automata cell, *Nanotechnology*, 10, 166, 1999.

[46] Amlani, I. et al., Digital logic gate using quantum-dot cellular automata, *Science*, 284, 289, 1999.

[47] Orlov, A.O. et al., Experimental demonstration of a binary wire for quantum-dot cellular automata, *Appl. Phys. Lett.*, 74, 2875, 1999.

[48] Amlani, I. et al., Experimental demonstration of a leadless quantum-dot cellular automata cell, *Appl. Phys. Lett.*, 77, 738, 2000.

[49] Orlov, A.O. et al., Experimental demonstration of a latch in clocked quantum-dot cellular automata, *Appl. Phys. Lett.*, 78, 1625, 2001.

[50] Tour, J.M. et al., Molecular scale electronics: a synthetic/computational approach to digital computing, *J. Am. Chem. Soc.*, 120, 8486, 1998.

[51] Lent, C.S., Molecular electronics: bypassing the transistor paradigm, *Science*, 288, 1597, 2000.

[52] Bandyopadhyay, S., Debate response: what can replace the transistor paradigm?, *Science*, 288, 29, June, 2000.

[53] Yu, J.–Y. et al., Silicon nanowires: preparation, devices fabrication, and transport properties, *J. Phys. Chem.B.*, 104, 11864, 2000.

[54] Ausman, K.D. et al., Roping and wrapping carbon nanotubes, *Proc. XV Intl. Winterschool Electron. Prop. Novel Mater.*, Euroconference Kirchberg, Tirol, Austria, 2000.

[55] Bahr, J.L. et al., Dissolution of small diameter single-wall carbon nanotubes in organic solvents?, *Chem. Commun.*, 2001, 193, 2001.

[56] Mickelson, E.T. et al., Solvation of fluorinated single-wall carbon nanotubes in alcohol solvents, *J. Phys. Chem. B.*, 103, 4318, 1999.

[57] Chen, J. et al., Dissolution of full-length single-walled carbon nanotubes, *J. Phys. Chem. B.*, 105, 2525, 2001.

[58] Bahr, J.L. et al., Functionalization of carbon nanotubes by electrochemical reduction of aryl diazonium salts: a bucky paper electrode, *J. Am. Chem. Soc.*, 123, 6536, 2001.

[59] Bahr, J.L. and Tour, J.M., Highly functionalized carbon nanotubes using in situ generated diazonium compounds, *Chem. Mater.*, 13, 3823, 2001.

[60] O'Connell, M.J. et al., Reversible water-solubilization of single-walled carbon nanotubes by polymer wrapping, *Chem. Phys. Lett.*, 342, 265, 2001.

[61] Pease, A.R. et al., Switching devices based on interlocked molecules, *Acc. Chem. Res.*, 34, 433, 2001.

[62] Collins, P.G. et al., Engineering carbon nanotubes and nanotubes circuits using electrical breakdown, *Science*, 292, 706, 2001.

[63] Chen, J. et al., Large on-off ratios and negative differential resistance in a molecular electronic device, *Science*, 286, 1550, 1999.

[64] Derosa, P.A. and Seminario, J.M., Electron transport through single molecules: scattering treatment using density functional and green function theories, *J. Phys. Chem. B.*, 105, 471, 2001.

[65] Seminario, J.M. et al., Theoretical analysis of complementary molecular memory devices, *J. Phys. Chem. A.*, 105, 791, 2001.

[66] Tour, J.M. et al., Synthesis and testing of potential molecular wires and devices, *Chem. Eur. J.*, 7, 5118, 2001.

[67] Kosynkin, D.V. and Tour, J.M., Phenylene ethynylene diazonium salts as potential self-assembling molecular devices, *Org. Lett.*, 3, 993, 2001.

[68] Chanteau, S. and Tour, J.M., Synthesis of potential molecular electronic devices containing pyridine units, *Tet. Lett.*, 42, 3057, 2001.

[69] Dirk, S.M. et al., Accoutrements of a molecular computer: switches, memory components, and alligator clips, *Tetrahedron*, 57, 5109, 2001.

[70] Seminario, J.M. et al., A theoretical analysis of metal–molecule contacts, *J. Am. Chem. Soc.*, 123, 5616, 2001.

[71] Ward, M.D., Chemistry and molecular electronics: new molecules as wires, switches, and logic gates, *J. Chem. Ed.*, 78, 321, 2001.

[72] Seminario, J.M. et al., Molecular current–voltage characteristics, *J. Phys. Chem.*, 103, 7883, 1999.

[73] Collier, C.P. et al., A [2]catenane-based solid-state electronically reconfigurable switch, *Science*, 289, 1172, 2000.

[74] Wong, E.W. et al., Fabrication and transport properties of single-molecule thick electrochemical junctions, *J. Am. Chem. Soc.*, 122, 5831, 2000.

[75] Collier, C.P., Molecular-based electronically switchable tunnel junction devices, *J. Am. Chem. Soc.*, 123, 12632, 2001.

[76] Fan, R.-F.F. et al., Determination of the molecular electrical properties of self-assembled monolayers of compounds of interest in molecular electronics, *J. Am. Chem. Soc.*, 123, 2424, 2001.

[77] Wold, D.J. and Frisbie, C.D., Fabrication and characterization of metal–molecule–metal junctions by conducting probe atomic force microscopy, *J. Am. Chem. Soc.*, 123, 5549, 2001.

[78] Donahauser, Z.J. et al., Conductance switching in single molecules through conformational changes, *Science*, 292, 2303, 2001.

[79] Ranganathan, S. et al., Covalently bonded organic monolayers on a carbon substrate: a new paradigm for molecular electronics, *Nano Lett.*, 1, 491, 2001.

[80] Huang, Y. et al., Logic gates and computation from assembled nanowire building blocks, *Science*, 294, 1313, 2001.

[81] Derycke, V. et al., Carbon nanotubes inter- and intramolecular logic gates, *Nano Lett.*, 1, 453, 2001.

[82] Bachtold, A. et al., Logic circuits with carbon nanotubes transistors, *Science*, 294, 1317, 2001.

[83] Schön, J.H. et al., Self-assembled monolayer organic field-effect transistors, *Nature*, 413, 713, 2001.

[84] Schön, J.H. et al., Field-effect modulation of the conductance of single molecules, *Science*, 294, 2138, 2001.

[85] Chau, R. et al., A 50 nm Depleted-Substrate CMOS Transistor (DST), International Electron Devices Meeting, Washington, D.C., December 2001.

[86] Barlage, D. et al., High-Frequency Response of 100 nm Integrated CMOS Transistors with High-K Gate Dielectrics, International Electron Devices Meeting, Washington, D.C., December 2001.

[87] Thompson, S. et al., An Enhanced 130 nm Generation Logic Technology Featuring 60 nm Transistors Optimized for High Performance and Low Power at 0.7–1.4 V, International Electron Devices Meeting, Washington, D.C., December 2001.

3

Unimolecular Electronics: Results and Prospects

Robert M. Metzger

3.1 Introduction

"Molecular electronics," "molecular-scale electronics," or "unimolecular electronics" (UE) [1] promises electronic devices with dimensions of 1 to 3 nm, useful for the ultimate miniaturization of electrical circuits. At least conceptually, well-designed electroactive molecules, with a large variation in electronic energy levels, should be able to perform whatever electronic functions inorganic solid-state devices can, with a component size that Si-based electronics will have trouble reaching. The design rule (DR), or the nearest distance between electronic components on an integrated circuit, was chronicled by Gordon Moore in the mid-1960s as dropping by a factor of two every two years using inorganic electronics; the circuit clock speed of the circuit could then be increased by this same factor of two [2]. This marvelous engineering progress has continued to this day: computers using integrated circuits with DR = 50 nm and GHz clock speeds are now available commercially, and DR = 30 nm is under study. The cost of erecting a new Si "foundry" grows exponentially with time (Moore's second law), but when production starts, the new, faster chips have a low unit cost. At the 3 to 5 nm level, heat dissipation is a huge problem in

Si-based electronics. Organic molecules in UE are less resistant to heat than inorganic compounds. Heat-resistant carbon nanotubes are very robust, but cannot yet be chemically selected by diameter, length, and electronic properties, so they remain research curiosities. If molecules can emit light rather than heat after excitation, that would be a major advantage for UE. Engineers in the Si industry are watching UE with interest and some bemusement: When will UE finally produce something useful? Of course, UE wants to be useful. It has grown in spurts; the chimera of nanotechnology has helped, but key experiments remain, and must remain, the driving force for progress. Other articles in this volume chronicle progress in other laboratories, or worry about how to use what is, or may be, produced. If UE becomes practical in time, ultra-fast molecule-based computing may be reached. The first serious device proposal of UE was made in 1974, when Aviram and Ratner (AR) proposed electrical rectification, or diode behavior, by a single molecule with suitable electronic asymmetry [3].

3.2 Donors and Acceptors; HOMOS and LUMOS

One simple way to understand how electroactive organic molecules can be used is to tabulate their first adiabatic ionization potentials I_D (for electron donors D) or their first adiabatic electron affinities A_A (for electron acceptors A), and compare them to the work-functions ϕ of inorganic metals that may be used to contact them (Figure 3.1). The match is not good, so positive or negative applied potentials are needed to bring the molecular energy levels into resonance with the wor function, or Fermi energy, of inorganic metal electrodes. Vertical approximations to I_D are easily measured; electron affinities A_A are difficult to measure. Usually, "good" or "strong" electron donors (relatively low I_D) are poor electron acceptors (have very small A_D), and conversely, "good" or "strong" acceptors (with large A_A) are difficult to oxidize (large I_A). The semimetal graphite, as the infinite two-dimensional extension of polycyclic

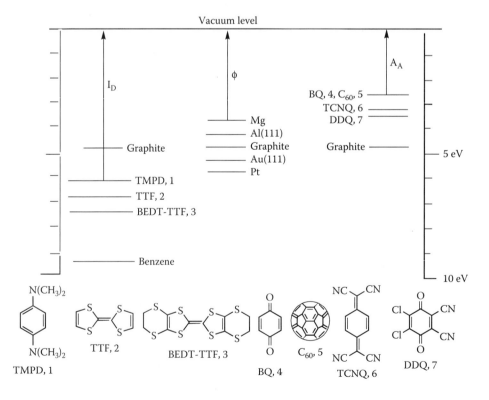

FIGURE 3.1   Representative one-electron donors, D, and their ionization potentials, I_D, one-electron acceptors, A, and their electron affinities, A_A, metals, and their approximate work functions ϕ.

aromatic hydrocarbons, is as good a donor as it is an acceptor. Theory yields estimates of I_D and A_A by Koopmans' theorem [4]: the HOMO (highest occupied molecular orbital) level is a vertical approximation to I_D, while the LUMO (lowest unoccupied molecular orbital) level is a vertical approximation to A_A. These approximations ignore electron correlation and Franck–Condon reorganization.

The practical range of I_D and A_A is limited, because the molecules and their cations (or anions) must be stable in ambient air or solvent: thus, very potent electron donors D or acceptors A can be desinged, but they are not stable enough for synthesis, analysis, or assembly. Also, we have found recently that the gas-phase I_D and A_A values are just a good guide; they probably change significantly (by 1 to 2 eV) for molecules in monolayers between metal electrodes, presumably because of image forces.

Tour [5] has avoided combining "strong" electron donors or "strong" electron acceptors in the same molecules. This strategy circumvented the difficulties of chemically bonding D and A molecules, which can create charge–transfer complexes instead of chemical linkages. Nevertheless, impressive molecular lengths and very interesting connectivity issues were addressed by the "Tour wires," and NDR was observed [6a] (if unforeseen).

3.3 Contacts

UE must physically "touch" a molecule to measure it. How? By using conducting polymers? By using metal electrodes? If a molecule gently "touches" a metal surface, then J. Willard Gibbs teaches us that the chemical potentials must become equal across the interface: the resultant band bending forms a surface dipole as the chemical potential or Fermi level of the metal and the HOMO of the molecule shift to become equal by partial electron transfer at the interface (this shift is the Schottky barrier [7]). A second, or third, or fourth electrode that must likewise interrogate the molecule (or monolayer of molecules) must "touch" the molecule without heating it, or compressing it. Thus, one seeks a contact that obeys Ohm's law [8] with as low a resistance as possible—i.e., avoiding an energy barrier to electron transfer across it. Such a "gentle" contact is not easily achieved, but the scanning probe methods help a lot when a point contact must be made (scanning tunneling microscopy (STM) [9], the atomic force microscopy (AFM) [10], and the conducting-tip AFM (CT-AFM) [11]).

Molecules deposited on surfaces by physisorption can move after deposition, either to reach a thermodynamic steady state on the surface, or in response to an externally applied field. If one puts a 1-Volt bias across a monolayer 1 nm thick, the electric field is large: 1 GV m^{-1}, probably large enough to move or reorient molecules in order to minimize the total energy.

Amphiphilic molecules can be transferred quantitatively from the Pockels–Langmuir (PL) monolayer at the air–water interface onto a metal or other solid substrate by the Langmuir–Blodgett (LB) or vertical transfer method [12, 13], or by the Langmuir–Schaefer (LS) or quasi-horizontal transfer [14]; the coverage of the surface is well quantified by the transfer ratio = [(area covered on the substrate) / (area lost from the PL film)]. To make the organic molecule amphiphilic, pendant alkyl groups (which yield a hydrophobic end) or pendant carboxylic acid groups (to make a hydrophilic end) are often necessary. After transfer, molecules may reorganize over time; thus, the kinetic packing of the PL monolayer may relax to a different thermodynamic order. LB and LS methods have the advantage of achieving full coverage kinetically, and the disadvantages that the films cover any adventitious impurities and may find a new, thermodynamically advantageous order over time.

Molecules can also be covalently bound to certain metal or semiconductor surfaces [15]: carboxylates onto oxide-covered aluminum, and thiols or thioesters to gold, and silanes onto hydroxylated silicon. These are called self assembled monolayers (SAMs), even though the term "self-assembly" is also used for a different purpose in biochemistry. The advantage of SAMs is that they are sturdily anchored at the right distance from the metal substrate, and they can displace adventitious physisorbed impurities. The disadvantage is that true perfect monolayer coverage is difficult to obtain or prove.

The best future strategy may be to achieve ordering at the air-water interface molecules with thiols at one end and long but detachable alkyl chains on the other, transfer the molecules by LB or LS methods onto Au, and finally remove the unwanted alkyl chains by gentle chemical or physical means.

Molecules with thiol terminations can be bound simultaneously to two, three, or more electrodes. So far, this has been done for two electrodes (break junctions), but as yet not for three.

Making 1- to 3-nm gaps between electrodes is difficult: electron beam lithography can make 50-nm gaps routinely, 20-nm gaps with considerable effort, and smaller gaps with even greater effort. Physicists have worked around these restrictions. Two-electrode break junctions were pioneered by Muller [16] and applied to 1,4-benzenedithiol by Reed and co-workers [17]. In the process, a thin Au wire is vapor-deposited onto a flexible substrate, with a narrower region in the center; the wider parts of the Au wire are anchored under two static supports on the top, then the thinner Au region is piezoelectrically pressed from below until the Au wire breaks, creating two Au shards with a narrow gap whose width can be controlled to within less than 0.1 nm by the piezo device. A 1,4-benzenedithiol solution in benzene, suspended above the break junction, will form many single one-end-only thiolate bonds to Au randomly along the wire, while in the 8 Å gap one (or two, or more) benzenedithiolates will bond simultaneously to both shards. The minimum conductance will be due to a single molecule in the gap [17]. Inspired by earlier work [18], Reed and co-workers developed a nanopore technique to study a small assembly of a few hundred to a few thousand molecules [19].

Nanogaps between electrodes can also be made by controlled electromigration; Au wires can be broken into very sharp tips, if a current is passed through them [20]. It helps a lot if the sample is held at 4.2 K, but this is not easy to control.

Chemists can bridge 50-nm gaps between electrodes by providing, for example, two 25-nm diameter nanoparticles of Au or Ag, coated with the usual "spinach" of bithiols, and then bonding them chemically to a 3-nm molecule squeezed between them.

Lindsay and co-workers established by CT AFM that the IV curves for octanedithiol, bonded to an Au(111) susbtrate and also bonded to an Au nanoparticle (to make contact easier), fell into several broad families, depending on the force used by the AFM cantilever, and estimated 900 ± 50 MΩ as the resistance per molecule [21].

So, can metal electrodes be deposited blithely atop an organic monolayer? No. If the metal atoms that impinge on an organic layer are too hot, they can "fry" the monolayer. Things are much better if (1) the organic layer is very thick (say, above 50 nm); (2) the metal has a low work function, e.g., Mg [22], or Ca; (3) if the substrate+organic monolayer are cryocooled to 77 K [23]; (4) if "cold gold," or thermalized Au atoms fall onto the cryocooled organic layer [24–26]. Systematic studies of (hot) metal deposition onto self-assembled monolayers on Au reveal interpenetration and even chemical reaction [27,28]. The cold gold method may prevent chemical damage and may retard, but not totally impede, atom penetration into the monolayer.

Careful attention has focused on the metal-molecule interface. Allara and co-workers established spectroscopically that Ti, when evaporated and deposited atop a SAM on Au, far from being a benign cover layer (potentially oxidized on the surface) actually interpenetrates within the monolayer [28]. Thus, careful ongoing studies seek to understand what does, or does not happen, at the metal–organic interface, both at zero bias and under applied voltage.

One example of indirect damage is the electromigration of oxide-free "cold" gold atoms into the monolayer under bias, forming stalagmites of gold, but not shorting the device [29]. This has also been reported elsewhere [30]. Another example was the report that Ti layers above an organic thiolate chemisorbed layer do form Ti-C bonds, as seen by infrared spectroscopy; Ti/TiO$_2$ adlayers are not inert [28].

3.4 Two-Probe, Three-Probe, and Four-Probe Electrical Measurements

Central to electronics is the IV measurement—that is, the measurement of the electrical current I through a device, as a function of the electrical potential, bias, or voltage V placed across it. Electrical devices are most often two-terminal devices (resistors, capacitors, inductors, rectifiers and diodes, and negative differential resistance [NDR] devices). Amplification is also possible with Esaki tunnel diodes and NDR devices (diode logic), because an input signal, applied across a load of R Ω, placed in series with an NDR device with negative resistance $-$R Ω, provides a zero net resistance at the output, and therefore large signal amplification across the sum of those two resistances. However, difficulties in controlling the tunneling resistance of Esaki diodes and, presumably, difficulties in using organic NDR thiolates at room temperature have prevented the commercialization of NDR devices as amplifiers. Commercial devices used for amplification are three-terminal devices (bipolar junction transistors [BJT], field-effect transistors [FET], vacuum triodes or four or five-terminal devices [vacuum tetrodes, vacuum pentodes]).

The best way to measure the resistance of a macroscopic device is to use four probes: the outermost two are employed to provide a current I from a constant current source, and the potential drop V between the inner two is measured: the resultant resistance R = V/I, after some corrections for geometry, is the true resistance of the device; the contact resistances at the probe–device interfaces cancel out. But for a device 3 nm × 3 nm × 3 nm, present technology cannot yet easily make four electrodes 3 nm apart. Electron-beam lithography can easily reach 20 nm × 20 nm × 20 nm, but going below that is difficult.

For two-probe measurements of a two-terminal device, all resistances (measuring instrument-to-first-electrode, first-electrode-to-molecule, molecule-to-second-electrode, and second-electrode-to-measuring-instrument) are additive. To minimize extraneous large resistances, droplets of wetting solders, Ag paint, Au paint, or Ga/In eutectic are used. To minimize Schottky barrier problems, the same metal is used on both sides of the molecule or monolayer. Most metals are covered by an oxide (impervious, or defect-ridden, such as Al). In contrast, gold has no oxide, but has another problem: Au atoms can migrate or creep after deposition, or under an electric field (electromigration) [31].

Three-electrode measurements have been made—where two electrodes are prepared beforehand, the molecule is placed between them by physisorption or chemisorption; the third "gate" electrode is an STM or CT-AFM tip. This technique has been used to measure FET behavior in a monolayer. The electric field for the FET can also be supplied from the gate conductor through the barrier oxide below the molecules being tested.

3.5   Resistors

Molecules can function as resistors. Of course, organic chemists tell us that saturated straight-chain alkanes conduct less well than unsaturated poly-alkenes or poly-conjugated aromatic hydrocarbons. In the 1960s, Henry Taube proved by kinetic studies that electron transfer rates between metal ions across alkane ligands occur more slowly than across unsaturated ligands [32,33]. Confirming this, in 1996 Weiss and co-workers studied the STM currents across a thioalkyl SAM on Au, and found a pronounced conjugation and molecular length dependence of the conductivity [34].

Ohm's law [8] indicates that the resistance R (Ohms, or Ω) and the conductance G (Siemens $=$ ohm^{-1}, or S) of a device is given by:

$$R = 1/G = V/I \tag{3.1}$$

where V is the applied potential (Volts), and I is the current (Amperes). This law is valid for macroscopic metals, or for semiconductors at any given temperature, where the resistance is mainly due to scattering off impurities and lattice defects in the material. In semiconductors, the current follows an Arrhenius-like temperature dependence,

$$I = I_0 \exp(-\Delta E/k_B T) \tag{3.2}$$

where ΔE is the activation energy for the dominant carriers (electrons or holes), T is the temperature, and k_B is Boltzmann's constant.

For nanoscopic objects, the current I is determined by Landauer's formula [35]:

$$I = (2e/h) \int_{-\infty}^{\infty} [f_L(\varepsilon) - f_R(\varepsilon)] Tr\{G^a(\varepsilon)\Gamma^R(\varepsilon)G^r(\varepsilon)\Gamma^L(\varepsilon)\}d\varepsilon \tag{3.3}$$

where e = the charge on one electron; h = Planck's constant; ε = energy; $f_L(\varepsilon)$ and $f_R(\varepsilon)$ = Fermi–Dirac distributions in the left and right electrodes, respectively; $G^a(\varepsilon)$ and $G^r(\varepsilon)$ = advanced (and retarded) Green's functions for the molecule; $\Gamma^R(\varepsilon)$ and $\Gamma^L(\varepsilon)$ = matrices that describe the coupling between molecule and the metal electrodes; and Tr{ } = trace operator. In this formula, the quantum of resistance, R_0, and its reciprocal, the quantum of conductance, G_0, are given by Landauer's constant (which is also called the von Klitzing constant of the fractional quantum Hall effect [36], now known to 1 part in 10^9):

$$R_0 = 1/G_0 = h/2e^2 = 12.813 k\Omega = 1/(7.75 \times 10^{-5} S) \tag{3.4}$$

This does not say that the intrinsic resistance of any molecule is 12.813 kΩ; it says that the resistance of that molecule *plus* the two metallic electrodes is 12.813 kΩ [35]. The *minimum overall* resistance of a molecular wire and its junctions to arbitrary metal electrodes is 25.626 kΩ (assuming two carriers). The conductance *within* the nanowire can be much higher, particularly if there is no scattering ("ballistic" conductance), but the overall conductance is no larger than R_0^{-1}.

The resistance of Equation (3.4) must be divided by a factor N, if N elementary one-dimensional wires, or N molecules, bridge the gap in parallel between the two metal contacts:

$$R_N = h/2e^2 N = (12.91/N) k\Omega \tag{3.5}$$

Bulk electrical conductivities range over 25 orders of magnitude (from 1.33×10^{-18} S m^{-1} for fused silica, to 1.56×10^{-3} S m^{-1} for silicon, to 5.5×10^{-5} S m^{-1} for ultra-pure "conductivity" water, to 20 S m^{-1} for the quasi-one-dimensional organic metal TTF TCNQ, to about 0.01 S m^{-1} for highly conducting organic polymers, to 6.3×10^7 S m^{-1} for Au, all at room temperature); the conductivity is essentially infinite for superconductors below their critical temperature.

For metal–insulator–metal (MIM) structures, where there is assumed to be a rectangular barrier of energy Φ_B and width d on both sides of the molecule, in the direct tunneling regime $V < \Phi_B e^{-1}$, the Simmons formula [37] can be used [38]:

$$I = e[2\pi hd^2)^{-1}\{(\Phi_B - eV/2)\exp[-4\pi(2m)^{1/2}h^{-1}\alpha(\Phi_B - eV/2)]$$
$$+ (\Phi_B + eV/2)\exp[-4\pi(2m)^{1/2}h^{-1}\alpha(\Phi_B + eV/2)]\} \tag{3.6}$$

where the dimensionless constant α corrects for a possible nonrectangular barrier, or for the effective mass in place of the true carrier (electron) rest mass, m. A fit to the experimental I versus V curves for a SAM of alkanethiols between Au electrodes in a very small pad of diameter 45 $\pm$ 7 nm at 300 K yielded $\Phi_B =$ 1.37 $\pm$ 0.06 Volts and $\alpha = 0.66 \pm 0.02$ for n-dodecanethiol, $C_{12}H_{25}SH$, and $\Phi_B = 1.40 \pm 0.04$ Volts and α = 0.68 $\pm$ 0.02 for n-hexadecanethiol, $C_{16}H_{33}SH$ [38]. Assuming a molecular cross-sectional area of 23 Å^2 (typical for LB monolayers of alkanes), the circular pad contains at most 300 molecules in parallel; I = 20 nA at V = 0.8 Volts for $C_{12}H_{25}SH$ yields an Ohm's law conductance of 2.5×10^{-8} S = (36 MΩ)$^{-1}$, and a specific conductance per molecule of 8.3×10^{-11} S molecule^{-1}. Using the SAM thickness of 14.4 Å for $C_{12}H_{25}S^-$ yields 2.5×10^{-8} S / 1.44×10^{-9} m = 19.4 S m^{-1} or 0.065 S m^{-1}molecule^{-1}: this may not be a fair use of the data, but it will give us some rough idea.

The range of conductivities between straight-chain hydrocarbons and aromatic hydrocarbons is much smaller than the 25 orders of magnitude mentioned for all bulk materials. Indeed, the "best" (Landauer formula) specific resistance of $2.5616 \times 10^4 \Omega$ molecule^{-1} is only six orders of magnitude smaller than the estimated 12 GΩ molecule^{-1} measured for n-dodecanethiol [38]. A good design for minimizing resistances suggests aromatic molecules whose LUMO is low enough to be reached with small biases (< 1 V).

Single-wall carbon nanotubes (SWCNT) [39] are very robust "molecules" of pure carbon, which behave either as electrical semiconductors or as quasi-metals, depending on the topology of folding [40,41]. Alas, the nanotubes are not yet fully chemically processable. For that we may need defect-free, differently end-derivatized SWCNT, e.g. A_n-SWCNT-B_m, with n polar or formally charged groups A and m polar or oppositely charged groups, B, such that the nanotubes can be chemically separated by chromatography by charge, dipole moment, and conductivity. If this can be achieved, then the A_n-SWCNT-B_m would become an ideal connector in UE.

There is also a quantum limit [42]: if an electron is confined to a small dot—i.e., a two-dimensional confined region, or quantum dot, of capacitance C (typically 1 fF)—then adding another electron will cost a "charging energy" e^2 / C. If (e^2 / 2 C) < k_B T (where k_B is Boltzmann's constant, and T is the absolute temperature), then a Coulomb blockade occurs [42]: no more charges can be added, for a threshold voltage V_{CB} < (k_B T / e). This causes a flat region of no current rise in the IV curve until V ≥ (k_B T / e) (at 300 K, V_{CB} = 0.026 Volts).

3.6   Rectifiers or Diodes

AR proposed [3] a D-σ-A molecular rectifier, with an electron donor moiety (D), bonded to an electron acceptor moiety (A) through an insulating saturated "σ" bridge; the current, small at negative bias, becomes large at and beyond a threshold positive bias, because at that bias HOMOs and LUMOs and Fermi levels of the two electrodes start to allow electron transfer to the electrodes, the first highly polar electronic excited state $D^+ - \sigma$-A^- gets populated, and will decay to the less polar ground state $D^0 - \sigma$-A^0 by inelastic tunneling through the molecule [3]. This decay may be enhanced by some intramolecular charge transfer (ICT) or intervalence transfer (IVT) mixing of the donor and acceptor states—i.e., the existence of an extra ICT or IVT absorption band. If the two moieties are too far apart (the σ bridge is too long), they will not communicate, and no rectification will occur. If they are too close, then a new single mixed ground-state will form, and the molecule will not rectify. What is the right length for σ? One might guess that σ should have between two and six C atoms or their equivalent.

Will the de-excitation of the excited molecule $D^+ - \sigma$-A^- occur without radiation (i.e., releasing heat), or with photon emission? The latter process is more desirable since a molecular device that must dispense with about 2 eV of heat (equivalent to a local "temperature" of 16,000 K) will likely melt or burn. Inorganic transistors will face a similar burden, at design rules below 10 nm, since their decay must be thermal: molecules will possess an advantage if and only if they can emit light as they relax. However, it is a truism that when an electric field is present, the decay by photon emission may be curtailed in favor of radiationless (i.e., thermal) processes. If this holds true for AR rectifiers, there will be no advantages for UE!

Three distinct processes exist for asymmetrical conduction (i.e., rectification) in metal–organic–metal (MOM) assemblies. The first is due to Schottky barriers [7] at the metal–organic interface(s): the "S" (for Schottky) rectifiers [22,43].

The second process arises if the "chromophore" (i.e., the part of the molecule whose molecular orbital must be accessed during conduction) is placed asymmetrically within a metal–molecule–metal sandwich, for example, because of the presence of a long alkyl "tail" [44,45]. We shall call molecules that rectify by this process "A" (for "asymmetric") rectifiers [46]. The inclusion of LB tails causes an A contribution.

The third process occurs when the current passing through a molecule, or monolayer of molecules, involves electron transfers between molecular orbitals, whose significant probability amplitudes are asymmetrically placed within the chromophore: this third process may be true "unimolecular rectification," or "U" (for unimolecular) rectification [44]; these U rectifiers are our goal.

The requirements for assembling organic molecules between two inorganic metal electrodes may result in a combination of A, S, and U effects. Pure U rectifiers are rare [45].

The electron transport from metal to organic material to metal has received theoretical attention [44,47]. First, asymmetries in current-voltage plots (often ascribed to rectification) also occur if a chromophore is placed asymmetrically within the electrode gap [44] (A rectifiers). This has been seen by STM [48]. Second, elastic electron transfer between a metal and a single molecular orbital of a molecule can be expressed by [47,49]:

$$I = I_0\{ \tan^{-1}[\theta(E_0 + peV)] - \tan^{-1}[\theta(E_0 - (1-p)eV)]\} \tag{3.7}$$

where E_0 is the molecular orbital energy (typically a LUMO or HOMO), V is the applied potential, and p is the fractional distance of the molecule from, say, the left electrode. If the molecule is centered in the gap, then p $= 1/2$. Tunneling across molecules is expected to be approximately exponential to some power of the potential, so a sigmoidal curve is usually seen symmetrical about I $= 0$ and V $= 0$.

Rectification has a figure of merit, the rectification ratio (RR), defined as the current at a positive bias V divided by the absolute value of the current at the corresponding negative bias $-V$:

$$RR(V) = I(V)/|I(-V)| \tag{3.8}$$

Commercial doped Si, Ge, or GaAs pn junction rectifiers have RR between 10 and 100.

Between 1982 and 1997, we studied many D-σ-A molecules as potential rectifiers [50–84], but could not reliably measure their IV properties. Between 1986 and 1993, Sambles developed reliable techniques for studying rectification by LB multilayers and even monolayers by sandwiching them between electrodes of different work functions: Mg on one side, to minimize damage to the film, and noble metals (Ag, Pt) on the other side [22,43,85]. To avoid difficulties with potentially asymmetric Schottky barriers, and to bypass the thorny issue of how electron transport occurs between adjacent layers in an LB multilayer, we studied almost exclusively single LB or LS monolayers, and used the same metal (first Al, then Au) on both sides of the monolayer. Since 1997, we have identified nine unimolecular rectifiers (structures 8–16 in Figure 3.2) as LB or LS monolayers, either between Al electrodes [23,87,88], or between Au electrodes [25,26,29], [87–96]. As shown in Figure 3.2, the structures have D and A moieties and all, except for 15 and 16 have pendant alkyl groups for organizing the molecules as monolayers. The evaporation of a metal electrode (Al or Au) onto glass, quartz, or very flat Si substrates was routine, as was the transfer of an LB monolayer atop the metal electrode. But depositing the second metal electrode atop the delicate LS or LB monolayer was not routine: a liquid-nitrogen-cooled sample stage in the vacuum evaporator was enough to cool Al vapor upon contact with the cold organic monolayer (at least 50% of the metal–organic–metal "pads" were not electrically shorted) [88]. For Au, this was not enough, so the "cold gold" technique was implemented (cooling the Au vapor atoms to room temperature by multiple collisions with Ar vapor) [24–26] (see Figure 3.3). Most molecules were studied at room temperature in a Faraday cage, but 8 was also studied for its rectification for 105 K < T < 370 K [87]. Characteristic IV curves for them at room temperature are shown in Figure 3.4. Most of these compounds were also studied for their spectroscopic properties (V-UV, IR, grazing-angle IR, spectroscopic ellipsometry, XPS, EPR of their radical ions, surface plasmon resonance, small-angle x-ray scattering) [23,97–99]. Efforts were made to identify the molecular mechanisms for the rectification, and to buttress them through theoretical calculations [23,44,47,100]. The direction of larger electron flow (forward direction) is shown by arrows in Figure 3.4, and is in the direction from the electron donor D to the electron acceptor A, as expected. Not all compounds tested rectified [101–103], because of their chemical structure and/or monolayer assembly. Table 3.1 summarizes

TABLE 3.1   Summary Data for Nine Unimolecular Rectifiers 8 thru 16

Str.	Transf. Type	Press. (mN/m)	LB or LS?	RR Eq. (9)	# pads	Survives cycling ?	U, A, or S?	Refs.
8	D$^+$ $-$ π-A$^-$	20	LB	2-27	16	no	U,A	[26]
9	D$^+$ $-$ π-A$^-$	28	LB	3-64	3	no	U,A	[95]
10	D$^+$ iodide	22	LB	8-60	24	no	A	[89]
11	D-σ-A	22	LB	2	1	no	A	[29]
12	D-σ-A	23	LS	2-16	9	yes	A	[91]
13	D-σ-A	32	LB	2-5	4	yes	U,A	[92]
14	D-σ-A	35	LB	28	1	yes	U,A	[92]
15	D-σ-A	35	LS	3	1	no	U	[94]
16	D-σ-A	20	LS	6	20	yes	U	[96]

All compounds were measured at room temperature in air between Au electrodes inside a Faraday cage (8 was also measured earlier between Al electrodes at 300 K [23], and also between 105 K and 370 K [87]). The column "# pads" lists how many independent typical MOM pads were discussed in each publication as rectifying (out of hundreds measured).

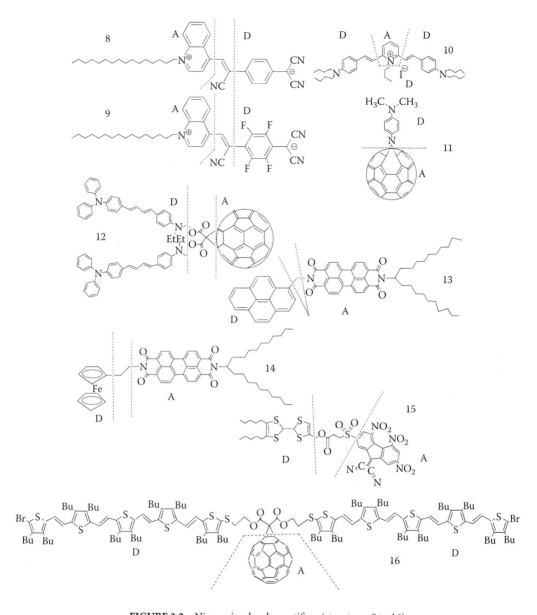

FIGURE 3.2 Nine unimolecular rectifiers (structures 8 to 16).

the characteristics of the measured rectifiers. By popular demand, this work has been reviewed almost too often [104–141].

Before discussing our results in detail, we mention some recent and very valuable contributions of other research groups: (1) Bryce, Petty, and co-workers studied a new DσA compound, inspired by our older effort [51,52] containing the D = TTF and the A = TCNQ. This TTFσTCNQ ester gave strong PL films at the air-water interface. But the TCNQ group lies flat, rather than end-on, on the water surface; the LB multilayers were Y-type, rather than Z-type, and rectification could not be observed [142]. (2) Bryce, Heath, and co-workers found a rectifier in an analog of 15 [143]; (3) Ashwell and co-workers studied several zwitterionic systems by scanning tunneling spectroscopy (STS), starting with 8 [144]. For 8, and for several other zwitterionic systems, the addition of acid could stop, or reverse, the rectification; many new impressive rectifiers were studied, including some bound to the electrode by ionic forces [145–155].

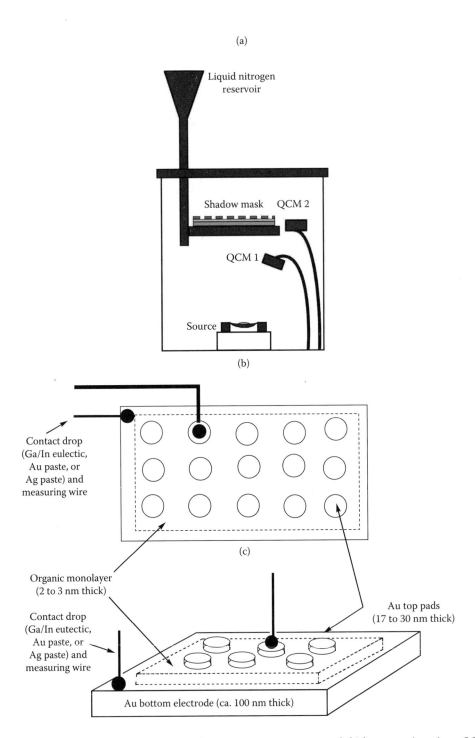

FIGURE 3.3 (a) Edwards E308 evaporator, with Au source, two quartz crystal thickness monitors (one, QCM1, pointed to an Au source to monitor Au vapor deposition on the chamber walls; the other, QCM2, monitored Au film thickness deposited through a shadow mask atop the organic layer). (b and c) The geometry of MOM Au–organic monolayer–Au pads.

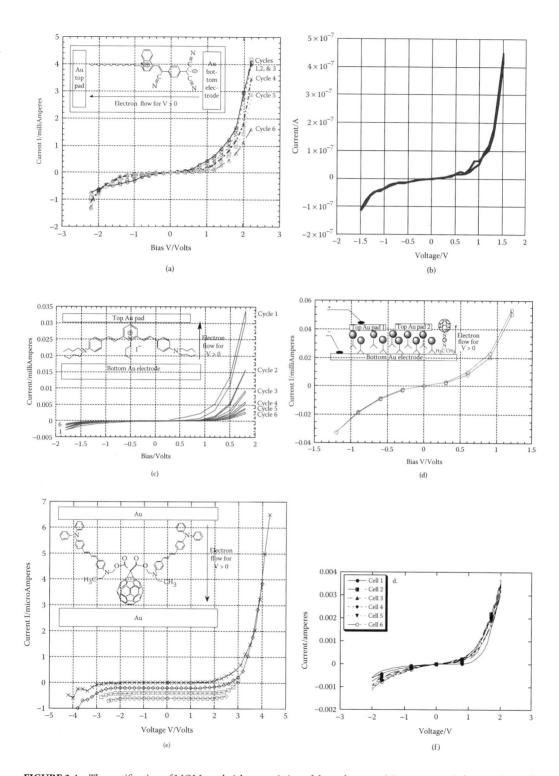

FIGURE 3.4 The rectification of MOM sandwiches consisting of three elements: (1) a macroscopic bottom Au or Al electrode, (2) a 0.3 mm² top Al or "cold Au" electrode pad, and, between them, (3) (a) an LB monolayer of 8 [26]; (b) an LB monolayer of 9 [95]; (c) an LB monolayer of 10 [89]; (d) an LB monolayer of 11 [29]; (e) an LS monolayer of 12 [91]; (f) an LB monolayer of 13 [92]; (g) an LS monolayer of 14 [92]; (h) an LS monolayer of 15 [94].

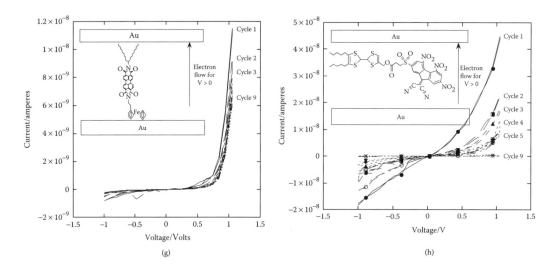

FIGURE 3.4 (Continued).

(4) Yu and co-workers reported rectification by STS, which could be reversed by protonation [156].
(5) Weber, Mayor, and co-workers linked bithiols between mechanically controlled Au break junctions: rectification (RRs scattered between 2 and 10) was seen if the molecule consisted of one tetrafluorophenyl group at one end, and a plain phenyl group at the other end, but no rectification was seen if the ends were chemically symmetrical [157].

The first confirmed rectifier, hexadecylquinolinium tricyanoquinodimethanide 8 [23,25,26,85–87] is a ground-state zwitterion D^+-π-A^-, connected by a twisted "pi" bridge (not a "sigma" bridge): it is sparingly soluble in polar solvents, and has a ground-state static electric dipole moment of 8 of μ_{GS} = 43 ± 8 Debyes at infinite dilution in CH_2Cl_2 [23]. The absorption spectrum in solution shows a hypsochromic band, peaked between 600 and 900 nm. This is an intervalence transfer (IVT) or internal charge–transfer band [23,97], which fluoresces in the near IR [97]. The excited-state dipole moment is μ_{ES} = 3 to 9 Debyes [97]. Since the molecule is hypsochromic, the ground state must be D^+-π-A^-, and the first electronically excited state must be D^0-π-A^0; the finite twist angle between the quinolinium ring and the tricyanoquinodimethanide ring allows for an intense IVT band between the D^+ and A^- ends of the molecule. There is an intermolecular aggregate peak at 570 nm [23] polarized in the plane of an LB multilayer [95] (which was believed to be the IVT peak [23]), and the "real" IVT peak, polarized perpendicular to the monolayer, at 535 nm [95]. A second, transient peak can appear at the air-water interface at 670 nm [158]. Molecule 8 forms amphiphilic Pockels–Langmuir monolayers at the air–water interface, with a collapse pressure of 34 mN m^{-1} and collapse areas of 50 Å^2 at 20°C [23], that transfers on the upstroke, with transfer ratios around 100% onto hydrophilic glass, quartz, aluminum [23,83], or fresh hydrophilic Au [25,26], but transfers poorly on the downstroke onto graphite, with a transfer ratio of about 50% [83]. The LB monolayer thickness of 8 is 23 Å [23] or 29 Å [26] by x-ray diffraction, 23 Å by spectroscopic ellipsometry [26], 22 Å by surface plasmon resonance [23,98], and 25 Å by x-ray photoelectron spectrometry (XPS) [98]. With an averaged monolayer thickness of 23 Å and a calculated molecular length of 33 Å, the molecule on Al or Au has a tilt angle of 46° from the surface normal [23]. The XPS spectrum of one monolayer of 8 on Au displays two N(1s) peaks [98]. An angle-resolved XPS spectrum shows that the N atoms of the CN group are closer to the Au substrate, than the quinolinium N [98]. The valence-band portion of the XPS spectrum agrees roughly with the density of molecular energy states [97]. The contact angle of a drop of water on fresh hydrophilic Au is 40° (it should be 0° if the gold is perfectly hydrophilic). This angle is 92° above a monolayer of 8 deposited on fresh hydrophilic Au (this exposes the nonpolar tail to water) [98]. The orientation of 8 is confirmed by a grazing-angle FTIR study of 8 on Al [23] or on Au [98].

LB monolayers and multilayers of 8 were sandwiched between macroscopic Al electrodes [23], and later, using the "cold gold" technique [24], between Au electrodes [25,26]. Between Al electrodes (with their inevitable patchy and defect-ridden covering of oxide), the monolayer has a dramatically asymmetric current. For 8, RR = 26 at 1.5 Volts [23]. Assuming a molecular area of 50 Å^2, the current at 1.5 Volts corresponds to 0.33 electrons molecule^{-1} s^{-1} [23]. The RRs vary from pad to pad, as does the current, because these are all two-probe measurements, with all electrical resistances (Al, Ga/In, or Ag paste, wires, etc.) in series. As high potentials are scanned repeatedly, the IV curves become less asymmetric; the RRs decrease gradually with repeated cycling of the bias across the monolayer. In the range 105 K < T < 390 K, the onset of rectification of 8 between Al electrodes showed no temperature dependence [87].

With oxide-free Au electrodes, the current through the "Au–monolayer of 8–Au" pads increased dramatically, as expected, but the asymmetry persisted: the highest current was 90,400 electrons molecule^{-1} s^{-1} [25,26]. The best RR was 27.53 at 2.2 Volts [26]. Figure 3.4(a) shows how the rectification ratio decreases from cycles 1 to 6. For some cells, the current increases until breakdown occurs, while in some cells this happens at 5.0 Volts (i.e., the cells suffer dielectric breakdown only at a field close to 2 GV m^{-1}) [26].

Ashwell and co-workers confirmed that Z-type 30-layer films of 8 rectify between Au electrodes [158]. The currents [158] were three orders of magnitude smaller than those reported for the monolayer [26].

A tetrafluoro analog of 8 (i.e., molecule 9) also rectifies, Figure 3.4(b) [95].

The unwelcome gradual decreases in the electrical conductivity and in the RR of an LB monolayer of 8 (from an initial value of 27 [23,26] to close to 1 upon repeated cycling) led to combining the LB and SAM techniques, by measuring thioacetyl variants of 8, which could bind strongly to Au electrodes [90,93]. These variants were synthesized [90,93] with the aim of preparing molecules that (1) form good Langmuir (or Pockels–Langmuir) monolayers at the air–water interface, then (2) bind covalently to an Au substrate after either LB or LS transfer. The good ordering, afforded by the LB technique, should combine with a very sturdy chemical bond to the Au substrate (SAM formation) after LB transfer. The variant of 8 [90] with an undecyl tail followed by a thioacetyl termination (C11 thioacetyl) gave disappointing results. The pressure-area isotherm indicated that the Pockels–Langmuir film collapsed at relatively low surface pressures, compared to 8, and yielded disordered LB monolayers, with competition between strong physisorption by the dicyanomethide end of the molecule and Au-to-thiolate chemisorption. The monolayer rectified in either direction, depending on where in the LB monolayer (i.e., on which molecule, "right side up" or "upside down") the STM tip was probing [90]. Longer variants (C14 and C16 thioacetyl derivatives) did much better [93].

2,6-Di[dibutylamino-phenylvinyl]-1-butylpyridinium iodide, 10, forms a Pockels–Langmuir film at the air–water interface, and transfers to hydrophilic substrates as a Z-type multilayer [89]. The monolayer thickness was 0.7 nm by spectroscopic ellipsometry, 1.3 nm by x-ray diffraction, and 1.15 nm or 1.18 nm by surface plasmon resonance at λ = 532 nm and 632.8 nm, respectively [89]. The films exhibit an absorption maximum at 490 nm (which is slightly hypsochromic in solution), attributable to iodide-to-pyridinium back-charge-transfer, and a second harmonic signal $\chi^{(2)}$ = 50 pm V^{-1} at normal incidence (λ = 1064 nm) and 150 pm V^{-1} at 45° [89]. The rectification shows a decrease of rectification upon successive cycles (Figure 3.4[c]). Some cells have initial RRs as high as 60. The favored direction of electron flow is from the gegenion to the pyridinium ion (i.e., in the direction of back-charge-transfer). The rectification in 2 may be attributed to an interionic electron transfer, or to an intramolecular electron transfer [89].

Dimethylaminophenylazafullerene, 11, is a moderate rectifier, but can also exhibit a tremendous but *spurious* apparent rectification ratio (as high as 20,000) [29], which is probably due to a partial penetration ("electromigration") of Au stalagmites [29]. The azafullerene 11 consists of a weak electron donor (dimethylaniline) bonded to a moderate electron acceptor (N-capped C$_{60}$), with an IVT peak at 720 nm [29]. The Langmuir film is very rigid—i.e., the slope of the isotherm is relatively large. However, the molecular areas are 70 Å^2 at extrapolated zero pressure, and 50 Å^2 at the chosen LB film transfer pressure of 22 mN m^{-1} [29], whereas the true molecular area of C$_{60}$ is close to 100 Å^2. Therefore, it is thought that the molecules 3, transferred onto Au on the upstroke, are somewhat staggered, as is shown in the insert of Figure 3.4(d), with the more hydrophilic dimethylamino group closer to the bottom Au electrode. The film thickness is 2.2 nm by XPS [29]. Angle-resolved N(1s) XPS spectra confirm that the two N atoms are closer to the bottom Au electrode than is the C_{60} cage [29]. One must ignore the IV plots that show large currents due to electromigration. Some cells show a much smaller current, which is slightly rectifying in the forward direction, with RR ≈ 2 (Figure 3.4[d]) [29].

Very sturdy rectification was seen in a Langmuir–Schaefer (LS monolayer of fullerene-bis-[4-diphenyl-amino-4]-[N-ethyl-N-2])-hydroxyethylamino-1,4-diphenyl-1,3-butadiene malonate 12 between Au electrodes [91]. Molecule 12 is based on two triphenylamines (two one-electron donors) and a single fullerene (weak one-electron acceptor): a Langmuir–Schaefer monolayer of 12 rectifies (Figure 3.4[e]); RR does not decrease at all upon successive cycling. The monolayer is probably very dense and stiff. So stiff, in fact, it cannot be transferred onto an Au substrate by the vertical LB process, and adheres to Au if it is transferred by the horizontal, or Langmuir–Schaefer, process [91].

N-(10-nonadecyl)-N-(1-pyrenylmethyl)perylene-3,4,9,10-bis(dicarboximide), 13, is a D-σ-A molecule, based on the moderate pyrene donor D, a one-carbon bridge, and the moderate perylenebisimide acceptor A [92]. It has a persistent RR (Figure 3.4[f]) [92].

N-(10-nonadecyl)-N-(2-ferrocenylethyl)perylene-3,4,9,10-bis(dicarboximide), 14, is a D-σ-A molecule, based on the moderate ferrocene donor D, a two-carbon bridge, and the moderate perylenebisimide acceptor A [92]. It has an IVT band that peaks at 595 nm [92]. Its Pockels–Langmuir isotherm shows that 14 can be transferred as a monolayer at the fairly high surface pressure of 35 mN m^{-1}, and forms a rectifier with RR between 25 and 35, which does not change much upon cycling (Figure 3.4[g]) [92].

4,5-Dipentyl-5'-methyltetrathiafulvalen-4'-methyloxy 2,4,5-trinitro-9-dicyanomethylene-fluorene-7-(3-sulfonylpropionate), 15, is also a D-σ-A molecule, based on D = tetrathiafulvalene, and A = di-cyanomethylenetrinitrofluorene. It has an IVT band maximum at 1220 nm, and was transferred to a fresh hydrophilic Au substrate at a surface pressure of 21 mN/m^{-1} [o57]: its IV curves (Figure 3.4[h]) show that the RRs decrease upon cycling, and become unity after about 9 cycles of measurement [94]. Similar results were reported for a very closely related molecule [143].

Fullerene-bis-[ethylthio-tetrakis(3,4-dibutyl-2-thiophene-5-ethenyl)-5-bromo-3,4-dibutyl-2-thio-phene] malonate, 16, when studied as an LS monolayer between Au electrodes at room temperature, has not one but two rectification regimes. At about 0.8 Volts it rectifies because the LUMO of 16 comes into resonance with the Fermi level of the Au electrode (Figure 3.5[a]) [96]. Beyond 2.5 Volts, however, it rectifies in the opposite direction because now both the HOMO and LUMO of the molecule become accessible from the electrodes (Figure 3.5[b])—i.e., 16 starts to behave as an Aviram–Ratner D-σ-A rectifier [96]. It also rectifies at low bias as an LS monolayer between Al and superconducting Pb electrodes at 4.2 K [96]

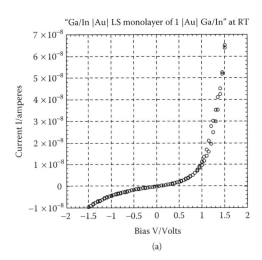

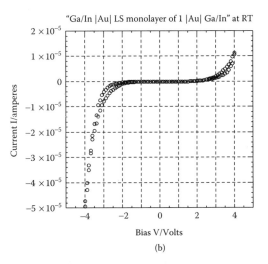

FIGURE 3.5 IV curves for an "Au–LS monolayer of 16–Au" junction: (a) low-bias range; (b) higher-bias range [96].

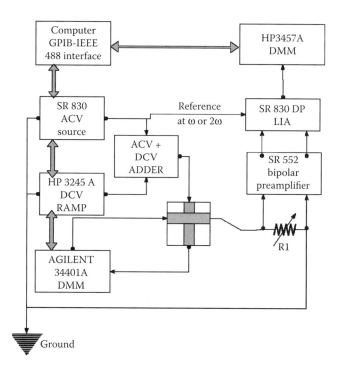

FIGURE 3.6 Diagram of the IETS spectrometer [96].

(IV curve not shown here). We developed an IETS spectrometer (Figure 3.6), and obtained for a cast film of benzoic acid between Pb and Al electrodes at 4.2 K the expected and often recorded IETS spectrum. We next measured the IETS spectrum of an LS monolayer of 16 between Pb and Al electrodes at 4.2 K for the bias range corresponding to molecular vibrations (Figure 3.7). The most prominent peak is that due

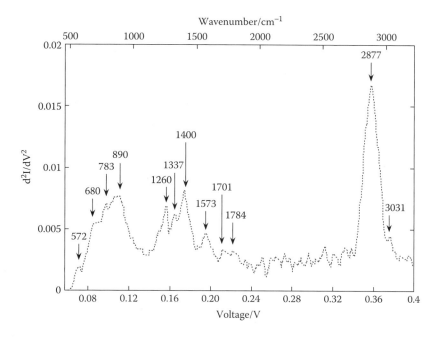

FIGURE 3.7 IETS vibrational spectrum of a "Pb–LS monolayer of 16–Al" junction at 4.2 K [96].

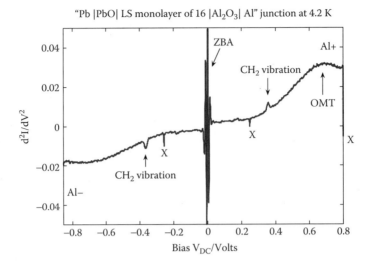

FIGURE 3.8   Wide-scan IETS spectrum of a "Pb–LS monolayer of 16–Al" junction at 4.2 K [96].

to the CH_2 vibration [96]. For a much wider bias range, the IETS spectrum (Figure 3.8) exhibited (1) the most prominent CH_2 vibrational peak at 0.36 Volts; (2) a well-known "zero-bias anomaly" (ZBA) close to 0 Volts, due to Al-O and Pb-O vibrations, and of no interest here; (3) some electronic artifacts labeled "X," and, most importantly, (4) a very broad peak at positive bias due to resonance between the LUMO of 16 and the Pb electrode, at the same bias (0.65 V) at which enhanced current was observed in the IV curve. This peak, previously dubbed "orbital mediated tunneling" [159,160] is broad because several crystallites of Pb, with different Miller indices for the exposed faces, have slightly different work functions (± 0.2 eV), and enter into resonance successively [96]. At negative bias, the OMT goes away because the LUMO is asymmetric in the gap [96]. This OMT peak is what we have sought for a long time: proof that the current indeed must be going through the molecule.

3.7   Switches

Switches require bi-stability. A crystal with bulk bi-stability is CuTCNQ, which is metastable between its neutral form Cu^0TCNQ^0 and its ionic form, Cu^+TCNQ^-: this allowed for high- and low-voltage conductivities [161], but despite much work, many publications and patents, it did not become a practical device. An LB monolayer of a bistable [3]catenane closed-loop molecule, with a naphthalene group as one "station," and tetrathiafulvalene as the second "station," and a tetracationic catenane hexafluorophospate salt traveling on the catenane, like a "train" on a closed track, was deposited on poly-Si as one electrode, and topped by a 5-nm Ti layer and a 100-nm Al electrode. The current-voltage plot is asymmetric as a function of bias (which may move the train on the track), and a succession of read-write cycles shows that the resistance changes stepwise, as the train(s) move from the lower-conductivity station(s) to the higher-conductivity station(s) [162]. Infrared spectroscopy showed that depositing Ti atop the catenane does indeed lead to chemical reaction of Ti with the "top" of the monolayer, but preserves the "working part" of the molecular switch [163].

3.8   Capacitors

Bi-stable molecules and unimolecular rectifiers could also be used as capacitors, but this possibility has not received much attention so far.

3.9 Future Flash Memories

A flash memory device has a middle electrode that is not in electrical contact with the outside circuit, so after a polarization pulse, charges can be stored for a long time (but not forever) on this middle electrode. Two monolayers of unimolecular rectifiers, separated by a middle "floating" electrode, could be used as flash memory devices, but this has never been tested.

3.10 Field-Effect Transistors

A field-effect transistor (FET) requires a semiconducting channel connecting source and drain electrodes whose "thickness" can be modified by an applied bias on the gate electrode. For this, any semiconductor will do. Present integrated circuits use FETs preferentially over BJTs because of their ease of fabrication. FET behavior was observed for LB monolayers and multilayers some time ago [164]. FET behavior has been observed by STM for a single-walled carbon nanotube curled over parallel Au lines, with the STM acting as a gate electrode [165], and much work has since been devoted to these FETs. The difficulty of ordering the nanotubes has so far prevented practical use.

3.11 Negative Differential Resistance Devices

Using a "nanopore" technique, molecules of 2'-amino-4-ethynylphenyl-4'-ethynylphenyl-5'-nitro-benzene-1-thiolate, attached to Au on one side and topped by a Ti electrode on the other, exhibit negative differential resistance (NDR) [166]. This molecule, when studied by STM, shows time-dependent oscillations in conductance, presumably due to a change in tilt angle of the organothiolate with respect to the Au substrate [167]. However, if the second "top" metal is deposited at room temperature, then evidence of chemical reactions at the open surface of alkoxyorganothiolates (Al, Cu, Ag, and especially the very reactive Ti) or of interpenetration to the bottom of the SAM close to the Au-S interface (Au) has been presented [27,168].

3.12 Coulomb Blockade Device and Single-Electron Transistor

The organometallic equivalent of a no-gain organometallic single-electron transistor (SET)—i.e., a Coulomb blockade device—was realized at 0.1 K with a Co(II) complex, using two electromigrated Au electrodes covalently bonded to the molecule, and a Si gate electrode at 30 nm from the molecule [169].

3.13 Future Unimolecular Amplifiers

A three-sided molecule, designed to control the current pathway within it by the judicious choice of three moieties with different electron affinities and/or ionization potentials, when covalently bonded to three metal electrodes 3 nm apart, should be the unimolecular equivalent of a bipolar junction transistor [127,129,135,137]. Many suitable molecules can be designed, with end-groups designed for SAM formation with two or three dissimilar metal electrodes, but at present it is highly nontrivial to fabricate, even by electromigration, three electrodes 3 nm apart.

3.14 Future Organic Interconnects

One a sufficient set of resistors, capacitors, rectifiers, and amplifiers have been demonstrated to work with conventional metal electrodes, one can initiate a new project, of assembling all-organic polymeric electrodes to replace the inorganic metals. This would lead to the all-organic computer! The controlled electrochemical growth of conducting oligomer filaments has already been demonstrated [170].

3.15   Acknowledgments

This work was made possible by the help, diligence, and insight of many colleagues, students, and post-doctoral fellows to whom I owe a large debt of gratitude. I hope they had fun working on these problems!

References

[1] Metzger, R.M., Prospects for truly unimolecular devices, in Metzger, R.M. et al., Eds., *Lower-Dimensional Systems and Molecular Electronics, NATO ASI Series*, B248, 659–666, 1991.

[2] Moore, G.E., Cramming more components onto integrated circuits, *Electronics*, 38, 8, 114–117, 1965.

[3] Aviram, A. and Ratner, M.A., Molecular rectifiers, *Chem. Phys. Lett.*, 29, 277, 1974.

[4] Koopmans, T., The classification of wave functions and eigen-values to the single electrons of an atom, *Physica*, 1, 104–113, 1934.

[5] Tour, J.M. and James, D.K., elsewhere in this volume.

[6] Reed, M.A. and Wang, W., elsewhere in this volume.

[7] Schottky, W., Simplified and extended theory of barrier–layer rectifiers, *Z. f. Phys.*, 118, 539–592, 1942.

[8] Ohm, G. S., *Die Galvanische Kette, Mathematisch Bearbeitet*, Riemann, Berlin, 1827.

[9] Binnig, G. et al., Surface studies by scanning tunneling microscopy, *Phys. Rev. Lett.*, 49, 57–61, 1982.

[10] Binnig, G. et al., Atomic force microscope, *Phys. Rev. Lett.*, 56, 930, 1986.

[11] D,rig, U. et al., Observation of metallic adhesion using the scanning tunneling microscope, *Phys. Rev. Lett.* 65, 349–352, 1990.

[12] Blodgett, K.B., Films built by depositing successive monomolecular layers on a solid surface, *J. Am. Chem. Soc.*, 57, 1007–1022, 1935.

[13] Blodgett, K.B. and Langmuir, I., Built-up films of barium stearate and their optical properties, *Phys. Rev.*, 51, 964–982, 1937.

[14] Langmuir, I. and Schaefer, V.J., Activity of urease and pepsin monolayers, *J. Am. Chem. Soc.* 60, 1351–1360, 1938.

[15] Bigelow, W.C. et al., Oleophobic monolayers. 1. Films adsorbed from solution in non-polar liquids, *J. Colloid Sci.* 1, 513–538, 1946.

[16] Muller, C.J., Ph.D. thesis, Univ. of Leiden, 1991.

[17] Reed, M.A. et al., Conductance of a molecular junction, *Science*, 278, 252–253, 1997.

[18] Ralls, K.S. et al., Fabrication of thin-metal nanobridges *Appl. Phys. Lett.* 55, 2459–2461, 1989.

[19] Zhou, C. et al., Nanoscale metal/self-assembled monolayer/metal heterostructures, *Appl. Phys. Lett.*, 71, 611–613, 1997.

[20] Park, H. et al., Fabrication of metallic electrodes with nanometer separation by electromigration, *Appl. Phys. Lett.*, 75, 1999.

[21] Cui, X.D. et al., Reproducible measurement of single-molecule conductivity, *Science* 294, 571–574, 2001.

[22] Geddes, N. J. et al., Fabrication and investigation of asymmetric current–voltage characteristics of a metal/Langmuir–Blodgett monolayer/metal structure, *Appl. Phys. Lett.* 56, 1916–1918, 1990.

[23] Metzger, R.M. et al., Unimolecular electrical rectification in hexadecylquinolinium tricyanoquinodimethanide, *J. Am. Chem. Soc.* 119, 10455–10466, 1997.

[24] Okazaki, N. and Sambles, J. R., In *Extended Abstracts of the International Symposium on Organic Molecular Electronics*, 66, Nagoya, Japan, 2000.

[25] Xu, T. et al., Rectification by a monolayer of hexadecylquinolinium tricyanoquinodimethanide between gold electrodes, *Angew. Chem. Intl. Ed.*, 40, 1749–1752, 2001.

[26] Metzger, R.M. et al., Electrical rectification by a monolayer of hexadecylquinolinium tricyanoquinodimethanide measured between macroscopic gold electrodes, *J. Phys. Chem.*, B105, 7280–7290, 2001.

[27] Walker, A.V. et al., The dynamics of noble metal atom penetration through methoxy-terminated alkanethiolate monolayers, *J. Am. Chem. Soc.*, 126, 3954–3963, 2004.

[28] Tighe, T. et al., Evolution of the interface and metal film morphology in the vapor deposition of Ti on hexadecanethiolate hydrocarbon monolayers on Au, *J. Phys. Chem.*, B109, 21006–21014, 2005.

[29] Metzger, R.M. et al., Large current asymmetries and potential device properties of a Langmuir–Blodgett monolayer of dimethyanilinoazafullerene sandwiched between gold electrodes, *J. Phys. Chem.*, B107, 1021–1027, 2003.

[30] Stewart, D.R. et al., Molecule-independent electrical switching in Pt/organic monolayer/Ti devices, *Nano Lett.*, 4, 133–136, 2004.

[31] Black, J.R., Electromigration – A brief survey and some recent results, *IEEE Trans. El. Dev.*, ED-16, 338–347, 1969.

[32] Taube, H. and Gould, E.S., Organic molecules as bridging groups in electron–transfer reactions, *Acc. Chem. Res.*, 2, 321–329, 1969.

[33] Taube, H., Electron transfer between metal–complexes — a retrospective view (Nobel Lecture), *Angew. Chem. Intl. Ed. Engl.*, 23, 329–339, 1984.

[34] Bumm, L.A. et al., Are single molecular wires conducting?, *Science*, 271, 1705, 1996.

[35] Landauer, R., Spatial variation of currents and fields due to localized scatterers in metallic conduction, *IBM J. Res. Dev.*, 1, 223–231, 1957.

[36] von Klitzing, K. et al., New method for high-accuracy determination of the fine-structure constant based on quantized hall resistance, *Phys. Rev. Lett.*, 45, 494–497, 1980.

[37] Simmons, J.G., Conduction in thin dielectric films *J. Phys.*, D4, 613, 1971.

[38] Chen, J. et al., Molecular electronic devices, in Reed, M.A. and Lee, T., Eds., *Molecular Nanoelectronics*, 39–114, 2003.

[39] Iijima, S. and Ichihashi, T., Single-shell carbon nanotubes of 1-nm diameter, *Nature*, 363, 603–605, 1993.

[40] Saito, R. et al., Electronic structure of carbon fibers based on C_{60}, in Chiang, L.Y. et al., Eds., *Electrical, Optical, and Magnetic Properties of Organic Solid-State Materials, MRS Symp. Proc.*, 247, 333–338, 1992.

[41] White, C.T. et al., Helical and rotational symmetries of nanoscale graphitic tubules, *Phys. Rev.*, B47, 5485–5488, 1993.

[42] Averin, D.V. and Likharev, K.K., Coulomb blockade of single-electron tunneling, and coherent oscillations in small tunnel junctions, *J. Low-Temp. Phys.*, 62, 345–373, 1986.

[43] Geddes, N.J. et al., The electrical properties of metal-sandwiched Langmuir–Blodgett multilayers and monolayers of a redox-active organic molecular compound, *J. Appl. Phys.*, 71, 756–768, 1992.

[44] Krzeminski, C. et al., Theory of rectification in a molecular monolayer, *Phys. Rev. B*, 64, # 085405, 2001.

[45] Mujica, V. et al., Molecular rectification: why is it so rare?, *Chem. Phys.*, 281, 147–150, 2002.

[46] Chabinyc, M.L. et al., Molecular rectification in a metal–insulator–metal junction based on self-assembled monolayers, *J. Am. Chem. Soc.*, 124, 11730–11736, 2002.

[47] Peterson, I.R. et al., Analytical model for molecular-scale charge transport, *J. Phys. Chem.*, A105, 4702–4707, 2001.

[48] Stabel, A. et al., Diodelike current–voltage curves for a single molecule–tunneling spectroscopy with submolecular resolution of an alkylated, *peri*-condensed hexabenzocoronene, *Angew. Chem. Int., Ed.* 34, 1609–1611, 1995.

[49] Hall, L.E. et al., Formalism, analytical model, and a-priori Green's function-based calculations on the current-voltage characteristics of molecular wires, *J. Chem. Phys.*, 112, 1510–1521, 2000.

[50] Metzger, R.M. and Panetta, C.A., Progress in cohesive energies and in building organic unimolecular rectifiers, *J. Phys. (Paris)*, 44 *Colloque* C3, 1605–1611, 1983.

[51] Metzger, R.M. and Panetta, C.A., Toward organic rectifiers, in F. L. Carter, Ed., *Molecular Electronic Devices*, Vol. 2, Dekker, New York, 1987.

[52] Panetta, C.A. et al., TTF-NHCO$_2$(CH$_2$)$_2$O- TCNQBr and TTF-CO$_2$(CH$_2$)$_2$-O-TCNQBr, two potential molecular rectifiers, *Mol. Cryst. Liq. Cryst.*, 107, 103–113, 1984.

[53] Metzger, R.M. et al., Toward organic rectifiers: Langmuir–Blodgett films and redox properties of the n-4-n-dodecyloxyphenyl and n-1-pyrenyl carbamates of 2-bromo,5-(2'–hydroxyethoxy)TCNQ, *J. Mol. Electronics*, 2, 119–124, 1986.

[54] Metzger, R.M. et al., Progress toward organic single-monolayer rectifiers, *Synth. Metals*, 18, 797–802, 1987.

[55] Metzger, R.M. and Panetta, C.A., Progress in molecular electronics: Langmuir–Blodgett films of donor-sigma-acceptor molecules, in J. L. Heiras and T. Akachi, Eds., *Proc. Eighth Winter Meeting on Low-Temperature Physics*, UNAM, Mexico City, 1987.

[56] Torres, E. et al., The preparation of 2-hydroxymethyl- 11,11,12,12-tetracyanoanthraquinodimethane and its carbamates with electron-donor moieties, *J. Org. Chem.* 52, 2944–2945, 1987.

[57] Metzger, R.M. and Panetta, C. A., Langmuir–Blodgett films of donor-sigma-acceptor molecules and prospects for organic rectifiers, in P. Delhaès and M. Drillon, Eds., *Organic and Inorganic Lower-Dimensional Materials*, NATO ASI Ser., B168, 271–286, Plenum, New York, 1988.

[58] Metzger, R.M., et al., Langmuir–Blodgett films of donor-urethane-tcnq and related molecules, *Langmuir* 4, 298–304, 1988.

[59] Miura, Y. et al., Electroactive organic materials. preparation and properties of 2-(2'-hydroxyethoxy)-7,7,8,8-tetracyanoquinodimethane, *J. Org. Chem.*, 53, 439–440, 1988.

[60] Miura, Y. et al., Crystal structure of 2-acetoxyethoxy-7,7,8,8-tetracyanoquinodimethan, AETCNQ, C$_{16}$H$_{10}$N$_4$O$_3$, *Acta Cryst.* C44, 2007–2009, 1988.

[61] Laidlaw, R.K. et al., Crystal structure of methyl 4-(N,N-dimethylamino)phenyl carbamate, DMAPCMe, C$_{10}$H$_{14}$N$_2$O$_2$, *Acta Cryst.* C44, 2009–2013, 1988.

[62] Laidlaw, R.K. et al., Crystal structure of 2-bromo-5-hydroxyethoxy-7,7,8,8-tetracyanoquino-dimethan, BHTCNQ, C$_{14}$H$_7$N$_4$O$_2$Br, *Acta Cryst.*, B44, 645–650, 1988.

[63] Miura, Y. et al., Preparative purification of 2-(2'- hydroxyethoxy)-terephthalic acid by countercurrent chromatography, *J. Liquid Chromatography*, 11, 245–250, 1988.

[64] Metzger, R.M. and Panetta, C.A., Langmuir–Blodgett films of potential donor-sigma-acceptor organic rectifiers, *J. Mol. Electronics*, 5, 1–17, 1989.

[65] Metzger, R.M. and Panetta, C.A., Rectification in Langmuir–Blodgett monolayers of organic D-σ-A molecules, *J. Chem. Phys.*, 85, 1125–1134, 1988.

[66] Metzger, R.M. and Panetta, C.A., Possible rectification in Langmuir–Blodgett monolayers of organic D-σ-A molecules, *Synth. Metals*, 28, C807–C814, 1989.

[67] Metzger, R.M. et al., Crystal structure of DMAP-C-HMTCAQ, N,N-dimethylaminophenyl-carbamate-2'hydroxymethoxy-11,11,12,12-tetracya-no-anthraquinodimethan, *J. Crystall. Spectro-scopic Res.*, 19, 475–482, 1989.

[68] Metzger, R.M. and Panetta, C.A., Langmuir–Blodgett films of potential organic rectifiers, in Aviram, A., Ed., *Molecular Electronics — Science and Technology*, New York Engineering Foundation, 1990.

[69] Metzger, R.M. et al., Monolayers and Z-type multilayers of donor-sigma-acceptor molecules with one, two, and three dodecoxy tails, *Langmuir* 6, 350–357, 1990.

[70] Metzger, R.M. and Panetta, C.A., Review of the organic rectifier project, Langmuir–Blodgett films of donor-sigma-acceptor molecules, in Metzger, R.M. et al., Eds., *Lower-Dimensional Systems and Molecular Electronics*, NATO ASI Ser. Ser., B248, 611–625, Plenum Press, New York, 1991.

[71] Metzger, R.M. and Panetta, C.A., Langmuir–Blodgett films of potential unidimensional organic rectifiers, in L.Y. Chiang et al., Eds., *Advanced Organic Solid State Materials*, Mater. Res. Soc. Symp. Proc. Ser., 173, 531–536, 1990.

[72] Metzger, R.M. and Panetta, C.A., The quest for unimolecular rectifiers, *New J. Chem.*, 15, 209–221, 1991.

[73] Metzger, R.M. and Panetta, C.A., Langmuir–Blodgett films of potential organic rectifiers, in J. L. Beeby, Ed., *Condensed Systems of Low Dimensionality NATO ASI Ser.*, B253, 779–793, Plenum Press, New York, 1991.

[74] Metzger, R.M. and Panetta, C.A., Langmuir–Blodgett films of potential organic rectifiers: new scanning tunneling microscopy and non-linear optical results, *Synth. Metals* 42, 1407–1413, 1991.

[75] Panetta, C.A. et al., Functionalized tetracyanoquinodimethane-type electron acceptors: suitable precursors for D-σ-A materials, *Synlett*, 301–309, 1991.

[76] Metzger, R.M., The search for organic unimolecular rectifiers, in A. Aviram, Ed., *Molecular Electronics — Science and Technology*, Am. Inst. Phys. Conf. Proc. 262, 85–92, 1992.

[77] Wu, X.-L. et al., Scanning tunneling microscopy and high-resolution transmission electron microscopy of $C_{16}H_{33}$-Q3CNQ, hexadecylquinolinium tricyanoquinodimethanide, *Synth. Metals*, 57, 3836–3841, 1993.

[78] Wang, P. et al., Scanning tunneling microscopy and transmission electron microscopy of Langmuir–Blodgett films of three donor-sigma-acceptor molecules: BDDAP-C-HETCNQ, BDDAP-C-HPTCNQ and BDDAP-C-HBTCNQ, *Synth. Metals*, 57, 3824–3829, 1993.

[79] Metzger, R.M., The long road towards organic unimolecular rectifiers, in Blank, M., Ed., *Electricity and Magnetism in Biology and Medicine*, San Francisco Press, San Francisco, 1993.

[80] Metzger, R.M., The quest for D-σ-A unimolecular rectifiers and related topics in molecular electronics, in Birge, R.R., Ed., *Molecular and Biomolecular Electronics*, Am. Chem. Soc. Adv. in Chem. Ser., 240, 81–129, American Chemical Society, Washington, DC, 1994.

[81] Nadizadeh, H. et al., Langmuir–Blodgett films of donor-bridge-acceptor (D-σ-A) compounds, where D = anilide donors with internal diyne or saturated lipid tails, σ = carbamate bridge, and A = 4-nitrophenyl or TCNQ acceptors, *Chem. Mater.*, 6, 268–277, 1994.

[82] Metzger, R.M., D-σ-A unimolecular rectifiers, *Matrls. Sci. Engrg.*, C3, 277–285, 1995.

[83] Metzger, R.M., et al., Is ashwell's zwitterion a molecular diode?, *Synth. Metals*, 85, 1359–1360, 1997.

[84] Metzger, R.M., The prospects for unimolecular rectification, in Sasabe, H., Ed., *Hyper-Structured Molecules I: Chemistry, Physics, and Applications*, Gordon & Breach Science Publishers, Amsterdam, 1999.

[85] Ashwell, G.J. et al., Rectifying characteristics of Mg | ($C_{16}H_{33}$-Q3CNQ LB film) | Pt structures, *J. Chem. Soc. Chem. Commun.*, 1374–1376, 1990.

[86] Martin, A.S. et al., Molecular rectifier, *Phys. Rev. Lett.*, 70, 218–221, 1993.

[87] Chen, B. and Metzger, R.M., Rectification between 370 K and 105 K in hexadecyl-quinolinium tricyanoquinodimethanide, *J. Phys. Chem.*, B103, 4447–4451, 1999.

[88] Vuillaume, D. et al., Electron transfer through a monolayer of hexadecylquinolinium tricyanoquinodimethanide, *Langmuir*, 15, 4011–4017, 1999.

[89] Baldwin, J.W. et al., Rectification and nonlinear optical properties of a Langmuir–Blodgett monolayer of a pyridinium dye, *J. Phys. Chem.*, B106, 12158–12164, 2002.

[90] Jaiswal, A. et al., Electrical rectification in a monolayer of zwitterions assembled by either physisorption or chemisorption, *Langmuir*, 19, 9043–9050, 2003.

[91] Honciuc, A. et al., Current rectification in a Langmuir–Schaefer monolayer of fullerene-bis-4-diphenylamino-4-(N-ethyl-N-2'''-ethyl)amino-1,4-diphenyl-1,3-butadiene malonate between Au electrodes, *J. Phys. Chem.*, B109, 857–871, 2005.

[92] Shumate, W.J. et al., Spectroscopic and rectification studies of three donor-sigma-acceptor compounds, consisting of a one-electron donor (pyrene or ferrocene), a one-electron acceptor (perylenebisimide), and a C19 swallowtail, *J. Phys. Chem.*, B110, 11146–11159, 2006.

[93] Jaiswal, A. et al., Comparison of unimolecular rectification in monolayers of $CH_3C(O)S$-$C_{14}H_{28}Q^+$-3CNQ$^-$ and $CH_3C(O)S$-$C_{16}H_{32}Q^+$-3CNQ$^-$ organized by self-assembly, Langmuir–Blodgett, and Langmuir–Schaefer techniques, unpublished.

[94] Shumate, W.J., Ph.D. thesis, Univ. of Alabama, 2005.

[95] Honciuc, A. et al., Polarization of charge-transfer bands and rectification in hexadecylquinolinium 7,7,8-tricyano-quinodimethanide and its tetrafluoro analog, *J. Phys. Chem*, B110, 15085–15093, 2006.

[96] Honciuc, A. et al., Electron tunneling spectroscopy of a rectifying monolayer, to be submitted.

[97]   Baldwin, J.W. et al., Spectro-scopic studies of hexadecylquinolinium tricyanoquinodimethanide, *J. Phys. Chem.*, B103, 4269–4277, 1999.

[98]   Xu, T. et al., A spectroscopic study of hexadecylquinolinium tricyanoquinodimethanide as a monolayer and in bulk, *J. Phys. Chem.*, B106, 10374–10381, 2002.

[99]   Terenziani, F. et al., From solution to Langmuir–Blodgett films: spectroscopic study of a zwitterionic dye, *J. Phys. Chem.*, B108, 10743–10750, 2004.

[100] Kwon, O. et al., Theoretical calculations of methyl-quinolinium tricyanoquinodimethanide (ch_3q-3cnq) using a solvation model, *Chem. Phys. Lett.*, 313, 321–331, 1999.

[101] Scheib, S. et al., In search of molecular rectifiers. The D-σ-A system derived from triptycenequinone and tetrathiafulvalene, *J. Org. Chem.*, 63, 1198–1204, 1998.

[102] Hughes, T.V. et al., Synthesis and Langmuir–Blodgett film formation of amphiphilic zwitterions based on benzothiazolium tricyanoquinodimethanide, *Langmuir*, 15, 6925–6930, 1999.

[103] Xu, T. et al., Current-voltage characteristics of an LB monolayer of di-decylammonium tricyanoquinodimethanide measured between macroscopic gold electrodes, *J. Mater. Chem.*, 12, 3167–3171, 2002.

[104] Metzger, R.M., et al., Electrical rectification by a molecule of hexadecylquinolinium tricyanoquinodimethanide, in L.Y. et al., Eds., *Electrical, Optical, and Magnetic Properties of Organic Solid-State Materials IV*, MRS Proceedings, 488, Materials Research Society, Pittsburgh, 1998.

[105] Metzger, R.M., et al., Observation of unimolecular electrical rectification in hexadecylquinolinium tricyanoquinodime-thanide, *Thin Solid Films*, 327–329, 326–330, 1998.

[106] Scheib, S. et al., In search of D-σ-A molecular rectifiers: The D-σ-A system derived from triptycenequinone and tetrathiafulvalene, *Thin Solid Films*, 327–329, 100–103, 1998.

[107] Metzger, R.M. and Cava, M.P. Rectification by a single molecule of hexadecylquinolinium tricyanoquinodimethanide, in *Molecular Electronics: Science and Technology*, Ann. N. Y. Acad. Sci., 852, 95–115, 1998.

[108] Metzger, R.M., Demonstration of unimolecular electrical rectification in hexadecyl-quinolinium tricyanoquinodimethanide, *Adv. Mater. Optics & Electronics*, 8, 229–245, 1998.

[109] Metzger, R.M., Unimolecular electrical rectification by hexadecylquinolinium tricyanoquinodimethanide, *Mol. Cryst. Liq. Cryst. Sci. Technol.*, A337, 37–42, 1999.

[110] Metzger, R.M., The unimolecular rectifier: Unimolecular electronic devices are coming, *J. Mater. Chem.*, 9, 2027–2036, 1999.

[111] Metzger, R.M., All about γ-hexadecylquinolinium tricyanoquinodimethanide: A unimolecular rectifier of electrical current, *J. Mater. Chem.*, 10, 55–62, 2000.

[112] Metzger, R.M., Unimolecular rectification down to 105 K and spectroscopy of hexadecylquinolinium tricyanoquinodimethanide, *Synth. Metals*, 109, 23–28, 2000.

[113] Metzger, R.M., Electrical rectification by a molecule: The advent of unimolecular electronic devices, *Acc. Chem. Res.*, 32, 950–957, 1999.

[114] Metzger, R.M. et al., Unimolecular rectification between 370 K and 105 K and spectroscopic properties of hexadecylquinolinium tricyanoquino-dimethanide, in Glaser, R. and Kaszinski, P., Eds., *Anisotropic Organic Materials — Approaches to Polar Order*, Am. Chem. Soc. Symp. Proc., 798, 50–65, 2001.

[115] Metzger, R.M., Unimolecular rectification down to 105 K and spectroscopic properties of hexadecylquinolinium tricyanoquinodimethanide, in Pantelides, S. K. et al., Eds., *Molecular Electronics*, Mater. Res. Soc. Symp. Proc., 582, paper H12.2, Materials Research Society, Warrendale, PA, 2001.

[116] Metzger, R.M., Hexadecylquinolinium tricyanoquinodimethanide, a unimolecular rectifier between 370 K and 105 K and its spectroscopic properties, *Adv. Mater. Optics & Electronics*, 9, 253–263, 1999.

[117] Metzger, R.M., Rectification by a single molecule, *Synth. Metals*, 124, 107–112, 2001; and in Ledoux-Rak, I. et al., Eds., *Molecular Photonics: From Macroscopic to Nanoscopic Applications, European Mater. Res. Soc. Proc.*, 96, Elsevier, Amsterdam, 2001.

[118] Metzger, R.M., Electrical rectification by a monolayer of hexadecylquinolinium tricyanoquinodimethanide sandwiched between gold electrodes, in Merhari, L., Eds., *Nonlithographic and*

Lithographic Methods of Nanofabrication – From Ultrahigh - Scale Integration to Photonics to Molecular Electronics, Mater. Res. Soc. Symp. Proc., 636 Materials Res. Soc., Warrendale, PA, 2001.

[119] Metzger, R.M., The quest of unimolecular rectification: from Oxford to Waltham to Exeter to Tuscaloosa, *J. Macromol. Sci.*, A38, 1499–1517, 2001.

[120] Metzger, R.M., Monolayer Rectifiers, *J. Solid St. Chem.* 168, 696–711, 2002.

[121] Metzger, R.M., Unimolecular Rectifiers, in Reed, M.A. and Lee, T. Eds., *Molecular Nanoelectronics*, American Scientific Publishers, Stevenson Ranch, CA, 2003.

[122] Metzger, R.M., Three Langmuir–Blodgett monolayer rectifiers, in *Structural and Electronic Properties of Molecular Nanostructures, AIP Conf. Proc.*, 663, AIP Conf. Proc., Melville, NY, 2002.

[123] Metzger, R.M., Electrical rectification by Langmuir–Blodgett Monolayers, *Nanotechnology*, 13, 585–591, 2003.

[124] Metzger, R.M., One-molecule-thick devices: Rectification of electrical current by three Langmuir–Blodgett monolayers, *Synth. Metals*, 137, 1499–1501, 2003.

[125] Adams, D. et al., Charge transfer on the nanoscale, *J. Phys. Chem.*, B107, 6668–6697, 2003.

[126] Metzger, R.M., Unimolecular electrical rectifiers, *Chem. Rev.* 103, 3803–3834, 2003.

[127] Metzger, R.M., Unimolecular rectifiers and proposed unimolecular amplifier, in Reimers, J.R. et al., Eds., *Molecular Electronics III, Ann. N. Y. Acad. Sci.*, 1006, 252–276, 2003.

[128] Metzger, R.M., Molecular rectifiers, in *Encyclopedia of Supramolecular Chemistry*, II, 1525–1537, Marcel Dekker, New York, NY, 2004.

[129] Metzger, R.M., Three unimolecular rectifiers and a proposed unimolecular amplifier, in Ouahab, L. and Yagubskii, E., Eds., *Organic Conductors, Superconductors and Magnets: from Synthesis to Molecular Electronics, NATO ASI Ser. II*, Kluwer, Dordrecht, The Netherlands, 2004.

[130] Allara, D.L. et al., The design, characterization and use of molecules in molecular devices, in Ouahab, L. and Yagubskii, E., Eds., *Organic Conductors, Superconductors and Magnets: from Synthesis to Molecular Electronics, NATO ASI Ser. II*, Kluwer, Dordrecht, The Netherlands, 2004.

[131] Metzger, R.M., Four examples of unimolecular electrical rectifiers, *The Electrochemical Society Interface*, 13, 40–44, 2004.

[132] Metzger, R.M., Unimolecular rectifiers and beyond, in Razeghi, M. and Brown, G.J., Eds., *Quantum Sensing and Nanophotonic Devices SPIE* 5359, 153–168, SPIE – The International Society for Optical Engineering, Bellingham, WA, 2004.

[133] Metzger, R.M., Unimolecular rectifiers and prospects for other unimolecular electronic devices, *Chem. Record*, 4, 291–304, 2004.

[134] Metzger, R.M., Electrical rectification by monolayers of three molecules, in Kahovec, J., Ed., *Electronic Phenomena in Organic Solids, Macromol. Symposia*, 212, 63–72, Wiley-VCH, Weinheim, Germany, 2004.

[135] Metzger, R.M., Four unimolecular rectifiers and what lies ahead, in *Proc. of the First Conference on Foundations of Nanoscience: Self-Assembled Architectures and Devices, Snowbird, UT*, Sciencetechnica, 2004.

[136] Peterson, I.R. and Metzger, R.M., Individual molecules as electronic components, *IEE Proc. Circ. Dev. Syst.*, 151, 452–456, 2004.

[137] Metzger, R.M., Six unimolecular rectifiers and what lies ahead, in Cuniberti, G. et al., Eds., *Introducing Molecular Electronics, Springer Lecture Notes on Physics*, 680, Springer, Berlin, 2005.

[138] Metzger, R.M., Unimolecular rectifiers and what lies ahead, *Colloids and Surfaces*, A285, 2–10, 2006.

[139] Metzger, R.M., Unimolecular rectifiers: methods and challenges, *Anal. Chim. Acta*, 568, 146–155, 2006.

[140] Metzger, R.M., Unimolecular rectifiers: Present status, *Chem. Physics*, 326, 176–187, 2006.

[141] Metzger, R.M., Eight unimolecular rectifiers, in Saito, G. and Maesato, M., Eds.*Multifunctional Conducting Molecular Materials*, Royal Society of Chemistry, London, accepted and in press.

[142] Perepichka, D.F. et al., A covalent tetrathiafulvalene-tetracyanoquinodimethane diad: extremely low HOMO–LUMO gap, thermoexcited electron transfer, and high-quality Langmuir–Blodgett films, *Angew. Chem. Int. Ed.*, 42, 4636–4639, 2003.

[143] Ho, G. et al., The first studies of a tetrathiafulvalene-σ-acceptor molecular rectifier, *Chem. Eur. J.*, 11, 2914–2922, 2005.

[144] Ashwell, G.J. et al., Molecular rectification: Self-assembled monolayers of a donor-(pi-bridge)-acceptor chromophore connected via a truncated Au-S-$(CH_2)_3$ bridge, *J. Mater. Chem.*, 13, 2855–2857, 2003.

[145] Ashwell, G.J. et al., Rectifying Au-S-CnH2n-P3CNQ derivatives, *J. Mater. Chem.*, 14, 2848–2851, 2004.

[146] Ashwell, G.J. and Berry, M., Hybrid SAM/LB device structures: Manipulation of the molecular orientation for nanoscale electronic applications, *J. Mater. Chem.*, 15, 108–110, 2005.

[147] Ashwell, G.J. et al., Molecular rectification: Self-assembled monolayers in which donor-(π-bridge)-acceptor moieties are centrally located and symmetrically coupled to both gold electrodes, *J. Am. Chem. Soc.*, 126, 7102–7110, 2005.

[148] Ashwell, G.J. et al., Orientation-induced molecular rectification and nonlinear optical properties of a squaraine derivative, *J. Mater. Chem.* 15, 1154–1159, 2006.

[149] Ashwell, G.J. et al., Induced rectification from self-assembled monolayers of sterically hindered pi-bridged chromophores, *J. Mater. Chem.*, 15, 1160–1166, 2005.

[150] Ashwell, G.J. and Mohib, A., Improved molecular rectification from self-assembled monolayers of a sterically hindered dye, *J. Am. Chem. Soc.*, 127, 16238–16244, 2005.

[151] Ashwell, G.J. et al., Dipole reversal in Langmuir–Blodgett films of an optically nonlinear dye and its effect on the polarity for molecular rectification, *J. Mater. Chem.*, 15, 4203–4305, 2005.

[152] Ashwell, G.J. et al., Molecular rectification: Stabilised alignment of chevron-shaped dyes in hybrid sam/lb structures in which the self-assembled monolayer is anionic and the Langmuir–Blodgett layer is cationic, *J. Chem. Soc. Faraday Disc.*, 131, 23–31, 2006.

[153] Ashwell, G.J. et al., Molecules that mimic Schottky diodes, *Phys. Chem. Chem. Phys.*, 8, 3314–3319, 2006.

[154] Ashwell, G.J. et al., Organic rectifying junctions fabricated by ionic coupling, *Chem. Commun.*, 618–620, 2006.

[155] Ashwell, G.J. et al., Organic rectifying junctions from an electron-accepting molecular wire and an electron-donating phthalocyanine, *Chem. Commun.*, 1640–1642, 2006.

[156] Morales, G.M. et al., Inversion of the rectifying effect in diblock molecular diodes by protonation, *J. Am. Chem. Soc.*, 127, 10456–10457, 2005.

[157] Elbing, M. et al., A single-molecule diode, *Proc. Natl. Acad. Sci. U.S.*, 102, 8815–8820, 2005.

[158] Ashwell, G.J. and Paxton G.A.N., Multifunctional properties of Z-beta-(N-hexadecyl-4-quinolinium)-alpha-cyano-4-styryldicyanomethanide: a molecular rectifier, optically non-linear dye, and ammonia sensor, *Austr. J. Chem.*, 55, 199–204, 2002.

[159] Mazur, U. and Hipps, K.W., Resonant tunneling bands and electrochemical reduction potentials, *J. Phys. Chem.*, 99, 6684–6688, 1995.

[160] Mazur, U. and Hipps, K.W. Orbital-mediated tunneling, inelastic electron tunneling, and electrochemical potentials for metal phthalocyanine thin films, *J. Phys. Chem.*, B103, 9721–9727, 1999.

[161] Potember, R.S. et al., Electrical switching and memory phenomena in Cu-TCNQ thin films, *Appl. Phys. Lett.*, 34, 405–407, 1979.

[162] Collier, C.P. et al., A 2.catenane based solid-state electronically reconfigurable switch, *Science*, 289, 1172–1175, 2000.

[163] DeIonno, E. et al., Infrared spectroscopic characterization of 2.rotaxane molecular switch tunnel junction devices, *J. Phys. Chem.*, B110, 7609–7612, 2006.

[164] Paloheimo, J. et al., Molecular field-effect transistors using conducting polymer Langmuir–Blodgett films, *Phys. Lett.*, 56, 1157–1159, 1990.

[165] Tans, S.J. et al., Individual single-wall carbon nanotubes as quantum wires, *Nature*, 386, 474–477, 1997.

[166] Chen, J. et al., Large on-off ratios and negative differential resistance in a molecular electronic device, *Science*, 286, 1550–1552, 1999.

[167] Donhauser, Z.J. et al., Conductance switching in single molecules through conformational changes, *Science*, 292, 2303–2307, 2001.

[168] Haynie, B.C. et al., Adventures in molecular electronics: how to attach wires to molecules, *Appl. Surf. Sci.*, 203–204, 433–436, 2003.

[169] Park, J. et al., Coulomb blockade and the kondo effect in single atom transistors, *Nature*, 417, 722–725, 2002.

[170] He, H. et al., A conducting polymer nanojunction switch, *J. Am. Chem. Soc.*, 123, 7730–7731, 2001.

4

Carbon Derivatives

Rikizo Hatakeyama

4.1 Introduction

Since carbon allotropes take on a diversity of structures and properties, carbon-based materials have attracted much interest in the areas of basic science and practical materials development. The carbon chemical bond is characterized by the s and p hybridized orbitals—carbyne with the sp hybridized orbital, graphite with the sp^2 hybridized orbital, and diamond with the sp^3 hybridized orbital are one-, two-, and three-dimensionally stretched, respectively. This leads to the generation of a vast variety of carbon derivatives given that 90% of 13 million kinds of terrestrial materials are carbon-filled organic compounds. The diamond (an insulator) is known to be the hardest material, while graphite (a conductor) is one of the softest crystals. On the other hand, newly discovered fullerenes [1,2], as well as carbon nanotubes [3,4], have both soccer ball–like and cylindrical structures with diameters of approximately 1 to 1-50 nm, which are zero- and one-dimensionally stretched, respectively. They are pure nanoscale allotropes of carbon formed in the sp^2 hybridized graphite structure, but that incorporate five-membered rings, as well as the normal hexagon rings, in order to bend the lattice and form a closed carbon cage. Their chemical/electric/magnetic/optical properties have been investigated in many different areas of chemistry, physics, biology, medicine, and engineering. The fullerenes and carbon nanotubes are expected to be candidates for key materials in bottom-up nanotechnology, given that top-down nanotechnology, which prevails mainly in the Si semiconductor industries, is approaching an inevitable limit regarding processing in nanoscale, because the electronic and mechanical properties combined with their small dimensions make them very suitable for constructing nanoscale electronic devices.

Furthermore, both of them can be modified internally and externally by incorporating other kinds of atoms and molecules, yielding a variety of unusually structured and functional carbon derivatives. Here, material processes, resultant derivatives, and their properties relating to nano- and molecular electronics are taken up, with an emphasis on inner nanospace control of carbon fullerenes and nanotubes.

4.2 Nano-Electronics Oriented Carbon Fullerenes

4.2.1 Cutting Edge Background

The encapsulation of atoms inside hollow cages of fullerenes has proved a fascinating task ever since its discovery. Endohedral fullerenes ("endo" meaning within) are molecules where one or more atoms are captured inside the carbon cage. These were given the appealing designation $M@C_n(n = 60, 70,$ and so on). Among the elements of the periodic table, electropositive metals such as cations, noble gas atoms, and group-V atoms have so far been encapsulated mainly by adding appropriate materials during the formation of the fullerenes in arc discharge and laser vaporization processes, applying high temperatures and high pressures to rare gas atoms, and other alternative methods such as atomic collisions in beams and ion implantation. However, group III-atoms (metals such as Sc, Y, La, Gd, etc.) have exclusively been encapsulated, not in C_{60}, but in higher fullerenes such as C_{82}, C_{84}, etc. [5,6]. Since C_{60} has the highest productivity among all the fullerenes, C_n, when using the normal fullerene-production methods, it is an urgent issue today to develop an efficient method for synthesizing the endohedral C_{60} ($M@C_{60}$) with high yields. When the production of $M@C_{60}$ in large quantities is realized, many more experiments will be feasible and exciting results can be expected, leading to the development of innovative applications relating to nano- and molecular electronics.

Before describing any experimental details, let's briefly discuss the historical development of electron-based electronics and information technology in order to understand the meaning of endofullerene-based nanoelectronics (see Figure 4.1). Electrons in atoms have both charge and spin, which have aided the technology of the semiconductor and magnetic materials fields, respectively. By simultaneously exploiting both, a key science was launched: semiconductor spintronics. Here, endofullerene-based nanoelectronics is claimed, where an atom-encapsulated fullerene is regarded as a pseudo atom. In the case of endohedral metallofullerene, charge transfer between atom and fullerene cage may result in the appearance of, for

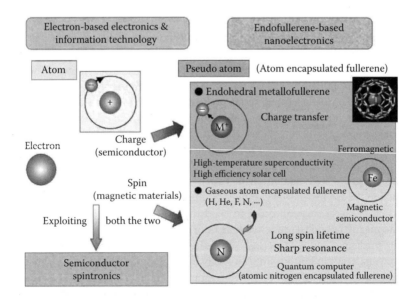

FIGURE 4.1 The historical meaning of endofullerene-based nanoelectronics.

example, high-temperature superconductivity and high efficiency of solar cells. Since the ionization energy of alkali metals such as Na, Cs, and so on is minimum and the most electropositive among all the elements, the alkali metal is predominantly taken up in the present context as a typical charge-transfer element in $M@C_{60}$ [7,8,9]. On the other hand, with a certain specified-atom encapsulated fullerene as one of the gaseous atom encapsulated fullerenes, phenomena such as a long spin lifetime and sharp resonance take place [10,11]. In this respect, the most unusual member of the various endohedral fullerenes is atomic nitrogen in C_{60} ($N@C_{60}$). The enclosed nitrogen atom is uncharged, unbound, and has a very long relaxation time in the quartet ground state ($^4S_{3/2}$) in the center of the fullerene at ambient conditions. Thus, potential applications, such as storing quantum information or quantum computations due to long spin lifetime and sharp resonance [12,13], have been proposed. Therefore, $N@C_{60}$ is also taken up in the present context as a spin-exploited endohedral fullerene. Furthermore, when a ferromagnetic element such as Fe is encapsulated, a magnetic semiconductor $M@C_{60}$ is expected to be synthesized, which may contribute to the development of nano semiconductor spintronics. However, it is not taken up in the present context.

4.2.2 Alkali – Metal Encapsulated Fullerenes

In order to produce macroscopic amounts of endohedral fullerenes. the arc discharge method for pristine (empty) fullerene production is usually employed by using metal or metal oxide doped graphite rods instead of pure ones, where the metal is trapped in the carbon cage, thus closing the fullerene network [6]. Another method of synthesizing endohedral fullerenes was developed by the author's group after gas phase ion-neutral collision experiments [7]. This method adopts a plasma consisting of positively and negatively charged particles, with an equal density to insert alkali ions into fullerenes [8]. A similar method, based on ion implantation in the absence of negatively charged particles was also developed by Cambell *et al.* [9].

Regarding the plasma method, the experiment is performed using a cylindrical vacuum chamber (10 cm diameter, 100 cm long), as shown in Figure 4.2(a). A plasma of electrons and positive-alkali ions (A^+) is produced by the contact ionization of alkali atoms on a 2.0 cm-diameter tungsten plate heated to $2000°C$, where Li, Na, K, and Cs are individually used as alkali metals (A). The background gas pressure

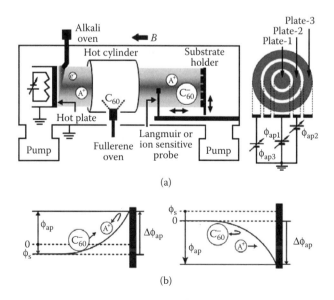

FIGURE 4.2 Experimental apparatus to produce alkali-metal encapsulated fullerenes by a plasma method (a), and potential structures depending on DC bias voltage $\phi_{ap}(\Delta\phi_{ap} = \phi_{ap} - \phi_s)$ (b).

is $P = (1 \sim 3) \times 10^{-4}$ Pa. The plasma, with a density of $n_p = (0.5 \sim 5) \times 10^9$ cm^{-3} and an electron temperature of $T_e \approx 0.2$ eV $\geq T_+$ (positive ion temperature), diffuses along a uniform magnetic field $B = 0.3$ T and passes through a copper cylinder (6.0 cm diameter, 20 cm long) near the tungsten plate. A copper oven is used for C_{60} sublimation, and has a small hole for C_{60} injection. The oven temperature, T_0, is set at 250 to 550°C, and the cylinder temperature is maintained around 450°C. C_{60}- ions appear as a result of electron attachment (e + $C_{60} \rightarrow C_{60}$-), generating an alkali-fullerene plasma which consists of positive alkali-metal ions (A$^+$) and negative ions (C_{60}-) [14]. Plasma parameters and spatial profiles of the C_{60}- ions are measured using Langmuir and ion-sensitive probes. As shown in Figure 4.2(a), an endplate consisting of three circular concentric segmented electrodes (plates 1, 2, 3) as substrates is mainly used, which are located at radial positions of $|r| = 0$ to 0.7, 0.9 to 1.6, and 1.8 to 2.5 cm, respectively. The radial density profile of C_{60}^- is hollow with a broad peak at $|r| = 1.0$ to 1.7 cm, whereas the A$^+$ density is higher at $|r| = 0$ to 0.3 cm, and decreases uniformly with increasing $|r|$. DC bias voltages of ϕ_{ap1}, ϕ_{ap2}, and ϕ_{ap3} are applied to each circular electrode with respect to the grounded tungsten plate. Here, $\Delta\phi_{ap}$ is defined as the difference ($\phi_{ap} - \phi_s$) between ϕ_{ap} and the plasma potential ϕ_s [Figure 4.2(b)]. Deposition time is 60 minutes, and the thickness of the thin films formed is between 1.0 to 1.5 μm. The mass analysis of the thin films is performed using a reflection-type laser-desorption time-of-flight mass spectrometer (LD-TOF MS).

According to typical mass analysis results of thin films accumulated on the substrate (plate–2) in the alkali-fullerene plasmas, the spectra show clear peaks corresponding to the Li-, Na-, and K-endohedral fullerenes Li@C_{60}, Li$_2$@C_{60}, Na@C_{60}, and K@C_{60} only when the DC bias voltages are slightly positive with respect to the plasma potential, i.e., $\Delta\phi_{ap} > 0$ (dimetallofullerene Li$_2$@C_{60} : fullerene containing two Li atoms in the cage). In this case, initially faster A$^+$ and slower C_{60}^- ions are decelerated and accelerated just in front of the substrate (in a plasma-sheath region), respectively, and their relative velocity approaches zero, resulting in a strong Coulomb interaction between them (see the left of Figure 4.2[b]). However, no such spectrum-peaks are observed when the negative DC bisases are applied ($\Delta\phi_{ap} < 0$), where C_{60}^- ions are reflected before arriving at the substrate and A$^+$ ions are accelerated by the sheath potential drop, impinging upon it with higher energies of $-e\Delta\phi_{ap}$ (see the right of Figure 4.2[b]). Figure 4.3 gives the bias dependence of the relative production ratios of Li@C_{60}, Na@C_{60}, K@C_{60}, and Cs@C_{60} to C_{60}, which are Li$^+$-C_{60}^-: ~ 0.6; Na$^+$-C_{60}^-: ~ 1.0; K$^+$-C_{60}^-: ~ 0.1; and Cs$^+$-C_{60}^-: 0, respectively. Here the production ratio of A@C_{60} to C_{60} is estimated by each intensity ratio of the mass spectrum-peaks. The production ratio of A@C_{60}/C_{60} is almost inversely proportional to the diameter of the alkali–metal ions except for the Li@C_{60} case, where the plasma density is lower than in the other alkali–fullerene plasmas due to a difference of Li ionization potential and boiling point from the other alkali metals. The bias voltage for the maximum production ratio increases in proportion to the diameter of the alkali–metal ions. Since the diameter (≈ 3.4 Å) of the Cs$^+$ ion is very large compared with other alkali ions (Li$^+$: $1.2 \sim 1.8$ Å; Na$^+$: $1.9 \sim 2.32$ Å [comparable to the average diameter, ~ 2.48 Å, of the C_{60} six-membered ring]; and K$^+$: $2.66 \sim 3.04$ Å), even a high DC bias voltage of $\Delta\phi_{ap} \approx 50$ V equivalent to the relative energy of C_{60}^- and Cs$^+$, ≈ 50 eV or greater, does not yield the production of Cs@C_{60} [15].

These results suggest that Coulomb interactions, originating from the acceleration and deceleration of A$^+$ and C_{60}^- ions controlled by the substrate bias voltage, have a significant effect on encapsulating atoms in C_{60} cages at extremely low energies. According to *ab initio* molecular dynamics simulations for collisions between C_{60}^- and Li$^+$ (or Na$^+$) ions [16], Li atom is easily encapsulated in C_{60} when Li$^+$ ions hit with kinetic energy of 5 eV to the center of the six-membered ring in the C_{60} cage. However, when Li$^+$ ions hit with 5 eV near the center of a C-C bond in C_{60}, the C_{60} cage distorts. Therefore, it is likely that an increase in approach probability due to Coulomb interactions causes distortion of the C_{60} cage.

In the case of ion implantation without background electrons, on the other hand, a schematic picture of the apparatus is shown in Figure 4.4 [17]. Alkali ions are generated from a thermal source, accelerated to the desired energy and deposited on a rotating metal cylinder. This situation is similar to the plasma experiment described in the right of Figure 4.2(b), i.e., $\Delta\phi_{ap} < 0$. Fullerenes are deposited simultaneously from an oven at a temperature of approximately 450°C. The oven temperature, fullerene deposition rate, and speed of rotation of the metal cylinder are controlled to ensure that one monolayer of fullerenes is

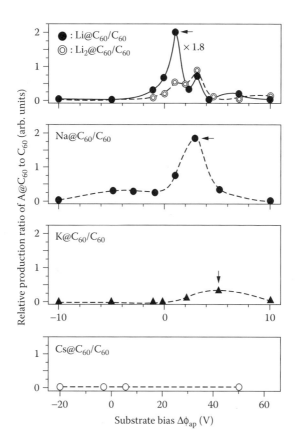

FIGURE 4.3 DC bias dependences of relative average production ratios of Li@C_{60}, Na@C_{60}, and K@C_{60}, and Cs@C_{60} to C_{60}.

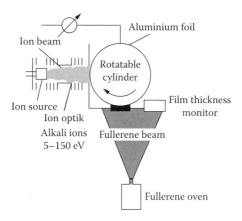

FIGURE 4.4 An apparatus to produce alkali–metal encapsulated fullerenes by an ion implantation method. (From Campbell, E. E. E. et al., *J. Phys. Chem. Solids*, 58, 1763, 1997.)

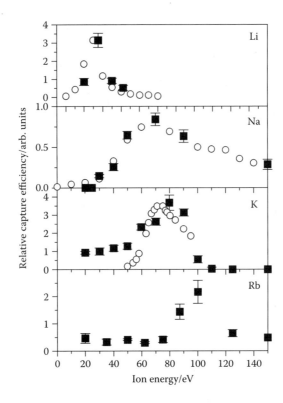

FIGURE 4.5 Comparison of the endohedral yield of material from the alkali ion-fullerene film collision experiments of Campbell *et al.* (squares) [17] and the sum of all endohedral product ions from the gas phase experiments of Anderson *et al.* (circles) [18]. (From Campbell, E.E.E. et al., *J. Phys. Chem. Solids*, 58, 1763, 1997.)

deposited on each full rotation. The deposition and irradiation is continued until films of typically up to a few hundred-nm thickness are formed. For ease of extraction of the endohedral fullerene material, the fullerenes are deposited on aluminum foil wrapped around the cylinder.

The yield of endohedral species, as a function of ion implantation energy, as determined by measuring the mass peaks in LD-TOF MS spectra, is plotted in Figure 4.5 for the different alkali metals [17]. The endohedral yield has been compared with the result in gas phase collision experiments [18]. The distributions coincide very nicely. In particular, the energetic thresholds are in good agreement, as well as the widths of the energetic window within which endohedral fullerene formation is possible.

It is to be noted that a conflicting A^+-acceleration condition for the endohedral–fullerene synthesis emerges in comparison with the experimental results of plasma and ion implantation methods. A^+ ions are not accelerated, but C_{60}^- ions are slightly accelerated (a few eV) in the former, while only A^+ ions are strongly accelerated (several tens of eV) in the latter. According to a recent experiment of plasma method, $A@C_{60}$ is also produced in the range of $\Delta\phi_{ap} < 0$ (a situation similar to the ion-implantation method) for certain experimental parameters. To solve this kind of experimental paradox, it's expected to achieve the goal of the high-yield production of alkali–metal endofullerenes.

4.2.3 Atomic-Nitrogen Encapsulated Fullerenes

$N@C_{60}$ belonging to the new category of endohedrals has so far been produced basically by ion implantation process—that is, ion bombardment where ions are prepared by a Kaufmann ion source [10] or simple glow discharge plasmas [11,19]. In any case, these methods parameters for the ion irradiation to C_{60} are not actively changed, and the effects of ion bombardment on the $N@C_{60}$ synthesis have been unclear.

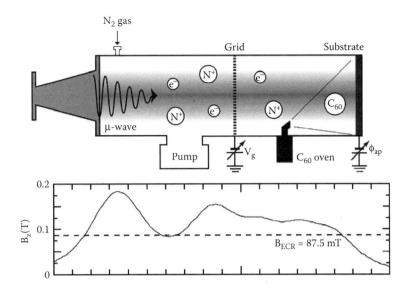

FIGURE 4.6 An experimental setup to produce atomic-nitrogen encapsulated fullerenes using a magnetic-mirror ECR plasma.

Thus, since the production ratio of N@C_{60} to C_{60} has been restricted to the order of 10^{-6} to 10^{-4}, further studies on N@C_{60} are forced to be delayed. Therefore, a new method is urgently needed that allows the production of N@C_{60} in large quantities. On the other hand, one of the other nitrogen-C_{60} compounds, the azaheterofullerene C_{59}N, has been investigated independently of N@C_{60} in conjunction with organic semiconductor devices in molecular electronics, in which a carbon atom of the cage is replaced with a nitrogen atom. It has mostly been produced by a complex and inefficient chemical reaction [20]. Therefore, it is of the utmost importance today to investigate the effects of irradiated energies on the formation of nitrogen-C_{60} compounds by generating atomic-nitrogen ions with controlled kinetic energies.

The synthesis of nitrogen-C_{60} compounds requires a plasma containing a considerable amount of N^+ as a nitrogen ion source. To generate N^+ from nitrogen gas (N_2), a desorption/ionization energy of 24.3 eV should be supplied to N_2 by energy transfer from electrons. Here, an electron cyclotron resonance (ECR) discharge in a mirror magnetic field is adopted, which leads to the energization of electrons and the effective resultant dissociation and ionization of N_2. The experimental setup is schematically shown in Figure 4.6, where a microwave is launched into a stainless steel chamber 11 cm in diameter through waveguides by a microwave generator (2.45 GHz, 800 W) [21]. N_2 (1×10^{-2} Pa) is fed to the chamber and ionized using the microwave. Since electrons in the ECR region are trapped in the mirror magnetic field and accelerated owing to ECR at the bottom of the mirror well ($B_z = 875$ G), a number of N^+ ions are expected to be generated as a result of the effective dissociation and ionization of N_2 gas, which is confirmed by optical emission spectroscopy. The ECR plasma containing the nitrogen ions diffuses toward the process region through the separation grid. C_{60} particles sublimated from an oven are deposited continuously on a DC-voltage-applied (ϕ_{ap}) substrate. The nitrogen ions arriving in front of the substrate are accelerated by the potential difference $\Delta\phi_{ap}(< 0)$ and irradiated to a C_{60} thin film with energy E_i ($= -e\Delta\phi_{ap}$) throughout the period of C_{60} deposition. The typical plasma density and electron temperature in the process region are 1×10^9 cm^{-3} and 0.5 eV, respectively.

The compound after ion irradiation is dissolved in toluene and filtered to divide it into a residue and a solution, which is analyzed by TOF-MS and electron spin resonance (ESR). Figure 4.7 shows an ESR spectrum, where the g value (magnitude of electron Zeeman factor for unpaired electron spin density) is found to be 2.00213 and hyperfine splitting constants are found to be 0.568 and 0.570 mT. Since the hyperfine splitting constants of free nitrogen and N@C_{60} are predicted to be 0.38 and 0.5665 mT,

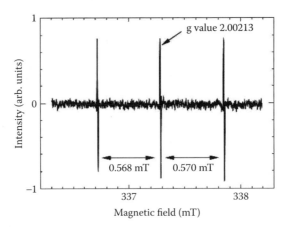

FIGURE 4.7   ESR spectrum of N@C$_{60}$ in toluene solution at room temperature. Since the nuclear spin of ^{14}N is 1/2, three lines of hyperfine split are observed.

respectively [22], the hyperfine splitting constant shown in Figure 4.7 agrees with that of N@C$_{60}$, demonstrating that N@C$_{60}$ exists in the solution. The amount of N@C$_{60}$ in the solution can be estimated from the ESR spectrum of the sample compared with that of a standard sample. When high-performance liquid chromatography (HPLC) is used in combination with ESR [23] to separate very low purity N@C$_{60}$ from C$_{60}$, the initial purity of N@C$_{60}$ in C$_{60}$ in the sample (about 10^{-3} to 10^{-2}%) is increased up to about 5%. Figure 4.8(a) gives a mass spectrum of the 5% purity sample analyzed by TOF-MS, the peak at the mass number 734 indicating N@C$_{60}$ is clearly observed, in addition to that at 720 of C$_{60}$, where the peaks at the mass numbers 721 and 722 are owing to the existence of isotopes. According to the TOF-MS analysis of residue including polymerized C$_{60}$ and various impurities, the azaheterofullerene C$_{59}$N is found to exist in large quantities as demonstrated by a distinct spectrum-peak at the mass number 722 in Figure 4.8(b). The results described previously show that N@C$_{60}$ and C$_{59}$N are produced under the same conditions and are successively separated by solvent extraction.

Figure 4.9 presents the dependences of the C$_{59}$N and N@C$_{60}$ syntheses on ion irradiation energy, where closed circles and open squares denote the mass peak ratio of C$_{59}$N (722) to C$_{60}$ (720) in the residue and the purity of N@C$_{60}$ in C$_{60}$ in the solution, respectively. The optimum ion energy is found to be about 40 to 50 eV for C$_{59}$N synthesis and almost no C$_{59}$N is synthesized for $E_i > 50$ eV. It is conjectured that the optimum ion energy for C$_{59}$N synthesis is determined by the desorption energy of a carbon atom (C) from C$_{60}$ (26.6 to 37.8 eV) [24] and the binding energy of C-N (3.03 eV). When a N$^+$ ion with a kinetic energy E_i higher than 26.6 eV collides with C$_{60}$ on the substrate, C is dissociated from C$_{60}$ and replaced with N, resulting in the formation of C$_{59}$N. However, when the energy of N remains higher than 9.09 eV after C$_{59}$N formation, a Natom with a C-N bond is considered to leave the C$_{60}$ cage again because the energy is higher than the binding energy of the N atom in C$_{59}$N. Thus, the optimum ion energy for C$_{59}$N synthesis is roughly estimated to be approximately 26.6 to 46.9 eV, which corresponds to the experimental result.

On the other hand, the N@C$_{60}$ purity is almost constant in a wide ion-irradiation energy range. However, the purity markedly decreases for an ion irradiation energy less than 20 eV. The energy barrier for the kinetic penetration of the N atom through the fullerene framework has been estimated to be about 20 eV [25]. Therefore, a N$^+$ ion with an energy larger than 20 eV appears to penetrate into the C$_{60}$ cage, leading to the synthesis of N@C$_{60}$. By taking into account the fact that the dependence of N@C$_{60}$ on ion energy is different from that of C$_{59}$N, both nitrogen-C$_{60}$ compounds can possibly be simultaneously or selectively produced under certain conditions by this method, which is important in realizing the application of C$_{60}$ derivatives to nano- and molecular electronics.

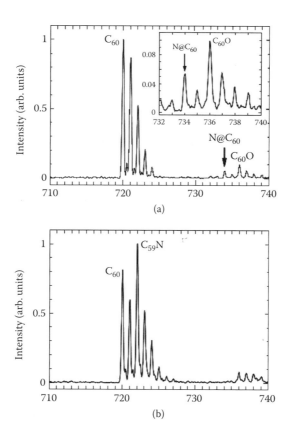

FIGURE 4.8 Mass spectra of (a) solution after solvent extraction using α-cyano-4-hydroxy cinnamic acid as a matrix (purity of N@C_{60} in C_{60}: about 5%), and (b) the residue after solvent extraction (C_{59}N at the mass number 722) obtained by TOF-MS analysis.

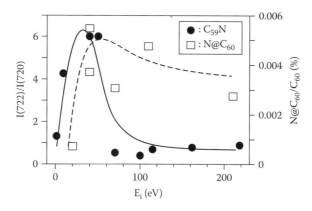

FIGURE 4.9 Dependences of C_{59}N and N@C_{60} syntheses on ion irradiation energy. Closed circles denote the spectrum-peak ratio of C_{59}N (722) to C_{60} (720) observed by TOF-MS. Open squares denote the purified ratio of N@C_{60} in C_{60} obtained by ESR and HPLC.

4.3   Alignment-Controlled Pristine Carbon Nanotubes

4.3.1   Motivation Background

Since the discovery of carbon nanotubes (CNTs) in 1991 by Iijima [3], they have attracted a great deal of attention in a wide range of application fields. Single-walled carbon nanotubes (SWNTs) [4], which consist of only one graphen layer, are especially expected to be a primal candidate for developing a nanoelectronics application due to their scaling advantage and unprecedented electrical characteristics. The electrical characteristics of SWNTs are fairly sensitive to an environment around the carbon shells, and hence the inner and outer spaces of SWNTs have huge potential to modify their electrical features, which is inevitable to construct sophisticated electrical circuits for future applications. Based on this background, several remarkable techniques for the diversification of SWNT electrical properties have been developed, such as C_{60} encapsulation with vacuum heating, alkali metal doping on the outer surface of SWNTs, and so on. The author's group also tackles this problem with a unique plasma method. Figure 4.10 shows an inclusive concept of nanoscopic plasma processing proposed by the author. A basic plasma technology is able to create a special situation, where sources of desired CNTs and materials to be encapsulated in them are independently prepared. In the latter source area, almost all of the atoms and molecules are positively or negatively charged to form different-polarity ion plasmas, and therefore can be selectively accelerated toward the former source area with an externally applied electrostatic force between the two sources inserted into the end-cap opened CNTs. In order to effectively realize inner space modifications of SWNTs with this method, well-organized SWNTs must be used as a target for accelerated ions, because all of the materials to be encapsulated are forced to pass through only from the edge of SWNTs into their inner hollow region—i.e., almost all of SWNTs have to stand individually.

As just mentioned, how to organize the as-grown state of individual SWNTs is one of the most important issues, not only for the modification of electrical characteristics but also the integration of them in electrical circuits for a practical application. In this chapter, several outstanding techniques for the alignment control

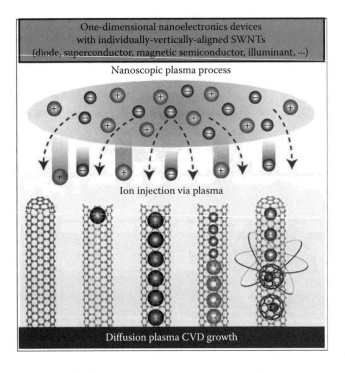

FIGURE 4.10   Nanoscopic plasma processing.

of individual SWNTs are introduced with a simple history of their development. Before touching ground it is convenient in this context to briefly note a significance of individual SWNTs in the following. Individual SWNTs have gathered much attention since the discovery of their outstanding electrical and optical characteristics, which are much superior to those of bundle-forming SWNTs. Their remarkable structures, with their high aspect ratio and extraordinary flexibility, can also provide a great number of opportunities to be used in wide application fields. For instance, field effect transistors, nanoprobes, sensors, wiring, field emission displays, and so on. Unfortunately, however, those structural features strongly limit a practical manipulation of the individual SWNTs, which cause a critical and huge barrier against fruitfully utilizing their potential abilities in industrial application fields. This background has motivated a large number of researchers to focus upon the ultimate aim, "perfect manipulation of the individual SWNTs" in the last decade.

4.3.2 Parallel-Direction Grown Carbon Nanotubes

The direct growth of individual SWNTs on a substrate was first reported by Cassell *et al.* in 1999 [26]. When the SWNTs' growth is performed using a thermal chemical vapor deposition (CVD) method with specially patterned substrates, they found that SWNTs are suspended bridges grown from a catalyst material placed on the top of regularly patterned silicon tower structures. In an area containing towers under a square configuration, a square of suspended nanotube bridges is obtained (Figure 4.11[a]). Directionality of the suspended tubes is simply a result of the designed substrate. During the CVD growth, nanotubes emanate from the top of the towers. Nanotubes growing toward adjacent towers become suspended, whereas tubes growing toward other directions fall onto the sidewalls of the towers.

After the results of suspending growth, a new concept, "control of the growth direction," was first introduced in the field of individual SWNTs. One of the most promising technique about the alignment control is the one applying an electrical force during the SWNTs' growth [27]. When the electric force is applied between two electrodes, SWNTs start to grow along with a line of electric force (Figure 4.11[b]).

A force of gas flow can also control the growth direction [28,29]. When SWNTs grow quite long using the thermal CVD method, it was found the growth direction corresponded completely with the flow direction of carbon source gas, as described in Figure 4.12(a). Interestingly, when the two-step SWNTs' growth is performed with the same substrate, it is possible to fabricate well-defined cross-network structures of SWNTs on a large scale by adjusting the gas flow direction. (Figure 4.12[b]).

The other method attracting strong attention as a novel technique to control the individual SWNTs growth direction is the one using a crystal structure of a substrate surface [30]. As shown in Figure 4.12(c),

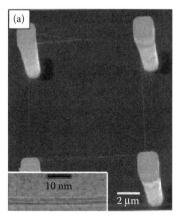 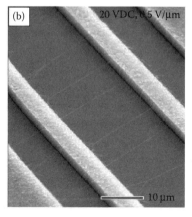

FIGURE 4.11 (a) A scanning electron microscope (SEM) image of a square of suspended SWNT bridges. Inset: A high magnification transmission electron microscope (TEM) image showing the structure of a synthesized SWNT. (From Cassell, A. M. et al. *J. Am. Chem. Soc.*, 121, 7975, 1999.) (b) SEM images of suspended SWNTs grown in electric fields. (From Zhang, Y. et al. *Appl. Phys. Lett.*, 79, 3155, 2001.)

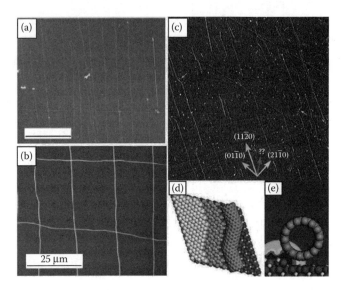

FIGURE 4.12   (See color insert following page **5**-6) (a) SEM image of horizontally aligned SWNTs compared with the direction of gas flow. (From S. Huang et al. *J. Am. Chem. Soc.*, 125, 5636, 2003.) (b) SEM image of the cross-network SWNTs fabricated by two-step growth process. The substrate angle is changed 90° after the first process. (From S. Huang et al. *Adv. Mater.*, 15, 1651, 2003.) (c) AFM amplitude image of kinked SWNTs growing along the $[112^-0]$ direction (blue) with short segments along the $[101^-0]$ direction (red), and occasionally $[21^-1^-0]$ (yellow) and $[011^-0]$ (green; image size: 5 mm). The short arrows in the respective color point to a few such segments. (d) Illustration of a (10,0)-(6,6)-(10,0) kinked nanotube along $[112^-0]$-$[101^-0]$-$[112^-0]$. (e) Model of a 1-nm-diameter SWNT along a $[112^-0]$ atomic step. The color gradient represents an estimated SWNT-step electrostatic interaction energy per unit of nanotube length as a function of SWNT axis position, $U(x,z)$. This was calculated from the force exerted on a polarizable body by an inhomogeneous field, $F = (\alpha E \Sigma \nabla)\,E$. Averaging the potential along the direction of the step and the SWNT (y) gives $U(x,z) = -1/2\alpha_{xx}E^2(x,z)$, where α_{xx} is the transverse polarizability of the SWNT per length and $E(x,z)$ is the local field. The latter was derived from the unreconstructed atomic step, by summation of Coulomb potentials from bulk Mulliken charges, averaged along the y axis and corrected for slab edge effects by subtracting a similar potential without the step. (The blue-to-red scale is 0–750 eVnm^{-1}.) (From A. Ismach et al. *Angew. Chem. Int. Ed.*, 43, 6140, 2004.)

the individual SWNTs grow in accordance with a certain crystal's orientation, something first developed by Ismach *et al.* in 2004. Several similar works have also been reported after their study [31,32]. These precise alignment control techniques could open the door to a wide range of industrial applications for individual SWNTs, including nano-electronics.

4.3.3   Individually Vertically-Aligned Carbon Nanotubes

As just described, these outstanding techniques enable individual SWNTs to be grown in optional directions. Noticeably, however, all of the reported methods have only achieved alignment control in directions parallel to substrate surfaces. To successfully integrate large arrays of devices—such as field effect transistors, nanoprobes, sensors, wiring, and field emission displays—the tube alignment control in both the parallel and perpendicular directions on a substrate surface must be effectively addressed.

Plasma-enhanced chemical vapor deposition (PECVD) is one of the well-known methods of forming nanotubes, and has outstanding benefits for the vertical growth of individual multiwalled carbon nanotubes [33–36]. Although this PECVD method has a strong potential to solve the aforementioned problem in one synthesis step for individual SWNTs (i.e., the vertical growth of isolated SWNTs), high-energy ions causing significant damage to tube structure have greatly restricted PECVD from being fruitfully applied to the nanotube growth. In order to utilize these advantages of plasma technology in industrial applications

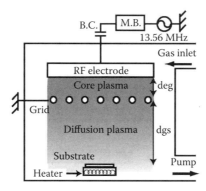

FIGURE 4.13 Experimental setup of the diffusion plasma process.

of SWNTs, a damage-free plasma process urgently needs to be developed. From this point of view, the author's group was greatly interested in a diffusion plasma process, which is considered to lead to the nanotube growth without significant damage even under a strong electric field in a plasma sheath. The diffusion plasma can be generated with a simple planer type radio frequency (RF, 13.56 MHz) plasma unit, as shown in Figure 4.13. An RF power (P_{RF}) is supplied to an upper planer electrode through an RF power supply system consisting of a power generator, a matching box (M.B.), and a blocking condenser (B.C.). A mesh-type grid is utilized as an anode on purpose to promote the diffusion of plasmas. In the diffusion area, electron temperature can be drastically decreased (~ 0.1 eV) with an increase in diffusion distance. Since the value of electron temperature is directly related to the energy of ions that flow into a substrate, ion-damage-free plasma processes can be realized using this diffusion PECVD method. With the diffusion plasma process, the author's group previously succeeded for the first time in forming SWNTs within the framework of the PECVD method, as well as synthesizing freestanding SWNTs based on plasma sheath effects [37–41].

Figure 4.14(a) is a low-magnification SEM image of produced materials with the diffusion PECVD method. The detailed SEM observations reveal that large amounts of filament shaped materials grow on the substrate surface. Surprisingly, these materials are often observed in a freestanding form, as described in Figure 4.14(b).

Based on meticulous and intensive TEM observations, it has been found that almost all of the materials grown are SWNTs. Furthermore, individually standing SWNTs can often be observed, as shown in Figure 4.14(c). Because the sample pre-treatment was carried out for TEM observations (softly peeling off from the substrate surface without any dispersion process), it is difficult to discuss the as-grown state of the produced materials from these results. However, when the results of SEM observations are combined with those of TEM observations, it can be concluded that the individually vertically aligned (freestanding) SWNTs are produced with the diffusion PECVD method.

Figure 4.14(d) describes a typical Raman scattering spectrum. A tangential stretching (TS) mode can clearly be observed to sharply split into the G^+ band (1590 cm^{-1}) and the G^- band (1557 cm^{-1}). This result also supports the belief that the material produced is SWNTs. Despite the sharp TS mode, any clear radial breathing mode (RBM) associated with tube diameters cannot be detected in the general RBM region (150 to 300 cm^{-1}). As shown in Figure 4.14(e), the results of TEM observations tell us that the main diameter of the produced SWNTs is about 3 nm. Following this result, it is possible to suppose that the reason for the absence of RBM in the 150 to 300 cm^{-1} region is related to the unique diameter distribution of the freestanding SWNTs produced here.

As demonstrated previously, our diffusion plasma method is capable of producing freestanding SWNTs on flat substrates, which are expected to contribute to the creation of a novel three-dimensional architecture in a nanoelectronics device field. Furthermore, these perfectly isolated SWNTs are extraordinarily suitable for inner- and outer-surface modifications, which can bring out the superior characteristics of SWNTs drastically.

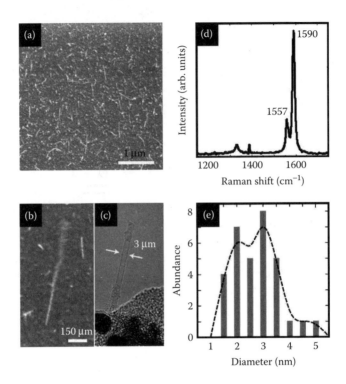

FIGURE 4.14 Typical low (a) and high (b) magnification SEM images of freestanding SWNTs, respectively. (c) A TEM image of freestanding individual SWNTs. (d) A Raman scattering spectrum of SWNTs. (e) A diameter distribution of SWNTs measured by TEM.

4.4 Nano-Electronic Oriented Carbon Nanotubes

4.4.1 Atomic/Molecule Encapsulated Carbon Nanotubes

As already mentioned, pristine CNTs have shown great potential as components of nanoscale electronic devices and sensors because of their novel structure, unique conducting properties, and high thermal capacity. Recently, modification of such pristine CNTs with other foreign materials is recognized to be an attractive ideal for the purpose of controlling their electronic properties through interaction with electron donors or acceptors. Although various methods have been explored up till now, compared with surface modification such as substitution and attachment, intercalation or encapsulation of CNTs with foreign atoms or molecules is considered to be a more fascinating way (in a relative sense) of easily and accurately controlling the physical properties of CNTs. Up until today, various atoms such as alkali–metals [42] and magnetic metals [43,44], molecules such as Br_2, I_2 [45,46], and various compounds [47,48] have been filled in multiwalled carbon nanotubes (MWNTs), SWNTs, or double-walled carbon nanotubes (DWNTs). The first doping reactions by K and Rb are performed on MWNTs prepared by an arc discharge method [49]. The first molecule ever reported in the case of SWNTs is fullerene C_{60} [50] and the insertion of C_{60} into their inner nanospace is realized by a pulse laser vapor method. On the other hand, the plasma ion-irradiation method described in Figure 4.10 (see Section 4.3.1) is found to be efficient for the insertion of metal atoms or molecules into nanotubes. When the alkali–fullerene plasma is exemplified for that purpose, different-polarity ions can selectively be accelerated toward a substrate coated with pristine SWNTs or DWNTs, in this case by adjusting its bias voltages, as previously shown in Figure 4.2 (see Section 4.2.2).

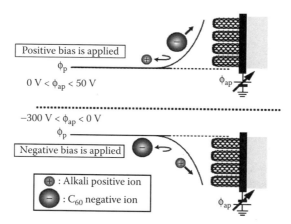

FIGURE 4.15 The plasma irradiation process.

In more detail, the controllable ion irradiation can be realized when largely positive or negative biases are applied to the substrate, as schematically described in Figure 4.15: $\phi_{ap} >> 0$ for C_{60}^-, $\phi_{ap} << 0$ for A^+. First, SWNTs or DWNTs start to be oriented along the plasma sheath electric field. Then, their structural modifications such as tube deflection, tube kink, tube cutting, and end-cap opening are observed to be caused, resulting in the formation of atom or molecule encapsulated carbon nanotubes [51–54].

Figure 4.16 gives the TEM images for (a) C_{60} encapsulated SWNT (C_{60}@SWNT) and (b) Cs encapsulated SWNT (Cs@SWNT), respectively [55]. The observations evidently reveal that the foreign materials have been filled inside SWNTs. In the case of DWNTs, owing to their large diameter only amorphous C_{60} molecules are observed inside DWNT, as shown in Figure 4.17(a), compared with the chain-like C_{60} molecules encapsulated SWNTs. In the case of Cs encapsulation, worm-like Cs clusters can clearly be observed within DWNT, as indicated in Figure 4.17(b) [56], which is somewhat similar to the morphologies of Cs@SWNTs. In addition to the C_{60} fullerene, metallofullerenes can also be filled inside SWNTs, which has been realized by Sato *et al.* [57] by means of a vapor reaction method. Figure 4.18 shows the TEM images of two Gd@C_{82} molecules aligned inside SWNTs, and black arrows indicate the Gd ions. Moreover, by using the vapor reaction method, organic compounds have been doped into SWNTs, and the charge transfer between SWNTs and various organic molecules have been investigated by Takenobu *et al.* [58]. Figure 4.19 shows the typical X-ray diffraction profiles for pristine and TCNQ-encapsulated SWNT materials. The most obvious difference between doped and undoped SWNTs is the strong reduction of peak intensity at approximately $Q \sim 0.4$ ($\overset{\circ}{A}^{-1}$). Such a behavior provides direct evidence for the encapsulation of organic molecules inside SWNTs.

According to a recent simulation work [59], one of the fascinating routes toward novel functional materials is to inject magnetic metals such as Fe, Co, or Ni into SWNTs or DWNTs, where both the

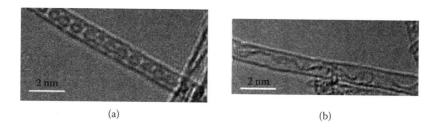

(a)   (b)

FIGURE 4.16 Typical TEM images of (a) C_{60} encapsulated SWNT and (b) Cs encapsulated SWNT.

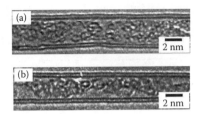

FIGURE 4.17   Typical TEM images of (a) C_{60} encapsulated DWNT and (b) Cs encapsulated DWNT.

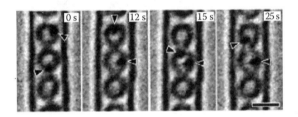

FIGURE 4.18   TEM images of a Gd@C_{82} encapsulated SWNT scale bar: 1 nm. (From Sato, Y. et al. *Phys. Rev. B*, 73 233409, 2006.)

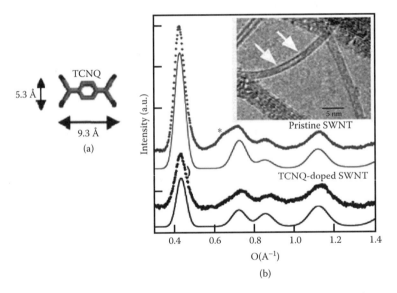

FIGURE 4.19   (a) A TCNQ molecule and (b) observed X-ray diffraction profiles of pristine and TCNQ encapsulated SWNTs. (From Takenobu, T. et al. *Nature Mater.*, 2, 683, 2003.)

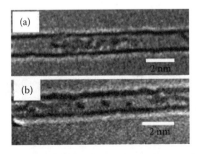

FIGURE 4.20   TEM images of (a) ferrocene encapsulated SWNT and (b) Fe encapsulated SWNT.

charge and spin of electrons may be allowed to be used for transport phenomena. Ferromagnetism and semiconducting properties are expected to coexist in a SWNTs-based semiconductor by encapsulating magnetic elements. One of the possible methods is carried out to fill magnetic metals in SWNTs (or DWNTs) using ferrocene as a starting substance. During the first step, ferrocene molecules are filled in SWNTs (or DWNTs) by a vapor reaction method [60,61]. Figure 4.20(a) shows the TEM image for ferrocene molecules encapsulated inside SWNT [61]. In the second step, samples of ferrocene-filled SWNTs (or DWNTs) are further annealed in a vacuum with an aim to release Fe atoms inside nanotubes. In Figure 4.20(b), several dark spots distributed in SWNT can clearly be identified. These dark spots with a diameter of about 0.5 to 0.7 nm suggest that each of them contains only several Fe atoms [61]. The preceding method is simple, and easily scalable for the mass production of Fe encapsulated SWNTs (Fe@SWNTs) and DWNTs (Fe@DWNTs).

4.4.2 Electronic Properties of Encapsulated Nanotubes

Due to their ideal one-dimensional structure and unique electronic properties, a lot of studies have been devoted to the transport properties of modified SWNTs and DWNTs, in which nano devices are fabricated using SWNTs and DWNTs as the channels of field-effect transistors (FETs). In the present case, an FET substrate typically consists of two Au electrodes (thickness of 150 nm) placed on a SiO_2 insulating layer (thickness of 500 nm), as schematically depicted in Figure 4.21(a). The Au electrodes are used as source and drain electrodes with a gap length of 500 nm between the two electrodes. A heavily Si-doped substrate is used as a back gate, which is prepared by Al evaporation. Pristine semiconducting SWNTs show only the well-known p-type semiconducting property at room temperature, as indicated by a source-drain current I_{DS} versus gate voltage V_G characteristic in Figure 4.21(b), while C_{60}@SWNTs are confirmed to maintain their p-type behavior due to the electron–acceptor quality of C_{60} [62]. By doping the alkali atom K around the outside surface of SWNTs, high-performance n-type semiconducting SWNTs have even been obtained by Dai *et al.* [63]. In their case, the on/off ratio for n-type SWNTs is about 10^6 with no ambipolar p-channel conductance.

On the other hand, in the case of encapsulating alkali atoms inside pristine SWNT or DWNTs, their electronic transport properties of SWNTs or DWNTs can be gradually modified by changing the doping time in the plasma irradiation method. Figure 4.22 presents an evolution of electronic properties of Cs encapsulated SWNTs created by varying the plasma ion-irradiation time from 15 min to 60 min, where (a), (b), (c), and (d) show the $I_{DS} - V_G$ characteristics of Cs@SWNTs measured with a source-drain voltage of $V_{DS} = 1$ V in a vacuum [64]. The samples are prepared by the Cs ion irradiation process, during which the concentration of Cs^+ is about 2.2×10^4, 4.3×10^4, 6.5×10^4, and 8.6×10^4 Cs ions per nm^2,

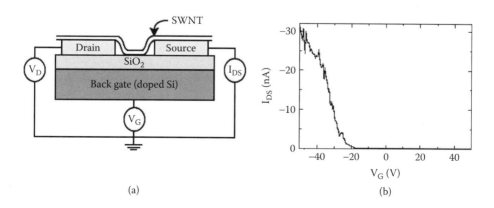

(a)

(b)

FIGURE 4.21 (a) A SWNT-FET device and (b) p-type semiconducting behavior of a pristine SWNT.

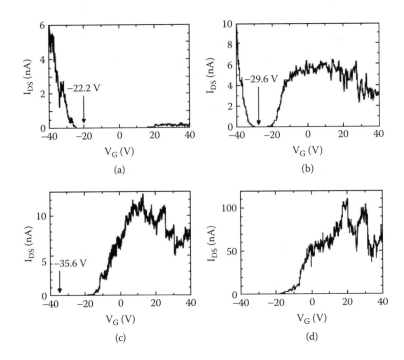

FIGURE 4.22 $I_{DS} - V_G$ characteristics of Cs@SWNTs measured at $V_{DS} = 1$ V in a vacuum by varying the Cs ion irradiation time: (a) p-type, (b) ambipolar, (c) and (d) n-type behavior.

respectively. Starting with the pure p-type conductance, ambipolar properties appear in the range of the low doping level. A threshold gate voltage for p-type conductance tends to shift more negatively from -22.2 to -35.6 V and the conductance region of n-type gradually increases as the total amount of irradiated Cs atoms increases. Finally, Cs@SWNTs change to show the complete n-type behavior, as described in Figure 4.22(d).

In general, the obtained nanotube FET transistors are very sensitive to the presence of air. Therefore, how to realize the air-stable nanotube-FET transistors is becoming an important topic. Recently, n-type FETs stabilized in air by polymer functionalization have been reported [65]. In addition, air stable n-type semiconducting SWNTs can be obtained by Cs encapsulation via the plasma method. Figure 4.23 shows an n-type Cs@SWNT-FET before (solid line) and after exposure to air (dashed line), and the characteristic indicates that the n-type behavior of Cs@SWNT remains even in air [64]. Thus, it is demonstrated that the reactive Cs atoms are protected against oxidation by the carbon layers because the Cs atoms are encapsulated inside SWNTs.

Compared with p-type pristine SWNTs, pristine semiconducting DWNTs exhibit amibipolar behavior due to their narrow bandgap of about 0.2 eV, as shown in Figure 4.24(a), which is similar to the semiconducting properties of SWNTs with a large diameter over 3 nm [66].

After the Cs encapsulation, Cs@DWNTs can clearly exhibit n-type transport behavior (as shown in Figure 4.24[b]) similar to the case of SWNTs, suggesting the occurrence of charge transfer. By comparison, it is found that n-type DWNTs exhibit high performance such as a high on/off ratio and mobility in contrast to those of Cs@SWNTs during the measurements [56]. The possible reason for the preceding difference is that the Cs encapsulation is accompanied by much more structural or chemical defects in the case of SWNTs, as compared with DWNTs and their novel structural merits.

Interestingly, strong resonant tunneling characteristics are found for C_{60}-encapsulated metallic DWNTs. Figure 4.25(a) illustrates a series of $I_{DS} - V_G$ curves when V_{DS} is increased from 0 to 6 V. The obtained data indicate that the measured current linearly increases at first for $V_{DS} \leq 3$ V, becoming gradually

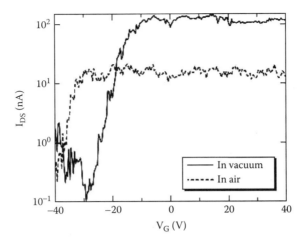

FIGURE 4.23 $I_{DS} - V_G$ characteristics of Cs@SWNTs in a vacuum (solid line) and in air (dashed line), measured at room temperature.

unstable when V_{DS} is larger than 3 V. At $V_{DS} = 6$ V, the current is encountered with an unexpected sharp decrease and the Coulomb blockade behavior appears. Figure 4.25(b) presents the source-drain current versus voltage curve measured at room temperature using this C_{60}@DWNT transistor. The measured curve at a gate voltage of $V_G = -20$ V exhibits unique negative differential resistance (NDR) characteristics at high bias voltages of both ~ 6 V and -6 V, namely, an initial rise of the current followed by a sharp decrease instead of the linear increase expected from Ohm's law when the voltage is progressively increased, indicating a typical feature for resonance tunneling devices. The distinct NDR feature with a peak-to-valley current ratio ($PVCR = I_P/I_V$) of up to 1300 is surprisingly observed at room temperature, providing strong evidence of the super resonance tunneling phenomenon [67].

Furthermore, SWNTs also hold considerable promise for applications in the field of electron spin, in addition to the aforementioned charge-dominated phenomena, due to their one-dimensional structure and ballistic transport properties. Spin-polarized transport properties of SWNTs have been investigated by connecting them with two ferromagnetic electrodes in magnetic fields [68]. By injecting ferromagnetic elements such as Fe, Co, or Ni into nonferromagnetic SWNTs, both the charge and spin of electrons

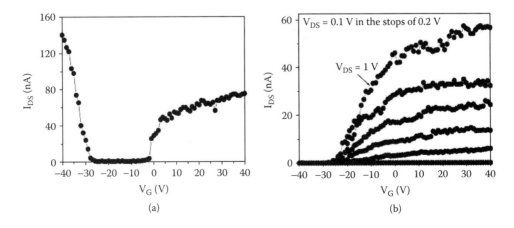

FIGURE 4.24 Electronic transport properties of (a) ambipolar behavior of pristine DWNTs and (b) *n*-type Cs@DWNTs.

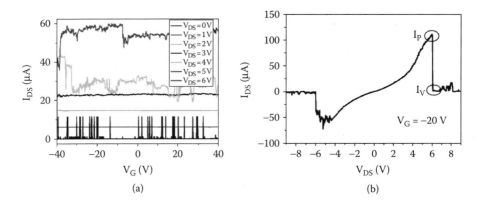

FIGURE 4.25   Negative differential resistance characteristics of metallic C_{60}@DWNT: (a) $I_{DS} - V_G$ curves with V_{DS} in the range 0 to 6 V, and (b) $I_{DS} - V_{DS}$ curve with $V_G = -20$ V.

can be used in SWNTs. Recent work demonstrates that high-performance unipolar n-type characteristics are observed for Fe encapsulated SWNTs. Figure 4.26(a) displays the current versus voltage $I_{DS} - V_G$ characteristics for n-type Fe@SWNTs measured at room temperature [69]. Figure 4.26(b) shows the magnetic characterization of Fe@SWNTs, suggesting they exhibit a ferromagnetic property [70]. Therefore, the present measurements demonstrate it is possible to create ferromagnetic semiconducting SWNTs by Fe encapsulation.

Additionally, in the case of the metallofullerene-encapsulated SWNTs, electronic properties are found to be significantly changed from those of pristine SWNTs, such as the shift of the Fermi level. Lee *et al.* have investigated the bandgap modulation of SWNTs by encapsulating a metal Gd in C_{82} (metallofullerene: Gd@C_{82}) [71]. Their studies using a scanning tunneling microscope (STM) demonstrate that the bandgap is narrowed due to encapsulation of metallofullerenes. Figure 4.27(a) shows the bandgap modulation of a Gd@C_{82} encapsulated SWNT. The original bandgap of 0.43 eV of SWNT is narrowed down to 0.17 eV when Gd@C_{82} molecules are encapsulated, which can be explained as the contribution of the elastic strain and the electron transfer to the Au substrate and to C_{82}. The modified conduction band of a different shape is shown in Figure 4.27(b). Various shapes of the modified energy band reflect the variation of local elastic strain and charge transfer or other physical quantities. Shimada *et al.* have reported related electronic properties of metallofullerene (Gd@C_{82}) encapsulated SWNTs [72]. According to their work given in Figure 4.28, the

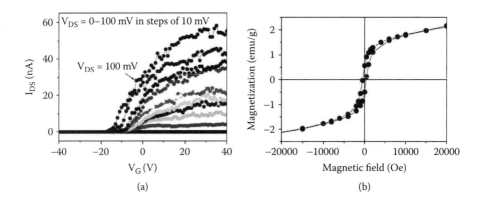

FIGURE 4.26   (see color insert) (a) $I_{DS} - V_G$ curves for n-type Fe@SWNTs measured at room temperature and (b) a magnetization curve of Fe@SWNTs measured at 5 K.

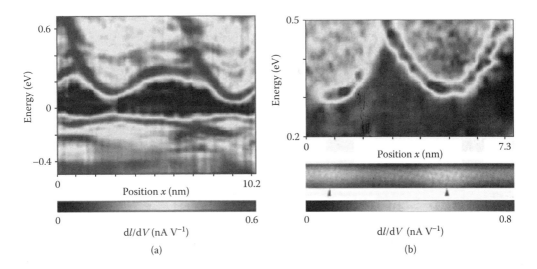

(a)

(b)

FIGURE 4.27 (see color insert) Bandgap modulation of (a) a 10.2-nm-long Gd@C_{82} encapsulated SWNT and (b) a 7.3-nm-long Gd@C_{82} encapsulated SWNT. (From Lee, J. et al. *Nature*, 415, 1005, 2002.)

metallofullerene encapsulated SWNTs show ambipolar, both p- and n-type, characteristics, being different from the C_{60} encapsulated SWNTs only exhibiting unipolar p-type characteristics as already mentioned in the beginning of this section. The difference in transport behavior can be explained in terms of a bandgap change depending on different dopants.

Since SWNTs belong to a microscopic system, their conduction is predicted to be suppressed at low temperatures because of the charging energy by a single electron, which leads to the appearance of some singular phenomena like Coulomb blockade at low temperatures. Here we can gain further insight into an encapsulated profile of alkali–metals inside SWNTs, because it is possible to estimate a quantum dot size made due to the potential barrier by observing Coulomb oscillations. Figure 4.29(a) shows the $I_{DS} - V_G$

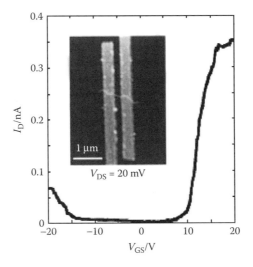

FIGURE 4.28 Electrical transport properties of Gd@C_{82} encapsulated SWNTs, indicating ambipolar properties. (From Shimada, T. et al. *Appl. Phys. Lett.*, 81, 4067, 2002.)

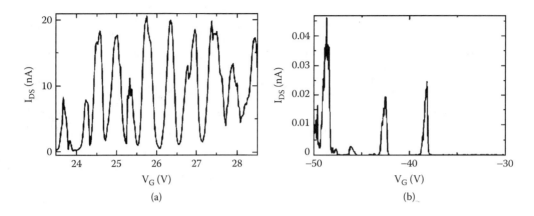

FIGURE 4.29   Coulomb oscillation characteristics measured at 11.5 K for (a) Cs@SWNTs and (b) C$_{60}$@SWNTs.

characteristic of Cs@SWNTs measured with $V_{DS} = 10$ mV at 11.5 K [64]. The Coulomb oscillation phenomenon is clearly observed, which indicates that the system evidently consists of multiple quantum dots. It is natural that Cs ion irradiation possibly causes some defects on SWNTs, which may lead to the forming of quantum dots between potential barriers. In order to elucidate which mechanism is dominant in the present situation, the same measurement is performed for the case of C$_{60}$ encapsulated SWNTs (C$_{60}$@SWNT-FET), as shown in Figure 4.29(b). A clear difference between their gate-voltage periods (ΔV_G) is observed even under the same condition concerned with ion irradiation effect on the defect of SWNTs. The observed ΔV_G for Cs@SWNTs is much smaller than that in the case of C$_{60}$@SWNTs, suggesting that the observed Coulomb oscillation transport behavior of SWNTs is possibly due to the encapsulation profile inside SWNTs.

4.4.3   The Possibility of Nano pn-Junction Formation

In an ambient environment, semiconducting SWNTs generally show unipolar p-type behavior, as repeatedly explained. By selectively doping on the outside surface of a semiconducting SWNT, as schematically shown in Figure 4.30(a), Dai *et al.* have realized the formation of a p-n junction in a nanotube [73]. In their work, the nanotube can be doped in the form of n-type for half of its length, and p-type for the other half. Figure 4.30(b) gives the electrical transport properties on gate voltage versus source-drain current for p-n junctions in SWNTs, being different from the results of simple p-type or n-type nanotubes, which should exhibit a monotonic decrease or increase of conductance by positively increasing the gate bias [73]. Instead, a hump feature with substantial conductance is surprisingly observed, which can be explained as a result of the formation of the p-n junction with electron and hole concentrations controlled by the backgate.

Instead of using the individual SWNT, Zhou *et al.* have fabricated p-n diodes based on the network of SWNTs [74]. When the channel is exposed to air or coated by poly methyl methacrylate (PMMA), the transistor operates in the unipolar p mode. By spin-coating the polymer polyethylenimine (PEI) on the channel region, the transistor can be switched to operate in the unipolar n mode. Patterning the exposure of a single channel to PMMA and PEI yields a p-n diode, as illustrated in Figure 4.31(a). This same coating procedure can be applied selectively to different parts of the channel of a single device to build complex devices. Figure 4.31(b) shows the output characteristics of the p-n diode made from an originally p-type device [74]. When a forward bias or positive bias voltage on the p-side is applied, the current increases rapidly as the voltage increases. However, for the reverse bias, no current flows until a breakdown voltage

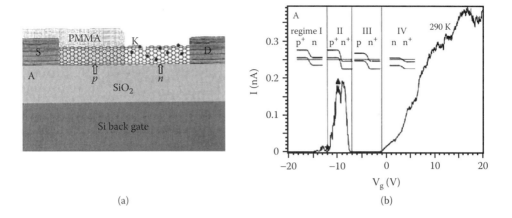

(a)　　　　　　　　　　　　　　　　　　　　　　　　　　　　(b)

FIGURE 4.30　(a) A SWNT *p-n* junction device and (b) electrical properties of the K-doped half of SWNT. (From Zhou, C. et al. *Science*, 290, 1552, 2000.)

of 9 V is reached. The current–voltage characteristics can qualitatively be explained by the *p-n* junction with series resistance effect.

On the other hand, the author's group has pointed out that a *p-n* junction can be formed by partially encapsulating Cs inside SWNTs through the plasma ion-irradiation method [53]. The Cs ions are evidently observed inside SWNTs by the *Z*-contrast method in scanning transmission electron microscopy (STEM). Figure 4.32 demonstrates the formation of alkali–metal encapsulated SWNTs, where Cs atoms (indicated by an arrow) are partially filled in an individual nanotube [53]. The preceding result can also be confirmed by the STM observation under the collaboration with Y. Kuk of Seoul National University, as shown in Figure 4.33(a). From the STM image, the bright and dark regions are found to reveal the parts where Cs is filled and unfilled, respectively. The density of state (DOS) for an unfilled and Cs-partially filled regime is measured using scanning tunneling spectroscopy (STS/STM), as illustrated in Figure 4.33(b). The DOS shift in the nanotube enables us to make an internally doped nanotube junction, which is predicated to behave as a nanodiode. Figure 4.33(c) shows the calculated DOS for a pure SWNT (10, 10), together with two cases of light and heavy doping of Cs atoms, respectively. The results indicate the formation of one and

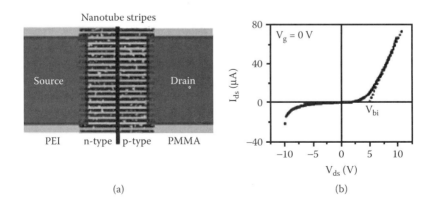

(a)　　　　　　　　　　　　　　　　　　　　　　　　　　　　(b)

FIGURE 4.31　(a) A *p-n* diode by patterning PEI on half of the channel region and (b) the output characteristic of a gated *p-n* diode at zero gate voltage. (From Zhou, C. et al. *Nano Lett.*, 4, 2031, 2004.)

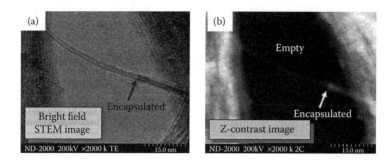

FIGURE 4.32   STEM image of (a) a SWNT partially filled with Cs, and a Z-contrast image of (b) a SWNT partially filled with Cs.

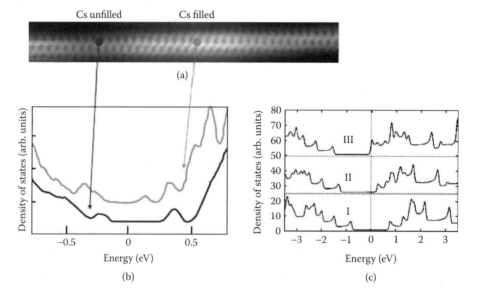

FIGURE 4.33   A STM image of (a) a SWNT partially filled with Cs, (b) a DOS of unfilled and Cs-filled SWNT (under collaboration of Y. Kuk of Seoul National University), and (c) simulated band structures for pristine (I), lightly Cs-doped (II), and heavily Cs-doped SWNT (III), respectively.

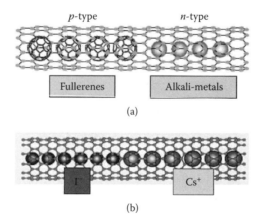

FIGURE 4.34   A SWNT *p-n* junction formation by filling (a) alkali–metals and C_{60} molecules, and (b) alkali–metals and halogen atoms.

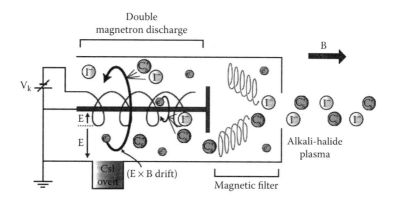

FIGURE 4.35 An alkali–halogen plasma source using double cathodes.

two nearly flat bands for the cases of light and heavy doping, respectively, within the conduction bands of the tube.

Compared with encapsulating only one kind of dopant in SWNTs, a better way to create the *p-n* junction is to fill both the electron acceptor and electron donor in a nanotube, as schematically shown in Figures 4.34(a) and (b), respectively. Alkali–metals like Cs are good electron donors [62], while C_{60} molecules and halogen atoms such as iodine (I) can operate as electron acceptors. In the case of the alkali–fullerene combination, a subsequent bias application with different polarity in the situation of Figure 4.15 is predicated to form a *p-n* junction, as shown in Figure 4.34(a). Besides, when the connection of rows of alkali–metal and halogen atoms inside SWNTs is attained, a resultant *p-n* junction is expected to yield rectifying properties predicted by the recent theoretical simulation [75]. With this aim, an alkali–halogen plasma source consisting of alkali positive ions and halogen negative ions is developed [76], which is suitable for a substrate-bias control in the encapsulation process. The alkali–halogen plasma is generated by a dc magnetron discharge under a uniform magnetic (B) field, as schematically shown in Figure 4.35. Spiral and linear cathodes of tungsten wire heated by resistive heating are set at the central axis of a grounded cylinder, and are negatively biased to form an electric field E perpendicular to B field lines. Alkali-salt vapor is introduced from an oven, filling the cylinder. Thermal electrons drift in the azimuthal (E × B) direction, and the electrons collide with alkali salt, dissociating and ionizing it. As a result of this process, alkali positive ions, halogen negative ions, and electrons are produced. The electrons can be removed from the processing region by a magnetic filter effect. The dissociation and the ionization can be controlled by only the E field under a constant B field. Therefore, different kinds of ions can selectively be filled inside SWNTs through the preceding procedure, leading to the formation of *p-n* junctions in SWNTs.

4.5 Molecular Electronics Oriented Carbon Nanotubes

4.5.1 Biomolecule Encapsulated Carbon Nanotubes

The inside modification methods of SWNTs have been mentioned in the previous section, where the thermal diffusion and plasma ion irradiation methods are mainly introduced and proved to provide materials encapsulated SWNTs with high yields. In these procedures, the encapsulation process is performed in the gas phase. Therefore, the encapsulated materials are limited to atoms and simple molecules, which have a high thermal tolerance and sublimation property. Structural-complicated molecules such as biomolecules have more various properties—in other words, functionalities, compared with atoms, and simple molecules in general. Since they cannot exist in a structural-stable state in the gas phase, a novel concept or method is required to achieve functional materials in combination with SWNTs, such as biomolecule encapsulated SWNTs. Biomolecules, including DNA, behave as ions in the presence of counter ions in solution (in optimized pH, ion concentration, and temperature), forming a sort of electrolyte plasma. Then, the ion irradiation is predicted to be possibly performed even in solution.

Here, a novel method for forming DNA encapsulated SWNTs is introduced. This method is an extension of concept of the plasma ion irradiation method in the gas phase. DNA consists of four kinds of base (adenine, thymine, cytosine, and guanine), sugar ring, and phosphoric acid part. The phosphoric acid has a negative charge and the other part is electrically neutral. Each base has a different electronic property [77], and a base sequence has come to be easily controlled recently. Since the bases are oriented at intervals of every 0.34 nm in the DNA molecule under the assumption that they form a helical conformation, the length of the DNA molecule can be estimated from the number of bases. Therefore, the control of electronic properties on a nanometer scale can be performed by selecting a base sequence. Since DNA is a typical polyelectrolyte and forms a negatively charged long molecule, the behavior of DNA negative ions can be understood by Debye and Huckel's theory on the electrolyte plasma [78] in the same manner as that in gaseous plasmas in consideration of valence of the DNA ion and its diameter. In general, the conformation of chain molecules takes on a random-coiled shape due to Brownian motion. When this ion irradiation method is applied, it is indispensable to take into account the apparent diameter of the ion in solution, which can be estimated by a three-dimensional random walk theory. This theory is represented by $R_N \propto N^{1/2}$ under the assumption that the volume of the components is not taken into account, where R_N is the apparent diameter and N is the number of bonds in a DNA molecule. According to the TEM observation, the average diameter of SWNTs used at present, which is produced by an arc discharge method and purified [79], is determined to be about 1.5 nm, while it is difficult to determine R_N exactly. Then R_N is roughly estimated to be larger than the diameter of SWNTs when more than 15 bases are included in one DNA molecule. In the performance of the ion irradiation method, the conformation of the DNA ion has to be changed in solution. A radio frequency (RF) electric field is known to be effective in changing the DNA conformation [80,81].

Figure 4.36 shows the schematic of an experimental apparatus for DNA negative-ion irradiation in solution. A direct current (DC) voltage V_{DC} is applied to a substrate covered with end-opened pristine SWNTs, and an RF voltage V_{RF} is superimposed on V_{DC}. The DNA negative-ion conformation can be changed from a random coiled to a stretched form by the RF electric field, and be moved to the substrate of anode electrode by the DC electric field generated for $V_{DC} > 0$ with respect to an oppositely situated cathode in this system [82]. A gap distance between the anode and cathode is set in the range of 1 mm.

After the DNA ion irradiation, intensive observations of SWNTs are performed by TEM (in which the acceleration voltage is 200 kV) to identify the conformation of the encapsulated DNA. The DNA-irradiated

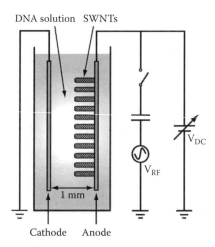

FIGURE 4.36 An experimental apparatus of DNA negative-ion irradiation used to form the DNA encapsulated SWNTs.

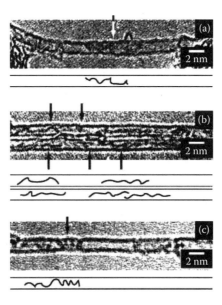

FIGURE 4.37 TEM images of SWNTs after DNA irradiation. (a) and (b): A_{15}, $V_{DC} = 10$ V, and $V_{RF} = 20$ V. (c): A_{30}, $V_{DC} = 10$ V and $V_{RF} = 150$ V. Bottom images indicate encapsulated DNA molecules schematically.

SWNTs are washed by water and dried in air. Then, they are sonicated in ethanol for several hours. The SWNT suspension is dropped onto a cupper grid, dried in air again, and its observation is started using this grid. The notation of DNA is as follows: four kinds of base are represented by A, C, T, and G, respectively. In addition, the numbers of bases are represented by subscripts of A, C, T, and G. For example, for DNA consisting of 15 adenines, the DNA molecule is represented by A_{15}. Figures 4.37(a) and (b) present the TEM images of SWNTs after DNA irradiation at $V_{DC} = 10$ V and $V_{RF} = 20$ V using A_{15}. The one-dimensional materials appear to be encapsulated in SWNTs; the bottom images are schematic ones. The length of all the encapsulated materials is about 5 nm, which corresponds to that of the DNA used, indicating that the formation of DNA encapsulated SWNTs (DNA@SWNTs) is realized for the first time. However, it is difficult to control the number of encapsulated DNA molecules, thus far, since its DNA-molecule number is different even under the same condition. For example, only one DNA molecule is observed to be encapsulated, such as that shown in Figure 4.37(a), and sometimes, the plural number of DNA can be inserted [Figure 4.37(b)]. It is believed an optimized condition, such as the orientation of DNA and SWNTs, allows high-yield insertion, and that the encapsulation process will be precisely controlled to some extent using highly oriented SWNTs in the near future. Individual vertically aligned SWNTs grown by the diffusion plasma CVD (see Section 4.3) could be a candidate for such high-quality pristine SWNTs. In the case of using A_{30} (Figure 4.37[c]), where the DNA irradiation conditions are $V_{DC} = 10$ V and $V_{RF} = 150$ V, the encapsulation is also verified and its length is estimated to be about 10 nm. In addition, A_{30} is found to form a helical conformation inside SWNTs and seems to be adsorbed onto the inner wall. On the basis of TEM observations, it is likely that the bases (the hydrophobic part of the DNA) taking up a certain DNA length tend to be adsorbed onto the inner wall and to form the helical conformation in SWNTs, respectively [83,84].

Since the DNA irradiation method described here utilizes not the base sequence but the negative charge of the DNA molecules and the stretching of the DNA molecules is caused by the interaction between the permittivity of DNA and the applied external RF electric field, the effectiveness of irradiation and stretching is independent of the length of the DNA molecules. Therefore, this formation method for DNA@SWNTs can be applied to not only a specific DNA base sequence—for example, that consisting of only adenines—but also other base sequences. Actually, in the case that guanine contained DNA is encapsulated, Raman spectra imply possibilities of the charge transfer between encapsulated DNA and the surrounding SWNT

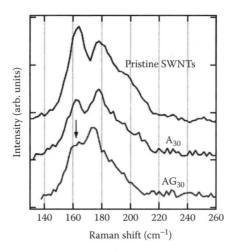

FIGURE 4.38   Raman spectra of DNA@SWNTs. Both the V_{DC} and V_{RF} are applied to the substrate in DNA solution.

or bandgap modulation of SWNT. Figure 4.38 shows Raman spectra of DNA@SWNTs in the region of radial breathing mode. The spectrum shape of AG_{30} ($AGAG...$) is different compared with the other case, especially the intensity of a peak around 160 cm^{-1} (indicated by the arrow) is largely decreased. This phenomenon indicates an indirect evidence of the electronic interaction between SWNT and guanine contained DNA to result in a resonance condition change of DNA@SWNTs.

Concerning DNA insertion into SWNTs, a simulation study is reported [85]. In this paper, Gao *et al.* mentioned that DNA molecules could be spontaneously inserted into SWNTs in a water solute environment, as illustrated in Figure 4.39. In addition, the encapsulated DNA is stable and the van der Waals interaction plays an important role in the insertion process. A similar situation can exist in the present experimental case, where the RF electric field affects DNA in the sense of not moving but stretching the DNA ion. Namely, when this stretched DNA ion is located near opened SWNTs for any reason, a part of DNA seems to start to be inhaled in any case [82–84]. Based on these two researches, it is attributed that DNA can be encapsulated into SWNTs without applying specified external forces whenever it is stretched to some extent, and DNA insertion can strongly be enhanced by applying the superimposed external forces such as stretching RF and accelerating DC electric fields.

Generally, the sizes of biomolecules such as amino acid, polysaccharide, and protein are too large to be encapsulated into SWNTs. Therefore, it is most likely impossible to encapsulate them into SWNTs. However, among them, one-dimensional materials such as carotenoids, which are one of important natural pigments, are possible to be encapsulated. They have also been used extensively as a model system in the

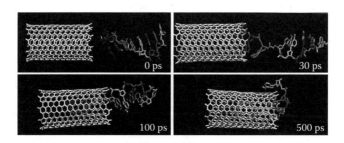

FIGURE 4.39   (see color insert) Simulation snapshots of a DNA insertion process at 0 to 500 psec. Water molecules are not displayed for clarity. (From Gao, H. et al. *Nano Lett.*, 3, 471, 2003.)

study of π-conjugated polyene molecules. Yanagi *et al.* focused on β-carotene as an encapsulated material, which holds the possibility of optical applications [86]. They reported that the encapsulation process is performed not in water but in a liquid state of organic solvents. As a result, although it takes several hours, they achieved a high encapsulation yield of β-carotene into SWNTs even in the absence of applying any external force.

4.5.2 Molecule-Wrapped Carbon Nanotubes

Since the van der Waals interaction of SWNTs is very strong, it is very difficult to obtain each individual SWNT in general. As mentioned earlier, although the individual SWNT can be obtained by the plasma CVD or other methods, these are still a little complicated and tricky. Actually, when SWNTs are prepared by an arc discharge or laser ablation methods, which can produce high quality ones, it has been difficult to obtain the individual SWNT. Thus, a certain modification method is required to obtain the individual SWNT independently of the SWNT-production methods.

Modification techniques of SWNTs are often used in order to obtain individual SWNTs, one of which is a wrapping method. Since various SWNTs with different chirality are mingled together in general because CNTs easily form bundles due to their strong van der Waals interaction, it has been difficult to investigate physical, chemical, and optical properties of specific SWNTs. Here, a SWNT dispersion method using surfactants is introduced from the point of view of outside functionalization—the so-called wrapping of SWNTs. It is reported that surfactants are good reagents to obtain isolated CNTs [87–89], and hence SWNTs are dissolved into sodium dodecyl sulfate (SDS) solution. After that, the dispersion is sonicated and centrifuged, yielding isolated SWNTs as shown in Figure 4.40, where the hydrophobic part of SDS is attached onto the sidewall of SWNTs and the hydrophilic part of SDS faces away from SWNTs. This isolation method plays an important role in the investigation of optical properties of SWNTs under the circumstance that the optical transition is limited to only the nonradiative process in the case of SWNT bundles. Eventually, the distinct electronic absorption and emission transition of semiconducting SWNTs have been revealed using the isolation method, as mapped in Figure 4.41 [87]. At pH less than 5, the absorption and emission spectra of individual nanotubes show evidence of bandgap–selective protonation of the sidewalls of the tube. This protonation is readily reversed by treatment with base or ultraviolet light. Of course, many kinds of surfactants are reported to be available to the SWNTs dispersion [90].

On the other hand, two groups, Zheng *et al.* and Nakashima *et al.*, have reported almost simultaneously that DNA is also an effective reagent to disperse SWNTs [91,92]. Since DNA has both the hydrophilic and hydrophobic parts in its structure, it is possible to dissolve SWNTs in water in the same way as SDS. Zheng *et al.* revealed the appearance of DNA wrapped SWNTs that are observed by atomic force microscopy (AFM). In the case of using single-stranded DNA, which contains both guanine and thymine, the AFM measurement shows that a single-stranded DNA-SWNT complex has a much more uniform periodic structure, with a regular pitch of about 18 nm. Furthermore, the dispersion of SWNTs is found

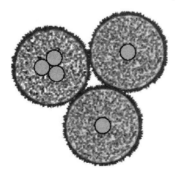

FIGURE 4.40 SWNT dispersion by SDS.

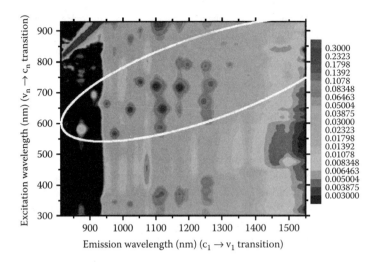

FIGURE 4.41   (see color insert) Photoluminescence of SWNTs, which are dispersed in SDS solution. (From Bachilo, S. M. et al. *Science*, 298, 2361, 2002.)

to depend on the pH and base sequence. Then, the separation of metallic and semiconductive SWNTs can be performed using anion exchange chromatography [91–94].

Understanding of the interaction between SWNTs and DNA develops the potential of their application to novel electronics and optics. Lu *et al.* studied a simulation related to the interaction between DNA and SWNTs by constructing the DNA-SWNT combined system and calculating the density of state. As a result, they suggest the possibilities of electron transfer in the system, as well as the application to not only electronic devices but also ultrafast DNA sequencing [95]. Furthermore, an interesting potential of the DNA-SWNT complex is delivered by Strano's group—that is, DNA hybridization could be detected from fluorescence of SWNTs [96]. They demonstrate the optical detection of DNA hybridization on the surface of SWNTs through fluorescence modulation, as shown in Figure 4.42. The energy shift of the fluorescence is modeled by correlating the surface coverage of wrapped DNA on SWNT to the exciton binding energy. In this paper, a new technique is used to suspend SWNTs using UV-Vis absorption spectroscopy and the subsequent removal of excess DNA from solution. They mentioned that optical detection of specific DNA sequences might have an application in the life science and medical fields as detectors of oligonucleotides.

4.5.3   Molecule-Attached Carbon Nanotubes

Since the sidewall of SWNTs is graphite, organic chemical synthetic techniques can be applied to the SWNTs. The reactivity of the edge of the graphite is so high that the position of the outside modification by chemical synthesis was mainly focused on the end or defect part of SWNTs in general. After attaching diazonium reagents, Strano *et al.* found that metallic SWNTs attain highly chemoselective reactions in contrast to the semiconducting nanotubes [97], which can be used to realize the separation of semiconducting SWNTs from metallic nanotubes. The change of band structure indicates that the extent of electron transfer is dependent on the density of states in that electron density near Fermi level.

The sidewall modification of SWNTs using biomolecules was reported by Williams *et al.* [98]. Peptide nucleotide acid (PNA) is covalently attached onto SWNT and this PNA-SWNT complex can recognize the DNA base sequence. Dwyer *et al.* reported a synthesis method of attaching DNA to the end of SWNTs. They opened the end of SWNTs by oxidation to make an active site and SWNTs tended to react with amine-terminated DNA [99]. Baker *et al.* reported another synthesis method, and also oxidized SWNTs at first to make an active site at not only the end but also at the sidewall of SWNTs. After several steps, they synthesized DNA linked to the end and sidewall of SWNTs by thiol termination [100]. Among biomolecules,

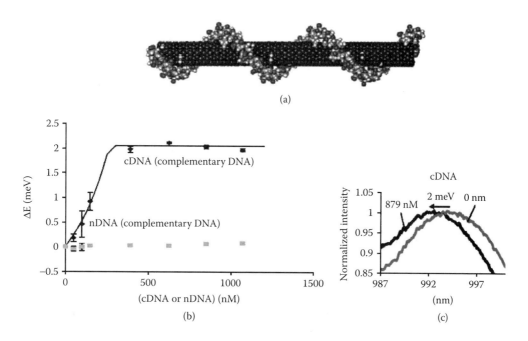

(a)

(b)

(c)

FIGURE 4.42 (see color insert) (a) SWNT decorated with ss-DNA as a sensor of the selective detection of DNA hybridization. (b) Addition of complementary DNA (cDNA) causes an increase in energy of the steady state (6,5) fluorescence peak while there is negligible energy change with noncomplementary DNA (nDNA). (c) Sample spectra of the fluorescence peak blue shift with cDNA addition. (From Jeng, E. S. et al. *Nano Lett.*, 6, 371, 2006.)

proteins can also be attached onto SWNTs. Huang *et al.* reported functionalization of SWNTs/MWNTs by proteins via diimide-activated amidation [101]. It is to be noted that complexes are shown to be highly water-soluble, and over 90% maintain bioactivity.

Next, we shall mention other adducts from the point of view of the solubilization of SWNTs. The improvement in solubility of SWNTs is important in performing various procedures in liquid phase and getting high yields. Although SDS, DNA, and other surfactants can dissolve SWNTs in water (as mentioned in the previous section), it is significant in the development of other functionalization methods for a variety of applications in the biomedical fields. Bianco *et al.* demonstrated the solubilization in aqueous media of sidewall modified CNTs and their derivatization with N-protected amino acid. Their functionalization method is based on the 1,3-dipolar cycloaddition reaction to the outside of CNTs. Then, it was demonstrated for the first time that functionalized CNTs are able to cross the cell membrane [102]. Figure 4.43 depicts the typical functionalized CNTs used in a series of their studies. According to their report, it clearly appears that CNTs are a very promising carrier system for future applications in drug delivery systems and targeting therapy. It is important to remember that CNTs can cross cell membranes and accumulate in the cytoplasm or reach the nucleus without being toxic even for primary cells belonging to the immune system [103] under the optimized condition, demonstrating the possibility of gene delivery systems using these functionalized SWNTs. The functionalized SWNTs have positive charges due to the end of the amino group, which is attached onto the sidewall of SWNTs. Based on several analyses, their results indicate that these cationic SWNTs are able to condense DNA to a varying degree, and both the surface area of SWNTs and charge density are critical parameters that determine the interaction and electrostatic complex formation between functionalized SWNTs and negatively charged DNA. Therefore, the functionalized SWNTs form supramolecular complexes with DNA through ionic interactions. Moreover, this complex can bind to, and also penetrate within, cells [104,105].

Thus, molecule-attached CNT derivatives have a broadening potential for applications ranging from molecular electronics to nanobiotechnology.

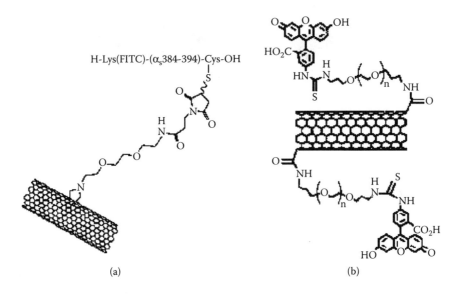

H-Lys(FITC)-(α_s384-394)-Cys-OH

(a)   (b)

FIGURE 4.43 Chemical structure of (a) a peptide-SWNTs complex and (b) an oxidized complex. (From Pantarotto, D. et al. *Chem. Comm.*, 16, 2004; Dumortier, H. et al. *Nano Lett.*, 6, 1522, 2006.)

4.6 Summary and Outlook

A special remark is focused from the viewpoint of nano- and molecular electronics on the development of novel-structured and new-functional carbon derivatives based on fullerenes and nanotubes among a variety of carbon allotropes, which is pioneered mainly using nanoscopic plasma processing technology.

First, the production of charge-exploited alkali-metal and spin-exploited atomic-nitrogen encapsulated C_{60} fullerenes in large quantities is challenged in order to sprout endofullerene-based nanoelectronics. Second, individually isolated and vertically aligned single-walled carbon nanotubes (SWNTs) are produced on a flat-surface substrate, making further inroads toward inner nanospace control.

Third, not only alkali-metal, fullerene, and ferromagnetic-atom encapsulated SWNTs but also double-walled carbon nanotubes (DWNTs) are created, and their electronic properties are found to be appealing as nanoelectronics elements in high-performance air-stable n-type semiconductors, magnetic semiconductor nanotubes, high-performance resonance tunneling nano FET, etc. Furthermore, the formation of nano pn-junction is challenged using "alkali–metal/fullerene" and "alkali–metal/halogen–atom" junctions encapsulated SWNTs or DWNTs.

Fourth, SWNTs encapsulating biomolecules such as DNA and carotene are created, while the modification of carbon nanotubes is extensively performed by externally wrapping and attaching various kinds of molecules through chemical reaction processes. Such a process is also available for DWNTs. Their electronic and biochemical properties are expected to be applied to electronic devices, biosensors, gene delivery, nano medical-tubing, etc. as nano- and molecular electronics.

4.7 Acknowledgments

The author would like to acknowledge the assistance of N. Sato, T. Mieno, K. Tohji, Y. Kawazoe, T. Hirata, G.–H. Jeong, T. Kaneko, W. Oohara, Y. F. Li, T. Okada, T. Kato, K. Baba, J. Shishido, Y. Neo, H. Mimura, K. Omote, Y. Kasama, Y. Kuk, and M. Takahashi.

References

[1] Kroto, H.W. et al., C_{60}: Buckminsterfullerene, *Nature*, 318, 162, 1985.

[2] Krätschmer, et al., Solid C_{60}: a new form of carbon, *Nature*, 347, 354, 1990.

[3] Iijima, S., Helical microtubules of graphitic carbon, *Nature*, 354, 56, 1991.

[4] Iijima, S. and Ichihashi, T., Single-shell carbon nanotubes of 1-nm diameter, *Nature*, 363, 603, 1993.

[5] Heath, J.R. et al., Lanthanum complexes of spheroidal carbon shells, *Journal of American Chemical Society*, 107, 7779, 1985.

[6] Shinohara, H. et al., Mass spectroscopic and ESR characterization of soluble yttrium-containing metallofullerenes YC_{82} and Y_2C_{82}, *Journal of Physical Chemistry*, 96, 3571, 1992.

[7] Wan, Z. et al., Collision of Li^+ and Na^+ with C_{60}: Insertion, fragmentation, and thermionic emission, *Physical Review Letters*, 69, 1352, 1992.

[8] Hirata, T. et al., Production and control of $K-C_{60}$ plasma for material processing, *Journal of Vacuum Science Technology A*, 14, 615, 1996.

[9] Tellgram, R. et al., Endohedral fullerene production, *Nature*, 382, 407, 1996.

[10] Murphy, T.A. et al., Nitrogen in nitrogen-implanted solid C_{60}, *Physical Review Letters*, 77, 1075, 1996.

[11] Pietzak, B. et al., Buckminsterfullerene C_{60}: A chemical faraday cage for atomic nitrogen, *Chemical Physics Letters*, 279, 259, 1997.

[12] Harneit, W., Fullerene-based electron-spin quantum computer, *Physical Review A*, 65, 032322, 2002.

[13] Suter, D. and Lim, K., Scalable architecture for spin-based quantum computers with a single type of gate, *Physical Review A*, 65, 052309, 2002.

[14] Sato, N. et al., Production of C_{60} plasma, *Physics of Plasmas*, 1, 3480, 1994.

[15] Hirata, T. et al., Correlation between fullerene-plasma production using different alkali metals and endohedral metallofullerene generation, *Proceedings of Plasma Science Symposium 2001/18th Symposium on Plasma Processing*, 741, 2001.

[16] Ohno, K. et al., *Ab Initio* molecular dynamics simulations for collision between C_{60}^- and alkali-metal ions: A possibility of $Li@C_{60}$, *Physical Review Letters*, 76, 3590, 1996.

[17] Campbell, E.E.B. et al., Production and LDMS characterisation of endohedral alkali-fullerene films, *Journal of Physics and Chemical of Solids*, 58, 1763, 1997.

[18] Wan, Z. et al., Collision of alkali ions with C_{60}/C_{70}: Insertion, thermionic emission, and fragmentation, *Jouranl of Chemical Physics*, 99, 5858, 1993.

[19] Huang, H. et al., $^{14}N@C_{60}$ formation in a nitrogen rf-plasma, *Chemical Communication*, 2076, 2002.

[20] Reuther, U. and Hirsh, A., Synthesis, properties and chemistry of Aza [60] fullerene, *Carbon*, 38, 1539, 2000.

[21] Abe, S. et al., Effects of ion energy control on production of nitrogen-C_{60} compounds by ion implantation, *Japanese Journal of Applied Physics*, 45, 8340, 2006.

[22] Heald, M.A. and Beringer, R., Hyperfine structure of nitrogen, *Physical Review*, 96, 645, 1954.

[23] Suetsuna, T. et al., Separation of $N_2@C_{60}$ and $N@C_{60}$, *Chemistry—A European Journal*, 22, 5079, 2002.

[24] Cui, F.Z. et al., Atomistic simulation of radiation damage to C_{60}, *Physical Review B.*, 49, 9962, 1994.

[25] Lips, K. et al., Atomic nitrogen encapsulated in fullerenes: realization of a chemical faraday cage, *Physical Status Solid*, 177, 81, 2000.

[26] Cassell A. M. et al., Directed growth of free-standing single-walled carbon nanotubes, *Journal of American Chemical Society*, 121 ,7975, 1999.

[27] Zhang, Y. et al., Electric-field-directed growth of aligned single-walled carbon nanotubes, *Applied Physics Letters*, 79, 3155, 2001.

[28] Huang, S. et al., Growth of millimeter-long and horizontally aligned single-walled carbon nanotubes on flat substrates, *Journal of American Chemical Society*, 125, 5636, 2003.

[29] Huang, S. et al., Ultralong, well-aligned single-walled carbon nanotube architectures on surface, *Advanced Materials*, 15, 1651, 2003.

[30] Ismach, A. et al., Atomic-step-templated formation of single-wall carbon nanotube patterns, *Ange-wandte Chemie International Edition*, 43, 6140, 2004.

[31] Han, S. et al., Template-free directional growth of single-walled carbon nanotubes on a- and r-plane sapphire, *Journal of American Chemical Society*, 127, 5294, 2005.

[32] Ago, H. et al., Aligned growth of isolated single-walled carbon nanotubes programmed by atomic arrangement of substrate surface, *Chemical Physics Letters*, 408, 433, 2005.

[33] Hirata, T. et al., Magnetron-type radio-frequency plasma control yielding vertically well-aligned carbon nanotube growth, *Applied Physics Letters*, 83, 1119, 2003.

[34] Jeong, G.–H. et al., Time evolution of nucleation and vertical growth of carbon nanotubes during plasma-enhanced chemical vapor deposition, *Japanese Journal of Applied Physics*, 42, L1340, 2003.

[35] Jeong, G.–H. et al., Simple methods for site-controlled carbon nanotube growth using radio-frequency plasma-enhanced chemical vapor deposition, *Applied Physics A*, 79, 85, 2004.

[36] Hatakeyama, R. et al., Effects of micro- and macro-plasma-sheath electric fields on carbon nanotube growth in a cross-field radio-frequency discharge, *Journal of Applied Physics*, 96, 6053, 2004.

[37] Kato, T. et al., Single-walled carbon nanotubes produced by plasma-enhanced chemical vapor deposition, *Chemical Physics Letters*, 381, 422, 2003.

[38] Kato, T. et al., Structure control of carbon nanotubes using radio-frequency plasma enhanced chemical vapor deposition, *Thin Sold Films*, 457, 2, 2004.

[39] Kato, T. et al., Freestanding individual single-walled carbon nanotube synthesis based on plasma sheath effects, *Japanese Journal of Applied Physics*, 43, L1278, 2004.

[40] Kato, T. et al., Diffusion plasma chemical vapor deposition yielding freestanding individual single-walled carbon nanotubes on a silicon-based flat substrate, *Nanotechnology*, 17, 2223, 2006.

[41] Kato, T. and Hatakeyama, R., Formation of freestanding single-walled carbon nanotubes by plasma-enhanced chemical vapor deposition, *Chemical Vapor Deposition*, 12, 345, 2006.

[42] Grigorian, L. et al., Transport properties of alkali-metal-doped single-walled carbon nanotubes, *Physical Review B*, 58, R4195, 1998.

[43] Karmakar, S. et al., Magnetic behavior of ion-filled multiwalled carbon nanotubes, *Journal of Applied Physics*, 97, 054306, 2005.

[44] Grobert, N. et al., Enhanced magnetic coercivities in Fe nanowires, *Applied Physics Letters*, 75, 3363, 1999.

[45] Chen, G. et al., Chemically doped double-walled carbon nanotubes: cylindrical molecular capacitors, *Physical Review Letters*, 90, 257403, 2004.

[46] Rao, A.M. et al., Evidence for charge transfer in doped carbon nanotube bundles from Raman scattering, *Nature*, 388, 257, 1997.

[47] Costa, P.M.F.J. et al., Imaging lattices defects and distortions in alkali-metal iodides encapsulated within double-walled carbon nanotubes, *Chemistry Materials*, 17, 3122, 2005.

[48] Sloan, J., et al., Integral atomic layer architectures of 1D crystals inserted into single-walled carbon nanotubes, *Chemical Communications*, 1319, 2002.

[49] Zhou, O. et al., Defects in carbon nanostructures, *Science*, 263, 1744, 1994.

[50] Smith, B.W. et al., Encapsulated C_{60} in carbon naotubes, *Nature*, 396, 323, 1998.

[51] Jeong, G.–H. et al., Structural deformation of single-walled carbon nanotubes and fullerene encapsulation due to magnetized-plasma ion irradiation, *Applied Physics Letters*, 79, 4213, 2001.

[52] Jeong, G.–H. et al., Formation and structural observation of cesium encapsulated single-walled carbon nanotubes, *Chemical Communications*, 152, 2003.

[53] Jeong, G.–H., et al., Cesium encapsulation in single-walled carbon nanotubes via plasma ion irradiation: Application to junction formation and ab initio investigation, *Physical Review B*, 68, 075410, 2003.

[54] Li, Y.F. et al., Synthesis of Cs-filled double-walled carbon nanotubes by a plasma process, *Carbon*, 44 1586, 2006.

[55] Hatakeyama, R. et al., Creation of novel structured nanocarbons based on plasma technology, *Journal of the Vacuum Society of Japan*, 48, 142, 2005.

[56] Li, Y.F. et al., Electronic transport properties of Cs-encapsulated double-walled carbon nanotubes, *Applied Physics Letters*, 89, 093110, 2006.

[57] Sato, Y. et al., Correlation between atomic rearrangement in defective fullerenes and migration behavior of encaged metal ions, *Physical Review B*, 73, 233409, 2006.

[58] Takenobu, T. et al., Stable and controlled amphoteric doping by encapsulation of organic molecules inside carbon nanotubes, *Nature Materials*, 2, 683, 2003.

[59] Kang, Y.J. et al., Electronic and magnetic properties of single-walled carbon nanotubes filled with iron atoms, *Physical Review B*, 71, 115441, 2005.

[60] Li, Y.F. et al., Synthesis and electrical transport measurement of functionalized double-walled carbon nanotubes by ferrocene encapsulation, *Nanotechnology*, 17, 4143, 2006.

[61] Li, Y.F. et al., Nano-sized magnetic particles with diameter less than 1 nm encapsulated in single-walled carbon nanotubes, *Japanese Journal of Applied Physics*, 45, L428, 2006.

[62] Izumida, T. et al., Measurement of electronic transport properties of single-walled carbon nanotubes encapsulating alkali-metals and C_{60} fullerene via plasma ion irradiation, *Japanese Journal of Applied Physics*, 44, 1606, 2005.

[63] Javey, A. et al., High performance n-type carbon nanotube field-effect transistor with chemically doped contacts, *Nano Letters*, 5, 345, 2005.

[64] Izumida, T. et al., Electronic transport properties of Cs encapsulated single-walled carbon nanotubes created by plasma ion irradiation, *Applied Physics Letters*, 89, 093121, 2006.

[65] Shim, M. et al., Polymer functionalization for air-stable *n*-type carbon nanotube field-effect transistors, *Journal of America Chemical Society*, 123, 11512, 2001.

[66] Javey, A. et al., Electrical properties and devices of large-diameter single-walled carbon nanotubes, *Applied Physics Letters*, 80, 1064, 2002.

[67] Li, Y.F. et al., Negative differential resistance in tunneling transport through C_{60} encapsulated double-walled carbon nanotubes, *Applied Physics Letters*, in press.

[68] Jensen, A. et al., Magnetoresistance in ferromagnetically contacted single-walled carbon nanotubes, *Physical Review B*, 72, 035419, 2005.

[69] Li, Y.F. et al., Electrical properties of ferromagnetic semiconducting single-walled carbon nanotubes, *Applied Physics Letters*, 89, 083117, 2006.

[70] Li, Y.F. et al., Magnetic characterization of Fe-nanoparticles encapsulated single-walled carbon nanotubes, *Chemical Communications*, 254, 2007.

[71] Lee, J. et al., Bandgap modulation of carbon nanotubes by encapsulated metallofullerenes, *Nature*, 415, 1005, 2002.

[72] Shimada, T. et al., Ambipolar field-effect transistor behavior of Gd@C_{82} metallofullerene peapods, *Applied Physics Letters*, 81, 4067, 2002.

[73] Zhou, C. et al., Modulated chemical doping of individual carbon nanotubes, *Science*, 290, 1552, 2000.

[74] Zhou, Y. et al., p-channel, n-channel thin film transistors and p-n diodes based on single wall carbon nanotube networks, *Nano Letters*, 4, 2031, 2004.

[75] Esfarjani, K. et al., Electronic and transport properties of N-P doped nanotubes, *Applied Physics Letters*, 74, 79, 1999.

[76] Oohara, W. et al., Alkali-halogen plasma generation by dc magnetron discharge, *Applied Physics Letters*, 88, 191501-1, 2006.

[77] Yoo, K.–H. et al., Electrical conduction through poly(dA)-poly(dT) and poly(dG)-poly(dC) DNA molecules, *Physical Review Letters*, 87, 198102, 2001.

[78] Debye, Von P. and Falkenhagen, H., Zur theorie der elektrolyte, *Physikalische Zeitschrift*, 29, 401, 1928.

[79] Tohji, K. et al., Purifying single-walled carbon nanotubes, *Nature*, 383, 679, 1996.

[80] Suzuki, S. et al., Quantitative analysis of DNA orientation in stationary AC electric fields using fluorescence anisotropy, *IEEE Transaction on Industry Application*, 34, 75, 1998.

[81] Washizu, M. and Kurosawa, O., Electrostatic manipulation of DNA in microfabricated structures, *IEEE Transaction on Industry Application*, 26, 1165, 1990.

[82] Okada, T. et al., Electrically triggered insertion of single-stranded DNA into single-walled carbon nanotubes, *Chemical Physics Letters*, 417, 289, 2006.

[83] Okada, T. et al., DNA negative ion irradiation toward carbon nanotubes in micro electrolyte plasmas, *Transaction of Material Research Society of Japan*, 31, 459, 2006.

[84] Okada, T. et al., Single-stranded DNA insertion into single-walled carbon nanotubes by ion irradiation in an electrolyte plasma, *Japanese Journal of Applied. Physics*, 45, 8335, 2006.

[85] Gao, H. et al., Spontaneous insertion of DNA oligonucleotides into carbon nanotubes, *Nano Letters*, 3, 471, 2003.

[86] Yanagi, K. et al., Highly stabilized β-carotene in carbon nanotubes, *Advanced Materials*, 18, 437, 2006.

[87] O'Connell, M.J. et al., Band gap fluorescence from individual single-walled carbon nanotubes, *Science*, 297, 593, 2002.

[88] Bachilo, S.M. et al., Structure-assisted optical spectra of single-walled carbon nanotubes, *Science*, 298, 2361, 2002.

[89] Fantini, C. et al., Optical transition energies for carbon nanotubes from resonant Raman spectroscopy: environment and temperature effects, *Physical Review Letters*, 93, 147406, 2004.

[90] Moore, V.C. et al., Individually suspended single-walled carbon nanotubes in various surfactants, *Nano Letters*, 3, 1379, 2003.

[91] Zheng, M. et al., Structure-based carbon nanotube sorting by sequence-dependent DNA assembly, *Science*, 302, 545, 2003.

[92] Nakashima, N. et al., DNA dissolves single-walled carbon nanotubes in water, *Chemistry Letters*, 32,456, 2003.

[93] Strano, M.S. et al., Understanding the nature of the DNA-assisted separation of single-walled carbon nanotubes using fluorescence and Raman spectroscopy, *Nano Letters*, 4, 543, 2004.

[94] Chou, S.G. et al., Optical characterization of DNA-wrapped carbon nanotube hybrids, *Chemical Physics Letters*, 397, 296, 2004.

[95] Lu, G. et al., Carbon nanotube interaction with DNA, *Nano Letters*, 5, 897, 2005.

[96] Jeng, E.S. et al., Detection of DNA hybridization using the near-infrared band-gap fluorescence of single-walled carbon nanotubes, *Nano Letters*, 6, 371, 2006.

[97] Strano, M.S. et al., Electronic structure control of single-walled carbon nanotube functionalization, *Science*, 301, 519, 2003.

[98] Williams, K.A. et al., Carbon nanotubes with DNA recognition, *Nature*, 420, 761, 2002.

[99] Dwyer, C. et al., DNA-functionalized single-walled carbon nanotubes, *Nanotechnology*, 13, 601, 2002.

[100] Baker, S.E. et al., Covalently bonded adducts of deoxyribonucleic acid (DNA) oligonucleotides with single-walled carbon nanotubes: Synthesis and hybridization, *Nano Letters*, 2, 1413, 2002.

[101] Huang, W. et al., Attaching proteins to carbon nanotubes via diimide-activated amidation, *Nano Letters*, 2, 311, 2002.

[102] Pantarotto, D. et al., Translocation of bioactive peptides across cell membranes by carbon nanotubes, *Chemical Communication*, 16, 2004.

[103] Dumortier, H. et al., Functionalized carbon nanotubes are non-cytotoxic and preserve the functionality of primary immune cells, *Nano Letters*, 6, 1522, 2006.

[104] Singh, R. et al., Binding and condensation of plasmid DNA onto functionalized carbon nanotubes: Toward the construction of nanotube-based gene delivery vectors, *Journal of American Chemical Society*, 127, 4388, 2005.

[105] Pantarotto, D. et al., Functionalized carbon nanotubes for plasmid DNA gene delivery, *Angewandte Chemie International Edition*, 43, 5242, 2004.

5

System-Level Design and Simulation of Nanomemories and Nanoprocessors

Shamik Das

Carl A. Picconatto

Garrett S. Rose

Matthew M. Ziegler

James C. Ellenbogen

Abstract

This chapter describes in detail system designs and system simulations for electronic nanocomputers that are integrated on the molecular scale. Here, these systems are considered as consisting primarily of the combination of two component subsystems: nanomemories and nanoprocessors. Challenges are enumerated for the design and development of both of these ultra-densely integrated components. Various system-level designs or architectures are presented that have been proposed to meet these challenges. Detailed consideration is given for both nanomemories and nanoprocessors to system designs based upon arrays of crossed nanowires. In each case, a system simulation is performed to assess and help optimize the prospective performance of the system component in advance of its fabrication. In the ongoing development of crossbar nanocomputer systems, these simulations have been integral to the refinement of designs because they assist in reducing the time and cost of such development.

5.1 Introduction

Much progress has been made recently in the field of molecular electronics. In particular, dramatic successes in the demonstration of nanoelectronic devices and simple molecular-scale circuits [1–11] suggest that soon we may be able to design, fabricate, and demonstrate an entire, ultra-dense nanoelectronic computer that is integrated on the molecular scale. In fact, development of such nanoelectronic computer systems already is underway. Despite significant challenges, this effort has produced functioning prototype nanomemories [12–14] and is likely to produce functioning prototype nanoprocessors within a few years [2,8,15–18].

In this chapter, we describe system designs [16,19,20] and system simulations [21,22] that have been and continue to be integral to those advances in nanoelectronic system hardware development. That design and simulation work has focused on approaches for novel, ultra-dense nanoelectronic circuits and systems that use crossed nanowire arrays [8,23–25] as their underlying circuit structures. Such arrays may be fabricated either from patterned nanowires [8,24,25] or from self-assembled nanowires [23]. It is expected that operational nanomemory and nanoprocessor systems based upon such crossed-nanowire array structures can achieve integration densities in excess of 10^{11} devices per square centimeter [26]. This is well beyond the densities presently envisioned [27] for electronic computer systems that use circuits based upon complementary metal-oxide-semiconductor (CMOS) devices, as do conventional microelectronic computers. Thus, higher-density nanomemory and nanoprocessor systems designed and fabricated from crossed nanowires might even be used to enhance CMOS-based electronics in a post-CMOS era. Here, however, we focus on the design and simulation of "pure" or "true" nanomemory and nanoprocessor systems that incorporate only nanometer-scale devices, such as crossed nanowires and molecules.

The process of developing such true nanocomputers opens up an entirely new frontier of systems objectives and issues that require research and development, beyond that in the much more numerous investigations that presently are being conducted upon isolated nanodevices and small nanocircuits [28–32]. Thus, in addition to describing our specific design and simulation investigations of crossed-nanowire nanocomputer systems, we also discuss the broader range of system issues encountered at this new frontier. Further, we survey some of the other device and system design approaches [33–46] being advanced to help address the problem of building entire nanomemory and nanoprocessor systems integrated on the molecular scale.

By integration on the molecular scale, we mean the basic switching devices, as well as the wire widths and pitch dimensions (i.e., spacing between the centers of neighboring wires), all will measure only a few nanometers — the size of a small molecule — in the computer systems of interest here. Such systems may function using only one or a few molecules within their basic devices [4,5,8,10,12,14,47]. On the other hand, the systems may not use molecules at all, employing instead solid-state quantum dots [35,38,48–50] and/or patterned or self-assembled nanowires [8,23,24,51,52], as mentioned above.

Consideration of the range of topics described previously in this section proceeds below as follows:

- In Section 5.2 of this work, we discuss the electrical behaviors required of molecular-scale devices in order to develop extended nanoelectronic systems. We describe the presently-available nanoelectronic devices that exhibit and yield the types of behavior necessary for computation.
- Section 5.3 considers the prospective performance of a crossbar-based nanomemory system that utilizes some of the nanoelectronic devices described in Section 5.2. An overview is given of the architecture and operational principles of this system. Then, metrics and a simulation methodology for the evaluation of system performance are described. This unique, bottom-up simulation methodology facilitates the detailed prediction of the performance of entire systems integrated on the molecular scale. Specific system simulation results are provided, followed by a discussion of the implications of these results for the construction of extended nanomemory systems.
- Building upon this analysis of nanomemories, Section 5.4 considers the more complex problem of nanoprocessor design. It begins with a review of the difficulties facing the design of nanoprocessor architectures and surveys the various system architecture approaches that have been proposed.

Detailed consideration is given to one promising system-level design approach, a crossed-nanowire approach by DeHon and Wilson [19]. Section 5.4 concludes by describing a detailed simulation of key circuits of a notional nanoprocessor based upon the DeHon–Wilson design approach.

The simulations described here are intended to illustrate in a very specific manner the types of issues that will be encountered in building and operating a nanocomputer. It is significant that these simulations can be and have been conducted well before an entire system of this type actually is fabricated and integrated on the molecular scale. As the research community attempts to move forward with detailed designs for an entire nanocomputer, system simulation can illuminate the detailed consequences of both the architecture-level design choices and the *a priori* device-level constraints. Still further, the results of the simulation serve to provide focus for nanodevice and nanofabrication research, showing where it may be necessary to push back on the limits of these technologies, and where such efforts can have the most benefit for the ultimate objective of building a nanocomputer.

5.2 Molecular Scale Devices in Device-Driven Nanocomputer Design[1]

Whether one considers the design, simulation, or fabrication of an entire computer system, there is a hierarchy of structure and function. In the usual approach of modern electrical engineering, this hierarchy is taken to start at the highest level of abstraction, the architecture level. Then it descends down to the level of its component circuits, and finally, proceeds down to the level of the component switch and interconnect devices [53]. To a great extent, this viewpoint mirrors the "top-down" approach used in the design and fabrication of microprocessors, in which the robust performance of the devices and the ability to tune precisely the structure and performance of those devices (i.e., microelectronic transistors) is somewhat taken for granted. Architectures often are optimized to suit first the high-level, system objectives, such as computational latency and throughput, then the circuits, and finally, the behavior of the devices may be adjusted to suit particular needs of the architecture.

At present, the situation is different when one sets out to design, simulate, or fabricate an entire nanocomputer system integrated on the molecular scale. The ability to tune the performance of nanodevices still is limited. This is partly because these molecular-scale devices are so new. Thus, the experiments [54–58] and the theory [59–66] necessary to understand them, design them, and make them to order still are very much in development. In addition, the ability to tune precisely the structure and performance of nanometer-scale devices may be limited inherently by the quantization of those structures and properties, which is ubiquitous on that tiny scale.

Further, designs for nanoelectronic circuits and systems are constrained by the very small size and small total currents associated with molecular-scale switches. This is coupled with the difficulty of making contact with them using structures and materials that are large and conductive enough to provide sufficient current and signal strength to serve an entire nanocomputer system. Such a system would be at least tens of square micrometers, if not tens of square millimeters, in extent, which is millions or trillions of times larger than the molecular-scale devices themselves.

Regardless of whether all these limitations are temporary or fundamental, for now they constrain both the circuits and architectures achievable in the relatively near term. Further, these limitations force us to begin consideration of the design and simulation of nanocomputer systems at the bottom-most level of the hierarchy, the device level.

As is true in most experiments on the electrical properties of molecules [3,55,56,67,68], for the purposes of discussing circuits and systems, a molecular-scale device consists of a junction between two metal

---

[1]Some of the material in this section has appeared previously in Das *et al.*, "Architectures and simulations for nanoprocessor systems integrated on the molecular scale," *Lect. Notes Phys.*, vol. 680, pp. 479–513, 2005.

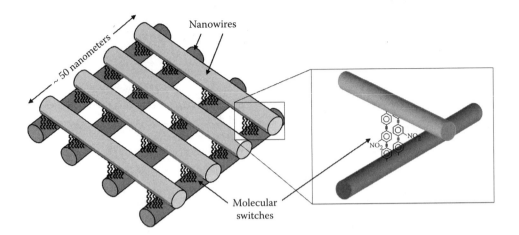

FIGURE 5.1   "Crossbar" array of nanowires with molecular devices at junctions.

or semiconductor surfaces with a molecular-scale structure sandwiched between. This molecular-scale structure may be one or a few molecules, as depicted in Figure 5.1. Or else, it may be a layer of molecules or atoms only a few nanometers thick, as in the nanowire junction diode depicted in Figure 5.2(a). While many electrical properties may be very important (especially capacitance), the electrical behavior of such junction nanoswitches is characterized primarily by the current response I to an applied voltage V, a so-called I-V curve, such as that shown in Figure 5.2(b).

I-V behaviors of such junctions include simple resistance at low voltage [69], rectification [57,70], negative differential resistance (NDR) [6], and hysteresis [69]. A variety of such junction nanodevices have been realized that might be useful for building extended nanoelectronic systems. The hysteretic behavior illustrated in Figure 5.2 is particularly valuable, as it allows the "programming" of a junction into one of two states. Such bistable switches are essential components of any computing system.

Development of molecular-scale switches with appropriate I-V behaviors is essential to constructing functional circuits that can be used to build up computer systems. For the logic components of such systems (i.e., nanoprocessors), it is of particular importance to have nanoscale switches that can be used to produce signal restoration and gain. These two features are essential in maintaining electrical signals as they move through multiple levels of logic. Nanoscale switches that produce signal restoration and gain

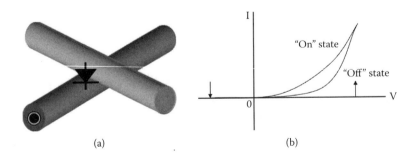

FIGURE 5.2   Illustrations of (a) a rectifying junction switch made of crossed nanowires that sandwich a molecule or layer of molecules or atoms and (b) a representative I-V characteristic for a hysteretic, rectifying device. Hysteresis is indicated by the multiple conductance states. The high-conductance "on" state and low-conductance "off" state are depicted, and the voltage thresholds at which the device switches between states are labeled with arrows. Rectification is indicated by the unequal responses to positive and negative voltages.

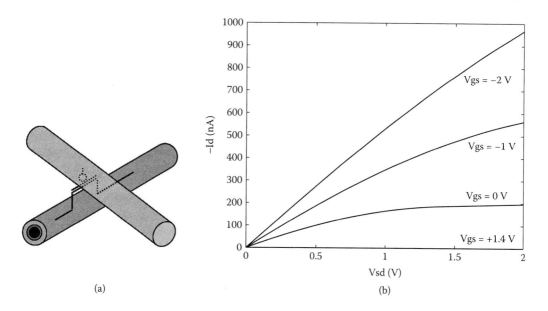

FIGURE 5.3 Illustrations of (a) a crossed-nanowire p-channel field effect transistor (PFET) and (b) a model of the I-V characteristic for this device. The experimental basis for this model was obtained from Huang *et al.* [7]. For this transistor, the threshold voltage, at which the device produces essentially zero current and turns "off," is observed to be approximately +1.4 V.

likely would be implemented using nanotransistors, although small circuits, e.g., latches incorporating molecular diodes, also can produce signal restoration [11]. Nanotransistors have been fabricated using carbon nanotubes (CNTs) [9,71–74], although it remains very difficult to use them in building extended systems. There also have been some suggestions for fabricating transistors from smaller molecules [57,75]. A few individual molecular transistors have been demonstrated based on small molecules, but only in very sensitive experiments under cryogenic conditions [76,77]. On the other hand, robust nanoscale transistors built from crossed nanowires have been demonstrated in a number of experiments at room temperature [7].[2] A diagram of such a nanowire nanotransistor is displayed alongside models of its I-V curves in Figure 5.3.

In addition to obtaining gain and signal restoration, other I-V behaviors, such as rectification from two-terminal nanodevices, are very important. Simulations show that even when using devices that provide good gain, rectification is necessary to ensure that signals do not take unintended and undesirable paths through circuits, especially in crossbar arrays. A strong rectifier can fulfill this role by permitting current to pass only in one direction in the circuit at the designed operating voltages.

The molecular-scale electronics community is just beginning to succeed in taking the key steps required for actually building and operating an extended nanocomputer system that integrates two-terminal junction nanodevices, such as rectifiers, as well as three-terminal nanotransistors. These steps form a hierarchy from the device to the system level, as follows: (a) development of nanofabrication approaches to build large numbers of the requisite junction nanodevices with precision and regularity, (b) development of interconnect and circuit design approaches that can incorporate such junction structures into extended circuit systems, and (c) determination of architectural approaches that include the aforementioned circuit

---

[2] Note that this transistor is not a junction nanoswitch since, ideally, no current flows between the nanowires. Rather, the top nanowire serves as a gate for the bottom "channel" nanowire, and the two are isolated from each other by a dielectric layer. This is in contrast to the nanowire diode shown in Figure 5.2, which is a junction nanoswitch.

designs and that can accommodate the limitations imposed by the constrained I-V behaviors available in present-day molecular electronic devices.

Challenges exist at each level of this hierarchy. However, these challenges may be mitigated by considering the development of a nanocomputer system separately for each of its two primary component systems, nanomemories and nanoprocessors. Nanomemory systems present fewer challenges by virtue of their less complex system architecture. Also, the lessons learned in designing and simulating nanomemory systems, as discussed in the next section, provide a foundation for addressing in Section 5.4 the more numerous and severe challenges of nanoprocessor design and simulation.

5.3   Crossbar-Based Design for Nanomemory Systems[3]

5.3.1   Overview

The crossbar architecture [10,12,14,16,25,69,78] is the most prevalent framework or approach now being employed for the design and fabrication of nanomemory systems integrated on the molecular scale. The basic crossbar architecture consists of the combination of planes of parallel wires laid out in orthogonal directions, such as is shown in Figure 5.1. The fundamental devices for memory storage are the molecular-scale junction switches formed at the crosspoints of the wires. For the purposes of this work, it is assumed that one bit is stored in each such fundamental crosspoint device.

A nanomemory system design based upon this architecture consists of three major subsystems: a nanowire crossbar memory array and two decoders, one for the array rows and one for the columns. Ultra-dense arrays of crossed nanowires are fabricated using specialized techniques such as nanoimprinting [8,24] and flow-based alignment [23]. Reasonably large memory arrays have been constructed using these techniques [12,14].

Figure 5.4 shows a system diagram and a corresponding circuit schematic of a notional 10×10 nanomemory based on the architectural design by DeHon [79]. In Figure 5.4, the nanowires forming the crossbar array are represented by thin black lines, as are the nanowires in the decoders. The decoders also contain much longer and much thicker micrometer-scale wires or "microwires" of the type used in conventional microelectronics. These are represented by thick gray lines. The crossbar array stores the data, whereas the decoders serve as an interface to this nanomemory array. The decoders permit an external microelectronic system to access a unique crossbar junction within the densely-integrated array. In addition, each decoder is connected to a microwire that supplies power to the system. These power supply lines are represented by thick black lines. They also serve to read or write a bit to the individual nanowire junction selected by the decoders, by imposing a voltage upon it.

Variations have been proposed for the nanomemory system design depicted in Figure 5.4. For example, Strukov and Likharev propose a "hybrid" nanomemory architecture [80] that utilizes nanowire crossbars for storage, but places these crossbars on top of a decoder structure that is fabricated entirely in conventional CMOS circuitry. Another example is provided by Nantero Corporation, which has demonstrated prototype nanomemories using an altogether different crossbar composition [13,81]. In the Nantero crossbar, one plane of wires is constructed using CMOS technology, while the other, orthogonal plane is created from a mesh of carbon nanotubes.

Several switch options have been proposed to store individual memory bits at the crosspoints of the ultra-dense nanowire arrays. For example, in the Lieber–DeHon nanomemory system [78], each nanowire-nanowire crosspoint in the crossbar array forms a bi-stable, nonvolatile nanowire (NVNW) diode, as depicted in Figure 5.2. In the Heath nanomemory system [10], a monolayer of bi-stable rotaxane molecules serves as an electronically rewritable memory bit [5]. Similarly, the system design of the Hewlett-Packard

---

[3]Some of the material in this section has appeared previously in Ziegler *et al.*, "Scalability simulations for nanomemory systems integrated on the molecular scale," *Ann. N.Y. Acad. Sci.*, vol. 1006, pp. 312–330, 2003.

Figure captions are available in the text.

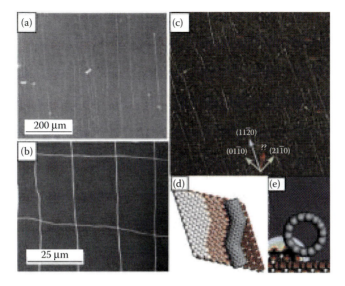

COLOR FIGURE 4.12

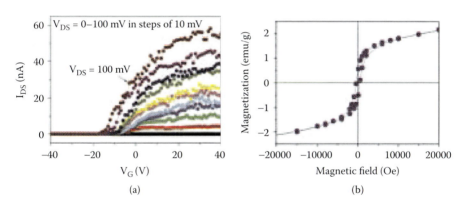

COLOR FIGURE 4.26

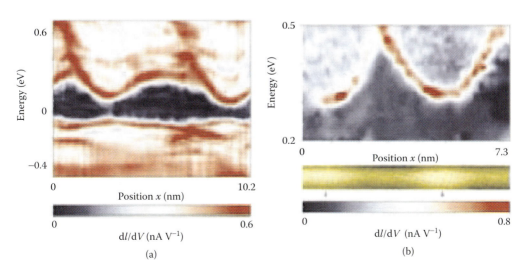

COLOR FIGURE 4.27

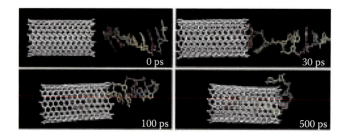

COLOR FIGURE 4.39

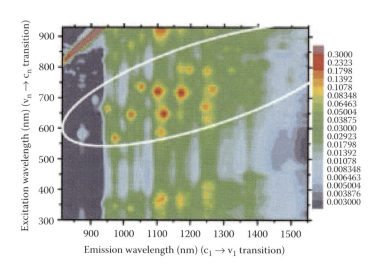

COLOR FIGURE 4.41

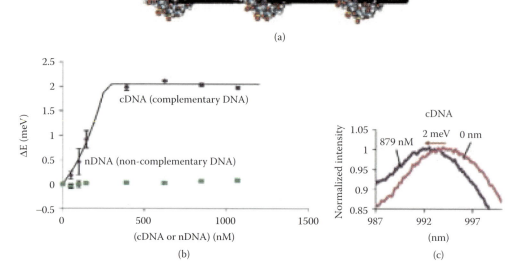

COLOR FIGURE 4.42

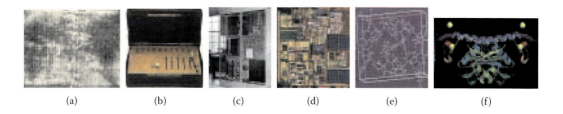

(a)   (b)   (c)   (d)   (e)   (f)

COLOR FIGURE 6.1

(a)   (b)

COLOR FIGURE 6.7

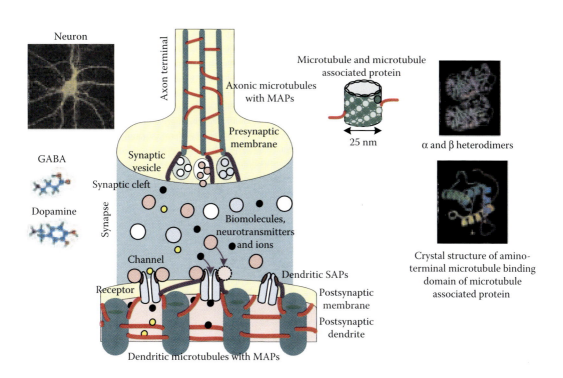

COLOR FIGURE 6.8

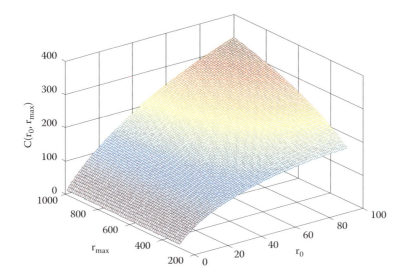

COLOR FIGURE 6.11

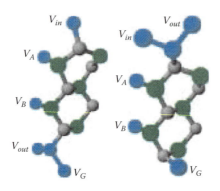

COLOR FIGURE 6.20

COLOR FIGURE 6.37

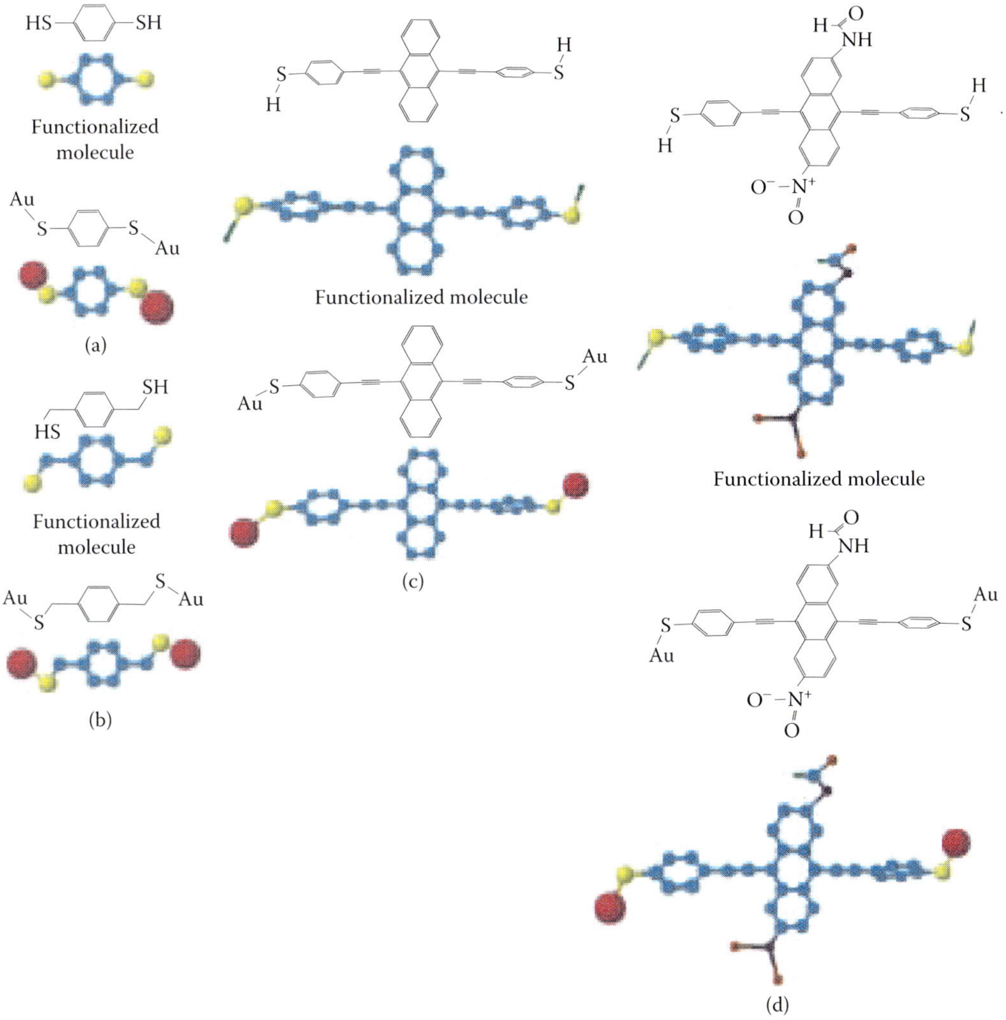

HS SH
Functionalized molecule
Au S S Au
(a)
SH
HS
Functionalized molecule
Au S S Au
(b)
S H
H S
Functionalized molecule
Au S S Au
(c)
H O
NH
S H
H S
O⁻ N⁺ O
Functionalized molecule
H O
NH
Au S
S Au
O⁻ N⁺ O
(d)

COLOR FIGURE 6.39

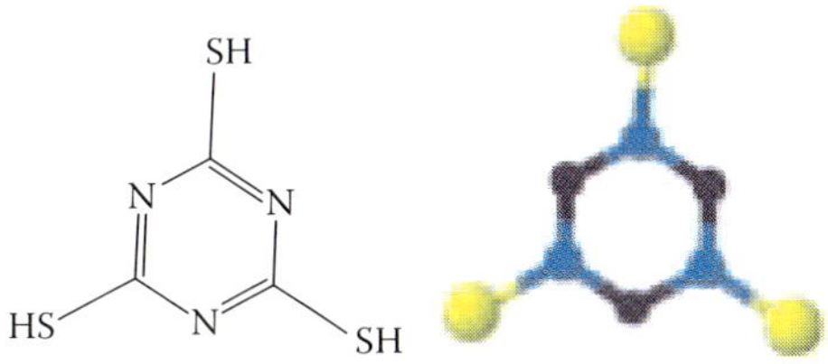

SH
N N
HS N SH

COLOR FIGURE 6.41

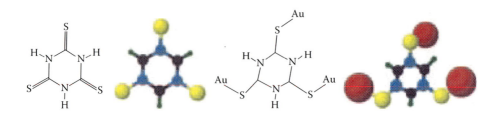

COLOR FIGURE 6.42

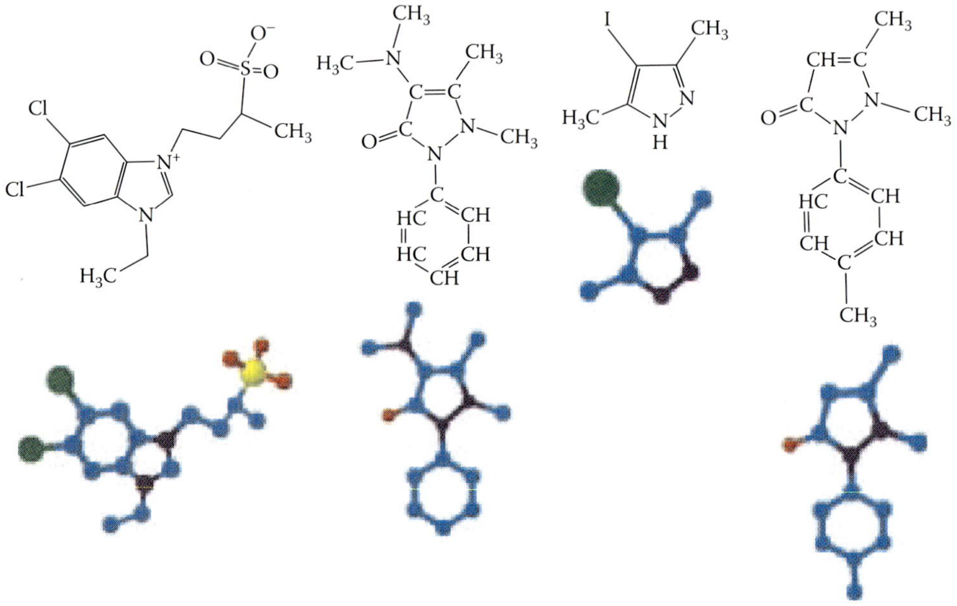

COLOR FIGURE 6.43

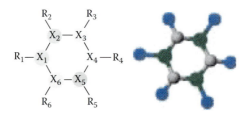

COLOR FIGURE 6.44

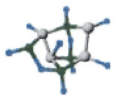

COLOR FIGURE 6.45

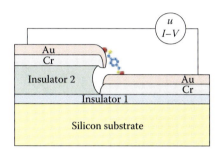

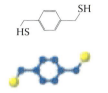

SH

HS

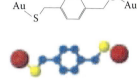

Au—S

S—Au

1, 4-phenylenedimethanethiol molecule with two thiol end groups

Functionalized 1, 4-phenylenedimethanethiol molecule with Au–S bonds

COLOR FIGURE 6.46

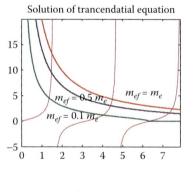

Solution of trancendatial equation

$m_{ef} = 0.5 \, m_e$

$m_{ef} = m_e$

$m_{ef} = 0.1 \, m_e$

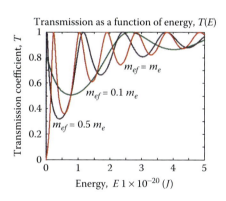

Transmission as a function of energy, $T(E)$

$m_{ef} = m_e$

$m_{ef} = 0.1 \, m_e$

$m_{ef} = 0.5 \, m_e$

Energy, $E \, 1 \times 10^{-20} \, (J)$

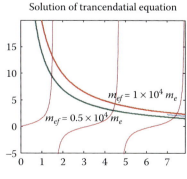

Solution of trancendatial equation

$m_{ef} = 1 \times 10^4 \, m_e$

$m_{ef} = 0.5 \times 10^4 \, m_e$

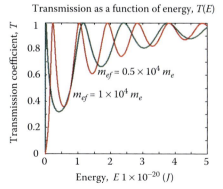

Transmission as a function of energy, $T(E)$

$m_{ef} = 0.5 \times 10^4 \, m_e$

$m_{ef} = 1 \times 10^4 \, m_e$

Energy, $E \, 1 \times 10^{-20} \, (J)$

COLOR FIGURE 6.51

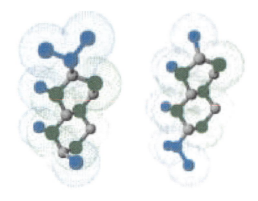

COLOR FIGURE 6.56

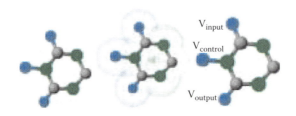

V_{input}

$V_{control}$

V_{output}

COLOR FIGURE 6.57

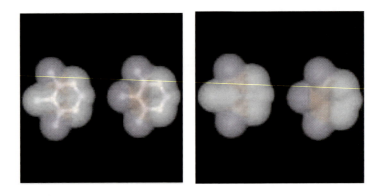

COLOR FIGURE 6.58

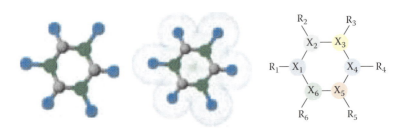

COLOR FIGURE 6.60

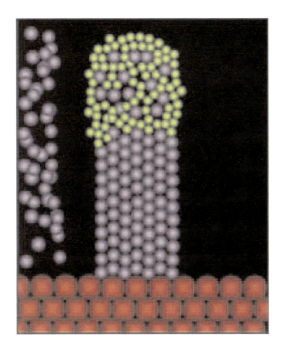

COLOR FIGURE 7.1

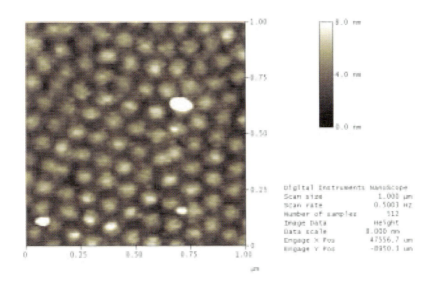

COLOR FIGURE 9.1

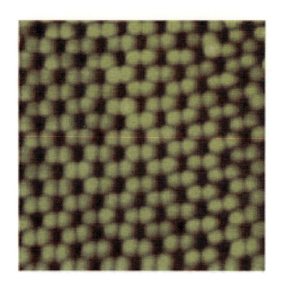

COLOR FIGURE 9.3

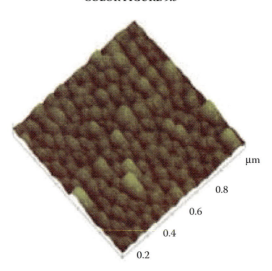

COLOR FIGURE 9.12

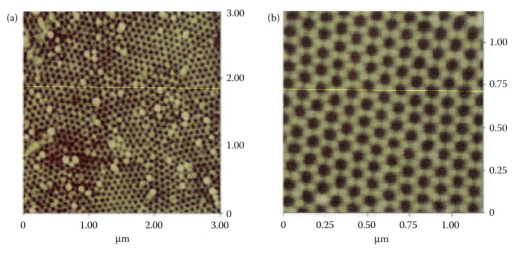

COLOR FIGURE 9.13

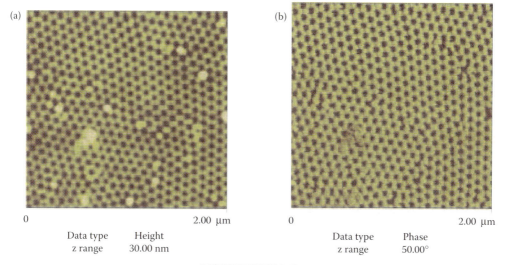

(a)

0   2.00 μm

Data type   Height
z range   30.00 nm

(b)

0   2.00 μm

Data type   Phase
z range   50.00°

COLOR FIGURE 9.18

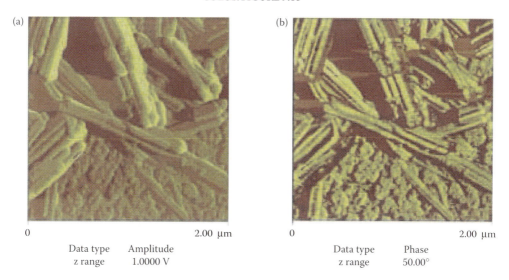

(a)

0   2.00 μm

Data type   Amplitude
z range   1.0000 V

(b)

0   2.00 μm

Data type   Phase
z range   50.00°

COLOR FIGURE 9.19

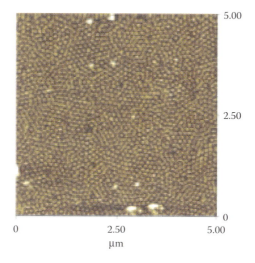

5.00

2.50

0

0   2.50   5.00
μm

COLOR FIGURE 9.30

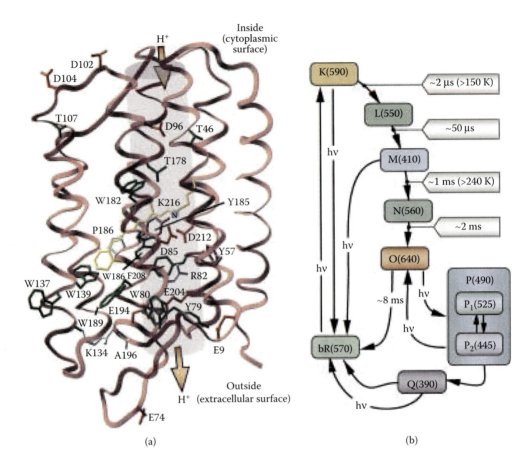

COLOR FIGURE 16.1

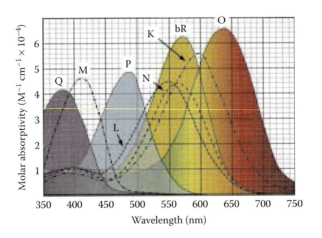

COLOR FIGURE 16.2

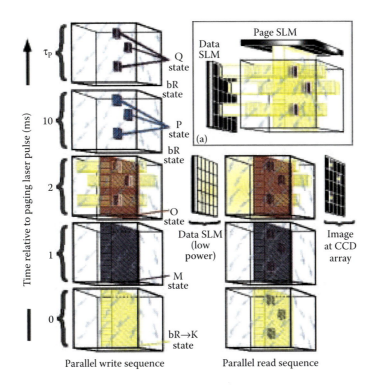

COLOR FIGURE 16.4

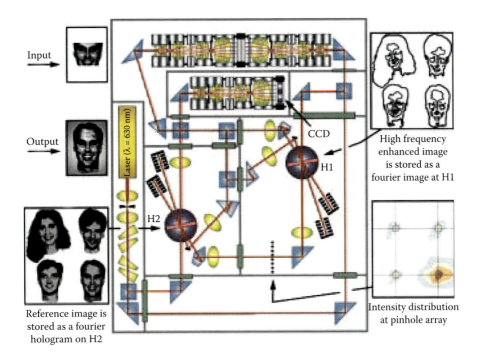

COLOR FIGURE 16.6

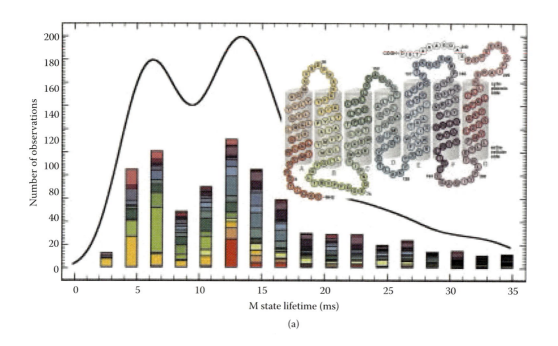

(a)

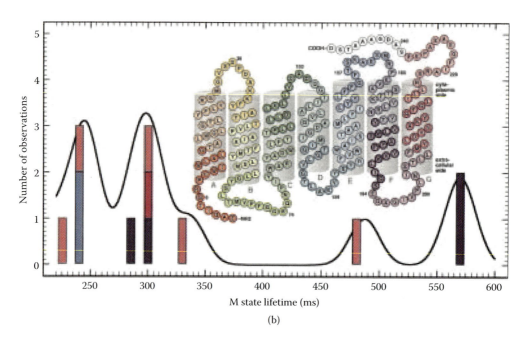

(b)

COLOR FIGURE 16.11

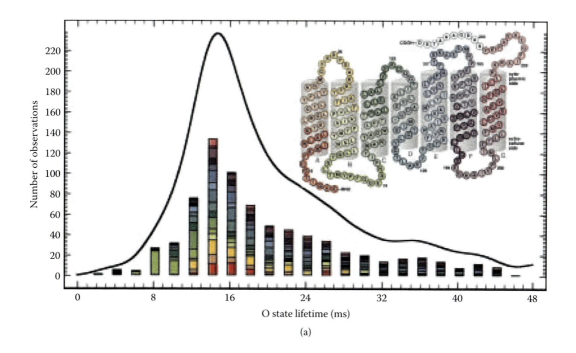

(a)

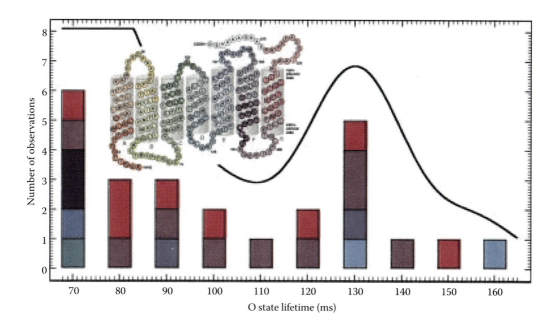

COLOR FIGURE 16.12

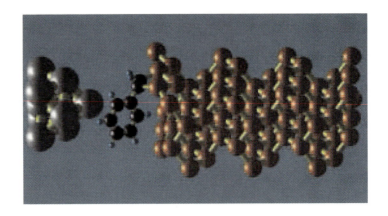

COLOR FIGURE 19.1

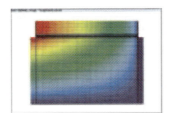 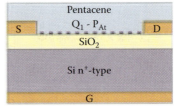

COLOR FIGURE 19.9

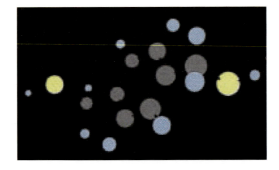

COLOR FIGURE 21.5

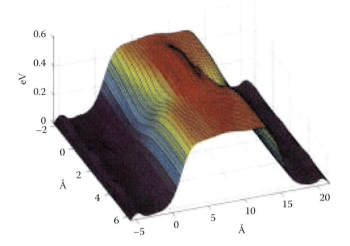

COLOR FIGURE 21.7

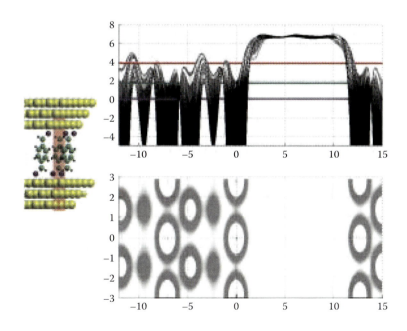

COLOR FIGURE 21.8

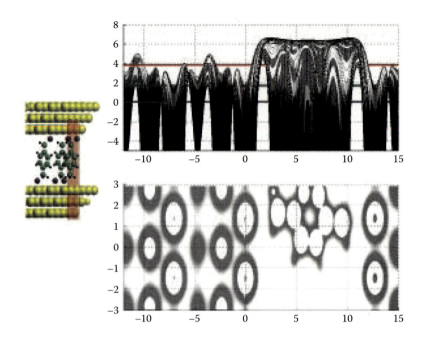

COLOR FIGURE 21.9

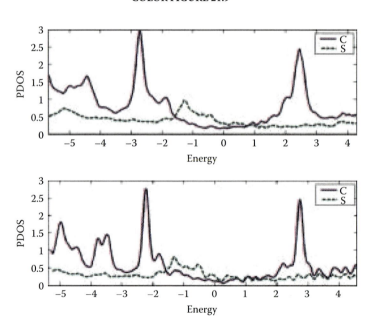

COLOR FIGURE 21.10

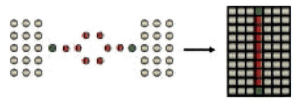

COLOR FIGURE 21.11

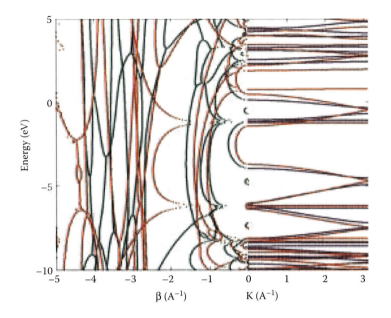

COLOR FIGURE 21.13

COLOR FIGURE 21.14

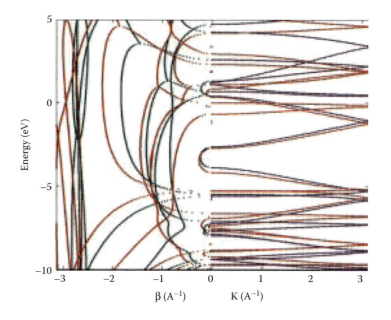

COLOR FIGURE 21.15

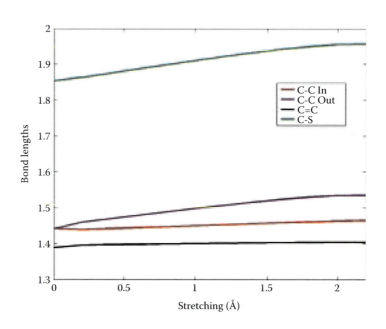

COLOR FIGURE 21.17

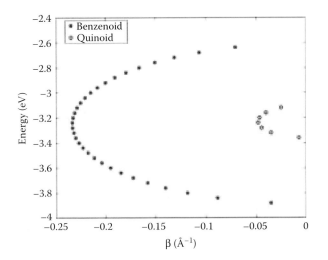

COLOR FIGURE 21.20

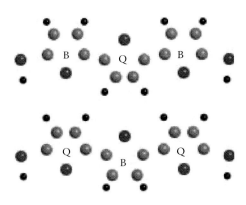

COLOR FIGURE 21.22

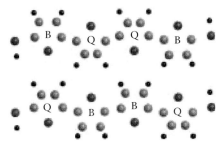

COLOR FIGURE 21.23

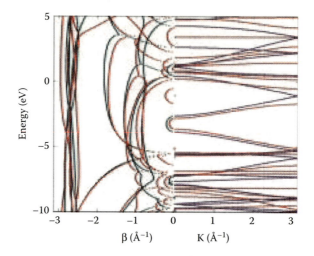

COLOR FIGURE 21.24

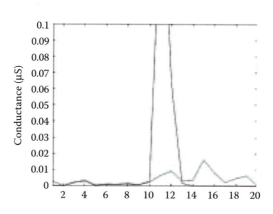

COLOR FIGURE 21.25

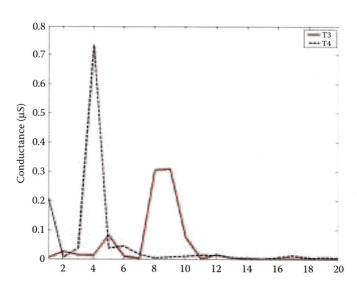

COLOR FIGURE 21.31

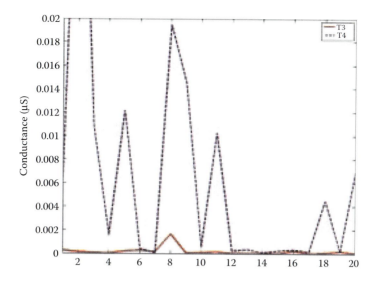

COLOR FIGURE 21.32

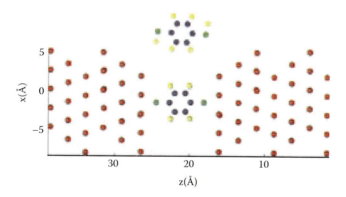

COLOR FIGURE 25.1

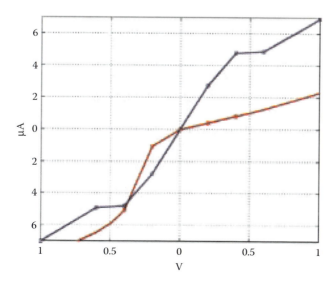

COLOR FIGURE 25.4

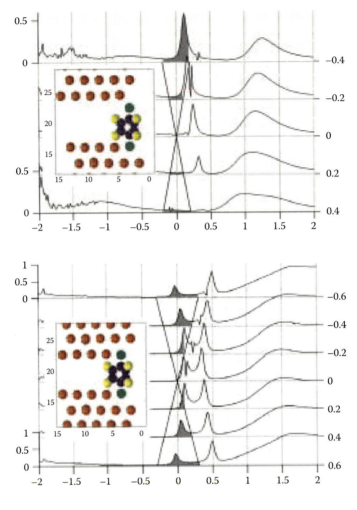

COLOR FIGURE 25.5

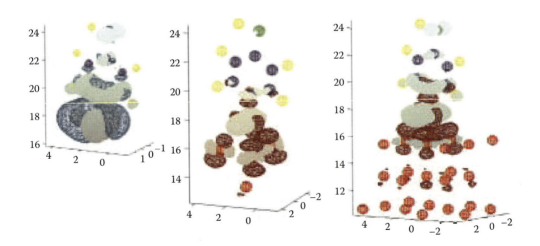

COLOR FIGURE 25.6

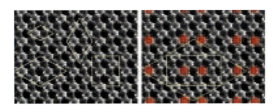

COLOR FIGURE 25.10

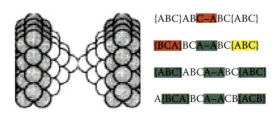

{ABC}ABC–ABC{ABC}

{BCA}BCA–ABC{ABC}

{ABC}ABCA–ABC{ABC}

A{BCA}BCA–ACB{ACB}

COLOR FIGURE 25.11

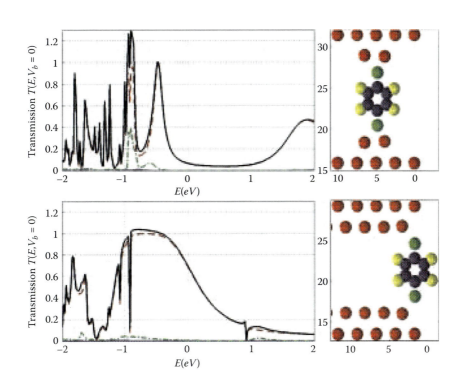

COLOR FIGURE 25.12

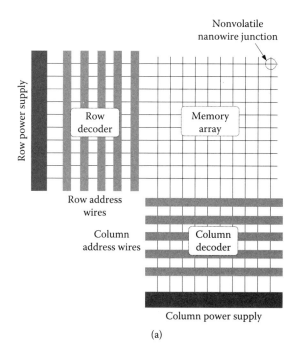

(a)

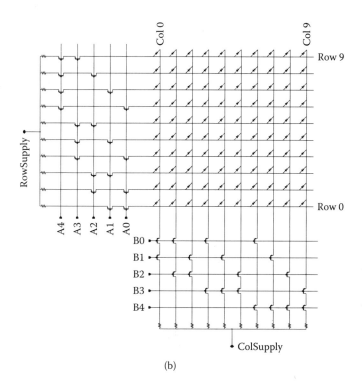

(b)

FIGURE 5.4 A sketch of (a) the structure and (b) a circuit schematic for a nanomemory design. This design consists of a crossbar nanowire memory array composed of nonvolatile nanowire diodes, plus two decoders composed of top-gated nanowire field-effect transistors.

corporation relies on a monolayer of 100 to 1000 organic molecules sandwiched between the inorganic contacts at each crosspoint [82]. The hybrid nanomemory proposed by Strukov and Likharev would employ single-electron latching switches [83]. In the Nantero nanomemory architecture, the carbon nanotubes that form part of the crossed-wire array also serve as the fundamental devices — individual memory bits are stored electromechanically by introducing reversible deflections or "kinks" at the desired crosspoints in the nanomemory.

A different set of molecular-scale devices is required to synthesize the decoder circuits that provide the interface to this storage array [84]. The Lieber–DeHon [85] and Heath [86] nanomemory architectures employ transistors that utilize semiconducting nanowires as their channels. Microwires, which gate these channels, are used to connect to the nanomemory from the microscale. Alternatively, Hewlett-Packard proposes a scheme in which the nanowire-microwire interface is generated stochastically by the random deposition of gold colloidal nanodots between the microwires and nanowires [87]. Through proper control of the deposition process, decoding of each of the individual nanowires in the nanomemory can be achieved with high probability. In contrast to the Lieber–DeHon and Hewlett-Packard approaches, the hybrid systems of Strukov and Likharev [80] and of Nantero Corporation [13,81] employ conventional CMOS in the decoder circuits.

For any of these nanomemory designs, it is costly, time-consuming, and difficult experimentally to determine whether such a nanomemory system will function correctly. In fact, for most of the architectures proposed for nanomemory systems, fabrication and physical testing have yet to be carried out. Thus, to shorten the design cycle and reduce costs, it is desirable to conduct full-system simulation of these nanomemory system designs before they are fabricated.

5.3.2   Simulation of an Example Nanomemory System

In this section, we describe simulations of a notional nanomemory system based upon the Lieber–DeHon architecture [79]. The fundamental devices [51,52,88] and small prototype circuits [23,85] of this nanomemory already have been demonstrated experimentally. Here, we utilize computer simulation to evaluate how extended system prototypes might perform if built using the same devices.

In the system simulation described here, the nanomemory storage array consists of nanowire diodes. Within the decoders, the microwire-nanowire crosspoints form field-effect transistors. These transistors permit the selection, or "addressing," of individual rows and columns in the memory array. The transistors are organized in a "2-hot" coding scheme [79]. The 2-hot scheme requires asserting a voltage on exactly two microwires in a decoder in order to select a unique nanomemory location, no matter the size of the storage array. This coding scheme differs from the binary schemes typically used in CMOS circuitry [53]. Binary coding would require asserting $\log_2 N$ microwires to select a unique wire from a set of N wires. The 2-hot addressing scheme is chosen for its additional defect tolerance. With 2-hot addressing, any failure of a single microwire impacts significantly fewer bits than a comparable failure in a binary scheme [79]. In addition, the 2-hot coding scheme requires the selected nanowire encounter exactly two transistors in series, regardless of the size of the array, whereas binary coding would meet $\log_2 N$ transistors in series. Reducing the number of transistors in series is beneficial because it ensures an ample amount of the supply voltage reaches the selected crossbar junction, rather than being dissipated by the decoders.

5.3.3   Nanomemory System Evaluation Metrics

First and foremost, we evaluate the ability of a nanomemory to read and write information accurately, with strong signals that are not easily lost in circuit noise or prone to other sources of error. To evaluate read operations, we focus on the output current differences (ΔI_{out}) between reading logic "1" and reading logic "0." This current difference is evaluated for the worst-case memory configuration (i.e., the worst-case pattern of "1"s and "0"s in the array) in order to ensure that a logic "1" can be distinguished from a logic "0" for each bit in every configuration of the entire nanomemory array. For write operations, we examine the voltage applied to the desired crossbar junction in order to verify that sufficient voltage is being applied. At the same time, we verify that no other junction receives enough voltage to alter its logical state.

Consideration also is given here to the nanomemory speed and power consumption. Output current switching times are analyzed, and methods are suggested to improve speed and reduce power.

5.3.4 Simulation Methodology and Device Modeling

The simulation of devices and complex circuit systems can be performed at a number of different levels of design abstraction [53]. The appropriate level of abstraction is determined by the goals of the simulation and by the ability of the simulation tools to handle complexity. Often, it is necessary to neglect levels involving fine details in order to capture the important overall behaviors of complex systems.

Three categories of electronic design abstraction exist: the device level, the circuit level, and the architectural level. The device level focuses on a single device (e.g., a diode or transistor) in great detail. Simulations at this level provide information about the operation and physics of individual devices, but generally do not consider the interactions among distinct devices in a circuit. In contrast, the architectural level considers very large systems, but typically does not include the physics or the behavior of individual devices. The circuit level bridges these two approaches and considers relatively large systems (on the order of tens of thousands of devices), while still retaining a connection to the underlying physical behavior. The simulations described here take place at this level.

Many concepts and techniques from conventional microelectronics are borrowed here for use in simulating nanoelectronic memories. For example, the commonly-utilized commercial Cadence Spectre VLSI CAD software tool [89] is our primary simulation program. One reason for applying such commercial off-the-shelf software tools from the microelectronics industry is the obvious timesaving and reliability associated with the use of readily available, well-tested software. This software also incorporates powerful features, such as modeling languages and graphics, developed specifically for the flexible modeling of extended circuitry. Finally, the use of conventional VLSI tools provides a seamless approach to the design and simulation of the nanomemory together with the peripheral microelectronic circuitry required for operation and communication with the outside world [90,91].

The work presented here also relies heavily on the conventional microelectronic concept of the device model, which captures the essential properties and response behavior of a circuit element. Models of experimentally observed behavior are required for all of the devices utilized in the nanomemory. In particular, the current–voltage transfer characteristics (I-V curves) are necessary for the simulation of steady-state behavior, and the capacitance–voltage transfer characteristics (C-V curves) are required for time-varying, or transient, simulation.

Typical models for microelectronic devices consist of compact equations based upon the well-understood, underlying physics of such devices. However, this physics-based approach is not workable, at present, for simulations involving molecular-scale devices, because the fundamental physics of most molecular-scale devices is not well understood. Thus, in this work, we utilize empirical models based on measured device characteristics.

Incorporating new models into conventional circuit simulators can be difficult. The addition of a new model often can require modifying proprietary source code. Open-source simulators do exist, such as SPICE3 [92], but adding new device models to these simulators is tedious [93]. Furthermore, these open-source simulators lack the robustness and simulation speed necessary to model large circuit systems and found in many commercial simulators.

Thus, to develop and simulate efficiently models for molecular-scale devices, we utilized the commercial Cadence Spectre simulator. This software permits the description of the empirical behavior of devices using the analog hardware description language (analog HDL) Verilog-A. This modeling approach is similar to one described elsewhere [94–96], except that the empirical equations derived in this work were tailored to the devices employed in the Lieber–DeHon nanomemory system. These empirical equations were incorporated into the Spectre circuit simulator, which supports co-simulation of both Verilog-A components and conventional SPICE-level devices.

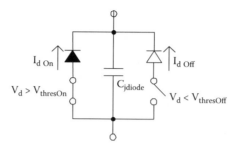

FIGURE 5.5   A sketch of a circuit schematic for the nonvolatile nanowire diode shown in Figure 5.2(a). The device model consists of two conventional diodes in parallel with a capacitor. The two individual diodes model the high current state (on-state) and low current state (off-state), respectively.

Simulations of the crossbar nanomemory system required three device models. The first two are models of nanowire devices: the nonvolatile nanowire (NVNW) diodes used in the storage array and the top-gated nanowire field-effect transistors (TGNW-FETs) used in the decoders. The third is a model of the nanowire interconnects of the nanomemory system.

5.3.4.1   Nonvolatile Nanowire (NVNW) Diode Model

Figure 5.5 shows a schematic diagram of a circuit that models the behavior of the NVNW diodes developed at Harvard University [78]. The model consists of two conventional, non-hysteretic diodes connected in parallel with a capacitor (C_{jdiode}). The model can be switched between a high current state (on-state) and a low current state (off-state) by switching which diode is connected to the circuit. This reproduces the hysteretic I-V behavior seen in the experimental device. The measured I-V characteristics of the actual NVNW diodes and the corresponding model I-V curve are shown in Figure 5.6. The measured I-V curves were fitted to empirical equations to produce the model. The apparatus used to collect the experimental data shown in Figure 5.6 was limited to measuring currents of up to 1000 nA, a limit that the device attains at a bias voltage of approximately 3 V. In the model, values for the current passing through the diode at bias voltages greater than 3 V were extrapolated from the available data.

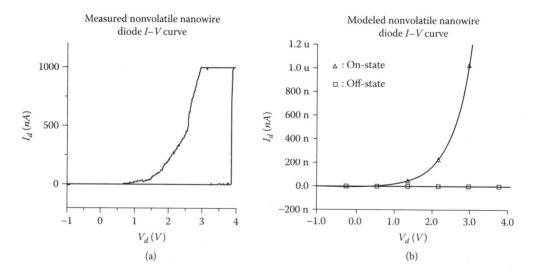

FIGURE 5.6   Hysteretic I-V curve for the nonvolatile nanowire diode. (a) The measured I-V curve for an experimentally fabricated nonvolatile nanowire diode. (b) The simulated I-V curve for the nonvolatile nanowire diode model.

The NVNW diode switches from the on-state to the off-state when a reverse bias voltage more negative than $V_{thres\,OFF}$ is applied across the device. In a similar fashion, a bias voltage greater (i.e., more positive) than $V_{thres\,ON}$ switches the device from the off-state to the on-state. Device threshold values for the experimental diodes are -2.75 V and 3.80 V for $V_{thres\,OFF}$ and $V_{thres\,ON}$, respectively. One issue for this simulation research is whether device characteristics, such as the threshold voltages, are optimal from the perspective of designing and building an extended memory circuit system, and whether this device behavior might be improved for that purpose. This question is addressed in Section 5.3.5.

Although this diode switch appears to exhibit relatively simple behavior, the hysteretic I-V curve creates a complicated device modeling task. A smooth transition between curves occurs when switching from the on-state to the off-state at $V_{thres\,OFF}$, but the device experiences an abrupt jump in current when switching from the off-state to the on-state at $V_{thres\,ON}$. This discontinuity in the current requires special provisions in the mathematical models used in the simulation. We avoid any possible difficulties at the discontinuity by simply recording when $V_{thres\,ON}$ has been surpassed, without actually changing the underlying state of the device. This is sufficient for the purposes of the work described here, because the memory array is simulated for only one configuration at a time. Thus, it is necessary only to determine which of its constituent diodes has crossed its switching threshold. Subsequent analysis of the nanomemory system with the diodes in the switched state is not required.

This technique is not suitable in all situations. Multiconfiguration simulations, such as those that calculate power consumption during write-read combinations, require simulation of diode transitions between the ON and OFF states. This cannot be modeled with the methodology described here. Nevertheless, the single-configuration simulations presented here are sufficient to determine whether the proposed memory system can be made to operate if constructed from presently available devices.

For time-varying simulation, information concerning the device capacitance is needed in addition to the I-V behavior. Ideally, we would obtain a transfer curve relating capacitance to voltage in a manner similar to that of obtaining the curve describing the I-V behavior. However, sufficiently detailed experimental data is not yet available to describe the change in capacitance versus voltage. Instead, we used a constant value of 1 aF for the NWNV diode junction capacitance (C_j) [78]. In the absence of detailed data, this first-order estimate must suffice for use in simulating overall memory performance. Nonetheless, the simulations developed in this work can incorporate more detailed capacitance characteristics as they are measured.

5.3.4.2 Top-Gated Nanowire Field Effect Transistor (TGNW-FET) Model

The decoders are composed of TGNW-FETs that are constructed by crossing a microscale wire over a nanowire covered with silicon dioxide. The silicon dioxide isolates the microwire from the nanowire and allows the device to behave like a field-effect transistor, with the microscale wire acting as the gate. Changing the voltage on the microwire gate controls the current flow through the nanowire channel. These field-effect devices are similar to the crossed nanowire FETs (cNWFETs) described by Huang *et al.* [7]. An illustration of a TGNW-FET and a circuit schematic of the device model are shown in Figure 5.7. The experimental I-V characteristics for p-type silicon nanowires coated with silicon dioxide and the

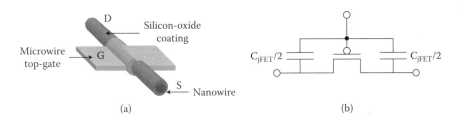

FIGURE 5.7 A sketch of (a) the structure and (b) a circuit schematic for a top-gated nanowire FET formed by depositing a microwire over a silicon-dioxide coated nanowire. The device model consists of a PFET transistor and two capacitors.

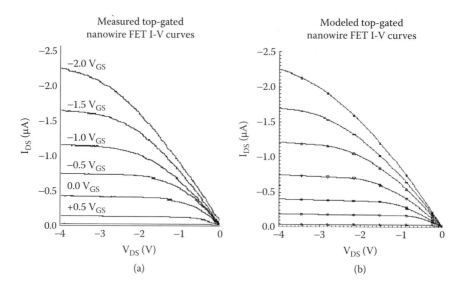

FIGURE 5.8 I-V curves for the top-gated nanowire FETs as a function of gate voltage. (a) Measured I-V curves for a p-silicon top-gated nanowire FET. (b) Simulated I-V curves for the top-gated nanowire transistor model.

corresponding TGNW-FET simulation model are shown in Figure 5.8. The device behaves as a p-channel MOSFET (PFET), where applying a positive voltage to the gate reduces the conductivity of the channel [53]. The I-V equations for the model are modified versions of first-order MOSFET I-V equations. The modifications to the MOSFET equations involve scaling the input voltages and adding an error correction term. These modifications are empirical in nature and remove any direct connection to the underlying physics. However, this is sufficient for the simulations presented here. It is not necessary to represent the underlying physics of the device, only to mimic its experimental behavior.

In addition, a capacitance between the nanowire and microwire is present in the model (C_{jFET}). We assume this capacitance is similar to that of the NVNW-diode junction (i.e., we set $C_{jFET} = C_{jdiode}$). This is a safe assumption, especially for large nanomemory arrays, because in these arrays C_{jFET} is dominated by C_{jdiode}.

5.3.4.3 Nanowire Interconnect Model

In conventional microelectronics, there is a clear-cut distinction between the devices and the wires that connect them. This distinction does not exist in the crossbar nanomemory considered here. Nanowires in the nanomemory form the devices and also connect these devices to one another. For simulation purposes, these two roles were divided artificially into separate models. The device behavior was captured in the models described in Sections 5.3.4.1 and 5.3.4.2. The interconnect behavior was captured in a third model.

Figure 5.9 shows an illustration and a circuit schematic of the interconnect model. This is a Π model [53] composed of a resistor and two capacitors. The figure details a unit crossbar (i.e., two crossing nanowires), each the length of the nanowire pitch. The resistance of the unit crossbar determines the values of the resistances (R_{NW}) in Figure 5.9, whereas the capacitors (C_{NWsub}) model the capacitances to the substrate below.

The interconnect model shown in Figure 5.9 optionally may incorporate a contact resistance R_c. This resistance models the contact between the microwire power supply lines and the nanowires. Its value is approximately 1 MΩ in present devices [52]. This value of R_c is dominant in comparison to the nanowire resistance R_{NW}. Thus, we assume that R_{NW} is negligible in our simulations. Although R_{NW} is not employed in the simulations presented in this paper, including it in the interconnect model provides the capability to account for the nanowire resistance when improvements in the fabrication techniques reduce R_c to a value where the two resistances are comparable.

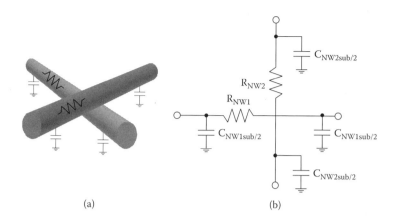

(a)   (b)

FIGURE 5.9   A sketch of (a) the structure and (b) a circuit schematic for the nanowire interconnect model, which consists of networks of resistors and capacitors.

The experimental value of the capacitance C_{NWsub} can be altered by changing the separation distance between the nanomemory array and the substrate or by changing the insulating dielectric between the array and the substrate. Thus, within simulations, C_{NWsub} is treated as a variable parameter. Its value has an important influence on system performance, as shown in Section 5.3.5.

Two additional parasitic influences were not included in the interconnect model, but may play a role in nanoscale systems. These are crosstalk capacitance between neighboring wires and parasitic inductances along the wires. These two effects may manifest themselves in systems with small wire pitches or in systems with long and narrow wires operating at high frequencies, respectively. However, these effects should not influence strongly the functionality of a low-speed, low-frequency prototype nanomemory, such as is considered here. That is, although these two parasitics may impact the speed and energy efficiency of the nanomemory, they will not affect whether or not the system can be made to operate.

5.3.5   Nanomemory Simulation and Analysis

The nanomemory is accessed by providing an address to the row and column decoders and then adjusting the supply voltages to force either a read or write operation. The decoders assert a row and a column by turning on the TGNW-FETs in the selected row and column, while turning off at least one TGNW-FET in each nonselected row and column. This procedure isolates a unique point or address in the nanomemory array.

When the TGNW-FETs are turned off, they create an open circuit and leave the voltage upon the non-selected rows and columns "floating," in the absence of a connection to a strong power supply. Allowing the rows and columns to float in this manner risks having nonselected diode junctions inadvertently reprogrammed if these diodes are subjected to voltages from elsewhere in the array that exceed programming thresholds. To help control the voltages across the nonselected rows and columns, a precharge signaling scheme is used. The precharge places a fixed charge on all of the nonselected diodes prior to evaluation. This limits the voltage difference across them.

Each operation is thereby divided into a precharge phase and an evaluation phase. Figure 5.10 shows the waveforms of the input and output signals of these two phases for a read operation on the 10×10 nanomemory shown in Figure 5.4. The simulation first reads diode (8,8) — that is, the diode in row 8 and column 8 — followed by a read of diode (9,9). It is of particular importance to be able to simulate the reading of diode (9,9) because it is the worst-case diode for both read and write operations — that is, it is the farthest from the power supplies. Simulation of the reading of diode (8,8) provides an example of the precharge scheme over successive memory accesses. In principle, any address location would do.

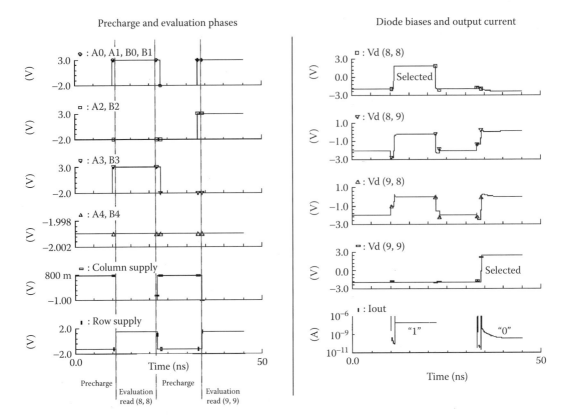

FIGURE 5.10 Input and output waveforms for the precharge and evaluation phases of two sequential read operations. The left half of the figure shows the input signals for a read of diode (8,8) followed by a read of diode (9,9). The voltage biases across the selected diodes, (8,8) and (9,9), their neighboring diodes, (8,9) and (9,8), and the memory's output current are shown in the right half of the figure.

The precharge phase asserts all address lines and places a voltage on all the rows and columns. Then, during the evaluation phase, only the selected row and column are asserted. The junction and parasitic capacitances on the nonselected lines hold the precharge voltage while they are isolated from the rest of the circuit. During the evaluation phase, at least one TGNW-FET in each nonselected row and column is turned off, leaving the only path between the row supply and column supply through the selected diode, enabling the reading or writing of a single bit.

When reading a bit from memory, voltages are placed on the row and column supplies such that the selected diode is forward biased, allowing the output current of the nanomemory to reflect the resistance of the selected diode. It is particularly important to choose operating voltages that forward bias only the selected diode. Forward biasing nonselected diodes will cause them to contribute, inadvertently, to the overall output current. In the worst-case memory configuration for reading a logic "0" bit (i.e., when the selected diode is in the off-state and the rest of the diodes are in the on-state), even a slight forward biasing of the nonselected junctions may make the state of the selected diode unreadable. This problem increases with the size of the array since there are more nonselected diodes that can contribute to the overall current.

To avoid this interference from nonselected diodes, we choose precharge and evaluation voltages for reading the memory that force nonselected diodes into a reverse bias or near zero bias. This strategy prevents nonselected diodes from contributing to the output current. The right half of Figure 5.10 shows simulation results for the strategy described here. The memory configuration is set to the worst case for reading logic "0". The worst-case diode, that is, diode (9,9), is set to logic "0" and the rest of the diodes

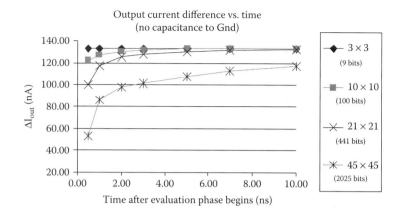

FIGURE 5.11 Plot of ΔI_{out} versus time. T = 0 corresponds to the beginning of the evaluation phase. The time it takes for ΔI_{out} to reach its maximum value has implications on the speed of the memory. The simulations here have zero capacitance between the nanowires and the substrate.

are set to logic "1". The top four waveforms are the voltage biases across the diodes being read and two neighboring diodes. The simulation results show that the diodes in nonselected rows and columns are either reverse biased or have a very small forward bias during the evaluation phase.

Although placing nonselected diodes under a reverse bias is effective for reducing unwanted current contributions to the output current, this scheme does run the risk of inadvertently programming on-state devices to off-state devices if the reverse bias exceeds $V_{thres\,OFF}$. Therefore, it is necessary to use supply voltages that are small enough to ensure $V_{thres\,OFF}$ is not surpassed. This, in turn, limits the bias that can be placed across the selected diode.

Nevertheless, in the simulation it is possible to achieve excellent ON/OFF current differences for a variety of different memory arrays, as is shown in Figure 5.11. Similarly, Table 5.1 provides details of the output currents I_{out} for worst-case read operations for both logic "1" and logic "0", as well as the current difference ΔI_{out} and "1"/"0" current ratio between them. These differences are sufficient to read each memory successfully. Furthermore, the high current ratios suggest that read operations can be performed successfully in memory arrays that have been scaled up to include even more rows and columns.

Data is written to the nanomemory by subjecting the selected diode to a bias exceeding the switching threshold. As discussed in Section 5.3.4.1, a diode in the on-state is switched to the off-state at $V_{thres\,OFF} \approx -2.75$ V and a diode in the off-state is switched to the on-state at $V_{thres\,ON} \approx 3.8$ V. As with the read operations, care must be taken to avoid inadvertently programming nonselected diodes. However,

TABLE 5.1 Simulation Results for a Read Operation Performed on the Nanomemory Shown in Figure 5.4

Nanomemory Array Size	I_{out} (nA) logic "1"	logic "0"	ΔI_{out} (nA)	"1"/"0" Current Ratio
3 × 3	134	0.8	133	168
10 × 10	134	0.9	133	149
15 × 15	134	1.1	133	122
21 × 21	134	1.6	132	84
45 × 45	134	16.7	117	8

Note: The simulations are performed with zero capacitance to ground and the reported values occur 10 nsec after the evaluation phase begins (see text for details).

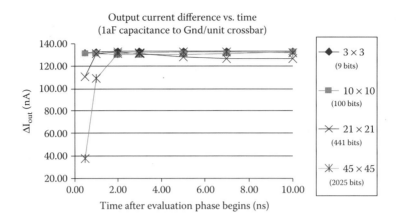

FIGURE 5.12   Same as Figure 5.11 except that the simulations have 1 aF of capacitance between the nanowires and the substrate. These simulations show that a small amount of capacitance produces shorter ΔI_{out} settling times.

simulations performed in this work suggest it is feasible to write either logic value to the memory. It was always possible to identify operating conditions that programmed the selected diode without subjecting nonselected diodes to voltages that exceeded thresholds.

The simulations shown in Figure 5.11 and Table 5.1 assume no capacitance between the nanowires and the substrate, that is, $C_{NWsub1} = C_{NWsub2} = 0$. This is a reasonable approximation that can be realized experimentally by raising the crossbar nanomemory sufficiently high above the substrate or by using a low-k dielectric between the nanomemory and substrate. Likewise, it should be possible to add a controlled amount of capacitance to the nanowires by reducing the height above the substrate or by employing an alternative dielectric. Recent experiments have shown that the capacitances between the memory cell of interest and the substrate may be estimated to be approximately 1aF. Thus, the simulations previously described above were repeated with this small capacitance to ground added to each unit crossbar in the nanowire interconnect model — that is, $C_{NWsub1} = C_{NWsub2} = 1$ aF. As shown in Figure 5.12, adding capacitance to ground reduces the ΔI_{out} settling times, particularly for the larger arrays. This reduction in settling times occurs because the capacitance to ground provides a better environment for holding the precharge. Without capacitance to ground, the junction capacitance dominates and capacitive coupling to crossing wires can reduce the effectiveness of the precharge.

The simulations developed in this work also can evaluate the effects of varying design parameters on specific aspects of nanomemory performance or evaluate the trade-offs between traditionally disparate design goals, such as high speed versus low power. For example, the output current difference ΔI_{out} can be improved either by shifting $V_{thresOFF}$ to a lower voltage (more negative voltage) or by increasing $V_{thresON}$. Increasing ΔI_{out} should lead to increased speed and array size. However, altering the programming threshold in this manner requires more energy during write operations. This, of course, increases power consumption. Simulation is an effective way to examine these trade-offs in a quantitative manner. It can be used to identify optimal operating parameters for specific design goals. For all of the simulations performed to date, the voltage swing for the input signals is relatively large, requiring the address lines to vary by 5 V, while the row supply and column supply vary by 2.75 V and 1.75 V, respectively. These large voltage swings most likely will consume significant dynamic power and require level shifting circuits to interface with conventional electronics. Thus, reducing the signal swing should be an experimental goal. This will reduce power consumption and ease integration with conventional circuits. However, achieving this goal may require smaller diode thresholds. This may reduce the memory speed and could affect functionality. Additional simulations that explicitly incorporate external CMOS circuits are required to explore this issue more fully.

Nevertheless, the simulation results, thus far, suggest that a 45 × 45 nanomemory would function correctly if built using the Lieber–DeHon architecture and devices. The general trends of these results suggest that larger memories will be functional as well. Furthermore, as shown in Section 5.3.6, the ability to assemble 45 × 45 nanomemories could be of considerable utility, because their use in a banked topology provides a route to realizing even larger nanomemories.

Although the simulations to date have suggested that the memory is scalable and will function under present device and design parameters, other factors should be considered in future simulations. For example, as the size of the nanomemory array grows, so does the capacitance and resistance of the rows and columns, which can hamper memory performance. Figure 5.11 shows the time dependence of ΔI_{out} for simulations of four different memory sizes. The figure shows that increasing the size of the memory also increases the time needed for ΔI_{out} to reach its maximum value. This settling time may reduce the speed of the memory. However, detailed information and models for the connection of the nanomemory to conventional microscale CMOS circuitry (in this case, signal amplifiers) are necessary for any realistic estimation of the memory speed.

5.3.6 Banking Topologies and Area Estimates

Increasing the size of a single nanomemory array may not be the most effective approach for producing memories with very high bit counts. As the size of a memory array increases, so do the resistances and capacitances associated with the array, which increase delay and power consumption. Ultimately, this may threaten functionality.

Further, large memory arrays are more susceptible to fabrication defects, since a single defect in a wire can render all the memory cells along it unusable. Reducing the vulnerability of nanomemories to defects is important. This is because, based on statistical and thermodynamic arguments, it is anticipated that the hierarchical self-assembly strategies being pursued for molecular-scale electronic circuits may produce a significant fraction of defective devices, or devices that are imprecisely positioned [97].

To increase defect and fault tolerance, instead of using a single large array to achieve a high bit count, banks of smaller memories might be employed. Figure 5.13 illustrates the notion of banking by showing how a one-kilobit memory array can be represented as a single 32 × 32 array or four 16 × 16 arrays. This strategy allows for the same level of defect tolerance with less redundancy, since any single defect impacts a smaller number of individual memory bits. Generally speaking, as the degree of banking increases, the amount of required redundancy should decrease, since smaller arrays pay a lower price per defect.

Adopting a banking strategy also increases the overall data throughput for the memory. First, the lower resistances and capacitances of the shorter nanowires in the smaller arrays allow faster access times. Second, banked arrays can be accessed in parallel (i.e., a bit can be accessed from each bank simultaneously)

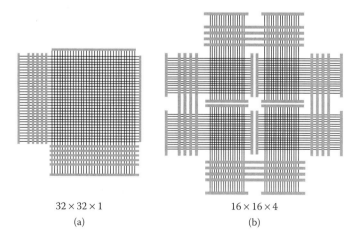

$$32 \times 32 \times 1$$

(a)

$$16 \times 16 \times 4$$

(b)

FIGURE 5.13 Illustration of two different topologies for realizing a 1-kilobit memory. (a) A single array. (b) A bank of four arrays with an equivalent number of bits.

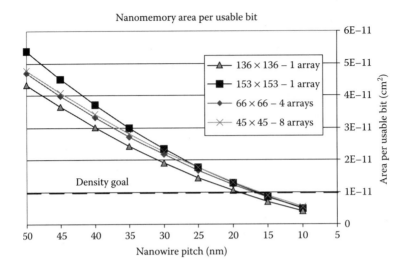

FIGURE 5.14 Plot of estimated area per usable bit versus the nanowire pitch for four memory arrangements. The calculations assume that 16 kilobits of data can be accessed and the remaining memory locations are reserved for redundancy. The microwire pitch is set at 100 nm for all four arrays.

significantly increasing memory performance. Although banked architectures can create more complex fabrication patterns, the regularity of the banks would seem to provide a feasible route to nanomemory assembly. For one example, Harvard University already has made significant progress in the parallel fabrication of multiple arrays in a tiled pattern [17].

The one significant trade-off generally associated with employing a banking strategy is an increase in area per usable bit. This occurs because each additional bank requires additional wires for encoding and decoding the memory array. Although some of these wires can be shared among the banks (see Figure 5.13), banking always results in an increase in the number of address wires. Thus, an optimal banking strategy will employ moderately sized arrays that not only take advantage of the coding scheme to increase density, but also achieve the requisite degree of defect tolerance, parallel access, and other design goals.

Despite the various banking topologies possible for producing a given extended nanomemory system, first-order area calculations suggest that the target nanowire pitch should be similar for a variety of topologies. Figure 5.14 shows the estimated bit density for three different banking strategies as a function of nanowire pitch. We also consider two different amounts of redundancy for a single array implementation. To compare these various strategies, these area estimations are premised on a goal of providing 16 kilobits of accessible memory, where any additional memory locations are assumed to be used only as replacements for faulty bits. In other words, the area per usable bit is calculated by dividing the total area for each topology by 16,000, regardless of the actual number of bits. The microwire pitch was set to 100 nm for all of the area calculations. Details of these four memory arrangements are given in Table 5.2.

TABLE 5.2 Estimated Area for Four Different Memory Arrangements Targeting a 16-Kilobit Nanomemory

Memory Arrangement	Total Locations	Percent Redundancy	Total Area (sq. μm)		
			20 nm pitch	15 nm pitch	10 nm pitch
136×136–1 array	18,496	15.6%	16.6	11.1	6.5
153×153–1 array	23,409	46.3%	20.4	13.5	7.8
66×66–4 arrays	17,424	8.9%	19.6	13.4	8.1
45×45–8 arrays	16,200	1.3%	20.9	14.4	8.8

The four different memory arrangements described in Figure 5.14 and in Table 5.2 all reach a density of 10^{11} bits/cm^2 when the nanowire pitch is approximately 15 nm. However, a more appropriate measure of the nanotechnology employed in the fabrication of the nanomemory might ignore the area occupied by the microwires and just consider the area occupied by the nanowires. In that case, a nanowire pitch of approximately 30 nm would suffice to achieve this density. Clearly, a variety of topological strategies will be viable to fabricate functional, extended nanomemory systems.

5.3.7 Summary of Nanomemory System Simulation

Simulations performed on a crossbar nanomemory system based upon the work of Lieber and DeHon [78,79] suggest that if such a system were built, it would operate. The simulation results suggest that a 45×45 nanomemory array would function properly if constructed from presently fabricated experimental devices. Furthermore, such arrays could be banked to build more extended nanomemory systems, such as a 16-kilobit molecular-scale electronic nanomemory with a bit density of 10^{11} bits/cm^2.

The favorable results from these simulations are encouraging for ongoing and future experiments in the fabrication and prototyping of post-CMOS, crossed-nanowire nanomemory systems. In addition, these results suggest that more complex, extended nanoprocessing systems also could be made to operate using crossed-nanowire architectures that build upon those described here for nanomemories. Thus, the next section of this paper addresses the additional challenges that must be faced in the design of nanoprocessor systems.

5.4 Beyond Nanomemories: Design of Nanoprocessors Integrated on the Molecular Scale[4]

5.4.1 Challenges for Developing Nanoprocessors

Many challenges must be faced at all levels of design and fabrication in order to utilize recent advances in molecular-scale devices and circuits to build extended nanoprocessor systems. Foremost, the structure and ultra-high density of novel molecular-scale devices make these devices difficult to employ in conventional microprocessor architectures. This motivates fundamental departures in the design of system architectures, which in turn necessitates the development of new circuits, interconnection strategies, and fabrication methods. The following sections discuss some of the challenges posed by the use of conventional electronic processor architectures, as well as the new difficulties that arise in using novel architectures.

5.4.1.1 Challenges Posed by the Use of Conventional Microprocessor Architectures

The principal challenge of using conventional architectures [98] for the development of nanoprocessor systems is that such architectures have too much heterogeneity and complexity for existing nanofabrication methods. Conventional processor architectures are heterogeneous at every level of the design hierarchy. At the top level, a modern microprocessor consists of logic, cache memory, and an input/output interface. In conventional microscale integration, these three architectural components may be designed using different circuit styles or even different fabrication methods. The logic component itself consists of arithmetic and control subcomponents, both of which require circuits that may be either combinational (e.g., AND, OR, XOR gates) or sequential (i.e., clocked elements such as registers) [98]. Further still, the synthesis of the aforementioned combinational logic gates requires multiple kinds of devices for optimal performance [53]. This differentiation into a wide variety of devices, circuits, and subsystems is an advantageous structural feature provided by the sophistication of modern microfabrication techniques. Providing such

---

[4]Some of the material in this section has appeared previously in Das *et al.*, "Architectures and simulations for nanoprocessor systems integrated on the molecular scale," *Lect. Notes Phys.*, vol. 680, pp. 479–513, 2005.

differentiation is beyond the reach of present nanofabrication techniques. As a result, nanoelectronics research has targeted the development of architectures for nanoprocessors that provide comparable function while avoiding as much as possible the introduction of structural heterogeneity at the hardware level.

5.4.1.2 Challenges in the Development of Novel Nanoprocessing Architectures

Most of the nanoprocessor architectures presently proposed [19,35–44,46,83,97,99–101] are essentially homogeneous at the hardware level and introduce diversification at the programming stage. In this way, they are able to do without the complexity of fabrication characteristic of conventional microprocessors.

Many of these nanoprocessor architectures inherit their design characteristics from microscale programmable logic [102], especially field-programmable gate arrays (FPGAs) [103] and programmable logic arrays (PLAs) [104]. As described in detail below in Section 5.4.3, FPGAs and PLAs are regular arrays of logic gates whose inter-gate wiring can be reconfigured. Software is used to configure FPGAs and PLAs to compute particular logic functions. In contrast, the logic functions in conventional microprocessors are hard-wired during construction. Thus, in FPGAs and PLAs, the use of software to "complete" the hardware construction allows the hardware design to be simplified to a homogeneous form.

Although these physically homogeneous architectures simplify fabrication, they do introduce a new set of challenges. For nanoprocessing, these challenges may be illustrated by considering the example of a nanoscale crossbar switch array. As discussed in Section 5.3, this is a homogeneous approach that combines a high degree of scalability with some of the smallest circuit structures demonstrated to date [8,10]. A number of architectural proposals for nanoprocessors have been put forth that involve the tiling of crossbar subarrays to form programmable fabrics, including the design shown in Figure 5.15 [18,19,43,79].

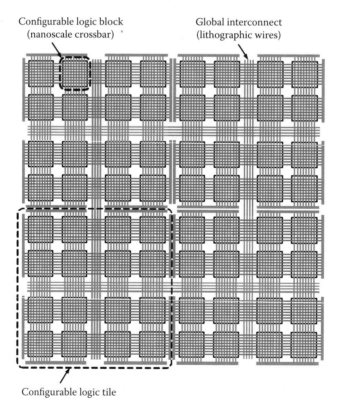

FIGURE 5.15 A programmable fabric incorporates molecular-scale devices into the crossbar structures shown in Figure 5.1. The fabric builds from them an extended structure of molecules or molecular devices, crossed nanowires, and microwires, such as is shown here. This can provide a platform for realizing a nanoprocessor [79].

Among the reasons why these regular crossbar structures are attractive is because it is possible to assemble them using presently available nanofabrication techniques. However, the structural regularity can increase the complexity of realizing logic at nearly every other level of the design hierarchy. One pays a penalty in the use of area and time in order to program topologically-irregular logic circuits into a physically homogeneous crossbar architecture. For example, programmable microscale circuits such as FPGAs incur approximately a 20 to 50-fold area penalty [105] and a 15-fold delay penalty [106] when compared to heterogeneous, custom-designed solutions. Thus, one significant challenge for nanoprocessing lies in developing programming algorithms that can produce area- and time-efficient realizations of heterogeneous logic using relatively homogeneous regular structures.

Furthermore, microscale PLAs and FPGAs are "mostly" regular, but some irregularity often is introduced at the lowest levels of the hardware hierarchy in order to promote more efficient utilization of physical resources [103]. Likewise, the ability to provide even a limited amount of irregularity with future nanofabrication methods might have a large, beneficial impact on the overall density and performance of a nanoprocessor.

In addition to the challenges enumerated earlier, the task of designing and developing novel nanoprocessor architectures must confront further difficulties in the circuit and device domains. Some of these challenges also are faced in the development of nanomemories, but for nanoprocessing, such issues are compounded. For example, in nanomemories, the use of two-terminal devices without gain imposes system-level constraints due to requirements for signal restoration. In nanoprocessors, requirements for signal restoration are more stringent, because the signals may need to traverse larger portions of nanoscale circuitry without the aid of the microscale amplifier circuits proposed for use with nanomemories [107]. Also, wires and the signals they carry must fan out in order to construct the complex logic required for processing, such as arithmetic functions. Still further, there are issues of signal integrity due to the signal coupling that arises when devices and interconnects are as densely packed as is proposed for nanoprocessors. The high density of devices also makes it difficult to maintain low enough power density that system temperature can be controlled [108].

A challenge for nanoprocessing that does not arise in nanomemories is that sequential (clocked) elements will be required. Such elements can be inefficient to realize using the combinational logic that is most readily available using crossbars that incorporate molecular-scale resistors and rectifiers. Specialized nanocircuits have been proposed to serve as sequential elements [44,109–112]. These circuits operate using Goto pairs [113] in implementations that were used previously in solid-state nanoelectronic circuit designs [114,115]. In crossbars, these circuits may be built by incorporating NDR molecules [6].

One virtue of using Goto-pair-based circuits for nanoelectronic systems is that they can provide restoration using only two-terminal devices. In effect, these circuits can provide some of the gain required to restore logic signals, thus reducing gain requirements for other circuits in the system. Such circuits might be able to limit, and possibly even eliminate, the need for nanotransistors. However, a potential drawback is that, unlike transistor-based circuits, Goto-pair circuits may require additional components in order to provide electrical isolation between logic stages. Such isolation might be provided by distinct nanodevices such as rectifiers. However, with or without such additional devices for isolation, localized insertion and placement of Goto-pair-based clocked elements into a crossbar array probably would require introducing a degree of heterogeneity into an otherwise regular nanofabric.

The need for heterogeneity might be reduced by using the crossbar latch designed by the Hewlett-Packard Corporation [11,116]. This latch has been demonstrated to produce signal restoration and inversion using only molecular two-terminal devices. It is a clocked element designed to be fabricated using junction molecular devices within the same homogeneous crossed-nanowire molecular-scale circuit systems (see Figure 5.15) that have been used to fabricate nanomemories [10,16,69,117]. Such latches could be introduced into nanoprocessor systems based on crossbars, without requiring a heterogeneous set of devices. Furthermore, as with the Goto-pair circuits, the use of these crossbar latches in a nanoelectronic system might reduce gain requirements for other circuits in the system, even to the point where nanotransistors may not be required. Nanoprocessor system architectures based on these latches are under development [118,119].

For all approaches to nanoprocessor system design based upon molecular switches, it is well understood that many device-level challenges also must be addressed [15,97]. Impedance matching between bulk solid contacts and molecular-scale devices, precise characterization of device behaviors, variability, and the yield of devices are among the chief examples. These challenges will be discussed further in connection with the nanoprocessor simulations described in Section 5.4.4. Such challenges must be managed either by improving fabrication capabilities or by introducing defect and variation tolerance into system architectures.

5.4.2 A Brief Survey of Nanoprocessor System Architectures

Section 5.4.1 discussed some of the challenges facing the design and fabrication of future nanoprocessors based on novel nanodevices and new nanofabrication techniques. In this section, we survey the major architectural approaches that have been proposed to address these challenges. Some of these approaches rely on new architectural paradigms that are very different from those applied in conventional microprocessors. Others borrow heavily from these microprocessor architectures. However, all of these nanoscale approaches attempt to harness molecules or molecular-scale structures to build up electronic circuits and systems. These approaches and the nanoelectronic systems that will be developed in accordance with them have the potential to utilize effectively the much higher device densities possible at the nanoscale. Further, because they take advantage of potentially inexpensive, novel nanofabrication techniques, it may be possible to address the issue of exponentially rising costs that presently plagues the microelectronics industry [120,121].

Substantial progress also continues to be made in the scaling of CMOS-based conventional microprocessors. Thus, some nanocomputer architects propose to leverage the substantial knowledge and infrastructure available in CMOS technology. Rather than devise new or modified architectures to accommodate the properties of novel nanodevices, these architects attempt to use them to augment the CMOS devices employed in conventional microprocessors. For the most part, such efforts retain conventional microprocessor architectural designs.

In the following sections, both the scaling of conventional architectures and the development of novel approaches are discussed. First, in Section 5.4.2.1, the aggressive miniaturization of conventional architectures to the molecular scale is described. Second, in Section 5.4.2.2, alternatives to conventional architectures are detailed for cases in which recent nanodevice and nanofabrication developments have made such architectures especially relevant.

5.4.2.1 Migration of Conventional Processor Architectures to the Molecular Scale

Virtually all conventional microprocessor architectures use CMOS to implement a basic architectural design originally due to von Neumann, Mauchly, and Eckert [122–124]. First described in the 1940s, this architecture divides a computer into four main "organs": arithmetic, control, memory, and input/output. Present examples of such CMOS-based processors include the well-known Intel Pentium® 4 and the AMD Opteron™ chips. As Figure 5.16 shows for the AMD Opteron, the organ structure still is evident.

Because of its long-term investment, industry places a high premium on maintaining these architectures as it seeks to achieve ultra-dense integration on the nanometer scale. The primary industry approach today to building nanoprocessors is the aggressive scaling of CMOS technology to nanometer dimensions.[5] However, for a number of years, industry investigators and others have examined the likely limits of CMOS technology [126–128,130,131] and the possibility that it might not be cost-effective to use it to build commercial systems with devices scaled down to a few tens of nanometers. This is one of the reasons new architectural ideas inspired by nanotechnology and molecular-scale electronics are so compelling.

---

[5]This topic has been reviewed and discussed extensively elsewhere [27,126–129]. We include a brief discussion of it here both for completeness and to provide a reference point for the other, more novel approaches we discuss.

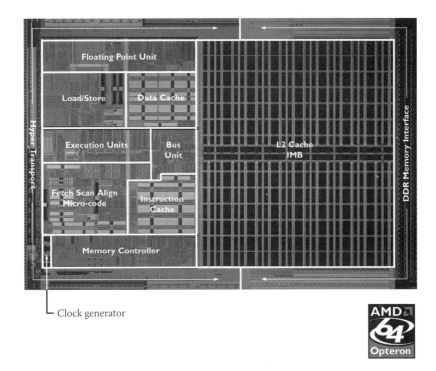

FIGURE 5.16 AMD Opteron™ die photo with annotated block structure [125].

An alternative to the straightforward, two-dimensional, aggressive scaling of CMOS is to expand silicon technology into a third dimension [130]. Three-dimensional integration, or 3D CMOS [132,133], refers to any of several methods that take conventional, "flat" CMOS wafers and stack them together with an inter-wafer interconnect [134–139]. For microprocessors, it has been shown that 3D integration allows for a substantial improvement in performance, and, furthermore, that this improvement increases as device and interconnect dimensions decrease [140]. Therefore, 3D architectures may have particular utility in combination with novel molecular-scale devices, such as might be implemented using a 3D crossbar array.

So-called "hybrid" approaches that incorporate novel nanostructures into CMOS devices constitute a third avenue by which conventional processor architectures may be migrated toward the molecular scale. Major industrial research laboratories have begun to explore how nanowires and CNTs might be employed to enhance CMOS and CMOS-like structures. For example, some of the Intel Corporation's designs for future transistors call for the incorporation of nanowire-like silicon channels to increase current density and control short-channel effects [141]. Similarly, work at IBM has examined the increased current that results from the use of CNTs in field-effect transistor channels [73,142].

Another hybrid approach involves the use of self-assembled monolayers (SAMs) of redox-active molecules to enhance the function of traditional silicon devices. Thresholds and conductances of the underlying silicon substrate can be altered by the incorporation of these monolayers. In addition, new and novel devices might be enabled. For example, the redox states of the molecules in the SAMs may be used to form multilevel bits (i.e., n-ary digits) [33,34]. Such so-called molecular FETs, or MoleFETs, which employ NDR molecules or charge-storage porphyrin molecules on silicon, might be used to implement multilevel memories or logic. It appears that molecules and molecular layers can be inserted into CMOS production processes for this purpose. For example, the porphyrin molecules proposed for some of these hybrid devices have been shown to be able to survive the $400°C$ processing temperature used for conventional CMOS components [143]. Also, as mentioned in Section 5.3, Nantero Corporation is succeeding in introducing novel carbon nanotube–based devices and circuits into a CMOS production line [13,81,144].

Hybridization also may be employed at the architectural level. An example of such a hybrid design is the CMOL architecture [83,145]. CMOL circuits combine CMOS with crossed nanowires and molecular devices. Specifically, CMOL circuits are to be fabricated in two layers, with one layer consisting of CMOS blocks, or "cells," and the other layer containing an array of crossed nanowires employed as interconnects between the CMOS cells. As with many other crossbar architectures, the nanowire crosspoints are designed to contain programmable molecular devices. These devices should permit reconfiguration of the nanowire-based connections between the CMOS cells. Therefore, if physical experiments confirm the designers' preliminary analyses [83,100,145], it is likely CMOL may be used to implement any architecture based upon programmable interconnects. Thus far, quantitative analyses of the CMOL designs seem promising, but no fabrication experiments have been completed to build and test CMOL circuits.

In general, hybridization at device, circuit, or architectural levels may allow the semiconductor industry to leverage the best features of both conventional CMOS and novel nanostructures. However, this combination does introduce additional challenges. One potential difficulty lies in designing the interface between CMOS and nanoscale components. For systems built solely from nanodevices, such an interface is required only at a relatively small number of points at the periphery of the nanoelectronic circuit system. In contrast, hybrid architectures necessitate many interfaces and problematic contacts to achieve tighter and denser integration of the many, many individual CMOS components and nanostructures *within* the circuit system.

For example, the CMOL approach proposes novel interface pins to accomplish this task [83]. However, such pins must be manufactured to tight, sublithographic tolerances. Also, to contact these pins, precise linear and angular alignment of the nanowire array is likely to be required.

A more fundamental difficulty introduced by combining CMOS with nanostructures is that overall scalability may be limited by the scalability of CMOS technology. Such technology is almost certain to hit physical barriers to further scaling. Thus, new processor architectures must be devised that can operate solely with novel nanodevices.

5.4.2.2 Overview of Novel Architectures for Nanoelectronics

A set of clever, yet profound architectural concepts underlies the prototype nanomemory and nano-processor circuit systems just now emerging [19,35,37,40,42,79,97]. These architectural innovations seek to take advantage of the strengths of novel nanodevices (especially, high device density and nonvolatile, low-power operation), as well as to ameliorate some of the limitations discussed in Section 5.4.1 in the techniques presently available for fabrication and assembly at the nanoscale (e.g., the inability to place nanostructures precisely or to make them readily with arbitrary shape or complexity). At the highest level, one may view these architectural innovations as falling into two classes, as discussed below.

1. **Radical Departures from Microelectronic Architectures** One broad class of architectures has been devised strictly by taking demonstrated nanodevices and considering how to combine them into circuits or circuit-like structures that may then be fashioned into complex systems. This bottom-up style of nanoprocessor design has resulted in a number of architectural approaches that differ drastically from conventional architectures. These novel approaches, which are considered in detail elsewhere, include quantum cellular automata (QCA) [35–39], nanoscale neural networks [40,83], nanocells [41,42,47], and biologically inspired electronic system structures such as the virus nanoblock (VNB) [146,147]. Each of these encompasses important ideas and has virtues either in ease of fabrication or in ultra-low power consumption.

   The QCA approach [35–39] seeks to use electric fields, rather than currents, to set bits and propagate signals by moving the charge distributions in arrays of multi-quantum-dot structures termed quantum-dot cells. The primary virtue of this approach is that it is predicted to have ultra-low power dissipation, which is highly desirable in a very dense array of nanostructures. Also, the tiny size of molecular quantum dots may permit this scheme to operate at room temperature, in contrast to solid-state QCA approaches that require cryogenic operation. However, a circuit employing a molecular QCA approach has not yet been demonstrated experimentally.

The nanocell architecture [41,42,47] employs an array of nanoparticles randomly distributed and randomly connected by self-assembled molecules that typically exhibit negative differential resistance and voltage-dependent switching. No attempt is made to control the placement of the molecules that make up the individual interconnects; rather, the designer takes advantage of the molecules' switching characteristics to program the nanocell after it has been assembled. Input and output connections are fabricated on the lithographic scale using conventional techniques. This permits relative ease in manufacturing nanocells, as well as in connecting them to form higher-order circuits. As such, high-level designs may be possible that are similar to today's Very Large Scale Integrated (VLSI) circuits [41].

The nanocell architecture avoids potential difficulties in precise nanoscale fabrication. Instead, the desired connectivity is established by intensive post-fabrication testing and programming. Because of its random assembly and post-fabrication programming, the nanocell approach is inherently defect and fault tolerant [41]. Experimental nanocell memories recently have been fabricated [47] and logic gates have been simulated, but not yet demonstrated.

These architectures, which depart significantly in their operational and organizational principles from those of present-day computers, may make important contributions over the long term. However, their differences from present industry architectures mean they cannot easily harness the significant infrastructure developed by the existing electronics industry. Thus, at the moment, they have more hurdles to overcome and appear to be further from being applied to build extended nanoprocessing systems than the regular array structures discussed below.

2. **Regular Array Architectures Derived from Microelectronics**   This second class of novel nano-electronic architectures is derived via the adaptation and ultra-miniaturization of microelectronic FPGAs and PLAs so they can be implemented with novel nanodevices and new nanofabrication techniques. For the purposes of achieving some near-term successes in developing and operating prototype nanoprocessors, these regular arrays occupy an important middle ground between the radical departures discussed above and the very inhomogeneous architectures used in conventional microprocessors. Nanoarray architectures have an appealing structural simplicity that takes advantage of a number of the strengths of novel nanodevices and nanofabrication techniques. Thus, physical prototypes of extended nanoarray processors are approaching realization based upon much systematic effort [8,10,17–19,25,69], including the detailed simulations described in Section 5.4.4.

There have been criticisms of the use of PLAs to develop nanoprocessors [83]. Some of these criticisms are premised on the assumption that nanoPLAs will not incorporate gain-producing or restoration-producing nanodevices. However, this is not necessarily the case. For example, the nanoPLA architecture due to DeHon and Wilson [19] does incorporate gain-producing nanowire-based nanotransistors, as is described in detail in Section 5.4.3.2. Other criticisms focus on the issue of heat dissipation. This is a valid concern, due to the high density of current-based devices. However, circuit techniques, such as the use of dynamic instead of static logic, may alleviate this problem [19].

Thus, because the path to the realization of these novel nanoelectronic architectures seems clearer and nearer at hand, the rest of this chapter will focus on a discussion of the operational principles, advantages, and trade-offs of FPGA- and PLA-type nanoarray processor architectures.

5.4.3   Principles of Nanoprocessor Architectures Based on FPGAs and PLAs

Having provided a brief survey earlier of various architectural approaches for nanoprocessors, we now focus our attention exclusively on regular arrays such as FPGAs and PLAs. Until recently, the use of such regular arrays in general-purpose, microscale computation has been disfavored relative to the use of conventional, heterogeneous architectures. Thus, to understand how regular arrays may be leveraged for nanoprocessing, it is important to review their use in conventional processing systems and to illustrate the benefits and challenges. Following this brief review, a specific regular array architecture for a nanoprocessor will be explored, the DeHon–Wilson PLA.

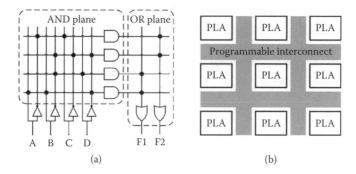

FIGURE 5.17 Schematic illustrations of (a) a single PLA and (b) an extended system architecture based on an array of PLAs. A single PLA consists of a plane of AND gates followed by a plane of OR gates. The interconnections between these gates are reconfigurable after fabrication. In this example, output F_1 is programmed to compute $(A \; AND \; B \; AND \; (NOT \; D)) \; OR \; ((NOT \; B) \; AND \; D)$, based on the configured connections shown by the black dots. More complex, hierarchical logic can be constructed using an array of PLAs like that shown in part (b). Here, outputs such as F_1 and F_2 can be used as inputs to other PLAs in the array.

5.4.3.1 Description of Regular Arrays, FPGAs, and PLAs: Advantages and Challenges

A regular array is a homogeneous two- or three-dimensional grid of configurable logic elements (such as four-input logic tables) interconnected by wires with embedded programmable switches (i.e., "programmable wires") [103]. The array is configured by programming the individual logic elements and switches to define a hardware implementation of a desired logic function. Thus, regular arrays attempt to eliminate heterogeneity at the hardware level, introducing it at the software level, instead. Present fabrication methods for nanoelectronics, which rely on bottom-up- self-assembly approaches, can produce such homogeneous systems of nanostructures [8,17,25].

In conventional microelectronics, regular structures are employed for special-purpose applications in the form of circuits such as FPGAs and PLAs. A schematic diagram of a PLA is given in Figure 5.17(a). Figure 5.17(b) shows an extended system architecture based on PLAs. This system structure is similar to that used for FPGAs. (See Section 5.4.1.2 for a brief description of FPGAs.)

Because of the underlying homogeneity of such structures, thus far they have been outperformed by classical microprocessor architectures in carrying out general-purpose computation. For example, in a given application, an FPGA may be programmed to outperform a general-purpose microprocessor. However, a key capability of general-purpose microprocessors is their ability to switch rapidly between various applications. If the FPGA is configured to provide an equal amount of so-called "context switching" capability, the FPGA implementation usually lags in performance [105].

This is because the general class of functions that can be computed by a conventional processor is quite large, and the best way to compute the whole class of functions on an FPGA has been to program the FPGA as a conventional processor. This is inefficient. However, this inefficiency is not believed to be fundamental. It may be the case that migration to the nanoscale will address this problem. At the nanoscale, it is conceivable that a system may operate with many trillions of devices per processor. With so many devices, it may be possible to implement simultaneously all the required functions that make up a given set of programs [148]. Similarly, the existence of programmable nanoscale interconnects may improve the efficiency of array-based implementations, since the area overhead of each switch can be reduced.

Thus, due to the large number of available devices and the inherent regularity produced by several nanofabrication methods, array architectures have become prominent in nanocomputation research. In the next section, we will describe one such promising architecture, due to DeHon and Wilson [19].

5.4.3.2 The DeHon–Wilson PLA Architecture

A detailed example of a nanoarray architecture that utilizes nanowires in readily realizable crossbar structures is the DeHon–Wilson PLA architecture [19,79,149,150]. A high-level diagram of this architecture is shown in Figure 5.15 and Figure 5.17(b), while Figure 5.18 provides a detailed view of the low-level implementation. As with microelectronic PLA-based designs [104], the large-scale architecture of this nanoprocessor combines a number of PLAs into still larger arrays.

In general, a PLA consists of a programmable AND plane (with a number of AND gates in parallel) followed by a programmable OR plane (with a number of OR gates in parallel), as shown in Figure 5.17(a). Inverters also are available for all inputs. Since any combinational logic function can be written as the OR of some number of AND terms, any such function can be synthesized using a PLA, assuming the PLA is large enough to contain all the logic terms [53].

In the DeHon–Wilson design, a crossbar subarray is used to provide the logical equivalents of the AND and OR planes of the PLA, as shown in Figure 5.18(a). The system is extended by tiling crossbar subarrays, as illustrated in Figure 5.15. Figure 5.18(a) shows the four major subsystems of the DeHon–Wilson PLA implementation: an array of crossed-nanowire diodes used as a programmable OR plane, one inverting subarray of crossed nanowire transistors, a similar buffering subarray, plus an input/output decoder. The inverting and buffering subarrays each are used to regenerate signals and maintain their strengths.

In this PLA scheme, the AND planes are replaced by logically-equivalent pairs of inverting subarrays and OR planes. Figure 5.18(b) shows a more detailed circuit-level characterization of the left-hand side of the system in Figure 5.18(a). In the bottom half of the subarray shown in Figure 5.18(b), all the crossed-wire junctions are taken to contain switchable or "programmable" diodes. By programmable, we mean that the diode can be set to either a high ("on") or low ("off") conductance state in the conductive direction. Where the diodes are not shown, they are taken to be always off, so that the block depicted produces the desired function.

The DeHon–Wilson architecture is notable because it is designed explicitly to tolerate shortcomings in present-day nanofabrication. Within the crossbars of the DeHon–Wilson architecture, redundant wires are used to overcome potential failures due to misalignment or physical defects. A stochastic scheme is used to connect to and thereby address specific wires so that unique addressing can be nearly guaranteed without the need to pick and place individual wires [149]. Also, the inverter and buffer arrays can function in two modes, static and dynamic [19]. In dynamic mode, static power consumption is reduced [53]. This ameliorates the potential problem [83] of heat dissipation in ultra-dense, current-based designs.

Efforts are underway to implement the DeHon–Wilson architecture. Prior to its actual fabrication, there are parameters that remain to be tuned and assumptions that remain to be verified. The most cost-effective method for doing this is the use of nanoprocessor system simulation, as has been demonstrated convincingly in the development of conventional microprocessors [151] and as is discussed further below.

5.4.4 Sample Simulation of a Circuit Architecture for a Nanowire-Based Programmable Logic Array

As stated earlier, system simulation can produce an integrated, multilevel view of the performance of a candidate nanocomputer architecture. This view considers optimization at the device level simultaneously with the problems of designing the system at the circuit and architecture levels. At this early stage of nanocomputer development, it is possible to provide useful insights and guidance to device developers, as well as system architects, by simulating even small component circuits and subsystems. Here, we describe such a simulation and analysis of the DeHon–Wilson PLA [19].

5.4.4.1 Device Models for System Simulation of the DeHon–Wilson NanoPLA

Construction of a nanoprocessor according to the DeHon–Wilson nanowire-based PLA architecture requires four distinct nanodevices, each of which requires a distinct I-V behavior model within the system simulation. All four of these devices are represented, for example, in the schematic in Figure 5.18(b).

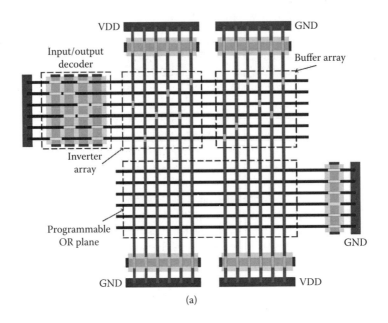

(a)

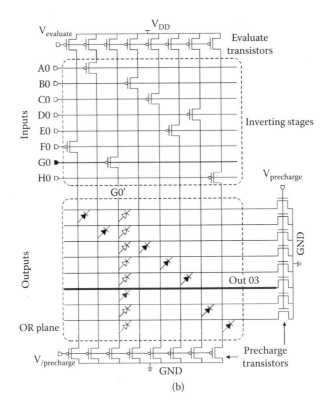

(b)

FIGURE 5.18 Illustrations of (a) the DeHon–Wilson PLA Architecture and (b) an 8 × 8 inverting block. The eight vertical wires shown in part (b) correspond approximately to the vertical wires in the left-hand side of the subarray in part (a).

Three of these devices also are employed in the construction of nanomemory prototypes and are described in Section 5.3.4. These are the nonvolatile nanowire (NVNW) diode, the microwire top-gated FET (TGNW-FET), and the nanowire interconnects.

The fourth device and device model required for the nanoPLA is the crossed-nanowire FET (cNWFET) [17,23,51,52], which acts as the input transistor for the restoration blocks. The cNWFETs are constructed by crossing a nanowire over another nanowire that is coated with silicon dioxide, as depicted in Figure 5.3(a) [23]. The oxide isolates the coated nanowire and allows it to act as the channel of a field-effect transistor, while the uncoated nanowire serves as the gate. Figure 5.3(b) shows an I-V behavior model that has been developed for this device and incorporated into the simulations. This model reproduces published experimental I-V characteristics [7], although some extrapolation beyond the measured voltages was necessary.

One important observation from the I-V characteristics of the cNWFETs is that the experimentally observed threshold voltage (V_T) of the p-channel FETs (PFETs) ranges into positive values. In contrast, conventional microelectronic circuits employ PFETs that have a negative threshold [53]. Some circuits, including the ones we explore here, can be made to function correctly using PFETs with positive thresholds. However, such operation is disadvantageous. In static mode, these circuits consume a great deal of power and usually are not capable of providing adequate signal restoration. Thus, dynamic-mode operation would be preferable. However, for the dynamic operation of the circuits we examine, the PFET V_T threshold must be negative.

Recent experimental results suggest that nanowire p-channel transistors can be fabricated with the desired negative thresholds [51] and that the value of this threshold can be controlled [52]. Based on these experimental results, we have extrapolated a cNWFET model with a reasonable negative value for the PFET threshold voltage. Use of this model permits simulation of these circuits in dynamic mode.

With the device models developed for all required devices, as described above, system simulations were conducted in accordance with the proposed architecture or system design shown in Figure 5.18. Parasitic behaviors of the nanowire arrays, such as coupling capacitance, also were incorporated.

5.4.4.2 Simulations and Analyses of the NanoPLA

The simulations described here consider primarily the performance of a 64-bit PLA. This is represented by an 8 × 8 OR plane driven by eight inverting stages, as shown in Figure 5.18(b). The PLA is programmed with the pattern of diodes depicted there and described in Section 5.4.3.2. The input vectors to the PLA are given in Table 5.3.

The generally accepted method for determining the viability of a circuit system is to assess its operation under the least favorable circumstances. Thus, analysis is performed here by examining the worst-case high and low output voltages. The signal OUT_{03}, which is labeled in Figure 5.18(b) and is the inversion of the G_0 input, is likely to produce the worst-case measurements. This is because, given the switch configuration shown, the length of wire traversed for this output is greatest, which results in the largest parasitic resistance and capacitances.

Functionality of the circuit can be determined by providing a specific input waveform and programmed function, then simulating the output waveform to determine if the function is realized. Such a simulation is illustrated in Figure 5.19, which shows an output waveform for OUT_{03} when the circuit in Figure 5.18(b) is programmed to implement the inversion of G_0. Also shown is the clocking scheme (i.e., the precharge and evaluate signals) for operating the inverting block in dynamic mode. To understand this scheme, it is

TABLE 5.3 PLA Input Vectors

A_0	B_0	C_0	D_0	E_0	F_0	G_0	H_0	
1	1	1	1	1	1	0	1	High Output
0	0	0	0	0	0	1	0	Low Output

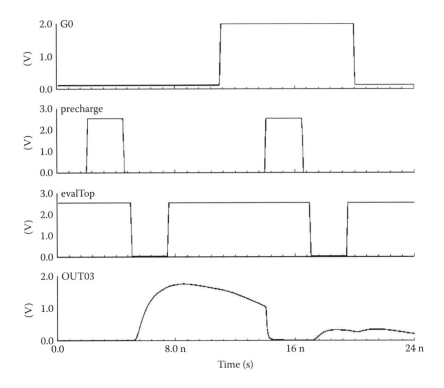

FIGURE 5.19 Waveforms describing how the circuit in Figure 5.18(b) inverts input signal G_0 to produce output signal OUT_{03}. See discussion in text.

first necessary to appreciate that the circuit operates in dynamic mode by storing charge on the wires and the terminals of the devices. Thus, the precharge signals serve to set the charge state of all these elements (e.g., to a charge state that produces a low voltage equivalent to logic "0"). Then, the evaluate signal is used to change the charge state appropriately on some of the wires and terminals (e.g., those for which the correct logic value would be "1").

The dynamic precharge-evaluate cycle first begins when the precharge signal goes high. This has the effect of switching on the n-channel FETs at the right of Figure 5.18(b) to discharge the outputs of the inverting block to a low voltage. After the precharge is completed, the evaluate signal transitions to a low voltage, which turns on the evaluate PFETs at the top of Figure 5.18(b) in order to produce the desired output signal on OUT_{03}. As can be seen in Figure 5.19, the OUT_{03} waveform will continue to be pulled to a high voltage until the evaluate signal is turned back high. After the evaluate transistors turn off, the signal begins to drop, due primarily to leakage through the transistors.

Analyses based upon simulations of this type allow the determination of system behavior and limits. For example, by setting *a priori* the levels for the minimum logic "1" voltage and maximum logic "0" voltage, a minimum operating frequency may be calculated from the signal decay data shown in the bottom graph of Figure 5.19. Thus, these simulations can help characterize how transistor leakage impacts the performance of the system.

Alternative simulations can examine still other effects. For example, diode loading can affect system operation. Simulations suggest there is a limit to the number of diodes that may be turned on and permitted to load a single input column of the inverting stage. For one such simulation, Figure 5.20 shows the output-voltage dependence of the number of diodes programmed in the "on" state along the G_0' column [see Figure 5.18(b)], which drives the OUT_{03} output row. The high output voltage, and thus the voltage swing, is reduced as more diodes are programmed "on" and load the driving column. This is a result of current being divided among multiple outputs.

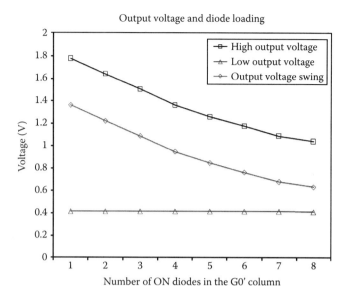

FIGURE 5.20 High and low output voltages and output voltage swing plotted against the number of diodes programmed ON in the G_0' column.

From another simulation for which results are plotted in the bottom curve of Figure 5.20, it is seen that the low or "0" output voltage signal remains relatively constant as the number of "on" diodes is increased. This is because the input vector shown in Table 5.3, and used in this simulation for the low output, drives all the row wires in Figure 5.18(b) except OUT_{03} to logic "1." This has the effect of reverse-biasing all the diodes on the G_0' column that connect to rows other than OUT_{03}. Thus, little current will flow through the diodes into those rows.

While these results show that the circuits can function correctly, they also suggest a limit to the number of "on" diodes that can load the restoring columns. The simulations suggest the maximum number of diodes that can load each column (i.e., the fan-out) is approximately five. Otherwise, it is found that the voltages representing "1" and "0" get so close together they cannot be distinguished by the gates in the downstream logic stages. Thus, there is a limit on the number of functions that may use the same input.

This limit can be increased in a number of ways. One way is to reduce leakage through the nanowire transistors. This requires that difficult experiments be carried out in order to alter device performance appropriately. Another way to increase the limit would be to increase the capacitance at each output. However, this increased capacitance, which takes longer to discharge, also takes longer to charge. This reduces the maximum operating speed of the system. Still a third way would be to introduce duplicate columns, where the input transistors are driven by the same row nanowire.

Also, the restoration-producing portions of the nanoPLA array are likely to be particularly sensitive to variability in the nanodevices. In simulations we have performed on the buffering subarrays, it is seen that a buffer can fail to restore signals adequately if the control signals that would derive from other logic subsystems vary outside of a small acceptable range. A likely source of control signal variation is variation in the structures of devices.

Specific results and design guidance, such as those described in the previous examples, illustrate that system simulation is an effective way to extrapolate from device experiments to consider and improve various nanoelectronic system design options.

5.4.5   Further Implications and Issues for System Simulations

Although the results shown earlier are derived from simulations of a particular nanoprocessor system design, the implications are significant for a wide variety of potential designs and architectures. Any system based on electronic currents flowing through densely-packed circuits must consider issues such as signal integrity, power density, fan-in, fan-out, and gain. For example, we have shown explicitly in Section 5.4.4.2 how the design of such systems must consider fan-out, which in the DeHon–Wilson architecture is the number of diode-connected rows a single inverting column can drive. Fan-out is an important issue for the design of any nanoscale architecture, in that greater fan-out capability aids in reducing the number of logic levels and the area required when implementing complex functions. Several of the nanoscale architectures proposed to date are based on PLAs, much like that envisioned in the DeHon–Wilson architecture [19,43,79,97,99]. As such architectures move toward realization, it will be up to device and circuit designers to find ways to address issues like fan-out for the purpose of optimizing system robustness.

It is important to note that the simulations presented here represent only the first steps toward detailed, extensive simulations of complete nanocomputer architectures. Further issues must be explored for the DeHon–Wilson architecture and other architectures. These issues include system impacts of crosstalk, transistor leakage, and power density. Crosstalk, the loss of signal through coupling capacitances between neighboring wires, can impair significantly the performance of any system consisting of closely-packed wires. Understanding the extent of crosstalk, and devising means for controlling it, can provide design flexibility to improve signal integrity, while possibly reducing power density. Leakage current is another factor that contributes to increased power consumption and to signal degradation. Preliminary experimental data suggest that leakage currents can be relatively large for many of the devices used in this architecture. This would result in increased static power consumption and decreased output voltage-level stability. While it probably will be feasible to reduce the leakage, this will require further careful experimentation.

5.5   Conclusion

In this chapter, we have examined potential approaches to the system-level design and simulation of an extended nanocomputer system that is integrated on the molecular scale. We have considered such systems to be the union of two component subsystems: nanomemories and nanoprocessors.

For each of these components, we have focused upon ultra-dense, array-based system-level design strategies or architectures that offer significant promise for the fabrication and demonstration of extended system prototypes in the near term. In the case of nanomemory systems, recent research in nanoelectronic devices and in the nanofabrication of prototype nanomemory arrays [12,14] has provided evidence of the efficacy of the crossbar array architecture. For nanoprocessor systems, we have surveyed a range of possible architectural approaches. Following this survey, we have focused upon crossbar-based architectures that occupy an important middle ground between conventional microelectronic architectures and a set of more radical nanoelectronic architectures.

To explore the prospective performance of nanocomputer systems based upon these crossbar-based architectures, we have adapted the simulation tools and techniques used widely by the microelectronics industry. In so doing, we are attempting to bridge the gap between the present realm of pure research in nanoelectronics and the application of the resultant innovations in functional, manufacturable systems.

Using detailed simulations of the circuits and subsystems embodied in these architectures for nanomemory and nanoprocessor systems, we have examined some of the trade-offs that affect nanoelectronic systems built from molecular-scale devices. Many of these trade-offs apply to almost any nanocomputer architecture that might be adopted to harness molecules or molecular-scale devices in ultra-dense electronic computing structures. System simulations such as we have described in this chapter can indicate the extent to which enhancements in devices might improve system performance. If such improvements are significant, it becomes worthwhile for experimentalists to invest in enhancing designs for nanodevices and techniques for fabricating them.

Thus, work of the type described here translates the hard-won results of difficult experiments upon nanodevices and small circuits into insights that illuminate the new frontier of nanocomputer systems development. Innovative system design and simulation strategies, coupled closely with device and system experiments, may both speed the realization and optimize the performance of ultra-dense electronic computers integrated on the molecular scale.

Acknowledgments

The authors thank Professors André DeHon and James Heath of the California Institute of Technology, Professor Charles Lieber of Harvard University, plus R. Stanley Williams, Phil Kuekes, Duncan Stewart, and Greg Snider of the Hewlett-Packard Corporation for their many generous discussions and for providing detailed information regarding their nanoscale devices and system designs. Thanks also are due to Professor Konstantin Likharev of Stony Brook University and Brent Segal of Nantero Corporation for so kindly providing us with copies of their forthcoming papers. This research was supported by the Intelligence Technology Innovation Center (ITIC) Nano-Enabled Technology Initiative (NETI).

References

[1] J. C. Ellenbogen and J. C. Love. Architectures for molecular electronic computers: 1. logic structures and an adder designed from molecular electronic diodes. *Proc. IEEE*, 88(3):386–426, 2000.

[2] K. S. Kwok and J. C. Ellenbogen. Moletronics: future electronics. *Materials Today*, 5(2):28–37, 2002.

[3] M. A. Reed, C. Zhou, C. J. Muller, T. P. Burgin, and J. M. Tour. Conductance of a molecular junction. *Science*, 278(5336):252–254, 1997.

[4] C. P. Collier, E. W. Wong, M. Belohradsky, F. M. Raymo, J. F. Stoddart, P. J. Kuekes, R. S. Williams, and J. R. Heath. Electronically configurable molecular-based logic gates. *Science*, 285:391–394, 1999.

[5] C. P. Collier, G. Mattersteig, E. W. Wong, Y. Luo, K. Beverly, J. Sampaio, F. M. Raymo, J. F. Stoddart, and J. R. Heath. A [2]catenane-based solid state electronically reconfigurable switch. *Science*, 289:1172–1175, 2000.

[6] J. Chen, M. A. Reed, A. M. Rawlett, and J. M. Tour. Large on-off ratios and negative differential resistance in a molecular electronic device. *Science*, 286:1550–1552, November 1999.

[7] Y. Huang, X. Duan, Y. Cui, L. J. Lauhon, K. H. Kim, and C. M. Lieber. Logic gates and computation from assembled nanowire building blocks. *Science*, 294:1313–1317, November 2001.

[8] N. A. Melosh, A. Boukai, F. Diana, B. Gerardot, A. Badolato, P. M. Petroff, and J. R. Heath. Ultrahigh-density nanowire lattices and circuits. *Science*, 300(5616):112–115, 2003.

[9] A. Bachtold, P. Hadley, T. Nakanishi, and C. Dekker. Logic circuits with carbon nanotube transistors. *Science*, 294:1317–1320, 2001.

[10] Y. Luo, C. P. Collier, J. O. Jeppesen, K. A. Nielson, E. Delonno, G. Ho, J. Perkins, H. Tseng, T. Yamamoto, J. F. Stoddart, and J. R. Heath. Two-dimensional molecular electronics circuits. *ChemPhysChem*, 3:519–525, 2002.

[11] P. J. Kuekes, D. R. Stewart, and R. S. Williams. The crossbar latch: Logic value storage, restoration, and inversion in crossbar circuits. *J. Appl. Phys.*, 97, 2005.

[12] W. Wu, G. Y. Jung, D. L. Olynick, J. Straznicky, Z. Li, X. Li, D. A. A. Ohlberg, Y. Chen, S. Y. Wang, J. A. Liddle, W. M. Tong, and R. Stanley Williams. One-kilobit cross-bar molecular memory circuits at 30-nm half-pitch fabricated by nanoimprint lithography. *Appl. Phys. A*, 80:1173–1178, 2005.

[13] D. K. Brock, J. W. Ward, C. Bertin, B. M. Segal, and T. Rueckes. Fabrication and applications of single-walled carbon nanotube fabrics. In A. Busnaina, Ed., *CRC Nanomanufacturing Handbook*, CRC Press, Boca Raton, FL, 2006, pp. 41–64.

[14] J. E. Green, J. W. Choi, A. Boukai, Y. Bunimovich, E. Johnston-Halperin, E. DeIonno, Y. Luo, B. A. Sheriff, K. Xu, Y. S. Shin, H.-R. Tseng, J. F. Stoddart, and J. R. Heath. A defect-tolerant molecular electronic memory. In press.

[15] J. R. Heath. Wires, switches, and wiring. A route toward a chemically assembled electronic nanocomputer. *Pure Appl. Chem.*, 72(1-2):11–20, 2000.

[16] P. J. Kuekes, J. R. Heath, and R. S. Williams. Molecular wire crossbar memory. United States Patent 6,128,214, October 2000.

[17] D. Whang, S. Jin, Y. Wu, and C. M. Lieber. Large-scale hierarchical organization of nanowire arrays for integrated nanosystems. *Nano Lett.*, 3(9):1255–1259, September 2003.

[18] G. Snider, P. Kuekes, and R. S. Williams. CMOS-like logic in defective, nanoscale crossbars. *Nanotechnology*, 15(8):881–891, August 2004.

[19] A. DeHon and M. J. Wilson. Nanowire-based sublithographic programmable logic arrays. In *Proc. ACM/SIGDA FPGA*, pp. 123–132, ACM Press, Monterey, CA, 2004.

[20] P. J. Kuekes, G. S. Snider, and R. S. Williams. Crossbar nanocomputers. *Scientific American*, pp. 72–80, November 2005.

[21] M. M. Ziegler, C. A. Picconatto, J. C. Ellenbogen, A. DeHon, D. Wang, Z. H. Zhong, and C. M. Lieber. Scalability simulations for nanomemory systems integrated on the molecular scale. In *Molecular Electronics III*, volume 1006 of *Ann. N.Y. Acad. Sci.*, pp. 312–330. 2003.

[22] S. Das, G. S. Rose, M. M. Ziegler, C. A. Picconatto, and J. C. Ellenbogen. Architectures and simulations for nanoprocessor systems integrated on the molecular scale. *Lect. Notes Phys.*, 680:479–513, 2005.

[23] Y. Huang, X. Duan, Q. Wei, and C. M. Lieber. Directed assembly of one-dimensional nanostructures into functional networks. *Science*, 291:630–633, January 2001.

[24] M. D. Austin, H. Ge, W. Wu, M. Li, Z. Yu, D. Wasserman, S. A. Lyon, and S. Y. Chou. Fabrication of 5 nm linewidth and 14 nm pitch features by nanoimprint lithography. *Appl. Phys. Lett.*, 84(26):5299–5301, June 2004.

[25] G. Y. Jung, S. Ganapathiappan, D. A. A. Ohlberg, D. L. Olynick, Y. Chen, W. M. Tong, and R. S. Williams. Fabrication of a 34 × 34 crossbar structure at 50 nm half-pitch by UV-based nanoimprint lithography. *Nano Lett.*, 4(7):1225–1229, 2004.

[26] A. DeHon, S. C. Goldstein, P. J. Kuekes, and P. Lincoln. Nonphotolithographic nanoscale memory density prospects. *IEEE Trans. Nano.*, 4(2):215–228, March 2005.

[27] International technology roadmap for semiconductors. Technical report, Semiconductor Industry Association, 2005.

[28] M. Ratner and V. Mujica. Molecular electronics from the bottom up. In S. E. Lyshevski, Ed., *CRC Handbook on Molecular and Nano Electronics*. CRC Press, Boca Raton, FL, 2007.

[29] W. Wang, T. Lee., and M.A. Reed. Electrical characterization of self-assembled monolayers. In S. E. Lyshevski, Ed., *CRC Handbook on Molecular and Nano Electronics*. CRC Press, Boca Raton, FL, 2007.

[30] R. Metzger. Unimolecular electronics: Results and prospects. In S. E. Lyshevski, Ed., *CRC Handbook on Molecular and Nano Electronics*. CRC Press, Boca Raton, FL, 2007.

[31] B. Yu and M. Meyyappan. Inorganic nanowires in electronics. In S. E. Lyshevski, Ed., *CRC Handbook on Molecular and Nano Electronics*. CRC Press, Boca Raton, FL, 2007.

[32] B. Kanchibotla, S. Pramanik, and S. Bandyopadhyay. Self assembly of nanostructures using nanoporous alumina templates. In S. E. Lyshevski, Ed., *CRC Handbook on Molecular and Nano Electronics*. CRC Press, Boca Raton, FL, 2007.

[33] Z. M. Liu, A. A. Yasser, J. S. Lindsey, and D. F. Bocian. Molecular memories that survive silicon device processing and real-world operation. *Science*, 302(5650):1543–1545, 2003.

[34] C. Li, W. Fan, B. Lei, D. Zhang, S. Han, T. Tang, X. Liu, Z. Liu, S. Asano, M. Meyyappan, J. Han, and C. Zhou. Multilevel memory based on molecular devices. *Appl. Phys. Lett.*, 84(11):1949–1951, March 2004.

[35] C. S. Lent and P. D. Tougaw. Device architecture for computing with quantum dots. *Proc. IEEE*, 85(4):541–557, 1997.

[36] G. L. Snider, A. O. Orlov, I. Amlani, G. H. Bernstein, C. S. Lent, J. L. Merz, and W. Porod. Quantum-dot cellular automata: Line and majority logic gate. *Japanese J. Appl. Phys. Part 1*, 38(12B):7227–7229, 1999.

[37] V. P. Roychowdhury, D. B. Janes, and S. Bandyopadhyay. Nanoelectronic architecture for Boolean logic. *Proc. IEEE*, 85(4):574–588, 1997.

[38] W. Porod, C. S. Lent, G. H. Bernstein, A. O. Orlov, I. Amlani, G. L. Snider, and J. L. Merz. Quantum-dot cellular automata: computing with coupled quantum dots. *Intl. J. Elect.*, 86(5):549–590, 1999.

[39] G. Toth, C. S. Lent, P. D. Tougaw, Y. Brazhnik, W. W. Weng, W. Porod, R. W. Liu, and Y. F. Huang. Quantum cellular neural networks. *Superlattices and Microstructures*, 20(4):473–478, 1996.

[40] O. Turel, J. H. Lee, X. Ma, and K. K. Likharev. Neuromorphic architectures for nanoelectronic circuits. *Int. J. Circ. Theor. Appl.*, 32:277–302, 2004.

[41] C. P. Husband, S. M. Husband, J. S. Daniels, and J. M. Tour. Logic and memory with nanocell circuits. *IEEE Trans. Elect. Dev.*, 50(9):1865–1875, 2003.

[42] J. M. Tour, W. L. van Zandt, C. P. Husband, S. M. Husband, L. S. Wilson, P. D. Franzon, and D. P. Nackashi. Nanocell logic gates for molecular computing. *IEEE Trans. Nano.*, 1(2):100–109, 2002.

[43] S. C. Goldstein and M. Budiu. Nanofabrics: Spatial computing using molecular electronics. In *Proc. Intl. Symp. Comp. Arch.*, June 2001.

[44] G. S. Rose and M. R. Stan. Programmable logic using molecular devices in a three-dimensional architecture. Presentation at the Engr. Intl. Conf. on Mol. Elect., San Diego, CA (unpublished), January 2005.

[45] J. Han and P. Jonker. A defect- and fault-tolerant architecture for nanocomputers. *Nanotechnology*, 14(2):224–230, 2003.

[46] V. Beiu. From unreliable nanodevices to low power nanoarchitectures. In S. E. Lyshevski, Ed., *CRC Handbook on Molecular and Nano Electronics*. CRC Press, Boca Raton, FL, 2007.

[47] J. M. Tour, L. Cheng, D. P. Nackashi, Y. X. Yao, A. K. Flatt, S. K. St. Angelo, T. E. Mallouk, and P. D. Franzon. Nanocell electronic memories. *J. Am. Chem. Soc.*, 125(43):13279–13283, 2003.

[48] A. Fijany and B. N. Toomarian. New design for quantum dots cellular automata to obtain fault tolerant logic gates. *J. Nanop. Res.*, 3(1):27–37, 2001.

[49] C. S. Lent and B. Isaksen. Clocked molecular quantum-dot cellular automata. *IEEE Trans. Elect. Dev.*, 50(9):1890–1896, 2003.

[50] C. S. Lent, B. Isaksen, and M. Lieberman. Molecular quantum-dot cellular automata. *J. Am. Chem. Soc.*, 125(4):1056–1063, 2003.

[51] A. B. Greytak, L. J. Lauhon, M. S. Gudiksen, and C. M. Lieber. Growth and transport properties of complementary germanium nanowire field-effect transistors. *Appl. Phys. Lett.*, 84(21):4176–4178, May 2004.

[52] Y. Cui, Z. Zhong, D. Wang, W. U. Wang, and C. M. Lieber. High performance silicon nanowire field effect transistors. *Nano Lett.*, 3(2):149–152, 2003.

[53] J. M. Rabaey, A. P. Chandrakasan, and B. Nikolic. *Digital Integrated Circuits*, 2nd ed. Prentice-Hall, Englewood Cliffs, NJ, 2002.

[54] J. van Ruitenbeek, E. Scheer, and H. B. Weber. Contacting individual molecules using mechanically controllable break junctions. *Lect. Notes Phys.*, 680:253–274, 2005.

[55] W. Wang, T. Lee, and M. Reed. Intrinsic electronic conduction mechanisms in self-assembled monolayers. *Lect. Notes Phys.*, 680:275–300, 2005.

[56] J. Tomfohr, G. Ramachandran, O. F. Sankey, and S. M. Lindsay. Making contacts to single molecules: Are we there yet? *Lect. Notes Phys.*, 680:301–312, 2005.

[57] R. M. Metzger. Six unimolecular rectifiers and what lies ahead. *Lect. Notes Phys.*, 680:313–350, 2005.

[58] E. Thune and C. Strunk. Quantum transport in carbon nanotubes. *Lect. Notes Phys.*, 680:351–380, 2005.

[59] J. Jortner, A. Nitzan, and M. A. Ratner. Foundation of molecular electronics — charge transport in molecular conduction junctions. *Lect. Notes Phys.*, 680:13–54, 2005.

[60] P. Hanggi, S. Kohler, J. Lehmann, and M. Strass. AC-driven transport through molecular wires. *Lect. Notes Phys.*, 680:55–76, 2005.

[61] R. DiFelice, A. Calzolari, D. Versano, and A. Rubio. Electronic structure calculations for nanomolecular systems. *Lect. Notes Phys.*, 680:77–116, 2005.

[62] K. Stokbro, J. Taylor, M. Brandbyge, and H. Guo. Ab-initio based nonequilibrium Green's function formalism for calculating electron transport in molecular devices. *Lect. Notes Phys.*, 680:117–152, 2005.

[63] A. DiCaarlo, A. Pecchia, L. Latessa, T. Frauenheim, and G. Seifert. Tight-binding DFT for molecular electronics (gDFTB). *Lect. Notes Phys.*, 680:153–184, 2005.

[64] N. Bushong and M. DiVentra. Current-induced effects in nanoscale conductors. *Lect. Notes Phys.*, 680:185–206, 2005.

[65] M. Wegewijs, M. H. Hettler, J. Konig, A. Thielmann, C. Romeike, and K. Nowack. Single electron tunneling in small molecules. *Lect. Notes Phys.*, 680:207–228, 2005.

[66] M. Thorwart, M. Grifoni, and R. Egger. Transport through intrinsic quantum dots in interacting carbon nanotubes. *Lect. Notes Phys.*, 680:229–249, 2005.

[67] L. A. Bumm, J. J. Arnold, M. T. Cygan, T. D. Dunbar, T. P. Burgin, L. Jones, D. L. Allara, J. M. Tour, and P. S. Weiss. Are single molecular wires conducting? *Science*, 271(5256):1705–1707, 1996.

[68] Y. Selzer, M. A. Cabassi, T. S. Mayer, and D. L. Allara. Temperature effects on conduction through a molecular junction. *Nanotechnology*, 15(7):S483–S488, 2004.

[69] Y. Chen, G. Jung, D. A. A. Ohlberg, X. Li, D. R. Stewart, J. O. Jeppesen, K. A. Nielsen, J. F. Stoddart, and R. S. Williams. Nanoscale molecular-switch crossbar circuits. *Nanotechnology*, 14:462–468, 2003.

[70] R. M. Metzger. All about (*N*-hexadecylquinolin-4-ium-1-yl) methylidenetricyanoquinodimethanide, a unimolecular rectifier of electrical current. *J. Mater. Chem.*, 10:55–62, 2000.

[71] S. J. Tans, A. R. M. Verschueren, and C. Dekker. Room-temperature transistor based on a single carbon nanotube. *Nature*, 393:49–52, May 1998.

[72] H. W. Ch. Postma, T. Teepen, Z. Yao, M. Grifoni, and C. Dekker. Carbon nanotube single-electron transistors at room temperature. *Science*, 293:76–79, July 2001.

[73] S. Heinze, J. Tersoff, and P. Avouris. Carbon nanotube electronics and optoelectronics. *Lect. Notes Phys.*, 680:381–410, 2005.

[74] A. Javey, R. Tu, D. B. Farmer, J. Guo, R. G. Gordon, and H. Dai. High performance n-type carbon nanotube field-effect transistors with chemically doped contacts. *Nano Lett.*, 5(2):345–348, 2005.

[75] J. C. Ellenbogen. Monomolecular electronic device. United States Patent 6,339,227, issued to the MITRE Corporation, 2002.

[76] J. Park, A. N. Pasupathy, J. I. Goldsmith, C. Chang, Y. Yaish, J. R. Petta, M. Rinkoski, J. P. Sethna, H. D. Abruna, P. L. McEuen, and D. C. Ralph. Coulomb blockade and the Kondo effect in single-atom transistors. *Nature*, 417:722–725, June 2002.

[77] W. Liang, M. P. Shores, M. Bockrath, J. R. Long, and H. Park. Kondo resonance in a single-molecule transistor. *Nature*, 417:725–729, June 2002.

[78] C. M. Lieber et al. Design and hierarchical assembly of nanowire-based moletronics. DARPA Moletronics PI Meeting, 2002.

[79] A. DeHon. Array-based architecture for FET-based, nanoscale electronics. *IEEE Trans. Nano.*, 2(1):23–32, 2003.

[80] D. B. Strukov and K. K. Likharev. Defect-tolerant architectures for nanoelectronic crossbar memories. *J. Nanosci. and Nanotech.*, August 2006.

[81] G. Stix. Nanotubes in the clean room. *Scientific American*, pp. 82–85, February 2005.

[82] D. R. Stewart, D. A. A. Ohlberg, P. A. Beck, Y. Chen, R. Stanley Williams, J. O. Jeppesen, K. A. Nielsen, and J. Fraser Stoddart. Molecule-independent electrical switching in Pt/organic monolayer/Ti devices. *Nano Lett.*, 4(1):133–136, 2004.

[83] K. K. Likharev and D. B. Strukov. CMOL: Devices, circuits, and architectures. *Lect. Notes Phys.*, 680:447–478, 2005.

[84] J. E. Savage, E. Rachlin, A. DeHon, C. M. Lieber, and Y. Wu. Radial addressing of nanowires. *ACM J. Emerg. Tech. in Comp. Sys.*, 2:129–154, 2006.

[85] Z. Zhong, D. Wang, Y. Cui, M. W. Bockrath, and C. M. Lieber. Nanowire crossbar arrays as address decoders for integrated nanosystems. *Science*, 302(5649):1377–1379, November 2003.

[86] R. Beckman, E. Johnston-Halperin, Y. Luo, J. E. Green, and J. R. Heath. Bridging dimensions: Demultiplexing ultrahigh-density nanowire circuits. *Science*, 310:465–468, 2005.

[87] T. Hogg, Y. Chen, and P. J. Kuekes. Assembling nanoscale circuits with randomized connections. *IEEE Trans. Nano.*, 5(2):110–122, March 2006.

[88] Y. Cui and C. M. Lieber. Functional nanoscale electronic devices assembled using silicon nanowire building blocks. *Science*, 291:851–853, 2001.

[89] Cadence Design Framework II, Version IC 5.0.33. Cadence Design Systems, Inc., San Jose, CA, 2004.

[90] M. M. Ziegler and M. R. Stan. A case for CMOS/nano co-design. In *Proc. IEEE/ACM Intl. Conf. Comp. Aid. Des.*, pp. 348–352, 2002.

[91] M. M. Ziegler and M. R. Stan. The CMOS/nano interface from a circuits perspective. In *Proc. Int'l Symp. Circuits and Systems*, volume 4, pp. 904–907, 2003.

[92] D. O. Pederson and A. Sangiovanni-Vincentelli. SPICE 3 version 3F5 user's manual. Department of EECS, University of California, Berkeley, CA.

[93] M. Bhattacharya and P. Mazumder. Augmentation of SPICE for simulation of circuits containing resonant tunneling diodes. *IEEE Trans. Computer-Aided Design Integrated Circuits Systems*, 20:39–50, 2001.

[94] M. M. Ziegler and M. R. Stan. Design and analysis of crossbar circuits for molecular nanoelectronics. In *Proc. IEEE Conf. on Nano.*, pp. 323–327, 2002.

[95] G. S. Rose, M. M. Ziegler, and M. R. Stan. Large-signal two-terminal device model for nanoelectronic circuit analysis. *IEEE Trans. VLSI*, 12(11):1201–1208, November 2004.

[96] G. S. Rose, A. C. Cabe, N. Gergel-Hackett, N. Majumdar, M. R. Stan, J. C. Bean, L. R. Harriott, Y. Yao, and J. M. Tour. Design approaches for hybrid CMOS/molecular memory based on experimental data. In *Proc. Great Lakes Symp. on VLSI*, pp. 2–7, Philadelphia, PA, May 2006.

[97] J. R. Heath, P. J. Kuekes, G. S. Snider, and R. S. Williams. A defect-tolerant computer architecture: Opportunities for nanotechnology. *Science*, 280(5370):1716–1721, 1998.

[98] J. L. Hennessy and D. A. Patterson. *Computer Architecture: A Quantitative Approach*, 3rd ed. Morgan Kaufmann, San Mateo, CA, 2002.

[99] M. R. Stan, P. D. Franzon, S. C. Goldstein, J. C. Lach, and M. M. Ziegler. Molecular electronics: From devices and interconnect to circuits and architecture. *Proc. IEEE*, 91(11):1940–1957, 2003.

[100] D. B. Strukov and K. K. Likharev. CMOL FPGA: A reconfigurable architecture for hybrid digital circuits with two-terminal nanodevices. *Nanotechnology*, 16:888–900, 2005.

[101] J. Han. Fault-Tolerant Architectures for Nanoelectronic and Quantum Devices. Ph.D. thesis, Delft University of Technology, Delft, The Netherlands, 2004.

[102] M. Barr. Programmable logic: What's it to ya? *Embedded Systems Programming magazine*, pp. 75–84, June 1999. Also available online at http://www.embedded.com/1999/9906/ 9906sr.htm.

[103] S. Trimberger, Ed. *Field Programmable Gate Array Technology*. Kluwer Academic Publishers, Boston, 1994.

[104] B. Zeidman. *Designing with FPGAs and CPLDs*. CMP Books, Lawrence, KS, 2002.

[105] A. DeHon. Reconfigurable architectures for general-purpose computing. Technical Report AITR-1586, Massachusetts Institute of Technology, 1996.

[106] P. S. Zuchowski, C. B. Reynolds, R. J. Grupp, S. G. Davis, B. Cremen, and B. Troxel. A hybrid ASIC and FPGA architecture. In *Proc. Intl. Conf. Comp. Aid. Des.*, pp. 187–194, 2002.

[107] M. M. Ziegler and M. R. Stan. CMOS/nano co-design for crossbar-based molecular electronic systems. *IEEE Trans. Nano.*, December 2003.

[108] M. Forshaw, R. Stadler, D. Crawley, and K. Nikolic. A short review of nanoelectronic architectures. *Nanotechnology*, 15(4):S220–S223, 2004.

[109] R. P. McConnell. Diode-based power gain for molecular-scale electronic digital computers. report MP 00W0000310, The MITRE Corporation, McLean, VA, 2000.

[110] R. P. McConnell, J. C. Ellenbogen, T. S. Mayer, T. E. Mallouk, and S. P. Goldstein. Requirements and designs for molecular computer architectures that incorporate gain-producing elements. Presentation at the Engr. Found. Conf. on Mol. Elect., Kona, HI (unpublished), December 2000.

[111] S. C. Goldstein and D. Rosewater. Digital logic using molecular electronics. In *Proc. Intl. Sol. St. Circ. Conf.*, 2002.

[112] G. S. Rose and M. R. Stan. Memory arrays based on molecular RTD devices. In *Proc. IEEE-NANO*, pp. 453–456, 2003.

[113] E. Goto, K. Murata, K. Nakazawa, K. Nakagawa, T. Moto-Oka, Y. Matsuoka, Y. Ishibashi, T. Soma, and E. Wada. Esaki diode high speed logical circuits. *IRE Trans. Elect. Comp.*, pp. 25–29, 1960.

[114] H. C. Liu and T. C. L. G. Sollner. High-frequency resonant-tunneling devices. In R. A. Kiehl and T. C. L. G. Sollner, editors, *Semiconductors and Semimetals*, volume 41, pp. 359–418. Academic Press, Boston, 1994.

[115] R. H. Mathews, J. P. Sage, T. C. L. G. Sollner, S. D. Calawa, C.-L. Chen, L. J. Mahoney, P. A. Maki, and K. M. Molvar. A new RTD-FET logic family. *Proc. IEEE*, 87(4):596–605, April 1999.

[116] P. J. Kuekes. Molecular crossbar latch. United States Patent 6,586,965, 2003.

[117] Y. Chen, D. A. A. Ohlberg, X. Li, D. R. Stewart, J. O. Jeppesen, K. A. Nielsen, J. F. Stoddart, D. L. Olynick, and E. Anderson. Nanoscale molecular-switch devices fabricated by imprint lithography. *Appl. Phys. Lett.*, 82(10):1610–1612, March 2003.

[118] G. Snider, P. Kuekes, T. Hogg, and R. Stanley Williams. Nanoelectronic architectures. *Appl. Phys. A*, 80:1183–1195, 2005.

[119] G. S. Snider and P. J. Kuekes. Nano state machines using hysteretic resistors and diode crossbars. *IEEE Trans. Nano.*, 5:129–137, 2006.

[120] M. van den Brink. Litho roadmap shows difficult terrain — part 2 — technology information. *Electronic News*, Jan. 17, 2000.

[121] Semi industry to reach $360 billion by 2010, says report. *Silicon Strategies*, Dec. 2, 2003.

[122] A. W. Burks, H. H. Goldstine, and J. von Neumann. Preliminary discussion of the logical design of an electronic computing instrument. In A. H. Taub, Ed., *John von Neumann Collected Works*, volume V, pp. 34–79. The Macmillan Co., New York, 1963.

[123] H. H. Goldstine and J. von Neumann. On the principles of large scale computing machines. In A. H. Taub, Ed., *John von Neumann Collected Works*, volume V, pp. 1–32. The Macmillan Co., New York, 1963.

[124] J. von Neumann. First draft of a report on the EDVAC. In N. Stern, Ed., *From ENIAC to Univac: An Appraisal of the Eckert-Mauchly Computers*. Digital Press, Bedford, MA, 1981.

[125] Reprinted from the AMD Virtual Pressroom at http://www.amd.com.

[126] J. D. Meindl, Q. Chen, and J. A. Davis. Limits on silicon nanoelectronics for terascale integration. *Science*, 293(5537):2044–2049, 2001.

[127] V. V. Zhirnov, R. K. Cavin, J. A. Hutchby, and G. I. Bourianoff. Limits to binary logic switch scaling — a gedanken model. *Proc. IEEE*, 91(11):1934–1939, November 2003.

[128] D. J. Frank, R. H. Dennard, E. Nowak, P. M. Solomon, Y. Taur, and H. S. P. Wong. Device scaling limits of Si MOSFETs and their application dependencies. *Proc. IEEE*, 89(3):259–288, 2001. This article is one of several that appeared in a Special Issue on Limits of Semiconductor Technology.

[129] M. T. Bohr. Nanotechnology goals and challenges for electronic applications. *IEEE Trans. Nano.*, 1(1):56–62, March 2002.

[130] J. A. Davis, R. Venkatesan, A. Kaloyeros, M. Beylansky, S. J. Souri, K. Banerjee, K. C. Saraswat, A. Rahman, R. Reif, and J. D. Meindl. Interconnect limits on gigascale integration (GSI) in the 21st century. *Proc. IEEE*, 89(3):305–324, March 2001. This article is one of several that appeared in a Special Issue on Limits of Semiconductor Technology.

[131] M. Ieong, B. Doris, J. Kedzierski, K. Rim, and M. Yang. Silicon device scaling to the sub-10-nm regime. *Science*, 306:2057–2060, 2004.

[132] A. Rahman. System-Level Performance Evaluation of Three-Dimensional Integrated Circuits. Ph.D. thesis, Massachusetts Institute of Technology, Cambridge, MA, 2001.

[133] R. Reif, A. Fan, K.-N. Chen, and S. Das. Fabrication technologies for three-dimensional integrated circuits. In *Proc. Intl. Symp. Qual. Elect. Des.*, pp. 33–37, 2002.

[134] S. F. Al-Sarawi, D. Abbott, and P. D. Franzon. A review of 3-D packaging technology. *IEEE Trans. CPMT B*, 21(1):2–14, 1998.

[135] A. Fan, A. Rahman, and R. Reif. Copper wafer bonding. *Elect. Sol. St. Lett.*, 2(10):534–536, 1999.

[136] J. A. Burns, C. Keast, K. Warner, P. Wyatt, and D. Yost. Fabrication of 3-dimensional integrated circuits by layer transfer of fully depleted SOI circuits. In *Proc. Mat. Res. Soc. Symp. G*, volume 768, April 2003.

[137] Y. Kwon, A. Jindal, J. J. McMahon, J.-Q. Lu, R. J. Gutmann, and T. S. Cale. Dielectric glue wafer bonding for 3-D ICs. In *Proc. Mat. Res. Soc.*, Spring 2003.

[138] L. Xue, C. C. Liu, H. S. Kim, S. Kim, and S. Tiwari. Three-dimensional integration: Technology, use, and issues for mixed-signal applications. *IEEE Trans. Elect. Dev.*, 50(3):601–609, 2003.

[139] V. Subramanian, P. Dankoski, L. Degertekin, B. T. Khuri-Yakub, and K. C. Saraswat. Controlled two-step solid-phase crystallization for high-performance polysilicon TFT's. *IEEE Elect. Dev. Lett.*, 18(8):378–381, 1997.

[140] S. Das. Design Automation and Analysis of Three-Dimensional Integrated Circuits. Ph.D. thesis, Massachusetts Institute of Technology, Cambridge, MA, 2004.

[141] R. Chau, B. Boyanov, B. Doyle, M. Doczy, S. Datta, S. Hareland, D. Jin, J. Kavalieros, and M. Metz. Silicon nanotransistors for logic applications. *Physica E: Low-dimensional systems and nanostructures*, 19(1-2):1–5, 2003.

[142] S. J. Wind, J. Appenzeller, R. Martel, V. Derycke, and P. Avouris. Vertical scaling of carbon nanotube field-effect transistors using top gate electrodes. *Appl. Phys. Lett.*, 80(20):3817–3819, 2002.

[143] Q. L. Li, S. Surthi, G. Mathur, S. Gowda, Q. Zhao, T. A. Sorenson, R. C. Tenent, K. Muthukumaran, J. S. Lindsey, and V. Misra. Multiple-bit storage properties of porphyrin monolayers on SiO_2. *Appl. Phys. Lett.*, 85(10):1829–1831, 2004.

[144] B. J. Feder. Nanotech memory chips might soon be a reality. *New York Times*, June 7, 2004.

[145] D. B. Strukov and K. K. Likharev. A reconfigurable architecture for hybrid CMOS/nanodevice circuits. In *Proc. ACM/SIGDA FPGA*, Monterey, CA, 2006. ACM Press.

[146] A. S. Blum, C. M. Soto, C. D. Wilson, J. D. Cole, M. Kim, B. Gnade, A. Chatterji, W. F. Ochoa, T. W. Lin, J. E. Johnson, and B. R. Ratna. Cowpea mosaic virus as a scaffold for 3-D patterning of gold nanoparticles. *Nano Lett.*, 4(5):867–870, 2004.

[147] J. Y. Fang, C. M. Soto, T. W. Lin, J. E. Johnson, and B. Ratna. Complex pattern formation by cowpea mosaic virus nanoparticles. *Langmuir*, 18(2):308–310, 2002.

[148] P. Beckett and A. Jennings. Towards nanocomputer architecture. In *Proc. ACS Conf. Res. Prac. Inf. Tech.*, volume 6, pp. 141–150, 2002.

[149] A. DeHon, P. Lincoln, and J. E. Savage. Stochastic assembly of sublithographic nanoscale interfaces. *IEEE Trans. Nano.*, 2(3):165–174, 2003.

[150] H. Naeimi and A. DeHon. A greedy algorithm for tolerating defective crosspoints in nano PLA design. In *Proc. IEEE Intl. Conf. Field Prog. Tech.*, 2004.

[151] N. H. E. Weste and D. Harris. *CMOS VLSI Design: A Circuits and Systems Perspective*, 3rd ed. Addison-Wesley, Reading, MA, 2004.

6

Three-Dimensional Molecular Electronics and Integrated Circuits for Signal and Information Processing Platforms

Sergey Edward Lyshevski

Abstract

Solid-state microelectronics and complementary metal–oxide–semiconductor (CMOS) technology are approaching the fundamental and technological limits. Innovative paradigms and technologies are emerging, promising to ensure revolutionary developments far beyond semiconductor devices and microelectronics scaling. Far-reaching developments are focused on three-dimensional (3D) biomolecular processing platforms (^{BM}PPs), as well as on *solid* and *fluidic* molecular electronics. Molecular electronics encompasses novel 3D-topology molecular devices (^{M}device), 3D organizations, innovative architectures, *bottom-up* fabrication, etc. The achievable volumetric dimensionality of *solid* molecular electronic devices (^{ME}device) is in the order of $1 \times 1 \times 1$ nm. These multiterminal ^{ME}devices can be synthesized and aggregated within neuronal hypercells ($^\aleph$hypercell) forming functional molecular integrated circuits (^{M}ICs). New device physics, innovative organizations, novel architectures, enabling capabilities/functionality, but not the dimensionality, are the most essential features of molecular and nanoelectronics. *Solid* and *fluidic* ^{M}devices, as compared to semiconductor devices, are based on new device physics, exhibit exclusive phenomena, provide enabling capabilities and possess unique functionality which should be utilized. From the system-level consideration, ^{M}ICs can be designed within novel 3D organizations and enabling architectures which guarantee superior performance. The aforementioned device physics and system innovations lead to unprecedented advantages and opportunities. At the same time, extraordinary fundamental and technological challenges emerge at the device, module, and system levels. In particular, significant challenges and unsolved problems exist in synthesis and design of ^{M}devices, molecular gates (^{M}gate), $^\aleph$hypercells, and ^{M}ICs. For ^{M}devices, a wide spectrum of fundamental, applied, and experimental issues related to device physics, phenomena, and functionality are not sufficiently examined. At the system level, design, optimization, aggregation, verification and other problems are formidable tasks. From fabrication viewpoints, one can synthesize 3D topology *solid* ^{ME}devices and aggregates, but the technology has not matured enough to evaluate and characterize complex ^{M}gates and $^\aleph$hypercells, not to mention ^{M}ICs. This chapter reports the fundamentals of molecular electronics and documents possible solutions to some of the aforementioned problems. The device physics of novel *solid* and *fluidic* ^{M}devices is examined, and ^{M}devices are studied in sufficient detail. Innovative concepts in the design of molecular processing platforms (^{M}PP), formed from ^{M}ICs, are researched, applying and advancing a *molecular architectronics* (^{M}architectronics) paradigm. These ^{M}PPs can be designed within enabling hierarchic architectures (neuronal, processor-and-memory, fused memory-in-processor, and others) utilizing the 3D organization of molecular processing and memory hardware. Neuromorphological reconfigurable *solid* and *fluidic* ^{M}PPs are devised by utilizing a ^{3D}networking-and-processing paradigm.

6.1 Introduction

Though a progress in various applications of nanotechnology has been announced, many of those declarations have been largely acquired from well-known theories and accomplished technologies of material science, biology, chemistry, and other matured areas established in olden times and utilized for centuries. Atoms and atomic structures were envisioned by Leucippus of Miletus and Democritus around 440 BC, and the basic atomic theory was developed by John Dalton in 1803. The Periodic Table of Elements was established by Dmitri Mendeleev in 1869, and the electron was discovered by Joseph Thomson in 1897. The composition of atoms was discovered by Ernest Rutherford in 1910 using the experiments conducted under his direction by Ernest Marsden in the scattering of α-particles. The quantum theory was largely developed by Niels Bohr, Louis de Broglie, Werner Heisenberg, Max Planck, and other scientists at the beginning of the 20th century. Those developments were advanced by Erwin Schrödinger in 1926. For many decades, comprehensive editions of chemistry and physics handbooks coherently reported thousands of organic and inorganic compounds, molecules, ring systems, purines, pyrimidines, nucleotides, oligonucleotides, organic magnets, organic polymers, atomic complexes, and molecules with dimensionality on the order of 1 nm. In the last 50 years, meaningful methods have been developed and commercially deployed to synthesize a great variety of nucleotides and oligonucleotides with various linkers and spacers, bioconjugated molecular aggregates, modified nucleosides, as well as other inorganic, organic, and bio molecules. The aforementioned fundamental, applied, experimental and technological accomplishments have provided essential foundations in many areas, including biochemistry, chemistry, physics, electronics, etc.

Microelectronics has achieved phenomenal accomplishments. For more than 50 years, the discovered microelectronic devices, integrated circuits (ICs), and high-yield technologies have matured and progressed, ensuring high-performance electronics. Many electronics-preceding processes and materials were advanced and fully utilized, as the following list of some past developments attests:

- Crystal growth, etching, thin-film deposition, coating, photolithography and other processes have been known and used for centuries.
- Etching was developed and performed by Daniel Hopfer from 1493 to 1536.
- Modern electroplating (electrodeposition) was invented by Luigi Brugnatelli in 1805.
- Photolithography was invented by Joseph Nicèphore Nièpce in 1822, and he made the first photograph in 1826.
- In 1837, Moritz Hermann von Jacobi introduced and demonstrated silver, copper, nickel, and chrome electroplating.
- In 1839, John Wright, George Elkington, and Henry Elkington discovered that potassium cyanide can be used as an electrolyte for gold and silver electroplating. They patented this process, receiving the British Patent 8447 in 1840.

In the fabrication of various art and jewelry products, as well as Christmas ornaments, these inventions and technologies have been used for many centuries.

By advancing microfabrication technology, the feature size has been significantly reduced. The structural features of solid-state semiconductor devices have been scaled down to tens of nanometers in dimension, and the thickness of deposited thin films can be less than 1 nm. The epitaxy fabrication process (invented in 1960 by J. J. Kleimack, H. H. Loar, I. M. Ross, and H. C. Theuerer) led to the growing layer after layer of silicon films identical in structure with the silicon wafer itself. Technological developments in epitaxy continued, resulting in the possibility to deposit uniform multilayered semiconductors and insulators with precise thicknesses in order to improve the ICs' performance. Molecular beam epitaxy is the deposition of one or more pure materials on a single crystal wafer, one layer of atoms at a time, under high vacuum, forming a single-crystal epitaxial layer. Molecular beam epitaxy was originally developed in 1969 by J. R. Arthur and A. Y. Cho. The thickness of the insulator layer (formed by silicon dioxide, silicon nitride, aluminum oxide, zirconium oxide, or other high-k dielectrics) in field-effect transistors (FETs) was gradually reduced from tens of nanometers to less than 1 nm.

The aforementioned—as well as other meaningful, fundamental, and technological developments—were not referred to as nanoscience, nanoengineering, and nanotechnology until just a few years ago. Indeed, the use of the prefix *nano* in many cases recently has become an excessive attempt to associate products, areas, technologies, and even theories with *nano*. Primarily focusing on atomic structures, examining atoms, and researching subatomic particles and studying molecules, biology, chemistry, physics, and other disciplines have been using the term *microscopic* even though they have dealt with the atomic theory of matter using pico- and femtometer atomic/subatomic dimensions, employing quantum physics, etc.

De Broglie's postulate provides a foundation of the Schrödinger theory, which describes the behavior of *microscopic* particles within the *microscopic* structure of matter composed from atoms. Atoms are composed from nuclei and electrons, and a nucleus consists of neutrons and protons. The *microscopic* theory has been used to examine *microscopic* systems (atoms and elementary particles), such as baryons, leptons, muons, mesons, partons, photons, quarks, etc. The electron and π-meson (pion) have masses 9.1×10^{-31} kg and 2×10^{-28} kg, while their radii are 2.8×10^{-15} m and 2×10^{-15} m. For these subatomic particles, the *microscopic* terminology has been used. The femtoscale dimensionality of subatomic particles has not been a justification to define them to be *"femtoscopic"* particles or to classify these *microscopic* systems to be *"femtoscopic."*

Molecular electronics centers on the developed science and engineering fundamentals, while the progress in chemistry and biotechnology can be utilized to accomplish *bottom-up* fabrication. The attempts to invent appealing terminology for well-established theories and technologies has sometimes led to a broad spectrum of newly originated terms and revised definitions. For example, well-established molecular, polymeric, supramolecular, and other motifs have been sometimes renamed to become such concepts as the directed nanostructured self-assembly, controlled biomolecular nanoassembling, etc. Designer-controlled self-replication (though performed in biosystems through complex and not fully comprehended mechanisms) is a long-term target, many decades away, that may eventually be accomplished utilizing biochemistry and biotechnology. Many recently announced and appealing declarations (molecular building blocks, molecular assembler, nanostructured synthesis, and others) are quite similar to *aromatic compounds, chemistry of coordination compounds, modern materials*, and other subjects covered in undergraduate biology, biochemistry, and chemistry textbooks published decades ago. In those texts, different organic compounds, ceramics, polymers, crystals, composites, and other materials, as well as distinct molecules, were discussed in light of corresponding synthesis processes that had been known for decades or even centuries.

With the current focus on electronics, one may be interested in analyzing major trends [1–4] in the field, as well as to define microelectronics and nanoelectronics. Microelectronics was well-established and matured with a more than 150-billion-dollar market per year. With the clarity on microelectronics [1], nanoelectronics should be defined stressing the underlined premises. The focus, objective, and major themes of nanoelectronics are defined as the following: *Nanoelectronics focuses on fundamental/applied/experimental research and technological developments in devising and implementing novel high-performance enhanced-functionality atomic/molecular devices, modules and platforms (systems), as well as high-yield bottom-up fabrication.*

Nano (molecular) electronics centers on:

1. Discovery of novel devices based on a new device physics
2. Utilization of exhibited unique phenomena and capabilities
3. The devising of enabling organizations and architectures
4. *Bottom-up* fabrication

Other features at the device, module, and system levels are emerging as subproducts of these four major themes. Compared with the solid-state semiconductor (microelectronic) devices, ^{M}devices exhibit new phenomena and offer unique capabilities that should be utilized at the module and system levels. In order to avoid discussions in terminology and definitions, the term molecular—and not the prefix *nano*—is mostly used in this chapter.

At the device level, IBM, Intel, Hewlett-Packard and other leading companies have been successfully conducting pioneering research and pursuing technological developments in *solid* ^{ME}devices, molecular wires, molecular interconnects, etc. Basic, applied, and experimental developments in *solid* molecular electronics are reported in [5–10]. Unfortunately, it seems limited progress has been made in molecular electronics, *bottom-up* fabrication, and technological developments. These revolutionary high-risk high-payoff areas recently emerged, requiring time, readiness, commitment, acceptance, investment, infrastructure, innovations, and market needs. Among the most promising directions, which will lead to revolutionary advances, are the devising and designing of:

- Molecular signal/data processing platforms
- Molecular memory platforms
- Integrated molecular processing-and-memory platforms
- Molecular information processing platforms

Our ultimate objective is to contribute to the developments of a viable ^{M}*architectronics* paradigm in order to radically increase the performance of processing (computing) and memory platforms. Molecular electronics should guarantee information processing preeminence, computing superiority, and memory supremacy.

In general, molecular electronics spans from new device physics to synthesis technologies and from unique phenomena/capabilities/functionality to novel organizations and system architectures. We present a unified synthesis taxonomy in the design of 3D ^{M}ICs, which are envisioned to be utilized in processing and memory platforms for a new generation of arrays, processors, computers, etc. The design of ^{M}ICs is accomplished by using a novel technology-centric concept based on the use of $^{\aleph}$hypercells consisting of ^{M}gates. These ^{M}gates are comprised from interconnected multiterminal ^{M}devices. Some promising ^{M}devices are examined in sufficient detail. Innovative approaches in the design of ^{M}PPs, formed from ^{M}ICs, are documented. Our major motivation is to further develop and apply a sound fundamental theory coherently supported by enabling solutions and technologies. We expand the basic and applied research towards technology-centric CAD-supported ^{M}ICs design theory and practice. The advancements and progress are ensured by using new sound solutions, and a need for a super-large-scale integration (SLSI) is emphasized. The fabrication aspects are covered. The results reported further expand the horizon of the molecular electronics theory and practice, information technology, the design of processing/memory platforms, as well as that of molecular technologies (nanotechnology).

6.2 Data and Signal Processing Platforms

We have touched on a wide spectrum of challenges and problems. It seems that the devising of ^{M}devices, *bottom-up* fabrication, the design of ^{M}ICs, and technology-centric CAD developments are among the most complex issues. However, before discussing ^{M}*architectronics* and its application, let's look at its past and then move on to its prospect and opportunities. The history of data retrieval and processing tools is traced back thousands of years ago. To enter the data, retain it, and perform calculations, people used a mechanical tool: the *abacus*. The early abacus, known as a counting board, was a piece of wood, stone, or metal with carved grooves or painted lines between which movable beads, pebbles, or wood/bone/stone/metal disks were arranged. When these beads were moved around, according to the "programming rules" memorized by the user, some recording and arithmetic problems were solved and documented. The abacus was used for counting, tracking data, and recording facts even before the concept of numbers was invented. The oldest counting board, found in 1899 on the island of Salamis, was used by the Babylonians around 300 BC. As shown in Figure 6.1(a), the Salamis abacus is a slab of marble marked with two sets of 11 vertical lines (ten columns), a blank space between them, a horizontal line crossing each set of lines, and Greek symbols along the top and bottom. Another important invention around the same time was the astrolabe, used for navigation.

In 1623, Wilhelm Schickard built his "calculating clock," a six-digit machine that can add, subtract, and indicate overflow by ringing a bell. Blaise Pascal is usually credited for building the first digital calculating machine. He created it in 1642 to assist his father who was a tax collector. This machine was able to add numbers entered with dials. Pascal also designed and built a "Pascaline" machine in 1644. These five- and

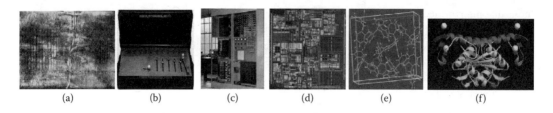

| (a) | (b) | (c) | (d) | (e) | (f) |

FIGURE 6.1   (see color insert following page 5-6) From the abacus (300 BC) to Thomas' *arithmometer* (1820), from the electronic numerical integrator and computer (1946) to the 1.5 × 1.5 cm 478-pin Intel Pentium 4 processor with 42 million transistors (2002), and onward toward 3D *solid* and *fluidic* molecular electronics and processing.

eight-digit machines used a different concept compared with the Schickard's "calculating clock," however. In particular, rising and falling weights instead of a gear drive were used. The "Pascalian" machine can be extended for more digits, but it cannot subtract. Pascal sold more than ten machines, and several of them still exist. In 1674, Gottfried Wilhelm von Leibniz introduced a "stepped reckoner" using a movable carriage to perform multiplications. Charles Xavier Thomas applied Leibniz's ideas and in 1820 made a mechanical calculator (see Figure 6.1[b]). In 1822, Charles Babbage built a six-digit calculator which performed mathematical operations using gears. For many years, from 1834 to 1871, Babbage carried out the Analytical Engine project. His design integrated the stored-program (memory) concept, envisioning the memory may hold more than 100 numbers. The proposed machine had a read-only memory in the form of punch cards. These cards were chained, and the motion of each chain could be reversed. Thus, the machine was able to perform conditional manipulations and integrated coding features. The instructions depended on the positioning of metal studs in a slotted barrel, called the *control barrel*. Babbage only partially implemented his ideas in designing a proof-of-concept programmable calculator because his innovative initiatives were far ahead of his era's technological capabilities and theoretical foundations. Nevertheless, the ideas and goals were set.

In 1926, Vannevar Bush proposed the *product integraph*, a semiautomatic machine for determining the characteristics of electric circuits. International Business Machines introduced the *IBM 601* in 1935, a punch-card machine with an arithmetic unit based on relays that could perform an advanced multiplication in one second. More than 1500 were made. In 1937, George Stibitz constructed a one-bit binary adder using relays. In the same year, Alan Turing published a paper reporting "computable numbers," wherein he solved mathematical problems and proposed a mathematical model of computing known as the *Turing machine*. The idea of electronic computers, however, can be traced back to the late 1920s. Major breakthroughs appeared later. In his master thesis in 1937, Claude Shannon outlined the application of relays, proposing an "electric adder to the base of two." George Stibitz developed a binary circuit based on Boolean algebra in the same year, eventually building and testing the proposed adding device in 1940. John Atanasoff completed a prototype of a 16-bit adder using diode vacuum tubes in 1939. The same year, Zuse and Schreyer examined the application of relay logic. Schreyer completed a prototype of the ten-bit adder using vacuum tubes in 1940, and built memory using neon lamps. Zuse demonstrated the first operational programmable calculator in 1940. The calculator had floating point numbers with a seven-bit exponent, 14-bit mantissa, sign bit, 64 words of memory with 1400 relays, and arithmetic and control units comprised of 1200 relays. Howard Aiken proposed a calculating machine which solved some problems of relativistic physics, and so built the Automatic Sequence Controlled Calculator Mark I. The project was finished in 1944, and the Mark I was used to calculate ballistics problems. This electromechanical machine was 15 m long, weighed 5 tons, and had 750,000 parts (72 accumulators with arithmetic units and mechanical registers with a capacity of 23 digits plus sign). The arithmetics were fixed-point, with a plug-board determining the number of decimal places. The input–output unit included card readers, a card puncher, paper tape readers, and typewriters. It had 60 sets of rotary switches, each of which could be used as a constant register (e.g., a mechanical read-only memory). The program was read from a paper tape, and data could be read from the other tapes, card readers, or constant registers. In 1943, the U.S. government contracted John Mauchly and Presper Eckert to design the Electronic Numerical Integrator and Computer which likely was the first electronic digital computer ever built. It was completed in 1946 (see Figure 6.1[c]). This machine performed 5000 additions or 400 multiplications per second, showing an enormous capability for that time. The Electronic Numerical Integrator and Computer weighed 30 tons, consumed 150 kW, and had 18,000 diode vacuum tubes. John von Neumann and his colleagues built the Electronic Discrete Variable Automatic Computer in 1945, using the so-called "von Neumann computer architecture."

Combinational and memory circuits are comprised from microelectronic devices, logic gates, and modules. Microelectronics textbooks coherently document the developments starting from the discoveries of semiconductor devices to the design of ICs. The major developments are reported next. Ferdinand Braun invented the solid-state rectifier in 1874. The silicon diode was created and demonstrated by Pickard in 1906. The field-effect devices were patented by von Julius Lilienfeld and Oskar Heil in 1926 and 1935, respectively. The functional solid-state bipolar junction transistor was built and tested on December 23, 1947 by John Bardeen and Walter Brattain. Gordon Teal made the first silicon transistor in 1948, and William Shockley invented the unipolar field-effect transistor in 1952. The first ICs were designed by Kilby and Moore in 1958.

Microelectronics has been utilized in signal processing and computing platforms. First-, second-, third-, and fourth-generation computers emerged, and tremendous progress has been achieved. The Intel Pentium 4 processor, illustrated in Figure 6.1(d), and Core^{TM} Duo processor families were built using advanced Intel microarchitectures. These high-performance processors are fabricated using 90- and 65-nm CMOS technology nodes. The CMOS technology was matured to fabricate high-yield high-performance ICs with trillions of transistors on a single die. The fifth generation of computers will utilize further scaled down microelectronic devices and enhanced architectures. However, even more progress and developments are needed. New solutions and novel enabling technologies are emerging.

The suggestion to utilize molecules as a molecular diode, which could be considered the simplest two-terminal *solid* ^{ME}device, was introduced by M. Ratner and A. Aviram in 1974 [11]. This visionary idea has been further expanded through meaningful theoretical, applied, and experimental developments [5–10]. Three-dimensional molecular electronics and ^{M}ICs, designed within a 3D organization, were proposed in [7]. These ^{M}ICs are designed as aggregated ^{N}hypercells, comprised from ^{M}gates engineered utilizing 3D-topology multi-terminal *solid* ^{ME}devices (see Figure 6.1[e]).

The United States Patent 6,430,511 "Molecular Computer" was issued in 2002 to J. M. Tour, M. A. Reed, J. M. Seminario, D. L. Allara, and P. S. Weiss. The inventors envisioned a molecular computer as formed by establishing arrays of input and output pins, "injecting moleware," and "allowing the moleware to bridge the input and output pins." The proposed "moleware includes molecular alligator clip-bearings 2-, 3-, and molecular 4-, or multiterminal wires, carbon nanotube wires, molecular resonant tunneling diodes, molecular switches, molecular controllers that can be modulated via external electrical or magnetic fields, massive interconnect stations based on single nanometer-sized particles, and dynamic and static random access memory (DRAM and SRAM) components composed of molecular controller/nanoparticle or fullerene hybrids." Overall, one may find a great deal of motivating conceptual ideas, expecting fundamental soundness and technological feasibility.

The questions regarding the feasibility of molecular electronics and ^{M}PPs arise. There does not exist conclusive evidence on the overall soundness of *solid* ^{M}ICs, as there was no analog for the solid-state microelectronics and ICs in the past. By contrast, ^{BM}PPs exist in nature. We briefly focus our attention on the most primitive biosystems. Prokaryotic cells (bacteria) lack extensive intracellular organization and do not have cytoplasmic organelles, while eukaryotic cells have well-defined nuclear membranes as well as a variety of intracellular structures and organelles. However, even a 2-μm long single-cell *Escherichia coli* (*E.coli*), *Salmonella typhimurium*, *Helicobacter pylori*, and other bacteria possess ^{BM}PPs, exhibiting superb information and signal/data processing. These bacteria also have molecular sensors $\sim$ 50 $\times$ 50 $\times$ 50 nm motors, as well as other numerous biomolecular devices and systems made from proteins. Though the bacterial motors (largest devices) have been studied for decades, baseline operating mechanisms are still unknown [12]. Biomolecular processing and memory mechanisms also have not been comprehended at the device and system levels. The fundamentals of biomolecular processing, memories, and device physics are not well understood even regarding single-cell bacteria. The information processing, memory storage, and memory retrieval are likely performed utilizing biophysical mechanisms involving ion (~ 0.2 nm) − biomolecule (~ 1 nm) − protein (~ 10 nm) electrochemomechanical interactions and transitions in response to stimuli. The *fluidic* molecular processing and ^{M}PPs, which mimic ^{BM}PPs, were first proposed in [7]. Figure 6.1(f) schematically illustrates the ion-biomolecule−protein complex. The electrochemomechanical interactions and transitions establish a possible device physics of biomolecular devices, ensuring feasibility and soundness of *synthetic* and *fluidic* molecular electronics.

Having emphasized the device levels, it should be stressed again that superb biomolecular 3D organizations and architectures are not comprehended. Assume that in prokaryotic cells and neurons, processing and memory storage are performed due to transitions in biomolecules such as folding transformations, induced potential, charge variations, bonding changes, etc. These electrochemomechanical changes are accomplished due to the binding/unbinding of ions and/or biomolecules, enzymatic activities, etc. The experimental and analytic results show that protein folding is accomplished within nanoseconds and requires $\sim 1 \times 10^{-19}$ to $\sim 1 \times 10^{-18}$ J of energy. Real-time 3D image processing is ordinarily accomplished even by primitive insects and vertebrates that have less than one million neurons. To perform these and other immense processing tasks, less than 1 μW is consumed. However, real-time 3D image processing cannot be performed by even envisioned processors with trillions of transistors, a device switching speed of ~ 1 THz, a circuit speed of ~ 10 GHz, a device switching energy of $\sim 1 \times 10^{-16}$ J, a writing energy of $\sim 1 \times 10^{-16}$ J/bit, a read time of ~ 10 nsec, etc. This is undisputable evidence of superb biomolecular processing that cannot be surpassed by any envisioned microelectronics enhancements and innovations.

Remark

The nomenclature "biomolecular electronics" may not be well founded because it may not reflect the baseline physics and phenomena. Thus, we introduced the term ^{BM}PP. *Fluidic*-centered molecular processing and ^{M}PPs provide the ability to mimic ^{BM}PPs. Our terminology does not imply that *fluidic* and biomolecular platforms are based or centered on the electron transport or only on electron-associated transitions. However, *synthetic* and *fluidic* molecular devices/modules exhibit electrochemomechanical transitions, and one uses electronic apparatuses to control these transitions. This provides the justification of the terminology used.

6.3 Microelectronics and Nanoelectronics: Retrospect and Prospect

To design and fabricate planar CMOS ICs, which consist of FETs and bipolar junction transistors (BJTs) as major microelectronic devices, processes and design rules have been defined. Taking note of the topological layout, the physical dimensions and area requirements can be estimated using the design rules, which center on: (1) minimal feature size and minimum allowable separation in terms of absolute dimensional constraints; (2) the lambda rule (defined using the *length unit* λ) which specifies the layout constraints, taking note of nonlinear scaling, geometrical constraints, and minimum allowable dimensions (e.g., width, spacing, separation, extension, overlap, width/length ratio, and so on). In general, λ is a function of exposure wavelength, image resolution, depth of focus, processes, materials, device physics, topology, etc. For different technology nodes, λ varies from $\sim 1/2$ to 1 of the minimal feature size. For the current front-edge 65-nm technology node, introduced in 2005 and deployed by some high-tech companies in 2006, the minimal feature size is 65 nm. It is expected that the feature size could decrease to 18 nm by 2018. For n-channel MOSFETs (physical cell size is $\sim 10\lambda \times 10\lambda$) and BJTs (physical cell size is $\sim 50\lambda \times 50\lambda$), the *effective* cell areas are in the range of hundreds and thousands of λ^2, respectively. For MOSFETs, the gate length is the distance between the active source and drain regions underneath the gate. This implies that if the channel length is 30 nm, it does not mean the gate width or λ is 30 nm. For FETs, the ratio between the *effective* cell size and minimum feature size will remain to be ~ 20.

One cannot define and classify electronic, optical, electrochemomechanical, and other devices or ICs only by taking note of their dimensions (length, area, or volume) or minimal feature size. The device dimensionality is an important feature primarily from the fabrication viewpoint. To classify devices and systems, one examines the device physics, system organization/architecture and fabrication technologies, assessing distinctive features, capabilities and phenomena utilized. Even if the dimension of CMOS transistors will be scaled down to achieve a 100×100 nm *effective* cell size for FETs by late 2020, these solid-state semiconductor devices may not be viewed as nanoelectronic devices because conventional phenomena and evolved technologies are utilized. The fundamental limits on microelectronics and solid-state semiconductor devices were known and reported for many years [1]. Though significant progress has been made, ensuring the high-yield fabrication of ICs, the basic physics of semiconductor devices has remained virtually unchanged for decades. Three editions (1969, 1981, and 2007) of a classical textbook *Physics of Semiconductor Devices* [13–15] coherently cover the device physics. The evolutionary technological developments will continue beyond the current 65-nm technology node. The 45-nm CMOS technology node is expected to emerge in 2007. Assume that by 2018, the 18-nm technology node will be deployed with the expected $\lambda =\sim 18$ nm and $\sim$ 7- to 8-nm effective channel length for FETs. This will lead to the estimated footprint area of the interconnected FET to be in the range of tens of thousands of nm^2 because the *effective* cell area is at least $\sim 10\lambda \times 10\lambda$. Sometimes, a questionable size-centered definition of nanotechnology surfaces, ambiguously picking the 100-nm dimensionality to be met. It is uncertain which dimensionality should be used. Also, it is unclear why 100 nm is declared. Why not 1 nm or 999 nm? On the other hand, why not use a volumetric measure such as 100 nm^3?

An electric current is a flow of charged particles. The current in conductors, semiconductors and insulators is due to the movement of electrons. In aqueous solutions, the current is due to the movement of charged particles (e.g., ions, molecules, etc.). The devices have not been classified using the dimension of the charged carriers (electrons, ions, or molecules). However, one may compare the device dimensionality with the size of the particle which causes the current flow or transitions. For example, considering a protein as a core component of a biomolecular device, and an ion as a charge carrier that affects the protein transitions, the device/carrier dimensionality ratio would be ~ 100. The *classical* electron radius r_0, called the Compton radius, is found by equating the electrostatic potential energy of a sphere with the charge e and radius r_0 to the relativistic rest energy of the electron which is $m_e c^2$. We have $e^2/(4\pi \varepsilon_0 r_0) = m_e c^2$, where e is the charge on the electron, $e = 1.6022 \times 10^{-19}$ C; ε_0 is the permittivity of free space, $\varepsilon_0 = 8.8542 \times 10^{-12}$ F/m; m_e is the mass of the electron, $m_e = 9.1095 \times 10^{-31}$ kg; and c is the speed of light in a vacuum: 299792458 m/sec. Thus, $r_0 = e^2/(4\pi \varepsilon_0 m_e c^2) = 2.81794 \times 10^{-15}$ m. With the achievable volumetric dimensionality of *solid* ^{ME}devices on the order of $1 \times 1 \times 1$ nm, one finds that the device is much larger than the carrier. Up to 1×10^{18} devices can be placed in 1 mm^3. This upper-limit device density may not be achieved due to synthesis constraints, technological challenges, expected inconsistency, aggregation/interconnect complexity, and other problems. The *effective* volumetric dimensionality of interconnected *solid* ^{ME}devices in ^{M}ICs is expected to be $\sim 10 \times 10 \times 10$ nm. For *solid* ^{ME}devices, quantum physics must be applied to examine the processes, functionality, performance, characteristics, etc. The device physics of *fluidic* and *solid* ^{M}devices is profoundly different. Emphasizing the major premises, nanoelectronics implies the use of:

1. Novel high-performance devices, devised and designed using new device physics, which exhibit unique phenomena and capabilities to be exclusively utilized at the gate and system levels.
2. Enabling 3D organizations and advanced architectures which ensure superb performance and superior capabilities. Those developments rely on the device-level solutions, technology-centric SLSI design, etc.
3. *Bottom-up* fabrication.

To design ^{M}ICs-comprised processing and memory platforms, one must apply novel paradigms and pioneering developments utilizing 3D-topology ^{M}devices, enabling organizations/architectures, sound *bottom-up* fabrication, etc. Tremendous progress has been accomplished within the last 60 years in microelectronics—e.g., from inventions and demonstrations of functional solid-state transistors to the fabrication of processors that comprise trillions of transistors on a single die. The current high-yield 65-nm CMOS technology node ensures minimal features ~ 65 nm, and FETs were scaled down to achieve the channel length below 30 nm. Using this technology for the SRAM cells, $\sim 500,000$ nm^2 foot-print area was achieved by Intel. Optimistic predictions foresee that within 15 years the minimal feature of planar (two-dimensional) solid-state CMOS-technology transistors may approach $\sim$ 10 nm, leading to the *effective* cell size for FET $\sim 20\lambda \times 20\lambda = 200 \times 200$ nm. However, the projected scaling trends are based on a number of assumptions and foreseen enhancements [1]. Though the FET cell dimension can reach 200 nm, the overall prospects in microelectronics (technology enhancements, device physics, device/circuits performance, design complexity, cost, and other features) are troubling [1–4]. The near-absolute limits of the CMOS-centered microelectronics can be reached by the next decade. The general trends, prospects and projections are reported in the *International Technology Roadmap for Semiconductors* [1].

The device size- and switching energy–centered version of Moore's first conjecture for high-yield room-temperature mass-produced microelectronics is reported in Figure 6.2 for past, current (90 and 65 nm), and foreseen (45 and 32 nm) CMOS technology nodes. For the switching energy, one uses eV or J, and 1 eV $=$ $1.602176462 \times 10^{-19}$ J. Intel expects to introduce 45-nm CMOS technology node in 2007. The envisioned 32-nm technology node is expected to emerge in 2010. The expected progress in the baseline characteristics, key performance metrics, and scaling abilities has already slowed down due to fundamental and techno-logical challenges and limitations. Correspondingly, new solutions and technologies have been sought and assessed [1]. The performance and functionality at the device, module, and system levels can be significantly improved by utilizing novel phenomena, employing innovative topological/organizational/architectural solutions, enhancing device functionality, increasing density, improving utilization, increasing switch-ing speed, etc. Molecular electronics (nanoelectronics) is expected to result in a departure from the first and second of Moore's conjectures. The second conjecture foresees that, in order to ensure the projected microelectronics scaling trends, the cost of microelectronics facilities can reach hundreds of billion dol-lars by 2020. High-yield affordable nanoelectronics is expected to ensure superior performance. Existing superb bimolecular processing/memory platforms and progress in molecular electronics are assured ev-idence of fundamental soundness and technological feasibility of molecular electronics and ^{M}PPs. Some data and expected developments, reported in Figure 6.2, are subject to adjustments because it is difficult to accurately foresee the fundamental developments and maturity of prospective technologies due to the impact of many factors. However, the overall trends are obvious and likely cannot be rejected. Having emphasized the emerging molecular (nano) electronics, it is obvious that solid-state microelectronics is a core 21st-century technology. CMOS technology will continue to be a viable technology for many decades, even as limits are reached and envisioned nanoelectronics mature. By 2030, core modules of super-high-performance processing (computing) platforms may be implemented using ^{M}ICs. However, microelectronics and molecular electronics are complementary technologies, and ^{M}ICs will not diminish the use of ICs. Molecular electronics and ^{M}PPs are impetuous revolutionary (not evolutionary) changes at the device, system, fundamental, and technological levels. The foreseen revolutionary changes towards ^{M}devices are analogous to abrupt changes from the vacuum tube to a solid-state transistor.

The fundamental and technological limits are also imposed on molecular electronics and ^{M}PPs. Those limits are defined by the device physics, circuit, system, CAD, and synthesis constraints. Some of these limitations are reported in this chapter. However, there is no end to the progress, and one will evolve beyond molecular electronics and processing. What lies beyond these innovations and frontiers? The hypothetical answer is provided next. In 1993, the Dutch theoretical physicist G. Hooft proposed the *Holographic Principle*, postulating that the information contained in some region of space can be represented as a *hologram* that gives the bounded region of space, at *most*, one degree of freedom per the Planck unit of area ($\lambda_p = 1.616 \times 10^{-35}$ m). In this chapter, we will utilize the so-called *standard model* (particles are considered to be points moving through space and coherently represented by mass, electric charge, interaction, spin, etc.). The *standard model* is consistent within quantum mechanics and the special theory of relativity. Other concepts have also been developed, some even utilizing *string theory*, with its various aspects like string vibration, distinct forces, multidimensionality, etc. It is difficult to theorize which far-reaching paradigms will emerge. Therefore, we will focus here on sound and practical paradigms.

Commercial high-yield high-performance molecular electronics is expected to come to the fore by 2015, as shown in Figure 6.2. Molecular devices and ^{M}ICs can operate due to electron tunneling, ion transport, photon interaction, biomolecular interactions, state transitions, etc. For distinct classes of ^{M}devices, basic

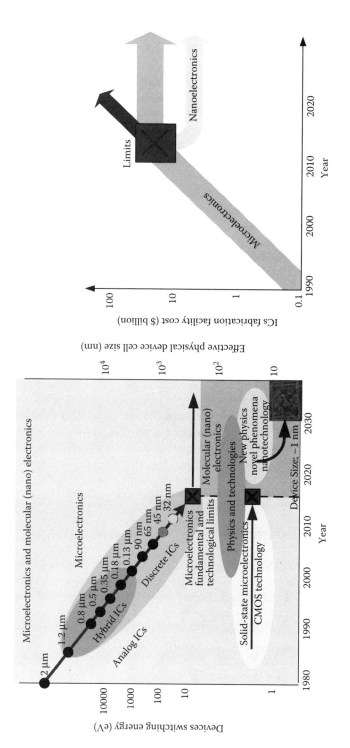

FIGURE 6.2 Envisioned molecular (nano) electronics advancements and microelectronics trends.

physics, phenomena exhibited, effects utilized, and fabrication technologies are profoundly different. Molecular electronics can be classified using the following four major classes:

- *Solid* organic/inorganic molecular electronics
- *Fluidic* molecular electronics
- *Synthetic* biomolecular electronics
- Hybrid molecular electronics

Distinct subclasses and classifiers can be developed, taking into account the device physics and system features. Biomolecular devices and platforms, which are not within the aforementioned classes, may be classified as well. The dominating premises of molecular (nano) electronics and ^{M}PPs have a solid bio association. There exists a great number of superb biomolecular systems and platforms. The device-level biophysics and system-level fundamentals of biomolecular processing are not fully comprehended, but, they are *fluidic* and molecule-centered.

For molecular electronics, theory, engineering practice, and technology are revolutionary advances compared with microelectronic theory and CMOS technology. From a 3D-centered topology/organization/architecture standpoint, *solid* and *fluidic* molecular electronics evolution mimics superb ^{BM}PPs. Information processing, memory storage, and other relevant tasks performed by biosystems are a sound proving ground of the proposed developments. Molecular electronics will lead to novel ^{M}PPs. Compared with the most advanced CMOS processors, molecular platforms will greatly enhance functionality and processing capabilities, radically decrease latency, power, and execution time, as well as drastically increase device density, utilization, and memory capacity. Many difficult problems at the device and system levels must be addressed, researched, and solved. For example, the following tasks should be carried-out: design, analysis, optimization, aggregation, routing, reconfiguration, verification, evaluation, etc. Many of the aforementioned problems have not even been addressed yet. Due to significant challenges, much effort must be devoted to solving these problems. We address and propose solutions to some previously mentioned fundamental and applied problems, thus establishing a ^{M}*architectronics* paradigm. A number of baseline problems are examined, progressing from the system level consideration to the module/device level and vice versa. Taking note of the diversity and magnitude of tasks under consideration, one cannot formulate, examine, and solve all challenging problems. A gradual step-by-step approach is pursued rather than attempting to solve abstract problems with a minimal chance of success. There is a need to stimulate further developments and foster advanced research focusing on well-defined existing fundamentals and future perspectives emphasizing the near-, medium- and long-term prospects, visions, problems, solutions, and technologies.

6.4   Performance Estimates

The combination and memory ^{M}ICs can be designed as aggregated ^{N}hypercells comprised from ^{M}gates and molecular memory cells [16]. At the device level, one examines functionality, studies characteristics, and estimates performance of 3D-topology ^{M}devices. Device- and system-level performance measures are of great interest. The experimental results indicate that protein folding is performed within 1×10^{-6} to 1×10^{-12} sec and requires $\sim 1 \times 10^{-19}$ to 1×10^{-18} J of energy. These transition times and energy estimates can be used for some *fluidic* and *synthetic* ^{M}devices. To analyze protein folding energetics, examine the switching energy in solid-state microelectronic devices, estimate *solid* ^{ME}device energetics, and perform other studies, distinct concepts have been applied. For solid-state microelectronic devices, the logic signal energy is expected to fall to $\sim 1 \times 10^{-16}$ J, and the energy dissipated is $E = Pt = IVt = I^2 Rt = Q^2 R/t$, where P is the power dissipation, I and V are the current and voltage along the discharge path, and R and Q are the resistance and charge.

The term $k_B T$ has been used to solve distinct problems. Here, k_B is the Boltzmann constant, $k_B = 1.3806 \times 10^{-23}$ J/K $= 8.6174 \times 10^{-5}$ eV/K. For example, expression $\gamma k_B T$ ($\gamma > 0$) has been used to perform energy estimates, and $k_B T \ln(2)$ was applied to assess the lowest energy bound for a binary switching. The applicability of distinct equations must be thoroughly examined and sound concepts must be applied.

Statistical mechanics and entropy analysis coherently utilize term $k_B T$ within a specific content as reported in the following, while for some other applications and problems, the use of $k_B T$ may be impractical.

Entropy and Its Application. For an ideal gas, the kinetic-molecular Newtonian model provides the average translational kinetic energy of a gas molecule. In particular, $\frac{1}{2}m(v^2)_{av} = \frac{3}{2}k_B T$. One concludes that the average translational kinetic energy per gas molecule depends only on the temperature. The most notable equation of statistical thermodynamics is the Boltzmann formula for entropy as a function only of the system's state—e.g.,

$S = k_B \ln w$, where w is the number of possible arrangements of atoms or molecules in the system.

Unlike energy, entropy is a quantitative measure of the system disorder in any specific state, and S is not related to each individual atom or particle. At any temperature above absolute zero, the atoms acquire energy, more arrangements become possible, and because $w > 1$, one has $S > 0$.

The entropy and energy are very different quantities. When the interaction between the system and environment involves only reversible processes, the total entropy is constant, and $\Delta S = 0$. When any irreversible process occurs, the total entropy increases, and $\Delta S > 0$. One may derive the entropy *difference* between two distinct states in a system that undergoes a thermodynamic process that takes the system from an initial *macroscopic* state 1 with w_1 possible *microscopic* states to a final *macroscopic* state 2 with w_2 associated *microscopic* states. The change in entropy is then $\Delta S = S_2 - S_1 = k_B \ln w_2 - k_B \ln w_1 = k_B \ln(w_2/w_1)$. Thus, the entropy *difference* between two *macroscopic* states depends on the ratio of the number of possible *microscopic* states. The entropy change for any reversible isothermal process is given using an infinitesimal quantity of heat ΔQ. For initial and final states 1 and 2, one has $\Delta S = \int_1^2 \frac{dQ}{T}$.

Example 6.4.1

To heat 1 ykg (1×10^{-24} kg) of silicon from $0°C$ to $100°C$ using the constant specific heat capacity $c = 702$ J/kg $\cdot$ K over the temperature range, the change of entropy is

$$\Delta S = S_2 - S_1 = \int_1^2 \frac{dQ}{T}$$

$$= \int_{T_1}^{T_2} mc \frac{dT}{T} = mc \ln \frac{T_2}{T_1} = 1 \times 10^{-24} \text{kg} \times 702 \frac{\text{J}}{\text{kg} \cdot \text{K}} \times \ln \frac{373.15\text{K}}{273.15\text{K}} = 2.19 \times 10^{-22} \text{ J/K}.$$

From $\Delta S = k_B \ln(w_2/w_1)$, one finds the ratio between *microscopic* states to be w_2/w_1. For the problem under consideration, $w_2/w_1 = 7.7078 \times 10^6$.

If $w_2/w_1 = 1$, the total entropy is constant, and $\Delta S = 0$.

The energy needed to heat 1×10^{-24} kg of silicon to $\Delta T = 100°C$ is $Q = mc\Delta T = 7.02 \times 10^{-20}$ J.

To heat 1 g of silicon from $0°C$ to $100°C$, one finds $\Delta S = S_2 - S_1 = mc \ln \frac{T_2}{T_1} = 0.219$ J/K and $Q = mc\Delta T = 70.2$ J.

Taking note of equation $\Delta S = k_B \ln(w_2/w_1)$, it is impossible to derive the numerical value for w_2/w_1.

For a silicon atom, the covalent, atomic, and van der Waals radii are 117, 117, and 200 pm. The Si-Si and Si-O covalent bonds are 232 and 151 pm, respectively. One can examine the thermodynamics using the enthalpy, Gibbs function, entropy, and heat capacity of silicon in its solid and gas states. The atomic weight of a silicon atom is 28.0855 amu, where amu stands for the atomic mass unit, 1 amu $= 1.66054 \times 10^{-27}$ kg. Hence, the mass of a single Si atom is 28.0855 amu $\times 1.66054 \times 10^{-27}$ kg/amu $= 4.6637 \times 10^{-26}$ kg. Therefore, the number of silicon atoms in 1×10^{-24} kg of silicon is $1 \times 10^{-24}/4.6637 \times 10^{-26} = 21.44$.

Consider two silicon atoms to be heated from $0°C$ to $100°C$. For $m = 9.3274 \times 10^{-26}$ kg, we have $\Delta S = S_2 - S_1 = mc \ln \frac{T_2}{T_1} = 2.04 \times 10^{-23}$ J/K. One obtains an obscure result $w_2/w_1 = 4.39$.

It should be emphasized again that the entropy and *macroscopic/microscopic* states analysis are performed for an ideal gas, assuming the accuracy of the kinetic-molecular Newtonian model. In general, to examine the particle and molecule energetics, quantum physics must be applied.

For particular problems, using the results reported one may carry out a similar analysis for other atomic complexes. For example, while carbon has not been widely used in microelectronics, organic molecular electronics is carbon-centered. Therefore, some useful information is reported. For a carbon atom, the covalent, atomic, and van der Waals radii are 77, 77, and 185 pm. Carbon can be in the solid (graphite or diamond) and gas states. Using the atomic weight of a carbon atom, which is 12.0107 amu, the mass of a single carbon atom is 12.0107 amu $\times$ 1.66054 $\times$ 10^{-27} kg/amu = 1.9944$\times$ 10^{-26} kg. ∎

Example 6.4.2

If $w = 2$, the entropy is found to be $S = k_B \ln 2 = 9.57 \times 10^{-24}$ J/K $= 5.97 \times 10^{-5}$ eV/K.

Having derived S, one cannot conclude that the minimal energy required to ensure the transition (*switching*) between two *microscopic* states or to erase a bit of information (energy dissipation) is $k_B T \ln 2$, which for $T = 300$K gives $k_B T \ln 2 = 2.87 \times 10^{-21}$ J $= 0.0179$ eV. In fact, under this reasoning, one assumes the validity of the *averaging* kinetic-molecular Newtonian model and applies the assumptions of distribution statistics at the same time, allowing only two distinct *microscopic* system's states. The energy estimates should be performed utilizing the quantum mechanics. ∎

6.4.1 Distribution Statistics

Statistical analysis is applicable only to systems with a large number of particles and energy states. The fundamental assumption of statistical mechanics is that in thermal equilibrium every distinct state with the same total energy is equally probable. Random thermal motions constantly change energy from one particle to another and from one form of energy to another (kinetic, rotational, vibrational, etc.) obeying the conservation of energy principle. The absolute temperature T has been used as a measure of the total energy of a system in thermal equilibrium.

In semiconductor devices, an enormous number of particles (electrons) are considered using the electrochemical potential $\mu(T)$. The Fermi–Dirac distribution function $f(E) = \dfrac{1}{1 + e^{\frac{E - \mu(T)}{k_B T}}}$ gives the average (probable) number of electrons of a system (device) in equilibrium at temperature T in a quantum state of energy E.

The electrochemical potential at absolute zero is the Fermi energy E_F, and $\mu(0) = E_F$. The occupation probability that a particle would have the specific energy is not related to quantum indeterminacy. Electrons in solids obey Fermi–Dirac statistics. The distribution of electrons, leptons, and baryons (*identical fermions*) over a range of allowable energy levels at thermal equilibrium is expressed as $f(E) = \dfrac{1}{1 + e^{\frac{E - E_F}{k_B T}}}$, where T is the equilibrium temperature of the system. Hence, the Fermi–Dirac distribution function $f(E)$ gives the probability that an allowable energy state at energy E will be occupied by an electron at temperature T.

For *distinguishable* particles, one applies the Maxwell–Boltzmann statistics with a distribution function $f(E) = e^{-\frac{E - E_F}{k_B T}}$.

The Bose–Einstein statistics are applied to *identical bosons* (photons, mesons, etc.). The Bose–Einstein distribution function is given as $f(E) = \dfrac{1}{e^{\frac{E - E_F}{k_B T}} - 1}$.

As was emphasized, the distribution statistics are applicable to electronic devices which consist of a great number of constituents where particle interactions can be simplified by deducing the system behavior from the statistical consideration. Depending on the device physics, one must coherently apply the appropriate baseline theories and concepts.

Example 6.4.3

For $T = 100$K and $T = 300$K, letting $E_F = 5$ eV, the Fermi–Dirac distribution functions are reported in Figure 6.3(a). Figure 6.3(b) documents the Maxwell–Boltzmann distribution functions $f(E)$. ∎

6.4.2 Energy Levels

In ^{M}devices, one can calculate the energy required to excite the electron, and the allowed energy levels are quantized. In contrast, solids are characterized by energy band structures that define electric characteristics. In semiconductors, the relatively small bandgaps allow excitation of electrons from the valence band to the conduction band by thermal or optical energy. The application of quantum mechanics allows one to derive the expression for the quantized energy.

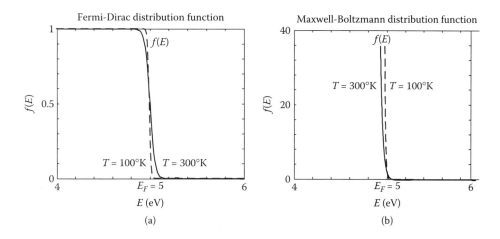

FIGURE 6.3 The Fermi–Dirac and Maxwell–Boltzmann distribution functions for $T = 100$ K and $T = 300$ K if $E_F = 5$ eV.

For a hydrogen atom, one has

$$E_n = -\frac{m_e e^4}{32\pi^2 \varepsilon_0^2 \hbar^2 n^2},$$

where $\hbar$ is the modified Planck constant,

$$\hbar = h/2\pi = 1.055 \times 10^{-34} \text{ J-sec} = 6.582 \times 10^{-16} \text{ eV} - \text{sec}.$$

The energy levels depend on the quantum number n. As n increases, the total energy of the quantum state becomes less negative, and $E_n \to 0$ if $n \to \infty$. The state of lowest total energy is the most stable state for the electron, and the normal state of the electron for a hydrogen (one-electron atom) is at $n = 1$.

Thus, for the hydrogen atom, in the absence of a magnetic field **B**, the energy $E_n = -\frac{m_e e^4}{32\pi^2 \varepsilon_0^2 \hbar^2 n^2}$ depends only on the principle quantum number n. The conversion $1 \text{ eV} = 1.602176462 \times 10^{-19}$ J is commonly used, and $E_{n=1} = -2.17 \times 10^{-18}$ J $= -13.6$ eV. For $n = 2$, $n = 3$, and $n = 4$, we have $E_{n=2} = -5.45 \times 10^{-19}$ J, $E_{n=3} = -2.42 \times 10^{-19}$ J and $E_{n=4} = -1.36 \times 10^{-19}$ J. When the electron and nucleus are separated by an infinite distance ($n \to \infty$), one has $E_{n \to \infty} \to 0$.

The energy difference between the quantum states n_1 and n_2 is $\Delta E = E_{n1} - E_{n2}$, and $\Delta E = E_{n1} - E_{n2} = \frac{m_e e^4}{32\pi^2 \varepsilon_0^2 \hbar^2} \left(\frac{1}{n_2^2} - \frac{1}{n_1^2} \right)$, where $\frac{m_e e^4}{32\pi^2 \varepsilon_0^2 \hbar^2} = 2.17 \times 10^{-18}$ J $= 13.6$ eV.

The excitation energy of an exited state n is the energy above the ground state—e.g., for the hydrogen atom one has ($E_n - E_{n=1}$). The first exited state ($n = 2$) has excitation energy $E_{n=2} - E_{n=1} = -3.4 + 13.6 = 10.2$ eV. In atoms, orbits are characterized by quantum numbers.

De Broglie conjecture relates the angular frequency v and energy E. In particular, $v = E/h$, where h is the Planck constant, $h = 6.626 \times 10^{-34}$ J-sec $= 4.136 \times 10^{-15}$ eV-sec. The frequency of a photon of electromagnetic radiation is found to be $v = \Delta E/h$.

Example 6.4.4

Examine the meaning of $\Delta \hat{E}$. The energy difference between the quantum states ΔE is not the energy uncertainty in the measurement of E, which is commonly denoted in the literature as ΔE. In this chapter, reporting the Heisenberg uncertainty principle, to ensure consistency, we use the notation $\Delta \hat{E}$. In partic- ular, Section 13.2 reports the energy-time uncertainty principle as $\sigma_E \sigma_t \geq \frac{1}{2}\hbar$ or $\Delta t \geq \frac{1}{2}\hbar$, where σ_E and σ_t are the standard deviations, and notations $\Delta \hat{E}$ and Δt are used to define the standard deviations as uncertainties, $\Delta \hat{E} = \sqrt{\langle \hat{E}^2 \rangle - \langle \hat{E} \rangle^2}$. ∎

For many-electron atoms, an atom in its normal (electrically neutral) state has Z electrons and Z protons. Here, Z is the atomic number, and for boron, carbon, and nitrogen, $Z = 5, 6$, and 7, respectively. The total electric charge of atoms is zero because the neutron has no charge while the proton and electron charges have the same magnitude but opposite sign. For the hydrogen atom, denoting the distance that separates the electron and proton by r, the Coulomb potential is $\Pi(r) = -e^2/(4\pi\varepsilon_0 r)$.

The radial attractive Coulomb potential felt by the single electron due to the nucleus having a charge Ze is

$$\Pi(r) = -Z(r)e^2/(4\pi\varepsilon_0 r),$$

where $Z(r) \to Z$ as $r \to 0$ and $Z(r) \to 1$ as $r \to \infty$.

By evaluating the average value for the radius of the shell, the effective nuclear charge Z_{eff} is found. The common approximation to calculate the total energy of an electron in the outermost populated shell is $E_n = -\frac{m_e Z_{eff}^2 e^4}{32\pi^2\varepsilon_0^2\hbar^2 n^2}$, and $E_n = -2.17 \times 10^{-18}\frac{Z_{eff}^2}{n^2}$ J.

The effective nuclear charge Z_{eff} is derived using the electron configuration. For boron, carbon, nitrogen, silicon, and phosphorus, three commonly used Slater, Clementi, and Froese–Fischer Z_{eff} are 2.6, 2.42, and 2.27 (for B); 3.25, 3.14, and 2.87 (for C); 3.9, 3.83, and 3.46 (for N); 4.13, 4.29, and 4.48 (for Si); 4.8, 4.89, and 5.28 (for P).

Taking note of the electron configurations for the earlier mentioned atoms, one concludes that ΔE could be on the order of $\sim 1 \times 10^{-19}$ to 1×10^{-18} J. If one supplies the energy greater than E_n to the electron, the energy excess will appear as kinetic energy of the free electron. The transition energy should be adequate to excite electrons. For different atoms and molecules with different exited states, as prospective *solid* ^{ME} devices, the transition (switching) energy can be estimated to be $\sim 1 \times 10^{-18}$ to 1×10^{-19} J. This energy estimate is valid for biomolecular and *fluidic* ^{M} devices.

The quantization of the orbital angular momentum of the electron leads to a quantization of the electron total energy. The space quantization permits only quantized values of the angular momentum component in a specific direction. The magnitude L_μ of the angular momentum of an electron in its orbital motion around the center of an atom and the z component L_z are

$$L_\mu = \sqrt{l(l+1)}\hbar \quad \text{and} \quad L_z = m_l\hbar,$$

where l is the orbital quantum number, and m_l is the magnetic quantum number, which is restricted to integer values $-l, -l+1, \ldots l-1, l$ (e.g., $|m_l| \le l$). If a magnetic field is applied, the energy of the atom will depend on the alignment of its magnetic moment with the external magnetic field.

In the presence of a magnetic field $\mathbf{B}$, the energy levels of the hydrogen atom are

$$E_n = -\frac{m_e e^4}{32\pi^2\varepsilon_0^2\hbar^2 n^2} - \boldsymbol{\mu}_L \cdot \mathbf{B},$$

where $\boldsymbol{\mu}_L$ is the orbital magnetic dipole moment, $\boldsymbol{\mu}_L = -\frac{e}{2m_e}\mathbf{L}, \mathbf{L} = \mathbf{r} \times \mathbf{p}$.

Let $\mathbf{B} = B_z \mathbf{z}$. One finds $E_n = -\frac{m_e e^4}{32\pi^2\varepsilon_0^2\hbar^2 n^2} + \frac{e}{2m_e}\mathbf{L}\cdot\mathbf{B} = -\frac{m_e e^4}{32\pi^2\varepsilon_0^2\hbar^2 n^2} + \frac{e}{2m_e}B_z L_z = -\frac{m_e e^4}{32\pi^2\varepsilon_0^2\hbar^2 n^2} + \frac{e}{2m_e}B_z m_l\hbar$.

If the electron is in an $l = 1$ orbit, the orbital magnetic dipole moment is $\mu_L = \frac{e\hbar}{2m_e} = 9.3 \times 10^{-24}$ J/T $= 5.8 \times 10^{-5}$ eV/T. Hence, if the magnetic field is changed by 1 T, an atomic energy level changes by $\sim 10^{-4}$ eV. The *switching* energy required to ensure the transitions between distinct *microscopic* states is straightforwardly derived using the wave function and allowed discrete energies.

Example 6.4.5

Consider a 1,3-butadiene molecule

The four delocalized π-electrons are assumed to move freely over the four-carbon-atom framework. Neglecting three-dimensional configuration, one may perform one-dimensional analysis. Solving the Schrödinger equation for a particle in the box $-\frac{\hbar^2}{2m_e}\nabla^2\Psi(x) + \Pi(x)\Psi(x) = E\Psi(x)$ with an infinite square well potential

$$\Pi(x) = \begin{cases} 0 & \text{for } 0 \le x \le L \\ \infty & \text{otherwise} \end{cases},$$

the wave function $\Psi_n(x)$ and allowed discrete energies are found [7]. In particular, $\Psi_n(x) = \sqrt{\frac{2}{L}}\sin\left(\frac{n\pi}{L}x\right)$ and $E_n = \frac{\hbar^2\pi^2}{2m_e L^2}n^2$. The state of the lowest energy is called the *ground state*.

The $C_1 = C_2$, $C_2 - C_3$, and $C_3 = C_4$ bond lengths are 0.1467, 0.1349, and 0.1467 nm, respectively. The electron wave function extends beyond the terminal carbons. We add 1/2 bond length at each end. Hence, $L = 0.575$ nm.

The π-electron density is concentrated between carbon atoms C_1 and C_2, as well as C_3 and C_4, because the predominant structure of butadiene has double bonds between these two pairs $C_1 = C_2$ and $C_3 = C_4$. Each double bond consists of a π-bond, in addition to the underlying σ-bond. One must also consider the residual π-electron density between C_2 and C_3. Thus, butadiene should be described as a resonance hybrid with two contributing structures $CH_2 = CH - CH = CH_2$ (dominant structure) and $^\circ CH_2 = CH - CH = CH_2^\circ$ (secondary structure). The lowest unoccupied molecular orbital (LUMO) in butadiene corresponds to the $n = 3$ particle-in-a-box state. Neglecting electron–electron interaction, the longest-wavelength (lowest-energy) electronic transition occurs at $n = 2$ which is the highest occupied molecular orbital (HOMO). This is visualized as

$$n = 3 \underline{\quad\quad} \text{LUMO} \underline{\;-\!\!\bullet\!-\;}$$
$$n = 2 \underline{\;-\!\!\bullet\!\!\bullet\!-\;} \text{HOMO} \underline{\;-\!\!\bullet\!\!\underline{\quad\quad}}$$
$$n = 1 \underline{\;-\!\!\bullet\!\!\bullet\!-\;} \quad\quad\quad \underline{\;-\!\!\bullet\!\!\bullet\!-\;}.$$

The HOMO→LUMO transition corresponds to $n \to (n+1)$. The energy difference between HOMO and LUMO is $\Delta E = E_3 - E_2 = \frac{\hbar^2\pi^2}{2m_e L^2}(3^2 - 2^2) = \frac{h^2}{8m_e L^2}(3^2 - 2^2) = 9.11 \times 10^{-19}$ J. From $\Delta E = hc/\lambda$, one finds the Compton wavelength $\lambda = 218$ nm. Performing the experiments, it is found that the maximum of the first electronic absorption band occurs at 210 nm. Hence, the use of quantum theory provides one with accurate results.

To enhance the accuracy, consider a rectangular $L_x \times L_y \times L_z$ 3D infinite-well box with

$$\Pi(x,y,z) = \begin{cases} 0 & \text{for } 0 \le x \le L_x,\ 0 \le y \le L_y,\ 0 \le z \le L_z \\ \infty & \text{otherwise} \end{cases}.$$

One solves a time-independent Schrödinger equation

$$-\frac{\hbar^2}{2m_e}\nabla^2\Psi(x,y,z) + \Pi(x,y,z)\Psi(x,y,z) = E\Psi(x,y,z).$$

We apply the separation of variables concept expressing the wave function as

$$\Psi(x,y,z) = X(x)Y(y)Z(z) \quad \text{and} \quad E = E_x + E_y + E_z.$$

One has $-\frac{\hbar^2}{2m_e}\frac{d^2X}{dx^2} = E_x X$, $-\frac{\hbar^2}{2m_e}\frac{d^2Y}{dy^2} = E_y Y$ and $-\frac{\hbar^2}{2m_e}\frac{d^2Z}{dz^2} = E_z Z$.

The general solutions are $X(x) = A_x\sin k_x x + B_x\cos k_x x$, $Y(y) = A_y\sin k_y y + B_y\cos k_y y$, and $Z(z) = A_z\sin k_z z + B_z\cos k_z z$. Here, $k_x^2 = \frac{2m_e}{\hbar^2}E_x$, $k_y^2 = \frac{2m_e}{\hbar^2}E_y$, and $k_z^2 = \frac{2m_e}{\hbar^2}E_z$.

Taking note of the boundary conditions, one finds $B_x = B_y = B_z = 0$ and $k_x L_x = n_x\pi$, $k_y L_y = n_y\pi$, and $k_z L_z = n_z\pi$. Normalizing the wave function, we obtain three-dimensional eigenfunctions as

$$\Psi_{n_x,n_y,n_z}(x,y,z) = \sqrt{\frac{8}{L_x L_y L_z}}\sin\left(\frac{n_x\pi}{L_x}x\right)\sin\left(\frac{n_y\pi}{L_y}y\right)\sin\left(\frac{n_z\pi}{L_z}z\right),\quad n_x, n_y, n_z = 1, 2, 3, \ldots$$

The allowed energies are found to be

$$E_{n_x,n_y,n_z} = \frac{h^2}{8m_e}\left(\frac{n_x^2}{L_x^2} + \frac{n_y^2}{L_y^2} + \frac{n_z^2}{L_z^2}\right) = \frac{\hbar^2\pi^2}{2m_e}\left(\frac{n_x^2}{L_x^2} + \frac{n_y^2}{L_y^2} + \frac{n_z^2}{L_z^2}\right) = \frac{\hbar^2 k^2}{2m_e},$$

where k is the magnitude of the wave vector $\mathbf{k}$, $\mathbf{k} = (k_x, k_y, k_z)$, $k_x = n_x\pi/L_x$, $k_y = n_y\pi/L_y$, and $k_z = n_z\pi/L_z$. Taking note of the wave vector, we conclude that each state occupies a volume $\pi^3/L_x L_y L_z = \pi^3/V$ of a k-space.

Suppose a system consists of N atoms, and each atom contributes M free electrons. The electrons are identical *fermions* that satisfy the Pauli exclusion principle. Thus, only two electrons can occupy any given state. Furthermore, electrons fill one octant of a sphere in k-space, whose radius is $k_R = (3NM\pi^2/V)^{1/3} = (3\rho\pi^2)^{1/3}$. The expression for k_R is derived taking note of $\frac{1}{8}\frac{4}{3}\pi k_R^3 = \frac{1}{2}NM\frac{\pi^3}{V}$. Here, ρ is the *free electron density* $\rho = NM/V$ (e.g., ρ is the number of free electrons per unit volume). The boundary separation of occupied and unoccupied states in k-space is called the Fermi surface, and the Fermi energy for a free electron gas is $E_F = \frac{\hbar^2}{2m_e}(3\rho\pi^2)^{2/3}$. The total energy of a free electron gas is

$$E_t = \frac{\hbar^2}{2m_e}\frac{V}{\pi^2}\int_0^{k_R} k^4 dk = \frac{\hbar^2 V}{10m_e\pi^2}k_R^5 = \frac{\hbar^2(3\pi^2 NM)^{5/3}}{10m_e\pi^2}V^{-2/3}.$$

The expression for E_t is found by taking note of the number of electron states in the shell

$$\frac{2\left(\frac{1}{2}\pi k^2 dk\right)}{\pi^3/V} = \frac{V}{\pi^2}k^2 dk$$

and the energy of the shell

$$dE = \frac{\hbar^2 k^2}{2m_e}\frac{V}{\pi^2}k^2 dk$$

(each state carries the energy $\frac{\hbar^2 k^2}{2m_e}$).

The analytic solutions exist for ellipsoidal, spherical, and other 3D wells for infinite, and some finite, potentials. The numerical solutions can be found for complex potential wells and barriers, as reported in Section 6.20. ∎

6.4.3 Device Switching Speed

The transition (switching) speed of ^{M} devices largely depends on the device physics, phenomena utilized, and other factors. One examines dynamic evolutions and transitions by applying the molecular dynamics theory, the Schrödinger equation, the time-dependent perturbation theory, numerical methods, and other concepts. The analysis of state transitions and interactions allows one to coherently study the controlled device behavior, evolution, and dynamics. The simplified steady-state analysis is also applied to obtain estimates. Considering the electron transport, one may assess the device features using the number of electrons. For example, for 1 nA current, the number of electrons that cross the molecule per second is $1 \times 10^{-9}/1.6022 \times 10^{-19} = 6.24 \times 10^9$, which is related to the device state transitions.

The maximum carrier velocity places an upper limit on the frequency response of semiconductor and molecular devices. The state transitions can be accomplished by a single photon or electron. Using the Bohr postulates, the average velocity of an optically exited electron is $v = \frac{Ze^2}{4\pi\varepsilon_0\hbar n}$. Taking note that for all atoms $Z/n \approx 1$s, one finds the orbital velocity of an optically exited electron to be $v = 2.2 \times 10^6$ m/sec, and $v/c \approx 0.01$. Considering an electron as a not relativistic particle, taking note of $E = mv^2/2$, we obtain the particle velocity as a function of energy as $v(E) = \sqrt{\frac{2E}{m}}$. Letting $E = 0.1$ eV $= 0.16 \times 10^{-19}$ J, one finds $v = 1.88 \times 10^5$ m/sec. Assuming 1 nm path length, the traversal (*transit*) time is $\tau = L/v = 5.33 \times 10^{-15}$ sec. Hence, ^{M} devices can operate at a high switching frequency. However, one may not conclude that the device switching frequency to be utilized is $f = 1/(2\pi\tau)$ due to device physics features (number of electrons, heating, interference, potential, energy, noise, etc.), system-level functionality, circuit specifications, etc.

Having estimated the $v(E)$ for ^{M}devices, the comparison to microelectronics devices is of interest. In silicon, the electron and hole velocities reach up to 1×10^5 m/sec at a very high electric field with the intensity 1×10^5 V/cm. The reported estimates indicate that particle velocity in ^{M}devices exceeds the carriers' saturated drift velocity in semiconductors.

6.4.4 Photon Absorption and Transition Energetics

Consider a rhodopsin, which is a highly specialized protein-coupled receptor that detects photons in the rod photoreceptor cell. The first event in the monochrome vision process, after photon (light) hits the rod cell, is the isomerization of the chromophore 11-*cis*-retinal to all-*trans*-retinal. When an atom or molecule absorbs a photon, its electron can move to the higher-energy orbital, and the atom or molecule makes a transition to a higher-energy state. In retinal, absorption of a photon promotes a π electron to a higher-energy orbital (e.g., there is a $\pi - \pi^*$ excitation). This excitation breaks the π component of the double bond allowing free rotation about the bond between carbon 11 and carbon 12. This isomerization, which corresponds to switching, occurs in a picoseconds range.

The energy of a single photon is found as $E = hc/\lambda$, where λ is the wavelength. The maximum absorbance for rhodopsin is 498 nm. For this wavelength, one finds $E = 4 \times 10^{-19}$ J. This energy is sufficient to ensure transitions and functionality.

It is important to emphasize that the photochemical reaction changes the shape of the retinal, causing a conformational change in the opsin protein, which consists of 348 amino acids covalently linked together to form a single chain. The sensitivity of the eye photoreceptor is one photon, and the energy of a single photon, which is $E = 4 \times 10^{-19}$ J, ensures the functionality of a molecular complex of $\sim$ 5000 atoms that constitute 348 amino acids. We derived the excitation energy (signal energy) which is sufficient to ensure state transitions and processing. This provides conclusive evidence that $\sim 1 \times 10^{-19}$ to 1×10^{-18} J of energy is required to guarantee the state transitions for complex molecular aggregates.

6.4.5 Processing Performance Estimates

Reporting the performance estimates, we focus on molecular electronics, basic physics and envisioned solutions. The 3D-centered topology/organization of envisioned *solid* and *fluidic* devices and systems are analogous to the topology/organization of ^{BM}PPs. Aggregated brain neurons perform superb information processing, perception, learning, robust reconfigurable networking, memory storage, and other functions. The number of neurons in the human brain is estimated to be $\sim$ 100 billions, mice and rats have $\sim$ 100 millions of neurons, while honeybees and ants have $\sim$ 1 million neurons. Bats use echolocation sensors for navigation, obstacle avoidance, and hunting. By processing the sensory data, bats can detect 0.1% frequency shifts caused by the Doppler effect. They distinguish echoes received $\sim$ 100 μsec apart. To accomplish these, as well as to perform shift compensation and transmitter/receiver isolation, real-time signal/data processing should be accomplished within at least microseconds. Flies accomplish a real-time precisely coordinated motion due to remarkable actuation and an incredible visual system which maps the relative motion using the retinal photodetector arrays. The information from the visual system and sensors is transmitted and processed within the nanoseconds range requiring μW of power. The dimension of the brain neuron is $\sim$ 10 μm, and the density of neurons is $\sim$ 100000 neurons/mm^3. The review of electrical excitability of neurons is reported in [17].

The biophysics and mechanisms of biomolecular information and signal/data processing are not fully comprehended. Biomolecular state transitions are accomplished with a different rate. The electrochemo-mechanical biomolecular transformations (propagation of biomolecules and ions through the synaptic cleft and membrane channels, protein folding, binding/unbinding, etc.) could require microseconds. In contrast, photon- and electron-induced transitions can be performed within femtoseconds. The energy estimates were documented obtaining the transition energy $\sim 1 \times 10^{-19}$ to 1×10^{-18} J.

Performing enormous information processing tasks with immense performance that are far beyond the foreseen capabilities of envisioned parallel vector processors (which perform signal/data processing), the human brain consumes merely $\sim$ 20 W. Only some of this power is required to accomplish information

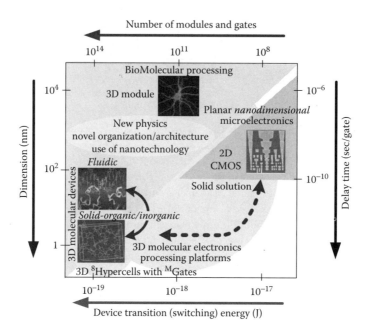

FIGURE 6.4 Toward molecular electronics and processing/memory platforms: (1) revolutionary advancements: from 2D microelectronics to 3D molecular electronics; (2) evolutionary developments: from [BM]PPs to *solid* and *fluidic* molecular electronics and processing.

and signal/data processing. This contradicts some postulates of slow processing, immense delays, high energy/power requirements, low switching speed, and other hypotheses reported in [18–21]. The human retina has 125 million rod cells and 6 million cone cells, and an enormous amount of data, among other tasks, is processed in real-time. Real-time 3D image processing, ordinarily accomplished even by primitive vertebrates and insects that consume less than 1 μW to perform information processing, cannot be performed by envisioned processors with trillions of transistors, a device switching speed of 1 THz, a circuit speed of 10 GHz, device switching energy of 1×10^{-16} J, a writing energy of 1×10^{-16} J/bit, a read time of 10 nsec, etc. Molecular devices can operate with the estimated transition energy $\sim 1 \times 10^{-19}$ to 1×10^{-18} J, discrete energy levels (ensuring multiple-valued logics) and femtosecond transition dynamicsguaranteeing exceptional device performance. These 3D-topology [M]devices result in the ability to design super-high-performance processing and memory platforms within 3D organizations, and enabling architectures ensuring unprecedented capabilities such as massive parallelism, robustness, reconfigurability, etc.

Distinct performance measures, estimates, and indexes are used. For profoundly different paradigms (microelectronics versus molecular electronics, distinguished by distinct topologies, organizations, and architectures), Figure 6.4 reports some baseline performance estimates—e.g., transition (switching) energy, delay time, dimension, and number of modules/gates. It was emphasized that the device physics and system organization/architecture are dominating features compared to the dimensionality or number of devices. Due to limited basic/applied/experimental results, as well as attempts to use four performance variables (reported in Figure 6.4), some performance measures and projected estimates are expected to be refined. Molecular electronics and [M]ICs can utilize diverse molecular primitives and devices that: (1) Operate due to different physics, such as electron transport, electrostatic transitions, photon emission, conformational changes, etc.; (2) Exhibit distinct phenomena and effects. Therefore, biomolecular systems and *fluidic* and *solid* [M]devices will exhibit distinct performance. As demonstrated in Figure 6.4, advancements are envisioned towards 3D *solid* molecular electronics departing from [BM]PPs by utilizing a familiar solid-state microelectronics solution. In Figure 6.4, a 3D-topology neuron is represented as a biomolecular information processing/memory module that may consist of [M]devices.

6.5 Synthesis Taxonomy in Design of ^{M}ICs and Processing Platforms

Molecular architectronics is a paradigm in the devising and designing of preeminent ^{M}ICs and ^{M}PPs. This paradigm is based on:

- The discovery of novel topological/organizational/architectural solutions, as well as the utilization of new phenomena and capabilities of 3D molecular electronics at the system and device levels.
- The development and implementation of sound methods, technology-centric CAD, and SLSI design concurrently associated with bottom-up fabrication.

In design of ^{M}ICs, one faces a number of challenging tasks such as analysis, optimization, aggregation, verification, reconfiguration, validation, evaluation, etc. Technology-centric synthesis and design at the device and system levels must be addressed, researched, and solved by making use of the CAD-supported SLSI design of super-complex ^{M}ICs. Molecular electronics provides a unique ability to implement signal/data processing hardware within 3D organizations and enabling architectures. This guarantees massive parallel distributed computations, reconfigurability and large-scale data manipulations, ensuring super-high-performance computing and processing. The combinational and memory ^{M}ICs should be designed as aggregated ^{N}hypercells and molecular memories [16]. The device physics is reported in this chapter for 3D-topology *solid* and *fluidic* molecular devices. Those ^{M}devices are aggregated as ^{M}gates which must guarantee the desired performance and functionality of ^{N}hypercells.

Various design tasks for 3D ^{M}ICs are not analogous to the CMOS-centered design, planar layout, placement, routing, interconnect, and other tasks that were successfully solved. Conventional VLSI/ULSI design flow is based on the well-established system specifications, functional design, conventional architecture, verification (functional, logic, circuit, and layout), as well as CMOS fabrication technology. The CMOS technology utilizes the two-dimensional topology of conventional gates with FETs and BJTs. For ^{M}ICs, device- and system-level technology-centric design must be performed using novel methods. Figure 6.4 illustrates the proposed 3D molecular electronics departing from two-dimensional multilayer CMOS-centered microelectronics. To synthesize ^{M}ICs, we propose to utilize a unified top-down (system level) and bottom-up (device/gate level) synthesis taxonomy within an *x*-domain flow map, as reported in Figure 6.5. The core 3D design themes are integrated within four domains:

- Devising with validation
- Analysis–evaluation
- Design–optimization
- Molecular fabrication

As reported in Figure 6.5, the synthesis and design of 3D ^{M}ICs and ^{M}PPs should be performed by utilizing a bidirectional flow-map. Novel design, analysis, and evaluation methods must be developed. Design in 3D space is radically different compared with VLSI/ULSI due to novel 3D topology/organization, enabling architectures, new phenomena utilized, enhanced functionality, enabling capabilities, complexity, technology-dependence, etc. The unified top-down/bottom-up synthesis taxonomy should be coherently supported by developing innovative solutions to carry out a number of major tasks such as:

- Devising and designing ^{M}devices, ^{M}gates, ^{N}hypercells, and networked ^{N}hypercells aggregates that form ^{M}ICs
- Developing new methods in design and verification of ^{M}ICs
- Analyzing and evaluating performance characteristics
- Developing technology-centric CAD to concurrently support design at the system and device/gate levels

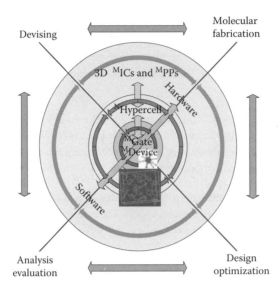

FIGURE 6.5   Top-down and bottom-up synthesis taxonomy within an x-domain flow-map.

The reported unified synthesis taxonomy integrates:

- *Top-Down Synthesis*: Devise super-high-performance molecular processing and memory platforms implemented by designed ^{M}ICs within 3D organizations and enabling architectures. These 3D ^{M}ICs are implemented as aggregated $^\aleph$hypercells composed from ^{M}gates engineered from ^{M}devices (see Figures 6.6[a] and 6.6[b]).
- *Bottom-Up Synthesis*: Engineer functional 3D-topology ^{M}devices that compose ^{M}gates in order to form $^\aleph$hypercells (for example, multiterminal *solid* ^{ME}devices are engineered as molecules arranged from atoms).

The proposed synthesis taxonomy utilizes a number of innovations at the system and device levels. In particular, (1) innovative architecture, organization, topology, aggregation and networking in 3D; (2) novel enhanced-functionality ^{M}devices that form ^{M}gates, $^\aleph$hypercells, and ^{M}ICs; (3) Unique phenomena, effects, and solutions (tunneling, parallelism, etc.); (4) bottom-up fabrication; (5) CAD-supported technology-centric SLSI design.

Super-high-performance molecular processing and memory platforms can be synthesized using $^\aleph$hypercells D_{ijk} within 3D topology/organization, which are analogous to the 3D topology/organization of biomolecules and their aggregates. A vertebrate brain is of the most interest. However, not only vertebrates, but also single-cell bacteria, possess superb 3D ^{BM}PPs. We focus major efforts on solid molecular electronics due to a limited knowledge of the baseline processes, effects, mechanisms, and functionality of ^{BM}PPs. Insufficient knowledge makes it virtually impossible to comprehend and prototype biomolecular devices that operate utilizing different phenomena and concepts, compared to *solid* ^{ME}devices. Performance and baseline characteristics of *solid* ^{ME}devices are drastically affected by the molecular structures, aggregation, bonds, atomic orbitals, electron affinity, ionization potential, arrangement, sequence, assembly, folding, side groups, and other features. Molecular devices and ^{M}gates must ensure desired transitions, switching, logics, electronic characteristics, performance, etc. Enhanced functionality, high switching frequency, superior density, expanded utilization, low power, low voltage, desired I–V characteristics, noise immunity, robustness, integration and other characteristics can be ensured through a coherent design. In ^{M}devices, performance and characteristics can be changed and optimized by utilizing and controlling distinct transitions, states and parameters. For solid ^{ME}devices, the number of quantum wells/barriers, their width, energy profile, tunneling length, dielectric constant, and other key features can be adjusted and optimized by engineering molecules with specific atomic sequences, bonds, side groups, etc. The goal is to ensure optimal achievable performance at the device, module, and system levels. The performance

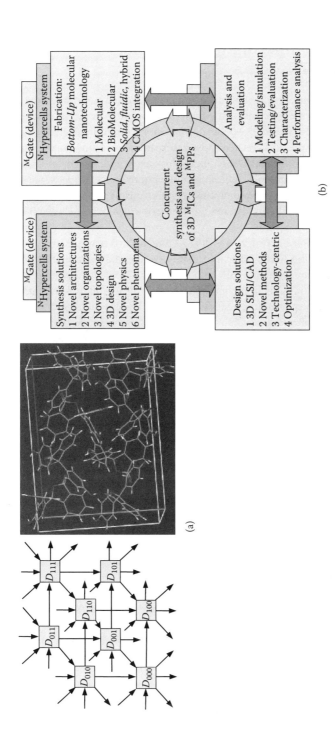

FIGURE 6.6 (a) Three-dimensional molecular electronics: aggregated ^{N}hypercells D_{ijk} composed from ^{M}gates that integrate multiterminal *solid* ^{ME} devices engineered from atomic complexes. (b) Concurrent synthesis and design at system, module, and gate (device) levels.

should be assessed by using the quantitative and qualitative performance measures, indexes, and metrics. The reported interactive synthesis taxonomy is coherently integrated within all tasks, including devising ^{M}devices, discovering 3D organization, synthesizing enabling architectures, designing ^{M}ICs, etc.

6.6 Neuroscience: Information Processing and Memory Postulates

Biosystems detect various stimuli, and the information is processed through complex electrochemo-mechanical phenomena and mechanisms at the molecular and cellular levels. Biosystems accomplish cognition, learning, perception, knowledge generation, storing, computing, coding, transmission, communication, adaptation, and other tasks related to the information processing. Appreciating neuroscience, neurophysiology, cellular biology, and other disciplines, this section addresses open-ended problems from engineering and technology standpoints, reflecting some author's inclinations. Due to a lack of conclusive evidence, no agreement exists regarding baseline mechanisms and phenomena (electrochemical, optochemical, electromechanical, thermodynamic, and other), which ultimately result in signal/data and information processing in biosystems.

The human brain is a complex network of $\sim 1 \times 10^{11}$ aggregated neurons with more than 1×10^{14} synapses. Action potentials, and likely, other information-containing signals, are transmitted to other neurons by means of very complex and not fully comprehended *axo-dendritic, dendro-axonic, axo-axonic,* and *dendro-dendritic* interactions utilizing axonic and dendritic structures. It is the authors' beliefs that a neuron, as a complex system, performs information processing, memory storage, and other tasks utilizing electrochemomechanically-induced interactions and transitions. For example, biomolecules (neurotransmitters and enzymes) and ions propagate in the synaptic cleft, membrane channels, and cytoplasm. This controlled propagation of *information carriers* result in charge distribution, interaction, release, binding, unbinding, bonding, switching, folding and other state transitions and events. The electrochemomechanical transitions and interactions of *information carriers* under electrostatic, magnetic, hydrodynamic, thermal and other fields (forces) were examined in [22]. Debates are ongoing concerning system- and device-level considerations, and neuronal aggregation, as well as fundamental phenomena observed, utilized, embedded, and exhibited by neurons and their organelles. There is no agreement on whether or not a neuron is a device (according to a conventional neuroscience postulate) or a system, or on how the information is processed, encoded, controlled, transmitted, routed, etc. The information processing and storage are far more complicated problems compared to data transmission, routing, communication, etc. Under these uncertainties, new theories, paradigms, and concepts have emerged.

By applying the possessed knowledge, it is a question of *whether it is possible to accomplish a coherent biomimetics (bioprototyping) and devise (discover - and - design) man-made bio-identical or bio-centered processing and memory platforms*. Unfortunately, even for signal/data processing platforms, it seems unlikely these objectives will be achieved in the near future. A great number of unsolved fundamental, applied, and technological problems remain. To some extent, a number of problems can be approached by examining and utilizing different biomolecular-centered processing postulates, concepts, and solutions. General and application-centric foundations are needed that do not rely on hypotheses, postulates, assumptions, and exclusive solutions that depend upon specific technologies, hardware, and fundamentals. Achievable technology-centric *solid* and *fluidic* molecular electronics are prioritized in this chapter due to not yet understood cellular phenomena and mechanisms in ^{BM}PPs. Some postulates, concepts, and new solutions are reported.

The anatomist Heinrich Wilhelm Gottfried Waldeyer-Hartz found that the nervous system consists of nerve cells in which there are no mechanical joints in between. In 1891, he used the word *neuron*. The cell body of a typical vertebrate neuron consists of the nucleus (soma) and other cellular organelles. Neuron branched projections (axons and dendrites) are packed with ~ 25 nm diameter microtubules which may play a significant role in signal/data transmission, communication, processing, and storage. The cylindrical wall of each microtubule is formed by 13 longitudinal protofilaments of tubuline molecules (e.g., altering α and β heterodimers). The cross-sectional representation of a microtubule is a ring of

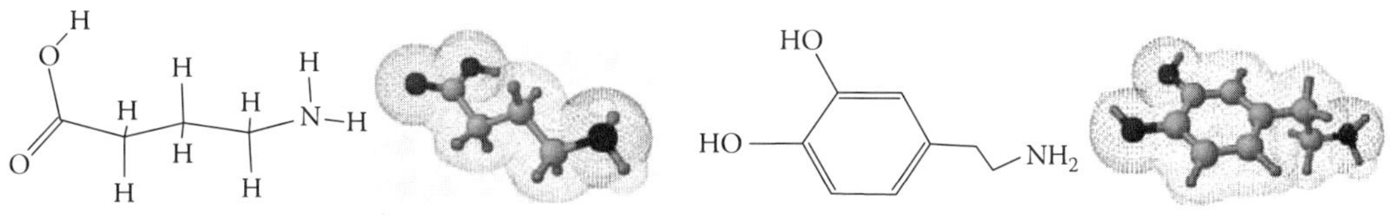

FIGURE 6.7 (see color insert) Gamma-aminobutyric acid and dopamine neurotransmitters.

13 distinct subunits. Numerous and extensively branched dendrite structures are believed to transmit information towards the cell body. The information is transmitted from the cell body through axon structures. The axon originates from the cell body and ends in numerous terminal branches. Each axon terminal branch may have thousands of synaptic axon terminals. These presynaptic axon terminals and postsynaptic dendrites establish the biomolecular-centered interface between neurons or between a neuron and target cells. Specifically, various neurotransmitters are released into the synaptic cleft and propagate to the postsynaptic membrane. It also should be emphasized that within a complex microtubule network are nucleus-associated-microtubules.

Neurotransmitter molecules are (1) synthesized (reprocessed) and stored into vesicles in the presynaptic cell; (2) released from the presynaptic cell, propagate, and bind to receptors on one or more postsynaptic cells; (3) removed and/or degraded. More than 100 known neurotransmitters were studied, while the total number of neurotransmitters is unknown. Neurotransmitters are classified as small-molecule neurotransmitters and neuropeptides (composed from 3 to 36 amino acids). It is reported that small-molecule neurotransmitters mediate rapid synaptic actions, while neuropeptides tend to modulate slower ongoing synaptic functions. As an example, the structure and 3D configuration of the gamma-aminobutyric acid (GABA) and dopamine neurotransmitters are illustrated in Figures 6.7(a) and (b).

Conventional neuroscience theory postulates that in neurons the information is transmitted by action potentials, which result due to ionic fluxes that are controlled by complex cellular mechanisms. The ionic channels are opened and closed by the binding and unbinding of neurotransmitters released from the synaptic vesicles (located at the presynaptic axon sites). Neurotransmitters propagate through the synaptic cleft to the receptors at the postsynaptic dendrite, see Figure 6.8. According to conventional theories, the binding/unbinding of neurotransmitters in multiple synaptic terminals results in the selective opening/closing of membrane ionic channels, and the flux of ions causes the action potential which is believed to contain and carry out information. At the cellular level, a wide spectrum of phenomena and mechanisms are not sufficiently studied or remain unknown. For example, the production, activation, reprocessing, binding, unbinding, and propagation of neurotransmitters, even though they have been studied for decades, are not adequately comprehended. Debates abound on the role of microtubules and microtubule associated proteins. With limited knowledge on signal transmission and communication in neurons, in addition to the action potential, other stimuli of different origin may exist and should be examined. Unfortunately, no sound explanation, justification, and validation exists for information processing, memory storage, and other related tasks.

The binding and unbinding of neurotransmitters and ions cause electrochemomechanically induced transitions at the molecular and cellular levels due to charge variation, force generation, moment transformation, potential change, orbital overlap variation, vibration, resonance, folding, and other effects. For neurons and envisioned *synthetic fluidic* devices/modules, these transitions ultimately can result in information processing (with other directly related tasks) and memory storage. For example, a biomolecule (protein) can be used as a *biomolecular electrochemomechanical switch* utilizing the conformational changes, or as a *biomolecular electronic switch* using the charge changes. *Axo-dendritic* organelles with microtubules and microtubule-associated proteins (MAPs), as well as the propagating ions and neurotransmitters in a synapse, are schematically depicted in Figure 6.8. There are axonic and dendritic microtubules, MAPs, synapse-associated proteins (SAPs), endocytic proteins, etc. Distinct pre- and post-synaptic SAPs have been identified and examined. Large multidomain scaffold proteins, including SAP and MAP families, form the framework of the presynaptic active zones (AZ), postsynaptic density (PSD), endocytic zone

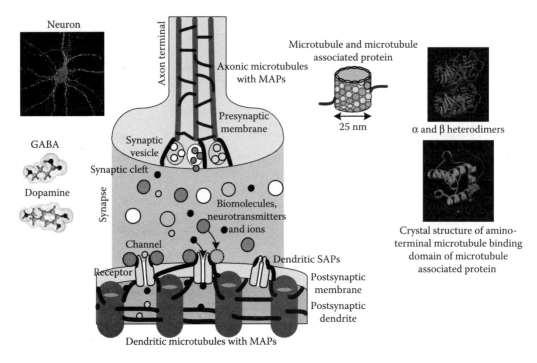

FIGURE 6.8 (see color insert) Schematic representation of the *axo-dendritic* organelles with AZ and PSD protein assemblies: (1) binding and unbinding of the *information* carriers (biomolecules, neurotransmitters, and ions) result in the state transitions leading to information processing and memory storage; (2) the 3D-topology lattice of SAPs and microtubules with MAPs ensures reconfigurable 3D organization utilizing *routing* carriers.

(EnZ), and exocytic zone (ExZ) assemblies. Numerous protein interactions occur between AZ, PSD, EnX, and ExZ proteins. With a high degree of confidence, one may conclude that these are the processing- and memory-associated state transitions in 3D extracellular and intracellular protein assemblies.

In a microtubule, each tubulin dimer ($\sim 8 \times 4 \times 4$ nm) consists of positively and negatively charged α-tubulin and β-tubulin (see Figure 6.8). Each heterodimer made from ~ 450 amino acids, and each amino acid contains ~ 15 to 20 atoms. Tubulin molecules exhibit different geometrical conformations (states). The tubulin dimer subunits are arranged in a hexagonal lattice with different chirality. The interacting negatively charged C–termini extend outward from each monomer (protrude perpendicularly to the microtubule surface), attracting positive ions from the cytoplasm. The intra-tubulin dielectric constant is $\varepsilon_r = 2$, while outside the microtubule $\varepsilon_r = 80$. The MAPs are proteins that interact with the microtubules of the cellular cytoskeleton. A large variety of MAPs have been identified. These MAPs accomplish different functions such as stabilizing/destabilizing microtubules, guiding microtubules towards specific cellular locations, interconnecting microtubules and proteins, etc. Microtubule-associated proteins bind directly to the tubulin monomers. Usually, the carboxyl-terminus -COOH (C-terminal domain) of the MAP interacts with tubulin, while the amine-terminus -NH$_2$ (N-terminal domain) binds to organelles, intermediate filaments, and other microtubules. Microtubule-MAPs binding is regulated by phosphorylation. This is accomplished through the function of the microtubule-affinity-regulating-kinase protein. Phosphorylation of the MAP by the microtubule-affinity-regulating-kinase protein causes the MAP to detach from any bound microtubules. MAP1a and MAP1b, found in axons and dendrites, bind to microtubules differently than other MAPs, utilizing the charge-induced interactions. While the C-terminals of MAPs bind the microtubules, the N-terminals bind other parts of the cytoskeleton or the plasma membrane. MAP2 is found mostly in dendrites, while the tau-MAP is located in the axon. These MAPs have a C-terminal microtubule-binding domain and variable N-terminal domains projecting outwards interacting with other proteins. In addition to MAPs, many other proteins affect microtubule behavior. These proteins are not considered to be MAPs, however, because they do not bind directly to tubulin monomers, but affect the functionality of microtubules and MAPs. The mechanism of the so-called synaptic plasticity and the role of proteins, neurotransmitters, and ions, which likely affect learning and memory, are not fully comprehended.

An innovative hypothesis of microtubule-assisted quantum information processing is reported in [23]. The authors consider microtubules as assemblies of oriented dipoles and postulate that [23]: (1) Conformational states of individual tubulins within neuronal microtubules are determined by mechanical London forces within the tubulin interiors, which can induce a conformational quantum superposition; (2) In superposition, tubulins communicate/compute with entangled tubulins in the same microtubule, with other microtubules in the same neuron, with microtubules in neighboring neurons, and through macroscopic regions of the brain by tunneling through gap junctions; (3) Quantum states of tubulins/microtubules are isolated from environmental decoherence by biological mechanisms, such as quantum isolation, ordered water, Debye layering, coherent pumping, and quantum error correction; (4) Microtubule quantum computations/superpositions are tuned by MAPs during a classical liquid phase which alternates with a quantum solid-state phase of actin gelation; (5) Following periods of preconscious quantum computation, tubulin superpositions reduce or collapse by Penrose quantum gravity *objective reduction*; (6) The output states which result from the *objective reduction* process is nonalgorithmic (noncomputable) and governs neural events such as the binding of MAPs, and regulating synapses and membrane functions; (7) The reduction or self-collapse in the *orchestrated objective reduction* model is a *conscious moment*, connected to Penrose's quantum gravity mechanism, which relates the process to fundamental space-time geometry. The results reported in [23] suggest that tubulins can exist in quantum superposition of two or more possible states until the threshold for quantum state reduction (quantum gravity mediated by *objective reduction*) is reached. A double-well potential, according to [23], enables the inter-well quantum tunneling of a single electron and spin states because the energy is greater than the thermal fluctuations. Debates continue on the soundness of this concept, examining the feasibility of utilization of quantum effects in tubulin dimers, the relatively high width of the well (the separation is ~ 1.5 nm), decoherence, noise, etc.

In neurons, biomolecules (neurotransmitters and enzymes) and ions can be the *information* (processing) and *routing* carriers. Publications [22,24] suggest that signal and data processing (computing, logics, coding, and other tasks), memory storage, memory retrieval, and information processing could (potentially) be accomplished by using neurotransmitters and ions as the *information carriers*. There are distinct *information* carriers e.g., *activating*, *regulating*, and *executing*.

Control of released specific neurotransmitters (*information carriers*) in a particular synapse and their binding to the receptors results in state transitions, ensuring cellular-level signal/data/information processing and memory mechanisms. The processing and memory may be robustly reconfigured utilizing *routing* carriers that potentially ensure networking. We thus state the following original major postulates:

- Certain biomolecules and ions are the *activating*, *regulating*, and *executing information* carriers that interact with SAPs, MAPs, and other cellular proteins. The controlled binding/unbinding of *information* carriers leads to biomolecular-assisted electrochemomechanical state transitions (folding, bonding, etc.), affecting the processing- and memory-associated transitions in protein assemblies. This ultimately results in processing and memory storage. Typifying examples include the following: (1) the binding/unbinding of *information* carriers ensures a combinational logics equivalent to *on* and *off* switching analogous to the AND- and OR-centered logics (see Figure 6.20); (2) charge change is analogous to the functionality of the molecular storage capacitor (see Figure 6.16).
- Specific biomolecules and ions are the *routing* carriers that interact with SAPs, MAPs, and other proteins. The binding and unbinding of *routing* carriers results in electromechanical state transitions, ensuring robust reconfiguration, networking, adaptation, and interconnect.
- Information processing and memories may be accomplished on a high radix by means of electromechanically-induced transitions/interactions/events in specific neuronal protein complexes.
- Presynaptic AZ and PSD (comprised from SAPs, MAPs, and other proteins), as well as microtubules, form a biomolecular 3D-assembly (organization) within a reconfigurable processing-and-memory neuronal architecture.

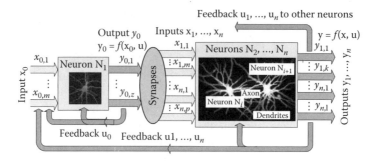

FIGURE 6.9   Input–output representation of $(n+1)$ aggregated neurons with *axo–dendritic* inputs and *dendro–axonic* outputs.

A biomolecular processing includes various tasks, such as communication, signaling, routing, reconfiguration, coding, etc. Consider biomolecular processing between neurons using the *axo-dendritic* inputs and *dendro-axonic* outputs. We do not specify the information-containing signals (action potential, polarization vector, phase shifting, folding modulation, vibration, switching, etc.) with possible corresponding cellular mechanisms which are due to complex biomolecular interactions and phenomena. The reported transitions can be examined using the *axo-dendritic* input vectors x_i (see Figure 6.9). For example, the inputs to neuron N_0 are $x_{0,1}, \ldots, x_{0,m}$, and $x_0 = [x_{0,1}, \ldots, x_{0,m}]$. The first neuron N_0 has m inputs (vector x_0) and z outputs (vector y_0). Spatially distributed $y_0 = [y_{0,1}, \ldots, y_{0,z}]$ furnish the inputs to neurons $N_1, N_2, \ldots, N_{n-1}, N_n$. The aggregated neurons $N_0, N_1, \ldots, N_{n-1}, N_n$ *process* the information by cellular transitions and mechanisms. The output vector y is $y = f(x)$, where f is the nonlinear function, and, for example, in the logic design of ICs, f is the *switching* function. To ensure robustness, reconfigurability, and adaptiveness, we consider the feedback vector u. Hence, the output of the neuron N_0 is a nonlinear function of the input vector x_0 and feedback vector $u = [u_0, u_1, \ldots, u_{n-1}, u_n]$, e.g., $y_0 = f(x_0, u)$. As the information is processed by N_0, it is fed to a neuronal aggregate $N_1, N_2, \ldots, N_{n-1}, N_n$. The neurotransmitters release—performed by all neurons—the *dendro-axonic* output y_i. As earlier emphasized, neurons have a branched dendritic tree with ending *axo-dendritic* synapses. The *fan-out* per neuron reaches 10,000. Figure 6.9 illustrates the 3D aggregation of $(n + 1)$ neurons with the resulting input–output maps $y_i = f(x_i, u)$. Dendrites may form *dendro–dendritic* interconnects, while in *axo–axonic* connects, one axon may terminate on the terminal of another axon and modify its neurotransmitter release.

Neurons, which perform various processing and memory tasks, are examined designing neuronal and integrated processor-and-memory bioinspired ^{M}PPs. Taking note of the *axo–dendritic* input and *dendro–axonic* output vectors x and y, the input–output mapping is schematically represented in Figure 6.10 for biomolecular, *fluidic* and *solid* molecular electronics. Taking note of the feedback vector u, which is a very important feature for robust reconfigurable (adaptive) processing, we have $y = f(x,u)$.

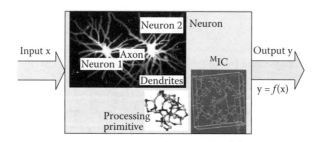

FIGURE 6.10   Input–output representation of aggregated neurons, processing molecular primitive and ^{M}IC.

6.7 Biomolecules and Ion Transport: Communication Energetics Estimates

Kinetic energy is the energy of motion, while the stored energy is called potential energy. Thermal energy is the energy associated with the random motion of molecules and ions, and therefore can be examined in terms of kinetic energy. Chemical reaction energy changes are expressed in calories, and 1 cal = 4.184 J. In cells, the directional motion of biomolecules and ions results in *active* and *passive* transport. For years, the analysis of neuronal activities has been largely focused on action potential.

Conventional neuroscience postulates that the neuronal communication is established by means of action potentials. A potential difference exists across the axonal membrane, and the resting potential is $V_0 = -0.07$ V. The voltage-gated sodium and potassium channels in the membrane result in the propagation of action potential with a speed of ~ 100 m/sec, and the membrane potential changes from $V_0 = -0.07$ V to $V_A = +0.03$ V. The ATP-driven pump restores the Na^+ and K^+ concentration to their initial values within $\sim 1 \times 10^{-3}$ sec, making the neuron ready to fire again, if triggered. Neurons can fire more than 1×10^3 times per second. Consider a membrane with the uniform thickness h. For the voltage differential, $\Delta V = (V_A - V_0)$ across the membrane, the surface charge density $\pm\rho_S$ inside/outside membrane is $\rho_S = \varepsilon E$. Here, ε is the membrane permittivity, $\varepsilon = \varepsilon_0 \varepsilon_r$, while E is the electric field intensity, $E = \Delta V/h$. The total active surface area is estimated as $A = \pi d L_A$, where d is the diameter, and L_A is the active length. The total number of ions that should propagate to ensure a single action potential is

$$n_I = \frac{A\rho_S}{q_I} = \frac{\pi d L_A \varepsilon_0 \varepsilon_r (V_A - V_0)}{q_I h},$$

where q_I is the ionic charge. Here, one recalls that for parallel-plate capacitors, the capacitance is $C = \varepsilon_0 \varepsilon_r A/h$, and the number of ions which flow per action potential is $n_I = Q/q_I$, where $Q = C\Delta V$.

Example 6.7.1

Let $d = 1 \times 10^{-5}$ m, $L_A = 1 \times 10^{-4}$ m, $\varepsilon_r = 2$, $V_A = 0.03$ V, $V_0 = -0.07$ V and $h = 8 \times 10^{-9}$ m. For Na^+ ions, $q_{Na} = e$. We have $n_{Na} = 4.34 \times 10^5$. Thus, 4.34×10^5 ions are needed to ensure $\Delta V = 0.1$ V. The synapses separation is $\sim 1\mu$m, and taking note that the single sodium pump maximum transport rate is ~ 200 Na^+ ions/sec and ~ 100 K^+ ions/sec, one may find that for the use L_A, the firing rate is ~ 1 spike/sec. ∎

The masses, diffusion coefficients (at 37^0C), and ionic radii of Na^+, Cl^-, K^+, and Ca^{2+} ions are $m_{Na} = 3.81 \times 10^{-26}$ kg, $m_{Cl} = 5.89 \times 10^{-26}$ kg, $m_K = 6.49 \times 10^{-26}$ kg, $m_{Ca} = 6.66 \times 10^{-26}$ kg, $r_{Na} = 0.95 \times 10^{-10}$ m, $r_{Cl} = 1.81 \times 10^{-10}$ m, $r_K = 1.33 \times 10^{-10}$ m, $r_{Ca} = 1 \times 10^{-10}$ m, $D_{Na} = 1.33 \times 10^{-9}$ m^2/s, $D_{Cl} = 2 \times 10^{-9}$ m^2/s, $D_K = 1.96 \times 10^{-9}$ m^2/s, and $D_{Ca} = 0.71 \times 10^{-9}$ m^2/s.

The instantaneous power is $P = dW/dt$, and using the force F and liner velocity v, one finds $P = Fv$. The output power can be found using the kinetic energy $\Gamma = 1/2 \, mv^2$, and $W = \Delta\Gamma$.

Consider a spherical particle with radius r that moves at velocity v in the liquid with viscosity μ. For the laminar flow, the Stokes's law gives the viscous friction (drag) force as $F_v = \eta v$, where η is the viscous friction (drag) coefficient, which is $\eta = 6\pi\mu r$. The inverse of the drag coefficient is called the mobility, $\mu_B = 1/\eta = 1/(6\pi\mu r)$.

The diffusion constant of a particle D is related to the mobility and the absolute temperature by the Einstein fluctuation dissipation theorem, which is given as $D = k_B T \mu_B$. Using the ionic radii of Na^+ and K^+ ions, for $\mu = 9.5 \times 10^{-4}$ N-s/m^2 at 37°C ($T = 310.15$ K), one calculates $D_{Na} = 2.52 \times 10^{-9}$ m^2/s and $D_K = 1.8 \times 10^{-9}$ m^2/s, which agree with the experimental values $D_{Na} = 1.33 \times 10^{-9}$ m^2/s and $D_K = 1.96 \times 10^{-9}$ m^2/s.

By regulating the ionic channels, the cell controls the ionic flow across the membrane. The membrane conductance g has been experimentally measured using an expression $I = gV$. It is found that for the sodium and potassium open channels, the conductance is in the range of $g = 2 \times 10^{-11}$ A/V. The current through the channel is estimated as $I = q_I J_I A_c$, where J_I is the ionic flux, $J_I = cv$; c is the ionic concentration; and A_c is the channel cross-sectional area.

The average velocity of an ion under the electrostatic field is estimated using the mobility and force as $v = \mu_B F = \mu_B q_I E = \mu_B q_I V/x$. Hence, one has $v = \frac{D}{k_B T} q_I \frac{V}{x}$. One finds the values for the velocity, force, and power. To transport a single ion, using the data reported, the estimated power is $\sim 1 \times 10^{-13}$ W. Taking note of the number of ions required to produce and amplify the action potential for the firing rate 100 spike/sec, and letting the instantaneous neuron utilization be 1%, hundreds of W are required to ensure communication only. It must be emphasized that binding/unbinding, production (reprocessing) of biomolecules, controlled propagation, and other cellular mechanisms require additional power.

The neuron energetics is covered in [25]. Using the longitudinal current, the intracellular longitudinal resistivity is found to be from 1×10^3 to 3×10^3 ohm-mm, while the channel conductance is 25 pS or $g = 2.5 \times 10^{-11}$ A/V. For a 100-μm segment with a 2-μm radius, the longitudinal resistance is found to be 8×10^6 ohm [25]. The membrane resistivity is 1×10^6 ohm-mm^2. To cross the ionic channel, the energy is $q_I V$. Taking note of the number of ions to generate a spike, the switching energy of a neuron is estimated to be $\sim 1 \times 10^{-14}$ J/spike. The cellular energetics is reported, taking note of the conventional consideration. The action potentials, ionic transport, spike generation, and other cellular mechanisms exist, guaranteeing the functionality and specificity of cellular processes. However, the role and specificity of some phenomena, effects, and mechanisms may be revisited and coherently examined from the communication energetics and other perspectives. Recently, the research in synaptic plasticity has culminated in meaningful results departing from the past oversimplified analysis. However, the complexity of the processes and mechanisms is overwhelming.

6.8   Applied Information Theory and Information Estimates with Applications to Biomolecular Processing and Communication

Considering a neuron as a switching device, which could be an oversimplified hypothesis, the interconnected neurons are postulated to be exited only by the action potentials I_i. Neurons are modeled as a spatio-temporal lattice of aggregated processing elements (neurons) by the second-order linear differential equation [19,20]

$$\frac{1}{ab}\left(\frac{d^2 x_i}{dt^2} + (a+b)\frac{dx_i}{dt} + abx_i\right) = \sum_{j \neq i}^{N}\left[w_{1ij} Q(x_j, q_j) + w_{2ij} f_j(t, Q(x_j, q_j))\right] + I_i(t),$$

$$i = 1, 2, \ldots, N-1, N,$$

$$Q(x, q) = \begin{cases} q\left(1 - e^{-\frac{e^x - 1}{q}}\right) & \text{if } x > \ln\left[1 - q\ln(1 + q^{-1})\right] \\ -1 & \text{if } x < \ln\left[1 - q\ln(1 + q^{-1})\right] \end{cases},$$

where a, b, and q are the constants; and w_1 and w_2 are the topological maps. This model, according to [19,20], is an extension of the results reported in [26,27] by taking into consideration the independent dynamics of the dendrites' wave density and the pulse density for the parallel axons' action.

Examining action potentials, synaptic transmission has been researched by studying the activity of the pre- and postsynaptic neurons [28–30] with attempts to study communication, learning, cognition, perception, knowledge generation, etc. Paper [31] proposes the learning equation for a synaptic adaptive weight $z(t)$ associated with a long-term memory as $\frac{dz}{dt} = f(x)[-Az + g(y)]$, where x is the activity of a presynaptic (postsynaptic) cell; y is the activity of a postsynaptic (presynaptic) cell; $f(x)$ and $g(y)$ are the nonlinear functions; and A is the matrix. Papers [28–30] suggest that matching the action potential generation in the pre- and postsynaptic neurons equivalent to the condition of associative (Hebbian) learning results in a dynamic change in synaptic efficacy. The excitatory postsynaptic potential results due to presynaptic action potentials. After matching, the excitatory postsynaptic potential changes. Neurons are firing irregularly at distinct frequencies. The changes in the dynamics of synaptic connections, resulting from Hebbian-type pairing, lead to significant modification of the temporal structure of excitatory postsynaptic potentials generated by irregular presynaptic action potentials [25]. The changes which occur in synaptic efficacy due to the Hebbian pairing of pre- and postsynaptic activity substantially change the dynamics of the synaptic connection. The long-term changes in synaptic efficacy (long-term potentiation or long-term depression) is believed to be dependent on the relative timing of the onset of the excitatory postsynaptic potential generated by the pre- and post-synaptic action potentials [28–30]. The previously reported, as well as other numerous concepts, have caused a lot of debates. The cellular mechanisms which are responsible for the induction of long-term potentiation or long-term depression are not known.

Analysis of distinct cellular mechanisms and even unverified hypotheses that exhibit sound merits have a direct application to molecular electronics, envisioned bioinspired processing, etc. For example, the design of processing and memory platforms may be performed by examining and comprehending baseline fundamentals at the device and system levels, making use of prototyping/mimicking cellular organization, phenomena, and mechanisms. Based upon the inherent phenomena and mechanisms, distinct networking and interconnect of the *fluidic* and *solid* electronics can be envisioned. This interconnect, however, most likely cannot be based on the semiconductor-centered interfacing reported in [32]. Biomolecular versus envisioned *solid/fluidic* ^{M}PPs can be profoundly different from the device and system-level standpoints.

Intelligent biosystems exhibit goal-driven behavior, evolutionary intelligence, learning, perception and knowledge generation functioning in a non-Gaussian, nonstationary rapidly changing dynamic environment. No generally accepted concept exists for a great number of key open problems such as biocentered processing, memory, coding, etc. Attempts have been pursued to perform bioinspired symbolic, analog, and both digital (discrete-state and discrete-time) and hybrid processing by applying stochastic and deterministic concepts. To date, those attempts have not been culminated in feasible and sound solutions. At the device/module level, utilizing biomolecules as the *information carriers*, novel devices and modules have been proposed for the envisioned *fluidic* molecular electronics [22,24]. The results were applied to control of the *information carriers* (intra- and outer-cellular ions and biomolecules) in cytoplasm, synaptic cleft, membrane channels, etc. The information processing platforms should be capable of mapping stimuli and capturing the goal-relevant information into the cognitive information processing, perception, learning, and knowledge generation [33]. For example, in bioinspired *fluidic* devices, to ensure processing one should control propagation, production, activation, and the binding/unbinding of biomolecules in *active*, *available*, *reprocessing*, and other states. Unfortunately, a significant gap exists between basic, applied, and experimental research, as well as consequent engineering practices and technologies. Due to technological and fundamental challenges and limits, this gap may not be overcome in the near future.

Neurons in the brain—among various information processing and memory tasks—code and generate signals (stimuli) that are transmitted to other neurons trough axon–synapse–dendrite *channels*. Unfortunately, we may not be able to coherently answer fundamental questions including how neurons process (compute, store, code, extract, filter, execute, retrieve, exchange, etc.) information. Even the communication in neurons is a disputed topic. The central assumption is that the information is transmitted and possibly processed by means of action potential—the spikes mechanism. Unsolved problems exist in other critical areas, including information theory. Consider a series connection of processing elements (an ^{ME}device, biomolecule, or protein). The input signal is denoted as x, while the outputs of the first and second processing elements are y_1 and y_2. Even simplifying the data processing to a Markov chain $x \rightarrow y_1(x) \rightarrow y_2(y_1(x))$, the information measures used in communication theory can be applied only to a very limited class of problems. One may not be able to explicitly, quantitatively, and qualitatively examine the information-theoretic measures beyond communication and coding problems. Furthermore, the information-theoretic estimates in neurons and molecular aggregates (shown in Figure 6.10) can be applied to the communication-centered analysis, assuming the availability of a great number of relevant data. Performing the communication and coding analysis, one examines the entropies of the variables x_i and y, denoted as $H(x_i)$ and $H(y)$. The probability distribution functions, conditional entropies $H(y|x_i)$ and $H(x_i|y)$, relative information $I(y|x_i)$ and $I(x_i|y)$, mutual information $I(y,x_i)$, as well as joint entropy $H(y,x_i)$ could be of interest.

In a neuron and its intracellular structures and organelles, baseline processes, mechanisms, and phenomena are not explicitly comprehended. The lack of ability to soundly examine and coherently explain the basic phenomena and processes has resulted in numerous hypotheses and postulates. From the signal/data processing standpoints, neurons are commonly studied as switching devices, while networked neuron ensembles have been considered, assuming *stimulus-induced, connection-induced, adaptive,* and other*correlations*. Conventional neuroscience postulates that networked neurons transmit data, perform information processing, accomplish communication as well as perform other functions by means of a sequence of spikes that are the propagating time-varying action potentials. Consider communication and coding in networked neurons assuming the validity of conventional hypotheses. Each neuron usually receives inputs from many neurons. Depending on whether input produces a spike (excitatory or inhibitory) and on how the neuron processes inputs determine the neuron's functionality. Excitatory inputs cause spikes, while inhibitory inputs suppress them. The rate at which spikes occur is believed to change due to stimulus variations. Though the spike waveform (magnitude, width, and profile) vary, these changes are usually considered to be irrelevant. In addition, the probability distribution function of the interspike intervals varies. Thus, input stimuli, as processed through a sequence of complex processes, result in outputs that are encoded as the pattern of action potentials (spikes). The spike duration is ~ 1 msec, and the spike rate varies from one to thousands of spikes per second. The premise that the spike occurrence, timing, frequency, and its probability distribution encode the information has been extensively studied. It is found that the same stimulus does not result in the same pattern, and debates continue with an alarming number of recently proposed hypotheses.

Let us discuss the relevant issues applying the information-theoretic approach. In general, one cannot determine if a signal (neuronal spike, voltage pulse in ICs, electromagnetic wave, etc.) is carrying information or not. There are no coherent information measures and concepts beyond communication- and coding-centered analysis. One of the open problems is to qualitatively and quantitatively define what the information is. It is not fully comprehended how neurons perform signal/data processing, not to mention information processing, but it is obvious that networked neurons are not analogous to combinational and memory ICs. Most importantly, by examining any signal, it is impossible to determine if it is carrying information or not, as well as to coherently assess the signal/data processing, information processing, coding, or communication features. It is evident that there exists a need to further develop the information theory. Those meaningful developments, as succeeded, can be applied in the analysis of neurophysiological signal/data and information processing.

The entropy, which is the Shannon quantity of information, measures the complexity of the set—e.g., sets having larger entropies require more bits to represent them. For M objects (symbols) X_i that have probability distribution functions $p(X_i)$, the entropy is given as

$$H(X) = -\sum_{i=1}^{M} p(X_i) \log_2 p(X_i), \quad i = 1.2, \dots, M-1, M.$$

Here, $H \geq 0$, and, hence, the number of bits required by the Source Coding Theorem is positive. Examining analog action potentials and considering spike trains, a *differential entropy* can be applied. For a continuous-time random variable X, the *differential entropy* is

$$H(X) = -\int p_X(x) \log_2 p_X(x) dx,$$

where $p_X(x)$ is a one-dimensional probability distribution function of x, $\int p_X(x) dx = 1$.

However, the *differential entropy* can be negative. For example, the *differential entropy* of a Gaussian random variable is $H(X) = 0.5 \ln(2\pi e \sigma^2)$, and $H(X)$ can be positive, negative, or zero depending on the variance. Furthermore, *differential entropy* depends on scaling. For example, if $Z = kX$, one has $H(Z) = H(X) + \log_2 |k|$, where k is the scaling constant. To avoid the aforementioned problems, from the entropy analysis standpoints, continuous signals are discretized. Let X_n denotes a discretized continuous random variable with a binwidth ΔT. Thus, we have $\lim_{\Delta T \to 0} H(X_n) + \log_2 \Delta T = H(X)$. The problem though is to identify the information carrying signals for which ΔT should be obtained.

One may use the a-order Renyi entropy measure as given by [34]

$$R^a(X) = \frac{1}{1-a} \log_2 \int p_X^a(x)dx,$$

where a is the integer, $a \geq 1$. The first-order Renyi information ($a = 1$) leads to the Shannon quantity of information. However, Shannon's and Renyi's quantities measure the complexity of the set, and, even for this specific problem, the unknown probability distribution function should be obtained.

The Fisher information $I_F = \int \frac{(dp(x)/dx)^2}{p(x)} dx$ is a metric for the estimations and measurements. In particular, I_F measures an adequate change in knowledge about the parameter of interest.

The entropy does not measure the complexity of a random variable which could be voltage pulses in ICs, neuron inputs or outputs (response) such as spikes, or any other signals. The entropy can be used to determine whether random variables are statistically independent or not. Having a set of random variables denoted by $X = \{X_1, X_2, \ldots, X_{M-1}, X_M\}$, the entropy of their joint probability function equals the sum of their individual entropies $H(\mathbf{X}) = \sum_{i=1}^{M} H(X_i)$ only if they are statistically independent.

One may examine the mutual information between the stimulus and the response in order to measure how similar the input and output are. We have

$$I(X,Y) = H(X) + H(Y) - H(X,Y)$$

$$I(X,Y) = \int p_{X,Y}(x,y) \log_2 \frac{p_{X,Y}(x,y)}{p_X(x)p_Y(y)} dxdy = \int p_{Y|X}(y|x)p_X(x) \log_2 \frac{p_{Y|X}(y|x)}{p_Y(y)} dxdy.$$

Thus, $I(X,Y) = 0$ when $p_{X,Y}(x,y) = p_X(x)p_Y(y)$ or $p_{Y|X}(y|x) = p_Y(y)$. For example, $I(X,Y) = 0$ when the input and output are statistically independent random variables of each other. When the output depends on the input, one has $I(X,Y) > 0$. The more the output reflects the input, the greater the mutual information. The maximum (infinity) occurs when $Y = X$. From a communications viewpoint, the mutual information expresses how much the output resembles the input. Taking note that for discrete random variables $I(X,Y) = H(X) + H(Y) - H(X,Y)$ or $I(X,Y) = H(Y) - H(Y|X)$, one may utilize the conditional entropy $H(Y|X) = -\sum_{x,y} p_{X,Y}(x,y) \log_2 p_{Y|X}(y|x)$. Here, $H(Y|X)$ measures how random the *conditional* probability distribution of the output is, on average, given a specific input. The more random it is, the larger the entropy, thus reducing the mutual information and $I(X,Y) \leq H(X)$, because $H(Y|X) \geq 0$. The less random it is, the smaller the entropy until it equals zero when $Y = X$. The maximum value of mutual information is the entropy of the input (stimulus).

The channel capacity is found by maximizing the mutual information subject to the input probabilities, e.g., $C = \max_{p_X(\cdot)} I(X,Y)$ [bit/symbol].

Thus, the analysis of mutual information results in the estimation of the channel capacity C which depends on $p_{Y|X}(y|x)$, which defines how the output changes with the input. In general, it is very difficult to obtain or estimate the probability distribution functions.

Using conventional neuroscience hypotheses, the neuronal communication to some extent is equivalent to the communication in the *point process channel* [35]. The *instantaneous* rate at which spikes occur cannot be lower than the r_{min} and greater than the r_{max} related to the discharge rate. Let the average sustainable spike rate be r_0. For a Poisson process, the channel capacity of the point processes, if $r_{min} \leq r \leq r_{max}$ is derived in [35] as

$$C = r_{min}\left[e^{-1}\left(1 + \frac{r_{max} - r_{min}}{r_{min}}\right)^{\frac{1+r_{min}}{r_{max}-r_{min}}} - \left(1 + \frac{r_{min}}{r_{max} - r_{min}}\right)\ln\left(1 + \frac{r_{max} - r_{min}}{r_{min}}\right)\right],$$

which can be expressed in the following form [36]:

$$C = \begin{cases} \frac{r_{min}}{\ln 2}\left(e^{-1}\left(\frac{r_{max}}{r_{min}}\right)^{\frac{r_{max}}{r_{max}-r_{min}}} - \ln\left(\frac{r_{max}}{r_{min}}\right)^{\frac{r_{max}}{r_{max}-r_{min}}}\right), & r_0 > e^{-1}r_{min}\left(\frac{r_{max}}{r_{min}}\right)^{\frac{r_{max}}{r_{max}-r_{min}}} \\ \frac{1}{\ln 2}\left((r_0 - r_{min})\ln\left(\frac{r_{max}}{r_{min}}\right)^{\frac{r_{max}}{r_{max}-r_{min}}} - r_0 \ln\left(\frac{r_0}{r_{min}}\right)\right), & r_0 < e^{-1}r_{min}\left(\frac{r_{max}}{r_{min}}\right)^{\frac{r_{max}}{r_{max}-r_{min}}} \end{cases}.$$

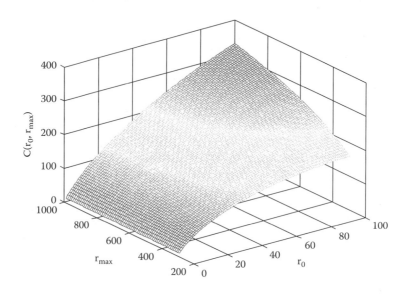

FIGURE 6.11   (see color insert) Channel capacity.

Let the minimum rate be zero. For $r_{\min} = 0$, the expression for a channel capacity is simplified to be

$$C = \begin{cases} \frac{r_{\max}}{e \ln 2}, & r_0 > \frac{r_{\max}}{e} \\ \frac{r_0}{\ln 2} \ln \left(\frac{r_{\max}}{r_0} \right), & r_0 < \frac{r_{\max}}{e} \end{cases}.$$

Example 6.8.2

Assume that the maximum rate varies from 300 to 1000 pulse/sec (or spike/sec), and the average rate changes from 1 to 100 pulse/sec. Taking note of $r_{\max}/e = 0.3679 r_{\max}$, one obtains $r_0 < r_{\max}/e$, and the channel capacity is given as $C = \frac{r_0}{\ln 2} \ln \left(\frac{r_{\max}}{r_0} \right)$. The channel capacitance $C(r_0, r_{\max})$ is documented in Figure 6.11. For $r_0 = 100$ and $r_{\max} = 1000$, one finds $C = 332.2$ bits, or $C = 3.32$ bits/pulse.

The entropy is a function of the window size T and the time binwidth ΔT. For $\Delta T = 3 \times 10^{-3}$ sec and $18 \times 10^{-3} < T < 60 \times 10^{-3}$ sec, the entropy limit is reported to be 157 ± 3 bit/sec [37]. For the spike rate, $r_0 = 40$ spike/sec, $\Delta T = 3 \times 10^{-3}$ sec and $T = 0.1$ sec, the entropy is 17.8 bits [38]. This data agrees with the previous calculations for the capacity of the *point process channel* (see Figure 6.11).

For $r_0 = 100$ and $r_{\max} = 1000$, one finds that $C = 332.2$ bits ($C = 3.32$ bit/pulse). However, this does not mean that each pulse (spike) represents 3.32 bits or any other number of bits of information. In fact, the capacity is derived for digital communication. In particular, for a Poisson process, using $r_{\min}, r_{\max}$, and r_0, we found specific rates with which digital signals (data) can be sent by a *point process channel* without incurring massive transmission errors.   ∎

For analog channels, the channel capacity is $C = \lim_{T \to \infty} \frac{1}{T} \max_{p_X(\cdot)} I(X, Y)$ [bit/sec], where T is the time interval during which communication occurs. In general, analog communication cannot be achieved through a noisy channel without incurring error. Furthermore, the probability distribution functions and the distortion function must be known to perform the analysis. Probability distributions and distortion functions are not available, and processes are non-Poisson. Correspondingly, only some estimates may be made using a great number of assumptions. The focus can be directed rather on the application of biomimetics using sound fundamentals and technologies gained.

Other critical assumptions commonly applied in the attempt to analyze bioprocessing features are a binary-centered hypothesis. Binary logics has a radix of two, meaning that it has two logic levels—e.g., 0 and 1. The radix r can be increased by utilizing r states (logic levels). Three- and four-valued logics are called ternary and quaternary [39]. The number of unique permutations of the truth table for r-valued logic is r^{r^2}. Hence for two-, three-, and four-valued logic, we have 2^4 (16), 3^9 (19,683), and 4^{16} (4,294,967,296) unique permutations, respectively. The use of multiple-valued logic significantly reduces circuitry complexity, device number, and power dissipation, and improves interconnect, efficiency, speed, latency, packaging, and other features. However, sensitivity, robustness, noise immunity, and other challenging problems arise. A r-valued system has r possible outputs for r possible input values, and one obtains r^r outputs of a single r-valued variable [39]. For the radix $r = 2$ (binary logic), the number of possible output functions is $2^2 = 4$ for a single variable x. In particular, for $x = 0$ or $x = 1$, the output f can be 0 or 1—e.g., the output can be the same as the input (identity function), reversed (complement) or constant (either 0 or 1). With a radix of $r = 4$ for quaternary logic, the number of output functions is $4^4 = 256$. The number of functions of two r-valued variables is r^{r^2}, and for the two-valued case $2^{2^2} = 16$. The larger the radix, the smaller the number of digits necessary to express a given quantity. The radix (base) number can be derived from optimization standpoints. For example, mechanical calculators, including Babbage's calculator, mainly utilize ten-valued design. Though the design of multiple-valued memories is similar to the binary systems, multistate elements are used. A T-gate can be viewed as a universal primitive. It has $(r + 1)$ inputs, one of which is an r-valued control input whose value determines which of the other r (r-valued) inputs is selected for output. Due to quantum phenomena in *solid* ^{ME} devices, or controlled release-and-binding/unbinding of specific *information carriers* in the *fluidic* ^{M} devices, it is possible to employ enabling multiple-valued logics and memories.

6.9   Fluidic Molecular Platforms

The activity of brain neurons has been extensively studied using single microelectrodes as well as microelectrode arrays to probe and attempt to influence the activity of a single neuron or assembly of neurons in brain and neural culture. The integration of neurons and microelectronics has been studied in [32,40–42]. Motivated by a biological-centered hypothesis that a neuron is a processing module (system) which processes and stores the information, we propose a *fluidic* molecular processing device/module. This module emulates a brain neuron [22], and cultured neurons can be potentially utilized in implementation of 3D processing and memory platforms. Signal/data processing and memory storage can be accomplished through release, propagation, and the binding/unbinding of molecules. The binding of molecules and ions results in the state transitions to be utilized. Due to fundamental complexity and technological limits, one may not coherently mimic and prototype bioinformation processing. Therefore, we propose to emulate 3D topologies and organizations of biosystems and utilize distinct molecules, thereby ensuring a multiple-valued hardware solution. These innovations imply novel synthesis, design, aggregation, utilization, functionalization, and other features. Using molecules and ions as *information* and *routing* carriers, we propose a novel solution to solve signal, and potentially, information processing problems. We utilize 3D topology/organization inherently exhibited by biomolecular platforms. The proposed *fluidic* molecular platforms can be designed within a processing-and-memory architecture. The *information* carriers are used as logic and memory inputs which lead to the state transitions. Utilizing *routing* carriers, persistent and robust morphology reconfiguration and reconfigurable networking are achieved. One may use distinct membranes and membrane lattices with highly selective channels, and different carriers can be employed. Computing, processing, and memory storage can be performed on the high radix. This ensures multiple-valued logics and memory.

Multiple *routing* carriers are steered in the fluidic cavity to the binding sites, resulting in the binding/unbinding of *routers* to the stationary molecules. The binding/unbinding events lead to a reconfigurable networking. Independent control of *information* and *routing* carriers cannot be accomplished through preassigned steady-state conditional logics, synchronization, timing protocols, and other conventional concepts. The motion and dynamics of the carrier release, propagation, binding/unbinding, and other events should be examined.

A 3D-topology *synthetic fluidic* device/module is illustrated in Figure 6.12. The silicon inner enclosure can be made of proteins, porous silicon, or polymers to form membranes with fluidic channels that should

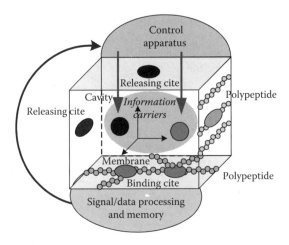

FIGURE 6.12   *A synthetic fluidic* molecular processing module.

ensure the selectivity. The *information* and *routing* carriers are encapsulated in the outer enclosure. The release and steering are controlled by the control apparatus.

The proposed device/module prototypes a neuron with synapses, membranes, channels, cytoplasm, and other components. Specific ions, molecules, and enzymes can pass through the porous membranes. These passed molecules (*information* and *routing* carriers) bind to the specific receptor sites, while enzymes free molecules from binding sites. The binding and unbinding of molecules result in the state transitions. The carriers that pass through selective fluidic channels and propagate through the cavity are controlled by changing the electrostatic potential or thermal gradient [22]. The goal is to achieve a controlled Brownian motion of carriers. Distinct control mechanisms (electrostatic, electromagnetic, thermal, hydrodynamic, etc.) allow one to uniquely utilize selective control ensuring super-high performance and enabling functionality.

The controlled Brownian dynamics of molecules and ions in the fluidic cavity and channels was examined in [22]. The nonlinear stochastic dynamics of Brownian particles is of particular importance in cellular transport, molecular assembling, etc. It is feasible to control the propagation (motion) of carriers by changing the force $F_n(t,\mathbf{r},\mathbf{u})$ or varying the asymmetric potential $V_k(\mathbf{r},\mathbf{u})$. The high-fidelity mathematical model is given as:

$$m_i \frac{d^2 \mathbf{r}_i}{dt^2} = -F_{v_i}\left(\frac{d\mathbf{r}_i}{dt}\right) + \sum_{i,j,n} F_n(t,\mathbf{r}_{ij},\mathbf{u}) + \sum_{i,k} q_i \frac{\partial V_k(\mathbf{r}_i,\mathbf{u})}{\partial \mathbf{r}_i} + \sum_{i,j,k} \frac{\partial V_k(\mathbf{r}_{ij},\mathbf{u})}{\partial \mathbf{r}_{ij}} + f_r(t,\mathbf{r},\mathbf{q}) + \xi_{ri},$$

$$\frac{d\mathbf{q}_i}{dt} = f_q(t,\mathbf{r},\mathbf{q}) + \xi_{qi}, \qquad i = 1,2,\ldots,N-1,N,$$

where $\mathbf{r}_i$ and $\mathbf{q}_i$ are the displacement and extended state vectors; $\mathbf{u}$ is the control vector; $\xi_r(t)$ and $\xi_q(t)$ are the Gaussian white noise vectors; F_v is the viscous friction force; m_i and q_i are the mass and charge; and $f_r(t,\mathbf{r},\mathbf{q})$ and $f_q(t,\mathbf{r},\mathbf{q})$ are the nonlinear maps.

The Brownian particle velocity vector $\mathbf{v}$ is $\mathbf{v} = d\mathbf{r}/dt$. The Lorenz force on a Brownian particle possessing the charge q is $\mathbf{F} = q(\mathbf{E}+\mathbf{v}\times\mathbf{B})$, while using the surface charge density ρ_v, one obtains $\mathbf{F} = \rho_v(\mathbf{E}+\mathbf{v}\times\mathbf{B})$. The released carriers propagate in the fluidic cavity and are controlled by a control apparatus varying $F_n(t,\mathbf{r},\mathbf{u})$ and $V_k(\mathbf{r},\mathbf{u})$ [22]. This apparatus is comprised of polypeptide or molecular circuits which change the temperature gradient or the electric field intensity. The state transitions occur in the anchored processing polypeptide as *information* and *routing* carriers bind and unbind. For example, conformational *switching*, charge changes, electron transport and other phenomena can be utilized. The settling time of electronic, photoelectric, and electrochemomechanical state transitions is from pico to microseconds.

In general, it is possible to design, and potentially synthesize, aggregated 3D networks of high-performance reconfigurable *fluidic* modules. These modules can be characterized in terms of input/output activity. The reported *fluidic* module, which emulates neurons, guarantees superior co-design features.

6.10 Neuromorphological Reconfigurable Molecular Processing Platforms

Consider a gate with binary inputs A and B. Given the outputs are generated by the universal logic gate, one has the following 16 functions: 0, 1, A, B, Ā, B̄, A + B, A+B̄, Ā + B, Ā + B̄, AB, AB̄, ĀB, ĀB̄, AB + ĀB, and AB + ĀB̄. The standard logic primitives (AND, NAND, NOT, OR, and other) can be implemented using a Fredkin gate which performs conditional permutations. Consider a gate with a *switched* input A and a *control* input B. As illustrated in Figure 6.13, the input A is routed to one of two outputs, conditional on the state of B. The routing events change the output switching function, which is AB or AB̄.

FIGURE 6.13 Gate schematic.

Utilizing the proposed *fluidic* molecular processing paradigm, *routable* molecular universal logic gates (^{M}ULG) can be designed and implemented. We define a ^{M}ULG as a reconfigurable combinational gate that can be reconfigured to realize specified functions of its input variables. The use of specific multi-input ^{M}ULGs is defined by the technology soundness, requirements and achievable performance. These ^{M}ULGs can realize logic functions using multi-input variables with the same delay as a two-input ^{M}gate. Logic functions can be efficiently factored and decomposed using ^{M}ULGs.

Figure 6.14 schematically depicts the proposed routing concepts for a reconfigurable logics. The typified 3D-topologically reconfigurable routing is accomplished through the binding/unbinding of *routing* carriers to the stationary molecules. For illustrative purposes, Figure 6.14 documents the reconfiguration of five ^{M}gates depicting a reconfigurable networking-and-processing in 3D. The *information carriers* are represented as the signals x_1, x_2, x_3, x_4, x_5, and x_6. The *routing* carriers ensure a reconfigurable routing and networking of ^{M} gates and hypercells uniquely enhancing and complementing the ^{N}hypercell design. In general, one may not be able to route any output of any gate/hypercell/module to any input of any other gate/hypercell/module. Synthesis constraints, selectivity limits, complexity to control the spatial motion of *routers*, and other limits should be integrated in the design.

It was documented that the proposed *fluidic* module can perform computations, implement complex logics, ensure memory storage, guarantee memory retrieval, etc. Sequences of conditional aggregation, carriers steering, 3D-directed routing, and spatial networking events form the basis of the logic gates and memory retrieval in the proposed neuromorphological reconfigurable *fluidic* ^{M}PPs. In Section 6.7, we documented how to integrate Brownian dynamics into the performance analysis and design. The *transit* time of *information* and *routing* carriers depends on the steering mechanism, control apparatus, particles

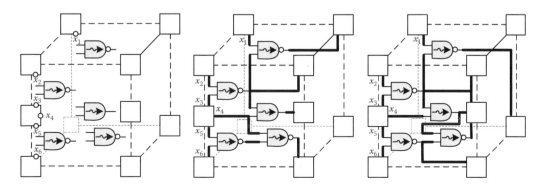

FIGURE 6.14 Reconfigurable routing and networking.

used, sizing features, etc. From the design prospective, one applies the state-space paradigm using the *processing* and *routing* transition functions F_p and F_r that map previous states to the resulting new states in $[t, t_+]$, $t_+ > t$. The output evolution is $\mathbf{y}(t_+) = F_i[t, \mathbf{x}(t), \mathbf{y}(t), \mathbf{u}(t)]$, where $\mathbf{x}$ and $\mathbf{u}$ are the state and control vectors. For example, $\mathbf{u}$ leads to the release and steering of the *routing* carriers with the resulting networking transitions. The reconfigurable system is modeled as $P \subset X \times Y \times U$, where X, Y, and U are the input, output, and control sets.

The proposed neuromorphological reconfigurable *fluidic* ^{M}PPs, which to some degree prototype ^{BM}PPs, can emulate any existing ICs surpassing the overall performance, functionality, and capabilities of envisioned microelectronic solutions. However, the theoretical and technological foundations of neuromorphological reconfigurable ^{3D}networking-processing-and-memory ^{M}PPs remain to be developed and implemented.

6.11   Toward Cognitive Information Processing Platforms

Information (I) causes changes either in the whole system (S) that receives information or in an information processing logical subsystem (S_I) of this system. Different types of information measures, estimates, and indexes exist. For example, potential or prospective measures of information should determine (reflect) what changes may be caused by I in S. Existential or synchronic measures of information should determine (reflect) what changes S experiences during a fixed time interval after receiving I. Actual or retrospective measures of information should determine (reflect) what changes were actually caused by I in S. For example, synchronic measures reflect changes in the short-term memory, while retrospective measures represent transformations in long-term memory. Consider the system mapping tuple (S, L, E), where E denotes the environment; L represents the linkages between S and E. The three structural types of information measurement are internal, integral, and external. The internal information measure should reflect the extent of inner changes in S caused by I. The integral information measure should reflect the extent of changes caused by I on S due to the L between S and E. Finally, the external information measure should reflect the extent of outer changes in E caused by I and S. the three constructive types of information measurement are abstract, realistic, and experiential. The abstract information measure should be determined theoretically under general assumptions, while a realistic information measure must be determined theoretically, subject to realistic conditions applying sound information-theoretic concepts. Finally, the experiential information measure should be obtained through experiments. The information can be measured, estimated, or evaluated only for simple systems examining a limited number of problems (communication and coding) for which the information measures exist. Any S has many quantities, parameters, stimuli, states, events, and outputs that evolve. In general, different measures are needed to be used in order to reflect variations, functionality, performance, capabilities, efficiency, etc. It seems that currently the prospect of finding and using a universal information measure is unrealistic. The structural-attributive interpretation of information does not represent information itself but may relate I to the information measures (for some problems), events, information carriers, and communication in S. In contrast, the functional-cybernetic consideration is aimed to explicitly or implicitly examine information from the functional viewpoint descriptively studying state transitions in systems that include information processing logical subsystems.

Cognitive systems are envisioned to be designed by accomplishing information processing, integrating knowledge generation, perception, learning, etc. By integrating interactive cognition tasks, there is a need to expand signal/data processing (primarily centered on binary computing, coding, manipulation, mining, and other tasks) to information processing. The information theory must be enhanced to explicitly evaluate knowledge generation, perception, and learning by developing an information-theoretic framework of information representation and processing. The information processing at the system and device levels must be evaluated using the cognition measures examining how systems represent and process the information. It is known that information processing depends on the statistical and deterministic structure of stimuli and data. These statistics may be utilized to attain statistical knowledge generation, learning, adaptation, robustness, and self-awareness. The information-theoretic measures, estimates and

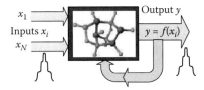

FIGURE 6.15 Cognitive information processing primitive $_P S$ Fredkin gate.

limits of cognition, knowledge generation, perception, and learning in S must be found and examined to approach fundamental limits and benchmarks. Cognizance has been widely studied from an artificial intelligence standpoint. However, limited progress has been achieved in basic theory, design, applications, and technology developments. New theoretical foundations, software, and hardware to support cognitive systems design must be developed. Simple increases in computational power and memory capacity will not result in cognizance and/or intelligence due to entirely distinct functionality, capabilities, measures, and design paradigms. From the fundamental, computational, and technological standpoints, the problems to be solved are far beyond conventional information theory, signal/data processing, and memory solutions.

Consider a data information set, which is a global knowledge with Σ states. By using the observed data D, the system gains and learns certain knowledge, but not all Σ. Before the observations, the system possesses some states from distribution $p(\Sigma)$ with the information measure $M(\Sigma)$. This $M(\Sigma)$ must be explicitly defined, which is an open problem. Once the system observes some particular data D, the enhanced perception of Σ is described by the reciprocal measure estimate $M(\Sigma|D)$, and $M(\Sigma|D) \leq M(\Sigma)$. The uncertainty about Σ reduces through observations, learning, perception, etc. We identify this process as the information gain that the system learned about Σ. Some data $D_< \in D$ will increase the uncertainty about Σs resulting in knowledge reduction. For this regret $D_<$, one finds $M(\Sigma|D_<)$, and the information reduction is expressed as $I_< = f[M(\Sigma), M(\Sigma|D_<)]$. With the goal to achieve cognition and learning by gaining the information (on average) $I_{D \to \Sigma}$, one should derive I using the information measures and estimates. By observing the data, the system cannot learn more about the global knowledge than $M(\Sigma)$. In particular, $M(\Sigma)$ may represent the number of possible states in which the knowledge is mapped, and $M(\Sigma)$ indicates the constrained system ability to gain knowledge due to the lack of possibilities in Σ. The system cannot learn more than the information measure that characterizes the data. In particular, the M of observations limits how much the system can learn. In general, M defines the capacity of the data D to provide or convey information. The information that the system can gain has upper and lower bounds defined by the M limits, while M bounds depend on the statistical properties, structure, and other characteristics of the observable data as well as the system S abilities. Consider a cognitive information processing primitive $_P S$ implemented as a multi-terminal molecule. Utilizing the continuous information-carrying inputs $\mathbf{x} = [x_1, x_2, x_3, x_4] \in X$, $_P S$ generates a continuous output $y(t)$, $y \in Y$ with distinguished states, as shown in Figure 6.15. Hence, the multiple-valued inputs $\mathbf{x}$ are observed and processed by $_P S$ with a transfer function $F(\mathbf{x}, y)$. The cognitive learning can be formulated as utilization and optimization of information measures through the S perception, knowledge generation, and reconfiguration. In general, S integrates subsystems $_S S$, modules $_M S$, and primitives (gate/devices level) $_P S$. The primitive $_P S$ statistical model can be described by $M(\mathbf{x}, \mathbf{y})$, as generated through learning and perception using observed $\mathbf{x} = [x_1, x_2, \ldots, x_{n-1}, x_n] \in X$ and $\mathbf{y} = [y_1, y_2, \ldots, y_{m-1}, y_m] \in Y$.

6.12 Molecular Electronics and Gates: Device and Circuits Prospective

Distinct ^{M}gates and ^{N}hypercells can be used to perform logic functions. To store the data, the memory cells are used. A systematic arrangement of memory cells and peripheral ^{M}ICs (to address and write the data into the cells as well as to delete data stored in the cells) constitute the memory. The ^{M}devices can be used to implement static and dynamic random access memory (RAM) as well as programmable and alterable read-only memory (ROM). Here, RAM is the read-write memory in which each individual molecular primitive

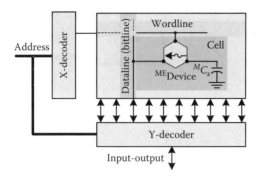

FIGURE 6.16   Dynamic RAM cell with ^{ME}device and storage molecular capacitor $^M C_s$.

can be addressed at any time, while ROM is commonly used to store instructions of a system's operating system. The static RAM may consist of a basic flip-flop ^{M}device with stable states (for example, 0 and 1). In contrast, the dynamic RAM, which can be implemented using one ^{M}device and a storage capacitor, stores one bit of information charging the capacitor. As an example, the dynamic RAM cell is documented in Figure 6.16. The binary information is stored as the charge on the molecular storage capacitor $^M C_s$ (logic 0 or 1). This RAM cell is addressed by switching *on* the access ^{ME}device via the worldline signal, resulting in the charge transferring into and out of $^M C_s$ on the dataline. The capacitor $^M C_s$ is isolated from the rest of the circuitry when the ^{ME}device is *off*. However, the leakage current through the ^{ME}device may require the RAM cell refreshment to restore the original signal. Dynamic shift registers can be implemented using transmission ^{M}gates and ^{M}inverters, flip-flops can be synthesized by cross-coupling NOR ^{M}gates, while delay flip-flops can be built using transmission ^{M}gates and feedback ^{M}inverters.

Among the specific characteristics under consideration are the read/write speed, memory density, power dissipation, volatility (data should be maintained in the memory array when the power is off), etc. The address, data, and control lines are connected to the memory array. The control lines define the function to be performed or the status of the memory system. The address and data lines ensure data manipulation and provide addresses into or out of the memory array. The address lines are connected to an address row decoder which selects a row of cells from an array of memory cells. A RAM organization, as documented in Figure 6.16, consists of an array (matrix) of storage cells arranged in 2^n columns (bitlines) and 2^m rows (wordlines). To read the data stored in the array, a row address is supplied to the row decoder, which selects a specific wordline. All cells along this wordline are activated and the contents of each cell are placed onto their corresponding bitlines. The storage cells can store one (or more) bit of information. The signal available on the bitlines is directed to a decoder. As reported in Figure 6.16, a binary (or high-radix) cell stores binary information utilizing a ^{ME}device at the intersection of the wordline and bitline. The ROM cell can be implemented as (1) a parallel molecular NOR (^{M}NOR) array of cells; (2) a series molecular NAND (^{M}NAND) array of cells requiring a single ^{M}device per storage cell. The ROM cell is programmed by either connecting or disconnecting the ^{M}device output (drain for FETs) from the bitline. Though a parallel ^{M}NOR array is faster, a series ^{M}NAND array ensures compacts and implementation feasibility.

In Figure 9.16, the multi-terminal ^{ME}device is denoted as ⟨↝⟩. There is a need to design ^{M}devices whose robustly controllable dynamics results in a sequence of quantum, quantum-induced, or not quantum state transitions that correspond to a sequence of computational, logic, or memory states. This is guaranteed even for quantum ^{M}devices because quantum dynamics is deterministic, and the nondeterminism of quantum mechanics arises when a device interacts with an uncontrolled outside environment or leaks information to an environment. In ^{M}devices, the *global* state evolutions (state transitions) should be deterministic, predictable, and controllable. The bounds posed by the Heisenberg uncertainty principle restrict observability and do not impose limits on both the device physics and device performance.

The logic device physics defines the mechanism of physical encoding of the logical states in the device. Quantum computing concepts emerged, proposing to utilize the quantum spins of electrons or atoms

FIGURE 6.17 Logical states and energy barriers.

to store information. In fact, a spin is a discrete two-state composition allowing a bit encoding. One can encode information using electromagnetic waves and cavity oscillations in optical devices. The information is encoded by DNA. The feasibility of different state encoding concepts depends on the ability to maintain the logical state for a required period. The stored information must be reliable—e.g., the probability of the spontaneous changing of the stored logical state to another value should be small. One can utilize energy barriers and wells in the controllable energy space for a set of physical states encoding a given logical state. In order for the device to change the logical state, it must pass the energy barrier. To prevent this, the quantum tunneling can be suppressed by using high and wide potential barriers, minimizing excitation and noise, etc. To change the logical state, one varies the energy barrier as illustrated in Figure 6.17. Examining the logical transition processes, the logical states can be retained reliably by potential energy barriers which separate the physical states. The logical state is changed by varying the energy surface barriers (as illustrated in Figure 6.17) for a one-dimensional case. The adiabatic transitions between logical states located at stable or meta-stable local energy minima result.

In VLSI design, resistor-transistor logic (RTL), diode-transistor logic (DTL), transistor-transistor logic (TTL), emitter-coupled logic (ECL), integrated-injection logic (IIL), merged-transistor logic (MTL), and other logic families have been used. All logic families and subfamilies (TTL includes Schottky, low-power Schottky, advanced Schottky, and others) have advantages and drawbacks. Molecular electronics offer unprecedented capabilities compared with microelectronics. Correspondingly, some logic families that ensure marginal performance using solid-state devices provide superior performance as ^{M}devices are utilized. The ^{M}NOR gate, realized using the molecular resistor-transistor logic (^{M}RTL), is documented in Figure 6.18(a). In electronics, NAND is one of the most important gates. The ^{M}NAND gate, designed by applying the molecular diode-transistor logic (^{M}DTL), is reported in Figure 6.18(b). In Figures 6.18(a) and 6.18(b), we use different symbols to designate molecular resistors $\text{-}\sqcap\sqcup\sqcap\text{-}$ ($^M r$), molecular diodes $\text{-}\!\!\!\blacktriangleright\!\!\!|\text{-}$ ($^M d$), and molecular transistors $\overset{\text{-}}{\text{-}\!\!\!\vdash}_\searrow$ ($^M T$). It will be documented below that a term $^M T$ may be used with great caution due to the distinct device physics of molecular and semiconductor devices. In order to introduce the subject, we use this incoherent terminology temporarily because $^M T$ may ensure characteristics similar to FETs and BJTs. However, the device physics of conventional three-, four-, and many-terminal FETs and BJTs is entirely different compared even with *solid* ^{ME}devices. Therefore, we depart from a conventional terminology. Even a three-terminal *solid* ^{ME}device with the controlled $I–V$

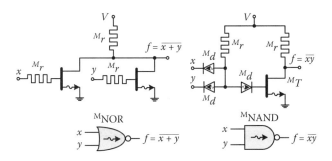

FIGURE 6.18 Circuit schematics: (a) a two-input ^{M}NOR gate; (b) two-input ^{M}NAND gates.

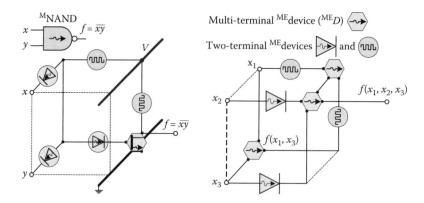

FIGURE 6.19 (a) Implementation of ^{M}NAND mapped by a ^{N}hypercell primitive; (b) a ^{N}hypercell primitive with two- and multiterminal ^{ME}devices.

characteristics may not be referenced as a transistor. New terminology can be developed in the observable future reflecting the device physics of ^{M}devices.

The ^{M}NAND gate, as implemented within a ^{M}DTL logic family, is illustrated within the ^{N}hypercell primitive schematics in Figure 6.19(a). We emphasized the need for developing a new symbols for molecular electronic devices. Quantum phenomena (quantum interaction, interference, tunneling, resonance, etc.) can be uniquely utilized. In Figure 6.19(b), a multi-terminal ^{ME}device (^{ME}D) is illustrated as ⟨↝⟩. Using the proposed ^{ME}D schematics, the illustrated ^{ME}D may have six *input, control,* and *output* terminals (ports) with corresponding molecular bonds for the interconnect. As an illustration, a 3D ^{N}hypercell primitive implementing a logic function $f(x_1,x_2,x_3)$ is shown in Figure 6.19(b). Two-terminal molecular devices (^Md ▷▶⊦ and ^Mr ⟨⅏⟩) are shown. The input signals (x_1, x_2, and x_3) and output switching function f are documented in Figure 6.19(b).

Molecular gates (^{M}AND and ^{M}NAND), designed within the molecular multi-terminal $^{ME}D–^{ME}D$ logic family, are documented in Figure 6.20. Here, three-terminal cyclic molecules are utilized as ^{ME}devices, the physics of which is based on the quantum interaction and controlled electron tunneling. The input signals are V_A and V_B, and are supplied to the *input* terminals, while the output signal is V_{out}. These ^{M}gates are designed using cyclic molecules within the carbon interconnecting framework, as shown in Figure 6.20. The details of synthesis, device physics, and phenomena utilized are reported in Sections 6.19, 6.20, and 6.21. A coherent design should be performed in order to ensure the desired performance, functionality, characteristics, aggregability, topology, and other features. Complex ^{M}gates can be synthesized implementing ^{N}hypercells, which form ^{M}ICs. The ^{M}AND and ^{M}NAND gates are documented in Figure 6.20, and Section 6.21 reports the device physics of the multi-terminal ^{ME}devices.

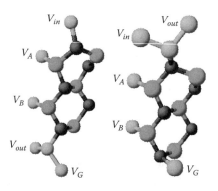

FIGURE 6.20 (see color insert) ^{M}AND and ^{M}NAND gates designed within the molecular $^{ME}D–^{ME}D$ logic family.

6.13 Decision Diagrams and Logic Design of ^{M}ICs

Innovative solutions to perform the system-level logic design for 3D ^{M}ICs should be examined. One needs to depart from 2D logic design (VLSI, ULSI, and postULSI) as well as from planar ICs topologies and organizations. We propose the SLSI design of ^{M}ICs which mimics hierarchical 3D bioprocessing platforms prototyping topologies and organizations observed in nature. This sound solution complies with the envisioned device-level outlook and fabrication technologies. In particular, when using ^{M}devices one may implement $^\aleph$hypercells that form ^{M}ICs. The use of $^\aleph$hypercells as baseline primitives in the design of ^{M}ICs and processing/memory platforms results in a technology-centric solution.

For 2D CMOS ICs, the decision diagram (unique canonical structure) is derived as a reduced decision tree by using topological operators. In contrast, for 3D ICs, a new class of decision diagrams and design methods must be developed to handle the complexity and 3D features. The design concept of a linear decision diagram, mapped by 3D $^\aleph$hypercells, was proposed in [43]. In general, hypercell (cube, pyramid, hexagonal, or other 3D topological aggregates) is a unique canonical structure that is a reduced decision tree. Hypercells are synthesized by using topological operators (deleting and splitting nodes). Optimal and suboptimal technology-centric topology mappings of complex switching functions can be accomplished and analyzed. The major optimization criteria are (1) the minimization of decision diagram nodes and circuit terminals; (2) the simplification of topological structures (linear arithmetic leads to the simple synthesis and straightforward embedding of linear decision diagrams into 3D topologies); (3) the minimization of pathlength in decision diagrams; (4) routing simplification; and (5) verification and evaluation. The optimal topology mapping results in power dissipation reduction, evaluation simplicity, testability enhancement, and other important features. For example, the switching power is not only a function of devices/gates/switches, but also a function of circuit topology, organization, design methods, routing, dynamics, switching activities, and other factors that can be optimized. In general, a novel CAD-supported SLSI should be developed to perform the optimal technology-centric design of high-performance molecular platforms. Through a concurrent design, the designer should be able to perform the following major tasks:

- The logic design of ^{M}ICs utilizing novel representations of data structures.
- The design and aggregation of $^\aleph$hypercells in functional ^{M}ICs.
- The design of multiple-valued and binary decision diagrams.
- CAD developments to concurrently support design tasks.

SLSI utilizes a coherent top-down/bottom-up synthesis taxonomy as an important part of a ^{M}architectronics paradigm. The design complexity should be emphasized. Current CAD-supported postULSI design does not allow one to design ICs with a number of gates more than 1,000,000. For ^{M}ICs, the design complexity significantly increases, and novel methods are sought. The binary decision diagrams (BDDs) for representing Boolean functions use state-of-the-art techniques in high-level logic design [43]. The reduced-order and optimized BDDs ensure large-scale data manipulations and are used to perform logic design and circuitry mapping utilizing hardware description languages. The design scheme is

$$\text{Function(Circuit)} \leftrightarrow \text{BDDModel} \leftrightarrow \text{Optimization} \leftrightarrow \text{Mapping} \leftrightarrow \text{Realization}.$$

The dimension of a decision diagram (the number of nodes) is a function of the number of variables and the variables' ordering. In general, the design complexity is $O(n^3)$. This enormous design complexity significantly limits the designer's abilities to design complex ICs without partitioning and decomposition. Commonly used word-level decision diagrams further increase the complexity due to the processing of data in word-level format. Therefore, novel and sound software-supported design approaches are needed. Innovative methods in data structure representation and data structure manipulation are developed and applied to ensure the design specifications and objectives. We synthesize 3D ^{M}ICs utilizing the linear word-level decision diagrams (LWDDs) that allow one to perform the compact representation of logic circuits using linear arithmetical polynomials (LP) [43,44]. The design complexity becomes $O(n)$. The proposed concept ensures compact representation of circuits compared with other formats and methods.

The following design algorithm guarantees a compact circuit representation:

$$\text{Function(Circuit)} \leftrightarrow \text{BDDModel} \leftrightarrow \text{LWDDModel} \leftrightarrow \text{Realization.}$$

The LWDD is embedded in 3D $^\aleph$hypercells that represent circuits in a 3D space. The polynomial representation of logical functions ensures the description of multi-output functions in a word-level format. The expression of a Boolean function f of n variables $(x_1, x_2, \ldots, x_{n-1}, x_n)$ is

$$LP = a_0 + a_1 x_1 + a_2 x_2 + \cdots + a_{n-1} x_{n-1} + a_n x_n = a_0 + \sum_{j=1}^{n} a_j x_j.$$

To perform a design in 3D, the mapping $\text{LWDD}(a_0, a_1, a_2, \ldots, a_{n-1}, a_n) \leftrightarrow \text{LP}$ is used. The nodes of LP correspond to a Davio expansion. The LWDD is used to represent any m-level circuit with levels L_i, $i = 1, 2, \ldots, m-1, m$ with elements of the molecular primitive library. Two data structures are defined in the algebraic form by a set of LPs as

$$L = \begin{cases} L_1: \text{ inputs } x_j; \text{ outputs } y_{1k} \\ L_2: \text{ inputs } y_{1k}; \text{ outputs } y_{2l} \\ \cdots\cdots\cdots\cdots\cdots\cdots\cdots \\ L_{m-1}: \text{ inputs } y_{m-2,t}; \text{ outputs } y_{m-1,w} \\ L_m: \text{ inputs } y_{m-1,w}; \text{ outputs } y_{m,n} \end{cases}$$

that corresponds to $L P_1 = a_0^1 + \sum_{j=1}^{n_1} a_j^1 x_j, \ldots, L P_m = a_0^n + \sum_{j=1}^{n_m} a_j^n y_{m-1,j}$, or in the graphic form by a set of LWDDs as

$$LWDD_1\left(a_0^1, \ldots, a_{n_1}^1\right) \leftrightarrow LP_1, \ldots, LWDD_m\left(a_0^n, \ldots, a_{n_m}^n\right) \leftrightarrow LP_m.$$

The use of LWDDs is a departure from existing logic design tools. This concept is compatible with the existing software, algorithms, and circuit representation formats. Circuit transformation, format transformation, modular organization/architecture, library functions over primitives, and other features can be accomplished. All combinational circuits can be represented by LWDDs. The format transformation can be performed for circuits defined in Electronic Data Interchange Format (EDIF), Berkeley Logic Interchange Format (BLIF), International Symposium on Circuits and Systems Format (ISCAS), Verilog, etc. The library functions may have a library of LWDDs for multi-input gates, as well as libraries of ^{M}devices and ^{M}gates. The important feature is that these primitives are realized (through logic design) and synthesized as primitive aggregates within $^\aleph$hypercells. The reported LWDD simplifies analysis, verification, evaluation, and other tasks.

Arithmetic expressions underlying the design of LWDDs are canonical representations of logic functions. They are alternatives of the sum-of-product, product-of-sum, Reed-Muller, and other forms of representation of Boolean functions. Linear word-level decision diagrams are obtained by mapping LPs, where the nodes correspond to the Davio expansion, and functionalizing vertices to the coefficients of the LPs. The design algorithms are given as

$$\text{Function(Circuit)} \leftrightarrow \text{LPModel} \leftrightarrow \text{LWDDModel} \leftrightarrow \text{Realization.}$$

Any m-level logic circuits with a fixed order of elements are uniquely represented by a system of mLWDDs. The proposed concept is verified by designing 3D ICs representing Boolean functions by hypercells. The CAD tools for logic design must be based on the principles of 3D realization of logic functions with a library of primitives. Linear word-level decision diagrams are extended by embedding the decision tree into the hypercell structure. For two graphs $G = (V, E)$ and $H = (W, F)$, we embed the graph G into the graph H. The information in the resulting $^\aleph$hypercells is subdivided according to the new structural properties of the cell and the type of the embedded tree. The embedding of a guest graph G into a host graph H is a one-to-one mapping $M_{GV}: V(G) \rightarrow V(H)$, along with the mapping M that maps an edge

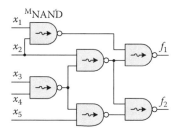

FIGURE 6.21   c17 with ^{M}NAND gates.

$(u;v) \in E(G)$ to a path between $M_{GV}(u)$ and $M_{GV}(v)$ in H. Thus, the embedding of G into H is a one-to-one mapping of the nodes in G to the nodes in H.

In SLSI design, decision diagrams and decision trees are used. The information estimates can be evaluated [43]. Decision trees are designed using the Shannon and Davio expansions. The best variable and expansion for any node of the decision tree in terms of information estimates must be found in order to optimize the design and synthesize optimal ^{M}ICs. The optimization algorithm should generate the *optimal paths* in a decision tree with respect to the design criteria. The decision tree is designed by arbitrarily choosing variables using either Shannon (S), positive Davio (pD), or negative Davio (nD) expansions for each node. The decision tree design process is a recursive decomposition of a switching function. This recursive decomposition corresponds to the expansion of switching function f with respect to the variable x. The variable x carries information that influences f. The initial and final state of the expansion $\sigma \in \{S, pD, nD\}$ can be characterized by the performance estimates. The information-centered optimization of ^{M}ICs design is performed in order to design optimal decision diagrams. A path in the decision tree starts from a node and finishes in a terminal node. Each path corresponds to a term in the final expression for f. For the benchmark c17 circuit, implemented using 3D NAND ^{M}gates (^{M}NAND) as reported in Figure 6.21, Davio expansions ensure optimal design as compared with the Shannon expansion [43].

The software-supported logic design of proof-of-concept 3D ^{M}ICs is successfully accomplished for complex benchmarking ICs in order to verify and examine the method proposed [43]. The size of LWDDs is compared with the best results received by other decision diagram packages developed for 2D VLSI design. Both the method reported and the software algorithms were tested and validated. The number of nodes, number of levels, and CPU time (in seconds) required to design decision diagrams for 3D ^{M}ICs are examined. In addition, volumetric size, topological parameters, and other performance variables are analyzed. We assume (1) a feedforward neural networked topology with no feedback; (2) threshold ^{M}gates as the processing primitives; (3) aggregated $^\aleph$hypercells comprised from ^{M}gates; and (4) multilevel combinational circuits over the library of NAND, NOR, and EXOR ^{M}gates implemented using three-terminal ^{ME}devices.

Experiments were conducted for a variety of ICs, and some results are reported in Table 6.1 [43]. The space size is given by X, Y, and Z that result in the volumetric quantity $V = X \times Y \times Z$. The topological characteristics are analyzed using the total number of terminals (N_T) and intermediate (N_I) nodes. For example, c880 is an eight-bit arithmetic logic unit (ALU). The core of this circuit is in the form of the eight-bit 74283 adder, which has 60 inputs and 26 outputs. A planar design leads to 383 gates. By contrast, 3D design results in 294 ^{M}gates. A 3D nine-bit ALU (c5315) with 178 inputs and 123 outputs is implemented using 1413 ^{M}gates, while a c6288 multiplier (32 inputs and 32 outputs) has 2327 ^{M}gates. Molecular gates are aggregated, networked, and grouped in 3D within $^\aleph$hypercell aggregates. The number of incompletely specified $^\aleph$hypercells was minimized. The $^\aleph$hypercells in the ith layer were connected to the corresponding $^\aleph$hypercells in $(i-1)$th and $(i+1)$th layers. The number of terminal nodes and intermediate nodes are 3750 and 2813 for a nine-bit ALU, while for a multiplier we have 9248 and 6916 nodes. To combine all layers, more than 10,000 connections were generated. The design in 3D was performed within 0.36 seconds for nine-bit ALU. The studied nine-bit ALU performs arithmetic and logic operations simultaneously on two

TABLE 6.1   Design Results for 3D ^{M}ICs

Circuit	I/O	Space Size				Nodes and Connections		
		#G	#X	#Y	#Z	#N_T	#N_I	CPU Time (sec)
c432	36/7	126	66	64	66	2022	1896	< 0.032
8-bit ALU c880	60/26	294	70	72	70	612	482	< 0.047
9-bit ALU c5315	178/123	1413	138	132	126	3750	2813	< 0.36
16 × 16 Multiplier								
c6288	32/32	2327	248	248	244	9246	6916	< 0.47

nine-bit input data words, as well as computes the parity of the results. Conventional 2D logic design for c5315 with 178 inputs and 123 outputs results in 2406 gates. In contrast, the proposed design, as performed using a proof-of-concept SLSI software, leads to 1413 ^{M}gates networked and aggregated in 3D. In addition to conventional parameters (diameter, dilation cost, expansion, load, etc.), we use the number of variables in the logic function described by $^\aleph$hypercells, the number of links, the fan-out of the intermediate nodes, statistics, and others to perform the evaluation. To ensure the similarity to 2D design, binary three-terminal ^{ME}devices were used. The use of multiple-valued multi-terminal ^{ME}devices results in superior performance.

The representative proof-of-concept CAD tools and software solutions were developed in order to demonstrate the 3D design feasibility for combinational ^{M}ICs. The compatibility with hardware description languages is important. Three netlist formats (EDIF, ISCAS, and BLIF) are used and embedded in a proof-of-concept SLSI software that features [43]:

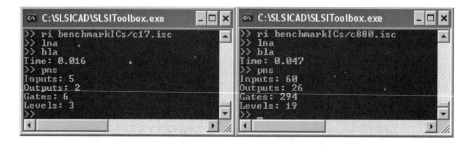

FIGURE 6.22   The design of 3D ^{M}ICs using a proof-of-concept SLSI software.

- A new design concept for 3D ^{M}ICs
- Synthesis and partitioning linear decision diagrams for given functions or circuits
- A spectral representation of logic functions
- Circuit testability and verification
- A compact format ensuring robustness and rapid-prototyping
- A compressed optimal representation of complex ^{M}ICs

For 3D ^{M}ICs, the results of the design are shown in Figure 6.22, which displays the data in the Command Window—in particular, the c17 circuit and eight-bit ALU (c880) designs.

The software and CAD developments in 3D logic design performed by Drs. S. Yanushkevich and V. Shmerko [43] are deeply appreciated and acknowledged.

6.14 Hypercell Design

The binary tree is a networked description that carries information about the dual connections of each node. The binary tree also carries information about the functionality of the logic circuit and its topology. The nodes of the binary tree are associated with the Shannon and Davio expansions, with respect to each variable and coordinate in 3D. A node in the binary decision tree realizes the Shannon decomposition $f = x_i f_0 \oplus x_i f_1$, where $f_0 = f|_{x_i=0}$ and $f_1 = f|_{x_i=1}$ for all variables in f. Thus, each node realizes the Shannon expansion, and the nodes are distributed over levels. The classical hypercube contains 2^n nodes, while the $^\aleph$hypercell has $2^n + \sum_{i=0}^{n} 2^{n-1} C_i^m$ nodes in order to ensure a technology-centric design of ^{M}ICs. The $^\aleph$hypercell consists of terminal nodes, intermediate nodes, and roots. This ensures a straightforward hypercell implementation, for example, by using the molecular multiplexer. The design steps are

Step 1: Connect the terminal node with the intermediate nodes.
Step 2: Connect the root with two intermediate nodes located symmetrically on the opposite faces.
Step 3: Pattern the terminal and intermediate nodes on the opposite faces and connect them through the root.

Figure 6.23(a) reports a 3D $^\aleph$hypercell implemented using two-to-one molecular multiplexers.

Several methods are used for representing logic functions, and a hypercell solution is utilized. In general, a $^\aleph$hypercell is a homogeneous aggregated assembly for massive super-high-performance parallel computing. We apply the enhanced switching theory integrated with a novel logic design concept. In the design, the graph-based data structures and 3D topology are utilized. The $^\aleph$hypercell is a topological representation of a switching function in an n-dimensional graph. In particular, the switching function

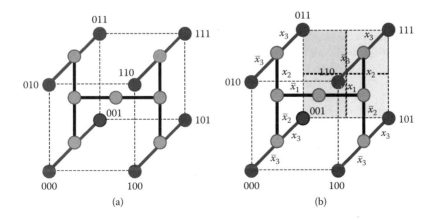

FIGURE 6.23 Multiplexer-based $^\aleph$hypercells: (a) a $^\aleph$hypercell with molecular multiplexers; (b) implementation of a switching function $f = \bar{x}_1 x_2 \vee x_1 \bar{x}_2 \vee x_1 x_2 x_3$.

f is given as

$$f^{\text{Switching Function}} \Rightarrow \underset{\underset{\text{Operation}}{\Uparrow}}{\mathbf{L}_{i=0}^{2^n-1}} \underset{\underset{\mathbf{K}_i}{\Downarrow}}{\text{Coefficient}} \left(x_1^{i_1} \ldots x_n^{i_n} \right) \Rightarrow f_F^{\text{Form of Switching Function}}.$$

The data structure is described in matrix form using the truth vector $\boldsymbol{F}$ of a given switching function f, as well as the vector of coefficients $\mathbf{K}$. The logic operations are represented by $\mathbf{L}$. $^\aleph$Hypercells compute f, and Figure 6.23(b) reports a $^\aleph$hypercell to implement $f = \bar{x}_1 x_2 \vee x_1 \bar{x}_2 \vee x_1 x_2 x_3$.

From the technology-centric viewpoints, we propose a concept that employs $^\text{M}$gates coherently mapping the device/module/system-level and data structure solutions by using $^\aleph$hypercells. Aggregated $^\aleph$hypercells can implement switching functions f of arbitrary complexity. The logic design in spatial dimensions is based on the advanced methods and enhanced data structures in order to satisfy the requirements of 3D topology. The appropriate data structure of logic functions and the methods of embedding this structure into $^\aleph$hypercells are developed. The algorithm in a logic functions manipulation needed to change the information carrier from the algebraic form (logic equation) to the hypercell structure consists of three steps:

Step 1: The logic function is transformed into the appropriate algebraic form (Reed-Muller, arithmetic or word-level in a matrix or algebraic representation).

Step 2: The derived algebraic form is converted to the graphical form (decision tree or decision diagram).

Step 3: The obtained graphical form is embedded and technologically implemented by $^\aleph$hypercells. The $^\aleph$hypercell aggregates form $^\text{M}$ICs.

The design is expressed as

$$\underset{\text{Step1}}{\text{LogicFunction}} \Leftrightarrow \underset{\text{Step2}}{\text{Graph}} \Leftrightarrow \underset{\text{Step3}}{\text{Hypercell}/^\text{M}\text{ICs}}.$$

The proposed procedure results in:

- Algebraic representations and robust manipulations of complex switching logic functions.
- Matrix representations and manipulations providing a consistency of logic relationships for variables and functions from the spectral theory viewpoint.
- Graph-based representations using decision trees.
- The direct mapping of decision diagrams into logical networks, as demonstrated for multiplexer-based $^\aleph$hypercells.
- The robust embedding of data structures into $^\aleph$hypercells.

From the synthesis viewpoint, the complexity of the molecular interconnect corresponds to the complexity of $^\text{ME}$devices. We introduce a 3D directly interconnected molecular electronics ($^\text{3D}$DIME) concept in order to reduce the synthesis complexity, minimize delays, ensure robustness, enhance reliability, etc. This solution minimizes the interconnect, utilizing a direct atomic bonding of *input*, *control*, and *output* $^\text{M}$devices terminals (ports) by means of direct device-to-device aggregation. We have documented that $^\text{ME}$devices and $^\text{M}$gates are engineered and implemented using cyclic molecules within a carbon framework, see Figure 6.20 and Section 12. For example, the *output* terminal of the $^\text{ME}$device is directly connected to the *input* terminal of another $^\text{ME}$device. This ensures synthesis feasibility, the compact implementation of $^\aleph$hypercells, the applicability of $^\text{M}$primitives, etc.

6.15   Three-Dimensional Molecular Signal/Data Processing and Memory Platforms

Advanced computer architectures (beyond von Neumann architecture) [43,45,46] can be devised and implemented to guarantee superior processing, communication, reconfigurability, robustness, networking, etc. In von Neumann computer architecture, the central processing unit (CPU) executes sequences of instructions and operands, which are fetched by the program control unit (PCU), executed by the data

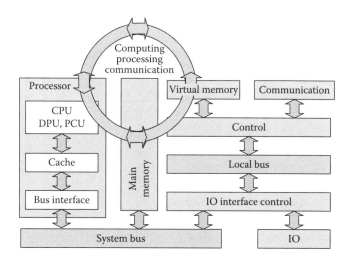

FIGURE 6.24 Computer architecture.

processing unit (DPU), and then placed in memory. In particular, caches (high-speed memory where data is copied when it is retrieved from random access memory, thus improving the overall performance by reducing the average memory access time) are used. The instructions and data form instruction and data streams which flow to and from the processor. The CPU may have more than one processor and coprocessor, with various execution units, multi-level instruction, and data caches. These processors can share, or have their own, caches. The *datapath* contains ICs to perform arithmetic and logical operations on words such as fixed- or floating-point numbers. The CPU design involves a trade-off between hardware, speed, and affordability. The CPU is usually partitioned on the control and *datapath* units. The control unit selects and sequences the data-processing operations. The core interface unit is a switch that can be implemented as autonomous cache controllers operating concurrently and feeding the specified number (64 or 128) of bytes of data per cycle. This core interface unit connects all controllers to the data or instruction caches of processors. Additionally, the core interface unit accepts and sequences information from the processors. A control unit is responsible for controlling the data flow between controllers, thus regulating *in* and *out* information flows. An interface to the input/output devices is also available. On-chip debuging, error detection, sequencing logic, self-test, monitoring, and other units must be integrated to control a pipelined computer. The computer performance depends on the architecture and hardware components, and Figure 6.24 illustrates a conventional computer architecture.

Consider signal/data and information processing between nerve cells. The key to understand processing, memory, learning, intelligence, adaptation, control, hierarchy, and other system-level basics lies in the ability to comprehend the phenomena exhibited, the organization utilized, and the architecture possessed by the central nervous system, neurons, and their organelles. Unfortunately, many problems have not been resolved. Each neuron in the brain that performs processing and memory storage has thousands of synapses with binding sites, membrane channels, microtubule- and synapse associated proteins, etc. The *information carriers* accomplish transitions performing and carrying out various information processing, memory, communication and other tasks. The information processing and memories are reconfigurable and constantly adapt. Neurons function within a 3D hierarchically distributed, robust, adaptive, parallel and networked organization. Making use of the existing knowledge, Figure 6.25 documents a 3D ^{M}PP. A processor executes sequences of instructions and operands, which are fetched (by the control unit) and placed in memory. The instructions and data form *instruction* and *data streams* which flow to and from the processor. The processor may have subprocessors with shared caches. The core interface unit concurrently controls operations and data retrieval. This interface unit interfaces all controllers to the data or processor instruction caches. The interface unit accepts and sequences information from the processors. A control unit is responsible for controlling data flow, thus regulating the *in* and *out* information flows.

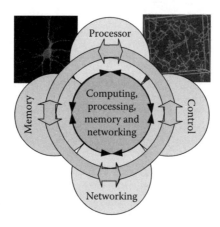

FIGURE 6.25 A molecular processing-and-memory platform.

The integrated processor-and-memory architecture which accomplishes the processing tasks is reported in Figure 6.25.

More study is needed of hierarchical distributed computing and nervous systems which process, store, code, compress, manipulate, route, and network the information in the optimal manner. Figure 6.26 shows the principle of organization of a nervous system similar to the ^{M}PP reported in Figure 6.25. The distributed central nervous system adaptively reconfigures based upon information processing, memory storage, communication, and control (instruction) parallelisms. This principle can be effectively used in the design of various processing and memory platforms within 3D organization and enabling architectures.

The envisioned implementation of ^{M}PPs primarily depends on the progress in device physics, system organization/architecture, molecular hardware and SLSI design. The critical problems in the design are the development, optimization, and utilization of hardware and software. The current status of fundamental and technology developments suggests that the ^{M}PPs will likely be designed utilizing a digital paradigm. Numbers in binary digital processors and memories are represented as a string of zeros and ones, and circuits perform Boolean operations. Arithmetic operations are performed based on a hierarchy built upon simple operations. The methods to compute, and the algorithms used, are different. Therefore, speed, robustness, accuracy, and other performance characteristics vary. The information is represented

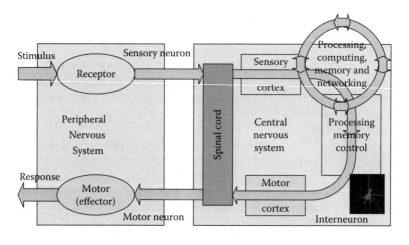

FIGURE 6.26 The vertebrate nervous system.

as a string of bits (zeros and ones). The number of bits depends on the length of the word (the quantity of bits on which the hardware is capable of operating). The operations are thus performed over the string of bits. Certain rules associate a numerical value X with the corresponding bit string $x = \{x_0, x_1, \ldots, x_{n-2}, x_{n-1}\}$, $x_i \in 0, 1$. The associated word (string of bits) is n bits long. If for every value X there exists one, and only one, corresponding bit string x, the number system is nonredundant. If it's possible to have more than one x representing the same value X, the number system is redundant. A *weighted* number system is used, and a numerical value is associated with the bit string x as $x = \sum_{i=0}^{n-1} x_i w_i$, $w_0 = 1, \ldots, w_i = (w_i - 1)(r_i - 1)$, where r_i is the *radix* integer. By making use of the multiplicity of instruction and data streams, the following classification can be applied:

1. Single instruction stream / single data stream — conventional word-sequential architecture including pipelined computing platforms with parallel arithmetic logic unit (ALU).
2. Single instruction stream / multiple data stream — multiple ALU architectures, e.g., parallel-array processor (ALU can be either bit-serial or bit-parallel).
3. Multiple instruction stream / single data stream.
4. Multiple instruction stream / multiple data stream — the multiprocessor system with multiple control units.

In biosystems, multiple instruction stream / multiple data stream are observed. No evidence exists that technology will provide the abilities to synthesize biomolecular processors, not to mention biocomputers, in the near future. Therefore, we shall concentrate here on computing platforms designed using *solid* molecular electronics that ensure soundness and technological feasibility. Performance estimates are reported in this chapter. Three-dimensional topologies and organizations significantly improve the performance of computing platforms guaranteeing, for example, massive parallelism and optimal utilization. Using the number of instructions executed (N), the number of cycles per instruction (C_{PI}), and clock frequency (f_{clock}), the program execution time is $T_{ex} = N C_{PI}/f_{clock}$. In general, circuit hardware determines the clock frequency f_{clock}, software affects the number of instructions executed N, while architecture defines the number of cycles per instruction C_{PI}. Computing platforms integrate functional controlled hardware units and systems which perform processing, storage, execution, etc. The ^{M}PP accepts digital or analog input information, processes and manipulates it according to a list of internally stored machine instructions, stores the information, and produces the resulting output. The list of instructions is called a program, and internal storage is called memory. A memory unit integrates different memories. The processor accesses (reads or loads) the data from the memory systems, performs computations, and stores (writes) the data back to memory. The memory system is a collection of storage locations. Each storage location (memory word) has an address. A collection of storage locations forms an address space. Figure 6.27 documents the data flow and its control, representing how a processor is connected to a memory system via address, control, and data interfaces. High-performance memory systems should be capable of serving multiple requests simultaneously, particularly for vector processors.

When a processor attempts to load or read the data from the memory location, the request is issued, and the processor stalls while the request returns. While ^{M}PPs can operate with overlapping memory requests,

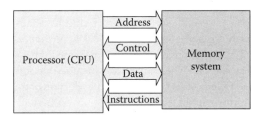

FIGURE 6.27 A memory–processor interface.

the data cannot be optimally manipulated if there are long memory delays. Therefore, a key performance parameter in the design is the effective memory speed. The following limitations are imposed on any memory systems: the memory cannot be infinitely large, cannot contain an arbitrarily large amount of information, and cannot operate infinitely fast. Hence, the major characteristics are speed and capacity. The memory system performance is characterized by the latency (τ_l) and bandwidth (B_w). The memory latency is the delay as the processor first requests a word from memory until that word arrives and is available for use by the processor. The bandwidth is the rate at which information can be transferred from the memory system. Taking note of the number of requests the memory can service concurrently $N_{request}$, we have $B_w = N_{request}/\tau_l$. Using 3D ^{M}ICs, it becomes feasible to design and build superior memory systems with superior capacity, low latency and high bandwidth approaching physical and technological limits. Furthermore, it becomes possible to match the memory and processor performance characteristics and capabilities.

Memory hierarchies ensure decreased latency and reduced bandwidth requirements, whereas parallel memories provide higher bandwidth. The ^{M}PP architectures can utilize a 3D-organization with a fast memory located in front of a large but relatively slow memory. This significantly improves speed and enhances memory capacity. However, this solution results in the application of registers in the processor unit, and the most commonly accessed variables should be allocated at registers. A variety of techniques, employing either hardware, software, or a combination of hardware and software, must be employed to ensure that most references to memory are fed by the faster memory. The locality principle is based on the fact that some memory locations are referenced more often than others. The implementation of spatial locality, due to the sequential access, provides one with the property that an access to a given memory location increases the probability that neighboring locations will soon be accessed. Making use of the frequency of program looping behavior, temporal locality ensures the access to a given memory location, thus increasing the probability that the same location will be accessed again soon. If a variable has not been referenced in a while, it is unlikely the variable will be needed in the near future.

The performance parameter, which can be used to quantitatively examine different memory systems, is the effective latency τ_{ef}. We have $\tau_{ef} = \tau_{hit} R_{hit} + \tau_{miss}(1 - R_{hit})$, where τ_{hit} and τ_{miss} are the hit and miss latencies; R_{hit} is the hit ratio, $R_{hit} < 1$. If the needed word is found in a level of the hierarchy, it is called a hit. Correspondingly, if a request must be sent to the next lower level, the request is said to be a miss. The miss ratio is given as $R_{miss} = (1 - R_{hit})$. These R_{hit} and R_{miss} are affected by the program being executed and influenced by the high/low-level memory capacity ratio. The access efficiency E_{ef} of multiple-level memory ($i - 1$ and i) is found using the access time, hit and miss ratios. In particular, $E_{ef} = \left(\frac{t_{access\,time\,i-1}}{t_{access\,time\,i}} R_{miss} + R_{hit} \right)^{-1}$.

The hardware can dynamically allocate parts of the cache memory for addresses likely to be accessed soon. The cache contains only redundant copies of the address space. The cache memory can be associative or content-addressable. In an associative memory, the address of a memory location is stored along with its content. Rather than reading data directly from a memory location, the cache is given an address and responds by providing data which might or might not be the data requested. When a cache miss occurs, the memory access is then performed from the main memory, and the cache is updated to include the new data. The cache should hold the most active portions of the memory, and the hardware dynamically selects portions of the main memory to store in the cache. When the cache is full, some data must be transferred to the main memory or deleted. A strategy for cache memory management is therefore needed. These cache management strategies are based on the locality principle. In particular, spatial (selection of what is brought into the cache) and temporal (selection of what must be removed) localities are embedded. When a cache miss occurs, hardware copies a contiguous block of memory into the cache, which includes the word requested. This fixed-size memory block can be small, medium, or large. Caches can require all fixed-size memory blocks to be aligned. When a fixed-size memory block is brought into the cache, it is likely another fixed-size memory block must be removed. The selection of the removed fixed-size memory block is based on efforts to capture temporal locality.

The cache can integrate the data memory and the tag memory. The address of each cache line contained in the data memory is stored in the tag memory. The state can also track which cache line is modified. Each line contained in the data memory is allocated by a corresponding entry in the tag memory to indicate the full address of the cache line. The requirement that the cache memory be associative (content-addressable) complicates the design because addressing data by content is more complex than by its address (all tags must be compared concurrently). The cache can be simplified by embedding a mapping of memory locations to cache cells. This mapping limits the number of possible cells in which a particular line may reside. Each memory location can be mapped to a single location in the cache through direct mapping. There is no choice of where the line resides and which line must be replaced, however, poor utilization results. In contrast, a two-way set-associative cache maps each memory location into either of two locations in the cache. Hence, this mapping can be viewed as two identical directly mapped caches. In fact, both caches must be searched at each memory access, and the appropriate data selected and multiplexed on a tag match-hit and on a miss. Then, a choice must be made between two possible cache lines as to which is to be replaced. A single least-recently-used bit can be saved for each such pair of lines in order to remember which line has been accessed more recently. This bit must be toggled to the current state each time. To this end, an M-way associative cache maps each memory location into M memory locations in the cache. Therefore, this cache map can be constructed from M identical direct-mapped caches. The problem of maintaining the least-recently-used ordering of M cache lines is primarily due to the fact that there are $M!$ possible orderings. In fact, it takes at least $\log_2 M!$ bits to store the ordering. In general, a multi-associative cache may be implemented.

Multiple memory *banks*, formed by ^{M}ICs, can be integrated together to form a parallel main memory system. Since each *bank* can service a request, a parallel main memory system with N_{mb} *banks* can service N_{mb} requests simultaneously, increasing the bandwidth of the memory system by N_{mb} times the bandwidth of a single *bank*. The number of *bank* is a power of two, e.g., $N_{mb} = 2^p$. An n-bit memory word address is partitioned into two parts: a p-bit *bank* number and an m-bit address of a word within a *bank*. The p bits used to select a *bank* number could be any p bits of the n-bit word address. Let us use the low-order p address bits to select the *bank* number. The higher order $m = (n - p)$ bits of the word address is used to access a word in the selected *bank*. Multiple memory *banks* can be connected using *simple paralleling* and *complex paralleling*. Figure 6.28 shows the structure of a simple parallel memory system where m address bits are simultaneously supplied to all memory *banks*. All *banks* are connected to the same read/write control line. For a read operation, the *banks* perform the read operation and accumulate the data in the latches. Data can then be read from the latches one by one by setting the switch appropriately. The *banks* can be accessed again to carry out another read or write operation. For a write operation, the latches are loaded one by one. When all latches have been written, their contents can be written into the memory *banks* by supplying m bits of address. In a simple parallel memory, all *banks* are cycled simultaneously. Each *bank* starts and completes its individual operations at the same time as every other *bank*, and a new memory cycle

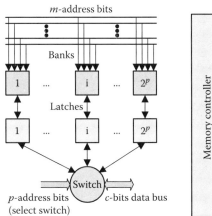 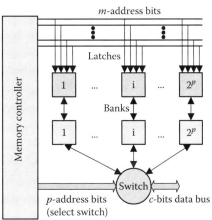

FIGURE 6.28 Simple and complex parallel main memory systems.

starts for all *banks* once the previous cycle is complete. A complex parallel memory system is documented in Figure 6.28. Each *bank* is set to operate on its own, independent of the operation of the other *banks*. For example, ith *bank* performs a read operation on a particular memory address, while $(i + 1)$th *bank* performs a write operation on a different and unrelated memory address. Complex paralleling is achieved using the address latch and a read/write command line for each *bank*. The *memory controller* handles the operation of the complex parallel memory. The processing unit submits the memory request to the memory controller, which determines which *bank* needs to be accessed. The controller then determines if the *bank* is busy by monitoring a busy line for each *bank*. The controller holds the request if the *bank* is busy, submitting it when the *bank* becomes available to accept the request. When the *bank* responds to a read request, the switch is set by the controller to accept the request from the *bank* and forward it to the processing unit. It can be foreseen that complex parallel main memory systems will be implemented as molecular vector processors. If consecutive elements of a vector are present in different memory *banks*, then the memory system can sustain a bandwidth of one element per clock cycle. Memory systems in ^{M}PPs can have thousands of *banks* with multiple memory controllers that allow multiple independent memory requests at every clock cycle.

Pipelining is a technique to increase the processor throughput with limited hardware in order to implement complex *datapath* (data processing) units (multipliers, floating-point adders, etc.). In general, a pipeline processor integrates a sequence of i data-processing molecular primitives which cooperatively perform a single operation on a stream of data operands passing through them. Design of pipelining ^{M}ICs involves deriving multistage balanced sequential algorithms to perform the given function. Fast buffer registers are placed between the primitives to ensure the transfer of data between them without interfering with one another. These buffers should be clocked at the maximum rate that still guarantees the reliable data transfer between primitives. As illustrated in Figure 6.29, ^{M}PPs must be designed to guarantee the robust execution of overlapped instructions using pipelining. Four basic steps (fetch F_i, decode D_i, operate O_i, and write W_i) and specific hardware units are needed to achieve these tasks. The execution of the instructions can be overlapped. When the execution of some instruction I_i depends on the results of a previous instruction I_{i-1} which is not yet completed, instruction I_i must be delayed. The pipeline is said to be stalled, waiting for the execution of instruction I_{i-1} to be completed. While it is not possible to eliminate such situations, it is important to minimize the probability of their occurrence. This is a key consideration in the design of the instruction set, as well as in the design of the compilers that translate high-level language programs into machine language.

The parallel execution capability (called superscalar processing), when added to pipelining of the individual instructions, means that more than one instruction can be executed per basic step. Thus, the execution rate can be increased. The rate R_T of performing basic steps in the processor depends on the processor clock rate. The use of multiprocessors speeds up the execution of large programs by executing subtasks in parallel. The main difficulty in achieving this is the decomposition of a given task into its parallel subtasks, and then ordering these subtasks to the individual processors in such a way that communication among the subtasks are performed efficiently and robustly. Figure 6.30 documents a block diagram of a multiprocessor system with the interconnection network needed for data sharing among the processors P_i. Parallel paths are needed in this network in to parallel activity to proceed in the processors as they access

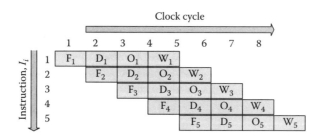

FIGURE 6.29   Pipelining of instruction execution.

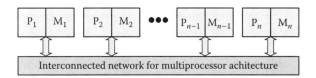

FIGURE 6.30   Multiprocessor architecture.

the global memory space—as represented by the multiple memory units M_i. This is performed utilizing 3D organization.

6.16   Hierarchical Finite-State Machines and Their Use in Hardware and Software Design

Simple register-level subsystems perform single data-processing operations—e.g., summation $X :=$ $x_1 + x_2$, subtraction $X := x_1 - x_2$, etc. To carry out complex data processing operations, multifunctional register-level subsystems should be designed. These register-level subsystems are partitioned as a data-processing unit (*datapath*) and a controlling unit (control unit). The control unit is responsible for collecting and controlling the data-processing operations (actions) of the *datapath*. To design the register-level subsystems, one studies a set of operations to be executed, and then designs ^{M}ICs using a set of register-level components that implement the desired functions. The ultimate goal is to achieve optimal achievable performance under the constraints. It is difficult to impose meaningful mathematical structures on register-level behavior using Boolean algebra and conventional gate-level design. Due to these difficulties, the heuristic synthesis is commonly accomplished using the following sequential algorithm:

1. Define the desired behavior as a set of sequences of register-transfer operations (each operation can be implemented using the available components) comprising the algorithm to be executed.
2. Examine the algorithm to determine the types of components, and their number, to ensure the required *datapath*.
3. Design a complete block diagram for the *datapath* using the components chosen.
4. Examine the algorithm and *datapath* in order to derive the control signals with the ultimate goal of synthesizing the control unit for the found *datapath* that meets the algorithm's requirements.
5. Test, verify, and evaluate the design performing analysis and simulation.

Let's now design virtual control units that ensure extensibility, flexibility, adaptability, robustness, and reusability. For the design, we shall use the hierarchic graphs (HGs). One significant problem is developing straightforward algorithms that ensure implementation (nonrecursive and recursive calls) and utilize hierarchical specifications. We will examine the behavior, perform logic design, and implement reusable control units modeled as hierarchical finite-state machines with virtual states. The goal is to attain the top-down sequential well-defined decomposition in order to develop a complex robust control algorithm step-by-step. We consider *datapath* and control units. The *datapath* unit consists of memory and combinational units. A control unit performs a set of instructions by generating the appropriate sequence of micro-instructions that depend on intermediate logic conditions or on intermediate states of the *datapath* unit. To describe the evolution of a control unit, behavioral models are developed. We use the direct-connected HGs containing nodes. Each HG has an entry (*Begin*) and an output (*End*). Rectangular nodes contain micro instructions, macro instructions, or both.

A micro instruction set U_i includes a subset of micro operations from the set $U = \{u_1, u_2, \ldots, u_{u-1}, u_u\}$. Micro-operations $\{u_1, u_2, \ldots, u_{u-1}, u_u\}$ control the specific actions in the *datapath*, as shown in Figures 6.31 and 6.28. For example, one can specify that u_1 sends the data in the local stack, u_2 sends the data in the output stack, u_3 forms the address, u_4 calculates the address, u_5 forwards the data from the local stack, u_6 stores the data from the local stack in the register, u_7 forwards the data from the

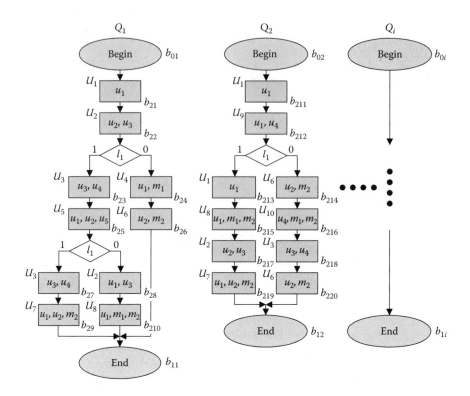

FIGURE 6.31   A control algorithm represented by HGs $Q_1, Q_2, \ldots, Q_{i-1}, Q_i$.

output stack to external output, etc. A micro-operation is the output causing an action in the *datapath*. Any macro-instruction incorporates macro-operations from the set $M = \{m_1, m_2, \ldots, m_{m-1}, m_m\}$. Each macro-operation is described by another lower-level HG. Assume that each macro instruction includes one macro operation. Each rhomboidal node contains one element from the set $L \cup G$. Here, $L = \{l_1, l_2, \ldots, l_{l-1}, l_l\}$ is the set of logic conditions, while $G = \{g_1, g_2, \ldots, g_{g-1}, g_g\}$ is the set of logic functions. Using logic conditions as inputs, they are derived by examining predefined sets of sequential steps described by a lower-level HG. Directed lines connect the inputs and outputs of the nodes. Consider a set $E = M \cup G, E = \{e_1, e_2, \ldots, e_{e-1}, e_e\}$. All elements $e_i \in E$ have HGs, and each e_i has a corresponding HG Q_i which specifies either an algorithm for performing e_i (if $e_i \in M$) or an algorithm for calculating e_i (if $e_i \in G$). Assume that $M(Q_i)$ is the subset of macro-operations and $G(Q_i)$ is the subset of logic functions that belong to the HG Q_i. If $M(Q_i) \cup G(Q_i) = \emptyset$, the well-known scheme results [45]. The application of HGs enables one to gradually and sequentially synthesize complex control algorithms, concentrating the efforts at each stage on a specified level of abstraction because specific elements of the set E are used. Each component of the set E is simple and can be checked and debugged independently. Figure 6.31 reports HGs $Q_1, Q_2, \ldots, Q_i$, which describe the control algorithm.

The execution of HGs is examined studying complex operations $e_i = m_j \in M$ and $e_i = g_j \in G$. Each complex operation e_i that is described by a HG Q_i must be replaced with a new subsequence of operators that produces the result executing Q_i. In the illustrative example, shown in Figure 6.32, Q_1 is the first HG at the first level Q^1, the second level Q^2 is formed by Q_2, Q_3, and Q_4, etc. We consider the following hierarchical sequence of HGs $Q_{1(\text{level }1)} \Rightarrow Q^2_{(\text{level }2)} \Rightarrow \cdots \Rightarrow Q^{q-1}_{(\text{level }q-1)} \Rightarrow Q^q_{(\text{level }q)}$. All $Q_{i(\text{level }i)}$ have the corresponding HGs. For example, Q^2 is a subset of the HGs used to describe elements from the set $M(Q_1) \cup G(Q_1) = \emptyset$, while Q^3 is a subset of the HGs employed to map elements from the sets $\cup_{q \in Q^2} M(q)$ and $\cup_{q \in Q^2} G(q)$. In Figure 6.32, $Q^1 = \{Q_1\}$, $Q^2 = \{Q_2, Q_3, Q_4\}$, $Q^3 = \{Q_2, Q_4, Q_5\}$, etc.

Micro-operations u^+ and u^- are used to increment and decrement the stack pointer. The problem of switching to various levels can be solved using a stack memory, see Figure 6.32. Consider an algorithm

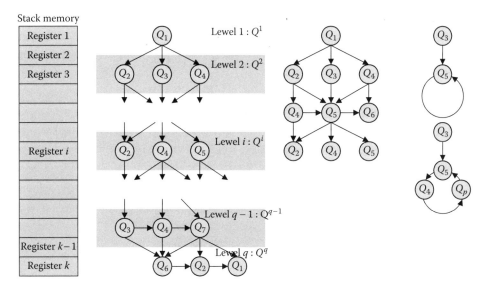

FIGURE 6.32 A stack memory with multiple-level sequential HGs, with an illustration of a recursive call.

for $e_i \in M(Q_1) \cup G(Q_1) = \emptyset$. The stack pointer is incremented by the micro operation u^+, and a new register of the stack memory is set as the current register. The previous register stores the state when it was interrupted. New Q_i becomes responsible for the control until terminated. After termination of Q_i, the micro operation u^- is generated to return to the interrupted state. As a result, control is passed to the state in which Q_f is called. The design algorithm is formulated as: For a given control algorithm A, described by the set of HGs, construct the finite-state machine that implements A. In general, the design includes the following steps: (1) the transformation of the HGs to the state transition table; (2) state encoding; (3) combinational logic optimization; (4) final structure design.

The first step is divided into three tasks as: (t1) Mark the HGs with labels b (see Figure 6.31); (t2) record transitions between the labels in the extended state transition table; (t3) convert the extended table to ordinary form. The labels b_{01} and b_{11} are assigned to the nodes *Begin* and *End* of the Q_1. The label $b_{02}, \ldots, b_{0i}$, and $b_{12}, \ldots, b_{1i}$ are assigned to nodes *Begin* and *End* of $Q_2, \ldots, Q_i$, respectively. The labels $b_{21}, b_{22}, \ldots, b_{2j}$ are assigned to other nodes of HGs, and inputs and outputs of nodes with logic conditions, etc. Repeating labels are not allowed. The labels are considered as the states. The extended state transition table is designed using the state evolutions due to inputs (logic conditions) and logic functions, which cause the transitions from $x(t)$ to $x(t + 1)$. All evolutions of the state vector $x(t)$ are recorded, and the state $x_k(t)$ has the label k. It should be emphasized that the table can be converted from the extended to the ordinary form. To program the Code Converter, as shown in Figure 6.32, one records the transition from the state x_1 assigned to the *Begin* node of the HG Q_1—e.g., $x_{01} \Rightarrow x_{21}(Q_1)$. The transitions between different HGs are recorded as $x_{ij} \Rightarrow x_{nm}(Q_j)$. For all transitions, the data-transfer instructions are derived. The hardware schematics are illustrated in Figure 6.33. Robust control algorithms are derived using the HGs, and employing both hierarchical behavior specifications and top-down decomposition. The reported method guarantees exceptional adaptation and reusability features through reconfigurable hardware and reprogrammable software for complex ICs and 3D ^{M}ICs.

6.17 Adaptive Defect-Tolerant Molecular Processing-and-Memory Platforms

Some molecular fabrication processes, such as organic synthesis, self-assembly and others, have been shown to be quite promising [5–10,47,48]. However, it is unlikely that near-future technologies will guarantee the reasonable repeatable characteristics, affordable high-quality high-yield, satisfactory uniformity, desired

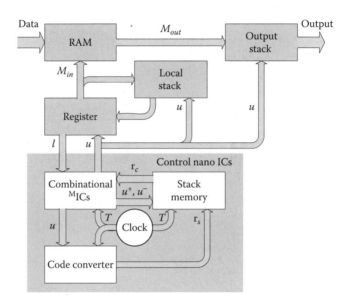

FIGURE 6.33   Hardware schematics.

failure tolerance, needed testability, and other important specifications imposed on ^{M}devices and ^{M}ICs. Therefore, the design of robust defect-tolerant adaptive (reconfigurable) architectures (hardware) and software to accommodate failures, inconsistencies, variations, nonuniformity, and defects is critical.

For conventional ICs, the programmable gate arrays (PGAs) have been developed and utilized. These PGAs lead one to the on-chip reconfigurable circuits. The reconfigurable logics can be utilized as a functional unit in the *datapath* of the processor, having access to the processor register file and to on-chip memory ports. Another approach is to integrate the reconfigurable part of the processor as a co-processor. For this solution, the reconfigurable logic operates concurrently with the processor. Optimal design and memory port assignments can guarantee the co-processor reconfigurability and concurrency. In general, the reconfigurable architecture synthesis emphasizes a high-level design, rapid prototyping, and reconfigurability in order to reduce time- and cost-improving performance. The goal is to design and fabricate affordable high-performance high-yield ^{M}ICs and application-specific ^{M}ICs. These ^{M}ICs should be testable to detect the defects and faults. The design of the application-specific ^{M}ICs involves mapping application requirements into specifications implemented by ^{M}ICs. The specifications are represented at every level of abstraction including the system, behavior, structure, physical, and process domains. The designer should be able to differently utilize ^{M}ICs to meet the application requirements.

Reconfigurable ^{M}PPs should use reprogrammable logic units, such as PGAs, to implement a specialized instruction set and arithmetic units to optimize the performance. Ideally, reconfigurable ^{M}PPs should be reconfigured in real-time (runtime), enabling the existing hardware to be reused depending on its interaction with external units, data dependencies, algorithm requirements, faults, etc. The basic PGAs architecture is built using the programmable logic blocks (PLBs) and programmable interconnect blocks (PIBs) (see Figure 6.34). The PLBs and PIBs then hold the current configuration setting until the adaptation is accomplished. The PGA is programmed by downloading the information in the file through a serial or parallel logic connection. The time required to configure a PGA is called the configuration time, and PGAs can be configured in series or in parallel. Figure 6.34 illustrates the basic architectures from which multiple PGAs architectures can be derived. For example, pipelined interfaced PGAs architecture fits for functions that have streaming data at specific intervals, while an arrayed PGAs architecture is appropriate for functions that require a systolic array. A hierarchy of configurability is different for the different PGAs architectures, and the specifics of ^{M}ICs impose emphasized constraints on the technology-centric SLSI.

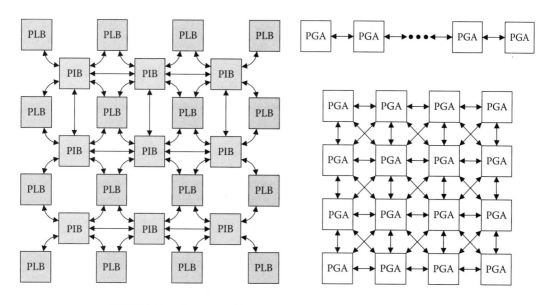

FIGURE 6.34 Programmable gate arrays and multiple PGAs organization.

The goal is to design reconfigurable ^{M}PP architectures with corresponding software to cope with less-than-perfect, entirely or partially defective and faulty ^{M}devices, ^{M}gates and ^{M}ICs used in arithmetic, logic, control, input-output, memory, and other units. To achieve our objectives, the redundant concept can be applied. The redundancy level is determined by the ^{M}ICs quality and software capabilities. Hardware and software evolutionary learning, adaptability, and reconfigurability can be achieved through decision-making, diagnostics, analysis and optimization of software, as well as the reconfiguring, pipelining, rerouting, switching, matching, controlling, and networking of hardware. Thus, one needs to design, optimize, build, test, and configure ^{M}PPs. The overall objective can be achieved by guaranteeing the evolution (behavior) matching between the ideal (C_I) and fabricated (C_F) molecular platform, its subsystems, or its components. The molecular compensator (C_{F1}) can be designed and implemented for a fabricated C_{F2} such that the response of the C_F will match the evolution of the C_I (see Figure 6.35). Both C_{F1} and C_{F2} represent ^{M}ICs hardware. The C_I gives the reference ideal evolving model which provides the ideal input–output behavior, and the compensator C_{F1} should modify the evolution of C_{F2} such that C_F, described by $C_F = C_{F1} \circ C_{F2}$ (series architecture), matches the C_I behavior and functionality. Figure 6.35

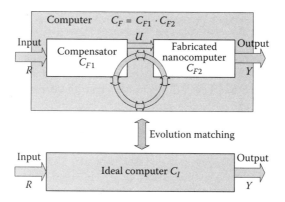

FIGURE 6.35 Molecular platform and evolution matching.

illustrates the concept. The necessary and sufficient conditions for strong and weak evolution matching based on C_I and C_{F2} must be derived.

To address analysis, control, diagnostics, optimization, and design problems, the explicit mathematical models of molecular platform or its units (subsystems) must be developed and applied. Different levels of abstraction in modeling, simulation, and analysis exist. High-level models can accept streams of instruction descriptions and memory references, while the low-level (device/gate-level) modeling can be performed by making use of streams of input and output signals examining nonlinear transient behavior and steady-state characteristics of devices. The subsystem/unit-level modeling (medium-level) also can be formulated and performed. A subsystem can contain billions of ^{M}devices, and may not be modeled as queuing networks, difference equations, Boolean models, polynomials, information-theoretic models, etc. Different mathematical modeling concepts exist and have been developed for each level. In this section, we concentrate on high-, medium-, and low-level systems modeling using the finite state machine concept.

Molecular processors and memories accept the input information, process it according to the stored instructions, and produce the output. Any mathematical model is the mathematical idealization based upon the abstractions, simplifications, and hypotheses made. It is virtually impossible to develop and apply the complete mathematical model due to complexity and uncertainties. It is possible to concurrently model a molecular platform by the six-tuple $\mathbf{C} = \{\mathbf{X}, \mathbf{E}, \mathbf{R}, \mathbf{Y}, \mathbf{F}, \mathbf{X}_0\}$, where $\mathbf{X}$ is the finite set of states with initial and final states $x_0 \in \mathbf{X}$ and $x_f \subseteq \mathbf{X}$; $\mathbf{E}$ is the finite set of events (concatenation of events forms a string of events); $\mathbf{R}$ and $\mathbf{Y}$ are the finite sets of the input and output symbols (alphabets); $\mathbf{F}$ is the transition functions mapping from $\mathbf{X} \times \mathbf{E} \times \mathbf{R} \times \mathbf{Y}$ to $\mathbf{X}$ (denoted as $\mathbf{F}_X$) to $\mathbf{E}$ (denoted as $\mathbf{F}_E$), or to $\mathbf{Y}$ (denoted as $\mathbf{F}_Y$), $\mathbf{F} \subseteq \mathbf{X} \times \mathbf{E} \times \mathbf{R} \times \mathbf{Y}$ (we assume that $\mathbf{F} = \mathbf{F}_X$—e.g., the transition function defines a new state to each quadruple of states, events, references and outputs, and $\mathbf{F}$ can be represented by a table listing the transitions, or by a state diagram).

The evolution of a molecular platform is due to inputs, events, state evolutions, parameter variations, etc. A vocabulary (or an alphabet) A is a finite nonempty set of symbols (elements). A world (or sentence) over A is a string of finite length of elements of A. The empty (null) string is the string which does not contain symbols. The set of all words over A is denoted as A_w. A language over A is a subset of A_w. A finite-state machine with output $\mathbf{C}_{FS} = \{\mathbf{X}, \mathbf{A}_R, \mathbf{A}_Y, \mathbf{F}_R, \mathbf{F}_Y, \mathbf{X}_0\}$ consists of a finite set of states S, a finite input alphabet A_R, a finite output alphabet A_Y, a transition function F_Y that assigns a new state to each state and input pair, an output function F_Y that assigns an output to each state and input pair, and an initial state X_0. Using the input–output map, the evolution of $\mathbf{C}$ can be expressed as $E_C \subseteq \mathbf{R} \times \mathbf{Y}$. That is, if $\mathbf{C}$ in state $x \in \mathbf{X}$ receives an input $r \in \mathbf{R}$, it moves to the next state $f(x,r)$, and produces the output $y(x,r)$. One can represent the molecular platform using the state tables, which describe the state and output functions. In addition, the state transition diagram (a direct graph whose vertices correspond to the states, and its edges correspond to the state transitions, where each edge is labeled with the input and output associated with the transition) is frequently used.

The quantum molecular platform is described by the seven-tuple $\mathbf{QC} = \{\mathbf{X}, \mathbf{E}, \mathbf{R}, \mathbf{Y}, \mathbf{H}, \mathbf{U}, \mathbf{X}_0\}$, where H is the Hilbert space, and U is the unitary operator in the Hilbert space that satisfies the specific conditions.

The parameters set $\mathbf{P}$ should be used. Designing reconfigurable fault-tolerant architectures, sets $\mathbf{P}$ and $\mathbf{P}_0$ are integrated, and $\mathbf{C} = \{\mathbf{X}, \mathbf{E}, \mathbf{R}, \mathbf{Y}, \mathbf{P}, \mathbf{F}, \mathbf{X}_0, \mathbf{P}_0\}$. It is evident the evolution of the $\mathbf{C}$ depends on $\mathbf{P}$ and $\mathbf{P}_0$. The optimal performance can be achieved through adaptive synthesis, reconfiguration, and diagnostics. For example, one can vary F and variable parameters P_v to attain the best possible performance. The evolution of states, events, outputs, and parameters is expressed as

$$(x_0, e_0, y_0, p_0) \overset{\text{evolution1}}{\Rightarrow} (x_1, e_1, y_1, p_1) \overset{\text{evolution2}}{\Rightarrow} \cdots \overset{\text{evolution} j-1}{\Rightarrow} (x_{j-1}, e_{j-1}, y_{j-1}, p_{j-1}) \overset{\text{evolution} j}{\Rightarrow} (x_j, e_j, y_j, p_j).$$

The input, states, outputs, events, and parameter sequences are aggregated within the model as given by $\mathbf{C} = \{\mathbf{X}, \mathbf{E}, \mathbf{R}, \mathbf{Y}, \mathbf{P}, \mathbf{F}, \mathbf{X}_0, \mathbf{P}_0\}$. The concept reported allows us to find and apply the minimal, but complete, functional description of molecular processing and memory platforms. The minimal subset of state, event, output, and parameter evolutions (transitions) can be used. That is, the partial description $\mathbf{C}_{partial} \subset \mathbf{C}$ results, and every essential quadruple (x_i, e_i, y_i, p_i) can be mapped by $(x_i, e_i, y_i, p_i)_{partial}$. This significantly reduces the complexity of modeling, simulation, analysis, and design problems.

Let the function $\mathbf{F}$ map from $\mathbf{X} \times \mathbf{E} \times \mathbf{R} \times \mathbf{Y} \times \mathbf{P}$ to $\mathbf{X}$ (e.g., $\mathbf{F} : \mathbf{X} \times \mathbf{E} \times \mathbf{R} \times \mathbf{Y} \times \mathbf{P} \rightarrow \mathbf{X}$, $\mathbf{F} \subseteq \mathbf{X} \times \mathbf{E} \times \mathbf{R} \times \mathbf{Y} \times \mathbf{P}$). Thus, the transfer function $\mathbf{F}$ defines a next state $x(t+1) \in \mathbf{X}$ based upon the current state $x(t) \in \mathbf{X}$, event $e(t) \in \mathbf{E}$, reference $r(t) \in \mathbf{R}$, output $y(t) \in \mathbf{Y}$, and parameter $p(t) \in \mathbf{P}$. Hence, $x(t+1) = F(x(t), e(t), r(t), y(t), p(t))$ for $x_0(t) \in \mathbf{X}_0$ and $p_0(t) \in \mathbf{P}_0$.

The robust adaptive algorithms must be developed. The control vector $u(t) \in \mathbf{U}$ is integrated into the model. We have $\mathbf{C} = \{\mathbf{X}, \mathbf{E}, \mathbf{R}, \mathbf{Y}, \mathbf{P}, \mathbf{U}, \mathbf{F}, \mathbf{X}_0, \mathbf{P}_0\}$, and the problem is to design the compensator. The strong evolutionary matching $C_F = C_{F1} \circ C_{F2} =_B C_I$ for given C_I and C_F is guaranteed if $E_{C_F} = E_{C_I}$. Here, $C_F =_B C_I$ means that the behaviors (evolution) of C_I and C_F are equivalent. The weak evolutionary matching $C_F = C_{F1} \circ C_{F2} \subseteq_B C_I$ for a given C_I and C_F is guaranteed if $E_{C_F} \subseteq E_{C_I}$. Here, $C_F \subseteq_B C_I$ means that the evolution of C_F is contained in the behavior C_I. The problem is to derive a compensator $C_{F1} = \{X_{F1}, E_{F1}, R_{F1}, Y_{F1}, F_{F1}, X_{F10}\}$ such that if $C_I = \{X_I, E_I, R_I, Y_I, F_I, X_{I0}\}$ and $C_{F2} = \{X_{F2}, E_{F2}, R_{F2}, Y_{F2}, F_{F2}, X_{F20}\}$, the following conditions

$\quad C_F = C_{F1} \circ C_{F2} =_B C_I$ (strong behavior matching)

or

$\quad C_F = C_{F1} \circ C_{F2} \subseteq_B C_I$ (weak behavior matching)

are satisfied. We assume that: (1) output sequences generated by C_I can be generated by C_{F2}; (2) the C_I inputs match the C_{F1} inputs.

The output sequences mean the state, event, output, and/or parameters vectors—e.g., we have (x,e,y,p). If there exists the state-modeling representation $\gamma \subseteq X_I \times X_F$ such that $C_I^{-1} \subseteq_B^{\gamma} C_{F2}^{-1}$ (if $C_I^{-1} \subseteq_B^{\gamma} C_{F2}^{-1}$, then $C_I C_B^{\gamma} C_{F2}$), then the evolution matching problem is solvable. The compensator C_{F1} solves the strong matching problem $C_F = C_{F1} \circ C_{F2} =_B C_I$ if there exists the state-modeling representations $\beta \subseteq X_I \times X_{F2}$, $(X_{I0}, X_{F20}) \in \beta$ and $\alpha \subseteq X_{F1} \times \beta$, $(X_{F10}, (X_{I0}, X_{F20})) \in \alpha$ such that $C_{F1} = {}_B^{\alpha} C_I^{\beta}$ for $\beta \in \Gamma = \{\gamma \,|\, C_I^{-1} \subseteq {}_B^{\gamma} C_{F2}^{-1}\}$. The strong matching problem is tractable if there exists C_I^{-1} and C_{F2}^{-1}. The C can be decomposed using an algebraic decomposition theory based on the closed partition lattice. For example, consider the fabricated C_{F2} represented as $C_{F2} = \{X_{F2}, E_{F2}, R_{F2}, Y_{F2}, F_{F2}, X_{F20}\}$. A partition on the state set for C_{F2} is a set $\{C_{F21}, C_{F22}, \ldots, C_{F2i}, \ldots, C_{F2k-1}, C_{F2k}\}$ of disjoint subsets of the state set X_{F2} whose union is X_{F2}—e.g., $\bigcup_{i=1}^{k} C_{F2i} = X_{F2}$ and $C_{F2i} \cap C_{F2j} = \emptyset$ for $i \neq j$. Hence, one designs and implements the compensators C_{F1i} for a given C_{F2i}.

6.18 Hardware–Software Design

Significant research activities have been focused on the synthesis of novel processing and memory platforms. The aforementioned activities must be supported by a broad spectrum of hardware-software co-design, including technology-centric CAD developments. The ${}^M architectronics$ paradigm can serve as the basis for the design and analysis of novel, efficient, robust, homogeneous, and redundant ^{M}PPs. Hardware and software co-design, integration, and verification are important problems to be addressed. The synthesis of concurrent architectures and their organization (a collection of functional hardware components, modules, subsystems, and systems that can be software programmable and adaptively reconfigurable) are among the most important issues. It is evident that software depends on hardware and vice versa. The concurrency may indicate hardware and software compliance and matching. It is impractical to fabricate high-yield ideal (perfect) complex ^{M}ICs. Furthermore, it is unlikely that the software can be developed for configurations not strictly defined, which must be adapted, reconfigured, and optimized. The not-perfect devices make diagnostics, reconfiguration, evaluation, testing, and other tasks to be implemented through robust software important. The systematic synthesis, analysis, optimization, and verification of hardware and software (as illustrated in Figure 6.36) are applied to advance the design and synthesis.

The performance analysis, verification, evaluation, characterization, and other tasks can be formulated and examined only as the molecular processing/memory platforms are devised, synthesized, and designed.

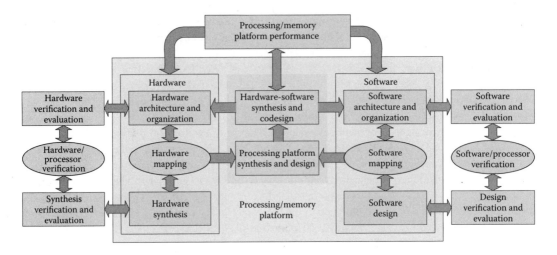

FIGURE 6.36   Hardware–software co-design for ^{M}PPs.

It is important to start the design process from a high-level, but explicitly defined, abstraction domain, which should:

- Coherently capture the functionality and performance at all levels.
- Examine and verify the proper functionality, behavior, and operation of devices, modules, subsystems, and systems.
- Depict the specification of different organizations and architectures, examining their adaptability, reconfigurability, optimality, etc.

System-level models describe processing and memory platforms as a hierarchical collection of modules, subsystems, and systems. For example, steady-state and the dynamics of gates and modules are studied to find out how these components perform and interact. The evolution of states, events, outputs, and parameters are of the designer interest. Different discrete events, process networks, Petri nets, and other methods have been applied to model computers. Models based on synchronous and asynchronous finite-state machine paradigms with some refinements ensure these meaningful features and map the essential behavior in different abstraction domains. Mixed control, data flow, data processing (encryption, filtering, and coding), and computing processes can be modeled.

A program is a set of instructions one writes to define what a computer should do. For example, if the ICs consists of *on* and *off* logic switches, one can assign that the first and second switches are *off*, while the third to eighth switches are *on* in order to receive the eight-bit signal 00111111. The program commands millions of switches, and should be written in the circuitry-level language. For ICs, software developments have progressed to the development of high-level programming languages. A high-level programming language allows you to use a vocabulary of terms—e.g., read, write, or do instead of creating the sequences of *on-off* switching which implements these functions. All high-level languages have their syntax, provide a specific vocabulary, and give an explicitly defined set of rules for using their vocabulary. A compiler is used to translate (interpret) the high-level language statements into machine code. The compiler issues the error messages if the programmer uses the programming language incorrectly. This allows one to correct the error and perform other translations by compiling the program again. The programming logic is an important issue because it involves executing various statements and procedures in the correct order to produce the desired results. One must use the syntax correctly and execute a logically constructed workable program. Two commonly used approaches to writing computer programs are procedural and object-oriented programming. Through procedural programming, one defines and executes computer memory locations (variables) to hold values, and writes sequential steps to manipulate these values. The object-oriented programming is the extension of the procedural programming because it involves generating objects (program components) and creating applications that use these objects. Objects are made up of states, and these states describe the characteristics of an object. For 3D ^{M}ICs, novel software environments must be developed that are organizationally/architecturally neutral or specific. A single software toolbox likely cannot be used, or will not be functional to all classes of ^{M}ICs that utilize different hardware solutions, and exhibit distinct phenomena, etc. For example, analog versus digital, binary versus multiple-valued, and so on.

Specific hardware and software solutions must be developed and implemented. For example, ICs are designed by making use of hardware description languages (HDLs), such as Very High Speed Integrated Circuit Hardware Description Language (VHDL) and Verilog. The design starts by interpreting the application requirements into architectural specifications. As the application requirements are examined, the designer translates the architectural specifications into behavior and structure domains. Behavior representation means the functionality required as well as the ordering of operations and the completion of tasks within specified times. A structural description consists of a set of ^{M}devices and their interconnection. Behavior and structure can be specified and studied using HDLs. These languages efficiently manage complex hierarchies, which can include millions of logic ^{M}gates. Another important feature is that HDLs are translated into net-lists of library components using synthesis software.

The structural or behavioral representations are meaningful ways of describing a model. In general, HDLs can be used for design, verification, simulation, analysis, optimization, documentation, etc. For conventional ICs, VHDL and Verilog are among the standard design tools. In VHDL, a design is typically partitioned into blocks. These blocks are then integrated to form a complete design using the schematic capture approach. This is performed using a block diagram editor or hierarchical drawings to represent block diagrams. In VHDL, every portion of a VHDL design is considered a block. Each block is analogous to an off-the-shelf IC, and is called an entity. The entity describes the interface to the block, schematics, and operation. The interface description is similar to a pin description and specifies the inputs and outputs to the block. A complete design is a collection of interconnected blocks. Consider a simple example of an entity declaration in VHDL. The first line indicates a definition of a new entity. The last line marks the end of the definition. The lines between, called the port clause, describe the interface to the design. The port clause provides a list of interface declarations. Each interface declaration defines one or more signals that are inputs or outputs to the design. Each interface declaration contains a list of names, a mode, and a type. While the interface declaration is accomplished, the architecture declaration is studied. As the basic building blocks using entities and their associated architectures are defined, one can combine them together to form other designs. The structural description of a design is a textual description of a schematic. A list of components and their connections is called a netlist. In the data flow domain, ICs are described by indicating how the inputs and outputs of built-in primitive components or pure combinational blocks are connected together. Thus, one describes how signals (data) flow through ICs. The architecture part describes the internal operation of the design. In the data flow domain, one specifies how data flows from the inputs to the outputs. In VHDL, this is accomplished with the signal assignment statement. The evaluation of the expression is performed, substituting the values of the signals in the expression and computing the result of each operator in the expression. The scheme used to model a VHDL design is called a discrete event time simulation. When the value of a signal changes, this means an event has occurred on that signal. The values of signals are only updated when discrete events occur. Since one event causes another, simulation proceeds in rounds. The simulator maintains a list of events that need to be processed. In each round, all events in a list are processed, any new events produced are placed in a separate list (scheduled) for processing in a later round. Each signal assignment is evaluated once, when simulation begins to determine the initial value of each signal to design ^{M}ICs. In general, one needs to develop new technology-centric HDLs coherently integrating 3D topologies/organization, enabling architectures, device physics, *bottom-up* fabrication, and other distinctive features of molecular electronics.

6.19   The Design and Synthesis of Molecular Electronic Devices: Toward Molecular Integrated Circuits

6.19.1   Synthesis of Molecular Electronic Devices

For ^{M}devices, the device-level research is a very important task. The performance, capabilities, and characteristics of ^{M}devices are defined by the phenomena exhibited, the effects utilized, etc. Fundamentals, high-fidelity modeling, heterogeneous simulation and the data-intensive analysis of ^{M}devices are addressed in Section 6.20. In particular, using quantum mechanics, we examine electron transport in atomic complexes in order to evaluate the soundness of the device physics of ^{ME}devices, assess device functionality, and analyze performance characteristics. These ^{M}devices must be synthesized, and sound high-yield *bottom-up* fabrication processes and technologies must be developed.

It was emphasized that ^{M}devices are comprised from functionalized aggregated molecules. All materials are composed from atoms and molecules. Lithographically defined microelectronic devices have been fabricated utilizing enhanced-functionality materials through photolithography, deposition, etching, doping, and other processes. In these solid-state microelectronic devices, individual atoms and molecules have not been examined and utilized from the device physics prospective. At the device level, the key differences between molecular and microelectronic devices are (1) phenomena exhibited, (2) effects utilized, (3) topologies, and (4) fabrication technologies. For solid-state devices using different composites, material science focuses largely on the top-down design in order to engineer enhanced-functionality materials (self-assembled thin films, templates, assemblies, etc.) with the overall goal of ensuring the desired characteristics of microelectronic devices [15]. The scaling down of microelectronic devices results in performance degradation due to quantum interference, discrete impurities, inelastic scattering, vortices, resonance, etc. [3]. The proposed ^{ME}devices exhibit the previously mentioned phenomena, and these effects are uniquely utilized, ensuring device functionality. This results in novel device physics.

One concludes that:

- For microelectronic devices, individual molecules and atoms do not depict the overall device physics and do not define device performance, functionality, and capabilities.
- For ^{M}devices, individual molecules and atoms depict the overall device physics and define device performance, functionality, and capabilities.

Focusing on ^{ME}devices and researching novel device solutions, the high-yield affordable fabrication technologies must be developed. One needs to synthesize not only *solid* or *fluidic* ^{M}devices, but complex ^{M}ICs. Those ^{M}ICs can be synthesized utilizing controlled molecular self-assembling and robust aggregation. The cyclic molecules fulfill the device physics and provide the desired synthesis capabilities. An aromatic hydrocarbon is a cyclic compound with the sp^2-hybridized atoms in the rings. This molecule, with a delocalized π-electron system, has free p-orbitals, thus ensuring the conduction of π-electrons. Some cyclic hydrocarbon molecules have $(4n + 2)\pi$-electrons, but they are not aromatic because at least one of the carbon atoms within the ring is not sp^2-hybridized. For example, cycloheptatriene has six π electrons; however, one of the seven carbon atoms is sp^3-hybridized, and the ring is not planar. The ring must be planar in order for the π-electrons to be delocalized in the ring. The planar structure ensures the stability and rigidity. Benzene is the most commonly known aromatic hydrocarbon having six π-electrons, with all six carbon atoms sp^2-hybridized; therefore, the ring is planar. In particular, the π-system of benzene is formed from six overlapping p-orbitals composing π-molecular orbitals with six π-electrons. In cyclic molecules, carbon atoms can be substituted. Figure 6.37 documents the structural and 3D-topological-view of pyridine, pyrrole, furan, and thiophene. The well-known heterocyclic biomolecules (purine and pyrimidine) contain nitrogen and oxygen (see Figure 6.37). It should be emphasized that the derivatives of purine and purimidine can be utilized to synthesize modified nucleotides. The colors are C – cyan, N – blue, O – red, and S – yellow.

Organic synthesis is the collection of procedures for the preparation of specific molecules and molecular aggregates. In planning the syntheses of desired molecules, the precursors must be selected. A great number

FIGURE 6.37 (see color insert) Pyridine, pyrrole, furan, thiophene, purine, and pyrimidine molecules.

of commercial and natural precursors are available. One carries out the retrosynthetic analysis as Target Molecule ⇒ Precursors, where the open arrow ⇒ means "is made from." More than one synthetic step is required—e.g., one has Target Molecule ⇒ Precursor 1 ⇒ ... ⇒ Precursor M ⇒ Starting Molecule. A linear synthesis, which is adequate for simple molecules, is a series of sequential steps to be performed, resulting in synthetic intermediates. For complex molecules under our consideration, convergent or divergent synthesis is required. Different procedures exist for the synthesis of synthetic intermediates. For new synthetic intermediates, the discovery, development, optimization, and implementation steps are required.

As an example, we report the Hantzsch pyridine (1,4-dihydropyridine dicarboxylate) synthesis as a multicomponent organic reaction between a formaldehyde (CH_2O), two molecules of an ethyl acetoacetate (Et denotes an ethyl C_2H_4), and an ammonium acetate (NH_4OAc) as a nitrogen donor. The initial reaction product is a dihydropyridine which can be oxidized in a subsequent step to a pyridine. The water is used as a reaction solvent, and the ferric chloride ($FeCl_3$) leads to aromatization in the second reaction step, as shown in Figure 6.38. A cyclic 1,4-dihydropyridine dicarboxylate molecule with side groups results.

Two-terminal molecular diodes and switches are reported in [5–11]. Devising multiterminal ^{ME}devices within a novel device physics is of great importance. In multiterminal *solid* ^{ME}device with the controlled $I–V$ and $G–V$ characteristics, distinct quantum phenomena could be used—for example, quantum interaction, quantum interference, quantum transition, vibration, Coulomb effect, electron spin, etc. The device physics, based on these phenomena, must be coherently complemented by the *bottom-up* synthesized molecular aggregates which will exhibit those phenomena. We consider a 3D-topology of two- and multiterminal *solid* ^{ME}devices for which one may utilize controlled tunneling, quantum interaction, and other effects.

Distinct *solid* ^{ME}devices have been proposed, ranging from resistors to multiterminal devices [5–11]. These ^{ME}devices are comprised of organic, inorganic, and bio molecules. The testing and characterization of some two-terminal ^{ME}devices are reported in [5,8,10,49]. Figure 6.39 documents different molecules, which were thiol-functionalized in order to perform the characterization measuring their $I–V$ and $G–V$ characteristics [5,8,10]. The sulfur binds to the gold cluster, usually consisting of four Au atoms (Figure 6.39 schematically illustrates only one Au atom at each binding site).

The density functional theory is used in [50] to examine the geometry, bonding, and energetics of thiol-functionalized molecules to the Au(111) surface. The gold electrodes comprise four-gold-atom clusters covalently bonded to S, and the schematics of the major structural motifs that are derived from energy minimization [50] are reported in Figure 6.40.

The aggregated molecules examined in [50] are $^{2s+1}[Au_4–X–Au_4]^q$, where **X** is the molecule, **X** = S–C_8H_8–S, **X** = S–CH_2–C_6H_4–CH_2–S (1,4-dithio-*p*-xylene), **X** = S–C_6H_4–S (1,4-phenyledithiol),

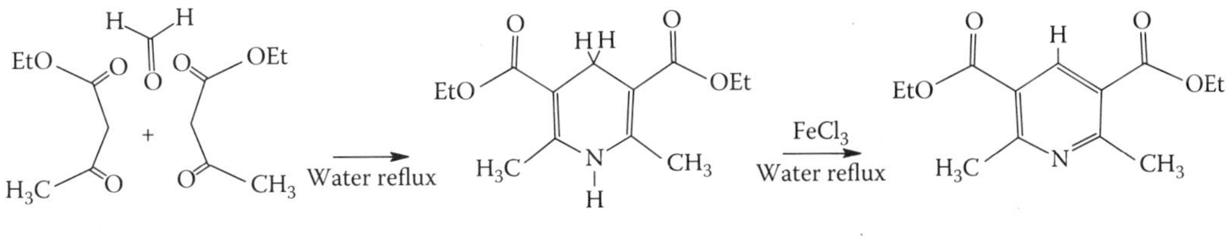

FIGURE 6.38 Synthesis of a cyclic 1,4-dihydropyridine dicarboxylate molecule.

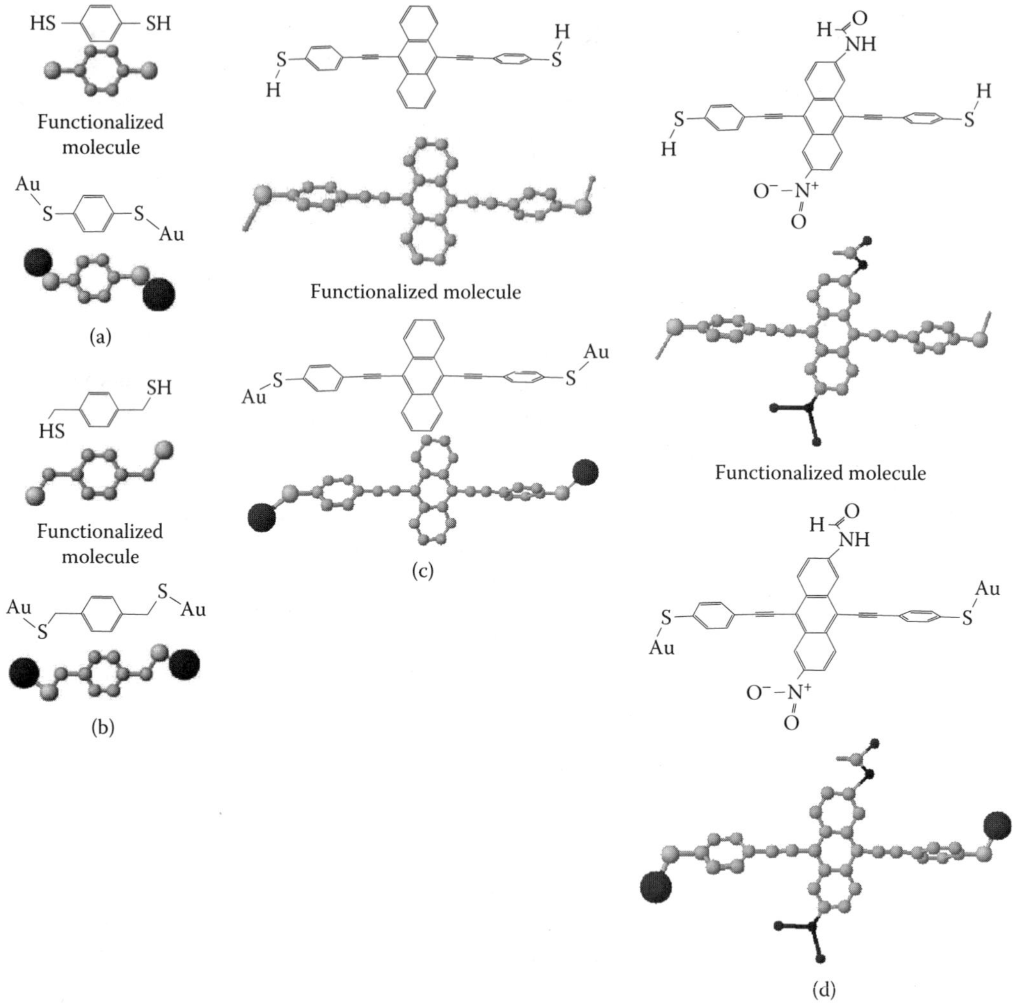

FIGURE 6.39 (see color insert) Molecules as potential two-terminal ^{ME} devices (atoms are colored as: H – green, C – cyan, N – blue, O – red, S – yellow, and Au – magenta). (a) 1,4-phenyledithiol molecule and functionalized 1,4-phenyledithiol molecule; (b) 1,4-phenylenedimethanethiol molecule; (c) 9,10-Bis([2'-para-mercaptophenyl]-ethinyl)-anthracene molecule; (d) 1,4-Bis([2'-paro-mercaptophenyl]-ethinyl)-2-acetyl-amino-5-nitro-benzene molecule.

X = S–C$_2$H$_4$–S (1,2-dithioethane), **X** = S–C$_2$H$_2$–S (1,2-dithioethylene), and **X** = S–C$_2$–S (dithioacetylene); Au$_4$ is the cluster of four Au atoms representing the electrode interconnect with the thiol-ended molecule **X**; s is the spin quantum number; amd q is the net charge on the complex. Here, $q = 0$ (neutral complex) with s = 1 and s = 3; $q = +1$ (cation) and $q = -1$ (anion) with s = 2. Different types of geometric structures (geometrical motifs) were found in [50]. The derived gold cluster geometries are

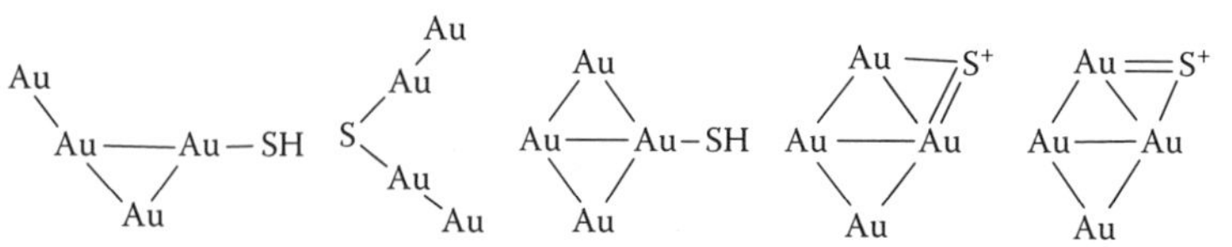

FIGURE 6.40 Gold cluster bonded to S motifs.

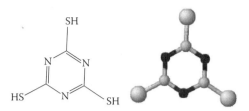

FIGURE 6.41   (see color insert) 1,3,5-triazine-2,4,6-trithiol molecules.

planar and developed from their initial tetrahedral Au$_4$ arrangement with the S atom on the three-fold axis equidistant from three Au atoms. The single Au–S and double Au $=$ S bonds result in the Au$_4$–S complexes. The bond distances are usually in the following range: Au–Au from 2.6 to 3.1Å; Au–S from 2.35 to 2.6Å; S–C from 1.6 to 1.9Å.

One faces significant challenges in fabrication of experimental test-beds (contact— $\sim$ 1 nm gap— contact) for multiterminal ^{ME}devices, as well as in their functionalization, testing, characterization, etc. Only a limited number of molecules have been tested and characterized as two-terminal devices. The major challenges are

1. Significant problems in functionalization of molecules and robust contact–molecule–contact interconnect. In fact, from the device characterization viewpoint, not all molecules of interest can be thiol-functionalized by a thiol end-group, which interacts with the Au(111) surface forming S-Au covalent bonds.
2. Current microelectronic fabrication technologies allow one to fabricate predominantly two-terminal test-beds with $\sim$ 1 nm gaps.
3. *I–V* and other baseline steady-state and dynamic (switching) characteristics are affected due to undesired effects. For example, variations of the contact–molecule–contact interconnect (bond length, interbond angle, orbital overlap, etc.), number of molecules functionalized, etc.

Fabrication challenges do not allow one to test and characterize multiterminal ^{ME}devices. As an illustration, Figure 6.41 documents a three-terminal 1,3,5-triazine-2,4,6-trithiol molecule.

To characterize a 1,3,5-triazinane-2,4,6-trione $(C_3N_3S_3^{3-})$ molecule (TMT), a H$_3$TMT molecule is utilized, as shown in Figure 6.42. A H$_3$TMT molecule can be prepared by treating the Na$_3$TMT 9H$_2$O compound with the concentrated hydrochloric acid in a 1:3 molar ratio. The 100 g of Na$_3$TMT 9H$_2$O is dissolved in 350 mL of DI water with subsequent filtering. Then 60 mL of concentrated hydrochloric acid (12.1 N) is added to the filtrate. Yellow precipitate is formed immediately, and the mixture is stirred briefly. The precipitate is isolated by filtration, washed by a copious amount of DI water, dried first at room temperature and then at 110°C. The typical yield is $\sim$ 40 g (91%), (mp 230°C). The H$_3$TMT molecule can be examined by IR spectroscopy, elemental analysis, and XRD pattern. All reagents should have 95% purity. As the molecule is synthesized for potential use as a ^{ME}device, the testing and determination of its electronic characteristics is of great importance. The molecule should be functionalized. The TMT

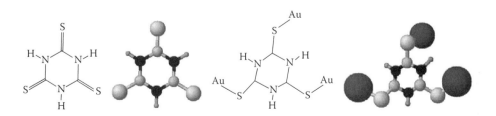

FIGURE 6.42   (see color insert) The H$_3$TMT molecule and H$_3$TMT molecule with three Au-S bonds.

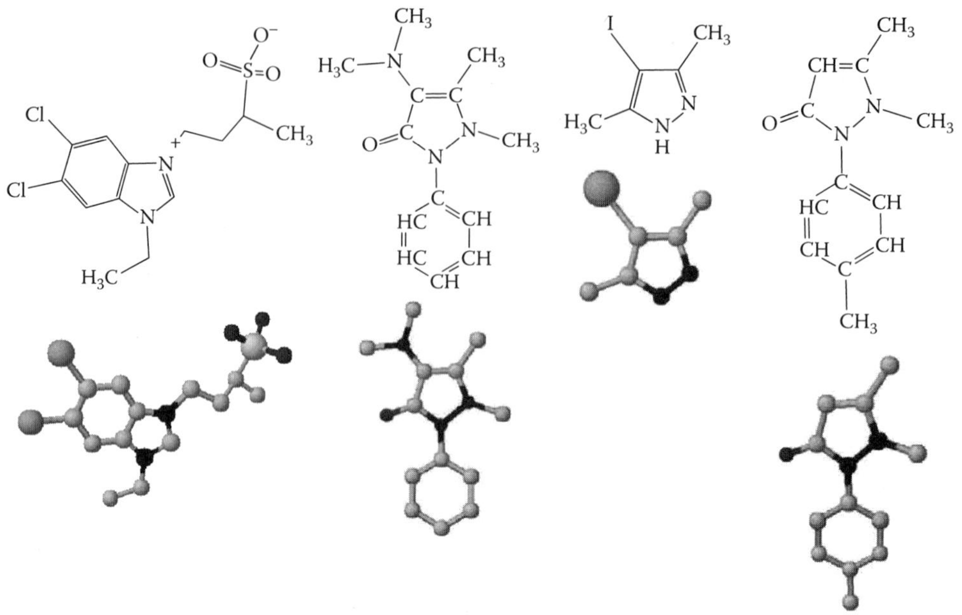

FIGURE 6.43 (see color insert) Multiterminal molecules with input, output, and control terminals: (a) (2*S*)-4-(5,6-dichloro-1-ethyl-1*H*-3,1-benzimidazol-3-ium-3-yl)butane-2-sulfonate molecule; (b) 4-(dimethylamino)-1,5-dimethyl-2-phenyl-1,2-dihydro-3*H*-pyrazol-3-one molecule; (c) 4-iodo-3,5-dimethyl-1*H*-pyrazole molecule; (d) 1,5-dimethyl-2-(4-methylphenyl)-1,2-dihydro-3*H*-pyrazol-3-one molecule.

molecule ensures stable complexes with transition metals. One can prepare the divalent molecule-metal aggregates containing ligands—e.g., TMT^{3-}, HTMT^{2-}, and H$_2$TMT$^-$. Using Au and utilizing the thiol end-group, one obtains the molecular complex, as shown in Figure 6.42.

The electronic characteristics of many organic molecules do not fully meet desired features due to insufficient controllability, symmetric I–V characteristics without desired current saturation region, thermodynamic sensitivity, etc. The electron tunneling, interactions, charge distributions, and other important features are modified by applying the potentials to the molecular terminals. However, the I – V characteristics of the functionalized monocyclic H$_3$TMT molecule are virtually symmetric without a desired saturation region. Devising, engineering, and analyzing new functional ^{ME}devices are extremely important. We depart from the symmetric organic ^{ME}devices proposing asymmetric multiterminal carbon-centered ^{ME}devices, comprised of B, N, O, P, S, I, and other atoms. To ensure synthesis feasibility and practicality, these ^{ME}devices are engineered from cyclic molecules and their derivatives. Molecular gates and ^{N}hypercells are formed from ^{ME}devices. We utilize cyclic molecules as a baseline concept. The ^{ME}devices are formed from cyclic molecules with side groups, ensuring device physics and aggregability. The proposed concept is documented in Figure 6.43. The reported molecules ensure the desired asymmetry of the I–V characteristics and saturation.

Consider a multiterminal *solid* ^{ME}device with the controlled electronic characteristics. Due to distinct device physics, one may find it unreasonable to employ the definitions and terminology of solid-state semiconductor transistors, where source-base-drain and emitter-base-collector terms are used for FETs and BJTs. To specify inputs, controls, and outputs, we propose to define the *input, output*, and *control* molecular terminals. By applying the voltage to the *control* terminal, one varies the potential, regulates the charge and electromagnetic field, and varies the interactions, as well as changes the tunneling affecting the electron transport. Hence, the input–output characteristics (I–V and G–V) can be controlled.

The monocyclic multiterminal molecule with side groups is illustrated in Figure 6.44. Here, X$_i$ denotes the specific atoms (B, C, N, O, Al, Si, P, S, Co, Br, and others); R$_i$ denotes the *input/control/output* terminals and/or side groups. The use of specific atoms and side groups is defined by the device physics, synthesis,

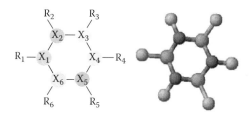

FIGURE 6.44 (see color insert) Monocyclic molecules as ^{M}devices.

aggregability, etc. The aggregation and interconnect of *input/control/output* terminals can be accomplished within the carbon framework. The reported ^{M}devices possess molecular-centered device physics, exhibiting and utilizing quantum phenomena and transitions. For example, (i) the electron transport is significantly affected by X_i and R_i; (ii) atomic structures of R_i can exhibit transitions under the external electromagnetic excitations and thermal gradient; and (iii) side groups R_i can be utilized as electron *donating* and electron *withdrawing* substituent groups, as well as interacting or interconnect groups, etc. The device aggregability and interconnect features are enhanced by utilizing side groups. The documented ^{M}devices can be employed in *solid*, *fluidic*, and hybrid molecular electronics. A three-terminal monocyclic molecule was used to design ^{M}gates within the ^{ME}D–^{ME}D logic family, as shown in Figure 6.20. A six-terminal monocyclic ^{M}device with a carbon interconnecting framework is documented in Figure 6.44.

Organometallic molecules, such as trimethylaluminum, $(CH_3)_3Al$; triethylborane $(CH_3CH_2)_3B$; tetra-ethylstannane, $(CH_3CH_2)_4Sn$; ethylmagnesium bromide, CH_3CH_2MgBr; and others can be potentially utilized in molecular electronics. As an alternative solution, a 3D-topology molecular cage with carbon interconnects, as a multiterminal ^{M}device, is shown in Figure 6.45.

Biomolecular processing platforms and molecular electronics provide indisputable evidence of super-high-performance and superb 3D-centered topology/organizations, far-surpassing any envisioned micro-electronics solutions. In general, *fluidic* ^{M}devices and ^{M}ICs offer a broader class of physics and phenomena for utilization, compared with *solid* ^{ME}devices and circuits. However, taking into account existing and prospective technologies, from a fabrication viewpoint it seems that *solid* molecular electronics may ensure a greater degree of feasibility in the near future. In biosystems, ^{BM}PPs are synthesized through robustly controlled molecular assembling, which is far beyond even envisioned comprehension and synthesis capabilities. Though the next sections primarily focus on *solid* ^{ME}devices, the *fluidic* and molecular solutions are of great importance, so we also take note of the steady progress in biomolecular technologies and fundamental advances.

6.19.2 Testing and Characterization of Proof-of-Concept Molecular Electronic Devices

To date, some proof-of-concept two-terminal ^{ME}devices have been characterized, and their I–V characteristics are measured [5,8,10,49,51,52]. To fabricate characterization test-beds, conventional microelectronic fabrication techniques, processes and materials are used. Horizontal and vertical gaps with separation between contacts in the range from ~ 1 nm to tens of nm were fabricated using photolithography, deposition,

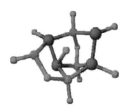

FIGURE 6.45 (see color insert) A molecular cage as a multiterminal ^{M}device.

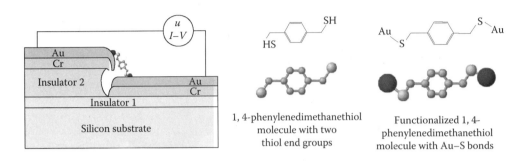

FIGURE 6.46   (see color insert) A characterization testbed with a 1,4-phenylenedimethanethiol molecule functionalized to Au contacts.

etching, and other processes. High-resolution photolithography defines planar (two-dimensional) patterns and profiles, thereby allowing one to achieve the specified patterns of insulator, metal, and other materials on the silicon wafer. Using photolithography, the mask pattern is transferred to a photoresist which is used to transfer the pattern to the substrate, as well as distinct layers on it, using sequential processes such as deposition and etching. Chemical and physical vapor deposition processes are used to deposit different insulators and conductors, while sputtering and evaporation are used to deposit Au, Pd, Ti, Cr, Al, and other metals. Wet chemical etching and dry etching are used to etch materials. Different etchants ensure desired vertical and lateral etching. Deep trenches and pits can be etched in a variety of materials, including silicon, silicon oxide, silicon nitride, etc. A combination of dry and wet etchings is integrated with materials, ensuring etching selectivity, vertical (planar) and lateral (wall) profile control, etch rate ratio control, uniformity, etc. The *anisotropic* etching uses etchants (potassium hydroxide, sodium hydroxide, ethylene-diamine-pyrocatecol, etc.) that etch different crystallographic directions at different etch rates. In contrast, the *isotropic* etching ensures the same (or close) etch rate in all directions. Different etch-stop materials are used, and these etch-stop layers can be sacrificial or structural. Shape, profile, thickness, and other features are controlled. The use of different materials, combined with etching and deposition processes, provides one with the opportunity to fabricate application-specific characterization testbeds. Molecules to be examined must be functionalized, with the metals forming robust contacts.

As a representative illustration, Figure 6.46 shows a cross-section view of a testbed, characterizing two-terminal ^{ME}devices. Chromium and gold are sequentially evaporated on the insulators. E-beam gold evaporation with adhesion layer (Cr or Ti) is a well-established process that deposits a gold layer with a specific thickness and uniformity. Through the lateral etching of insulator 2, a nanogap is engineered. If needed, unwanted Cr (near nanogap) can be removed using Cr etchants. Figure 6.46 does not reflect the dimensionality or thickness of insulators (silicon oxide, silicon dioxide, silicon nitride, aluminum oxide, zirconium oxide, or other high-k dielectrics), and adhesive (Cr or Ti) and contact/pad (Au) layers. Distinct molecules, to be characterized as ^{ME}devices, can be functionalized to the evaporated gold or titanium layers using the thiol end-group. A functionalized 1,4-phenylenedimethanethiol molecule is shown in Figure 6.46. The gap separation can be controlled by varying the processes (deposition and etching time, concentration, density, temperature, etc.). The separation between Au must match the functionalized molecule geometry and Au-**X**-Au length. For 1,4-benzenedimethane-thiols, the separation should be $\leq \sim$ 1.2 nm to form a contact–molecule–contact assembly. After fabrication, a testbed is cleaned in the Ar/O$_2$ plasma, rinsed with ethanol, and then stored in a glovebox to avoid the oxidation of Au. The molecular deposition (functionalization) involves immersion of a testbed in a 1,4-phenylenedimethanethiol solution (1 to 10 mM in ethanol) and soaking for 20 to 30 hours. Following an ethanol rinse, the *I–V* and *G–V* characteristics are measured [51]. Many variables significantly affect the electronic characteristics—e.g., the attachment of multiple functionalized molecules to contacts, variations of the contact–S bonds, tunneling, leakage, electrostatic phenomena, etc. Molecules are attached by thiol (−SH) groups, which adsorbs to the gold lattice. The thiol group ensures conduction between metal and molecule. Though thiol is the most common end-group for attaching molecules to metals, it may not form the desired coherence for testing and characterization of ^{ME}devices. For example, the geometry of the S orbitals may not ensure the conjugated π-orbitals from the molecule to interact strongly with the conduction orbitals of metal. The orbitals' mismatch creates an energy barrier at each bonding terminal, significantly impacting the electron transport. One also should avoid or minimize surface oxidation and any side effects.

Other processes to fabricate the so-called step junctions were reported in [51] with the positive slope formed using the AZ-1518 photoresist. The chromium was used as a sacrificial layer. The electrodes were formed using Ti and Au.

Electromigration-induced mechanical and electrical break nanogaps have been used. Alternative solutions are nanopore, nanoimprint, crossed wire, etc. The step and electromigration-induced gaps are relatively easy to fabricate and characterize using microscopy. An electromigration-induced break-junction technique at room temperature was introduced in [52]. Photolithographically defined Au electrodes are evaporated on the oxide-coated silicon substrate silanized with (3-Mercaptopropyl)trimethoxysilane. The subsequent electromigration procedure is carried out at room temperature to create a $\sim$ 1- to 2-nm gap between the two Au electrodes. The electromigration process is affected by the local Joule heating, melting, surface tension, migrating ions, electron forces, etc. The dissipated power per volume is estimated as $J^2\rho$, where J is the current density, which is a function of the cross-section area; ρ is the resistivity. The threshold current density is found to be from 1 to 2.5 A/μm^2 [52]. The testbed with the electromigration-induced break junction is cleaned in the Ar/O$_2$ plasma and rinsed with ethanol to remove the oxides from Au. Then, the substrate is immersed in a 1-mM solution of 1,4-phenylenedimethanethiol in ethanol. The Au–molecule contacts are formed through chemisorbed Au–S coupling, which forms contacts at both ends of the molecule. Then, the $I-V$ characteristics are measured and reported in [52].

The application of different biomolecules and modified biomolecules for molecular electronics was considered in [7,53,54]. The $I-V$ characteristics of DNA were reported in [55]. Three different short ($\sim$ 5.4 nm) double-stranded DNA (dsDNA), functionalized using short oligonucleotide linkers and thiol end-groups, were examined in [55]. In particular, paper [55] documents the experimental results when 15 base-pair single-stranded oligonucleotides **X** 3'-(CCGCGCGCCCGCCCG)-5' with a complementary **X'**, **Y** 3'-(CCGCGTTTTTGCCCG)-5' with **Y'**, and **Z** 3'-(GCCTCTCAACTCGTA)-5' with **Z'**, were hybridized to form dsDNA. They were immobilized and functionalized to the gold electrodes using the -(CH$_2$)$_3$SH and -(CH$_2$)$_6$SH oligonucleotide linkers to their 3' and 5' ends. The electromigration-induced-break-gap testbed with its $\sim$ 10-nm gap was used to test the functionalized dsDNA. The uncertainties in the testing and characterization were emphasized. The quantitative results indicate that for the applied voltage $\pm$1.2 V, the current in the **X-X'** dsDNA is $\pm$0.35 nA, while the current in **Y-Y'** dsDNA is $\pm$0.065 nA. No current was measured in the random paired **Z-Z'** dsDNA. Though it is difficult to make conclusive assertions, one may engineer ^{M}devices to ensure desired characteristics.

6.19.3   Molecular Integrated Circuits

The synthesis of multiterminal carbon-centered ^{ME}devices was covered in this section. The *bottom-up* fabrication provides techniques to engineer not only stand-alone ^{ME}devices, but also ^{M}gates and ^{M}IC. In particular, combinational logics can be implemented using molecular multiplexers or ^{M}XOR gates, while the memories can be realized applying ^{M}NAND gates. It was illustrated that there exist procedures to synthesize complex molecular aggregates progressing from ^{M}gates to ^{N}hypercell aggregates. The ^{M}PPs are envisioned to be fabricated aggregating molecular primitives (^{N}hypercells) through the robust controlled synthesis and assembly. Three-dimensional ^{M}ICs can be synthesized and implemented as cross-bar fabrics which guarantee the desired reconfigurability. The promising molecular interconnecting and interfacing solutions have been developed. The integration of molecular electronic and microelectronics is very important to ensure ICs-^{M}ICs-ICs interconnect and interface, as well as to test and characterize devices, modules, and systems. However, this does not imply that envisioned ^{M}ICs must employ microelectronics-centered interconnect solutions and technologies. There is no need to utilize the thiol and other end-groups and/or linkers to accomplish the contact–molecule–contact interconnect, as reported for proof-of-concept device testing, characterization, and evaluation. For envisioned ^{M}ICs, at the module and system levels, optical, electromagnetic, quantum, and other high-end I/O interfacing paradigms and technologies are under development.

Molecular chemistry allows one to synthesize a wide range of complex molecules from atoms linked by covalent bonds. Utilizing noncovalent and covalent intermolecular interactions as well as precisely controlling spatial (structural) and temporal (dynamic) features, supramolecular chemistry provides methods to synthesize even more complex atomic aggregates, resulting in the ability to implement ^{N}hypercells. The molecular recognition is based on well-defined interaction patterns (hydrogen bonding arrays, sequences of donor and acceptor groups, ion coordination sites, etc.). One can design preorganized molecular receptors capable of binding specific substrates with high efficiency and selectivity. The major features of the supramolecular noncovalent synthesis are (1) molecular recognition based on molecular reactivity, catalysis, and transport; (2) templating; (3) controllable robust self-assembly; (4) adaptive hierarchical self-organization (generation of well-defined, organized, and functional supramolecules by self-assembly); (5) adaptation and evolution; (6) accurate entity positioning with postassembly modification through covalent bond formation; (7) synthesis of interlocked molecular aggregates; (8) programmable preorganization; (9) recognition based on specific interaction patterns; (10) self- and complementary-selection with self-recognition, etc. A self-organization process involves three main steps: molecular recognition for the selective binding of the basic components; growth through the sequential and hierarchical binding of multiple components in the correct relative disposition; and termination of the process using a built-in feature. The self-organization should be stable towards interfering interactions (metal coordination, van der Waals stacking, etc.) and robust towards modifications of parameters (concentrations and stoichiometries of the components, presence of other species, etc.). Multimode coordinated self-organization provides additional features. The ^{M}ICs and ^{M}PPs can be synthesized utilizing the *bottom-up* fabrication as reported.

6.20   Modeling and Analysis of Molecular Electronic Devices

6.20.1   Introduction to Modeling Concepts

A great variety of molecules have been synthesized and examined for applications other than electronics. This section is devoted to the analysis of electron transport in ^{ME}devices that should ensure functionality, desired characteristics, and specified performance. These ^{ME}devices, composed from atomic aggregates ensuring chemical synthesis soundness, exhibit quantum phenomena which should be utilized. Molecular electronics devices should be examined by applying quantum mechanics. Coherent high-fidelity mathematical models are needed to carry out data-intensive analysis and to examine electron transport in molecular complexes. Mathematical models should accurately describe the basic phenomena, be computationally tractable, and suit heterogeneous simulations as needed to carry out data-intensive analysis. The modeling and analysis of electronic devices are based on the Schrödinger equation, Green's function, and other methods [7,56–58]. The kinetic energy, potentials, Fermi energy E_F, energy level broadening E_B, charge density, and other quantities, variables, and parameters are used. Figure 6.47(a) schematically illustrates 3D-topology multiterminal and two-terminal ^{ME}devices.

It has been shown that by using quantum mechanics, one can derive the dimensionless transmission probability of electron tunneling $T(E)$, which is a function of energy E, and $0 \leq T(E) \leq 1$. The conductance of molecular wires and some two-terminal ^{ME}devices was examined in [7,56–58]. A linear conductance that neglects thermal relaxation and other effects can be estimated by applying the so-called Landauer [59] or Landauer–Buttiker [60] expression $g(E) = \frac{e^2}{\pi\hbar}T(E)$. Here, the total transmission coefficient $T(E)$ is evaluated at the energy E, which is equal to the Fermi energy E_F at zero voltage bias. The so-called quantum conductance is defined to be $g_0 = \frac{e^2}{\pi\hbar} = 7.75 \times 10^{-5}\,\Omega^{-1}$. The constant $\frac{e^2}{3\pi^2\hbar^2}$ in defining the expression for conductance was originally reported in [59], where the electron transport was studied in the electric field. By making use of the acceleration of electrons $\frac{d\mathbf{k}}{dt} = -\frac{e\mathbf{E}}{\hbar}$, the expression for conductivity was provided. In particular, assuming the equilibrium condition, paper [59] states: "For our isotropic band structure and isotropic background scattering the conductivity . . . is given by $\sigma_B = \frac{\tau_B}{3\pi^2}\frac{e^2}{\hbar^2}k^2\frac{dU}{dk}$." Here, τ_B is the relaxation time, and k is the wave number.

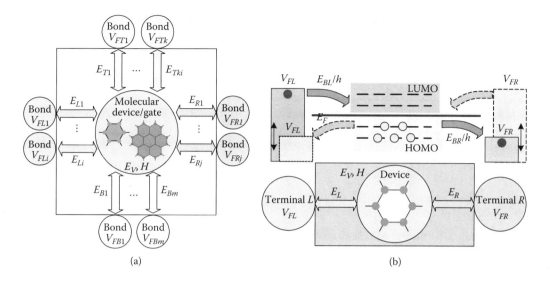

FIGURE 6.47 Molecular electronic devices: (a) a multiterminal ^{ME}device with the left (L), right (R), top (T), and bottom (B) bonds forming *input*, *control*, and *output* terminals; (b) a two-terminal ^{ME}device with Hamiltonian H, single energy potential E_V, and varying left/right potentials V_{FL} and V_{FR}.

Assuming the applicability of the Fermi–Dirac distribution, the current–voltage ($I - V$) characteristics for two-terminal ^{ME}devices (see Figure 6.47[b]) are commonly found by applying the following equation [58]

$$I(E) = \frac{2e}{h} \int_{-\infty}^{+\infty} T(E)\left[f(E_V, V_{FL}) - f(E_V, V_{FR})\right] dE,$$

where $f(E_V, V_{FL})$ and $f(E_V, V_{FR})$ are the Fermi–Dirac distribution functions, $f(E_V, V_{FL}) = (1 + e^{\frac{E_V - V_{FL}}{kT}})^{-1}$ and $f(E_V, V_{FR}) = (1 + e^{\frac{E_V - V_{FR}}{kT}})^{-1}$; E_V is the single energy potential that depends on the charge density $\rho(E)$ or the number of electrons N, $E_V = E_{V0} + V_{SC}$; V_{SC} is the self-consistent potential to be determined by solving the Poisson equation using the charge density, $V_{SC} = f_\rho(\rho)$ or $V_{SC} = V(N - N_0)$; N is the electron concentration; N_0 is the number of electrons at the equilibrium, $N_0 = 2f(E_{V0}, E_F) = 2(1 + e^{\frac{E_{V0} - E_F}{kT}})^{-1}$; and V_{FL} and V_{FR} are the left and right electrochemical potentials related to the Fermi levels.

The electrochemical potentials V_{FL} and V_{FR} vary, and there is no electron transport if $V_{FL} = V_{FR}$. The HOMO and LUMO orbitals, as well as the Fermi level, are documented in Figure 6.47(b). Depending on the HOMO and LUMO levels, as well as E_F, the electron transport takes place trough particular orbitals. The electron transport rates E_{BL}/h and E_{BR}/h are functions of the broadening energies E_{BL} and E_{BR}. One estimates the number of electrons and current as [58]

$$N = 2\frac{E_{BL} f(E_V, V_{FL}) + E_{BR} f(E_V, V_{FR})}{E_{BL} + E_{BR}} \quad \text{and} \quad I = \frac{eNE_{BR}}{h} = \frac{2e E_{BL} E_{BR}\left[f(E_V, V_{FL}) - f(E_V, V_{FR})\right]}{h(E_{BL} + E_{BR})}.$$

The approach reported previously is well-suited for semiconductor microelectronic devices. For ^{ME}devices, many assumptions and postulates made may not be ensured. Correspondingly, other methods have been applied, as was reported in Section 6.4. The application of quantum theory will be reported to examine the performance and baseline characteristics of ^{ME}devices. The wave function $[\Psi(t,\mathbf{r})]$ allowed energies, potentials, and other quantities must be studied to qualitatively and quantitatively examine the time and spatial evolution of quantum system (^{M}device) states. This ensures a coherent analysis of behavior and phenomena including electron transport. For example, the transmission coefficient, the expectation values of system variables and other quantities are derived using the wave function, which is obtained by solving the Schrödinger equation.

6.20.2   The Heisenberg Uncertainty Principle

We apply quantum theory and perform some analysis from an experimental prospective by employing the Heisenberg uncertainty principle The Heisenberg uncertainty principle specifies that no experiments can be performed to furnish uncertainties below the limits defined by the uncertainty relationship. For a perturbed particle, using complementary observable variables A and B, the generalized uncertainty principle is given as:

$$\sigma_A^2 \sigma_B^2 \geq \left(\frac{1}{2i} \langle [\hat{A}, \hat{B}] \rangle \right)^2$$

where σ_A and σ_B are the standard deviations, and $[\hat{A}, \hat{B}]$ is the commutator of two Hermitian operators $\hat{A}$ and $\hat{B}$, $[\hat{A}, \hat{B}] = \hat{A}\hat{B} - \hat{B}\hat{A}$.

We conclude that it is impossible to measure simultaneously two complementary observable variables with arbitrary accuracy. One may use the observable position x, for which $\hat{A} = x$, and the momentum p has the corresponding operator $\hat{B} = -i\hbar\frac{\partial}{\partial x}$. By taking note of the canonical commutation relation $[\hat{x}, \hat{p}] = i\hbar$, we obtain the position–momentum uncertainty principle as:

$$\sigma_x^2 \sigma_p^2 \geq \left(\frac{1}{2i} i\hbar \right)^2 = \left(\tfrac{1}{2}\hbar \right)^2 \quad \text{or} \quad \sigma_x \sigma_p \geq \tfrac{1}{2}\hbar.$$

The energy-time uncertainty principle is

$$\sigma_E \sigma_t \geq \tfrac{1}{2}\hbar.$$

Notations Δx, Δp, ΔE, and Δt are frequently used to define standard deviations as uncertainties. In Section 6.4, and in quantum mechanics books, ΔE gives the energy difference between the quantum states. Hence, covering the Heisenberg uncertainty principle, we use the notation Δ which is not ΔE.

One defines the uncertainties ΔA and ΔB in the measurement of A and B by their dispersion—for example:

$$(\Delta A)^2 = \left\langle \left(\hat{A} - \langle \hat{A} \rangle \right)^2 \right\rangle = \langle \hat{A}^2 \rangle - \langle \hat{A} \rangle^2 \quad \text{and} \quad (\Delta B)^2 = \left\langle \left(\hat{B} - \langle \hat{B} \rangle \right)^2 \right\rangle = \langle \hat{B}^2 \rangle - \langle \hat{B} \rangle^2,$$

or

$$\Delta A = \sqrt{\langle \hat{A}^2 \rangle - \langle \hat{A} \rangle^2} \quad \text{and} \quad \Delta B = \sqrt{\langle \hat{B}^2 \rangle - \langle \hat{B} \rangle^2}.$$

The uncertainty relation is $\Delta A \Delta B \geq \tfrac{1}{2} \left| \langle [\hat{A}, \hat{B}] \rangle \right|$.

The position–momentum and energy–time uncertainty principles are rewritten as

$$\Delta x \Delta p_x \geq \tfrac{1}{2}\hbar, \quad \Delta y \Delta p_y \geq \tfrac{1}{2}\hbar, \quad \Delta z \Delta p_z \geq \tfrac{1}{2}\hbar$$

and

$$\Delta \hat{E} \Delta t \geq \tfrac{1}{2}\hbar.$$

Example 6.20.3

Consider in detail the position–momentum uncertainty relation $\Delta x \Delta p_x \geq \tfrac{1}{2}\hbar$. The subscript x is used for the momentum p_x to indicate that $\Delta x \Delta p_x \geq \tfrac{1}{2}\hbar$ applies to the motion of a particle in a given direction, and relates the uncertainties in position x and momentum p_x in that direction only. The relationship $\Delta x \Delta p_x \geq \tfrac{1}{2}\hbar$ gives an estimate (one cannot do better) of the minimum uncertainty that can result from any experiment, and the measurement of the position and momentum of a particle give uncertainties Δx and Δp_x. Hence, the Heisenberg uncertainty principle indicates that if the x−component of the momentum of a particle is measured with uncertainty Δp_x, then its x−position cannot be measured more accurately than $\Delta x \geq \frac{\hbar}{2\Delta p_x}$. Thus, it is impossible to simultaneously measure two observable variables with an arbitrary accuracy. ∎

Hence, accuracy is limited. One cannot perform experiments better than conditions imposed by $\Delta x \Delta p_x \geq \frac{1}{2}\hbar$, $\Delta y \Delta p_y \geq \frac{1}{2}\hbar$, $\Delta z \Delta p_z \geq \frac{1}{2}\hbar$, and $\Delta \hat{E} \Delta t \geq \frac{1}{2}\hbar$, no matter which measuring hardware is used. It must be emphasized that the particle position, momentum, and energy are dynamic variables (measurable characteristics of the system or device) at any given time. In contrast, time is the independent variable of which the dynamic quantities are functions. That is, in $\Delta \hat{E} \Delta t \geq \frac{1}{2}\hbar$, Δt is the time it takes the system to change substantially. For example, Δt represents the amount of time it takes the expectation value of E to change by one standard deviation in order to ensure the observability of E.

The reported results impose constraints and limits on the testing, evaluation, and characterization of quantum systems, including ^{M}devices. The ability to conduct measurements for particular devices depends on the device physics, functionality, phenomena, carriers (photon, electron, or ion), etc. The uncertainty principle does not define or imply the dimensionality, switching time, power dissipation, switching energy, and other device characteristics. Those quantities must be found coherently by applying other concepts reported in this section.

Example 6.20.4

For a single photon of energy E, the momentum is $p = E/c$. The de Broglie formula relates the momentum and the wavelength λ as $p = h/\lambda$. The rest energy of the electron $m_e c^2$ is 5.1×10^5 eV. For the electron with the kinetic energy Γ, if $\Gamma \ll m_e c^2$, one may use nonrelativistic formalism to find the momentum as $p = \sqrt{2m_e \Gamma}$. Letting $\Gamma = 1$ eV, we have $p = 5.4 \times 10^{-25}$ kg-m/sec which gives $\lambda = 1.2$ nm. The frequency of radiation is $v = \frac{c}{\lambda}$. ∎

Example 6.20.5

Derive the position uncertainties Δx for a 9.1×10^{-31} kg electron (microscopic particles) and a 9.1×10^{-3} kg bullet (macroscopic particles). Let their speed be 1000 m/sec, measured with an uncertainty of 0.001%. Using $p = mv$, one finds $\Delta p = m\Delta v$. Hence, from $\Delta x \geq \frac{\hbar}{2\Delta p_x}$, for an electron one obtains $\Delta x \geq 0.00577$ m, while for a bullet we have $\Delta x \geq 5.77 \times 10^{-31}$ m. For the electron, taking note of the atomic radius of the silicon atom, which is 117 pm, one concludes that the position uncertainty Δx is 2.47×10^7 larger than the diameter of a Si atom. In contrast, the dimension of a 1-cm bullet is 1.73×10^7 times larger than Δx, thus guaranteeing no restrictions on measurements for a bullet. ∎

6.20.3 Particle Velocity

For ^{ME}devices, it is important to examine how wave packets evolve in time and space thus providing an answer regarding the motion of quantum particles in space. The velocity of the group of matter waves is equal to the particle velocity whose motion they are governing. For the wave packets propagating in the x-direction, in order to examine the time evolution, we apply the following equation

$$\Psi(t, x) = \frac{1}{\sqrt{2\pi}} \int_{-\infty}^{\infty} \phi(k) e^{i(kx - \omega t)} dk,$$

where $\phi(k)$ is the magnitude of the wave packet; k is the wave number; and ω is the angular frequency.

Examining the time evolution of the wave packet, the group and phase velocities are given as

$$v_g = \frac{d\omega(k)}{dk} = v_{ph} + k\frac{dv_{ph}}{dk} = v_{ph} + p\frac{dv_{ph}}{dp}$$

and

$$v_{ph} = \frac{\omega(k)}{k}.$$

The group velocity represents the velocity of motion of the group of propagating waves that compose the wave packet. The phase velocity is the velocity of propagation of the phase of a single mth harmonic wave $e^{ik_m(x-v_{ph}t)}$. The wave packet travels with the group velocity. Taking note of $E = \hbar\omega$ and $p = \hbar k$, one obtains

$$v_{gb} = dE(p)/dp \quad \text{and} \quad v_{ph} = E(p)/p.$$

From $E = \frac{p^2}{2m} + \Pi$, assuming that $\Pi = \text{const}$, we have

$$v_g = dE(p)/dp = p/m = v \quad \text{and} \quad v_{ph} = E(p)/p = p/2m + \Pi/p.$$

Thus, the group velocity of the wave packet is equal to the particle velocity v.

For a free electron, the energy is $E = \frac{p^2}{2m} = \frac{\hbar^2 k^2}{2m} = \hbar\omega$. Therefore, one finds $v_g = \frac{d\omega}{dk} = \frac{\hbar k}{m} = \frac{p}{m} = v$. Consider a free electron in the electric field with the intensity E_E. We have

$$dE = eE_E dx = eE_E \frac{dx}{dt}dt = eE_E v dt \quad \text{and} \quad dE = \hbar d\omega = \hbar\frac{d\omega}{dk}dk = \hbar v dk.$$

Thus, one finds $qE_E = \hbar\frac{dk}{dt}$.

The time derivative of the electron velocity $v = \frac{d\omega}{dk} = \frac{1}{\hbar}\frac{dE}{dk}$ gives the acceleration of the electron, and $a = \frac{dv}{dt} = \frac{1}{\hbar}\frac{d^2E}{dkdt} = \frac{1}{\hbar}\frac{d^2E}{dk^2}\frac{dk}{dt} = \frac{1}{\hbar^2}\frac{d^2E}{dk^2}eE_E$. The force acting on the electron is $F = \frac{dp}{dt} = \hbar\frac{dk}{dt}$ or $F = eE_E$. Hence, $a = \frac{1}{\hbar^2}\frac{d^2E}{dk^2}F$. The expression $F = \hbar^2\left(\frac{d^2E}{dk^2}\right)^{-1}\frac{dv}{dt}$ is used in solid-state semiconductor devices to introduce the so-called *effective* mass of an electron, which is $m_{eff} = \hbar^2\left(\frac{d^2E}{dk^2}\right)^{-1}$.

In solid ^{ME}devices, the device physics and 3D-topology must be coherently integrated. The derived expressions for the particle velocity can be used to obtain the $I-V$ and $G-V$ characteristics, estimate propagation delays, analyze the switching speed, and examine other characteristics of ^{ME}devices.

Example 6.20.6

Consider a wave packet corresponding to a relativistic particle. The energy and momentum are $E = mc^2 = \frac{m_0 c^2}{\sqrt{1-v^2/c^2}}$ and $p = mv = \frac{m_0 v}{\sqrt{1-v^2/c^2}}$, where m_0 is the rest mass of the particle. From $E = c\sqrt{p^2 + m_0^2 c^2}$, one obtains $v_g = \frac{dE}{dp} = \frac{d\left(c\sqrt{p^2+m_0^2 c^2}\right)}{dp} = \frac{pc}{\sqrt{p^2+m_0^2 c^2}} = v$ and $v_{ph} = \frac{E}{p} = \frac{c^2}{v}$. ∎

Example 6.20.7

Considering an electron as a non-relativistic particle, from $E = mv^2/2$, one has $v = \sqrt{\frac{2E}{m}}$. Let $E = 0.1$ eV $= 0.1602176462 \times 10^{-19}$ J. For a non-relativistic electron, we find $v = 1.88 \times 10^5$ m/sec. The time it takes an electron to travel a 1-nm distance is thus $t = L/v = 5.33 \times 10^{-15}$ sec. ∎

The particle (electron) traversal time is of interest to analyze the device performance [7,62]. In a one-dimensional case, for a particle with an energy E in $\Pi(x)$, one has $\tau(E) = \int_{x_0}^{x_f}\sqrt{\frac{m}{2[\Pi(x)-E]}}dx$. For a one-dimensional rectangular barrier with Π_0 and width L, the equation is $\tau(E) = \sqrt{\frac{m}{2(\Pi_0-E)}}L$. By using the transmission probabilities of two particle states $T_1(E)$ and $T_2(E)$, we have [63] $\tau(E) = \lim_{\lambda \to 0}\left(\frac{\hbar}{|\lambda|}\sqrt{\frac{T_2(E)}{T_1(E)}}\right)$.

Example 6.20.8

If $(\Pi_0 - E) = 0.1$ eV $= 0.16 \times 10^{-19}$ J and $L = 1$ nm, one finds $\tau = 5.33 \times 10^{-15}$ sec. The estimated τ agrees with the results reported for $\tau(E)$ in [63], where the transmission probabilities are used. As will be

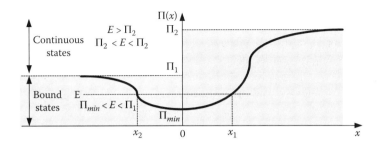

FIGURE 6.48 One-dimensional potential and motion.

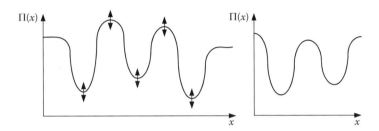

FIGURE 6.49 One-dimensional metastable potentials.

documented in Section 6.13.5, when using the wave function, one may derive the expected value for the momentum to obtain $\tau(E)$. ∎

6.20.4 Particle and Potentials

Consider a particle in a finite one-dimensional potential $\Pi(x)$ for $x \to \pm\infty$ with $\Pi(-\infty) = \Pi_1$ and $\Pi(+\infty) = \Pi_2$, as shown in Figure 6.48. If the potential has one minimum $\Pi_{min} < \Pi_1 < \Pi_2$, bound states (states whose wave functions are finite or zero at $x \to \pm\infty$) occur because the particle with energy $\Pi_{min} < E < \Pi_1$ cannot move to infinity—e.g., the particle is confined (bound) at all energies to move within a finite and limited region. The Schrödinger equation admits only a discrete solution—for example, infinite square well and harmonic oscillator problems. Unbound states (a continuous spectrum) occur when the motion of a particle is not confined—e.g., a particle is free. In particular, if $\Pi_1 < E < \Pi_2$, the particle moves towards $-\infty$—e.g., the particle moves between x_1 and $-\infty$. The energy spectrum is continuous, and none of the energy eigenvalues is degenerate. If $E > \Pi_2$, the energy spectrum is continuous, and particle motion is infinite in $\pm\infty$. The energy levels are doubly degenerate. It should be emphasized that the mixed spectrum corresponds to potentials that confine the particle for some energies only—for example, Coulomb and molecular potentials.

Let a particle be trapped in a metastable potential well [61], as shown in Figure 6.49. Due to thermodynamic fluctuations and electromagnetic fields, the particle can gain the energy from the environment or control apparatus to escape, transmit, or tunnel. Theoretical results reported in the literature provide one with various details and contradictory results. Quantum theory can be applied to ^{ME}devices emphasizing the engineering solutions that are based on solid theoretical fundamentals. Taking note of $\Pi(x)$, the Wentzel–Kramers–Brillouin approximation for the transmission $T(E)$ is given as

$$T(E) \cong e^{-\frac{2}{\hbar} \int_{x_0}^{x_f} \sqrt{2m[\Pi(x) - E]}dx} .$$

The fundamentals and applications of quantum theory are reported next.

6.20.5   The Schrödinger Equation

The time-invariant (time-independent) Schrödinger equation for a particle in the Cartesian coordinate system is given as

$$-\frac{\hbar^2}{2m}\nabla^2\Psi(x, y, z) + \Pi(x, y, z)\Psi(x, y, z) = E(x, y, z)\Psi(x, y, z),$$

where ∇^2 is the Laplacian, $\nabla^2 = \frac{\partial^2}{\partial x^2} + \frac{\partial^2}{\partial y^2} + \frac{\partial^2}{\partial z^2}$; $\Pi(x,y,z)$ is the potential energy function; and $E(x,y,z)$ is the total energy.

The Hamiltonian is $H = -\frac{\hbar^2}{2m}\nabla^2 + \Pi$.

Hence, $H(x, y, z)\Psi(x, y, z) = E(x, y, z)\Psi(x, y, z)$ or $H(\mathbf{r})\Psi(\mathbf{r}) = E(\mathbf{r})\Psi(\mathbf{r})$.

The time-dependent Schrödinger equation is

$$-\frac{\hbar^2}{2m}\nabla^2\Psi(t, x, y, z) + \Pi(t, x, y, z)\Psi(t, x, y, z) = i\hbar\frac{\partial\Psi(t, x, y, z)}{\partial t}$$

or

$$-\frac{\hbar^2}{2m}\nabla^2\Psi(t, \mathbf{r}) + \Pi(t, \mathbf{r})\Psi(t, \mathbf{r}) = i\hbar\frac{\partial\Psi(t, \mathbf{r})}{\partial t}.$$

The Schrödinger equation is (1) consistent with the de Broglie–Einstein postulates $p = h/\lambda$ and $v = E/h$; (2) consistent with total, kinetic, and potential energies—e.g., $E = p^2/2m + \Pi$; and (3) linear in $\Psi(t,\mathbf{r})$.

The Schrödinger equation should be solved using normalizing, boundary, and continuity conditions in order to find the wave function. In general, $\Psi(t,\mathbf{r})$ is a nonlinear function of energy, mass, etc. The probability of finding a particle within a volume V is $\int_V \Psi^*(t, \mathbf{r})\Psi(t, \mathbf{r})\, dV$, where $\Psi^*(t,\mathbf{r})$ is the complex conjugate of $\Psi(t,\mathbf{r})$. The wave function is normalized as $\int_{-\infty}^{\infty}\Psi^*(t, \mathbf{r})\Psi(t, \mathbf{r})\, dV = 1$, where in the Cartesian coordinate system $dV = dxdydz$.

The time evolution of the system's states is defined by the wave function. The basic connection between the properties of $\Psi(t,\mathbf{r})$ and the behavior of the associated particle is expressed by the probability density $P(t,\mathbf{r})$. For example, the quantity $P(t,x)$ specifies the probability, per unit length, of finding the particle near x at time t. Thus,

$$P(t, x) = \Psi^*(t, x)\Psi(t, x).$$

For a physical observable C that has an associated operator $\hat{C}$, the average expectation value of the observable is $\langle C \rangle = \int \Psi^*(t, \mathbf{r})\hat{C}\,\Psi(t, \mathbf{r})\, dV$. The following momentum and energy operators $p \leftrightarrow -i\hbar\frac{\partial}{\partial x}$ and $E \leftrightarrow i\hbar\frac{\partial}{\partial t}$ are applied. In general, for a momentum one has $p \leftrightarrow -i\hbar\nabla$.

For a given probability density $P(t,x)$, the expected values of any function of x can be derived. In particular,

$$\langle f(x) \rangle = \int_{-\infty}^{\infty} f(x)P(t, x)\, dx = \int_{-\infty}^{\infty}\Psi^*(t, x)f(x)\,\Psi(t, x)\, dx.$$

For example, the expectation values of x and x^2 are

$$\langle x \rangle = \int_{-\infty}^{\infty} xP(t, x)\, dx = \int_{-\infty}^{\infty}\Psi^*(t, x)x\,\Psi(t, x)\, dx$$

and

$$\langle x^2 \rangle = \int_{-\infty}^{\infty} x^2 P(t, x)\, dx = \int_{-\infty}^{\infty}\Psi^*(t, x)x^2\,\Psi(t, x)\, dx.$$

For a one-dimensional case, the expectation values of the momentum and total energy are

$$\langle p \rangle = \int_{-\infty}^{\infty} \Psi^*(t,x) \left(-i\hbar \frac{\partial}{\partial x} \right) \Psi(t,x)\, dx = -i\hbar \int_{-\infty}^{\infty} \Psi^*(t,x)\, \frac{\partial \Psi(t,x)}{\partial x}\, dx$$

and

$$\langle E \rangle = \int_{-\infty}^{\infty} \Psi^*(t,x) \left(i\hbar \frac{\partial}{\partial t} \right) \Psi(t,x)\, dx = i\hbar \int_{-\infty}^{\infty} \Psi^*(t,x)\, \frac{\partial \Psi(t,x)}{\partial t}\, dx$$

$$= \int_{-\infty}^{\infty} \Psi^*(t,x) \left(-\frac{\hbar^2}{2m} \frac{\partial^2}{\partial x^2} + \Pi(t,x) \right) \Psi(t,x)\, dx.$$

For $f(p)$, we have $\langle f(p) \rangle = \int_{-\infty}^{\infty} \Psi^*(t,x) f\left(-i\hbar \frac{\partial}{\partial x} \right) \Psi(t,x)\, dx.$

For example, one finds $\langle p^2 \rangle = \int_{-\infty}^{\infty} \Psi^*(t,x) \left(-i\hbar \frac{\partial}{\partial x} \right)^2 \Psi(t,x)\, dx = -\hbar^2 \int_{-\infty}^{\infty} \Psi^*(t,x) \frac{\partial^2 \Psi(t,x)}{\partial x^2}\, dx.$

For any dynamic quantity which is a function of x and p—for instance, $f(t,x,p)$—the expectation value is

$$\langle f(t,x,p) \rangle = \int_{-\infty}^{\infty} \Psi^*(t,x) f\left(t, x, -i\hbar \frac{\partial}{\partial x} \right) \Psi(t,x)\, dx.$$

As an illustration, for a potential $\Pi(t,x)$, we have $\langle \Pi(t,x) \rangle = \int_{-\infty}^{\infty} \Psi^*(t,x) \Pi(t,x)\, \Psi(t,x)\, dx.$

Example 6.20.9

Let the wave function for the lowest energy state of a free particle be

$$\Psi(t,x) = \begin{cases} A \cos \frac{\pi x}{L} e^{-\frac{iE}{\hbar} t} & \text{for } -\frac{1}{2}L < x < \frac{1}{2}L \\ 0 & \text{for } x \leq -\frac{1}{2}L,\ x \geq \frac{1}{2}L \end{cases}.$$

As will be documented later, we consider a particle in a one-dimensional potential well with $\Pi(x) = 0$ to be $-L/2 < x < L/2$, and $\Pi(x) = \infty$ otherwise.

One finds the total energy E by using the Schrödinger equation, which is

$$-\frac{\hbar^2}{2m} \frac{\partial^2 \Psi}{\partial x^2} = i\hbar \frac{\partial \Psi}{\partial t} \quad \text{for } -L/2 < x < L/2.$$

The expressions for the spatial and time derivatives are

$$\frac{\partial \Psi}{\partial x} = -\frac{\pi}{L} A \sin \frac{\pi x}{L} e^{-\frac{iE}{\hbar} t}, \frac{\partial^2 \Psi}{\partial x^2}$$

$$= -\frac{\pi^2}{L^2} A \cos \frac{\pi x}{L} e^{-\frac{iE}{\hbar} t} = -\frac{\pi^2}{L^2} \Psi \quad \text{and} \quad \frac{\partial \Psi}{\partial t} = -\frac{iE}{\hbar} A \cos \frac{\pi x}{L} e^{-\frac{iE}{\hbar} t} = -\frac{iE}{\hbar} \Psi.$$

Thus, the Schrödinger equation gives

$$\frac{\hbar^2}{2m} \frac{\pi^2}{L^2} \Psi = -i\hbar \frac{iE}{\hbar} \Psi.$$

Therefore, $E = \frac{\pi^2 \hbar^2}{2mL^2}.$

The expectation values of x and x^2 are found by making use of

$$\langle x \rangle = \int_{-\infty}^{\infty} x P(t,x)\, dx = \int_{-\infty}^{\infty} \Psi^*(t,x) x\, \Psi(t,x)\, dx \quad \text{and}$$

$$\langle x^2 \rangle = \int_{-\infty}^{\infty} x^2 P(t,x)\, dx = \int_{-\infty}^{\infty} \Psi^*(t,x) x^2\, \Psi(t,x)\, dx.$$

Taking note of $\Psi(t, x)$, we have

$$\langle x \rangle = \int_{-\frac{1}{2}L}^{\frac{1}{2}L} A \cos \frac{\pi x}{L} e^{\frac{iE}{\hbar}t} x A \cos \frac{\pi x}{L} e^{-\frac{iE}{\hbar}t} dx = A^2 \int_{-\frac{1}{2}L}^{\frac{1}{2}L} x \cos^2 \frac{\pi x}{L} dx = 0,$$

and

$$\langle x^2 \rangle = \int_{-\frac{1}{2}L}^{\frac{1}{2}L} A \cos \frac{\pi x}{L} e^{\frac{iE}{\hbar}t} x^2 A \cos \frac{\pi x}{L} e^{-\frac{iE}{\hbar}t} dx = A^2 \int_{-\frac{1}{2}L}^{\frac{1}{2}L} x^2 \cos^2 \frac{\pi x}{L} dx$$

$$= 2A^2 \int_0^{\frac{1}{2}L} x^2 \cos^2 \frac{\pi x}{L} dx = 2A^2 \frac{L^3}{\pi^3} \int_{-\frac{1}{2}L}^{\frac{1}{2}\pi} \left(\frac{\pi x}{L}\right)^2 \cos^2 \frac{\pi x}{L} d\frac{\pi x}{L} = A^2 \frac{L^3}{24\pi^2} (\pi^2 - 6).$$

The wave function should be normalized, and the amplitude A can thus be found. One has

$$\int_{-\infty}^{\infty} \Psi^*(t, x)\Psi(t, x)\, dx = A^2 \int_{-\frac{1}{2}L}^{\frac{1}{2}L} \cos^2 \frac{\pi x}{L} dx = 2A^2 \frac{L}{\pi} \int_0^{\frac{1}{2}\pi} \cos^2 \frac{\pi x}{L} d\frac{\pi x}{L} = 2A^2 \frac{L}{\pi} \frac{\pi}{4}.$$

By normalizing the wave function as $\int_{-\infty}^{\infty} \Psi^*(t, x)\Psi(t, x)\, dx = 1$, we obtain $A = \sqrt{\frac{2}{L}}$. Hence, $\langle x^2 \rangle = \frac{2}{L} \frac{L^3}{24\pi^2} (\pi^2 - 6) = \frac{L^2}{12\pi^2} (\pi^2 - 6)$, which gives the fluctuations of the particle about the average, and the root-mean-square value is $\sqrt{\langle x^2 \rangle}$.

From $\langle p^2 \rangle = -\hbar^2 \int_{-\infty}^{\infty} \Psi^*(t, x)\frac{\partial^2 \Psi(t,x)}{\partial x^2}\, dx$, one has $\langle p^2 \rangle = \hbar^2 \frac{\pi^2}{L^2} \int_{-\infty}^{\infty} \Psi^*(t, x)\Psi(t, x)\, dx = \frac{\hbar^2 \pi^2}{L^2}$. Thus, the root-mean-square momentum is $\sqrt{\langle p^2 \rangle} = \frac{\pi \hbar}{L}$, and $\sqrt{\langle p^2 \rangle}$ represents the average momentum fluctuations about the average $\langle p \rangle = 0$. By making use of $E = \frac{\pi^2 \hbar^2}{2mL^2}$, from $p = \pm\sqrt{2mE}$ one concludes that the magnitude of momentum is $\frac{\pi \hbar}{L}$. ∎

Example 6.20.10

Let

$$\Psi(x) = \begin{cases} 2a\sqrt{a}xe^{-ax} & \text{for } x \geq 0 \\ 0 & \text{for } x < 0 \end{cases}.$$

The peak of $P(x) = |\Psi(x)|^2$ occurs at $\frac{dP(x)}{dx} = 4a^3 \frac{d(x^2 e^{-2ax})}{dx} = 0$. By making use of $x(1 - ax)e^{-2ax} = 0$, we have $x = 1/a$. The expected values for x and x^2 are $\langle x \rangle = \int_0^{\infty} x(4a^3 x^2 e^{-2ax})dx = \frac{1}{4a} \int_0^{\infty} y^3 e^{-y}dy = \frac{3!}{4a} = \frac{3}{2a}$ and $\langle x \rangle^2 = \int_0^{\infty} x^2 (4a^3 x^2 e^{-2ax})dx = \frac{4!}{8a^2} = \frac{3}{a^2}$. ∎

For a one-dimensional problem, the probability current density $J(t, x)$ is given as

$$J(t, x) = \frac{i\hbar}{2m} \left(\Psi(t, x)\frac{\partial \Psi^*(t, x)}{\partial x} - \Psi^*(t, x)\frac{\partial \Psi(t, x)}{\partial x} \right).$$

The probability of finding a particle in the region $a < x < b$ at time t is $P_{ab}(t) = \int_a^b \Psi^*(t, x)\Psi(t, x)dx$ and $\frac{dP_{ab}}{dt} = J(t, a) - J(t, b)$. For the probability density $P(t, x) = \Psi^*(t, x)\Psi(t, x)$, one finds $\frac{\partial P(t,x)}{\partial t} + \frac{\partial J(t,x)}{\partial x} = 0$. Let the solution of the Schrödinger equation be $\Psi(t, x) = e^{-i\frac{E}{\hbar}t}\Psi(x)$. The probability density does not depend on time, $dP_{ab}/dt = 0$, and $J(t, x) = $ const. For example, if $\Psi(x) = Ae^{ikx}$, we have $P_{ab} = |A|^2(b - a)$ and $P|A|^2$. Hence, $J = \frac{\hbar k}{m}|A|^2 = \frac{\hbar k}{m}P$.

For a three-dimensional problem, we have $\frac{\partial P(t,\mathbf{r})}{\partial t} + \nabla \cdot \mathbf{J}(t, \mathbf{r}) = 0$. Here, the probability density and probability current density are $P(t, \mathbf{r}) = \Psi^*(t, \mathbf{r})\Psi(t, \mathbf{r})$ and $\mathbf{J}(t, \mathbf{r}) = \frac{i\hbar}{2m} [\Psi(t, \mathbf{r})\nabla\Psi^*(t, \mathbf{r}) - \Psi^*(t, \mathbf{r})\nabla\Psi(t, \mathbf{r})]$.

Example 6.20.11 Discussion on Meaning of Probability Current Density and Current Density

It must be emphasized that the probability current density $\mathbf{J}(t, \mathbf{r})$ and the current density $\mathbf{j}$ are entirely different variables. In semiconductor devices, one of the basic equations is $\mathbf{j} = Q\mathbf{v}$, where Q is the charge density; $\mathbf{v}$ is the velocity of the charge carrier (electron or hole), which is found by making use of the applied potential, electric field, and other quantities. Taking note of the volume charge density ρ_V, one has $\mathbf{j} = \rho_V \mathbf{v}$. Electric charges in motion constitute a current. As charged particles move from one region to another within a *conducting* path, electric potential energy is transformed. The current through the closed surface is $I = \oint_S \mathbf{j} \cdot d\mathbf{s}$, and $I = dQ/dt$. The current density in electronic devices is the number of electrons crossing a unit area per unit of time $N_s \bar{v}_x$ (the unit for N_s is [electrons/cm^2]) multiplied by the electron charge. For a one-dimensional case $j_x = -eN\bar{v}_x$ or $j_x = -e\sum_i \bar{v}_{xi}$. Here, the average net velocity is found using the average momentum per electron, $\bar{v}_x = \bar{p}_x/m$. By contrast, in quantum mechanics, $\mathbf{J}(t,\mathbf{r})$ represents the rate of probability changes, allowing one to estimate $\langle p \rangle$, which is found using $\Psi(t,\mathbf{r})$. ∎

Example 6.20.12

Reference [15] thoroughly reports the device physics and application of the basic laws to straightforwardly obtain and examine the steady-state and dynamic characteristics of FETs, BJTs, and other solid-state electronic devices. The deviations are straightforward, and some well-known basics are briefly reported next. For FETs, one may find the total charge in the channel Q and the *transit* time t, which gives the time it takes an electron to pass between the source and the drain. Thus, the drain-to-source current is $I_{DS} = Q/t$. The electron velocity is $\mathbf{v} = -\mu_n \mathbf{E}_E$, where μ_n is the electron mobility; $\mathbf{E}_E$ is the electric field intensity. One also has $\mathbf{v} = \mu_p \mathbf{E}_E$, where μ_p is the hole mobility. At room temperature for intrinsic silicon, μ_n and μ_p reach ~ 1400 cm^2/V-s and ~ 450 cm^2/V-s, respectively. It should be emphasized that μ_n and μ_p are functions of the field intensity, voltages, and other quantities, therefore the *effective* μ_{neff} and μ_{peff} are used. Using the x component of the electric field, we have $E_{Ex} = -V_{DS}/L$, where L is the channel length. Thus, $v_x = -\mu_n E_{Ex}$, and $t = L/v_x = L^2/\mu_n V_{DS}$. The channel and the gate form a parallel capacitor with plates separated by an insulator (gate oxide). From $Q = CV$, taking note that the charge appears when the voltage between the gate and the channel V_{GC} exceeds the n-channel threshold voltage V_t, one has $Q = C(V_{GC} - V_t)$. Using the equation for parallel-plate capacitors with length L, width W, and a plate separation equal to the gate-oxide thickness T_{ox}, the gate capacitance is $C = WL\varepsilon_{ox}/T_{ox}$, where ε_{ox} is the gate-oxide dielectric permittivity, and for silicon dioxide (SiO$_2$), ε_{ox} is $\sim 3.5 \times 10^{-11}$ F/m. We briefly reported the baseline equations in deriving size-dependant quantities, such as current, capacitance, velocity, *transit* time, etc. Furthermore, the analytic equations for the I–V characteristics for FETs and BJTs are straightforwardly obtained and reported in [15]. The derived expressions for the so-called Level 1 model of nFETs in the *linear* and *saturation* regions are

$$I_D = \mu_n \frac{\varepsilon_{ox}}{T_{ox}} \frac{W_c}{L_c - 2L_{GD}} \left[(V_{GS} - V_t)V_{DS} - \frac{1}{2}V_{DS}^2 \right] (1 + \lambda V_{DS}) \qquad \text{for } V_{GS} \geq V_t, \quad V_{DS} < V_{GS} - V_t$$

and

$$I_D = \frac{1}{2}\mu_n \frac{\varepsilon_{ox}}{T_{ox}} \frac{W_c}{L_c - 2L_{GD}} (V_{GS} - V_t)^2 (1 + \lambda V_{DS}) \qquad \text{for } V_{GS} \geq V_t, \quad V_{DS} \geq V_{GS} - V_t.$$

Here, I_D is the drain current; V_{GS}, V_{DS} are the gate source and drain source voltages; L_c and W_c are the channel length and width; L_{GD} is the gate-drain overlap; the device physics; and L_{GD} is the channel length modulation coefficient. For pFETs, in the equations for I_D, one uses μ_p. The coefficients and parameters used to calculate the characteristics of nFETs and pFETs are different. Due to distinct device physics, phenomena exhibited, and effects utilized, the foundations of semiconductor devices are not applicable to ^{ME}devices. For example, the electron velocity and I–V characteristics can be found using $\Psi(t,\mathbf{r})$, which depends on the three-dimensional $\mathbf{E}(\mathbf{r})$, as documented in Section 6.13. ∎

Example 6.20.13

If $\Psi(x) = Ae^{ikx} + Be^{-ikx}$, we have $\Psi^*(x) = A^*e^{-ikx} + B^*e^{ikx}$ and $\frac{d\Psi(x)}{dx} = ik(Ae^{ikx} - Be^{-ikx})$. Thus, $J(x) = \frac{\hbar k}{m}\left(|A|^2 - |B|^2\right) = \frac{p}{m}\left(|A|^2 - |B|^2\right)$. For $\Psi(x) = Ae^{\frac{iD(x)}{\hbar}}$, one finds $J(x) = \frac{1}{m}A^2\frac{dD(x)}{dx}$. ∎

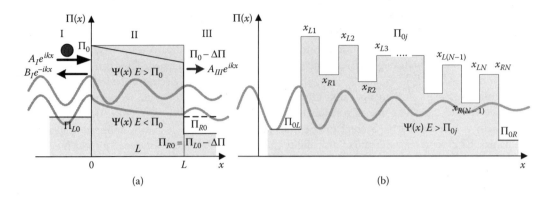

FIGURE 6.50 Electron tunneling through finite potential barriers: (a) single potential barrier; (b) multiple potential barriers.

For the potential barriers documented in Figures 6.50(a) and 6.50(b), one studies a tunneling problem examining the incident and reflected wave function amplitudes. As shown, for $\Psi(x) = Ae^{ikx} + Be^{-ikx}$, one has $J = \frac{\hbar k}{m}\left(|A|^2 - |B|^2\right)$, which can be defined as the difference between incident and reflected probability current densities—e.g., $J = J_I - J_R$. The reflection coefficient is $R = J_R/J_I = |B|^2/|A|^2$. One may find the velocity and probability density of incoming, injected, and backward electrons. The potential can vary as a result of the applied voltage (voltage bias is $\Delta V = V_L - V_R$), electric field, transitions, and other factors. Using the potential difference $\Delta\Pi$, the variation of a piecewise continuous energy potential barrier $\Pi(x)$ is shown in Figure 6.50(a). The analysis of the wave function and current (if $E < \Pi$ or $E > \Pi$) is of specific interest. One may examine electrons that move from the region of negative values of coordinate x to the region of positive values of x. At x_{Lj} and x_{Rj}, electrons encounter intermediate finite potentials Π_{0j} with width L_j (see Figures 6.50[a] and [b]). At the left and right (x_{L1} and x_{RN}), the finite potentials are denoted as Π_{0L} and Π_{0R}. There is a finite probability for transmission and reflection. The electrons on the left side that occupy the energy levels E_n can tunnel through the barrier to occupy empty energy levels E_n on the right side. The currents have contributions from all electrons.

Example 6.20.14

Consider a particle in the following infinite potential well:

$$\Pi(x) = \begin{cases} 0 & \text{for } 0 \leq x \leq L \\ \infty & \text{for } x < 0 \text{ and } x > L \end{cases}$$

The solution of the time-independent Schrödinger equation

$$-\frac{\hbar^2}{2m}\frac{\partial^2\Psi(x)}{\partial x^2} + \Pi(x)\Psi(x) = E(x)\Psi(x) \quad \text{in} \quad 0 \leq x \leq L$$

is a standing wave $\Psi(x) = A_1 e^{ikx} + B_1 e^{-ikx}$, or $\Psi(x) = A\sin kx + B\cos kx$, where $k^2 = \frac{2mE}{\hbar^2}$.

Inside the well, the wave function is found by solving $\frac{d^2\Psi(x)}{dx^2} + k^2\Psi(x) = 0$. It is evident that $\Psi(x) = 0$ at $x = 0$ and $x = L$—that is, the standing wave has nodes at the walls. From $\Psi(0) = 0$, one concludes that $B = 0$. While $\Psi(L) = 0$ results in $k_n L = n\pi$, $n = 1, 2, 3, \ldots$ By normalizing the wave function as $1 = \int_{-\infty}^{\infty}\Psi^*(x)\Psi(x)\,dx = A^2\int_0^L\sin^2\frac{\pi x}{L}dx = \frac{1}{2}A^2 L$, the amplitude A is found to be $A = \sqrt{\frac{2}{L}}$.

Thus, we have $\Psi_n(x) = \sqrt{\frac{2}{L}}\sin\frac{n\pi}{L}x = \frac{1}{i\sqrt{2L}}\left(e^{i\frac{n\pi}{L}x} - e^{-i\frac{n\pi}{L}x}\right), 0 \leq x \leq L$.

The energy levels are quantized as $E_n = \frac{\hbar^2\pi^2}{2mL^2}n^2$. Here, the integer n designates the allowed energy level (n is called the quantum number), $n = 1, 2, 3, \ldots$

The solution $\Psi_n(x) = \sqrt{\frac{2}{L}} \sin \frac{n\pi}{L} x$ is real, and we have

$$\langle p \rangle = \int_{-\infty}^{\infty} \Psi_n^*(x) \left(-i\hbar \frac{\partial}{\partial x} \right) \Psi_n(x) dx = -\frac{i\hbar}{2} \int_0^L \frac{d\,(\Psi_n(x))^2}{dx} dx = -\frac{i\hbar}{2} \left(\Psi_n^2(L) - \Psi_n^2(0) \right) = 0.$$

Taking note of $p^2 = 2mE$, for a particle in the infinite well we have $\langle p^2 \rangle = 2mE_n = \frac{\hbar^2\pi^2}{L^2}n^2$. The energy at the ground state (state of lowest energy) is $E_{n=1} = \frac{\hbar^2\pi^2}{2mL^2}$, which is different when compared to a classical particle at rest with $p = 0$ and $\Pi(x) = 0$—e.g., the sum of the kinetic and potential energy is zero.

The potential examined and a wave function derived for the standing waves are not directly related to the electron transport problem due to the infinite potential studied. To study electron transport, finite potentials which correspond to realistic potentials in atomic complexes should be examined. However, the considered example associates with the insulation and immunity problems important in ^{ME}devices. For an infinite potential, the difference between the energy levels ($E_n - E_{n-1}$) is proportional to $1/L^2$. That is, a small width L leads to high ($E_n - E_{n-1}$) distinguishing molecular (nano) electronics, for which the width is in the range of Å, and microelectronics. ∎

Example 6.20.15

Consider a finite square well of length L with three regions (I, II, and III) similar to the potential barrier as documented in Figure 6.50(a). Let

$$\Pi(x) = \begin{cases} 0 & \text{for } x < \frac{1}{2}L \\ -\Pi_0 & \text{for } -\frac{1}{2}L \leq x \leq \frac{1}{2}L. \\ 0 & \text{for } x > \frac{1}{2}L \end{cases}$$

The potential admits bound states ($E < 0$), and scattering states with $E > 0$.

Outside (if $|x| > \frac{1}{2}L$) and inside (for $-\frac{1}{2}L \leq x \leq \frac{1}{2}L$) the quantum well, the Schrödinger equations are

$$-\frac{\hbar^2}{2m}\frac{d^2\Psi}{dx^2} = E\Psi \quad \text{or} \quad \frac{d^2\Psi}{dx^2} + k^2\Psi = 0, \quad \text{where} \quad k^2 = \frac{2m}{\hbar^2}E,$$

and

$$-\frac{\hbar^2}{2m}\frac{d^2\Psi}{dx^2} - \Pi_0\Psi = E\Psi \quad \text{or} \quad \frac{d^2\Psi}{dx^2} + \kappa^2\Psi = 0, \quad \text{where} \quad \kappa^2 = \frac{2m}{\hbar^2}(E + \Pi_0).$$

The general solutions are thus:

$$\Psi_I(x) = A_I e^{ikx} + B_I e^{-ikx}, \quad x < -\frac{1}{2}L,$$
$$\Psi_{II}(x) = A_{II} e^{i\kappa x} + B_{II} e^{-i\kappa x}, \quad -\frac{1}{2}L \leq x \leq \frac{1}{2}L,$$
$$\Psi_{III}(x) = A_{III} e^{ikx} + B_{III} e^{-ikx}, \quad x > \frac{1}{2}L,$$

where A_i and B_i are the constants that can be derived using the boundary conditions, and $B_{III} = 0$.

Using boundary and continuity conditions, one finds the unknown coefficients A_j and B_j. The Schrödinger differential equations are valid in all three regions. In order to simplify the solution, we are using the particular solution, taking into account the convergence (decaying) of $\Psi(x)$—e.g., real exponentials can be used instead of complex exponentials.

Taking note of $\Psi_{II}(x) = A_{II} \sin(\kappa x) + B_{II} \cos(\kappa x)$, the continuity of $\Psi(x)$ and $d\Psi(x)/dx$ at $x = -L/2$ gives

$$A_I e^{-\frac{1}{2}ikL} + B_I e^{\frac{1}{2}ikL} = -A_{II} \sin\left(\frac{1}{2}\kappa L\right) + B_{II} \cos\left(\frac{1}{2}\kappa L\right)$$

and

$$ik\left(A_I e^{-\frac{1}{2}ikL} - B_I e^{\frac{1}{2}ikL}\right) = \kappa \left[A_{II} \cos\left(\frac{1}{2}\kappa L\right) + B_{II} \sin\left(\frac{1}{2}\kappa L\right)\right],$$

while at $x = L/2$, one obtains

$$A_{II} \sin\left(\frac{1}{2}\kappa L\right) + B_{II} \cos\left(\frac{1}{2}\kappa L\right) = A_{III} e^{\frac{1}{2}ikL}$$

and

$$\kappa\left[A_{II} \cos\left(\frac{1}{2}\kappa L\right) - B_{II} \sin\left(\frac{1}{2}\kappa L\right)\right] = ik A_{III} e^{\frac{1}{2}ikL}.$$

Here, A_I, B_I, and A_{III} are the incident, reflected, and transmitted amplitudes. With the ultimate objective to study the transmission coefficient, we have $A_{III} = \frac{e^{-ikL}}{\cos(\kappa L) - i\frac{k^2+\kappa^2}{2k\kappa}\sin(\kappa L)} A_I$. Correspondingly, the transmission coefficient is

$$T(E) = \frac{|A_{III}|^2}{|A_I|^2} = \left[1 + \frac{\Pi_0^2}{4E(E+\Pi_0)}\sin^2\left(\frac{L}{\hbar}\sqrt{2m(E+\Pi_0)}\right)\right]^{-1}.$$

The transmission coefficient is a periodic function. The maximum achievable transmission—e.g., the *total* transmission $T(E) = 1$, is guaranteed if $\frac{L}{\hbar}\sqrt{2m(E_n + \Pi_0)} = n\pi$, $n = 1, 2, 3, \ldots$

The energies for a *total* transmission are related as $E_n + \Pi_0 = \frac{n^2\pi^2\hbar^2}{2mL^2}$. Denoting $K = \frac{1}{2}L\kappa$ and $K_0 = \frac{L}{\sqrt{2}\hbar}\sqrt{m\Pi_0}$, the transcendental equation that defines K and E is $\tan K = \sqrt{\frac{K_0^2}{K^2} - 1}$. This equation can be solved analytically and numerically.

For shallow and narrow quantum wells, there is a limited number of bound states. There is always one bound state, and for $K_0 < \frac{1}{2}\pi$, only one state remains. Having solved the transcendental equation, one uses E to find the transmission coefficient which is a function of energy. The potential $\Pi(x)$, mass and well width, result in variations of $T(E)$.

We will now examine two quantum wells. If $\Pi_0 = 0.3$ eV, $L = 14$ nm, and $L = 0.14$ nm, effective masses are $0.1 m_e$, $0.5 m_e$, and m_e (semiconducting heterogeneous structure) as well as $0.5 \times 10^4 m_e$ and $1 \times 10^4 m_e$ (electron transport in organic molecules for which the bond lengths C–C and C–N are approximately 0.14 nm). Figures 6.51(a) and (b) document the numerical solutions for the studied heterogeneous structure and atomic complex. ∎

Example 6.20.16

Consider a one-dimensional scattering problem for a particle of mass m that moves from the left to the potential barrier

$$\Pi(x) = \begin{cases} 0 & \text{for } x < 0 \\ \Pi_0 & \text{for } 0 \leq x \leq L. \\ 0 & \text{for } x > L \end{cases}$$

As represented in Figure 6.50(a), we consider two cases when $E > \Pi_0$ and $E < \Pi_0$. The Schrödinger equation results in two differential equations for two distinct regions when $\Pi(x) = 0$ or $\Pi(x) = \Pi_0 \neq 0$.

Consider

$$-\frac{\hbar^2}{2m}\frac{d^2\Psi(x)}{dx^2} = E\Psi(x) \quad \text{or} \quad \frac{d^2\Psi(x)}{dx^2} + k^2\Psi(x) = 0, \quad \text{where } k^2 = \frac{2m}{\hbar^2}E,$$

and

$$-\frac{\hbar^2}{2m}\frac{d^2\Psi(x)}{dx^2} + \Pi_0\Psi(x) = E\Psi(x) \quad \text{or} \quad \frac{d^2\Psi(x)}{dx^2} + \kappa_i^2\Psi(x) = 0,$$

where κ_i (κ_1 or κ_2) depend on the amplitudes of the incident particle energy E and potential Π_0.

For $E > \Pi_0$, the general solutions in three regions are

$$\Psi_I(x) = A_I e^{ikx} + B_I e^{-ikx}, \; x < 0$$
$$\Psi_{II}(x) = A_{II} e^{i\kappa_1 x} + B_{II} e^{-i\kappa_1 x}, 0 \leq x \leq L,$$
$$\Psi_{III}(x) = A_{III} e^{ikx} + B_{III} e^{-ikx}, \; x > L$$

where $\kappa_1^2 = \frac{2m}{\hbar^2}(E - \Pi_0)$.

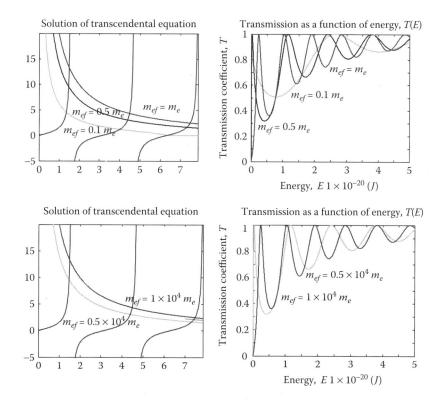

FIGURE 6.51 (see color figure) Solution for the transcendental equation and transmission coefficient $T(E)$ for finite quantum wells: (a) $\Pi_0 = 0.3$ eV and $L = 14$ nm; (b) $\Pi_0 = 0.3$ eV and $L = 0.14$ nm.

While, for $E < \Pi_0$, one defines $\kappa_2^2 = \frac{2m}{\hbar^2}(\Pi_0 - E)$, and

$$\Psi_I(x) = A_I e^{ikx} + B_I e^{-ikx}, \ x < 0$$
$$\Psi_{II}(x) = A_{II} e^{-\kappa_2 x} + B_{II} e^{\kappa_2 x}, 0 \le x \le L$$
$$\Psi_{III}(x) = A_{III} e^{ikx} + B_{III} e^{-ikx}, \ x > L.$$

For $E > \Pi_0$, applying a classical consideration, the particle with a momentum $p_1 = (2mE)^{1/2}$ entering the potential slows to a momentum $p_2 = [2m(E - \Pi_0)]^{1/2}$, and gains momentum as $x = L$ resuming p_1 at $x = L$ and keeping p_1 for $x > L$. For example, in regions $x < 0$ and $x > L$ we have a *total* transmission.

The application of quantum mechanics leads to the solution of the Schrödinger equation considering three distinct regions. For $E > \Pi_0$, one obtains

$$T(E) = \frac{|A_{III}|^2}{|A_I|^2} = \left[1 + \frac{\Pi_0^2}{4E(E - \Pi_0)}\sin^2\left(\frac{L}{\hbar}\sqrt{2m(E - \Pi_0)}\right)\right]^{-1}$$

which is usually written as

$$T(E) = \left[1 + \frac{1}{4\frac{E}{\Pi_0}\left(\frac{E}{\Pi_0} - 1\right)}\sin^2\left(\frac{L}{\hbar}\sqrt{2m\Pi_0\left(\frac{E}{\Pi_0} - 1\right)}\right)\right]^{-1}.$$

The *total* transmission occurs when the incident energy of a particle is $E_n = \Pi_0 + \frac{n^2\pi^2\hbar^2}{2mL^2}, n = 1, 2, 3, \dots$ The maxima of the $T(E)$ coincide with the energy eigenvalues of the infinite square well potential known as resonances which do not appear in classical physics consideration. This resonance phenomenon is due to interference between the incident and reflected waves observed in the atomic structures studying low energy ($E \sim 0.1$ eV) scattering electrons such as Ramsauer–Townsend and other effects.

If $E >> \Pi_0$, $T(E) \approx 1$, and $R(E) \approx 0$.

The tunneling problem is focused on analyzing the propagation of particles through regions (barrier) where the particle energy is smaller than the potential energy—e.g., $E < \Pi(x)$. For tunneling, $E < \Pi_0$, and one has

$$T(E) = \left[1 + \frac{\Pi_0^2}{4E(\Pi_0 - E)} \sinh^2\left(\frac{L}{\hbar}\sqrt{2m(\Pi_0 - E)}\right)\right]^{-1}$$

and

$$R(E) = \frac{\Pi_0^2}{4E(\Pi_0 - E)} \sinh^2\left(\frac{L}{\hbar}\sqrt{2m(\Pi_0 - E)}\right) T(E).$$

For $E << \Pi_0$, taking note of the following approximation $\sin h(z) \approx e^z/2$, one finds

$$T(E) \approx \frac{16E}{\Pi_0}\left(1 - \frac{E}{\Pi_0}\right) e^{-\frac{2L}{\hbar}\sqrt{2m(\Pi_0 - E)}}.$$

If $E \approx \Pi_0$, we obtain $T(E) = \left(1 + \frac{mL^2\Pi_0}{2\hbar^2}\right)^{-1}$ and $R(E) = \left(1 + \frac{2\hbar^2}{mL^2\Pi_0}\right)^{-1}$. ∎

Example 6.20.17

In the analysis, the scattering and transfer matrices can be used. Let

$$\Pi(x) = \begin{cases} 0 & \text{for } x < -L \\ \Pi_0 & \text{for } -L \leq x \leq L. \\ 0 & \text{for } x > L \end{cases}$$

The Schrödinger equation when $\Pi(x) = 0$ is

$$-\frac{\hbar^2}{2m}\frac{d^2\Psi(x)}{dx^2} = E\Psi(x) \quad \text{or} \quad \frac{d^2\Psi(x)}{dx^2} + k^2\Psi(x) = 0, \quad \text{where } k^2 = \frac{2m}{\hbar^2}E.$$

In the region where $\Pi(x) = \Pi_0 \neq 0$, one has

$$-\frac{\hbar^2}{2m}\frac{d^2\Psi(x)}{dx^2} + \Pi_0\Psi(x) = E\Psi(x) \quad \text{or} \quad \frac{d^2\Psi(x)}{dx^2} + \kappa^2\Psi(x) = 0,$$

$$\text{where } \kappa^2 = \frac{2m}{\hbar^2}(\Pi_0 - E) \quad \text{for } E < \Pi_0.$$

The expressions for wave functions are

$$\Psi_I(x) = A_I e^{ikx} + B_I e^{-ikx}, \; x < -L,$$
$$\Psi_{II}(x) = A_{II} e^{-\kappa x} + B_{II} e^{\kappa x}, \; -L \leq x \leq L,$$
$$\Psi_{III}(x) = A_{III} e^{ikx} + B_{III} e^{-ikx}, \; x > L.$$

Using the continuity of $\Psi(x)$ at $x = -L$, one obtains $A_I e^{-ikL} + B_I e^{ikL} = A_{II} e^{-\kappa L} + B_{II} e^{\kappa L}$, while continuity of $d\Psi/dx$ at $x = -L$ gives $ik A_I e^{-ikL} - ik B_I e^{ikL} = -\kappa A_{II} e^{\kappa L} + \kappa B_{II} e^{-\kappa L}$. Hence, we have

$$\begin{bmatrix} A_I \\ B_I \end{bmatrix} = M_1 \begin{bmatrix} A_{II} \\ B_{II} \end{bmatrix},$$

where $M_1 \in \mathbb{R}^{2 \times 2}$ is the matrix,

$$M_1 = \begin{bmatrix} \frac{ik-\kappa}{2ik} e^{-(ik+\kappa)L} & \frac{ik+\kappa}{2ik} e^{-(ik-\kappa)L} \\ \frac{ik+\kappa}{2ik} e^{(ik-\kappa)L} & \frac{ik-\kappa}{2ik} e^{(ik+\kappa)L} \end{bmatrix}.$$

Furthermore, continuity conditions for $\Psi(x)$ and $d\Psi/dx$ at boundary $x = L$ result in

$$\begin{bmatrix} A_{II} \\ B_{II} \end{bmatrix} = M_2 \begin{bmatrix} A_{III} \\ B_{III} \end{bmatrix},$$

where $M_2 \in \mathbb{R}^{2 \times 2}$ is the matrix. The transfer matrix which relates the amplitudes of wave functions in the regions I, II, and III is $M = M_1 M_2$. This transfer matrix, which provides the relationship between the incident, reflected, and transmitted wave functions, is straightforwardly applied to derive $T(E)$.

The results in deriving $T(E)$ are enhanced by applying the Wentzel–Kramers–Brillouin approximation. For a continuous slow-varying potential $\Pi(x)$, one has

$$T(E) \cong e^{-\frac{2}{\hbar} \int_{x_0}^{x_f} \sqrt{2m[\Pi(x) - E]}\, dx}.$$

This equation is obtained by making use of the Schrödinger equation $-\frac{\hbar^2}{2m} \frac{d^2\Psi}{dx^2} + \Pi(x)\Psi = E\Psi$, which is rewritten as $\frac{d^2\Psi(x)}{dx^2} + \frac{p^2(x)}{\hbar^2}\Psi(x) = 0$, where $p^2(x) = 2m[E - \Pi(x)]$. The general approximate solution is

$$\Psi(x) \cong \frac{A}{\sqrt{p(x)}} e^{\pm \frac{1}{\hbar} \int |p(x)| dx}, \quad \text{and} \quad |\Psi(x)|^2 \cong \frac{|A|^2}{p(x)}.$$

In the *classical* region with $E > \Pi(x)$, $p(x)$ is real, while for the tunneling problem, $p(x)$ is imaginary because $E < \Pi(x)$. The $\Psi(x)$ amplitudes are used to derive the Wentzel–Kramers–Brillouin expression for $T(E)$. As an alternative, the potential $\Pi(x)$ can be approximated using a number of steps $\Pi_j(x)$. ∎

Example 6.20.18

Consider an electron with mass m and energy E under the external time-invariant electric field E_F. The potential barrier that corresponds to the scattering of electrons (the cold emission of electrons from metal with the work function Π_0) is

$$\Pi(x) = \begin{cases} 0 & \text{for } x \leq 0 \\ (\Pi_0 - eE_F x) = (\Pi_0 - fx) & \text{for } x > 0 \end{cases}, \quad \text{where } f = eE_F.$$

One finds the transmission coefficient of tunneling as $T(E) \cong e^{-\frac{2}{\hbar} \int_0^{(\Pi_0 - E)/f} \sqrt{2m(\Pi_0 - ax - E)}\, dx} = e^{-\frac{4\sqrt{2m}}{3\hbar f}(\Pi_0 - E)^{\frac{3}{2}}}$. Here, the x_f is found by taking note of $(\Pi_0 - fx) = E$ at x_f. ∎

Example 6.20.19

A proton of energy E is incident from the right to a nucleus of charge Ze. To estimate the transmission coefficient that provides one with the perception of how a proton penetrates the nucleus, one considers the repulsive Coulomb force of the nucleus. The radial Coulomb potential barrier is $\Pi(r) = -Z(r)e^2/(4\pi\varepsilon_0 r)$ or $\Pi(r) = -Z_{eff} e^2/(4\pi\varepsilon_0 r)$. To simplify the resulting expression for $T(E)$, without a loss of generality, let $\Pi(r) = Ze^2/r$. Taking note of E, one finds $E = \Pi(r)|_{\text{at } b=r}$, and $b = Ze^2/E$. Thus, we have

$T(E) \propto e^{-\frac{2}{\hbar} \int_b^0 \sqrt{2m\left(\frac{Ze^2}{r} - E\right)}\, dr} = e^{-\frac{2\sqrt{2mE}}{\hbar} \int_{Ze^2/E}^0 \sqrt{\frac{Ze^2}{Er} - 1}\, dr}$. By using a new variable $y = E/(Ze^2 r)$, we have $\frac{2\sqrt{2mE}}{\hbar} \int_{Ze^2/E}^0 \sqrt{\frac{Ze^2}{Er} - 1}\, dr = \frac{2Ze^2}{\hbar} \sqrt{\frac{2m}{E}} \int_0^1 \sqrt{\frac{1}{y} - 1}\, dy = \frac{Ze^2 \pi}{\hbar} \sqrt{\frac{2m}{E}}$ because $\int_0^1 \sqrt{\frac{1}{y} - 1}\, dy = \frac{1}{2}\pi$.

Hence, $T(E) \cong e^{-\frac{\sqrt{2m} Ze^2 \pi}{\hbar} \frac{1}{\sqrt{E}}}$. ∎

Example 6.20.20

The tunneling of a particle through the rectangular double barrier with the same potentials ($\Pi_0 = \Pi_{01} = \Pi_{02}$) is considered. Let $E < \Pi_0$. Denote the barrier width as $L (L = L_1 = L_2)$ and the barriers spacing as l. By making use of the Schrödinger equation and having derived $\Psi_i(x)$, the analytic expression for the

Transmission as a function of energy, $T(E)$

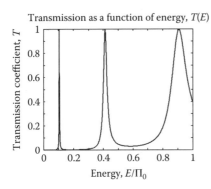

Transmission as a function of energy, $T(E)$

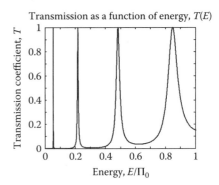

FIGURE 6.52 Tunneling as a function of energy for an electron in a rectangular double barrier with spacings $l = 4L$ and $l = 6L$.

transmission coefficient is found to be $T(E) = \left| \frac{4A^2 B^2}{C} \right|^2$, where $A = \sqrt{\frac{2m}{\hbar^2}(\Pi_0 - E)}$, $B = \sqrt{\frac{2m}{\hbar^2} E}$ and

$$C = e^{iB(l+2L)}[(e^{i2lB}(A^2 + B^2)^2 - A^4 - B^4)\sinh^2 LA$$
$$+ A^2 B^2(1 + 3\cosh 2LA) + i2AB(A^2 - B^2)\sinh 2LA].$$

Let $m = m_e = 9.11 \times 10^{-31}$ kg, $\Pi_0 = 7$ eV $= 7 \times 1.6 \times 10^{-19}$ J $= 1.12 \times 10^{-18}$ J and $L = 0.14$ nm. For two distinct l values ($l = 4L = 0.56$ nm, and $l = 6L = 0.84$ nm), the plots for $T(E)$ with three and four resonant states at different energies are documented in Figure 6.52. Significant changes of $T(E)$ are observed. ∎

Consider a finite multiple potential $\Pi(x)$, as illustrated in Figure 6.50(b). In all regions, using the Schrödinger equation, one obtains a set of $(2N + 2)$ second-order differential equations

$$-\frac{\hbar^2}{2m}\frac{d^2\Psi_j}{dx^2} + \Pi_{0j}\Psi_j = E\Psi_j \quad \text{or} \quad \frac{d^2\Psi_j}{dx^2} + \kappa_{nj}^2\Psi_j = 0, \qquad j = 0, 1, \ldots, 2N, 2N+1,$$

where κ_{nj} (κ_{1j} or κ_{2j}) depend on the particle energy E and potentials Π_{0L}, Π_{0j}, and Π_{0R}.

For $E > \Pi_{0L}$, $E > \Pi_{0j}$, and $E > \Pi_{0R}$, the general solutions are

$$\Psi_I(x) = A_I e^{i\kappa_{10}x} + B_I e^{-i\kappa_{10}x}, x < x_{L1},$$
$$\Psi_{II_j}(x) = A_{II_j} e^{i\kappa_{1j}x} + B_{II_j} e^{-i\kappa_{1j}x}, \ x_{L1} \leq x < x_{R1}, \ x_{R1} \leq x < x_{L2}, \ldots,$$
$$x_{R(N-1)} \leq x < x_{LN}, \ x_{LN} \leq x \leq x_{RN}, j = 1, 2, \ldots, N-1, N,$$
$$\Psi_{III}(x) = A_{III} e^{i\kappa_{12N+1}x} + B_{III} e^{-i\kappa_{12N+1}x}, x > x_{RN}.$$

where $\kappa_{1_0}^2 = \frac{2m}{\hbar^2}(E - \Pi_{0L})$, $\kappa_{1j}^2 = \frac{2m}{\hbar^2}(E - \Pi_{0j})$, and $\kappa_{12N+1}^2 = \frac{2m}{\hbar^2}(E - \Pi_{0R})$.

If $E < \Pi_{0L}$, $E < \Pi_{0j}$, and $E < \Pi_{0R}$, we have

$$\Psi_I(x) = A_I e^{-\kappa_{20}x} + B_I e^{\kappa_{20}x}, x < x_{L1},$$
$$\Psi_{II_j}(x) = A_{II_j} e^{-\kappa_{2j}x} + B_{II_j} e^{\kappa_{2j}x}, \ x_{L1} \leq x < x_{R1}, \ x_{R1} \leq x < x_{L2}, \ldots,$$
$$x_{R(N-1)} \leq x < x_{LN}, \ x_{LN} \leq x \leq x_{RN}, j = 1, 2, \ldots, N-1, N,$$
$$\Psi_{III}(x) = A_{III} e^{-\kappa_{22N+1}x} + B_{III} e^{\kappa_{22N+1}x}, x > x_{RN},$$

where $\kappa_{2_0}^2 = \frac{2m}{\hbar^2}(\Pi_{0L} - E)$, $\kappa_{2j}^2 = \frac{2m}{\hbar^2}(\Pi_{0j} - E)$, and $\kappa_{22N+1}^2 = \frac{2m}{\hbar^2}(\Pi_{0R} - E)$.

One may simply modify the preceding solutions, taking note of other possible relationships between potentials (Π_{0L}, Π_{0j}, and Π_{0R}) and E. The boundary and continuity conditions, as well as normalization, are used to obtain the wave functions and unknown A_I, A_{IIj}, A_{III}, B_I, B_{IIj}, and B_{III}. The interatomic

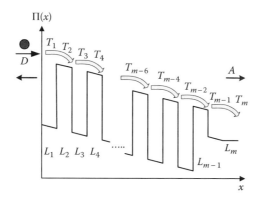

FIGURE 6.53 The energy profile.

bond lengths in various organic molecular aggregates are usually from 1 to 2 Å. For example, in fullerenes, the C–C, C–N, and C–B interatomic bond lengths are from 1.4 to 1.45 Å. Assuming that $L = (x_{Rj} - x_{Lj}) = $ const, the procedure for deriving $\Psi(x)$ and $T(E)$ can be simplified.

Molecular aggregates exhibit a complex energy profile. The Schrödinger equation is

$$\left(\frac{\hbar^2}{2m_j} \frac{d^2}{dx^2} - \Pi_j(x) + E_j(x) + E_{aj}(x) \right) \Psi_j(x) = 0$$

where $E_{aj}(x)$ is the applied external energy. The boundary and continuity conditions to be used are

$$\Psi_j(x_j) = \Psi_{j+1}(x_j) \quad \text{and} \quad \frac{1}{m_j} \left. \frac{\partial \Psi_j(x)}{\partial x} \right|_{x=x_j} = \frac{1}{m_{j+1}} \left. \frac{\partial \Psi_{j+1}(x)}{\partial x} \right|_{x=x_j}.$$

The general solutions were reported for $E_{aj} = 0$ and $\Pi_j(x) = $ const. For the energy profile, illustrated in Figure 6.53, the analytic solution is derived using Airy's functions Ai and Bi. In particular,

$$\Psi_j(x) = A_j \text{Ai}[C_j(x)] + B_j \text{Ai}[C_j(x)], \quad C_j(x) = \frac{\hbar^2 k_j^2 - 2m_j E_{aj} x}{(2m_j E_{aj})^{2/3}}.$$

For the scattering state, we have

$$M_j \begin{bmatrix} A_j \\ B_j \end{bmatrix} = M_{j+1} \begin{bmatrix} A_{j+1} \\ B_{j+1} \end{bmatrix}.$$

The transfer matrix is $M_{1 \to m} = M_{1 \to 2} M_{2 \to 3}, \ldots, M_{(m-2) \to (m-1)} M_{(m-1) \to m}$.

The analytic solution of the Schrödinger equation has been emphasized. For practical problems, including electron transport, one may depart from some assumptions and simplifications made in order to derive analytic solutions. Though the explicit expressions for wave functions, incident/reflected/transmitted amplitudes, and other quantities are of significant interest, those results are difficult to obtain for complex energy profiles. Therefore, numerical solutions, computational algorithms, and numerical methods are emphasized. Consider the Schrödinger equation

$$-\frac{\hbar^2}{2m} \frac{d^2 \Psi(x)}{dx^2} + \Pi(x)\Psi(x) = E\Psi(x),$$

which is given as a second-order differential equation

$$\frac{d^2 \Psi(x)}{dx^2} = -k^2(x)\Psi(x)$$

to be numerically solved. Here, $k^2(x) = \frac{2m}{\hbar^2} [E - \Pi(x)]$.

The Euler approximation is used to represent the first spatial derivative as a first difference—e.g.,

$$\frac{d\Psi(x)}{dx} \approx \frac{\Psi_{n+1} - \Psi_n}{\Delta_h},$$

where Δ_h is the spatial discretization spacing. Thus, the Schrödinger equation can be numerically solved through discretization, applying high-performance software. For example, MATLAB provides one with distinct application-specific differential equation solvers. Various discretization formulas and methods can be utilized.

The Numerov three-point-difference expression is $\frac{d^2\Psi(x)}{dx^2} \approx \frac{\Psi_{n+1} - 2\Psi_n + \Psi_{n-1}}{\Delta_h^2}$. From $\frac{d^2\Psi(x)}{dx^2} = -k^2(x)\Psi(x)$, one obtains a simple recursive equation

$$\Psi_{n+1} = \frac{2\left(1 - \frac{5}{12}k_n^2\Delta_h^2\right)\Psi_n - \left(1 + \frac{1}{12}k_{n-1}^2\Delta_h^2\right)\Psi_{n-1}}{1 + \frac{1}{12}k_{n+1}^2\Delta_h^2}.$$

Assigning initial values for Ψ_{n-1} and Ψ_n (for example, Ψ_0 and Ψ_1), the value of Ψ_{n+1} is derived. The *forward* or *backward* calculations of Ψ_i are performed with the accuracy $0(\Delta_h^6)$. The initial values of Ψ_{n-1} and Ψ_n can be assigned using the boundary conditions. One assigns and refines a *trial* energy E_n, guaranteeing a stability and convergence of the solution.

Using the Numerov three-point-difference expression, the Schrödinger equation is discretized as

$$\frac{\hbar^2}{2m}\left(\frac{(\Psi_{n+1} - \Psi_n) - (\Psi_n - \Psi_{n-1})}{\Delta_h^2}\right) - \Pi_n\Psi_n + E_n\Psi_n = 0.$$

Using the Hamiltonian matrix $\mathbf{H} \in \mathbb{R}^{(N+2)\times(N+2)}$, vector $\boldsymbol{\Psi} \in \mathbb{R}^{N+2}$ that contains Ψ_i, and the source vector $\mathbf{Q} \in \mathbb{R}^{N+2}$, the following matrix equation $(E\mathbf{I} - \mathbf{H})\boldsymbol{\Psi} = \mathbf{Q}$ should be solved. Here, $\mathbf{I} \in \mathbb{R}^{(N+2)\times(N+2)}$ is the identity matrix. For a two-terminal ^{ME}device, the entities of the diagonal matrix $\mathbf{H}$ are $H_{n,n} = -\frac{\hbar^2}{2m\Delta_h^2} + \Pi_n$, except $H_{0,0}$ and $H_{(N+1)(N+1)}$, which depend on the self-energies that account for the interconnect interactions. By taking note of notations used for the incoming wave function $\Psi(x) = Ae^{ik_Lx} + Be^{-ik_Lx}$, which leads to $\Psi_{-1} = Ae^{-ik_L\Delta_h} + Be^{ik_L\Delta_h} = Ae^{-ik_L\Delta_h} + (\Psi_0 - A)e^{ik_L\Delta_h}$ and $\Psi_{N+2} = \Psi_{N+1}e^{ik_R\Delta_h}$, one has $H_{0,0} = -\frac{\hbar^2}{m\Delta_h^2}(1 + \frac{1}{2}e^{ik_L\Delta_h}) + \Pi_0$ and $H_{(N+1),(N+1)} = -\frac{\hbar^2}{m\Delta_h^2}(1 + \frac{1}{2}e^{ik_R\Delta_h}) + \Pi_{N+1}$. Hence, the solution of the Schrödinger equation is reduced to the solution of a linear algebraic equation. The probability current density is

$$J = \frac{i\hbar}{2m}\left(\Psi_n\frac{\Psi_{n+1}^* - \Psi_n^*}{\Delta_h} - \Psi_n^*\frac{\Psi_{n+1} - \Psi_n}{\Delta_h}\right).$$

6.20.6 Quantum Mechanics and Molecular Electronic Devices: Three-Dimensional Problems

The electron transport in ^{ME}devices must be examined in 3D and applying quantum mechanics. The time-independent Schrödinger equation $-\frac{\hbar^2}{2m}\nabla^2\Psi(\mathbf{r}) + \Pi(\mathbf{r})\Psi(\mathbf{r}) = E(\mathbf{r})\Psi(\mathbf{r})$ can be solved in different coordinate systems depending on the problem under consideration. In the Cartesian system, we have

$$\nabla^2\Psi(\mathbf{r}) = \nabla^2\Psi(x, y, z) = \frac{\partial^2\Psi}{\partial x^2} + \frac{\partial^2\Psi}{\partial y^2} + \frac{\partial^2\Psi}{\partial z^2},$$

while in the cylindrical and spherical systems, one solves

$$\nabla^2\Psi(\mathbf{r}) = \nabla^2\Psi(r, \phi, z) = \frac{1}{r}\frac{\partial}{\partial r}\left(r\frac{\partial\Psi}{\partial r}\right) + \frac{1}{r^2}\frac{\partial^2\Psi}{\partial\phi^2} + \frac{\partial^2\Psi}{\partial z^2}$$

and

$$\nabla^2\Psi(\mathbf{r}) = \nabla^2\Psi(r, \theta, \phi) = \frac{1}{r^2}\frac{\partial}{\partial r}\left(r^2\frac{\partial\Psi}{\partial r}\right) + \frac{1}{r^2\sin\theta}\frac{\partial}{\partial\theta}\left(\sin\theta\frac{\partial\Psi}{\partial\theta}\right) + \frac{1}{r^2\sin^2\theta}\frac{\partial^2\Psi}{\partial\phi^2}.$$

The solution of the Schrödinger equation is obtained by using different analytical and numerical methods. The analytical solution can be found by using the separation of variables. For example, if the potential is $\Pi(x,y,z) = \Pi_x(x) + \Pi_y(y) + \Pi_z(z)$, one has

$$[H_x(x) + H_y(y) + H_z(z)]\Psi(x,y,z) = E\,\Psi(x,y,z),$$

where the Hamiltonians are

$$H_x(x) = -\frac{\hbar^2}{2m}\frac{\partial^2}{\partial x^2} + \Pi_x(x),\ H_y(y) = -\frac{\hbar^2}{2m}\frac{\partial^2}{\partial y^2} + \Pi_y(y),\quad \text{and}\quad H_z(z) = -\frac{\hbar^2}{2m}\frac{\partial^2}{\partial z^2} + \Pi_z(z).$$

The wave function is given as a product of three functions $\Psi(x,y,z) = X(x)Y(y)Z(z)$. This results in

$$\left[-\frac{\hbar^2}{2m}\frac{1}{X(x)}\frac{d^2X(x)}{dx^2} + \Pi_x(x)\right] + \left[-\frac{\hbar^2}{2m}\frac{1}{Y(y)}\frac{d^2Y(y)}{dy^2} + \Pi_y(y)\right] + \left[-\frac{\hbar^2}{2m}\frac{1}{Z(z)}\frac{d^2Z(z)}{dz^2} + \Pi_z(z)\right] = E,$$

where the constant total energy is $E = E_x + E_y + E_z$. The separation of variables technique results in a reduction of the three-dimensional Schrödinger equation to three independent one-dimensional equations—e.g.,

$$\left[-\frac{\hbar^2}{2m}\frac{d^2}{dx^2} + \Pi_x(x)\right]X(x) = E_x X(x),\ \left[-\frac{\hbar^2}{2m}\frac{d^2}{dy^2} + \Pi_y(y)\right]Y(y) = E_y Y(y),$$

and

$$\left[-\frac{\hbar^2}{2m}\frac{d^2}{dz^2} + \Pi_z(z)\right]Z(z) = E_z Z(z).$$

The cylindrical and spherical systems can be effectively used to reduce the complexity and make the problem tractable. In the spherical system, one uses $\Psi(r,\theta,\phi) = R(r)Y(\theta,\phi)$. The Schrödinger partial differential equation is solved using the continuity and boundary conditions, and the wave function is normalized as $\int_V \Psi^*(\mathbf{r})\,\Psi(\mathbf{r})\,dV = 1$.

Example 6.20.21

For an infinite spherical potential well, let

$$\Pi(r) = \begin{cases} 0 & \text{for } r \le a \\ \infty & \text{for } r > a \end{cases}.$$

For a particle in $\Pi(r)$ (as shown in Figure 6.54), the Schrödinger equation is

$$-\frac{\hbar^2}{2m}\left[\frac{1}{r^2}\frac{\partial}{\partial r}\left(r^2\frac{\partial\Psi}{\partial r}\right) + \frac{1}{r^2\sin\theta}\frac{\partial}{\partial\theta}\left(\sin\theta\frac{\partial\Psi}{\partial\theta}\right) + \frac{1}{r^2\sin^2\theta}\frac{\partial^2\Psi}{\partial\phi^2}\right] + \Pi(r,\theta,\phi)\Psi(r,\theta,\phi) = E\,\Psi(r,\theta,\phi).$$

We apply the separation of variables concept. The wave function is given as $\Psi(r,\theta,\phi) = R(r)Y(\theta,\phi)$. Outside the well, when $r > a$, the wave function is zero. The stationary states are labeled using three quantum

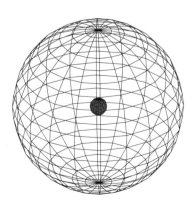

FIGURE 6.54 A particle in an infinite spherical potential well $\Pi(r)$.

numbers n, l, and m_l. Our goal is to derive the expression for $\Psi_{nlm_l}(r,\theta,\phi)$. The energy depends only on n and l—e.g., E_{nl}. In general, $\Psi_{nlm_l}(r,\theta,\phi) = A_{nl} S_{BL}(s_{nl}r/a) Y_l^{m_l}(\theta,\phi)$, where A_{nl} is the constant which must be found through the normalization of the wave function; S_{BL} is the spherical Bessel function of order l, $S_{BL}(x) = (-x)^l \left(\frac{1}{x}\frac{d}{dx}\right)^l \frac{\sin x}{x}$, and for $l = 0$ and $l = 1$, we have $S_{B0} = \sin x/x$ and $S_{B1} = \sin x/x^2 - \cos x/x$; s_{nl} is the nth zero of the lth spherical Bessel function.

Inside the well, the radial equation is

$$\frac{d^2 u}{dr^2} = \left(\frac{l(l+1)}{r^2} - k^2\right) u, \quad k^2 = \frac{2mE}{\hbar^2}.$$

The general solution of this equation for an arbitrary integer l is

$$u(r) = Ar\, S_{BL}(kr) + Br\, S_{Nl}(kr),$$

where S_N is the spherical Neumann function of order l, $S_{Nl}(x) = -(-x)^l \left(\frac{1}{x}\frac{d}{dx}\right)^l \frac{\cos x}{x}$, and for $l = 0$ and $l = 1$, one finds $S_{N0} = -\cos x/x$ and $S_{N1} = -\cos x/x^2 - \sin x/x$.

The radial wave function is $R(r) = u(r)/r$. We use the boundary condition $u(a) = 0$. For $l = 0$, from $\frac{d^2 u}{\partial r^2} = -k^2 u$, we have $u(r) = A\sin kr + B\cos kr$, where $B = 0$. Taking note of the boundary condition, from $\sin ka = 0$, one obtains $ka = n\pi$. The normalization of $u(r)$ gives $A = \sqrt{2/a}$.

The angular equation is

$$\sin\theta \frac{\partial}{\partial\theta}\left(\sin\theta \frac{\partial Y}{\partial\theta}\right) + \frac{\partial^2 Y}{\partial\phi^2} = -l(l+1)\sin^2\theta\, Y.$$

By applying $Y(\theta,\phi) = \Theta(\theta)\Phi(\phi)$, the normalized angular wave function (spherical harmonics) is known to be $Y_l^{m_l}(\theta,\phi) = \gamma\sqrt{\frac{2l+1}{4\pi}\frac{(l-|m_l|)!}{(l+|m_l|)!}} e^{im_l\phi} L_l^{m_l}(\cos\theta)$, where $\gamma = (-1)^{m_l}$ for $m_l \geq 0$ and $\gamma = 1$ for $m_l \leq 0$; $L_l^{m_l}(x)$ is the Legendre function, $L_l^{m_l}(x) = (1-x^2)^{\frac{1}{2}|m_l|}\left(\frac{d}{dx}\right)^{|m_l|} L_l(x)$; and $L_l(x)$ is the lth Legendre polynomial, $L_l(x) = \frac{1}{2^l l!}\left(\frac{d}{dx}\right)^l (x^2-1)^l$.

Thus, the angular component of the wave function for $l = 0$ and $m_l = 0$ is $Y_0^0(\theta,\phi) = \frac{1}{\sqrt{4\pi}}$.

Hence, $\Psi_{n00} = \frac{1}{\sqrt{2\pi a}}\frac{1}{r}\sin\frac{n\pi r}{a}$, and the allowed energies are $E_{n0} = \frac{\pi^2\hbar^2}{2ma^2}n^2$, $n = 1, 2, 3, \ldots$ Using the nth order of the lth spherical Bessel function S_{Bnl}, the allowed energies are $E_{nl} = \frac{\pi^2\hbar^2}{2ma^2}S_{Bnl}^2$. ∎

The Schrödinger differential equation is numerically solved in all regions for the specified potentials, energies, potential widths, boundaries, etc. For 3D-topology ^{ME}devices, using potentials, tunneling paths, interatomic bond lengths, and other data, having found $\Psi(t,\mathbf{r})$, one obtains, $P(t,\mathbf{r})$, $T(t,E)$, expected values of variables, and other quantities of interest. For example, having determined the velocity (or momentum) of a charged particle as a function of control variables (time-varying external electric or magnetic field) and parameters (mass, interatomic lengths, permittivity, etc.), the electric current is derived. As documented, the particle momentum, velocity, transmission coefficient, traversal time, and other variables change as functions of the time-varying external electromagnetic field. Therefore, depending on the device physics varying—for example, $\mathbf{E}(tr)$ or $\mathbf{B}(tr)$—one controls the electron transport. Different dynamic and steady-state characteristics are examined. For example, the steady-state experimental I–V and G–V characteristics emphasized in Section 6.12 are derived using the theoretical fundamentals reported.

For the planar solid-state semiconductor devices, to derive the transmission coefficient $T(E)$, Green's function $G(E)$ has been used. In particular, we have

$$T(E) = \text{tr}[E_{BL}G(E)E_{BR}G^*(E)].$$

To obtain the I–V characteristics, one self-consistently solves the coupled transport and Poisson's equations [7,58].

The Poisson equation $\nabla\cdot(\varepsilon(\mathbf{r})\nabla V(\mathbf{r})) = -\rho(\mathbf{r})$ is used to find the electric field intensity and electrostatic potential. Here, $\rho(\mathbf{r})$ is the charge density, which is not a probability current density $\rho(t,\mathbf{r})$; and $\varepsilon(\mathbf{r})$ is the dielectric tensor.

For example, letting $\rho_x = \rho_0 \text{sech} \frac{x}{L} \tanh \frac{x}{L}$, we solve $\nabla^2 V_x = -\frac{\rho_x}{\varepsilon}$, obtaining the following expressions: $E_x = -\frac{\rho_0}{\varepsilon} L \text{sech} \frac{x}{L}$ and $V_x = 2\frac{\rho_0}{\varepsilon} L^2 (\tan^{-1} e^{\frac{x}{L}} - \frac{1}{4}\pi)$.

For 3D-topology ^{ME}devices, the Poisson equation is of great importance in attaining a self-consistent solution. The Schrödinger and Poisson equations are solved utilizing robust numerical methods using the difference expressions for the Laplacian, integration–differentiation concepts, etc. It is possible to solve differential equations in 3D using a finite difference method that gives lattices. Generalizing the results reported for the one-dimensional problem, for the Laplace equation one has $\frac{\partial^2 V(i,j,k)}{\partial^2 r} = \frac{V(i+1,j,k)-2V(i,j,k)+V(i-1,j,k)}{\Delta_h^2}$, where (i, j, k) gives a grid point; Δ_h is the spatial discretization spacing in the x, y, or z directions.

For Poisson's equation, we have

$$\nabla \cdot (\varepsilon(\mathbf{r})\nabla V(\mathbf{r})) = \frac{C_{i,j,k}^{i+1,j,k}(V_{i+1,j,k} - V_{i,j,k}) - C_{i-1,j,k}^{i,j,k}(V_{i,j,k} - V_{i-1,j,k})}{\Delta_x^2}$$

$$+ \frac{C_{i,j,k}^{i,j+1,k}(V_{i,j+1,k} - V_{i,j,k}) - C_{i,j-1,k}^{i,j,k}(V_{i,j,k} - V_{i,j-1,k})}{\Delta_y^2}$$

$$+ \frac{C_{i,j,k}^{i,j,k+1}(V_{i,j,k+1} - V_{i,j,k}) - C_{i,j,k-1}^{i,j,k}(V_{i,j,k} - V_{i,j,k-1})}{\Delta_z^2}, \quad C_{l,m,n}^{i,j,k} = \frac{2\varepsilon_{i,j,k}\varepsilon_{l,m,n}}{\varepsilon_{i,j,k} + \varepsilon_{l,m,n}}.$$

Thus, using the number of grid points, equation $\nabla \cdot (\varepsilon(\mathbf{r})\nabla V(\mathbf{r})) = -\rho(\mathbf{r})$ is represented and solved as $\mathbf{AV} = \mathbf{B}$, where $\mathbf{A} \in \mathbb{R}^{N \times N}$ is the matrix; and $\mathbf{B} \in \mathbb{R}^N$ is the vector of the boundary conditions. The self-consistent problem that integrates the solution of the Schrödinger (gives the wave function, energy, etc.) and Poisson (provides the potential) equations is solved in updating the potentials and other variables obtained through iterations. The convergence is enforced and specified accuracy is guaranteed by applying robust numerical methods.

6.20.7 Electromagnetic Field and Control of Particle Motion

For a free particle in the Cartesian coordinate system, $E(\mathbf{r}) = \frac{\mathbf{p}^2}{2m}, \mathbf{p}^2 = p_x^2 + p_y^2 + p_z^2$. Taking into account a potential, one uses $E(\mathbf{r}) = \frac{\mathbf{p}^2}{2m} + \Pi(\mathbf{r})$. In a magnetic field, the interaction of a magnetic moment μ with a magnetic field $\mathbf{B}$ changes the energy by $-\mu \cdot \mathbf{B}$.

Consider a particle with a charge q and mass m in a one-dimensional potential $\Pi(x)$. Let a particle propagate under an external time-varying electric field $E_E(t\mathbf{r})$, where the particle Hamiltonian is $H = \frac{1}{2m}p^2 + \Pi(x) + qE_E(t)x$. For example, $E_E(t) = E_{E0}\sin\omega t$, where E_{E0} is the amplitude of the electrostatic field. It should be emphasized that the operators are commonly used in deriving the expressions for the Hamiltonian, which can be time-invariant or time-dependent.

The external electromagnetic field, which can be controlled, affects the Hamiltonian. In general, for a particle with a charge q in a uniform magnetic field $\mathbf{B}$, one has

$$H = \frac{1}{2\mu}\mathbf{p}^2 + \Pi(\mathbf{r}) - \frac{q}{2\mu c}\mathbf{B} \cdot \mathbf{L} + \frac{q^2}{8\mu c^2}[B^2 r^2 - (\mathbf{B} \cdot \mathbf{r})^2],$$

where μ is the angular momentum, and $\mathbf{L}$ is the orbital angular momentum.

In H, the term $-\frac{q}{2\mu c}\mathbf{B} \cdot \mathbf{L} = -\mu_L \cdot \mathbf{B}$ represents the energy resulting from the interaction between the particle orbital magnetic moment $\mu_L = q\mathbf{L}/(2\mu c)$ and the magnetic field $\mathbf{B}$. If the charge q has an intrinsic spin $\mathbf{S}$, the spinning motion results in the magnetic dipole moment $\mu_S = q\mathbf{S}/(2\mu c)$, which interacts with an external magnetic field generating the energy $-\mu_S \cdot \mathbf{B}$. Thus, we have

$$H = \frac{1}{2\mu}\mathbf{p}^2 + \Pi(\mathbf{r}) - \mu_L \cdot \mathbf{B} - \mu_S \cdot \mathbf{B} + \frac{q^2}{8\mu c^2}[B^2 r^2 - (\mathbf{B} \cdot \mathbf{r})^2].$$

Consider the hydrogen atom under an external uniform magnetic field $\mathbf{B}$. The atom energy levels are shifted as shown in Section 6.4. This energy shift is known as the Zeeman effect. Considering the normal Zeeman effect, neglecting the electron spin (the anomalous Zeeman effect takes into consideration the spin of the electron utilizing the perturbation theory), we assume that $\mathbf{B} = B_z\mathbf{z}$—e.g., $\mathbf{B} = [0, 0, B_z]$.

The Hamiltonian is

$$H = \frac{1}{2\mu}\mathbf{p}^2 - \frac{e^2}{4\pi\varepsilon_0 r} + \frac{e}{2\mu c}B_z L_z + \frac{e^2 B_z^2}{8\mu c^2}(x^2 + y^2),$$

where $H_0 = \frac{1}{2\mu}\mathbf{p}^2 - \frac{e^2}{4\pi\varepsilon_0 r}$ is the atom Hamiltonian in the absence of the magnetic field; and L_z is the orbital angular momentum.

The third term of H is usually rewritten as $\frac{e}{2\mu c}B_z L_z = \frac{\mu_B}{\hbar}B_z L_z$, where μ_B is the Bohr magneton, $\mu_B = \frac{e\hbar}{2\mu c} = \frac{e\hbar}{2m_e} = 9.274 \times 10^{-24}$ J/T $= 5.7884 \times 10^{-5}$ eV/T.

The electron's orbital magnetic dipole moment, resulting from the orbital motion of the electron about the proton, is $\boldsymbol{\mu}_L = -e\mathbf{B}/(2\mu c)$.

The term $\frac{e^2 B_z^2}{8\mu c^2}(x^2 + y^2)$ may be small, and usually is neglected. The spherical and Cartesian coordinates are related as $x = r\sin\theta\cos\varphi$, $y = r\sin\theta\sin\varphi$, and $z = r\cos\theta$.

One concludes that the propagation of electrons can be effectively controlled by changing the electromagnetic field in ^{ME}devices. The control variables are time-varying. One examines a time-dependent Schrödinger equation $H(t, \mathbf{r})\Psi(t, \mathbf{r}) = i\hbar\frac{\partial\Psi(t,\mathbf{r})}{\partial t}$.

Consider a time-varying one-dimensional potential $\Pi(t,x)$ as given by $\Pi(t,x) = \Pi_t(t,x) + \Pi_0(x)$. If $\Pi(t,x) = \Pi_0(x)$, the solution of the Schrödinger equation is

$$\Psi_n(t, x) = \Psi_n(t)\Psi_n(x) = e^{-\frac{iE_n}{\hbar}t}\Psi_n(x),$$

where E_n and $\Psi_n(x)$ are the unperturbed eigenvalues and eigenfunctions.

Taking note of a time-varying $\Pi(t,x)$, the solution is

$$\Psi(t, x) = \sum_n a_n(t)\Psi_n(t, x),$$

where $a_n(t)$ is the time-varying function found by solving a set of differential equations depending on the problem under consideration.

The transition probability is related to $a_n(t)$, as $P_m = \sum_{n,\, n\neq m} a_n^*(t)a_n(t)$.

Our goal is to study how the quantum state, given by $\Psi(t)$, evolves over time. In particular, for a given initial state $\Psi(t_0)$ the system's dynamic behavior, governed by the Schrödinger equation, to the following (intermediate or final) state with $\Psi(t_f)$ is of interest. We have

$$\Psi(t) = U(t_0, t)\Psi(t_0), t > t_0,$$

where $U(t_0, t)$ is the unitary operator, which gives the finite time transition.

To find the *time-evolution* operator $U(t_0, t)$, one substitutes $\Psi(t) = U(t_0, t)\Psi(t_0)$ into the time-dependent Schrödinger equation, yielding $\frac{\partial U(t_0, t)}{\partial t} = -\frac{i}{\hbar}HU(t_0, t)$. If the Hamiltonian H is not a function of time, using the unit initial condition $U(t_0, t_0) = I$, we have

$$U(t_0, t) = e^{-\frac{iH(t-t_0)}{\hbar}},$$

and

$$\Psi(t) = \Psi(t_0)e^{-\frac{iH(t-t_0)}{\hbar}}.$$

To find a solution for a time-varying potential $\Pi(t,x) = \Pi_t(t,x) + \Pi_0(x)$, let $\Pi_0(x) \gg \Pi_t(t,x)$.

Assume

$$\Pi(t) = \begin{cases} \Pi(t) & \text{for } 0 \leq t \leq \tau \\ 0 & \text{for } t < 0,\ t > \tau \end{cases}.$$

The solution of the Schrödinger equation in $0 \leq t \leq \tau$ gives $\Psi(t) = U_H(t_0, t)\Psi(t_0)$, where $U_H(t_0, t) = e^{\frac{iH_0}{\hbar}t}U(t_0, t)e^{-\frac{iH_0}{\hbar}t}$; H_0 is the time-independent part of the Hamiltonian, $H_0 > \Pi(t)$. From the time-dependent Schrödinger equation, one obtains

$$i\hbar\frac{\partial U_H(t_0, t)}{\partial t} = e^{\frac{iH_0}{\hbar}t}\Pi(t)e^{-\frac{iH_0}{\hbar}t}U_H(t_0, t).$$

The solution of this equation is

$$U_H(t_0, t) = I - \frac{i}{\hbar} \int_{t_0}^{t} e^{\frac{iH_0}{\hbar}t} \Pi(t) e^{-\frac{iH_0}{\hbar}t} U_H(t_0, t) dt.$$

The time-dependant perturbation theory provides the first-, second-, third-, and other high-order approximations. The first-order approximation is derived substituting $U_H(t_0,t) = I$. Thus, $U_H^{(1)}(t_0, t) = I - \frac{i}{\hbar} \int_{t_0}^{t} e^{\frac{iH_0}{\hbar}t} \Pi(t) e^{-\frac{iH_0}{\hbar}t} dt$. Having found the initial and final states defined by Ψ_i and Ψ_f, the transition probability is $P_{if}(t) = |\Psi_f U_H(t_0, t)\Psi_i|^2$, and the second-order approximation as $P_{if}(t) = |-\frac{i}{\hbar} \int_0^t \Psi_f \Pi(t')$ $\Psi_i e^{i\omega_t t'} dt'|^2$, where ω_f is the transition frequency between the initial and final system's states, $\omega_t = \frac{E_f - E_i}{\hbar} = \frac{\Psi_f H_0 \Psi_f - \Psi_i H_0 \Psi_i}{\hbar}$.

For practical engineering problems, the time-dependent problem can be solved numerically. In general, the numerical formulation and solution relax the complexity of analytic results, which are usually based on a number of assumptions and approximations of the time-dependent perturbation theory.

6.20.8 Green's Function Formalism

Electronic devices can be modeled using the Green's function method [7,56–58].

The time-independent Schrödinger equation $-\frac{\hbar^2}{2m}\nabla^2\Psi(\mathbf{r}) + \Pi(\mathbf{r})\Psi(\mathbf{r}) = E(\mathbf{r})\Psi(\mathbf{r})$ is written as the Helmholtz equation by using the inhomogeneous term $Q(\Psi)$. In particular, we have

$$(\nabla^2 + k^2)\Psi = Q,$$

where $k^2 = \frac{2mE}{\hbar^2}$ and $Q = \frac{2m}{\hbar^2}\Pi\Psi$.

Our goal is to find a function $G(\mathbf{r})$, called Green's function, that solves the Helmholtz equation with a delta-function *source*, which is given as $(\nabla^2 + k^2)G(\mathbf{r}) = \delta^3(\mathbf{r})$. The wave function $\Psi(\mathbf{r}) = \int G(\mathbf{r} - \mathbf{r}_0)Q(\mathbf{r}_0)d^3\mathbf{r}_0$ satisfies the Schrödinger equation

$$(\nabla^2 + k^2)\Psi(\mathbf{r}) = \int [(\nabla^2 + k^2)G(\mathbf{r} - \mathbf{r}_0)]Q(\mathbf{r}_0)d^3\mathbf{r}_0 = \int \delta^3(\mathbf{r} - \mathbf{r}_0)Q(\mathbf{r}_0)d^3\mathbf{r}_0 = Q(\mathbf{r}).$$

The general solution of the Schrödinger equation is

$$\Psi(\mathbf{r}) = \Psi_0(\mathbf{r}) - \frac{m}{2\pi\hbar^2}\int \frac{e^{ik|\mathbf{r}-\mathbf{r}_0|}}{|\mathbf{r} - \mathbf{r}_0|}\Pi(\mathbf{r}_0)\Psi(\mathbf{r}_0)d^3\mathbf{r}_0,$$

where $\Psi_0(\mathbf{r})$ satisfies the homogeneous equation $(\nabla^2 + k^2)\Psi_0 = 0$.

It should be emphasized that to solve the integral Schrödinger equation derived, one must know the solution, because $\Psi(\mathbf{r}_0)$ is under the integral sign.

Using the Hamiltonian H, one obtains the equation $(E - H)G(\mathbf{r}, \mathbf{r}_E) = \delta(\mathbf{r} - \mathbf{r}_E)$.

Studying electron–electron interactions in the π-conjugated molecules, one may apply the semiempirical Hamiltonian [64–66]. For a molecule, one has

$$H_M = \sum_{i,\sigma} E_i a_{i\sigma}^+ a_{i\sigma} - \sum_{\langle ij \rangle, \sigma} \left(t_{ij} a_{i\sigma}^+ a_{j\sigma} + t_{ij}^* a_{j\sigma}^+ a_{i\sigma} \right)$$

$$+ U\sum_i n_{i,\uparrow} n_{i,\downarrow} + \frac{1}{2}\sum_{i,j,i\neq j} U_{ij}\left(\sum_\sigma a_{i\sigma}^+ a_{i\sigma} - 1\right)\left(\sum_\sigma a_{j\sigma}^+ a_{j\sigma} - 1\right),$$

where E_i are the orbital energies; $a_{i\sigma}^+$ and $a_{i\sigma}$ are the creation and annihilation operators for the π-electron of ith atom with spin $\sigma(\uparrow, \downarrow)$; t_{ij} are the tight-binding hopping matrix entities for the p_z orbitals of the nearest neighbor atoms; $\langle ij \rangle$ denotes the sum of the nearest neighbor sites i and j; U is the onsite Coulomb repulsion between two electrons occupying the same atom p_z orbital; n_i is the total number of π-electrons on site i, $n_i = \sum_\sigma a_{i\sigma}^+ a_{i\sigma}$; and U_{ij} is the intersite Coulomb interaction.

In H_M, the third and fourth terms represent the electron–electron interactions, which depend on the distance. For the interaction energies, the parametrization equation is $U_{ij} = \frac{U}{\sqrt{1 + k_r r_{ij}^2}}$, where r_{ij} is the distance between sites i and j; and k_r is the screening constant. The parameters can thus be obtained. In particular, t_{ij} varies from 2 to 3 eV for orbitals depending on atom placement, bonds (single, double, or triple), while $U \sim 10$ eV and $k_r \sim 50$ (for r_{ij} given in nm).

The Hamiltonian for the molecular complex is constructed by using molecule H_M, tunneling H_T, terminal H_C, and external H_E Hamiltonians. That is, we have

$$H = H_M + H_T + H_C + H_E.$$

For the oscillator with mass m, momentum p, and a resonant angular frequency of radial vibrations ω_0, one obtains $H_0 = \frac{p^2}{2m} + \frac{1}{2}m\omega_0^2 x^2$. However, H_M integrates the core energy-based single-electron Hamiltonian and electron–electron interaction Hamiltonian—e.g., $H_M = \sum_i \left(\frac{1}{2m}\mathbf{p}_i^2 + \frac{1}{2}m\omega_{0i}^2\mathbf{r}_i^2 \right) + \sum_{i,j} B_{ij}a_i^+ a_j$, where $B_{ij}(t)$ are the time-varying amplitudes; a and a^+ are the electron annihilation and creation operators, and the steady-state number of electrons is $N = \sum_{i,j} \langle a_i^+ a_j \rangle$; and i and j are indices that run over the molecule. Using the molecular orbital indices n and l, the last term in the equation for H_M can be expressed as $\sum_{(in)(jl)} B_{(in)(jl)}a_{(in)}^+a_{(jl)}$. The current can be estimated as $I = -e\frac{d}{dt}\sum_{i,j} \langle a_i^+(t) a_j(t) \rangle$, where the equations of motion for $a(t)$ and $a^+(t)$ are derived by using the Hamiltonian.

The tunneling Hamiltonian that describes the electrons' transport to and from the molecule is $H_T = -\sum_{i,j \in \text{Terminals}} e^{-\frac{r}{\lambda_j}} \left(T_{ij}a_i^+ b_j + T_{ij}^* a_i^+ b_j \right)$, where $T_{ij}(t)$ represents the time-varying tunneling amplitudes; b and b^+ are the electron annihilation and creation operators at the input (L) and output (R) terminals; λ_j represents the tunneling lengths between the molecule's conducting and terminal atoms; h.c. denotes the Hamiltonian conjugate; T_{ij_L} and T_{ij_R} are the amplitudes of the electron transfer—for example, from the j_Lth occupied orbital to the molecule's lowest unoccupied molecular orbital $|i_{LUMO}\rangle$, and to the j_Rth unoccupied orbital; $|j_L\rangle$ and $|j_R\rangle$ are the contacts' orbitals; and $|i_{LUMO}\rangle$ is the molecule's lowest unoccupied molecular orbital.

The terminal Hamiltonian is expressed as $H_C = \sum_{j \in L,R} C_j b_j^+ b_j$, where $C_j(t)$ represents the time-varying energy amplitudes.

The external Hamiltonian depends on the device physics, as was documented in Section 6.13.7. For example, $H_E = -\sum_{i,j} E_{ij} \cdot \mathbf{m}_{ij}a_i^+ a_j$, where $E_{ij}(t)$ is the function controlled by varying the electrostatic potential; and $\mathbf{m}_{ij}$ is the electron dipole moment vector.

Taking note of the Hamiltonians derived, one finds a total Hamiltonian H. To examine the functionality and characteristics of ^{ME} devices (input/control bonds–molecule–output bonds), the wave function should be derived by solving the Schrödinger equation. Alternatively, the Keldysh nonequilibrium Green function concept can be applied. Green's function is a wave function of energies at $\mathbf{r}$ resulting from an excitation applied at $\mathbf{r}_E$. We study the retarded Green's function G that represents the behavior of the aggregated molecule, and the equation

$$(E - H)G(\mathbf{r}, \mathbf{r}_E) = \delta(\mathbf{r} - \mathbf{r}_E)$$

is used. The boundary conditions must be satisfied for the transport and Poisson equations.

To examine a finite molecular complex, the molecular Hamiltonian of the isolated system and the complex self-energy functions are used instead of the single energy potential and broadening energies. In the matrix notations, using the overlap matrix S, one has $[G(E)] = (E[S] - [H] - [V_{SC}] - \sum_i [E_i])^{-1}$, where $[E_i] = [S_i][G_i][S_i^*]$; and S_i is the geometry-dependent terminal coupling matrix between the molecular terminals. The imaginary non-Hermitian self-energy functions of the input and output electron reservoirs E_L and E_R are $[E_L] = [S_L][G_L][S_L^*]$ and $[E_R] = [S_R][G_R][S_R^*]$, where G_L and G_R are the surface Green's functions found by applying the recursive methods.

By taking note of Green's function, the density of state $D(E)$ is found as $D(E) = -\frac{1}{\pi}\text{Im}\{G(E)\}$. The spectral function $A(E)$ is the anti-Hermitian term of Green's function, and $A(E) = i[G(E) - G^*(E)] = -2Im[G(E)]$. One obtains $D(E) = \frac{1}{2\pi}\text{tr}[A(E)S]$, where tr is the trace operator, and S is the overlap matrix for orthogonal basis functions. Using the broadening energy functions E_{BL} and E_{BR}, one obtains the spectral functions

$$[A_L(E)] = [G(E)][E_{BL}(E)][G^*(E)] \quad \text{and} \quad [A_R(E)] = [G(E)][E_{BR}(E)][G^*(E)].$$

Multiterminal ^{ME}devices attain equilibrium at the Fermi level, and the nonequilibrium charge density matrix is

$$[\rho(E)] = \frac{1}{2\pi} \int_{-\infty}^{\infty} \sum_{k,i \in L, j \in R} f(E_{Vk}, V_{Fi.j})[A_{i,j}(E)]dE$$

$$= \frac{1}{2\pi} \int_{-\infty}^{\infty} \sum_{k,i \in L, j \in R} f(E_{Vk}, V_{Fi.j})[G(E)][E_{Bi,j}(E)][G^*(E)]dE,$$

where $V_{Fi,j}$ are the potentials, and $f(E_V, V_{Fi,j})$ are the distribution functions.

Utilizing the transmission matrix $T(E) = \text{tr}[E_{BL}G(E)E_{BR}G^*(E)]$ and taking note of the broadening, the current between terminals is found as

$$I_k = \frac{2e}{h} \int_{-\infty}^{+\infty} \text{tr}[E_{BL}G(E)E_{BR}G^*(E)] \sum_{k,i \in L, j \in R} f(E_{Vk}, V_{Fi,j})dE.$$

For a two-terminal ^{ME}device,

$$\rho(E) = \frac{1}{2\pi} \int_{-\infty}^{\infty} [f(E_V, V_{FL})G(E)E_{BL}G^*(E) + f(E_V, V_{FR})G(E)E_{BR}G^*(E)]dE$$

and

$$I = \frac{2e}{h} \int_{-\infty}^{+\infty} \text{tr}[E_{BL}G(E)E_{BR}G^*(E)][f(E_V, V_{FL}) - f(E_V, V_{FR})]dE.$$

As emphasized in Section 6.13.1, one may apply these equations using the applicable distribution functions if the assumptions of the statistical mechanics are valid for the electronic device under consideration.

6.21 Multiterminal Quantum-Effect ^{ME}Devices

Quantum-well resonant tunneling diodes and FETs, Schottky-gated resonant tunneling, heterojunction bipolar, and resonant tunneling bipolar, and other transistors have been introduced to enhance the microelectronic device performance. The tunneling barriers are formed using AlAs, AlGaAs, AlInAs, AlSb, GaAs, GaSb, GaAsSb, GaInAs, InP, InAs, InGaP, and other composites and spacers with a thickness in the range of 1 nm to tens of nm. The CMOS-technology high-speed double-heterojunction bipolar transistors ensure the cut-off frequency ~ 300 GHz, the breakdown voltage is ~ 5 V, and the current density is $\sim 1 \times 10^5$ A/cm^2. The one-dimensional potential energy profile, shown in Figure 6.55, schematically depicts the first barrier (L_1, L_2), the well region (L_2, L_3), and the second barrier (L_3, L_4), with the quasi-Fermi levels E_{F1}, E_{F23}, and E_{F2}. The device physics of these transistors is reported in [15], and the electron transport in

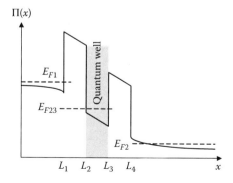

FIGURE 6.55 A one-dimensional potential energy profile and quasi-Fermi levels in the double-barrier single-well heterojunction transistors.

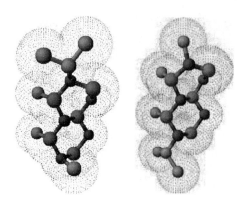

FIGURE 6.56   (see color insert) ^{M}NAND and ^{M}AND gates comprised from cyclic molecules.

double-barrier single-quantum-well is straightforwardly examined by applying a self-consistent approach and numerically solving the one- or two-dimensional Schrödinger and Poisson equations.

The ^{M}AND and ^{M}NAND gates were documented in Figure 6.16 utilizing multiterminal ^{ME}devices to form ^{M}gates. Figure 6.56 illustrates the overlapping molecular orbitals for cyclic molecules used to implement these ^{M}gates.

In Section 6.19, we reported 3D-topology multiterminal ^{ME}devices formed using cyclic molecules with a carbon interconnecting framework (see Figure 6.44). In this section, consider a three-terminal ^{ME}device with the *input, control,* and *output* terminals (as shown in Figure 6.57). The device physics of the proposed ^{ME}device is based on the quantum interaction and controlled electron tunneling. The applied $V_{\text{control}}(t)$ changes the charge distribution $\rho(t,\mathbf{r})$ and $E_E(t,\mathbf{r})$, affecting the electron transport. This ^{ME}device operates in the controlled electron-exchangeable environment due to quantum interactions. Thus, controlled super-fast potential-assisted tunneling is achieved. The electron-exchangeable environment interactions qualitatively and quantitatively modify the device behavior and its characteristics. Consider the electron transport in the time- and spatial-varying metastable potentials $\Pi(t,\mathbf{r})$. From the quantum theory viewpoints, it is evident that the changes in the Hamiltonian result in: (i) changes of tunneling $T(E)$, and (ii) quantum interactions due to variations of $\rho(t,\mathbf{r})$, $E_E(t,\mathbf{r})$, and $\Pi(t,\mathbf{r})$. The device controllability is ensured by varying $V_{\text{control}}(t)$, which affects the device switching, I–V, and other characteristics.

We solve high-fidelity modeling and data-intensive analysis problems for the studied ^{ME}device. For heterojunction microelectronic devices, one usually solves the one-dimensional Schrödinger and Poisson equations by applying the Fermi–Dirac distribution function. In contrast, for the devised ^{ME}devices, a 3D problem arises which cannot be simplified. Furthermore, the distribution functions and statistical mechanics postulates may not be straightforwardly applied.

For the studied cyclic molecule which forms an interconnected ^{ME}device, we consider nine atoms with motionless protons of charges q_i. The radial Coulomb potentials are $\Pi_i(r) = -\frac{Z_{eff\,i} q_i^2}{4\pi\varepsilon_0 r}$. For example, for

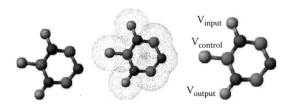

FIGURE 6.57   (see color insert) A three-terminal ^{ME}device comprised from a cyclic molecule with a carbon interconnecting framework.

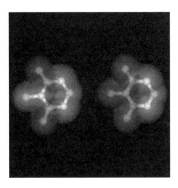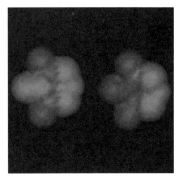

FIGURE 6.58   (see color insert) Charge distribution $\rho(\mathbf{r})$.

carbon, $Z_{effC} = 3.14$. Using the spherical coordinate system, the Schrödinger equation

$$-\frac{\hbar^2}{2m}\left[\frac{1}{r^2}\frac{\partial}{\partial r}\left(r^2\frac{\partial\Psi}{\partial r}\right) + \frac{1}{r^2\sin\theta}\frac{\partial}{\partial\theta}\left(\sin\theta\frac{\partial\Psi}{\partial\theta}\right) + \frac{1}{r^2\sin^2\theta}\frac{\partial^2\Psi}{\partial\phi^2}\right]$$
$$+ \Pi(r,\theta,\phi)\Psi(r,\theta,\phi) = E\,\Psi(r,\theta,\phi)$$

should be solved. For the problem under consideration, it is impractical to find the analytic solution as obtained in Example 6.20.21 by using the separation of variables concept. We represented the wave function as $\Psi(r,\theta,\phi) = R(r)Y(\theta,\phi)$ in order to derive and solve the radial and angular equations. In contrast, we discretize the Schrödinger and Poisson equations, as reported in this section, with the ultimate objective of numerically solving these differential equations. The magnitude of the time-varying potential applied to the control terminal is bounded due to the thermal stability of the molecule—e.g., $|V_{control}| \leq V_{controlmax}$. In particular, we let $|V_{control}| \leq 0.25$ V. The charge distribution is of particular interest. Figure 6.58 documents a three-dimensional charge distribution in the molecule if $V_{control} = 0.1$ V and $V_{control} = 0.2$ V. The total molecular charge distribution is found by summing the individual orbital densities.

The Schrödinger and Poisson equations are solved using a self-consistent algorithm in order to verify the device physics soundness and examine the baseline performance characteristics. To obtain the current density j and current in the ^{ME}device, the velocity and momentum of the electrons are obtained by making

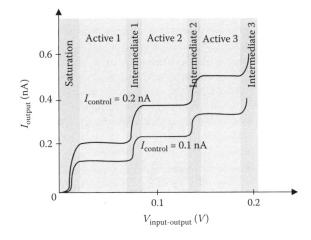

FIGURE 6.59   Multiple-valued $I-V$ characteristics.

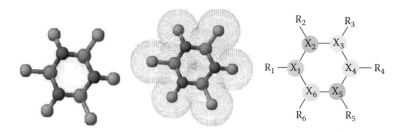

FIGURE 6.60   (see color insert) Six-terminal ^{ME}devices.

use of $\langle p \rangle = \int_{-\infty}^{\infty} \Psi^*(t,\mathbf{r})\left(-i\hbar\frac{\partial}{\partial \mathbf{r}}\right)\Psi(t,\mathbf{r})\,d\mathbf{r}$. The wave function $\Psi(t,\mathbf{r})$ is derived for distinct values of V_{control}. The $I\text{--}V$ characteristics of the studied ^{ME}device for two different control currents (0.1 and 0.2 nA) are reported in Figure 6.59. The results documented imply that the proposed ^{ME}device may be effectively used as a multiple-valued or symbolic ^{M}primitive in order to design enabling multiple-valued or symbolic logics and memories.

The traversal time of electron tunneling is derived from the expression $\tau(E) = \int_{\mathbf{r}_0}^{\mathbf{r}_f}\sqrt{\frac{m}{2[\Pi(\mathbf{r})-E]}}\,d\mathbf{r}$. It is found that τ varies from 2.4×10^{-15} to 5×10^{-15} sec. Hence, the proposed ^{ME}device ensures super-fast switching.

The reported monocyclic molecule can be used as a six-terminal ^{ME}device, as illustrated in Figure 6.60. The proposed carbon-centered molecular hardware solution, in general:

- Ensures a sound *bottom-up* synthesis at the device, gate, and module levels
- Guarantees aggregability to form complex ^{M}ICs
- Results in the experimentally characterizable ^{ME}devices and ^{M}gates.

The use of the side groups R_i, shown in Figure 6.60, ensures the variations of the energy barriers and wells potential surfaces $\Pi(t,\mathbf{r})$. This results in the controlled electron transport and varying quantum interactions. As reported, the studied ^{ME}devices can be utilized in combinational and memory ^{M}ICs. In addition, those devices can be used as routers. Hence, one achieves a reconfigurable networking-processing-and-memory, as covered in Section 6.5 regarding *fluidic* platforms. We conclude that neuromorphological reconfigurable *solid* ^{M}PPs can be designed.

A *generic* modeling concept is reported next. A ^{M}device may have two or more terminals. The interconnected ^{M}devices are well defined in the sense of their time-varying variables—e.g., input $\mathbf{r}(t)$, control $\mathbf{u}(t)$, output $\mathbf{y}(t)$, state $\mathbf{x}(t)$, disturbance $\mathbf{d}(t)$, and noise $\xi(t)$ vectors. The electric or magnetic field intensities can be considered as $\mathbf{u}(t)$, while velocity, displacement, current, and voltage can be the state and/or output variables. For example, for controllable *solid* ^{ME}devices examined, the voltage at any terminal is well defined with respect to a common datum node (ground). Figure 6.61 documents a multitermi-

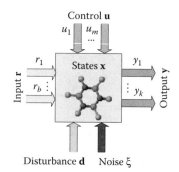

FIGURE 6.61   A molecular device with time-varying variables that characterize dynamic and steady-state device performance.

nal ^{M}device with *input*, *control*, and *output* terminals, with the corresponding of time-varying variables. The disturbances and noise vectors are also documented. The phenomena exhibited, and the effects utilized, are defined by the device physics. As illustrated, the proposed ^{ME}device is modeled by using the Schrödinger and Poisson equations. The ^{M}devices for admissible inputs and disturbances are completely characterized by the differential equations and constitutive relations that describe transient dynamics and steady-state behavior. In particular, neglecting disturbances, unmodelled phenomena, and noise, the ^{M}device is described by the quadruple $(\mathbf{r,x,y,u})$, where $\mathbf{r} \in \mathbb{R}^b$, $\mathbf{x} \in \mathbb{R}^n$, $\mathbf{y} \in \mathbb{R}^k$, and $\mathbf{u} \in \mathbb{R}^m$. We denote by R, X, Y,and U the universal sets of achievable values for each of these vector variables. Thus, the ^{M}device response (behavioral) quadruple is $(\mathbf{r,x,y,u}) \in R \times X \times Y \times V$. The measurement set is given as $M = \{(\mathbf{r,x,y,u}) \in R \times X \times Y \times U, \forall t \in T\}$. The electrochemomechanical state transitions (electron transport, conformation, etc.) are controlled by changing $\mathbf{u}$ to meet the optimal transient achievable performance and thus guarantee the desired steady-state performance characteristics.

6.22 Conclusions

We reported innovative developments in 3D *solid* and *fluidic* molecular electronics utilizing novel ^{M}devices, 3D organizations, enabling architectures, *bottom-up* fabrication, etc. The proposed 3D-topology ^{M}devices are based on a new device physics. Enabling 3D organizations and novel architectures are utilized at the module and system levels. A wide spectrum of fundamental, applied, and experimental issues, related to the device physics, phenomena, functionality, performance, and capabilities, were researched. Therefore, we advanced a ^{M}*architectronics* paradigm. It was found and demonstrated that *solid* and *fluidic* ^{M}devices exhibit novel phenomena and possess unique functionality, thus providing enabling capabilities. These ^{M}devices were aggregated within ^{N}hypercells, which form ^{M}ICs. Innovative concepts in the SLSI design of ^{M}ICs and ^{M}PPs were also reported and examined. The proposed ^{M}PPs were designed within 3D organizations and enabling architectures, guaranteeing superior performance. The *bottom-up* fabrication issues were covered for the carbon-centered molecular electronics. Biomolecular processing platforms were examined, researching baseline fundamentals, cognition, and information processing. Though a great number of problems remain to be solved, it was demonstrated that the proposed 3D-centered topologies/organizations and novel architectures guarantee overall supremacy. The reported solutions led to the envisioned ^{M}PPs and cognizant information processing. Neuromorphological reconfigurable *solid* and *fluidic* ^{M}PPs were devised utilizing a ^{3D}networking-and-processing paradigm. The proposed design and analysis concepts at the device and system levels were coherently studied and illustrated through numerous examples.

Acknowledgments

The author sincerely acknowledges partial support from the *Microsystems and Nanotechnologies* under the U.S. Department of Defense, Department of the Air Force (Air Force Research Laboratory) contracts 8750024 and 8750058. *Disclaimer*: Any opinions, findings, conclusions, or recommendations expressed in this chapter are those of the author and do not necessarily reflect The U.S. Department of Defense, Department of the Air Force views.

The software support from the MathWorks, Inc. and Accelrys Software Inc. are sincerely acknowledged.

References

[1] *International Technology Roadmap for Semiconductors*, 2005 ed., Semiconductor Industry Association, Austin, TX, 2006.
[2] Brewer, J.E. et al., Memory technology for the post CMOS era, *IEEE Circuits and Devices Magazine*, 21, 2, 13, 2005.
[3] Ferry, D.K. et al., Semiconductor device scaling: Physics, transport, and the role of nanowires, *Proc. IEEE Conf. Nanotechnology*, Cincinnati, OH, 2006.

[4] Zhirnov, V.V. et al., Emerging research memory and logic technologies, *IEEE Circuits and Devices Magazine*, 21, 3, 2005.

[5] Ellenbogen, J.C. and Love, J.C., Architectures for molecular electronic computers: Logic structures and an adder designed from molecular electronic diodes, *Proc. IEEE*, 88, 3, 2000.

[6] Heath, J.R. and Ratner, M.A., Molecular electronics, *Physics Today*, 1, 2003.

[7] Lyshevski, S.E., *NEMS and MEMS: Fundamentals of Nano- and Microengineering*, 2nd ed., CRC Press, Boca Raton, FL, 2005.

[8] Chen, J. et al., Molecular electronic devices, in *Handbook Molecular Nanoelectronics*, Reed, M.A. and Lee, L. Eds., American Science Publishers, 2003.

[9] Tour, J.M. and James, D.K., Molecular electronic computing architectures, in *Handbook of Nanoscience, Engineering and Technology*, Goddard, W.A. et al., Eds., CRC Press, Boca Raton, FL, 2003.

[10] Wang, W. et al., Inelastic electron tunneling spectroscopy of an alkanedithiol self-assembled monolayer, *Nano Lett.*, 4, 4, 2004.

[11] Aviram, A. and Ratner, M.A., Molecular rectifiers, *Chem. Phys. Letters*, 29, 1974.

[12] Berg, H.C., The rotary motor of bacterial flagella, *J. Ann. Rev. Biochem.*, 72, pp. 19–54, 2003.

[13] Sze, S.M., *Physics of Semiconductor Devices*, Wiley, NJ, 1969.

[14] ———, *Physics of Semiconductor Devices*, Wiley, NJ, 1981.

[15] Sze, S.M. and Ng, K.K., *Physics of Semiconductor Devices*, Wiley, NJ, 2007.

[16] Lyshevski, S.E., Design of three-dimensional molecular integrated circuits and molecular architectronics, *Proc. IEEE Conf. Nanotechnology*, Cincinnati, OH, 2006.

[17] Kaupp, U.B. and Baumann, A., Neurons –the molecular basis of their electrical excitability, in *Handbook of Nanoelectronics and Information Technology*, Waser, R., Ed., Wiley-VCH, Darmstadt, Germany, 2005.

[18] Churchland, P.S. and Sejnowski, T.J., *The Computational Brain*, MIT Press, Cambridge, MA, 1992.

[19] Freeman, W., *Mass Action in the Nervous System*, Academic, New York, 1975.

[20] Freeman, W., Tutorial on neurobiology from single neurons to brain chaos, *Int. J. Biforcation Chaos*, 2, 3, 1992.

[21] Laughlin, S. et al., The metabolic cost of neural computation, *Nature Neurosci.*, 1, 1998.

[22] Lyshevski, M.A. and Lyshevski, S.A., Fluidic nanoelectronics and Brownian dynamics, *Proc. NSTI Nanotechnology Conf.*, Boston, MA, 3, 2006.

[23] Hameroff, S.R. and Tuszynski, J., Search for quantum and classical modes of information processing in microtubules: implications for the living atate, in *Handbook on Bioenergetic Organization in Living Systems*, Musumeci, F. and Ho, M.-W., Eds., World Scientific, Singapore, 2003.

[24] Lyshevski, M.A., Fluidic molecular electronics, *Proc. IEEE Conf. Nanotechnology*, Cincinnati, OH, 2006.

[25] Dayan, P. and Abbott, L.F., *Theoretical Neuroscience: Computational and Mathematical Modeling of Neural Systems*, MIT Press, Cambridge, MA, 2001.

[26] Grossberg, S., On the production and release of chemical transmitters and related topics in cellular control, *J. Theoretical Biol.*, 22, 1969.

[27] ———, *Studies of Mind and Brain*, Reidel, Amsterdam, The Netherlands, 1982.

[28] Abbott, L.F. and Regehr, W.G., Synaptic computation, *Nature*, 431, 2004.

[29] Markram, H. et al., Regulation of synaptic efficacy by coincidence of postsynaptic APs and EPSPs, *Science*, 275, 1997.

[30] Rumsey, C.C. and Abbott, L.F., Equalization of synaptic efficacy by activity- and timing-dependent synaptic plasticity, *J. Neurophysiol.*, 91, 5, 2004.

[31] Grossberg, S., Birth of a learning law, *Neural Networks*, 11, 1, 1968.

[32] Frantherz, F., Neuroelectronics interfacing: semiconductor chips wth ion channels, nerve cells, and brain, in *Handbook of Nanoelectronics and Information Technology*, Waser, R., Ed., Wiley-VCH, Darmstadt, Germany, 2005.

[33] Lyshevski, S.E., Molecular cognitive information-processing and computing platforms, *Proc. IEEE Conf. Nanotechnology*, Cincinnati, OH, 2006.

[34] Renyi, A., On measure of entropy and information, *Proc. Berkeley Symp. Math. Stat. Prob.*, 1 , 1961.

[35] Kabanov, Yu.M., The capacity of a channel of the Poisson type, *Theory Prob. Appl.*, 23, 1978.

[36] Johnson, D., Point process models of single-neuron discharges, *J. Comp. Neurosci.*, 3, 1996.

[37] Strong, S.P. et al., Entropy and information in neuronal spike trains, *Phys. Rev. Lett.*, 80, 1, 1998.

[38] Rieke, F. et al., *Spikes: Exploring the Neural Code*, MIT Press, Cambridge, MA, 1997.

[39] Smith, K.C., Multiple-valued logic: A tutorial and appreciation, *Computer*, 21, 4, 1998.

[40] Buitenweg, J.R. et al., Modeled channel distributions explain extracellular recordings from cultured neurons sealed to microelectrodes, *IEEE Trans. Biomed. Eng.*, 49, 11, 2002.

[41] Sigworth, F.J. and Klemic, K.G., Microchip technology in ion-channel research, *IEEE Trans. Nanobioscience*, 4, 1, 2005.

[42] Suzuki, H. et al., Planar lipid membrane array for membrane protein chip, *Proc. Conf. on MEMS*, 2004.

[43] Yanushkevich, S., *Logic Design of Nano ICs*, CRC Press, Boca Raton, FL, 2005.

[44] Malyugin, V.D., Realization of corteges of Boolean functions by linear arithmetical polynomials, *Automica and Telemekhica*, 2, 1984.

[45] Lyshevski, S.E., Nanocomputers and nanoarchitectronics, in *Handbook of Nanoscience, Engineering and Technology*, Goddard, W., Ed., CRC Press, Boca Raton, FL, 2002.

[46] Porod, W., Nanoelectronic circuit architectures, in *Handbook of Nanoscience, Engineering and Technology*, Goddard, W. A., et al., Ed., 2003.

[47] Williams, S.R. and Kuekes, P.J., Molecular nanoelectronics, *Proc. Int. Symp. Circuits and Systems*, Geneva, Switzerland, 1, 2000.

[48] Kamins, T. I. et al., Ti-catalyzed Si nanowires by chemical vapor deposition: Microscopy and growth mechanism, *J. Appl. Phys.*, 89, 2001.

[49] Reichert, J. et al., Driving current through single organic molecules, *Phys. Rev. Lett.*, 88, 17, 2002.

[50] Basch H. and Ratner, M.A., Binding at molecule/gold transport interfaces. V. Comparison of different metals and molecular bridges, *J. Chem. Phys.*, 119, 22, 2003.

[51] Lee, K., Measurement of $I-V$ characteristic of organic molecules using step junction, *Proc. IEEE Conf. Nanotechnology*, Munich, Germany, 2004.

[52] Mahapatro, A.K. et al., Nanometer scale electrode separation (nanogap) using electromigration at room temperature, *Proc. IEEE Trans. Nanotechnology*, 5, 3, 2006.

[53] Carbone, A. and Seeman, N.C., Circuits and programmable self-assembling DNA structures, *Proc. Nat. Acad. Science*, 99, 20, 2002.

[54] Porath, D. et al., Charge transport in DNA-based devices, *Top. Curr. Chem.*, 237, 2004.

[55] Mahapatro, A.K. et al., Electrical behavior of nano-scale junctions with well engineered double stranded DNA molecules, *Proc. IEEE Conf. Nanotechnology*, Cincinnati, OH, 2006.

[56] Galperin. M. and Nitzan, A., NEGF-HF method in molecular junction property calculations, *Ann. NY Acad. Sci.*, 1006, 2003.

[57] Galperin, M. et al., Resonant inelastic tunneling in molecular junctions, *Phys. Rev.*, B, 73, 045314, 2006.

[58] Paulsson, M. et al., Resistance of a molecule, in *Handbook of Nanoscience, Engineering and Technology*, Goddard, W. Ed., CRC Press, Boca Raton, FL, 2002.

[59] Landauer, R., Spatial variation of current and fields due to localized scatterers in metallic conduction, *IBM J.*, 1, 3, 1957. Reprinted in *IBM J. Res. Develop.*, 44, 1/2, 2000.

[60] Büttiker, M., Quantuized transmission of a saddle-point constriction, *Phys. Rev. B*, 41, 11, 1990.

[61] Büttiker, M. and Landauer, R., Escape-energy distribution for particles in an extremely underdamped potential well, *Phys. Rev. B*, 30, 3, 1984.

[62] ———, Traversal time for tunneling, *Phys. Rev. Lett.*, 49, 23, 1982.

[63] Galperin, M. and Nitzan, A., Traversal time for electron tunneling in water, *J. Chem. Phys.*, 114, 21, 2001.

[64] Pariser, R., and Parr, R.G., A semiempirical theory of the electronic spectra and electronic structure of complex unsaturated molecules I, *J. Chem. Phys.*, 21, 1953.

[65] ———, A semiempirical theory of the electronic spectra and electronic structure of complex unsaturated molecules II, *J. Chem. Phys.*, 21, 1953.

[66] Pople, J.A., Electron interaction in unsaturated hydrocarbons, *Trans. Faraday Soc.*, 49, 1953.

II

Nanoscaled Electronics

7

Inorganic Nanowires in Electronics

Bin Yu

M. Meyyappan

7.1 Introduction

One of the areas nanotechnology is expected to have a significant impact on in terms of paradigm change and large scale economy is nanoelectronics. The concern that the downscaling of silicon CMOS according to Moore's law will come to an end has been around for quite some time. At present, it is not clear what the feature scale of the last CMOS generation will be (perhaps 10 nm? even smaller?). The current speculation is that the end of downscaling will happen in a decade or so. For this reason, tremendous interest has risen in exploring alternative technologies that can give higher performance and integration density than the presumed last-generation silicon CMOS device. The alternatives include molecular electronics, carbon nanotube–based nanoelectronics, single electron transistors, quantum computing architecture, and so on—all of which are topics covered in this book. The inorganic nanowire-based approach to electronics does not really belong with the previous class of revolutionary technologies. The material system here will still be silicon or germanium with one-dimensional nanowires replacing conventional 2D thin films. This could possibly provide some flexibilities in fabrication and/or performance advantages. Certainly, the possibility of vertical transistors (instead of the traditional planar CMOS device) exists, which can help attain a higher device density and possibly even lead to a 3D architecture. Investigation of nanowires in electronics is at its very early stages. Indeed, the volume of work in the literature to date is far smaller than the efforts on molecular and carbon nanotube–based electronics.

This chapter will provide a review of the current status on using semiconducting nanowires in device fabrication. First, the impetus for using nanowires will be established. Then, a discussion of growth techniques will be provided, followed by growth results and the properties of nanowires. Finally, a summary of device fabrication efforts to date will be presented.

7.2 Why Nanowires?

In the last two to three decades, a variety of semiconducting, metallic, dielectric and other materials—both elemental and compound materials—have been grown as thin films using chemical vapor deposition

(CVD), metal organic chemical vapor deposition (MOCVD), molecular beam epitaxy (MBE), and other techniques. These materials covered the wavelength range from UV to far IR, and correspondingly the bandgap from about 3.6 eV down to 0.4 eV. Extraordinary dimensional control has been demonstrated in growing multiple quantum well layers where the layer thickness is just 1 nm. These achievements in the controlled growth of epitaxial layers have resulted in numerous advances in electronics (logic and memory) and optoelectronics (lasers, detectors, etc.). In the last few years, most of these materials have been grown as one-dimensional nanowires. The nanowires are single crystal with very well defined surface structural properties. Their one-dimensionality offers the lowest dimension transport channel for the best field effect transistor (FET) scalability. The bandgap of semiconducting nanowires varies inversely with the diameter. One-dimensional quantum confinement in the radial direction, as well as reduced phonon scattering, provide interesting physics for logic devices. The electronic properties can be altered by doping. All of these make nanowires an interesting candidate for electronics applications.

7.3   Nanowire Growth

There are several approaches reported in the literature for growing various nanowires. One of the earliest methods described is to use a nanoporous template to guide the nanowire growth by selective deposition in the openings of the template [1]. The most common template is the anodized aluminum oxide (AAO) thin film, which is stable, insulating, and characterized by reasonable density and a uniformity of pores. The porous film is prepared by the anodizing of 99.999% Al on a desired substrate, where the pores self-organize into a highly ordered hexagonal array of parallel vertical pores. To date, several types of nanowires—for example, Cd, Cds, CdS_xSe_{x-1}, and $Zn_xCd_{1-x}S$—have been reported by electrodeposition. The challenges facing this approach include the ease of removal of the template during device fabrication and the uniformity of pore size on large areas.

The most widely used method to grow nanowires is the vapor–liquid–solid (VLS) approach pioneered by Wagner and Ellis [2]. This technique uses a metal catalyst film which, at normal growth temperatures, is in a molten form, consisting of tiny droplets. The source vapor for a given nanowire—generated either by sublimation of a metal, laser ablation of a target, or from chemical reactions of the feedstock gases—dissolves into the catalyst droplets (see Figure 7.1), and when supersaturation is reached, precipitation of the nanowire occurs from the catalyst particle. Typically, the particle is carried at the head of the growing nanowire. The VLS approach is amenable for growth on patterned substrates where lithography can be used to specify desired patterns. The approach is also amenable for integration with device fabrication schemes.

For growth of silicon nanowires, either SiH_4 or $SiCl_4$, mixed with H_2, can be used as a source gas [3,4]. The silane route is a lower temperature process (520°C versus 850°C). The research scale reactor usually consists of a quartz tube inserted in a two-zone furnace (Figure 7.2). The upstream section can be used to generate the source vapor by maintaining the zone at the appropriate temperature mentioned previously. For example, the source material in the form of powder, pellets, foil, or an equivalent can be placed in a boat and the source vapor can be generated by sublimation. An example would be ZnO powder by itself or ZnO powder mixed with graphite powder for carbothermal reduction to generate the source vapor [5]. This type of sublimation approach is not common in silicon nanowire preparation; instead, the source feedstock, such as silane or $SiCl_4$, is decomposed in Zone 1. In the downstream section, a wafer containing a thin film of the catalyst metal, gold being the most common, is placed, and the local temperature corresponds to the melting point of the catalyst metal. In practice, very thin films (~ 1 nm) melt at a temperature lower than the bulk melting point. Several alternatives to gold, particularly those with lower melting points such as Ga or In, are also possible choices [6]. The catalyst, sputtered or evaporated as a thin film (1 to 10 nm), breaks into droplets facilitating VLS growth. The thickness of the film can be changed to match various wire thicknesses; however, for each layer thickness, there will always be a distribution of the catalyst particle size and, hence, the nanowire diameter. Growth temperature and feedstock composition also influence the nanowire diameter and the morphology. Absolute control on the diameter uniformity across a full scale wafer (of any size) has not been demonstrated and this may only be possible if each catalyst

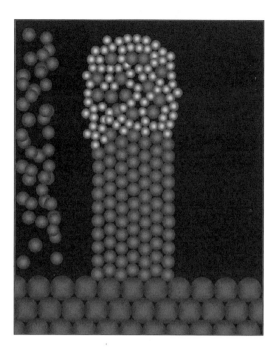

FIGURE 7.1 (Please see color insert following page 5-6) Nanowire growth by vapor–liquid–solid (VLS) mechanism. The orange-colored atoms denote the substrate atoms. The purple spheres represent the source vapor molecules, and the yellow spheres denote the metal catalysts.

particle is patterned at a prescribed location and further prevented from coalescing with a neighboring droplet.

Though a variety of oxide, nitride, metal, and other nanowires have been reported in the literature, the following discussion concerns only the growth literature pertaining to silicon and germanium due to the electronics applications. After the early works of Wagner and Ellis [2] on VLS growth, Westwater *et al.* [3] provided the first comprehensive report on silicon nanowires from silane at temperatures from 320 to 600°C and silane partial pressures of 0.01 to 1 Torr. They found that nanowires as thin as 10 nm can be grown by keeping the partial pressure high and temperature low. These wires are single crystal but tend to have defects of kinks and bends. Lowering the partial pressure yielded thicker wires but with reduced defects. More recently, Mao *et al.* [7] conducted a similar study for the $SiCl_4 + H_2$ system and found that only a narrow set of conditions yielded vertical nanowires. Using the same $SiCl_4 + H_2$ system and gold colloids, Hochbaum *et al.* [8] grew vertical silicon nanowires with a narrow size distribution and an

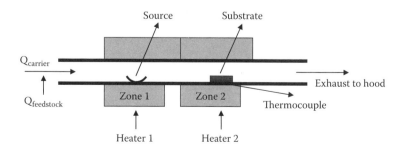

FIGURE 7.2 Schematic of a VLS growth reactor setup.

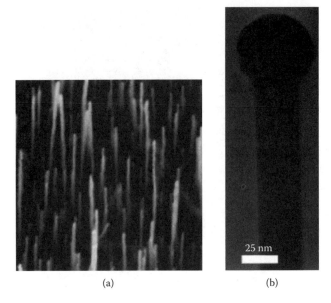

FIGURE 7.3   (a) Scanning electron microscopy (SEM) image of germanium nanowires. (b) Transmission electron microscopy (TEM) image of a germanium wire with the gold catalyst at the top.

average diameter of 39 nm. They were able to vary the density from 0.1 to 1.8 wires/μm^2. Sunkara *et al.* [9] used a microwave plasma of N$_2$ + H$_2$ with gallium as the catalyst. They avoided a silicon containing precursor gas; instead, the hydrogen plasma etched the exposed silicon wafer to provide the silyl radicals. The VLS growth in their work provided silicon nanowires with diameters of 6 nm at temperatures less than 400°C. Using silane in the same plasma setup works as well by providing a variety of reactive precursors and yields nanowires of 10 to 100 μm long [10]. In summary, it is now possible to grow silicon and germanium nanowires of various diameters (a few nm to about 100) and 1 to 100 μm long. The nanowire morphology is also controllable, and it is possible to obtain vertical nanowires on patterned locations. Tan *et al.* [11] indicate there is no thermodynamic limit on the attainable minimum size in the VLS growth, and arbitrarily small nanowires can be grown until reaching some sort of kinetic limit.

Figure 7.3 shows vertical germanium wires grown on a Ge substrate using the VLS process described earlier [12]. The source here consists of a 1:1 weight ratio of germanium powder and graphite powder. The latter provides enhanced surface area for the evaporation of germanium and control of the germanium vapor partial pressure for a given gas flow rate. The source temperature is 1020°C and the substrate temperature is 470°C, with a carrier gas of Ar + H$_2$ mixture. The Ge nanowires are of uniform diameter (42 ± 10 nm) and length distribution (1.0 ± 0.2 μm). Analysis by transmission electron microscopy (TEM) indicates that the nanowire elongates in the [111] direction. The TEM image in Figure 7.3b shows a hemispherical gold catalyst head on top of the wire. High resolution TEM (not shown here) reveals well-defined lattice fringes in the (111) and $(11\bar{1})$ planes, along with a smooth surface. The lattice fringe is 3.26 Å which matches well with x-ray powder diffraction data for bulk germanium. Typically, a thin oxide sheath (1 to 2 nm) is always observed on the nanowires. Selected area diffraction patterns confirm that the nanowires are composed of highly crystalline germanium with the cubic diamond structure and a preferred growth direction of [111].

Figure 7.4 shows vertical silicon nanowires grown using a SiCl$_4$ + H$_2$ system. These are grown at 925°C and atmospheric pressure with 0.04% SiCl$_4$ and 4% H$_2$ in argon. The wires are well aligned and crystalline. The diameter ranges from 68 to 96 nm and the growth density is about 4 wires/μm^2. The TEM image (not shown here) shows regular crystal lattices throughout the wire body with the exception of a thin outer layer consisting of the native oxide.

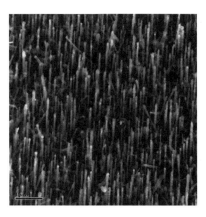

FIGURE 7.4 SEM image of silicon nanowires grown using silicon tetrachloride and hydrogen in argon. (Image courtesy of A. Mao.)

7.4 Nanowires: Morphology and Properties

Ma and co-authors [13] measured the bandgap of silicon nanowires using scanning tunneling spectroscopy. First, they carefully removed the native oxide surrounding the nanowires to obtain reliable measurements. At 7 nm, the bandgap of silicon nanowire is found to be close to the bulk value of 1.1 eV. Then, it gradually increases with decreasing diameter, eventually to 3.5 eV at 1.3 nm. The thermal conductivity of nanowires also depends on the diameter as found from measurements and theoretical calculations [14, 15]. In general, thermal conductivity of the nanowire is significantly smaller than the bulk value [14]. For silicon nanowires, this reduction is by two orders of magnitude, and thermal conductivity decreases with wire diameter. The diameter dependence was ascribed to the increased phonon-boundary scattering [14]. Figure 7.5 shows the computed thermal conductivity of germanium wires of various diameters as a function of temperature. Typically, germanium wires exhibit a lower thermal conductivity than the silicon wires of the same size.

The properties of nanowires are significantly modified by doping, something which has been studied extensively by Lieber and co-workers [16]. Cui *et al.* synthesized boron-doped (p-type) and phosphorous-doped (n-type) silicon nanowires and estimated the mobilities from gate-dependent transport measurements. Their studies showed the possibility of heavy doping of the silicon wires, approaching metallic characteristics.

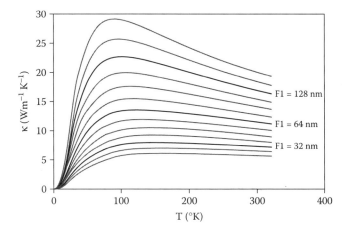

FIGURE 7.5 Thermal conductivity of germanium nanowires as a function of diameter and temperature. (Image courtesy of N. Mingo.)

7.5   Device Fabrication

Compared to the efforts on carbon nanotube and molecular electronics, nanowire device fabrication has received less attention to date. For this reason, devices using materials other than silicon and germanium are also included in the following discussion though their chances of making to the logic device family in the Moore's law paradigm and beyond are relatively small.

Kim *et al.* [17] fabricated Schottky diodes on single GaN nanowires and studied their transport properties. GaN has a direct bandgap of 3.4 eV and therefore is useful for UV detectors and emitters and high-temperature electronics. Ga metal and GaN power were used along with flowing ammonia at 1000°C to grow the nanowires in a VLS process using nickel catalyst. The diameter distribution was found to be in the 30 to 70 nm range with a single crystal hexagonal worzite morphology. An Al contact provided a Schottky junction with the GaN wire, and Ti/Au formed an ohmic contact. The I-V characteristics showed a clear rectifying behavior with no reverse-bias breakdown till −5 V. Wang *et al.* [18] observed a rectifying effect in boron nanowire devices. In their case, nickel formed ohmic contacts, and Ti formed Schottky contacts to boron wires. The breakdown of the device happened at a reverse bias of −20 V. Sun and Sirringhaus [19] reported high-performance thin film transistors (TFTs) using solution-processable ZnO nanorods. ZnO also exhibits a large bandgap of 3.37 eV. The as-prepared nanorods, about 10 nm wide and 65 nm long, were processed using spin coating to make the TFTs. The devices exhibited a mobility of 0.61 cm^2/vs at 230°C and an on-off ratio at 3×10^5.

There have been a few device fabrication efforts reported in the literature [20–28] using silicon. Lui and Lieber [21] reported n^+− p − n silicon nanowire bipolar transistors. These early devices showed a common emitter current gain of 16 when the corresponding base width is 15 μm. The same group also reported [22] a variety of logic-gate structures (OR, AND, NOR, etc.) with substantial gain and showed their implementation in basic computation. SiNW FETs have been fabricated with Ti for making the source and drain contacts [23]. After thermal annealing of the contacts, the device performance improved substantially with peak values of transconductance of 2000 nS and mobility 1350 cm^2/V.S. These values are substantially higher than planar silicon devices of comparable feature size.

The early device fabrication efforts used a planar geometry (as in conventional silicon CMOS processing) with a nanowire replacing a two-dimensional epitaxial layer. Ng *et al.* provided generic processing schemes for bottom-up integration of nanowire devices and discussed both top gate [29] and surround gate [5] structures using metal oxide channels. Figure 7.6 shows a schematic of an In_2O_2-based vertical FET; the process flow to fabricate this device is shown in Figure 7.7. The nanowire is grown on an a-sapphire substrate, and a thin continuous layer of In_2O_3 at the bottom of the wire serves as the source electrode. The In_2O_3 nanowire grown through the VLS process is the conducting channel. The gold catalyst at the top of

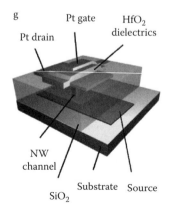

FIGURE 7.6   Schematic of a vertical top-gate nanowire transistor. (Image courtesy of H.T. Ng.)

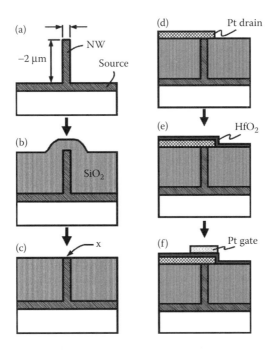

FIGURE 7.7 Process flow for the fabrication of device in Figure 7.6 [29].

the wire provides evidence of the VLS mechanism. Next, the structure is covered with SiO_2 by chemical vapor deposition using tetraethoxy silane (TEOS) as the source. The well-known TEOS process provides a conformal coating over the nanowire. This is followed by a chemical mechanical polishing (CMP) step to provide a smooth top surface and remove the gold particle. The next step involves the selective patterning of an HfO_2 dielectric as gate oxide. This is followed by the deposition of a lithographically patterned 15-nm thick platinum electrode for the drain contact. This device showed an on-off ratio of 2.8×10^3 and an effective mobility of 6.9 cm^2/V.S. in the channel. These authors also fabricated a vertical surround gate transistor as shown in Figure 7.8.

The concept of vertical transistors and surround gates has been around for some time in the Semiconductor Industry Association Roadmap. There have even been a few reports on the fabrication of such transistors, but they involved etching nanopillars from epitaxial silicon to create the active channel. The plasma etching invariably leaves too poor a surface quality to be of use for commercial logic devices to fit

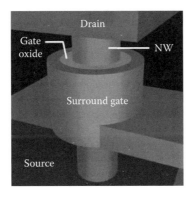

FIGURE 7.8 Schematic of a vertical surround gate nanowire transistor. (Image courtesy of H.T. Ng.)

into the Moore's Law scaling curve. With high-quality single-crystal nanowires with well-defined surface properties, it is entirely possible to realize the vertical transistor concept. The most important advantages of vertical transistors include the following: (1) lithography-free approach to define the source-drain separation, (2) increasing density of devices, and (3) three-dimensional architectures. After the early demonstrations of vertical transistors by Ng *et al.* [5,29] using ZnO and indium oxide as channel materials, Schmidt *et al.* [27] described a similar effort using silicon nanowires. However, the gold catalyst was left intact creating a Schottky contact at the source. This device showed poor characteristics with an on-off ratio of only 6. A more realistic and successful effort with silicon nanowire was presented by Goldberger *et al.* [28]. They reported on-off ratios from 10^4 to 10^6 and a normalized transconductance of 0.65 to 7.4 $\mu S/\mu m$. By connecting a 200-MΩ resistor to the p-type device, they also demonstrated inverter logic.

Other device configurations using silicon nanowires include a dual-gated FET by Koo *et al.* [25] which had both a top gate and bottom gate. In this case, the back gate accumulates or inverts the channel while the top gate modulates the energy band picture. With this dual control, an on-off ratio of 10^6 was reported along with suppression of the ambipolar behavior commonly seen in nanowire and carbon nanotube transistors.

7.6 Summary

Growth of one-dimensional inorganic nanowires has been an active area of research recently. Metal, semiconductor, oxide, and nitride nanowires have been reported using a variety of growth techniques. Applications for these structures include sensors, lasers, and other optoelectronic devices, field emitters, electronics, and several others. In this chapter, a review has been presented with a focus on electronics applications. Silicon and germanium nanowires of varying diameters can now be grown successfully. The bandgap varies inversely with the wire diameter for small-diameter nanowires. The ability to control the diameter precisely and other critical requirements for device fabrication, such as positional control and patterned growth, are yet to happen. Early device fabrication efforts show the promise of these structures for future nanoelectronics. Both conventional planar transistors and vertical (top gate or surround gate) transistors have been successfully fabricated. Application of semiconducting nanowires in electronics is in its very early stages and, indeed, the amount of work detailed in literature is very limited. At present, nanowires in electronics as a topic is overshadowed by carbon nanotubes and molecular electronics. This is expected to change in the coming years since it is likely that the industry may feel more comfortable with silicon and germanium with just a change in dimension (2D versus 1D) or structure (thin film versus wire).

References

[1] Routkevitch, D. et al., *IEEE Trans. Elec. Dev.*, 43, 1646, 1996.
[2] Wagner, R.S. and Allis, W.C., *Appl. Phys. Lett.*, 4, 89, 1964.
[3] Westwater, J. et al., *Sci. Technol. B.*, 15, 554, 1997.
[4] Mao, A. et al., *Nanotechnology*, 5, 831, 2005.
[5] Ng, H.T. et al., *Nano Lett.*, 4, 1247, 2004.
[6] Nguyen, P. et al., *Adv. Mat.*, 17, 1773, 2005.
[7] Mao, A., M.S. thesis, San Jose State University, 2005.
[8] Hochbaum, A.I. et al., *Nano Lett.*, 5, 457, 2005.
[9] Sunkara, M.K. et al., *Appl. Phys. Lett.*, 79, 1546, 2001.
[10] Sharma, S. and Sunkara, M.K., *Nanotechnology*, 15, 130, 2004.
[11] Tan, T.Y. et al., *Appl. Phys. Lett.*, 83, 1199, 2003.
[12] Nguyen, P. et al., *Adv. Mat.*, 17, 549, 2005.
[13] Ma, D.D.D. et al., *Science*, 299, 1874, 2003.
[14] Li, D. et al., *Appl. Phys. Lett.*, 83, 2934, 2003.
[15] Mingo, N. et al., *Nano Lett.*, 3, 1713, 2003.

[16] Cui, Y. et al., *J. Phys. Chem. B.*, 104, 5213, 2000.

[17] Kim, J.R. et al., *Nanotechnology*, 13, 701, 2002.

[18] Wang, D. et al., *IEEE Trans. Nanotechnol.*, 3, 328, 2004.

[19] Sun, B. and Sirringhaus, H., *Nano Lett.*, 5, 2408, 2005.

[20] Yu, J.Y. et al., *J. Phys. Chem. B.*, 104, 11864, 2000.

[21] Cui, Y. and Lieber, C.M., *Science*, 291, 851, 2001.

[22] Huang, Y. et al., *Science*, 294, 1313, 2001.

[23] Cui, Y. et al., *Nano Lett.*, 3, 149, 2003.

[24] Ecoffey, S. et al., *IEDM Digest*, 05-277, 2005.

[25] Koo, S.M. et al., *Nano Lett.*, 5, 2519, 2005.

[26] Wang, D. et al., *Nano Lett.*, 6, 1096, 2006.

[27] Schmidt, V. et al., *Small*, 2, 85, 2006.

[28] Goldberger, J. et al., *Nano Lett.*, 6, 973, 2006.

[29] Nguyen, P. et al., *Nano Lett.*, 4, 651, 2004.

8

Quantum Dots in Nanoelectronic Devices

Gregory L. Snider

Alexei O. Orlov

Craig S. Lent

8.1 Introduction

The electronics industry has enjoyed unparalleled progress over the last 45 years. However, as the end of CMOS scaling nears, the industry faces some difficult questions. Should research into devices continue, or should we merely accept the maturity of silicon devices and place the responsibility for future development onto the shoulders of circuit designers? If device development beyond CMOS is possible, what sort of devices will be used? Will transistors be abandoned? Transistors have been a very successful paradigm for the processing of information. Will moving beyond transistors require a dramatic change in the circuits and architectures used? Will charge be used as the state variable (the quantity holding the information), or will another quantity such as spin be used? Will the new devices interface easily to silicon CMOS, or will sophisticated transitional structures be needed to marry the two?

Alternative devices have been investigated for many years, and low-dimensional structures have received particular attention for use in devices. In these structures, the carriers are confined to small sizes in one or more directions. At these sizes, the quantum mechanical nature of holes and electrons begins to play a significant role. In MOSFETs, these typically lead to undesirable effects such as gate or source-drain leakage due to tunneling. In low-dimensional devices, the wave nature of the carriers can be used to advantage to give performance that cannot be achieved with conventional devices. Quantum dots are structures where carriers are confined in all three dimensions to form a zero-dimensional "dot." The most common example of a quantum dot is a small volume of a narrow band-gap semiconductor surrounded on all sides by a wide band-gap semiconductor. In such cases, the carriers behave like the well-known particle in a box, with finite barriers defined by the band offset between the two semiconductors. If the volume of the dot is large, the separation between allowed states will be small. As the volume is decreased, the state separation increases. The separation between states is an important quantity. If it is greater than $k_B T$, where k_B is the Boltzmann constant, it is possible to observe effects due to the wavelike properties of the carriers, and to manipulate them one at a time. Similarly, single-electron effects can also be observed in small metallic islands where the energy to add an electron is greater than $k_B T$. Thus, single electron devices can be implemented in a wide variety of materials systems.

Even if devices can be scaled to the point of operating with single electrons, will this be a useful device? Will it solve the problems of power dissipation? This leads to the more basic question: what are the fundamental limits of computing? In this chapter, we will examine these issues beginning with an introduction to single-electron devices, followed by an introduction to a binary computing paradigm that addresses the issues of power dissipation, and finishing with a discussion on the nature of information and the limits of binary logic.

8.2   Single-Electron Devices

Devices that manipulate single electrons represent the ultimate in device scaling. Such devices have been investigated both theoretically and experimentally for a number of years. Theoretical investigations began long ago [1,2] with the recognition that on a very small island the charge already on the island could affect further charging. As the name "single electron" implies, this family of devices controls individual electrons. Intuitively, to accomplish this the device must be small. But how small? What must we do to control and manipulate individual electrons? By the 1980s, fabrication techniques had developed to the point that devices showing single-electron effects could possibly produce single-electron devices in both metallic and semiconductor systems [3,4], and the theoretical framework to understand them was developed [5,6]. Single-electron devices must meet several conditions, but the overall goal is to design a structure that separates electrons so they no longer behave as a continuous fluid (as in MOSFETs) but like separable particles. To understand how this is done, consider a capacitor. When charge is added to a capacitor, a voltage is required to force the charge onto the plates of the capacitor. The energy of the system is described by the well-known equation:

$$E = \frac{1}{2}CV^2 = \frac{Q^2}{2C} \tag{8.1}$$

where E is the energy, C is the capacitance, V is the voltage, Q is the charge, and $Q = CV$. Now consider the case in which the charge you want to add to the capacitor is just one electron. How much energy is required? Replace Q in Equation (8.1) with the charge of an electron, e, and the energy is now the charging energy (E_C), the energy required to add a single electron to the capacitor. When the value of the capacitance is in the range of commonly available capacitors— say, 1 pF—the energy required to add one electron is very small: only 8×10^{-8} eV—far less than $k_B T$ at room temperature. This means that electrons can enter and leave the capacitor using only the thermal energy they possess. Thus, the exact number of electrons on the capacitor cannot be determined. The applied voltage determines the average number of electrons, but the exact number at any time will fluctuate as electrons enter and leave due to thermal energy. If the capacitance is made very small, the charging energy becomes larger, and if it is greater than $k_B T$, electrons can no longer enter and leave the capacitor due only to thermal energy. Now the number of electrons on the capacitor is quantized. This argument is over-simplified since the capacitor is connected to the voltage source by wires so it is not clear where the capacitor begins and ends. In addition, the wires add to the overall capacitance, reducing the charging energy. We must therefore spatially define the capacitor where the number of electrons will be quantized. The problem with a macroscopic wire is that there are many conduction channels open [7], and the electrons can move from one place to another in a continuous fashion. To localize the electrons, we require that the electrons be at one place or another without a continuous connection. This means we must squeeze down the wire until there are no open conductance channels, which implies the electron cannot move classically—it must tunnel. Following the Landauer formalism, each conductance channel contributes $2e^2/h$ of conductance, including spin, corresponding to a resistance of $1/2R_Q$ where R_Q is the quantum resistance (~ 25 kΩ). If the resistance is higher than the quantum resistance, any electron transport is by tunneling.

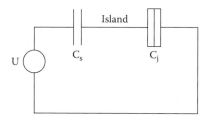

FIGURE 8.1 A schematic diagram of a single-electron box.

To summarize, control and manipulation of single electrons requires that the capacitance of the device be small enough that the charging energy is larger than $k_B T$, and that the transport through the device is by tunneling. This requires that the resistance be greater than the quantum resistance. The simplest device where single-electron effects are demonstrated is the single-electron box, shown in Figure 8.1 [8]. Here, a single tunnel junction capacitor (C_j) is coupled to an isolated island, which can be viewed as an extension of one plate of the tunnel junction, and the island is also connected to a nontunneling capacitor (C_s). To observe single electron effects, the total capacitance of the island, $C_j + C_s$, must give a charging energy greater than $k_B T$. A voltage supply, U, is applied to C_s, and when a sufficient potential is applied, a single electron tunnels onto the island. To see when this occurs, consider the potential on the island. When there are n electrons on the island, the free energy of the system is given by:

$$E(n) = \frac{(ne - Q)^2}{2(C_s + C_j)} \tag{8.2}$$

where $Q = C_s U$, the charge on C_s. The energy of the configuration where $n + 1$ electrons reside on the island can be calculated by replacing n by $n + 1$ in Equation (8.2). The energy of each configuration is a parabola as a function of U. Assume that the island contains n electrons at $U = 0$, as in Figure 8.2(a). At this voltage, the configuration n has the lowest energy and is thus the ground state. No additional electrons can be added or removed without adding energy to the system equal to the charging energy. This is called the Coulomb blockade. As the voltage U is increased, the energy of the configuration n increases while that of $n + 1$ decreases. When $Q = 1/2\,e$, the energies of the two configurations are the same, and if U is increased further, the configuration $n + 1$ becomes the lowest energy configuration. To stay in the ground state, one electron will tunnel through the capacitor C_j, bringing the island population to $n + 1$. Thus, as the voltage is swept, configurations with different numbers of electrons become the ground state, and electrons tunnel through C_j as needed to stay in the ground state. The population of the island as a function of U is therefore a staircase, as shown in Figure 8.2(b).

The single-electron box demonstrates the control of single electrons but is of limited use. To experimentally track the tunneling events onto and off of the box, another device such as an electrometer must be connected to the island. The ideas of the single-electron box can be extended to produce the most basic useful single-electron device, the single-electron transistor (SET). An SET uses an island, as in the box, but now a second tunnel junction is connected to the island, as shown in Figure 8.3. This structure resembles a field effect transistor in that there are two current leads and one voltage lead, but the operation is quite different. To operate, the SET must be in a Coulomb blockade, which requires that the total capacitance of the island $C_l + C_r + C_g$ must be such that the charging energy is greater than $k_B T$. In practice, the charging energy must be at least $4k_B T$ to obtain acceptable performance. As in the box, the gate voltage U is used to control the charge on the nontunneling capacitor C_g. The two tunnel junctions are connected to voltage supplies. As shown in Figure 8.3, a small differential bias ($V \ll E_C/e$) is applied across the device with $V/2$ applied to C_l, and $-V/2$ applied to C_r. As for the box, the energy of the configuration with n

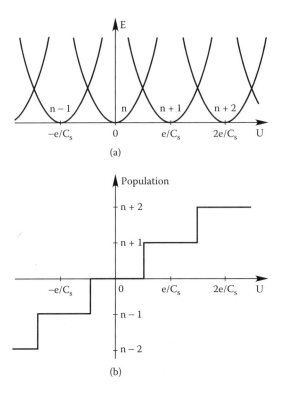

(a)

(b)

FIGURE 8.2 (a) The energy of different charge configurations plotted as a function of the applied voltage U. (b) The electron population of the island of a function of U.

electrons on the island can be expressed as:

$$E(n) = \frac{(ne - Q)^2}{2(C_g + C_l + C_r)} \tag{8.3}$$

where $Q = C_g U$, the charge on C_g. As before, the energy for each population of electrons on the island is a parabola, and as U is increased, the energy of the configuration n increases, while that of the configuration $n + 1$ decreases. At the point when they cross, the two configurations are equally probable, and something interesting happens. As with the box, the population of the island increases by one electron, which tunnels onto the island through the junction C_r, connected to the potential, $-V/2$. However, at the point when the two configurations have equal energy, the electron can tunnel off the island to return to the n configuration.

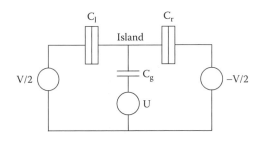

FIGURE 8.3 A schematic diagram of a single-electron transistor (SET). The central island contains a fixed number of electrons when the device is biased between Coulomb blockade peaks.

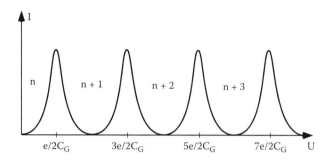

FIGURE 8.4 A rendition of the current through the SET showing the Coulomb blockade peaks.

Here is where the second junction of the SET comes into play. Since the junction C_l is connected to the potential V/2, it is energetically favorable for the electron to tunnel through the junction C_l rather than return through C_r. Once the electron has tunneled off the island and the population is again n, another electron can tunnel onto the island since the configuration $n + 1$ is equally probable. The cycle repeats itself with electrons tunneling, one at a time, onto the island through C_r, and off of the island through C_l. This produces a net current, and if the tunneling occurs at a fast rate, the current can be measured. Thus, as the voltage is swept, a series of current peaks through the device will be observed, as shown in Figure 8.4. Since these peaks are seen in an island in a Coulomb blockade, they are commonly referred to as "Coulomb blockade peaks," or "Coulomb blockade oscillations" (CBOs). Each of the current peaks corresponds to a gate voltage where two population configurations are equally probable. Between the current peaks, the population of the island is stable with a quantized number of electrons residing on the island. It is important to note that the current peaks do not represent the current of one electron entering the island, but the current of a large number of electrons streaming through the island one at a time. As the gate voltage moves through the peak into the stable region, one electron is captured and remains on the island.

One important application of SETs is in the sensing of charge [8]. In fact, SETs make the most sensitive electrometers known, with demonstrated sensitivities of 3×10^{-6} electrons/sqrt (Hz) [9]. These electrometers are extremely sensitive because the current through the SET is strongly influenced by the potential of the island. Any change in the potential of the island, caused by a small change in a nearby potential, such as a dot gaining or losing an electron, produces a large change in the current through the SET. Figure 8.5(a)

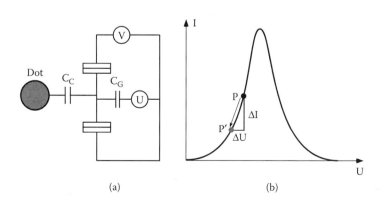

FIGURE 8.5 (a) Schematic diagram of a dot coupled to an SET electrometer. (b) Operation of the electrometer. A change of potential on the island is coupled to the SET island where it acts analogously to a change in the gate voltage U. Since the peaks are sharp, a small voltage change gives a large change in the current.

shows a schematic of a dot capacitively coupled to an SET, and Figure 8.5(b) shows one of the Coulomb blockade peaks to demonstrate the operation of the electrometer. A voltage is applied to the gate electrode so that the SET is biased to point P on the left side of the peak (a convention used by most experimental groups), and held constant. Any change in the electrostatic environment of the electrometer is analogous to a change in the gate voltage, and since the peak is sharp, a small change in the gate voltage leads to a large change in the current. For instance, an electron entering the measured dot will lower its own potential, as well as that of the SET through the capacitance C_c, reducing the current to point P'. Calculating the charge change is simply a matter of measuring the current change in the electrometer, ΔI, and finding the corresponding change in the gate voltage, ΔU, from the Coulomb blockade peak, Figure 8.5(b). The charge change on the dot being measured is then:

$$\Delta Q_{dot} = \frac{e \Delta U}{V_{Period}} \frac{C_G}{C_c} \tag{8.4}$$

where V_{Period} is the voltage change of U required to add one electron, and ΔU is the equivalent change of gate voltage caused by the dot potential. A sharp peak gives greater sensitivity since a given change of charge on the dot will yield a greater change in the current.

A few words are necessary about the effect of temperature on the Coulomb blockade peaks. Current can flow through the SET island only when two (or more) population configurations are energetically accessible. Accessible means there is a significant probability for the system to be in that configuration. Since the occupation probability depends on the temperature, the occupation probability—and hence the current through the SET—should be temperature dependent. At $T = 0$, the only voltage where two configurations are accessible are at the crossing points, which implies that current peaks should resemble delta functions. As the temperature increases, the current peaks widen because two configurations are accessible when they are within a few $k_B T$ of each other, with the lower occupation probability of the higher-energy configuration limiting the current. Thus, the current is highest at the voltage where the configuration energies cross, and then falls to either side as the occupation of the higher configuration drops. As the temperature increases, the peaks get wider and wider, reducing the region between the peaks where the current is zero and the island population is stable. At some temperature, the two peaks will begin to overlap and the current does not go to zero. This means that the population of the island is never stable. At this temperature, there are not just two but three configurations accessible to the electrons, and there is no gate voltage at which electrons are not able to tunnel from the source onto the island and back off into the drain. As the temperature is further increased, the current in the valleys of the CBOs increases, reducing the peak-to-valley ratio and making the CBOs less distinct. Eventually, as more population configurations become accessible, the CBOs wash out completely. This highlights an important aspect of single electron devices: In order to control and manipulate single electrons, there must be an operating point where the electron population is quantized. This means that at that point only one population configuration is energetically accessible. If the energy separation between configurations is small, the temperature must be reduced, but if higher temperature operation is required, a large energy separation is needed. To obtain a large energy separation between population configurations, the device must be small. An extreme example is an atom, which controls single electrons at temperatures far beyond room temperature by confining the electrons very strongly at very small dimensions. Achieving room temperature operation is challenging for a device fabricated by conventional means such as electron beam lithography, because the dimensions needed for the device are so small. For example, room temperature operation of an SET requires that the total capacitance of the island be on the order of 1 aF ($E_C \sim 3$ to $4 \, k_B T$). We can obtain a feel for the required size of this device by considering just the self-capacitance of the island. If we assume the island to be a sphere surrounded by free space, the diameter of the sphere must be less than 18 nm in diameter. Gate and junction capacitors add to the total capacitance, so a realistic estimate of the required size of the island of a room temperature SET is on the order of 5 nm, which is beyond the resolution of direct lithography. Room temperature SETs have been demonstrated [10–12], but they have relied on fabrication techniques to reduce the size of the device after lithography. Room temperature single-electron devices are a significant challenge, but are possible with molecular or molecular-sized electronics.

Single-electron devices appear to be ideal candidates for digital logic. An SET, as the name implies, is a three-terminal device, and over a certain range of voltage will operate much like a normal transistor, providing power gain. SETs can be used to implement logic gates [13,14]; however, these circuits fail to address a serious problem that faces all logic devices: power dissipation. As mentioned earlier, in the region of high conductance many electrons flow through the device. Since this current comes from a supply voltage, there will be power dissipation, and in a highly integrated system this dissipation will be excessive. A simple calculation illustrates the problem. The power dissipated per unit area is given by:

$$P = \frac{N}{A} V Q f \tag{8.5}$$

where V is the supply voltage, Q is the charge moved from the supply voltage to ground in each period, f is the frequency, N is the number of devices, and A is the area. Let's examine an extreme case where $V = 0.25$ V, $Q = 1$ e, $f = 1$ THz, and $N/A = 5 \times 10^{11}$ devices/cm^2. The clock frequency, supply voltage, and device density numbers are not unreasonable based on where the electronics industry would like to be in 10 to 15 years, but the charge flow through the device is the extreme situation of just a single electron flowing through the device. In this case, the power dissipation would be approximately 20,000 W/cm^2! For reference, a nuclear reactor typically generates 200 W/cm^2. Today's chips are already straining the cooling limits of packages at 100 W/cm^2.

The situation facing the electronics industry is quite dire. Our simple calculation shows that even single-electron transistors cannot solve the problem of power dissipation. Any device used in the conventional switching paradigm, taking charge from a supply voltage and moving it to ground, will dissipate too much power to be implemented at full speed with ultimate scaling densities. Now the power dissipation can be reduced from our estimated value by considering that not every device conducts current at every clock cycle. Aggressive power management techniques can be applied, such as turning off certain areas of the chip when not used. More of the chip's area can be devoted to cache memories that dissipate less power, or multicore processors can be implemented where each runs at a lower clock rate. However, these techniques under-utilize the true capabilities of the chip.

Power dissipation and the associated heat are the true limiters of scaling. If the conventional FET switching paradigm is used, it does not matter what technology is employed to implement the FET: FinFET, carbon nanotube, or nanowire. However, this conclusion applies only to conventional current-switching paradigms. As will be shown later, this does not imply that charge-based computing faces fundamental limits, as has been suggested by Zhirnov *et al.* [15,16]. Charge-based computations can be done at extremely low levels of power dissipation, but a break must be made with the current switching paradigms that have served so well in the past.

8.3 Quantum-Dot Cellular Automata

Quantum-Dot Cellular Automata (QCA), proposed by Lent *et al.* [17,18], is a new paradigm that holds the promise of overcoming the limits of power dissipation. The key is to encode information in a charge configuration where the charge is not moved from a supply voltage to ground. QCA employs arrays of coupled quantum dots to implement Boolean logic functions, and its advantage lies in the extremely high packing densities possible due to the small size of the dots, the simplified interconnection, and the extremely low power-delay product.

A basic QCA cell consists of four quantum dots in a square array coupled by tunnel barriers. Here, a quantum dot can be anything that localizes an electron. Electrons are able to tunnel between the dots, but cannot leave the cell. If two excess electrons are placed in the cell, Coulomb repulsion forces the electrons to dots on opposite corners. Thus, two energetically equivalent ground state polarizations exist, as shown in Figure 8.6, which can be labeled logic "0" and "1." If two cells are brought close together, Coulombic interactions between the electrons cause the cells to take on the same polarization. If the polarization of one of the cells is gradually changed from one state to the other, the second cell exhibits a highly abrupt, bi-stable switching of its polarization.

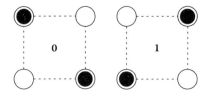

FIGURE 8.6   Basic four-dot QCA cell showing the two possible ground-state polarizations.

The simplest QCA array is a line of cells, shown in Figure 8.7(a). Since the cells are capacitively coupled to their neighbors, the ground state of the line is for all cells to have the same polarization. In this state, the electrons are as widely separated as possible, giving the lowest possible energy. To illustrate the operation of QCA devices, we'll start with the simplest case, abrupt switching, and then explain QCA clocking. In abrupt switching, an input is applied at the left end of the line, breaking the degeneracy of the ground state of the first cell and forcing it to one polarization. Since the first and second cells are now of opposite polarization, with two electrons close together, the line is in a higher energy state. The energy difference between this state and the ground state is called the kink energy, and is the characteristic energy of a QCA system. In short, it represents the energy required to place two adjacent cells in the opposite polarization. This occurs when an input is applied to the line, but it also occurs if a mistake occurs in the line. If external energy sources such as thermal energy approach the kink energy, errors appear in the QCA system, so the kink energy must be greater than $k_B T$. Since the kink energy is less than the charging energy, a QCA system requires lower temperatures than an SET implemented in the same technology. Returning to abrupt switching of the QCA line, after the first cell of the line is switched, all subsequent cells in the line must flip their polarization to reach the new ground state of the line. An inverter, or NOT, is shown in Figure 8.7(b). In this inverter, the input is first split into two lines of cells, and then brought back together at a cell that is displaced by 45° from the two lines, as shown. The 45° placement of the cell produces a polarization that is opposite that in the two lines, as required in an inverter. AND and OR gates are implemented using the topology shown in Figure 8.7(c), called a majority gate. In this gate, the three inputs "vote" on the polarization of the central cell, and the majority wins. The polarization of the central cell is then propagated as the output. One of the inputs can be used as a programming input to select the AND or OR function. If the programming input is a logic 1, then the gate is an OR, but if it is 0, the gate is an AND. Thus, with majority gates and inverters, it is possible to implement all combinational logic functions.

We began with a presentation of the abrupt switching of QCA systems because it is instructive to the understanding of QCA operation. In real QCA systems, a slightly different mode of operation would be used: quasi-adiabatic switching [18,19]. Quasi-adiabatic switching is based on the early work of Keyes and Landauer [20]. In this scheme, an electron is in one of two wells, separated by an energy barrier, as shown

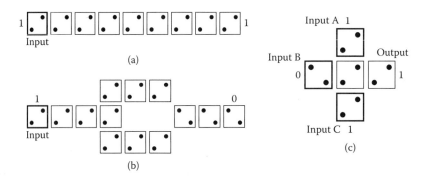

FIGURE 8.7   (a) Line of QCA cells; (b) a QCA inverter; (c) a QCA majority gate.

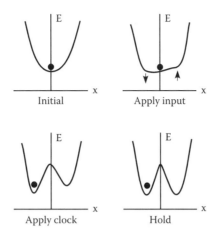

FIGURE 8.8 Schematic representation of the QCA clocking scheme. The black dot represents an electron. The input breaks the symmetry and when the clock is applied the electron localizes on the side indicated by the input. When the clock raises the barrier high, the electron is held on one side.

in Figure 8.8. To quasi-adiabatically switch the electron to the other well, the barrier between the wells is lowered so the electron can access both wells, an input is applied which nudges the electron to the other well, and finally the barrier is raised forcing the electron into the selected well. With the barrier raised high, the electron is locked into its well and the input can be removed. Thus, the device acts as a latch. By performing these switching operations slowly, relative to the settling time of the electron, the energy of the switching can be lowered below $k_B T \ln 2$. More details on quasi-adiabatic switching will be given later in the chapter.

In addition, the potential that modulates the barrier can do work on the system, and provide power gain. Power gain is extremely important in digital systems because it is needed to restore logic levels as a signal progresses through the system. Without power gain the signal level will decay at each element until it is lost in the noise. In a QCA cell, the barrier is modulated by a clock signal, and the input can be that of an adjacent cell. If the coupling between cells is weak, power gain can be achieved since the input merely nudges the electron toward the proper dot, while the clock does the work of forcing the electron to that dot. Clocking of the QCA system also allows one to control the flow of information in the system and to implement latching, memory, and pipelining. Clocking can be implemented in both semiconductor and metallic QCA systems.

8.3.1 Experimental Demonstrations

QCA cells can be implemented in a number of ways. To date, working cells have been demonstrated using aluminum islands with aluminum-oxide tunnel junctions [21], molecules [22], and doped islands in silicon [23]. The most extensive experiments are based on QCA cells using aluminum islands and aluminum-oxide tunnel junctions, fabricated on an oxidized silicon wafer. The fabrication uses standard electron beam lithography and dual shadow evaporations to form the islands and tunnel junctions [3]. A completed device is shown in the SEM micrograph of Figure 8.9. The area of the tunnel junctions is an important quantity since this dominates island capacitance, determining the charging energy of the island, and hence the operating temperature of the device. For our typical devices, the area is approximately 60 by 60 nm, giving a junction capacitance of 200 to 300 aF. These metal islands stretch the definition of a quantum dot, but we will refer to them as such because the electron population of the island is quantized and can be changed only by quantum mechanical tunneling of electrons.

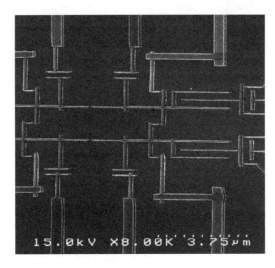

FIGURE 8.9   Scanning electron micrograph of an aluminum dot QCA cell, along with the associated electrometers.

8.3.2   QCA Cell

The first step in the development of QCA systems is a functional QCA cell where we can switch the polarization of the cell. This confirms the basic premise of the QCA paradigm: that the switching of a single electron between coupled quantum dots can control the position of a single electron in another set of dots [21]. A simplified schematic diagram of our latest QCA system is shown in Figure 8.10(a). For clarity, not shown are the single-electron transistors (SETs) coupled to D3 and D4. The four-dot QCA cell is formed by dots D1 to D4, which are coupled in a ring by tunnel junctions. A tunnel junction source or drain is connected to each dot in the cell. The device is mounted on the cold finger of a dilution refrigerator that has a base temperature of 10 mK, and characterized by measuring the conductance through various branches of the circuit using standard ac lock-in techniques. A magnetic field of 1 T was applied to suppress the superconductivity of the aluminum metal. Full details of the experimental measurements are described elsewhere [24].

QCA operation is demonstrated by biasing the cell, using the gate voltages so that an excess electron is on the point of switching between dots D1 and D2, and a second electron is on the point of switching between D3 and D4. A differential voltage is then applied to the input gates V_1 and V_2 ($V_2 = -V_1$), while all other gate voltages are kept constant. As the differential input voltage is swept from negative to positive, the electron starts on D1, and then moves from D1 to D2. This forces the other electron to move from D4 to D3. The experimental measurements confirm this behavior. Using the electrometer signals, we can calculate the differential potential in the output half-cell, $V_{D3} - V_{D4}$, as a function of the input differential voltage. This is plotted in the top panel of Figure 8.10(b), along with the theoretically calculated potential at a temperature of 70 mK. Although at a temperature of $0°$K the potential changes are abrupt, the observed potential shows the effects of thermal smearing, and theory at 70 mK shows good agreement with experiment. The middle and bottom panels of Figure 8.10(b) plot the theoretical excess charge on each of the dots in the input and output half-cells, at 70 mK. This shows an 80% polarization switch of the QCA cell, and confirms the polarization change required for QCA operation.

8.3.3   QCA Shift Register

Clocking in any digital system brings many advantages, and the QCA paradigm is no different. Clocking allows us to greatly reduce the power dissipation, control the flow of information, and implement pipelining. In a QCA device, clocking is accomplished by modulating the barriers between the dots. In a

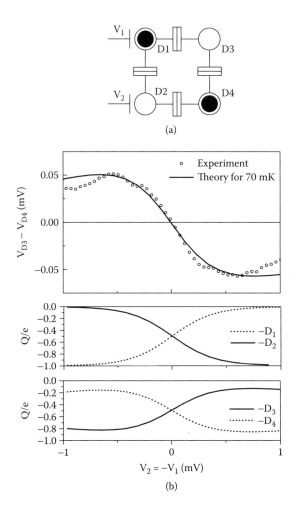

FIGURE 8.10 (a) Simplified schematic of a single QCA cell. Electrometers coupled to D3 and D4 are not shown. (b) Data from the measurement of the cell. The top panel shows the differential potential of the right half of the cell, while the bottom two panels show the calculated charge of the dots.

semiconductor dot system, this is easily accomplished by using gates to directly change the barrier between dots. However, in the metal-dot QCA cells that we use, the barrier between the dots is aluminum oxide, and hence cannot be modulated. The variable barrier in this case is formed by adding two additional dots to each cell, as shown in Figure 8.11(a). The potential of these additional dots is set by the clock line to control the tunneling of the electron between the top and bottom dots on the left and right halves of the cell. In this case, it is possible to apply a different clock to each set of three dots to create a shift register of two three-dot cells, as shown.

The operation of the QCA is shown in Figure 8.11(b). At the starting point, t = 0 ms, both latch 1 (L1) and latch 2 (L2) are set to the monostable, or "null" state. First, latch L1 is activated (i.e., switched from the null to a bi-stable state), while L2 is kept in the null state. To activate L1, first a small differential signal VIN corresponding to logical "0" is applied to the inputs at 50 ms. L1 remains in the null state until CLK1 is set HIGH at 100 ms (note that clock HIGH is actually negative voltage). When CLK1 is set high, L1 becomes active, and an electron is transferred to the bottom dot. The Coulomb barrier separating the end dots is now high, so the electron is thus locked in the bottom dot. Once L1 is locked, the signal input is removed at 150 ms and the state of L1 no longer depends on the input signal for as long as CLK1 remains HIGH.

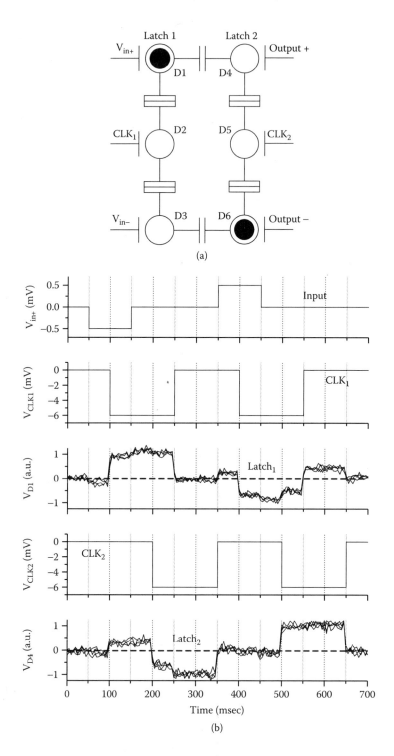

FIGURE 8.11 (a) Simplified schematic of a clocked QCA cell. Separate clocks are applied to the two halves, forming a two-stage shift register. (b) Operation of the two-stage shift register.

The dipole electric moment created by locking an electron L1 acts as the input signal for L2 in that the bottom dot of L2 is biased negatively relative to the top dot. Next, L2 is activated (CLK2 is set HIGH) at 200 ms. As a result, an electron in L2 switches to the top dot. L2 holds the bit after CLK1 is removed at 250 ms, for as long as CLK2 is high (until 350 ms). At this point in time, the shift register returns to its initial null state and is ready to receive new binary input. From $t = 400$ ms onwards, the sequence is repeated for an opposite input. In the output data shown for latches 1 and 2 in Figure 8.11(b), three successive traces are shown to demonstrate the reproducibility of the experiment [25].

One of the crucial parameters for every logic device is the speed of switching for binary operations. The operational speed of the QCA latch is determined primarily by the tunneling time of the electron ($t \approx R_J C_J \approx 10^{-10}$ sec, where $R_J \approx 3 \times 10^5 \Omega$ and $C_J \approx 3 \times 10^{-16}$ F are the resistance and the capacitance of the junction, respectively). For quasi-adiabatic operation, this gives the switching "speed limit" of the order of 1 ns for this Al/AlO$_x$ prototype. Due to much lower total capacitance ($C \sim 10^{-19}$ F), the expected switching speed is on the order of picoseconds for future molecular QCA cells. Note that the clock speed in our current experiment is limited not by the switching speed in the latch, but by the bandwidth of the electrometer circuits. Since the temporal resolution of the electrometer readout is about 0.2 ms, the detector simply cannot resolve any events occurring at a higher rate.

8.3.4   QCA Power Gain

Power gain is an important requirement for any practical electronic device. Power gain allows logic elements to restore signal levels and to overcome noise in a system. Without power gain, the signal energy put into the system by the inputs is quickly lost to the environment. Power gain in digital logic devices is different than in linear amplifiers because logic devices are saturating amplifiers. In a system of saturating amplifiers, net power gain occurs only when a weak signal is applied to the input. If a strong signal is applied, the output is equal to the input and the power gain is unity. In conventional digital logic, if a weak signal is applied to a gate, power is drawn from the voltage supply to produce an output signal with the full logic voltage. In a QCA cell, the clock line plays the role of the voltage supply in providing the power to restore the signal. In a QCA system, a weak signal could be caused by loss of energy in the system, an abnormal capacitor, or by a latch with a low charging energy. However, a small input signal is sufficient to decide the direction of switching while the clock provides most of the energy for the switching. Thus, when a weak input occurs, the clock provides the energy required to switch the latch and restore the logic level. Notice that the capability of power gain does not imply large power dissipation. Power is drawn only when needed to restore logic levels. In QCA, only this amount of power is drawn, and no more. Thus, a QCA system has power gain when needed, but low power dissipation overall.

In our demonstration of QCA power gain, we must first make clear our definition of power gain, which is the ratio of the power delivered by a cell to the power applied to that cell, as shown in Equation (8.6), which also relates the power gain to the work done by and on the cell divided by one clock period.

$$\text{Power Gain} = \frac{P_{out}}{P_{in}} = \frac{W_{out}/T}{W_{in}/T} \tag{8.6}$$

where P_{out} is the power delivered by the cell, P_{in} is the power applied to the cell, W_{out} is the work performed by the cell, W_{in} is the work performed on the cell, and T is the period of the input signal. Using this definition, we will demonstrate power gain by measuring the work done on a cell by the input over one clock cycle, as well as the work done by the cell on the next cell, and then compare the two. The work done is defined by Equation (8.7):

$$W = \int V dQ \tag{8.7}$$

where V is the voltage at a cell lead, and Q is the charge at that connection. In the experiment, the work is measured by measuring the lead voltage, and since the leads to the cell connect through capacitors,

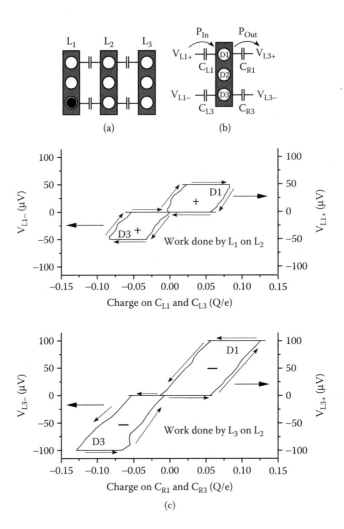

FIGURE 8.12 (a) Simplified schematic of a QCA shift register. (b) Simplified schematic illustrating the power gain experiment. (c) Experimental measurement of the work done by and on latch L2, showing a power gain of 2.07.

the charge is calculated through the voltage on each side of the capacitor and the value of the capacitor. Over one clock cycle, a plot of V versus Q forms a loop whose area is the work done, and the direction of traversing the loop denotes whether work is done by the cell or on the cell. Clockwise denotes work done on the cell, and counterclockwise is the work done by the cell.

In the experiment, a single latching cell will be used (L2), and power supplies will be used to simulate a shift register with an input cell to the left (L1), and an output cell to the right (L3), as shown in Figure 8.12(a). Voltages V_{L1+}, V_{L1-}, V_{L3+}, and V_{L3-} (input and output, in short) are applied to capacitors C_{L1}, C_{L3}, C_{R1}, and C_{R3}, respectively, to simulate latches L1 and L3 as shown in Figure 8.12(b). To determine the magnitudes of V_{L1} and V_{L3}, the dot potential swing in latch L2 is first measured, and then a smaller signal is used to simulate a weak latch L1 while a signal of the same magnitude is used to simulate a normal latch L3. Figure 8.12(c) shows the Q-V plots for the signals V_{L1} and V_{L3} over one clock period, simulating bit motion in the shift register, as in Section 8.2.2. Full details of the experiment are given elsewhere [26]. The Q-V plot for VL1 (the top panel of Figure 8.12[c]) shows a clockwise direction indicating that latch L1 performs work on latch L2. However, the Q-V plot for V_{L3} (bottom panel of Figure 8.12[c]) displays a counterclockwise direction, indicating that latch L2 performs work on latch L3. Hence, work is being performed in the same direction as the transfer of the bit. The ratio of the area enclosed by the plot in the top panel to the area enclosed by the plot in the bottom panel gives the power gain. The ratio calculated from Figure 8.12(c) is 2.07 and demonstrates that clocked QCA cells can provide true power gain.

8.3.5 Molecular QCA

A great limitation of QCA devices demonstrated so far is the low temperature of operation. To raise the operating temperature, the QCA cell must be made smaller. Unlike conventional FET-based logic, QCA devices can be scaled to molecular dimensions, and the performance actually improves.

The first step in implementing molecular QCA is to demonstrate external electric field-induced switching of an electron within a molecule. To accomplish this task, we have created an assembly of biased, vertically oriented two-dot cells sandwiched between two electrodes, and have measured the capacitance of the parallel plate device as a function of applied voltage across the plates. The schematic outline of the experiment is shown in Figure 8.13(a). The cells are covalently bound to a highly doped Si substrate in a monolayer film. In the active state, where there is a mobile electron within the molecule, the molecules are charged, and counter ions are incorporated in the film to compensate the charge. To avoid degradation of the two-dot cells, the top electrode is elemental Hg [22]. Mercury electrodes are versatile and cause little perturbation of the molecular layer. The capacitance is measured using a conventional high-frequency capacitance meter, consisting of a dc bias voltage in series with small-magnitude 1 MHz ac voltage.

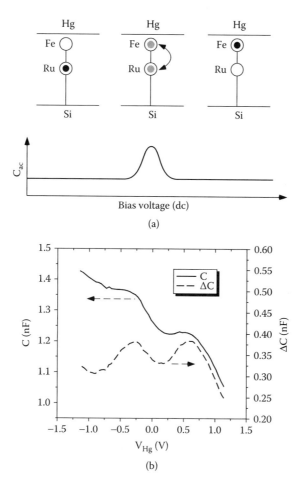

FIGURE 8.13 (a) Simplified diagram of the experimental setup for the molecular measurement, showing the physical configuration of the system and the expected ac capacitance. (b) Capacitance measurement exhibiting peaks demonstrating electron switching within the molecule.

The molecules contain one electron that can switch between a Ru atom and an Fe atom within the molecule. If there is no bias voltage, the electron should be localized on the lower Ru atom, and the capacitance should have a constant value. As the dc bias voltage is increased, the electric field raises the potential of the Fe atom, and at some bias value it becomes energetically possible for the electron to tunnel from the Re to the Fe atom, creating a dipole between the moving electron and the stationary counter ion. This increases the measured capacitance as the dipole is created and destroyed in response to the small ac voltage. As the applied dc voltage is further increased, the stable dipole no longer contributes to the capacitance. Thus, the expected response is a peak in capacitance as a function of voltage where the peak corresponds to the voltage at which the electron switches from the Ru to the Fe atom. Figure 8.13(b) shows the result of the measurement at room temperature. The top curve shows the raw capacitance data, and the lower curve the data where the sloping background, as measured on a control sample with inactive molecules, has been subtracted. The sloping background is due to depletion in the highly doped p-type substrate. Instead of the expected one peak, two peaks are observed in the capacitance data. We believe that the two peaks are associated with two types of counter-ions in the film. In the mixed valence state, the QCA molecules are charged +2, and to achieve charge neutrality, negatively charged counter ions are required. There are two types of counter-ions associated with the QCA film. The first is a [PF_6]– molecule incorporated into the film, while the second is an Si-O complex at the surface. The QCA molecules close to each of these counter-ions will switch at different applied voltages, leading to two distinct peaks in the C-V signal. We verify this by changing the fabrication procedure to vary the ratio of the two types of counter-ions. For instance, a longer wash of the molecular film with solvents will remove [PF_6]– and increase the number of Si-O complexes. As expected, the relative heights of the peaks changes. Full details of this experiment are given elsewhere [22].

A subsequent experiment confirmed the switching of an electron within a molecule and the role of the counter ions [27]. Another molecule was synthesized that had a longer linker between the two dots (in this case, two Ru atoms), and a preparation method was used that produced only one type of counter ion. As expected, the capacitance signal was stronger due to the greater distance between the dots, and only one capacitance peak was observed. This demonstrates field controlled switching of single electrons within molecules.

8.4   Limits to Binary Computing

Since QCA cells can be scaled down to the molecular level, device densities of 10^{11} to 10^{12} cm^{-2} are possible, and at these integration densities it is imperative to understand where and when dissipation occurs in computation. Information is physical [28], and binary information can be represented physically in a system with two stable states. The energy landscape of an isolated two-state system is shown schematically in Figure 8.14. The two energy valleys represent stable states of the system which we associate with bit values of "1" and "0," respectively. The energy barrier between the two states plays an important role here. It must be large enough that thermal fluctuations from the environment do not result in random flipping of the state. The system must be able to hold bit information for a reasonable period of time and enable us to distinguish a 1 from a 0.

The fundamental energetics of binary switching are independent of the physical degree of freedom used to represent the bit. We can encode the information with the spatial position of electronic charge, or with the spin of a single electron, the aggregate spin of several electrons with nuclear spins or aggregated nuclear spins (i.e., magnetization), or with the dipole moment of a molecule or assembly of molecules. Fundamentally, each yields a multi-well energy landscape that we must be able to controllably alter in order to switch the state. The notion that there is some fundamental advantage to spin systems over, for example, charge-based systems is simply naïve (see [16,29] and the critique in [30]).

Figure 8.15 illustrates three ways in which the bi-stable state can flip from the 0 to the 1 state in this energy landscape. Thermal fluctuations can excite the system over the energy barrier as shown in Figure 8.15(a). Note that the initial and final states have the same energy, so this kind of event does not dissipate energy. By energy dissipation, we mean a net transfer of energy from the system into the thermodynamically large

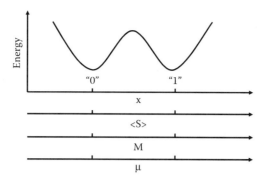

FIGURE 8.14 The energy landscape for an isolated physical system with two stable states used to represent a binary 1 or 0. The energy discussion is the same whether the dynamic degree of freedom used to encode the information is position, spin, magnetization, or dipole moment.

number of degrees of freedom of the environment. Energy came into the system from the environment to excite it up to the barrier top, but then energy flowed from the system back out to the environment as the system relaxed to its new local energy minimum. A quantum system could also tunnel from one side of the barrier to the other as illustrated in Figure 8.15(b). Again, while this is an undesirable result (assuming the goal is to hold the bit in place), it is not a dissipative event. Figure 8.15(c) shows a driven switching event in which the energy landscape has been altered by driving one side of the well to a higher energy. This lifts the system into an excited state ΔE above the ground state. The system then de-excites through either thermal excitation over the reduced barrier, or tunneling to the lower energy state (here, the 1 state on the right). In this case, the system must dissipate an amount of energy equal to ΔE to the environment in the form of heat. Standard CMOS employs precisely this technique. The charge stored on the CMOS node is discharged to ground through a transistor channel.

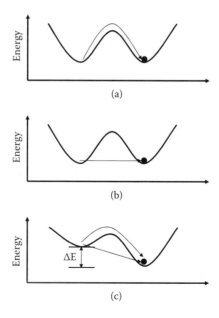

FIGURE 8.15 Switching of a bi-stable system. Neither (a) thermal excitation over the barrier nor (b) quantum tunneling through the barrier dissipate energy to the environment. By contrast falling "down-hill," as illustrated in (c), dissipates energy ΔE.

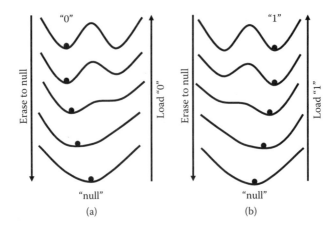

FIGURE 8.16 Adiabatic switching of a bi-stable system. A bi-stable system can be switched from the 0 state and the null state; the "erase to null" operation as shown in (a). The reverse process loads the 0 bit into the system. Adiabatic erasure and the loading of a 1 bit are shown in (b). Importantly, to adiabatically erase the bit, a copy of the bit information must exist outside the system so the system can be initially biased into the state it is already in. This is illustrated in the second figure from the top in both (a) and (b). Without this, unavoidable energy dissipation of at least $k_B T \log(2)$ occurs. With the copy, the bit can be erased from the system with no fundamental minimum energy dissipation.

Landauer pointed out that there is another way to switch a bit that keeps the system always close to its instantaneous energy minimum and therefore allows the system to dissipate only arbitrarily small amounts of heat [20,31]. If the energy landscape is continuously altered, the system can be moved with minimal excitation. This is illustrated easiest by adopting a third "null" state which is not bi-stable but monostable. Figure 8.16(a) shows the process of changing a 0 state to a null state. First, the energy of the right well is raised, then the barrier is slowly lowered until the system is in the single energy minimum, as shown at the bottom of the figure. To load a 0, the process is simply reversed. Loading a 1 into the null cell is shown in Figure 8.16(b). First, the energy on the left is raised so the landscape is tilted. Then, the barrier is raised and the system moves into the energy minimum on the right, corresponding to a 1 bit. Finally, the bias can be removed when the barrier height is large enough to assure the system is held on one side. This process is termed "adiabatic" or quasi-adiabatic switching because the system is held close to its instantaneous (local) minimum energy state.

The process of storing a 1 or 0 bit presents no fundamental problem. The system begins in the null state as shown at the bottom of Figure 8.16(a) and (b). It is biased in the appropriate direction, corresponding to the incoming information, which is then locked into the system by raising the energy barrier sufficiently to suppress thermal and quantum mechanisms. The biasing field can then be removed and the bit has been written, as shown in the top part of Figure 8.16(a) and (b). More subtle considerations attend the "erase to null" operation.

Consider now the erasure process labeled "erase to null" in Figure 8.16. To perform this process adiabatically requires that the system be biased into the state in which it is already. This is illustrated in the second figures from the top of Figure 8.16(a) and (b). To bias the system into the state, a copy of the bit must exist outside the system itself. This could be the result of measuring the state of the bit, for example. Measurement means making a copy. This simple observation—that to adiabatically erase the bit a copy of the bit information must exist—has fundamental consequences.

The thermodynamics of bit erasure are illustrated in Figure 8.17 and Figure 8.18. We plot the time evolution of the physical configuration space of a system. Note that what is being plotted represents physical degrees of freedom, not simply logical space. The system can initially exist in two possible physical configurations, which we associate with bit values 0 and 1. In Figure 8.17, the number of configurations

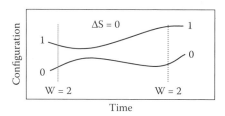

FIGURE 8.17 A physically reversible process. Since the microscopic laws of physics are physically reversible, a system with two possible initial configurations has two possible configurations at a later time. An isolated system like this has no change in entropy.

denoted that W has the value 2 for both the initial and final state. This corresponds to simple time evolution under the microscopically physically reversible laws of physics. Physical reversibility refers to the fact that, except for exotic situations involving the weak nuclear force, the microscopic laws of physics are the same for both forward and reverse processes. That is, if we take a movie of a physical process and play it backwards, no physical laws are violated. If the system moves from state A to B, we could reverse all velocities and the system would move from B to A. Stated another way, the microscopic laws of physics make it possible to determine if we find the system in state B at time t_2, which state A it was in at an earlier time t_1. The entropy is defined as the Boltzmann constant times the number of accessible states in the system, $S = k_B$ log(W) (here, log denotes the natural logarithm). For Figure 8.17, both the initial and final entropy is k_B log(2) and the change in entropy is zero.

Figure 8.18 corresponds to the situation of an "erase to null" event. Initially, the system can be in either the (physical) 1 or 0 state, but the final state in either case is the null state. This collapse results in a change of entropy $\Delta S = k_B$ log(1) $- k_B$ log(2) $= -k_B$ log(2). The situation shown in Figure 8.18 cannot occur under the time evolution of the microscopically reversible laws of physics. The change in entropy means there must be another system involved. Two possibilities exist:

1. Another finite system is coupled to this one and it now has a copy of the information. That is, the configuration space is actually larger than the one shown. The apparent degeneracy of the final state would be broken if we included the second system in the description (the line on the right would split into two branches in the perpendicular plane, for example). The microscopic time evolution is still one-to-one and thus physically and logically reversible.
2. The system is coupled to a thermodynamically large system with many degrees of freedom (the environment or heat bath) at temperature T. A transfer of free energy ($F = TS$) equal to $\Delta F =$

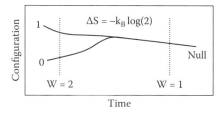

FIGURE 8.18 A physically irreversible process. The system with two possible initial configurations evolves into a single state. Information has left the system and the entropy has decreased. One of two things must have occurred: (1) Information has been copied into another finite system whose degrees of freedom are not shown here. Were the complete configuration of both systems shown, there would be no collapse and states would stay reversible. (2) Entropy and free energy have been transferred into the thermodynamically large manifold of states in the environment. In this case, we say energy has been dissipated and a bit has been irreversibly erased.

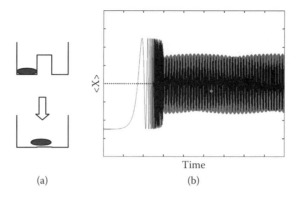

(a)   (b)

FIGURE 8.19 Attempt to erase without a copy. The time-dependent Schrödinger equation is solved for a double well system. In this case, the barrier between the two wells is slowly lowered, though without any bias applied to the system. The state is initially localized in the left well but oscillates as the barrier is lowered. These oscillations characterize an excited state which would relax dissipatively to the new ground state.

$T\Delta S = k_B T \log(2)$ has occurred from the system to the environment. This increased the entropy of the environment by $\Delta S = +k_B \log(2)$, exactly making up for the entropy lost by the system. The process results in the true loss of information (into the dense manifold of environmental states) and dissipation of energy in the form of heat in the amount of $k_B T \log(2)$.

This makes it clear why it is the erasure operation that is fundamentally linked to heat generation. Some amount of heat will be generated in any physical process representing a logical switching event. But there is a minimum amount of heat ($k_B T \log(2)$) that *must* be generated when a bit of information is erased. The adiabatic switching approach (Figure 8.16) can lower the amount of heat generated for logically reversible processes (which do not entail information loss) to an arbitrary small level (at the cost of speed—see the following), but cannot change the fundamental connection between energy dissipated as heat and information erasure.

To adiabatically erase a bit, we need a copy of the bit so we can bias the system into the state it is already in, as we have seen in Figure 8.16. What happens if we try to erase the bit without a copy? To see this, we examine a very simple quantum system, the double potential well. Figure 8.19 shows the results of solving the time-dependent Schrödinger equation for a double-well system as the well barrier is slowly lowered. In this case, no bias is applied, so the potential is always symmetric. The result of lowering the barrier is that the wave function, initially localized in one well, begins to oscillate. The oscillations will persist forever if (as here) there is no way for the energy to dissipate. Figure 8.20 shows exactly the same situation, but with a field which biases the system toward its initial state. Now when the barrier is lowered, the system relaxes smoothly to its new ground state. We emphasize that in order to bias the system into the state it is already in, we must have a copy of the information outside the system itself. Without the copy, we lose control of the system, and oscillations—which in reality would be damped by dissipative processes—result. The classical version of this loss of control as the system is relaxed was discussed by Landauer [31].

Erasure of information causes unavoidable heat dissipation. However, we can design around this, again with some trade-offs. Bennett showed [32–35] that any logically irreversible function could be embedded within a reversible calculation provided intermediate results were all kept. The stored results could then be used to "decompute" the result so that at the end one has only a copy of the inputs and the desired output. This comes at the cost of the computing process taking at least twice as long (compute + decompute). In addition, in most implementations, the number of intermediate results can explode geometrically and force large amounts of overhead in temporary storage. In QCA, this Bennett style computation may be much more tractable because intermediate results are simply stored in place as the state of QCA cells. Bennett-clocking of QCA [30] can be pipelined and offers the possibility of tuning the amount of dissipation within the heat budget, making the trade-off between throughput and heat generation genuinely tunable.

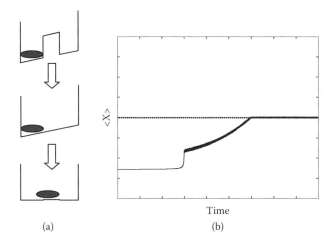

(a)   (b)

FIGURE 8.20   Erasure with a copy. The time-dependent Schrödinger equation is solved for a double well system, as in the case shown here. The difference is that now the initial state of the system is known (a copy exists) so the system can be biased into the state it is already in. The result is a smooth erasure to the new ground state. The tiny visible oscillations can be made arbitrarily small by switching slowly.

Adiabatic switching must be done "slowly" so the system remains close to its instantaneous ground state. A simple analogy is a person walking with a mass suspended on the end of a string. If the person walks gently and slowly from one side of the room to another, the mass barely sways. If the person accelerates quickly, walks to the other side, and halts abruptly, the mass oscillates (an excited state) and then damps gradually down to the new ground state, dissipating energy through friction. What does it mean then to switch slowly? It means slowly compared to the natural excitation times of the system. For CMOS circuitry, this means slow compared to the RC charging times, which alas can be quite long. For molecular QCA [36], this means slowly compared to the tunneling time for an electron to move from one part of a molecule to another (which could be femtoseconds). Adiabatic switching therefore becomes practical in systems with very fast natural response times.

Finally, let us emphasize again that these considerations of the energetics of switching are entirely independent of the nature of the physical degree of freedom used to represent the information. Landauer used magnetization in his original arguments. Figures 8.14 through 8.20 could apply equally to spin systems or charge-transfer molecules. There is no magic system. Basic choices for information processing are analog/digital and reversible/irreversible. Quantum computing represents the extreme requirements of being both analog, because quantum degrees of freedom are continuous, and irreversible, because unitary operators are always reversible. QCA represents a modest choice of digital information and the ability to choose the amount of irreversibility included through circuit design, and so manage the associated heat dissipation.

8.5   Summary

As the scaling of CMOS comes to its end, there is a real need to address the issue of power dissipation coupled with device density. Single-electron devices such as SETs and pumps have niche applications in sensing and metrology, but even single-electron devices fail to provide the small power dissipation needed for large-scale computing. A break with the FET-based paradigm is needed, and while more conventional single-electron devices fail to give the needed performance, others can. A device paradigm based on QCA cells offers the opportunity to break away from FET-based logic, and to exploit the quantum effects that come with small size. In this new paradigm, the basic logic element is no longer a current switch but a small array of quantum dots, and the logic state is encoded as the position of electrons within a quantum dot cell. We have demonstrated the operation of a QCA cell fabricated in aluminum islands with an aluminum-oxide tunnel where the polarization of the cell can be switched by applied bias voltages, along with a QCA shift register, and true power gain. QCA cells are scalable to molecular dimensions, and since the performance improves as the size shrinks, a molecular QCA cell should operate at room temperature with a switching energy of a few yJ.

As devices are scaled down to the molecular size range, the issue of how power dissipation occurs is no longer just a topic for academic discussions. To attain integration densities of 10^{11} cm^{-2} the power dissipation of each device must be miniscule. This will require a shift away from transistors, which are extremely wasteful of power. However, at high integration levels every source of power dissipation must be considered. Since dissipation occurs whenever information is erased, the high-performance computing systems of the future will need to carefully husband information. Erasure can be avoided using reversible techniques such as Bennett-clocking, but these require some system overhead. Future systems will likely be broken into pipelined blocks where reversible methods are used within the blocks to avoid dissipation, while some erasure with its associated dissipation is allowed at the edges of the blocks. In this way, the requirements of both power dissipation and system performance can be met.

References

[1] Gorter, C.J., *Physica*, 17, 777, 1951.

[2] Neugebauer, C.A. and Webb, M.B., *J. Appl. Phys.*, 33, 74, 1962.

[3] Fulton, T.A. and Dolan, G.H., Observation of single-electron charging effects in small tunnel junctions, *Phys. Rev. Lett.*, 59, 109–112, 1987.

[4] Scott-Thomas, J.H.F. et al., Conductance oscillations periodic in the density of a one-dimensional electron-gas, *Phys. Rev. Lett.*, 62, 583–586, 1989.

[5] Likharev, K.K., Correlated discrete transfer of single electrons in ultrasmall tunnel junctions, *IBM J. Res. Dev.*, 32 (January), 144–158, 1988.

[6] Averin, D.V. and Odintsov, A.A., Macroscopic quantum tunneling of the electric charge in small tunnel junctions, *Phys. Lett. A*, 140, 251–257, 1989.

[7] Landauer, R., Electrical transport in open and closed systems, *Z. Phys. B*, 68, 217–228, 1987.

[8] Lafarge, P. et al., Direct observation of macroscopic charge quantization, *Z. Phys. B*, 85, 327–332, 1991.

[9] Aassime, A. et al., Radio-frequency single-electron transistor: Toward the shot-noise limit, *Appl. Phys. Lett.*, 79, 4031–4033, 2001.

[10] Matsumoto, K., Room temperature–operated single electron transistor made by STM/AFM nano-oxidation process, *Phys. B*, 227 (1–4), 92–4, 1996.

[11] Takahashi, Y. et al., Fabrication technique for Si single-electron transistor operating at room temperature, *Electron. Lett.*, 31, 136–137, 1995.

[12] Saitoh, M. et al., Room-temperature demonstration of low-voltage and tunable static memory based on negative differential conductance in silicon single-electron transistors, *Appl. Phys. Lett.*, 85, 6233–5, 2004.

[13] Chen, R.H. et al., Single-electron transistor logic, *Appl. Phys. Lett.*, 68, 1954–1956, 1996.

[14] Gerousis, C.P. et al., Nanoelectronic single-electron transistor circuits and architectures, *Intl. J. Circuit Theory Appl.*, 32, 323–338, 2004.

[15] Zhirnov, V.V. et al., Emerging research logic devices, *IEEE Circuits & Devices*, 21, 37–46, 2005.

[16] Zhirnov, V.V. et al., Limits to binary logic switch scaling – A Gedanken model, *Proc. IEEE*, 91, 1934–1939, 2003.

[17] Lent, C.S. et al., Quantum cellular automata, *Nanotechnology*, 4, 49–57, 1993.

[18] Lent, C.S. and Tougaw, P.D., A device architecture for computing with quantum dots, *Proc. IEEE*, 85, 541–557, 1997.

[19] Tougaw, P.D. and Lent, C.S., Dynamic behavior of quantum cellular automata, *J. Appl. Phys.*, 80, 4722–4736, 1996.

[20] Keyes, R.W. and Landauer, R., Minimal energy dissipation in logic, *IBM J. Res. Dev.*, 14, 152–157, 1970.

[21] Orlov, A.O. et al., Realization of a functional cell for quantum-dot cellular automata, *Science*, 277, 928–930, 1997.

[22] Qi, H. et al., Molecular quantum cellular automata cells. Electric field driven switching of a silicon surface bound array of vertically oriented two-dot molecular quantum cellular automata, *J. Am. Chem. Soci.*, 125, 15250–15259, 2003.

[23] Mitic, M. et al., Demonstration of a silicon-based quantum cellular automata cell, *Appl. Phys. Lett.*, 89, 13503–13, 2006.

[24] Amlani, I. et al., Demonstration of a six-dot quantum cellular automata system, *Appl. Phys. Lett.*, 72, 2179–2181, 1998.

[25] Orlov, A.O. et al., Clocked quantum-dot cellular automata shift register, *Surface Sci.*, 532–535, 1193–1198, 2003.

[26] Kummamuru, R.K. et al., Power gain in quantum-dot cellular automata, *Appl. Phys. Lett.*, 81, 1331–1334, 2002.

[27] Qi, H. et al., Dependence of field switched ordered arrays of dinuclear mixed-valence complexes on the distance between the redox centers and the size of the counterions, *J. Am. Chem. Soci.*, 127, 15218–15227, 2005.

[28] Landauer, R., The physical nature of information, *Phys. Lett. A*, 217, 188–193, 1996.

[29] Cavin, R.K. et al., Energy barriers, demons, and minimum energy operation of electronic devices, *Fluctuation and Noise Lett.*, 5, C29–C38, 2005.

[30] Lent, C.S. et al., Bennett clocking of quantum-dot cellular automata and the limits to binary logic scaling, *Nanotechnology*, 17, 4240–4251, 2006.

[31] Landauer, R., Irreversibility and heat generation in the computing process, *IBM J. Res. Dev.*, 14, 183–191, 1988.

[32] Bennett, C.H., Notes on the history of reversible computation, *IBM J. Res. Dev.*, 32, 16–23, 1988.

[33] ———, Notes on Landauer's principle, reversible computation, and Maxwell's Demon, *Studies in History and Philosophy of Modern Physics*, 34, 501–510, 2003.

[34] ———, Logical reversibility of computation, *IBM J. Res. Dev.*, 17, 525–532, 1973.

[35] Leff, H. and Rex, A., *Maxwell's Demon 2*, IOP, Bristol, 2003.

[36] Lent, C.S. et al., Molecular quantum-dot cellular automata, *J. Am. Chem. Soc.*, 125, 1056–1063, 2003.

9

Self Assembly of Nanostructures Using Nanoporous Alumina Templates

Bhargava Kanchibotla

Sandipan Pramanik

Supriyo Bandyopadhyay

9.1 Introduction

Nanotechnology, the science and engineering of nanometer-sized objects, is a major research endeavor owing to the innumerable possibilities—and profit—the field has to offer. Nanostructures exhibit unique physical and chemical properties compared to "bulk" structures. They have important applications in developing cutting-edge technology in such diverse areas as medicine, computing, communication, sensing, defense, homeland security, informatics, and the life sciences [1–5].

Fabrication or synthesis is the first step in producing nanostructures. A plethora of nanofabrication techniques has been developed in recent years, ranging from the very expensive to the inexpensive. *Self assembly* is a nanofabrication technique that has captured the imagination of many in the field, primarily because it is easy, inexpensive, has a rapid throughput, and often requires minimal investment in equipment and infrastructure. Self assembly is any process where nanostructures spontaneously nucleate or assemble in a suitable medium under the proper conditions. Depending on the medium and the external environment, the nanostructures can sometimes *self order*, meaning they automatically arrange themselves into a regimented (usually periodic) array in space. This technique is *parallel*, unlike direct write lithography, which is *serial*. In direct writing, nanostructures are delineated serially, one at a time, leading to an unacceptably slow throughput. In self assembly, several nanostructures are produced simultaneously (in "parallel"), leading to a fast throughput. Furthermore, self assembly is usually a gentle process, causing little or no processing damage to the finished nanostructures. This is usually not the case with more conventional nanofabrication techniques such as direct write lithography, or x-ray lithography, where the patterning or exposing beam (often an electron or ion beam, or x-ray) causes severe processing damage. This is particularly debilitating in the case of semiconductor nanostructures since surface states caused by the damages can pin the Fermi level in the middle of the bandgap of the semiconductor and deplete the entire structure of mobile charge carriers.

A multitude of self-assembly techniques have been reported in the literature. Some of them are "physical," such as the Stranski–Krastanow mode growth of quantum dots on a lattice-mismatched substrate where the nanostructures nucleate because of strain (mechanical forces). Others are "chemical" in that chemical reactions orchestrate the spontaneous nucleation of nanostructures. In this chapter, we will focus on a chemical technique that is rapidly emerging as the workhorse for making well-regimented quasi-periodic arrays of nanostructures of virtually any material on arbitrary substrates. This technique utilizes nanoporous alumina membranes that have a quasi-periodic array of pores with nearly uniform diameter in the sub-10 to few hundred nm range. It is produced by the anodization of aluminum in a suitable acid with a suitable voltage. Since the pores develop spontaneously because of a process known as field-assisted dissolution, the template is "self assembled." Under appropriate conditions, the pores form a nearly periodic hexagonal close packed array. Each pore is a cylinder of uniform diameter. The aspect ratio (length-to-diameter ratio) can be very large without causing the diameter to vary along the length. Aspect ratios of 200:1 are routinely realized in our laboratory. Pore diameter can be made to vary from 8 nm to 200 nm by changing the nature of the acid and the anodizing voltage. These pores can be *selectively* filled up with metals, semiconductors, insulators, and even superconductors by using electrodeposition. That results in a vertically standing, quasi-periodic array of nanowires or quantum dots of these chosen materials housed in a ceramic matrix. The finished product is a self-assembled spatially ordered array of nanostructures.

Nanoporous alumina membranes are also useful as excellent templates for pattern transfer on any generic substrate. *Reactive ion etching* is a standard etching technique used for such pattern transfer. It is also discussed in this chapter.

9.2 Nanoporous Alumina Membranes

Anodized aluminum oxide (alumina) nanoporous membranes have been a subject of considerable interest for many years, detailed studies having been undertaken for more than half a century [6–8]. Since the early 1990s, several techniques to fabricate these membranes have been developed. Their salient features include (1) a highly ordered arrays of pores, (2) a pore diameter that can be controllably varied from 8 nm to a few hundred nm, and (3) pore density ranges from 10^{10}–10^{12} pores/cm^2. These membranes are a convenient template for synthesizing highly ordered, vertically standing nanowires in a ceramic matrix. The wires are electrically isolated from each other by the intervening alumina, which is an insulator. The alumina matrix can be gated with a metallic pad and a potential is applied to this pad to lower the barrier between neighboring nanowires, thereby introducing weak electrical coupling (tunneling). This has intriguing device applications. Alumina is also quite effective for optical isolation since it has a relative dielectric constant between 2.5 and 3.0 at optical wavelengths [9]. Nanowires embedded in alumina will act as optical waveguides if the nanowire material has a larger relative dielectric constant than what alumina has. Most semiconductors used as optical waveguides have relative dielectric constants larger than 3.0. Therefore, nanowire waveguides embedded in alumina can be effective waveguides, if the optical wavelength is comparable to, or smaller than, the nanowire diameter.

9.2.1 Electrochemical Synthesis of Alumina Membranes: Electropolishing of Aluminum Foils

Fabrication of alumina membranes with an ordered array of nanopores involves several critical processes and steps. The starting material is high-purity aluminum foils. We use 99.998% pure, 100-μm thick aluminum foils consisting of polycrystalline aluminum (single crystal aluminum is considerably more expensive). Square coupons of area 1 in. × 1 in. are cut from the aluminum foils for ease of handling. The surface condition of the aluminum foil is an important factor that affects the ordering of the pores and the uniformity of the pore diameter. The surface roughness of Al foils, purchased off the shelf, is typically on the order of a few μm. In order to reduce that to a few tens of nanometers, we first degrease the coupons in acetone and then electropolish them in a Leeco electrolyte consisting of a solution of perchloric acid, butyl cellusolve, ethanol, and distilled water for ~ 10 sec using 45 V DC. Electropolishing creates a shiny, mirror-like surface. Closer inspection (using atomic force microscopy) has revealed that electropolishing actually creates intriguing patterns on the surface of the aluminum [10]. Depending on the voltage and duration of electropolishing, periodic ridges (with heights of a few nanometer) can be produced on the Al surface, as shown in Figure 9.1. The chemical dynamics governing this process are quite complex and have been explored in detail in a number of publications [11–14]. In this chapter, we will not dwell on this

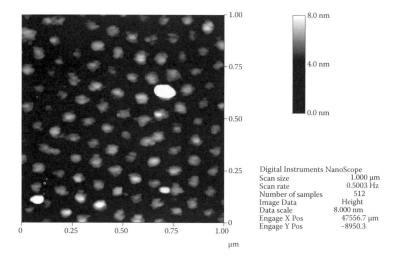

b_ec60v30s.001

FIGURE 9.1   (see color insert following page **5-6**) Atomic force micrograph of quasi periodic bumps formed on an Al foil electropolished at 60V for 30 seconds.

aspect since these patterns, while highly interesting from the perspective of basic chemical reactions and kinetics, have no major device applications.

Some groups have investigated the effect of the polycrystallinity of the aluminum foil on the regularity of pore formation. The pores form a nearly periodic hexagonal close packed array, but only within a domain of size of around a few μm. Some reports indicate the domain size can be increased if the foils are annealed prior to electropolishing to increase the grain size in the polycrystalline film. It is not clear if there is any correlation between the grain size and the domain size. To date, there is no widespread consensus about this matter and it remains an open question.

In our group, electropolishing is carried out in an Electromet-4 (Buehler Ltd.) electro polisher, with the Al foil placed at the anode and in physical contact with the Leeco electrolyte that acts as the etchant for electropolishing [10]. The most commonly used Leeco electrolyte is the L1 electrolyte, comprising 93 cc perchloric acid, 150 cc butyl cellusolve, 1050 cc ethyl alcohol, and 205 cc distilled water. The perchloric acid is the etchant for aluminum, while ethanol and butyl cellusolve are polarizable molecules that sheath regions (crests) on the surface, which have the highest electric fields. As a result, troughs on the surface dissolve faster. This effect is counter-balanced by the fact that mass transport of Al^{3+} ions from the crests is more efficient. The interplay of these two effects is responsible for the formation of ridges on the surface as shown in Figure 9.1.

9.2.2 Anodization of Aluminum Foils

Anodization of the electropolished Al foils in a suitable acid leads to the growth of the porous alumina film on the foil surface. An electropolished Al coupon is anodized in an electrolytic cell. Figure 9.2 is the schematic diagram of the experimental step up we use to perform electrochemical anodization. The electrolytic cell is a commercial flat cell. A regulated power supply provides a highly stable DC voltage for anodization. The aluminum foil acts as the anode, and a platinum grid is used as a cathode since it does not react with the electrolytes used for anodization. The electrolytes employed are a moderately strong acid (pH < 4). Acids typically used for anodization are 3% oxalic acid, 15% sulphuric acid, and 10% phosphoric acid. Anodization using AC current is not feasible because anodic alumina allows current to flow in one direction only. Figure 9.3 is a top-view AFM image of Al foil anodized in 3% oxalic acid at 40 V. The darker areas are the pores, and the lighter areas surrounding the pores are the cell walls. The diameter of the pores is about 50 nm, and the interpore separation is about 100 nm[8].

9.2.3 Growth Kinetics of Nanoporous Alumina Membranes

Many theoretical models have been proposed to explain pore growth [15–19]. O'Sullivan *et al.* [16] proposed a model based on inhomogeneous electric field distribution at pore tips and field-assisted

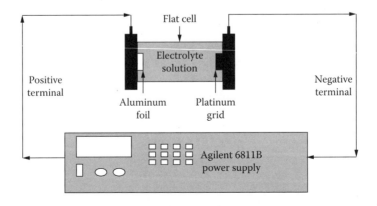

FIGURE 9.2   Schematic diagram of the experimental set up used for anodization.

FIGURE 9.3 (see color insert) Top view AFM image of the sample electro polished and anodized in 3% oxalic acid at 40 V.

dissolution, which explains why pores grow and what determines the size distribution of the pores. The pore growth is perpendicular to the surface. It was shown that parameters like the pore diameter, the interpore separation, and the steady-state barrier layer thickness are directly proportional to the applied voltage. Shershulsky *et al.* [17] gave a microscopic explanation for the dependence of pore parameters on the applied voltage or the electrolyte used. Later, Jessensky *et al.* [18] developed a model to explain the densely packed hexagonal pore structure developed under special anodization conditions. The forces between the neighboring pores in the hexagonal structure are due to the mechanical stress associated with the volume expansion that occurs during the conversion of aluminum to alumina. Expansion of the material takes place only in the vertical direction because oxidation occurs over the entire pore bottom simultaneously, and not all of the alumina contributes to the oxide formation since some Al^{3+} ions remain mobile in the oxide under the applied voltage. According to Jessensky *et al.*, the voltage and the electrolyte composition influence the relative thickness of the porous alumina layer compared to the consumed aluminum. The most ordered pores are obtained using 0.3M solution oxalic acid at an anodizing voltage of about 40 V. Li *et al.* [19] claimed that the best ordered periodic arrangements of pores are obtained when the relative volume expansion of the aluminum during oxidation is about 1.4, independent of the electrolyte. If the anodization voltage is small, the pores become disordered as the volume of the alumina formed is smaller and the relative thickness ratio is less than 1.4. On the other hand, if the voltage is large, the volume expansion will be too large and the ratio will be greater than 1.4. Thus, producing ordered periodic pores with small or large voltages is difficult.

A typical anodization current–versus-time characteristic for a sample anodized in 3% oxalic acid at 40 V at room temperature is shown in Figure 9.4. During the first few seconds of anodization, the current decreases rapidly until a minimum is reached. This is followed by a slow rise in the anodization current until the steady state is achieved. This process is illustrated in a sequence of steps shown in Figure 9.5.

Initially, when the voltage is applied, an aluminum oxide layer begins to form due to oxidation of aluminum by oxygen containing ions [20]. During this process, the anodization current gradually drops as indicated in region A of the anodization current curve (Stage A). In the region B, fine-featured pores begin to develop, because of local dissolution of the oxide (Stage B) by hydrogen-containing ions in the acid. The rate of oxidation is, however, greater than the rate of oxide dissolution. Pore formation is gradually stabilized with the further passage of time. This corresponds to the region C (Stage C) in the current curve. Finally, the steady state region D is reached when the pore structure stabilizes completely

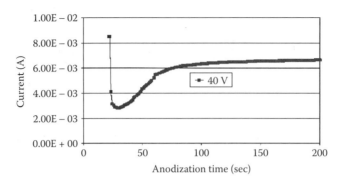

FIGURE 9.4 Anodization current versus time curve obtained experimentally for a sample anodized in 3% oxalic acid at 40 V, at room temperature.

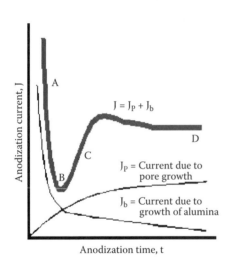

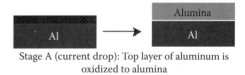

Stage A (current drop): Top layer of aluminum is oxidized to alumina

Stage B (minimum current): Aluminum is continuously oxidized to alumina and pores begin to develop due to dissolution of alumina
Rate of oxidation > Rate of dissolution

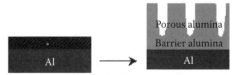

Stage C (rise in current): Aluminum is continuously oxidized to alumina and pores continue to grow in alumina
Rate of oxidation > Rate of dissolution

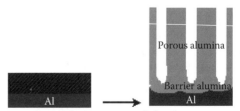

Stage D (constant current): Steady state
Rate of oxidation = Rate of dissolution

FIGURE 9.5 Sequence of steps explaining the growth kinetics of nanoporous alumina membrane.

and the corresponding anodization current reaches a constant value (Stage D). This corresponds to the equilibrium state where the rate of oxide growth is balanced by the rate of oxide dissolution. At this stage, the pores grow perpendicular to the surface and are parallel to each other.

The chemical reactions occurring during anodization are as follows:

At the Anode

Oxidation of Aluminum: $2Al + 3H_2O = Al_2O_3 + 6H^+ + 6e^-$

Dissolution of Alumina: $Al_2O_3 + 6H^+ = 2Al^{3+} + 3H_2O$

At the Cathode

Hydrogen Evolution: $6H^+ + 6e^- = 3H_2$

As shown by O'Sullivan *et al.*, the growth of porous aluminum is a consequence of two competing mechanisms: oxide growth and the partial dissolution of aluminum oxide by the hydrogen ions. Oxide growth is due to the migration of oxygen containing ions O^{2-}/OH^- from the electrolyte through the oxide layer to the pore bottom, and the dissolution of the oxide is due to the migration of Al^{3+} ions, which drift through the oxide layer and are ejected into the electrolyte. The net current density J can thus be thought of as being a sum of two current contributions, namely, J_b, due to the growth of the barrier alumina (which decreases with time), and J_p, due to pore growth (which increases with time and finally reaches a steady state, as shown in Figure 9.5).

9.2.4 Factors Controlling the Self Ordering of Pores

The electrolyte used for anodization, the voltage applied, and the duration of anodization are key factors known to influence the porous structure. It has been reported in the literature that the temperature of the electrolyte during anodization does influence the pore structure [21], but this influence is minor.

Figure 9.6 is the schematic cross section of the nanoporous structure after anodization. In the figure, the top layer is porous consisting of an array of vertical pores aligned parallel to each other. Underneath lies the U-shaped barrier layer. At the very bottom is the remaining aluminum layer left behind after anodization.

In Figure 9.6, d is the diameter of the pore, L is the length of the porous membrane, B is the thickness of the barrier layer, D is the interpore separation, and A is the thickness of the unreacted aluminum. The diameter of the pores, the thickness of the barrier layer, and the interpore separation depend on the electrolyte used and the applied voltage. The length of the pores depends on the duration of anodization and the starting thickness of the aluminum foil. In the following, we discuss how these parameters are influenced by the anodization conditions.

9.2.4.1 The Effect of Acid Strength

Weak or quasi-neutral acidic solutions (pH 5–7) are not good for anodization because they lead to the formation of flat nonporous aluminum oxide membranes, commonly called a barrier-type layer

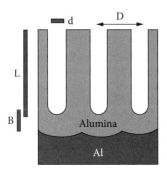

FIGURE 9.6 Schematic diagram of the cross-section of nanoporous alumina after anodization.

or film [22]. Anodization in a strong acid (pH < 4) results in porous aluminum oxide film. Based on the theoretical model proposed by Shershulsky *et al.*, the pore radius varies linearly with the pH according to the relationship:

$$R_e \sim (1 + C\Phi)$$

where R_e is the pore radius, C is a constant, and Φ is the pH value.

Hence, anodization using 5% phosphoric acid produces the largest pore diameter at high voltages (high pH value). The smallest pore diameter is obtained using 15% sulphuric acid since it is a strong acid (small pH value). Recently, Ono *et al.* [23] showed self-ordering porous alumina having cell diameters of 300 nm and 500 nm, using organic acid electrolytes such as malonic and tartaric acid solutions, respectively. These acids have large pH values, yielding large pore diameters.

9.2.4.2   The Effect of Anodizing Voltage

The DC voltage applied between the Al foil and the platinum electrode, also known as the "anodization voltage," is a critical parameter that greatly influences the porous structure. Any arbitrary anodization voltage does not result in the formation of porous membranes. It is well known that the pore diameter and the barrier layer thickness vary with the applied voltage. The model proposed by O'Sullivan *et al.* showed a linear relation between the pore diameter and the anodization voltage. For samples anodized in 3% oxalic acid, the optimum voltage for obtaining a well-ordered hexagonal pattern is about 40 V, which yields a pore diameter of 50 nm. The optimum anodization voltage in the case of samples anodized using 15% sulphuric acid is around 10 V, and the resulting pore diameter is around 10 nm. In the case of samples anodized in 5% phosphoric acid, a pore diameter as large as 350 nm can be obtained using high anodization voltages, while retaining some semblance of spatial ordering. Few groups have recently reported self organization of porous alumina arrays using a mixture of acids as electrolyte [24]. Figure 9.7 summarizes the self-ordering voltages reported up to now and the corresponding pore diameters.

The pore density varies inversely as the square of the anodizing voltage. Exponential dependence of pore density as a function of anodization voltage was reported by Palibroda *et al.* [25]. Almawlawi *et al.* [26] showed that the pore density ρ varies with the anodization voltage V according to:

$$\rho = \frac{\alpha}{(d + \beta V)^2}$$

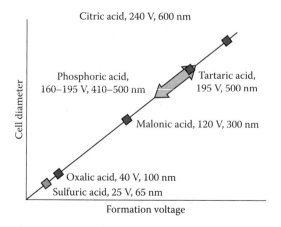

FIGURE 9.7   Cell diameter (center-to-center separation between neighboring pores) versus anodizing voltage for various acids. (Reproduced from ref. [23] with permission from Elsevier.)

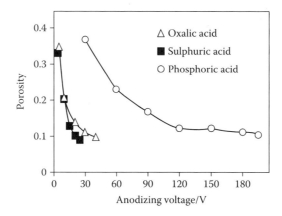

FIGURE 9.8 Porosity of anodic films with anodization voltages measured for three different electrolytes. (Reproduced from ref. [27] with permission from the Electrochemical Society.)

where α is a constant ~ 1.15, β is a constant that depends on the acid and temperature and d is another constant. Using 3% oxalic acid, pore densities of the order 10^{10} pores/cm^2 can be obtained. Pore densities $> 10^{11}$ pores/cm^2 can be achieved using 15% sulphuric acid. Recently, reports have arisen regarding changes in the porosity of anodic alumina with increasing anodization voltages [27]. Figure 9.8 shows the porosity versus anodization voltage curve for different electrolytes. The porosity values decrease as the anodization voltage is increased, irrespective of the electrolyte used.

9.2.4.3 The Effect of Temperature

It has been reported in the literature that temperature affects the ordering of the pore structure but does not have any significant influence on pore diameter or the interpore separation when the anodization is carried out at a constant anodization voltage and electrolyte concentration [28]. Pore structures formed at low temperatures are not as regular when compared to that formed at room temperature. They are twisted or bent if produced at low temperatures. Sang Li *et al.* [29] reported the effect of temperature on the ordering of the pore structure. The regularity of the pore structure was analyzed by applying fast Fourier transformation (FFT) to the SEM images.

9.2.5 Current Density During Pore Growth

The values of current densities obtained during porous film growth are a key factor controlling the self ordering of pores. The pore growth mechanism based on current density was discussed in Section 9.2.3. Recently, Ono *et al.* reported a systematic study of pore growth in different electrolytes, focusing on the influence of current density. Figure 9.9 shows the anodization current versus time characteristics obtained for samples anodized using sulfuric acid. The anodization voltage ranges from 5 to 27 V. For voltages ranging from 5 to 25 V, the *I-t* curves obtained are similar to the curves discussed in section 2.3, indicating porous membrane growth. At about 27 V, a high current density value was observed along with intense gas evolution at the sample surface. Interestingly, no porous growth was seen at this stage. The absence of pores at this stage is probably because the anodization voltage exceeded a certain critical value.

Similar studies were performed using oxalic and phosphoric acids. Figure 9.10 shows the anodization voltage versus current density curves for three different electrolytes. Extreme high current was observed for voltages higher than the individual self-ordering voltages for the three electrolytes. It was concluded that self ordering occurs only for voltages below the individual critical voltages for the electrolytes used. Beyond the critical voltage, no film growth takes place.

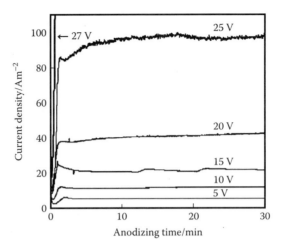

FIGURE 9.9   Current time curves of anodizing in sulphuric acid in the voltage range from 5 V to 27 V. (Reproduced from ref. [27] with permission from the Electrochemical Society.)

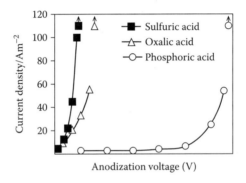

FIGURE 9.10   Changes in steady-state current density values at anodization voltages for three different electrolytes. (Reproduced from ref. [27] with permission from the Electrochemical Society.)

9.3   Fabrication of Highly Ordered Porous Templates Using Multistep Anodization

Porous membranes, obtained using a one-step anodization, do not contain very well-ordered pores. This issue limited the use of porous templates in nanofabrication. Long range ordering is absent in these structures, even though short range ordering was present. Annealing the Al foils prior to electropolishing and anodization was suggested as a remedy, but proved unsatisfactory. The first satisfactory solution to the problem was revealed in a landmark paper published by Masuda *et al.* [30] in 1995. They developed a technique called *multi step anodization* that significantly improved the periodicity of the pores produced in both oxalic and sulphuric acids. A highly ordered honeycomb pore structure over large areas was achieved by this technique.

Figure 9.11 is a sequence of steps that illustrates multi step anodization. First, the Al foil is electropolished and then anodized for a short time (5 to 10 min) which results in the formation of a thin alumina film on the surface. Now the sample is treated using a mixture of 0.2M chromic acid and 0.4M phosphoric acid in order to remove the film. Then the sample is again anodized for a long time (10 to 12 hours or overnight). The long period of anodization significantly reduces the number of defects and dislocations

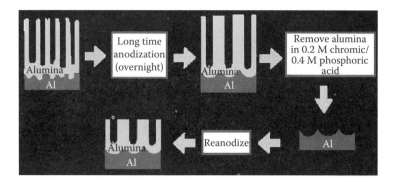

FIGURE 9.11   Schematic diagram showing the sequence of steps during the multistep anodization process.

in the sample. The thickness of the film at this stage is several microns. The thick film is then etched away in a mixed acid (0.2M chromic/0.4M phosphoric acid) at 60°C. This leaves behind a hexagonally ordered array of scallops on the surface of the aluminum. Figure 9.12 is the AFM image of the sample after being treated with the chromic/phosphoric acid mixture. The presence of the aluminum scallops corresponds with the U-shaped barrier layer at the base of the pores. The sample is then re-anodized for a few minutes to acquire the required pore length. At this stage, the ordered array of scallops acts as nucleation centers for the re-growth of the nanoporous alumina. This results in ordered pore growth.

The significant improvement in the ordering of the pores can be gleaned from the images. The AFM image after the first step anodization shows the defects Figure 9.13 (left). The highly ordered hexagonal structure with no defects can be seen after the multi step anodization process from Figure 9.13 (right). Recently, Vrublevsky *et al.* [31] carried out a detailed study of the growth of porous membranes during the re anodization process, details of which are beyond the scope of the present work. Interested readers are referred to the article for more intricate details.

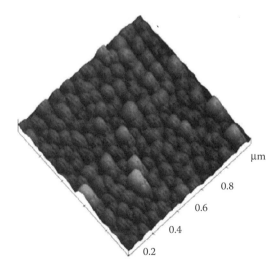

FIGURE 9.12   (see color insert) Top-view AFM image of the aluminum scallops after removing the porous membrane with chromic/phosphoric acid mixture.

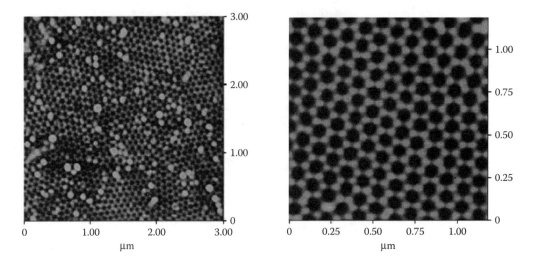

FIGURE 9.13   (see color insert) (left) Top-view AFM image of porous membrane showing the defects. (right) Top-view AFM image of highly ordered defect-free porous membrane obtained after multistep anodization.

9.4   Nanostructure Synthesis Using Porous Alumina Membranes

Nanoporous alumina membranes are popular templates for the fabrication of regimented nanowire arrays, as well as arrays of quantum dots of various materials like metals, semiconductors, polymers, alloys, etc. Selective electrodeposition of the desired material within the pores is commonly used to fabricate these structures. Electrodeposition can be performed with either an AC or a DC voltage. Electrodeposition of magnetic and semiconductor nanowires are discussed next.

9.4.1   Techniques for the Removal of the Barrier Alumina Layer

The thickness of the U-shaped barrier layer formed at the porous membrane and Al interface is about 10 to 30 nm (Figure 9.6, Section 9.2.4). This thickness depends on the electrolyte used, as well as the anodization voltage. The barrier layer is a spoiler. This layer is nonconducting in nature, which prevents the passage of *conduction current* along the length of the pores, and hence the DC electrodeposition of materials inside the pores is impossible unless this layer is removed. AC electrodeposition is another matter. There, atomic species are transported to within the pores via the *displacement current*, as opposed to conduction current. Since the barrier layer does not completely impede the displacement current, it is possible to selectively electrodeposit materials within the pores using AC electrodeposition, even if the barrier layer is present. However, another disadvantage of having the barrier layer is that it is hard to make electrical contacts to the materials inside the pores from the bottom since the barrier layer intervenes. Recently, techniques have been developed to either remove or thin the barrier layer.

9.4.1.1   Thinning or Dissolution of the Barrier Layer Using Phosphoric Acid:

Ansermet *et al.* [32] developed a method for thinning the barrier layer by soaking the anodized aluminum membrane in an aqueous solution of 5% phosphoric acid for 50 minutes. This process thins the barrier layer at the pore bottom but also widens the pores since the etching in phosphoric acid is more or less isotropic. Once the barrier layer is sufficiently thinned, materials can be AC or DC electrodeposited into the porous membrane and electrical measurements can be carried out by making contacts at the top and bottom of the membrane.

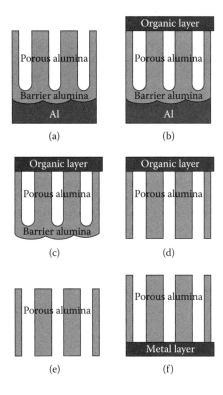

FIGURE 9.14 Sequence of steps showing the method to remove the barrier layer to facilitate DC electrodeposition. (Reproduced from ref. [7] with permission from Marcel–Dekker publishers.)

Another technique widely used to facilitate DC electrodeposition is based on complete dissolution of the barrier layer. This method can also be used to make template for pattern transfer purposes. Figure 9.14 shows the sequence of steps describing this well-known technique. First, the top layer of the porous membrane is coated with a thick organic layer to provide mechanical stability to the film. Then, the sample is treated with 3% $HgCl_2$ solution which dissolves the unreacted Al layer from the bottom (Step C) to expose the barrier layer from the bottom. The sample is then soaked in a 5% phosphoric acid solution, which removes the exposed barrier layer. The top organic layer is then etched by treating the sample with ethyl alcohol solution (Step E). After this stage, just the thin porous membrane (a "see through" film) is left behind. This delicate film can be fished out of the ethyl alcohol solution with tweezers and carefully placed on a conducting substrate for DC electrodeposition (this method calls for some operator skill since careless handling will make the film wrinkle and render it useless).

9.4.1.2 Removal of the Barrier Layer Using "Reverse Polarity Etching"

In a technique developed by Rabin *et al.* [33], the barrier layer was etched in 5% phosphoric acid, not by mere soaking but by applying a negative bias (the negative terminal is connected to Al foil). Typically, a negative bias of 2.25 V is used. The phosphoric acid can be replaced by a dilute solution of KCl, which has a faster etching rate. In our labs, we perform the reverse polarity etching using 1% phosphoric acid [34]. Figure 9.15(a) is the top-view SEM image of a sample anodized in 3% Oxalic acid at 40 V. Figure 9.15(b) is the bottom-view SEM image of the same sample after performing reverse polarity etching. The see-through pores can be clearly seen from this image.

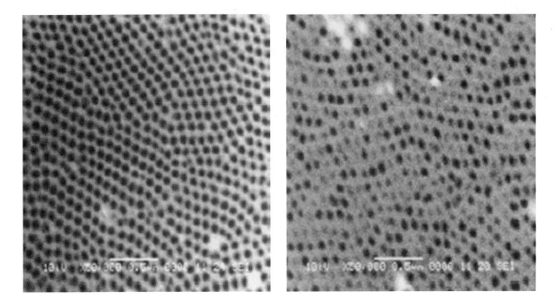

FIGURE 9.15   (left) Top-view SEM image of the porous membrane anodized in 3% oxalic acid at 40 V before reverse polarity etch. (right) Bottom-view SEM image showing through pores after reverse polarity etch. (Reproduced from Ref. [34] with permission from American Scientific Publishers.)

9.5   AC Electrodeposition

AC electrodeposition is the standard technique used for depositing various materials inside the pores with the barrier layer intact. Metal nanowires like Ni, Fe, Au, Cu, Co, etc. can be fabricated using this process. Alloys (FeCo, FeNi, CoCu, and CoNi, etc.) and semiconducting nanowires like CdS, InP, and ZnTe, etc. can also be synthesized using this technique.

9.5.1   Magnetic Nanowires

High-density arrays of magnetic nanostructures are of significant technological importance because this geometry offers considerable advantages in storing data [6]. It has been estimated that a real density of the order of $\sim$ 300 Gbit/in^2 can be realized using the ordered array structure. AC electrodeposition of magnetic materials inside an anodized aluminum oxide (AAO) template offers a cheap and easy method of achieving such high-density arrays. In our lab, we fabricated ordered nanowire arrays of various magnetic materials. The experimental setup is similar to the one shown in Figure 9.2 of Section 9.2. An AC voltage is applied across the Pt mesh and the porous alumina, which act as the two electrodes. The electrolyte used is an aqueous solution of the salt of the material to be deposited. For example, for deposition of Fe nanowires, the electrolyte used is 0.1 M FeSO$_4$ solution. Boric acid is added to increase the conductivity of the solution. The AC voltage and frequency are optimized. For electrodeposition inside pores with diameter 50 nm, the appropriate voltage and frequency values are 40 V and 200 Hz. During every negative half-cycle, metal ions in the solution (example, Fe^{2+}, Co^{2+}) migrate toward alumina. They are reduced to zero-valent metal atoms such as Fe (after combining with 2e$^-$) at the porous alumina, and then collect inside the pores because the pores *offer the least impedance paths for the displacement current to flow.* The following reactions occur during electrodeposition.

At anode:
$$Fe^{2+} + 2e^- = Fe$$
$$2H^+ + 2e^- = H_2$$

FIGURE 9.16 Cross-section SEM image of an alumina template containing electrodeposited Fe nanowires. (Courtesy of Prof. Latika Menon.)

At cathode:

$$2H_2O = O_2 + 4H^+ + 4e^-$$

During the positive half-cycles, the zero valent atoms are not reoxidized to the divalent cations since alumina is a valve metal oxide that allows current to flow only in one direction.

The length of the magnetic nanowires formed within the pores depends on the kind of electrolyte, the concentration of the electrolyte, and the time of deposition. Figure 9.16 shows a cross-section SEM image of an alumina template containing electrodeposited Fe nanowires. The diameter of the wires is about 50 nm and inter-wire separation is around 100 nm. To prevent the surface oxidation of the metal nanowires after deposition inside the pores, the template is immersed in boiling water for about 10 minutes, which seals the pores, thus inhibiting oxidation.

As mentioned before, one of the factors that influence the length of the wires is the time of deposition. For certain applications it is critical the pores are not over-filled with the material. Hence, having control

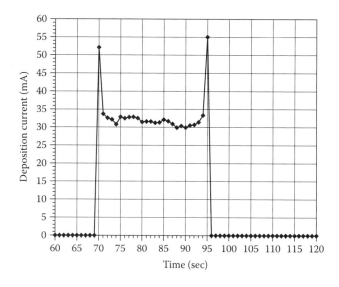

FIGURE 9.17 Deposition current versus deposition time curve of Fe nanowires inside the porous membrane.

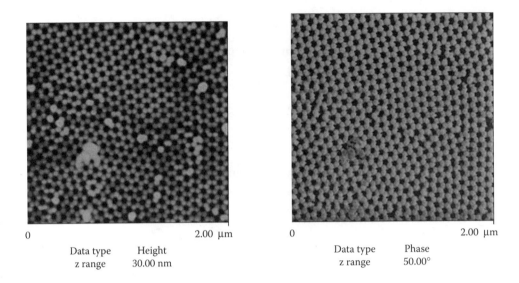

Data type	Height
z range	30.00 nm

Data type	Phase
z range	50.00°

0   2.00 μm   0   2.00 μm

FIGURE 9.18 (see color insert) (left) Top-view AFM image of the Fe nanowires inside the porous membrane. (right) MFM image showing the magnetic signal of the Fe nanowires inside the pores. Time of electrodeposition of Fe is about 50 sec.

of the electrodeposition time is of utmost importance. Figure 9.17 is the I-t curve for deposition of Fe nanowires inside the pores.

From the curve, one can notice that the current density rises up sharply after 95 sec. This indicates the pores are over-filled. Ideally, by stopping the deposition between the two transition regions, over-filling of the pores can be avoided and nanowires of the desired length can be fabricated. Figure 9.18 is the AFM (left) and MFM (right) image of the Fe nanowires deposited inside the porous alumina template. The sample is anodized in 3% oxalic acid at 40 V and the deposition is carried out at 40 V, by applying a frequency of 250 Hz. The deposition time for Fe is around 50 sec.

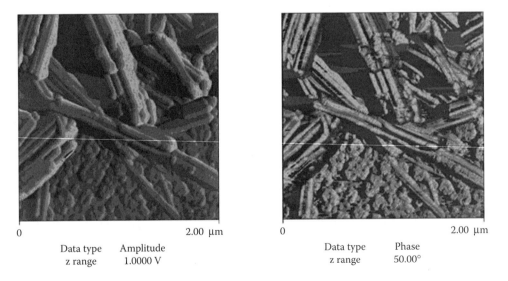

Data type	Amplitude
z range	1.0000 V

Data type	Phase
z range	50.00°

0   2.00 μm   0   2.00 μm

FIGURE 9.19 (see color insert) (left) AFM image of Fe nanowires released from an AAO template onto a substrate. (right) MFM image of Fe nanowires released from an AAO template onto a substrate.

FIGURE 9.20 TEM image of 8-nm cobalt nanowire. (Courtesy of Dr. Yi Liu.)

Figure 9.19 shows the AFM (left) and MFM (right) image of Fe nanowires released from an AAO template onto a substrate. The sample is soaked in 0.2 M chromic acid/0.4M phosphoric acid mixture at 60°C. This solution etches the aluminum oxide, releasing the wires into the solution. The sample is then rinsed in distilled water in a container with the substrate placed at the bottom. In order to prevent the wires from being rinsed away, a magnet was held below the substrate. The diameter of the wire is $\sim$ 50 nm, and the length of the wire is $\sim$ 1 μm.

Using the procedure described earlier, various metal nanowires of desired diameter and length can be synthesized by choosing the appropriate aqueous solution of the salt of the metal to be deposited. To fabricate Ni nanowires, 0.1M $NiSO_4$ aqueous solution can be used. Similarly, 0.1M $CoSO_4$ aqueous is used for depositing Co nanowires. In general, the aqueous solution is of the type MSO_4, where M is assumed to be a divalent metal element. A TEM image of 8-nm Co nanowires is shown in Figure 9.20. This was produced by electrodeposition of Co in pores formed by anodization in sulfuric acid.

The drawback of using aqueous solutions for electrodeposition is that it limits the choice of materials that can be deposited inside the pores. The limit is imposed by the fact that the metal salt *must dissolve in water*. Sometimes the normal sulfate salt does not dissolve in water. An example is lead sulfate. However, lead acetate dissolves in water and therefore can be used for the electrodeposition of lead.

9.5.2 Nanowires of Magnetic Alloys

Binary [FeCo, CoNi, FeNi, CoPt] or ternary [FeCoNi] magnetic alloy nanowires, which are of significant interest in the area of magnetic recording because of their giant coercivities, can be easily synthesized by electrochemical deposition—for example, $Fe_{1-x}Co_x$, where $0 \leq x \leq 1$ can be deposited using an aqueous solution consisting of a mixture of 0.1 M $FeSO_4$ and 0.1M $CoSO_4$. Electrodeposition is carried out using an AC voltage of 20 V at 250 Hz [35]. Different crystal phases of Fe and Co can be obtained by varying the content of the material added into the mixture. Details about the magnetic properties of these alloy nanowires are discussed in ref [6]. Ferromagnetic metal–nonmagnetic metal (FM-NM) heterogeneous alloys nanowires can also be fabricated. Recently, Coey *et al.* [36] deposited a Co-Cr alloy from an aqueous bath containing $CrCl_3$:$6H_2O$, $CoCl_2$:$6H_2O$, H_3BO_3, $HCOONH_4$, and H_2NCH_2COOH. Electrodeposition was carried out at a constant potential in a conventional three electrode configuration. Fabrications of other heterogeneous alloys like Ni_xZn_{1-x}, Co-Cu have also been reported [37].

9.5.3   Semiconductor Nanowires

II-VI group semiconductors have potential applications in quantum devices, photo detectors, ultraviolet light sources, etc. Using the AC electrodeposition technique, various kinds of II-VI semi-conductor nanowires can be synthesized [6]. In our laboratory, we have successfully deposited materials like CdS, CdSe, and CdTe and fabricated other semiconductor nanowires like ZnS, ZnSe, and ZnTe. The non-aqueous solution of dimethlysulfoxide (DMSO) is used as the electrolyte to deposit semiconductors having a composition of the form MX where M can be Cd or Zn, and X stands for S, Se, or Te. For example, in order to electrodeposit CdS, 10.5 grams of cadmium per chlorate and 1 gm of sulfur powder are mixed with 250 gm of DMSO. Then, 2.5 grams of lithium per chlorate is added to this mixture to improve the conductivity. An AC voltage of about 25 volts rms is applied and the deposition frequency is maintained at 250 Hz. The deposition is carried out at a temperature of about 70°C. Porous alumina membrane and a platinum mesh act as the electrodes. The Cd^{+2} ions in the solution are reduced to zero valent Cd during the negative cycle and are selectively deposited inside the pores where they react with sulfur in the heated solution, forming cadmium sulfide nanowires.

9.6   DC Electrodeposition

Multilayered nanowires like Co/Cu, Co/Ni, and "spin valve structures" can be easily fabricated using DC electrodeposition. A spin valve structure is basically a tri-layered nanowire where a paramagnetic layer is sandwiched between two ferromagnetic layers. In our lab, we have successfully deposited various metals like Ni, Co, etc. individually and tri-layers (Co-Ge-Ni or Co-Cu-Ni) forming spin valve structures. The main advantage of DC electrodeposition is that it allows well-controlled deposition of the material inside the pores, which is difficult to achieve using the AC deposition technique. The critical issue with DC electrodeposition is to estimate correctly and accurately the deposition rate of the material being deposited. The ratio of the pore length to the pore fill time gives a reasonable estimate of the deposition rate for the given material. For example, we calibrate the deposition rate of Ni as follows. Initially, a porous template with a pore length of 1 μm and a pore diameter of 50 nm is synthesized using 3% oxalic acid. Reverse polarity etching is used to remove the barrier layer. The electrolyte employed for fabricating Ni nanowires is the aqueous solution of the salt of the material to be deposited. The applied DC voltage is about 3 V. After 10 seconds, the deposition current rises drastically, indicating the pores are being over-filled. The deposition is stopped at this point. Based on the preceding experiment, the deposition rate of Ni is calculated to be 100 nm per second. Similar calibration studies were done to determine the deposition rates of Co, Fe, and Cu for a known pore diameter. Recently, we have successfully electrodeposited Ge nanowires inside

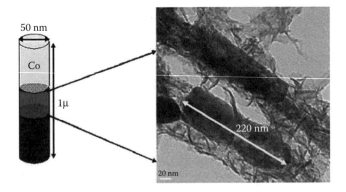

FIGURE 9.21   Schematic diagram of nanowire-based spin valve structure consisting of Ni, Ge, and Co. Enlarged image is the TEM image of 220-nm Ge nanowire. (The figure on the right is reproduced from Ref. [38] with permission from the American Institute of Physics.)

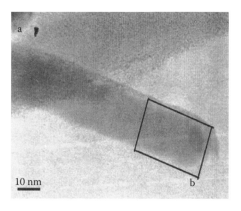 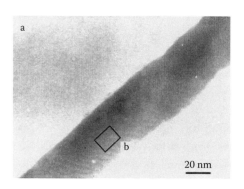

FIGURE 9.22 (left) TEM image of a 20-nm CdS nanowire obtained using DC deposition. (Reproduced from ref. [40] with permission from the International Union of Pure and Applied Chemistry.) (right) TEM image of 20-nm CdTe nanowire fabricated using DC deposition. (Reproduced from ref. [40] with permission from the International Union of Pure and Applied Chemistry.)

porous alumina templates [38]. The electrolyte consisted of a mixture of an ionic liquid 1-butyl-3-methyl-imidazolium-hexafluorophosphate ($BMIPF_6$), GeI_4crystals, and 250 ml of dimethyl sulfoxide (DMSO). The electrodeposition is carried out under appropriate DC bias and the deposition rate is approximately estimated. Knowledge of the deposition rates of the constituent materials allows us to fabricate tri-layered structures (spin valve structures) that exhibit the spin valve effect. Figure 9.21 is a schematic diagram of a nanowires-based spin valve structure.

Another standard technique used for DC electrodeposition is to have a porous alumina template with just the through holes, synthesized using the technique outlined in Section 9.4.2. Conducting materials like Au are coated on one side of the template (usually by resistive or e-beam evaporation) to facilitate DC deposition. The deposition is realized using a potentiostat and a three-terminal setup where a porous alumina membrane is used as the working electrode, Pt mesh as the counter electrode, and a reference electrode as the third terminal. Multilayer nanowires can be electrodeposited inside the pores using a single electrolyte or separate electrolytes [39]. II-VI compound semiconductor nanowires like CdTe, CdS, and CdSe can also be prepared by DC electrodeposition using non-aqueous electrolytes. Figure 9.22 (left) is the TEM images of 20-nm CdS, and (right) CdTe nanowires fabricated using DC electrodeposition [40].

9.7 Oxide Nanowires

Oxide nanowires have applications in optoelectronics because of the novel properties exhibited by these nanostructures. Synthesis of oxide nanowires like ZnO, In_2O_3, MnO_2, etc. have been reported by several groups. Anodized Aluminum Oxide (AAO) templates can be used to fabricate oxide nanowires.

ZnO, with a bandgap of 3.2 eV, has applications in the area of optoelectronic devices like LEDs and diode lasers operating in the ultraviolet range [41–43]. Starting with an AAO template, the unreacted Al layer and the barrier layer are removed and the bottom side of the template coated with Au to facilitate deposition of ZnO. DC electrodeposition of Zn is carried out at 1.25 V in a standard three-terminal cell with an electrolyte containing 80g/l $ZnSO_4.7H_2O$ and 20g/l H_3BO_3. After electrodepositing Zn nanowires inside the pores, the sample is annealed at 300°C in air, resulting in the formation of ZnO nanowires. Figure 9.23 is a TEM image of the ZnO nanowires inside the porous template.

ZnO nanowires can also be synthesized using the AC electrodeposition method [44]. A porous alumina template of thickness ~ 1 micron is fabricated using 3% oxalic acid solution. The solution used for depositing Zn consists of 250 ml of DMSO and 10.5 gm of $ZnClO_4$. 2.5 gm of lithium perchlorate are added to improve the conductivity of the solution. The deposition is carried out with an AC voltage of about 25 V rms and an AC frequency of 250 Hz. The sample is then treated with H_2O_2 for about five hours,

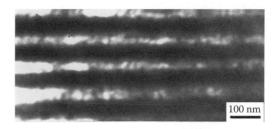

FIGURE 9.23   TEM image of the ZnO nanowires inside the porous template. (Reproduced from ref. [43] with permission from the American Institute of Physics. © 2000 American Institute of Physics.)

which oxidizes Zn, and thus ZnO nanowires with diameters of 50 nm were obtained. ZnO nanowires of diameter 10 nm and 25 nm were also synthesized. The photoluminescence measurements of 25-nm diameter ZnO nanowires showed strong photoluminescence due to exciton recombination. Figure 9.24 is a TEM image of single ZnO nanowire that was released from the AAO template and captured on a TEM grid for imaging.

Recently, West *et al.* [45] synthesized amorphous MnO_2 nanowires inside porous alumina membranes. They have applications in high energy and high power density Li ion batteries. Commercially available porous templates with a pore diameter of 200 nm were used. One side of the membrane was coated with Au to enable the deposition of the material inside the pores using the potentiostatic method. Two different approaches were followed in depositing the nanowires. In the first technique, an electrolyte bath consisting of an aqueous solution of 0.01M $MnSO_4$ and 0.03M $(NH_4)_2SO_4$ was used. A slight amount of H_2SO_4 was added to maintain the pH of the solution around 8. The deposition was carried out using a potentiostatic control by varying the applied voltage between 0.2 and 1.2 V. The second deposition method used an electrolyte with 1M $MnSO_4$ and 0.5M H_2SO_4 with a galvanostatic control set to 10 mA/cm^2 at room temperature. After the electrodeposition, the alumina membrane was removed using NaOH solution and

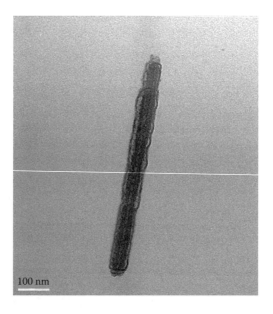

FIGURE 9.24   TEM image of single ZnO nanowire released from the AAO template. (Reproduced from Ref [44] with permission from Springer Science and Business Media.)

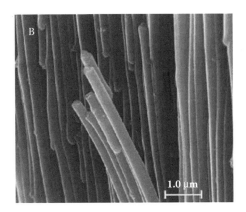

FIGURE 9.25 (a) Cross-section SEM image of amorphous MnO$_2$ nanowires deposited using potentiostatic method. Reproduced from ref. [45] with permission from Elsevier. (b) Cross-section SEM image of MnO$_2$ nanowires deposited using galvanostatic control. (Reproduced from ref. [45] with permission from Elsevier.)

the sample was then dried in air for about 24 hours, forming amorphous MnO$_2$ nanowires. Figure 9.25 (left) is a SEM image of amorphous MnO$_2$ nanowires deposited using the potentiostatic method, while Figure 9.25 (right) is the SEM image of MnO$_2$ nanowires fabricated using galvanostatic control.

9.8 Half Cell Method for Fabricating Inorganic Nanowires Using Porous Alumina

Recently, Yuanzhe Piao [46] and co-workers developed a new and simple technique for fabricating inorganic nanowires called the "paired cell" method. Figure 9.26 is a schematic diagram of the paired cell.

A 1.5-cm diameter and 40-μm thick porous membrane separates Teflon half cells. Solutions A and B with the same volume and proper concentration are then poured into the cell and left for a certain time at room temperature, resulting in nanowires of the form A$_m$B$_n$ inside the porous membrane. Using this method, inorganic nanowires of AgI, CuS, and Ag$_2$S were synthesized. For example, in synthesizing AgI nanowires, an aqueous solution of 0.01M AgNO$_3$ and 0.01M KI were poured into the cell and allowed to react for about 48 hours at room temperature. AgI/Al$_2$O$_3$ nanowires were then formed inside the porous membrane. The AAO membrane is then etched away leaving the wires behind, which are rinsed in DI water several times. Figure 9.27 is an SEM image of AgI nanowires released from the porous membrane. Figures 9.28 and 9.29 are SEM images of CuS and Ag$_2$S nanowires, synthesized using the paired cell method.

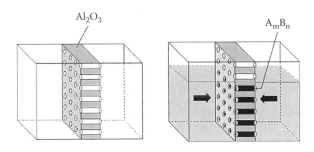

FIGURE 9.26 Schematic diagram of the paired cell. (Reproduced from ref. [46] with permission from Elsevier.)

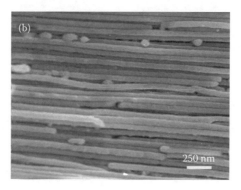

FIGURE 9.27   SEM image of AgI nanowires released from the porous template. (Reproduced from ref. [46] with permission from Elsevier.)

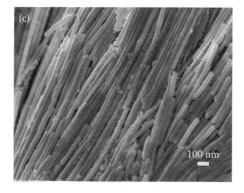

FIGURE 9.28   SEM image of CuS nanowires synthesized using paired cell method. (Reproduced from ref. [46] with permission from Elsevier.)

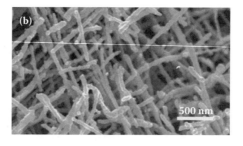

FIGURE 9.29   SEM image of Ag_2S nanowires synthesized using the paired cell method. (Reproduced from ref. [46] with permission from Elsevier.)

9.9 Pattern Transfer Using Anodic Alumina Templates

In spite of its many attractive features, realizing a stand-alone functional device (particularly a three-terminal device) using exclusively AAO membranes is currently impossible. Ideally, conventional semiconductors like Si, GaN, etc. with well-ordered nanostructures similar to the porous membrane are desired since they will enhance the performance of the present-day devices and also open up new possibilities in device fabrication technology. One possible application is in quantum dot flash memories [47] which can benefit from the use of AAO templates to embed well-ordered, size-controlled quantum dots in the gate insulator of the metal oxide semiconductor transistor used as a memory element. Using a porous alumina membrane as a deposition mask, uniform-sized nanodot arrays of any material can be synthesized on any substrate at minimal cost. Over the next few paragraphs, we discuss this methodology.

First, a thin through-hole porous membrane of desired pore diameter is fabricated using the method outlined in Section 9.3 of this chapter [6]. The template is then physically transferred onto a substrate and dried. Using e-beam evaporation, the material of interest is evaporated inside the template. E-beam evaporation is predominantly "line-of-sight" and is therefore ideal for transporting the evaporated atoms directly into the pores. The evaporation should be carried out under optimum deposition conditions such as in high vacuum (10^{-7} Torr or better) with a well-controlled rate of deposition. A faster rate of deposition results in blocking the material from depositing inside the pores since the material would form clusters and block the pore entrance. After depositing the material, the porous template is etched away in NaOH. Figure 9.30 is a top-view AFM image of Co nanodots with a diameter of 70 nm and a height of 40 nm on a Si substrate. Various metal nanodots like Au, Fe, and GaN can be deposited using this method. The advantage of these kinds of ordered nanodots arrays is their usefulness as catalysts for growing nanowires and fabricating nanopillars.

Xu *et al.* [48] synthesized highly ordered ZnO nanorods using Au nanodots as seeds for growth. The Au nanodots were fabricated using alumina template on a GaN substrate. The diameter of the Au nanodots is 50 nm, with center-to-center spacing of 110 nm. ZnO nanorods of an average length of 400 nm and a mean diameter of 60 nm were synthesized using the VLS growth method. Figure 9.31(a) is a top-view SEM image of the highly ordered ZnO nanorods, with traces of Au on top, while 9.31(b) shows the oblique-view SEM image of the quasi-ordered nanorod array.

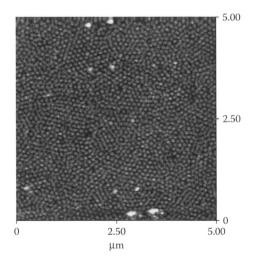

FIGURE 9.30 (see color insert) Top-view AFM image of Co nanodots with diameter of 70 nm and height of 40 nm on an Si substrate.

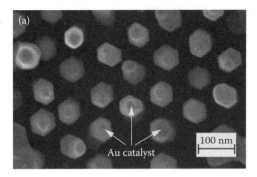

FIGURE 9.31   (a) Top-view SEM image of the highly ordered ZnO nanorods with traces of Au on top. (Reproduced from ref. [8] with permission from Elsevier.) (b) Oblique-view SEM image of the quasi-ordered nanorod array. (Reproduced from ref. [8] with permission from Elsevier.)

9.10   Semiconductor Nanoporous Structures

A thin membrane (0.5 to 1μm thickness) with a highly ordered array of pores is physically transferred onto a Si substrate and allowed to dry, whereupon it sticks to the Si surface due to surface tension. It is important that the template make excellent physical contact with the substrate. Figure 9.32 is a cross-section SEM image of the template physically transferred onto Si wafer [20].

The Si wafer with the template on top is then subjected to a plasma etch, carried out in a commercial inductively coupled plasma (ICP) system with reactive ion etching (RIE). The main reason for using plasma etch is that it is highly anisotropic, unlike the isotropic wet etch processing [49]. Chlorine gas diluted with Ar gas is used as an etchant. The flow rates are 20 sccm and 5 sccm, respectively. The anodized sample is held in a vacuum inside the reactor and mounted on a quartz wafer. Etching is carried out under optimized conditions, which are as follows: ICP Power $\sim$ 300W; RIE power $\sim$ 200W; Pressure $\sim$ 8 to 9M torr. The etching duration is about 3 minutes. This duration determines the depth of the pores obtained. After

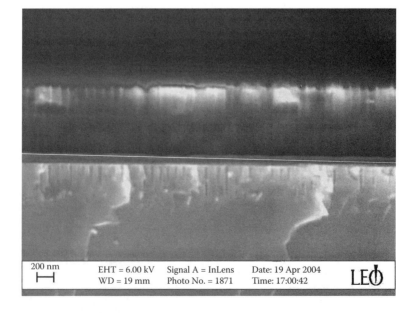

FIGURE 9.32   Cross-section SEM image of the template physically transferred onto an Si wafer.

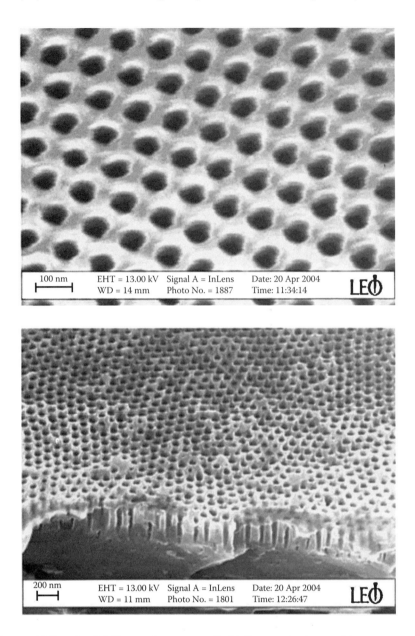

FIGURE 9.33   (a) Top-view SEM image of nanoporous Si. (b) Cross-section image of the nanopore array of Si.

etching, the traces of the template are removed by treating the wafer with a mixed chromic/phosphoric acid solution at 600°C. The wafer is cleaned in methanol and DI water, and then dried in air. This leaves behind the Si wafer, consisting of ordered arrays of pores. Figure 9.33 (left, right) is the top-view and cross-section image of the nanopore array of Si.

9.11   Nanopore Arrays on Generic Substrate

Nanopores can be fabricated on any generic substrate by a plasma etch of the porous membrane directly deposited on the given substrate [50]. Figure 9.34 is a schematic diagram outlining the fabrication of nanopores and nanopillars of the desired material.

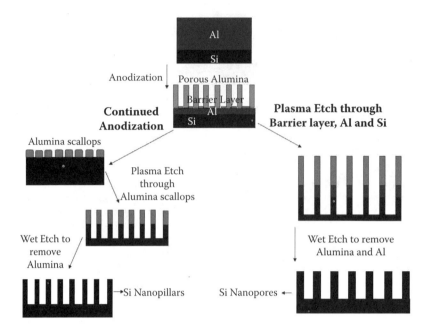

FIGURE 9.34 Schematic diagram showing the procedure for fabrication nanopores and nanopillars on a generic substrate.

9.11.1   Integration of Nanoporous Films with Arbitrary Substrates

In this method, a nanoporous alumina membrane is *directly* formed on an arbitrary substrate. The initial starting substrate could be a Si wafer cleaned in methanol and DI water. A 0.5-μm thick Al layer is deposited using thermal evaporation. The base pressure used for the deposition is $\sim 10^{-7}$ Torr. After the deposition of the Al layer, the sample is anodized in 3% oxalic acid at 40 V. The Al film atop the Si wafer is anodized for a few minutes. This results in the growth of a thin layer of porous alumina membrane over the Al. Figure 9.35 (left) shows the cross-section SEM image of the template after a 2-min anodization. The alumina layer is $\sim 0.6\ \mu$m thick and the pores have a diameter of ~ 50 nm. There is also a thin Ti layer below the aluminum layer, which is deposited in order to improve adhesion to the Si surface. A magnified image showing the U-shaped barrier layer is seen in Figure 9.35 (right).

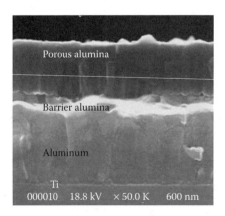
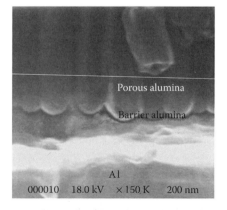

FIGURE 9.35   (a) Cross-section SEM image of the template after a 2-min anodization. (b) Magnified image showing the U-shaped barrier layer. (Reproduced from Ref [50] with permission from the Electrochemical Society, Inc.)

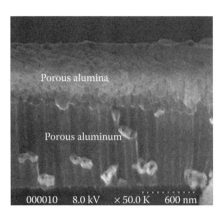 

FIGURE 9.36 (a) Cross-section SEM image of the sample after the plasma etching step. (b) Top-view SEM image showing the nanoporous Al membrane.

Plasma etching is employed to transfer the porous pattern by etching through the barrier alumina layer and the Al. Cl_2/Ar gas is used as an etchant. The etch parameters are as follows: ICP Power $\sim$ 300 W; RIE power $\sim$ 200 W; Cl_2/Ar flow rate $\sim$ 20/5 sccm; Pressure $\sim$ 8 to 9 mTorr. The etching was performed for about 1 minute. Figure 9.36 (left) shows the SEM image of the sample after the plasma etching step. One can clearly see the top nanoporous layer. The etch has penetrated the Al layer after etching through the barrier layer. Thus, we are left with a porous alumina layer, followed by the porous aluminum layer. The porous alumina layer can be removed by means of a wet etch—that is, by treating the sample with a 0.2M chromic/0.4M phosphoric acid mixture, resulting in a pore pattern on the aluminum layer as shown in Figure 9.36 (right) (a top-view SEM image showing the nanoporous Al membrane). By continuing etching into Si, an array of pores can be created in Si.

9.12 Fabrication of Nanopillars

A 1-μm thick Al layer is deposited on top of a Si substrate. As discussed in Section 9.2.3, once the anodization process begins, the current begins to decrease, followed by a rise, and finally the steady state, indicating both oxide growth and pore formation. This steady-state process continues until the interface of the porous membrane with the Si substrate is reached. At this point, no aluminum layer is left. All the aluminum has been consumed to produce alumina. Hence, no further oxide growth is possible. Oxidation of the substrate is still a possibility, however. Once the interface region is reached, the anodization current drops sharply, indicating the entire Al layer has been converted to porous alumina. The anodization process is thus complete. We now have alumina scallops (or pillars) on the surface of the wafer. This is because the dissolution rate of alumina is very slow. In fact, it is the slow rate of alumina dissolution that is responsible for the growth of pores inside alumina rather than the complete dissolution of the alumina by the hydrogen ions in the acid.

The thickness of the porous membrane obtained at this stage is about 500 nm. The sample is then subjected to plasma etching in Cl/Ar. Chlorine begins to etch the scalloped-alumina layer. The regions around the scallop where the alumina layer is thinnest are etched faster than the tops of each scallop. Hence, the Si interface is reached faster around the scallop. Once the interface is reached, it begins to etch Si very rapidly [20]. An *array of Si pillars with traces of porous membrane on top of the pillars* results. The traces of the porous membrane are etched away using the chromic/phosphoric mixture. Figures 9.37(a) and 9.37(b) show the cross-section and top-view SEM images of the sample after the plasma etch.

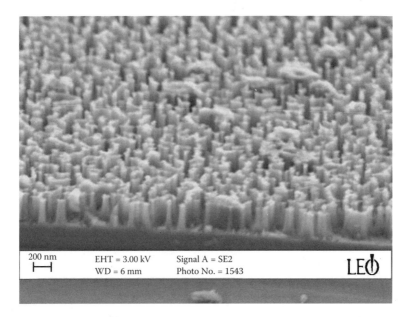

FIGURE 9.37   (a) Cross-section SEM image of Si nanopillars array. (b) Top-view SEM image of the nanopillars array.

9.13   Conclusion

In this chapter, we have described various routes to the synthesis of nanostructures using anodic alumina membrane technology. Many of these nanostructures, fabricated in our laboratories and elsewhere, exhibit intriguing device functionality [51]. Thus, the anodic alumina template technology has made nanoscience and nanoengineering accessible to a large community of researchers.

Acknowledgment

Much of this work was carried out under the sponsorship of multiple agencies, including the U.S. National Science Foundation, the U.S. Army Research Office, the U.S. Air Force Office of Scientific Research, the U.S. Office of Naval Research, the Civilian Research and Development Foundation, and the U.S. Defense

Advanced Research Projects Agency. The authors are grateful to many past and present collaborators: Prof. Latika Menon, Prof. Yi Liu, Mr. James Moore, Mr. Sridhar Patibandla, and Prof. Nikolai Kouklin.

References

[1] Duan, X. et al., Single-nanowire electrically driven lasers, *Nature*, 421, 241,2003.

[2] Alivisatos, P., The use of nanocrystals in biological detection, *Nature Biotechnology*, 22, 47, 2004.

[3] Michalet, X. et al., Quantum dots for live cells, in vivo imaging, and diagnostics, *Science*, 307, 538, 2005.

[4] Yeh, H.C. et al., Quantum dot-mediated biosensing assays for specific nucleic acid detection, *Nanomedicine: Nanotechnology, Biology, and Medicine*, 1, 115, 2005.

[5] Mikulskas, I. et al., Aluminum oxide photonic crystals grown by new hybrid method, *Adv. Mater.*, 13, 1574, 2001.

[6] Menon, L., Synthesis of nanowires using porous alumina, in *Quantum Dots and Nanowires*, Bandyopadhyay, S and Nalwa, H.S., Eds., American Scientific, Stevenson Ranch, CA, 2003, Chapter 4, and references within Menon, L., Nanoarrays from porous alumina, in *Dekker Encyclopedia of Nanoscience and Nanotechnology*, Schwarz, J.A et al., Eds., Marcel-Dekker Publishers, New York, 2004.

[7] See, for example, Thomson, G.E. et al., *Nature* (London), 272, 433, 1978; Furneaux, R.C. et al., *Nature* (London), 337, 147, 1989, Martini, C.R., *Science*, 266, 1961, 1994.

[8] Chik, H. and Xu, J.M., Nanometric superlattice: non-lithographic fabrication, materials, and prospects, *Materials Science and Engineering*, R 43, 103, 2004.

[9] Thompson, D.W. et al., Optical characterization of porous alumina from vacuum ultraviolet to mid infrared, *J. Appl. Phys.*, 97, 113511, 2005.

[10] Bandyopadhyay, S. et al., Electrochemically assembled quasi-periodic quantum dot arrays, *Nanotechnology*, 7, 360, 1996.

[11] Kolb, D.M., Reconstruction phenomena at metal-electrolyte interfaces, *Prog. Surf. Sci.*, 51,109, 1996.

[12] Thamida, S. and Chang, H.C., Nanoscale pore formation dynamics during aluminum anodization, *Chaos*, 12, 240, 2002.

[13] Yuzhakov, V.V. et al., Pattern selection during electropolishing due to double-layer effects, *Chaos*, 9, 62, 1999.

[14] Yuzhakov, V.V. et al., Pattern formation during electropolishing, *Phys. Rev. B.*, 56, 12608, 1997.

[15] Siejka, J. and Ortega, C., An O 18 study of field-assisted pore formation in compact oxide films on aluminum, *J. Electrochem. Soc.*, 124, 883, 1977.

[16] O' Sullivan, J.P. and Wood, G.C., The morphology and the mechanism of formation of porous anodic films on aluminum, *Proc. Roy. Soc. London Ser A.*, 317, 511, 1970.

[17] Shershulsky, V.I. and Parkhutik, V.P., Theoretical modeling of porous oxide growth on aluminum, *J. Phys. D.*, 25, 1258, 1992.

[18] Jessensky, O. et al., Self-organized formation of hexagonal pore arrays in anodic alumina, *Appl. Phys. Lett.*, 72, 1173,1998.

[19] Li, A.P. et al., Polycrystalline and monocrystalline pore arrays with large interpore distance in anodic alumina, *Electrochem Solid-State Lett.*, 3, 131, 2000.

[20] Kanchibotla, B., Magnetic properties of Iron nanostructures, Master's thesis, Texas Tech University, 2005.

[21] Csokan, P., Structure deformation in the anodic oxide coating on aluminum caused by internal stress, *Trans. Inst. Metal Finishing.*, 52 ,92, 1974.

[22] Lohrengel, M.M., Thin anodic oxide layers on aluminium and other valve metals: high-field regime, *Mater. Sci. Eng.*, R11, 243, 1993.

[23] Ono, S., et al., Self-ordering of anodic porous alumina formed in organic acid electrolytes, *Electrochimica Acta.*, 51, 82, 2005.

[24] Shingubara, S. et al., Self-Organization of a porous alumina nanohole array using a sulfuric/oxalic acid mixture as electrolyte, *Electrochemical and Solid State Lett.*, 7, E15–E17, 2004.

[25] Palibroda, E., Aluminum porous oxide growth—II. On the rate determining step, *Electrochim. Acta.*, 40, 1051 ,1995.

[26] Almawlawi, D. et al., Magnetic properties of Fe deposited into anodic aluminum oxide pores as a function of particle size, *J. Appl. Phys.*, 70, 4421, 1991.

[27] Ono, S. et al., Controlling factor of self-ordering of anodic porous alumina, *Electrochem. Soc.*, 151, B473, 2004.

[28] Ramazani, A. et al., The effect of temperature and concentration on the self-organized pore formation in anodic alumina, *J Phys D: Appl Phys.*, 38, 2396, 2005.

[29] Li, W.S. et al., Influence of anodizing conditions on the ordered pore formation in anodic alumina, *J Phys D: Appl Phys.*, 33, 2527, 2000.

[30] Masuda, H. and Fukuda, K., Ordered metal nanohole arrays made by a two-step replication of honeycomb structures of anodic alumina, *Science*, 268, 1466, 1995.

[31] Vrublevsky, V. et al., Study of chemical dissolution of the barrier oxide layer of porous alumina films formed in oxalic acid using a re-anodizing technique, *Appl. Surf. Science*, 236, 270, 2004.

[32] Ansermet, J.P et al., Bridging the gap between template synthesis and microelectronics: spin-valves and multilayers in self-organized anodized aluminium nanopores, *Nanotechnology*, 14, 978, 2004.

[33] Rabin, O. et al., Nanofabrication using self assembled alumina templates, *Mat. Res. Soc. Symp.*, 636, D.4.7.1, 2001.

[34] Pramanik, S. et al., Spin transport in self-assembled all-metal nanowire spin valves: A study of the pure Elliott–Yafet mechanism, *J. Nanosci. Nanotech.* 6, 7, 2006.

[35] Menon, L. et al., Magnetic and structural properties of electrochemically self-assembled $Fe_{1-x}Co_x$ nanowires, *J. Nanosci. Nanotech.*, 1, 149, 2001.

[36] Chaure, N.B. and Coey, J., M.D., Fabrication and characterization of electrodeposited $Co_{1-x}Cr_x$ nanowires, *J. Magnetism Magnetic Mater.*, 303, 236, 2006.

[37] Wang, G. et al., Electrochemical fabrication and structure of Ni_xZn_{1-x} alloy nanowires, *Nanotechnology*, 17, 19, 2006.

[38] Patibandla, S. et al., Spin relaxation in a germanium nanowire *J. Appl. Phys.*, 100, 044303. 2006.

[39] Schwarzacher, W., Metal Nanostructures: A New Class of Electronic Devices, *Electrochem. Soc. Interface*, 8, 20, 1999.

[40] Xu, D. et al., Preparation of II-VI group semiconductor nanowire arrays by dc electrochemical deposition in porous aluminum oxide templates, *Pure Appl. Chem.*, 72, 127, 2000.

[41] Meng, G. et al., Ordered Ni nanowire tip arrays sticking out of the anodic aluminum oxide template, *J. Appl. Phys.*, 97, 064303, 2005.

[42] Chik, H. et al., Periodic array of uniform ZnO nanorods by second-order self- assembly, *Appl. Phys. Lett.*, 84, 3376, 2004.

[43] Li, Y. et al., Ordered semiconductor ZnO nanowire arrays and their photoluminescence properties *Appl. Phys. Lett.*, 76, 2011, 2000.

[44] Ramanathan, S. et al., Fluorescence and infrared spectroscopy of electrochemically self-assembled ZnO nanowires, *J. Mat. Sci: Mater. Electron.* 17, 651, 2006.

[45] West, W.C. et al., Electrodeposited amorphous manganese oxide nanowire arrays for high energy and power density electrodes, *J. Power Sources*, 126, 203, 2004.

[46] Piao, Y. et al., Nanostructured materials prepared by use of ordered porous alumina membranes, *Electrochimica Acta.*, 50, 2997, 2005.

[47] Tiwari, S. et al., A silicon nanocrystals based memory, *Appl. Phys. Lett.*, 68, 1377, 1996.

[48] Xu, J.M. et al, Periodic array of uniform ZnO nanorods by second-order self-assembly, *Appl. Phys. Lett.*, 84, 3376, 2004.

[49] Tian, L. et al., Optical properties of a nanoporous array in silicon, *J. App. Phys.*, 97, 026101, 2005.

[50] Menon, L. et al., Anodization and plasma etching of nanoporous pattern on a generic substrate, *J. Electrochem. Soc.*, 151, C492, 2004.

[51] Bandyopadhyay, S. et al., in *Quantum Dots and Nanowires*, Bandyopadhyay, S. and Nalwa, H.S., Eds., American Scientific Publishers, Stevenson Ranch, CA, 2003.

10

Neuromorphic Networks of Spiking Neurons

Giacomo Indiveri

Rodney Douglas

Recent advances in neuroscience are revealing the principles of neural computation used by the mammalian brain [7], while modeling studies show how neural architectures composed of diversified and unreliable computing elements (neurons and synapses) can support robust and reliable computation using computational primitives of both analog and digital nature [10]. It is now clear that the principles of computation used by nervous systems are radically different from those generally used in current computers. Unlike computers, neuronal networks process information using energy-efficient asynchronous, event-driven, methods. They are self-constructing and repairing, self-programming, and they are able to flexibly compose complex behaviors from simpler elements. These biological abilities offer an attractive alternative to conventional computing technology, and could have enormous consequences for future generations of artificial information processing and behaving systems.

The quest to design and fabricate electronic neural systems composed of, for example, retinas, cochles, and neuronal networks whose architecture and design principles are based on those of biological nervous systems is known as *neuromorphic engineering* [4, 18]. In this chapter, we describe some general properties of neuromorphic systems, and also how neurons and synapses can be implemented in the CMOS (complimentary metal-oxide semiconductor) electronic medium using hybrid analog/digital VLSI (very large scale integrated) technology.

10.1 Neuromorphic vs. Conventional Use of VLSI Technology

The primary feature of the vast majority of conventional VLSI circuits is that they are purely digital: They use transistors as on-off switches, and they represent numbers as collections of binary digits. Because these circuits use only a binary encoding, it is possible to reduce the performance of their component transistors to the extent that they only reliably determine a single bit. These simple bits can then be combined to encode variables of arbitrary high precision. It is these two features—simplicity and reliability—combined with the Turing Machine concept of how to encode an algorithm on simple symbols, that has enabled the long and hugely successful growth of digital electronics.

This success has propelled a dramatic development of VLSI fabrication technology. However, this industry has focused almost exclusively on producing circuits consistent with the digital—Turing—von Neumann method, with the result that modern computer processors now contain millions of transistors organized into stereotyped processor, communication, and memory architectures that manipulate large fixed length variables via serial algorithms. As a consequence, present computers are obliged to use deterministic, very high precision methods to deal even with real-world tasks whose natural characteristics are usually asynchronous, stochastic, parallel, and have very low precision that could be encoded with a few bits. For many such problems, particularly those in which the input data are ill-conditioned and the computation can be specified in a relative manner, biological solutions are many orders of magnitude more effective than those using digital methods.

Because digital computation requires that many binary nodes be combined to encode one variable, the operation of the nodes of any one variable as well as its interactions with other variables, must be carefully synchronized. This, together, with the need for serial implementation of algorithms means that digital systems require global coordination supported by precise global clocking. Unlike these digital circuits, the continuous variables of analog circuits interact with one concurrently, and in real-time. So, if their time constants are set appropriately, their processing is inherently synchronized with real-world events.

Digital variables have no meaning in and of themselves. In order to express a function, the binary components of digital variables must be set according to an externally imposed encoding scheme, and then combined algorithmically to obtain a required result, which is then decoded according to the same scheme. For example, to compute an exponential requires that the value of an argument be loaded into a register. Then, many successive binary shifts and additions must be applied to multibit variables using circuits that involve hundreds of transistors. Finally, the result must be read from the output register.

By contrast, analog processing is more compact than its digital counterpart: A single analog variable is continuous and thus able to represent many bits of information; and computational primitives are expressed directly in terms of the physical properties of the analog devices. For example, using the analog approach, an exponential can be generated by just a single properly configured transistor operating in its sub-threshold analog domain [15]. The transistor generates a drain current that is exponential in its gate voltage. Moreover, this drain current does not *encode* the exponential result, it *is* the result itself; available directly as a measurable, physical quantity. Not all functions can be computed so efficiently by analog circuits. However, by exploiting the physics of silicon, it is rather straightforward to perform operations such as invert, add, differentiate, integrate, and correlate; and to generate exponentials, logarithms, tanh, and the like. Indeed, these are exactly the kind of operations and functions required for emulating the electrophysiological behavior of neurons.

Although straightforward in principle, analog computing is difficult to implement in practice, because the physical properties of the material used to construct the machine plays an important role in the solution of the problem. Biology is successful in implementing analog computation because it is able to directly grow adaptive structures. But we are still obliged to use feed-forward manufacturing procedures for constructing VLSI circuits. Using this approach, it is difficult to control the physical properties of micron-sized devices so their analog characteristics are well matched. The adaptive techniques used in neuromorphic VLSI design, inspired by biology, play a key role in compensating for the effects of device mismatch due to component differences. These adaptation techniques, used at the basic circuit level, lead naturally to the design of systems that, at a fundamental level, can learn about their environment.

Adaptation may also become relevant for conventional technology, because as VLSI technology progresses and transistors become smaller, the individual processing components consume proportionately more power, and become less reliable. Reliability is a serious problem for advanced digital computing systems whose component counts are ever increasing, because if even only a single transistor is defective, the functionality of the whole large circuit is compromised. Most neuromorphic circuits are designed to emulate populations of spiking neurons and so comprise massively parallel arrays of silicon synapses and neurons. Consequently, these circuits are able to exploit their redundancy to achieve fault tolerance, and compensate for the large variations of transistor properties.

10.2 Simulation vs. Emulation

Neuronal systems are composed of large numbers of nonlinear elements, and have a wide range of time constants. Consequently, their mathematical behavior can rarely be solved analytically. Instead, simulation methods must be used to explore their quantitative behavior [13]. Simulation is a strong tool for examining the detailed, precise behavior of neuronal elements. However, the speed of simulation is limited by the shortest time constant of the problem, and performance slows dramatically as the number and degree of coupling of elements increases. Hybrid VLSI offers a medium in which neuronal networks can be emulated directly in hardware rather than simply simulated on a general-purpose computer. Here, the silicon neurons operate in real-time, and the speed of the network is independent of the number of neurons or their coupling. However, analog circuits provide only a good qualitative approximation to the exact performance of simulated neurons. Moreover, the design of special-purpose hardware is a significant investment, particularly if it is analog hardware, since analog VLSI (aVLSI) design remains very much an art form. Designing robust novel circuits depends on engineers with considerable experience, preferably working in small groups with sufficient collective knowledge to monitor and criticize one another's design assumptions and circuit layouts. So, in their present form, neuromorphic circuits are not a substitute for quantitative simulation. On the other hand, the electronic neural circuits do offer an advantage in the investigation of questions concerning the strict real-time interaction of a neuronal system with its environment; possibly accepting sensory input from neuromorphic sensors such as silicon retinas [14] or silicon cochleas [20]. Emulation also has a strong role in exploring the relationship between algorithm and architecture; understanding asynchronous, event-based methods; and understanding how to compute with enormous numbers of imprecise and variable elements.

10.3 Action Potentials and the Address-Event Representation

Biological neurons communicate with one another through dedicated point-to-point axons that must ramify widely to make the necessary connections between a source neuron and its many target synapses. It is impractical to wire all VLSI neurons in this way: The connecting "wires" occupy too much space; the general routing problem cannot be solved; the many axonal wires cannot be routed between chips; and hard-wired connection schemes can not be modified by use and learning. Fortunately, the event-driven nature of action potential generation provides a convenient solution for the VLSI emulations. Neurons process only very slowly: Their shortest time constant is on the order of 1 millisecond, and their action potentials occur relatively infrequently, and so the output of many different neurons can be multiplexed on a single high-speed digital bus. For example, using a 10-MHz digital bus, one million other action potentials can be transmitted during the inter-spike interval of a particular neuron firing at 100-Hz.

The binary (all-or-nothing) nature of action potentials makes them robust against computational and intrinsic noise, and permits reliably transmission of these spike-events between multiple chips [16]. One popular communication protocol used to transmit spikes between neuromorphic chips is the *address-event-representation* (AER). In this protocol, each source node (e.g., the soma of a silicon neuron) on a sending device is assigned a unique address. When a neuron generates an action potential its address is written to a digital bus using asynchronous logic (see Figure 10.1). The AER is asynchronous and in real-time, which means that the distributed processing amongst the connected neurons is inherently synchronized by time, and the analog signals of the source somata are encoded by the intervals between their successive action potentials.

In the case of single-sender/single-receiver communication, a simple handshaking mechanism ensures that all events generated at the sender side arrive at the receiver. The address of the sending element is

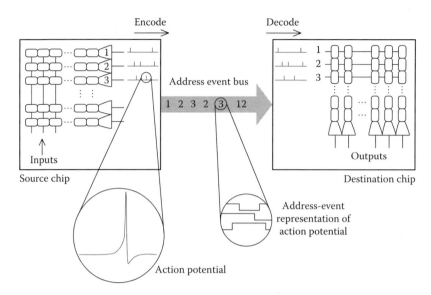

FIGURE 10.1 Asynchronous communication scheme between two chips using the address-event representation (AER). When a neuron on the source chip generates an action potential, its address is placed on a common digital bus. The receiving chip decodes the address events and routes them to the appropriate synapses.

conveyed as a parallel word of sufficient length, while the handshaking control signals require only two lines. Event "collisions" (cases in which sending nodes attempt to transmit their addresses at exactly the same time) are managed by on-chip arbitration schemes. Systems containing more than two AER chips can be constructed by implementing additional special purpose off-chip arbitration schemes. Address-events detected by the target chip are conveyed to the local synapses which initiate their local synaptic action (see Figure 10.1). In this way, those target synapses that are "connected" to the source neuron are activated whenever the source neuron spikes. These multiplexing strategies are optimal when, like their biological counterparts, the neurons of a network are sparsely active.

AER facilitates network reconfigurability: the source address-event can be translated from a post-synaptic bus onto a pre-synaptic bus through a programmable look-up table, which maps the addresses of source neurons to (lists of) destination synapse addresses. In this way, arbitrary network topologies can be programmed, and it is possible to construct large-scale multichip reconfigurable systems.

10.4 Silicon Neurons

The last half a century of intensive investigations in neuronal processing has revealed the rich membrane biophysics of synapses, dendrites, somata, and axons. Nevertheless, we still do not have a clear understanding of the formal processing characteristics of neurons and their networks. That is, we do not have a more sophisticated abstract model for the neuron that could replace the McCulloch–Pitts type simplicity with one that has both richer biophysical verisimilitude and a clear computational specification. Such a new neuronal model would provide the necessary foundation for understanding the global nature of spike-based computations supported by neuronal circuits. In addition to its relevance for neuroscience, this model would also be relevant for future technology. In their contribution to this quest, the relatively small neuromorphic engineering community has, during the last decade, steadily developed a robust infrastructure for studying event-driven computation in networks of neuron-like elements. They have developed circuits for neurons, dendrites, and synapses, as well as a general method for event-driven communication between neurons distributed over possibly many chips. It is now possible to assemble quite complex systems of such neurons [19].

A cornerstone of these systems is the integrate and fire (I&F) neuron. I&F neurons integrate pre-synaptic input currents and generate a voltage pulse analogous to an action potential when the integrated voltage reaches a threshold. Many variants of these circuits were built during the 1950s and 1960s using discrete electronic components. The first simple VLSI version was probably the *Axon–Hillock* circuit, built by Mead in the late 1980s [18]. In this circuit, a capacitor that represents the neuron's membrane capacitance integrates current input to the neuron. When the capacitor (membrane) potential crosses the spiking threshold, a pulse is generated by a positive feedback loop and the membrane potential is reset. This circuit captures the basic principle of operation of biological neurons, but cannot faithfully reproduce all of the dynamic behaviors observed in real neurons.

More elaborate models of neurons, that take into account the biophysical properties of the voltage dependent conductances and currents present in real neurons have also been proposed. The first and most influential conductance-based silicon neuron is perhaps that of Douglas and Mahowald [17]. This silicon neuron is composed of connected compartments, each of which is populated by modular subcircuits that emulate particular ionic conductances. This circuit, as well as other conductance-based neurons that have since been proposed [1], can reproduce in great detail many of the behaviors observed in real neurons, but their overall size and circuit complexity is significantly larger than the one of the simpler Axon–Hillock circuit.

A compromise between the preceding two approaches is provided by more elaborate models of I&F neurons that include additional neural characteristics, such as spike–frequency adaptation properties and refractory period mechanisms [3]. An example of such a circuit is shown in Figure 10.2. In addition to

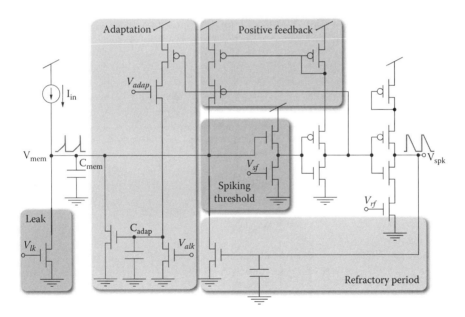

FIGURE 10.2 Schematic diagram of an integrate-and-fire neuron. The input current I_{in} is integrated onto the neuron's membrane capacitor C_{mem} until the spiking threshold is reached. At that point, the output signal V_{spk} goes from zero to the power supply rail, signaling the occurrence of a spike, and the membrane capacitor is reset to zero. The "leak" module implements a current leak on the membrane. The "spiking threshold" module controls the voltage at which the neuron spikes. The "adaptation" module subtracts a firing rate dependent current from the input node. The amplitude of this current increases with each output spike and decreases exponentially with time. The "refractory period" module sets a maximum firing rate for the neuron. The "positive feedback" module is activated when the neuron begins to spike, and is used to reduce the transition period in which the inverters switch polarity, dramatically reducing power consumption. The circuit's biases (V_{lk}, V_{adap}, V_{alk}, V_{sf}, and V_{rf}) are all subthreshold voltages that determine the neuron's properties.

implementing the basic behavior of integrating input currents and producing output pulses at a rate that is proportional to the amplitude of its input, this low-power I&F neuron [11] implements a *leak* mechanism (as in leaky I&F neuron models); an adjustable *spiking threshold* mechanism for adapting or modulating the neuron's spiking threshold; a *refractory period* mechanism for limiting the maximum possible firing rate of the neuron; and a spike–frequency *adaptation* mechanism, for modeling some of the adaptation mechanisms observed in real neurons.

10.5   Silicon Synapses

Synapses are highly specialized structures that, by means of complex chemical reactions, allow neurons to transmit signals to target neurons. When an action potential generated by the soma of a source neuron reaches its pre-synaptic terminal, a cascade of events leads to the release of neurotransmitters that give rise to a flow of ionic currents into or out of the post-synaptic neuron's membrane. These excitatory or inhibitory post-synaptic currents (EPSC or IPSC, respectively) have temporal dynamics with a characteristic time course that can last up to several hundreds of milliseconds [12]. In most simulations of pulse-based neural models, synaptic currents are often reduced to simple instantaneous charge impulses, so as to reduce computational load. This simplification can be a critical problem, because the precise timing of spikes and the dynamics of the neuron's transfer function play an important computational role for learning neural codes, and for encoding spatio-temporal patterns of spikes [9]. In neuromorphic chips, the detailed dynamics of post-synaptic currents can be easily modeled using dedicated subthreshold analog circuits. An example of a compact circuit that reproduces the temporal properties of synaptic transmission and accounts for the linear summation property of post-synaptic currents is shown in Figure 10.3. This circuit implements a log-domain linear temporal filter and supports a wide range of

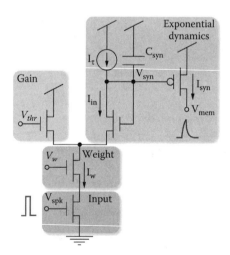

FIGURE 10.3   A compact synaptic circuit that exhibits the exponential dynamics observed in real synapses. The circuit's analog output current I_{syn} encodes the frequency of the input spikes, represented as digital voltage pulses arriving at the V_{spk} node. The circuit's time constant is set by adjusting the value of the I_τ current. The V_{thr} bias in the "gain" module can be used to set the circuit's gain: lower values increase the gain, while higher ones decrease it. Similarly, the synaptic efficacy is controlled by the V_w bias voltage in the "weight" module. In this case, higher values of V_w lead to higher output currents. Additional circuits can be connected to the V_w node to implement short- and long-term plasticity mechanisms, and to locally store or refresh the value of the weight.

synaptic properties, ranging from short-term depression to conductance-based EPSC generation, to synaptic plasticity.

10.5.1 Plasticity and Learning

One of the key properties of biological synapses is their ability to exhibit short- and long-term plasticity. Short-term plasticity produces dynamic modulation of the synaptic strength by the timing of the input spikes alone, while long-term plasticity produces sustained changes in synaptic strength that are induced by the correlations in the spiking activity of the pre- and post-synaptic processes. Circuits have been developed that emulate the short time-scale synaptic depression [5]. Various learning circuits that implement long-term plasticity of synapses use spike timing information and/or rate information [8, 11]. Spike-driven plasticity mechanisms are particularly well suited to VLSI implementation. In these mechanisms, the precise timing of spikes generated by the pre-synaptic neurons, and the state of the post-synaptic neuron at those instants play an important role in shaping the synaptic efficacy.

Several modeling studies have developed learning algorithms based on spike-driven plasticity, and demonstrated how systems that use these types of algorithms can carry out complex information processing tasks. In the past few years, several neuromorphic synaptic circuits that implement both long- and short-term plasticity have been proposed. These circuits are now being integrated into large arrays in VLSI technology to build large-scale *learning* neural networks.

10.5.2 Memory and Synaptic Weight Storage

Fundamental differences in architecture exist between conventional computers and neural networks. Conventional computers have a single, or small number of processors connected to a random access memory. This global access means that the state of the machine can be conveniently loaded and examined. By contrast, in neuronal networks the memory and processing is massively distributed, and co-localized at the synapses. In this case, direct loading and inspection of the memory element (i.e., the synaptic weight) is not possible, except at the huge cost of providing duplicate access lines. Long-term plasticity circuits alleviate these problems slightly (for both biology and neuromorphic VLSI) by allowing the weights of synapses to be set automatically, without requiring dedicated access to individual synapses. But the difficulty of experimental observation and control remains for these circuits.

Basic considerations concerning the relation between precision and the area of silicon required to support it, suggest that biologically realistic low-resolution weights could be stored as analog voltages across capacitors. Unfortunately, CMOS capacitors are subject to leakage, and so their voltages will decay over a few seconds if not refreshed. Alternatively, the analog values can be stored using non-volatile technology similar to that used in electrically erasable programmable memory (EEPROM). In this case, the analog value is stored as charge on a "floating gate" which is the gate of a FET isolated between two layers of (perfect) oxide insulation. Charge is added or removed from the floating gate by Fowler–Nordheim tunneling and impact-ionized hot-electron injection. The uncontrolled decay of charge from floating gates is negligible, so that learned synaptic weights can be retained for decades, even when the power to the circuits is off. This exciting technology is still under development, but simple structures such as single transistor synapses, and examples of networks that implement learning algorithms based on these circuits have been demonstrated [6].

The problem of synaptic weight storage would be simplified considerably if only one binary bit, rather than an analog value, needs to be stored. In fact, it has been demonstrated that networks of sufficiently large numbers of only *binary* synapses are adequate for any memory task [8]. Circuits for such binary synapses have been proposed. They bypass the need for having specialized structures for nonvolatile analog memory within each synapse, as long as the circuit's power supply is active. These binary synaptic circuits have been successfully applied to learning tasks, and systems using them have been capable of classifying complex patterns [2, 11].

10.6 Multi-Chip Neural Networks

Several examples of successful multi-chip networks of spiking neurons have been demonstrated during recent years. Using present CMOS technology, it is possible to implement on the order of hundreds of neurons and thousands of synapses per square millimeter of silicon (see, for example, Figure 10.4).

In principle, networks of this type can be scaled up to any arbitrary size, but in practice the size is limited by the maximum silicon area and AER bandwidth available. Given the current speed and specifications of the AER interfacing circuits [3] and the availability of present silicon VLSI technology, the network size could be increased by at least two orders of magnitude. It is likely that large neuronal networks, such as those of the neocortex, are dominated by local connectivity, with only a relatively small fraction of long range connections. In this case, it may be possible to make similarly large-scale VLSI networks, in which multiple regional AER busses with the same address spaces carry local event traffic, and are interconnected by sparser long range traffic between local domains. However, before testing those connectivity limits, there is more immediate interesting work to be done with these multichip networks. For example, existing methods can be used for investigating complex spike-based learning algorithms in real-time. These studies are all the more interesting when considering problems of adaptation and learning in neuronal networks interfaced to neuromorphic AER sensors such as silicon retinas or silicon cochleas.

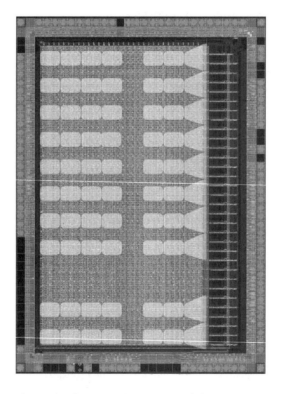

FIGURE 10.4 Layout of a multi-neuron chip comprising an array of silicon neurons connected to plastic synapses. The chip was fabricated using a standard 0.35-μm CMOS technology, and occupies an area of 12^2 mm. It implements an array of 32 neurons and 8192 synapses. Each of the 32 neurons in the array is connected to 256 synaptic circuits. Superimposed on the layout is a diagram of neurons and synapses similar to the one shown in Figure 10.1. Neural networks of arbitrary topology can be implemented by simply rerouting output spikes to input synapses of the same chip, or of other multineuron chips, using the AER communication protocol.

10.7   Acknowledgments

This work was supported by the EU Grants ALAVLSI (IST-2001-38099) and DAISY (FP6-2005-015803), and by the Zurich ETH TH0-20174-04 grant.

References

[1] Alvado, L. et al., Hardware computation of conductance-based neuron models, *Neurocomputing*, 58–60,109–115, 2004.

[2] Arthur, J. and Boahen, K., Learning in silicon: Timing is everything, in Weiss, Y. et al., Eds, *Advances in Neural Information Processing Systems*, 18, MIT Press, Cambridge, MA, 2006.

[3] Boahen, K.A., Communicating neuronal ensembles between neuromorphic chips, in Lande, T.S., Ed., *Neuromorphic Systems Engineering*, Kluwer Academic, Norwell, MA, 1998.

[4] Boahen, K.A., Neuromorphic microchips, *Scientific American*, 2005.

[5] Boegerhausen, M. et al., Modeling short-term synaptic depression in silicon, *Neural Computation*, 15, 2, 2003.

[6] Cauwenberghs, G. and Bayoumi, M.A., Eds., *Learning on Silicon: Adaptive VLSI Neural Systems*, Kluwer, Boston, 1999.

[7] Douglas, R.J. and Martin, K.A.C., Neural circuits of the neocortex, *Annual Review of Neuroscience*, 27, 2004.

[8] Fusi, S. et al., Spike–driven synaptic plasticity: theory, simulation, VLSI implementation, *Neural Computation*, 12, 2000.

[9] R. Gütig, R. and Sompolinsky, H., The tempotron: a neuron that learns spike timing–based decisions, *Nature Neuroscience*, 9, 2006.

[10] Hahnloser, R. et al., Digital selection and analog amplification co-exist in an electronic circuit inspired by neocortex, *Nature*, 405, 6789, 2000.

[11] Indiveri, G. et al., A VLSI array of low-power spiking neurons and bistable synapses with spike–timing dependent plasticity, *IEEE Transactions on Neural Networks*, 17, 1, 2006.

[12] Koch, C., *Biophysics of Computation: Information Processing in Single Neurons*, Oxford University Press, Oxford, 1999.

[13] Koch, C. and Segev, I., *Methods in Neural Modeling*, 2nd ed., MIT Press, Cambridge, MA, 2001.

[14] Lichtsteiner, P. et al., A 128 × 128 120dB 30mW asynchronous vision sensor that responds to relative intensity change, in *2006 IEEE ISSCC Digest of Technical Papers*, IEEE, 2006.

[15] Liu, S.-C. et al., *Analog VLSI: Circuits and Principles*, MIT Press, Cambridge, MA, 2002.

[16] Mahowald, M., *An Analog VLSI System for Stereoscopic Vision*, Kluwer, Boston, 1994.

[17] Mahowald, M. and Douglas, R., A silicon neuron, *Nature*, 354, 1991.

[18] Mead, C.A., *Analog VLSI and Neural Systems*, Addison-Wesley, Reading, MA, 1989.

[19] Serrano-Gotarredona, R. et al., AER building blocks for multi-layer multi-chip neuromorphic vision systems, in Becker, S. et al., Eds., *Advances in Neural Information Processing Systems*, 15, MIT Press, Cambridge, MA, 2005.

[20] van Schaik, A. and Liu, S.-C., AER EAR: A matched silicon cochlea pair with address event representation interface, in *IEEE International Symposium on Circuits and Systems*, 5, 2005.

11

Allowing Electronics to Face the TSI Era—Molecular Electronics and Beyond

G. F. Cerofolini

Abstract

If single atoms at solid surfaces are used for storing the information, the chemical composition of the surface is fixed, and the information is coded as a conformation or an electronic state (between two possible ones) of any given surface atom (the Feynman dream), then the maximum bit density would be of the order 10^{15} cm^{-2}—the peta scale integration (PSI). The manipulation of atoms, however, requires the use of macroscopic-scale apparatuses that may be operated at a negligible rate. Information coded by electrons, that can be lodged with densities on the order of 10^{12} bit cm^{-2} (the tera scale integration, TSI) at most, can instead be managed and felt by already existing mesoscopic-scale apparatuses in the giga scale integration (GSI). Even though there is no idea for the full exploitation of the performances of TSI devices, this density is within reach of present technology. Rather than scaling down conventional CMOS (complementary metal-oxide-semiconductor) circuits, the TSI may almost be achieved via a hybrid architecture where a silicon-based CMOS circuitry controls a nanoscopic crossbar structure hosting in each cross-point a collection of reprogrammable molecules able to mimic by themselves the behavior of a nonvolatile memory cell. Solutions to the severe problems posed by this hybrid architecture are proposed.

11.1 Introduction

The race toward larger and larger scale integration of silicon integrated circuits (ICs) has always met some "physical limits." Their impact, however, has never been so strong as to slow down the exponential growth of microelectronics (see Table 11.1). That was achieved more with the sophistication of the production technology than with the use of materials with better intrinsic properties.[1] An inspection of the major changes associated with the evolution of microelectronics (see Table 11.2 [1]) shows, however, that the situation has changed in recent years.

The latest difficulties met in the scaling down of the basic constituent of ICs, the metal-oxide-semiconductor (MOS) field-effect transistor (FET), have been

- The non scalability of the gate SiO_2 thickness to the values required by gate insulators of next-generation circuits [6,7].
- The adverse scaling behavior of interconnects, producing the prevalence of delay times due to interconnection parasitics on the intrinsic switching time of transistors when horizontal dimensions are shrunk on going from a scale integration to a larger one [8].

Both problems are now attacked with the use of tailored materials: the first via the substitution of an oxide (HfO_2) with high dielectric constant (high-κ insulator) for SiO_2, and the second (initially overcome uniquely with the use of more and more interconnection layers [9]) via the substitution of copper for aluminum (reduction of resistivity by a factor of 2 [10]) and of materials with low dielectric constant (low-κ insulators) for SiO_2 [11] (reduction of κ by a factor close to 4, when very porous materials will succeed in replacing SiO_2 [12]).

The bit density ρ currently achievable with the silicon technology is in the giga scale of integration (GSI); although 10-Gbit ICs have been announced, it is a widespread opinion that circuits with a complexity of 10^2 Gbit will not be producible within the current technological paradigm [13]. The limits of the silicon technology have been identified by the International Technology Roadmap for Semiconductors

TABLE 11.1 The Evolution of Semiconductor Integrated Circuits

					Size			
				Transistor [μm]		Chip [mm^2]		
Year	Class		Bits					
1960 − 1968	SSI	2	−	128	> 10		1	−	15
1965 − 1975	MSI	64	−	4K	10 → 3		10	−	25
1972 − 1983	LSI	2K	−	128K	4 → 1.5		15	−	50
1980 − 1988	VLSI	64K	−	4M	2 → 0.75		25	−	75
1985 − 1993	ULSI	2M	−	128M	1 → 0.5		50	−	200
1990 − 2006	GSI	64M	−	4G	0.5 → 0.09		100	−	1000

The acronyms SSI, MSI, LSI, VLSI, ULSI, and GSI mean small, medium, large, very large, ultra-large, and giga-scale integration, respectively.

---

[1]The major progresses in IC integration were actually obtained with the invention of techniques that allow the *self-alignment* of a mask with respect to an underlying pattern. Previous analyses of the major changes associated with the evolution of microelectronics have indeed shown that the passage from one generation to another was not associated with the use of a material or device with better intrinsic performance [1]. In the author's view, the milestones characterizing the major progresses in IC integration are the silicon-gate technology [2], the local oxidation of silicon (LOCOS) [3,4], and the spacer-patterning technique (SPT) [5].

TABLE 11.2 Change of Technology and Physical Properties Associated with the Change of Device

Device	Technology	Physical Property
tube → transistor	vacuum → solid state	v_{sat} [cm/s]: $3 \times 10^{10} \to 10^7$
Ge → Si	mesa → planar	v_{sat}: unchanged; $\mu_{Si} < \mu_{Ge}$
bipolar → pMOS		l [μm]: $0.1 - 1 \to 10 - 100$
pMOS	Al gate → Si gate	ϱ [Ω cm]: $10^{-6} \to 10^{-3}$
pMOS → nMOS	ion implantation; LOCOS; $(1\,1\,1) \to (1\,0\,0)$	μ [cm^2/V s]: $300(p) \to 1000(n)$
nMOS → CMOS		*transistors:* P [μW/gate]: $300 \to 10 - 100$; μ [cm^2/V s]: $1000(n) \to 300(p)$
	oxidation → CVD	$SiO_2 \to$ high-κ *interconnections:* Al → Cu $SiO_2 \to$ low-κ

Symbols and acronyms: v_{sat}, saturation velocity; μ, mobility; l, device length (base in bipolar transistor, gate in MOS transistor); ϱ, resistivity; P, dissipated power; low-κ, dielectric with low dielectric constant ($\kappa < 2$); high-κ, dielectric with high dielectric constant ($\kappa \gg 10$); CVD, chemical vapour deposition.

(the "Roadmap") [14]. They are discussed at length by many papers devoted to the practical limits of circuit integration [15–21].

The Roadmap has been constructed under the assumption that the scaling down of MOS devices is not limited by fundamental physical reasons. Going beyond the Roadmap poses the problem of identifying which principles may be exploited for the construction of useful devices and which physical laws will eventually impact on the scaling down.

In recent years, numerous efforts have been made to remove the constraint of using current technology or devices: the latest experiments on condensed-matter–based systems (with characteristic size on the super-nanometer length scale) for quantum-information processing have shown that they also can carry information, for instance, in macroscopic quantum states (superconducting charge or flux qubits) or in charge or spin states of quantum dots. It is possible that overcoming the classical Turing machine by new paradigms (not only with intrinsically higher parallelism [quantum computers] but also with learning [neural networks] or fault tolerance [bio-inspired systems][2]) will render vain the current efforts toward larger and larger integration; this work, however, is based on the stipulation of the persistence of the current computational paradigm.

Within this *conservative* approach (for which computation is essentially a dissipative irreversible process), I shall also try to be

- *Application-driven*, having in mind devices operating at room temperature
- *Concrete*, concentrating my attention on a device which, though highly ideal, has features that resemble those of a real device—the nonvolatile memory cell
- *Technology-biased*, limiting the analysis to memory cells arranged in planar configurations

---

[2] For a recent review of this topic, see Ref. [22].

Of course, removing one or the other of the preceding constraints (allowing, for instance, operations close to 0 K, three-dimensional arrangements, etc.) would lead to different conclusions.

11.2   The Physical Limits of Computation

The figure of merit of a planar memory is the flop rate per unit area, $\dot{f}$, which should be as high as possible:

$$\dot{f} = \text{max.}$$

Since $\dot{f} = \nu\rho$, where ν is the cell switching frequency and ρ is the cell density, the problem of maximizing f is reduced to the problem of finding the combination of ν and ρ, satisfying the following condition:

$$\nu\rho = \text{max.}$$

Needless to say, computation must satisfy the underlying physical laws, that are specialized according to the size of the cell, its material structure and environment, and how it is linked to the rest of the world. If the attention is posed on the dynamics, the limits are presumably dictated by the uncertainty principle and the ballistic motion of the information carrier; specifying the materials of which the cell is formed will introduce constraints related to material science; if the cell is embedded in a thermal reservoir at temperature T, decoherence phenomena and statistical properties ultimately related to the Second Law of thermodynamics must be taken into account; and considering the phenomena (semiconductivity, superconductivity, etc.) exploited for computation will result in limits given by the macroscopic laws governing those phenomena. At the current level of development, the efforts toward larger and larger integration have mainly been limited by the last kind of constraints. However, they are related to the particular choice of materials and technology, and will henceforth be ignored.

Computers can be thought of as engines for transforming free energy into mathematical work and waste heat. The thermodynamics of computing is relatively well established and does not depend appreciably on the assumed model of the computer [23,24]. How fast the computer is performing the calculation, however, is totally ignored by thermodynamics, and its identification requires a minimal dynamical model of the computing act.

The present analysis is based on the fundamental assumption that the cell is constituted by a single object (the "information carrier") of mass m; the dynamics of this object are described by the Schrödinger equation in a potential with two minima (coding the states $|0\rangle$ and $|1\rangle$) separated by a barrier of height ΔE_{bit}. Moreover, to simplify the calculations, all potentials are assumed to be sufficiently smooth to allow the use of classical mechanics to describe the ballistic motion of the carrier (Ehrenfest theorem), provided that the quantum expectation values are substituted for the corresponding classical variables (in this situation, tunnelling from one minimum to the other is disallowed).

The system may be prepared in one or the other of two[3] orthogonal states ($|0\rangle$ or $|1\rangle$), where the carrier is in one or the other of the potential minima.[4] In the absence of any other phenomenon, the information is gradually lost via spreading of the wavefunction over the entire cell. Under certain conditions, however, spontaneous decoherence phenomena reduce the quantum position eigenstates to classical states, 0 or 1, which in the absence of interactions with the neighborhood do not vary with time, at least for a significantly shorter time than that required for the thermal migration of the carrier from one minimum to the other. The cell is assumed to undergo two possible operations: *programming* (i.e., the confirmation of the original

---

[3]Actually, even this assumption is a strong restriction. The use of three or more logic levels is by itself a way to increase the *bit* density more than the *cell* density.

[4]States $|0\rangle$ or $|1\rangle$ are orthogonal, thus defining distinguishable states, only if their supports in the position representation do not overlap; they may be regarded as position eigenstates in a coarse grain description where all points around the minimum of each potential well are nearly equivalent.

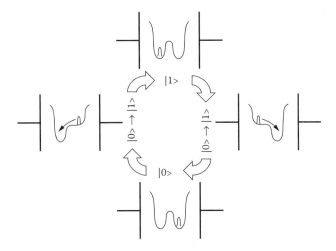

FIGURE 11.1   Programming the cell: from $|0\rangle$ to $|1\rangle$ and vice versa.

state, or the switching to the other one) or *sensing* (i.e., the detection of the actual state of the particle). In this way, the memory cell resembles the cell of nonvolatile memories.[5] Even though this model is highly ideal and real computation requires certainly more complicated arrangements, it is nonetheless often considered, and is at the basis of several fundamental investigations [25,26]. In this work, without anyway specifying the structure of the ultimate memory cell, I shall admit that it is characterized by the same operation modes that characterize nonvolatile memories: PROGRAM and READ.

Programming is achieved by decreasing the potential energy of one well, in such a way as either to stabilize further the carrier therein if already there, or to produce the drift of the carrier from the other well. This application is assumed to last a time sufficiently long to allow the dissipation of the original particle energy as heat. In this way, any programming act implies the dissipation of an energy of about $\Delta E_{\rm bit}$. Figure 11.1 sketches the situation.

Sensing is essentially a measurement process where the state of the carrier is detected. The theoretical description of sensing is a difficult point because it is intrinsically associated with the measurement of a microscopic state—a problem that has hitherto gotten only partial and unsatisfactory solutions. However, under certain circumstances (discussed in Section 11.2.1.2) the microscopic state may be regarded as classical, thus allowing its determination without running into the difficulties of the measurement problem.

11.2.1   Programming

11.2.1.1   The Limits Imposed by the Particle Dynamics

In order that the particle remains localized in one or the other of the energetically equivalent potential wells, the position eigenstate must be characterized by a certain energy spread ΔE. This spread is ultimately responsible for a loss of information. The minimum time required by the system to switch from $|0\rangle$ and

---

[5]Of course, the similarity must not be taken in a strict sense. In complementary MOS (CMOS) nonvolatile memories, states 0 and 1 are classical macroscopic states; programming is achieved either via the injection of $10^2 - 10^4$ electrons into a floating gate from the underlying channel ($1 \rightarrow 0$: WRITE) or via their field-assisted tunneling to the source ($0 \rightarrow 1$: ERASE). Moreover, whereas erasing is a relatively efficient process involving the dissipation of approximately 1 eV per electron, the efficiency of writing depends on memory architecture: for NOR memories writing is an inefficient process (yield on the order of 0.1%) involving "hot" electrons (with kinetic energy of $3 - 4$ eV), whereas for NAND memories writing is a relatively efficient process occurring via field emission from the channel.

$|1\rangle$ or vice versa, is given by

$$\Delta t \geq \hbar / \Delta E, \tag{11.1}$$

where $\hbar$ is the reduced Planck constant (states $|0\rangle$ and $|1\rangle$ are supposedly orthogonal) [27]. The higher ΔE, the shorter the switching time Δt; on another side, in order to prevent a spontaneous loss of information ΔE must be lower than the energy ΔE_{bit} required for switching from the corresponding logical state to another. Inserting condition $\Delta E = \Delta E_{bit}$ into Equation (11.1) gives the minimum switching time Δt_{min} as:

$$\Delta t_{min} = \hbar / \Delta E_{bit}.$$

A complete switching cycle is obtained when two transitions, $|0\rangle \rightarrow |1\rangle \rightarrow |0\rangle$ or $|1\rangle \rightarrow |0\rangle \rightarrow |1\rangle$, have occurred. Thus, the switching frequency of a single memory cell satisfies the following condition

$$\nu \leq 1/2\Delta t_{min}$$
$$= \Delta E_{bit}/2\hbar. \tag{11.2}$$

A similar argument may be applied to the minimum size of the cell. Let Δq be the measure of the support of the wave function in each of the states $|0\rangle$ and $|1\rangle$. Because of a well-known theorem of quantum mechanics [28] (see also Chapter 6), the momentum dispersion associated with each of them is given by

$$\Delta p \Delta q \geq \tfrac{1}{2}|\langle [p,q] \rangle| = \tfrac{1}{2}\hbar.$$

The carrier momentum p cannot be smaller than its dispersion; condition $p \geq \Delta p$ provides a lower limit to the kinetic energy E_{kin} of the information carrier in one or the other of states $|0\rangle$ or $|1\rangle$:

$$E_{kin} = p^2/2m \geq (\Delta p)^2/2m \geq \hbar^2/2m(\Delta q)^2. \tag{11.3}$$

The condition for localization to be possible is $E_{kin} < \Delta E_{bit}$, so that combining this inequality with (11.3) one has

$$\Delta q \geq \hbar / \sqrt{2m\Delta E_{bit}} \tag{11.4}$$
$$=: \Delta q_{min}$$

A closer inspection of the preceding derivation shows it assumes that the transition from one state to another does not involve any material motion. This assumption, however, is unphysical: Whichever process one can hypothesize for modifying the information (redox process, conformational change, change of electronic state, etc.), it is inherently associated with some material motion (redox processes imply electron transfer; conformational or electronic-state changes imply nuclear motions, etc.) in a region which defines the cell size.[6] For that, one has to consider the minimum distance ℓ that the information carrier must travel to allow the cell to switch from one state to another (ℓ may be thought of as the distance separating the minima defining the memory cell). If one wants the switching to be externally controlled, thermally activated or tunnel transitions from one state to another must be negligible, so that one can limit the attention to motions with energy higher than ΔE_{bit}. Describing them classically, the carrier average

---

[6]Actually, situations which allow information to be stored, sensed, and modified without appreciable motion, at least on the atomic scale, can be hypothesized. For instance, that could be achieved exciting, sensing, and de-exciting internal degrees of freedom of nuclei. Since the energy involved in these processes is higher than the thermal one by so many orders of magnitude, its thermalization is highly improbable—suggesting use of this route for non-dissipative computation. However, in line with my conservative approach and in the absence of any idea for the exploitation of this route, I shall ignore it.

velocity $\bar{v}$ is given by $\bar{v} = \chi \sqrt{2\Delta E_{bit}/m}$, where χ is a quantity depending on the actual energy of the carrier and on the potential energy inside the cell. Having assumed that the carrier attains its equilibrium configuration in the second cell via dissipation of its kinetic energy, this implies the dissipation of the maximum kinetic energy $\frac{1}{2}mv_{max}^2$ per switching. If one wants to minimize this energy, it must be just above ΔE_{bit}, which gives $\chi \simeq \frac{1}{2}$.

The assumption that the time required for the ballistic motion, $\ell/\bar{v}$, combines quadratically with the time involved by the uncertainty principle enhances the minimum switching time to a new value Δt_* given by

$$\Delta t_* \geq \sqrt{(\Delta t_{min})^2 + (\ell/\bar{v})^2}$$

$$= \frac{\hbar}{\Delta E_{bit}} \sqrt{1 + \frac{s}{4\chi^2} \left(\frac{\ell}{a_\circ}\right)^2}, \tag{11.5}$$

where

$$s = \frac{m}{m_0} \frac{\Delta E_{bit}}{E_\circ}, \tag{11.6}$$

$a_\circ$, m_0, and $2E_\circ$ are the atomic units of distance, mass and energy: $a_\circ = 1$ bohr $= 0.53$ Å, $m_0 = 9.1 \times 10^{-28}$ g—the electron mass, and $E_\circ = 1$ rydberg $= 13.6$ eV.

The assumption that Δq_{min} combines quadratically with ℓ enhances the minimum spatial localization to a new value Δq_* given by

$$\Delta q_* \geq \sqrt{(\Delta q_{min})^2 + \ell^2}$$

$$= \sqrt{\frac{\hbar^2}{2m\Delta E_{bit}} + \ell^2}$$

$$= a_\circ \sqrt{\frac{1}{s} + \left(\frac{\ell}{a_\circ}\right)^2} \tag{11.7}$$

There is no special reason for assuming quadratic combinations; other (equally arbitrary) choices, like linear combinations, would however produce nearly the same result.

Chapter 6 of this handbook (S. E. Lyshevski, "Three-Dimensional Molecular Electronics and Integrated Circuits For Signal and Information Processing Platforms") emphasizes that the Heisenberg uncertainty principle provides the limits on the measurements, accuracy, and experimental device characterization. It reports that the cornerstone equations of the uncertainty principle, e.g., the *position-momentum* and *energy-time* uncertainties $\Delta x \Delta p_x \geq \frac{1}{2}\hbar$ and $\Delta \hat{E} \Delta t \geq \frac{1}{2}\hbar$ are not related to the device physics as well as to device dynamics and steady-state characteristics. In particular, the energetics, device sizing, switching energy, switching time, as well as other "limits" and estimates, cannot be derived. To assess the device characteristics, other concepts should be applied and coherently utilized.

11.2.1.2 Limits Imposed by the Thermal Embedding

The thermal environment has contrasting effects on the information content. On one side it produces thermal migration of the information carrier over the potential barrier separating the wells; on another side it produces a decoherence of the quantum state of the information carrier.

The first effect, eventually responsible for the loss of the information, is detrimental. In fact, the time τ_T required to migrate from one well to the neighbouring one is given by

$$\tau_T = \tau_\infty \exp(\Delta E_{bit}/k_B T) \tag{11.8}$$

with τ_∞ a characteristic period in the ground state. Assume τ_∞ in the interval $10^{-14} - 10^{-12}$ s (corresponding to electronic motion in weakly bound states in the lower regime, or to hindered atomic vibrations in the higher regime); at 300 K, values of ΔE_{bit} around 0.8 eV would produce lifetimes consistent with dynamic memories (for which the information must be continuously refreshed) whereas ΔE_{bit} around or greater than 1.3 eV would produce lifetimes consistent with nonvolatile memories:

$$\Delta E_{\text{bit}} = 0.8 \text{ eV} \Longrightarrow \tau_T = 2 \times 10^{-1} - 2 \times 10^{1} \text{ s},$$
$$\Delta E_{\text{bit}} = 1.3 \text{ eV} \Longrightarrow \tau_T = 5 \times 10^{7} - 5 \times 10^{9} \text{ s}.$$

The second effect is beneficial because it prevents the carrier from tunneling from one well to the other. The quantum state is indeed expected to be reduced to a classical state (thus justifying the classical description given in the previous part) when the time t_{spread} occurring for the wavefunction spreading over the entire cell ($t_{\text{spread}} \approx 2m\ell^2/\hbar$) is much longer than t_{deco}, the decoherence time:

$$2m\ell^2/\hbar \gg t_{\text{deco}} \qquad (11.9)$$

Though a complete understanding of the emergence of classical properties from the underlying quantum behavior is still missing, it is generally believed that the appearance of the classical behavior is due to interaction with the environment [29,30]. In this interpretation, if the decoherence is ascribed to the erratic collisions of the information carrier with the particles of the thermal reservoir at temperature T, t_{deco} decreases with T. Thus, it seems not unrealistic to assume

$$t_{\text{deco}} \approx \tau_{\text{vib}}, \qquad (11.10)$$

where τ_{vib} is a time of the order of the period of atomic vibrations. Combining Equations (11.9) and (11.10) one has

$$\left(\frac{m}{m_0}\right)\left(\frac{\ell}{a_\circ}\right)^2 \gg \frac{E_\circ \tau_{\text{vib}}}{\hbar}. \qquad (11.11)$$

The decoherence length ℓ determines not only the distance between the regions of the cell coding 0 and 1, but also between adjacent cells. In fact, if that distance were shorter than ℓ, the information would be destroyed via spreading over (quantum diffusion to) nearby cells. In a planar arrangement the maximum bit density ρ per unit area cannot be larger than $1/(\Delta q_*)^2$; even more, the consideration that wires and contacts are required to allow the cell to be power supplied, addressed, and sensed leads to the introduction of a packing factor k ($k < 1$) specifying the fraction of space linearly occupied by the cell along the x or y direction:

$$\rho \le (k^2/\Delta q_*)^2.$$

From Equations (11.7) and (11.5), one eventually has

$$\rho \le \frac{1}{a_\circ^2} \frac{k^2 s}{1 + s(\ell/a_\circ)^2} \qquad (11.12)$$

$$v \le \frac{\Delta E_{\text{bit}}}{2\hbar} \frac{1}{\sqrt{1 + \frac{1}{4\chi^2} s(\ell/a_\circ)^2}} \qquad (11.13)$$

$$f \le \frac{\Delta E_{\text{bit}}}{2\hbar a_\circ^2} \frac{k^2 s}{[1 + s(\ell/a_\circ)^2]\sqrt{1 + \frac{1}{4\chi^2} s(\ell/a_\circ)^2}}. \qquad (11.14)$$

If each cycle involves the dissipation of an energy $2\Delta E_{\text{bit}}$, the maximum dissipated power per unit area is thus given by

$$\dot{W} \le \frac{(\Delta E_{\text{bit}})^2}{\hbar a_\circ^2} \frac{k^2 s}{[1 + s(\ell/a_\circ)^2]\sqrt{1 + \frac{1}{4\chi^2} s(\ell/a_\circ)^2}}. \qquad (11.15)$$

For electrons ($m = m_0$) and $\tau_{vib} \approx 10^{-13}$ s, Equation (11.11) gives $\ell > 2.4$ nm; assuming that

- The cell has the corresponding minimum size ($\ell/a_\circ = 40 - 50$)
- $\chi = 0.5$
- $k = 0.5$ (a very aggressive packing factor),
- $\Delta E_{bit} = 1$ eV
- The maximum rate is limited by the switching of each cell from one state to another only,

the maximum bit density, flop rate, and dissipated power are given by

$$\rho \leq 4.3 \times 10^{12} \text{ cm}^{-2},$$
$$\dot{f} \leq 1.7 \times 10^{26} \text{ cm}^{-2}\text{s}^{-1}$$
$$\dot{W} \leq 5.4 \times 10^{14} \text{ erg cm}^{-2}\text{s}^{-1}$$

The upper limit of ρ is not dramatically far from what has already been achieved: The maximum reported bit densities, obtained with crossbar structures produced transforming "vertical sizes" (which can be controlled on the nanometer length scale) into "horizontal sizes," are indeed in the range 10^{10} to 10^{11} cm^{-2} [30–36] (this matter will be discussed in Section 11.4.1). On the contrary, the upper value of dissipated power (5.4×10^7 W cm^{-2}!) is manifestly inconsistent with the properties of matter. Let $\dot{W}_{max}$ be the maximum allowed power dissipation. The constraint $\dot{W} \leq \dot{W}_{max}$ may thus be satisfied limiting the flop rate to a value lower than $\dot{W}_{max}/2\Delta E_{bit}$. For $\Delta E_{bit} = 1$ eV and $\dot{W}_{max} = 1$ W cm^{-2} (a value that may assumed to be representative for silicon at 300 K) this would limit the maximum flop rate to 3×10^{18} cm^{-2}s^{-1} and determines a maximum parallelism of 8×10^4.

11.2.2 Reading

As far as decoherence destroys the quantum nature of the state (transforming the quantum states $|0\rangle$ and $|1\rangle$ into the classical states 0 and 1) the determination of the memory state does not involve the characteristic problems associated with the measurement of a quantum system.

11.2.2.1 Coupling the Carrier with the External World

The classical nature of the information carrier allows the specification of the most convenient quantity for the management of the information carrier.

Excluding spin (because intrinsically it's a quantum property, disappearing in the classical limit) the attention is focused on the material properties that characterize matter: mass, charge, electric dipole, magnetic dipole, refractive index, etc. Of them, the most convenient choice is thus dictated by the ease of actuating the $0 \rightleftarrows 1$ transitions (PROGRAM) and sensing the corresponding states (READ).

Excluding mass (that can be actuated and sensed through gravitational fields: in general, extremely weak with respect to the Earth one), one must focus the attention on electromagnetic quantities—and among them the electric charge is the quantity that allows the easiest manipulation.

If the information carrier is charged, the state can be measured by embedding one region of the memory cell (for instance, that coding state 1) in an electrometer and connecting it to the gate of an MOS transistor. To be a little bit more concrete, assume the electron as an information carrier and imagine that in the absence of the electron (electron in the zone coding 0) the transistor is in the OFF condition, whereas in the presence of the electron the transistor is switched to ON.[7] In this arrangement, the macroscopic output will be at ground potential (OUT = 0) when the electron is in the region coding 0 (responsible for the transistor in the OFF condition), or it will have a positive voltage (OUT = 1) when the electron is in the region coding 1 (responsible for the transistor in ON condition).

---

[7]That would occur for a hypothetical p-channel MOS transistor with gate capacity such that the addition of a unitary charge produces an increase of surface potential that is able to bring the transistor from accumulation to inversion—about 1 V for silicon.

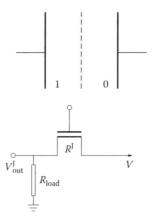

FIGURE 11.2 The cell is reading the current flowing through a resistance whose value is controlled by the charge in one half of the cell.

Of course, the switching of an MOS transistor requires more than one electron (actually, $10^2 - 10^4$ electrons are currently required for that). However, experiments involving the conduction along molecules have demonstrated that certain molecules are able to store a single electron in a metastable state and that the conductance in this state is higher than that in the neutral state by a measurable amount [37]. This matter is discussed in detail in Section 11.4.4. In the following we are instead interested in the maximum reading frequency.

11.2.2.2 Physical Limits in READ Operations

As far as the measurement of the conduction state of the cell state is essentially a classical measurement, it can be attacked with elementary considerations. It is particularly easy if the sensing element of the memory cell can be regarded simply as a reconfigurable resistance R^J (with one or the other of two possible values, $R^J = R^0$ when in the OFF condition or $R^J = R^1$ when in the ON condition, with $R^1 \ll R^0$).

Denoting, as sketched in Figure 11.2, the supply voltage with V, the output voltage with V_{out}^J, the load resistance with R_{load}, the current flowing through the sensing element with I^J, and the dissipated power with $\dot{W}^J$, one has

$$I^J = \frac{V - V_{out}^J}{R^J} \tag{11.16}$$

$$V_{out}^J = \frac{R_{load}}{R^J}(V - V_{out}^J) \tag{11.17}$$

$$\dot{W}^J = \frac{R_{load} + R^J}{(R^J)^2}(V - V_{out}^J)^2 \tag{11.18}$$

with e being the positive charge.

The minimum time required for reading the cell state is immediately obtained from Equation (11.16): Whichever is the resistance state J, the electron flux Φ^J flowing through the resistance is given by

$$\Phi^J = (V - V_{out}^J)/eR^J. \tag{11.19}$$

Assume first that the resistance is in the ON state, $J = 1$. The measurement is actually able to sense the ON state only if the duration lasts a time Δt sufficiently long enough to allow the passage of at least one electron, that happens when $\Phi^1 \Delta t > 1$, i.e.:

$$\Delta t \gtrsim eR^1/(V - V_{out}^1). \tag{11.20}$$

On another side, the system will detect the passage of an electron even when the molecule is in the OFF state, provided that the cell is observed for a time longer than $e R^0/(V - V_{\text{out}}^0)$. Since $V_{\text{out}}^0 < V_{\text{out}}^1$, the detection of the passage of one or more electrons in the time interval

$$\left\{ \Delta t : e R^1/(V - V_{\text{out}}^1) \ll \Delta t \ll e R^0/(V - V_{\text{out}}^1) \right\} \tag{11.21}$$

may be considered to provide evidence for resistance in the ON state; missing any evidence for flowing electrons in the time interval (11.21) may instead be considered evidence for resistance in the OFF state. Let

$$\Delta t^{\text{READ}} := e R^1/(V - V_{\text{out}}^1) \tag{11.22}$$

the quantity Δt^{READ} is thus the minimum time[8] required for the detection of the state, 0 or 1, of the cell.

The time sensitivity can be improved, thus increasing the potential difference $V - V_{\text{out}}^J$ (of course, maintaining it below the breakdown potential). In so doing, however, the dissipated power increases quadratically because of Equation (11.18). Many believe that *the* ultimate physical limit to integration is power dissipation [37–40]. In this line of thought, one can try to minimize the energy ΔE_{diss} dissipated in the switching. Actually, although non-dissipative computational schemes have been developed [41–46], as far as computation is an irreversible physical process, it necessarily involves some dissipation. Moreover, if dissipation is certainly detrimental, irreversibility is useful to decouple the outcomes of the computational process from the intermediate states, thus rendering the results independent of the back propagation of noise. The minimum dissipation required for computation in a thermal reservoir at temperature T, where the computing device is embedded, is given by the famous formula

$$\Delta E_{\text{diss}} \geq k_B T \ln(2) \tag{11.23}$$
$$=: \Delta E_{\text{min}}$$

with k_B the Boltzmann constant. This formula, first asserted (without justification) by von Neumann and confirmed by Landauer analysis [25], is fully supported by the discussion of Meindl and Davis [39].

Since dissipation decreases with the total resistance ($R_{\text{load}} + R^J$), the attention can be limited to the case of resistance in the ON condition; assuming that even reading is limited by power dissipation, $V - V_{\text{out}}^J$ must thus be as close as possible to the minimum voltage involved in the computation:

$$e(V - V_{\text{out}}^1) \simeq k_B T \ln(2). \tag{11.24}$$

For assigned R^1 and $V - V_{\text{out}}^1 [= k_B T \ln(2)/e,]$ Equation (11.18) requires that R_{load} is as small as possible. On another side, if one requires that even V_{out}^1 is capable of computation [that is, $V_{\text{out}}^1 \geq k_B T \ln(2)/e$], Equation (11.17) gives $R_{\text{load}} \geq R^1$. Combining this constraint with the need for minimizing dissipation, one gets the optimum combination

$$R_{\text{load}} = R^1 \tag{11.25}$$

The insertion of Equation (11.24) into Equation (11.20) gives

$$\Delta t_{\text{min}}^{\text{READ}} \simeq \frac{e^2 R^1}{k_B T \ln(2)}; \tag{11.26}$$

the minimum power dissipation in reading a single cell for the optimum circuit is instead obtained, inserting Equation (11.24) into Equation (11.18):

$$\dot{W}_{\text{min}} = \frac{2e^2}{R^1 [k_B T \ln(2)]^2} \tag{11.27}$$

---

[8]Actually, assuming that the electron flow is a Poisson process, Δt^{READ} is just the time at which the standard deviation of the number of detected electrons coincides with the average number. Statistical significance is obtained only for $\Phi^1 \Delta t^{\text{READ}} > r\sqrt{\Phi^1 \Delta t^{\text{READ}}}$, with $r \simeq 3$. This condition reads $\Phi^1 \Delta t^{\text{READ}} \gtrsim 10$: i.e., $\Delta t^{\text{READ}} \gtrsim 10 e R^1/(V - V_{\text{out}}^1)$.

There remains to discuss R^1, the sensing resistance. In view of the many mechanisms responsible for the transport through molecules (tunnel, field emission from the contacts, Poole or Poole–Frenkel conduction, etc.) [46], it is extremely difficult to assert anything about molecular conductance in general. However, the discussion is greatly simplified if the attention is focused on carrier conduction from one contact to another occurring through a classically forbidden region. In this case, indeed, one has

$$\frac{1}{R^J} = \frac{e^2}{h} T^J \tag{11.28}$$

where $h = 2\pi\hbar$, $h/2e^2$ is Landauer's quantum resistance (a kind of minimum contact resistance associated with any classically forbidden region), and T^J is a quantity associated with the transparency of the barrier separating the two contacts and with the density of states in the arrival region. This case is particularly interesting because T^J can be modified easily by electric fields in the vicinity of the tunneling region [generated, for instance, by the presence (J = 1) of absence (J = 0) of an electron in its vicinity—that allows in principle an easy implementation; see Section 11.4.4].

The insertion of Equation (11.28) into Equations (11.26) and (11.27) gives

$$\Delta t_{min}^{READ} \simeq \frac{h}{k_B T \ln(2)} \frac{1}{T^1} \tag{11.29}$$

and

$$\dot{W}_{min}^J = 4 \frac{[k_B T \ln(2)]^2}{h} T^J. \tag{11.30}$$

Equations (11.29) and (11.30) specify how Δt_{min}^{READ} and $\dot{W}_{min}^J$ depend on the environment (through T) and on internal properties of the resistance (through T^J).

As already mentioned, it is difficult to assert anything about R^J. In general, for the molecules which have been considered of potential interest, resistances in the ON condition of the order of 10^9 Ω (thus, much higher than the Landauer resistance, $h/2e^2 = 12.9$ kΩ) have been found; the assumption that conduction is limited by tunnelling would give $T^1 \approx 10^{-5}$. The typical current flowing through a molecule in ON is thus on the order of 1 nA for a voltage of 1 V, or of 10 pA for the minimum voltage allowing computation at room temperature.

Assuming the condition of minimum dissipation, for $T^1 \approx 10^{-5}$ Equations (11.29) and (11.30) give $\Delta t_{min}^{READ} \simeq 2.4 \times 10^{-8}$ s and $\dot{W}_{min}^1 \simeq 4.7 \times 10^{-13}$ W—that would allow the simultaneous reading of all memory cells if they are packed at the maximum density, and the allowed dissipation is of the order of 1 W cm^{-2}.

Up to now, we have assumed that the condition $T^{OFF} \ll T^{ON}$ may be achieved quite irrespective of the applied signal. Actually, the sketch of Figure 11.2 would suggest that the minimum variation of input potential required to switch the transistor from OFF to ON is that required to bring the transistor from accumulation to inversion—roughly 1 V for silicon (E_g/e, with E_g the energy gap). For $T^{ON} \approx 10^{-5}$ Equations (11.26) and (11.27) give access time and dissipated power per cell on the order of 10^{-9} s and 10^{-10} W, respectively. The last datum shows that in this situation only a part of the total allowed cells could be simultaneously managed to be consistent with an allowed dissipation of 1 W cm^{-2}.

An important quantity not considered yet is the current flowing through the reconfigurable resistance in ON: of the order of 1 nA for a supply voltage of 1 V or about 20 pA for a supply voltage of $k_B T \ln(2)/e$. Both these currents are easily measurable, but with an external amplifier only. If one needs to integrate it into the control circuitry (this is mandatory if a number of cells must be managed in parallel) this value would be reduced on the 1 μA current scale [47]. The effect of this technological constraint is considered in Section 11.4.2.

How the signal of PROGRAM and READ are imparted to the cell has not been considered yet. Irrespective of the detailed modes used to vary cell potentials or to measure flowing currents, the correspondingly involved apparatuses are macroscopic, with a size in general much larger than the cell. If these apparatuses must be arranged in a planar configuration too, *only an architecture allowing the sensing of all cells by relatively few measuring apparatuses permits tera-scale integration.* The integration of the ultimate memory cells with such apparatuses is the final goal of nanoelectronics.

11.3 How a Hybrid Technology Can Approach the Physical Limits—Molecular Electronics

Let N_{cell} denote the number of memory cells forming the circuit. The simplest way for sensing them with relatively few measuring apparatuses is to organize the memory in a matrix where each cell is defined by two crossing lines.[9] The idea of using crossbar structures for next-generation circuits has been subjected to systematic investigation since the early proposals [48]. The crossbar structure has actually been used in exploratory memories exploiting reconfigurable molecules as memory cells [49]; though envisaged by several years [50], the idea of integrating molecular devices in the existing CMOS technology has been explored more recently [50–54].

Among the several architectures proposed for circuits with a complexity larger than 0.1 Tbit [55], I shall focus on an architecture where a conventional, silicon-based, microelectronic part controls a nanoscopic crossbar structure, in turn hosting in each cross-point a nanoscopic device behaving as a memory cell [48]:

TSI IC = micro IC ∪
nano crossbar ∪
nanoscopic cells.

This architecture shifts the problem of preparing TSI ICs to that of producing nanoscopic memory cells and inserting them into the cross-points of a crossbar structure. It would be a mere declaration of will were it not for the fact that molecules by themselves are able to behave as memory cells and have been not only designed [55–60], but also synthesized [49,61]. This fact opens immediately the possibility of a *hybrid route* to TSI ICs:

hybrid TSI IC = micro IC ∪
nano crossbar ∪
grafted reconfigurable molecules

This approach, in which the electronic-transport properties of programmable molecules are exploited for the preparation of externally accessible circuits, is usually referred to as *molecular electronics.* The hybrid molecular nonvolatile memory is thus formed by a nanometer-sized kernel (the functionalized crossbar) linked to a conventional submicrometer-sized control circuitry.

The implementation of hybrid architecture requires the solution of a number of problems. The major ones seem to be

1. The set up of an economically sustainable technology[10] for the preparation of 10^{11} cross-points cm^{-2}
2. The linkage of the addressing lines to the microelectronic circuit for writing and sensing

---

[9]The dimensionality d of the cell arrangement is critical not only for the possibility of linking a cell to its neighbors (for balls of the same radius, the maximum kissing number being 2 for $d = 1$, 6 for $d = 2$, and 8 for $d = 3$), but also for the number N^*_{cell} of externally accessible cells, $N^*_{cell} \propto N_{cell}^{1-1/d}$. Examples of memories with $d = 1, 2$, or 3 are the shift register (N^*_{cell} independent of N_{cell}), static random access memory (RAM, with $N^*_{cell} \propto N_{cell}^{1/2}$), and brain cortex (with $N^*_{cell} \propto N_{cell}^{2/3}$). A three dimensional (3D) arrangement is manifestly required when a large amount of information must be simultaneously managed; the crossbar memory has a 2D arrangement; dynamic RAMs and NAND NVM may be viewed as organized in a multi-1D arrangement.

[10]Here "economically sustainable" means "without the extended use of electron beam lithography (EBL)." Of course, this exclusion does not mean that no layer can be defined via EBL, but rather that this technique is limited to the definition of a few selected layers only.

3. The design, synthesis, and electrical characterization of the reconfigurable molecules

4. The grafting of the functional molecules to those cross-points via batch processing

All the preceding problems have been attacked, and for all of them there are now clues suggesting the possibility of their solution:

1′. Ways for the preparation of simple geometries (line arrays) with pitch on the nanometer length scale (NLS) are known [30–35].

2′. Solutions have been found to the problem of demultiplexing [61–69].

3′. Molecules admitting two conduction states are known from the literature [48,55–60].

4′. Hydrogen-terminated silicon can be functionalized by simple exposure to alkene-terminated molecules with the formation of chemically robust, environmentally stable, $Si-C$ bonds [34].

11.4   Beating the Limits of Conventional and Electron-Beam Lithography

It is possible to define wire arrays with pitch on the NLS without a massive use of EBL. Their preparation exploits

- The possibility of transforming vertical features into horizontal features
- The fact that the thickness (a "vertical" feature) of many films can be controlled to the sub-NLS.

For instance, a contact mask with nanometer-sized features for *imprint lithography* (IL) [70] may be prepared via a process involving the following steps [32]:

i. Growing on a substrate a quantum well via molecular beam epitaxy

ii. Cutting the sample perpendicularly to the surface

iii. Polishing the newly exposed surface

iv. Etching selectively the different strata of the well

Another way is the multi-spacer patterning technique (S^nPT). In this technique an array of n bars is directly defined onto a substrate via a sequence of n conformal depositions and anisotropic etchings [32–34]. The basic idea of the SPT is shown on the left in Figure 11.3: the upper part sketches the process, while the lower part shows the cross-section of a wire produced via this technique. The basic idea of the S^nPT is shown on the right in Figure 11.3: the upper right part sketches the process; the lower part shows instead how poly-silicon arrays separated by SiO_2 dielectrics with sub-lithographic pitch (35 nm) can indeed be produced.

In the approach based on the imprint lithography, the entire pattern is defined in a single step; the S^nPT, instead, requires the repetition of n (or $n/2$, if the process is carried out in a trench, like the structure shown in Figure 11.3) cycles of conformal depositions and anisotropic etchings. This implies that the first approach is by far more convenient when the array is formed by numerous lines. On the contrary, the second approach is more convenient for the preparation of test structures with few devices on the NLS (however, see Section 11.4.2.2).

Although the producible geometries are quite simple both for imprint lithography and multispacer patterning, the minimum achievable pitches (16 nm [32] and 20 nm [71,72], respectively) are remarkably smaller than the ones currently producible with the the most advanced lithography (around 40 nm for EBL [71]), because both techniques exploit the self-alignment of a wire with respect to the adjacent ones (see note 1).

The ability to produce wire arrays with pitch on the 10-nm length scale does not allow by itself the preparation of transistors with such a feature length because the masks for the transistor involve a design more complex than simple arrays, and masks at different levels must not only have *definitions* but also

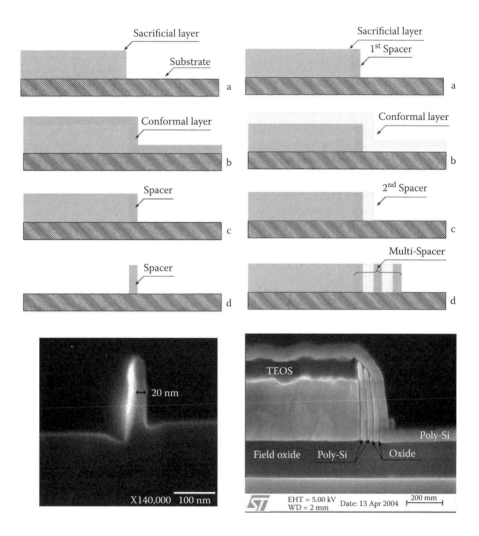

FIGURE 11.3 *Up on the left:* a sequence for obtaining sublithographic features via the spacer patterning technique: (a) deposition of a sacrificial layer and definition of a vertical step by means of lithography and an anisotropic etching; (b) deposition of a conformal layer; (c) anisotropic etching; and (d) a selective chemical etching of the sacrificial layer. *Down, on the left:* an example of structure produced via SPT (from Choi, Y.-K., et al., *J. Phys. Chem. B*, 107, 3340, 2003). *Up on the right*, the S^nPT: (a) fabrication of a first spacer by means of the SPT; (b) deposition of a novel conformal layer; (c) anisotropic etching and fabrication of a second array of another material; (d) iteration of the sequence (b)–(c). *Down, on the right:* an example of a S^3PT multi spacer (with a pitch of 35 nm and formed by a double layer poly-Si|SiO$_2$) resulting after three repetitions of the SPT (from Cerofolini, G. F., et al., *Nanotechnology*, 16, 1040, 2005).

alignments on the NLS. Both these factors limit the scaling down of MOSFETs to the mask definition limit. They, however, disappear using the crossbar structure, in which two perpendicularly oriented arrays (orientation not being critical) are defined on two different levels, and each cross-point between lower and upper arrays is the site containing the active memory element.

11.4.1 The Crossbar Structure

Seemingly, the simplest way to prepare a crossbar circuit consists of the following fundamental steps:

(XB1), definition of the first-level wire array

(XB2), deposition of the active (reconfigurable) element, working also as vertical spacer separating lower and upper arrays

(XB3), deposition and definition of the second-level array

This approach was used by a collaboration between Hewlett-Packard (HP) and the University of California at Los Angeles (UCLA) to prepare the kernel of a non-volatile memory [49]. The HP–UCLA collaboration employed as an active element a complex bi-stable organic molecule transferred to the substrate via the Langmuir–Blodgett technique. The HP–UCLA approach has however a serious limit: the organic active element is incompatible with high-temperature processing, so the counter-electrode layer must be deposited at room, or slightly higher, temperature. This need implied a preparation based on physical vapour deposition,[11] where the metallic electrode results from the condensation of metal *atoms* on the outer surface of the deposited organic films. This process, however, poses severe problems of compatibility, because isolated metal atoms, quite irrespective of their chemical nature, are mobile and decorate the molecule, rather than being held at its outer extremity [72–76].

Assuming that the reconfigurable is formed by organic molecules, *the only way to overcome the difficulties met by the HP–UCLA collaboration seems to consist in the preparation of the two arrays defining the crossbar matrix* before *the insertion of the organic element* [51,54].

That objective may be achieved by modifying the sequence (XB1)–(XB3) as follows:

(XB*1), preparation of a first-floor array of wires

(XB*2), deposition of a sacrificial layer as vertical spacer separating lower and upper arrays

(XB*3), preparation of a second-floor array crossing the first-floor array

(XB*4), selective chemical etching of the spacer

(XB*5), insertion of the reprogrammable moles in a way to link upper and lower wires in each cross-point.

A simplified scheme is sketched in the upper part of Figure 11.4.

Of course, both the definition of a sacrificial layer with thickness adapted to the molecular size and the grafting in recessed regions with typical dimension on the NLS are not trivial. The first problem, however, seems to have a relatively easy solution using poly-silicon wires and growing a thin layer of SiO_2 (whose thickness is easily tuned to match the molecule length), and constructing a second-level array of wires oriented perpendicularly to the underlying one. A selective chemical etch is finally performed to remove the interlayer oxide and to form the hosting sites for the molecular guests.

In the case of arrays produced via $S^n PT$, in order to have an oxide of the same thickness at each stage of the second-floor $S^n PT$, the process is specified as follows: Let D_X, E_Y and O denote the operations of conformal deposition of layer X, anisotropic etching of layer Y, and silicon oxidation to a certain thickness; if $B^\frown A$ means that operations A and B are carried out in that order (operation $^\frown$ is not commutative: $B^\frown A \neq B^\frown A$), then the first- and second-floor $S^n PT$ may be summarized with the following symbols:

$$S^n PT = S^1 PT ^\frown S^{n-1} PT$$

$$S^1 PT = \begin{cases} E_{SiO_2} ^\frown D_{SiO_2} ^\frown E_{Si} ^\frown D_{Si} \\ \qquad \text{for first floor} \\ \\ E_{SiO_2} ^\frown D_{SiO_2} ^\frown E_{Si} ^\frown D_{Si} ^\frown O \\ \text{for second or upper floor} \end{cases} \qquad (11.31)$$

---

[11] Actually, alternative methods can be envisioned. Imagine, for instance, that the upper layer is formed by a suspension of metallic nanoclusters (stabilized by their interaction with the medium) in a volatile liquid—the evaporation of the liquid would result in a nearly compact layer. Another method could be the spin deposition of an organic conductor. No such methods have been explored yet.

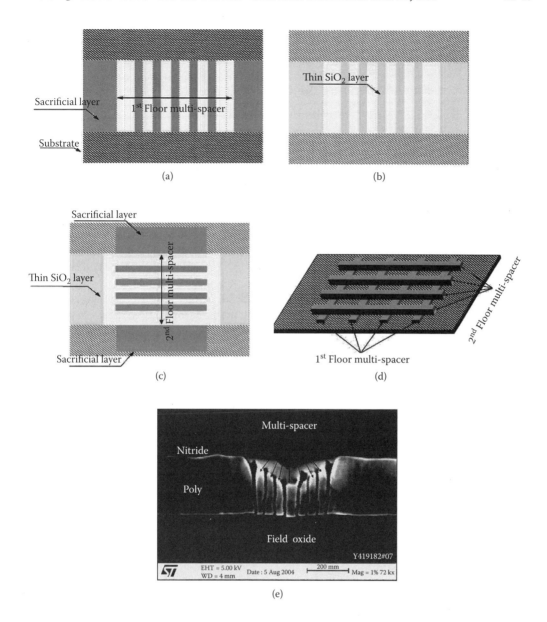

FIGURE 11.4 *Up*; crossbar fabrication steps: (a) fabrication of a first-floor array of double spacers by means of S^{n}PT; (b) thermal growth of a thin film of SiO$_2$; (c) fabrication of a second-floor array of double spacers crossing the first-floor array; (d) selective chemical etching of the intralayer SiO$_2$. *Down*: image at the scanning electron microscope of the cross-section of a seven-spacer crossbar structure obtained with the S^{n}PT using undoped poly-silicon and thermally grown SiO$_2$.

[33–35]. This method has the additional advantage that its simplest implementation results in poly-silicon arrays, whose surface has a modest reactivity but still allows its controlled functionalization with organic molecules [53].

The lower part of Figure 11.4 shows an image at the scanning electron microscope of a cross-section of a crossbar structure obtained with the S^{n}PT using undoped poly-silicon and thermally grown SiO$_2$. This picture is of certain interest because it shows how crossing two overlapped arrays is able to arrange, without special care, over a region with area 350×350 nm^2, 49 cross-points at a density of 4×10^{10} cm^{-2}.

According to Ref. [71], this technique would allow the preparation of cross-points at densities of 1.2×10^{11} cm^{-2}.

11.4.2 Addressing and Sensing the Cross-Points

The motivations for the industrial development of hybrid nano ICs are twofold:

- They allow the preparation of memories with the highest integration (denoting with F the minimum achievable feature size, the crossbar memory cell area is about F^2 versus $4.5F^2$ for NAND or $10F^2$ for NOR nonvolatile memories [47]).[12]
- In principle, the nanoscopic crossbar is producible without a massive use of EBL with definitions F_X much better than the feature size F_{litho} allowed by optical lithography.

For $F_X = 0.3F_{\text{litho}}$ the combination of the preceding factors gives for crossbar memories a higher density than that of current NAND–NOR memories by a factor of 40–100. This advantage is paid with the difficulties of connecting features with sublithographic pitch to lithographic features.

The development of strategies able to address ultra-high dense architecture on the NLS is under systematic investigation [61–68]. The major strategy for attacking this problem has been the stochastic linking of nanoscopic lines to the external circuitry with structured wires embodying in themselves the ability to address single cells [62,63,65,67]. In this way, a large amount of cells are left out of control (being either not addressed or having its different cells identified by a unique code) and a special logic must be employed to manage this situation, eliminating unaddressed cells and possibly exploiting redundancy (TERAMAC approach) [68,76]. Needless to say, this deal is convenient only if the non-functioning cells are a minority and the circuitry required to manage the stochasticity remains relatively simple.

That these conditions may really be satisfied in practice is not clear, especially if the expected cross-point density, 10^{11} cm^{-2}, is compared with the currently achievable density (10^{10} cm^{-2}) for NAND memories, whose control is deterministic.

In the following, the possibility of a deterministic access to all cross-points shall be explored. In particular, three methods shall be discussed: one addressed to all kinds of nanoscopic crossbars (irrespective of their preparation procedure) and the other two specific to crossbars prepared by the S^{n}PT. Two of them are based on the geometric separation of wires, while the other separates one wire from the other via a kind of energetic filtering.

11.4.2.1 Geometric Separation

The horizontal bevelling technique   The first technique, originally proposed with the name of CMOL circuit by Likharev (see Ref. [69] for a recent review) is reminiscent of the bevelling technique once used, together with micro-sectioning and staining, for the measurement of junction depth (and, more recently, of carrier concentration profiles using spreading resistance techniques); because of this, it will henceforth be referred to as horizontal beveling technique (HBT). Consider, as described earlier, an array of n conductive parallel wires aligned in the x direction with width ℓ_y and pitch L_y, so densely arranged they cannot be singularly accessed by EBL and thus cannot be linked to the hosting microelectronic circuit. A kind of hardware demultiplexing is possible via the following process: After the deposition of a protective insulating cap, define (for instance, but not necessarily, by EBL — SPT could also be used) a bar with width b, oriented in a direction tilted by a small angle α with respect to the x direction, and so long as to cross all wires, and use this pattern to etch the underlying insulating cap. This process will result in

---

[12] Actually, the different estimates for crossbar and NAND memories are essentially due to the fact that lithographic F (which can, in principle, be reduced well below the lithographic limit simply by over-etching) is conveniently set to $F = 0.5L$, with L the pitch, while in S^{n}PT the pitch may be very close to F.

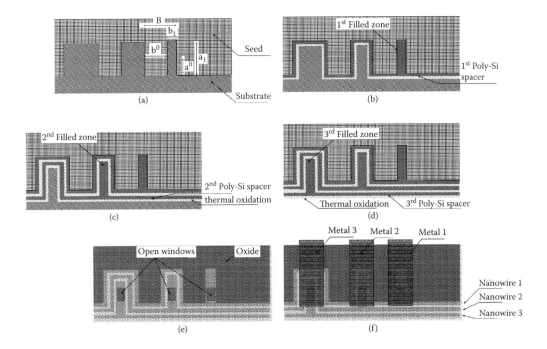

FIGURE 11.5 Plan view of the mask defining the wires and the trick adopted to allow the separate contact of each wire using electron-beam lithography.

the exposition of N rhomboids with pitch Δ_x and major side δ_x in the x direction given by

$$\Delta_x = L_y / \tan \alpha \tag{11.32}$$

$$\delta_x = b / \sin \alpha \tag{11.33}$$

the area of each open contact being thus $b\ell_y / \sin \alpha$. The orientation is chosen in such a way that both δ_x and Δ_x are in the reach of (not necessarily electron-beam) lithography. Because of Equations (11.32) and (11.33), it is certainly possible for a sufficiently small α. Since the method produces from an array with pitch L_y another array with pitch Δ_x [controlled uniquely by the orientation angle α of the cutting bar, see Equation (11.32)], it may viewed as a kind of *pitch converter* with a geometric gain of about α^{-1}.

Fusing adjacent lines in S^nPT The S^nPT offers a major advantage over other techniques—the possibility to "separate" the wires by a distance which allows their connection to the microelectronic world via EBL or even conventional lithography. Imagine indeed that the mask defining the sacrificial layer has the shape shown in Figure 11.5 (a) where the kth indentation extends along, and perpendicularly to, the wire by segments of length b^k (with pitch B^k) and a^k, respectively (the upper index k runs from 1 to n and defines the wire in the array). Let these quantities satisfy the following conditions:

$$b^k = 2t_{Si} + 2(k-1)t_{sp} \tag{11.34}$$

$$B^k = b_0 + b^k \tag{11.35}$$

$$a^k = (n-1)t_{sp} + a_0 = a \text{ (independent of } k) \tag{11.36}$$

where t_{Si} is the thickness of the deposited poly-silicon (transformed by SPT in poly-silicon width), t_{sp} is the width of each double spacer, b_0 and $a_\circ$ are sizes that allow the alignment of contacts (defined by EBL) in the zone where a single poly-silicon deposition fills the underlying region. Figures 11.5 (b), (c), (d), etc. show

how the structure evolves after the deposition of 1, 2, 3, etc. spacers, provided that Equations (11.1–11.3) are really satisfied; the whole sequence shows how it is possible to contact individually each wire with lithographic resolution. Once lithographically accessible regions have been defined, the process continues depositing on the surface a dielectric film and etching the insulator over those regions.

In both cases (S^nPT and HBT), the process continues with the deposition of a uniform metallic film and by the lithographic definition of wires connecting the nano-wires to conventional micro-contacts.

This number would be reasonable (on the order of 10^6 cm^{-2}) even for N in the terabit range, but such an arrangement imposes extremely strong constraints on the current–voltage characteristics of the memory cell ($I_{ON}/I_{OFF} \gg 10^6$, where I_{ON} and I_{OFF} denote the currents in the ON and OFF condition, respectively).

It is thus possible that addressing in TSI devices should be obtained not only via geometric selection but also, for instance, via energetic filtering.

11.4.2.2 Energetic Filtering

Addressing lines poses serious architectural problems. For instance, if the memory is organized as a rectangular matrix of $n_x \times N_y$ (with $n_x \ll N_y$) cells and is addressed via $n_x + N_y$ externally accessible bit and word lines, the number of lines would increase as $n_x(1 + N_{cell}/n_x^2)$, where N_{cell} is the number of cells. This number has a minimum when $n_x = N_y$, in which case the number of lines is reduced to $2N$, where $N = \sqrt{N_{cell}}$.

Addressing $N \times N$ memory elements organized in a crossbar of $N + N$ rows and columns (collectively referred to as lines) requires that one arbitrary row and one arbitrary column are simultaneously enabled while all others lines are disabled. The conceptually simplest way to enable or disable ("control") a line is to put it as the output of an AND (or NAND) gate whose inputs are all the signals coding the line. Since a binary coding of $2N$ inputs requires $\log_2(2N)$ bits, addressing the N^2 cells of the crossbar requires in principle $2N$ gates each with $\log_2(2N)$ inputs.

A hybrid silicon-organic memory is manifestly of practical interest only if it has a complexity much larger than the ones already available (of the order of 10^9 bits). Therefore, the example of a memory with $2^{34}(\simeq 1.7 \times 10^{10})$ bits is relevant. The control of a crossbar with $N^2 = 2^{34}$ cells would require $2N = 2^{18}(\simeq 2.6 \times 10^5)$ gates, each with $\log_2(2N) = 18$ inputs. Clearly enough, gates with so large fan-in are inconsistent with the current technology; assuming a fan-in of 2, a single decoder would instead require a maximum of $\log_2(N) = 9$ dual-input AND (or NAND) gates per line, and the total number of gates would be $2N \log_2(N) \simeq 2.3 \times 10^6$. The example shows that in principle a very large TSI crossbar memory can be controlled by a silicon circuitry with, at most, GSI complexity.

The hypothesized architecture is, however, practically impossible for the following reason: A single line, indeed, has width on the NLS and height of the order of 10^2 nm, but a length of the order of 1 cm. The line resistance, R_{line}, is thus expected to become so high as to make it impossible to recognize if the molecules linking the two sides of the cross-points are in the ON or OFF state (this matter will be discussed in the following). This difficulty can be made up by organizing the entire memory in K smaller sub memories, each with n^2 cross-points, with $N^2 = Kn^2$ (and presumably $n \ll N$). Assume for this arrangement the same solutions hypothesized earlier for the complete memory; the addressing of the entire memory would therefore require from a minimum of $2nK = 2N(N/n)$ gates with fan-in of $\log_2(2n)$ to a maximum of $2nK \log_2(2n) = 2N(N/n)\log_2(2n)$ dual-input gates. The complexity of the silicon circuitry therefore increases, by a factor ranging from N/n, for a fan-in of $\log_2(2n)$, to $(N/n)\log_2(2n)/\log_2(2N)$, for a fan-in of 2. To give an idea of the increase of complexity resulting from this organization, observe that an architecture with sub-memories with $n = 2^5$ (1024 bits) would require 1.1×10^9 gates with a fan-in of 6 (or 3.1×10^9 gates with a fan-in of 2). This simple example shows that the organization in sub-memories of 1 kbits would require a silicon circuitry with almost the same complexity as the molecular part, thus making useless the crossbar approach. The complexity of the addressing circuitry decreases with n. The preceding example shows that TSI crossbar memory can be controlled by a silicon circuitry with GSI complexity only for

$$n \gg 30. \tag{11.37}$$

This need, however, is in conflict with the need for integrating the sense amplifier in the silicon control circuitry. The minimum cross-point area is indeed controlled by the ability of sensing the current flowing through the memory cells.

- The integrated sense amplifier has current sensitivity of the order of 1 μA (one order lower than the value currently used [77,78], but presumably not an unrealistic forecasting [47]); and
- For an applied voltage of 1 V, each molecule carries a current in ON condition of the order of 1 nA, $I^{ON} \approx 1$ nA [59].

If these conditions are satisfied, the current flowing through the cross-point in ON condition is detectable when it contains 10^3 molecules (additivity of effects is assumed). If each molecule occupies an area of approximately 1 nm^2 and all are closely packed,[13] the cross-point area is about 10^3 nm^2, corresponding to a square with sides around 30 nm. If the dielectric separation between adjacent cross-points has a negligible thickness (this is the situation achieved in the demonstrator on the right hand side of Figure 11.3) the resistance of an array of n cross-point is given by $R_{line} = R_\square n$, where $R_\square$ is the sheet resistance of the poly-silicon layer. For heavily doped poly-silicon of thickness of the order of 10^2 nm, $R_\square$ may be assumed of the order of 10^3 $\Omega/\square$, $R_\square \approx 10^3$ $\Omega/\square$. The ON state can be detected only if $R_{ON} \gg 2R_\square n$ (otherwise, the current would be smaller than the sensitivity limit of the amplifier; the factor of 2 comes from the fact that the cross-point is accessed by *two* lines). This condition gives

$$n \ll R_{ON}/2R_\square \approx 5 \times 10^2. \tag{11.38}$$

Even though perhaps there is a narrow range of n where inequalities (11.37) and (11.38) are simultaneously satisfied, they still show that a robust design of a hybrid silicon-molecular memory actually requires a strategy for accessing the cross-points in a more sophisticated fashion than a simple geometric separation.

Whereas in IL all wires are defined collectively, in S^{n}PT they are constructed sequentially. The batch fabrication would make preferable IL over S^{n}PT unless one were able to *use the sequential array deposition for the external recognition of single wires*.

An inspection of the right side of Figure 11.3 shows that multispacer patterning results in wires whose height is progressively reduced with the order of preparation (readers will easily convince themselves that the amount of this decrease can be magnified in a controlled way via slight changes in the S^{n}PT conditions).

Let r_i and c_j denote the i-th row and j-th column, respectively, and (i, j) be the cross-point they define. Imagine now that all rows r_i (columns c_j) go to a unique photolithographically defined contact C_r (C_c) running below a photolithographically defined electrode GE_r (GE_c) and dielectrically separated from it by an insulator of fixed thickness. Figure 11.6 shows a plan view and two cross-sections of the considered arrangement.

Considering the wires are formed of poly-silicon, electrodes GE_r and GE_c may be operated as control gates for the underlying wires. Imagine now that for assigned silicon doping and insulator thickness, the voltages of control gates GE_r and GE_c are so arranged that only the tallest wires are not totally depleted. In this case, the current $I_{(1,1)}$ flowing from C_r to C_c is controlled by the conduction state of cross-point $(1,1)$ alone. Assume now that voltage of GE_c is modified in such a way that even c_2 is enabled: the current flowing from C_r to C_c is thus the sum of the current $I_{(1,1)\cup(1,2)}$ flowing through $(1,1)$ and $(1,2)$. If $I_{(1,1)}$ was memorized, the current $I_{(1,2)}$ flowing through $(1,2)$ (giving its state) is obtained as the difference $I_{(1,1)\cup(1,2)} - I_{(1,1)}$. In this way, it is easy to convince oneself that, reiterating the argument to all cross-points, the states of each of them can be determined provided the information contents of all (256 or 1024) cross-points are memorized in a buffer memory. This implies that only a small part of the crossbar can

---

[13] In Section 11.2.1, the minimum separation between different memory cells was estimated to be larger than 2 nm. In the present case, it's assumed instead that molecules are spaced by about 1 nm. There is no inconsistency, because the previous estimate applies to the distance between different memory cells, while the present one applies to the molecular distance within the same cell.

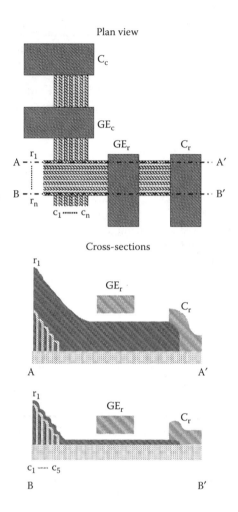

FIGURE 11.6 *Top*: Plan view of the crossbar and lithographically defined contacts. *Middle*: View projected on plane AA′ perpendicular to the surface and crossing the tallest row line. *Bottom*: View projected on plane BB′ perpendicular to the surface and crossing the smallest row line.

be read in parallel (otherwise, the silicon circuitry should have at least the same number of bits as the nanoscopic memory).

However, it is noted that this difficulty would disappear if the molecules operating as memory elements behaved as Schmitt triggers, with a non-negligible conductance only in a certain narrow voltage interval. Actually several molecules considered as candidates for molecular electronics are reported to have this behavior [79–82].

Remember now that the S^nPT is more conveniently carried out in a cave, because the repetition of n SPT produces $2n$, rather than n wires. A single control electrode covering all wires would be unable to resolve those with the same height, this difficulty my be circumvented by assigning to the cave a size allowing the growth of just $2n - 1$ wires (the central one having twice the width of the others, and using *two* control-gate electrodes (GE_l^u and GE_l^d per line, $l = r, c$), each covering totally $n - 1$ outer electrodes and partially the central electrode. Reading is achieved by first enabling GE_l^u and disabling GE_l^d (reading sequentially the current through upper wires), then enabling GE_l^d and disabling GE_l^u (reading sequentially the current through lower wires), and eventually enabling both GE_l^d and GE_l^u (reading the current through the middle wire). Needless to say, other solutions of the same kind are possible.

11.4.3   Grafting Reconfigurable Molecules to Poly-Silicon Cross-Points

Several additional modes of organic molecules to silicon are known. Among them are

1. The *cycloaddition* of molecules with alkene or alkyne functions to double-bonded silicon dimers [83,84]
2. The *alkyl/arylation* with Grignard reagents of halogen-terminated silicon surfaces [85], and of hydrogen-terminated silicon dimers [86]
3. The *condensation* of molecules containing alcoholic groups at hydroxyl-terminated silicon surfaces [87]
4. The *silanization* of hydroxyl-terminated surfaces [88] as well as of hydrogen-terminated surfaces [89] via reaction with chlorosilane-terminated molecules and the elimination of HCl
5. The *silanization* of hydrogen-terminated surfaces via reaction with azanes and the elimination of NH_3 [89]
6. The *arylation* of hydrogen-terminated silicon via its reaction with diazonium salts and the elimination of HBF_4 and N_2 [90]
7. The *hydrosilation* of alkenes at hydrogen-terminated surfaces [83,84].

If compatibility with poly-silicon is a major issue, cycloaddition is immediately discarded from these routes. In fact, cycloaddition is limited to the extremely reactive, *clean, 2 × 1 reconstructed, (1 0 0) surface of single crystalline silicon*, which is hardly producible at the poly-silicon surfaces — poly Si, indeed, is usually producible with (1 1 1) orientation. (Actually, the adoption of the method described in [72] for silicon-on-insulator substrates would allow the use of cycloaddition for molecular grafting on the lower silicon-wire array.)

Grignard reagents XMgR (with X a halogen and R an alkyl or aryl group) are known to be very effective in the derivatization of halogen-terminated silicon SiX'_n [85]. However, both the difficulties of preparing well-defined halogen-terminated silicon surfaces and the nature of the waste reaction products of this reaction, MgXX′, suggest that this route is poorly compatible with silicon device processing. Grignard reagents are known to derivatize the silicon surface via the cleavage of the weak Si–Si bond of $(SiH)_2$ dimers. Such a pathway is thus effective for *porous silicon* [86] and expectedly for the *monohydrogen-terminated 2 × 1 (1 0 0) surface of single crystalline silicon*, but does not seem suitable for prevailing (1 1 1) poly-silicon.[14]

Since there may be no process yet known for the controlled preparation of *hydroxyl-terminated silicon surfaces*, condensation and silanization of these surfaces seem difficult to implement too. (An additional reason for discarding condensation is due to the environmental weakness of the Si$-$O$-$C group, which undergoes hydrolysis in the presence of humidity.)

On the contrary, device quality, relatively stable, *hydrogen-terminated silicon surfaces* may be prepared by etching with an aqueous solution of HF (HF_{aq}) a sacrificial film of SiO_2 thermally grown on the surface [92]. Moreover, the quality of HF_{aq}-etched surfaces can greatly be improved by their exposure to H_2 at moderate (700 to 800°C [93]) or high (850 to 1100°C [94]) temperature. This suggests the use of such terminations for the functionalization. Highly homogeneous hydrogen-terminated surfaces are especially useful if the reprogrammable molecules contain polar or ionic groups (as it happens for bipyridinium and its derivatives) that could interact strongly with polar surface sites like silanols.

Of the considered processes, the ones involving the diazonium salts and chlorosilanes seem of difficult implementation. Integration issues, indeed, locate the derivatization in the back-end stage of the process [33], so that the waste reaction products (HCl and BF_3) could react, even vivaciously, with the circuit metallizations.

---

[14]Actually, even the hydrogen-terminated (1 1 1) surface, 1 × 1 (1 1 1)SiH, has been known to react in the absence of any catalyst with Grignard reagents with the formation of Si–C bonds [91], but the actual pathway likely involves first a surface substitution of halogen X for H due to XR contamination of the Grignard reagent, (1 1 1)SiH + XR $\longrightarrow$ (1 1 1)SiX + HR, followed by the reaction of the resulting surface with XMgR to produce silicon derivatization, (1 1 1)SiX + XMgR $\longrightarrow$ (1 1 1)SiR + MgX$_2$.

Commercial availability of precursors of functional molecules (larger for alkene-terminated molecules than for azanes) thus makes *hydrosilation of unsaturated hydrocarbons at the hydrogen-terminated silicon surface* the most serious candidate process for silicon functionalization.

This conclusion is further supported by an additional reason: while the silylation and arylation with diazonium salts of hydrogen-terminated surface occurs at room temperature, the hydrocarbon hydrosilation is effective at higher temperatures (say, $\simeq 170°C$) only. Though this difference may seem a disadvantage for hydrosilation, actually it is convenient because the functionalization via arylation or silanization would result in an immediate reaction at the periphery of the cross-point, thus inhibiting the entrance of other molecules in the cross-point region; rather, the derivatization via hydrosilation can be conducted in two steps: the first at lower temperatures (say, $< 100°C$) to allow the functional molecules to flow and fill the cross-point region, and the second at higher temperatures, (say, $\simeq 170\ °C$) to allow the grafting of the functional molecules to the poly-silicon surfaces.

Even though for the hypothesized application one should be mainly involved in the hydrosilation at hydrogen-terminated poly-silicon, the process is better understood at the surface of single crystalline silicon. Most studies have been carried out functionalizing the hydrogen-terminated (1 1 1) or (1 0 0) silicon surfaces with 1-alkenes [94–100] or 1-alkynes [101–107].

Having in mind the hydrosilation at poly-silicon wires, the best model surface is certainly the (1 1 1) surface. Actually, though this surface has been the subject of most studies, relatively little is known about the stability in air of the Si-C interface. Rather, this property was mainly considered for the (1 0 0) surface. Extended investigation based on x-ray photoemission spectroscopy (XPS) has demonstrated the possibility of derivatizing the hydrogen-terminated (1 0 0) Si surface via hydrosilation with 1-alkyne and the long-term environmental stability of the derivatized surfaces [102–107].

11.4.4   Crossbar Functionalization

The hydrogen-terminated crossbar structure is characterized by a fixed distance between upper and lower wires. The crossbar structure is eventually derivatized by its immersion at appropriate temperatures in a solution of an unreactive solvent containing the functional molecules. If they are suitably terminated (for instance, with terminal alkene groups) and their length is adapted to the inter-array distance, they are preferentially grafted to the silicon (hydrogen terminated, in the considered example) linking upper and lower wires.

Reprogrammable molecules able to mimic the structure and functioning of a flash memory cell can be hypothesized, as shown in the following example.

Consider a molecule connecting two heavily doped p-type poly-silicon electrodes. The molecule is formed by three moieties: a *wire* (for instance, a π-conjugated chain), covalently bonded at its extremes to the silicon electrodes; a side *redox center* R, whose state charge may be varied via electron transfer from, or to, the neighbouring electrodes; and an *arm*, linking materially the redox center to the wire. The arm is supposed to decouple the redox center from the wire, in such a way that the state of the latter is affected by the state of the former only electrostatically.

Figure 11.7 sketches a hypothetical molecule with the preceding characteristics.

The redox center admits two states: neutral R^0 or reduced R^-. The reduced state is more stable with respect to R^0 plus a free electron by a positive electron affinity A_{el}:

$$R^0 + e^- \longrightarrow R^- - A_{el}$$

The energy level $-A_{el}$ of R^- is supposed to be appreciably lower than the energy E^a_{LUMO} of the lowest unoccupied molecular orbital of the arm ($A_{el} > |E^a_{LUMO}|$). In this case, the reduced state of the redox center is metastable and may decay to the neutral state via electron tunnelling through the vacuum to the electrode (with energy barrier A_{el}) or through the arm to the wire and then to the electrode, with energy barrier $A_{el} - |E^a_{LUMO}|$. (The decay has been supposed not to be thermally activated. If the process were thermally activated, the lifetime should be given by $\tau_0^- \exp[(A_{el} - |E^a_{LUMO}|)/k_B T)]$, where τ_0^- is the oscillation period of the outer-electron in R^-).

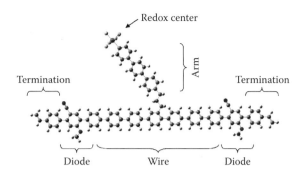

FIGURE 11.7   A candidate molecule for mimicking a flash memory cell, where the wire is formed by a π-conjugated poly-*para*phenylene chain, the arm is a rigid chain linked to the wire by a group discontinuing the π conjugation, and the redox centre, SbF_5, is a group with high electron affinity. The chain contains two groups (an acceptor CN and donor OCH_3 group on the same benzene ring) that allows it to behave as a rectifier.

In the absence of negative charge on the redox center, the wire is supposed to be an insulator, which means (for hole conduction) that the energy E_{HOMO}^{W} of the highest occupied molecular orbital of the wire is lower than the valence band edge E_{VB} ($|E_{VB}| < |E_{HOMO}^{W}|$). When a negative charge is injected into the redox center, however, its effect is to increase (for purely Coulombic interaction) the HOMO energy of the wire. If the arm length is short enough and the wire is suitably chosen, it may thus happen that E_{HOMO}^{W} will become higher than the valence band edge ($|E_{VB}| > |E_{HOMO}^{W}|$). This situation allows the injection of holes from the electrode into the wire and, in the ultimate analysis, brings it into a conductive state. Thus, according to the 0/− charge state of the redox center, the wire will be in different insulating/conductive states—this allows the use of the described molecular arrangement to code a bit and to sense it by the application of a weak electric field between the poly-silicon contacts. Since the charge distribution in the redox center is pointlike, it may happen that condition $|E_{VB}| > |E_{HOMO}^{W}|$ is satisfied only in the region of the wire closest to the redox center. In this case, the conduction mechanism will change from direct tunneling to superexchange-mediated tunneling, thus producing an increase of conductance by probably several orders of magnitude.

It remains to be explained how the redox center may be modified. Assuming the neutral state as the equilibrium one, the reduced state may be obtained via the application between the contacts of a potential difference so strong as to allow an avalanche injection of electrons into the region containing R^0. Once an electron is captured by the center, the current will then flow mainly through the wire, allowing an external circuitry to switch off the writing potential. Erasing the R^- state may be achieved via tunneling of the trapped electron into the writing electrode, sustained by the application of a large as reverse potential difference between the contacts (sufficiently large to increase the tunneling rate to a poly-silicon conduction band but not not so as large as to produce an avalanche injection). This can be achieved without an unwanted flow of current through the wire via the insertion along it of a rectifying moiety (like the one hypothesized by Aviram and Ratner [56]).

The reason for considering hole conduction through the wire (rather than electron conduction, thus destroying the complete analogy with conventional flash memories) is that transition $R^0 + e^- \longrightarrow R^-$ usually involves an energy in the electronvolt energy scale, while transition $R^0 \longrightarrow R^+ + e^-$ involves an energy in the energy scale of 10 eV.

Figure 11.8 shows the electronic configurations of the molecule in the considered states.

Let 0 and 1 denote the neutral and negative states, respectively, of the redox center. As discussed earlier, the 0 state imparts to the wire a low conductance, while the 1 state imparts a high conductance, allowing determination of the charge state of the redox center by a measurement of the wire conductance under static or quasi-static conditions— a quantity which seems accessible to current apparatuses.

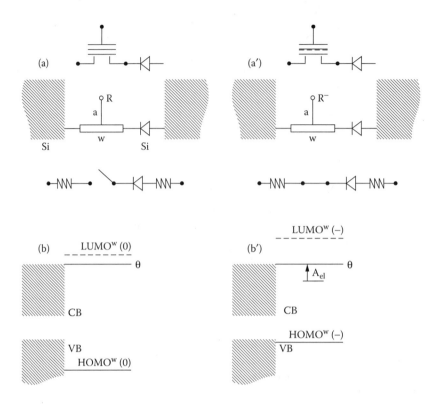

FIGURE 11.8 Energy configurations for the molecule in the neutral and charged states, and their analogy with the flash memory cell.

If the molecule admits two electronic states characterized by largely different conductance (as for the HP–UCLA collaboration) the whole structure may be viewed as a hybrid molecular IC with achievable bit density of the order of 10^{11} cm^{-2}.

11.5 How the Hybrid Technology Can Attack Physical Problems—Beyond Molecular Electronics

The analysis of Section 11.4.1 has led to identify in the hybrid approach a suitable route to nanoelectronics. The hybrid device is formed by the following parts: a conventional, silicon-based, circuitry; a non-lithographic, silicon-based, crossbar; and reprogrammable organic molecules providing the function of a nonvolatile memory cell. This structure offers two advantages: the problem of nanotechnology (definition of wires and their demultiplexing, left to the silicon technology) is decoupled from that of the design and production of the nonvolatile memory cell (left to organic chemistry), and the actuation and sensing of individual molecules is possible with macroscopically accessible nanoscopic apparatuses rather with macroscopic apparatuses (like scanning tunneling microscopes).

However, the hybrid approach requires the insertion of organic molecules in regions of size adapted to the molecule length. Needless to say, this is a difficult task, and is unavoidable if the reprogrammable elements are organic in nature. The substitution of inorganic reprogrammable materials for the organic molecules would allow easier processing (with the deposition of the inorganic functional material *before* the preparation of the second-floor array) and, likely, better reliability. Since that substitution seems possible in principle, it is expected that in the long run, nanoelectronics will no longer be hybrid but rather fully inorganic.

Along this line, one might argue about the usefulness of developing a hybrid technology. In the author's view, such a technology deserves to be developed not only for its usefulness in nanoelectronics but also for its potential regarding the manipulation and sensing of single molecules. In fact, it seems relatively easy to prepare crossbar structures, embedding in each cross-point a number of molecules of the order of unity. This situation would allow the testing, at the scale of one molecule and with a minimum of perturbation, of fundamental properties otherwise known only via collective behavior [109].

To be concrete and show how the accessibility of single molecules to the macroscopic world may bring a new sap to the body of basic science, let's consider three examples, related to (i) the effect of repeated measurements, (ii) the appearance of interference phenomena in electronic conduction, and (iii) the possibility of using Brownian motion as the engine for charge separation under a static electric potential.

11.5.1 Dynamic Memories and the Quantum Zeno Effect

The first case to consider is similar to that sketched in Figures 11.1 and 11.2 but characterized by a spatial separation between states $|0\rangle$ and $|1\rangle$ smaller than the one consistent with the appearance of the classical behavior.

In that case, decoherence phenomena are unable to produce a stable localization so that the information stored in the quantum state ($|0\rangle$ or $|1\rangle$) is gradually lost unless the datum is refreshed (i.e., projected onto the original position eigenvector) within a time much shorter than the spreading time of the wavefunction. According to a fundamental principle of quantum mechanics (the quantum Zeno effect [109–111] the refresh is achieved by repeated measurements of the state of the system with a period much shorter than $\hbar/2m\Delta q_*^2$. Since the measurement of state $|1\rangle$ is nothing but the passage of a current I satisfying condition (11.22), this implies that a system prepared in state $|1\rangle$ persists indefinitely in that state, provided that a current higher than $e\hbar/2m\Delta q_*^2$ flows through the detector.

The preceding description, however, is formal and advocates the unproved property that the setup sketched in Figure 11.2, an apparatus for the measurement of a *classical, though microscopic, state*, is actually a measuring apparatus for the quantum state, too. A dynamic, rather than axiomatic, analysis of the situation previously considered, suggests that the electron quantum state $|1\rangle$ may be stabilized by its Coulombic interactions with the holes flowing through the sensing resistance. That the seldom and erratically acting Coulombic interactions really succeed in refreshing the position eigenstate, however, is extremely difficult to demonstrate, but can be attacked on experimental grounds. In fact, if the considered arrangement is a measuring apparatus for quantum mechanical states, a current of only $10\mu A$ should be sufficient to stabilize the information. In that case the information could be stored even at densities higher than that allowed by the classical limit, thus allowing a kind of dynamic memory.

Actually, the current allowing for localization is too high for the molecular systems currently produced (a resistance of $10^9 \, \Omega$ would require a voltage of about 10^4 V applied over a region of few nanometers!); a molecule with the minimum resistance, the Landauer limit $3 \times 10^4 \, \Omega$, would instead require a voltage of 0.3 V, low enough for not producing any breakdown. The availability of molecular wires with resistance very close to the Landauer limit would thus allow study of how a classical measuring apparatus may behave as a quantum measuring apparatus.[15]

11.5.2 Interference Among Molecular Wires

Exploring the idea of a practical exploitation of molecules in electronic devices, let's consider a collection of them, assuming a linear superposition of their observable effects:

$$\text{conductance} = \text{the sum of molecular conductances.} \tag{11.39}$$

This condition, however, ignores the interference phenomena in the conduction along molecular wires.

---

[15]This situation is the counterpart of the textbook problem of studying the conditions, specified by Ehrenfest's theorem, which allows a quantum system to be described in classical terms.

FIGURE 11.9   Two-dimensional sketch of a three-wire molecule possibly displaying interference effects.

Consider an electron impacting a surface with n funnels through which the particle may move from one electrode to the other. As far as the motion is truly ballistic, the particle wavefunction—supposedly localized at one funnel—undergoes a lateral spreading to the surrounding channels, thus giving rise to interference phenomena and violating the assumption (11.39) of the superposition of classical effects. Of course, this consideration applies only to motions preserving the phase relationships (ballistic motion); it does not apply to situations that destroy those relationships (molecular resistance controlled by erratic scattering). The amount of the interference effects depends critically on the mutual distances between pairs of wires. In view of the statistical nature of the grafting process of the molecule to the surface (and of the defects it contains) the overall effect is expected to be negligible.

A regular distribution of wire distance can only be achieved *within* a unique molecule, (for instance, arranging therein two or more equivalent moieties departing from a unique atom and arriving to another unique atom). Figure 11.9 shows an example of such an arrangement.

The chemical equivalence of the various wires does not necessarily imply their conformational equivalence, however. A possible, but not unique, reason for their non-equivalence is due to the thermal reservoir, which modifies dynamically the conformation of the various wires, thus partially destroying the interference effects. The maximum of such effects is achieved using rigid wires (well represented in practice by π-conjugated moieties) and operating at the lowest possible temperature. The theoretical description of this situation (given, for instance, by Magoga and Joachim), predicts that they combine in a nonlinear fashion [113]. The experimental verification of this prediction is, however, still far away.

Limit now the attention to a system composed by two wires: their equivalence is destroyed by the presence on one of them of a charged site (like in the molecule sketched in Figure 11.7) whose effect on the wire is controlled by the potential difference applied to the contacts. This arrangement, sketched in Figure 11.10, is somewhat similar to the one hypothesized by Aharonov and Bohm for predicting their celebrated effect [114]. The availability of many-wire molecules and nanometre-sized contacts in such an arrangement would thus allow a study of the Aharonov–Bohm effect [115] on a length scale otherwise inaccessible.

11.5.3   A Molecular Van de Graaff Machine Exploiting the Brownian Motion

Among the molecular systems considered for their ability to mimic the behaviour of macroscopic machines, rotaxanes play a central role. A rotaxane (A, R) is a supramolecular system formed by a cyclic molecule R ("ring") threaded on a linear molecule A ("axle") with bulky caps on the heads to prevent the ring from falling off. Chemically speaking, the ring is usually a rigid, nearly flat, arene (a famous example being the "paraquat"—two doubly charged bipyridinium units linked by two benzene rings, with the net

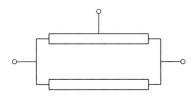

FIGURE 11.10   Two-dimensional sketch of a molecule able to display the Aharonov–Bohm effect.

positive charge balanced by four PF_6^- anions), whereas the axle is formed by n sites (for instance, benzene units) typically connected by chains of ethoxy groups. See [115–118] for topochemistry, [120] for supramolecular chemistry, [121,122] for molecular machines, and [37,50,57,58] for applications in electronics.

Up to now, rotaxanes have mainly (or perhaps uniquely) been hypothesized or used because of the possibility of functionalizing the axle via the insertion of two or more positions where the ring is preferentially hosted [37]. However, there is another possibility that has likely not been considered yet, but that could be exploited: the functionalization of the ring. Assume that

- The axle A is an unstructured, rigid, nearly linear, insulating wire connecting two electrodes C_1 and C_f.
- C_f is floating and geometrically arranged between C_1 and another electrode C_2.
- A potential difference may be applied between C_1 and C_2 so that C_f attains an intermediate potential.
- The electron affinity of the ring is intermediate between the Fermi energy of C_1 and C_f, so that R spontaneously ionizes in close proximity to C_1, whereas the negatively charged state R^- spontaneously undergoes oxidation, ceding its extra electron to C_f when in its close proximity.
- No other reaction is thermodynamically allowed.

An idea for obtaining a redox centre with the required features is to insert covalently in the ring a transition metal M admitting two oxidation states $M^{(n)}$ and $M^{(n+1)}$, where M is covalently bonded to n and $n + 1$ atoms, respectively. When in oxidation state $(n + 1)$, the ring R is globally neutral and all valencies of M are fully saturated; in oxidation state (n), the ring is globally negatively charged and the unsaturated valencies of M are stabilized by suitable ligands possibly undergoing conformational changes after the redox reaction.

Assume that at time $t = 0$ the ring is neutral and in an intermediate position between C_1 and C_f. Since R is in the neutral state, it does not feel the electric field applied between the electrodes and is, therefore, subjected only to a Brownian motion due its erratic collisions against the gas-phase molecules. The time required by the ring to dock either C_1 or C_f is of the order of $(\Delta x)^2/8D$, where Δx is the distance separating C_1 from C_f and D is the diffusion coefficient of the ring along the axle; let τ_{s1} and τ_{sf} be the sojourn times of R in close vicinity of electrodes C_1 and C_f, respectively (tentatively, $\tau_{s1} \approx \tau_{sf}$). Consider first the docking on C_f. The thermodynamics of the system inhibits any reaction so that the particle will leave C_f, eventually diffusing to C_1 (after a time, on the order of $(\Delta x)^2/2$). Let $\tau_{0/-1}$ and $\tau_{-1/0}$ be the mean times required for the ionization and neutralization of the ring:

$$\tau_{0/-1} : R + e^- \xrightarrow{C_1} R^-,$$
$$\tau_{-1/0} : \quad R^- \xrightarrow{C_f} R + e^-.$$

Once in contact with C_1, if the ionization of R is a Poisson process, the probability of its occurrence is given by $1 - \exp\left(\tau_{s1}/\tau_{0/1}\right)$. If there is no electron transfer from C_1 within the sojourn time, the system will evolve by a sequence involving the desorption of the neutral ring and its docking to the electrode until, after the repetition of $\left[1 - \exp\left(\tau_{s1}/\tau_{0/1}\right)\right]^{-1}$ cycles on the mean, the ring will undergo ionization. This will produce a coupling of the ring with the applied electric field and its drift with average velocity $\overline{v}$ from C_1 to C_f. If the mean free path in the medium is smaller than Δx, $\overline{v}$ is controlled by the mobility μ in the medium: $\overline{v} = \mu \mathcal{E}$, with μ given by the Einstein relation $\mu = (k_B T/e)D$. Once in contact with C_f, the ring will stay thereon (stabilized by the electric field) until it undergoes spontaneous neutralization (in a time $\tau_{-1/0}$).

Once the ring regains its neutral state, the initial condition is restored, thus allowing the repetition of the cycle, as sketched in Figure 11.12. The cycle can be repeated until the increase of Fermi energy, produced by the accumulated electrons, thermodynamically disallows further electron transfer to the floating electrode. The electron transfer, indeed, increases the voltage (and thus Fermi energy) of the floating electrode by an amount e/C_{C_f}, where C_{C_f} is its capacity.

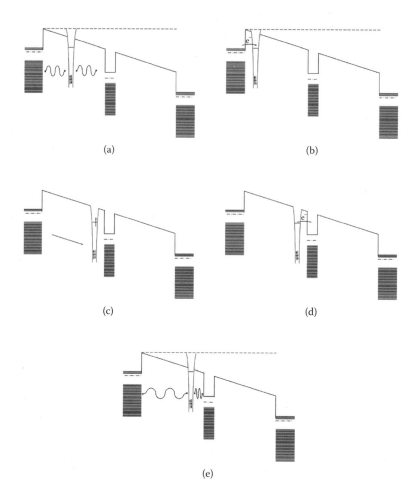

FIGURE 11.11   Various states of the nano Van de Graaff (from top to bottom): (a) the neutral ring in Brownian motion does not feel the applied electric field; (b) once in the vicinity of contact C_1, the ring undergoes ionization; (c) the ionized ring feels the applied electric field and shifts to electrode C_f; (d) the ionized ring is kept by the electric field in the vicinity of electrode C_f until it is neutralized, ceding its extra electrode to the floating electrode; (e) the neutral ring does not feel any longer the applied field and is subjected to the Brownian motion only.

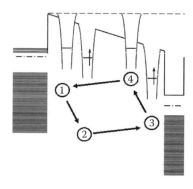

FIGURE 11.12   The cycle allowing the rotaxane to behave as a van de Graaff generator.

The described device is reminiscent of the ratchet-and-pawl machine sketched by Feynman in his famous *Lectures* [123], the pawl being provided by the ionization and by the applied electric field. Rather than trying to use this device for extracting energy from the medium, the device is operated as a nanoscopic Van de Graaff generator, where the drift of the electron carrier is provided by the applied electric field while its return to the low-voltage electrode is provided by Brownian motion.

What is of uppermost interest in this arrangement is the fact that many of the quantities that affect the observable increase of voltage of C_f are experimentally accessible (for instance: Δx, by choosing molecules with different axle lengths; $\mathcal{E}$, by applying different voltages; D, by varying the gas pressure in the medium), thus allowing a study of the involved phenomena (electron transfer, Brownian motion, drift) at the scale of a few (or possibly one) molecules.

11.6 Conclusions

Planar arrangements allowing bit density of the order of 0.1 Tbit cm^{-2} can be achieved by exploiting electrically reconfigurable molecules in concert with a silicon-based circuitries for supplying power, addressing, and sensing.

Hybrid architectures allowing the large-scale production of ICs on the 0.1 Tbit integration scale have been proposed. Solutions have been found for each of the putative factors (the definition of cross-points, demultiplexing of lines, choice of functional molecules, and surface derivatization) that are expected to limit the development of such an architecture. All of them require the adaptation of already existing technologies to the proposed architecture.

Of course, this is only a minor step in the road going from GSI to TSI, not only because almost all technological details that are really necessary to the goal have been ignored, but also because the complexity of the arrangement impacting on memory architecture (fault tolerance, intrinsic statistical behavior, management of the information, and certainly others still to be identified) have not been considered yet.

Identification of the scaling possibilities of the silicon technology is nonetheless a fundamental step to address the research and eventually to allow it to face the TSI era. Moreover, the setup of a technology for molecular electronics will allow phenomena like the quantum Zeno effect or the Aharonov–Bohm effect, currently studied on a length scale that requires the use of macroscopic quantum states, to be studied on a length scale (the molecular one) where they are expected to occur spontaneously.

References

[1] Cerofolini, G.F. et al., Substrates: The material bases of microelectronics and nanoelectronics, *Progr. Quant. Electr.*, 17, 273, 1993.

[2] Vadasz, L.L. et al., Silicon-gate technology, *IEEE Spectrum*, 6, 28, 1969.

[3] Appels, J.A. and Paffen, M.M., Local oxidation of silicon; new technological aspects, *Philips Res. Repts.*, 6, 157, 1971.

[4] Kooi, E. et al., LOCOS devices, *Philips Res. Repts.*, 26, 166, 1971.

[5] Hunter, W.R. et al., New edge-defined approach for submicrometer MOSFET fabrication, *IEEE Electron Device Lett.*, EDL 2, 4, 1981.

[6] Plummer, J.D. and Griffin, P.B., Material and process limits in silicon VLSI technology, *Proc. IEEE*, 89, 240, 2001.

[7] Kingon, A.I. et al., Alternative dielectrics to silicon dioxide for memory and logic devices, *Nature*, 406, 1032, 2000.

[8] Bohr, M.T., Interconnect scaling—the real limiter to high performance ULSI, *Solid St. Technol.*, 39, 9, 105, 1996.

[9] Keyes, R.W., Electronics in large systems, in *Molecular Electronics and Molecular Electronic Devices*, Sienicki, K., Ed., CRC Press, Boca Raton, FL, 1993, chap. 1.

[10] Murarka, S.P., Multilevel interconnections for ULSI and GSI era, *Mater. Sci. Eng. R*, 19, 87, 1997.

[11] Homma, T., Low dielectric constant materials and methods for interlayer dielectric films in ultralarge-scale integrated circuit multilevel interconnections, *Mater. Sci. Eng. R*, 23, 243, 1998.

[12] Cerofolini, G.F., Strategies for low-κ dielectrics, *Mater. Sci. Semicond. Process*, 5, 265, 2003.

[13] Brewer, J.E. et al., Memory technology for the post CMOS era, *IEEE Circuits & Devices Mag.*, 21, 2, 13, 2005.

[14] Semiconductor Industry Association (SIA), *International Technology Roadmap for Semiconductors*, 2005 ed.; available at http://public.itrs.net.

[15] Chiabrera, A. et al., Physical limits of integration and information processing in molecular systems, *J. Phys. D: Appl. Phys.*, 22, 1571, 1989.

[16] Packan, P.A., Pushing the limits, *Science*, 285, 2079, 1999.

[17] Peercy, P.S., The drive to miniaturization, *Nature*, 406, 1023, 2000.

[18] Ito, T. and Okazaki, S., Pushing the limits of lithography, *Nature*, 406, 1027, 2000.

[19] Keyes, R.W., Fundamental limits of silicon technology, *Proc. IEEE*, 89, 227, 2001.

[20] Frank, D.J. et al., Device scaling limits of Si MOSFETs and their application dependencies, *Proc. IEEE*, 89, 259, 2001.

[21] Harriott, L.R., Limits of lithography, *Proc. IEEE*, 89, 366, 2001.

[22] Gramß, T. et al., *Non-Standard Computation*, Wiley-VCH, Weinheim, 1998.

[23] Bennett, C.H., The thermodynamics of computation—a review, 21, 905, 1982.

[24] Zurek, W.H., Thermodynamic cost of computation, algorithmic complexity and the information metric, *Nature*, 341, 119, 1989.

[25] Landauer, R.W., Irreversibility and heat generation in the computing process, *IBM J. Res. Develop.*, 5, 183, 1961.

[26] Feynman, R.P., *Lectures on Computation*, Hey, J.G. and Allen, R.W., Eds., Addison-Wesley, Reading, MA, 1995, chap. 5.

[27] Aharonov, Y. and Bohm, D., Time in quantum theory and the uncertainty relation for the time and energy domain, *Phys. Rev.*, 122, 1649, 1961.

[28] Robertson, H.P., The uncertainty principle, *Phys. Rev.*, 34, 163, 1929.

[29] Joos, E. and Zeh, H. D., The emergence of classical properties through interaction with the environment, *Z. Phys. B - Condensed Matter*, 59, 223, 1985.

[30] O'Connell, R.F., Decoherence in quantum systems, *IEEE Trans. Nanotechnol.*, 4, 77, 2005.

[31] Natelson, D. et al., Fabrication of extremely narrow metal wires, *Appl. Phys. Lett.*, 77, 1991, 2000.

[32] Melosh, N.A et al., Ultrahigh density nanowire lattices and circuits, *Science*, 300, 112, 2003.

[33] Cerofolini, G.F. et al., Strategies for nanoelectronics, *Microelectr. Eng.*, 81, 405, 2005.

[34] Cerofolini, G.F. et al., A hybrid route for molecular electronics, *Nanotechnology*, 16, 1040, 2005.

[35] Cerofolini, G.F., An extension of microelectronic technology to nanoelectronics, *Nanotechnol. E-Newslett.*, 7, 3, 5, 2005.

[36] Cerofolini, G.F. and Mascolo, D., A hybrid micro-nano-molecular route for nonvolatile memories, *Semicond. Sci. Technol.*, 21, 1315, 2006.

[37] Mendes, P.M. et al., Nanoelectronic devices from self-organized molecular switches, *Appl. Phys. A*, 80, 1197, 2005.

[38] Keyes, R.W., Physical limits in semiconductor electronics, *Science*, 195, 1230, 1977.

[39] Meindl, J.D. and Davis, J.A., The fundamental limit on binary switching energy for terascale integration (TSI), *IEEE J. Solid-St. Circuits*, 35, 1515, 2000.

[40] Davis, J.A. et al., Interconnect limits on gigascale integration (GSI) in the 21st century, *Proc. IEEE*, 89, 305, 2001.

[41] Zhirnov, V.V. et al., Limits to binary logic switch scaling—a *gedanken* model, *Proc. IEEE*, 91, 1934, 2003.

[42] Bennett, C.H., Logical reversibility of computation, *IBM J. Res. Develop.*, 17, 525, 1973.

[43] Fredkin, E. and Toffoli, T., Conservative logic, *Int. J. Theor. Phys.*, 21, 219, 1982.

[44] Likharev, K.K., Classical and quantum limitations on energy consumption in computation, *Int. J. Theor. Phys.*, 21, 311, 1982.

[45] Likharev, K.K. and Korotkov, A.N., Reversible computation in a system with discrete eigenstates, *Science*, 273, 763, 1996.

[46] McCreery, R.L., Molecular electronic junctions, *Chem. Mater.*, 16, 4477, 2004.

[47] Bez, R. and Cappelletti, P., Flash memory and beyond, *IEEE-VLSI-TSA Symposium on VLSI Technology*, 84, 2005.

[48] Heath, J.R. et al., A defect-tolerant computer architecture: Opportunities for nanotechnology, *Science*, 280, 1716, 1998.

[49] Luo, Y. et al., Electronically configurable molecular-based logic gates, *Chem. Phys. Chem.*, 3, 519, 2002.

[50] Tour, J.M., Molecular electronics. Synthesis and testing of components, *Acc. Chem. Res.*, 33, 791, 2000.

[51] Cerofolini, G.F. and Ferla, G., Toward a hybrid micro-nanoelectronics, *J. Nanoparticle Res.*, 4, 185, 2002.

[52] Cerofolini, G.F., Nano-mesoscopic interface: hybrid devices, in *Encyclopedia of Nanoscience and Nanotechnology*, Schwarz, Ed., Dekker, New York, 2004.

[53] Cerofolini, G.F. et al., The hydrosilation of 1-alkene and 1-alkyne at terraced, dihydrogen terminated, 1×1 (100) silicon, in *Trends in Semiconductor Research*, Elliot, T. B., Ed., Nova Science, Hauppauge, NY, 2005.

[54] Howell, S.W. et al., Mass-fabricated one-dimensional silicon nanogaps for hybrid organic/nanoparticle arrays, *Nanotechnology* 16, 754, 2005.

[55] Stan, M.R. et al., Molecular electronics: From devices and interconnect to circuits and architecture, *Proc. IEEE*, 91, 1940, 2003.

[56] Aviram, A. and Ratner, M., Molecular rectifiers, *Chem. Phys. Lett.*, 29, 277, 1974.

[57] Reed, M.A., Molecular-scale electronics, *Proc. IEEE*, 87, 652, 1999.

[58] Joachim, C. et al., Electronics using hybrid-molecular and mono-molecular devices, *Nature*, 408, 541, 2000.

[59] Joachim, C. and Ratner, M.A., Molecular electronics: Some views on transport junctions and beyond, *Proc. Natl. Acad. Sci.*, 102, 8801, 2005.

[60] Elbing, M. et al., A single-molecule diode, *Proc. Natl. Acad. Sci.*, 102, 8815, 2005.

[61] Ashwell, G.J. et al., Rectifying characteristics of $Mg|(C_{16}H_{33}$-Q3CNQ LB film)$|Pt$ structures, *J. Chem. Soc. Chem. Commun.*, 1374, 1990.

[62] Huang, Y. et al., Logic gates and computation from assembled nanowire building blocks, *Science*, 294, 1313, 2001.

[63] Zhong, Z. et al., Nanowire crossbar arrays as address decoders for integrated nanosystems, *Science*, 302, 1377, 2003.

[64] Ziegler, M.M. and Stan, M.R., CMOS/nano co-design for crossbar-based molecular electronic systems, *IEEE Trans. Nanotechnol.*, 2, 217, 2003.

[65] DeHon, A., Array-based architecture for FET-based, nanoscale electronics, *IEEE Trans. Nanotechnol.*, 2, 23, 2003.

[66] Forshaw, M. et al., A short review of nanoelectronic architectures, *Nanotechnology*, 15, S220, 2004.

[67] Beckman, R. et al., Bridging dimensions: demultiplexing ultrahigh-density nanowire circuits, *Science*, 310, 465, 2005.

[68] Wu, W. et al., One-kilobit cross-bar molecular memory circuits at 30-nm half-pitch fabricated by nanoimprint lithography, *Appl. Phys. A*, 80, 1173, 2005.

[69] Likharev, K. and Strukov, D.B., CMOL: A silicon-based bottom-up approach to nanoelectronics, in *Introducing Molecular Electronics*, Cuniberti, G., Fagas, G., and Richter, K., Eds., Springer, Berlin, 2005, ch. 16.

[70] Guo, L.J., Recent progress in nanoimprint technology and its applications, *J. Phys. D: Appl. Phys.*, 37, R123, 2004.

[71] Choi, Y.-K. et al., A spacer patterning technology for nanoscale CMOS, *IEEE Trans. Electron Devices*, 49, 436, 2002.

[72] Choi, Y.-K. et al., Fabrication of sub-10-nm silicon nanowire arrays by size reduction lithography, *J. Phys. Chem. B*, 107, 3340, 2003.

[73] Service, R.F., Next-generation technology hits an early midlife crisis, *Science*, 302, 556, 2003.

[74] Stewart, D.R. et al., Molecule-independent electrical switching in Pt/organic monolayer/Ti devices, *Nano Lett.*, 4, 133, 2004.

[75] Lau, C.N. et al., Direct observation of nanoscale switching centers in metal/molecule/metal structures, *Nano Lett.*, 4, 569, 2004.

[76] Snider, G. et al., Nanoelectronic architectures, *Appl. Phys. A*, 80, 1183, 2005.

[77] Seevinck, E. et al., Current-mode techniques for high-speed VLSI circuits with application to current sense amplifier for CMOS SRAM's, *IEEE J. Solid St. Circuits*, 26, 525, 1991.

[78] Torelli, G. et al., Charge-and-split sense amplifier for multilevel nonvolatile memories, *Electronics Lett.*, 35, 796, 1999.

[79] Gittins, D.I. et al., A nanometre-scale electronic switch consisting of a metal cluster and redox-addressable groups, *Nature*, 408, 67, 2000.

[80] Chen, J. et al., Observation of a large on-off ratio and negative differential resistance in an electronic molecular switch, *Science*, 286, 1550, 1999.

[81] Galperin, M. et al., Hysteresis, switching, and negative differential resistance in molecular junctions: A polaron model, *Nano Lett.*, 5, 125, 2005.

[82] Blum, A.S. et al., Molecularly inherent voltage-controlled conductance switching, *Nature Mater.*, 4, 167, 2005.

[83] Bent, S.F., Organic functionalization of group IV semiconductor surfaces: Principles, examples, applications, and prospects, *Surf. Sci.*, 500, 879, 2002.

[84] Buriak, J.M., Organometallic chemistry on silicon and germanium surfaces, *Chem. Rev.*, 102, 1271, 2002.

[85] Bansal, A. et al., Alkylation of Si surfaces using a two-step halogenation/Grignard route, *J. Am. Chem. Soc.*, 118, 7225, 1996.

[86] Kim, N.Y. and Laibinis, P.E., Derivatization of porous silicon by Grignard reagents at room temperature, *J. Am. Chem. Soc.*, 120, 4516, 1998.

[87] Cleland, G. et al., Direct functionalization of silicon *via* the self-assembly of alcohols, *J. Chem. Soc. Faraday Trans.*, 91, 4001, 1995.

[88] Ulman, A., Self-assembled monolayers of alkyltrichlorosilanes: Building block for future organic materials, *Adv. Mater.*, 2, 573, 1990.

[89] Roth, C.A., Silylation of organic chemicals, *Ind. Eng. Chem. Prod. Res. Develop.*, 11, 134, 1972.

[90] Stewart, M.P. et al., Direct covalent grafting of conjugated molecules onto Si, GaAs, and Pd surfaces from aryldiazonium salts, *J. Am. Chem. Soc.*, 126, 370, 2004.

[91] Fellah, S. et al., Hidden electrochemistry in the thermal grafting of silicon surfaces from Grignard reagents, *Langmuir*, 20, 6359, 2004.

[92] Ubara, H. et al., Formation of Si–H bonds on the surface of microcrystalline silicon covered with SiO_x by HF treatment, *Solid St. Comm.*, 50, 673, 1984.

[93] Aoyama, T. et al., Silicon (0 0 1) surface after annealing in hydrogen ambient, *J. Vac. Sci. Technol. A*, 14, 2909, 1996.

[94] Cerofolini, G.F. et al., The formation of terraced, nearly flat, hydrogen terminated, (1 0 0) Si surfaces after high-temperature treatments in H_2 of single crystalline silicon, *Phys. Rev. B*, 72, 125431, 2005.

[95] Hovis, J.S. and Hamers, R.J., Structure and bonding of ordered organic monolayers of 1,5-cyclooctadiene on the silicon(1 0 0) surface, *J. Phys. Chem. B*, 101, 9581, 1997.

[96] Liu, H. and Hamers, R.J., An x-ray photoelectron spectroscopy study of the bonding of unsaturated organic molecules to the Si(1 0 0) surface, *Surf. Sci.*, 416, 354, 1998.

[97] Terry, J. et al., Electronic structure of alkyl monolayers on Si(1 1 1), *J. Appl. Phys.*, 85, 213, 1999.

[98] Lehner, A. et al., Hydrosilylation of crystalline silicon (1 1 1) and hydrogenated amorphous silicon surfaces: A comparative x-ray photoelectron spectroscopy study, *J. Appl. Phys.*, 94, 2289, 2003.

[99] Sieval, A.B. et al., Highly stable Si–C linked functionalized monolayers on the silicon (1 0 0) surface, *Langmuir*, 14, 1759, 1998.

[100] Sieval, A.B. et al., An improved method for the preparation of organic monolayers of 1-alkenes on hydrogen-terminated silicon surfaces, *Langmuir*, 15, 8288, 1999.

[101] Kosuri, M. et al., Vapor-phase adsorption kinetics of 1-decene on H-terminated Si(1 0 0), *Langmuir*, 19, 9315, 2003.

[102] Sieval, A.B. et al., Monolayers of 1-alkynes on the H-terminated Si(1 0 0) surface, *Langmuir*, 16, 10359, 2000.

[103] Cerofolini, G.F. et al., Functionalization of the (1 0 0) surface of hydrogen-terminated silicon via hydrosilation of 1-alkyne, *Mater. Sci. Eng. C*, 23, 253, 2003.

[104] Cerofolini, G.F. et al., The addition of organic functional groups to silicon via hydrosilation of 1-alkynes to hydrogen-terminated, 1×1 reconstructed (1 0 0) silicon surfaces, *Semicond. Sci. Technol*, 18 423, 2003.

[105] Cerofolini, G.F. et al., X-ray photoemission spectroscopy study at different takeoff angles of hydrosilation of 1-alkynes at hydrogen-terminated 1×1-reconstructed (1 0 0)-oriented silicon, *Mater. Sci. Eng. C*, 23, 989, 2003.

[106] Cerofolini, G.F. et al., Grafting of 1-alkynes to hydrogen-terminated (1 0 0) silicon surfaces, *Appl. Phys. A*, 80, 161, 2004.

[107] ———— Hydrosilation of 1-alkyne at nearly flat, terraced, homogeneously hydrogen-terminated (1 0 0) silicon surfaces, *Surf. Interface Anal*, 36, 71, 2004.

[108] Cerofolini, G.F. et al., Functionalization of atomically flat, dihydrogen terminated, 1×1 (100) silicon via reaction with 1-alkyne, *Appl. Surf. Sci.*, 246, 52, 2005.

[109] Clemente-León, M. et al., Towards organization of molecular machines at interfaces: Langmuir films and Langmuir–Blodgett multilayers of an acid-base switchable rotaxane, *Adv. Mater.*, 18, 1291, 2006.

[110] Misra, B. and Sundarshan, E.G., The Zeno's paradox in quantum theory, *J. Math. Phys.*, 18, 756, 1977.

[111] Peres, A., Zeno paradox in quantum theory, *Am. J. Phys.*, 48, 931, 1980.

[112] Itano, W.M. et al., Quantum Zeno effect, *Phys. Rev. A*, 41, 2295, 1990.

[113] Magoga, M. and Joachim, C., Conductance of molecular wires connected or bonded in parallel, *Phys. Rev. B*, 59, 16011, 1999.

[114] Aharonov, Y. and Bohm, D., Significance of electromagnetic potentials in the quantum theory, *Phys. Rev.*, 115, 485, 1959.

[115] Hamilton, J., *Aharonov–Bohm and other Cyclic Phenomena*, Springer, Berlin–Heidelberg, 1997.

[116] van Gulick, N., Theoretical aspects of the linked ring problem, *New J. Chem.*, 17, 619, 1993; originally presented at the *Reaction Mechanisms Conference* held in Princeton, NJ, 1960.

[117] Frisch, H.L. and Wasserman, E., Chemical topology, *J. Am. Chem. Soc.*, 83, 3789, 1961.

[118] Walba, D.M., Topological chemistry, *Tetrahedron*, 41, 3161, 1985.

[119] Sauvage, J.-P., Interlacing molecular threads on transition metals: Catenands, catenates, and knots, *Acc. Chem. Res.*, 23, 319, 1990.

[120] Lehn, J.M. and Ball, P., Supramolecular chemistry, in *The New Chemistry*, Hall, N., Ed., Cambridge University Press, Cambridge, 2000, ch. 12.

[121] Credi, A. et al., Molecular logic. An XOR gate based on a mechanical molecular machine, *J. Am. Chem. Soc.*, 119, 2679, 1997.

[122] Balzani, V. et al., Molecular machines, *Acc. Chem. Res.*, 31, 405, 1998.

[123] Feynman, R.P. et al., *The Feynman Lectures on Physics*, Addison-Wesley, Reading, MA, 1989, ch. 46.

12

On Computing Nano-Architectures Using Unreliable Nano-Devices

Valeriu Beiu

Walid Ibrahim

Abstract

This chapter will start with a brief review of nanoelectronic challenges while focusing on the reliability challenge. One of the most recent call-to-arms [1] raises two fundamental questions: *what is the meaning of reliability for systems that use nano or new technologies, and how do we interpret this meaning in practice?* The chapter has a hands-on approach for implementing computations (logic). The reader will be walked through a series of simple CMOS-based examples starting from the device (e.g., transistor) level and moving up to the gate, the circuit, the block, and only pointing towards the system level. Reliable memory design will not be dealt with in this chapter, as viable solutions (based on detection and errors correcting codes, stand-by spare rows and columns, reconfiguration, etc.) have long been used in industry, and are well known (the only question being how much will these be needed). The choice of CMOS for most of the examples is due to the broad design base available, but the ideas presented here can be translated to other nanotechnologies (as a few single electron technology set examples will show). The design approach will constantly be geared towards enhancing reliability as much as possible at all the levels. Unexpectedly, the final solution will not be a very power hungry one, but on the contrary, it will be quite low power for one that incorporates so much redundancy (at all the levels). Possible explanations can be found in some neural computations (and communications) articles, suggesting that unreliable "devices" (i.e., neurons and synapses) can lead to reliable computations (and communications) by cleverly combining redundancy and encoding, while simultaneously minimizing energy. The main conclusions of this chapter are that:

- Reliable designs should not be weighted with respect to their redundancy factors (as commonly done in the literature), but with respect to power, energy, and/or area (as it is customary in the VLSI community).
- Reliable designs for implementing computations using spatial redundancy are possible, and can even be low power. [Remark: For memory and communications, error detection and correction codes—maybe in combination with novel data encoding techniques—will most certainly prevail.]
- Defects and faults manifest themselves in different ways, one of them being by increasing currents (i.e., also power and heat). Hence adaptive local detection at reasonably low-levels could be designed based on current sensors (built-in I_{DD} testing), which could automatically trigger reconfiguration at the higher levels (when a predefined threshold current is exceeded), leading to self-healing systems.
- A lot of effort has to be invested for the development of EDA tools for quickly and precisely estimating the overall reliability at the system level, and for their integration with the EDA tools currently used for estimating area, delay and power. This aspect was also very strongly vouched for at the IEEE/ACM International Conference on Computer-Aided Design [2] where chip designers spoke of their difficulties in coping with reliability, power, clocking, statistical timing, verification and analog/mixed signal design, mentioning that *"there are a number of major holes in the IC design flow, and more research and development are urgently needed to fill them"* and that *"design-for-reliability needs help."*

12.1   Introduction

12.1.1   Challenges Due to Scaling

CMOS scaling has been the mean by which the semiconductor industry has achieved its historically unprecedented gains in productivity and performance quantified by Moore's Law [3]. Scaling CMOS technology to the next generation (*"a mad descent into the infinitesimal"*) has always increased transistor densities, improved performance, and reduced power consumption. The most recent International Technology Roadmap for Semiconductors (ITRS) report [4] predicts that the semiconductor industry will still continue its success in downscaling CMOS for a few more generations. It also predicts that the scaling of CMOS devices and process technology, as it is known today, will become very difficult as the industry approaches the 16 nm technology node (which is only a few generations away). Scaling might continue further, but it is expected that alternative nano-devices will start to be integrated with CMOS onto a silicon platform. The alternative nano-devices can be classified into solid-state (e.g., rapid single flux quanta, 1D-structures like nanowires and carbon nano tubes, resonant tunneling devices, SET, ferromagnetic devices, spin devices, etc.) and molecular ones. However, there are many fundamental and technical challenges that must be resolved to continue the scaling CMOS technologies deep into the nanometer regime [5]–[12]. The three greatest challenges are: *power and heat* dissipation, *reliability*, and the overall *complexity* (of fabrication, design, test, etc.) [13], [14]. Other great challenges include: connectivity and communication, verification, as well as logic encoding and hybrid integration (see [13]–[16]). The most difficult aspect is that all of these challenges are intricately (but hopefully not intractably) intertwined.

   Power dissipation (and the associated heat) is strongly affected by the increasing leakage currents. With the advent of the sub-100 nm CMOS technology (i.e., the "nanoera"), leakage currents have reached a level that cannot be ignored anymore. Leakage will continue to increase the static power dissipation exponentially (about 5× at each generation at 30°C), till new high-k dielectrics and/or multi-gate transistors will become mainstream. These are expected to reduce leakage currents down to less than 10% (multi-gate transistors), and even down to 1% (new high-k dielectrics). Nevertheless, with forecasted 10^{12} devices per chip [17], [18] (for comparison the human body has roughly 10^{13} cells, while the Brain has about 10^{12} neurons), even SET—which is obviously advocated as an ultra-low power technology—is going to become power constrained [4], [19].

With device geometries scaling below the 65 nm range, the *available reliability margins* are drastically being reduced [13]–[16], [20]. As a result, the reliability community will be forced to thoroughly investigate what exactly is determining these margins, and how we can change our reliability assessment methodology to gain new reliability space for the most advanced technologies [1], [4], [14–17]. From the chip designers' perspective, reliability manifests itself more and more as time-dependent uncertainties in electrical parameters. In the sub-65 nm era, these device-level parametric uncertainties will be too high to handle with prevailing worst-case design techniques—without incurring significant penalty in terms of area, delay, power, and energy. Additionally, with continued scaling, the copper (Cu) resistivity is starting to increase sharply due to interfacial and grain boundary scattering, while the introduction date for newer ultra low-k dielectrics has been slower than ITRS predictions. Besides the performance issues associated with interconnect scaling, several interconnect-related reliability issues are very troublesome (electromigration, stress migration, and heating), while, equally important, others are increasing with scaling (poor pattern definition, line-edge roughness, nano-scale corrosion, low-k dielectric cracks, post-chemical-mechanical polishing residues, etc.). The global picture is that the reliability issue looks like one of the greatest threats to the design of future integrated computing systems. For emerging nano-devices and their associated interconnects [21]–[23], the expected higher probabilities of failures, as well as the higher sensitivities to noise and variations, could make future chips prohibitively unreliable. The result is that the current IC design approach based on the conventional zero-defect foundation might simply not work. Therefore, fault- and defect-tolerance techniques that allow the system to recover from manufacturing and operational errors will have to be considered from the very early design phases.

Finally, *complexity* will certainly be the name-of-the-game for nanoelectronics. Miniaturization will increase the device density, which will subsequently increase the complexity of every aspect related to the design and manufacturing of future chips. Future EDA tools will have to consider quantum behaviors [24], while currently the modeling complexity of a multilevel interconnect network in a Gigascle chip is of the order of 10^{17} coupling inductances and capacitances throughout a nine-to-ten-level metal stack [11]. These will aggravate many other problems, like testing and verification [25], as well as integration and packaging [11], [17].

It is important to highlight here once again that all the challenges enumerated above are strongly and intimately entangled. On one hand, increasing the design complexity will increase the power and heat dissipation and also reduce the reliability margins. On the other hand, adding more devices (e.g., spatial redundancy) for improving the design's reliability will also increase the connectivity, and lead to extra power and heat. Furthermore, the increasing design complexity will also exacerbate connectivity both as the total number of wires as well as the overall length of these wires, and so on.

12.1.2 The Reliability Challenge

The reasons chip reliability is becoming the major hurdle are on the one hand due to the continuous increase in internal electrical fields and current densities, and on the other hand due to the introduction of new materials and devices with unknown reliability behavior—let alone the increase in the number of devices and interconnects. Another reason for the increased importance of reliability is that, in spite of these scaling trends, the market has continuously been demanding higher reliability levels. In the past, the technological reliability margins have always been sufficiently high, and have been guaranteed at the technology level (e.g., based on accelerated stress tests). Currently, the semiconductor industry approach is to extensively test the fabricated circuits and abandon those not operating correctly. Still, in 1994 Intel had to start the first ever chip recall campaign that cost US$ 475 millions when it was discovered that the Pentium processor generated slightly incorrect results for some floating-point operations. In the future, larger number of devices will be deployed in many applications and embedded systems, and reliability could turn out to be a showstopper for economically viable technology scaling: the cost to perform a similar recall—especially in the realm of failure-sensitive and energy-conscious real-time embedded systems—will be exponentially larger. Thus, there is very high pressure to make sure that future nanoelectronic systems will be functioning correctly over their expected lifetime—even if not free of faults and defects!

There is also an increasing concern that the massive scaling of the CMOS devices deep into the nanometer regime will introduce extreme static and dynamic parameter fluctuations at material, device, and circuit levels [8], [10], [17], [26]. Extreme parameter variations are a major barrier to achieving reliable and predictable system implementations [27]. At the material level, one of the potentially significant sources of fluctuation is the randomness in the exact location of doping atoms [8], [26]. Although the average concentration of doping is well controlled by ion implantation and annealing processes, it was shown in [17] that the standard deviation of the number of doping atoms per device could increase without bound as the doping volume decreases. This will lead to a device-to-device fluctuation in key device parameters, including the threshold voltage (V_{th}), as device dimensions scale down. The probability of failure of a CMOS inverter due to such fluctuations of V_{th} [15] can be calculated as $p_{INV} = \exp[-2V_{DD}/\sigma(V_{th})]$ (mentioned as early as 1980 by Mead and Conway [28]) where $\sigma(V_{th})$ is the variance of V_{th}, and V_{DD} is the supply voltage. As a result, some devices will not perform correctly. Fluctuations in manufacturing tolerances, on top of temperature variations and V_{DD} changes at the circuit level, will prevent circuit performances from reaching those defined by nominal physical limits. Typical increases in propagation delay and power dissipation due to such fluctuations are expected to be in the 30% to 50% above nominal for the 45 nm generation CMOS logic circuits [17].

Additionally, soft errors will occur for example when the radioactive atoms in the chip's material decay and release alpha particles that contain positive charges and kinetic energy. Similar effects are due to electromagnetic interference or electrostatic discharge. Since capacitance and voltages will decrease in future technologies to tiny fractions of what they are today, very small charges will be needed to flip a bit in memory, the output of a gate, or the value on a wire. Although such an event is highly unlikely per single device, soft errors are becoming a major reliability concern for future systems based on nanodevices [29]−[33], due to the expected massive number of devices the system will have. For instance, a future design having one Terabyte of memory and a probability of soft error per single bit of one per million years, will experience a soft error about every 3.94 seconds. Typically, data in memory is protected using error-correcting codes (on top of stand-by spare rows and columns, and off-line reconfiguration), or might be even self-testable [34], while fresh research results can be found in [35]. However, mechanisms to protect latches and flip-flops (that store state) and random logic have only recently started to appear [30]−[33].

According to classical scaling theory, the gate insulator thickness must shrink with the other transistor dimensions in order for the newly scaled transistor to function similarly as its predecessor. Shrinking the gate insulator thickness increases the gate leakage current and the power consumption exponentially. At the same time, there is an increasing concern about the ability of the very thin layer of insulation material to maintain its insulating properties while subjected to high electric fields and high temperatures for many years. The current flowing through the gate oxide causes reliability problems by leading to long-term parameter shifts and eventually to oxide breakdown. Frank *et al.* argue in [8] that a range of different device designs (for different applications), utilizing a range of gate insulator thicknesses could alleviate the problem (but these will not simplify the fabrication process), while ITRS predicts the use of high-*k* dielectrics starting from the 45 nm node [4].

Interconnect scaling raises another reliability concern for future nano-devices. In [11], Davis *et al.* mentioned that the miniaturization of interconnects, unlike transistors, does not enhance their performance (see also [36]). Interconnect scaling will significantly affect the circuit reliability due to increased crosstalk and latency. Reliability will be also negatively affected by the electrical, thermal, and mechanical stresses in a multilevel wire stack. Sakurai was among the very first to conclude in 2000 that the interconnects—rather than transistors—will be the major factor in determining the cost, delay, power, reliability and turn-around time of the future semiconductor industry [7].

Besides electrical and thermal stresses, as well as structural uncertainties (e.g., line edge roughness and oxide thickness), the introduction of new materials could sharply decrease reliability margins. Lack of thorough understanding of degradation mechanisms and their interactions with new materials and ultra-scaled structures demands much higher reliability margins to maintain confidence levels in reliability projections, and *"we are likely to see many no-fault-found failures"* [1]. These conflicting trends are rendering technologists unable to meet—as per prevailing reliability assessment methodology—failure rate targets and impose the delegation of reliability qualification to designers, using EDA tools based on appropriate reliability models.

Unfortunately, reliability problems for beyond CMOS technologies [37], [38] are expected to get even worse, as device failure rates are predicted to be as high as 10% (e.g., background charge for SET [21]) going up to 30% (e.g., self-assembled DNA [22], [23]). Clearly, achieving 100% correctness at the system level using such devices (and their associated interconnects) will be not only outrageously expensive, but might be plainly impossible! Therefore, relaxing the requirement of 100% correctness for devices (and interconnects) might significantly reduce the costs of manufacturing, verification, and test [25]. Still, this will lead to more transient and permanent failures, and most (if not all) of these failures will have to be compensated by architectural/design level techniques [13]–[16], [39]–[44].

Previously fault-tolerance has been an issue only for high profile high-cost designs, but it looks like it will be here to stay, i.e., it will become part of any future design [14], [16], [42], [45]–[47]. From the system design perspective errors fall into one of the following three classes: permanent (also called defects), intermittent, and transient [20]. The origins of these errors can be found in the manufacturing process (spot defects, bridging faults, metal silvers, cracks, lithography and process variations, edge and striation effects), the physical changes appearing during operation (electromigration and self-heating— amplified by the existence of mousebites or hillocks—soft breakdown, temperature effects, hot carriers, relative mismatch), internal (crosstalk noise, antenna effects, skin effects, hot spots, IR drops, ground bounce, timing inaccuracies), and external noise (radiation, electromagnetic interference, electrostatic discharge). The oldest and most commonly used model is the stuck-at fault model. It is not clear if emerging technologies will not require new fault models [43], [46], or if multiple errors might have to be dealt with [32]. In fact Kuo mentions in [1] that *"we are unsure as to whether much of the knowledge that is based on past technologies is still valid for reliability analysis."*

The well-known approach for developing fault-tolerant architectures in the face of uncertainties (both permanent and transient faults) is to incorporate redundancy [48]–[50]. Redundancy can be either static (in space, time, or information) or dynamic (requiring fault detection, location, containment, and recovery). Space (hardware) redundancy relies on voters (e.g., generic, inexact, mid-value, median, weighted average, analog, hybrid, etc.) and includes among others the well-known: modular redundancy, cascaded modular redundancy, and multiplexing (including von Neumann multiplexing [48], enhanced von Neuman multiplexing [42], [51], [52], and parallel restitution [53]). Time redundancy is trading space for time (e.g., alternating logic, recomputing with shifted operands, recomputing with swapped operands, etc.), while information redundancy is based on error detection and error correction codes. Hybrid approaches are also known, e.g., time-shared triple modular redundancy, recomputing with triplication with voting, hardware partitioning in time redundancy, recomputing with partitioning and voting, quadruple time redundancy, and many fresh ones [54]–[59]. For dynamic fault tolerance one should rely on periodic tests, self-checking circuits, and watchdog timers. Examples of fault detection include: duplication with comparison, recomputing with duplication with comparison, error detection codes, checkers, and so on. The fault recovery relies either on repeating calculation(s) based on an automatic 'repeat request', or on the use of extra spare modules and reconfiguration. Examples here include: hot and cold spares, standby spares, reconfigurable duplication, self-purging systems, shift-out modular redundancy, triple-duplex architectures, and reconfigurable arrays [39].

Many of the reliability-enhanced schemes enumerated above can be implemented at different levels: device, gate, circuit, block, and system. All of them have in common that improved reliability is traded off for increased area and higher connectivity, normally leading to higher power consumptions, and/or slower computations [60]. As an early example for nanoelectronics, Roychowdhury *et al.* [61] showed that a Quantum-dot Cellular Automata (QCA) circuit implemented with a redundancy factor $R = 16$ (i.e., the number of replicated identical copies) will be able to perform correctly even if 15% of the devices failed. However, very serious manufacturing concerns were raised later [62]. That is why it was strongly argued that, since manufacturing processes will not be able to produce perfect devices at a sustainable cost, future designs will have to be defect- and fault-tolerant, reconfigurable, functioning despite imperfections, and able to diagnose and repair failures dynamically (from the very early [63], followed recently by [64], [43], [42], and currently by many-many more). Any architecture that will disregard the fact that the underneath devices and interconnects are unreliable is anticipated to be impractical.

Still, implementing effective defect- and fault-tolerant architectures intimately depends on the existence of accurate fault models, and both device and gate levels modes have been left behind. Currently there are no models that can be used to precisely estimate the manufacturing defects and transient error rates of (future) nano-devices [39], [43], [64]. In 2003, Fortes even mentioned that existing fault modes need to be reevaluated and maybe new ones will have to be devised [45] (see also [1], [47]). However, it is possible to make estimates of the amount of redundancy that is required to ensure reliable system operation, given gate failure probabilities. Recently, even the assumption of how effective such an approach is was challenged [65]–[67], by showing that approximating the probabilities of failure of the gates by a (bounding) constant introduces non-negligible errors. It is only very recently that we were able to find the explanation for this unexpected behavior [68]. Innovative EDA tools, which incorporate the newly developed fault models for different nano-devices, have to be developed [1], [43], [69]. In addition to power consumption and delay, future EDA tools should be able to estimate—reasonably quickly and precisely—the reliability of circuits and architectures starting from the device levels. Using such EDA tools will allow the designers to evaluate the effectiveness of different defect- and fault-tolerant nano-architectures and select the most suitable one for the application at hand.

Till now, VLSI designers did not have to care about reliability, which was characterized at the technology level (i.e., it was considered to be the job of the technology/process engineer). In the future only material and device engineering together with logic design will not suffice anymore for tackling reliability, which will become a key design concern at all levels. No 'silver bullet' will be able to cope with all the types of (new) faults in nanoscale circuits and systems, and the blending of several techniques will certainly be needed. Boosting reliability will require a cooperative approach, where high-level techniques will rely upon lower levels' support.

In the following, we shall walk the reader through a series of examples starting from the device level and going towards the system level by trying to emphasize the synergy between the design and the technology levels and qualitatively and quantitatively analyzing the benefits such an approach will bring.

12.2 Reliability Simulation Tools

Trying to determine the reliability of a large (i.e., complex) logic circuit through simulations would allow not only verifying the many theoretical results [48]–[53], [70]–[85], but will also help in the design/selection of better/optimal nano-architectures. The methods used for simulating stochastic systems can be divided into experimental and numerical. In the case of *experimental methods*, the analysis is performed implicitly by observing the results obtained from many experiment runs. The most popular experimental method is Monte Carlo simulation, which reproduces the behavior of the system. For doing this, Monte Carlo relies on random number generators that sample the random activities of the system being analyzed. Once the model is built, the computer performs as many sample runs from the model as necessary to draw meaningful conclusions about the model's behavior. It follows that the Monte Carlo analysis is conducted indirectly, based on the observation of many sample runs. The biggest advantages of Monte Carlo are its intuitiveness and its ability of simulating models for which deterministic solutions are intractable. The Monte Carlo method is widespread in the VLSI community. It will have to be used in the future for thoroughly analyzing the behavior of (novel) devices, gates, and small sub-circuits. Being (very) time consuming its use appears to be limited, but the reliability results obtained should be collected and used at the higher design/circuit levels (e.g., as parameters in future libraries of gates). Different implementations of the same gate will have different reliabilities, hence could be used for optimizing circuit reliability in a similar way multiple V_{th} designs are used for reducing power consumption.

Numerical methods are designed for analyzing stochastic models without incorporating the random behavior. The simulation results that they deliver are the same for the same model parameters. These methods work by describing the flow of probabilities within the system—usually using differential equations and numerical methods for solving them. Markov chains can be used for describing and analyzing models that contain exclusively exponentially distributed state changes. Depending on the character of the time domain, there are discrete-time Markov chains (DTMCs) and continuous-time Markov chains (CTMCs). In the following we shall provide a brief review of some of the numerical tools and techniques with emphasis on those that have been developed lately. The interested reader can find many earlier results including: REL70, RELCOMP, CARE, CARSRA, CAST, CARE-III, ARIES-82, SAVE, MARK, HARP, SHARPE, GRAMP, SURF, SURE, SUPER, ASSIST, SPADE, METASAN, METFAC, ARM, and SUPER, in the excellent review of Johnson and Malek [86]. These were followed by ASSURE, CAREL, SIDECAR, and MEASURE+.

The Hybrid Automated Reliability Predictor (*HARP*) tool was pioneered in 1981 at Duke (www.ee.duke.edu/~kst/software_packages.html) and Clemson University. HARP uses a fault-tree analysis technique for describing the failure behavior of complex systems. Fault tree diagrams are logical block diagrams that display the state of a system in terms of its components. The basic elements of the fault tree are usually failures of different components of one system. The combination of these failures determines the failure of the system as a whole. Further developments have led to Symbolic Hierarchical Automated Reliability and Performance Evaluator (*SHARPE*) [87] (Duke University) and Monte Carlo Integrated HARP (*MCI-HARP*) [88] (developed at Northeastern University).

In the early 90s a few other tools providing numerical analyses have been developed: *TimeNET* at the Technical University of Berlin (pdv.cs.tu-berlin.de/~timenet/), *UltraSUN* (and later *Möbius*) at the University of Illinois at Urbana-Champaign (www.mobius.uiuc.edu/index.html), and *SMART* at the University of California at Riverside (www.cs.ucr.edu/~ciardo/SMART/). These were followed in the mid-90s by Dynamic Innovative Fault Tree (*DIFTree*) [89], and *Galileo* [90] (www.cs.virginia.edu/~ftree/), both from the University of Virginia. Galileo extended the earlier work on HARP, MCI-HARP and DIFTree using a combination of binary decision diagrams (BDD) and Markov methods.

In 1999 a team from the University of Birmingham introduced the Probabilistic Symbolic Model Checker (*PRISM*) [91], [92] (www.cs.bham.ac.uk/~dxp/prism/). PRISM relies on a probabilistic model checking for determining if a given probabilistic system satisfies given probabilistic specifications. The circuit is described as a state transition system with probabilities attached to each of the transitions. It applies algorithmic techniques to analyze the state space and calculate performance measures associated to the probabilistic model. PRISM supports the analysis of three types of probabilistic models: DTMCs, CTMCs, and Markov decision processes (MDPs).

The *proxel* based method was introduced in 2002 [93] as an alternative to simulating discrete stochastic models. For want of a better phrase, and borrowing from pixel, proxel is the abbreviation for *probability element*. It describes every probabilistic configuration of the model in a minimal and complete way. Each proxel carries enough information for generating its successor proxels, i.e., for determining probabilistically how the model will behave [94]. This transforms a non-Markovian model into a Markovian one. The approach analyzes models in a deterministic manner, avoiding the typical Monte Carlo problems (e.g., finding good-quality pseudo-random-number generators) and partial differential equations (difficult to set-up and solve). The underlying stochastic process is a DTMC, which is constructed on-the-fly by inspecting all the possible behaviors of the model.

The *probabilistic transfer matrices (PTMs)* framework was first presented in 2003 [95], but the underlying concept can be traced back to the 60's [96]. The PTMs can be used to evaluate the circuit overall reliability by combining the PTMs of elementary gates or sub-circuits [97], [98]. It performs simultaneous computation over all possible input combinations, and *calculates the exact probabilities of errors*. Another advantage (beside accuracy) is that it is trivial to have different probabilities of failures for different gates (see e.g. [66]). PTM however has a major memory bottleneck: for a circuit with n inputs and m outputs, the straightforward PTM representation requires $O(2^{n+m})$ memory space. This limits the size of the circuits that can be simulated to about 16 I/O signals. The use of the algebraic decision diagrams compression method can reduce the memory requirements, and circuits with about 40 I/O signals have been evaluated [97].

Probabilistic gate models (PGM) [99], [100] is another analytical approach for estimating reliabilities of circuits. This method can be used for any type of gate and fault models. A circuit is divided into many smaller modules (i.e., gates), and I/O signals are assumed to be statistically independent (uncorrelated). The overall reliability for a circuit is obtained by multiplying the individual reliabilities for each output. In a circuit that includes fan-outs, signals are correlated, hence, in general PGMs lead to approximate results, which are traded off for PGM's higher speeds and reduced computational resources.

Recent work has also been done in modeling signal dependencies using *Bayesian Networks* (BNs) [101]–[103]. The relation between circuit signals and Markov random fields was presented in the context of probabilistic computations. The conditional probability of output(s) given input signals determines how errors are propagated through a circuit. Using this theoretical model, it is possible to estimate the probability of output error given the gate errors. Kuo, the Editor-in-Chief of the *IEEE Transactions on Reliability*, mentions in his opening article from December 2006 [1] that *"we predict that the Bayesian approach will be even more frequently utilized in the nano era as product life cycles based on new technologies become even shorter, and it is becoming impossible to obtain sufficient data before a new product requires reliability assessment."*

These sustained efforts over almost four decades have led to only a few commercial products:

- *Galileo* from Exelix (www.exelix.com), which was already mentioned;
- *Predict* from ReliaSoft (www.reliasoft.com/predict /index.htm);
- *ProTesT* from Doble (www.doble.com/products/protest.html); and
- the products from Reliass (www.reliability-safety-software.com/), including PRISM® (which is completely different from the PRISM developed at the University of Birmingham and mentioned above).

12.3 More Reliable Gates

The design of more reliable gates has been of high interest when vacuum tubes were the elementary devices [16], [48]–[50]. By that time, threshold logic (including majority and ternary logic) was an active topic of research, even used in building some of the early computers. During 1957 and 1958, Rosenblatt together with Charles Wightman and others constructed the Mark I Perceptron having 512 adjustable weights [104]. Shortly afterwards, Bernard Widrow together with his students developed the ADALINE (ADAptive LINear Element) [105]. The next threshold logic computer was DONUT [106], followed later by Setun [107], [108]. Due to the low reliability of vacuum tubes, fast elements on miniature ferrite cores and semiconductor diodes were designed for implementing ternary logic. Brousentsov stated in [107] that: *"Ternary threshold logic gates, as compared with the binary ones, provide more speed and reliability, and required less equipment and power. These were the reasons to design a ternary computer."*

With the advent of MOS and CMOS integrated circuits, such topics were quickly forgotten. Still, radiation hardened [109] has been constantly in demand for special applications. Some of the designs that we shall present in this sub-chapter belong to the so-called *rad-hard by design* class, while other CMOS designs specifically targeting ultra deep sub-micron technologies are only now starting to appear. Such designs can be classified into either *layout-level* (i.e., based on modifying the layout) or *switch-level* (i.e., based on modifying the circuit at the transistor level) techniques, while hybrid and even adaptive techniques are emerging.

Most circuit designers have used *the switch-level approach*. A CMOS fault tolerant architecture based on encoding the circuit outputs with Berger codes (error correcting codes) requires the introduction of additional networks to provide tolerance to single stuck-on faults, as well as to a number of multiple faults, while also reducing unidirectional faults [110]. Another approach is quadruplication applied to combinatorial CMOS gates, both at the net (the *p*-stack and the *n*-stack are quadruplicated separately) and at the transistor level (every transistor is quadruplicated preserving the interconnection topology of

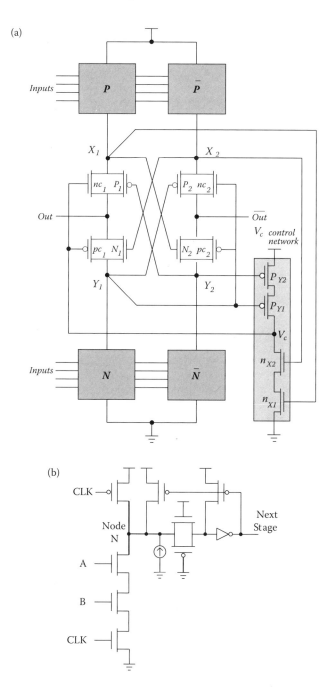

FIGURE 12.1 Radiation hardened by design: (a) differential fault-tolerant architecture at the CMOS level [112]; (b) soft error suppression technique for domino logic [113] using isolation device and dual keeper; (c) single event upset hardened inverter in silicon-on-insulator [114]; (d) classical output-wired-inverter solution [117] for a hardened majority voter [114].

the net) [111]. An alternate switch-level solution can be seen in Figure 12.1 (a). It duplicates the *n*-stack and the *p*-stack and adds cross-coupled transistors, achieving fault tolerance with almost no speed degradation [112]. A low-power soft error suppression technique for dynamic logic can be seen in Figure 12.1 (b). It adds pass transistor device(s) as isolation and weak keeper(s) to standard domino logic [113]. Optimizing the size of the keeper (layout-level) can also help. As can be seen from these few examples, the solutions

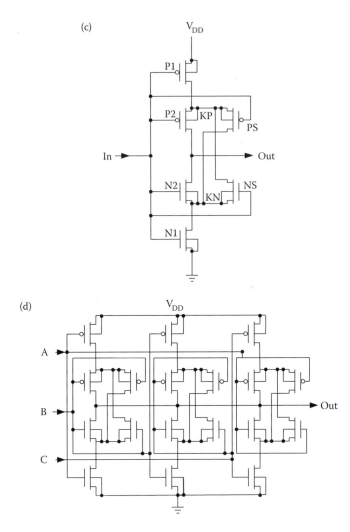

FIGURE 12.1   (Continued).

developed are quite involved. Under switch-level redundancy, we could also include active biasing and isolated well transistors [114] (see Figure 12.1 (c)). These prevent transients (in combinational logic) from reaching the output node. Such approaches complement noise-immune designs, like e.g., those detailed in [115], [116], and can be combined with hardwired voting (see Figure 12.1 (d)), e.g., based on the old output-wired-inverters idea [117].

In parallel, the analysis of the failure mechanisms of integrated circuits, has led to defining layout rules enabling to improve the testability of circuits [118]. Later layout-level design for testability rules were used for avoiding some hard to detect faults or even undetectable faults [119]. Such layout rules include: redundancy of contacts, ring-shaped or close loop conductive layers, and duplication of interconnections and I/O conductive paths. These avoid some open faults, or reduce their appearance probabilities. Another recently advocated layout technique—borrowing from analog designs—is high matching (Figure 12.2 (a)). This has been used in a combined switch- and layout-level approach for enhancing the noise immunity of threshold logic gates [116]. Having their roots in hardwired voting [117], hybrid solutions (bridging the switch- and layout-level) have been detailed for CMOS [120] (see Figure 12.2 (b)), as well as for SET [69], [52] (Figure 12.2 (c)), and for nano-PLA nano-circuits [121], [122] (Figure 12.2 (d)). These rely on built-in transistor-level redundancy, with redundant signals added on top, and can be combined at higher levels with system-level voting and/or with error correcting codes.

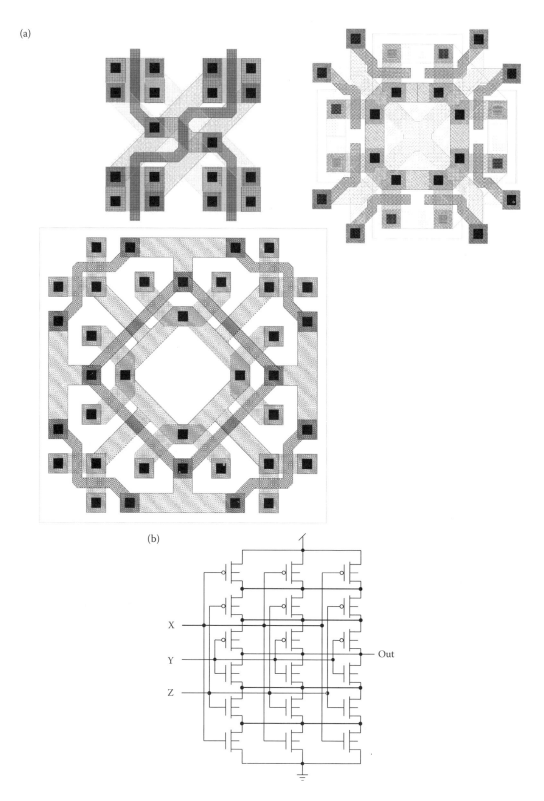

FIGURE 12.2 Built-in redundancy at the device level: (a) high matching techniques (used in analog designs) applied to digital circuits [116]; (b) an extension of the output-wired (inverter) principle [120]; (c) the high matching technique applied to SET circuits [69], [52]; (d) PLA style circuit (NASIC circuit [121], [122]).

(c)

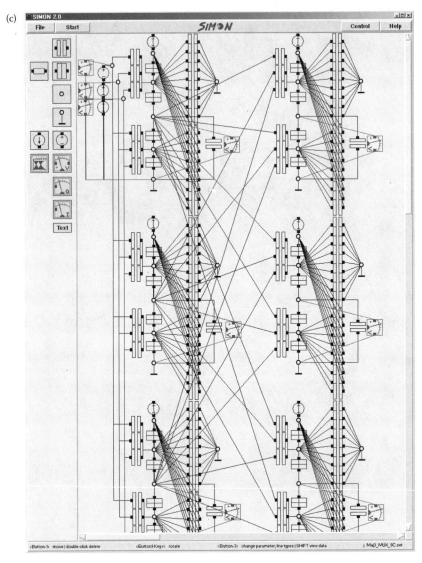

(d)

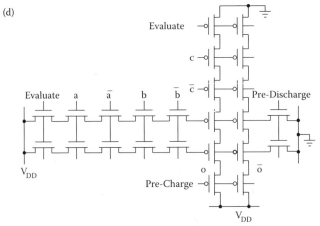

FIGURE 12.2   (Continued).

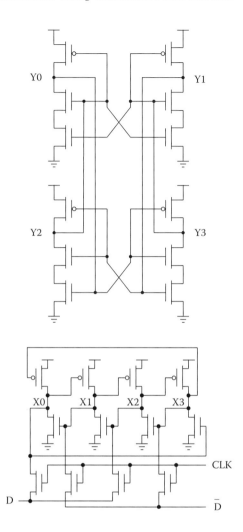

FIGURE 12.3 Radiation tolerant latches: (a) single event resistant latch structure [123] based on asynchronous sequential circuits; (b) soft error rate tolerant latch [124] utilizing local redundancy (see [125])—i.e., similar to the designs presented in Figure 12.2.

All these solutions are for logic gates, but original designs have been detailed for latches too. One of them is based on an asynchronous design [123] (Figure 12.3 (a)), while another hardened feedback topology, which does not incur speed penalties, is presented in Figure 12.3 (b) [124], [125].

Among the few other robust circuit and system design methodologies specifically intended for nanometer-scale devices we can mention the use of threshold logic gates for evaluating an analog average [126], [127], while [128] advocates for real time reconfigurable threshold logic gates (see also [129]). Finally, process variation compensation techniques should also be mentioned. They range from trying to optimally adjust the strength of the keepers (for domino logic gates) based on an on-die leakage current sensors [130], to the use of active body bias and transient noise attenuation via voltage division [56].

All these solutions are highly innovative, but there is a clear need for comparing such non-standard designs. Only a few preliminary results based on detailed Monte Carlo simulations for CMOS have started to be reported (e.g., [131]), while there are almost none for beyond CMOS nanotechnologies (for SET see [69], as well as [65] and [67]).

We conclude this sub-chapter by mentioning that a growing number of publications are dealing with such problems. This is encouraging, while there is a tendency to move too quickly to higher levels [29],

[32], [56]–[58], [132]. This should be done at a slower pace, as the expectation is that the highest reliability rewards will be at the lowest level, so we should first of all take advantage of all the 'low lying fruits'. These imply that many versions of the same gate, having different reliability performances, should be designed and tested. These will be equivalent to multiple V_{th} designs of the same gate, which have long been used in the bid for low power. Future EDA tools should use such (libraries of) gates for optimizing circuits' reliability—in the same way current EDA tools are using multiple V_{th}, and multiple V_{DD}, for optimizing power consumption. This means that *future EDA tools will have to optimize in four dimensions: timing, area, power, and reliability.*

12.4   Reliable Full Adders

12.4.1   Full Adder Cells

Addition is by far the most common arithmetic operation in a wide variety of applications including digital signal processors (DSPs), microprocessors, microcontrollers, and memories [133], [134]. Adders are crucial in arithmetic logic units (ALUs), floating-point units (FPUs), and for cache or memory address calculations. They are also easy to compare and benchmark due to the wealth of results already published.

A full-adder (FA) is a basic Boolean logic circuit that takes three binary inputs: A, B, and C_{in}, and provides two binary outputs: the Sum bit (S) and the Carry-out bit (C_{out}). The 2-bit binary number (C_{out}, S) is the binary representation of the arithmetic sum of the three inputs bits. The Boolean logic equations are as follows:

$$S = A \oplus B \oplus C_{in} = (A \oplus B) \oplus C_{in} \tag{12.1}$$

$$C_{out} = A \cdot B + A \cdot C_{in} + B \cdot C_{in} = \text{MAJ}(A, B, C) \tag{12.2}$$

or

$$C_{out} = A \cdot B + C_{in} \cdot (A \oplus B) \tag{12.3}$$

The increasing demand for mobile electronic devices such as cellular phones, PDA's, and portable computers has raised the demand for power efficient VLSI circuits. The FA cells usually lie in a critical path that determine the system overall performance. Therefore, designing a new FA cell that is both faster and more power efficient than a standard FA cell was the driving force behind many research papers during the last decade [135]–[145]. However, all of these FA designs assumed that the circuits are implemented using reliable gates. Therefore, reliability was not considered as one of the FAs' optimization criteria. It is only very recently that reliability has started receiving revived attention in the nanotechnology community [20], [30], [146], [147].

In this sub-chapter we will use the PTM technique (mentioned in subchapter 12.2) to numerically evaluate the reliability of four different FA implementations. We shall also try to link such PTM gate-level simulations with device-level estimates. For each FA a reliability model is constructed. These models are then used to study the effect of the gates (and devices) malfunctions on the reliability of each FA. The first FA implementation uses seven NAND-2 gates and five inverters (used as a PGM example in [99]), and will be called NAND-FA (Figure 12.4 (a)). The second implementation uses three NAND-2 gates and two Exclusive-OR (XOR) gates and will be called XOR-FA (Figure 12.4 (b)). This FA is the normal implementation when using direct synthesis (see e.g. [144]). The third implementation uses five majority (MAJ-3) gates and three inverters and will be called MAJ-FA (Figure 12.4 (c)). This solution was advocated by early QCA designs [148]. The fourth implementation uses three minority (MIN-3) gates and two inverters and will be called MIN-FA (Figure 12.4 (d)). Solutions having only three MAJ (and/or MIN) gates have been advocated as the simplest ones in the threshold logic community since 1961 [149]. This solution has been adopted later by many SET designs [150]–[152], as well as by recent QCA designs [153], while some authors do not seem to be aware of the very early threshold logic results (for reviews about threshold logic see [154], [155]).

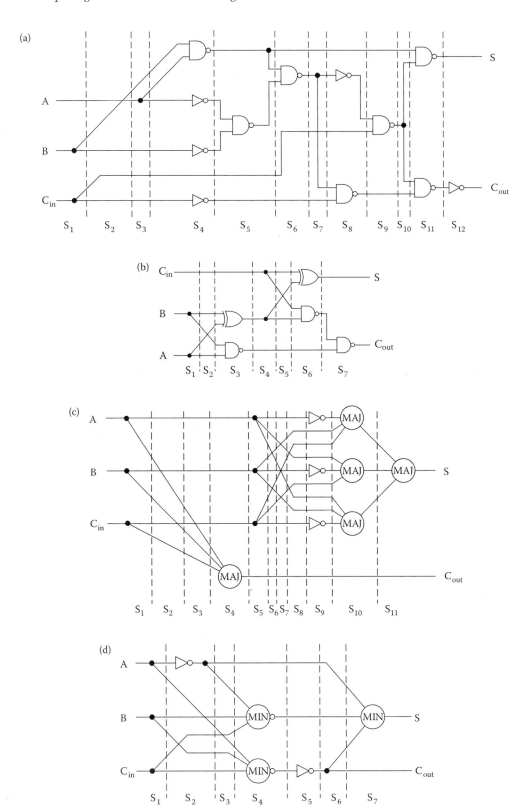

FIGURE 12.4 Four full adder (FA) implementations: (a) NAND-FA; (b) XOR-FA; (c) MAJ-FA; (d) MIN-FA.

12.4.2   Probability Transfer Matrix Calculations

In an error-free operation, the function of a combinational logic circuit or gate can be represented by a truth table, which shows how a logic circuit's output responds to various combinations of the inputs, using logic '1' for true and logic '0' for false. For example, the truth table for a NAND-2 (NAND gate of fan-in $= 2$) maps the input value '01' to the output value '1'. However, in the presence of (transient) faults or (manufacturing) defects this input may (occasionally) lead to a '0' at the output. If how often this is likely to happen is known, the gate behavior can be modeled using PTM. In this model, column indices represent input values, and row indices represent output values, while matrix elements capture the transition probabilities. For instance, consider the PTM for a standard NAND-2 gate with a probability of failure of $p_{\text{NAND}-2}$. The PTM elements corresponding to the input value '01' are $p_{\text{NAND}-2}$ and $1 - p_{\text{NAND}-2}$ for output '0' and '1' respectively.

Patel *et al.* in [95] showed that the PTM concept is not restricted to gates. It can also be used to model the functionality of (unreliable) circuits. In general, the cell a_{ij} in the truth table/matrix represents the probability that the ith output value is observed when the jth input value is given to the circuit. In order to calculate the circuit PTM, it is assumed that gate errors occur independently (i.e., uncorrelated) and the PTM of each gate in the circuit is already known. The PTM of a larger circuit is constructed by combining the PTMs of the individual gates into sub-circuits' PTMs. The sub-circuits' PTMs are then combined to generate the circuit's PTM. Gates and sub-circuits are connected using wires. Wires can be treated either as a special class of gates, or—if they make no errors—as Ideal Transfer Matrixes (ITMs), i.e., they are equivalent to ideal gates.

Gates and sub-circuits are combined using the following three rules:

1. If two gates (or sub-circuits) with PTM_1 and PTM_2 are connected in series, then their combined PTM is the product of the individual gates (or sub-circuits) PTMs:

$$\text{PTM}_{12} = \text{PTM}_1 \times \text{PTM}_2. \qquad (12.4)$$

2. If two gates (or sub-circuits) are connected in parallel, then their combined PTM is the tensor product of the individual gates (or sub-circuits) PTMs:

$$\text{PTM}_{12} = \text{PTM}_1 \otimes \text{PTM}_2. \qquad (12.5)$$

3. If two gates (or sub-circuits) are connected with fan-out (an output of one gate/sub-circuit is connected to more than one input of the following gates/sub-circuits), the combined PTM is constructed by calculating the tensor product of the two gates/sub-circuits and eliminating all the columns that have different values for the fan-out inputs.

To construct the PTM associated with each FA, the FAs were divided into several serially connected stages/sub-circuits (see Figure 12.4). The above second and third combinations rules were used to construct the PTM associated with each stage. The PTMs were then combined together using the first combination rule.

Stages are classified into either *a computing stage* or *a connecting stage*. A *connecting stage* contains only a group of connecting wires (or simply connections). Connecting stages are preparing the inputs for the following computing stages. Connecting stages can be used to augment the number of connections using the fan-out rule #3. For example, the connecting stage S_1 of the NAND-FA (see Figure 12.4 (a)) uses the fan-out rule to increase the number of wires form three (input wires) to five (output wires). [Remark: The wires for each stage are always counted from top to bottom.] A connecting stage can also be used to swap any two wires (when required) using a wire swap matrix (e.g., like the ones shown in Figure 12.5). For example, the 2-wire swap matrix is used to swap wires number 2 and 3 as well as wires number 7 and 8 in stage S_7 of the MAJ-FA (see Figure 12.4 (c)). At the same time, the 3-wire swap matrix is used to swap wires number 4 and 6 (over wire number 5) in the same S_7 stage of the MAJ-FA (Figure 12.4 (c)).

Unlike a connecting stage, *a computing stage* must contain at least one gate. A computing stage may also contain a group of connecting wires (e.g., S_2, S_4, S_5, and S_7 of the MIN-FA in Figure 12.4 (d)). The MAJ-FA (Figure 12.4 (c)) consists of four computing stages (S_4, S_9, S_{10}, and S_{11}) and seven connecting stages

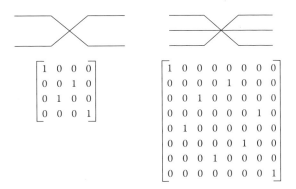

$$\begin{bmatrix} 1 & 0 & 0 & 0 \\ 0 & 0 & 1 & 0 \\ 0 & 1 & 0 & 0 \\ 0 & 0 & 0 & 1 \end{bmatrix} \qquad \begin{bmatrix} 1 & 0 & 0 & 0 & 0 & 0 & 0 & 0 \\ 0 & 0 & 0 & 0 & 1 & 0 & 0 & 0 \\ 0 & 0 & 1 & 0 & 0 & 0 & 0 & 0 \\ 0 & 0 & 0 & 0 & 0 & 0 & 1 & 0 \\ 0 & 1 & 0 & 0 & 0 & 0 & 0 & 0 \\ 0 & 0 & 0 & 0 & 0 & 1 & 0 & 0 \\ 0 & 0 & 0 & 1 & 0 & 0 & 0 & 0 \\ 0 & 0 & 0 & 0 & 0 & 0 & 0 & 1 \end{bmatrix}$$

FIGURE 12.5   Wire swap matrices: (a) 2-wire swap matrix; (b) 3-wire swap matrix.

(S_1, S_2, S_3, S_5, S_6, S_7, and S_8). The first three stages (S_1, S_2, and S_3) are used to fan-out a copy of the three input signals and prepare the connections for the computation of the C_{out} (in stage S_4). The following connecting stages S_5 to S_8 create three copies of the input signals and swap the wires to prepare the inputs for the first computing stage of the sum calculation (S_9). The summation of the three digits is performed by the computing stages S_9, S_{10}, and S_{11}.

After calculating the PTMs for each stage independently, these are combined for obtaining the overall PTM. The probability of failure of each FA is then calculated as the average probability of failure for all the input combinations. Since all the stages are serially connected, the overall PTM is constructed using the first combination rule as follow:

$$\mathrm{PTM}_{S1-2} = \mathrm{PTM}_{S2} \times \mathrm{PTM}_{S1}, \tag{12.6}$$

$$\mathrm{PTM}_{S1-3} = \mathrm{PTM}_{S3} \times \mathrm{PTM}_{S1-2}, \ldots, \tag{12.7}$$

$$\mathrm{PTM}_{S1-n} = \mathrm{PTM}_{Sn} \times \mathrm{PTM}_{Sn-1}. = \text{the product (PI) from } i = 1 \text{ to } n \text{ of PTM Sub } si \tag{12.8}$$

12.4.3   Full Adders Reliability Simulations

Numerical simulations have been conducted to assess the reliability of the four FAs. In the first experiment, the probability of failure of each FA was measured against the probability of failure pGATE of the elementary gates. For each elementary gate, the numerical simulations were performed assuming that only a single type of gate is unreliable while all the other types of gates are ideal ($p_{GATE} = 0$). This was done for estimating the influence of each type of gate on the FAs' reliabilities.

In a second set of simulations, the reliability of each FA was estimated assuming that all the elementary gates are unreliable with the same p_{GATE} (as it is customary, see e.g., [40], [41], [46]–[53], [95]–[103]). Figure 12.6 presents all these simulation results. They point to the NAND-FA as the least reliable of the four FAs. $P_{NAND-FA}$ goes almost up to 0.4 when $p_{NAND-2} = p_{INV} = 0.06$ (Figure 12.6 (a)). For the same conditions both P_{XOR-FA} and P_{MAJ-FE} are slightly below 0.25 (Figure 12.6 (b)), while $P_{MIN-FA} \approx 0.225$ (Figure 12.6 (d)). The simulation results also show that the elementary gates influence the FAs' reliabilities. For instance, using ideal MAJ-3 gates ($p_{MAJ-3} = 0$) and unreliable inverters ($p_{INV} = 0.06$) leads to $P_{MAJ-FA} = 0.09$. However, using unreliable MAJ-3 gates with $p_{MAJ-3} = 0.06$ and ideal inverters ($p_{INV} = 0$) leads to quite a different value, namely $P_{MAJ-FA} \approx 0.19$ (see Figure 12.6(c)). These results suggest that the MAJ-3 gates have a stronger influence on the probability of failure of MAJ-FA than the inverters, hence one should try to enhance first the reliability of the MAJ-3 gates. Furthermore, the simulation results show that the influence of the elementary gates on the circuits' reliabilities intimately depends on the circuit's design. Although the MIN-FA and the MAJ-FA have different number of inverters (2 and respectively 3) in their design, using unreliable inverters $p_{INV} = 0.06$ causes the two FAs to have the same probability of failure ($P_{MAJ-FA} = P_{MIN-FA} = 0.09$). Since gates are implemented together using similar nano-devices, assuming that only one type of gate is unreliable while (all) the others are ideal is unrealistic. However, this assumption has been used in the previous experiments only for studying the effect of each different type of gate on the FAs' reliability.

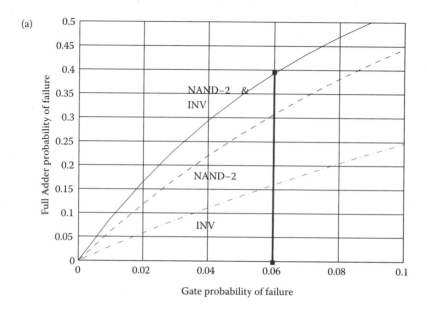

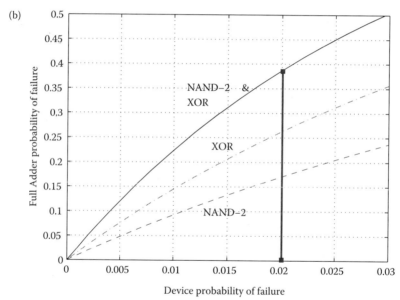

FIGURE 12.6 Probability of failure of four full adders (FAs) versus those of their elementary gates: (a) NAND-FA; (b) XOR-FA; (c) MAJ-FA; and (d) MIN-FA.

In a third series of experiments, the probability of failure of each FA was calculated assuming that the gates have different probabilities of failures. The simulation results from these experiments (Figure 12.7 and Figure 12.8) can be used to determine the probability of failure of the FAs corresponding to any combinations of p_{GATE} values. For example, if the NAND-FA is implemented with unreliable gates where $p_{INV} = 0.02$ and $p_{NAND-2} = 0.08$, Figure 12.8 (a) shows that $P_{NAND-FA} = 0.4$. These simulation results can also be used to determine the maximum allowable p_{GATE} for each type of gate in order to guarantee a predefined allowable P_{FA}. For instance, to implement a NAND-FA satisfying $P_{NAND-FA} \leq 0.1$, the elementary gates cannot exceed $p_{INV} \leq 0.035$ and $p_{NAND-2} \leq 0.018$ respectively (Figure 12.8 (a)). For more simulation results see [156].

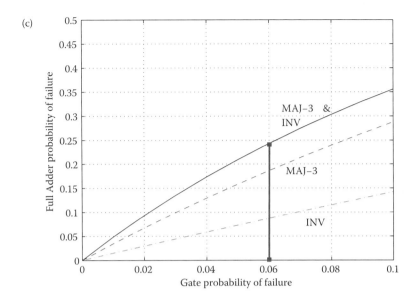

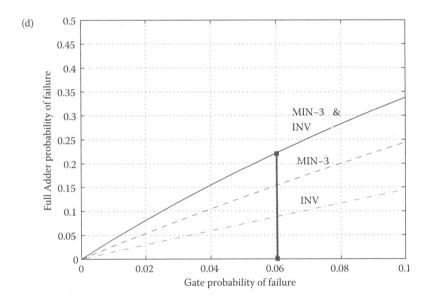

FIGURE 12.6 (Continued).

As mentioned earlier, different types of gates are implemented using similar nano-devices. However, different gates are built using different numbers of devices, not mentioning various logic styles and maybe atypical materials. While a standard CMOS inverter is implemented using two transistors, the NAND-2 and the MAJ-3 gates are implemented in standard CMOS using 4 and respectively 6 transistors. In [146] Forshaw *et al.* suggested an old and simple possible estimation of the probability of failure of a gate:

$$p_{\text{GATE}} \approx 1 - (1 - \varepsilon)^n \tag{12.9}$$

where ε denotes the probability of failure of a nano-device (e.g., CMOS transistor, SET junction, etc.), and n is the number of devices per gate.

(a)

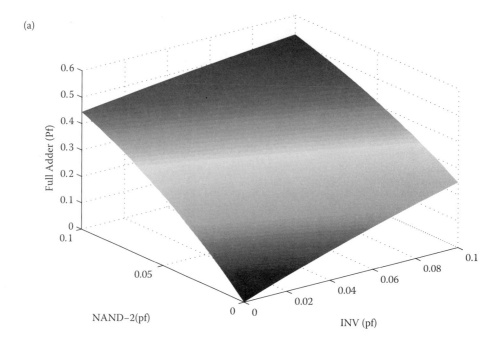

(b)

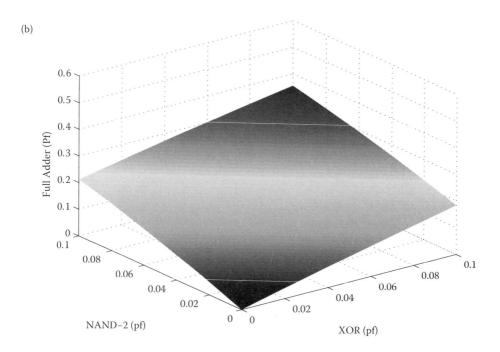

FIGURE 12.7   3D views of the full adders' (FAs) probabilities of failure versus those of their elementary gates: (a) NAND-FA; (b) XOR-FA; (c) MAJ-FA; (d) MIN-FA.

(c)

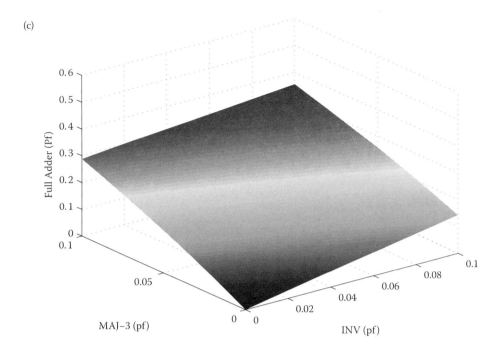

(d)

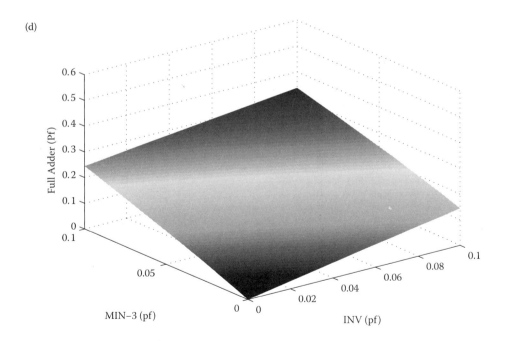

FIGURE 12.7 (Continued).

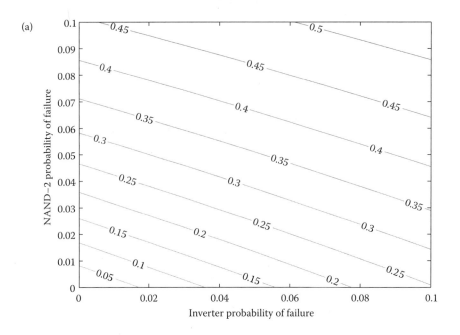

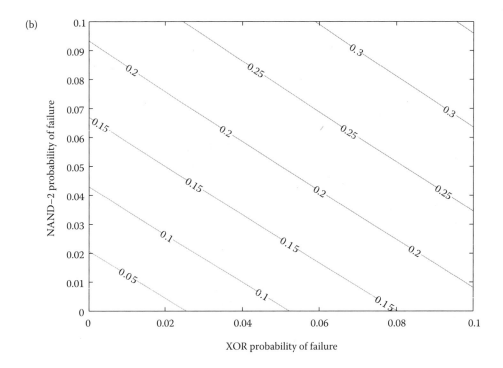

FIGURE 12.8 Contour plots of the full adders' (FAs) probabilities of failure versus those of their elementary gates: (a) NAND-FA; (b) XOR-FA; (c) MAJ-FA; (d) MIN-FA.

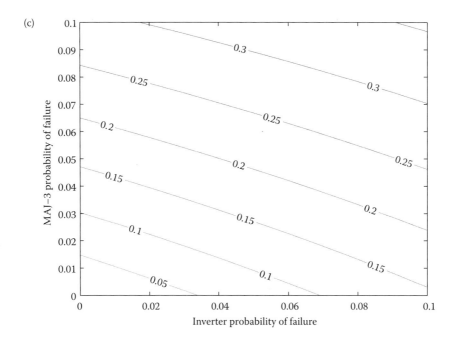

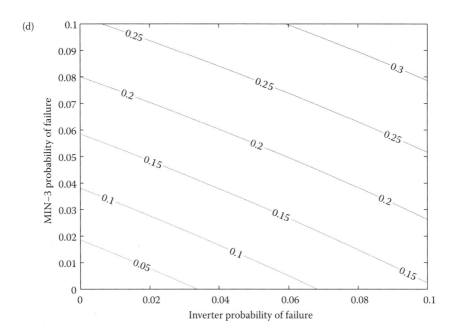

FIGURE 12.8 (Continued).

Figure 12.9 shows the P_{FA} as a function of ε when the gates are implemented using standard CMOS and p_{GATE} is estimated using eq. (12.9). These simulation results reveal that NAND-FA is still the least reliable FA from the ones considered, followed by XOR-FA. However, the difference amongst NAND-FA, XOR-FA and MAJ-FA is reduced when ε is used to estimate P_{FA} (compare Figure 12.9 with Figure 12.6). The simulation results also show that the effect of the inverters' malfunctions on the reliability of the FAs (which use it) is reduced when p_{GATE} is estimated using eq. (12.9). For instance, if $p_{GATE} = 0.04$, the MAJ-FA will fail with $P_{MAJ-FA} = 0.175$ (Figure 12.6 (c)). However, using MAJ-3 gates with $p_{MAJ-3} = 0.04$ implies that $\varepsilon \approx 0.0068$ (using eq. (12.9) for $n = 6$). Hence, $p_{INV} \approx 0.0135$ (using eq. (12.9) for $n = 2$) and the P_{MAJ-FA} reduces to about 0.15 (see Figure 12.9 (c)).

The PTM simulation results reported here show that different FAs have quite different reliabilities, which should be considered (in addition to speed, area, and power consumption) when a FA design is evaluated for implementation using (novel) nanoelectronic devices. The simulation results also show that the reliability of a FA does not only depend on the number of gates it has, but also on the way these gates are implemented and interconnected. Moreover, the accuracy of the FA reliability estimates depends on how accurate the reliabilities of the individual gates are evaluated. An accurate evaluation of the gate's reliability cannot be obtained only by considering the number of devices per gate (like in eq. (12.9)) and should rely on exhaustive Monte Carlo simulations [65], [67] based on better experimental data and fault/defect models [157]–[159]. Unfortunately, preliminary results on these lines point to the fact that bounding the probabilities of failure (with constants) will not be good enough [65]–[69]. Detailed simulations as well as preliminary theoretical explanations are expected to appear in [68]. Last but not least, the time dependence of p_{GATE} should not be forgotten [157]–[159].

12.5 Reliable Adders

12.5.1 A Theoretical Analysis

Binary addition has been studied extensively, starting with the classical (serial) ripple carry (RC) adder and going towards parallel implementations [160]–[164]. It is commonly accepted that RC is the slowest, while Kogge-Stone (KS) [161] is, theoretically, the fastest, but requires about 5× more transistors (i.e., larger area and power—both dynamic due to longer wires, and leakage due to more gates). Still, only a few recent studies have analyzed the reliability of such adders [165]–[170].

To get rough reliability estimates, four different adders will be analyzed theoretically in this sub-chapter. The four adders under investigation are RC, Brent-Kung (BK) [163] (Figure 12.10 (a)), Han-Carlson (HC) [164] (Figure 12.10 (b)), and KS (Figure 12.10 (c)) [161]. All four adders have been characterized by their number of layers, their number of nodes (i.e., blocks), their number of gates, and the length of their wires on the longest path (for more details see [171]).

- The number of layers has been estimated as $\text{Layers}_{RC} = n$, $\text{Layers}_{BK} = 2\log_2(n)$, $\text{Layers}_{HC} = 3 + \log_2(n)$, and $\text{Layers}_{KS} = 2 + \log_2(n)$.
- The number of nodes has been estimated as $\text{Nodes}_{RC} = n$, $\text{Nodes}_{BK} = 4n - \log_2(n) - 2$, $\text{Nodes}_{HC} = 2n + n\log_2(n)/2$, and $\text{Nodes}_{KS} = 2n + n\log_2(n) - (n-1)$.
- The number of gates has been estimated as $\text{Gates}_{RC} = 3n$, $\text{Gates}_{BK} = 4n + 3[2n - \log_2(n) - 2]$, $\text{Gates}_{HC} = 4n + 3n\log_2(n)/2$, and $\text{Gates}_{KS} = 4n + 3[n\log_2(n) - (n-1)]$.
- Finally, the length of the wires on the critical path was estimated geometrically as $\text{Length}_{RC} = n$, $\text{Length}_{BK} = n + 2\text{Layers}_{BK}$, $\text{Length}_{HC} = n + 2\text{Layers}_{HC}$, $\text{Length}_{KS} = n + 2\text{Layers}_{KS}$.

Here, the factor 2 (used for estimating the Length for BK, HC, and KS) is conservative, and accounts for the height of the cells and for the routing space between adjacent layers. For a better understanding compare the layout of a 64-bit KS adder shown in Figure 12.10 (d)—a few other layouts can be seen in [172]–[175]—with the schematic of a 16-bit KS adder presented in Figure 12.10 (c) drawn at a similar

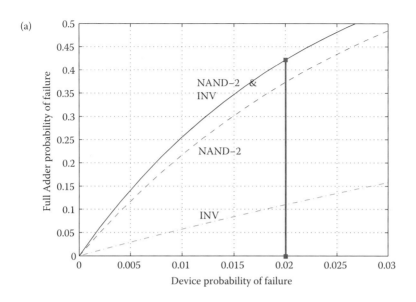

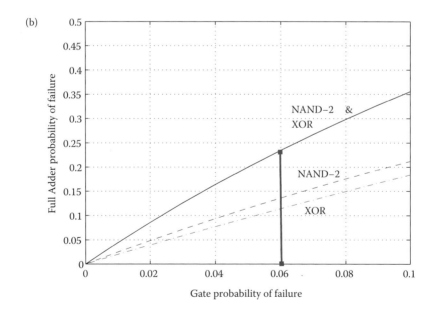

FIGURE 12.9 Probability of failure of four full adders versus the device probability of failure: (a) NAND-FA; (b) XOR-FA; (c) MAJ-FA; (d) MIN-FA.

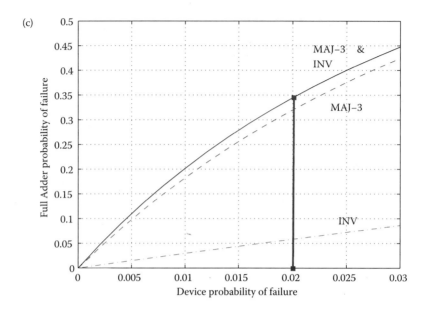

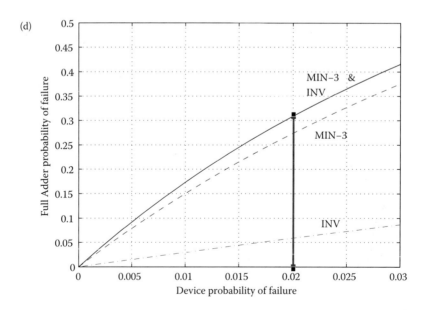

FIGURE 12.9   (Continued).

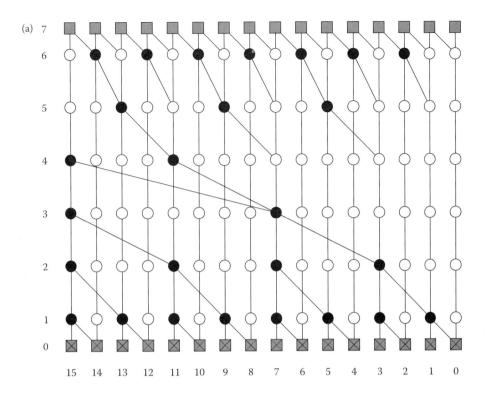

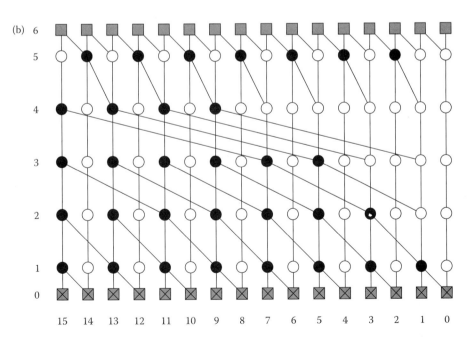

FIGURE 12.10 Interconnect pattern for parallel adders: (a) Brent and Kung (BK); (b) Han and Carlson (HC); (c) Kogge and Stone (KS); and (d) layout of a 64-bit Kogge-Stone adder.

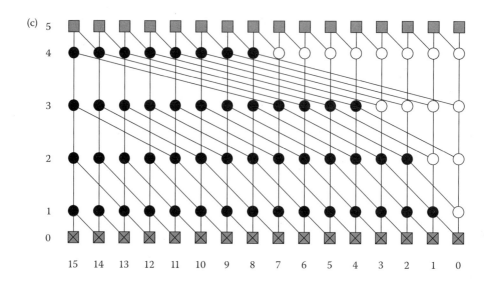

FIGURE 12.10   (Continued).

scale (see also [176]). Remark: All the above equations have been refined using ceilings and floor when appropriate, while even more precise estimates are possible [177], [178].

The reliability of these four adders was quickly (but crudely) estimated based on the number of gates as $P_{ADDER} = 1 - (1 - p_{GATE})^{\#Gates}$ (i.e., using eq. (12.9) at the gate level). [Remark: An alternate estimate could have started from the device failure rate ε, as done in the previous sub-chapter, and use the total number of devices.] The results of these simulations can be seen in Figure 12.11, and are supporting the intuition that a simpler structure is more reliable. [Remember Albert Einstein's quote: *"Everything should be made as simple as possible but not simpler."*] The interested reader can find a few other probabilistic simulation results for RC, BK, and KS in [166] and [167], as well as some theoretical ones for RC in [170]. Additionally, any redundancy scheme is much easier to integrate with RC [179], [180] than with a parallel adder. This claim is in fact valid for any locally connected architecture like e.g., cellular automata, systolic arrays, cellular neural networks, quantum cellular automata, etc.

12.5.2   Reliability Enhancements

Multiplexing has been advocated as a powerful solution for enhancing reliability, unfortunately one that requires large redundancy factors [48], [53], [146]. An exact performance evaluation of multiplexing schemes (MUX) was reported in [51], [52], with simulation results detailed in [67], [69], [65], and [68]. All of these confirm that there is a maximum threshold for p_{GATE} up to which multiplexing schemes improve on the reliability of the individual gates. For SET and $R = 6$ the values obtained from Monte Carlo simulations [65] are ≈ 0.13 for NAND-2 MUX and ≈ 0.225 for MAJ-3 MUX (when implemented using capacitive SET). If the gate failure probability becomes higher than this threshold, the multiplexed schemes are not able to improve anymore, leading in fact to worsening the overall reliability. A simplistic solution would be to increase the redundancy factor to $R = 10$ (i.e, use MAJ-5 gates), but more efficient approaches seem possible.

The integration of MAJ-3 MUX with KS (as an example of parallel adders) was shown in [179], with the conclusion that the connectivity pattern is nontrivial (and the longer wires contribute both to increasing the delay and the switching power). Based on all of the above factors, and on the simulation results from [181], we decided to focus our attention on multiplexed RC (see Figure 12.12). The main block of the RC is the well-known FA. The FA we will consider here is the standard CMOS implementation and is represented by a square. For enhancing reliability, a multiplexed RC with a redundancy factor $R = 3$ is

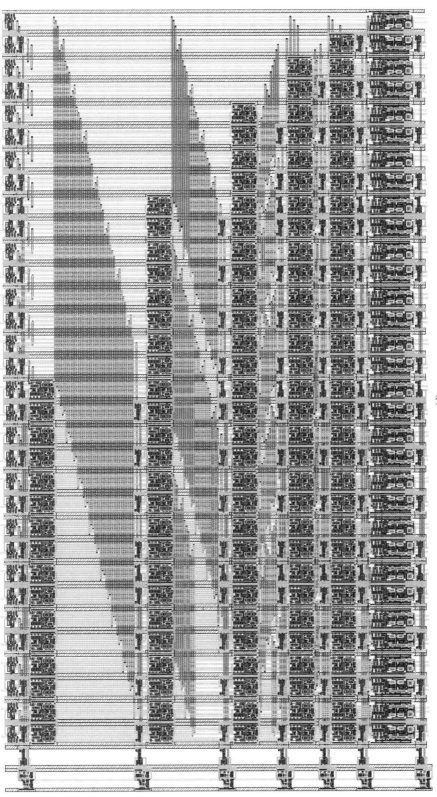

(d)

FIGURE 12.10 (Continued).

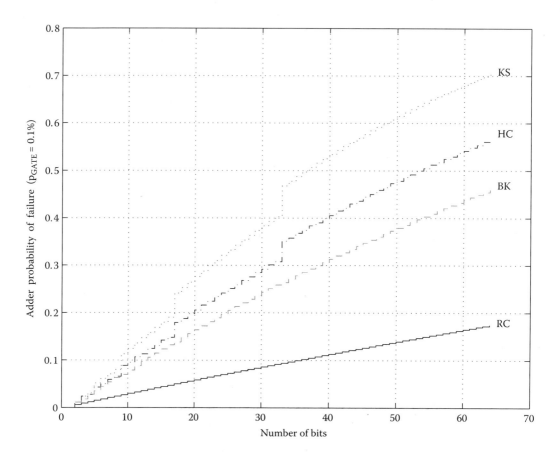

FIGURE 12.11   Estimated probability of failure of four different adders versus the number of input bits: RC, BK, HC, and KS, for a gate failure probability of pGATE = 0.1%.

used (3-RC). [Remark: Here the redundancy factor 3 is being counted at the block level, i.e., there are 3 adders in parallel. This corresponds in fact to a MAJ-3 vN-MUX with $R = 6$, while as we shall see below, it can really be $R = 3$, depending on the way some of the MAJ-3 gates are implemented.] Figure 12.12 (a) presents a block diagram of a standard RC. A multiplexed RC (3-RC) has three FAs (squares) per stage as follows:

- three RCs (used in parallel);
- three MAJ-3 gates (circles) 'vote' on the C_{out} coming from these three FAs (at position i);
- the output of each of these three MAJ-3 is used to drive the three C_{in} of the next three FAs (at bit position $i + 1$).

These ideas can be implemented by several different structures out of which three are presented here (see Figure 12.12 (b), (c), (d)), with each subsequent configuration being 'simpler' than the previous one. The first of the three structures (Figure 12.12 (b)) properly implements three MAJ-3 gates for the restorative stages (circles). This solution doubles the delay and increases power. The second solution is simpler, as the outputs of the FAs are fed to restorative inverters (triangles in Figure 12.12 (c)). The MAJ-3 gates have now been replaced by inverters (similar to the output-wired-inverters [117]). This solution will be faster, and will dissipate less than the previous one, as long as there are no faults/defects. In case of faults/defects, there will be fighting, which will increase the power consumption, while the inverters will try to restore the logic levels. The simplest structure (Figure 12.12 (d)) eliminates even the restorative

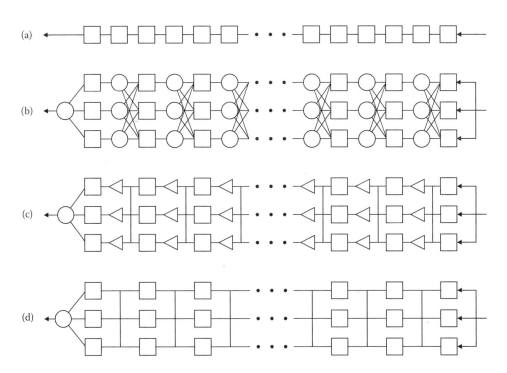

FIGURE 12.12 (a) Classical ripple carry (RC) adder where the square blocks represent full adders (FAs). Three different multiplexed RCs: (b) using MAJ-3 gates (circles) in between FAs; (c) short-circuiting the outputs of three FAs and using three inverters (triangles) to recover the voltage level; and (d) short-circuiting the outputs of three FAs (the voltage is recovered by the next three FAs).

inverters and relies upon the next stage of FAs for providing the needed signal restoration. The MAJ-3 gates have now completely vanished [180]. This solution will be the fastest, as long as there are no faults/defects. In case of faults/defects the shorting of the outputs will result in fighting, increasing the current and the signal propagation delay. These three structures have been tested for stuck-at faults, which is a simplistic scenario, as in practice a fault/defect could manifest itself as an analog/nondeterministic value (see [46], [126], [127], [179]–[181]).

Figure 12.13 compares the results of the multiplexed RC (3-RC) adder with the standard RC, BK, HC, and KS adders (compare Figure 12.13 with Figure 12.11). We have used $p_{GATE} = 0.01$ in Figure 12.13 (a), and $p_{GATE} = 0.1$ in Figure 12.13 (b), i.e., 1% and 10% respectively. These two values have been used in $P_{ADDER} = 1 - (1 - p_{GATE})^{\#Gates}$ for RC, BK, HC and KS. For 3-RC we used a hybrid approach. We have used PTM to calculate the reliability of one multiplexed block, and have used $P_{3-RC} = 1 - (1 - p_{BLOCK})^{\#Blocks}$ (where #Blocks is equal to the number of bits n) to estimate the probability of failure of the 3-RC as a whole. PTM has calculated $p_{BLOCK} = 0.000298$ for $p_{GATE} = 0.01$, and $p_{BLOCK} = 0.028$ for $p_{GATE} = 0.1$ (see [65], [66], [68]). These simulation results show that multiplexing (even at the smallest redundancy factor) can significantly improve reliability. The case when $p_{GATE} = 0.01$ (i.e., 1%) is presented in Figure 12.13 (a). For 16-bit adders the multiplexing scheme was able to reduce the probability of failure of the RC adder from 0.35 to ≈ 0.01 (3-RC), i.e., 35 times. The simulation results for $p_{GATE} = 0.01$ show that none of the adders will work (!), while 3-RC is the only one which might be used, but only for very small length addition. When $p_{GATE} = 0.1$ (i.e., 10% errors), the simulation results from Figure 12.13 (b) show that even the 3-RC adder is not reliable enough anymore. This suggests that more redundancy is needed. One solution is to increase R (long advocated by von Neumann [48]), see 5-RC, 7-RC, 9-RC, and 11-RC in Figure 12.13 (b). We advocate for an alternate solution, namely to keep $R = 3$ and use more reliable gates (like the ones presented in subchapter 12.3) in the design of the elementary FAs, or to design novel highly reliable FAs. Such an approach would reduce p_{BLOCK}, hence could make the combined approach—reliable gates

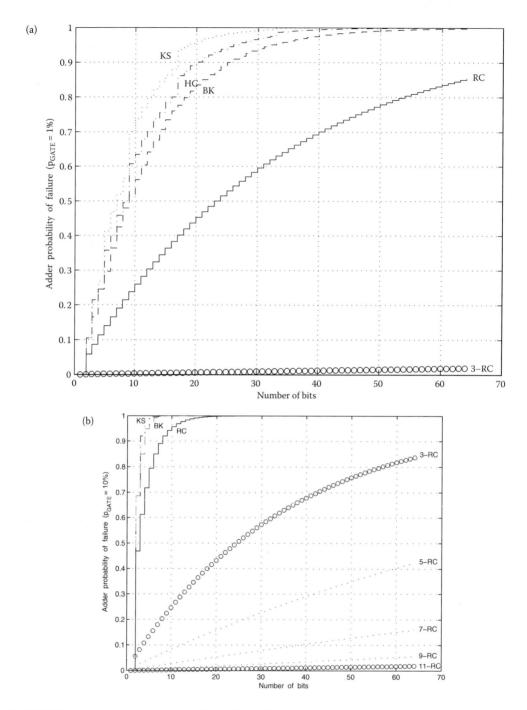

FIGURE 12.13 Estimated probability of failure of four different adders and the multiplexed RC versus the number of input bits: 3-RC, RC, BK, HC, and KS for: (a) gate failure probability of 1%; and (b) gate failure probability of 10%.

by design and low redundancy multiplexing—perform similarly to the 3-RC shown in Figure 12.13 (a). Unfortunately, simulation results for backing such claims require massive Monte Carlo simulations for multiplexing with the hardened gates. Supercomputing or grid computing power is required, as only one simulation for p_{BLOCK} with MAJ-11 takes about 6 hours (on an Intel Xeon 3.73 GHz processor with 16 GB RAM), while it takes over 9 days, for MAJ-19. Still, preliminary results [16] suggest that such a hybrid approach should be the most rewarding one, confirming that redundancy is more efficiently when used at the lowest possible level, i.e., at the device/component level.

The last two solutions (Figure 12.12 (c) and (d)) have one more advantage, namely the fact that an error will cause a fighting. Apparently this would seem to be a disadvantage, as it is going to increase both the current and the power consumption. This is true, and detailed simulations for several technology nodes can be found in [180]. Those simulations clearly show that each error (defect or fault) translates into a current increment. In case of a transient fault, the current will increase, but will return to the nominal value once the fault disappears. In case of a defect, the current will remain higher. This means that we have a very simple way to detect errors: by locally monitoring the current [130], [180], [182]. This will allow us to log faults, and also to identify the case when a fault becomes a defect. Knowing that the circuit can tolerate a few defects, we could set a certain threshold value for a local current sensor (built-in I_{DD} testing). If the current becomes larger than the threshold the local sensing circuit could automatically send a request to the higher level. A control scheme at the higher level could automatically reconfigure the I/O connections, power up a spare unit, and shut down the defective circuit (hot swap). Such an approach combines several reliability schemes:

- highly reliable gates at the lowest level (maybe including multiplexing at the device level);
- low redundancy designs at the gate/circuit level;
- automatic (local current-based) detection at the block level;
- reconfiguration at the block/system level.

This would span not only all the design levels, but also three different schemes for enhancing reliability, leading to a self-healing system. One last remark is that such an approach is not possible for the classical solution presented in Figure 12.12 (b) which includes MAJ-3 gates (circles), as it does not lead to fighting, so the I_{DD} current cannot be used for sensing (i.e., faults and defects cannot be detected by monitoring a current; simulations results can be seen in [180]).

12.6 Power-Performance Considerations

Reliability by itself cannot be the one and only design goal, but it should certainly support performance as represented by the well-known speed-power or energy-delay trade-offs. Many results concerning adders have been reported over the years, with recent ones including both simulations and measurements [176]–[178], [183]–[185]. On top of these, many other results have been reported using threshold logic instead of Boolean logic [155], or mixed Boolean-threshold logic solutions (advocated in [186]), and even capacitive SET threshold logic [19]. Originally, speed was the one and only goal, with speed-power and energy-delay optimizations emerging later [178], [187].

One approach in the bid to lower power dissipation is the reduction of V_{DD}—obviously a very effective way of reducing all the components of power (dynamic, static, and leakage) [188]. The aggressive scaling of V_{DD} to below V_{th} (known as subthreshold operation) has been known and used in ultra-low power designs [189]–[197]. The major disadvantage is its very slow speed. Therefore, subthreshold operation has been considered a poor approach where the much-needed speed is sacrificed for ultra low power (thus limiting its application range). Unfortunately, while reducing the voltage supply might save the day for power consumption, it is detrimentally affecting reliability. This is because in subthreshold noise plays a significant role [191], [29], [30], [188], let alone the higher sensitivity to variations [26]. In spite of these, many subthreshold results have been reported recently [198]–[203] (see also [188]). Falling in this trend, an RC and a KS operated in subthreshold have been compared in [181]. The main conclusions were that:

- the wires reduce the speed advantage of KS over RC to half (from 4.5× to 2.2×); other results showing that wire delays in parallel adders are significant have been presented in [178]);
- the higher speed of KS at a given V_{DD} can be matched by an RC powered at a V_{DD} which is only 10% to 20% higher (in subthreshold);
- at equal speeds the RC still maintains both its power and energy advantages.

Obviously, wires are playing an important role in subthreshold, both affecting the delay and influencing the dynamic power. The simulation results reported in [181] were used to fine-tune our estimates for the Delay of the four adders analyzed in the previous sub-chapter. The Delay of the adders was evaluated taking into account both the number of gates on the longest path (which is $Gates_{Longest\_path} = 2 \times Layers - 2$ for KS, HC, and BK, while being only $Gates_{Longest\_path} = Layers$ for RC), and the length of the wires on the longest path (Length):

$$Delay = Gates_{Longest\_path} + \alpha \times Length. \tag{12.10}$$

When $\alpha = 0$, only the gates are contributing to the delay, while the wires are not. By increasing α, the wires (Length) start affecting the delay more and more. A value of $\alpha \approx 0.25$ was determined by fitting the results of the simulations obtained for RC and KS in [181] (i.e., when operating in subthreshold).

For estimating power, both the length on the longest path (Length) and the total number of gates (Gates) were considered. These account for part of the switching capacitance (Length), hence dynamic power is sub-estimated for parallel adders (BK, HC, and KS), and for the total leakage power (Gates). A factor β was used for specifying indirectly the ratio between dynamic and leakage power leading to the following estimate:

$$Power = (1 - \beta) \times Length + \beta \times Gates. \tag{12.11}$$

Closer estimates of power are possible (see [177], [178], [183]). Preliminary estimations indicate that these do not lead to significant changes.

Finally, the power-delay-product (PDP) and the energy-delay-product (EDP) have been estimated in a straightforward manner as:

$$PDP = Power \times Delay = Energy \tag{12.12}$$
$$EDP = PDP \times Delay. \tag{12.13}$$

The results of these approximations can be seen in Figure 12.14, where Delays, PDPs, and EDPs are shown for two cases:

- Without wire delays and with no leakage power, i.e., for $\alpha = 0$ and $\beta = 0$ (see Figure 12.14 (a), (c), (e)).
- With a delay comparable with that of subthreshold operation at 65 nm, and an estimated leakage power of 33% of the estimated dynamic power, i.e., for $\alpha = 0.25$ and $\beta = 0.25$ ($\beta/(1 - \beta) = 0.25/0.75 = 1/3$) (Figure 12.14 (b), (d), (f)).

Obviously, when scaling down CMOS, α will increase, so the delays of the KS, HC, and BK adders will also increase, but KS, HC, and BK are always going to be faster than RC (as $\alpha < 1$). The more interesting results are the ones showing PDPs and EDPs. For $\alpha = 0$, RC has the best PDP for $n < 58$ (Figure 12.14 (c)), and the best EDP for $n < 26$ (Figure 12.14 (e)). When wires are properly accounted for (i.e., for $\alpha = 0.25$), the RC gets the best PDP always (Figure 12.14 (e)), while achieving the best EDP for any $n < 24$ (Figure 12.14 (f)). These results should get even better for a real implementation, as the power for KS, HC, and BK were underestimated, while being reasonably correct for RC. The plots in Figure 12.14 support the intuitive claim that serial adders could achieve better PDP and EDP than parallel adders in general, but clearly have more weight when one or more of the following are true:

- wires are introducing substantial delays;
- leakage power represents a significant percentage of the power consumption;
- circuits are operated in subthreshold;
- elementary devices have small gain.

These four different adders show very different power-delay tradeoffs both when working correctly and when faulty. The somehow unexpected result is that a very reliable adder (RC) can also be a low-power

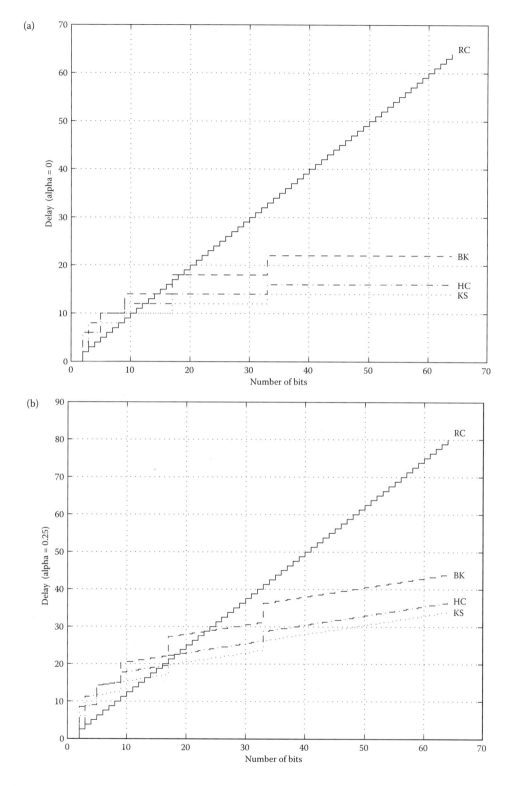

FIGURE 12.14 Estimates of the delay ((a) and (b)), the power-delay-product ((c) and (d)), and the energy-delay-product ((e) and (f)) for the four different adders (RC, BK, HC, and KS) with ((a), (c), and (e)) and without wires ((b), (d), and (f)).

(c)

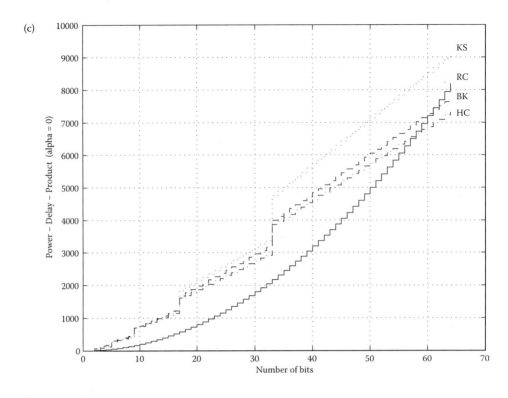

(d)

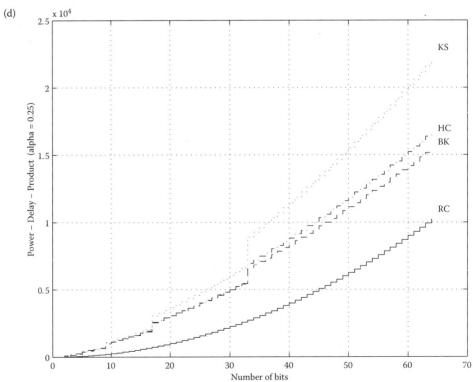

FIGURE 12.14   (Continued).

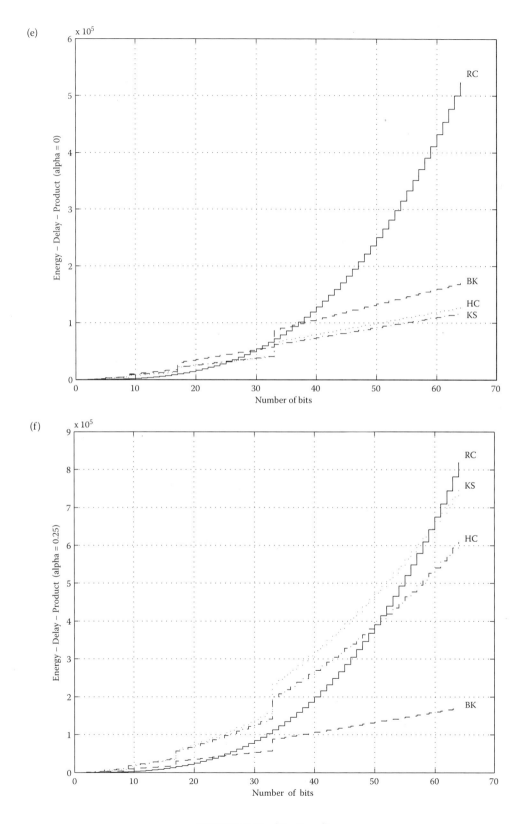

FIGURE 12.14 (Continued).

one, but it is clearly not the fastest one. Still, it also looks like the speed advantage of parallel adders will reduce when gain will become very small (e.g., SET, molecular, DNA).

Finally, subthreshold operation might become an interesting design approach particularly because the operation speeds are improving as scaling proceeds towards smaller technology nodes, but also because it might allow for an easier hybridization with ultra low-voltage technologies (SET, molecular, DNA). It is not difficult to envision a situation in which designs in older technology nodes operating at standard power supply voltages (nominal V_{DD}) would have comparable speeds to those in advanced technologies but operating in subthreshold. This would mean for example that a microprocessor designed to run at 1 GHz in 0.18 μm (or 0.13 μm) at nominal V_{DD} could be redesigned to operate at 1 GHz in subthreshold in say 32 nm (or 22 nm). The main advantage would be *a power reduction of one to two orders of magnitude.* Even incorporating spatial redundancy, such a solution (1 GHz, 22 nm, subthreshold) could be significantly lower power than the one from which we have started (1 GHz, 0.18 μm, nominal V_{DD}), while also being reliable enough. The research path is being opened for portable applications that could enjoy extended operation times (see e.g. [188], [204]), and also towards low-power analog computations [179], [205]–[208].

12.7   Conclusions

Although we have not discussed memory and communications, one could expect that these will be non-volatile and respectively (at least partly) optical and asynchronous, but it is not clear how future computations will be implemented. Irrespective of the technology, reliability will be a concern at all levels. Based on what we have presented in this chapter we should expect the following.

- Optimization will become even more difficult, as there are many options for trading power and speed for reliability ("complexity strikes again").
- Serial solutions (or more generally regular ones like e.g., cellular automata, systolic arrays, cellular neural networks, quantum cellular automata, etc.) will be a good bet from the power-reliability perspective [181] — in addition to the fabrication/manufacturing point of view.
- The probabilities of failure should be represented as functions [65], [157] and not as bounding constants (as they are now), and novel EDA tools will have to take advantage of these.
- Algorithms for quickly and closely estimating the reliability of large (complex) systems will have to be developed. Systematic approaches for estimating reliability will have to take advantage of the results obtained at lower levels, e.g., experimental and Monte Carlo results from the device and gate levels should be incorporated into reliability tools at higher levels.
- Subthreshold operation might be a simple and practical way to test novel reliability ideas, as a subthreshold design in 65 nm might be "as sensitive as" a CMOS design in 22 nm at nominal V_{DD} (or a molecular, or a SET one).
- Redundant designs should not be weighted with respect to (their) redundancy factors (as commonly done in the literature), but with respect to power, energy, and area, as it is customary in the VLSI community. This also implies that circuit optimization will have to be done in four dimensions: timing, area, power, and reliability.
- Reliable designs for implementing computations can be low-power. Ad hoc reliable designs within a low power budget have already been reported [123], [130], [209]–[212]. There is no simple explanation for this, but high level approaches suggest that working at the reliability limit—like what the Brain seems to be doing—optimizes energy [213]–[221], [14]. Moreover, encoding could help both for reliable communications and for reducing power.
- For certain designs, defects and faults could manifest themselves as reasonably well defined current steps. If this is the case, current sensors (for limits see [222]) could automatically trigger reconfiguration at the higher levels [180].
- A lot of effort has to be invested for the development of EDA tools for quickly and precisely estimating reliability at the system level, and for the integration with current EDA tools (used for estimating area, delay, and power) [1].

Finally, one more aspect that certainly deserves a lot of attention—and was not discussed in this chapter—is that of asynchronous or self-timed circuits. MAJ gates are well positioned here again, as a C-Muller element is nothing more than a MAJ gate with feedback. Such designs will be needed for two reasons: the fact that delay variations will have to be compensated (and it might be cheaper to use self-timed circuits than hardwired redundancy, while it seems that nobody knows a good/detailed answer to this tradeoff), and the fact that the power required for clock distribution is already prohibitive. Unfortunately, we are not aware of optimal energy-delay organizations for asynchronous designs, while clock distribution is already limiting the maximum frequency. It follows that quite a number of items should be added to an already busy reliability-related research agenda.

References

[1] W. Kuo, "Challenges related to reliability in nano electronics," *IEEE Trans. Reliability*, vol. 55, no. 4, Dec. 2006, pp. 569–570.

[2] *Proc. IEEE/ACM Intl. Conf. Computer-Aided Design ICCAD'06*, San Jose, CA, USA, Nov. 2006.

[3] G.E. Moore, "Cramming more components onto integrated circuits," *Electronics*, vol. 38, no. 8, Apr. 19, 1965, pp. 114–117.

[4] Semiconductor Industry Association (SIA), International Technology Roadmap for Semiconductors (ITRS), 2005 Edition and 2006 Update, Intl. SEMATECH, Austin, TX, USA. Available at: http://public.itrs.net/

[5] K.F. Goser, C. Pacha, A. Kanstein, and M.L. Rossmann, "Aspects of systems and circuits for nanoelectronics," *Proc. IEEE*, vol. 85, Apr. 1997, pp. 558–573.

[6] S. Borkar, "Design challenges of technology scaling," *IEEE Micro*, vol. 19, Jul.-Aug. 1999, pp. 23–29.

[7] T. Sakurai, "Design challenges for $0.1\mu m$ and beyond," *Proc. Asia & South Pacific Design Autom. Conf. ASP-DAC'00*, Tokyo, Japan, Jan. 2000, pp. 553–558.

[8] D.J. Frank, R.H. Dennard, E. Nowak, P.M. Solomon, Y. Taur, and H.-S.P. Wong, "Device scaling limits of Si MOSFETs and their application dependencies," *Proc. IEEE*, vol. 89, Mar. 2001, pp. 259–288.

[9] R.W. Keyes, "Fundamental limits of silicon technology," *Proc. IEEE*, vol. 89, Mar. 2001, pp. 227–239.

[10] R.E. Bryant, K.-T. Cheng, A.B. Kahng, K. Keutzer, W. Maly, R. Newton, L. Pillegi, J.M. Rabaey, and A. Sangiovanni-Vincentelli, "Limitations and challenges for computer-aided design technology for CMOS VLSI," *Proc. IEEE*, vol. 89, Mar. 2001, pp. 341–365.

[11] J.A. Davis, R. Venkatesan, A. Kaloyeros, M. Beylansky, S.J. Souri, K. Banerjee, K.C. Saraswat, A. Rahman, R. Reif, and J.D. Meindl, "Interconnect limits on Gigascale integration (GSI) in the 21st century," *Proc. IEEE*, vol. 89, Mar. 2001, pp. 305–324.

[12] Q. Chen, and J.D. Meindl, "Nanoscale metal-oxide-semiconductor field-effect transistors: Scaling limits and opportunities," *Nanotechnology*, vol. 15, Jul. 2004, pp. S549–S555.

[13] V. Beiu, U. Rückert, S. Roy, and J. Nyathi, "On nanoelectronic architectural challenges and solutions," *Proc. IEEE Conf. Nanotech. IEEE-NANO'04*, Munich, Germany, Aug. 2004, pp. 628–631.

[14] V. Beiu, "Limits, challenges, and issues in nano-scale and bio-inspired computing," in M.M. Eshaghian-Wilner (Ed.), *Bio-inspired and Nano-scale Integrated Computing*, John Wiley & Sons, 2007/8 (in progress).

[15] S. Tiwari, A. Kumar, C.C. Liu, H. Lin, S.K. Kim, and H. Silva, "Electronics at nanoscale: Fundamental and practical challenges, and emerging directions," *Proc. Intl. Conf. Emerging Tech. – Nanolelectr. NanoSingapore'06*, Singapore, Jan. 2006, pp. 481–486.

[16] V. Beiu, and U. Rückert (Eds.), *Emerging Brain Inspired Nano Architectures*, World Scientific Press, Singapore, 2007/8 (in progress).

[17] J.D. Meindl, Q. Chen, and J.A. Davis, "Limits on silicon nanoelectronics for Terascale integration," *Science*, vol. 293, 14 Sep. 2001, pp. 2044–2049.

[18] S.E. Lyshevski, "Nanotechnology and super high-density three-dimensional nanoelectronics and nanoICs," *Proc. IEEE Conf. Nanotech. IEEE NANO'03*, San Francisco, CA, USA, Aug. 2003, pp. 655–658. [For more details see Chapter 6 in this book.]

[19] M.H. Sulieman, V. Beiu, "Characterization of a 16-bit threshold logic single electron adder," *Proc. Intl. Symp. Circ. & Sys. ISCAS'04*, Vancouver, Canada, May 2004, pp. 681–684.

[20] C. Constantinescu, "Trends and challenges in VLSI circuit reliability," *IEEE Micro*, vol. 23, Jul.-Aug. 2003, pp. 14–19.

[21] K.K. Likharev, "Single-electron devices and their applications," *Proc. IEEE*, vol. 87, Apr. 1999, pp. 606–632.

[22] U. Feldkamp, and C.M. Niemeyer, "Rational design of DNA nanoarchitectures," *Angew. Chem. Intl. Ed.*, vol. 45, 13 Mar. 2006, pp. 1856–1876.

[23] C. Lin, Y. Liu, S. Rinker, and H. Yan, "DNA tile based self-assembly: Building complex nanoarchitectures," *ChemPhysChem*, vol. 7, 11 Aug. 2006, pp. 1641–1647.

[24] V. Sverdlov, H. Kosina, and S. Selberherr, "Current transport in nanoelectronic semiconductor devices," *Proc. Intl. Conf. Emerging Tech. – Nanolelectr. NanoSingapore'06*, Singapore, Jan. 2006, pp. 490–495.

[25] S.K. Shukla, R. Karri, S.C. Goldstein, F. Brewer, K. Banerjee, and S. Basu, "Nano, quantum, and molecular computing: Are we ready for the validation and test challenges?," *Proc. Intl. High-Level Design Validation & Test Workshop HLDVT'03*, San Francisco, CA, USA, Nov. 2003, pp. 3–7.

[26] A.R. Brown, A. Asenov, and J.R. Watling, "Intrinsic fluctuation in sub 10-nm double-gate MOSFETs introduced by discreteness of charge and matter," *IEEE Trans. Nanotech.*, vol. 1, Dec. 2002, pp. 195–200.

[27] S. Borkar, "Designing reliable systems from unreliable components: The challenges of transistor variability and degradation," *IEEE Micro*, vol. 25, no. 6, Nov.-Dec. 2005, pp. 10–16.

[28] C. Mead, and L. Conway, *Introduction to VLSI Systems*, Addision-Wesley, 1980.

[29] V. Degalahal, R. Ramanarayanan, N. Vijaykrishnan, Y. Xie, and M. J. Irwin, "The effect of threshold voltages on the soft error rate," *Proc. Intl. Symp. Quality Electr. Design ISQED'04*, San Jose, CA, USA, Mar. 2004, 503–508.

[30] P. Sivakumar, M. Kistler, S.W. Keckler, D. Burger, and L. Alvisi, "Modeling the effect of technology trends on soft error rate of combinatorial logic," *Proc. Intl. Conf. Dependable Sys. & Networks DSN'02*, Washington DC, USA, Jun. 2002, pp. 389–398.

[31] M. Nicolaidis, "Design for soft error mitigation," *IEEE Trans. Dev. & Material Reliability*, vol. 5, Sep. 2005, pp. 405–418.

[32] D. Rossi, M. Omaña, F. Toma, and C. Metra, "Multiple transient faults in logic: An issue for next generation ICs?," *Proc. Intl. Symp. Defect & Fault Tolerant VLSI Sys. DFT'05*, Monterey, CA, USA, Oct. 2005, pp. 352–360.

[33] K. Constantinides, S. Plaza, J. Blome, B. Zhang, V. Bertacco, S. Mahlke, T. Austin, and M. Orshansky, "Assessing SEU vulnerability via circuit-level timing analysis," *Proc. Workshop Arch. Reliability WAR-1*, Barcelona, Spain, Nov. 2005. Available at: http://cccp.eecs.umich.edu/papers/kypros-war05.pdf

[34] V. Beiu, "VLSI implementation of self-testable real content addressable memory," *Proc. Intl. Conf. Control Sys. & Comp. Sci. CSCS'84*, Bucharest, Romania, May 1984, pp. 400–405.

[35] L. Anghel, N. Achouri, and M. Nicolaidis, "Evaluation of memory built-in self repair techniques for high defect density technologies," *Proc. Pacific Rim Intl. Symp. Dependable Comp. PRDC'04*, Papeete, Tahiti, French Polynesia, Mar. 2004, pp. 315–320.

[36] W. Burleson, and A. Maheshwari, *VLSI Interconnects: A Design Perspective*, Elsevier/Morgan Kaufman, San Francisco, CA, USA, 2007, in press.

[37] J.A. Hutchby, G.I. Bourianoff, V.V. Zhirnov, J.E. Brewer, "Extending the road beyond CMOS," *IEEE Circ. & Dev. Mag.*, vol. 18, Mar. 2002, pp. 28–41.

[38] R. Waser (Ed.), *Nanoelectronics and Information Technology*, 2nd Edition, Wiley-VCH, 2005.

[39] J.R. Heath, P.J. Kuekes, G.S. Snider, and R.S. Williams, "A defect-tolerant computer architecture: Opportunities for nanotechnology," *Science*, vol. 280, 12 Jun. 1998, pp. 1716–1721.

[40] K. Nikolić, A. Sadek, and M. Forshaw, "Architectures for reliable computing with unreliable nanodevices," *Proc. IEEE Conf. Nanotech. IEEE-NANO'01*, Maui, HI, USA, Oct. 2001, pp. 254–259.

[41] K. Nikolić, A. Sadek, and M. Forshaw, "Fault-tolerant techniques for nanocomputers," *Nanotechnology*, vol. 13, May 2002, pp. 357–362.

[42] V. Beiu, "Neural inspired architectures for nanoelectronics: Highly reliable, ultra low-power, reconfigurable, asynchronous," *Special Session, Neural Information Processing Systems NIPS'03*, Whistler, Canada, Dec. 2003. Available at: http://www.eecs.wsu.edu/~vbeiu/workshop_nips03/

[43] M. Forshaw, R. Stadler, D. Crawley, and K. Nikolić, "A short review of nanoelectronic architectures," *Nanotechnology*, vol. 15, Feb. 2004, pp. S220–S223.

[44] M. Forshaw, D. Crawley, P. Jonker, J. Han, and C. Sotomayor Torres, "A review of the status of research and training into architectures for nanoelectronic and nanophotonic systems in the European research area," *Tech. Rep. FP6/2002/IST/1 Contract #507519*, Jul. 2004. Available at: http://www.ph.tn.tudelft.nl/People/albert/papers/NanoArchRev_finalV2.pdf

[45] J.A.B. Fortes, "Future challenges in VLSI system design," *Proc. Intl. Symp. VLSI ISVLSI'03*, Tampa, FL, USA, Feb. 2003, pp. 5–7.

[46] T. Lehtonen, J. Plosila, and J. Isoaho, "On fault tolerance techniques towards nanoscale circuits and systems," *Turku Center for CS (TUCS) Tech. Rep. No. 708*, Univ. of Turku, Dept. of IT, Turku, Finland, Aug. 2005. Available at http://www.tucs.fi/publications/attachnet.php?fname=TR708.pdf.

[47] M.T. Niemeir, M. Crocker, X. Sharon Hu, and M. Lieberman, "Using CAD to shape experiments in molecular QCA," *Proc. Intl. Conf. Comp.-Aided Design ICCAD'06*, San Jose, CA, USA, Nov. 2006, pp. 907–914.

[48] J. von Neumann, "Probabilistic logics and the synthesis of reliable organisms from unreliable components," in C.E. Shannon, and J. McCarthy (Eds.): *Automata Studies*, Princeton Univ. Press, Princeton, NJ, USA, 1956, pp. 43–98.

[49] E.F. Moore, and C.E. Shannon, "Reliable circuits using less reliable relays," *J. Franklin Inst.*, vol. 262, 1956, pp. 191–208.

[50] S. Winograd, and J.D. Cowan, *Reliable Computation in the Presence of Noise*, MIT Press, Cambridge, MA, USA, 1963.

[51] S. Roy, and V. Beiu, "Multiplexing schemes for cost-effective fault-tolerance," *Proc. IEEE Conf. Nanotech. IEEE-NANO'04*, Munich, Germany, Aug. 2004, pp. 589–592.

[52] S. Roy, and V. Beiu, "Majority multiplexing—economical redundant fault-tolerant designs for nanoarchitectures," *IEEE Trans. Nanotech.*, vol. 4, no. 4, Jul. 2005, pp. 441–451.

[53] A.S. Sadek, K. Nikolić, and M. Forshaw, "Parallel information and computation with restitution for noise-tolerant nanoscale logic networks," *Nanotechnology*, vol. 15, Jan. 2004, pp. 192–210.

[54] A.J. KleinOsowski, and D.J. Lilja, "The NanoBox project: Exploring fabrics of self-correcting logic blocks for high defect rate molecular device technologies," *Proc. IEEE Annual Symp. VLSI ISVLSI'04*, Lafayette, LA, USA, Feb. 2004, pp. 19–24.

[55] A.J. KleinOsowski, K. KleinOsowski, V. Rangarajan, P. Ranganath, and D.J. Lilja, "The recursive NanoBox processor grid: A reliable system architecture for unreliable nanotechnology devices," *Proc. Intl. Conf. Dependable Sys. & Networks DSN'04*, Florence, Italy, Jun. 2004, pp. 167–176.

[56] M. Zhang, and N.R. Shanbhag, "A CMOS design style for logic circuit hardening," *Proc. Intl. Reliability Physics Symp. IRPS'05*, San Jose, CA, USA, Apr. 2005, pp. 223–229.

[57] K. Constantinides, S. Plaza, J. Blome, B. Zhang, V. Bertacco, S. Mahlke, T. Austin, and M. Orshansky, "BulletProof: A defect-tolerant CMP switch architecture," *Proc. Intl. Symp. High-Perf. Comp. Arch. HPCA'06*, Austin, TX, USA, Feb. 2006, pp. 5–16.

[58] S. Shyam, K. Constantinides, S. Phadke, V. Bertacco, and T. Austin, "Ultra low-cost defect protection for microprocessor pipelines," *Proc. Intl. Conf. Arch. Support Prog. Lang. & Op. Sys. ASPLOS'06*, San Jose, CA, USA, Oct. 2006, pp. 73–82.

[59] S. Mitra, M. Zhang, N. Seifert, T.M. Mak, and K.S. Kim, "Soft error resilient system design through error correction," *Proc. IFIP Intl. Conf. VLSI VLSI-SoC'06*, Nice, France, Oct. 2006, pp. 332–337. Available at: www.gigascale.org/pubs/893/ifip06%5B1%5D.final.v7.pdf

[60] V. Beiu, "The quest for practical redundant computations" (Invited Plenary Talk), *Proc. Intl. Conf. Microelectronics ICM'05*, Islamabad, Pakistan, Dec. 2005, p. xix.

[61] V.P. Roychowdhury, D.B. Janes, and S. Bandyopadhyay, "Nanoelectronic architectures for Boolean logic," *Proc. IEEE*, vol. 85, Apr. 1997, pp. 574–588.

[62] M. Macucci, G. Iannaccone, M. Governale, C. Ungarelli, S. Francaviglia, M. Girlanda, L. Bonci, and M. Gattobigio, "Critical assessment of the QCA architecture as a viable alternative to large scale integration," in H. Nakashima (Ed.): *Mesoscopic Tunneling Devices*, Research Signpost, Kerala, India, 2004, pp. 161–202.

[63] V. Beiu, "Self-testable and self-repairable units: A must for VLSI structures," *Proc. Annual Conf. Electr. Telecom. Autom. Ctrl. CNETAC'84*, Bucharest, Romania, Nov. 1984 (in Romanian).

[64] J.E. Harlow III, "Toward design technology in 2020: Trends, issues, and challenges," *Proc. Intl. Symp. VLSI ISVLSI'03*, Tampa, FL, USA, Feb. 2003, pp. 3–4.

[65] V. Beiu, W. Ibrahim, Y.A. Alkhawwar, and M.H. Sulieman, "Gate failures effectively shape multiplexing," *Proc. Intl. Symp. Defect & Fault Tolerance VLSI Sys. DFT'06*, Arlington/Washington DC, USA, Oct. 2006, pp. 29–40.

[66] W. Ibrahim, V. Beiu, and Y.A. Alkhawwar, "On the reliability of four full adder cells," *Proc. Intl. Design & Test Workshop IDT'06*, Dubai, UAE, Nov. 2006, in press.

[67] V. Beiu, and M.H. Sulieman "On practical multiplexing issues," *Proc. IEEE Conf. Nanotech. IEEE-NANO'06*, Cincinnati, OH, USA, Jul. 2006, pp. 310–313.

[68] V. Beiu, W. Ibrahim, and S. Lazarova-Molnar, "What von Neumann did not say about multiplexing: Beyond gate failures — The gory details," *Intl. Work-conf. Artif. Neural Networks IWANN'07*, San Sebastián, Spain, Jun. 2007, in press.

[69] M.H. Sulieman, and V. Beiu, "Design and analysis of SET circuits: Using MATLAB modules and Simon," *Proc. IEEE Conf. Nanotech. IEEE-NANO'04*, Munich, Germany, Aug. 2004, pp. 618–621.

[70] H.A. Haus, and R.B. Adler, *Circuit Theory of Linear Noisy Networks*, MIT Press, Cambridge, MA, USA, 1959.

[71] G.I. Kirienko, "On self-correcting schemes from functional elements," *Prob. Kibern.*, vol. 12, 1964, pp. 29–37.

[72] W.H. Pierce, *Failure-Tolerant Computer Design*, Academic Press, New York, NY, USA, 1965.

[73] J.D. Cowan, "Synthesis of reliable automata from unreliable components," in E.R. Caianello (Ed.), *Automata Theory*, Academic Press, New York, NY, USA, 1966, pp. 131–145.

[74] G.I. Kirienko, "Synthesis of self-correcting schemes from functional elements for the case of growing number of faults in the scheme," *Diskret. Anal.*, vol. 16, 1970, pp. 38–43.

[75] D. Uhlig, "On the synthesis of self-correcting schemes from functional elements with a small number of reliable elements," *Math. Notes Acad. Sci. USSR*, vol. 15, 1974, pp. 558–562.

[76] R.L. Dobrushin and S.I. Ortyukov, "Lower bound for the redundancy of self-correcting arrangements of unreliable functional elements," *Prob. Inform. Transm.*, vol. 13, no. 1, Jan.–Mar. 1977, pp. 59–65.

[77] R.L. Dobrushin and S.I. Ortyukov, "Upper bound for the redundancy of self-correcting arrangements of unreliable functional elements," *Prob. Inform. Transm.*, vol. 13, no. 3, Jul.–Sep. 1977, pp. 203–218.

[78] N. Pippenger, "On networks of noisy gates," *Proc. Intl. Symp. Foundations Comp. Sci. FOCS'85*, Portland, OR, USA, Oct. 1985, pp. 30–38.

[79] D. Uhlig, "Reliable networks from unreliable gates with almost minimal complexity," *Proc. Fundamentals Comp. Theory*, Springer LNCS, vol. 278, 1987, pp. 462–469.

[80] R. Reischuk, and B. Schmeltz, "Area efficient methods to increase the reliability of combinatorial circuits," *Proc. Intl. Symp. Theoretical Aspects of Comp. Sci. STACS'89*, Paderborn, Germany, Feb. 1989, Springer, LNCS, vol. 349, pp. 314–326.

[81] N. Pippenger, "Reliable computation by formulas in the presence of noise," *IEEE Trans. Info. Theory*, vol. 34, no. 2, Mar. 1988, pp. 194–197.

[82] T. Feder, "Reliable computation by networks in the presence of noise," *IEEE Trans. Info. Theory*, vol. 35, no. 3, May 1989, pp. 596–571.

[83] N. Pippenger, "Developments in 'The synthesis of reliable organism from unreliable components'," *Proc. Intl. Symp. Pure Maths.*, vol. 50, 1990, pp. 311–324.

[84] N. Pippenger, G.D. Stamoulis, and J.N. Tsitsiklis, "On a lower bound for the redundancy of reliable networks with noisy gates," *IEEE Trans. Info. Theory*, vol. 37, no. 3, May 1991, pp. 639–643.

[85] J. Han, and P. Jonker, "A system architecture solution for unreliable nanoelectronic devices," *IEEE Trans. Nanotech.*, vol. 1, no. 4, Dec. 2002, pp. 201–208.

[86] A.M. Johnson Jr., and M. Malek, "Survey of software tools for evaluating reliability, availability, and serviceability," *ACM Comp. Surveys*, vol. 20, no. 4, Dec. 1988, pp. 227–269.

[87] R.A. Sahner, and K.S. Trivedi, "Reliability modeling using SHARPE," *IEEE Trans. Reliability*, vol. 36, no. 2, Jun. 1987, pp. 186–193.

[88] M.A. Boyd, and S.J. Bavuso, "Simulation modeling for long duration spacecraft control systems," *Proc. Annual Reliability & Maintainability Symp. RAMS'93*, Atlanta, GA, USA, Jan. 1993, pp. 106–113.

[89] J.B. Dugan, B. Venkataraman, and R. Gulati, "DIFTree: A software package for the analysis of dynamic fault tree models," *Proc. Annual Reliability & Maintainability Symp. RAMS'97*, Philadelphia, PA, USA, Jan. 1997, pp. 64–70.

[90] D. Coppit, and K.J. Sullivan, "Galileo: A tool built for mass-market applications," *Proc. Intl. Conf. Software Eng.*, Limerick, Ireland, Jun. 2000, pp. 273–282.

[91] M. Kwiatkowska, G. Norman, D. Parker, and R. Segala, "Symbolic model checking of concurrent probabilistic systems using MTBDDs and simplex," *Tech. Rep. CSR-99-01*, School of Comp. Sci., Univ. of Birmingham, Birmingham, UK, Jan. 22, 1999. Available at: http://www.cs.bham.ac.uk/~dxp/papers/CSR-99-01.pdf

[92] M. Kwiatkowska, G. Norman, and D. Parker, "Verifying randomized distributed algorithms with PRISM," *Proc. Workshop Adv. Verif. WAVe'00*, Chicago, IL, USA, Jul. 2000. Available at: http://www.cs.bham.ac.uk/~dxp/ papers/wave00-prism.pdf

[93] G. Horton, "A new paradigm for the numerical simulation of stochastic Petri nets with general firing times," *Proc. European Simulation Symp. ESS'02*, Dresden, Germany, Verlag, Oct. 2002. Available at: http://www.scs-europe.net/conf/ess2002/meth-20.pdf

[94] S. Lazarova-Molnar, "The proxel-based method: Formalisation, analysis and applications," *PhD dissertation*, Faculty of Informatics, Otto-von-Guericke-Universität, Magdeburg, Germany, Nov. 2005. Available at: http://diglib.uni-magdeburg.de/Dissertationen/2005/sanlazarova.pdf

[95] K.N. Patel, I.L. Markov, and J.P. Hayes, "Evaluating circuit reliability under probabilistic gate-level fault models," *Proc. Intl. Workshop Logic Synthesis IWLS'03*, Laguna Beach, CA, USA, May 2003, pp. 59–64.

[96] V.L. Levin, "Probability analysis of combination systems and their reliability," *Eng. Cyber.*, vol. 6, Nov-Dec. 1964, pp. 78–84.

[97] S. Krishnaswamy, G.F. Viamontes, I.L. Markov, and J.P. Hayes, "Accurate reliability evaluation and enhancements via probabilistic transfer matrices," *Proc. Design Autom. & Test Europe DATE'05*, Munich, Germany, Mar. 2005, pp. 282–287.

[98] S. Krishnaswamy, I.L. Markov, and J.P. Hayes, "Logic circuit testing and transient faults," *Proc. European Test Symp. ETS'05*, Tallin, Estonia, May 2005, pp. 102–107.

[99] J. Han, E.R. Taylor, J.B. Gao, and J.A.B. Fortes, "Faults, error bounds and reliability of nanoelectronic circuits," *Proc. Intl. Conf. Appl.-Specific Sys., Arch. & Processors ASAP'05*, Samos, Greece, Jul. 2005, pp. 247–253.

[100] E.R. Taylor, J. Han, and J.A.B. Fortes, "Towards accurate and efficient reliability modeling of nanoelectronic circuits," *Proc. IEEE Conf. Nanotech. IEEE-NANO'06*, Cincinnati, OH, USA, Jul. 2006, pp. 395–398.

[101] T. Rejimon, and S. Bhanja, "An accurate probabilistic model for error detection," *Proc. Intl. Conf. VLSI Design VLSID'05*, Kolkata, India, Jan. 2005, pp. 717–722.

[102] T. Rejimon, and S. Bhanja, "Time and space efficient method for accurate computation of error detection probabilities in VLSI circuits," *IEE Proc. Comp. & Digital Tech.*, vol. 152, Sep. 2005, pp. 679–685.

[103] T. Rejimon, and S. Bhanja, "Probabilistic error model for unreliable nano-logic gates," *Proc. Intl. Conf. Nanotech. IEEE-NANO'06*, Cincinnati, OH, USA, Jul. 2006, pp. 47–50.

[104] N.J. Nilsson, *Leraning Machines*, McGraw-Hill, New York, NY, USA, 1965.

[105] B. Widrow, and M.E. Hoff, "Adaptive switching circuits," IRE Wescon Conv., Rec. 4, 1960, pp. 96–104.

[106] C. Coates, and P. Lewis, "DONUT: A threshold gate computer," *IRE Trans. Electr. Comput.*, vol. 13, 1964, pp. 240–247.

[107] N.P. Brousentsov, "Computing machine Setun of Moscow State University" (in Russian), *New Develop. on Comp. Tech.*, 1960, pp. 226–234 [see http://www.computer-museum.ru/english/setun.htm].

[108] N.P. Brousentsov, "Threshold realization of three-valued logic on electromagnetic elements" (in Russian), *Comp. Pbls. Cyber.*, vol. 9, 1972, pp. 3–35.

[109] H.L. Hughes, and J.M. Benedetto, "Radiation effects and hardening of MOS technology: Devices and circuits," *IEEE Trans. Nuclear Sci.*, vol. 50, no. 3, Jun. 2003, pp. 500–521.

[110] C. Bolchini, G. Buonanno, D. Sciuto, and R. Stefanelli, "A CMOS fault tolerant architecture for switch-level faults," *Proc. Intl. Workshop Defect & Fault Tolerance VLSI Sys. DFT'94*, Montreal, Canada, Oct. 1994, pp. 10–18.

[111] C. Bolchini, G. Buonanno, D. Sciuto, and R. Stefanelli, "Static redundancy techniques for CMOS gates," *Proc. Intl. Symp. Circ. & Sys. ISCAS'96*, Atlanta, GA, USA, vol. 4, May 1996, pp. 576–579.

[112] C. Bolchini, G. Buonanno, D. Sciuto, and R. Stefanelli, "An improved fault tolerant architecture at CMOS level," *Proc. Intl. Symp. Circ. & Sys. ISCAS'97*, Kowloon, Hong Kong, Jun. 1997, pp. 2737–2740.

[113] J. Kumar, and M.B. Tahoori, "A low power soft error suppression technique for dynamic logic," *Proc. Intl. Symp. Defect & Fault Tolerant VLSI Sys. DFT'05*, Monterey, CA, USA, Oct. 2005, pp. 454–462.

[114] M.P. Baze, S.P. Buchner, and D. McMorrow, "A digital CMOS design technique for SEU hardening," *IEEE Trans. Nuclear Sci.*, vol. 47, no. 6, Dec. 2000, pp. 2603–2608.

[115] V. Beiu, "Ultra-fast noise immune CMOS threshold gates," *Proc. Intl. Midwest Symp. Circ. & Sys. MWSCAS'00*, Lansing, MI, USA, Aug. 2000, pp. 1310–1313.

[116] S. Tatapudi, and V. Beiu, "Split-precharge differential noise-immune threshold logic gate (SPD-NTL)," *Proc. Intl. Work-conf. Artif. Neural Networks IWANN'03*, Menorca, Spain, Springer, LNCS vol. 2687, Jun. 2003, pp. 49–56.

[117] J.B. Lerch, "Threshold gate circuits employing field-effect transistors," *U.S. Patent 3715603*, Feb. 6, 1973.

[118] J. Galiay, Y. Crouzet, and M. Vergniault, "Physical versus logical fault models MOS LSI circuits: Impact on their testability," *IEEE Trans. Comp.*, vol. C-29, Jun. 1980, pp. 527–531.

[119] F.C. Blom, J. Oliver, M. Rullán, and C. Ferrer, "Layout level design for testability strategy applied to a CMOS cell library," *Proc. Intl. Workshop Defect & Fault Tolerance VLSI Sys. DFT'93*, Venice, Italy, Oct 1993, pp. 199–206.

[120] S. Aunet, Y. Berg, and V. Beiu, "Ultra low power redundant logic based on majority-3 gates," *Proc. IFIP Intl. Conf. VLSI Sys.-on-Chip VLSI-SoC'05*, Perth, Australia, Oct. 2005, pp. 553–558.

[121] C.A. Moritz, and T. Wang, "Towards defect-tolerant nanoscale architectures," *Proc. IEEE Conf. Nanotech. IEEE-NANO'06*, Cincinnati, OH, USA, Jul. 2006, pp. 331–334.

[122] T. Wang, M. Bennaser, Y. Guo, and C.A. Moritz, "Combining circuit level and system level techniques for defect-tolerant architectures," *Intl. Workshop on Defect and Fault Tolerant Nanoscale Arch. NanoArch'06*, Boston, MA, USA, Jun. 2006. Available at: http://www.ecs.umass.edu/ece/ssa/papers/NanoArch06.pdf

[123] J. Gambles, L. Miles, J. Hass, W. Smith, and S. Whitaker, "An ultra-low-power, radiation-tolerant Reed Solomon encoder for space applications," *Proc. Custom IC Conf. CICC'03*, San Jose, CA, USA, Sep. 2003, pp. 631–634.

[124] T. Calin, M. Nicolaidis, and R. Velazco, "Upset hardened memory design for submicron CMOS technology," *IEEE Trans. Nuclear Sci.*, vol. 43, Dec. 1996, pp. 2874–2878.

[125] P. Hazucha, T. Karnik, S. Walstra, B.A. Bloechel, J.W. Tschanz, J. Maiz, K. Soumyanath, G.E. Dermer, S. Narendra, V. De, and S. Borkar, "Measurements and analysis of SER-tolerant latch in a 90-nm dual-V_T CMOS process," *IEEE J. Solid-State Circ.*, vol. 39, Sep. 2004, pp. 1356–1543.

[126] A. Schmid, and Y. Leblebici, "Robust circuit and system design methodologies for nanometer-scale devices and single-electron transistors," *IEEE Trans. VLSI Sys.*, vol. 12, Nov. 2004, 1156–1166.

[127] A. Schmid, and Y. Leblebici, "Robust circuit and system design methodologies for nanometer-scale devices and single-electron transistors," *Proc. IEEE Conf. Nanotech. IEEE-NANO'03*, San Francisco, CA, USA, Aug. 2003, vol. 2, 516–519.

[128] S. Aunet, and M. Hartmann, "Real-time reconfigurable linear threshold elements and some applications to neural hardware," *Proc. Intl. Conf. Evolvable Sys. ICES'03*, Trondheim, Norway, Mar. 2003, pp. 365–376.

[129] S. Aunet, and V. Beiu, "Ultra low power fault tolerant neural inspired CMOS logic," *Proc. Intl. Joint Conf. Neural Networks IJCNN'05*, Montreal, Canada, Aug. 2005, pp. 2843–2848.

[130] C.H. Kim, K. Roy, S. Hsu, R.K. Krishnamurthy, and S. Borkar, "A process variation compensation technique with an on-die leakage current sensor for nanometer scale dynamic circuits," *IEEE Trans. VLSI Sys.*, vol. 14, Jun. 2005, pp. 646–649.

[131] K. Granhaug, and S. Aunet, "Improving yield and defect tolerance in multifunction subthreshold CMOS gates," *Proc. Intl. Symp. Defect & Fault-Tolerance VLSI Sys. DFT'06*, Arlington/Washington, DC, USA, Oct. 2006, pp. 20–28.

[132] Y. Cao, H. Qin, R. Wang, P. Friedberg, A. Vladimirescu, and J.M. Rabaey, "Yield optimization with energy-delay constraints in low-power digital circuits," *Proc. Intl. Conf. Electron Dev. & Solid-State Circ. EDSSC'03*, Kowloon, Hong Kong, Dec. 2003, 285–288.

[133] M.D. Ercegovac, and T. Lang, *Digital Arithmetic*, Morgan Kaufmann, San Francisco, CA, USA, 2004.

[134] I. Koren, *Computer Arithmetic Algorithms*, A. K. Peters, Ltd., Natick, MA, USA, 2002.

[135] A.M. Shams, and M.A. Bayoumi, "A new full adder cell for low-power applications," *Proc. Great Lakes Symp. VLSI GLSVLSI'98*, Lafayette, LA, USA, Feb. 1998, pp. 45–49.

[136] A.M. Shams, and M.A. Bayoumi, "A framework for fair performance evaluation of 1-bit full adder cells," *Proc. Midwest Symp. Circ. & Sys. MWSCAS'99*, Las Cruces, NM, USA, Aug. 1999, pp. 6–9.

[137] A.M. Shams, and M.A. Bayoumi, "A novel high-performance CMOS 1-bit full-adder cell," *IEEE Trans. Circ. & Syst. II*, vol. 47, May 2000, pp. 478–481.

[138] J.M. Quintana, M.J. Avedillo, R. Jiménez, and E. Rodríguez-Villegas, "Low-power logic style for full adder circuits," *Proc. Intl. Conf. Electr. Circ. & Sys. ICECS'01*, Malta, Sep. 2001, pp. 1417–1420.

[139] A.M. Shams, T.K. Darwish, and M.A. Bayoumi, "Performance analysis of low-power 1-bit CMOS full adder cells," *IEEE Trans. VLSI Sys.*, vol. 10, Feb. 2002, pp. 20–29.

[140] Y. Berg, S. Aunet, Ø. Næss, O. Hagen, and M. Høvin, "A novel floating-gate multiple-valued CMOS full-adder," *Proc. Intl. Conf. Circ. & Sys. ISCAS'02*, Scottsdale, AZ, USA, May 2002, pp. 877–880.

[141] M. Alioto, and G. Palumbo, "Analysis and comparison on full adder block in submicron technology," *IEEE Trans. VLSI Sys.*, vol. 10, Dec. 2002, pp. 806–823.

[142] S. Aunet, B. Oelmann, T.S. Lande, and Y. Berg, "Multifunction subthreshold gate used for a low power full adder," *Proc. Norchip Conf.*, Oslo, Norway, Nov. 2004, pp. 44–47.

[143] S. Aunet, and Y. Berg, "Three sub-fJ power-delay-product subthreshold CMOS gates," *Proc. IFIP Intl. Conf. VLSI Sys.-on-Chip VLSI-SoC'05*, Perth, Australia, Oct. 2005, pp. 465–470.

[144] K. Granhaug, and S. Aunet, "Six subthreshold full adder cells characterized in 90 nm CMOS technology," *Proc. Intl. Workshop Design & Diagnostics Electr. Circ. & Sys. DDECS'06*, Prague, Czech Republic, Apr. 2006, pp. 25–30.

[145] T. Vigneswaran, B. Mukundhan, and P.S. Subbarami Reddy, "A novel low power, high speed 14 transistor CMOS full adder cell with 50% improvement in threshold loss problem," *Enformatika Trans. Eng. Comp. & Tech.*, vol. 13, May 2006, pp. 81–85.

[146] M. Forshaw, K. Nikolić, and A.S. Sadek, "ANSWERS: Autonomous Nanoelectronic Systems With Extended Replication and Signaling," *MEL-ARI #28667*, 3rd Year Report, 2001, pp. 1–32. Available at: http://ipga.phys.ucl.ac.uk/research/answers/reports/3rd_year_U CL.pdf

[147] P. Shivakumar, S.W. Keckler, C.R. Moore, and D. Burger, "Exploiting microarchitectural redundancy for defect tolerance," *Proc. Intl. Conf. Comp. Design ICCD'03*, San Jose, CA, USA, Oct. 2003, pp. 481–488.

[148] C.S. Lent, P.D. Tougaw, W. Porod, and G.H. Bernstein, "Quantum cellular automata," *Nanotechnology*, vol. 4, no. 1, Jan. 1993, pp. 49–57.

[149] W.S. Curry, Jr., "Transistor majority logic adder," *U.S. Patent 2999637*, Sep. 12, 1961.

[150] Y. Ono, H. Inokawa, and Y. Takahashi, "Binary adders of multi-gate single-electron transistor: Specific design using pass-transistor logic," *IEEE Trans. Nanotech.*, vol. 1, Jun. 2002, pp. 93–99.

[151] T. Oya, T. Asai, T. Fukui, and Y. Amemiya, "A majority logic device using an irreversible single-electron box," *IEEE Trans. Nanotech.*, vol. 2, Mar. 2003, pp. 15–22.

[152] M.H. Sulieman, and V. Beiu, "On single-electron technology full adders," *IEEE Trans. Nanotech.*, vol. 4, Nov. 2005, pp. 669–680.

[153] R. Zhang, K. Walus, W. Wang, and G.A. Jullien, "A method of majority logic reduction for quantum cellular automata," *IEEE Trans. Nanotech.*, vol. 3, Dec. 2004, pp. 443–450.

[154] V. Beiu, J.M. Quintana, M.J. Avedillo, "VLSI implementations of threshold logic: A comprehensive survey," *IEEE Trans. Neural Networks*, no. 14, Sep. 2003, pp. 1217–1243.

[155] V. Beiu, "Constructive threshold logic addition: A synopsis of the last decade," *Proc. Intl. Conf. Neural Networks ICANN'03*, Istanbul, Turkey, Jul. 2003, pp. 745–752.

[156] W. Ibrahim, V. Beiu, and M.H. Sulieman "On the reliability of majority gates full adders," under review. Available at http://facultx.uaeu.ac.ae/walidibr/publications.htm

[157] J. Srinivasan, S.V. Adve, P. Bose, and J.A. Rivers, "The impact of technology scaling on lifetime reliability," *Proc. Intl. Conf. Dependable Sys. & Networks DSN'04*, Florence, Italy, Jun. 2004, pp. 177–186.

[158] J. Srinivasan, S.V. Adve, P. Bose, and J.A. Rivers, "Lifetime reliability: Toward and architectural solution," *IEEE Micro*, vol. 25, May-Jun. 2005, pp. 2–12.

[159] J. Srinivasan, "Lifetime reliability aware microprocessors," *PhD dissertation*, Dept. CS, Univ. Illinois at Urbana-Champaign, USA, May 2006. Available at: http://rsim.cs.uiuc.edu/Pubs/srinivsn-phd-thesis.pdf

[160] A. Weinberger, and J.L. Smith, "A logic for high-speed addition," *Natl. Bur. Stand. Circ. 591*, 1958, pp. 3–12.

[161] P.M. Kogge, and H. Stone, "A parallel algorithm for the efficient solution of a general class of recurrence equations," *IEEE Trans. Comp.*, vol. 22, Aug. 1973, pp. 786–793.

[162] R.E. Ladner, and M.J. Fischer, "Parallel prefix computations," *J. ACM*, vol. 27, Oct. 1980, pp. 831–838.

[163] R.P. Brent, and H.T. Kung, "A regular layout for parallel adders," *IEEE Trans. Comp.*, vol. 31, Mar. 1982, pp. 260–264.

[164] T. Han, and D.A. Carlson, "Fast area-efficient VLSI adders," *Proc. Intl. Symp. Comp. Arithmetic ARITH'87*, Como, Italy, May 1987, pp. 49–56.

[165] M. Nicolaidis, "Carry checking/parity prediction adders and ALUs," *IEEE Trans. VLSI Sys.*, vol. 11, no. 1, Feb. 2003, pp. 121–128.

[166] R. Ramanarayanan, N. Vijaykrishnan, Y. Xie, and M.J. Irwin, "Soft errors in adder circuits," *Proc. Military & Aerospace Appls. of Programable Logic Devs. & Tech. MAPLD'04*, Washington, DC, USA, Sep. 2004. Available at: http://klabs.org/mapld04/abstracts/ramanarayanan_a.pdf

[167] S. Tosun, O. Ozturk, N. Mansouri, E. Arvas, M. Kandemir, Y. Xie, and W.-L. Hung, "An ILP formulation for reliability-oriented high-level synthesis," *Proc. Intl. Symp. Quality Electr. Design ISQED'05*, San Jose, CA, USA, Mar. 2005, pp. 364–369.

[168] S. Peng, and R. Manohar, "Fault tolerant asynchronous adder through dynamic self-reconfiguration," *Proc. Intl. Conf. Comp. Design: VLSI in Computers & Processors ICCD'05*, San Jose, CA, USA, Oct. 2005, pp. 171–178.

[169] F. Worm, P. Thiran, and P. Ienne, "Designing robust checkers in the presence of massive timing errors," *Proc. Intl. On-Line Test Symp. IOLTS'06*, Como, Italy, Jul. 2006, pp. 281–286.

[170] J.P. Hayes, I. Polian, and B. Becker, "A model for transient faults in logic circuits," *Proc. Intl. Design & Test Workshop IDT'06*, Dubai, UAE, Nov. 2006, in press. Available at: http://www.eecs.umich.edu/~jhayes/JPH_DubaiPaperV3_Nov7-06.pdf

[171] R. Zimmermann, "Binary adder architectures for cell-based VLSI and their synthesis," *PhD dissertation*, Diss. ETH No. 12480, Swiss Federal Inst. Tech., Zurich, Switzerland, 1997. Available at: http://www.iis.ee.ethz.ch/~zimmi/publications/adder_arch.pdf

[172] A. Glodovsky, H.R. Srinivas, R. Kolagotla, and R. Hengst, "A folded 32-bit prefix tree adder in 0.16-μm static CMOS," *Proc. Midwest Symp. Circ. & Sys. MWSCAS'00*, Lansing, MI, USA, Aug. 2000, pp. 238–373.

[173] Y. Shimazaki, R. Zlatanovici, and B. Nikolić, "A shared-well dual-supply-voltage 64-bit ALU," *Proc. Intl. Solid-State Circ. Conf. ISSCC'03*, San Francisco, CA, USA, Feb. 2003, pp. 104–105. [Also in *IEEE J. Solid-State Circ.*, vol. 39, no. 3, Mar. 2004, pp. 494–500.]

[174] Q.-W. Kuo, V. Sharma, and C.C.-P. Chen, "Substrate-bias optimized 0.18μm 2.5GHz 32-bit adder with post-manufacture tunable clock," *Proc. Intl. Symp. VLSI Design, Autom. & Test VLSI-TSA-DAT'05*, Hsinchu, Taiwan, Apr. 2005, pp. 341–344.

[175] G. Yang, S.-O. Jung, K.-H. Baek, S.H. Kim, S. Kim, and S.-M. Kang, "A 32-bit carry lookahead adder using dual-path all-N logic," *IEEE Trans. VLSI Sys.*, vol. 13, no. 8, Aug. 2005, pp. 992–996.

[176] M. Ziegler, and M.R. Stan, "Optimal logarithmic adder structures with a fan-out of two for minimizing the area-delay product," *Proc. Intl. Symp. Circ. & Sys. ISCAS'01*, Sydney, Australia, May 2001, pp. 657–660.

[177] R.A. Freking, and K.K. Parhi, "Theoretical estimation of power consumption in binary adders," *Proc. Intl. Symp. Circ. & Sys. ISCAS'98*, Monterey, CA, USA, Jun. 1998, pp. 453–457.

[178] V.G. Oklobdzija, and R. Krishnamurthy, "Design of power efficient VLSI arithmetic: Speed and power trade-offs," *Proc. Intl. Symp. Comp. Arithmetic ARITH'03*, Santiago de Compostela, Spain, Jun. 2003, pp. 280 (tutorial). Available at: http://www.acsel-lab.com/Presentations/ARITH-Tutorial-Vojin.pps

[179] V. Beiu, "A novel highly reliable low-power nano architecture: When von Neumann augments Kolmogorov," *Proc. Intl. Conf. App.-specific Sys., Arch. & Processors ASAP'04*, Galveston, TX, USA, Sep. 2004, 167–177.

[180] V. Beiu, S. Aunet, R.R. Rydberg III, A. Djupdal, and J. Nyathi, "The vanishing majority gate: Trading power and speed for reliability," *Intl. Workshop Defect & Fault Tolerant Nanoscale Architectures. NanoArch'05*, Palm Springs, CA, USA, May 2005. Available at: http://www.eecs.wsu.edu/~vbeiu/Publications/2005%20NanoArch.pdf

[181] V. Beiu, A. Djupdal, and S. Aunet, "Ultra low power neural inspired addition: When serial might outperform parallel architectures," *Proc. Intl. Work-conf. Artif. Neural Networks IWANN'05*, Barcelona, Spain, Jun. 2005, pp. 486–493.

[182] D.F. Hepner, and A.D. Walls, "Predictive failure analysis and failure isolation using current sensing," *U.S. Patent 7003409*, Feb. 21, 2006.

[183] K. Johansson, O. Gustafsson, and L. Wanhammar, "Power estimation for ripple-carry adders with correlated input data," *Proc. Intl. Workshop Power & Timing Modeling, Optimization & Simulation PATMOS'04*, Springer, LNCS vol. 3254, Santorini, Greece, Sep. 2004, pp. 662–674.

[184] S. Kao, R. Zlatanovici, B. Nikolić, "A 240ps 64b carry-lookahead adder in 90nm CMOS," *Proc. Intl. Solid-State Circ. Conf. ISSCC'06*, San Francisco, CA, USA, Feb. 2006, pp. 1735–1744.

[185] D.M. Markovic, "A power/area optimal approach to VLSI signal processing," *Tech. Rep. UCB/EECS-2006-65*, EE&CS, Berkeley, May 18, 2006. Available at: http://www.eecs.berkeley.edu/Pubs/TechRpts/2006/EECS-2006-65.html

[186] P. Celinski, S.F. Al-Sarawi, D. Abbott, S.D. Cotofana, and S. Vassiliadis, "Logical effort based design exploration of 64-bit adders using a mixed dynamic-CMOS/threshold-logic approach," *Proc. Annual Symp. VLSI ISVLSI'04*, Lafayette, LA, USA, Feb. 2004, pp.127–132.

[187] K. Ishibashi, T. Yamashita, Y. Arima, I. Minematsu, and T. Fujimoto, "A 9μW 50MHz 32b adder using a self-adjusted forward body bias in SoCs," *Proc. Intl. Solid-State Circ. Conf. ISSCC'03*, San Francisco, CA, USA, Feb. 2003, pp. 116–117.

[188] V. Beiu, J. Nyathi, S. Aunet, and M.H. Sulieman, "Femto Joule switching for nano electronics" (Tutorial), *Proc. ACS/IEEE Intl. Conf. Comp. Sys. & Appls. AICCSA'06*, Sharjah, UAE, Mar. 2006, pp. 415–423.

[189] R.M. Swanson, and J.D. Meindl, "Ion-implanted complementary MOS transistors in low-voltage circuits," *IEEE J. Solid-State Circ.*, vol. 7, Apr. 1972, pp. 146–153 (preliminary version in *Proc. Intl. Solid-State Circ. Conf. ISSCC'72*, San Francisco, CA, USA, Feb. 1972, pp. 192–193).

[190] E.A. Vittoz, and J. Fellrath, "CMOS analog integrated circuits based on weak inversion operation," *IEEE J. Solid-State Circ.*, vol. 12, Jun. 1977, pp. 224–231.

[191] C.A. Mead, "Neuromorphic electronic systems," *Proc. IEEE*, vol. 78, Oct. 1990, pp. 1629–1636.

[192] E.A. Vittoz, "Very low power circuit design: Fundamentals and limits," *Proc. Intl. Symp. Circ. & Sys. ISCAS'93*, Chicago, IL, USA, May 1993, vol. 2, pp. 1439–1442.

[193] E.A. Vittoz, "Low-power design: Ways to approach the limits" (Invited Plenary Talk), *Proc. Intl. Solid-State Circ. Conf. ISSCC'94*, San Francisco, CA, USA, vol. 1, Feb. 1994, pp. 14–18.

[194] J.B Burr, and J. Shott, "A 200mV self-testing encoder/decoder using Stanford ultra low power CMOS," *Proc. Intl. Solid-State Circ. Conf. ISSCC'94*, San Francisco, CA, USA, Feb. 1994, pp. 84–85.

[195] G. Schrom, and S. Selberherr, "Ultra low-power CMOS technologies," *Proc. Intl. Annual Semicond. Conf. CAS'96*, Sinaia, Romania, Oct. 1996, vol. 1, pp. 237–246.

[196] T.S. Lande, D.T. Wisland, T. Sœther, and Y. Berg, "FLOGIC – Floating-gate logic for low-power operation," *Proc. Intl. Conf. Electr., Circ. & Sys. ICECS'96*, Rhodes, Greece, Oct. 1996, vol. 2, pp. 1041–1044.

[197] C.H. Kim, H. Soeleman, and K. Roy, "Ultra-low-power DLMS adaptive filter for hearing aid applications," *IEEE Trans. VLSI Sys.*, vol. 11, Dec. 2003, pp. 1058–1067.

[198] A.P. Chandrakasan, S. Sheng, and R.W. Brodersen, "Low-power CMOS digital design," *IEEE J. Solid-State Circ.*, vol. 27, Apr. 1992, pp. 473–484.

[199] H. Soeleman, K. Roy, and B. Paul, "Robust subthreshold logic for ultra-low power operation," *IEEE Trans. VLSI Sys.*, no. 9, Feb. 2001, pp. 90–99.

[200] A. Wang, A.P. Chandrakasan, and S.V. Kosonocky, "Optimal supply and threshold scaling for sub-threshold CMOS circuits," *Proc. Annual Symp. VLSI ISVLSI'02*, Pittsburgh, PA, USA, Apr. 2002, pp. 5–9.

[201] D.D. Wentzloff, B.H. Calhoun, R. Min, A. Wang, N. Ickes, and A.P. Chandrakasan, "Design considerations for next generation wireless power-aware microsensor nodes," *Proc. Intl. Conf. VLSI Design VLSID'04*, Mumbai, India, Jan. 2004, pp. 361–367.

[202] S. Aunet, B. Oelmann, S. Abdalla, Y. Berg, "Reconfigurable subthreshold CMOS perceptron," *Proc. Intl. Joint Conf. Neural Networks IJCNN'04*, Budapest, Hungary, Jul. 2004, pp. 1983–1988.

[203] B.H. Calhoun, A. Wang, and A.P. Chandrakasan, "Device sizing for minimum energy operation in subthreshold circuits," *Proc. Custom IC Conf. CICC'04*, Orlando, FL, USA, Oct. 2004, 95–98.

[204] V. Beiu, J. Nyathi, and S. Aunet, "Sub-pico Joule switching: High-speed reliable CMOS circuits are feasible," *Proc. Innovations in Info. Tech. IIT'05*, Dubai, UAE, Sep. 2005. Available at: http://www.it-innovations.ae/iit05/proceedings/articles/E_5_IIT05_Beiu.pdf

[205] R. Sarpeshkar, "Analog versus digital: Extrapolating from electronics to neurobiology," *Neural Computation*, vol. 10, 1998, 1601–1638.

[206] R. Sarpeshkar, "Brain power: Borrowing from biology makes for low-power computing," *IEEE Spectrum*, vol. 43, no. 5, May 2006, pp. 24–29.

[207] G.E.R. Cowan, "A VLSI analog computer/math co-processor for a digital computer," *PhD dissertation*, Columbia University, Nov. 2005. Available at: http://digitalcommons.libraries.columbia.edu/dissertations/AAI3174769/

[208] G.E.R. Cowan, R.C. Melville, and Y.P. Tsividis, "A VLSI analog computer/digital computer accelerator", *IEEE J. Solid-State Circ.*, vol. 41, Jan. 2006, pp. 42–53.

[209] J. Donald, and M. Martonosi, "Power efficiency for variation-tolerant multicore processors," *Proc. Intl. Symp. Low Power Electr. & Design ISPLED'06*, Tegernsee, Germany, Oct. 2006, pp. 304–309.

[210] A. Datta, S. Bhunia, S. Mukhopadhyay, and K. Roy, "Delay modeling and statistical design of pipelined circuit under process variation," *IEEE Trans. CAD of IC & Sys.*, vol. 25, Nov. 2006, pp. 2427–2436.

[211] J.K. McIver III, and L.T. Clark, "Reducing radiation-hardened digital circuit power consumption," *IEEE Trans. Nuclear Sci.*, vol. 52, Dec. 2005, pp. 2503–2509.

[212] K. Ishibashi, T. Fujimoto, T. Yamashita, H. Okada, Y. Arima, Y. Hashimoto, K. Sakata, I. Minematsu, Y. Itoh, H. Toda, M. Ichihashi, Y. Komatsu, M. Hagiwara, and T. Tsukada, "Low-voltage and low-power logic, memory, and analog circuit techniques for SoCs using 90 nm technology and beyond," *IEICE Trans. Electron.*, vol. E89-C, Mar. 2006, pp. 250–262.

[213] W. Bialek, and F. Rieke, "Reliability and information transmission in spiking neurons," *Trends Neurosci.*, vol. 15, 1992, pp. 428–434.

[214] C.F. Stevens, "Cooperativity of unreliable neurons," *Curr. Biol.*, vol. 4, 1994, pp. 268–269.

[215] D.K. Smetters, and A. Zador, "Synaptic transmission: Noisy synapses and noisy neurons," *Curr. Biol.*, vol. 6, no. 10, Oct. 1996, pp. 1217–1218.

[216] J.E. Lisman, "Bursts as a unit of neural information: Making unreliable synapses reliable," *Trends Neurosci.*, vol. 20, 1997, pp. 38–43.

[217] A. Zador, "Impact of synaptic unreliability on the information transmitted by spiking neurons," *J. Neurophysiol.*, vol. 79, 1998, pp. 1219–1229.

[218] A. Manwani, and C. Koh, "Detecting and estimating signals over noisy and unreliable synapses: Information-theoretic analysis," *Neural Computation*, vol. 13, Jan. 2001, pp. 1–33.

[219] W.B. Levy, and R.A. Baxter, "Energy-efficient neuronal computation via quantal synaptic failures," *J. Neurosci.*, vol. 22, Jun. 1, 2002, pp. 4746–4755.

[220] D.B. Chklovskii, "Exact solution for the optimal neuronal layout problem," *Neural Computation*, vol. 16, no. 10, Oct. 2004, pp. 2067–2078.

[221] Q. Wen and D.B. Chklovskii, "Segregation of the brain into gray and white matter: A design minimizing conduction delays," *PLoS Comp. Biol.*, vol. 1, no. 7 (e78), Dec. 2005, pp. 617–630.

[222] B. Linares-Barranco, T. Serrano-Gotarredona, R. Serrano-Gotarredona, and C. Serrano-Gotarredona, "Current mode techniques for sub-pico-ampere circuit design," *Analog Integr. Circ. & Signal Proc.*, vol. 38, no. 2–3, Feb.–Mar. 2004, pp. 103–119.

III

Biomolecular Electronics and Processing

13

Properties of "G-Wire" DNA

Thomas Marsh

James Vesenka

Abstract

The observed properties of guanine-quadruplex "G-wire" DNA are summarized. These observations include the evaluation of the growth kinetics, along with structural and electronic properties of the dry self-assembled $G_4T_2G_4$ polymer. The primary investigative tool for these studies was scanning probe microscopy. Growth kinetics studies indicate the self-assembly process is diffusion limited and provides Poisson-like distribution of G-wire lengths upon reaching equilibrium. This evidence suggests that self-assembly is driven by thermodynamic processes. The average lengths of these molecules are around 100 nm after 24 hours of growth. Longer G-wire DNA molecules (many micrometers) have been found both in flexible and crystalline forms following many months of growth. The latter structures are extremely interesting candidates for molecular templates.

Hydration layer scanning tunneling microscopy (HLSTM) of G-wires on mica was carried out under controlled humidity conditions. The HLSTM images were similar to those measured by atomic force microscopy (AFM) in dry air. The G-wire height above the mica substrate, interpreted as quadruplex diameter, and the G-wire width appeared to decrease slightly with increasing humidity. Though much of the lateral broadening is likely a result of shielding by residual cations and the lower resolving ability of HLSTM, the dependence of the DNA height and width on humidity suggests a simple explanation in terms of a hydration layer. An increased thickness of the hydration layer of up to 0.6 nm was observed.

The electrical conductivity of G-wire DNA adsorbed to the surface of mica was examined with the assistance of silicon shadow masks. Four-point probe masks were fabricated in silicon using photolithographic patterning and dry reactive ion etching. The silicon "stencils" were designed specifically for use in generating shadow-deposited metal contacts on top of G-wire DNA samples adsorbed to the atomically flat surface of mica. Two types of metal contacts were employed in these experiments: electron beam evaporated gold using a high vacuum system, and argon sputtered gold using a low vacuum scanning electron microscopy sample coating apparatus. The metal electrode patterning was characterized through AFM imaging. The conductivity of the G-wire DNA samples was analyzed using a high impedance multimeter. The lower limit of resistance of the G-wire DNA networks was determined to be in excess of 1 GΩ (sheet resistance of 10^5 GΩ-m^2) indicating from these experiments that "dry" G-wire DNA is an insulator.

13.1 Introduction

The electrical conductivity of duplex DNA has been extensively characterized. Conductivity through short DNA sequences ($\sim$ 15 base pairs) has been established by photo-induced and flash-quench techniques[1]. Experiments on single-stranded DNA have suggested that short segments of DNA ($\sim$ 30 bp, about 10 nm) are electrically conducting [2–6]. However, the majority of experiments on longer strands (> 100 nm) indicate that duplex DNA is an insulator [7–10]. Those experiments that have indicated conductivity in bundles of DNA suggest the conduction mechanism may be related to charge migration through hydrated samples [9]. In this paper, we examine the growth kinetics, structure, and electrical properties of four-stranded "G-wire" DNA, primarily via scanning probe microscopy.

G-wires belong to a polymorphic family of quadruple helical nucleic acids collectively known as G-DNA. The common structural motif of G-DNA is the guanine tetrad (G-quartet) [11,12], in which four guanine bases are H-bonded in a cyclic planar array (Figure 13.1[a]). The formation and stability of G-DNA is greatly enhanced by the coordination of monovalent and/or divalent metal cations [13–17]. G-DNA is known to occur in a variety of functional settings within the genomes of living cells [18–24]. The important biological role of guanine self-recognition may also be viewed as a useful materials property for the development of self-assembling supramolecular structures. Guanine-rich oligonucleotides (GROs) spontaneously form G-DNA under appropriate conditions [12,25,26]. A number of GRO have lead to the development of self-assembling supramolecular G-DNA structures [27–30] from a single sequence component. One supramolecular G-DNA in particular, the G-wire, has been extensively characterized by scanning probe microscopy (SPM) [31,32]. The fundamental building block of a G-wire is a 10mer oligonucleotide with the telomere-like sequence d(GGGGTTGGGG) designated "Tet1.5" (Figure 13.1[b]). Self-assembly occurs in the presence of group I and group II metal cations over a broad range of temperatures to produce a supramolecular G-DNA that appears as rigid linear fibers when examined by SPM (Figure 13.1[c]). Whereas duplex DNA tends to collapse on the surface of mica, G-wire DNA appears to hold its cylindrical shape (Figures 13.2 and 13.3) [33]. This report summarizes baseline conditions for the growth of G-wire DNA, as well as substrate deposition, and estimates the equilibrium constant for multimers. These and future efforts may define conditions for the controlled growth and manipulation of "superstructures" (many micrometers long).

Tapping Mode™ atomic force microscopy (TMAFM) and hydration layer scanning tunneling microscopy (HLSTM) [34,35], high-resolution near-field three-dimensional imaging techniques, were used to characterize G-wires prepared in different growth media and imaged under controlled humidity. HLSTM suggested semiconductivity of hydrated G-wires [36]. The conduction mechanism may be the result of the base stacking of G-quartets and caged monovalent cations·[37] (Figures 13.1[a] and 13.1[b]). HLSTM measurements indicate G-wires could be reproducibly imaged at tunneling currents above a picoampere at moderate relative humidity (Figure 13.4) [36]. The contrast mechanism for nonconductive molecules imaged by HLSTM stems in part from the hydration layer on top of a hygroscopic substrate, such as mica in humid air, depending only upon the applied "bias" voltage. Under low bias, high-resolution imaging is maintained by conduction through the hydration layer. At high voltages, ballistic tunneling takes place through the air gap into the hydration layer, at the expense of resolution. The HLSTM observations were interesting because of a 10 to 100 *increase* factor in tunneling current attained in the presence of G-wire DNA, compared to double-stranded DNA. The implication was that the G-wires were assisting conductivity over the substrate. The caged metal cations integrated into a hydrated G-wire molecule may have some mobility. In addition, the metal cations are sufficiently close to each other (estimated between 0.3 and 0.7 nm) to support electron tunneling. Lastly, the π-bonding of adjacent Guanine-quartets may overlap enough to enhance electron "hopping." G-wire DNA inherent stability, uniformity, and long lengths make them candidates for molecular wiring [38,39]. This report describes humidity dependence of G-wire HLSTM images and estimates the thickness of the hydration layer. We also summarize efforts for direct measurement of G-wire DNA conductivity under dried conditions.

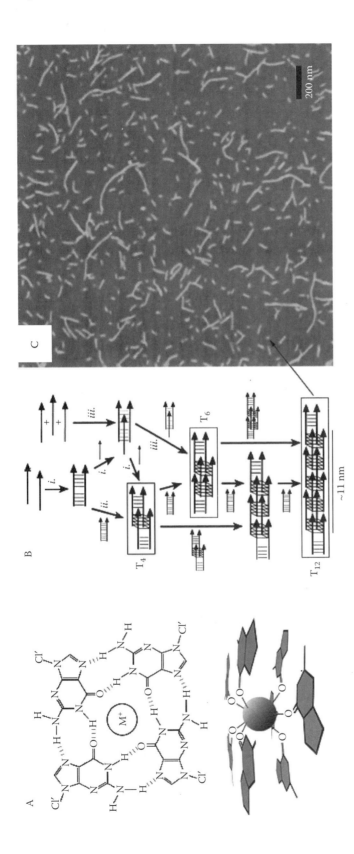

FIGURE 13.1 (a) Structure of a G-quartet showing cation coordination. (b) Hypothetical pathway of G-wire self-assembly. The thin arrows represent the sequence $d(G_4T_2G_4)$, and parallelograms represent G. Initial formation of a G-DNA nucleating structure may occur through multiple paths: The italic *i*, *ii*, and *iii* represent sequential stepwise assembly, dimerization, and triplex disproportionation, respectively, as potential models for the initial G-DNA structure formation [40]. (c) A TMAFM image of G-wires in a Na^+/Mg^{2+} buffer, deposited on a mica substrate. Cation species such as potassium, sodium, or magnesium are thought to help stabilize the G-wires in the base-stacked core of the structure as seen in (a). The thymine groups may act as flexible links that can "bunch up" in solution, or after adsorption onto a substrate.

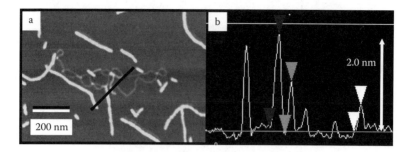

FIGURE 13.2 Topographic comparison between duplex and quadruplex G-wire DNA co-adsorbed on the same substrate (a). The duplex DNA collapses on the surface of mica to a height of 0.5 nm above the surface, as seen in the cross-section of (b). Even the supercoiled segments of the double-stranded DNA measure only about 1.0 nm above the mica substrate. This is about half the diameter expected from Watson–Crick duplex DNA in solution. However, the quadruplex DNA is uniformly about 2.2 nm in diameter, very close to the NMR and x-ray spectra of G-quartet DNA (2.4 nm in diameter). The vertical height range is 10 nm from black to white.

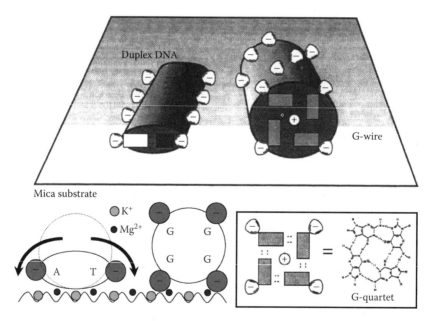

FIGURE 13.3 The "pinning model" suggests the greater internal attractive forces of G-wire DNA, comprised of guanine-quartet building blocks, four in a row, enable it to retain its solution state structure when exposed to tether cations (shown here as magnesium). However, the stronger tethering force exerted on the unsupported phosphate backbone of duplex DNA, illustrated here with an example A-T base pair, pins the duplex DNA flat to the mica substrate.

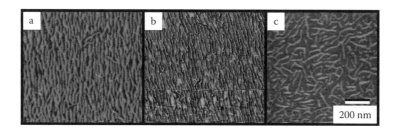

200 nm

FIGURE 13.4   G-wires freshly adsorbed onto mica imaged via Tapping Mode (a) and the same sample imaged by the same tip 24 hours later after drying in an oven at 37°C (b). Note the preferential orientation is NOT a sample preparation artifact—e.g., due to rinsing. G-wires appear to align with the underlying potassium vacancy sites of the freshly cleaved mica substrate. Note the broadening of the G-wire structure due to hydrated residual buffer salts. After drying in an oven, the G-wires appear much narrower and the buffer salts appear to distribute themselves in between the DNA strands. It is the freshly made, hydrated form of the G-wires that are essential for hydration layer scanning tunneling microscopy (HLSTM) imaging seen in (c). This HLCSTM image of G-wires freshly adsorbed on mica is recorded at 1pA tunneling current, −7 V bias and 80% relative humidity in a sealed imaging chamber. Successful images of G-wires on mica occur most frequently on freshly prepared samples, high G-wire densities, but NOT G-wire "networks" as seen in (a) and (b), and high concentrations of residual buffer cations. Note in this image sample there are pathways of tunneling current that do not require travel over the G-wires. These pathways are important for providing electron conduction from a bias voltage source. The vertical height range is 10 nm from black to white.

13.2   Materials and Methods

Quadruplex G-wire DNA was prepared according to the procedure outlined by Marsh *et al.*[31]. Melting of the $G_4T_2G_4$ monomers (Tet1.5) was maintained with a PCR Thermocylcer (Thermo Hybaid, U.K.)— that is, the growth cocktail and Tet1.5 were raised to 95°C for ten minutes to promote the melting of fortuitous G-4 structures (i.e., to ensure a monomeric concentration of G-wires). Samples of concentrated G-wires (monomer concentration 1.0 mM) were diluted to a factor of 10 to 100 in a buffer consisting of 10 mM Tris (pH 7.6), and 1 mM MCl_x, where M = sodium, potassium, magnesium, or zinc. G-wire networks: The oligonucleotide Tet1.5 d(GGGGTTGGGG) were purchased from integrated DNA technologies (Corralville, IA). Samples of pure oligonucleotide were dissolved to a final concentration of 100 μM in a buffer consisting of 50 mM NaCl, 10 mM $MgCl_2$, 10 mM Tris-HCl pH 7.5, and 1 μM spermidine, and then incubated at 37°C for three days. These samples were allowed to sit on Parafilm® for ten minutes at room temperature and adsorbed onto freshly cleaved muscovite mica. The samples were then incubated on the mica between 0 to 10 minutes, rinsed with 1 mL deionized water, dried in a stream of dry nitrogen, and allowed to stabilize in a 37°C oven before imaging. Samples were imaged with 125 μm × 20 μm silicon nitride probes using a Nanoscope E controller (Digital Instruments: Santa Barbara, CA) in contact mode in dry air. Fresh samples were imaged under ambient humidity using a Nanoscope IIIa controller in Tapping Mode and 75-μm long Tapping Tips. Freshly prepared G-wire DNA was also imaged with a PicoSPM (Molecular Imaging: Tempe, AZ) low-current STM. G-wires were imaged at RH ≈ 75 to 85%, bias voltages of −5 to −10 V, and a tunneling current of ≈ 1 to 3.0 picoamperes. A Nanoscope IIIa controller and Multimode AFM operated in Tapping Mode were used to image freshly prepared and dessicated samples. Silicon shadow masks, or stencils, were made by lithographic patterning and dry reactive-ion etching. The stencil pattern creates four metal contacts, separated by 600 microns, with which a four-point probe measurement can be made to determine the sheet resistance of the film. Metal contacts were made by thermal electron beam evaporation of gold onto the DNA samples at high vacuum (Temescal, 10^{-9} Torr) or by argon sputter coating at low vacuum (Polaron, 10^{-1} Torr). A Nanoscope IIIa controller and Multimode AFM operated in Tapping Mode were used to image the G-wire DNA networks and the integrated metalized contacts. Conductivity measurements were made using both two- and four-point probes and an HP 3258A Multimeter, with a maximum resistance measurement of 1.2 GΩ at a potential of 500 V and maximum current capacity of 500 nA throughout the sample.

13.3   Results and Discussion

Figure 13.4 is an example of a concentrated G-wire network that was created by depositing a sample containing a concentrated 24-hour-old G-wire solution onto freshly cleaved mica. The sample is rinsed and immediately imaged in TappingMode, revealing an oriented network of G-wire strands over the surface of mica, as shown in Figure 13.4(a). The orientation affect is due to the G-wire alignment with potassium vacancy sites on the surface of freshly cleaved mica [32]. The density of the G-wire DNA appears to depend upon local variations of the mica surface. For example, imaging a region of the mica surface a millimeter away can provide results similar to Figure 13.4(c) in which the G-wires are clearly separated. In Figure 13.4(b), the same sample from Figure 13.4(a) had been dried for 24 hours. Note that the G-wire DNA appears much narrower because of the dehydration of the hydration layer over the surface. The hydration layer is absolutely essential for imaging of the molecules via low current scanning tunneling microscopy (LCSTM), seen in Figure 13.4(c).

Monovalent cations (Na^+ and K^+) are assumed to stabilize a G-wire internally by residing in the coordination pocket provided by the four O6 groups of the G-quartet (Figure 13.1[a]). However, self-assembly into relatively long supramolecular structures (Figure 13.1[c]) requires Mg^{2+} [27]. Compared to monovalent cations such as Na^+ and K^+, Mg^{2+} is known to stabilize a nucleic acid structure to a 100-fold or more effectively, primarily functioning as a counter ion for the negatively charged phosphates in the backbone. The manner in which divalent cations interact with G-DNA to stabilize a quadruplex is more complex. Studies of divalent cation stabilization of various G-DNAs have shown a general trend for divalent metal cations to stabilize G-DNA structures at lower concentrations (M^{2+} below 10 mM) and destabilize G-DNA at higher M^{2+} concentrations [40,41]. Oligoucleotides such as d(GGGGTTGGGG) [27] and d(GGGGTTTTGGGG) are an apparent exception to this general trend. In the case of the hairpin dimer forming $dG_4T_4G_4$, increasing Ca^{2+} concentration induces a switch in the conformation of a d($G_4T_4G_4$) G-DNA from an anti-parallel hairpin dimer to a parallel tetramer G-DNA leading to a supramolecular assembly similar to the Tet1.5 oligonucleotide [42]. The model presented in Figure 13.1(b) suggests a slipped or out-of register strand(s) promotes the further growth of G-wires once a stable G-DNA is formed. In this scheme, combination monovalent and divalent cation (G-DNA stabilizing and destabilizing) effectors promote longer G-wires by reducing the number of G-DNA intermediate species exposed to a solvent. The initial assembly of a slipped strand G-DNA may occur through several paths, as indicated in Figure 13.1(b). Molecular dynamic simulations of precursor G-DNA structures predict stable slipped strand structures stabilized by Na^+ or K^+ occur in possible assembly pathways [43,44] and lend support to this model. The precise role divalent cations play in G-DNA assembly and, by extension, in G-wire self-assembly is not known at this time.

This study outlines a simple model based on a duplex dimerization [45] model for the formation of an out-of-register G-wire nucleating G4-DNA. The kinetics of G-DNA assembly explored through spectroscopic methods [42] indicate it is an initially slow process that is greatly enhanced by the presence of monovalent cations, and once formed, remains stable for extended periods of time [20]. Here, the rate of G-wire formation was tracked by SPM and considers the overall rate of G-wire formation. Two monomers combine to form the ladder structure L_1, or the staggered structure $L_{1'}$, with associated equilibrium constants K_1 or $K_{1'}$. The rate-determining step in the growth process of the G-wires is the generation of L_2, an essential structure required for the growth of extensible G-quartet multimers. After the construction of L_2, the growth of longer wires is possible. Initially, the concentration increase of L_2 is driven thermodynamically:

$$\Delta G = \Delta G^o + RTln(Q) \tag{13.1}$$

where the reaction quotient "Q" is defined by

$$Q = a_{L2}/a_{L1}^2 \approx [L_2]/[L_1]^2 \tag{13.2}$$

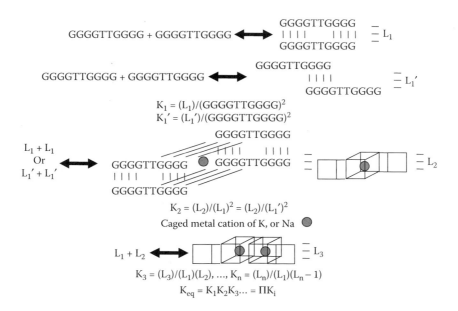

FIGURE 13.5 The self-assembly process initially is thermodynamically driven by the excess of ladder monomers L_1 and L_1'. As the concentration of dimers (L_2) reaches equilibrium with the monomers, the process of growing larger G-wire structures is driven by the excess numbers of dimers and monomers compared to the smaller concentrations of multimers (L_n). Characterization of the equilibrium constant for different length molecules involves the product of the equilibrium constant of the different length multimers. Complicating the growth process even more are the caged metal ions: sodium and potassium. Incorporation of potassium leads to a more uniform spread of smaller G-wire DNA, whereas incorporation of sodium leads to a greater distribution in G-wire lengths.

Here, "a_{L2}" is the activity of the dimer and "a_{L1}" is the activity of the monomer. The ratio of the activities is approximated by the ratio of their respective concentrations.

The free energy initially drives the formation of the dimers because the concentration of L_2, and thus Q, is zero. Consequently, ΔG is large and negative (spontaneous assembly). The system rapidly seeks to reach equilibrium between dimers and monomers. Eventually, the concentration of L_2 increases sufficiently enough that the system approaches equilibrium and the growth of longer wires is thermodynamically driven by the incorporation of dimers into the lower concentrations of multimers (Figure 13.5).

The formation of dimers is always spontaneous because of the competition between polymerization into multimers, with decomposition into L_1 or $L_{1'}$. The total self-assembly equilibrium constant, K_{eq}, can be expressed as a product of all the individual equilibrium constants for a given length G-wire multimer L_n:

$$K_{eq} = \prod K_n = [L_n]/[G_4T_2G_4]^2[L_1]^{n-1} \approx [L_n]/[G_4T_2G_4]^{n+1} \tag{13.3}$$

where the last approximation is the result of assuming the concentrations of Tet1.5 and L_1 are about the same.

If the model is accurate, we would expect to see a Poisson-like distribution of G-wire DNA lengths, with a greater number of smaller length G-wires compared to longer structures. Furthermore, if the process is diffusion limited by the concentration of monomers interacting with multimers, we would expect a comparison of length versus time to behave in a square root of time dependence. Both of these features can be seen in Figure 13.6. In the time study presented in Figure 13.6, the mean length $<L>$ (in nm) of the empirically determined growth rate is:

$$<L> = 28 \text{ nm} + [11 \text{ nm/(day)}^{1/2}](t)^{1/2} \tag{13.4}$$

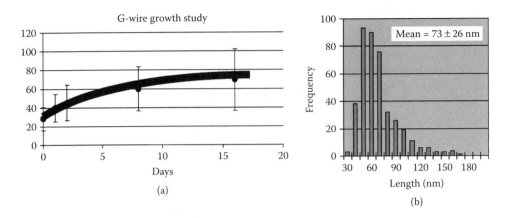

FIGURE 13.6   A measurement of the distribution of average G-wire lengths indicates a greater number of smaller wires (median equilibrium value of about 50 nm) over longer wires in (a). N = 400 from an image taken of G-wires grown for over a month. Error bars represent standard deviation. Note the presence of 30 nm in (b) wires even after growth for about a month. These short wires are present from essentially the very first measurement during the growth process. Increasing the concentration appears to reduce the growth time almost linearly.

where "t" is time in days. The appearance of short wires happens almost immediately at the concentrations undertaken in this study—i.e., 1.0 mM Tet 1.5 monomer (Figure 13.6[a]). Attempts to resolve mean lengths at smaller time scales (seconds) always yield short length G-wire DNA. At 5 to 10 nm resolutions with our best SPM probes, the initially measured average of 28 nm is well within resolution limits. As mentioned in the sample preparation procedure, the melting of fortuitous G-quartet structures through initial heating does not appear to generate a zero elapsed time, zero G-wire length. We speculate that growth of the multimers takes place extremely fast in a variety of growth conditions. Samples of the Tet 1.5 monomer taken from melting temperatures, and imaged by the SPM, have yielded no G-wire structures.

In Figure 13.6[b], we find the common Poisson-like distribution of G-wire lengths from samples allowed to self assemble for several weeks. Under these conditions, the mean length is approximately 73 nm in a potassium-rich mixture. With each L_2 half-length of about 1.55 nm, this corresponds to an average length multimer L_n of $n = 47$. Depending on growth conditions, as much as 25% of the mixture remains as Tet1.5 oligonucleotide. Each Tet1.5 oligo has a molecular weight of 3180 g/mole. The frequency distribution of G-wire lengths in Figure 13.6(b) provides us with a relative concentration of average G-wire lengths of 23%. If we assume this concentration reflects the relative amount of monomers tied up in the mean length multimer, then we can estimate the concentration of L_n for $n = 47$:

$$[L_{47}] = (0.23)(0.75 \text{ mM})(3180g/\text{mole})/(286200g/\text{mole}) = 0.0019 \text{ mM} \qquad (13.5)$$

Thus the equilibrium constant is

$$K_{eq} = [L_{47}]/[G_4T_2G_4]^{48} \approx 10^{26} \qquad (13.6)$$

The estimated equilibrium constant greatly favors the growth of the wires.

The fact that G-wires appear to self-assemble into smaller average lengths is challenging for the purposes of determining their macroscopic electronic characteristics—that is, large wires facilitate macroscopic electronic characterization. Curiously, diluting G-wire DNA samples with either growth or imaging buffers does not appear to greatly affect the average length of the molecules. From a thermodynamic growth standpoint, dilution should result in disassembly of the G-wires. However, dilution of samples to 1/1000 of their initial concentrations only reduces the density over the mica surface. The reason for this behavior is still being investigated.

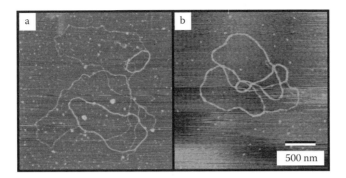

FIGURE 13.7 Interesting structures sometimes emerge, including these examples of loops of single- and multiple-stranded G-wires. These results are found with numerous smaller G-wires surrounding the immediate vicinity as seen in the two examples provided here. We speculate that looped G-wire structures are stable against disassembly after dilution and rinsing by virtue of continuity. G-wires tend to maintain an average length much smaller than the contour lengths of the loops shown here. Multiple-stranding of G-wire DNA is common, as shown in and Figures 13.8 and 13.9, a result that is surprising in view of the strong electrostatic repulsion expected from the four strands of phosphate backbones on each G-wire. The vertical height range is 10 nm from black to white.

13.3.1 G-Wire Superstructures

Occasionally, as shown in Figures 13.7 through 13.10, G-wire superstructures are observed. They have appeared in three different forms: helices, ribbons and tubes (G-wire DNA), and intentionally cross-linked. Figure 13.7 includes two examples of flexible G-wire loops, similar in shape and size to plasmid DNA. If not for the fact these structures retain a minimum diameter of 2.2 nm when imaged in dry air—i.e., the diameter of G-wire DNA—contamination of the sample might be suspect. Double-stranded DNA collapses on the surface of mica due to interactions between the substrate and the DNA. The four-stranded G-wire DNA has always maintained its integrity on the surface of mica in dry air. The G-wire loops might be kinetically stable because, by self-assembling into a closed structure, disassembly is discouraged. There is no "end" on a closed loop for the ladder building blocks to vacate (in the absence of cross-linkers).

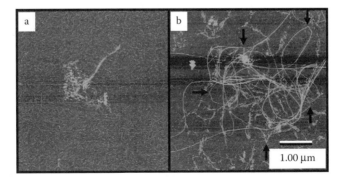

FIGURE 13.8 Two successive images from the same sample can be seen baring G-wires of the typical 200 nm length in the left image, and extremely long (hundreds of micrometers), spaghetti-like structures in the right image. Note the arrows in the right image indicating kinked regions of the spaghetti-like structures, and that the alignment of the straight lengths all are in the same direction. This orientation effect is a common feature of G-wire DNA. The diameter of these long structures (as measured by the height above the surface) is measured to be integrals of 2.2 nm. The latter is the diameter of the G-wire DNA as measured by SPM and NMR, and x-ray crystallography. This combination of evidence suggests that G-wire DNA under the right conditions can form one-dimensional crystals.

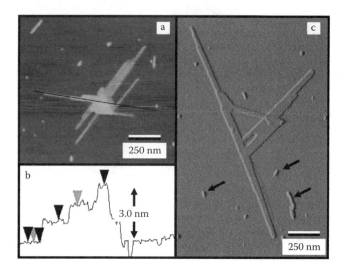

FIGURE 13.9   (a) One-dimensional crystals observed at 60° intervals, orienting themselves with the underlying potassium vacancies in the surface of mica. A contact AFM image taken in dry air with a vertical height scale of 10 nm from dark to light. The black line through the crystal is a cross-section described in (b). The cross-section indicates the different heights of the crystals, which form three distinct layers of heights—1.4 nm, 2.2 nm, and 3.6 nm tall (each ± 0.2nm)—incremental steps exactly one half the diameter of G-wire DNA. (c) Deflection image example of the purported crystalline of G-wires, indicating the micrometer lengths that these structures can attain. Note the arrowed flexible G-wire DNA segments in (c) and their alignment with the crystals, evidence of the lattice match with the underlying potassium vacancies of mica, and indirect evidence of the similar nature of the crystalline and flexible structures on the same surface.

The conditions under which such long structures develop, when the surrounding solution contains only smaller 100-nm length G-wire DNA, is still under investigation.

Flexible 1-D crystals have been found on occasion, as shown in Figure 13.8. Figure 13.8(a) provides a glimpse of a typical surface of mica inundated with G-wires approximately 100 nm long. Only the bright patch near the center of the image provides a hint of the longer structures found on the same surface a few hundred micrometers away. These spaghetti-like structures are several micrometers long, the kind

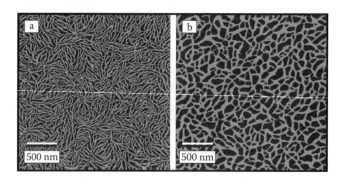

FIGURE 13.10   G-wire networks revealed by slow deposition (a) or rapid deposition (b) of self-assembled Tet1.5 on mica. Slow deposition involves a five-minute adsorption period prior to rinsing the sample with water and drying under a stream of nitrogen. Rapid deposition involves drying the sample immediately after spreading, followed by a water rinse and a second drying step. Note the greater condensation of G-wires upon slow deposition, a process that might enable the G-wires to aggregate.

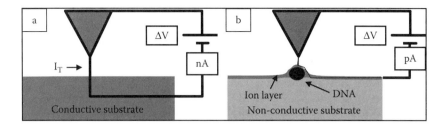

FIGURE 13.11 In this cross-section cartoon of the electron conduction path between the tip and sample, the traditional STM tunneling current (a) is typically 1000 times greater than the hydration layer scanning tunneling microscopy (HLSTM) current (b). The substantial reduction in HLSTM current is due to the greater resistance of the conduction pathway. The current passes through a thin hydration layer atop the hygroscopic mica surface coated with residual buffer ions in the presence of high humidity. In STM, tunneling currents are typically in the nanoampere range, whereas in HLSTM the tunneling currents are in the picoampere range.

of lengths of interest for electronic characterization. Two forms of evidence that these are G-wires are diameters over 2.2 nm or larger (larger presumably due to multiple wraps of the G-wires) and alignment of some segments of the structures with the mica substrate (the arrowed regions in Figure 13.8[b]). This alignment has been associated with the potassium vacancy-sites of the mica substrate. The sharp bends in these structures (arrowed regions) are suggestive of flexible one-dimensional crystals.

Evidence for more rigid structures is seen in Figure 13.9(a) and 13.9(c). Alignment with the potassium vacancies in the mica substrate is strongly evident in these pictures. The hexagonal arrangement of the potassium vacancies demands orientation at 60° intervals, as seen in these images. Unlike Figure 13.8, the structures in Figure 13.9 are rigid. Furthermore, they appear to be multilayered (Figure 13.9[b]), as seen in the cross-section image of Figure 13.9(a). The layers are integrals of half diameter of G-wires (about 1.1 nm). These structures might be examples of ladder G-DNA (L_1) crystals, as shown in Figure 13.5. Since these structures are much wider than the artificial broadening generated by finite tip geometry, the greater width could be explained by parallel packing of the ladder structures. A lattice match with the ladders and the potassium vacancy sites of the mica substrate (1.04 nm) would create a surface that the ladder DNA could also adsorb to—i.e., create multiple layers of ladder G-DNA. Note in Figure 13.9(c) that flexible strands of G-wire DNA are also seen aligning in the same direction as the crystalline structures.

Finally, networks of G-wires have been observed with samples deposited directly from the self-assembly cocktail onto a freshly cleaved mica substrate. Analysis of G-wire self-assembly by gel electrophoresis often shows material that appears at limit mobility even at very low gel density or that remains adsorbed at the origin of the gel [27,31]. It is possible that higher-order aggregation or network formation accounts for this phenomenon. The standard protocol used for the analysis of G-wires by SPM involves diluting the self-assembly cocktail to secure well-resolved structures. This process is intended to disrupt aggregates and may also disrupt any weak bifurcation point along the G-wire. A small aliquot (~0.5 μL or less) of G-wire cocktail is gently spread across freshly cleaved mica with buffer, and is allowed to adsorb to the mica substrate prior to rinsing and drying. This method yields a random, highly bifurcated network of G-wires (Figure 13.10[a]). All of the structures are of uniform height and width, and often adopt a preferred alignment with the mica in a manner consistent with previous observations. Alternatively, a rapid drying of the sample after spreading the G-wires results in the fortuitous crowding of concentrated G-wires on the substrate. Rapid solvent evaporation reproducibly generates networks of G-wires that trace a continuous path across the entire scan range of the image (> 2.5 μm). An example of such a network on mica is shown in Figure 13.10(b). This rapid method for deposition of G-wires may enable a much broader range of substrate materials to be used in device construction.

Figure 13.12 is a panel representing the humidity dependence of G-wire contrast. Most notable is the apparent reduction in width as the humidity of the imaging chamber was increased, even over a very small range. Figure 13.13 reflects the quantitative measurements of the images in Figure 13.12, clearly

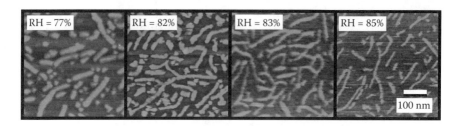

FIGURE 13.12   Freshly made G-wires imaged by HLSTM at 3pA tunneling current, −7 V bias, and four different relative humidities in a sealed chamber with different Pt-Ir tips. N.B. the apparent decrease in width as the humidity increases. The vertical height range is 5 nm from black to white.

indicating decreasing trend lines. The error bars are so large in the average width measurements, and the relative humidity range so small, it is not possible to glean the exact type of relationship between the two, though it appears the reduction in width is significant. Height information from the same samples indicate a decrease of 0.6 nm as the humidity increases, but the measurement error bars are so large that we are unable to determine if this decrease is significant. In our hands, the relative humidity range at which images can be intermittently collected is between 65 and 90%, with stable imaging between 75 and 85%. Crashing tips is a common casualty as the humidity changes, and all these measurements involved different tips. Consequently, the trend lines could represent fortuitous tip broadening since HLSTM *can* involve a slightly different contrast mechanism compared to regular STM—i.e., through ballistic tunneling. In ballistic tunneling, many apical atoms contribute to the tunneling current, reducing the resolution of the image.

The diagram in Figure 13.14 speculates about a possible contrast mechanism that would explain the data in Figure 13.13, namely a reduction in height and width of the G-wire DNA as the humidity in the chamber rises. With increased humidity comes greater adsorption of water into the hygroscopic hydration layer on the surface of the sample. If the DNA is firmly anchored onto the surface of the mica, the rising hydration layer would slowly submerge the DNA. The decrease in the height of the DNA is thus reflective of the increase in the thickness of the hydration layer, about 0.6 nm over the range indicated. This thickness

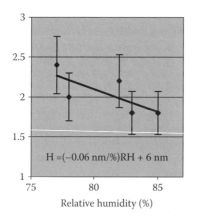
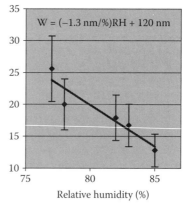

FIGURE 13.13   Height and width information plotted as a function of relative humidity. The trend lines decrease in both with increasing humidity, but the exact relationship with relative humidity can not be established because of the large error bars and limited range of the humidity measurements. Unlike the decrease in width, the slight reduction in height—about 0.6 nm as the humidity increases—is within the measurement error and cannot be established as being significant.

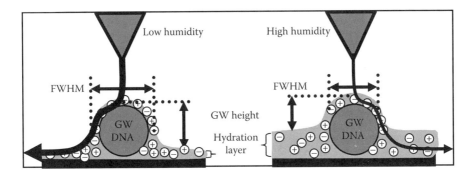

FIGURE 13.14 This cross-section cartoon speculates about the possible contrast mechanism observed in Figure 13.13. The full width half maximum (FWHM) of the G-wire DNA imaged at low humidity will be greater when the hydration layer is thinner than at higher humidity. Also the height of the DNA may appear diminished—assuming it is firmly anchored to the substrate—by the rising hydration layer. The conduction may pass through the dissolved ions from the buffered media (indicated by positive and negative circles shown above) or through electron hopping.

is substantially larger than thicknesses measured for water on mica by scanning polarization microscopy [46]. However, the latter results are for pure water on mica without any ions in the solution.

The metalized contacts were made by evaporating or sputtering gold through shadow masks, such as those shown in Figure 13.15(a) and (b). The contacts were deposited (150 nm—Figure 13.16[a]) over a variety of conductive and nonconductive surfaces, including G-wire DNA networks (Figure 13.16[b]). The samples stayed very nearly at room temperature during the coating procedure to ensure no degradation due to heating. The impact of extreme dehydration of the samples under vacuum conditions was not examined. Table 13.1 details average conductivity measurements from three samples each of three different types of conductive thin films, control surfaces (mica, mica rinsed and dried with imaging buffer) and six different samples of G-wire DNA networks. The thin film thicknesses were measured with the AFM in tapping mode. No visible surface deposits were imaged on either the mica or buffer treated mica surfaces. G-wire DNA networks were found to be in excellent contact with the gold coatings (Figure 13.16[b]). The calculated values of the resistance are in the same range as the measured resistances found with the multimeter. No observed conductivity was found on the mica and buffer-treated mica samples, as expected. Also no measurable conductivity was found over G-wire DNA network samples in either dry

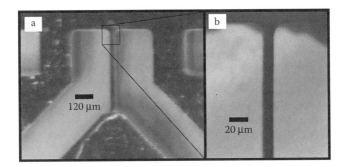

FIGURE 13.15 (a) A low-resolution bright-field optical image of a four-point probe configuration. The two arms on the wings are for setting up large currents at high voltages in conductive samples. (b) A high-resolution optical image of gold contacts made by evaporating gold through the shadow mask at left. The gap between the two center probes is about 15 μm and can be used for traditional two-point probe experiments. N.B.: the shadow on the top right of the gold contacts is a result of a thinner gold coating due to incomplete removal of the silicon oxide layer during the dry etching process.

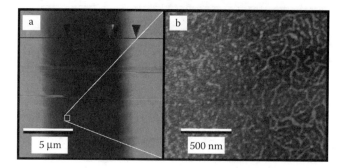

FIGURE 13.16   (a) Gold contacts on either side of this image are 150 nm above the G-wire DNA networks observed in the background in (b). The transition region between gold contacts and the G-wire DNA networks is shown in (b), where the height scale is 10 nm from dark to bright.

TABLE 13.1   Sample Conductivities

Sample	Depth (nm)	$\rho\ (\Omega\text{-m}^2)$	$R_{calculated}(\Omega)$	$R_{measured}(\Omega)$
Gold	13 ± 1	2.5×10^{-9}	1.8	18 ± 5
Chromium	13 ± 1	1.3×10^{-8}	10	40 ± 5
Carbon	12 ± 1	10×10^{-5}	8300	8700 ± 1000
Freshly cleaved mica	—	10^9	$> 1.2 \times 10^9$	$> 1.2 \times 10^9$
Mica rinsed w/buffer, water, and dried	—	—	$> 1.2 \times 10^9$	$> 1.2 \times 10^9$
G-wire DNA in either wet or dry air	2.0 ± 0.1	$> 10^5$	?	$> 1.2 \times 10^9$

or hydrated states. Rehydration of the G-wires was undertaken by exposing the samples to moist air over a saturated salt solution of potassium chloride (85% relative humidity at 20°C) over an hour. The resistance was measured with the probes already connected and inside the enclosed humidity chamber. The time used in this experiment may very well be insufficient to rehydrate the G-wires sufficiently to return the biomolecules to their native solution state. The results appear to be inconsistent with the HLCSTM data, though the reasons for the difference are probably due to structural differences between dried and hydrated DNA. In summary, the lower limit of G-wire resistance is in excess of 1 GΩ ($10^5\Omega$-m^2) under dry conditions.

Though not conductive under dry conditions, G-wire DNA still is an interesting candidate for molecular templates because of the long lengths and narrow dimensions. Rinaldi *et al.* [38] undertook experimental investigation in which ribbon-like guanine structures (for an example, see Figure 13.8) were part of a metal–semiconductor–metal photodetector. I-V curves indicated "striking" semiconductor behavior. Calzolari *et al.* [39] subsequently made first principle theoretical calculations that tube-like G-wire DNA should have semiconductor-like behavior. There results assume idealized configurations that are not achieved under dry conditions. Future research in our lab will focus on the electrical properties of hydrated G-wires.

Acknowledgments

The authors gratefully acknowledge the undergraduate research assistance of Tamieka Armstrong, David Bagg, Geoffrey Champagne, Nicholas Demers, Kristin Eccleston, Matthew Fletcher, Brandon Goblirsch, Mellissa Holden, Peter Hulsey, Marci Luhrs, Bethany Rioux, Joe Skaja and graduate students Robert Baron, Robert Kretschmer, Patrick Spinney, and Matthias Urban. Drs. Joe Brom and Rosemary Smith provided helpful comments in preparing the manuscript. We acknowledge the financial support from the University of New England, Research Corporation Cottrel College Science Award, and the David and Lucille Packard Foundation Interdisciplinary Science. This work was supported in part by the MRSEC Program of the National Science Foundation under award number DMR-0212302 and NSF Major Research Instrumentation Initiative DMR-0116398.

References

[1] Meade, T.J. and Kayyem, J.F., *Angew. Chem. Int. Engl.*, 34, 352–354, 1995.

[2] Fink, H.-W. and Schönenberger, C., *Nature*, 398, 407–410, 1999.

[3] Porath, D. et al., *Nature*, 43, 635–638, 2000.

[4] Yoo, K.–H. et al., *Phys. Rev. Lett.*, 87, 198102-1-198102-4, 2001.

[5] Cohen, H. et al., *Proc. Nat. Acad. Sci. U.S.A.*, 102, 11589–11593, 2005.

[6] Mahapatro, A.K. et al., *IEEE 2006—Nano Conference Proceedings*, 1-4244-0078, 2006.

[7] Zhou, Y.X. et al., *Nano Lett.*, 3, 1371, 2000.

[8] de Pablo, P.J. et al., *Adv. Mater.*, 12, 573–576, 2000.

[9] de Pablo, P.J. et al., *Phys. Rev. Lett.*, 85, 4992–4995, 2000.

[10] Storm, A.J. et al., *Appl. Phys. Lett.*, 79, 3881, 2001.

[11] Williamson, J.R. et al., *Cell.* 59:871–880, 1989.

[12] Williamson, J.R., *Proc. Natl. Acad. Sci. U.S.A.*, 90, 3124–3124, 1993.

[13] Sen D. and Gilbert, W., *Nature*, 344, 410–414, 1990.

[14] Hardin, C. et al., *Biochemistry*, 30, 4460–4472, 1991.

[15] Balagurumoorthy, P. and Brahmachari, S.K., *J. Biol. Chem.*, 269, 21858–21869, 1994.

[16] Keniry, M.A., *Biopolymers*, 56, 123–146, 2001.

[17] Parkinson, G.N. et al., *Nature*, 417, 876–880, 2002.

[18] Henderson, E. et al., *Cell*, 51, 899–908, 1987.

[19] Sen, D. and Gilbert, W., *Nature*, 334, 364–366, 1988.

[20] Williamson, J.R., *Annu. Rev. Biophys. Biomol. Struct.*, 23, 703–730, 1994.

[21] Siddiqui-Jain, A. et al., *Proc. Natl. Acad. Sci.*, 99, 11593–11598, 2002.

[22] Huppert, J.L. and Balasubramanian, S., *Nucleic Acids Res.*, 33, 2908–2916, 2005.

[23] Todd, A.K. et al., *Nucleic Acids Res.*, 33, 2901–2907, 2005.

[24] Rawal, P. et al., *Genome Research*, 16, 644–655, 2006.

[25] Sen, D. and Gilbert, W., *Nature*, 344, 410–414, 1990.

[26] Smith, F.W. et al., *Structure*, 3, 997–1008, 1995.

[27] Marsh, T.C. and Henderson, E., *Biochem.*, 33, 10718–1072, 1994.

[28] Sen, D. and Gilbert, W., *Biochem.*, 31, 65–70, 1992.

[29] Dai, T.Y. et al., *Biochem.*, 34, 3655–3662, 1995.

[30] Protozanova E., and Macgregor Jr., R.B., *Biochem.*, 35,16638–16645, 1996.

[31] Marsh, T.C., et al., *Nucleic Acids Res.*, 23, 696–700, 1995.

[32] Vesenka, J. et al., *AIP Conference Proceedings*, 640, 109–122, 2002.

[33] Muir, T. et al.,*J. Vac. Sci. Technol. A.*, 16, 1172–1177, 1998.

[34] Heim, M. et al., *Vac. Sci. Technol. B.*, 14, 1498, 1996.

[35] Heim, M. et al., *J. Structural Bio.*, 119, 212–221, 1997.

[36] Armstrong, T. et al., *AIP Conference Proceedings*, 725, 59–64, 2004.

[37] Gottarelli, G. et al., in *Comprehensive Supramolecular Chemistry*, Atwood, J.L. et al., Eds. Pergamon, New York, 9, 1996.

[38] Rinaldi, R. et al., *App. Phys. Lett.*, 78, 3541–3543, 2001.

[39] Calzolari, A.R. et al., *Appl. Phys. Letters*, 80, 3331–3334, 2002.

[40] Hardin, C.C.et al., *Biochemistry*, 32, 5870–5880, 1993.

[41] Hardin, C.C. et al., *Biopolymers*, 56, 147–194, 2001.

[42] Miyoshi, D. et al., *Nucleic Acids Res.*, 31, 1156–63, 2003.

[43] Spackova, N. et al., *J. Am. Chem. Soc.*, 121:5519-553, 1999.

[44] Stefl, R. et al., *Biophys J.*, 85, 1787–1804, 2003.

[45] Wyatt, J.R. et al., *Biochemistry*, 35, 8002–8008, 1996.

[46] Hu, J. et al., *Surf. Sci.*, 355, 255, 1996.

14

Metalloprotein Electronics

Andrea Alessandrini

Paolo Facci

14.1 Introduction: Metalloprotein Electronics

The term *metalloprotein electronics* refers to a particular branch of molecular electronics that aims to exploit the peculiar electron transport characteristics of redox metalloproteins for implementing hybrid electronic devices. Metalloprotein electronics shares with molecular electronics the bottom-up approach to the assembly of devices, as well as the use of intrinsically functional units. In both these cases, the functionality to be instilled in a device is already intrinsically present in the molecules rather than achieved by implementing a particular design or by exploiting a specific material for building a device.

This characteristic makes it possible to identify a single molecule with a device, and is why so much is expected of bioelectronics (and not just in regards to device miniaturization). Moreover, at variance with organic, synthetic molecules, the use of particular biomolecules such as redox metalloproteins (see Section 14.2) bears a number of potential advantages. As we will see, redox metalloproteins are biomolecules that have evolved over billions of years to perform the task of transferring electrons between molecular partners. Therefore, their effectiveness is usually so high it is virtually impossible to further enhance it by molecular biology approaches such as protein engineering. Furthermore, a protein's chemical nature endows its external surface with a remarkable chemical richness that can be exploited for anchoring molecules in desired positions. Protein's surface chemistry is also responsible for its (possible) self-assembling ability (i.e., the possibility of self-organizing in complex supra-molecular edifices—for example, 2D crystals). The ability to transfer electrons by changing their oxidation state (see Section 14.3) may also be connected with special protein functional characteristics (i.e., catalytic in general), which could open the door to novel device functionality, something that is unachievable with conventional electronic devices.

Using proteins for technological scopes counts on another important feature of these biomolecules connected with their production: the possibility of their recombinant, massive expression in suitable systems (e.g., bacteria, yeast, other eukaryotic cells). This aspect is of paramount importance if large high-purity amounts of a certain native or specifically mutated protein has to be obtained. Indeed, high-yield expression levels can in general be achieved, optimizing the expression procedure at the level of the chosen expression vector, as well as at that of the expression system (e.g., a particular bacterial strain), and of course at that of protein purification. Notwithstanding the aforementioned difficulty of improving artificial electron transport performance of natural proteins, protein engineering is a very important resource for the production of proteins for bioelectronics purposes. In fact, point-directed mutagenesis can be readily exploited to insert special functional groups on the protein surface that can be used for specific molecular immobilization at surfaces. Moreover, a similar approach can also be employed for inserting a chemical functionality that can facilitate protein purification (e.g., Hys tag).

Among the most diffused protein-based electronic devices, one should recall hybrid structures used for sensing biointeractions of different kinds. These are usually electronic devices (e.g., solid-state field effect transistors) coupled with a biomolecular layer (e.g., a layer of antibodies or a self assembled monolayer of functionalized ssDNA) [1], which imparts biospecificity and biorecognition sensitivity to the electronic transducer. Many examples of these kinds of devices have been reported in the literature, and all of them share the common feature of limiting the role of the electronic device to that of a transducer, and of exploiting the specific features of the biomolecular layer to sense some interaction (e.g., immunological interaction, hybridization, etc.).

Only recently, the idea of exploiting molecules of biological origin for assembling electronic devices has come into being. Indeed, biomolecules such as redox metalloproteins could be used to assemble a functional part of an electronic device such as the channel of a field effect biotransistor. The implementation of such a device requires, of course, a deep knowledge of the mechanisms ruling the transfer of electrons in single metalloproteins as well as an understanding of intermolecular electron flows in strong electric fields. Two- and three-terminal protein hybrid planar devices have been already demonstrated and characterized, and models for device functioning have been reported (see Section 14.9.2).

The intrinsic functionality present and demonstrated in each metalloprotein suggests the charming implementation of single-molecule devices, which requires a number of technological and biophysical problems be solved. Among the most relevant ones, we recall the ultimate lithographic resolution needed to fabricate a nanometer gap suitable for locating a single metalloprotein in it, as well as the need for effective approaches for gating the current via the molecule. Some solutions to these problems have been recently proposed, giving rise to the first single-metalloprotein transistor operating in a liquid environment and endowed with an electrochemical gate. Such a recent achievement has shown how the concepts of protein bioelectronics can differ from those typical of today's solid-state electronics. Namely, the idea that bioelectronics, rather than competing with conventional electronics, could be fruitfully used to implement novel functionalities and generate devices operating in unconventional environments, thus promising charming potentialities for this novel field.

14.2 Proteins and Redox Metalloproteins

Proteins are biological macromolecules that represent the main functional units in living systems. As such, they are deputed to perform an extremely wide range of tasks in every biologically relevant event. In particular, the transfer of electrons by biological macromolecules represents the means by which a relevant number of complex functions is partially or fully accomplished [2]. Key phenomena such as respiration, photosynthesis, catalytic reactions, etc. involve, as single or multiple crucial steps, the transfer of one or more electrons between molecular partners along free energy cascades. A peculiar role in biological electron transfer is played by special classes of biomolecules called redox metalloproteins.

These molecules belong to the larger family of metalloproteins, biomolecules containing one or more metal ions or clusters, which represent about 25 to 30% of the known proteins [3]. Generally speaking, the set of biological metals (i.e., those metals important for biological functions) includes vanadium, magnesium, manganese, iron, cobalt, nickel, copper, zinc, molybdenum, and tungsten [4]. These metals and their ligands configure prosthetic groups that are usually covalently bound to the polypeptide backbone by endogenous ligands brought about by amino-acid side-chains. Redox or electron transfer metalloproteins are characterized by bearing one or more metal ions in their active sites, whose oxidation state can reversibly change, being the basis of their ability to exchange electrons. These metalloproteins can be typically subdivided into three families: blue copper proteins (or cuprodoxins), cytochromes (containing haem groups), and iron–sulphur complexes, to which one can add chlorophyll-based photosynthetic complexes (e.g., reaction centers), which are redox metalloproteins where a photoionization event triggers the electron transfer cascade [5].

Due to their peculiar functional activity and extreme biological relevance, redox metalloproteins have been the object of intense study using computational (theoretical) and experimental approaches [6]. These different approaches range from classical and semiclassical (QM/MM) molecular dynamics to spectroscopy (UV-Vis and IR absorption, Raman, EXAFS, neutrons, fluorescence, NMR, EPR, XPS), diffraction (x-rays, electrons), pulsed radiolysis, electrochemistry, and, more recently, scanning probe microscopy (STM, EC-STM, AFM), often assisted by protein engineering to help unravel proteins' fine structural and functional details by site-specific mutagenesis.

The generalized scientific effort towards metalloproteins characterization, has resulted in a deep understanding of the structure and function of several redox metalloproteins and, in many cases, it has helped in clarifying to a good extent the mechanisms by which these biomolecules shuttle electrons between molecular partners.

The latter aspect is, of course, very important as far as the technological exploitation of metalloproteins electron transfer is concerned.

The recent advent of nanoscience has fostered the possibility of attaining a so far unmatched level of comprehension of the electron transfer mechanisms ruling the functional activity of redox metalloproteins: that of the single molecule. Indeed, at such a level, one can get rid of the statistical average that tends to smear out the specific features inherent to the different ET mechanisms, hoping to be able to attain a quantitative, mechanistic description of the observed phenomena.

The efficiency with which redox metalloproteins perform the task of transferring electrons, along with their intrinsic nanometer size, optimized by natural evolution along billions of years, makes them ideally suitable candidates for technological aims [7], namely for implementing bioelectronic devices.

Of course, the way to fully attaining such results is still far off since many data and technological approaches are still lacking. For instance, it is necessary to rely on very powerful experimental tools, capable of direct-space investigation and, of course, of achieving molecular or submolecular resolution. Moreover, the technological exploitation of (single) redox metalloproteins requires the development of methods for the robust anchoring of these molecules onto surfaces (e.g., metal electrodes, oxides) in locations of choice and to carry out a full characterization of their behavior in such a particular environment.

Ultimate lithographic approaches along with current-sensitive scanning probe techniques, performed in a liquid, physiologic-like environment, are among the experimental tools most needed for the implementation and the characterization of hybrid bioelectronic devices.

14.3 Biological Electron Transfer

At the root of the marked interest in bioelectronics for redox metalloproteins is their unique ability to perform electron transfer reaction (i.e., to shuttle electrons between molecular partners as crucial steps of a large pool of different biological events).

Electron transfer reactions are characteristic features of a variety of fundamental biological processes that include energy metabolism (photosynthesis, respiration, nitrogen fixation), hormone biosynthesis, and xenobiotic detoxification. For most of the proteins involved in these processes, the active site is comprised of a metal center, although organic cofactors (e.g., quinones, flavins) may also accomplish this function. Irrespective of the specific features of each electron transfer reaction, the basic principles that rule the electron transfer rate are common to any reaction. In order to understand the basis of this phenomenon and to define many of the mechanistic issues related to biological electron transfer reactions, it is useful to consider the impulse from pioneering inorganic chemistry studies to the current biological research activities.

The genesis of the nowadays approach to biological electron transfer reactions dates back to the late 1940s, when coordination compounds with radio-labeled transition metals were used to study inorganic electron transfer reactions [8]. These initial studies allowed for the first time the determination of the rate constants for the transfer of an electron from the reduced to the oxidized form of a coordination compound (self-exchange rate constant). The availability of reliable experimental information for self-exchange reactions combined with the chemical simplicity of electron transfer reactions (no bonds are formed or destroyed and the free energy of the products is identical to that of the reactants) attracted the attention of both experimentalists and theoreticians.

We can divide inorganic electron transfer into two different general types. The simplest mechanism is that called outer-sphere electron transfer in which the coordination shells of the reactants stay intact. It involves three steps:

$$A + D \leftrightarrow A \parallel D \tag{14.1}$$

$$A \parallel D \rightarrow A^- \parallel D^+ \tag{14.2}$$

$$A^- \parallel D^+ \leftrightarrow A^- + D^+ \tag{14.3}$$

Step (1) corresponds to the formation of the so-called precursor complex (A||D), in which the electron acceptor (A) and the donor (D) interact through the ligands coordinated to the central metal atom. Following the formation of the precursor complex, the electron is transferred from the donor to the acceptor, with the rate constant k_{et}, to form the successor complex (A$^-$||D$^+$), step (2). The reaction is completed with the dissociation of the successor complex to the reaction products, step (3). Here, the rate of intramolecular electron transfer will be influenced only by factors affecting k_{et}, whereas the intermolecular reaction will also be a function of those factors influencing the mutual interaction of the reactants (diffusion enabling the formation of the precursor complex, electrostatic interaction between reactants).

A different scenario is that characterizing the inner-sphere electron transfer reactions. Here, a reaction intermediate characterized by an inner coordination sphere of donor and acceptor sharing a common ligand of a metal ion is transiently formed.

Generally speaking, coordination complexes that are relatively inert to ligand substitution employ outer-sphere electron transfer mechanisms, whereas those complexes in which coordinating ligands are more labile, are more prone to undergoing inner-sphere reactions. The requirement for formation of a ligand-bridged intermediate prior to electron transfer indicates that the energetic barrier to formation of the transition state is significantly greater in the inner-sphere electron transfer case. This difference in activation barriers between the two mechanisms reflects the Franck–Condon principle, which states that nuclear rearrangements are slower than electronic ones.

The relative simplicity of the outer-sphere electron transfer reaction promoted the development of theoretical efforts towards the prediction of electron transfer rate constants. In what follows, we will recall only the basic ideas referring the interested reader to one of the numerous works reviewing the topic [8–12].

Let us start by considering the reaction described in Equation (14.2) in the framework of the transition-state theory (see Figure 14.1). In this panel, the potential energies of the precursor complex (A||D) and the successor one (A$^-$||D$^+$) are depicted in 2D as a function of the nuclear coordinate, representative of the

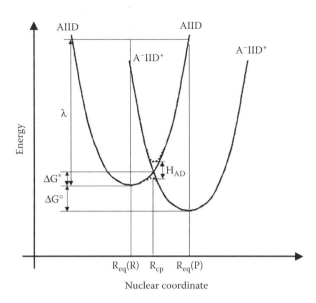

FIGURE 14.1 Energy diagram for an electron-transfer process between A and D. The potential energy of the nuclei for both the reactants and the products is represented by two parabolas (A||D and A$^-$||D$^+$) with equilibrium configuration R$_{eq}$(R) and R$_{eq}$(P). The passage from one parabola to the other represents the electron transfer process, which can occur in the proximity of the crossing point R$_{cp}$. The activation energy for the reaction is given by ΔG^*, whereas the thermodynamic driving force is given by ΔG°. The reorganization energy λ is the energy required to reorganize all the atoms from their equilibrium position in the reactant state to the equilibrium position in the product state without transferring the electron (staying on the reactant curve). In case of strong interaction between the donor and the acceptor, at the crossing point the two parabolas split, giving two curves separated by H$_{AD}$. This is referred to as an adiabatic transition, with a smooth transition from the reactants to the products (no quantum jump).

nuclear motion. The Franck–Condon principle states that electronic transitions take place in times that are very short compared to the time required for nuclei to move significantly. This implies that electron transitions take place at the crossing point R$_{cp}$. Mutual proximity of the acceptor and donor centers within the A||D complex causes electronic interaction between the centers that yield mixing and splitting of the two potential energy surfaces in the region of their intersection (the crossing point R$_{cp}$). The greater this electronic coupling (H$_{AD}$), the greater the separation between the upper and the lower curves of Figure 14.1, and the greater is the probability that the reactants will proceed from the A||D curve to the A$^-$||D$^+$ one at R$_{cp}$. Reactions characterized by a strong electronic coupling are called "adiabatic" (i.e., no electron quantum jumps); in this case, the probability that the activated complex will proceed to products is nearly unity. For complexes in which this coupling is poor (as in the case where distances separating centers are large), the separation between upper and lower curves is small and there is an increased probability that reactants will not progress to the product curve at the crossing point. These types of reactions are referred to as "non-adiabatic." The active sites of redox metalloproteins are quite insulating in nature and their metal ions are usually separated by relatively large distances from each other. As a result, transfer of electrons in proteins are usually non-adiabatic in nature, and mechanistic analysis of such reactions involves quantum-mechanical electron tunneling.

In order to compare experimental electron-transfer rate constants for non-adiabatic reactions with theoretical predictions, one needs to consider the following contributions:

1. The magnitude of the electronic coupling between donor and acceptor as described by the matrix element H$_{AD}$ (the separation between the A||D and the A$^-$||D$^+$ curve at the crossing point)
2. The contribution of the Franck–Condon factor (FC)

All these aspects are accounted for by the fundamental equation of non-adiabatic electron transfer theory, according to Fermi's golden rule:

$$\kappa_{et} = (2\pi\hbar)|H_{DA}|^2 FC \qquad (14.4)$$

Equation (14.4) highlights that it is possible to decouple the electronic and the nuclear contribution in the electron transfer rate. This is due to the fact that electron transfer occurs in a time too short for nuclei to change their position.

The magnitude of the matrix element H_{AD} depends on the nature of the donor and acceptor centers and the distance between them, and can be described by the relationship of Gamow:

$$|H_{DA}|^2 = |H_M|^2 \exp(-\beta R) \qquad (14.5)$$

where R is the distance separating the centers, β is the decay length of electronic coupling with R, and H_M is the value of the matrix element for maximum coupling.

Marcus [13] suggested that the FC factor in Equation (14.4) depends on:

1. The energy λ associated with the rearrangement of the atomic nuclei of the reactants in their configuration in the products
2. The thermodynamic driving force for the reaction, $-\Delta G^0$, which is derived from the difference in the midpoint redox potential of donor and acceptor.

According to Marcus, the classical expression for *FC* factor is

$$FC = (4\pi\lambda k_B T)^{\frac{1}{2}} \exp\left[-(\lambda + \Delta G^0)^2/4\lambda k_B T\right] \qquad (14.6)$$

where λ is the reorganization energy, k_B is Boltzmann's constant and $-\Delta G^0$ is the thermodynamic driving force of the reaction. $\ln(k_{et})$ varies parabolically with $-\Delta G^0$ such that k_{et} increases with the driving force until $-\Delta G^0 = \lambda$. As the driving force increases beyond this point, k_{et} decreases (if λ is constant). This critical prediction configures the so-called Marcus inverted region.

Therefore, the two dominant factors contributing to the electron transfer rate are the distance between the donor and acceptor and the thermodynamic driving force of the reaction. A consequence of these considerations, first noted by Marcus, is that the cross reaction rate constant (k_{12}) for electron transfer between two substitutionally inert complexes (i.e., λ is nearly constant) is a function of the self-exchange rate constants for each of the reactants (k_{11} and k_{22}) and of the equilibrium constant of the reaction K_{12} which is related to the difference in their reduction potentials (i.e., the thermodynamic driving force of the reaction):

$$k_{12} = (k_{11}k_{22}K_{12}f)^{\frac{1}{2}} \qquad (14.7)$$

where

$$\log f = (\log K_{12})^2/4\log(k_{11}k_{22}/z^2) \qquad (14.8)$$

and z is the collisional frequency between neutral molecules in aqueous solution.

The simplicity of the Marcus theory, along with the availability of the self-exchange rate constants for a variety of inorganic complexes (since the 1950s and early 1960s), made it amenable to experimental corroboration. Particularly, it was possible to predict cross-reaction rates between couples of coordination complexes from their self-exchange rate constants and the thermodynamic driving force of the reaction. An increasing amount of experimental results provided compelling evidence for the validity of this theory.

Electron transfer proteins in which the active sites are metal centers have many features in common with simple coordination complexes. Metalloproteins that function solely in electron transfer reactions possess metal centers that undergo minimal changes upon varying their oxidation state. On the other hand, metalloproteins that perform electron-transfer reactions as a part of catalytic cycles can exhibit more extensive structural rearrangements upon changing their oxidation state. In general, the location of the metal center within the protein and, hence, its accessibility by a redox partner, allows a treatment of the electron transfer behavior of these molecules according to the Marcus theory. Particularly, the electron transfer rate constants measured for numerous redox metalloproteins revealed that fast, long-range electron transfer takes place at distances of up to 14 Å, involving inverse electron transfer decay lengths of up to about $1\,\text{Å}^{-1}$ [14].

At any rate, the Marcus theory represents the starting point for any further refinement, and still configures the most suitable framework for the interpretation of data involving the transport of electrons by proteins.

14.4 Two Empirical Models: Pathways and Average Packing Density

To account for the electron transfer rate between two redox centers through the intervening protein matrix, two empirical models have been proposed.

In order to get a good estimate for the k_{et} in intermolecular electron transfer between metalloproteins, a single average value for β might not be appropriate to describe the intervening medium, which is known to modulate significantly the electron transfer phenomenon. Electron tunneling between covalently bridged redox centers in synthetic systems ($\beta \sim 0.9\text{Å}^{-1}$) has proven to be much faster than tunneling through a vacuum ($2.8\text{Å}^{-1} < \beta < 3.5\text{Å}^{-1}$). That is why the last years have witnessed the development of increasingly successful theoretical approaches for predicting k_{et} in proteins taking secondary and tertiary structure into account [15]. In 1991, Beratan and co-workers proposed an empirical model [16] named "Pathways for estimating $\mathbf{H}_{DA}$." Considering discrete steps in a chain between the donor and the acceptor, their model treats the intervening medium as different segments that have their own decay factors (i.e., separating the factors determining the coupling into bonded, nonbonded, and hydrogen-bonded steps). From this approach, results:

$$|\mathbf{H}_{DA}|^2 = \mathbf{A}^2 \left(\prod_i \varepsilon_{bond}(\mathbf{i}) \right)^2 \left(\prod_j \varepsilon_{space}(\mathbf{j}) \right)^2 \left(\prod_k \varepsilon_{H-bond}(\mathbf{k}) \right)^2 \tag{14.9}$$

where $\mathbf{A}$ is the value corresponding to the maximum coupling. As such, the Pathway model reflects the huge difference in through-bond and through-space electronic propagation. The unit-less decay factors corresponding to the different connection types are defined as:

$$\varepsilon_{bond} = 0.6 \tag{14.10a}$$

$$\varepsilon_{space} = 0.5\varepsilon_{bond}\exp[-1.7(R_{DA} - 1.4)] \tag{14.10b}$$

$$\varepsilon_{H-bond} = \varepsilon_{bond}^2\exp[-1.7(R_{DA} - 2.8)] \tag{14.10c}$$

where R_{DA} is the distance between the donor and the acceptor. An alternative empirical model was suggested by Dutton and co-workers, the so-called average packing density tunneling model [14]. Their formulation similarly balances the through-bond and through-space decay of electronic interactions in proteins. The model writes:

$$|H_{DA}|^2 = A^2\exp\{-[0.9\rho + 2.8(1 - \rho)][R_{DA} - 3.6]\} \tag{14.11}$$

Within this approach, the overall packing density ρ of protein atoms in the volume between redox centers, rather than the pathway connectivity, defines the coupling. ρ is defined as the fraction of the volume between redox cofactors that is within the van der Waals radius of intervening atoms. In principle, ρ can range from 1, corresponding to a fully packed medium, to a value of 0. Dutton surveyed a large set of proteins with known structure involved in electron transfer reactions, finally affirming that although long-range electron couplings depend on protein structure, there has been no necessity for proteins to evolve optimized routes between redox centers, provided the distance between the redox centers is within a critical value (1.4 nm), thus denying the role of preferred pathways in accompanying protein-mediated electron transfer. The critical role of the protein structure in determining the electronic coupling still remains a subject of debate in the literature. However, the differences in the predictions of the average packing density and pathway approaches are generally not significant when through-bond interactions dominate the coupling, given the mathematical isomorphism and the simplicity of the models.

14.5   Azurin

Among the electron transfer metalloproteins, an important family is that of the so-called blue copper proteins. This family is comprised of redox biomolecules bearing a single Cu ion in their active site, whose particular distorted coordination geometry imparts an intense blue color to them. They can be found in bacteria and plants and are usually involved in early steps of respiration or photosynthesis [17]. Their family is comprised of azurin, plastocyanin, rusticyanin, amicyanin, caeruloplasmin, etc. Azurin, in particular, has been the object of extensive investigations by electrochemically assisted scanning probe techniques, owing to a number of peculiar reasons that will be discussed later.

Azurin from *Pseudomonas aeruginosa* is a soluble protein (molecular mass 14600) involved in the oxidative phosphorylation of its expressing bacterium. It is believed to accomplish the function of shuttling electrons between two molecular partners, cytochrome c551 and nitrite reductase, which act as the primary donor and acceptor, respectively [18]. Its functional behavior is guaranteed by the reversible oxidation of Cu^{+1} to Cu^{+2}. The peculiar electronic properties of this molecule are connected with the special structural features of its redox active site. Indeed, it contains a copper ion ligated to five aminoacids (two hystidines and a cysteine strongly bound to copper, and two weaker axial ligands: a methionine and a main chain carbonyl oxygen) [19] according to a peculiar ligand-field symmetry that provides the center with unusual spectroscopic and electrochemical properties. Among them, an intense electron absorption band at 628 nm (due to the S(Cys)-Cu bonding to anti-bonding transition), a small hyperfine splitting in the electron paramagnetic spectrum [20] and an unusually large standard potential (+116mV versus SCE) [21] in comparison to the Cu(II/I) aqua couple (−89 mV versus SCE) [22].

Its structure is characterized by a globular shape, endowed mainly with a β-barrel conformation (Figure 14.2). The active redox site is located at $\sim$ 8 Å below the globule surface, and an exposed S-S bridge is formed between the Cys3 and Cys26 side chain thiol moieties.

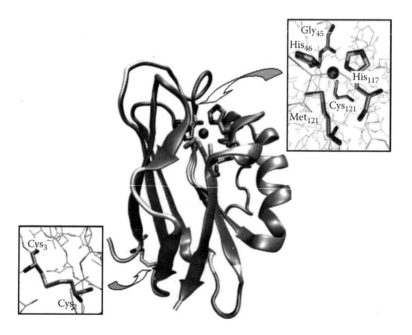

FIGURE 14.2   Cartoon representation of azurin. In the insets, the active site and the disulphide bridge are highlighted. Structural information from PDB file 1E5Y.

Both these characteristics make azurin an ideally suited candidate for single molecule investigation by scanning probe techniques. Particularly, the presence of the aforementioned exposed disulfide offers a very convenient means for anchoring this molecule to the surface of electronically soft metals (Au, Pt, Ag, Cu, etc.) by the formation of one or two S-Me bonds [23], whereas its β-barrel conformation endows the protein with an excellent resistance to repeated probe scans.

A number of different characterization techniques (FTIR in ATR configuration, QCM, SFM, XPS, CV, etc.) confirm that azurin chemisorbs on Au and that it retains its redox activity once arranged in submonolayer [24].

14.6   Cytochrome *c*

Cytochrome *c* is a small monomeric water-soluble protein that belongs to the haemproteins family. It is located in the intermembrane mitochondrial space where its role is to shuttle electrons between ubiquinol/cytochrome *c* oxidoreductase (QCR, cytochrome bc1 complex) and cytochrome *c* oxidase [25,26].

Cytochrome *c* (Cyt *c*) is one of the most extensively studied proteins because of its central role in electron transfer in living organisms. Its diverse functional roles and the availability of high-resolution crystallographic data since the early 1970s have helped make this protein a paradigm in the study of electron transfer processes [27].

Cyt *c* contains a c-type haem prosthetic group covalently bound to the polypeptide chain through two thioether bonds with the side chain of two cysteines, Cys14 and Cys17 (horse heart numbering) (Figure 14.3). The haem contains an iron atom, which is coordinated by the four pyrrole nitrogens of the haem moiety, by the $N\varepsilon 1$ of His18 and by the $S\gamma$ of Met80 [28,29]. The iron has two physiologically relevant oxidation states: Fe (II) and Fe (III) [28].

Alignment of mitochondrial cyt *c* sequences from different eukaryotes showed that the primary sequence is well preserved during evolution, the most conserved region being the one around the axial Met80 ligand [29]. The polypeptide chain wraps the prosthetic group so that only 7 to 10% (depending on the different cyt *c*) of the haem surface area is exposed at the molecular surface [30].

Naturally, most of the hydrophobic residues are buried. All the internal residues in the van der Waals contact with the haem macrocycle are bulky hydrophobic groups, which help stabilize the overall protein

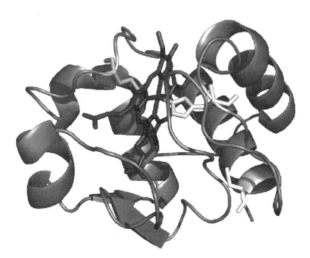

FIGURE 14.3   The three-dimensional fold of *S. cerevisiae* iso-1-cytochrome *c*. Haem inside the protein is highlighted together with axial ligands His18 and Met80. Cys102 is highlighted in the right-bottom part of the structure (PDB code: 1YCC).

structure. Further stabilization of the native structure is conferred by an extensive array of hydrogen bonds. Most of the charged groups are located on the protein surface, where they are distributed in a highly asymmetric manner [31]. Lysines are clustered predominantly around the exposed haem edge and on the right and left sides, while carboxylates are located primarily on the top and back of the molecule. Comparative analysis of amino acidic sequences of more than 100 mitochondrial cytochromes *c* shows that exposed residues are significantly more variable than buried ones, which are constrained by the requirements to provide a stable folded protein structure.

Therefore, a high degree of conservation of specific exposed amino acids implies an important functional role in intermolecular recognition and electron transfer: indeed, residue-specific interactions within a framework of electrostatic complementarity are important for the biological role of cyt *c* in electron transfer.

Within the family of cytochromes *c*, iso-1-cyt *c* from *Saccharomyces cerevisiae* is unique in containing a single surface-exposed cysteine that may be used for chemisorption on gold or other soft metals.

14.7 Tools for Studying the Functional Behavior of Redox Metalloproteins

Bioelectronics approaches must rely on a multitude of strategies to link biological molecules with electronic circuits. After the biomolecules and an inorganic structure have been interfaced, experimental techniques able to investigate the quality of the coupling are needed. The following is a brief survey of the main techniques used to investigate this fundamental issue in the case of metalloproteins.

14.7.1 Protein Film Voltammetry

Once transferred onto a surface, the retention of the functional behavior of proteins must be investigated, this last being of course a critical step towards the implementation of functional hybrid devices. For redox proteins on a solid substrate, the natural sign of a successfully assembled protein film without denaturation is the retention of the native redox electrochemical features with no significant differences from what is measured in solution. For example, in the case of cytochrome *c*, when adsorption on a surface is accomplished at the expense of the native conformation, changes in the ligation to the haem, or increased exposure of the redox active site to the solvent can greatly modify the redox potential.

Redox film voltammetry is one of the methods of choice in the study of the functional properties of redox proteins immobilized on a substrate [32,33]. Voltammetric techniques are dynamic electrochemical methods in which controlled potentials are applied to the electrodes, inducing electron transfer reactions measured through the current response. A schematic setup for protein film voltammetry is shown in Figure 14.4. In this configuration, the electroactive molecules are confined to the working electrode surface of a three-electrode electrochemical cell in which electron transfer to and from the proteins is induced by altering the electrode potentials. One advantage of protein film voltammetry over conventional solution voltammetry is that diffusion effects do not complicate the interpretation of the voltammetric signals. Figure 14.5 reports a typical cyclic voltammogram (the voltage is continuously varied at a constant rate in a cyclic loop) obtained in the case of a protein film undergoing a reversible oxidation-reduction cycle upon voltage sweeping. In this case, the reductive and oxidative waves are symmetrical, one being the mirror image of the other with respect to the potential axis, while the reduction potential $E^{\circ\prime}$ of the reaction (14.12) corresponds to the position of the maxima of the two waves.

$$\text{Protein}^{Ox} + ne^- \mathrel{\substack{\longrightarrow \\ \longleftarrow}} \text{Protein}^{Red} \qquad (14.12)$$

Integration of the area under the oxidative or reductive wave allows one to estimate the number of exchanged electrons and, consequently, the number of electroactive molecules on the surface. In fact,

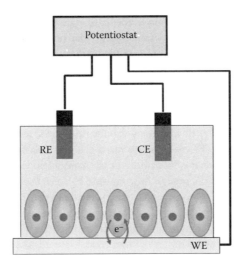

FIGURE 14.4 Schematic representation of a configuration employed for protein film voltammetry. The proteins are arranged on a conductive substrate either directly or via a "soft" self assembled monolayer. The substrate represents the working electrode of an electrochemical cell endowed with a reference electrode and a counter electrode driven by a potentiostat. Electrons are exchanged between the redox active center in the proteins and the working electrode.

the peak current, after subtraction of the capacitive contribution baseline to the overall current, is given by:

$$i_p = \frac{n^2 F^2 v A \Gamma}{RT} \tag{14.13}$$

where n is the number of exchanged electrons, F is the Faraday constant, v is the scan rate, A is the electrode area, and Γ is the surface coverage. A comparison of the surface coverage obtained from protein voltammetry analysis with data obtained from other experimental techniques, such as scanning probe microscopy (see Section 14.7.2), that are not directly sensitive to the electric charge exchanged with the substrate, may allow one to establish the amount of proteins which, upon adsorption to the electrode, preserve their functional behavior.

Protein film voltammetry offers an easy way to establish whether the obtained electrochemical signal is really ascribable to an immobilized protein layer. Due to the relation between peak current and voltage scan rate, a linear relation should be found between the scan rate and the current maximum, instead of the usual square-root dependence observed for solution experiments due to diffusion effects [32].

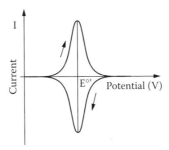

FIGURE 14.5 Scheme of an ideal cyclic voltammetry response for a monolayer of adsorbed proteins. The dependence of the current as a function of the working electrode potential is shown. The reductive and the oxidative waves are symmetrical and the current reaches a maximum at the formal reduction potential E°.

In the ideal case of reversible electron transfer, the half-height peak width is expected to be 3.53 RT/nF (90.6 mV at 25°C for $n = 1$). Deviation from this value may be caused by a spread of redox potentials indicating a nonhomogeneous distribution of the modes of molecular interactions with the electrode [34]. If the electron transfer process is followed by a chemical reaction that leads the species into a more stable electroinactive configuration, or to a species with a different electronic coupling with the electrode, the voltammetric response will be highly asymmetric. However, by using different potential scan rates, it is possible to investigate this kind of "gated" redox reaction [35].

Protein film cyclic voltammetry allows one to also measure the rate of electron transfer between the redox site and the electrode. In this case, the standard electron transfer rate can be derived from the behavior of the anodic and cathodic peak positions as the scan rate increases. Increasing the voltage scan rate, if the electron transfer is a slow process, a separation between the oxidation and reduction peak is observed. Fitting an appropriate analytical model of the electron transfer kinetics to the experimental peak positions as a function of voltage scan rate allows one to extrapolate, as a fitting parameter, the standard electron transfer rate [36], which represents the electron transfer rate at a zero free energy reaction. Usually, a relation different from the Butler–Volmer model is used to account for the fact that all the energy levels below the Fermi level in the metal electrodes are to be considered and that the potential energy varies parabolically along the reaction coordinate. This technique is particularly useful to establish the degree of coupling of the redox active site to the metal electrode, when coupled with protein engineering experiments to find the best efficiency of electronic coupling by modifying the immobilization strategy. Moreover, the same technique may be exploited to investigate the electronic coupling between proteins and the electrode when a self assembled monolayer (SAM) is used to immobilize the biomolecules (see Section 14.8). Using SAMs of molecules having different lengths, the degree of electronic coupling can be varied. For example, the behavior of cytochrome c immobilized on SAMs composed of alkanethiols terminating with carboxylic acid head groups has been studied as a function of the number of methylene units (n)[37]. It has been found that at large reactant-electrode separations $(n > 9)$, the electron transfer rate has an exponential dependence on the charge-transfer distance with a decay constant of about 1Å^{-1} per methylene unit analogous to the tunneling transport through saturated hydrocarbon chains. Decreasing the cytochrome c/electrode distance $(n < 9)$, the electron transfer rate reaches a plateau, highlighting the occurrence of a different electron transfer mechanism between the protein and the metal electrode. This behavior has been interpreted either as the transition from a regime of weak electronic coupling for $n > 9$ (non-adiabatic regime) to an adiabatic charge transfer mechanism $(n < 9)$ as a consequence of a stronger coupling between the protein and the electrode [37] or as a conformational variation gated electron transfer process that, when the electron transfer is fast $(n < 9)$, starts to limit the observed electron transfer rate [38].

14.7.2   Scanning Probe Microscopy

Scanning probe microscopy includes a variety of techniques that allows the investigation of molecular adsorbates on a solid substrate all the way down to the level of the single molecule. Using these techniques, information on single molecule behavior can be gained, at variance with the voltammetric methods, which are limited to ensemble behaviors. Scanning probe microscopy techniques are based on measuring an interaction between a tiny probe and a sample in close, mutual proximity. The tiny probe scans the sample surface while measuring the interaction locally; thus, an image representing the two-dimensional map of the sensed physical quantity is obtained. Usually, the measured physical quantity has a strong dependence on the mutual probe-sample distance and a feedback system is used to keep the interaction constant during scanning. The feedback operates on the probe's "vertical distance" and the voltage signals imparted to a piezo-translator in order to maintain the interaction at the setpoint value represent the third dimension, which adds to the 2D scan of the probe on the sample. The strong dependence of the interaction on the probe/sample distance is the keystone to the high spatial resolution capabilities of these microscopes. The main elements constituting a scanning probe microscope are then the probe (which is able to sense locally a certain property of the sample), a system enabling the measurement of the interaction, a feedback circuit, and the piezo-translators enabling both the 2D scan of the sample and the vertical movements. The great breakthrough of the scanning probe techniques is not limited to the high spatial resolution exceeding the optical diffraction limits, rather it includes the possibility of working in a number of different conditions, including aqueous solutions. This is particularly relevant for the case of proteins, which usually require a physiological environment to be characterized from a functional point of view. Moreover, all these techniques are endowed with spectroscopic potentialities that add to the more conventional surface topography information. In what follows, we will concentrate mainly on two techniques, atomic force microscopy (AFM) and electrochemical scanning tunneling microscopy (ECSTM). It will become evident that, apart from the potentiality in characterizing the morphology of molecular adsorbates, scanning probe techniques represent an optimal tool to establish a molecular junction in which a single metalloprotein is sandwiched between two electrodes. When the techniques used are sensitive to current, they allow a straightforward characterization of the current flowing through a single metalloprotein.

14.7.2.1 Atomic Force Microscope

Atomic force microscopy [39] generates a three-dimensional image of a sample surface by scanning a tiny tip mounted at the end of a flexible cantilever in contact with the sample (Figure 14.6). Extensive reviews can be found in literature on issues related to the instrumental aspects of AFM [40,41], and we refer the interested reader to them. Briefly, as the tip in contact with the sample scans the surface, variations in the sample topography are translated into variations of the vertical deflections of the cantilever. The spring constant of the cantilever can be so small (~ 0.01 N/m) that damage to the sample is minimal. In the framework of metalloproteins, investigation using AFM offers the possibility of characterizing the molecules on the surface from a morphological point of view, enabling an estimate of surface coverage independently of the retention of their functionality. In the following, some basic aspects regarding the operating principles and imaging modes are given.

Cantilever deflections are measured with high accuracy by optical means [42]. A laser beam is focused on the back of the cantilever and reflected towards a Position Sensitive Detector (PSD). Deflections of the cantilever are amplified by the optical systems and are translated in voltage signals. In constant force operation mode, the feedback system operates by regulating the tip-sample distance in order to maintain

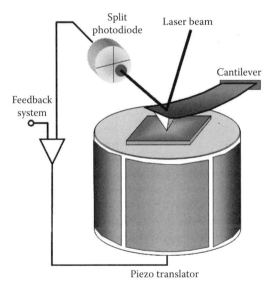

FIGURE 14.6 Schematic representation of an atomic force microscope (AFM). The main elements are highlighted.

the cantilever deflection (which can be translated directly in tip-sample force) constant. Different imaging techniques have been developed which can be grouped mainly into three categories: contact mode, non-contact mode, intermittent-contact mode. The difference between them derives from the different regions of the tip-sample force-distance curve exploited to acquire the image. In the contact mode, the tip gently touches the sample surface throughout the scan. In the noncontact imaging mode, the tip never touches the sample, exploiting long-range attractive forces to map the sample. In the intermittent-contact mode the cantilever oscillates over the sample surface, making the tip experience both the repulsive and attractive region of the force–distance curve [43]. The tip touches the sample only intermittently, allowing a strong reduction of the dragging forces which can limit resolution in the case of highly deformable and softly bound samples.

If both AFM probe and substrate are conductive, it is possible to use the two surfaces as two metal electrodes contacting a sample and to measure the current carried, for example, by a single sandwiched molecule. This approach configures the Conductive Atomic Force Microscopy (C-AFM) [44,45] in which it is possible to both image the topography and measure the current between the tip and the substrate through the sample under an applied bias. The Holy Grail of C-AFM applications in the framework of metalloprotein transport characterization would be the extension of this technique to physiologic-like environments in order to measure current across metalloproteins while controlling the probe-sample distance by the conventional AFM apparatus.

14.7.2.2 Electrochemical Scanning Tunneling Microscopy

Electrochemical scanning tunneling microscopy (EC-STM) is a particular implementation of the pro-genitor of all the scanning probe microscopy family: the scanning tunneling microscope (STM) [46]. An STM consists of a conductive probe that is brought near a conductive sample, or a sample positioned on a conductive substrate, until, under the presence of a small bias voltage (in the order of hundreds of millivolts), electrons can tunnel between the two electrodes (Figure 14.7). Different operation modes can be implemented. In the constant-current mode of operation, the tunneling current is kept constant via the feedback system by varying the vertical distance between the tip and the sample at each measurement point. The strong dependence of the tunneling current on tip/sample separation allows images to be acquired at high lateral resolution. Atomic resolution is routinely achieved by STM due to the fact that only atoms at the very apex of the tip contribute to the measured tunneling current.

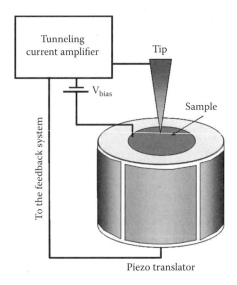

FIGURE 14.7   Schematic representation of an STM setup. The main components are highlighted.

The STM, like other members of the SPM family, has enlarged the idea of a microscope, coupling the possibility of performing high lateral resolution images with the ability to perform various types of spectroscopies with high lateral resolution. In such a way, analytical information can be added to the imaging capabilities of the microscopes. In the case of STM, scanning tunneling spectroscopy (STS) [47] allows the local electronic structure of a sample to be studied. Different methods can be used to collect STS information: imaging the same sample area at a different bias voltage and comparing the acquired images; positioning the tip at a fixed position with respect to the sample and ramping the bias voltage while recording the tunneling current (I-V curves); positioning the tip at a fixed lateral position and modulating the vertical tip-sample distance while recording the tunneling current (I-z curves). As a general comment, it is important to realize that STM does not map the true topography of the sample; rather, due to its sensitivity to the local electronic structure of the sample, it maps a surface of constant tunneling probability [48]. As a consequence, information about sample topography can not be obtained straightforwardly from STM images.

The electrochemical scanning tunneling microscope is an extension of the STM, and is endowed with the possibility of working in liquid and in an electrochemical cell. Provided the current which enters the feedback circuit arises mainly from the tunneling region between the tip and the sample, it was shown that the STM could work also in liquid [49]. Moreover, the conductive substrate could be configured as the working electrode of an electrochemical cell comprising a reference electrode and a counter electrode [50,51]. With respect to a conventional three-electrode electrochemical cell, the conductive STM tip can be considered a polarizable electrode, configuring a four-electrode cell (two working electrodes: the substrate and the tip) [52] (Figure 14.8). This configuration allows one to keep the tip and the substrate under potentiostatic control while imaging by the tunneling current. Non-tunneling currents have to be reduced to a small percentage of the tunneling setpoint current by both reducing the exposed tip-electrolyte interfacial area and by controlling the electrochemical potentials of the tip and substrate. The tip is coated by an electrical insulating material such as Apiezon wax [53] or electropolymerizable paints [54], leaving only the very apex of the tip, which has been previously electrochemically sharpened, exposed to the solution. The current that enters the feedback control at a large tip-sample separation can be characterized as a function of both tip and substrate potential in the electrochemical cell allowing it to establish a potential window in which non-tunneling currents can be kept to a very small contribution, so as not to prevent the

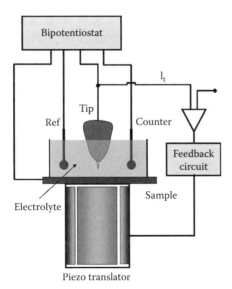

FIGURE 14.8 Schematic representation of an EC-STM setup with bipotentiostatic control. I_t represents the tunneling current that feeds the feedback system in constant current imaging.

imaging capabilities [55]. In ECSTM experiments, the reference electrode is usually a Ag wire in solution, representing a "quasi reference" whose potential can be measured with respect to other common reference electrodes such as a Saturated Calomel Electrode (SCE) or Ag/AgCl. The counter electrode is usually made of Pt and is the one through which the cell current is passed, playing the role of a sacrificial electrode. In the bipotentiostatic approach, the tip and substrate potential are independently controlled with respect to the reference electrode in solution, and their difference represents the tunneling bias voltage.

ECSTM represents the tool of choice to study the electron transfer properties of metalloproteins immobilized on a conductive substrate. The conductive substrate to be used in EC-STM experiments has to fulfil mainly two requirements. It has to allow the immobilization of the proteins (and not prevent their functional behavior), and it has to be atomically flat. The first requirement is usually fulfilled by using electrodes onto which proteins can directly bind via chemical functional groups present at their surface (for example, exploiting cysteine on gold) or by functionalizing electrodes with self assembled monolayers terminating with chemical groups able to immobilize the proteins (for example, ω-functionalized alkanethiol on gold) (see Section 14.8). The second requirement is a mandatory condition for having high spatial resolution in SPM experiments. If small adsorbates are to be visualized by SPM, the roughness of the substrate surface has to be as low as possible and lower than the typical dimensions of the samples. Usually, Highly Oriented Pyrolitic Graphite (HOPG) or a well-defined crystallographic metal surface is used as a substrate for EC-STM studies. Among the second type of substrates, Au(111) is one of the most often used because of the ease in its preparation [56], its resistance to oxidation, and the possibility of exploiting the sulphydril reactivity for developing immobilization strategies.

In the framework of metalloprotein investigation for biomolecular electronics applications, the use of EC-STM allows one to study the behavior of the electron transfer molecules as a function of the potential of the electrode onto which the molecules are immobilized [57]. As a common rule, when interpreting EC-STM experiments, it has to be considered that EC-STM data contain information on both the electronic structure of water and of the molecules interposed between the tip and the substrate.

14.8   Strategies for Protein Immobilization on Surfaces

A bioelectronic device in general requires the wiring of biomolecules on a conductive or semi-conductive substrate—a critical issue being the retention of the functional properties of the molecules [58]. Metalloproteins' functional properties involve the ability to exchange electrons with the substrate. In the following, a description of the main strategies adopted to immobilize proteins on surfaces will be given. Although biomolecules could be immobilized on surfaces by means of physisorption—as in cases where hydrophobic effects or electrostatic interactions are exploited—only the strategies that yield a chemisorbed protein layer will be treated, because only those approaches provide the surface stability usually required in bioelectronic devices.

The assembled protein layers can be investigated by different techniques [59] such as surface plasmon resonance (SPR) scanning probe microscopies (SPMs), infrared spectroscopy, x-ray photoelectron spectroscopy (XPS), ellipsometry, UV-vis spectroscopy, surface enhanced raman spectroscopy, quartz crystal microbalance (QCM), and electrochemical techniques. Some of these techniques are also able to provide a functional assay of the assembled protein (see the following).

The immobilization approaches can be divided into two main strategies: covalent bonding by means of heterobifunctional crosslinkers or direct immobilization of the proteins on the surface. Which of the two is to be used depends on the surface material and on the specific requirements on the final protein layer.

The general approach for covalent immobilization of proteins by means of heterobifunctional crosslinkers is sketched in Figure 14.9. A surface exposed reactive group is exploited to bind a short molecule endowed with two functional groups at the ends. In some cases, the affinity of the surface material for the chemical groups of the heterobifunctional linker can also be exploited. We will refer to this first step as the "surface chemical modification." The presence of two different functional groups assures that only one of the two is able to covalently bind to the surface. However, under particular circumstances, even a crosslinker with two identical terminal functional groups could be used, provided that one gets rid of the possibility that

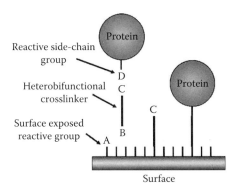

FIGURE 14.9 Schematic of the protein immobilization approach based on a surface chemical modification by a self-assembled monolayer of heterobifunctional crosslinkers. Surface-exposed reactive groups (A) react with one functional group of the crosslinkers (B) endowing the surface with the other functional group (C), which is then exploited for the immobilization with protein surface functional groups (D).

they both bind to the surface, yielding a surface-lying molecule. The assembled layer provides an exposed functional group which is then exploited to covalently bind a chemical group of the exposed protein aminoacids. From this scheme, it is evident that the final configuration of the assembled layer strictly depends on the number of exposed side-chain groups available to react with the pre-assembled surface layer. This aspect influences the orientation and the homogeneity of the immobilized biomolecules.

We will deal with two main types of surface chemical modifications: chemical modification based on the formation of siloxane monolayers and on sulfhydril group reactivity.

The first type of chemical modification is exploited for substrates which expose hydroxyl groups. This is the case, for example, with metal oxides [60], mica [61], glass [62], and silicon oxide SiO_x, resulting both from oxidation in air (native oxide) and from chemical or thermal oxidation [63]. The surface density of hydroxyl groups can be controlled by the use of O_2-plasma, through the use of strong oxidants, or by electrochemical oxidation. The surface endowed with hydroxyl groups is then exposed to organosilanes. Organosilanes are based on a Si atom which is tetrahedrally bound to three similar hydrolyzable functional groups and to a non-hydrolyzable one by means of a short methylene chain. Typically, chlorosilane or alkyloxysilane are used. The driving force for the surface functionalization is the formation of siloxane monolayers, which are connected to the surface by means of Si-O-Si bonds, as shown in Figure 14.10. The assembled layer results are thermally stable, thus affording a very robust system due to surface cross-polymerization.

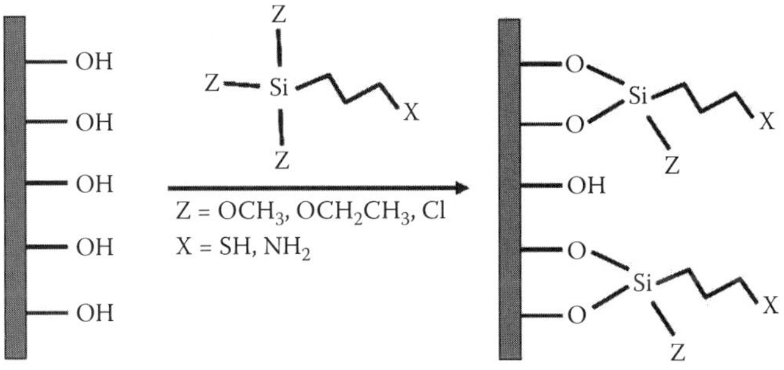

FIGURE 14.10 Scheme of the immobilization strategy based on silanization of surfaces endowed with hydroxyl groups.

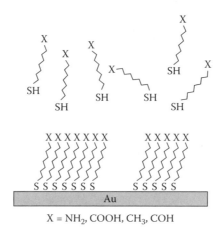

X = NH₂, COOH, CH₃, COH

FIGURE 14.11   Scheme of the surface chemical modification based on SAM of alkanethiols.

The main difficulty in the formation of silane monolayers is the presence of water in the reaction environment, which is required for the hydrolization of the alkyloxy groups, but which may also lead to polymerization of the silane molecules before bonding to the surface [64,65]. Great care should be paid to the pre-treatment of the surface before silanization to reduce contamination and also increase the density of reactive hydroxyl groups [66]. The level of order in the obtained silane monolayer depends on the roughness of the functionalized substrate, and on the underlying structure of the polysiloxane cross-polymerization, which, in its turn, depends on the distribution of hydroxyl groups on the surface, the presence of water molecules, and surface contamination. Different silanization methods can be used based on liquid phase or gas phase strategies. In the first case, surfaces to be silanized are immersed in a solution of silane molecules in an organic solvent followed by extensive wash to remove nonbonded silane molecules [66]. In the case of gas phase silanization, the reaction is performed in a vacuum, but it usually requires a long reaction time, and heating of the environment to enhance silane evaporation [67,68].

Biomolecule immobilization based on the exploitation of surface chemical functionalization by thiol-based molecules makes use of the strong affinity between sulphydryl groups and metals such as gold, silver, platinum, and copper (Figure 14.11). Very extensive reviews can be found in the literature about self assembled monolayers (SAMs) of thiolates [69,70], so only a short introduction to the technique will be given here. Briefly, sulphydryl groups have a strong affinity to transition metal surfaces and sulphur–metal bonds are formed by immersing the substrate in a solution containing the desired thiols by a reaction that can be considered an oxidative addition of the SH bond to gold as in reaction 14.14

$$R - S - H + Au^0 \rightarrow R - S - Au^+ \bullet Au^0 + 1/2H_2 \tag{14.14}$$

Initial adsorption is fast and is followed by a slow organization phase that is mainly driven by van der Waals forces (intermolecular forces). The length of the alkyl chain is a determinant in the properties of the SAM. The most studied system is the SAM of thiolates on gold because gold is easy to obtain even in thin films. It gives relatively simply single-crystal surfaces and is reasonably inert under standard conditions. Moreover, the thiol–gold affinity is so high that if a physisorbed contamination layer is present on the surface to be functionalized, it will be displaced by the thiols. In the process, both primary thiols and molecules containing a disulfide bond can be used, given the tendency of the disulfide bond to reduce to form two S-Au bonds [71]. Both thiol molecules with straight alkyl chains, and with aromatic rings, can be used to obtain well-ordered SAMs.

Using different ω-functionalized alkanethiol molecules, the functionalized surface can be endowed with chemical functionalities exploitable to immobilize biomolecules. The most relevant ω-functionalizations

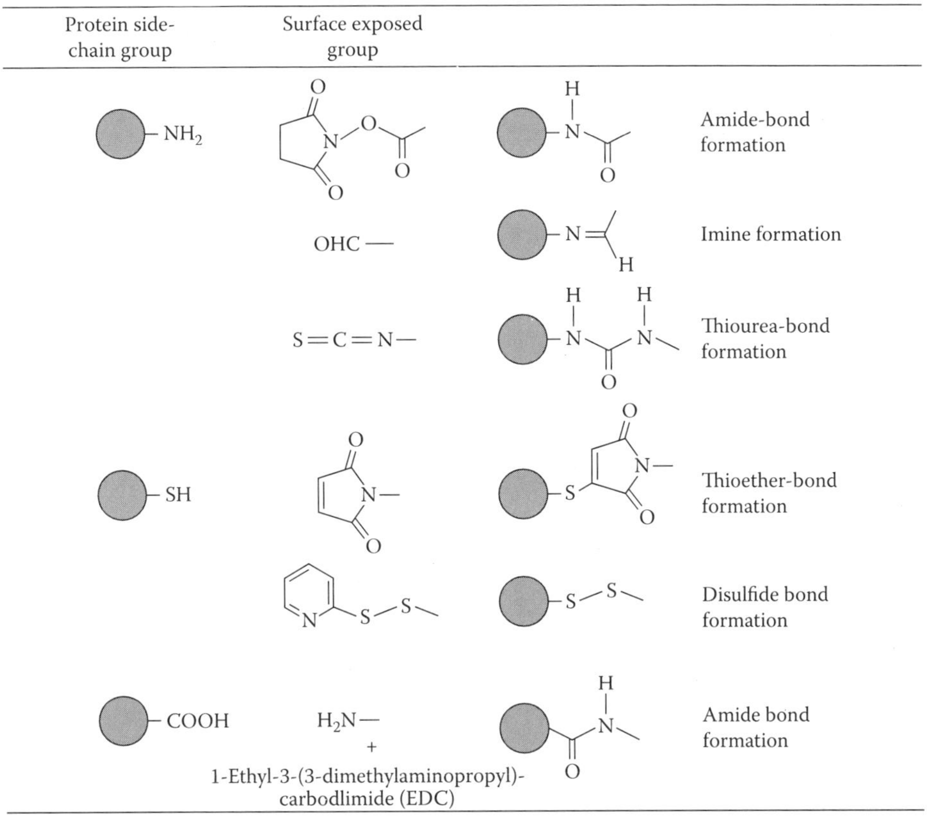

FIGURE 14.12 Immobilization strategies exploiting different protein side-chain groups.

are SH, COOH, NH$_2$, and CHO. Alternatively, SAMs can be modified to incorporate other chemical groups exploiting nucleophilic substitutions, esterification, and nucleophilic addition. By using this last strategy, chemical groups that are impossible to introduce into the alkanethiol chain due to synthetic difficulties can also be inserted.

Once a functionalized surface endowed with chemical reactive groups has been obtained by means of silane molecules or alkanethiols, the biomolecules of interest can be immobilized [72]. In the case of proteins, the typical side-chain chemical functionalities exploited are NH$_2$ groups from lysines, SH groups from cysteines, and COOH groups from aspartic acid and glutamic acid [73]. Figure 14.12 summarizes the most common reactions used to immobilize proteins on a SAM. Attention should be paid to the possibility that side reactions involving other amino acid residues occur.

When amine functions of lysine residues are present at the surface of proteins, reactive esters on the surface, such as N-hydroxysuccinimide can be used to end with an amide-bond formation. An alternative route to the immobilization of protein-exploiting NH$_2$ groups is the coupling with aldehyde groups to the creation of imine bonds (Schiff bases). Otherwise, isothiocyanate on the surface can result, in a basic environment, in the formation of a thiourea bond with exposed amines. In most proteins, several lysine residues are present at the surface and many bonds with the surface groups can be established. This fact would increase the orientational heterogeneity in the immobilized proteins.

If the proteins have exposed SH groups, the reaction with maleinimides can lead to the formation of thioether bonds. Otherwise, a thiol disulfide exchange reaction using pyridyldisulfides can result in disulfide bond formation. Thiol groups can result also from the reduction of cystines. Cystines, which result from a disulfide bond between two cysteines, usually contribute to the protein stability—in some cases, they can be reduced and the two thiols can bind the surface without affecting protein integrity. Cysteines are less abundant then lysine and, in the case where a single free cysteine is accessible, a more controllable immobilization can take place.

The carboxylic group of aspartic/glutaric acid residues can be exploited by using 1-ethyl-3-(3-dimethylaminopropyl)-carbodiimide (EDC) to convert the carboxylic acid into a reactive intermediate. This is susceptible to attack by amines on the surface, yielding an amide bond.

Under some circumstances, proteins can be immobilized on a surface directly, without the use of a soft SAM. An example of this immobilization strategy is that occurring when exposed cysteines are present and immobilization is performed on a gold surface. This approach is exploitable also when cystines are available. In this latter case, the immobilization proceeds by a reduction of the cystine to give two sulphydril groups that bind to the gold surface. The main drawback of a direct protein immobilization on a substrate consists in the possible denaturation of the molecules, due to the high surface energy of a solid-gas interface [74]. Nevertheless, it has been demonstrated that in several cases proteins can bind directly to gold, preserving their functionality.

An alternative route to those previously described consists in the tagging of the proteins in solution with a molecule that also provides a reactive group that is subsequently exploited for the immobilization of the construct on the substrate.

Genetic engineering techniques represent a major means towards the task of efficient protein immobilization on surfaces. Site-directed mutagenesis has been exploited to introduce specific functional groups on the protein surface in order to immobilize them according to specific demands. If, for example, protein immobilization strategy is based on sulfhydryl reactivity, introducing a single cysteine on a protein surface, which is naturally devoid of this type of residue, at a specific position, can lead to a uniformly oriented protein film [75]. Moreover, in the case of a metalloprotein, an appropriate choice of the position in which the cysteine is inserted can lead to an enhanced coupling of the active site to the substrate, allowing it to establish a more efficient electron transfer. In a reversed approach, native cysteines can be replaced by other amino acid residues if the cysteines are not necessary for the biological activity of the protein, and a new cysteine can be inserted in a location of choice. Site-specific mutagenesis can be exploited both in the case where immobilization is performed, exploiting a heterobifunctional linker and in that of direct immobilization on the substrate.

Another approach to protein immobilization consists of the exploitation of a polyhistidine tail towards a metal support in a fashion resembling affinity chromatography. Proteins can be expressed with a specific affinity tag that may consist of a sequence of histidines to help purification, but that can also be used to immobilize the proteins on a substrate [76]. The immobilization results from the strong affinity between histidines and metal ions such as Zn^{2+}, Cu^{2+}, Ni^{2+}, and Co^{2+}.

A well-controlled immobilization strategy allows one to avoid the formation of multilayers and agglomerates on the surface and, by a controlled orientation of the immobilized proteins, the protein–protein interaction between an immobilized protein and its partner in solution can be controlled. Also the surface density of the immobilized proteins can be controlled by using a suitable approach. In cases where a different immobilization scheme also leads to a different protein function, it could be possible to assemble patterned protein films where the protein function is controlled by the molecular architecture exploited for immobilization.

The possibility of controlling immobilization with spatial resolution, and of introducing dynamical immobilization and release properties to a surface is of great appeal in the framework of molecular bioelectronics. A control on these aspects would allow one to selectively immobilize molecules only on a specific electrode out of a full set. Approaches based on surface chemical modification by redox active molecules such as quinones [77–81] or on the reductive desorption technique (see the following) are being explored to pursue this aim. The first strategy relies on the reactivity of quinones towards functional groups such as thiols and primary amines only in their oxidized state. It is, indeed, possible to drive electrochemically the binding of biomolecules at surfaces and to address an individual electrode in a matrix. The second strategy relies on the electrochemically driven detachment of preassembled SAM on gold to expose an electrode to a solution containing other molecules to be immobilized.

14.8.1 Metalloproteins Surface Immobilization

In the framework of metalloprotein bioelectronics, different strategies have been exploited to come up with well-defined protein film on various types of surfaces. Most of the works concentrated on the immobilization of two well-known types of metalloproteins: haem-based metalloproteins, such as cytochrome c; and blue copper proteins, such as azurin.

Cytochromes *c* are haem proteins which immediately appeared as good candidates for bioelectronic applications. Different species of cytochrome *c* may be found in nature and, among them, iso-1-ncytochrome *c* from *Saccharomyces cerevisiae* (YCC) is endowed with a feature that has attracted much interest for its immobilization on substrates. In fact, YCC has a unique surface cysteine residue at position 102 (Cys102), which can be exploited for its immobilization. In principle, the immobilization could be obtained directly on gold, exploiting the affinity of the thiol for this metal, but, due to its limited accessibility [82], only under particular conditions this residue can be exploited directly, preventing protein denaturation (see the following). Immobilization of this molecule on different substrates has been obtained by the formation of disulfide bonds with extrinsic thiols provided by a surface chemical modification [83–85]. Depending on the nature of the substrate for the protein immobilization, silane chemistry (on glass or silicon oxide), or a double-step chemistry (on gold), has been exploited. In the first case, monolayers of 11-(trichlorosilyl)undecyl thioacetate or 3-mercaptopropyl-(trimethoxysilane) (3-MPTS) [85] have been assembled on the substrate and the cytochrome *c* has been made reacting with the exposed SH moieties through the formation of an intermolecular disulfide bond. The chemistry used in the case of 3-MPTS silane SAM is sketched in Figure 14.13.

To demonstrate that the protein immobilization effectively took place via the Cys102 residue, a mutant in which the Cys102 residue was substituted for a threonin (C102T) was expressed. This mutant was not expected to react with the surface SH moieties, and this was confirmed by imaging the assembled layers by atomic force microscopy (AFM). Comparing the morphology of the assembled layers on mica (mica can be silanized and at the same time provides a low surface roughness essential in atomic force microscopy imaging of small features such as proteins), it was possible to attribute bumps on the surface to single proteins in case of native cytochrome *c* (Figure 14.14), whereas no bumps were visible for the C102T mutant. The monolayers obtained with the native protein and the mutant were also characterized

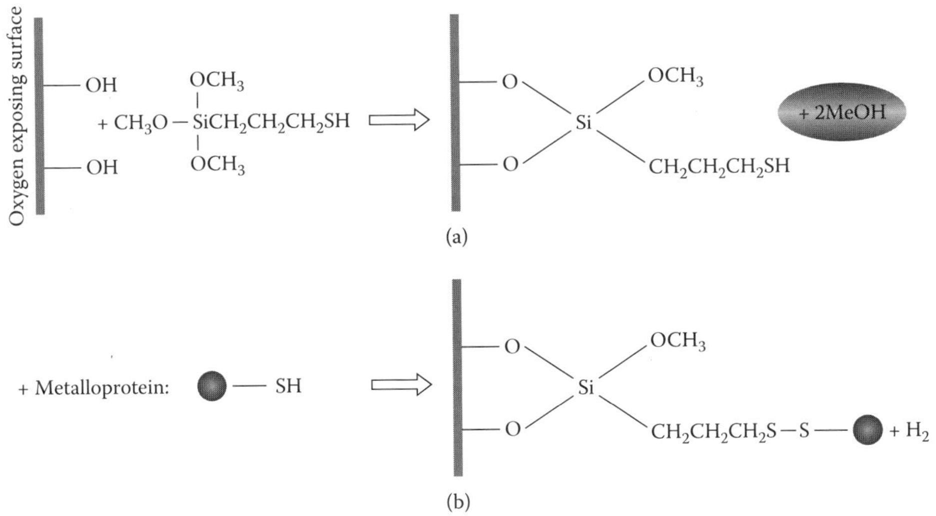

FIGURE 14.13 Protein immobilization scheme based on silanization of surfaces endowed with hydroxyl groups (a) and an intermolecular disulfide bridge between the thiol group of the silanes and the thiol group of a protein surface cysteine residue (b). (From Gerunda, M. et al., *Langmuir*, 20, 8812, 2004. With permission.)

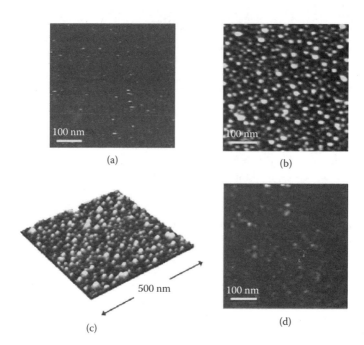

FIGURE 14.14   (a) Atomic force microscopy image of 3-MPTS silylated mica; (b) 3-MPTS silylated mica after exposure to native iso-1-cytochrome *c*; (c) 3D view of (b); (d) 3-MPTS silylated mica after exposure to the C102T mutant as a comparison. All the images are reported with the same z scale of 7 nm. (From Gerunda, M. et al., *Langmuir*, 20, 8812, 2004. With permission.)

by optical absorption spectroscopy. Cytochrome *c* from yeast was chemisorbed on a glass slide exploiting silanization by 3-MPTS and the corresponding absorption spectrum showed typical features of the protein, whereas upon substituting it with the C102T mutant, no spectroscopic features were observed. To verify the redox properties retention upon this immobilization strategy, voltammetric techniques were applied using Indium Tin Oxide (ITO) as the working electrode and substrate. A linear sweep voltammetry of a monolayer of native cytochrome *c* showed a clear Faradaic signal from which the surface coverage of functional molecules was estimated. A value of 1.8×10^{-11} mol/cm^2 was obtained, which is consistent with a near-monolayer coverage. Interestingly, the surface coverage obtained from the functional signal was consistent with the value obtained from absorption spectroscopy, highlighting that the fraction of proteins which lost their functional activity upon chemisorption was negligible. The oxidation peak of the cytochrome *c* turned out to be 168 mV (versus SHE), pointing to a lower reduction potential of the protein with respect to the freely diffusing protein in solution. Since the optical absorption spectrum did not show a variation of the spin state of the haem iron, this shift could not be attributed to the detachment of coordinating ligands that occurs upon denaturation. This shift is probably due to a variation in the solvation properties of the proteins and to intra- and intermolecular electrostatic and hydrophobic interactions that occur upon protein chemisorption.

The immobilization strategy based on disulfide bond formation with a thiol capped SAM on the surface should, in principle, lead to uniformly oriented protein layers. The orientation distribution can be assessed by absorbance linear dichroism measurements which probe the orientation of the haem plane with respect to the substrate [84,86]. It has been found that the assembled protein monolayer has a certain degree of order but, the overall angular distribution of the haem plane around the average value could be made narrower if aspecific adsorption of cytochrome *c* could be kept to a negligible level.

Direct YCC immobilization on gold by the C102 residue can be accomplished if the protein is immobilized in the reduced form [87,88]. Under these circumstances, the immobilized protein layer is still electrochemically active, having a very fast interfacial electron transfer with the substrate. Direct

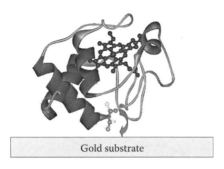

Gold substrate

FIGURE 14.15 Scheme of the direct immobilization of YCC on bare gold exploiting the C102 residue. The SH group of the cysteine in position 102, which is positioned in a hydrophobic pocket in the native form, unwinds to allow the formation of a sulphur–gold bond.

immobilization on bare gold allows the buried edge of the haem to be nearer to the gold electrode with respect to the case in which the immobilization is performed exploiting a heterobifunctional crosslinker (Figure 14.15). This is the reason why the direct immobilization assures a high electron transfer rate to the gold substrate, whereas the other immobilization strategies are severely rate limited in electron transfer due to the additional distance from the surface. In the framework of metalloprotein bioelectronics, it is to be stressed that direct immobilization on gold may present some advantages with respect to other immobilization strategies besides the increased electron transfer rate. In particular, some metalloproteins undergo conformational changes upon environmental variations (e.g., pH), which are reflected in a modification of the redox midpoint. The use of a SAM for protein immobilization usually prevents the possibility of exploring suitable pH values due to stability problems of the SAM to strong pH variations. For example, a direct immobilization on gold has allowed the investigation of the YCC low pH transition, whose study is usually prevented in diffusion experiments due to the detachment from the surface of the promoter layer used in freely diffusing electrochemistry of cytochrome *c* [88].

Another possible strategy for the immobilization of YCC on gold is the exploitation of a double-step chemical surface modification. In this case, a first layer of cysteamine (2-Mercaptoethylamine) is assembled on the gold surface and it is subsequently coupled to a glutaric dialdehyde layer. Glutaric dialdehyde is a symmetric molecule endowed with two terminal aldheyde groups able to bind to the NH_2 group of cysteamine from one side, and to lysine NH_2 groups on the protein from the other side.

A great deal of work has been dedicated to immobilization strategies for blue copper metalloproteins such as azurin. Azurin, from *Pseudomonas aeruginosa*, is endowed with a surface disulfide bridge between Cys3 and Cys26 which can be exploited for direct immobilization on gold. Evidence that azurin chemisorption on gold takes place by the reduction of the disulfide bridge and the formation of two thiol gold bonds has been obtained by different techniques [89]. Among the techniques used, reductive desorption is one of those more frequently used. It concerns measurement of the Faradaic signal that occurs when S is detached from the gold substrate at negative potentials. Each S atom contributes one negative charge to the total Faradaic peak, which is measured by conventional linear or cyclic voltammetry. Experiments are usually performed at a high pH in order to shift to lower potentials the onset of hydrogen evolution which could hide the desorption peak. Reductive desorption of an azurin monolayer immobilized on gold showed a desorption peak at -900 ± 30 mV versus SCE, which is attributable to S desorption [89]. Integration in time of the current under the desorption peak allows one to establish that the proteins were assembled in a quasi close-packed monolayer. Another technique that permits one to retrieve information on the immobilization details is x-ray photoelectron spectroscopy (XPS). From the binding energy of S(2p), the typical features of S bound to Au were identified, plus a contribution that may be due to other S atoms in the protein.

An azurin layer immobilized on a bare gold surface via the disulfide bridge is still endowed with a stable electrochemical activity (see Figure 14.16) [90–93]. The electron transfer rate is around 300 to 400 s^{-1} due

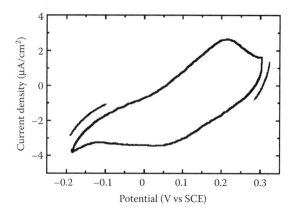

FIGURE 14.16 Cyclic voltammogram of an azurin (sub)monolayer chemisorbed on a Au(111) electrode in NH$_4$Ac 50 mM, pH 4.6 at a sweeping rate of 5 mV/s. Redox midpoint + 120 mV versus SCE, peak separation 160 mV. (From Facci, P., Alliata, D., and Cannistraro, S., *Ultramicroscopy*, 189, 291, 2000. With permission.)

to the high distance between the redox center and the surface (about 3 nm). Moreover, the redox potential measured on immobilized azurin films is very similar to those obtained in solution diffusion experiments, pointing to negligible conformational changes of the proteins upon immobilization.

Azurin mutants in which cysteines have been inserted in specific positions on the protein surface with the aim of obtaining a different orientation of the immobilized proteins on gold have been expressed. The mutant S118C, in which a serine in position 118 has been substituted for a cysteine, was immobilized on gold electrodes and the obtained protein film was imaged by STM and characterized by cyclic voltammetry [94]. The position at which the cysteine was inserted is adjacent to the residue 117, which is thought to reside on the electron transfer pathway between the copper ion and the natural electron transfer partners of azurin. The electrochemical response of the protein layer was stable to repeated scans pointing out the retention of protein structure upon immobilization. Also, the K27C azurin mutant, in which the lysine in position 27 has been substituted by a cysteine, has been expressed, and its electrochemical activity upon immobilization on gold has been confirmed by cyclic voltammetry [93].

Azurin molecules have also been immobilized on oxygen-exposing surfaces (SiO$_2$, mica, glass, etc.) by different strategies [24,95]. On oxygen-exposing surfaces, the method of choice for protein covalent immobilization is based on silanization. Using 3-Aminopropyltriethoxysilane (3-APTS), a silane layer exposing primary amines groups has been assembled on silicon, glass, and mica. Exposing the silylated sample to glutaric dialdehyde (GD), an imine bond is created, endowing the surface with exposed aldehyde groups. Exposing this surface to an azurin solution, protein chemisorption via formation of imine bonds between the aldehyde groups and protein primary amines (especially from lysines) is realized. A scheme of the immobilization steps is reported in Figure 14.17. The various stages of azurin film formation have been monitored by XPS and ellipsometry in order to measure the average thickness of the layers. The data obtained by both techniques agreed very well. According to x-ray crystallography, 12 amino groups (including the terminal NH$_2$ group) are present on the azurin surface, so no particular uniform orientation is imparted on the protein layer. The protein film has also been characterized by atomic force microscopy in liquid. Images clearly showed the presence of bumps with a lateral extent of about 10 nm, which can be attributed to single proteins (Figure 14.18).

The possibility of tuning the orientation of immobilized azurin molecules both on surfaces exposing hydroxyl groups and on gold has been demonstrated [96]. Whereas the previous immobilization strategy for azurin relied on the presence of surface amino groups, which amount to a total of 12, resulting in a non-uniform protein orientation, a new immobilization strategy was used, resulting in a more uniform orientation. Exploiting 3-mercaptopropyl-(trimethoxysilane) (3-MPTS), a surface chemical modification resulting in a surface endowed with thiol groups was obtained. Azurin immobilization was achieved by

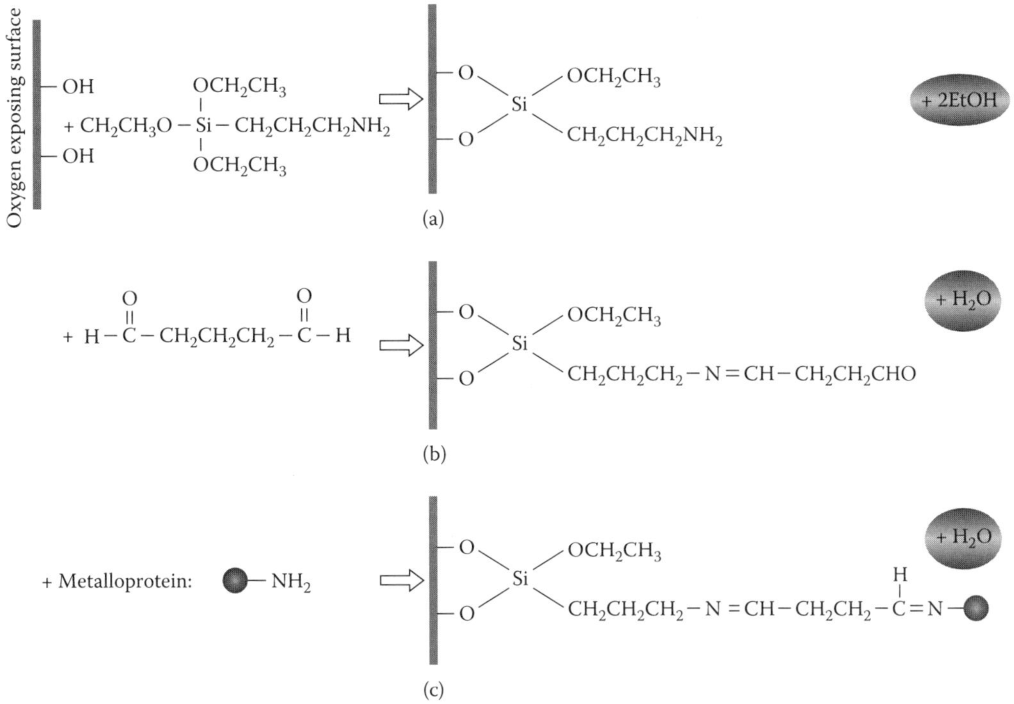

(a)

$+ 2EtOH$

(b)

$+ H_2O$

(c)

$+ H_2O$

FIGURE 14.17 Protein immobilization scheme based on a two-step surface chemical modification by 3-APTS and GD (a, b) and on an imine bond formation between the aldehyde group of GD and the primary amines on the protein surface (c).

exploiting the disulfide bridge between Cys3 and Cys26 and a disulfide thiol exchange reaction with the surface thiols. The different uniformity in the protein immobilization has been assessed both by XPS and cyclic voltammetry. XPS analysis of the Cu2p3/2 signal from the protein active site shows a different intensity depending on the exploited immobilization strategy. This behavior can be attributed to the asymmetric position of the Cu ion in the azurin molecule (see Figure 14.2). In fact, the Cu ion is buried

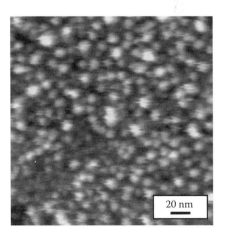

20 nm

FIGURE 14.18 Atomic force microscopy image of an azurin monolayer on functionalized mica. The image was acquired in contact mode under NH_4Ac buffer. z scale 3.9 nm.

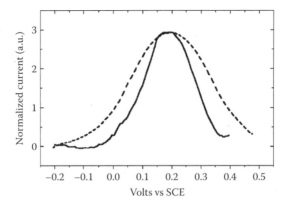

FIGURE 14.19 Comparison of the oxidation peaks of linear voltammetries of 2-MEA+GD+azurin (dashed line) and 1,4-benzenedimethanethiol+azurin (solid line) on gold. (From Alessandrini, A. et al., *Surf. Sci.*, 542, 64, 2003. With permission.)

0.8 nm under the globule surface, almost at the opposite side of the disulfide bridge from which it is 3 nm distant. The signal probed by XPS reflects the particular attenuation due to the location of the Cu ion in the protein layer. In the case of azurin molecules immobilized by the disulfide bridge, if aspecific binding is avoided, a unique orientation of the proteins is obtained and the average distance of the Cu ion from the surface will be higher with respect to the case of immobilization with random orientation. According to this prediction, the obtained Cu/N ratio was higher in the case of chemisorption via 3-MPTS with respect to chemisorption via 3-APTS+GD pointing to a higher distance from the surface of the Cu ion. The different orientation distribution of the two immobilization strategies were also studied by cyclic voltammetry. In order to obtain a voltammetric signal, similar immobilization schemes were developed for gold. To obtain a random orientation distribution, a first layer of 2-mercaptoethylamine (2-MEA) was chemisorbed on the gold surface, followed by a layer of GD, whereas, for a uniformly oriented protein layer, a SAM of 1,4-benzene dimethanethiol was assembled on the gold surface. Molecules endowed with sulphydril groups at both ends are prone to chemisorb on a gold yielding lying down molecules. However, it has been established that in the case of short or rigid molecules, such as 1,4-benzene dimethanethiol, the molecules on gold adopt a configuration in which a sulphydryl group is available off the surface for further reaction [97]. Using this configuration, a uniformly oriented protein layer is obtained as in the case of 3-MPTS. In both cases, a stable electrochemical signal was obtained and the FWHM of the oxidation waves were compared (see Figure 14.19). The FWHM varied from 0.17 V (in the case of the oriented layer) to 0.26 V (in the randomly oriented layer). Being that the FWHM is an indication of the distribution of formal potentials, the obtained behavior could be interpreted as the sign of a larger distribution of formal potentials in the case of the immobilization via 2-MEA+GD, due to a random orientation of the immobilized proteins.

14.9   Characterization of the Transport Properties of Immobilized Metalloproteins

In the framework of molecular bioelectronics, metalloproteins' transport properties have been investigated both in a configuration in which they are positioned between two planar electrodes, possibly using a third electrode to modify the energy levels in the molecules, and in a configuration in which they are sandwiched between two electrodes represented by the probe and the substrate of a scanning probe technique. Both types of investigation will be the subject of specific sections in which the potentiality and the main results so far achieved by each approach will be illustrated.

14.9.1 Scanning Probe Techniques in Metalloprotein Transport Properties Characterization

Scanning probe techniques are the most suitable methods for characterizing the behavior of proteins interacting with inorganic surfaces at the level of the single molecule in real space. Specifically, the two most important techniques that have been exploited in the context of metalloprotein bioelectronics are the STM (in EC-STM configuration) and the C-AFM. Here we will summarize the most important results so far obtained by these techniques.

The EC-STM allows one to characterize (sub)monolayers of metalloproteins adsorbed on a substrate and to shed light on the long range electron transfer mechanism across a single molecule. Starting from the characterization of the tunneling current between the tip and the substrate through a single metalloprotein, at the end of this section we will show how the EC-STM setup can be thought as a device that can implement a single molecule transistor in a physiological environment. Moreover, we will complement experimental results with theoretical interpretations in light of the developed models for the electron transfer mechanism in the EC-STM setup.

The potentiality of EC-STM to perform a sort of molecular spectroscopy of the density of states of the redox adsorbates was theoretically discussed by Schmickler in the early 1990s in trying to explain the conductivity of redox adsorbates in an STM setup [98,99]. Later on, the first experimental data appeared and more efforts were also concentrated on the theoretical models that could account for the observed behaviors.

Even if it did not concentrate on a biomolecular adsorbate, it is worthwhile introducing the experimental results related to EC-STM starting from the work of Tao on Fe-protoporphyrins [57]. In this work, Tao investigated Fe(III)-protoporphyrin (FePP) with respect to protoporphyrin (PP) adsorbed on a Highly Oriented Pyrolytic Graphite (HOPG) electrode. FePP is able to change reversibly its oxidation state from Fe(III) to Fe(II) at a substrate potential value available to the EC-STM setup, whereas the reference sample (PP), without the metal center, is electrochemically inactive. Imaging a sample composed of a mixture of FePP and PP at a substrate potential near the redox midpoint of FePP (−0.48 V versus SCE), the molecular species bearing Fe appeared brighter than the reference one (see Figure 14.20). The enhanced contrast for the electroactive molecules comes from the coupling between the Fe ion density of oxidized states and the Fermi levels of the tip and the sample. The presence of these molecular orbitals supports the electron tunneling between the tip and the substrate. Imaging in the constant current mode implies that molecular adsorbates which elicit higher tunneling current appear brighter in the image, corresponding to a virtual height increment. Tao measured the apparent height increment of FePP with respect to PP as a function of substrate potential at constant bias between the tip and the substrate, obtaining a resonance-like behavior peaked at a substrate potential (−0.42 versus SCE) very near to the redox potential of FePP.

One of the possible interpretations which has been invoked for explaining the obtained behavior of the tunneling current as a function of substrate potential is the resonant tunneling scheme, in which the electrons transfer between the tip and the substrate without being trapped in the molecule between them. As a general rule, all the proposed theories for the interpretation of the EC-STM contrast mechanism on redox adsorbates extrapolate an expression of the current as a function of the applied bias and the substrate potential involving microscopic quantities related to the molecule and to the molecule–environment interactions. In the resonant tunneling model, electrons tunnel from the tip to the substrate, exploiting the empty redox level of the molecule. From the electron energy point of view, the resonant tunneling mechanism implies that the starting level in the tip, the intermediate level on the molecule, and the final empty level in the substrate all have the same energy. Even if the electron never localizes on the molecular adsorbate, a significant role in establishing the electron current is played by the reorganization energy of the redox molecule, λ. This quantity can be interpreted as the difference between the center of the Gaussian distribution of the density of oxidized states and the redox energy level of the molecule derived from the standard reduction potential [100]. The latter quantity has to be considered in the particular configuration of the EC-STM setup, in which the redox adsorbate is under the influence of the electrostatic potentials produced by the charged tip and substrate in a nanometer-sized gap. Under the assumptions

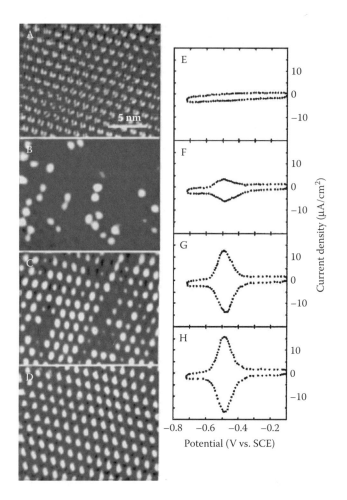

FIGURE 14.20 EC-STM images of FePP/PP adsorbed on a HOPG substrate from solutions containing FePP and PP at the ratio of 0:1 (a), 1:4 (b), 4:1 (c), and 1:0 (d). All the images were acquired at a substrate potential of –0.41 V versus SCE, with a tunneling current of 30 pA and a tip-substrate bias of –0.1 V. The percentage of brighter spots increases from A to D accordingly with the increase in the presence of FePP in the solution. From (E) to (H), the corresponding cyclic voltammograms of the adsorbed samples are reported. Clearly, the redox current increases with the presence of FePP. (From Tao, N.J., *Phys. Rev. Lett.*, 76, 4066, 1996. With permission.)

of a resonant tunneling transfer mechanism, the maximum in the tunneling current is expected when the maximum of the density of oxidized states is half way between the tip and substrate potential. The width of the resonance-like behavior of the tunneling current depends on its turn on the reorganization energy of the molecular species. This allows one to obtain an estimate for this molecular parameter of 0.2 eV, which should lead to a maximum in the tunneling current for a substrate potential equal to −0.73 V versus SCE instead of the experimentally obtained value of −0.42 V versus SCE. In order to predict a value for the potential corresponding to the tunneling current maximum, the quantities involved should be quantified. These quantities include both the reorganization energy and the redox energy level in the EC-STM setup. Dealing with the former, it should be pointed out that the value to be considered refers not to that obtained in homogeneous electron transfer reactions in solution, but to the specific configuration of EC-STM experiments, where the redox active molecules experience both a reduced solvent exposure and the electrostatic potentials of both electrodes. These considerations explain why the reorganization energy is a difficult quantity to be measured independently in this configuration. Theoretical calculations allows one to establish that it is not straightforward to derive the reorganization energy in the EC-STM setup from that in solution [101]. However, as a general behavior, λ increases with the tip-substrate distance. The other quantity involved in the relationship between tunneling current and substrate potential is the redox energy level in the EC-STM setup, which shows how the potential felt by the redox center is influenced by the electrostatic field in the nanometric-spaced tip–substrate gap. Also, this quantity is difficult to be extrapolated since it depends on the particular behavior of the voltage drop in the gap. All these considerations explain why the road to establish which is the electron transfer mechanism underlying the experimental data is not devoid of problems and could explain the apparent discrepancies found in Tao's experiment.

FePP studied by Tao is the prosthetic group of many haem-based metalloproteins such as cytochrome *c* and haemoglobin. One of the first EC-STM investigations performed on a metalloprotein was that by Friis *et al.* [102], in which they investigated the behavior of the well-known molecule, azurin, chemisorbed on Au (111) under double potentiostatic control in an aqueous solution.

The aspects related to azurin chemisorption on gold and its characterization have already been discussed in Section 14.8.1. The most important feature in the framework of molecular bioelectronics is that azurin, once chemisorbed on gold, retains its functional property—i.e., its ability to exchange electrons. Friis *et al.* [102] demonstrated that azurin molecules formed a submonolayer on gold that could sustain repeated scans by the STM tip. Moreover, they were able to identify a submolecular feature in the EC-STM images represented by a central brighter spot on each molecule, which was attributed to electron transfer enhancement due to the presence of the copper active site.

In parallel to experimental data from this first work and other data which followed, a theoretical effort was devoted to better understand the mechanism of long-range electron transfer in EC-STM on metalloproteins. These efforts lead to the theoretical development of other possible scenarios apart from the already described resonant tunneling mechanism.

In the regime in which electron transfer is much slower than the typical redox energy fluctuations of the redox adsorbates, the overall electron transfer between the tip and the substrate (or vice versa, depending on the sign of the bias) can be seen as a sequential combination of two processes: first, an electron transfers from the tip to the molecule, and another electron transfer event occurs from the molecule to the substrate. The overall tunneling current can be described in terms of electron transfer rates between the metal electrodes and the molecule. In the case of equally small electronic coupling between the substrate and tip to the redox molecule, the electron transfer mechanism has been defined as a "sequential two-steps model" [103]. In this scenario, after the first electron transfer process has taken place, the molecule relaxes to its reduced equilibrium configuration and a second thermally activated transition is necessary to transfer the electron to the second electrode. The maximum in the tunneling current in this model occurs when the redox level of the molecule is midway between the Fermi levels of the tip and substrate [103]. An alternative to the sequential two-step model is represented by the "vibrationally coherent model" in which the second electron transfer process takes place before the full relaxation of the molecule [104]. In this case, the second step must not be thermally activated because the transfer occurs when the energy of the molecular level is higher then the Fermi level of the second electrode. So, even in this case, similar to the resonant tunneling scheme, the electron never gets trapped in the molecular redox level. The substrate potential of the maximum in tunneling current corresponds to the maximum in the density of oxidized (or reduced) states of the molecule. Another scenario applies when the molecule–electrodes coupling is strong, configuring the system in a full-adiabatic regime. In this case, an initially oxidized molecule, after receiving an electron from the more negative electrodes, starts to relax to the reduced equilibrium configuration, opening a multitude of channels for the electron transfer between the two electrodes [103,105]. This phenomenon leads to an amplified electron current with its maximum at a substrate potential value equal to that of the sequential two-step process.

It is clear from all the previous discussion on the mechanism of electron transfer that all the theories foresee a dependence of the tunneling current on the substrate potential value in EC-STM experiments on redox adsorbates. This dependence has indeed been observed in Tao's experiment, but it has also been observed on metalloproteins. The first example of a substrate potential dependent contrast in an EC-STM

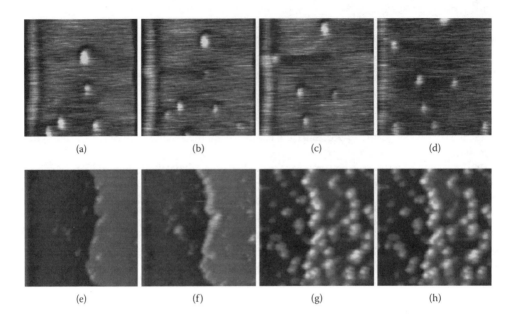

FIGURE 14.21 Upper set: EC-scanning force microscopy images obtained in contact mode at different values of substrate potential with respect to SCE: –225 mV (a), –125 mV (b), –25 mV (c), and +75 mV (d). Scan area 200 × 200 nm^2. Vertical range: 1 nm. Lower set: EC-STM images of the same sample recorded at the same substrate potentials as for EC-SFM: –225 mV (e), –125 mV (f), –25 mV (g), and +75 mV (h). Imaging conditions: tunneling current 2 nA, bias 400 mV tip positive. Scan area 70 × 70 nm^2. (From Facci, P., Alliata, D., and Cannistraro, S., *Ultramicroscopy*, 189, 291, 2000. With permission.)

experiment on metalloproteins is that by Facci *et al.* [90]. In their experiment, Facci and co-workers concentrated on azurin molecules immobilized on a Au(111) substrate. Freshly prepared substrates were incubated in a 10^{-4} M azurin solution in 50 mM NH$_4$Ac buffer, pH 4.6, which corresponds to the isoelectric point for azurin. Incubation time was long (three days) for the cyclic voltammetry characterization, whereas it was on the order of 20 to 40 minutes for the EC-STM characterization. This allowed a strong signal in CV and the possibility of concentrating on single molecules in EC-STM images. Cyclic voltammetry was used to ascertain the retention of the functionality of azurin molecules once chemisorbed on the surface.

The voltammetry highlighted a redox midpoint not significantly different from the solution value (+116 mV vs. SCE) pointing out that no significant conformational variations occurred upon chemisorption despite the involvement of the disulfide bridge between Cys3 and Cys26, which is considered structurally relevant for azurin. The most striking result of this investigation was the substrate potential contrast dependence of the azurin molecules on gold. When imaging was performed at –225 mV and –125 mV versus SCE, the azurin molecules were almost invisible on gold, whereas, imaging at −25 mV and +75 mV versus SCE resulted in bright spots on gold attributable to azurin molecules (Figure 14.21). It is to be stressed that AFM imaging, which is sensitive only to topographical features, performed on a similar sample under potentiostatic control of the substrate revealed no significant difference for the same potential values used in the EC-STM investigation. This feature assures that the contrast difference observed in EC-STM is due to a different behavior of azurin molecules for the different values of substrate potential and not to a removal of the sample by the STM tip. This last aspect is further confirmed by the fact that it was possible to step back and forth with the substrate potential value and switch on and off the visible spots accordingly, as shown in Figure 14.22.

The possibility of tuning substrate potential represents, therefore, a sort of control mechanism on the tunneling current elicited by azurin molecules, enabling a higher or lower value of the current. The trend of the tunneling current did not reveal a resonant-like behavior characterized by a rising of the current,

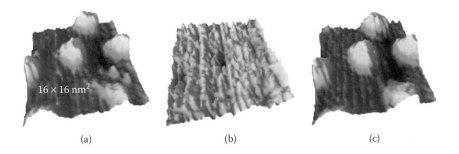

(a) (b) (c)

FIGURE 14.22 Sequence of consecutive EC-STM images obtained on the same sample area for three different values of the substrate potential with respect to SCE: –25 mV (a), –125 mV (b), and –25 mV (c). (Other imaging conditions: tunneling current setpoint 2 nA, bias 400 mV tip positive.) Azurin molecules are clearly visible only at –25 mV. Even if at –125 mV, the molecules appear as darker spots, they again reappear if the substrate potential is brought back to –25 mV, highlighting that the molecules are not removed by the tip.

followed by a decrease upon changing the substrate potential at constant bias, as would be expected on the basis of the previously described theories. This particular feature of the tunneling current, characterized by a rising step followed by a plateau, can be understood in light of the particular experimental conditions used in this work. As has been stressed before, the details of the electron transfer mechanism depend on different quantities of the redox adsorbates and their coupling with the electrodes which can be modulated by varying the experimental conditions. The high value of the tunneling current setpoint used in the work by Facci and co-workers [90] (2nA) implies a low tip-sample distance. Reducing the tip-sample distance causes a decrease of the reorganization energy of the immobilized azurin molecules [101]. The system may find itself in a configuration where the voltage bias (times e) is higher than twice the molecular reorganization energy. In light of the two-step electron transfer model, this consideration implies that once the electron transfers from the first electrode to the molecule, the latter relaxes to its equilibrium reduced configuration, but the energy level of the molecular configuration is still higher than the second electrode's Fermi level (see the scheme in Figure 14.23). This means that the second electron transfer takes place without the need for a thermally activated transition. Until both the energy levels of the redox adsorbates, which differ by twice the reorganization energy, are within the Fermi level difference of the electrodes imposed by the bias, no sensible variation of the tunneling current upon substrate potential is expected. In addition, when the tip-surface distance is lowered, the electrostatic coupling of the molecular redox level with the electrodes increases, and the effective overpotential is, consequently, decreased. This means that a larger overpotential must be applied to reach the point where the current decreases, leading to a longer plateau.

On the basis of these considerations, if the setpoint current were lowered maintaining the same applied bias, it would be possible to switch to a configuration in which the tunneling current shows a resonance-like behavior. This has been indeed demonstrated in another work by EC-STM on azurin molecules performed with a lower setpoint current (1 nA) [91]. The latter work concentrated on establishing the role of the redox active center on the tunneling current observed in the EC-STM imaging of azurin. This aim was pursued comparing the behavior of azurin molecules with Cu^{2+} as the ion in the redox center with azurins having, instead Zn^{2+}. Zn-azurin is structurally very similar to Cu-azurin, but it is not electroactive in the potential window accessible to EC-STM experiments using gold as the substrate. EC-STM imaging performed on a defined mixture of Cu- and Zn-azurin molecules at a substrate potential value of -250 mV versus SCE revealed the presence of spots of different brightness with a bimodal distribution of the apparent peak-height. This is consistent with the presence of two classes of adsorbates with a different efficiency in eliciting tunneling current through them. Imaging the mixture at different values of substrate potential showed that some spots in the images varied their intensity in a resonant-like fashion, whereas some spots remained unchanged (see Figure 14.24). Concentrating on a single spot which varies its intensity

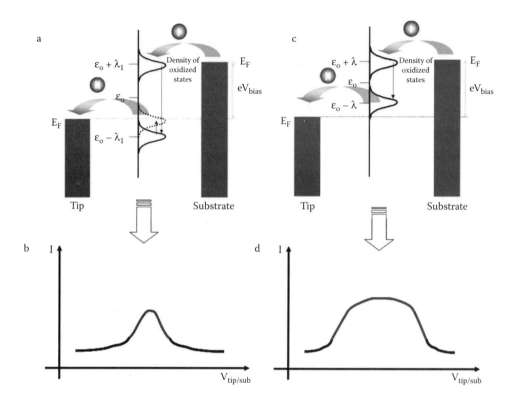

FIGURE 14.23 Energy level schemes for the two-step sequential transport mechanism in the case of small bias with respect to the reorganization energy ($2\lambda > eV_{bias}$) (a) and in the case of large bias ($2\lambda < eV_{bias}$) (c). The expected behavior of the tunneling current as a function of the tip or substrate potential versus the reference electrode is reported in (b) and in (d) for the (a) and (c) cases, respectively.

upon substrate potential (see Figure 14.24[h]), it is possible to obtain the potential corresponding to the maximum in the tunneling current. The obtained value is –210 mV versus SCE which is sensibly different from the redox potential of 120 mV versus SCE. The same value of the substrate potential corresponding to the maximum current is obtained when the apparent height variation is averaged over more molecules. To finally confirm that spots in the images which did not change their apparent height were attributable to Zn-azurin molecules, a control experiment with a sample consisting of only Zn-azurin was executed. This analysis showed that no dependence of the apparent height on the substrate potential could be found, corroborating the aforementioned interpretation of the experiment with the mixed sample. The position of the current maximum in principles provides a mean of establishing which is the electron transfer mechanism ruling the experimental data. This opportunity can be fully exploited if the reorganization energy of azurin in this particular configuration is known, along with the local electrostatic potential sensed by the redox active center. A theoretical investigation performed to evaluate the reorganization energy for azurin in the EC-STM setup reported a value of 430 meV, which makes the two-step electron transfer mechanism for the process involved more plausible than resonant tunneling.

14.9.2 Planar Biomolecular Devices

The amount of data retrieved from surface-immobilized redox metalloproteins samples by scanning probe techniques operated in an electrochemical environment suggests clearly that these molecules behave like molecular switches, being able to allow or prevent electrons flowing through them according to the availability of molecular electronic levels in between the Fermi levels of tip and substrate. Such particular

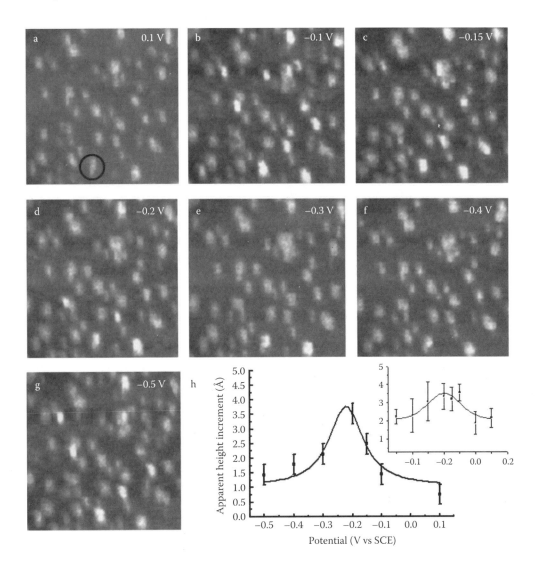

FIGURE 14.24 (a)–(g) EC-STM spectroscopy-like-imaging of a mixture of Cu- and Zn-azurin molecules adsorbed on a Au(111) substrate. The value of the substrate potential with respect to SCE is reported in each image. Some molecules do not change their apparent height upon substrate potential, whereas others do. In (h), the apparent height increment of a spot (in the black circle in [a]) with respect to the non-varying ones is reported. The inset to (h) reports the apparent height increment averaged over six molecules. (Other imaging conditions: tunneling current setpoint 1 nA, bias 400 mV tip positive, scan range 130 × 130 nm^2.) (From Alessandrini, A. et al., *Chem. Phys. Lett.*, 376, 625, 2003. With permission.)

feature matches, at best, the general requirement of molecular electronics, which needs elementary units (building blocks) endowed with intrinsic functionality. The possibility of switching the conduction state of an object such as a molecule or a semiconductor cluster by gating the flow of electrons through it, indeed represents the basic feature of a molecular (nano)transistor. Therefore, the idea of trying to implement a single (few) molecule(s) transistor exploiting the electronic conduction properties of metalloproteins (e.g., azurin) arises naturally [7]. Redox metalloproteins arranged in monomolecular film (or even single molecules) could thus act as the channel of a field-effect three-terminal device in the gap defined by a couple of planar electrodes. Of course, the problems to be faced for implementing such devices seem serious

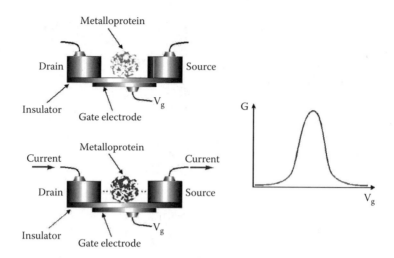

FIGURE 14.25   Possible scheme for a single metalloprotein planar nanotransistor. The gate electrode acts on the electronic levels brought about by the protein. When these levels are tuned to the Fermi levels of the leads, conduction via the molecule is enabled.

and numerous. Among them are the two most relevant issues: (1) the implementation and theoretical description of effective electrical contacts between metalloproteins and metal electrodes, and (2) the understanding and optimization of the mechanisms and conditions of intermolecular electron transfer in 2D ensembles of metalloproteins. Albeit, a final solution for these classes of problems is still lacking, but the practical realization of working devices appears possible and has already been achieved.

The implementation of a FET- or single particle transistor-like protein device, should likely take advantage of state-of-the-art lithographic techniques for the definition of planar (nano)electrodes. Figure 14.25 depicts the operating principle of a generic single metalloprotein planar transistor. Here a redox metalloprotein is located in the nanometric gap between two planar electrodes and is electrostatically coupled to a gate electrode. The coupling is responsible for shifting the electronic levels of the molecule with respect to the Fermi levels of the metal leads, enabling or hindering the electronic flow via the aforementioned levels in a process that usually involves electron tunnelling.

Interestingly, the physics behind the operating principles of such a single molecule transistor is the same that describes the electrochemical gating of the current in EC-STM experiments on surface-immobilized azurin. In the latter case, however, the gate is somehow "diffused," being electrochemical in nature.

To date, a limited number of different planar nanoelectronic devices have been implemented. All of them make use of the redox metalloproteins that have so far shown the most robust performance in withstanding large electric fields without denaturation, and to survive a nonphysiological, waterless environment such as that on the surface of a hybrid electronic device.

The implementations reported so far have been those of two- and three-terminal devices. They have had the scope of demonstrating the feasibility of such hybrid devices and have been used especially to characterize those particular systems, trying to shed light on the role of molecular organization, as well as on that of the particular metal ion in the protein.

The first implementation of a hybrid nanoelectronic device was a solid-state molecular rectifier based on a (sub)monolayer of azurins [106]. A monomolecular carpet of metalloproteins has been chemisorbed on the surface of thermally grown SiO_2 in between two gold nanoelectrodes (Cr-Au) defined by e-beam lithography (EBL). The chemistry which was used for the chemisorption of the protein carpet was a silane-based one. Particularly, by functionalizing the SiO_2 surface with 3-MPTS (see Section 14.8), it was possible to achieve a surface endowed with active thiols. Hence, exploiting the disulfide bond present at the surface of azurin (see Section 14.8.1), intermolecular disulfide bonds were formed between azurin and 3-MPTS by

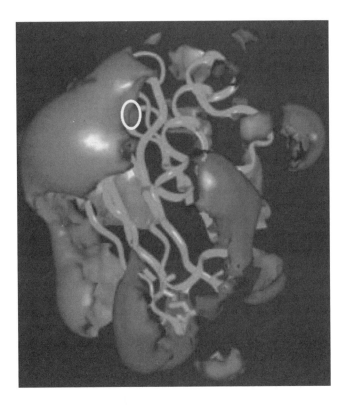

FIGURE 14.26 MEP of an oxidized azurin. The white circle identifies the Cu ion in the active site. Iso-potential surfaces corresponding to −0.5 (dark grey) and 0.5 kcal/mol/e (light grey) are highlighted. (From Rinaldi, R. et al., *Adv. Mater.*, 14, 1449, 2002. With permission.)

means of a thiol-disulfide exchange reaction. This construct guaranteed a covalent linkage of the molecule to the surface as well as a defined, uniform orientation of the proteins at the surface, considering the aforementioned disulfide was the only one available for the described reaction. Control experiments on non-uniformly oriented azurin hybrid devices were performed using 3-APTS and glutaric dialdehyde, exploiting the protein surface amine, as described in Section 14.8.1. The achievement of a uniformly oriented protein film in between nanoelectrodes is of extreme importance for the performance of a hybrid device for two broad reasons: (1) it enables a higher surface coverage, which can facilitate intermolecular electron transfer between neighboring metalloproteins; and (2) for a given protein coverage, it endows the molecular carpet with a uniformity in the location and orientation of the Cu-based redox active sites, increasing electron transfer probability. There is, moreover, another element that has to be taken into account as far as protein orientation is concerned: protein electrostatics. Proteins, being complex zwitterionic molecules, possess on their surface a variety of charged and polar groups whose global effect is described by MEPs (molecular electrostatic potentials). MEPs are known to play a major role in protein interaction and biorecognition in solution and drive the phenomena of self-assembling at surfaces and adsorption kinetics. MEPs for azurins have been computed (see Figure 14.26), in a range of different conditions involving different protein oxidation states, as well as solution pH, and ionic strength. Calculations have revealed that azurin in the experimental conditions used in this work possesses a strong dipolar moment (150 D) that can sum up in an oriented, 2D molecular assembly, influencing the overall field sensed by electrons flowing in the mono-molecular layer.

The electrical characterization of a two-electrode device with a gap of 60 nm for the uniformly oriented and random immobilization of the proteins is reported in Figure 14.27. Some important features and differences are evident: (1) the uniformly oriented sample shows a marked rectifying behavior (rectification

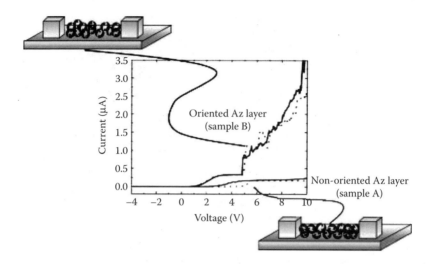

FIGURE 14.27 Current–voltage characterization for the randomly oriented azurin sample (sample A) and for the uniformly oriented azurin sample (sample B) in a 60-nm gap. Schemes of the possible orientation of azurin molecules in the two configurations are shown. (From Rinaldi, R. et al., *Adv. Mater.*, 14, 1449, 2002. With permission.)

ratio of 175 at 1.5 V versus just a figure of 10 for the randomly oriented counterpart); (2) the uniformly oriented sample shows a tenfold higher current with respect to the randomly oriented one; this effect, along with that of rectification, can be attributed to the molecular dipoles summing up in the self-assembled edifice and providing an electrostatic field superimposing to the external bias; (3) marked steps in the I/V curves, tentatively ascribable to the involvement in the conduction mechanism of protein redox levels, are present in both samples, albeit differently in intensity. This first demonstration of a metalloproteins-based molecular rectifier showed poor stability against aging and STP environmental conditions, suggesting that the nature of the observed phenomena was biomolecular in nature.

Another work addressed several open issues related to protein sample purity, and the role of the Cu ion in azurin active site [107]. In nanogaps 50 to 100 nm wide, it turned out that recombinant native azurin performed better than the wild type since it showed a rectifying ratio of 500 at 10 V. Moreover, an extensive study with different molecules (recombinant native, Zn-azurin, and Apo-azurin—without the metal ion), all immobilized with a uniform orientation, was undertaken in order to understand the role of the particular metal ion in the active site, in a similar fashion to what was previously studied by EC-STM at the level of the single molecule.

Figure 14.28 shows I/V curves for the three different samples. These results confirm the key role of Cu ions and that of the electronic levels brought about by its presence in assisting current flow through the molecular carpet. Zn-azurin and Apo-azurin, albeit almost identical to Cu-azurin in molecular structure, were in fact unable to let flowing any appreciable current into the device. Finally, the role of relative humidity in preserving over time the performances of the hybrid device were pointed out. A relative humidity of around 50% was found to be optimal towards device performance and lifetime.

As a further step, a three-electrode device was implemented [108]—i.e., a device where the current flow through the molecular carpet was controlled by a gate electrode in a FET-like implementation. The experimental configuration for such a device was pretty much similar to the two-electrode one (EBL defined Cr/Au arrow-shaped nanoelectrodes facing in a 100-nm gap on a Si/SiO$_2$ substrate). However, the main difference was the presence of a Ag electrode on the back of the structure, making an ohmic contact with the p-doped Si substrate. This electrode was used as a gate in a FET configuration. The main results of this work were the presence of a marked resonance in the measured transcharacteristics (see Figure 14.29), which displayed a peak-to-valley ratio of 2 at V$_g$ = 1.25 V. Only devices assembled

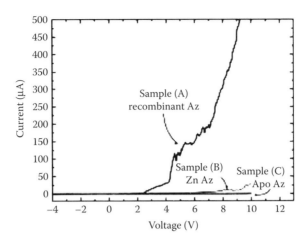

FIGURE 14.28 Current–voltage curves for samples of recombinant azurin (sample A), Zn-azurin (sample B) and Apo-azurin (sample C). The gap between the gold electrodes was 60 nm. (From Rinaldi, R. et al., *Appl. Phys. Lett.*, 82, 472, 2003. With permission.)

from Cu-azurin displayed such a feature, at variance with Zn- and apo-azurin. The realized devices were unable to withstand more than a dozen cycles. Their aging was attributed to the degradation of the metal leads, subject to strong electric fields rather than to protein denaturation in virtue of the results of a autofluorescence-based investigation devoted to assessing the effect of strong electric fields (of the order of $10^4 \div 10^5$ V/cm) on the native structure of azurin.

Interestingly, the switching behavior of this protein-based FET was also modeled upon a simplified frame of protein chains of alternated reduced and oxidized molecules, invoking a hopping mechanism for intermolecular electron transfer. Within this framework, it was possible to account for the main experimental features arising in the implemented device, namely: (1) the appearance of the described resonance in transcharacteristics, and (2) the marked onset in the I/V curves, as already described in the implementation of the two-terminal devices.

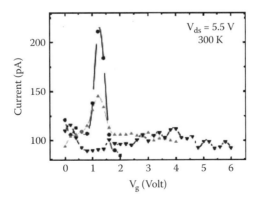

FIGURE 14.29 Three transfer characteristics obtained in sequence (dark-grey circles, light grey triangles, and black inverted triangles) on a 100-nm gap with azurin molecules immobilized in a uniform orientation. In the resonance region, the transconductance changes from positive to negative values. The resonance feature gradually disappears. (From Maruccio, G. et al., *Adv. Mat.*, 17, 816, 2005. With permission.)

14.9.3   A Single Protein Transistor in a Wet Environment

The described planar devices demonstrate the active role played by a metalloprotein monolayer in eliciting a current in between metal electrodes. The particular configuration of the immobilized protein layer is able to influence the electric current properties, imparting a directionality when a uniformly oriented molecular assembly is realized. However, the planar devices so far implemented make use of dried protein layers and of electrical characterization performed in air. Although the preservation of a non-denatured protein conformation upon exposure to a high electric field has been demonstrated, the possibility to take advantage of the protein functionalities in a physiologic-like environment would lead to a better exploitation of the protein properties that nature has evolved and selected over billions of years. Moreover, in the framework of metalloproteins bioelectronics, the possibility to work in an electrochemical environment allows one to relate the electrode voltages to the redox midpoint of the proteins via the presence of a reference electrode. The presence of a reference voltage allows one to establish a correlation between the transport properties of a bioelectronic device and the electrochemical properties of the biomolecules, which is not easy to establish in a "dry" device. For example, the bias voltage at which a sudden variation of the current occurs, or the gate voltage corresponding to the appearance of a resonance in the I-V curve are not easily related to the protein redox midpoint. Furthermore, it is difficult to explore single molecule properties in planar devices, due to difficulties in defining nanometer gaps. For this reason, the nanometer gap which naturally forms between the probe and the substrate of a scanning probe microscope appears as a valuable platform to implement a single biomolecule electronic device. Moreover, scanning probe techniques are easily operated in a physiologic-like environment even in an electrochemical cell. All these considerations prompted the idea of implementing a single-metalloprotein wet transistor exploiting an EC-STM based setup.

We already discussed the capability of EC-STM to give access to the density of states of the redox adsorbates exploiting a sort of spectroscopy-like imaging. This means that different images taken with different values of the substrate potential or bias voltage on the same sample area are compared. The usual tunneling spectroscopy technique based on positioning the tip at a fixed position in three dimensions and ramping the bias voltage is, in this case, prevented both by lateral drift and by the capacitive currents that arise when the bias voltage is swept in an aqueous environment. Moreover, all the electron transfer theories are formulated in terms of electron transfer rates, meaning electron current. When EC-STM imaging is performed in the constant current mode, variations in the tunneling current are converted to tip-sample distance differences. Extrapolating the corresponding variations in tunneling current can be done only if the decay factor of the current is known. Since it is not easy to estimate this quantity, direct access to the tunneling current variation is not possible in this experimental configuration. All these problems (measurement stability and the direct measurement of the tunneling current) have been circumvented by a reversed approach in which a metalloprotein has been immobilized on an EC-STM tip and brought into tunneling distance to a conductive substrate in a work by Alessandrini *et al.* [109]. The overall setup on the experiment is sketched out in Figure 14.30. In this work, a gold tip was electrochemically sharpened, insulated by molten Apiezon wax in order to reduce faradaic currents, and then incubated with a 10^{-4} M azurin solution in 50 mM NH$_4$Ac pH 4.6. Tip functionalization by azurin molecules has been characterized by transmission electron microscopy images at high resolution, which highlighted the presence of blobs protruding from the tip surface with a dimension compatible with that of azurin molecules (see Figure 14.31[a]). Protein functionality upon immobilization was confirmed by cyclic voltammetry directly on a non-insulated tip (see Figure 14.31[b]).

The metalloprotein-coated tip was then brought into the tunneling distance with a Au(111) substrate in an electrochemical cell under bipotentiostatic control. Tunneling current intensities not exceeding a few tens of picoAmps were adopted for the setpoint in order to achieve large enough tunneling gaps to preserve azurin integrity. Once the current setpoint had been reached, the system was left to stabilize over several minutes. Then, the feedback system was switched off and the tip potential was swept (keeping the bias constant) recording the tunneling current. At the end of each voltage sweeping cycle, the feedback system was again re-established. In such a configuration, it is plausible that the current flows through a

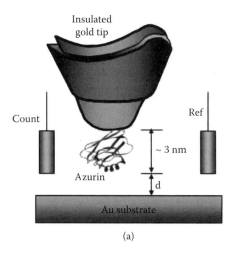

(a)

FIGURE 14.30 Schematic representation of the setup used to implement the single metalloprotein wet transistor. Azurin molecules were immobilized on a gold tip and faced to a gold substrate in an electrochemical cell endowed with a reference and a counter electrode. (From Alessandrini, A. et al., *Appl. Phys. Lett.*, 86, 133902, 2005. With permission.)

single azurin molecule in light of the exponential dependence of the tunneling current on distance since there will be one molecule nearest to the substrate.

Repeated tip potential sweeps resulted in tunneling current showing a resonance-like behavior, as shown in Figure 14.32(a). The same trend was not observed when the tip was functionalized with Zn-azurin instead of Cu-azurin, or the potential was swept when the tip was out of tunneling distance from the substrate. Moreover, the measured behavior of the tunneling current was reproducible and reversible, ensuring that the phenomenon was not related to irreversible structural modifications of azurin. Interestingly, in some curves reporting the tunneling current as a function of the tip potential, a bump was seen at higher potential with respect to the current maximum. This feature can be attributed to the faradaic contribution that enters the tunneling current circuit due to azurin molecules immobilized on the tip but out of tunneling distance. This aspect further confirms the nature of the observed resonance-like behavior, which is different from a faradaic effect.

The difficulty of obtaining I-V characteristics in an aqueous environment was in this case circumvented by performing repeated measurements of the same kind just described, but with different bias voltages

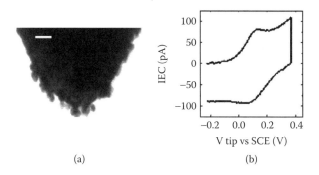

(a) (b)

FIGURE 14.31 Characterization of the azurin functionalized tip. (a) TEM image of an azurin-coated gold tip. Scale bar = 20 nm. Each blob protruding from the tip surface is attributable to a single protein. (b) Cyclic voltammetry of an azurin-coated tip. The presence of both the oxidation and reduction peak assures the preserved electron transfer functionality of the immobilized azurin molecules. (From Alessandrini, A. et al., *Appl. Phys. Lett.*, 86, 133902, 2005. With permission.)

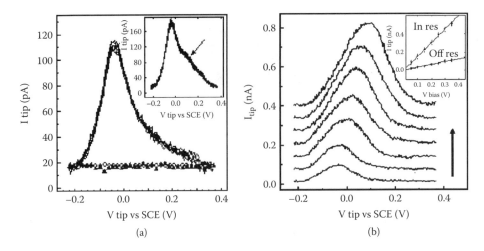

FIGURE 14.32 (a) Six repeated tip potential sweeps (initial conditions: $V_{bias} = +100$ mV, tunneling current = 15 pA, tip potential = -225 mV) showing tunneling current peaks compared to the behavior of a bare insulated gold stylus (hollow circles) and to that of a Zn azurin-coated tip (filled triangles). Inset: eight repeated sweeps (initial conditions: $V_{bias} = +100$ mV, $I_t = 20$ pA, tip potential = -225 mV) on another sample also showing an oxidation wave (outlined by the arrow). (b) Bias voltage dependence of the trans-characteristics. Curves (displayed with an offset of 50 pA each to avoid overlap) were obtained at different bias voltages stepped by 0.05 V in the range 0.05 to 0.35 V, starting from the lowest curve (see the arrow), using a sweep rate of 2.4 V/s. Inset: I_{tip}–V_{bias} characteristics measured from the set of curves at $V_{tip} = -0.2$ V ("off res") and at each current maximum position ("in res"), showing a different tunneling barrier transparency. (From Alessandrini, A. et al., *Appl. Phys. Lett.*, 86, 133902, 2005. With permission.)

across the hybrid nano-junction. The obtained trend is reported in Figure 14.32(b). The resonance intensity and width increase, and the tip potential corresponding to the maximum shifts towards positive tip potential values upon a bias voltage increase. This behavior allows one to further shed light onto the involved mechanism of electron transfer. In fact, the experimental $I(V_{tip})$ curves are nicely fitted by the two-step electron transfer models, whereas, a pure resonant tunneling mechanism appears unlikely in light of the mean residence time of the electrons on the protein.

From the obtained set of data, it was possible to estimate the tunneling resistance in the two limiting cases of "in" and "off resonance." This can be done by measuring the maximum tunneling current corresponding to each applied bias, and the tunneling current measured at a point far from resonance. The data reported a tunneling resistance of 660 ± 30 MΩ for the "in resonance" case, and 3.28 ± 0.25 GΩ for the "off resonance" case. A fivefold decrease of the tunneling barrier in resonant conditions was observed due to the temporary population of azurin redox states brought about by the redox active site and aligned with the Fermi levels of the gold electrodes.

This experiment showed for the first time that metalloprotein redox properties can be exploited to vary the electronic transparency of a wet bio-inorganic junction comprising even a single molecule between two metal electrodes. The great breakthrough of this demonstration is the possibility of exploiting the electron transfer properties of a metalloprotein in the framework of molecular electronics in its physiological environment. The described setup can be thought of as configuring a wet single-metalloprotein-based transistor, where the gating effect is provided by the electrochemical control of the solution potential.

14.9.4 Further SPM-Based Implementations

Using the C-AFM, it is possible to probe both the sample conductivity and the topography. If a metalloprotein is immobilized on a conductive substrate, which may be biased with respect to the conductive tip, the conductibility properties of the protein sandwiched between the two electrodes can be investigated (see

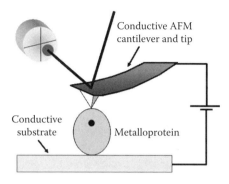

FIGURE 14.33 Schematic representation of a C-AFM experiment in which the transport properties of a single metalloprotein are investigated.

Figure 14.33). At the moment, the main shortcoming of this investigation technique is that it cannot be implemented in an aqueous environment where the full functional properties of the metalloproteins are retained. This limitation rises mainly from the faradaic currents, which can not be limited as easily as in EC-STM tips. High faradaic currents may obscure the signal coming only from the current passing from one electrode to the other through the protein. In principle, if instrumental developments could bring to our disposal a C-AFM able to work in an aqueous environment, we would have an investigation technique of improved potentiality with respect to EC-STM. In fact, the position of the tunneling tip with respect to the protein is essentially unknown in EC-STM experiments, and also in STM experiments aimed at studying molecular transport in general. The independent control on the applied force would provide a valuable means to work with well-defined electrodes–molecule contacts. The C-AFM technique, used in nonconducting solvents, has so far allowed one to obtain impressive results on the conduction properties at the level of single molecules for alkanethiols and other organic molecules [45,110,111].

Nevertheless, important information can be gained by studying the transport properties of metalloproteins by C-AFM in air. Davis *et al.* measured the dependence of the tunneling barrier of azurin molecules sandwiched between a conductive substrate and a conductive atomic force microscope tip on the applied force [112–114]. This has been accomplished by immobilizing azurin molecules either on a gold substrate or on a gold coated tip faced to a conductive substrate (usually HOPG). It has been found that under low force conditions (< 2 to 3 nN) azurin molecules are highly insulating due to the lack of a reliable protein–AFM tip electrical contact. Increasing the applied force (> 3 nN) current through azurin increases linearly with bias for a low bias regime, while the entire range of the current–voltage characteristic is well described by a Simmons tunneling model or a modified one [115], where the protein resembles a tunneling barrier characterized by a specific width and a mean height. The high reproducibility of the data suggests that the transport properties of a single molecule are being measured. The zero bias resistance obtained in this condition is about 45 to 60 GΩ, which is one order of magnitude higher than the tunneling resistance observed in the off-resonance status in the experiment in Ref. [103] under potentiostatic control of the metal electrodes. Upon a further increase in the applied force (5 to 50 nN) the current is enhanced till a breakdown occurs, pointing out a facile direct tunneling or even a direct contact between the tip and the sample. The enhanced current is a consequence of an increase in the atom packing density, which causes a decrease in the β tunneling coefficient and in the barrier height. However, these experiments usually lack the requirement of reversibility of the measurements. After a cycle in which the force is increased to the breakdown limit, upon releasing the force, electron transport does not take itself back to the initial value. Interestingly, the variation of the transported current upon the applied force can be deconvoluted into contributions due to the variation of the barrier height and to the variation of the tunneling distance. In all the previously described experiments, the protein matrix has been considered the same as an insulating film. More relevant to the presence of an active redox-site and to the redox properties' accessibility is the case in which a negative differential resistance (NDR) has been observed by C-AFM for low applied force and high bias ($\pm$ 3 to 4 V) [116]. The presence of an NDR feature is relevant in the framework of molecular electronics in general, because it would configure the possibility of obtaining a two-electrodes switchable device. In the literature, a correlation between an NDR feature and the electroactivity of the investigated molecules in a nanoscale junction have been already established [117]. This correlation is relevant to the case in which an NDR phenomenon is observed for azurin molecules. In fact, this behavior is strictly related to the presence of the copper ion as evidenced by a control experiment with Zn substituted azurin molecules for which no such phenomenon is observed. The experimental observation that in the case of high applied force no NDR effect is found seems to point out the relevance of having two non-negligible tunneling barriers in the junction. However, a clear correlation between the potentials at which NDR happens and the potentials typically encountered in electrochemical investigations of azurin has not been established, and the phenomenon is still poorly understood. Moreover, as usual in the case of NDR effects, the occurrence of this feature requires a high applied bias voltage. Recently [118], an NDR effect has been observed for a single molecule I(V) characterization of an STM-based junction with electrochemical potential control of the electrodes. In this case, the decrease of current upon the bias voltage increase has been observed for very low biases with respect to usual experiments performed without potentiostatic control and has been attributed to a bias-induced variation of the molecules' oxidation state. However, the details of the NDR mechanism, even in this case, are still much debated.

14.10   Future Trends in Metalloprotein Electronics

These first 15 years of activity on metalloproteins characterization towards their exploitation in hybrid electronic devices, along with the achieved relevant successes, has drawn the attention of scientists active in the field, wondering if such an approach is worthy of pursuit. The particular features of metalloproteins— i.e., the need for operating in a wet environment and the relative slowness of their intrinsic electron transfer mechanism, suggest that biomolecular electronics will hardly be able to act as surrogate to conventional solid-state electronics.

Rather than seeing such things as negative remarks, these considerations have inspired scientists to look for alternate routes to make biomolecular electronics a technological reality.

In fact, both the need of a wet environment and the richness and diversity of functional activity displayed by these molecules are now the basis of the most advanced and innovative ideas concerning their exploitation.

This means that instead of trying to emulate solid-state electronics by forcing a biomolecular system to undergo their rules, scientists now tend to find out the peculiarities of these molecular systems that might allow them to implement devices endowed with unique functionalities.

Among these possibilities are those connected with the characteristics of biomolecular-specific recognition and binding. In general, these features are, or can be made, dependent on the oxidation state of a redox biomolecule. Therefore, it is plausible to foresee redox state-dependent molecular binders or releasers, whose oxidation state can be tuned by an external electronic source. This general framework should open the way to a series of so far unmatched applications connected with the possibility of modulating biological phenomena at the molecular level by an electric source. Furthermore, it is even possible to imagine scenarios in which important key biomolecules, such as transcription factors in gene expression, will be functionally modulated by an external potential source.

Moreover, the general demonstration of the possibility of varying the transparency to electrons of a bio-inorganic single-molecule junction represents in itself a paradigm of the implementation of effective ways for interfacing biological systems with the inorganic world of solid-state electronics and technology.

In this direction goes also another future trend of metalloprotein electronics: that of exploiting special enzymes for hydrogen production. The possibility of interfacing in an effective and optimized way with enzymes (possibly engineered) belonging to the family of hydrogenases with solid-state metal electrodes represents a charming possibility of taking advantage of the functional features of special metalloproteins and of the amount of knowledge developed during these years in the complex field of metalloprotein electronics.

These few examples set the future standard in this fascinating and promising discipline.

References

[1] Grattarola, M., and Massobrio, G., *Bioelectronics Handbook: MOSFETs, Biosensors, and Neurons,* McGraw-Hill, Columbus, 1998.

[2] Gray, H.B. and Ellis, W., in *Electron Transport Metalloproteins in Bioinorganic Chemistry,* Bertini, I. et al., Eds., University Science Books, Sausalito, CA, 1994.

[3] Cowan, J.A., *Inorganic Biochemistry: An Introduction,* Wiley-VCH, New York, 1997.

[4] Holm, R.H. et al., Structural and functional aspects of metal sites in biology, *Chem. Rev.,* 96, 2239, 1996.

[5] Devault, D., *Quantum Mechanical Tunnelling in Biological Systems,* 2nd ed., Cambridge University Press, Cambridge, 1984.

[6] Harrison, P.M., *Metalloproteins Parts I and II,* Academic Press, New York, 1985.

[7] Facci, P., Single metalloprotein at work: towards a single-protein transistor, in *Nano-Physics & Bio-Electronics: A New Odyssey,* Chakraborty, T. et al., Eds., Elsevier, Amsterdam, 2002, 323.

[8] Marcus, R.A., Electron transfer reactions in chemistry and biology. Theory and experiments, in *Protein Electron Transfer,* Bendall, D.S., Ed., BIOS Scientific Publishers, Oxford, 1996, 249.

[9] Marcus, R.A., and Sutin, N, Electron transfer in chemistry and biology, *Biochim. Biophys. Acta,* 811, 265, 1985.

[10] Devault D., *Quantum Mechanical Tunnelling in Biological Systems,* Cambridge University Press, Cambridge, 1984.

[11] Moser, C.C. et al., Nature of biological electron transfer, *Nature,* 335, 796, 1992.

[12] Moser, C.C. and Dutton, P.L., Outline of theory of protein electron transfer, in *Protein Electron Transfer,* Bendall, D.S., Ed., BIOS Scientific Publisher, Oxford, 1996.

[13] Marcus, R.A., On the theory of oxidation-reduction reactions involving electron transfer: I, *J. Chem. Phys.,* 24, 966, 1956.

[14] Page, C.C. et al., Natural engineering principles of electron tunnelling in biological oxidation-reduction, *Nature,* 402, 47, 1999.

[15] Jones, M.L. et al., The nature of tunneling pathway and average packing density models for protein-mediated electron transfer, *J. Phys. Chem. A,* 106, 2002, 2002.

[16] Beratan, D.N. et al., Protein electron-transfer rates set by the bridging secondary and tertiary structure, *Science,* 252, 1285, 1991.

[17] Adman, E.T., Copper protein structures, *Adv. Protein Chem.,* 42, 145, 1991.

[18] Bendall, D.S., Interprotein electron transfer, in *Protein Electron Transfer,* Bendall D.S., Ed., BIOS Scientific Publisher, Oxford, 1996.

[19] Nar, H. et al., Crystal structure analysis of oxidized *Pseudomonas aeruginosa* azurin at pH 5.5 and pH 9.0. A pH-induced conformational transition involves a peptide bond flip, *J. Mol. Biol.,* 221, 765, 1991.

[20] Brill, A.S., *Transition Metals in Biochemistry,* Springer, Berlin, 1977.

[21] Chi, Q. et al., Electrochemistry of self-assembled monolayers of the blue copper protein *Pseudomonas aeruginosa* azurin on Au(111), *Electrochem. Commun.,* 1, 91, 1999.

[22] Lide, D.R., Ed., *CRC Handbook of Chemistry and Physics,* 74th ed., CRC Press, Boca Raton, FL, 1993.

[23] Di Felice, R. et al., DFT study of cysteine adsorption on Au(111), *J. Phys. Chem. B,* 107, 1151, 2003.

[24] Schnyder, B. et al., Comparison of the self-chemisorption of azurin on gold and on functionalized oxide surfaces, *Surf. Interface Anal.,* 34, 40, 2002.

[25] Lange, C. and Hunte, C., Crystal structure of the yeast cytochrome bc(1) complex with its bound substrate cytochrome *c, Proc. Natl. Acad. Sci. USA,* 99, 2800, 2002.

[26] Saraste, M., Oxidative phosphorylation at the fin de siecle, *Science,* 283, 1488, 1999.

[27] Pascher, T. et al., Protein folding triggered by electron transfer, *Science,* 271, 1558, 1996.

[28] Banci, L. and Assfalg, M., in *Handbook of Metalloproteins,* vol. 1, Messerschmidt, A. et al., Eds., Wiley, Chichester, 2001.

[29] Banci, L. et al., Mitochondrial cytochromes *c*: a comparative analysis, *J. Biol. Inorg. Chem.,* 4, 824, 1999.

[30] Brayer, G.D. and Murphy, M.E.P., in *Cytochrome c: A Multidisciplinary Approach*, Scott, R.A., and Mauk, A.G., Eds., University Science Book, Sausalito, CA, 1996; part II, 103.

[31] Koppenol, W.H. and Margoliash, E., The asymmetric distribution of charges on the surface of horse cytochrome *c*. Functional implications, *J. Biol. Chem.*, 257, 4426, 1982.

[32] Bard, A.J. and Faulkner, L.R., *Electrochemical Methods—Fundamentals and Applications*, 2nd ed., John Wiley & Sons, New York, 2000.

[33] Armstrong, F.A. et al., Reactions of complex metalloproteins studied by protein-film voltammetry, *Chem. Soc. Rev.*, 26, 169, 1997.

[34] Nahir, T.M., and Bowden, E.F., The distribution of standard rate constants for electron transfer between thiol-modified gold electrodes and adsorbed cytochrome *c*, *J. Electroanal. Chem.*, 410, 9, 1996.

[35] Brunschwig, B.S. and Sutin., N., Directional electron transfer: conformational interconversions and their effects on observed electron-transfer rate constants, *J. Am. Chem. Soc.*, 111, 7454, 1989.

[36] Weber, K., and Creager, S.E., Voltammetry of redox-active groups irreversibly adsorbed onto electrodes. Treatment using the Marcus relation between rate and overpotential, *Anal. Chem.*, 66, 3164, 1994.

[37] Khoshtariya, D.E. et al., Charge-transfer mechanism for cytochrome *c* adsorbed on nanometer thick films. Distinguishing frictional control from conformational gating, *J. Am. Chem. Soc.*, 125, 7704, 2003.

[38] Avila, A. et al., An electrochemical approach to investigate gated electron transfer using a physiological model system: Cytochrome *c* immobilized on carboxylic acid–terminated alkanethiol self-assembled monolayers on gold electrodes, *J. Phys. Chem. B*, 104, 2759, 2000.

[39] Binnig, G. et al., Atomic force microscope, *Phys. Rev. Lett.*, 56, 930, 1986.

[40] Colton, R.J. et al., Eds., *Procedures in Scanning Probe Microscopy*, Wiley, New York, 1998.

[41] Giessibl, F.J., Advances in atomic force microscopy, *Rev. Mod. Phys.*, 75, 949, 2003.

[42] Meyer, G. and Amer, N.M., Novel optical approach to atomic force microscopy, *Appl. Phys. Lett.*, 53, 1045, 1988.

[43] Zhong, Q. et al., Fractured polymer/silica fiber surface studied by tapping mode atomic force microscope, *Surf. Sci. Lett.*, 290, L688, 1993.

[44] De Wolf, P. et al., Two-dimensional carrier profiling of InP structures using scanning spreading resistance microscopy, *App. Phys. Lett.*, 73, 2155, 1998.

[45] Cui, X.D. et al., Reproducible measurement of single-molecule conductivity, *Science*, 294, 571, 2001.

[46] Binnig, G. et al., Surface studies by scanning tunneling microscopy, *Phys. Rev. Lett.*, 49, 57, 1982.

[47] Bonnell, D.A., Ed., *Scanning Tunneling Microscopy and Spectroscopy: Theory, Techniques, and Applications*, Weinheim, VCH, Verlagsgesellschaft, 1993.

[48] Chen, C.J, *Introduction to Scanning Tunneling Microscopy*, Oxford University Press, Oxford, 1993.

[49] Sonnenfeld, R. and Hansma, P.K., Atomic-resolution microscopy in water, *Science*, 232, 211, 1986.

[50] Lustenberger, P. et al., Scanning tunneling microscopy at potential controlled electrode surfaces in electrolytic environment, *J. Electroanal. Chem.*, 248, 451, 1988.

[51] Itaya, K. and Tomita, E., Scanning tunneling microscope for electrochemistry — a new concept for the in situ scanning tunneling microscope in electrolyte solutions, *Surf. Sci.*, 201, L507, 1988.

[52] Siegenthaler, H. and Christoph, R., in *Scanning Tunneling Microscopy and Related Methods*, Behm, R.J. et al., Eds., NATO ASI Series E, Vol 184, Kluwer, Dordrecht, 1990.

[53] Thundat, T. et al., Modification of tantalum surfaces by scanning tunneling microscopy in an electrochemical cell, *J. Vac. Sci. Technol. A*, 8, 3537, 1990.

[54] Bach, C.E. et al., Effective insulation of scanning-tunneling-microscopy tips for electrochemical studies using an electropainting method, *J. Electrochem. Soc.*, 140, 1281, 1993.

[55] Christoph, R. et al., *In situ* scanning tunneling microscopy at potential controlled Ag(100) substrates, *Electrochim. Acta*, 34, 1011, 1989.

[56] Kolb, D.M., Reconstruction phenomena at metal–electrolyte interfaces, *Prog. Surf. Sci.*, 51, 109, 1996.

[57] Tao, N.J., Probing potential-tuned resonant tunneling through redox molecules with scanning tunneling microscopy, *Phys. Rev. Lett.*, 76, 4066, 1996.

[58] Willner, I. and Katz., E., Integration of layered redox proteins and conductive supports for bioelectronic applications, *Angew. Chem. Int. Ed.*, 39, 1180, 2000.

[59] Lösche, M., Protein monolayers at interfaces, *Curr. Opin. Solid State Mater. Sci.*, 2, 546, 1997.

[60] Gun, J. et al., On the formation and structure of self-assembling monolayers : II. A comparative study of Langmuir–Blodgett and adsorbed films using ellipsometry and IR reflection–absorption spectroscopy, *J. Colloid Interface Sci.*, 101, 201, 1984.

[61] Kessel, C.R. and Granick, S., Formation and characterization of a highly ordered and well anchored alkylsilane monolayer on mica by self-assembly, *Langmuir*, 7, 532, 1991.

[62] Gun, J. and Sagiv, J., On the formation and structure of self-assembling monolayers: III. Time of formation, solvent retention, and release, *J. Colloid Interface Sci.*, 112, 457, 1986.

[63] Sagiv, J., Organized monolayers by adsorption: Formation and structure of oleophobic mixed monolayers on solid surfaces, *J. Am. Chem. Soc.*, 102, 92, 1980.

[64] Angst, D.L. and Simmons, G.W., Moisture absorption characteristics of organosiloxane self-assembled monolayers, *Langmuir*, 7, 2236, 1991.

[65] Le Grange, J.D. et al., Effects of surface hydration on the deposition of silane monolayers on silica, *Langmuir*, 9, 1749, 1993.

[66] Cras, J.J. et al., Comparison of chemical cleaning methods of glass in preparation for silanization, *Biosens. Bioelectron.*, 14, 683, 1999.

[67] Hong, H.G. et al., Cysteine-specific surface tethering of genetically engineered cytochromes for fabrication of metalloprotein nanostructures, *Langmuir*, 10, 153, 1994.

[68] Ledung, G. et al., A novel method for preparation of disulfides on silicon, *Langmuir*, 17, 6056, 2001.

[69] Love, J.C. et al., Self-assembled monolayers of thiolates on metals as a form of nanotechnology, *Chem. Rev.*, 105, 1103, 2005.

[70] Ulman, A., Formation and structure of self-assembled monolayers, *Chem. Rev.*, 96, 1533, 1996.

[71] Biebuyck, H.A. et al., Comparison of organic monolayers on polycrystalline gold spontaneously assembled from solutions containing dialkyl disulfides or alkanethiols, *Langmuir*, 10, 1825, 1994.

[72] Rao, S.V. et al., Oriented immobilization of proteins, *Mikrochim. Acta*, 128, 127, 1998.

[73] Ferretti, S. et al., Self-assembled monolayers: a versatile tool for the formulation of bio-surfaces, *Trends Anal. Chem.*, 19, 530, 2000.

[74] Gaines Jr, G.L., *Insoluble Monolayers at Liquid-Gas Interfaces*, John Wiley & Sons, New York, 1966.

[75] Stayton, P.S. et al., Genetic engineering of surface attachment sites yields oriented protein monolayers, *J. Am. Chem. Soc.*, 114, 9298, 1992.

[76] Johnson, D.L. and Martin, L.L., Controlling protein orientation at interfaces using histidine tags: An alternative to Ni/NTA, *J. Am. Chem. Soc.*, 127, 2018, 2005.

[77] Kim, K. et al., Electrochemically induced and controlled one-step covalent coupling reaction on self-assembled monolayers, *Langmuir*, 20, 3821, 2004.

[78] Bunimovich, Y.L. et al., Electrochemically programmed, spatially selective biofunctionalization of silicon wires, *Langmuir*, 20, 10630, 2004.

[79] Kim, K. H. et al., Protein patterning based on electrochemical activation of bioinactive surfaces with hydroquinone-caged biotin, *J. Am. Chem. Soc.*, 126, 15368, 2004.

[80] Yeo, W.-S. and Mrksich, M., Electroactive substrates that reveal aldehyde groups for bio-immobilization, *Adv. Mater.*, 16, 1352, 2004.

[81] Devaraj, N.K. et al., Selective functionalization of independently addressed microelectrodes by electrochemical activation and deactivation of a coupling catalyst, *J. Am. Chem. Soc.*, 128, 1794, 2006.

[82] Scott, R.A. and Mauk, A.G., *Cytochrome c: A Multidisciplinary Approach*, University Science Book, Sausalito, CA, 1995.

[83] Amador, S.M. et al., Use of self-assembled monolayers to covalently tether protein monolayers to the surface of solid substrates, *Langmuir*, 9, 812, 1993.

[84] Wood, L.L. et al., Molecular orientation distributions in protein films: Site-directed immobilization of yeast cytochrome *c* on thiol-capped self-assembled monolayers, *J. Am. Chem. Soc.*, 119, 571, 1997.

[85] Gerunda, M. et al., Grabbing yeast iso-1-cytochrome *c* by Cys102: An effective approach for the assembly of functionally active metalloprotein carpets, *Langmuir*, 20, 8812, 2004.

[86] Stayton, P.S. et al., Genetic engineering of surface attachment sites yields oriented protein monolayers, *J. Am. Chem. Soc.*, 114, 9298, 1992.

[87] Heering, H.A. et al., Direct immobilization of native yeast iso-1 cytochrome *c* on bare gold: Fast electron relay to redox enzymes and zeptomole protein-film voltammetry, *J. Am. Chem. Soc.*, 126, 11103, 2004.

[88] Bortolotti, C.A., et al., The redox chemistry of the covalently immobilized native and low-pH forms of yeast iso-1-cytochrome *c*, *J. Am. Chem. Soc.*, 128, 5444, 2006.

[89] Chi, Q. et al., Molecular monolayers and interfacial electron transfer of *Pseudomonas aeruginosa* azurin on Au(111), *J. Am. Chem. Soc.*, 122, 4047, 2000.

[90] Facci, P. et al., Potential-induced resonant tunneling through a redox metalloprotein investigated by electrochemical scanning probe microscopy, *Ultramicroscopy*, 189, 291, 2000.

[91] Alessandrini, A. et al., Electron tunnelling through azurin is mediated by the active site Cu ion, *Chem. Phys. Lett.*, 376, 625, 2003.

[92] Andolfi, L. et al., The electrochemical characteristics of blue copper protein monolayers on gold, *J. Electroanal. Chem.*, 565, 21, 2004.

[93] Davis, J.J. et al., Genetic modulation of metalloprotein electron transfer at bare gold, *Chem. Commun.*, 5, 576, 2003.

[94] Davis, J.J. et al., Protein adsorption at a gold electrode studied by *in situ* scanning tunneling microscopy, *New J. Chem.*, 1119, 1998.

[95] Facci, P. et al., Formation and characterization of protein monolayers on oxygen-exposing surfaces by multiple-step self-chemisorption, *Surf. Sci.*, 504, 282, 2002.

[96] Alessandrini, A. et al., Tuning molecular orientation in protein films, *Surf. Sci.*, 542, 64, 2003.

[97] Rieley, H. et al., X-ray studies of self-assembled monolayers on coinage metals: Alignment and photooxidation in 1,8-octanedithiol and 1-octanethiol on Au, *Langmuir*, 14, 5147, 1998.

[98] Schmickler, W., Investigation of electrochemical electron-transfer reactions with a scanning tunneling microscope—A theoretical study, *Surf. Sci.*, 295, 43, 1993.

[99] Schmickler, W. and Widrig, C., The investigation of redox reactions with a scanning tunneling microscope—Experimental and theoretical aspects, *J. Electroanal. Chem.*, 336, 213, 1992.

[100] Gerischer, H., Uber den Ablauf von Redoxreaktionen an Metallen und an Halbleitern II. Metall-Elektroden, *Z. Phys. Chem. NF*, 26, 325, 1960.

[101] Corni, S., The reorganization energy of azurin in bulk solution and in the electrochemical scanning tunneling microscopy setup, *J. Phys. Chem. B*, 109, 3423, 2005.

[102] Friis, E.P. et al., An approach to long-range electron transfer mechanisms in metalloproteins: *In situ* scanning tunneling microscopy with submolecular resolution, *Proc. Natl. Acad Sci. USA*, 96, 1379, 1999.

[103] Zhang, J. et al., Electronic properties of functional biomolecules at metal/aqueous solution interfaces, *J. Phys. Chem. B*, 106, 1131, 2002.

[104] Friis, E.P. et al., *In situ* scanning tunneling microscopy of a redox molecule as a vibrationally coherent electronic three-level process, *J. Phys. Chem. A*, 102, 7851, 1998.

[105] Zhang, J. et al., Electrochemistry and bioelectrochemistry towards the single-molecule level: Theoretical notions and systems, *Electrochim. Acta*, 50, 3143, 2005.

[106] Rinaldi, R. et al., Solid-state molecular rectifier based on self-organized metalloproteins, *Adv. Mater.*, 14, 1449, 2002.

[107] Rinaldi, R. et al., Electronic rectification in protein devices, *Appl. Phys. Lett.*, 82, 472, 2003.

[108] Maruccio, G. et al., Towards protein field-effect transistors: Report and model of a prototype, *Adv. Mat.*, 17, 816, 2005.

[109] Alessandrini, A. et al., Single-metalloprotein wet biotransistor, *Appl. Phys. Lett.*, 86, 133902, 2005.

[110] Wold, D.J. and Frisbie, C.D., Formation of metal–molecule–metal tunnel junctions: Microcontacts to alkanethiol monolayers with a conducting AFM tip, *J. Am. Chem. Soc.*, 122, 2970, 2000.

[111] Xu, B. et al., Measurements of single-molecule electromechanical properties, *J. Am. Chem. Soc.*, 125, 16164, 2003.

[112] Zhao, J. and Davis, J.J., Force dependent metalloprotein conductance by conducting atomic force microscopy, *Nanotechnology*, 14, 1023, 2003.

[113] Zhao, J. et al., Exploring the electronic and mechanical properties of protein using conducting atomic force microscopy, *J. Am. Chem. Soc.*, 126, 5601, 2004.

[114] Zhao, J. and Davis, J.J., Molecular electron transfer of protein junctions characterised by conducting atomic force microscopy, *Colloids Surf. B*, 40, 189, 2005.

[115] Simmons, J.G., Generalized formula for the electric tunnel effect between similar electrodes separated by a thin insulating film, *J. Appl. Phys.*, 34, 1793, 1963.

[116] Davis, J.J. et al., Metalloprotein tunnel junctions: Compressional modulation of barrier height and transport mechanism, *Faraday Discuss.*, 131, 167, 2006.

[117] Gorman, C.B. et al., Negative differential resistance in patterned electroactive self-assembled mono-layers, *Langmuir*, 17, 6923–6930, 2001

[118] Chen, F. et al., A molecular switch based on potential-induced changes of oxidation state, *Nano Lett.*, 5, 503, 2005.

15

Localization and Transport of Charge by Nonlinearity and Spatial Discreteness in Biomolecules and Semiconductor Nanorings. Aharonov–Bohm Effect for Neutral Excitons

F. Palmero

J. Cuevas

F.R. Romero

J.C. Eilbeck

R.A. Römer

J. Dorignac

Abstract

In this review, we describe some of our recent contributions related to the nonlinear localization of vibrational energy and electric charge in spatially discrete systems. We present some results concerning a polaronic charge transport mechanism in a DNA model, as well as on the existence and stability of *polarobreathers* solutions in the Peyrard–Bishop–Holstein model. Also, we review a lattice model for a neutral *exciton* (bound state of an electron and a hole) in a nanoring structure, and show that this neutral entity exhibits Aharonov–Bohm (AB) oscillations when the system is immersed in a magnetic field. Our results indicate a route to enhancing the possibility of experimental verification of the excitonic AB effect.

15.1   Introduction

Nonlinear localization of vibrational energy, electric charge and magnetic flux in spatially discrete systems has been widely investigated in during recent years (for a review of theoretical, experimental, and possible applications, see [1–5]).

In this review, we summarize some of our results obtained in this field, in particular our research concerning the polaronic charge transport mechanism in DNA, the existence and stability of *polarobreathers* in the Peyrard–Bishop–Holstein model, and some properties of electrons and holes in a nanoring using a 2D attractive fermionic Hubbard model. In particular, in this model, we demonstrate the existence of bound states of an electron-hole pair *exciton* and its sensitivity to variations of magnetic flux. Our results indicates route to enhance the possibility of experimentally verifying the excitonic AB effect.

This review is organized as follows: In Section 15.2, we present the results about the polaronic charge transport mechanism in DNA chains by means of a nonlinear, three-dimensional, semi-classical, tight-binding model. In Section 15.3, our findings related to the existence and stability of *polarobreathers* in the Peyrard–Bishop–Holstein model are given. Finally, in Section 15.4, we describe our research on the existence of Aharonov–Bohm oscillations for an exciton in a nanoring using a 2D attractive fermionic Hubbard model.

15.2   Charge Transport by Polarons in a DNA Molecule

Since the discovery of the DNA structure by James Watson and Francis Crick in 1953, research into DNA and the unraveling of the genetic code has given a great understanding of many processes of life. This knowledge has permitted the implementation of many useful applications in the sciences of life and biotechnology. Recently, scientists have become increasingly interested in other potential technological applications of DNA not directly related to the coding for functional proteins that is the expressed form of genetic information. All these potential applications imply the development of new disciplines, which are highly interdisciplinary, requiring biology, physics, engineering, chemistry, mathematics, computer science, and so on. Between others, we specifically mention the construction of nanostructures, DNA computing, DNA memory, and DNA electronics.

One of the most interesting applications of DNA is related to the construction of nanostructures of high complexity, as in the creation of "crystal lattices", structures where biomolecules of pharmacologic interest could be located periodically and then studied using standard crystallographic methods, or they could be used as "nanobreadboard," to which diverse components of a nanoelectronic device could be added. Also, molecular precision structures made of, or with, DNA could be used to build sensors, or, in general, different nanomachines [6]. At present, in our knowledge, it has been possible to create two-dimensional arrays with less than 20 nm of spacing [7], three-dimensional shapes [8], folding long single-stranded DNA molecules, arbitrary two-dimensional shapes [9], and structures which offer a wide range of potential applications.

Other possible technological applications based on DNA are related to DNA computers and DNA computing. The similarity between the way that DNA works and the operations of a Turing machine, a theoretical device that processes and stores information, suggests the possibility of using DNA to perform computations. In general, DNA computing is similar to parallel computing, taking advantage of different DNA molecules to simultaneously perform different calculations. In this field, the role of information processing in evolution and the possibility to reproduce these issues in a controlled environment was first addressed by Adleman's experiment, where the so-called Hamiltonian Path problem was set out [10] and, since this experiment, some Turing machines have proven to be constructible; even DNA computers with different input and output modules capable of fighting diseases have been developed [11]. On the other hand, molecular logic gates are crucial for the development of nanocomputers. Logical gates perform basic logic operations, (such as AND, NOT, OR), and in this field DNA have proven to be useful in building them [12, 13]. For a review about DNA computing, see [14, 15].

On the other hand, a strand of DNA is encoded with four bases (known as nucleotides), adenine (A), guanine (G), cytosine (C), and thymine (T) spaced every 0.35 nm along the molecule, giving a data density as approximately one bit per cubic nanometer, or a potentially exabyte (10^{18}) amount of information in a gram of DNA [16]. This characteristic can be used to build an inexpensive, long-lasting medium for information storage. In fact, it has been created in encoded DNA strand, hidden behind a dot in a printed document, sealed and mailed through the U.S. Postal Service, and, finally, recovered via the message [17]. Recent advances in genetic engineering allows us to introduce foreign DNA into living cells, and it has been shown that information can be encoded as an artificial DNA strand, and safely and permanently stored in a living host. This living organism could grow and multiply, and eventually it could be possible to extract this information back from the organism [18]. The true potential of the organic data memory based on DNA is still under research, and it is a challenging field.

In the field of nano-technology, molecular electronics based on electronic transport through bio-molecules has recently attracted great interest. Material scientists have argued that DNA is of fundamental interest to the development of DNA-based molecular technologies, as it possesses ideal structural and molecular recognition properties for use in self-assembling nanodevices with a definite molecular architecture [19]. Also, the robust, malleable one-dimensional structure of DNA can be used to design electronic devices [20–25], serving as a wire, transistor switch, or rectifier depending on its electronic properties [23, 26, 27]. For example, some type of sequences have p-type properties, while others have n-type ones [28]. Combinations of such sequences could form powerful logic elements, because only a short sequence of DNA base-pairs would be necessary to create all the combined n-p type properties. The conductivity properties of DNA have made it a candidate for the design of DNA-based electronic devices that could be much faster, smaller, and energy efficient than semiconductor-based electronic devices.

On the other hand, electronic transport can play an important role for some biological functions of DNA, such as biosynthesis and DNA repair after radiation damage. This is of great importance since some mutations in living systems and radical migrations are critical issues in carcinogenesis studies, and may yield insight into damage prevention or repair processes [29–31]. Also, charge transport has been demonstrated to proceed within HeLa cell nuclei [32] as well as in the nucleosome core particles, and can provide a practical method of genetic screening for known gene sequences and an alternative method for hybridization-based arrays [33].

At present, a clear picture about the electronic properties of DNA does not exist. Experimental results are controversial and probably the different base pairs sequences, layouts, and environmental conditions impact upon its conductivity properties [20, 34]. DNA can behave as a well conducting one-dimensional molecular wire [21, 22], an insulator [24], or even a semiconductor [25]. In general, charge carriers can hop along the DNA over distances of at least a few nanometers, but the evidence for electronic transport over larger distances is not so clear.

It has been proposed that charge transference through DNA proceeds along a one-dimensional pathway constituted by the overlap between *p*-orbitals in neighboring base pairs [35], and can be viewed as a "molecular wire," a one-dimensional twisted chain of stacked base pairs with a somewhat flexible structure [36]. Also, in order to explain the different experimental results, some different conduction mechanisms have been considered, such as coherent tunneling [35], incoherent phonon-assisted hopping [37], classical diffusion under thermal fluctuations [38], variable range hopping between localized states [39], charge carriers assisted by solitons [40], and polarons [41, 42].

In this review, we focus on some results obtained recently, related to polaronic charge transport mechanisms in DNA chains, by means of a simple tight-binding model.

15.2.1   Model for Polaron-Like Charge Transport Along a DNA Chain

In order to describe the dynamics of a DNA chain, we consider a variant of the *twist-opening* model [43–45], which itself is a modification of the Peyrard–Bishop model, taking into account the helical structure of the molecule and the torsional deformations induced by the opening of the base pairs. Each nucleotide and base is considered as a single non-deformable object, with mass m, which is an averaged estimate of the nucleotide mass. As we are interested in base pair vibrations, and not in acoustic motions, we fix the center of mass of each base pair (*i.e.*, the two bases in a base pair are constrained to move symmetrically with respect to the molecular axis). Moreover, the distances between two neighboring base pair planes is treated as fixed, because in the axial direction DNA seems less deformable than within the base pair planes [43].

In this model, all bond potentials are considered harmonic. This can be justified because charge transport is related to small deformations of the double chain. Furthermore, as the angular twist and the radial vibrational motion evolve on two different time scales, they can be considered as decoupled degrees of freedom in the harmonic approximation [46]. Thus, the position of the nth base pair is represented by the variables (r_n, ϕ_n), where r_n represents the radial displacement of the base pair from the equilibrium value R_0, and ϕ_n is angular displacement from equilibrium angles with respect to a fixed external reference frame. A sketch of the model is shown in Figure 15.1.

The Hamiltonian of the system is decomposed into three parts $\widehat{H} = \widehat{H}_{el} + \widehat{H}_{rad} + \widehat{H}_{twist}$. The term $\widehat{H}_{el}$ corresponds to the part related to the charged particle described by a tight-binding system: $\widehat{H}_{el} = \sum_n E_n |n\rangle\langle n| - V_{n-1,n}|n-1\rangle\langle n| - V_{n+1,n}|n+1\rangle\langle n|$, where $|n\rangle$ represents a localized state of the charge carrier at the nth base pair. The quantities $\{V_{n,n-1}\}$ are the nearest-neighbor transfer integrals along base pairs, and $\{E_n\}$ are the energy onsite matrix elements. A general electronic state is given by $|\Psi\rangle = \sum_n c_n(t)|n\rangle$, where $c_n(t)$ is the probability amplitude of finding the charged particle in the state $|n\rangle$. The time evolution of the $\{c_n(t)\}$ is obtained from the Schrödinger equation $i\hbar(\partial\Psi/\partial t) = \widehat{H}_{el}|\Psi\rangle$.

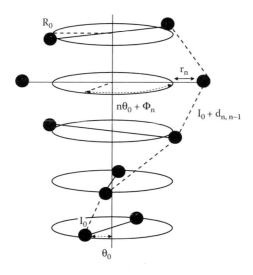

FIGURE 15.1   Sketch of the model. Filled circles represent bases. The variables used in the text are displayed.

The nucleotides are large molecules and their molecular motions are slow compared to those of a charged particle; thus, the lattice oscillators may be treated classically and $\widehat{H}_{rad}$ and $\widehat{H}_{twist}$, describing the radial and torsional contributions to $\widehat{H}$ are *de facto* classical Hamiltonians. They are given by (omitting the hat symbol):

$$H_{rad} = \sum_n \left[\frac{1}{2M} \left(p_n^r \right)^2 + \frac{M\Omega_{r_n}^2}{2} r_n^2 \right] \tag{15.1}$$

$$H_{twist} = \sum_n \left[\frac{1}{2J} \left(p_n^\phi \right)^2 + \frac{J\Omega_\phi^2}{2} (\phi_n - \phi_{n-1})^2 \right] \tag{15.2}$$

where p_n^r and p_n^ϕ are the conjugate momenta of the radial and angular coordinates. M is the reduced mass and the inertia moment of each base pair, respectively, m is an average estimate of the nucleotide mass, and Ω_{r_n} is the linear radial frequency, which is proportional to the strength of the hydrogen bonds.

The study of charge transport in this review is restricted to homogeneous DNA strands, where all base pairs are of the same type, or homogeneous DNA chains where all bases are of the same type, except for one of them which is of a different type. It can be supposed that the ratio between the elastic constants of bonds in a C-G base pair and an A-T base pair is 3/2 because the first involves three hydrogen bonds and the second involves only two. Then, we take $b_n = 0.8$ for an A-T base pair, and $b_n = 1.2$ for a C-G base pair [47]. The linear twist frequency is Ω_ϕ, and we represent the deviation of the relative angle between two adjacent base pairs from its equilibrium value θ_0 by $\theta_{n,n-1} = (\phi_n - \phi_{n-1})$.

In general, the ionization potential of different nucleotides differs by around 0.2 to 1.0 eV [48]. This could imply different values of the onsite energies E_n^0 for each base pair. If the study is focused only on geometrical effects due to the stretching of the chain over the charge, E_n^0 can be considered independent of the base type.

The interactions between the electronic variables and the structure variables, r_n and ϕ_n, arise from the dependence of the matrix elements E_n and $V_{n,n-1}$ on them. The first are given by $E_n = E_n^0 + kr_n$ [49], expressing the variation of the onsite electronic energies $\{E_n^0\}$ with the radial deformations. It is assumed the transfer matrix elements $V_{n,n-1}$ depend on the distances between two consecutive bases along a strand $d_{n,n-1}$, as $V_{n,n-1} = V_0(1 - \alpha d_{n,n-1})$, where

$$d_{n,n-1} = [a^2 + (R_0 + r_n)^2 + (R_0 + r_{n-1})^2 - 2(R_0 + r_n)(R_0 + r_{n-1})\cos(\theta_0 + \theta_{n,n-1})]^{1/2} - l_0 \tag{15.3}$$

with $l_0 = (a^2 + 4R_0^2 \sin^2(\theta_0/2))^{1/2}$, and a is the vertical distance between base pairs. The parameter α describes the influence of the distances between nucleotides on the transfer integrals. Note that, if the perturbations are small enough, a linear approximation of the distances $d_{n,n-1}$ can be considered [42, 50, 51].

There exists a general agreement about the range of values of some parameters for a DNA molecule [44, 52]. It is considered that $a = 3.4$ Å, $M = 4.982 \times 10^{-25}$ kg, $R_0 = 10$ Å, $\Omega_r = [6 - 8] \times 10^{12}$ s^{-1}, $\Omega_\phi = [0.5 - 1] \times 10^{10}$ s^{-1}, and $\theta_0 = 36^0$. Also, we have assumed that $V_0 = 0.1$ eV, a supposition widely used and found to reproduce ab initio results and experiments [36, 53].

No reliable data are available for the parameters α and k, and as in some other studies, the value of k can be fixed, and α can be taken as an adjustable parameter. In later studies, we have performed some quantum–chemical calculations in order to determine the values of parameters α and k [54], and the quantum–chemical estimates for these coupling parameters differs from the previous one. The results obtained using the parameter k as fixed and parameter α as adjustable are shown in the next three subsections, and the most relevant results obtained by computing credible values for the electron–mode coupling strengths k and α are shown in the last subsection.

We scale the time according to $t \rightarrow \Omega_r t$, and introduce the dimensionless quantities: $\tilde{r}_n = r_n (M\Omega_r^2/V_0)^{1/2}$, $\tilde{k}_n = k_n/(M\Omega_r^2 V_0)^{1/2}$, $\tilde{E}_n = E_n/V_0$, $\tilde{\Omega} = \Omega_\phi/\Omega_r$, $\tilde{V} = V_0/(J\,\Omega_r^2)$, $\tilde{\alpha} = \alpha\,(V_0/M\,\Omega_r^2)^{1/2}$, $\tilde{R}_0 = R_0\,(M\,\Omega_r^2/V_0)^{1/2}$. The scaled dynamical equations of the system, from which the tildes have been omitted, are

$$i\,\tau\dot{c}_n = (E_n + k\,r_n)\,c_n - (1 - \alpha\,d_{n+1,n})\,c_{n+1} - (1 - \alpha\,d_{n,n-1})\,c_{n-1} \tag{15.4}$$

$$\ddot{r}_n = -b_n r_n - k\,|c_n|^2 - \alpha\left[\frac{\partial d_{n,n-1}}{\partial r_n}(c_n^* c_{n-1} + c_n c_{n-1}^*) + \frac{\partial d_{n+1,n}}{\partial r_n}(c_{n+1}^* c_n + c_{n+1} c_n^*)\right] \tag{15.5}$$

$$\ddot{\phi}_n = -\Omega^2\,(2\phi_n - \phi_{n-1} - \phi_{n+1})$$
$$-\alpha\,V\left[\frac{\partial d_{n,n-1}}{\partial \phi_n}(c_n^* c_{n-1} + c_n c_{n-1}^*) + \frac{\partial d_{n+1,n}}{\partial \phi_n}(c_{n+1}^* c_n + c_{n+1} c_n^*)\right] \tag{15.6}$$

and the quantity $\tau = \hbar\,\Omega_r/V_0$ determines the time-scale separation between the fast electron motion and the slow bond vibrations. Observe that, in the case of $\alpha = 0$ and uniform $E_n = E_0$, the set of coupled equations represents the Holstein system, widely used in studies of polaron dynamics in one-dimensional lattices; for $\alpha = k = 0$, with random values of E_n, the Anderson model is obtained. In all cases, $E_0 c_n$ can be eliminated by a gauge transformation $c_n \rightarrow \exp(-i\,E_0 l/\tau)c_n$.

Using the expectation value for the electronic contribution to the Hamiltonian, the new Hamiltonian $\overline{H} = \langle\phi|\widehat{H}|\phi\rangle/V_0$ is given by

$$\overline{H} = \sum_n \left\{ \frac{1}{2}\left(\dot{r}_n^2 + b_n r_n^2\right) + \frac{R_0^2}{2}\left[\dot{\phi}_n^2 + \Omega^2(\phi_n - \phi_{n-1})^2\right] \right.$$
$$\left. + \left(E_n^0 + k r_n\right)|c_n|^2 - (1 - \alpha d_{n,n-1})(c_n^* c_{n-1} + c_n c_{n-1}^*) \right\} \tag{15.7}$$

15.2.2   Stationary Polaron-Like States

The study of nonlinear charge transport in this DNA model requires us to focus on the localized stationary solutions of Equations (15.4–15.6). Since the adiabaticity parameter τ is small, the fastest variables are the $\{c_n\}$, with a characteristic frequency (the linear frequency of the uncoupled system) of order $1/\tau \sim 19$. Next come the $\{r_n\}$, with frequency unity, and finally the $\{\phi_n\}$, with $\Omega_\phi \sim 0.11$. Using the Born–Oppenheimer approximation, it can be supposed initially that r_n and ϕ_n are constant in order to obtain the stationary localized solutions. For this purpose, it is necessary to use a modification of the numerical method outlined in Refs. [49, 55]. Substituting $c_n = \Phi_n \exp(-i\,E\,t/\tau)$ in Equation (15.4), with time-independent Φ_n's, results in a nonlinear difference system $E\Phi = \widehat{A}\Phi$, with $\Phi = (\Phi_1, ..., \Phi_N)$, from which a map $\Phi' = \widehat{A}\Phi/\|\widehat{A}\Phi\|$ is constructed, $\|.\|$ being the quadratic norm.

Thus, using Equations (15.4–15.6), the stationary solutions must be attractors of the map:

$$r_n' = -\frac{k}{b_n}|c_n|^2 \tag{15.8}$$

$$-\frac{\alpha}{b_n}\left[\frac{\partial d_{n,n-1}}{\partial r_n}(c_n^* c_{n-1} + c_n c_{n-1}^*) + \frac{\partial d_{n+1,n}}{\partial r_n}(c_{n+1}^* c_n + c_{n+1} c_n^*)\right] \tag{15.9}$$

$$\phi_n' = \frac{1}{2}(\phi_{n+1} + \phi_{n-1}) - \frac{\alpha V}{2\Omega^2}\left[\frac{\partial d_{n,n-1}}{\partial \phi_n}(c_n^* c_{n-1} + c_n c_{n-1}^*) + \frac{\partial d_{n+1,n}}{\partial \phi_n}(c_{n+1}^* c_n + c_{n+1} c_n^*)\right] \tag{15.10}$$

$$c_n' = \frac{[(E_n + k r_n')c_n - (1 - \alpha d_{n+1,n}')c_{n+1} - (1 - \alpha d_{n,n-1}')c_{n-1}]}{\|\{(E_n + k r_n')c_n - (1 - \alpha d_{n+1,n}')c_{n+1} - (1 - \alpha d_{n,n-1}')c_{n-1}\}\|} \tag{15.11}$$

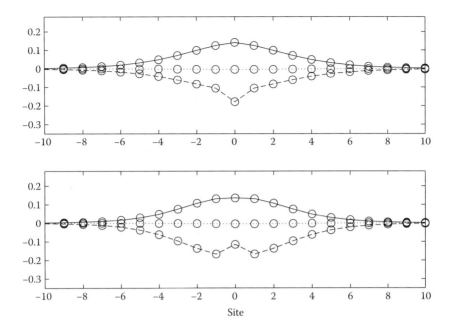

FIGURE 15.2   Profiles of the ground state in a homogeneous DNA chain with a different base pair at site 0. $a = 3.4$ Å, $M = 4.982$ kg, $R_0 = 10$ Å, $\Omega_r = 8 \times 10^{12}$ s^{-1}, $\Omega_\phi = 9 \times 10^{11}$ s^{-1}, $\theta_0 = 36^0$, $V_0 = 0.1\,eV$, $E_n^0 = 0$, $k = 1$, and $\alpha = 0.05$. Homogeneous C-G chain with an A-T base-pair inhomogeneity (top) and homogeneous A-T chain with a C-G base-pair inhomogeneity (bottom). The continuous line represents the wavefunction amplitude $|C_n|^2$, dashes are the static radial displacements r_n, and dots are the twist elongations $\theta_n = \phi_n - \phi_{n-1}$.

where $d' = d(r', \phi')$. The starting point is a completely localized state given by $c_n = \delta_{n,0}$, $r_n = 0$, and $\phi_n = 0$, $\forall n$. The map is applied until convergence is achieved. In this way, both stationary solutions and their energies E are obtained.

We have studied both the homogeneous case, where $E_n^0 = E_0$, and the random case, where $E_n^0 \in [-\Delta E, \Delta E]$ is a random parameter with mean value zero. Different interval sizes ΔE can be introduced in order to reproduce a certain degree of parametric disorder in the onsite electronic energy. A typical ground state is shown in Figure 15.2, it can be appreciated that the charge is fairly localized at only a few sites, and the amplitude decays as the distance increases from the central site. In the random case, if ΔE is not too large, the patterns do not change qualitatively.

15.2.3 Charge Transport

It can be expected the the polarons become mobile under perturbations. In order to move a polaron, the discrete gradient method [56] is an easier alternative to the more systematic *pinning mode* approach [57]. By means of the discrete gradient method, we perturb the (zero) velocities of a stationary state $\{r_n(0)\}$, $\{\dot{\phi}_n(0)\}$ in a direction parallel to the vectors $(\nabla r)_n = (r_{n+1} - r_{n-1})$, and/or $(\nabla \phi)_n = (\phi_{n+1} - \phi_{n-1})$. A typical profile of the moving polaron is shown in Figure 15.3.

If an amount of disorder in the onsite energies E_n^0 is introduced, it is found that mobile polarons exist below a critical value, ΔE_{crit}. Beyond this value, polarons are static. In general, the mobility induced by angular activation is more robust with respect to parametric disorder, the polaron has lower velocity, and the activation energy is higher than in the radial mobility regime.

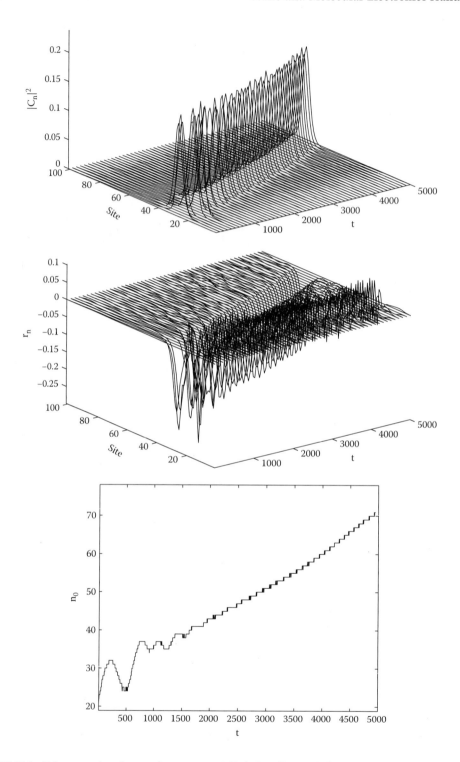

FIGURE 15.3 Polaron motion along an homogenous A-T chain. All magnitudes are in dimensionless units, and time is measured in units of the natural period of the unperturbed polaron $T = 2\pi\tau/E$, E being the energy of the unperturbed system. The moving polaron is generated by perturbing a static polaron located at site 21 by means of a radial perturbation of magnitude $\Delta E = 0.002$. $a = 3.4\,\text{Å}$, $b_n = 0.8$, $M = 4.982$ kg, $R_0 = 10\,\text{Å}$, $\Omega_r = 8 \times 10^{12}\,\text{s}^{-1}$, $\Omega_\phi = 9 \times 10^{11}\,\text{s}^{-1}$, $\theta_0 = 36^0$, $V_0 = 0.1\,eV$, $E_n^0 = 0$, $k = 1$, and $\alpha = 0.01$. Electronic breather (top), radial vibrational breather (center), and position of center of the electron breather n_0 (bottom).

15.2.4 Twist and Radial Polarons in a DNA Chain with a Base-Pair Inhomogeneity

In order to study the interaction of a mobile polaron with a single inhomogeneity, a mobile polaron induced far enough from the local inhomogeneity can be launched toward it, perturbing radial variables (radial mobility regime), angular variables (angular mobility regime), or both (mixed regime). Different kinds of behavior can be observed in Figure 15.4.

In the radial mobility regime, we have not observed the transmission of the charge through the local inhomogeneity. In the case of the homogeneous G-C chain with an A-T base pair inhomogeneity, when the polaron reaches the A-T base pair, it is trapped by it. This is caused by resonances between the polaron and the stationary state centered at the inhomogeneity. In the case of an A-T chain with a G-C base pair, when the polaron approaches the inhomogeneity, it is reflected by it. This phenomenon is due to the low probability of a resonance, due to the different shapes of the radial mode of the polaron and the ground state centered at the local inhomogeneity. These phenomena of trapping or reflection have been found in a great variety of situations, such as the interaction between a moving breather and an impurity in a Klein–Gordon chain [58]. A detailed study of these phenomena can be found in [59].

Some recent experiments on electron transfer in the DNA molecule [60] shows that an electron can migrate over long distances between triplet C-G base pairs and a C-G base pair separated by a number n of A-T base pairs. Also, the triplet C-G base pairs acts as a sink for holes in the chain. In our model, by decreasing slightly the ionization energy of the C-G base pair, as occurs in real DNA, we are able to reproduce (qualitatively) the experimental results.

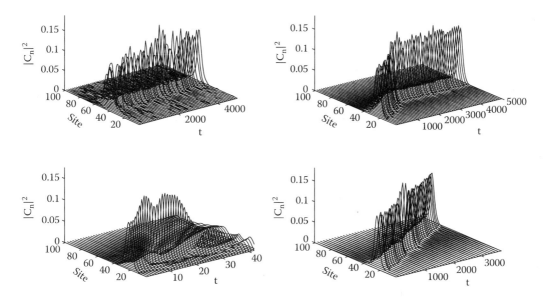

FIGURE 15.4 Polaron motion along a C-G chain with an A-T base pair located at site 51. All magnitudes are in dimensionless units, $\alpha = 0.01$, and time is measured in units of the natural period of the unperturbed polaron $T = 2\pi\tau/E$, E being the energy of the unperturbed system. The moving polaron is generated by perturbing a static polaron located at site 21 by means of a perturbation of magnitude ΔE. $a = 3.4$ Å, $M = 4.982$ kg, $R_0 = 10$ Å, $\Omega_r = 8 \times 10^{12}$ s^{-1}, $\Omega_\phi = 9 \times 10^{11}$ s^{-1}, $\theta_0 = 36^0$, $V_0 = 0.1eV$, $E_n^0 = 0$, and $k = 1$. (Top left) Trapping phenomenon due to the interaction between a radial moving polaron and the impurity. The moving polaron has been obtained by perturbing a static polaron with a radial perturbation of magnitude $\Delta E = 0.002$. (Top right) Trapping phenomenon due to the interaction between a twist polaron and the impurity. The moving polaron has been generated by perturbing a static polaron with an angular perturbation of magnitude $\Delta E = 0.002$. (Bottom left) Destruction of a static polaron due to a radial perturbation of magnitude $\Delta E = 0.10$. (Bottom right) Transmission phenomenon through the impurity of a twist moving polaron. The moving polaron has been obtained by perturbing a static polaron with an angular perturbation of magnitude $\Delta E = 0.10$.

15.2.5 Charge Transport in DNA Polymers Using Computational Values for the Electron-Mode Coupling Parameters

The study of charge transport by polarons in synthetic DNA polymers, built from a single type of base pairs can be carried out using a simplified version of the previous model. This is done by assuming that the impact of other vibrational degrees of freedom are negligible in comparison with the transverse vibrations.

Within this simplified model, some quantum–chemical calculations on symmetrical homodimers have allowed us to calculate reasonable values for the electron-mode coupling strengths k and α [42, 54].

Using the set of consistent parameter values for a DNA chain, $a = 3.4$ Å, $M = 4.982 \times 10^{-25}$ kg, $R_0 = 10$ Å, $\Omega_r = 6.25 \times 10^{12}$ s^{-1}, $\theta_0 = 36^0$, and $V_0 = 0.1$ eV for the coupling parameters of the poly(dA)-poly(dT) DNA polymer, we find $k = 0.0778917\,eV/$Å and $\alpha = 0.053835$ Å^{-1}. The corresponding values for the poly(dG)-poly(dC) DNA polymer are $k = -0.090325\,eV/$Å and $\alpha = 0.383333$ Å^{-1}. The quantum–chemical results for the coupling parameters in this simplified model differ from the ones previously used. These facts, the simplification of the model and different coupling parameters, implies that, in general, some different behaviour related to the mobility of the polaron can be expected.

Using the method described in previous sections, we have studied the existence of polarons and their mobility, observing that in the (dG)-(dC) case, unlimited electron propagation proceeds with uniform velocity. In comparison, the (dA)-(dT) electron travels with a smaller velocity, and can even stop and become trapped eventually, after having traversed a few lattice sites.

15.2.6 Experimental Results

An open question in our model is whether the polarons survive at ambient temperature. The values of radial variables are small, but the transfer integral elements are larger than $k_B T$, and this suggests that they would survive. In fact, our results are in agreement with experimental data [61] which show that poly(dG)-poly(dC) DNA polymers form a better conductor than their poly(dA)-poly(dT) counterparts.

For example, in these works, direct measurements of electrical transport through poly(dG)-poly(dC) and poly(dA)-poly(dT) DNA molecules containing identical base pairs have been performed. Some results are shown in Figure 15.5, where conductance at null tension has been numerically calculated from experimental $I - V$ curves. It is easy to appreciate that poly(dG)-poly(dC) DNA polymers are much better conductors than the poly(dA)-poly(dT) polymer.

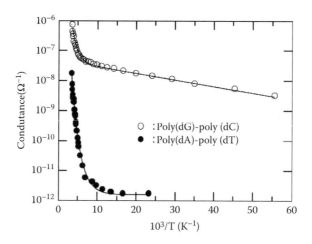

FIGURE 15.5 Conductance versus inverse temperature for poly(dA)-poly(dT) (●) and poly(dG)-poly(dC) (○). (Reprinted with permission from K.-H. Yoo, D.H. Ha, J.-O. Lee, J.W. Park, Jinhee Kim, J.J. Kim, H.-Y. Lee, T. Kawai, and Han Yong Choi. *Phys. Rev. Lett.*. 87, 19810, 2001. Copyright (2001) by the American Physical Society.)

15.3 Polarobreathers in the Peyrard–Bishop–Holstein Model

In some nonlinear systems, such as the Holstein model, the study of the conditions for the existence of intrinsic localization due to interacting degrees of freedom, even in a linear regime, is very interesting [49, 62].

It has been shown that a great variety of nonlinear systems sustain single- and multi humped polaronic and excitonic solutions [63]. The notion of effective nonlinearity in the setting of coupled excitonic and vibrational degrees of freedom was central to Davydov's suggestion of solitonic excitations arising in biomolecules [64, 65]. Also, a new aspect has been added to this type of problem [66, 67], by considering the interplay of the linear self-trapping with a soft nonlinear potential, such as a Morse potential. This has been proposed as a more general model relevant to many soft-matter applications.

The main stationary solutions obtained in the earlier work of [66] (namely, the single-site polarons) are dynamically stable within their region of existence. Furthermore, a different class of solutions which are genuinely "breathing" in time was termed *polarobreathers*. Such solutions were first obtained and discussed in the context of the Holstein model in [49, 68]. The domain of existence of such solutions was also obtained, together with their stability, by performing the corresponding Floquet spectral analysis. In this section, we show the most relevant results obtained regarding this type of model (a detailed study can be found in [69]).

15.3.1 The Model

The coupled charge/excitation-lattice model introduced in [66], describes the competition between linear polaronic self-trapping and self-focusing effects of a soft nonlinear potential. The Hamiltonian is given by:

$$H = \sum_n \left[\frac{1}{2} m\dot{u}_n^2 + V(u_n) + \frac{k}{2}(u_n - u_{n+1})^2 \right] - \sum_n \left[\chi(|\Psi_n|^2 u_n) + J(\Psi_n \Psi_{n+1}^* + \text{c.c.}) \right]. \quad (15.12)$$

where dots denote time derivatives, and the lattice index n runs from 1 to N (the total number of lattice sites). $\Psi_n(t)$ represents the "electronic" degrees of freedom, u_n corresponds to the lattice displacements (i.e., "vibrational" degrees of freedom), while the parameters J, k, and χ denote, respectively, the transfer integral, the lattice spring constant, and the coupling constant between the interacting fields. Finally, $V(u_n)$ is a Morse onsite potential:

$$V(u) = D[\exp(-bu) - 1]^2. \quad (15.13)$$

The lattice parameter values are chosen the same as in [70]: $D = 0.04$ eV, $b = 4.45$ Å, and $m = 300$ a.m.u. We have also chosen $J = 0.4$ meV, $k = 0.21$ eV/Å^2, and χ is a free parameter.

In order to get the lattice and electronic coordinates vibrating with the same frequency, the following transformation is introduced:

$$\Xi_n(t) = \Psi_n(t)e^{-iE_0t/\hbar}. \quad (15.14)$$

Then, the dynamical equations take the following form:

$$i\dot{\Xi}_n - \frac{E_0}{\hbar}\Xi_n + J(\Xi_{n+1} + \Xi_{n-1}) + \chi u_n \Xi_n = 0. \quad (15.15)$$

$$\ddot{u}_n + V'(u_n) - \chi|\Xi_n|^2 - k(u_{n+1} - 2u_n + u_{n-1}) = 0. \quad (15.16)$$

Stationary solutions can be obtained when all time derivatives in Equations (15.15) and (15.16) are set to zero. On the other hand, time-periodic solutions are characterized by a frequency of oscillation ν_b (or period $T_b = 2\pi/\nu_b$) in the time dependence of both the electronic wavefunction Ξ_n and of the lattice displacements u_n. Our aim is to study the existence and stability of such stationary and breathing polaron solutions; the latter, adopting the terminology of [49, 68], will be henceforth called "polarobreathers."

The results presented next have been obtained with fixed values of the transfer integral and lattice spring constant, and by letting the coupling constant χ and the polarobreather frequency ν_b vary.

In order to find either stationary or time-periodic solutions of the dynamical equations (15.15) and (15.16), we have used the methods based on the anti-continuous limit [71–73]. Once the numerically exact solutions are obtained, linear stability analysis should be performed to examine the dynamical stability of the solutions. In the case of stationary solutions, this has been implemented by a normal modes analysis [67]. On the other hand, in the case of (time-periodic) polarobreather solutions, a Floquet analysis has been performed [68].

15.3.2 Numerical Results

Apart from static polarons, we have considered the existence of stable, time-periodic, and spatially localized solutions within the context of the model. Figure 15.6 shows an example of these solutions and its relevant

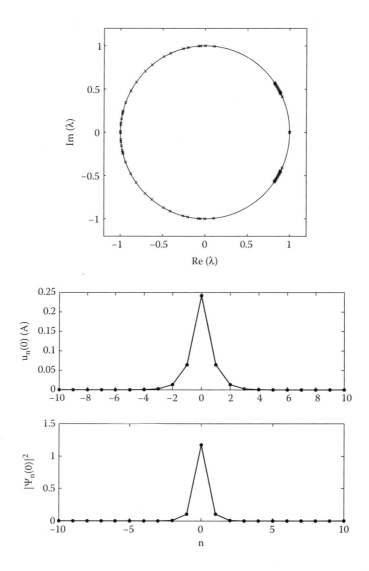

FIGURE 15.6 Floquet spectrum of a l-site polarobreather with $\nu_b = 0.91$ THz and $\chi = 0.53$ eV/ Å^2 (top). Profiles of the lattice (u_n, center) and electronic component ($|\Psi_n|^2$, bottom) corresponding to this solution.

Floquet spectrum. Polarobreathers exist as long as the conditions of the MacKay–Aubry's theorem are fulfilled [72]. A necessary condition is that none of the harmonics of the polarobreather frequency resonate with the linear modes. Based on this criterion, an analysis of the linear modes can provide the range of existence of polarobreathers for a given frequency. Figure 15.7 shows the real part of the normal mode frequencies and indicates the existence of a continuum band of (extended) linear modes together with several localized modes.

As the system is chosen to have an onsite soft potential, the polarobreathers must stem from the linear mode which is at the bottom of the band of extended modes. Then, the polarobreather frequency must be smaller than this localized-mode frequency. This fact is indicated in the top panel of Figure 15.7. It can be the observed that the bifurcation points correspond to the values of χ, for which the frequency of the localized mode coincides with the frequency of the polarobreather branch. The existence of polarobreathers is also limited by second-harmonic resonances.

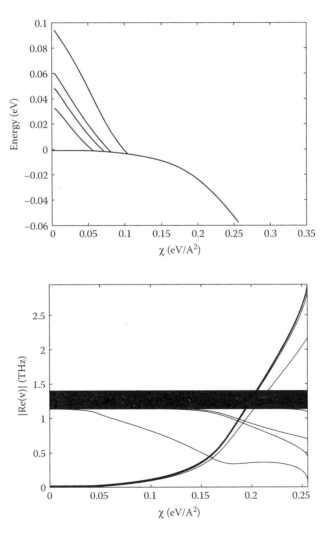

FIGURE 15.7 The top panel shows the dependence of the polarobreather energy as a function of χ for different values of the frequency (from left to right): $\nu_b = 1.02, 0.95, 0.91, 0.79$ THz. The line at the bottom corresponds to the static polaron. The bottom panel shows the real part of the linear mode spectrum.

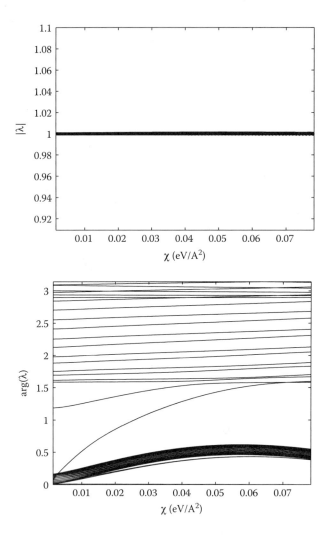

FIGURE 15.8 The modulus (top) and argument (bottom) of the Floquet eigenvalues as a function of the coupling constant χ for one-site polarobreathers.

An analysis of the stability of polarobreathers is shown in Figure 15.8. It is observed that, for the branch in question, all the solutions are linearly stable. It is worth noticing that for every polarobreather, two pairs of phase/growth modes exist.

In addition to one-site (site-centered) polarons, we have also considered the existence and stability of two-site (bond-centered) static polarons and polarobreathers. Whereas one-site polarons are stable, two-site ones are unstable. An example is shown in Figure 15.9, where the real and imaginary parts of the linear mode spectrum for two-site polarons are illustrated.

Contrary to their static counterparts, two-site polarobreathers are unstable. This instability is exponential (i.e., the Floquet eigenvalues responsible for the instability have zero phase). In Figure 15.10, an example of two-site polarobreather and its Floquet spectrum is shown. A full stability analysis for a branch of solutions is shown in Figure 15.11.

Two-site polarobreathers can be continued as a function of χ for J and k fixed. As in the one-site case, the branches for a fixed frequency merge with a static solution branch. The domains of existence of the one-site and two-site polarobreathers are shown in Figure 15.12. It is interesting to note that, similar to the case of their static counterparts [66], one-site polarobreathers have a narrower domain of existence (the grey region in Figure 15.12) than their two-site counterparts (the grey and black region in Figure 15.12).

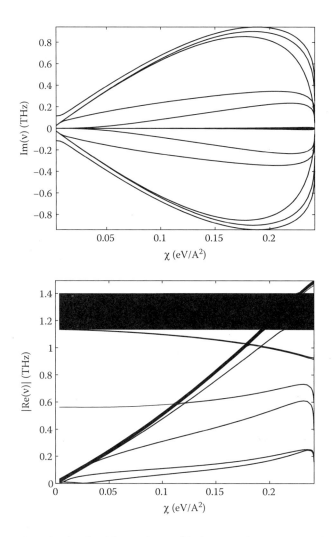

FIGURE 15.9 The imaginary (top) and real (bottom) part of the linear mode spectrum as a function of the coupling constant χ for two-site polarons.

Given that the termination of such branches occurs upon their collision with the stationary branch of solutions, this trait is natural to expect in the present setting. A wider domain of existence for two-site breathers was also observed in a model with competing attractive and repulsive interaction [74].

15.4   Aharonov–Bohm Effect for an Exciton in a Hubbard Ring

The Aharonov–Bohm effect (AB effect) is a quantum effect of a charged particle interacting with a electromagnetic vector potential, even when the field is zero [75]. The wave function of a charged particle in a ring geometry subject to a magnetic field acquires a phase proportional to the magnetic flux Φ through the ring [76, 77], and all observable phenomena depend periodically upon that flux, with period $\Phi_0 = hc/e$, the universal flux quantum [77–80]. Measurement of the original AB effect has been performed [80] and investigations and applications of this effect have been studied in different materials and devices [77, 78, 81–84].

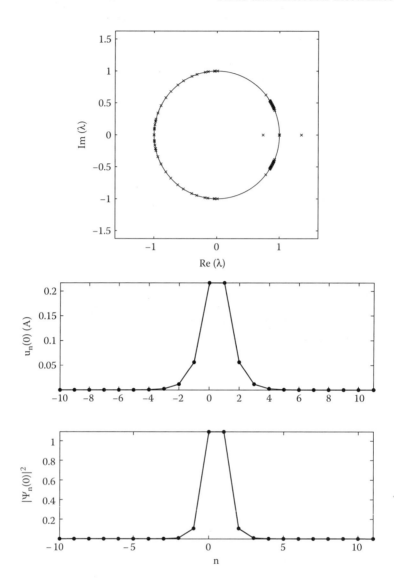

FIGURE 15.10   Floquet spectrum of a two-site polarobreather with $\nu_b = 0.91$ THz and $\chi = 0.053$ eV/ Å^2 (top). Profiles of the lattice (u_n, center) and electronic component ($|\Psi_n|^2$, bottom) corresponding to this solution.

It has been proposed that a neutral excitation, as in an electron–hole pair (exciton), can also exhibit AB oscillations [85–88]. However, this effect might be very weak, because when the ring becomes large enough, the effect is suppressed. On the other hand, progress in micro-structuring technology allows the construction of nano-sized InGaAs rings by self-assembly [89–91] or by using lithographic techniques [92, 93]. Similarly, PbSe-based nano-rectangles have been synthesized in solution through oriented attachment of nanocrystal building blocks [94]. In these systems, electrons and holes can propagate coherently and the AB effect can be explored. The AB effect for charged excitons was observed in optical experiments on quantum rings [93] and AB oscillations have been detected in Type–II quantum dot ensembles for a neutral exciton [95].

We present the main results obtained in a lattice model of a nanoring with an electron and hole subjected to a perpendicular and uniform magnetic flux. This model allows a finite number of transport channels,

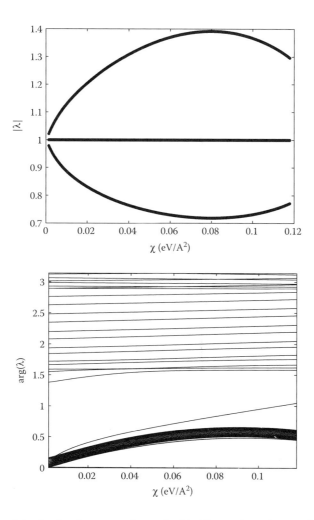

FIGURE 15.11   The modulus (top) and argument (bottom) of the Floquet eigenvalues as a function of the coupling constant χ for two-site polarobreathers.

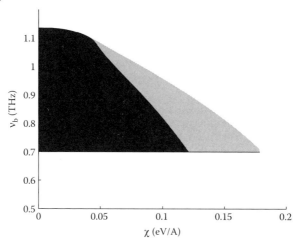

FIGURE 15.12   Domain of existence of one-site (black region) and two-site (black *and* grey region) polarobreathers in the χ-ν_b parameter plane.

either arranged annularly (coupled rings laid out in a 2D plane with increasing radius), or stacked vertically. Therefore, we can study in a controlled way how the excitonic AB oscillations evolve when additional rings are added, either in-plane or vertically.

15.4.1 The Model

In order to describe the dynamics of the electron/hole pair, we consider a simple two-band model with spinless particles, which is suitable to study some interesting physical situations, such as the dynamics of excitons in some semiconductor nanostructures [96].

The electron and the hole are located in an anharmonic lattice of N rings, each ring having M sites. The system is subject to a uniform magnetic field **B**, perpendicular to the lattice, with magnetic flux Φ, as shown in Figure 15.13.

We assume a short ranged interaction potential and describe the system by an attractive fermionic Hubbard model where only hopping between nearest-neighbors sites is possible. Thus, the Hamiltonian is

$$
\begin{aligned}
\hat{H} = &-\gamma \sum_{n=1}^{N} \sum_{m=1}^{M} a_{n,m}^{\dagger} a_{n,m} b_{n,m}^{\dagger} b_{n,m} \\
&-\sum_{n=1}^{N-1} \sum_{m=1}^{M} t_{n}^{\perp} [a_{n,m}^{\dagger} a_{n+1,m} + a_{n+1,m}^{\dagger} a_{n,m} + \mu(b_{n,m}^{\dagger} b_{n+1,m} + b_{n+1,m}^{\dagger} b_{n,m})] \\
&-\sum_{n=1}^{N} \sum_{m=1}^{M} t_{n}^{\|} [e^{2\pi i \varphi_n / M}(a_{n,m}^{\dagger} a_{n,m+1} + \mu b_{n,m+1}^{\dagger} b_{n,m}) + e^{-2\pi i \varphi_n / M}(a_{n,m+1}^{\dagger} a_{n,m} + \mu b_{n,m}^{\dagger} b_{n,m+1})]
\end{aligned}
\tag{15.17}
$$

where the operator $a_{n,m}^{\dagger}$ ($a_{n,m}$) creates (destroys) an electron in the conduction band at site m and ring n, and $b_{n,m}^{\dagger}$ ($b_{n,m}$) creates (destroys) a hole in the valence band at the same place, and satisfies the standard fermionic anticommutation rules. We assume periodic boundary conditions ($M + 1 = 1$).

The parameters $t_n^{\|}$ represent the hopping coefficient between neighboring sites along the nth ring, and $t_n^{\perp}$ the hopping coefficient between neighboring sites, one corresponding to the nth ring and the other to the $(n + 1)$th ring. In general, they will be a function of the distance between sites. In a single ring with no interaction, the effective mass of electron and holes at the extrema of conduction and valence band, respectively, are proportional to the inverse of the square of the distance between sites [96]. Hence, we

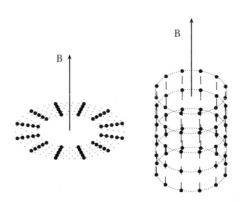

FIGURE 15.13   Sketch of the model. (Left) The lattice is a 2D system of N rings and M sites in each ring. (Right) The lattice is a system of N stacked rings and M sites in each ring. Each system is subject to a uniform magnetic field **B** perpendicular to the rings, with Φ the magnetic flux.

consider a similar dependence for the intra-ring and inter-ring hopping coefficient

$$t_n^{\perp} = \frac{\epsilon}{(r_{n+1} - r_n)^2}, \quad t_n^{\|} = \frac{\epsilon}{4r_n^2 \sin^2(\pi/M)}, \tag{15.18}$$

where r_n is the radius of the nth ring, and $d_n = 2r_n \sin(\pi/M)$ is the chord distance between two neighboring lattice points of the nth ring. Also, the formal continuum limit can be obtained, making this distance zero. It is supposed that the distance between rings $\Delta r = r_{n+1} - r_n$ is constant, $t_n^{\perp} = t^{\perp}$, and thus independent of the ring.

The parameter γ represents the short-range repulsive interaction between the electron and the hole, γ/ϵ being the ratio between anharmonicity and the nearest neighbor hopping energy. The parameter μ is the ratio of the effective masses of electron and holes at the bottom and top of the conduction and valence band, respectively, and $\varphi_n = \Phi_n/\Phi_0$, Φ_n the magnetic flux through the nth ring, Φ_0 being the flux quantum. In general, we consider $\epsilon = 1$ and $\mu = 0.2$, as in Ref. [97].

In the case of a stacked array of single rings with identical radius located along the magnetic field direction, $t_n^{\|} = t^{\|}$ and $\varphi_n = \varphi$.

The Hubbard model has been widely studied. In particular, the 1D Hubbard model with nearest-neighbor hopping terms has been solved exactly in the seminal papers by Lieb and Wu [98] and with flux by Shastry and Sutherland [99, 100]. Recent results are reviewed in Ref. [101]. Similar exact results for extensions of the model are rare. Some approximations to calculate the spectrum of the Hubbard model have been proposed, (see, e.g., [102]), where the exact spectrum and thermodynamics for a long-range hopping Hubbard chain with linear dispersion is calculated. Also, an exact solution for the two-electron case in a d-dimensional Hubbard model has been obtained [103].

In a recent work, we have used the number–state method [104] to calculate the eigenvalues and eigenvectors of the Hamiltonian operator (15.17) with a single electron and a single hole [88]. In this section, a different approach is used [105].

The standard creation operators can be transformed to the new set of operators

$$\mathbf{a}_{n,l}^{\dagger} = \frac{1}{\sqrt{M}} \sum_{r=1}^{M} \tau_l^{-r} a_{n,r}^{\dagger}, \quad \mathbf{b}_{n,l}^{\dagger} = \frac{1}{\sqrt{M}} \sum_{r=1}^{M} \tau_l^{-r} b_{n,r}^{\dagger} \tag{15.19}$$

where $\tau_l = e^{ik_l}$, and $k_l = 2\pi l/M$, with l integer.

Thus, the original operators can be expressed as

$$a_{n,r}^{\dagger} = \frac{1}{\sqrt{M}} \sum_{l=1}^{M} \tau_l^{r} \mathbf{a}_{n,l}^{\dagger}, \quad b_{n,r}^{\dagger} = \frac{1}{\sqrt{M}} \sum_{l=1}^{M} \tau_l^{r} \mathbf{b}_{n,l}^{\dagger} \tag{15.20}$$

In a similar way, a new set of lowering operators can be defined, and they satisfy the standard anticommutation rules. In terms of these operators, the original Hamiltonian (15.17) becomes

$$\widehat{H} = -\frac{\gamma}{M} \sum_{n=1}^{N} \sum_{r=1}^{M} \sum_{r'=1}^{M} \sum_{t=1}^{M} \sum_{t'=1}^{M} \delta_{r+pt,r'+t'} \mathbf{a}_{n,r}^{\dagger} \mathbf{a}_{n,r'} \mathbf{b}_{n,t}^{\dagger} \mathbf{b}_{n,t'} \tag{15.21}$$

$$-2 \sum_{n=1}^{N} \sum_{l=1}^{M} t_n^{\|} \cos(k_l - 2\pi \varphi_n/M) \mathbf{a}_{n,l}^{\dagger} \mathbf{a}_{n,l} - t^{\perp} \sum_{n=1}^{N-1} \sum_{l=1}^{M} (\mathbf{a}_{n,l}^{\dagger} \mathbf{a}_{n+1,l} + \mathbf{a}_{n+1,l}^{\dagger} \mathbf{a}_{n,l})$$

$$-2\mu \sum_{n=1}^{N} \sum_{l=1}^{M} t_n^{\|} \cos(k_l + 2\pi \varphi_n/M) \mathbf{b}_{n,l}^{\dagger} \mathbf{b}_{n,l} - \mu t^{\perp} \sum_{n=1}^{N-1} \sum_{l=1}^{M} (\mathbf{b}_{n,l}^{\dagger} \mathbf{b}_{n+1,l} + \mathbf{b}_{n+1,l}^{\dagger} \mathbf{b}_{n,l})$$

15.4.2 A Single Electron or Hole

Consider first the case where there is a single electron in the conduction band and the valence band is full. It is possible to block-diagonalize the Hamiltonian operator using eigenfunctions of the periodic translation (or rotation) operator $\hat{T}$ defined as $\hat{T} a_{n,m}^{\dagger} = a_{n,m+1}^{\dagger} \hat{T}$. In each block, the eigenfunctions have a fixed value of the momentum k_l, with $\tau_l = \exp(ik_l)$ being an eigenvalue of $\hat{T}$ such that $k_l = 2l\pi/M$ and l integer ($l = 1, \ldots, M$).

For a given value of the momentum k_l, a basis for this relevant subspace is the set

$$|a_{n,l}^{\dagger}> = a_{n,l}^{\dagger}|0> \quad (n = 1, \ldots, N) \tag{15.22}$$

where $|0>$ denotes a state where the valence band is full and the conduction band is empty. This state satisfies the condition $\mathbf{a}_{n,l}|0> = \mathbf{b}_{n,l}|0> = 0$. We denote the dual basis

$$< a_{n,l}^{\dagger}| = < 0|a_{n,l} \tag{15.23}$$

For the single electron case, it can be shown that

$$\widehat{H}|a_{n,l}^{\dagger}> = -2t_n^{\parallel} \cos(k_l - 2\pi\varphi_n/M)|a_{n,l}^{\dagger}> -t^{\perp}(|a_{n-1,l}^{\dagger}> +|a_{n+1,l}^{\dagger}>,$$

and the $N \times N$ tridiagonal Hamiltonian matrix in each momentum subspace is given by

$$H_l^e = - \begin{bmatrix} q_1 & t^{\perp} & 0 & . & . & 0 \\ t^{\perp} & q_2 & t^{\perp} & 0 & . & 0 \\ 0 & . & . & . & . & . \\ . & . & . & . & . & t^{\perp} \\ 0 & . & . & 0 & t^{\perp} & q_N \end{bmatrix}. \tag{15.24}$$

where $q_n = 2t_n^{\parallel} \cos(k_l - 2\pi\varphi_n/M)$.

For each value of k_l, we denote the N eigenvalues of this matrix as $E_l^e(p)$ ($p = 1 \ldots N$), and the corresponding normalized eigenvectors as $|\phi_l^e(p)>$, which can be written as a linear combination of the N basis states $|a_{n,l}^{\dagger}>$ ($n = 1 \ldots N$). Since only states with momentum k_l contribute, the eigenvectors can be expressed as

$$|\phi_l^e(p)> = \sum_{n=1}^{N} c_{n,l}^e(p)|a_{n,l}^{\dagger}>, \tag{15.25}$$

where $c_{n,l}^e(p) = < a_{n,l}^{\dagger}|\phi_l^e(p)>$. We denote the dual eigenvector of $|\phi_l^e(p)>$ as $< \phi_l^e(p)|$.

In a similar fashion, in the case of a single hole, the Hamiltonian matrix can be calculated: $H_l^h = \mu H_l^e$, but now $q_n = 2t_n^{\parallel} \cos(k_l + 2\pi\varphi_n/M)$.

We denote the N eigenvalues and eigenstates of this matrix as $E_l^h(p)(p = 1 \ldots N)$ and $|b_{n,l}^{\dagger}>$, respectively. Also, these eigenvectors can be expressed as a linear combination of the corresponding N basis states

$$|\phi_l^h(p)> = \sum_{n=1}^{N} c_{n,l}^h(p)|b_{n,l}^{\dagger}>, \tag{15.26}$$

where $c_{n,l}^h(p) = < b_{n,l}^{\dagger}|\phi_l^h(p)>$. We denote the dual eigenvector of $|\phi_l^h(p)>$ as $< \phi_l^h(p)|$.

15.4.3 An Electron and a Hole Together

In the trivial case of no interaction between the electron and the hole ($\gamma = 0$), the eigenstates of $\hat{H}$ are

$$|\phi_l^e(p)\phi_{l'}^h(p') >= \sum_{n=1}^{N}\sum_{n'=1}^{N} c_{n,l}^e(p)c_{n',l'}^h(q)|a_{n,l}^{\dagger}b_{n',l'}^{\dagger} >, \qquad (15.27)$$

where $(l,l') \in \{1\ldots M\}, (p,p') \in 1\ldots N$, and $|a_{n,l}^{\dagger}b_{n',l'}^{\dagger} >= a_{n,l}^{\dagger}b_{n',l'}^{\dagger}|0 >$. The energy of the system is given by $E_l^e(p) + E_{l'}^h(p')$.

In the simplest case of a single ring ($N = 1$), the energy is given by

$$E_l^e(1) + E_{l'}^h(1) = -2t_1^{\parallel}\cos(k_l - 2\pi\varphi_1/M) - 2\mu t_1^{\parallel}\cos(k_{l'} + 2\pi\varphi_1/M), \qquad (15.28)$$

where the momentum of each single particle is $k_l = 2\pi l/M$ for the electron and $k_{l'}' = 2\pi l'/M$ for the hole are the single particle wave vectors in absence of a magnetic field [96]. The continuous limit is archived when $M \to \infty, d \to 0$ and the radius of the ring is fixed. Considering the energy of the ground state in the absence of the field for the electron and hole as zero, Equation (15.28) reduces to the standard parabolic expressions for charges in a continuous ring threaded by a magnetic flux and no interaction [86].

In the case of interaction between the electron and the hole ($\gamma \neq 0$), the preceding states are not eigenstates of $\hat{H}$, but this set can be used as a basis. For a given total momentum of the system $K_Q = k_l + k_{l'}', Q \in \{1\ldots M\}$, and due to the translational (rotational) invariance of the system, it is possible to block-diagonalize the Hamiltonian operator for this given value and to write an eigenstate $|\psi^Q >$ with energy E as

$$|\psi^Q >= \sum_{l}\sum_{p}\sum_{p'} g_l(p,p')|\phi_l^e(p)\phi_{Q-l}^h(p') >, \qquad (15.29)$$

The eigenvalue equation $(\hat{H} - E)|\psi^Q >= 0$, can be written as

$$\sum_{l}\sum_{p}\sum_{p'} g_l(p,p')(W^e + E_l^e(p) + E_{Q-l}^h(p') - E)|\phi_l^e(p)\phi_{Q-l}^h(p') >= 0, \qquad (15.30)$$

where

$$W = -\frac{\gamma}{M}\sum_{n=1}^{N}\sum_{r=1}^{M}\sum_{r'=1}^{M}\sum_{t=1}^{M}\sum_{t'=1}^{M} \delta_{r+t,r'+t'} a_{n,r}^{\dagger} a_{n,r'} b_{n,t}^{\dagger} a_{n,t'} \qquad (15.31)$$

Multiplying by an element of the dual basis $< \phi_{l''}^e(p'')\phi_{Q-l''}^h(p''')|$, and using the orthonormal property of the eigenstates of the non-interactive electron–hole system, we obtain

$$\left[\sum_{l}\sum_{p}\sum_{p'} g_l(p,p') < \phi_{l''}^e(p'')\phi_{Q-l''}^h(p''')|W|\phi_l^e(p)\phi_{Q-l}^h(p') >\right] \qquad (15.32)$$
$$+\left\{E_l^e(p) + E_{Q-l}^h(p') - E\right\}g_{l'',Q-l''}(p'',p''') = 0.$$

We can write $W = -(\gamma/M)\sum_{n=1}^{N} V_n$, where

$$V_n = \sum_{r=1}^{M}\sum_{r'=1}^{M}\sum_{t=1}^{M}\sum_{t'=1}^{M} \delta_{r+t,r'+t'} a_{n,r}^{\dagger} a_{n,r'} b_{n,t}^{\dagger} b_{n,t'}, \qquad (15.33)$$

and the matrix elements of V_n are

$$< \phi_{l''}^e(p'')\phi_{Q-l''}^h(p''')|V_n|\phi_l^e(p)\phi_{Q-l}^h(p') >=$$

$$\sum_{r=1}^{M}\sum_{r'=1}^{M}\sum_{t=1}^{M}\sum_{t'=1}^{M}\delta_{r+t,r'+t'} < \phi_{l''}^e(p'')\phi_{Q-l''}^h(p''')|a_{n,r}^\dagger a_{n,r'} b_{n,t}^\dagger b_{n,t'}|\phi_l^e(p)\phi_{Q-l}^h(p') >=$$

$$\sum_{r=1}^{M}\sum_{r'=1}^{M}\sum_{t=1}^{M}\sum_{t'=1}^{M}\delta_{r+t,r'+t'} < \phi_{l''}^e(p'')\phi_{Q-l''}^h(p''')|b_{n,t}^\dagger a_{n,r}^\dagger a_{n,r'} b_{n,t'}|\phi_l^e(p)\phi_{Q-l}^h(p') >=$$

$$< \phi_{l''}^e(p'')|a_{n,l''}^\dagger >< \phi_{Q-l''}^h(p''')|b_{n,Q-l''}^\dagger >< a_{n,l}|\phi_l^e(p) >< b_{n,Q-l}|\phi_{Q-l}^h(p') >=$$

$$c_{n,l}^e(p)c_{n,Q-l}^h(p')c_{n,l''}^{e*}(p'')c_{n,Q-l''}^{h*}(p'''), \quad (15.34)$$

where an asterisk denotes a complex conjugate. Thus, Equation (15.33) becomes

$$-\gamma M^{-1}\sum_{n=1}^{N} c_{n,l''}^{e*}(p'')c_{n,Q-l''}^{h*}(p''')\sum_{l}\sum_{p}\sum_{p'} g_l(p,p')c_{n,l}^e(p)c_{n,Q-l}^h(p') \quad (15.35)$$

$$+\left\{ E_{l''}^e(p'') + E_{Q-l''}^h(p''') - E \right\}g_{l''}(p'',p''') = 0.$$

Defining

$$G_n = \sum_{l}\sum_{p}\sum_{p'} g_l(p,p')c_{n,l}^e(p)c_{n,Q-l}^h(p') \quad (15.36)$$

we can write

$$g_{l''}(p'',p''') = \gamma M^{-1}\frac{\sum_{n=1}^{N} c_{n,l''}^{e*}(p'')c_{n,Q-l''}^{h*}(p''')G_n}{E_{l''}^e(p'') + E_{Q-l''}^h(p''') - E} \quad (15.37)$$

and substituting in Equation (15.36),

$$G_n = \gamma M^{-1}\sum_{l}\sum_{p}\sum_{p'} c_{n,l}^e(p)c_{n,Q-l}^h(p')\frac{\sum_{n'=1}^{N} c_{n',l}^{e*}(p)c_{n',Q-l}^{h*}(p')G_{n'}}{E_l^e(p) + E_{Q-l}^h(p') - E} \quad (15.38)$$

$$= \gamma M^{-1}\sum_{n'=1}^{N} F_{n,n'}(E)G_{n'},$$

where

$$F_{n,n'}(E) = \sum_{l}\sum_{p}\sum_{p'}\frac{c_{n,l}^e(p)c_{n',l}^{e*}(p)c_{n,Q-l}^h(p')c_{n',Q-l}^{h*}(p')}{E_l^e(p) + E_{Q-l}^h(p') - E} \quad (15.39)$$

The condition to have no trivial solutions is

$$\det(I - \gamma M^{-1}F(E)) = 0 \quad (15.40)$$

where $F(E)$ is a $N \times N$ matrix with elements $F_{ij}(E)$. From this equation, the eigenvalues E of the system can be obtained.

15.4.4 A Single Ring

In the single ring case ($N = 1$), some analytical expressions can be obtained. The characteristics equation, Equation (15.40), is similar to the expression obtained in the continuous case [88], and can be written as

$$1 + \frac{\gamma}{2M|q|} \sum_{l=1}^{M} \frac{1}{\cos(2\pi l/M + \theta) - \cos \nu} = 0, \tag{15.41}$$

where $q = t_1^{\|} e^{2\pi i \varphi/M}(\mu + e^{-iK_Q})$, $\theta = \arg(q)$, and $\cos(\nu) = -E/2|q|$.

This sum can be calculated exactly, and the characteristic equation is thus

$$\frac{\tan(M\nu)}{\sin(\nu)} = \frac{2|q|}{\gamma} \left[1 - \frac{\cos(M\theta)}{\cos(M\nu)} \right]. \tag{15.42}$$

The solutions of this equation provide the energy spectrum. The eigenvectors are given by

$$|\psi^Q> = \sum_l g_l(1, 1)|\phi_l^e(1)\phi_{Q-l}^h(1) >, \tag{15.43}$$

with

$$g_l(1, 1) = \frac{C}{E_l^e(1) + E_{Q-l}^h(1) - E}, \tag{15.44}$$

C being a normalization constant.

Solutions of Equation (15.42) on the real axis ($\nu \in \mathbb{R}$) represent extended states, where the probability of finding the electron and the hole together at the same site is small. Also, the unique solution on the imaginary axis, if it exists, represents the bound state, where there is a high probability of finding the electron and the hole at the same site. Also, bound state solutions exist if

$$\gamma > \frac{2|q|}{M}(1 - \cos M\theta). \tag{15.45}$$

These solutions correspond to the bound electron–hole pair, the exciton. In Figure 15.14 we show the typical band structure, with an isolated band below the continuum band corresponding to the bound

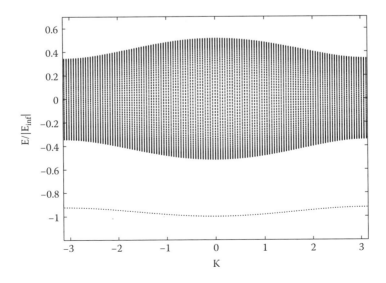

FIGURE 15.14 Band structure of the eigenvalues E (in units of $|E_{inf}|$) as function of the total wave vector K. $M = 125$, $N = 1$, $\gamma = 4$, $\varphi_1 = 0$, $d_1 = 1$, $\epsilon = 1$, and $\mu = 0.2$.

states. Due to the rotational invariance of the system, there is the same probability of finding the two fermions at the same place.

In the limit $M \to \infty$ and d constant, the energy of the exciton reduces to $E_\infty = -\sqrt{\gamma^2 + 4|q|^2}$, which is independent of the value of the magnetic field. This is the binding energy of an exciton in a straight wire [86]. Also, the anharmonic parameter γ must scale as $1/d$ to assure the existence of bound states, as occurs in systems with a point–like δ-function interaction [106].

If M is large, it is possible to derive an explicit approximate expression for the energy of a bound state, similar to the perturbative exciton energy obtained for a large ring radius in the continuum model [86]. This expression is given by

$$E = E_\infty \left[1 + \frac{2\gamma^2}{E_\infty^2} \cos(M\theta) \, e^{-\text{Marcsinh}(\frac{\gamma}{2|q|})} \right]. \tag{15.46}$$

Also, if d is small enough, condition (15.45) can be approximated by $2\pi r\beta > 1 - \cos(2\pi \Phi / \Phi_0)$, where $\beta = \text{arcsinh}(\gamma/2|q|)/d$ is the inverse decay length of energy oscillations given by Equation (15.46).

In general, the energy of an eigenstate depends on the magnetic flux. If we analyze the dependence of the energy of the ground state with the magnetic flux, despite the high probability of finding the electron and the hole in a bound state, where the electric charge is null, it is possible to observe the AB effect, as shown in Figure 15.15. The maximum of the amplitude corresponds to the integer or half-integer values of Φ / Φ_0 and this amplitude decreases rapidly with increasing ring radius. On the other hand, the energies corresponding to nonbound states, where the probability of finding the electron and the hole in different sites of the ring increases, are more sensitive to the magnetic flux, and the AB oscillations do not decrease as the radius of the ring increases.

In general, in a bound state, the amplitude of the AB oscillations decreases exponentially with the length of the ring, as shown in Figure 15.16 (top). Also, the characteristic decay length is a decreasing function of the interaction parameter γ, as shown in Figure 15.16 (bottom).

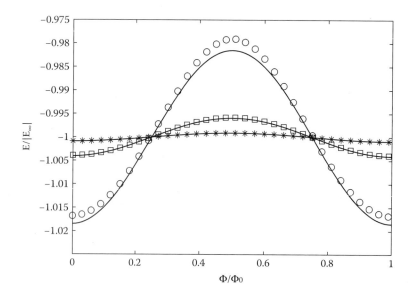

FIGURE 15.15 The Aharonov–Bohm oscillations of exciton energy (in units of $|E_{inf}|$) corresponding to three values of the (single) ring circumference ρ (in units of d) obtained by solving Equation (15.42) numerically for $\gamma = 2, d = 1$, $\epsilon = 1$, $\mu = 0.2$, and $k = 0$. (○) $M = 5$ ($\rho = 5.34$), (□) $M = 7$ ($\rho = 7.24$), and (*) $M = 9$ ($\rho = 9.19$). The lines denote the approximate solutions given by Equation (15.46).

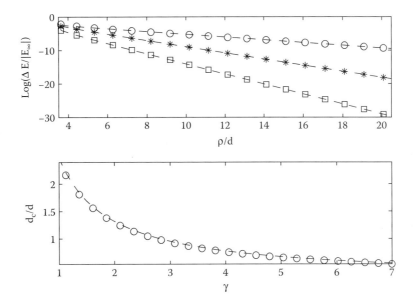

FIGURE 15.16 (Top) Energy oscillations amplitude of AB oscillations (in units of $|E_\infty|$ and on a semi-logarithm scale) as a function of the circumference of the (single) ring ρ (in units of circumferential lattice spacing d), obtained by solving Equation (15.42) numerically for $\gamma = 1$, corresponding to a weakly bound state ($\circ$), $\gamma = 2.5$, a typical bound state ($*$), and $\gamma = 5$, corresponding to a strongly bound state ($\square$). $\epsilon = 1$, $\mu = 0.2$, and $k = 0$. The dashed lines are approximate solutions given by Equation (15.46). (Bottom) Decay length d_c of AB oscillations (in units of the circumferential lattice spacing d) for a single ring as a function of the anharmonic parameter γ. $\epsilon = 1$, $\mu = 0.2$, and $k = 0$, obtained by the numerical solution of Equation (15.42) ($*$). The dashed line is the approximate solution given by Equation (15.46).

15.4.5 General Case

In the general case of M sites and N rings, we observe isolated bands corresponding to bound states of electron and holes around each ring if the interaction parameter γ is high enough, as shown in Figure 15.17.

For strong attraction, and due to the difference between hopping terms in each ring, the ground state is mainly a bound state localized around the inner ring. However, if the distance between rings is small, there exists a significant probability to find the exciton in the others rings. Thus, the variation of the energy with the magnetic flux is a superposition of different oscillations, the main oscillation with period Φ_1/Φ_0 and minor contributions of the other oscillations with periods Φ_n/Φ_0 ($n = 2, \ldots, N$).

Also, there exist contributions corresponding to states where the electron and the hole are in contiguous sites but on different rings. These oscillations persist when the circumference of the ring increases, the main period of this oscillation is given by the phase difference between the magnetic field contribution in each ring. Essentially, its period is $T = 2M/(\Phi_2/\Phi_1 - 1) = 2M/(r_2/r_1 - 1) = 2Mr_1/\Delta r$, corresponding to the phase difference when $\gamma = 0$. These oscillations will be dominant if the distance between rings is small (large values of the inter-ring hopping coefficients), and decreases with the circumference of the ring. A model in which the hole is located more towards the center of the ring has been put forward [107] in the context of the self-assembled nanorings [91].

If distance between the rings is small, the ground state of the system is mainly a bound state localized around the inner ring, and presents AB oscillations with period Φ_1/Φ_0. The first excited states correspond to bound states localized around each ring and oscillate with a period Φ_n/Φ_0 ($n = 2, \ldots, N$). If the distance between rings decreases, the inter-ring hopping coefficient increases and additional oscillations appear due to the contribution of nonlocalized components in different rings.

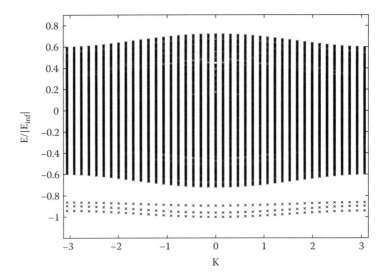

FIGURE 15.17 Band structure of the eigenvalues E (in units of $|E_{inf}|$) as function of the total wave vector K. $M = 41$, $N = 3$, $\gamma = 4$, $\varphi_1 = 0$, $d_1 = 1$, $\epsilon = 1$, $\mu = 0.2$, and $\Delta r = 1$.

In general, the amplitude of the oscillations of the energy depends weakly on the number of rings, as shown in Figure 15.18. This dependence becomes negligible when the number of rings is more than a certain number, which is unity if distance between rings is large enough. Thus, except in the cases where distance between rings is very small, the AB oscillations are essentially a one-dimensional phenomenon.

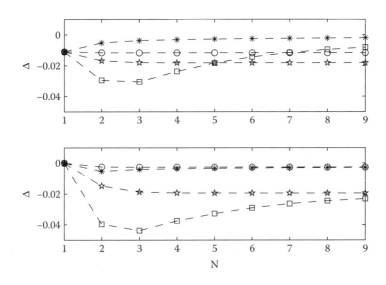

FIGURE 15.18 Oscillations of the energy of the ground state as a function of the number of rings N. $\Delta = (E_{max} - E_{min})/(2E)$, where E is the exciton energy for $B = 0$. $f = 5$, corresponding to $\rho_1 = 5.34$ (top) and $f = 11$, corresponding to $\rho_1 = 11.15$ (bottom). (*) $\Delta r = 0.1$, ($\square$) $\Delta r = 0.7$, ($\star$) $\Delta r = 1.0$, and ($\circ$) $\Delta r = 1.5$. $\gamma = 2.5$, $d_1 = 1$, $\epsilon = 1$, and $\mu = 0.2$.

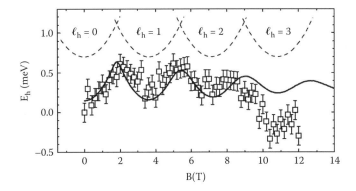

FIGURE 15.19 Hole energy dependence on the magnetic field (open squares), showing the Aharonov–Bohm oscillations with period ϕ_0. The error bars are the uncertainties on the peak position after the Gaussian fits. The dashed curves are parabolas following a theoretical model. (Reprinted with permission from E. Ribeiro, A.O. Govorov, W. Carvalho, Jr. and G. Medeiros-Ribeiro, *Phys. Rev. Lett.*, 92, 126402, 2004. Copyright (2004) by the American Physical Society.)

15.4.6 A System of Stacked Rings

In a system of stacked rings, as shown in Figure 15.13, where they are all threaded by the same magnetic flux Φ_0, an analogous behaviour to the single ring case is observed. Only oscillations of period $T = \Phi/\Phi_0$ are observed.

Analyzing the decay of the amplitude oscillations with the size of the ring, we rederive an approximate exponential decay, and it is possible to calculate the decay length d_c. If Δr is large, the behavior is similar to the one-ring case, but if this distance decreases, the decay length increases. This suggests that the exciton uses the additional rings as further channels along which to tunnel when prompted by the increasing magnetic flux. This effect indicates a route for enhancing the possibility of experimental detection of the AB effect for a neutral exciton.

15.4.7 Experimental Results

As we have mentioned before, AB oscillations have been detected in Type-II quantum dot ensembles for a neutral exciton [95]. In Figure 15.19, experimental AB oscillations, with period ϕ_0, for a neutral exciton are shown.

15.5 Acknowledgments

We are most grateful to our colleague Oliver Penrose for valuable suggestions and discussions during the stages of our research concerning the existence and properties of AB oscillations in a finite-width ring. F. Palmero thanks Heriot-Watt University for their hospitality, and the HPC-Europa program, funded under the European Commission's Research Infrastructures activity of the Structuring the European Research Area program, contract number RII3-CT-2003-506079, for financial support. This work has been supported by the MECD–FEDER project FIS2004–01183.

References

[1] Flach, S. and Willis, C.R., Discrete breathers, *Phys. Rep.*, 295, 181, 1998; Physica D, 119, special volume, Flach, S. and Mackay, R.S., Eds., 1999: Kevrekidis, P.G., et al., The discrete nonlinear Schrödinger equation, *Int. J. Mod. Phys. B*, 15, 2833, 2001; focus issued, Kivshar, Y.S. and Flach, S., Eds., *Chaos*, 13, 586, 2003; *Localization and Energy Transfer in Nonlinear Systems*, Vázquez, L., et al., Eds. World Scientific, Singapore, 2003.

[2] Scott, A.C., et al., Quantum lattice solitons, *Physica D*, 78, 194, 1994.

[3] Fleurov, V., Discrete quantum breathers: What do we know about them?, *Chaos*, 13, 676, 2003; MacKay, R.S., Discrete breathers: classical and quantum, *Physica A*, 288, 174, 2000.

[4] Fillaux, F. et al., Inelastic-neutron-scattering study of the sine-Gordon breather interactions in isotopic mixtures of 4-methyl-pyridine, *Phys. Rev. B*, 58, 11416, 1998; Swanson, B.I. et al., Observation of intrinsically localized modes in a discrete low-dimensional material, *Phys. Rev. Lett.*, 82, 3288, 1999; Asano, T. et al., ESR investigation on the breather mode and the Spinon-Breather dynamical crossover in Cu Benzoate, *Phys. Rev. Lett.*, 84, 5880, 2000; Schulman, L.S. et. al., Slow relaxation, confinement, and solitons, *Phys. Rev. Lett.*, 8, 224101, 2002.

[5] Hamm, P. and Edler, J., Quantum vibrational polarons: Crystalline acetanilide revisited, *Phys. Rev. B*, 73, 094302, 2006.

[6] Seeman, N.C., Nanotechnology and the double helix. *Scientific American,* June, 2004.

[7] Seeman, N.C., Nucleic–acid junctions and lattices. *J. Theor. Biol.*, 99, 237, 1982; Seeman, N.C. and Lukeman, P.S., Nucleic acid nanostructures: Bottom–up control of geometry on the nanoscale, *Rep. Prog. Phys.*, 68, 237, 2005; Chworos, A. et al., Building programmable jugsaw puzzles with RNA, *Science*, 306, 2068, 2004.; Park, S.H. et al. Finite-Size, fully-addressable DNA tile lattices formed by hierarchical assembly procedures, *Angew. Chem.*, 118, 749, 2006.

[8] Chen, J. and Seeman, N.C., The synthesis from DNA of a molecule with the connectivity of a cube. *Nature*, 350, 631, 1991; Zhang, Y. and Seeman, N.C., The construction of a DNA truncated octaedron, *J. Am. Chem. Soc.*, 116, 1661, 1994.

[9] Rothemund, P.W.K., Folding DNA to create nanoscale shapes and patterns, *Nature*, 440, 297, 2006.

[10] Adleman, L., Molecular computation of solutions to combinatorail problems, *Science*, 266, 1021, 1994.

[11] Benenson et al., An autonomous molecular computer for logical control of gene expression, *Nature*, 429, 423, 2004.

[12] Okamoto, A. et al., DNA logic gates, *J. Am. Chem. Soc.*, 126, 9458, 2004.

[13] Chen, X. et al., Construction of molecular logic gates with a DNA-cleaving deoxyribozyme, *Angew. Chem. Int. Ed.*, 45, 1759, 2006.

[14] Amos, M., *Theoretical and Experimental DNA Computation*, Springer, 2005.

[15] Ezziane, Z., DNA computing: applications and challenges, *Nanotechnology*, 17, R27, 2006.

[16] Chen, J. et al., Large-scale genomic monitoring of profiling using a DNA-based memory and microarrays, *24th Army Science Conf. Orlando FL*, 2004.

[17] Clelland Taylor, C. et al., Hiding messages in DNA microdots, *Nature*, 399, 533, 1999.

[18] Wong, P.C. et al., Organic data memory, *Commun. ACM*, 46, 95, 2003.

[19] Mao, C. et al., Logical computation using algorithmic self-assembly of DNA triple-crossover molecules, *Nature*, 407, 493, 2000.

[20] Ratner, M., Photochemistry: Electronic motion in DNA, *Nature*, 397, 480, 1999.

[21] Fink, H.W. and Schönenberger, C., Electrical conduction through DNA molecules, *Nature*, 398, 407, 1999.

[22] Tran, P. et al., Charge transport along the λ-DNA double helix, *Phys. Rev. Lett.*, 85, 1564, 2000.

[23] Bhalla, V. et al., DNA electronics, *Eur. Mol. Biol. Rep.*, 4, 442, 2003.

[24] Braun, E. et al., DNA templated assembly and electrode attachment of conducting silver wire, *Nature*, 391, 775, 1998.

[25] Porath, D. et al., Direct measurements of electrical transport through DNA molecules, *Nature*, 403, 635, 2000.

[26] Endres, R.G. et al., The quest for high-conductance DNA, *Rev. Mod. Phys.*, 76, 195, 2004.

[27] Porath, D. et al., Charge transport in DNA-based devices, *Top. Curr. Chem.*, 237, 183, 2004.

[28] Lee, H.Y. et al., Control of electrical conduction in DNA using oxygen hole doping, *Appl. Phys. Lett.*, 80, 1670, 2002.

[29] Boiteux, S. et al., *Free Radical Biol. Med.*, 32, 1244, 2002.

[30] Burrows, C.J. and Muller, J.G., Oxidative nucleobase modifications leading to strand scission, *Chem. Rev.*, 98, 1109, 1998.

[31] Friedman, K.A. and Heller, A., On the non-uniform distribution of guanine in introns of human genes: Possible protection of exons against oxidation by proximal intron poly-G sequences, *J. Phys. Chem. B*, 105, 11859, 2001.

[32] Núñez, M.E. et al., Dendritic cells transduced with HIV nef express normal levels of HLA-A and HLA-B class I molecules, *Biochem.*, 40, 12465, 2001.

[33] Boon, E.M., Electrochemical sensors based on DNA–mediated charge transport, PhD thesis, California Institute of Technology, Pasadena, 1998.

[34] Klotsa, D. et al., Electronic transport in DNA, *Biophys. J.*, 89, 2187, 2005; Wang, H. et al., Electronic transport and localization in short and long DNA, in *Biopolymers*, Starikov, J. Ed., Elsevier, Amsterdam, 2006 (to appear).

[35] Eley, D.D. and Spivey, D.I., Semiconductivity of organic substances, Part 9, Nucleic acid in the dry state, *Trans. Faraday Soc.*, 58, 411, 1962.

[36] Berlin, Y.B. et al., DNA as a molecular wire, *Superlattices Microstruct*, 28, 241, 2000.

[37] Jortner, J. et al., Charge transfer and transport in DNA, *Proc. Nat. Acad. Sci. USA*, 95, 12759, 1998.

[38] Bruinsma, R. et al., Fluctuation-facilitated charge migration along DNA, *Phys. Rev. Lett.*, 85, 4393, 2000.

[39] Yu, Z.G. and Song, X. Variable range hopping and electrical conductivity along the DNA double helix, *Phys. Rev. Lett.*, 86, 6018, 2001.

[40] Hermon, Z. et al., Prediction of charge and dipole solitons in DNA molecules based on the behaviour of phosphate bridges as tunnel elements, *Europhys. Lett.*, 43, 482, 1998.

[41] Conwell, E. and Rakhmanova, S.V., Polarons in DNA, *Proc. Nat. Acad. Sci.*, 97, 4556, 2000; Rakhmanova, S.V. and Conwell, E., Polaron motion in DNA, *J. Phys. Chem. B*, 105, 2056, 2001; Ly, D. et al., Mechanism of charge transport in DNA: Internally linked anthraquinone conjugates support phonon-assisted polaron hopping, *J. Am. Chem. Soc.*, 121, 9400, 1999; Komineas, S. et al., Effects of intrinsic base-pair fluctuations on charge transport in DNA, *Phys. Rev. E*, 65, 061905, 2002.

[42] Hennig, D. and Archilla, J.F.R., *Recent Developments in Biopolymers*, Elsevier, Amsterdam, 2006 (to appear).

[43] Barbi, M. Localized Solutions in a Model of DNA Helicoidal Structure, PhD thesis, Universitát degli Studi di Firenze, Italy, 1998.

[44] Barbi, M. et al., Helicoidal model for DNA opening, *Phys. Lett. A*, 253, 358, 1999.

[45] Barbi, M. et al., A twist opening model for DNA, *J. Biol. Phys.*, 24, 97, 1999.

[46] Cocco, S. and Monasson, R., Theoretical study of collective modes in DNA at ambient temperature, *J. Chem. Phys.*, 112, 10017, 2000.

[47] Salerno, M., Discrete model for DNA-promoter dynamics, *Phys. Rev. A*, 44, 5292, 1991.

[48] Brunaud G. et al., An effective Hamiltonian for hole transfer along B-DNA double strands, *Phys. Chem. Chem. Phys.*, 4, 6072, 2002.

[49] Kalosakas, G. et al., Polaron solutions and normal-mode analysis in the semiclassical Holstein model, *Phys. Rev. B*, 58, 3094, 1998.

[50] Hennig, D. et al., Nonlinear charge transport mechanism in periodic and disordered DNA, *Physica D*, 180, 256, 2003.

[51] Archilla, J.F.R. et al., in *Localization and Energy Transfer in Nonlinear Systems*, Vázquez, L. et al., Eds., World Scientific, 2003.

[52] Stryer, L., *Biochemistry*, Freeman, New York, 1995.

[53] Cuniberti G. et al., Backbone-induced semiconducting behavior in short DNA wires, *Phys. Rev. B*, 65, 241314, 2002.

[54] D. Hennig, E.B. et al., Charge transport in poly(dG)-poly(dC) and poly(dA)-poly(dT) DNA polymers, *J. Biol. Phys.*, 30, 227, 2004.

[55] Voulgarakis, N.K. and Tsironis, G.P., Stationary and dynamical properties of polarons in the anharmonic Holstein model, *Phys. Rev. B*, 63, 14302, 2001.

[56] Ibañes, M. et al., Dynamical properties of discrete breathers in curved chains with first and second neighbor interactions, *Phys. Rev. E*, 65, 041902, 2002.

[57] Chen, D. et al., Breather mobility in discrete φ 4 lattices, *Phys. Rev. Lett.*, 112, 4776, 1996.

[58] Cuevas, J. et al., Moving discrete breathers in a Klein–Gordon chain with an impurity, *J. Phys. A: Math. Gen.*, 35, 10519, 2002.

[59] Palmero, F. et al., Effect of base-pair inhomogeneities on charge transport along the DNA molecule, medited by twist and radial polarons, *New J. Phys.*, 6, 13, 2004.

[60] Giese, B. et al., Direct observation of hole transfer through DNA by hopping between adenine bases and by tunnelling, *Nature*, 412, 318, 2001.

[61] Cai, L. et al., Self-assembled DNA networks and their electrical conductivity, *Appl. Phys. Lett*, 77, 3105, 2000; Yoo, K.H. et al., Electrical conduction through poly(dA)-poly(dT) and poly(dG)-poly(dC) DNA molecules, *Phys. Rev. Lett*, 87, 198102, 2001; Lee, H.Y. et al., Control of electrical conduction in DNA using oxygen hole doping, *Appl. Phys. Lett*, 80, 1670, 2002.

[62] Holstein, T., Studies of polaron motion, *Ann. Phys. (N.Y.)*, 8, 325, 1959.

[63] Aubry, S. et al., Chaotic polaronic and bipolaronic states in the adiabatic Holstein model, *J. Stat. Phys*, 67, 675, 1992.

[64] Davydov, A.S., The theory of contraction of proteins under their excitation, *J. Theor. Bio.*, 38, 559, 1973; Davydov, A.S., Solitons and energy transfer along protein molecules, *J. Theor. Bio.*, 66, 379, 1977.

[65] Scott, A.C., Davydov's soliton, *Phys. Rep*, 217, 1, 1992.

[66] Fuentes, M.A. et al., Multipeaked polarons in soft potentials, *Phys. Rev. E*, 70, 025601(R), 2004.

[67] Maniadis, P. et al., Polaron normal modes in the Peyrard-Bishop-Holstein model, *Phys.Rev. B*, 68, 174304, 2003.

[68] Aubry, S., Breathers in nonlinear lattices: Existence, linear stability and quantization, *Physica D*, 103, 201, 1997.

[69] Cuevas, J. et al., ArXiv:cond-mat/0512695.

[70] Dauxois, T. et al., Dynamics and thermodynamics of a nonlinear model for DNA denaturation, *Phys. Rev. E*, 47, 684, 1993.

[71] Eilbeck, J.C. et al., Soliton structure in crystalline acetanilide, *Phys. Rev. B*, 30, 4703, 1984.

[72] MacKay, R.S. and Aubry, S., Proof of existence of breathers for time-reversible or Hamiltonian networks of weakly coupled oscillators, *Nonlinearity*, 7, 1623, 1994.

[73] Marín, J.L. and Aubry, S., Breathers in nonlinear lattices: Numerical calculation from the anticontinuous limit, *Nonlinearity*, 9, 1501, 1996.

[74] Cuevas, J. et al., Moving breathers in a DNA model with competing short and long range dispersive interactions, *Physica D*, 163, 106, 2002.

[75] Aharonov Y. and Bohm, D., Significance of electromagnetic potentials in the quantum theory, *Phys. Rev.*, 115, 485, 1959.

[76] Berry, M.V., Quantal phase factor accompanying adiabatic interaction, *Proc. R. Soc. London, Ser. A*, 392, 45, 1984.

[77] Byers, N. and Yang, C.N., Theoretical considerations concerning quantized magnetic flux in superconducting cylinders, *Phys. Rev. Lett.*, 7, 46, 1961.

[78] Bloch, F., Simple interpretation of the Josephson effect, *Phys. Rev. Lett.*, 21, 1241, 1968.

[79] Büttiker, M. et al., Josephson behavior in small normal one-dimensional rings, *Phys. Lett. A*, 96, 365, 1983.

[80] Peshkin M. and Tonomura, A., The Aharonov-Bohm Effect, in *Lecture Notes in Physics*, Vol. 340. Springer Verlag, Berlin, 1989.

[81] Washburn S. and Webb, R.A., Quantum transport in small disordered samples from the diffusive to the ballistic regime, *Rep. Prog. Phys.*, 55, 1311, 1992.

[82] Parks, R.D. and Little, W.A., Fluxoid quantization in a multiply-connected superconductor, *Phys. Rev.*, 133, A97, 1964.

[83] Hod, O. et al., Feasible nanometric magnetoresistance devices, *J. Phys. Chem. B*, 108, 14807, 2004.

[84] Shen, J.Q. and He, S., Geometric phases of electrons due to spin-rotation coupling in rotating C_{60} molecules, *Phys. Rev. B*, 68, 195421, 2003; Shen, J.Q. et al., Aharonov-Carmi effect and energy shift of valence electrons in rotating C_{60} molecules, *Eur. Phys. J. D*, 33, 35, 2005.

[85] Chaplik, A., Magnetoexcitons in quantum rings and in antidots, *Pis'ma Zh. Eksp. Teor. Fiz.*, 62, 885, 1995 [*JETP Lett.* 62, 900, 1995].

[86] Römer, R.A. and Raikh, M.E., Aharonov-Bohm effect for an exciton, *Phys. Rev. B*, 62, 7045, 2000.

[87] Ando, T., Excitons in carbon nanotubes revisited: Dependence on diameter, Aharonov-Bohm flux, and strain, *J. Phys. Soc. Jpn.*, 73, 3351, 2004.

[88] Palmero, F. et al., Aharonov-Bohm effect for an exciton in a finite width nano-ring, *Phys. Rev. B*, 72, 075343, 2005.

[89] Lorke, A. et al., Spectroscopy of nanoscopic semiconductor rings, *Phys. Rev. Lett.*, 84, 2223 2000.

[90] Emperador, A. et al., Far-infrared spectroscopy of nanoscopic InAs rings, *Phys. Rev. B*, 62, 4573, 2000.

[91] Lorke, A. et al., Electronic structure of nanometer-size quantum dots and quantum rings, *Microelectronic Eng.*, 47, 95, 1999.

[92] Bayer, M. et al., Hidden symmetries in the energy levels of excitonic 'artificial atoms', *Nature*, 405, 923, 2000.

[93] Bayer, M. et al., Optical detection of the Aharonov-Bohm effect on a charged particle in a nanoscale quantum ring, *Phys. Rev. Lett.*, 90, 186801, 2003.

[94] Cho, K.S. et al., Designing PbSe nanowires and nanorings through oriented attachment of nanoparticles, *J. Am. Chem. Soc.*, 127, 7140, 2005.

[95] Ribeiro, E. et al., Aharonov-Bohm signature for neutral polarized excitons in type-II quantum dot ensembles, *Phys. Rev. Lett.*, 92, 126402, 2004.

[96] K. Maschke et al., Coherent dynamics of magnetoexcitons in semiconductor nanorings, *Eur. Phys. J. B*, 19, 599, 2001.

[97] Hu, H. et al., Aharonov-Bohm effect of excitons in nanorings, *Phys. Rev. B*, 63, 195307, 2001.

[98] Lieb, E.H. and Wu, F.Y., Absence of mott transition in an exact solution of the short-range, one-band model in one dimension, *Phys. Rev. Lett.*, 20, 1445, 1968.

[99] B. S. Shastry and B. Sutherland, *Phys. Rev. Lett.*, 65, 243, 1990.

[100] Sutherland, B. and Shastry, B.S., Twisted boundary conditions and effective mass in Heisenberg-Ising and Hubbard rings, *Phys. Rev. Lett.*, 65, 1833, 1990.

[101] Essler, F. et al., *The One-Dimensional Hubbard Model*, Cambridge University Press, Cambridge, 2005.

[102] Gebhard F., and Ruckenstein, A.E., Exact results for a Hubbard chain with long-range hopping, *Phys. Rev. Lett.*, 68, 244, 1992.

[103] Caffarel M. and Mosseri R., Hubbard model on d-dimensional hypercubes: Exact solution for the two-electron case, *Phys. Rev. B*, 57, R12651, 1998.

[104] Scott, A.C. *Nonlinear Science*, 2nd ed., OUP, Oxford, 2003.

[105] Penrose, O., Private communication.

[106] Göhmann, F. and Korepin, V.E., Universal correlations of one-dimensional interacting electrons in the gas phase, *Phys. Lett. A*, 260, 516, 1999.

[107] Govorov, A.O. et al., Excitons in nanorings and the optical Aharonov-Bohm effect, *Phys. Rev. B*, 66, 081309R, 2002.

16

Protein-Based Optical Memories

Jeffrey A. Stuart

Robert R. Birge

Mark P. Krebs

Bangwei Xi

William Tetley

Duane L. Marcy

Jeremy F. Koscielecki

Jason R. Hillebrecht

16.1   Introduction

Nanotechnology has been quick to embrace the realm of biology, and is redefining our understanding of how biology functions at the cellular level and below. The techniques developed for the manipulation of molecular systems are now being applied to the problem of examining biological molecules, thereby facilitating an understanding of fundamental biological processes at the molecular scale. The central goal of bionanotechnology is the need to understand how biology works at the molecular level, and how to probe—and manipulate—biological events at this scale.

The field of biomolecular electronics can be considered a subdiscipline of bionanotechnology. Broadly speaking, this field is defined as the evaluation and utilization of molecules of biological origin for artificial applications. Numerous examples of such novel technologies are currently under development, including architectures for chemical and biological sensors, protein-based optical memories and holographic media, the integration of biological molecules into microelectronic devices, and DNA-based computational techniques. The natural question that emerges when considering such technologies concerns the rationale behind turning to molecules of biological origin for device architectures—what advantages are to be gained?

The answer to this question is found in the vast array of responses and behaviors that is possible—the process of evolution has allowed nature to solve a tremendous number of problems living organisms have faced throughout the history of life on earth. Biological molecules are unique in that they have benefited from eon upon eon of natural selection and evolution. The result is an expansive set of highly optimized properties well suited to specific biological functions—the toolbox offered by biological molecules performing their natural functions far exceeds what man has achieved through synthetic means. Nature has found unique solutions to structure and support (bone, connective tissue), metabolism (anabolism and catabolism, oxidative phosphorylation, photosynthesis), and information processing (senses, cognition, memory, chromosomes). And these abilities are facilitated by the complex molecules that define life, specifically enzymes, proteins, and DNA/RNA. Adaptation to a wide range of environmental conditions has resulted in functional biomolecules with efficiencies often unmatched by nonbiological or synthetic methods. Furthermore, many of the functions facilitated by biological molecules cannot be duplicated by other means. It stands to reason that if nature has succeeded in creating highly complex, diverse, and efficient molecules capable of a wide range of functions, the researcher should be able to exploit preexisting natural functions, or modify them toward a given application.

The goal of biomolecular electronics, therefore, is to utilize the unique set of properties characterized by molecules that originate from life, to enhance manmade technologies. Intelligent selection by the researcher can bring their novel capabilities to device applications. Often, these properties offer comparative advantages over more traditional approaches to device design and construction. In fact, proteins and other biomolecules pose a unique solution to the problem of size versus function in that they represent a class of extremely sophisticated molecules that perform specific functions under what can be a fairly broad range of conditions. Sensors employing DNA or RNA can detect pathogenic organisms much more quickly and specifically than conventional laboratory techniques. Enzymes can catalyze chemical reactions that might be of industrial interest, or can be worked into detection schemes (e.g., utilization of glucose oxidase in a sensor might aid in monitoring glucose levels in a diabetic patient). And, as will be the focus of this chapter, some proteins that respond to light can be used as the basis of device architectures, such as optical computer memories and holographic recording media.

Key to the utilization of biological molecules in many device architectures is the ability to interrogate them; the ability to determine the state of a molecule presents one of the largest challenges to biomolecular electronics today. For this reason, photoactive proteins are of special interest. All photoactive proteins have two things in common: the ability to respond to light in a reproducible manner, and the presence of an internal chromophore that mediates light absorption. The chromophore acts as a convenient and constant probe of protein structure and function, and is often a highly conjugated molecule that reacts to light either by isomerization around a double bond (the rhodopsins, photoactive yellow protein [PYP], phytochrome) or redox activity that results in electron transfer (photosynthetic reaction centers, flavoproteins). In any case, the internal chromophore facilitates light to chemical energy transduction, and is a most convenient tool for interrogation, whereby light or voltage can be used to monitor the state of the protein. Photochromic proteins are well suited for device architectures from a number of standpoints. A common property among many proteins in this group is the presence of multiple states accessed upon exposure to light, which often can be interrogated by both optical and electronic means, thereby providing a direct reading of functional activity. Furthermore, responses from many photoactive proteins can be elicited by exposure to either light or voltage, and are reproducible. The remainder of this chapter will focus on bacteriorhodopsin, a photoactive protein with potential in optical and holographic computer memory architectures, and efforts to modify the protein for this specific goal. As will be discussed next, the ability to modify bacteriorhodopsin (and proteins in general) has proven to be paramount to this protein's usefulness to the applied sciences.

A question central to the field of biomolecular electronics deals with efficiency. Although the processes of natural selection and evolution have produced biological molecules highly optimized for their natural function, it remains to be seen whether these molecules can be utilized toward non-natural applications—at least in their native form. Indeed, optimizing protein function has been in practice for over a decade; enzymes of interest to industrial processes or commercial product lines (e.g., laundry detergents) have been optimized with respect to thermal stability, chemical stability, and substrate specificity [1–5]. Generally speaking, libraries of randomly generated mutant proteins are screened for properties of interest. The organisms responsible for generating proteins with favorably enhanced properties are used as the parent to the next generation of random progeny. This process is accomplished in an iterative cycle until a highly optimized class of proteins is produced, specific to a given application. However, in examples of this sort, the optimization is of a preexisting property, toward greater efficiency in non-native conditions. The larger question is whether the researcher can enhance non-native functions, specifically for device applications. A protein incorporated into a device is likely to experience a very different environment than it would under normal circumstances, including nonaqueous conditions and large temperature fluctuations. And as indicated earlier, the protein may be required to perform a non-natural function. The potential of employing proteins as the active elements of device architectures is made possible through the significant advances made in molecular biology and genetic/protein engineering over the past decade. Two such advances are the techniques of random mutagenesis and directed evolution.

16.2 Bacteriorhodopsin Biochemistry and Biophysics

Bacteriorhodopsin (BR) occupies a unique place in the history of protein-based applications—it is not only one of the more thoroughly examined proteins with respect to structure and function, it has also been evaluated for a number of device architectures. Early examples included use of the protein as the active element in reusable holographic materials [6–16]. Eventually, architectures were devised for random access memories, nonvolatile computer memories, photodiodes, and applications as active elements in microelectronic circuits [9,17–34]. The consistent level of interest in bacteriorhodopsin as a photonic material stems from its stability, availability, and, finally, the way it interacts with light.

Bacteriorhodopsin (Figure 16.1) is the primary integral membrane protein produced by the halophilic archaea, *Halobacterium salinarum* (syn. *Halobacterium halobium, salinarium*). This organism thrives in salt marshes where the temperature can exceed 60°C, and the concentration of sodium chloride can reach 4 M, roughly six times that of seawater. BR has 248 amino acids with a molecular weight of 26 kD, and shares a common structural motif with G-protein coupled receptors (GPCRs), consisting of a seven transmembrane alpha helical bundle. The similarities may be serendipitous, however, as BR functions as a photosynthetic protein, and GPCRs play an important role in intercellular communication. An all-*trans* retinal cofactor is coupled with the BR apoprotein, bound to Lys-216 *via* a protonated Schiff base bond. The retinal prosthetic group mediates visible light absorption and is responsible for the protein's characteristic deep purple color. Furthermore, the retinal chromophore enables light to chemical energy transduction by the protein.

The regions of the cell membrane that contain bacteriorhodopsin have unique properties, and occur as discrete membrane patches consisting of approximately 75% BR and 25% lipid. Typically referred to as the purple membrane, these patches are stable in aqueous suspension, and are characterized by a very high level of organization. As revealed by high-resolution AFM images, the protein monomers are arranged in a hexagonal lattice of trimers, and the resulting patches are semicrystalline [35,36]. The retinal chromophores mirror the trimer arrangement, oriented with an angular separation between monomers of 60°. This arrangement ensures that the purple membrane captures all polarizations of light with equal efficiency. Once removed from the purple membrane, BR's stability drops from years to days. For this reason, researchers seldom utilize the solubilized protein—the ease of isolation, ability to process into different forms, and the rugged stability of purple membrane patches make them nearly ideal for use in device applications.

16.2.1 The Bacteriorhodopsin Photocycle

In its native organism, BR acts as a light-driven proton pump; proton translocation across the cell membrane is facilitated by the protein, and the resulting electrochemical gradient (ΔpH ≈ 1) is used by the cell to do work. The proton gradient fuels ATP synthesis through a standard F_0F_1 ATPase, and served as one of the first examples of chemiosmosis [37–39]. As mentioned earlier, the retinal chromophore mediates

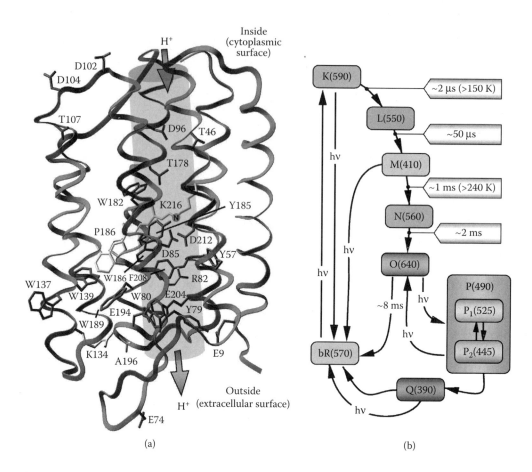

FIGURE 16.1   (see color insert following page **5**-6) (a) Schematic of bacteriorhodopsin illustrating certain key amino acids and the purported path of the proton pump. The all-*trans* retinal chromophore traverses the binding pocket roughly perpendicular to the membrane normal and the alpha helices. (b) The bacteriorhodopsin photocycle, including the branched photocycle originating at the O-state. Absorption maxima in nm are shown in parentheses for each intermediate.

light to chemical energy transduction by the protein. A light-induced isomerization of the chromophore initiates a series of events in the protein that ultimately results in the translocation of a proton across the cell membrane. This process is described by the protein's photocycle, a series of spectrally distinct intermediates (Figure 16.2) that characterizes the step-wise process of passing a proton across the membrane. The intermediates are produced sequentially, and are denoted as the K, L, M, N, and O states (see Figure 16.1). A branch off of the O-state accesses the P and Q states, which will be discussed later. Historically, the observable intermediates were defined by thermal trapping [40]. Each intermediate is characterized by a combination of factors, including the wavelength of maximum absorption, lifetime, chromophore configuration (isomerization state and overall conformation), the electronic environment of the binding site, the protonation state of the retinal Schiff base, and both the protonation state and orientation of key amino acids in the proton channel (reflecting the progress of a proton across the membrane). Light-induced chromophore isomerization from all-*trans* to 13-*cis*, often referred to as the primary event, results in a shift in electron density toward the protonated Schiff base during the formation of the K-state, the only photochemically generated state in the native photocycle. The bR→K transition proceeds with an impressive quantum efficiency of ∼ 65%. The redistribution of charge changes the electrostatic nature of the binding site and provides the motive force that drives the formation of subsequent thermal

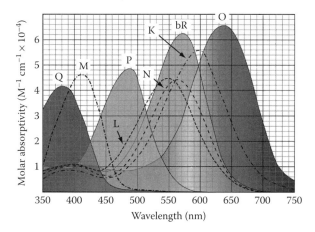

FIGURE 16.2 (see color insert) Absorption spectra of the various intermediate states associated with the bacteriorhodopsin photocycle. The bR, M, O, and Q states are of particular interest to BR-based memory operations.

intermediates (i.e., L, M, N, and O), and, ultimately, proton translocation. A branch off of the O-state has two additional intermediates, P and Q. The blue-shifted M and Q states are of the most interest to optical device applications. On average, the protein can go through this cycle $> 10^6$ times (a benchmark known as the cyclicity). The proton-pump mechanism is well described in the literature [41–52].

16.3 Bacteriorhodopsin as a Photochromic Material

Bacteriorhodopsin occupies an unprecedented position with respect to continuing efforts to incorporate it into a variety of optical technologies. It was discovered early on that BR's photocycle could be manipulated through chemical additives that would modulate the M-state lifetime [10,11,53]: any conditions that inhibited proton mobility prolonged the M-state, but the addition of various amine compounds were found to be particularly effective. The bR-M binary pair used as an optical switch: 570-nm light would drive the protein to the M-state, and 410-nm light would drive M back to bR. The binary pair formed by these two states has become the basis for a number of architectures, including binary optical and holographic memories and processors, and spatial light modulators [6,8,12,15,54–60]. Unfortunately, there are two drawbacks to this approach: the addition of chemicals that enhance M-state formation inevitably decreases cyclicity, and it was not possible to produce a truly permanent bi-stable system (i.e., the M-state cannot be made permanent). As a result, information encoded optically using the bR-M pair lasts only as long as the lifetime of the M-state. However, the wide spectral separation between the two states makes them an ideal pair for transient optical or holographic memory operations, and interest remained focused on efforts along these lines for many years, especially in the former Soviet Union, Project Rhodopsin was a military research program with the goal of developing polymer films containing bacteriorhodopsin that could be used in optical memory and processing applications. A wide variety of chemical additives were explored as a means by which the M-state lifetime could be prolonged, resulting in a BR-based material referred to as Biochrome. This real-time photochromic and holographic material could be written, read, and erased by application of the appropriate wavelength [11]. But its inability to produce a highly prolonged or permanent M-state at room temperature precluded practical applications. Regardless, several architectures have been proposed based on M-state photochromism, including the only commercially viable BR-based technology thus far developed, albeit utilizing a genetically engineered BR variant, D96N. MIB Biochemicals in Marburg, Germany markets a real-time interferometer, Fringemaker-Plus[TM]. The primary application of this system is nondestructive testing of manufactured components through microscopic deformation and interferometric analysis [61]. Real-time holographic interference patterns recorded in BR D96N films are digitized for computer analysis. The technology allows the detection of microscopic flaws that cannot be seen based on visual inspection, alone. Such technology could conceivably facilitate quality control during the manufacturing process at a level not previously possible—each and every manufactured component could be tested for defects in real time. The use of the bacteriorhodopsin variant D96N in holographic media remains an active area of research [62].

Were it not for two developments, the further progress toward BR-based device applications might not have continued. Despite two decades of effort, it has not been possible to develop a BR-based material with a permanent M-state, either through chemical or genetic means. Genetic-engineering provided a wealth of information on the protein's structure and function, but could not prolong the M-state beyond a few minutes. And for many years, the goal of a permanent bi-stable state that could be used for optical memory and processing operations remained elusive.

However, in 1993, a branch off of the main photocycle was described as leading to a pair of states, one highly blue-shifted, that could be accessed by illuminating the O-state with red light [63]. The O-state is the last intermediate in the normal bacteriorhodopsin photocycle, and represents the last step toward resetting the BR resting state, thereby re-priming the proton pump. The chromophore has returned to the all-*trans* state, but the resting state electronic environment of the binding site has not been fully restored (D85 has not yet become deprotonated). Excitation of the protein in this state produces the P-state (absorbing at ∼ 490 nm), which decays to the blue-shifted Q through thermal decay (380 nm). Chromophore extraction and spectroscopic studies of these states reveals that both are characterized by a 9-*cis* chromophore, which accounts for their unusual properties, and models of the binding site indicate that a 9-*cis* conformation cannot be accommodated comfortably [64]. Formation of the P-state results in a strained binding site that only finds relief through hydrolysis of the Schiff base bond—once this bond is broken, the chromophore can reposition to a less sterically challenged arrangement, thereby forming Q. The Q-state is for all intents and purposes permanent, with its 9-*cis* chromophore being unbound, but trapped, in the binding site. Only absorption of blue light will reset the bR resting state by driving 9-*cis* retinal back to all-*trans*, coupled with spontaneous reformation of the Schiff base bond.

Although a permanent bi-stable system has been described for BR, it suffers from low quantum efficiency (∼ 10^{-3}). Because most of the strain associated with the generation of the 9-*cis* state is due to the methyl group on C-13, a BR variant incorporating the 13-desmethyl chromophore was examined [65]. The variant was able to produce 9-*cis* photoproducts far more easily than the unaltered protein. However, the absence of the C-13 methyl group precluded the formation of the Q-state, and the barrier between 9-*cis* and all-*trans* retinal configurations was low enough to allow the interconversion to occur easily. In essence, the C-13 methyl group is not only responsible for the low quantum efficiency of Q-state formation, but is the primary factor allowing Q to be permanent.

BR-based optical memory architectures are therefore faced with a dilemma: the M-state can be produced with high quantum efficiency (0.65), but is only transient, and although the Q-state is permanent, it can only be produced with low quantum efficiency. Despite this trade-off, the branched photocycle has proven invaluable for bacteriorhodopsin-based memory research efforts. The Q-state lifetime has been demonstrated to be at least seven to ten years ([22, 64] and in-house observations). It also has a slightly better holographic diffraction efficiency than the M-state (8.5% versus 6.4%), due to the wider spectral separation between the 570 nm bR and 380 nm Q states, as illustrated in Figure 16.2. As such, the M-state is the best option for short-term holographic applications (e.g., Fringemaker-Plus), while the Q-state should be used for long-term optical data storage systems.

Discovery and characterization of the Q-state was the first development to reenergize BR-based device applications. The second development is advances in molecular biology and genetic engineering that are revolutionizing biomolecular electronics. Techniques such as random and semi-random mutagenesis, and directed evolution, have made possible the development of highly optimized proteins for specific applications. For bacteriorhodopsin, these methods have paved the way toward variants that produce the Q-state in much higher yields with improved efficiency. Research efforts along these lines are described in the following section.

16.4 Progress Toward Bacteriorhodopsin-Based Optical Memories

Optical computer memories offer comparative advantages to traditional disk and platter architectures in the form of greater storage density, speed, mechanical simplicity, and data throughput. Unfortunately, materials capable of operating in such architectures are rare, and are often plagued by unwanted photochemistry and the need for powerful lasers (e.g., two-photon architectures [66, 67]). Furthermore, many of the proposed (nonbiological) materials are not easily reusable. Bacteriorhodopsin-based memory applications have been a goal ever since the Soviet Union launched Project Rhodopsin in the 1970s. Over three decades later, BR-based optical memories remain a goal, due, no doubt, to BR's remarkable biophysics and the novel concept of a computer memory based on a protein. The discovery and characterization of the branched photocycle brings researchers one step closer. In this section, two current optical memory architectures are discussed, and in the section that follows, an introduction will be given to the advanced genetic engineering techniques that are responsible for a renaissance in this field.

16.4.1 The Binary Optical Branched Photocycle Architecture

A number of approaches to BR-based memory architectures have been explored over the last decade, all exploiting the protein's photochromic abilities. The earliest designs were based on the M-state, but could only work at reduced temperatures where the M-state can be trapped (< 240K). Most were platter-type assemblies, although one architecture relied on two-photon transitions in order to operate in three-dimensions [6,17,19,26,68,69]. Both approaches define the bR state as a binary zero, and the M-state as a binary one. But low-temperature architectures are not practical, and these approaches were abandoned. The branched photocycle enabled new architectures employing the Q-state. In the binary optical approach (i.e., non-holographic), the protein is suspended in a three-dimensional polymer matrix in a one-centimeter cuvette, and data is written/read using orthogonal laser irradiation [22].

The bR and Q-states are defined as binary zeros and ones, respectively, as illustrated in Figure 16.3. Writing is accomplished in three dimensions by sequential two-photon techniques: a "paging" operation initiates the photocycle in an optically defined slice (page) of the BR-based memory cube. Because the branched photocycle is accessed from the O-state, a writing operation must wait until the bacteriorhodopsin in the paged region reaches the O-state ($\sim$ 2 ms). At this point, data is encoded spatially into the writing (actinic) laser with a spatial light modulator, and the previously illuminated page is irradiated—because the writing laser is oriented orthogonally with respect to the paging laser, the full plane defined by the page can be illuminated simultaneously, allowing for parallel writing operations. It is important to note that

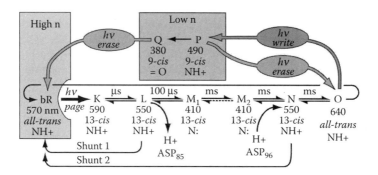

FIGURE 16.3 BR-based holographic memory operations utilizing the Q and bR states rely in the difference in the refractive indices of the two states for diffraction efficiency. In the case of the binary photonic architecture, the bR state is defined as a binary "0" and the Q state as a binary "1."

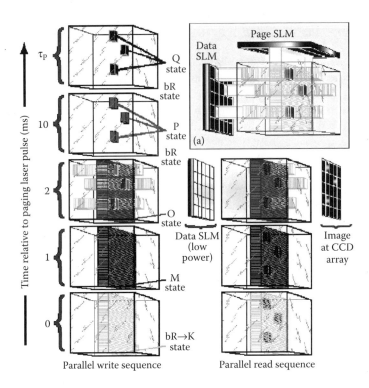

FIGURE 16.4 (see color insert) Parallel write sequence (left) and parallel read sequence (right) for the hypothetical storage and retrieval of three binary ones (22 binary zeroes) in a page of memory using the branched-photocycle architecture (see text). The inset details the concept of orthogonal accessing, a method by which any volume in a three-dimensional medium can be accessed.

branched photochemistry (Q-state formation) only occurs in the doubly irradiated regions of the page ("voxels").

A parallel reading operation is done similarly, in that a paging operation is performed to initiate photochemistry, followed by exposure of the page by the writing beam, but at a much lower intensity and with no spatially encoded data. The intensity is such that it will be absorbed by the O-state without causing appreciable photochemistry. The light will be absorbed by regions of the page in the O-state (binary zeros), but will be transmitted through the areas of the page where the Q-state was previously produced. The transmitted light can be imaged by a CCD array, thereby registering as binary ones (the absence of imaged light at the array is interpreted as zeroes). This process is illustrated in Figure 16.4. Ideally, the paging wavelength would be 570 nm, were it not for overlap with the P-state (which goes back to O upon illumination). So in practice, both the paging and writing wavelengths are done in the red. A drawback to this approach is that the architecture is path-dependent; it depends upon production of the Q-state through a specific route, driving bR→O and O→Q using specific wavelengths of light, in order to operate in three dimensions. However, several competitive routes to the Q-state have been discovered which degrade the efficiency of this architecture[64]; such routes introduce unwanted photochemistry and make it more difficult to distinguish between binary zeroes (bR) and binary ones (Q) (see Figure 16.5). Holography may offer a better alternative for BR-based memory storage in that it depends only on photo-converting bR to Q, without concern for the route. Thus, it is path independent, and photo-generation of Q by any available route is utilized.

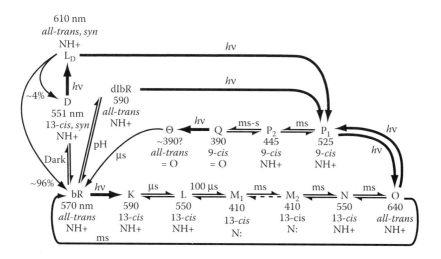

FIGURE 16.5 The bacteriorhodopsin photocycle with emphasis on the competitive routes to the Q state [64].

16.4.2 Bacteriorhodopsin-Based Holographic Memories

Holographic memories and associative processors are technologies with a wide variety of applications, offering higher storage densities than conventional media, fewer moving parts, and higher resistance to damage (mechanical or otherwise). A figure often cited is that a terabyte of information theoretically could be stored in a medium the size of a sugar cube. Because information is stored and retrieved with light, mechanical heads for handling data are no longer necessary, and the nature of holography allows information stored in a homogeneous medium to be retrieved from any location in the medium—if one part suffers damage, all previously stored data can still be retrieved from undamaged sections. Similar to the binary optical system described earlier, data is written and read in parallel, allowing higher throughput. Holographic associative memories offer a means through which object identification by the human brain can be simulated, thus retrieving information in parallel through association, rather than through file name and location. Object identification is an excellent example of a task that is handled quite efficiently by a holographic associative memory, and inefficiently by conventional computer memories. The source for this discrepancy is that holography, like the human brain, operates in parallel. An input object is compared simultaneously to all items stored in the memory bank, and the one with the highest correlation is retrieved. Conventional technologies accomplish this task sequentially, making it a much more time- and computationally-intensive process. The concept of a holographic associative loop was first described by Paek and Psaltis in 1987 [70], and applications have been proposed in neural networks, adaptive optical networks, and adaptive image correlators [71–74]. Examples include fingerprint and counterfeit currency identification, autonomous general pattern recognition, intelligence, and weather satellites (e.g., data reduction), and ultimately, artificial intelligence architectures. The Paek and Psaltis design is illustrated in Figure 16.6, as modified to incorporate bacteriorhodopsin-base holographic media [70].

16.4.3 Bacteriorhodopsin as a Holographic Material

Certainly, the technologies described previously are not limited to holographic media based on bacteri-orhodopsin. And yet neither has become commercially viable. The primary reason for this observation is the lack of real-time write-read-erasable holographic media. One popular material, lithium niobate, can be expensive to produce and cannot be erased in real time. Photorefractive polymers are under develop-ment and have potential, but it is not yet clear whether they will prove to be fully reusable. Most such materials require electrical or thermal fixing to make them semi-permanent and to prevent degradation

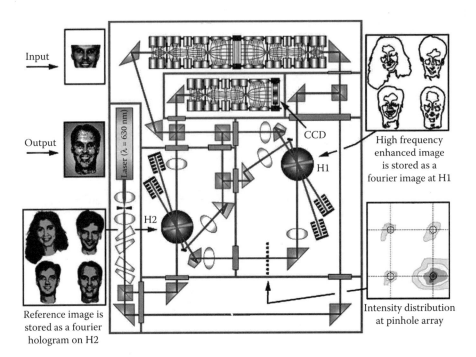

FIGURE 16.6 (see color insert) The Fourier transform holographic associative memory uses a thin film of BR as a medium for real-time holographic storage. This device can select from thousands of images simultaneously and requires only a partial input image to select and regenerate the entire image from a stored hologram.

of diffraction efficiency over time (e.g., lithium niobate)—real-time holography is essentially impossible [75]. Conceivable applications include any technology that requires search capabilities for large-scale databases. Bacteriorhodopsin has been shown to be reusable, and may succeed where other materials have so far failed—it offers not only real-time write-read-erase capabilities without processing steps, but also dynamic capabilities, high cyclicity, increased light sensitivity, optical resolution > 5000 lines/mm, can be produced easily and inexpensively, and can be processed into a number of different forms. A basic associative memory, a holographic autocorrelator, employing bacteriorhodopsin has already been demonstrated [58].

An angularly multiplexed holographic memory based on bacteriorhodopsin has recently been demonstrated by researchers at Syracuse University and the University of Connecticut. In this architecture, a cuvette containing BR suspended in a polymer matrix (the recording material) was placed on a precision rotational base, allowing multiple holograms to be stored in the same volume, separated by small angular intervals. Assuming the protein is randomly oriented within the gel, each angle accesses a largely independent population of BR molecules. Simultaneous exposure with two red laser beams both initiates the photocycle and drives the formation of the Q-state (Figure 16.7). A typical read-out is accomplished by rotating the cuvette back through the previously recorded holograms, while exposing the recording medium only to the reference beam—Figure 16.8 shows a series of four diffraction patterns stored with a 1° angular separation. The variation in diffraction efficiency (the peak height in Figure 16.8) results from a preprogrammed recording schedule to minimize cross-talk. The ability to gauge the dynamic range of a volumetric holographic medium is evaluated by determining the M/# of the material; the ability of a recording medium to store holograms in three dimensions is limited, and a point will be reached where successive recordings will degrade previously stored diffraction patterns. This benchmark is defined as the sum of the square root of the diffraction efficiencies of all the holograms stored in a particular volume [76], and facilitates comparison between different materials. Research is still underway to determine if BR is competitive with conventional materials, despite its other advantages.

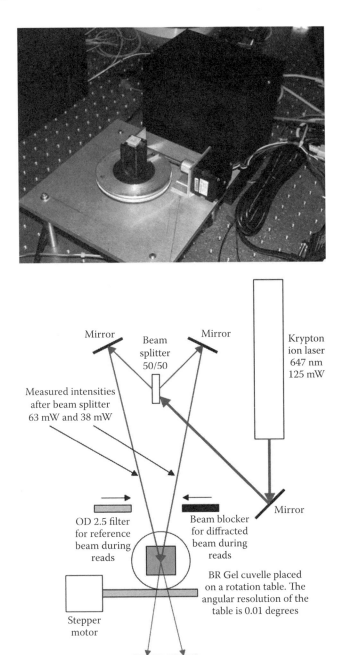

FIGURE 16.7 Below: Schematic of optical layout used for determining M/# and angular multiplexing. Above: Photo of BR sample sitting on the rotational state. The BR cuvette is housed in the center of the platform.

One major problem that remains to be overcome is efficiency. Although demonstration prototypes have been developed for both of the branched-photocycle technologies described earlier, the efficiency with which the Q-state is formed in the wild-type protein is rather poor. For the binary photonic optical memory, photoconversion requires high light intensities and prolonged exposures, both of which preclude

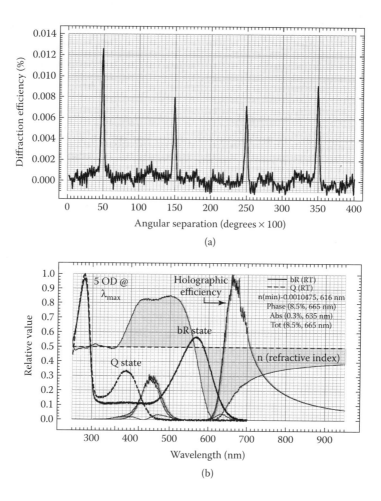

FIGURE 16.8　(a) Q-based diffraction peaks generated from diffraction patterns with one degree angular separation in a 1-cm BR poly(acrylamide) cube with an optical density of 1.5. (b) A Kramers–Kronig analysis of wavelength dependence of the diffraction efficiency associated with the formation of the Q-state.

commercial development. The holographic architectures have the same problem, although not quite as pronounced because writing is done using simultaneous exposure to two laser beams. Three methods are available for bacteriorhodopsin modification that can be used to modulate the photocycle with the goal of optimized efficiency: chemical modification, synthetic chromophores, and genetic engineering. Examples of both chemical modification and the use of a synthetic chromophore have been detailed earlier—amines are often used to extend M-state lifetime, and the ability to enhance Q formation was explored with the 13-desmethyl retinal chromophore. Neither approach is capable of providing the optimization necessary for commercial viability, meaning that more sophisticated techniques are required. The third method, genetic engineering, offers the most powerful solutions.

16.5　Optimization of Bacteriorhodopsin Through Genetic Engineering

Perhaps no technique has had a larger impact upon understanding the structure and function of bacterio-rhodopsin than site-directed mutagenesis. Site-directed mutagenesis (SDM) is the classical approach to genetic engineering, and has proven to be a powerful technique for elucidating BR's biophysical properties, especially the proton pump mechanism. Even crystallographic studies on the protein depended on genetically engineered variants for the purpose of examining photocycle intermediates [49,51]. SDM offers control of the protein at the genetic level, by enabling individual replacement of specific amino acids. The resulting mutant protein can be analyzed for new properties, which can provide tremendous insight into structure-function studies. With respect to BR, the majority of mutants examined are in regions of the protein known to be important for structure and function. Amino acids far removed from the chromophore binding site or the proton channel are seldom examined, because they are assumed to play no direct role in protein function, apart from structural. Given the often-synergistic interactions present in most proteins, such assumptions are unwarranted. Despite the extensive body of knowledge that exists for bacteriorhodopsin, researchers lack the ability to predict the actual active contribution of any given amino acid that is far removed from portions of the protein that are well understood. Therefore, the researchers' ability to design proteins *de novo* with a specific goal in mind is very limited. To date, D96N is the only BR mutant that has both provided valuable insight into how the protein works, and found application in device architectures.

Given the number of mutants possible, only a small area in the vast mutational landscape has been explored. With 248 amino acids, and 19 possible substitutions at each location, a total of nearly 4700 single mutants is possible in bacteriorhodopsin. And that figure does not take multiple mutations into account—the number of potential mutations is mind-boggling. So the prospect of comprehensively exploring the mutational landscape using SDM is clearly unrealistic. Furthermore, the researcher's perceived *a priori* knowledge of the protein might bias the direction of studies away from areas with high potential. Fortunately, the advent of advanced techniques in molecular biology has opened up new possibilities—researchers can now explore global diversification techniques coupled with high-throughput screens to identify candidate proteins with properties of interest [77].

Random mutagenesis and directed evolution allow characterization of the mutational landscape in an unbiased manner—the success of these techniques depends on a completely random approach to generating genetically engineered variants that otherwise might be overlooked. Whereas site-directed mutagenesis allows the researcher to make specific changes to proteins by design, random and semi-random methods make unpredictable changes to the genome, resulting in a library of strains expressing proteins with a wide range of properties. The resulting proteins are isolated and screened for properties of interest (increased efficiency, etc.). A compromised approach, semi-random mutagenesis, allows a more directed exploration of specific researcher-defined areas of the protein. A variety of *in vitro* genetic recombination methods have also been developed, which work by shuffling multiple genes or multiple variations of a single gene in order to increase the diversity of the resulting library of mutant proteins; examples include DNA shuffling, the staggered extension process (StEP), and randomly primed recombination [1,78–82]. Taken together, these methodologies provide a means by which proteins can be optimized toward specific, well-defined, properties. Indeed, this new capability is proving critical to the development of competitive protein-based technologies. The primary challenge for the researcher is that these approaches work too well, producing more proteins than can be screened for properties efficiently.

16.5.1 Semi-Random Mutagenesis, Random Mutagenesis, and Directed Evolution

At present, the most powerful tool in the protein engineer's genetic arsenal is directed evolution, which involves an iterative approach to building upon improvements in protein properties using random or semi-random mutagenesis (SRM). Random mutagenesis probes the entire mutational landscape of the protein by introducing a random mutagenic factor to the gene which generates mutations throughout the resultant protein with equal probability. One method for introducing random mutations into the gene is through the use of an error-prone polymerase chain reaction (PCR) that fails to reproduce DNA faithfully by using a polymerase with a high error rate. Errors are introduced at a frequency from 1 to 20 mutations per 1 Kb [83]. The advantages to random mutagenic methods have already been discussed—i.e., the lack of bias based on perceived knowledge of the protein. The disadvantage is that the large number of resulting variant proteins needs to be screened for properties, a task that can be both tedious and time-consuming. Properties like improved thermal stability or enzyme activity are relatively easy to identify, but looking for improvements to more complex photoactive or nonnative properties can be complicated.

Semi-random mutagenesis works by focusing genetic modifications to a specific region of the protein, often identified through previous structure-function studies. Each amino acid residue in the region can be mutated with an equal probability, while leaving the rest of the protein unchanged. As such, this is an excellent method for exploring the contributions made to a protein from a specific region. For example, it is well documented that the sequence including the stretch from BR E194 through F208 plays an important role in proton release and the O-state lifetime. Probing that region with SRM will result in BR variants with a range of modified properties that can provide insight into this stage of the proton pump mechanism. With respect to device applications, the O-state is the gateway to the branched photocycle—therefore, any mutations with enhanced yields of O will be beneficial to architectures that depend upon formation of the Q-state [84]. SRM is preferable over purely random mutagenic methods because of the reduced scale of the endeavor, especially when preexisting knowledge of the protein can define regions with a high probability of success.

Directed evolution (DE) is a genetic strategy based on an iterative approach to random mutagenic methods, where screening is used to identify the most promising mutations, which will then be used as the genetic starting point for the next round of random mutagenesis. Screening once again is used to identify improved proteins, and the process is repeated. Iterative rounds of genetic manipulation and differential screening are used to introduce optimized properties to the protein. In this way, each improved protein becomes a parent to the next generation of random progeny, thereby building upon improvements until a desired target is reached (Figure 16.9) [78,85–88]. Directed evolution mimics the process of natural selection on a much shorter time scale in a laboratory environment, and is the most efficient method of producing proteins optimized toward a desired target property. DE has been used widely in industry to optimize thermal stability, substrate specificity, enzyme kinetics, and chemical stability of proteins and

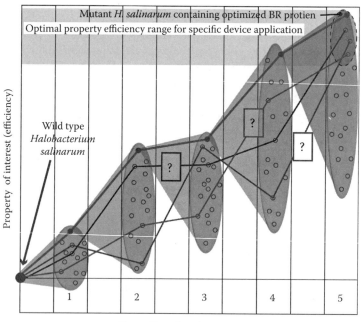

Generation of random progeny

FIGURE 16.9 Schematic illustration of the directed evolution of a bacterial protein toward a predefined goal. The bold line indicates the route selected by the researcher, where, based on screening, the strain containing the most efficient mutant protein of each generation was chosen to parent the next. However, as shown by the question marks, other synergistic routes may be possible, but never discovered.

enzymes to operate—to perform native function—in nonnative conditions [1–5]. As stated earlier, the major limitation to this technique is often evaluating the properties of the resulting library of genetically engineered proteins, a process that is only possible by defining a series of appropriate screening systems; *in vivo* approaches provide the best route toward a streamlined, high-throughput system [84,89,90].

Optimization of bacteriorhodopsin for device applications through directed evolution presents the researcher with another unique challenge. Whereas all of the examples of successful applications of directed evolution cited earlier enhanced naturally occurring protein functions (i.e., natural functions in nonnative conditions), the optimization of bacteriorhodopsin for device applications requires the modification of nonnative functions—no protein has evolved through natural selection as a more efficient holographic or optical memory medium. In order to enhance bacteriorhodopsin's function for permanent data storage, the yield and efficiency of Q formation must be improved. No concrete evidence exists supporting a physiological role for the branched portion of the BR photocycle—the ability to enhance a property that does not offer a comparative advantage in BR's natural role as a proton pump is unique, and may very well be the true strength of directed evolution.

Enhancing Q-state production is best approached from two directions that are pursued simultaneously: optimization of both O-state yield and efficiency of the O→P conversion. If more O-state is formed in the main photocycle, it stands to reason that larger amounts of Q can be produced. Formation of the O-state is fairly well understood, making this part of the optimization at least somewhat straightforward. However, optimizing the quantum efficiency of the branched photocycle is more complex and cannot be accomplished based upon the current understanding of P & Q formation alone [63, 64]. Furthermore, screening for Q-state formation is complicated because it is not part of the main BR photocycle. Full descriptions of the approaches to both the genetic engineering and screening of BR variants taken thus far can be found in the literature [77,84,89–92], but some salient results are shown below to illustrate the versatility of these techniques.

The first attempts to apply semi-random mutagenic techniques to bacteriorhodopsin resulted in better than 800 variants [89], all of which were screened for photokinetic properties of the M, O, and Q states. O-state photokinetic traces for an assortment of variants are shown in Figure 16.10, indicating the wide variety of responses possible. The M and O data are summarized in Figures 16.11 and 16.12, wherein the contribution to an M or O lifetime by a particular amino acid mutation is correlated to its location in the protein. The resulting histograms are extraordinarily valuable for mapping out the structure–function relationships related to these two states, and thereby providing a direction for future mutational studies. Close inspection of the data reveals several interesting trends, both from the standpoint of protein biophysics and device studies.

As would be expected, the vast majority of mutations provide no enhancement of either the M or O-states; the processes of natural selection and evolution have already provided an efficient protein, and the fact that the majority of changes introduce no significant change in the properties can essentially be viewed as a protective mechanism that allows continued function in the face of naturally occurring mutations. However, in both the M and O histograms, a select number of mutants exhibit prolonged lifetimes, as compared to the wild type proteins. For the M-state (natural lifetime of about 15 ms), the majority of the mutations are for the most part silent (i.e., not significantly different from wild type), and no clear trend is observable for lifetimes less than 35 ms. But for M lifetimes longer than 200 ms, it is clear that the majority of those exhibiting extended lifetimes have mutations confined to helices F and G. A similar trend is seen for the O-state. For lifetimes less than 50 ms, the mutations are randomly distributed throughout the protein. But for mutations with extended O-state lifetimes, the mutations are confined to helices E, F, and G, especially between residues E194 and F208 [90]. Q-state formation was improved by as much as two orders of magnitude relative to wild type BR through the first rounds of mutagenesis.

16.6 Discussion and Future Directions

Testament to bacteriorhodopsin's remarkable qualities is the fact that this protein remains a strong candidate for device applications 30 years after its discovery. The promise of a fully write-read-erasable optical

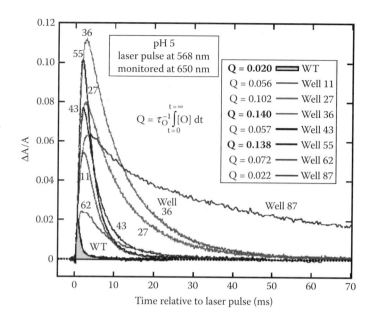

FIGURE 16.10 Photokinetic traces of the O-state of a selection of mutant BR proteins, compared with wild type (smallest trace). Most of these mutants were in the region of the F and G helices, which play an important role in O-state biophysics.

recording medium may finally be realized. However, it is clear that very few technologies are possible based on the wild-type protein, which lacks the efficiency to be competitive with conventional materials. Genetic engineering and directed evolution offer the only viable paths toward optimized proteins for device architectures, in the form of advanced methods that enable the researcher to custom-tailor biomaterials toward specific goals.

Critical to this endeavor for bacteriorhodopsin is the optimization of the Q-state yield, either through better production of the O-state or by improving the quantum efficiency of the O→P transition. Although a powerful technique in itself, site-directed mutagenesis is limited by the lack of a full understanding of BR biophysics. Random and semi-random mutagenic approaches provide a superior alternative for examining the roles of specific amino acids or well-defined regions of the protein. A particularly powerful aspect of these techniques is the ability to produce multiple mutations (e.g., double, triple, quadruple mutants, etc.) with unforeseen properties resulting from synergistic interactions. The properties of these variants are impossible to predict. For example, a sextuple mutant resulted from the first round of random mutagenesis with an O-state lifetime of several seconds. A pair of triple mutants based on the sextuple mutant was constructed, and the resulting proteins were evaluated for O-state lifetime; neither had O-states as long-lived as the parent sextuple mutant. And the sum of the two lifetimes still fell short, indicating that the effect is not additive (unpublished data). Several similar examples have resulted from these studies.

A number of amino acid residues have been identified in the literature (and through the studies discussed earlier) that govern the lifetime and yield of the O state, primarily in the F and G loop region (residues 194–208). The two most critical residues appear to be Glu194 and Glu204 [93–99], and efforts are ongoing to target these residues with saturation mutagenesis (i.e., to make all 19 possible mutations at those two points). E194Q is particularly promising, producing more O-state than all others examined thus far. Semi-random studies using this mutant as a template for further optimization are being pursued. However, an additional five to six rounds of optimization may be required before a viable candidate for optical memories is found.

Screening is still a problem, and requires that a minimal amount of protein be isolated for each mutant. *In vivo* methods would streamline the process considerably, and preliminary results indicate that the

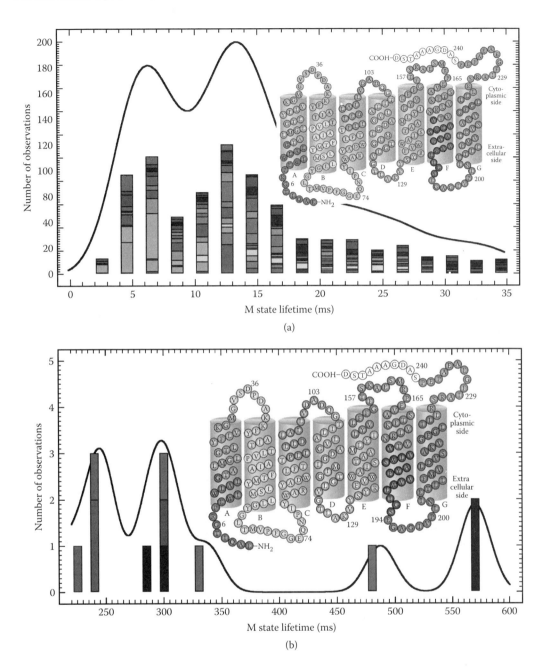

FIGURE 16.11 (see color insert) Histograms showing the correlation between the position of mutated residues and the M-state lifetime. Colors in the bars correspond to similarly colored regions of the secondary structural model of BR (inset).

production of Q can be followed indirectly through monitoring the magnitude of the O-state produced upon illumination of a cell colony [77]; the amount of O will decrease as the amount of Q increases. A design for a three-stage cell-sorting system has been proposed that could identify proteins *in vivo* with improved properties based upon sequential illumination [84]. Cells could be screened based on protein expression (stage one), O→P conversion efficiency (stage two), and Q→bR conversion efficiency (stage three).

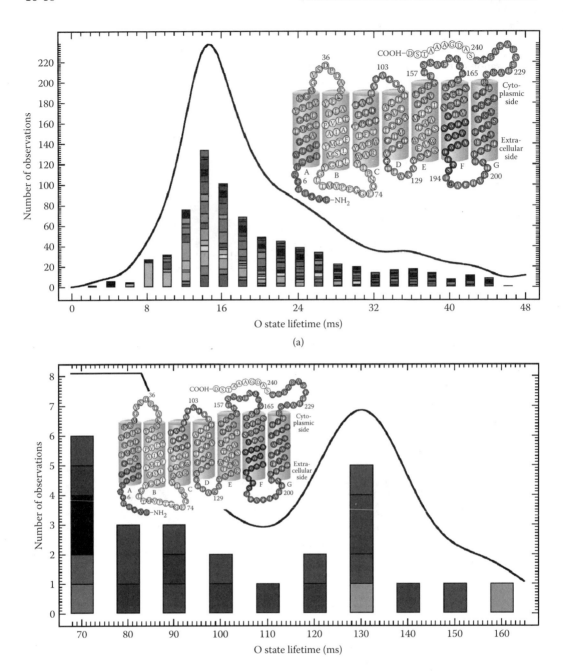

FIGURE 16.12   (see color insert) Histogram analysis of the O-state lifetime of mutants. The same color scheme is used as in Figure 16.11.

As more mutants are produced through the genetic engineering effort, they are evaluated for photo-kinetics, and ultimately, holographic benchmark parameters. The latter includes diffraction efficiency and the multiplexing parameter (M/#). A parallel research effort seeks to optimize the lifetime of the M-state for long-term data storage. In addition, ongoing studies are evaluating a close homologue of bacteriorhodopsin, the marine protein proteorhodopsin. This protein was first described in 2000 [100], and has already proven capable of transient holography (unpublished results; from a manuscript in preparation).

16.7 Conclusions

Utilization of proteins and other biomolecules has great potential for introducing novel functionalities to preexisting conventional technologies, as well as facilitating new architectures. Biological molecules have benefited from the stringent processes of natural selection and evolution, resulting in proteins highly optimized for a specific function. The challenge faced by the applied scientist is to exploit their properties for device applications. Bacteriorhodopsin is one such protein that performs its natural function with an extraordinary level of efficiency. Although BR has long been considered for a variety of device architectures, it has lacked the efficiency needed for binary photonic and holographic optical computer memories. Fortunately, molecular biology and genetic engineering techniques have been developed that address this issue, resulting in variant proteins with highly improved properties. Random mutagenic techniques and directed evolution are realizing genetically engineered bacteriorhodopsin variants optimized for binary photonic and holographic optical memories. The ability to manipulate structure and function at the genetic level, allowing the researcher to custom-tailor biomaterials toward specific functions, is critical not only to the development of protein-based optical memories, but also to the ultimate success of bionanotechnology.

Acknowledgments

The authors gratefully acknowledge the following funding organizations for their generous support of the research efforts described herein: The Syracuse University Keck Center for Molecular Electronics, The New York State Center for Excellence in Infotonics UC-004-0007, Nation Science Foundation grants #NSF-0432151, including REU support (JAS, RRB) and # NSF-0412387 (RRB), and the Nation Institutes of Health, NIH-GM34548 (RRB).

References

[1] Dalby, P.A., Optimising enzyme function by directed evolution, *Curr. Opin. Struct. Biol.*, 13(4): 500–505, 2003.

[2] Kirk, O., T.V. Borchert, and C.C. Fuglsang, Industrial enzyme applications, *Cur. Op. Biotech.*, 13: 345–351, 2002.

[3] Morawski, B., S. Quan, and F. Arnold, Functional expression and stabilization of horseradish peroxidase by directed evolution in Saccharomyces cerevisiae, *Biotechnol. Bioeng.*, 76(2): 99–107, 2001.

[4] Sterner, R. and W. Liebl, Thermophillic adaptation of proteins, *Cri. Rev. Biochem. Mol. Bio.*, 2001. 36(1): 39–106.

[5] Whaley, S.R., D.S. English, E.L. Hu, P.F. Barbara, and A.M. Belcher, Selection of peptides with semiconductor binding specificity for directed nanocrystal assembly. *Nature*, 2000. 405(6787): 665–668.

[6] Birge, R.R., P.A. Fleitz, R.B. Gross, J.C. Izgi, A.F. Lawrence, J.A. Stuart, and J.R. Tallent, Spatial light modulators and optical associative memories based on bacteriorhodopsin. *Proc. IEEE EMBS*, 1990. 12: 1788–1789.

[7] Birge, R.R., K.C. Izgi, J.A. Stuart, and J.R. Tallent, Wavelength dependence of the photorefractive and photodiffractive properties of holographic thin films based on bacteriorhodopsin. *Proc. Matl. Res. Soc.*, 1991. 218: 131–140.

[8] Hampp, N., A. Popp, C. Bräuchle, and D. Oesterhelt, Diffraction efficiency of bacteriorhodopsin films for holography containing bacteriorhodopsin wildtype BR_{WT} and its variants BR_{D85E} and BR_{D96N}. *J. Phys. Chem.*, 1992. 96(11): 4679–4685.

[9] Birge, B., P. Fleitz, R. Gross, J. Izgi, A. Lawrence, J. Stuart, and J. Tallent. Spatial light modulators and optical associative memories based on bacteriorhodopsin. In *Materials Research Society*. Boston, MA, 1990.

[10] Vsevolodov, N.N., A.B. Druzhko, and T.V. Djukova, Actual possibilities of bacteriorhodopsin application in optoelectronics. In *Molecular Electronics: Biosensors and Biocomputers*, F.T. Hong, Ed., Plenum Press, New York, 1989, 381–384.

[11] Vsevolodov, N.N. and V.A. Poltoratskii, Holograms in biochrome, a biological photochromic material. *Sov. Phys. Tech. Phys.*, 1985. 30: 1235.

[12] Hampp, N., C. Bräuchle, and D. Oesterhelt, Bacteriorhodopsin wildtype and variant aspartate-96 Ø asparagine as reversible holographic media. *Biophys J.*, 1990. 58: 83–93.

[13] Hampp, N., C. Bräuchle, and D. Oesterhelt, Mutated bacteriorhodopsins: competitive materials for optical information processing? *MRS Bulletin*, 1992. XVII(11): 56–60.

[14] Hampp, N., C. Bräuchle, and D. Oesterhelt. Mutated bacteriorhodopsins: new materials for optical storage and information processing. In *4th International Symposium on Bioelectronic and Molecular Electronic Devices*. 1992. Miyazaki, Japan: R&D Association for Future Electron Devices.

[15] Oesterhelt, D., C. Bräuchle, and N. Hampp, Bacteriorhodopsin: a biological material for information processing. *Quart. Rev. Biophys.*, 1991. 24: 425–478.

[16] Druzhko, A.B. and S.K. Zharmukhamedov, Biochrome film based on some analogues of bacteriorhodopsin, in *Photosensitive Biological Complexes and Optical Recording of Information*, G.R. Ivanitskiy and N.N. Vsevolodov, Eds., USSR Academy of Sciences, Biological Research Center, Institute of Biological Physics, Pushchino, 1985, 119–125.

[17] Birge, R.R., Protein based optical computing and memories. *IEEE Comp.*, 1992. 25(November): 56–67.

[18] Birge, R.R., Three-dimensional optical memories. *Am. Sci.*, 1994. 82: 349–355.

[19] Birge, R.R., Protein-based three-dimensional memory. *Am. Sci.*, 1994. 82(July–August): 348–355.

[20] Birge, R.R., Protein-based computers. *Sci. Am.*, 1995. 272(3): pp. 90–95.

[21] Birge, R.R., Z. Chen, R.B. Gross, S.B. Hom, K.C. Izgi, J.A. Stuart, J.R. Tallent, and B.W. Vought, Biomolecular photonics based on bacteriorhodopsin. In *CRC Handbook of Organic Photochemistry and Photobiology*, W. M. Horspool, Ed., CRC Press: Boca Raton, FL, 1995, 1568–1585.

[22] Birge, R.R., N.B. Gillespie, E.W. Izaguirre, A. Kusnetzow, A.F. Lawrence, D. Singh, Q.W. Song, E. Schmidt, J.A. Stuart, S. Seetharaman, and K.J. Wise, Biomolecular Electronics: Protein-based associative processors and volumetric memories. *J. Phys. Chem. B*, 1999. 103: 10746–10766.

[23] Birge, R.R. and R.B. Gross, Biomolecular optoelectronics. In *Introduction to Molecular Electronics*, M.C. Petty, M.R. Bryce, and D. Bloor, Eds. Edward Arnold, London, 1995, 315–344.

[24] Birge, R.R., D.S.K. Govender, R.B. Gross, A.F. Lawrence, J.A. Stuart, J.R. Tallent, E. Tan, and B.W. Vought, Bioelectronics, three-dimensional memories and hybrid computers. IEEE IEDM *Tech. Dig.*, 1994. 94: 3–6.

[25] Birge, R.R., B. Parsons, Q.W. Song, and J.R. Tallent, Protein-based three-dimensional memories and associative processors. In *Molecular Electronics*, M.A. Ratner and J. Jortner, Eds., Blackweel Science Ltd., Oxford, 1997, 439–471.

[26] Birge, R.R., C. Zhang, and A.F. Lawrence, Optical random access memory based on bacteriorhodopsin. In *Molecular Electronics: Biosensors and Biocomputers*, F.T. Hong, Ed., Plenum Press, New York, 1989, 369–379.

[27] Stuart, J.A., D.L. Marcy, and R.R. Birge. Photonic and optoelectronic applications of bacteriorhodopsin. in *Bioelectronic Applications of Photochromic Pigments*. IOS Press, Szeged, Hungary: 2000.

[28] Stuart, J.A., D.L. Marcy, and R.R. Birge. Bacteriorhodopsin-based three-dimensional optical memory. in *Bioelectronic Applications of Photochromic Pigments*. IOS Press, Szeged, Hungary: 2000.

[29] Stuart, J.A., D.L. Marcy, K.J. Wise, and R.R. Birge, Volumetric optical memory based on bacteriorhodopsin. *Synth. Met.*, 2002. 127: pp. 3–15.

[30] Stuart, J.A., D.L. Marcy, K.J. Wise, and R.R. Birge, Biomolecular electronic device applications of bacteriorhodopsin. In *Molecular Electronics: Bio-Sensors and Bio-Computers*, L.E.A. Barasanti, Ed., Kluwer Academic Publishers, 2003, 265–299.

[31] Stuart, J.A., J.R. Tallent, E.H.L. Tan, and R.R. Birge, Protein-based volumetric memory. *Proc. IEEE Nonvol. Mem. Tech.* (INVMTC), 1996. 6: 45–51.

[32] Bhattacharya, P., J. Xu, G. Váró, D.L. Marcy, and R.R. Birge, Monolithically integrated bacteriorhodopsin-GaAs field-effect transistor photoreciever. *Opt. Lett.*, 2002. 27(10): 839–841.

[33] Xu, J., P. Bhattacharya, and G. Váró, Photo-induced anisotropic photoelectric response in oriented bacteriorhodopsin films. *Optic. Mater.*, 2003. 22: 321–326.

[34] Rayfield, G., Ultra high speed bacteriorhodopsin photodetectors. In *Molecular Electronics*, F.T. Hong, Ed., Plenum, New York, 1989, 361–368.

[35] Müller, D.J., F.A. Schabert, G. Büldt, and A. Engel, Imaging purple membranes in aqueous solutions at sub-nanometer resolution by atomic force microscopy. *Biophys. J.*, 1995. 68(May): 1681–1686.

[36] Müller, D.J., D. Fotiadis, S. Scheuring, S.A. Müller, and A. Engel, Electrostatically balanced sub-nanometer imaging of biological specimens by atomic force microscope. *Biophys. J.*, , 1999. 76: 1101–1111.

[37] Danon, A. and W. Stoeckenius, Photophosphorylation in Halobacterium halobium. *Proc. Natl. Acad. Sci. USA*, 1974. 71(No. 4): 1234–1238.

[38] Stoeckenius, W., Purple membrane of halobacteria: A new light-energy converter. *Acc. Chem. Res.*, 1980. 13: 337–344.

[39] Oesterhelt, D. and W. Stoeckenius, Functions of a new photoreceptor membrane. *Proc. Natl. Acad. Sci. USA*, 1973. 70(No. 10): 2853–2857.

[40] Balashov, S. and T. Ebrey, Trapping and spectroscopic identification of the photointermediates of bacteriorhodopsin at low temperatures. *Photochem. Photobiol.*, 2001. 73(5): 453–462.

[41] Ebrey, T.G., Light energy transduction in bacteriorhodopsin. In *Thermodynamics of Membrane Receptors and Channels*, M.B. Jackson, Ed., CRC Press, Boca Raton, 1993, 353–387.

[42] Stuart, J.A. and R.R. Birge, Characterization of the primary photochemical events in bacteriorhodopsin and rhodopsin. In *Biomembranes*, A.G. Lee, Ed. JAI Press, London, 1996, 33–140.

[43] Birge, R.R., Nature of the primary photochemical events in rhodopsin and bacteriorhodopsin. *Biochim. Biophys. Acta*, 1990. 1016: 293–327.

[44] Lanyi, J.K., Bacteriorhodopsin. *Intl. Rev. Cytol.*, 1999. 187: 161–202.

[45] Lanyi, J.K. and H. Luecke, Bacteriorhodopsin. *Curr. Opin. Struct. Biol.*, 2001. 11(4): 415–9.

[46] Lanyi, J.K. and G. Váró, The photocycles of bacteriorhodopsin. *Isr. J. Chem.*, 1995. 35: 365–385.

[47] Lanyi, J., Bacteriorhodopsin: a paradigm for proton pumps? *Biophys. Chem.*, 1995. 56: 143–151.

[48] Luecke, H., H.T. Richter, and J.K. Lanyi, Proton transfer pathways in bacteriorhodopsin at 2.3 angstrom resolution. *Science*, 1998. 280(5371): 1934–7.

[49] Luecke, H., B. Schobert, H.-T. Richter, J.-P. Cartailler, and J.K. Lanyi, Structural changes in bacteriorhodopsin during ion transport at 2 angstrom resolution. *Science*, 1999. 286: 255–260.

[50] Luecke, H., B. Schobert, H.-T. Richter, J.-P. Cartailler, and J.K. Lanyi, Structure of bacteriorhodopsin at 1.55 Å resolution. *J. Mol. Biol.*, 1999. 291: 899–911.

[51] Luecke, H., B. Schobert, J.-P. Cartailler, H.-T. Richter, A. Rosengarth, R. Needleman, and J.K. Lanyi, Coupling photoisomerization of retinal to directional transport in bacteriorhodopsin. *J. Mol. Biol.*, 2000. 300: 1237–1255.

[52] Spassov, V.Z., H. Luecke, K. Gerwert, and D. Bashford, pKa calculations suggest storage of an excess proton in a hydrogen-bonded water network in bacteriorhodopsin. *J. Mol. Biol.*, 2001. 312: 203–219.

[53] Yoshida, M., K. Ohno, Y. Takeuchi, and Y. Kagawa, Prolonged lifetime of the 410-nm intermediate of bacteriorhodopsin in the presence of guanidine hydrochloride. *Biochem. Biophys. Res. Comm.*, 1977. 75(4): 1111–1116.

[54] Renner, T. and N. Hampp, Bacteriorhodopsin-films for dynamic time averaging interferometry. *Optics Comm.*, 1993. 96: 142–149.

[55] Thoma, R., N. Hampp, C. Bräuchle, and D. Oesterhelt, Bacteriorhodopsin films as spatial light modulators for nonlinear-optical filtering. *Opt. Lett.*, 1991. 16(9): 651–653.

[56] Thoma, R. and N. Hampp, Real-time holographic correlation of two video signals by using bacteriorhodopsin films. *Opt. Lett.*, 1992. 17(16): 1158–1160.

[57] Wolperdinger, M. and N. Hampp, Bacteriorhodopsin variants as versatile media in optical processing. *Biophys. Chem.*, 1995. 56: 189–192.

[58] Gross, R.B., K.C. Izgi, and R.R. Birge, Holographic thin films, spatial light modulators and optical associative memories based on bacteriorhodopsin. *Proc. SPIE*, 1992. 1662: 186–196.

[59] Gross, R.B., Z. Chen, and R.R. Birge, Observtion and analysis of self-diffraction in bacteriorhodopsin films. *Mat. Res. Soc. Symp. Proc.*, 1994. 330: 251–256.

[60] Song, Q.W., C. Zhang, R. Blumer, R.B. Gross, Z.C. Chen, and R.R. Birge, Chemically enhanced bacteriorhodopsin thin-film spatial light modulator. *Opt. Lett.*, 1993. 18(16): 1373–1375.

[61] Hampp, N. and T. Juchem. Fringemaker — the first technical system based on bacteriorhodopsin. In *Bioelectronic Applications of Photochromic Pigments*. IOS Press, Szeged, Hungary, 2000.

[62] Yao, B., Z. Ren, N. Menke, Y. Wang, Y. Zheng, M. Lei, G. Chen, and N. Hampp, Polarization holographic high-density optical data storage in bacteriorhodopsin film. *Appl. Opt.*, 2005. 44(34): 7344–8.

[63] Popp, A., M. Wolperdinger, N. Hampp, C. Bräuchle, and D. Oesterhelt, Photochemical conversion of the O-intermediate to 9-cis-retinal-containing products in bacteriorhodopsin films. *Biophys. J.*, 1993. 65(October): 1449–1459.

[64] Gillespie, N.B., K.J. Wise, L. Ren, J.A. Stuart, D.L. Marcy, J. Hillebrecht, Q. Li, L. Ramos, K. Jordan, S. Fyvie, and R.R. Birge, Characterization of the branched-photocycle intermediates P and Q of bacteriorhodopsin. *J. Phy. Chem. B*, 2002. 106: 13352–13361.

[65] Gillespie, N.B., L. Ren, L. Ramos, H. Daniell, D. Dews, K.A. Utzat, J.A. Stuart, C.H. Buck, and R.R. Birge, Characterization and photochemistry of 13-desmethyl bacteriorhodopsin. *J. Phys. Chem. B*, 2005. 109(33): 16142–52.

[66] Liang, Y.C., A.S. Dvornikov, and P.M. Rentzepis, Nonvolatile read-out molecular memory. *Proc. Natl. Acad. Sci. USA*, 2003. 100(14): 8109–8112.

[67] Dvornikov, A.S. and P.M. Rentzepis, 3D optical memory devices. System and materials characteristics. *Proc. IEEE Nonvol. Mem. Tech.* (INVMTC), 1996. 6: 40–44.

[68] Birge, R.R., R.B. Gross, M.B. Masthay, J.A. Stuart, J.R. Tallent, and C.F. Zhang, Nonlinear optical properties of bacteriorhodopsin and protein based two-photon three-dimensional memories. *Mol. Cryst. Liq. Cryst. Sci. Technol. Sec. B. Nonlinear Opt.*, 1992. 3: 133–147.

[69] Birge, R.R. and D.S.K. Govender, Three-Dimensional Optical Memory. Syracuse University Press, Ithaca, NY, 1993.

[70] Paek, E.G. and D. Psaltis, Optical associative memory using Fourier transform holograms. *Opt. Eng.*, 1987. 26: 428–433.

[71] Psaltis, D., D. Brady, X. Guang, and S. Lin, Holography in artificial neural networks. *Science*, 1990. 343: 325–330.

[72] Psaltis, D., D. Brady, and K. Wagner, Adaptive optical networks using photorefractive crystals. *Appl. Opt.*, 1988. 27(9): 1752–9.

[73] Psaltis, D., M.A. Neifeld, and A. Yamamura, Image correlators using optical memory disks. *Opt. Lett.*, 1989. 14(9): 429–431.

[74] Psaltis, D., M.A. Neifeld, A. Yamamura, and S. Kobayashi, Optical memory disks in optical information processing. *Appl. Opt.*, 1990. 29(14): 2038–2057.

[75] Psaltis, D. and G.W. Burr, Holographic data storage, in *Computer*. 1998. 52–60.

[76] Burr, G.W. and D. Psaltis, System metric for holographic memory systems. *Opt. Lett.*, 1996. 21(12): 896–898.

[77] Hillebrecht, J.R., J.F. Koscielecki, K.J. Wise, M.P. Krebs, J.A. Stuart, and R.R. Birge, Optimizing photoactive proteins for optoelectronic environments by using directed evolution, in *Smart Biosensor Technology*. 2006 (in press).

[78] Arnold, F., P.L. Wintrode, K. Miyazaki, and A. Gershenson, How enzymes adapt: lessons from directed evolution. *Trends Biochem. Sci.*, 2001. 26(2): 100–106.

[79] Crameri, A., S. Raillart, E. Bermudez, and W.P.C. Stemmer, DNA shuffling of a family of genes from diverse species accelerates directed evolution. *Nature*, 1998. 391: 288–291.

[80] Graddis, T.J., R.L.J. Remmele, and J.T. McGrew, Designing proteins that work using recombinant technologies. *Cur. Pharm. Biotechnol.*, 2002. 3(4): 285–297.

[81] Georgescu, R., G. Bandara, and L. Sun, Saturation mutagenesis. *Meth. Mol. Biol.*, 2003. 231: 75–83.

[82] Zhao, H., L. Giver, Z. Shao, J.A. Affholter, and F.H. Arnold, Molecular evolution by staggered extention process (StEP) in vitro recombination. *Nature Biotech.*, 1998. 16: 258–261.

[83] Arnold, F.H. and G. Georgiou, eds. Directed Evolution Library Creation. *Meth. Mol. Biol.*, 2003. 231.

[84] Hillebrecht, J.R., J.F. Koscielecki, K.J. Wise, D.L. Marcy, W. Tetley, R. Rangarajan, J. Sullivan, M. Brideau, M.P. Krebs, J.A. Stuart, and R.R. Birge, Optimization of protein-based volumetric optical memories and associative processors using directed evolution. *Nanobiotechnology*, 2005. 1(2): 141–152.

[85] Kuchner, O. and F. Arnold, Directed evolution of enzyme catalysts. *Trends Biotech.*, 1997. 15: 523–530.

[86] Arnold, F. and J.C. Moore, Optimizing industrial enzymes by directed evolution. *Adv. Biochem. Eng.*, 1997. 58: 1–14.

[87] Arnold, F.H., Design by directed evolution. *Acc. Chem. Res.*, 1998. 31: 125–131.

[88] Arnold, F.H. and A.A. Volkov, Directed evolution of biocatalysts. *Curr. Opin. Chem. Biol.*, 1999. 3(1): 54–9.

[89] Wise, K.J., N.B. Gillespie, J.A. Stuart, M.P. Krebs, and R.R. Birge, Optimization of bacteriorhodopsin for bioelectronic devices. *Trends Biotechnol.*, 2002. 20(9): 387–394.

[90] Koscielecki, J.F., J.R. Hillebrecht, and R.R. Birge, Directed evolution of proteins for device applications. *The Biomedical Engineering Handbook*, 3rd ed., J.D. Bronzino, Ed., Taylor & Francis, Boca Raton, FL, 21: 1–5.

[91] Wise, K.J., J.R. Hillebrecht, J.F. Koscielecki, J.A. Stuart, and R.R. Birge, Optimization of proteins for molecular and biomolecular electronic devices, in *Encyclopedia of Molecular Cell Biology and Molecular Medicine*, R.A. Meyers, Ed., John Wiley & Sons, New York, 2005.

[92] Hillebrecht, J.R., K.J. Wise, J.F. Koscielecki, and R.R. Birge, Directed evolution of bacteriorhodopsin for device applications. *Meth. Enzymol.*, 2004. 388: 333–47.

[93] Brown, L.S., J. Sasaki, H. Kandori, A. Maeda, R. Needleman, and J.K. Lanyi, Glutamic acid 204 is the terminal proton release group at the extracellular surface of bacteriorhodopsin. *J. Biol. Chem.*, 1995. 270(45): 27122–27126.

[94] Kandori, H., Y. Yamazaki, M. Hatanaka, R. Needleman, L.S. Brown, H.-T. Richter, J.K. Lanyi, and A. Maeda, Time-resolved Fourier transform infrared study of structural changes in the last steps of the photocycles of Glu-204 and Leu-93 mutants of bacteriorhodopsin. *Biochemistry*, 1997. 36(17): 5134–5141.

[95] Misra, S., R. Govindjee, T.G. Ebrey, N. Chen, J.-X. Ma, and R.K. Crouch, Proton uptake and release are rate-limiting steps in the photocycle of the bacteriorhodopsin mutant E204Q. *Biochemistry*, 1997. 36(16): 4875–4883.

[96] Sampogna, R. and B. Honig, Electrostatic coupling between retinal isomerization and the ionization state of Glu-204: A general mechanism for proton release in bacteriorhodopsin. *Biophys. J.*, 1996. 71: 1165–1171.

[97] Lazarova, T., C. Sanz, E. Querol, and E. Padros, Fourier transform infrared evidence for early deprotonation of Asp(85) at alkaline pH in the photocycle of bacteriorhodopsin mutants containing E194Q. *Biophys. J.*, 2000. 78(4): 2022–30.

[98] Zscherp, C., R. Schlesinger, and J. Heberle, Time-resolved FT-IR spectroscopic investigation of the pH-dependent proton transfer reactions in the E194Q mutant of bacteriorhodopsin. *Biochem. Biophys. Res. Commun.*, 2001. 283: 57–63.

[99] Balashov, S., E. Imasheva, T. Ebrey, N. Chen, D. Menick, and R. Crouch, Glutamate 194 to cysteine mutation inhibits fast light-induced proton release in bacteriorhodopsin. *Biochemistry*, 1997. 36: 8671–8676.

[100] Beja, O., E.N. Spudich, J.L. Spudich, M. Leclerc, and E.F. DeLong, Proteorhodopsin phototrophy in the ocean. *Nature*, 2001. 411(6839): 786–9.

17

Subneuronal Processing of Information by Solitary Waves and Stochastic Processes

Danko D. Georgiev

James F. Glazebrook

Abstract

We discuss a framework for subneuronal processing of information in terms of certain biophysical principles subject to the dynamics of solitary waves and stochastic processes. A particular focus concerns the propagation of electromagnetic solitons within neurons resulting from the interaction between the cytosolic water electric dipole field and the quantized electromagnetic field induced by transmembrane neuronal currents. We show that soliton collisions may be viewed as a type of logical gate application where the resulting output is ensured by interaction of the soliton with C–terminal tails (CTTs) projecting from the cytoskeletal microtubules. The CTT energase action by vibrationally assisted tunneling may influence the conformational dynamics of the neuronal cytoskeletal protein network by releasing energy and providing efficiency for mechanisms leading to neuronal neurite outgrowth, synaptogenesis, and membrane fusion. The nanomolecular neurobiology discussion utilizes quantum mechanical effects such as quantum tunneling within molecular processes and leads to comparisons with features of nonlinear optical networks and quantum entanglement. The described function of the synapse as a sandwich-like array modeled on a Josephson junction affords a similarity with nanomolecular devices as applied in nanoelectronics and quantum computers. The biological processes discussed may be viewed as a basis for a neuro-informatic/cybernetic network within the cytoskeleton of neurons.

17.1   Introduction

This chapter surveys certain bioenergetic mechanisims that can be modeled by soliton-type (solitary and travelling) waves and certain stochastic processes. Such models are related to various fundamental neurobiological/physiological phenomena occurring within the neuronal cytoskeleton and which may function as regulators of a cybernetic system. The foundation for this investigation firstly concerns *energases*, which are enzymes that have the ability to make or break noncovalent chemical bonds. One can regard every protein–protein interaction as an enzymatic process in which each protein functions as an energase [41,87] and determines a conformational change in the relevant constituents of its partner. The second concerns *vibrationally assisted quantum tunneling* which is introduced as a factor influencing protein conformational dynamics and proton transfer in enzymatic action (see e.g., [50,75,103–106]). From the biological background, we describe how such phenomena are realized in the case of the conformational dynamics of neuronal microtubules within the cytoskeleton of neurons leading to the activation of allied processes involving membrane and vesicle fusion (such as exocytosis). A notable feature here is that microtubules are considered in view of their superficial *C–terminal tubulin tails* (CTTs), a subject of intense experimental investigation in recent years. We provide an exposition of the biological framework for the CTTs and describe how their dynamics may be influenced by solitonic effects induced by the local electromagnetic field around neuronal microtubules. In this way, the CTT dynamics as modeled on systems of Klein–Gordon type (such as the sine/sinh–Gordon equations and their double-well companions) may be influential on attachment sites and motor–protein functions within the influence of a water electric dipole field. These are processes which we conceive as linked to recent models of dendritic Ca^{2+} signaling and ion passage through membranes [22,23].

The mechanisms discussed here are seen as cooperating with certain vital neurophysiologal procedures exhibiting stochastic properties to which we devote part of the exposition. Ultimately, one hopes to understand how the conformational dynamics of microtubules through protein (enzymatic) tunneling, the operative functioning of exocytosis, neuronal plasticity in relationship to *presynaptic active zones* and the SNARE complex, along with other bioenergetic processes, can contribute to an understanding of the neurobiological basis for mind and memory. We propose that the influence of solitary waves and stochastic resonance within the environment of the cytoskeleton may be essential for tunneling through energetic barriers that facilitate synaptic pore opening and membrane fusion.

Time and space do not permit a complete survey of the vast amount of literature available on the subject (both specialized and interdisciplinary). Electrobiological sensors, protein catalysis, and electron tunneling within molecular processes are just some facets of current nanomolecular research linked to biophysics and biomedical engineering (see e.g., [88]). However, let us address the "information" aspect. It is clearly not implied in the sense of a library database, and neither do we make an excursion into quantifying information relative to conditional entropy and automata, about which much has been written. Recall that a neuron comprises three basic components consisting of dendrites, the soma, and the axon forming, in their respective order, an input–output system in which the soma may be viewed as a kind of central processor (see e.g., [55]). In broad terms, this is the context for information as considered, vis-a-vis, the study of sensory receptors leading up to neural networks. Our particular approach includes the underlying (biological) dynamic processes, not only restricted to those enabling communication via dendritic signaling, phase-locking, etc. (as for neural networks), but with the purpose of reaching a broader level of events by including protein catalysis, solitonic effects, and the electrochemistry of the cytoplasm as members of a combined work force that could engender transition states contributing to such a mode of information. Consequently, we are able to develop an equation-oriented empirical model based on select neurobiophysical phenomena as recently investigated and supported by hypotheses that are proven or are strongly plausible.

17.2 Solitons and Bioenergetic Systems

17.2.1 Background to Solitons

Solitons can be viewed as solitary, pulsating waves that retain their form and velocity upon collision. They are dissipationless waves whose theory and applications are fundamental to areas of nonlinear science such as quantum physics, statistical mechanics, atmospherics, oceanography, telecommunications, cellular automata, and molecular–biological systems. Some well-known examples appearing in the wealth of literature on the subject include the equations of Korteweg–de Vries, Boussinesq, Klein–Gordon, and the nonlinear Schrödinger equation, besides the related mathematical models such as Lax systems, the Toda lattice, and the KP–hierarchy [1,17,27,35,95]. These robust, often bell-shaped, waves can propagate in a pulsating manner while retaining their shape and velocity in undergoing collisions. So in a sense they can be compared with systems of interacting particles. Their universality as a nonlinear phenomenon suggests they are essential to understanding the processes of biological and physiological information within a unified framework.

There is a very significant class of solitons for which the governing equations constitute part of a hierarchy of integrable, or "solvable", systems admitting high degrees of symmetry and for which the techniques of spectral transform and inverse scattering are frequently employed [1,17,35]. For suitable functions $u \equiv u(x, t)$, the prototypes considered mainly in this chapter include the *nonlinear Schrödinger equation* (NLS)

$$\iota\frac{\partial u}{\partial t} + \frac{\partial^2 u}{\partial x^2} + 2u|u|^2 = 0 \tag{17.1}$$

and the *sine–Gordon equation* (SG)

$$\frac{\partial^2 u}{\partial t^2} - \frac{\partial^2 u}{\partial x^2} = \pm \sin u \tag{17.2}$$

The NLS equation, and the SG equation play an important role throughout theoretical physics. The SG equation models spin propagation, electromagnetic fields within plasma, wave packets in dynamical media, resonant optical pulsations, and the propagation of fluxons in Josephson junctions. One master system comprises the general nonlinear *Klein–Gordon equation*, which can be elegantly expressed in terms of characteristic coordinates $\xi = \frac{1}{2}(x + t)$, $\tau = \frac{1}{2}(x - t)$ and appropriate functions Φ, by

$$\frac{\partial^2 u}{\partial \xi \, \partial \tau} = \Phi(u) \qquad (17.3)$$

From Equation (17.3), the SG, along with a collection of other important equations serving as model systems for various phenomena, can be derived (see e.g., [35]), and we will outline several of these next.

Regarded as solutions to nonlinear wave equations, solitons do not normally obey the superposition principle, so much so that when two solutions are combined, a complicated wave is formed. Eventually however, the two waves are seen to actually pass through each other, revealing an unusual characteristic that has far-reaching applications. Of specific interest here are *kink* and *antikink* solutions, which are common to a number of solvable systems with prescribed boundary conditions where spatial derivatives are localized, and typically, where resulting waves pulsate in a twisting fashion under certain asymptotic properties. Besides kink and antikink solutions, there may also be oscillatory solutions known as *breathers* which will play an instrumental role in the development of ideas since these waves admit more structure compared to a usual traveling wave, owing to their internal oscillations, and furthermore, they can evolve without energy activation. Detailed analysis of SG systems and a means of obtaining their solutions, can be found in, for example, [2,34,35,81].

17.2.2   Bioenergetics and Molecular Chains

Significant to the biological perspective is a wide class of nonlinear equations which can be derived or transformed from either the NLS and SG equations, or other equations of solitary wave type. The corresponding models feature in various aspects of molecular theory, nonlinear optics, bioenergetics and neurophysiology, as realized for instance in the quanta of collective excitations in quasiperiodic molecular chains, electron transport, ATP (adenosine triphosphate)–hydrolysis, proton pumps in cellular membranes, and neuromuscular contractions. Many of these phenomena have been given comprehensive biophysical treatment in the works of Davydov [26,27]. We also recommend the reader to the paper of Scott [95] for a masterly exposition of Davydov's ideas.

Particular attention is paid to the role of solitons in α–helical protein molecules. Davydov's solitons propagating through the backbone of a protein, mediate conformational transitions that could be utilized not only for specific biological functions such as actin–myosin contraction [28] or enzymatic catalysis, but may be essential for folding of the protein into its native state [18]. Hence the solitons should not be simply viewed as exotic events explaining a particular biological function, but instead should be viewed as indispensable for the proper definition of the living state since they are actively engaged in sustaining the natural biological structure of proteins.

Regarding the protein constitution of the cytoskeleton, protein molecules (such as myosin–actin complexes, tubulin, etc.) once incorporated, proceed to create transduction energy and intracellular couplings, all of which assist and determine energy release of hydrolysis of ATP or GTP (guanosine triphosphate) molecules, while at the same time the molecule's excited states induce a resonant interaction between peptide groups located within distinct chains. According to Davydov [26,27], a class of solitons evolve at the origin of each chain and so can be created within short intervals of α–helix proteins. Such solitons are often referred to as *polarons* with characteristic properties such as a valid continuum limit approximation, they have an "acoustic" nature since their self-localization is induced by interactions with acoustic modes of a corresponding lattice, and they are weakly coupled since the anharmonic energy is small compared to the phonon bandwidth [95]. At a later stage, we will further discuss the role of Davydov solitons in regard to molecular chains, excitations of the amide–I group and their relevance to anesthesia.

17.3   Klein–Gordon Equations

17.3.1   Slowly Varying Envelope and Phase Approximations

We commence by recalling a basic framework for studying the interaction of electromagnetic and dipolar fields. If $E(x, t)$ denotes an electric field, $P(x, t)$ an electric dipole field, and n a (small) number of 'atomic' densities, then the *Reduced Maxwell–Bloch* (RMB) equation in $(1 + 1)$–dimensions is given by

$$\frac{\partial E(x,t)}{\partial x} + \frac{1}{c}\frac{\partial E(x,t)}{\partial t} = -\frac{2\pi n}{c}\frac{\partial P(x,t)}{\partial x} \tag{17.4}$$

As derived from the Maxwell equation (classical) and the Bloch equation (quantum) for a spin-$\frac{1}{2}$ magnet, the RMB equation is semiclassical, and moreover leads to a completely integrable system. As seen by a sequence of *slowly varying envelope and phase approximations* (SVEPAs), this property of integrability, in relationship to the SG and NLS equations, descends (in order) to the phenomenon of *self-induced transparency* (SIT). The SIT phenomenon arises via short coherent optical impulses (on the order of 10^{-9} sec), which enter into a resonant medium and reshape themselves in "sech" wave forms before reemerging without loss of energy [16].

The SVEPA is governed by resonant optical impulses, nano-second envelopes with carriers at 10^{15} Hz, and exploits ratios of time-scales between 10^{-15} and 10^{-9} sec. In SIT, there is the following form of the electric field

$$E(x,t) = \mathcal{E}(x,t)\cos[kx - \omega t + p(x,t)] \tag{17.5}$$

where $\mathcal{E}(x,t)$ is an envelope function, $p(x,t)$ is a phase function, k denotes inverse wavelength, and ω is the frequency. For the "sharp line" case at exact resonance, the SIT equations are seen to reduce to the system

$$\frac{\partial \mathcal{E}}{\partial x} + c^{-1}\frac{\partial \mathcal{E}}{\partial t} = \alpha P$$

$$\frac{\partial P}{\partial t} = \mathcal{E}N, \quad \frac{\partial N}{\partial t} = -\mathcal{E}P \tag{17.6}$$

where $N = N(x, t, \omega_0)$ and $a = 2\pi n p^2 \omega \hbar^{-1} c^{-1}$. This system can be solved by utilizing the function

$$\bar{u} = \int_{-\infty}^{t} \mathcal{E}(x, \tau)\, d\tau \tag{17.7}$$

together with $P = \sin \bar{u}$ and $N = \cos \bar{u}$. Then, from the preceding equations it can be deduced that

$$\frac{\partial^2 \bar{u}}{\partial x \partial t} + \frac{1}{c}\frac{\partial^2 \bar{u}}{\partial t^2} = \alpha \, \sin \bar{u} \tag{17.8}$$

So, on setting $c = 1, \alpha = m^2$, and implementing a change in the independent variables, we obtain the SG equation

$$\frac{\partial^2 u}{\partial x^2} - \frac{\partial^2 u}{\partial t^2} = m^2 \sin u \tag{17.9}$$

Derived from the latter are kink and antikink solutions, as given by

$$u(x,t) = 4\tan^{-1}\exp\left[\frac{\pm m(x - vt)}{\sqrt{1 - v^2}}\right] \tag{17.10}$$

Here, the "$+$" corresponds to a 2π–kink which takes $u(x,t) = 0$ as $x \to -\infty$, to $u(x,t) = 2\pi$ for $x \to \infty$. Whereas the "$-$" corresponds to the 2π–antikink, which takes $u(x,t) = 0$ as $x \to -\infty$, to $u(x,t) = -2\pi$ for $x \to \infty$.

A breather solution is given by

$$u(x,t) = 4 \tan^{-1}[\tan \mu \, \sin \Theta_I \, \mathrm{sech}\Theta_R] \qquad (17.11)$$

where for $0 < \mu < \frac{\pi}{2}$, we have set

$$\Theta_R = (m \sin \mu)(x - vt)(1 - v^2)^{-\frac{1}{2}}$$
$$\Theta_I = (m \cos \mu)(t - vx)(1 - v^2)^{-\frac{1}{2}} \qquad (17.12)$$

The 2π–kink "sech pulse" equation

$$\frac{\partial u}{\partial x} = \frac{2m}{\sqrt{1 - v^2}} \, \mathrm{sech}\left[\frac{m(x - vt)}{\sqrt{1 - v^2}}\right] \qquad (17.13)$$

describes an electric field in analogy with how electric signals may propagate in an optical fiber. In the absence of any resonant frequency in the latter, the (nonlinear) 'sine' term in the SG equation can, via a SVEPA, be replaced by the (nonlinear) term u^3, yielding a complex field u. A little more work leads to the completely integrable NLS equation

$$-\iota\frac{\partial u}{\partial t} = \frac{\partial^2 u}{\partial x^2} - 2c|u|^2 u \qquad (17.14)$$

Since an envelope modulates a (complex) oscillatory term, this system with parameter $c < 0$ comprises breather-type solutions derived directly from the breather solutions of the SG via an appropriate SVEPA. These breather solutions of the preceding NLS are of the form

$$u = |c|^{-\frac{1}{2}}q \, \mathrm{sech}[q(x - vt)] \exp \iota\left[\left(q^2 - \frac{1}{4}v^2\right)t + \frac{1}{2}vx\right] \qquad (17.15)$$

with $u \to 0$, $\frac{\partial u}{\partial x} \to 0$, as $x \to \pm\infty$, and $q, v \in \mathbb{R}$ (see [16] and references therein).

17.3.2 The Sine–Gordon Equation and Its Double

We have just noted how optical solitons of the Maxwell–Bloch family of equations include both the SG and NLS equations as completely integrable Hamiltonian systems. For suitable functions $u \equiv u(x,t)$, we recall the SG equation in (17.9) (for $m = 1$) in its laboratory coordinates:

$$\frac{\partial^2 u}{\partial t^2} - \frac{\partial^2 u}{\partial x^2} = \sin u \qquad (17.16)$$

The SG equation is integrable by the inverse transform method and corresponds to the Hamiltonian

$$H[u] \equiv \gamma_0^{-1} \int \left\{ \frac{1}{2}\left(\frac{\partial u}{\partial t}\right)^2 + \frac{1}{2}\left(\frac{\partial u}{\partial x}\right)^2 + (1 - \cos u) \right\} \qquad (17.17)$$

where $\gamma_0 > 0$ is the dimension coupling constant.

The partition function (functional integral) in the classical limit is given by

$$Z = \int D\Pi \, Du \exp[-\nu H[u]] \qquad (17.18)$$

where $\Pi = \gamma_0^{-1}u_t$ and the term ν^{-1} corresponds to temperature. The classical integrability of this system involves action–angle variables with Poisson brackets commencing from $\{\Pi(x,t), u(x',t)\} = \delta(x - x')$, which provide an alternative form of the Hamiltonian suitable for starting a quantization procedure involving the functional integral

$$Z = \int D\mu \exp[-\nu H[u]] \tag{17.19}$$

where $D\mu$ represents a certain measure.

By exploiting the analytic properties of the transmission coefficient of the spectral transform, it is shown in [108] that in terms of the action–angle variables, there are kink Δ_k (antikink $\Delta_{\bar{k}}$) phonon phase shifts

$$\Delta(k, k') = \frac{1}{4}\gamma_0[k\omega(k') - k'\omega(k)] \quad \text{[Bose shift]}$$

$$\Delta(k, p) = 2\tan^{-1}\left[\frac{1}{\gamma}[k - v\,\omega(k)]\right] \quad \text{[Fermi shift]} \tag{17.20}$$

where $\omega(k)$ denotes an oscillatory function of phase parameters k, $\gamma \equiv (1 - v^2)^{-\frac{1}{2}}$, and $\Delta_{\bar{k}}(k, p) = \Delta_k(k, p)$, etc. Bose shifts $\Delta_b(k, k')$ and Fermi shifts $\Delta_f(k, k')$ are related by a Bose–Fermi correspondence:

$$\frac{d\Delta_b}{dk} = \frac{d\Delta_f}{dk} - 2\pi\delta(k - k') \tag{17.21}$$

We refer to [108] for complete details (for convenience, we have set $m = 1$ here).

The procedure involving action–angle coordinates and the functional integration method amounts to introducing quantum parameters, thus converting the SG equation to the *quantum sine–Gordon* (QSG) equation in $(1 + 1)$–dimensions where the latter can be described as a Bose system. In [108], the associated kinks and antikinks are postulated as "classical fermions." In strictly mathematical terms, the Bethe ansatz, in relationship to the QSG, provides (bijective) correspondences with certain classes of solvable lattice models such as the quantum spin$\frac{1}{2}$ – XYZ model, Baxter's 8–vertex model of statistical mechanics (which contains the two-dimensional Ising model [9]), together with the quantum massive Thirring model (see [16,77]). In a broad sense, the quasi–relativistic effects and supersonic propagation of solitons in relationship to the SG model, are discussed in [82], demonstrating analogies with acoustical and optical systems, a feature of our discussion that will be explored at a later stage.

The *double sine–Gordon* (DSG) equation as given by

$$\frac{\partial^2 u}{\partial x^2} - \frac{\partial^2 u}{\partial t^2} = \pm\left(\sin u + \frac{1}{2}\lambda\sin\frac{1}{2}u\right) \tag{17.22}$$

in contrast to the SG equation (integrable), comprises a non-integrable system. Relevant here is that by taking the "+" sign and $\lambda > 0$ sufficiently small on the right side, the DSG is exhibited as a perturbation of the SG–equation. The DSG equation is applicable to studying the propagation of intermittent optical impulses, charge density in polymer chains, and the condensate theories of organic linear conduction. Because the DSG leads to a non-integrable system, there are diffirent criteria for kinks–antikinks and radiating breather oscillations (see [15]).

Another perturbed SG–equation

$$\frac{\partial^2 u}{\partial x^2} - \frac{\partial^2 u}{\partial t^2} = \sin u + a_1\frac{\partial u}{\partial t} - a_2\frac{\partial^3 u}{\partial x^2\,\partial t} - a_3 \tag{17.23}$$

models *Josephson Junctions* [35]. Here a_1 denotes a dissipation term, a_2 a damping term due to surface impedance effects, and a_3 a driving term representing a bias current within the system. These terms account for the slowing down of kinks (dissipation) as well as their speeding up (driving), and so they represent the nature of "bunched fluxon" modes for the Josephson effect, something we will return to in Section 17.8.

17.3.3   The Sinh–Gordon Equation and Its Double

The *sinh–Gordon* (SHG) equation is given by

$$\frac{\partial^2 u}{\partial x^2} - \frac{\partial^2 u}{\partial t^2} = \sinh u \tag{17.24}$$

Like the SG equation, it leads to an integrable system solvable by the inverse scattering (spectral transform) method. Both the SG and SHG equations are Hamiltonian systems, but with differing features: whereas the SG admits soliton kink, antikink, and breather solutions on $-\infty < x < \infty$, the SHG does not admit soliton solutions in the usual sense.

The SHG transforms to the SG equation via $u \mapsto -\iota u$, as can be easily seen. The SHG Hamiltonian is given by

$$H[u] \equiv \gamma_0^{-1} \int \left\{ \frac{1}{2}\gamma_0^2 \Pi^2 + \frac{1}{2}\left(\frac{\partial u}{\partial x}\right)^2 + (\cosh u - 1) \right\} \tag{17.25}$$

where again $\Pi = \gamma_0^{-1}\frac{\partial u}{\partial t}$. Quantization and the statistical mechanics of the SHG follow in an analogous fashion to that of the SG equation [14,15]. There are corresponding Bose–Fermi shifts, where the Bose shift Δ_b is a classical propagator of phonons of arbitrary amplitude (but not strictly solitonic in nature).

Following [53], the *double sinh–Gordon* (DSHG) equation is a quasi–exactly solvable system specified by the double-well SHG–potential

$$V(u) = (\zeta \cosh 2u - n)^2 \tag{17.26}$$

where $\zeta > 0$ is a parameter and n a natural number. A particular property of this potential is that the n least eigenvalues and wave functions can be determined. The DSHG–Hamiltonian is given by

$$H[u] = \int \frac{dx}{\ell} \left[\frac{m}{2}\left(\frac{\partial u}{\partial t}\right)^2 + V(u) + \frac{mc_0^2}{2}\left(\frac{\partial u}{\partial x}\right)^2 \right] \tag{17.27}$$

where ℓ denotes the lattice spacing, m the mass of particles (ions), and c_0 the velocity of low amplitude sound waves. The DSHG equation is given by

$$m\frac{\partial^2 u}{\partial t^2} - \frac{\partial^2 u}{\partial x^2} + 2\zeta(\zeta \sinh 4u - 2n \sinh 2u) = 0 \tag{17.28}$$

that is

$$m\frac{\partial^2 u}{\partial t^2} - \frac{\partial^2 u}{\partial x^2} = V'(u) \tag{17.29}$$

It is an equation governing higher and lower phonons and phonon dispersion around the minima $\pm u_0$. Because a double-well potential is involved, then over certain temperature ranges there are meaningful analogs of kink and anti kink (lattice) solutions. One such single kink solution for the DSHG follows as a solution to the equation

$$\frac{1}{2}g\left(\frac{\partial u}{\partial x}\right)^2 = V(u), \quad g = mc_0^2 \tag{17.30}$$

Traveling kinks are obtained by a boost in velocity v, as given by

$$x \mapsto (1 - v^2)^{-\frac{1}{2}}(x - vt) \tag{17.31}$$

17.3.4   The ϕ^4–Equation

The ϕ^4–equation

$$m\frac{\partial^2 u}{\partial x^2} - \frac{\partial^2 u}{\partial t^2} = V_4'(u) \tag{17.32}$$

corresponding to the double-well potential $V_4(u) = [(n+\zeta)u^2 - (n-\zeta)]^2$, transforms under $u \mapsto \tanh u$ to the DSHG and under $u \mapsto \tan u$ to the DSG equation. In [53], numerical analysis of the kink statistical mechanics used simulations derived from an associated Langevin equation to compute the kink density and transportation. Taking η to denote viscosity and $\hat{F}(x,t)$ Gaussian white noise, the Langevin equation for the DSHG is

$$\frac{\partial^2 u}{\partial t^2} = \frac{\partial^2 u}{\partial x^2} - \eta\frac{\partial u}{\partial t} - 4\zeta(\zeta\cosh 2u - n)\sinh 2u + \hat{F}(x,t) \tag{17.33}$$

subject to the statistical averages of the fluctuation–dissipation theorem

$$\langle \hat{F}(x,t), \hat{F}(x',t')\rangle = 2nB^{-1}\delta(x-x')\delta(t-t') \tag{17.34}$$

where B denotes a temperature dependent parameter. The probability distribution function calculated from the Langevin dynamics corresponds to $|\psi|^2$, namely that of the Schrödinger wave function ψ.

Brownian motion in a double-well potential of the generic (and more elementary) ϕ^4–type

$$V_4(x) = \frac{1}{2}ax^2 + \frac{1}{4}bx^4 \tag{17.35}$$

for $-\infty < x < \infty$ (a, b constants) describes noise-driven motion in many models of (bio)physical and chemical systems. A one-dimensional translational Brownian motion of a particle in the well (17.35) with position $x = x(t)$, is governed by a Langevin equation of the form

$$m\frac{\partial^2 x}{\partial t^2} + \eta\frac{\partial x}{\partial t} + ax + bx^3 = F(t) \tag{17.36}$$

where as before, $\eta\frac{\partial x}{\partial t}$ is the viscous drag and $F(t)$ is the white noise driving force subject to statistical averages

$$\langle F(t)\rangle = 0, \quad \langle F(t), F(\tau)\rangle = 2kT\eta\,\delta(t-\tau) \tag{17.37}$$

Ignoring mass, Equation (17.36) becomes simply

$$\eta\frac{\partial x}{\partial t} + ax + bx^3 = F(t) \tag{17.38}$$

The corresponding Fokker–Planck equation for the probability density function $W = W(x,t)$ of the position is given by

$$\eta\frac{\partial W}{\partial t} = \frac{\partial}{\partial x}[(ax+bx^3)W] + kT\frac{\partial^2 W}{\partial x^2} \tag{17.39}$$

(see [21]).

Remark 17.1   Much more can be said concerning the Klein–Gordon systems. For instance, there is the Liouville equation

$$\frac{\partial^2 u}{\partial x\,\partial t} = e^u - e^{-2u} \tag{17.40}$$

which features in the theory of molecular systems subject to laser induced vibrations. Let us also remark, as a further point of motivation, that several of the preceding models are instrumental in the study of optical solitons (see the survey article [16]) and thus bear some relevance to the study of brain models proposed by Pribram [85], which are based upon certain principles of quantum optics and holography (see also [66]).

Remark 17.2   As we have pointed out, a number of the equations considered admit perturbations or transformations to other types as may be realized by environmental effects within the cytoplasm caused by such factors as thermal fluctuation and levels of resonance (or chaos). Such transformations/perturbations should then be seen within the context of a mathematical framework that describes the links between the corresponding neurobiological/physiological processes.

17.3.5   The Fröhlich Dielectric Theory

Besides the work of Davydov (part of which will be discussed in Section 17.9), the relevance of soliton dynamics to biophysics can be traced back to the studies of a number of researchers, in particular, to those of Fröhlich [38,39] who considered one-dimensional electron systems occurring in biological phenomena. When such systems admit holes of some kind, it was conjectured that electron–hole pairing leads to the existence of intra-cellular solitonic dynamics, inducing dissipationless energy transfer. Fröhlich claimed the existence of unusual protein dipole moments and wave frequencies, which were exhibited by cell membranes and certain enzymes. These dielectric systems were considered as producing longitudinal electric oscillations across matter. At suitable levels, energy can be channeled into a single mode and sufficiently ordered so as to sustain coherent electric waves, an ordering suggestive of long-range quantum–coherence that may be comparable to Bose–Einstein (BE) condensation whereby particles unite into a condensate regulated by a single wave function, but particles exterior to the condensate disperse erratically.

Later research revealed molecules beneath the cell membrane as exhibiting dipolar vibrational activity where thin layers appeared to function like biological superconductors from which the resulting (wave) propagation leads to Fröhlich waves possessing a frequency of order 10^{11} to 10^{12} sec^{-1} (see e.g., [51]). In [91] are considered small interactions between neighboring protofilaments in microtubules (see the following) whereby dipoles within a given protofilament comprise a system of quasi-torsion oscillators where the nonlinear dynamics of dipoles in a single protofilament are modeled upon the ϕ^4–potential (17.35) in relationship to Equation (17.32). It was suggested that protein dipoles in a common electromagnetic field exhibit resonating effects when energy is supplied. Such waves are realized by dipolar oscillations maintained by hydrogen bonds and non–localized electrons within the hydrophobic regions of protein molecules. However, subsequent computations in [109] revealed the ferroelectric model of microtubule lattices to be insensitive to external electric fields unless the field strength exceeds 10^4 to 10^5 Vm^{-1}, while complete ordering of all tubulin dipoles is observed for electric fields exceeding 10^7 Vm^{-1}. Such electric fields are not sustainable in neuronal cytoplasm where the field strength is just on the order 1 to 10 Vm^{-1} [47]. In order for the external field to be effective, it must overcome the dipole–dipole interaction energies [109], so it appears that a typical ferroelectric model is unsuitable for subneuronal computation. More recently, we have shown that microtubules may be sensitive to external (cytosolic) electric fields via their C–terminal tails [41–43,47] since the coupling between individual tails is exceeded by the thermal energy. In fact, recent simulations in [43] have revealed the tail–tail coupling is exceeded by eight orders of magnitude by the thermal energy.

17.4   Microtubules and C–Terminal Tubulin Tails

17.4.1   Neurobiological Background to Microtubules

The neuronal cytoskeleton is the major internal structure that defines the external shape and polarity of the neuron and organizes its cytoplasm to perform motile and metabolic activities essential to life.

The constituents of the cytoskeleton consist of microtubules, intermediary filaments, actin, microtubule-associated proteins (MAPs), motor proteins, cross–linking proteins (e.g., the plakin superfamily), cytoskeletal scaffold proteins, along with various cytoskeletal-bound phosphatases or kinases.

Microtubules form the major component of the cytoskeleton and are composed from tubulin, a heterodimer of two 50-kDa subunits, types α and β. The microtubule wall is a 2D polymer of such tubulin dimers, each of which is connected by two types of bonds. Longitudinal bonds connect dimers into protofilaments (PFs) and lateral bonds connect dimers in adjacent PFs. The PFs consist of alternating α and β subunits that are spaced 4.0 to 4.2 nm apart. *In vivo*, assembled microtubules have 13 PFs and form a cylinder-like structure with wall thickness of 4 nm and a diameter of 24 nm.

Microtubules organize the cellular shape, which is vital for neurons in stabilizing their cable-like projections called *neurites* (dendrites or axons). They also provide anchors for compartmentalization of various enzymes, including some of those involved in *glycolysis*, different *phosphatases* or *kinases*, etc., and serve as tracks for regulating the functions of motor proteins (such as kinesin, dynein, etc.) which drive cargo vesicles that transport various proteins and structural components of the plasma membrane of neurons.

Let us outline a basic principle concerning the assembly of α/β tubulin dimers. This is a process requiring nucleotide GTP to be bound to both α and β tubulins prior to assembly. The α–bound GTP never hydrolyzes, whereas the GTP–molecule tied to the β–tubulin is hydrolyzed to nucleotide GDP (guanosine diphosphate) soon after the dimer is incorporated into the growing microtubule lattice. We recall that in forming the protofilament there is the intermediate unstable state

$-\alpha$–tubulin–GTP–β–tubulin–GTP–α–tubulin– GTP–β–tubulin–GTP-

that is converted into

$-\alpha$–tubulin–GTP–β–tubulin–GDP–α–tubulin– GTP–β–tubulin–GDP-

The released energy from the GTP hydrolysis is then stored in the microtubule wall as an elastic strain, and the β–tubulin bound GDP cannot be further phosphorylated or exchanged for GTP because the successive α–tubulin in the protofilament occludes the preceding β–tubulin nucleotide binding pocket [60]. There remains only at the microtubule plus end a single layer of β–tubulin bound GTP molecules that form a protective microtubule "cap," which if hydrolysed, triggers fast (microtubule) disassembly.

17.4.2   C–Terminal Tubulin Tails

Part of the vital functions performed by microtubules can be attributed to the tiny C–*terminal tubulin tails* projecting from each tubulin monomer. These were revealed by Sackett [90] who suggested that microtubules are not smooth cylinders but in fact possess such superficial, tiny hairy-like projections of 4 to 5 nm length extending from each tubulin (see Figure 17.1). Owing to the high flexibility of these projections, the PDB structure was revealed only recently in [68] where the helicity of each of the α (404–451) and β (394–445) tubulin C–terminal recombinant peptides was determined by using nuclear magnetic resonance (NMR). It was shown that the C–terminal domain of tubulins has a different length and structure in α– and β–tubulin. In general, the C–terminal domain has a C–terminal helix H12 and a random coil C–terminal tubulin tail. In α–tubulin molecules, amino acid residues 418–432 form the C–terminal helix H12 and amino acid residues 433–451 form the α–tubulin tail. The α–tubulin tail is 19 amino acids long and possesses 10 negatively charged residues. The situation in the β–tubulin C–terminal domain is more interesting. In [68] was exhibited a 9 amino acid longer helix H12 of the β–tubulin compared to previous PDB models (cf [83]). This suggests an extension in the protein, supporting the possibility of a functional coil–to–helix transition at the C–terminal zone.

The β–tubulin C–terminal helix H12 is formed by amino acid residues 408–431, but it seems that the reversible transition between the coil and helix of the last nine amino acid residues 423–431 could either decrease or increase the length of the helix H12, at the same time increasing or decreasing (respectively) the β–tubulin tail length. The β–tubulin tail has 14 amino acids and nine negatively charged residues, but depending on the conformational status of the residues 423–431, the β–tubulin tail random coil can extend to 23 amino acid residues, bearing 11 negative charges. Following the C–terminal helices α–H12

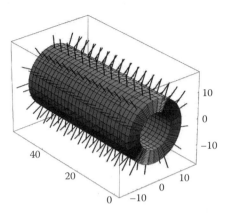

FIGURE 17.1   Computer simulation of a segment of 13-protofilament, 3-start helical left-handed "haired" micro-tubule. The outer microtubule diameter is 24 nm, the inner diameter is 14 nm, the microtubule wall thickness is 5 nm, and with a CTT length of 5 nm. The CTTs extending from each protofilament are visualized as rows of hairy projections.

and β–H12, the C–terminal residues of the respective α– and β–tubulin tails are observed to be disordered by NMR. It is a dynamical disordering and is effectively the manifestation of the extreme sensitivity of the tubulin tails to environmental conditions and local electric fields, thus giving rise to a wide range of metastable conformations [47].

17.4.3   Dynamics of Motor Proteins

Located within dendrites and axonal projections, microtubules serve as tracks for the transportation of post-Golgi vesicles by microtubule bound motor proteins (such as kinesin and dynein). Microtubules, however, are not passive elements in the vesicle transport, and results of experiments have shown that the tubulin C–terminal tails modulate the kinesin function. The results of [96] have revealed the β–tubulin tail as interacting with the kinesin switch II domain, while the α–tubulin tail possibly interacts with kinesin α 7–helix in such a way that after the kinesin bound ATP is hydrolyzed, the kinesin takes a stroll along the microtubule surface. Native microtubules that possess tubulin tails cannot be decorated by ADP (adenosine diphosphate)–kinesin molecules because of the weak ADP–kinesin/tubulin tail binding, while subtilisin treated microtubules that lack tubulin tails bind stably ADP–kinesin, thus blocking the kinesin walk. From such an experiment, one may conclude that the tubulin tails catalyze the detachment of the kinesin–ADP complex from the microtubule surface, thus permitting the kinesin dimer to take a "step" along the microtubule protofilament.

Microtubules do not only regulate motor protein function but also attach with their C–terminal tubulin tails different MAPs and protein kinases and phosphatases, thus organizing the intraneuronal space. The proper attachment/detachment of these proteins could regulate their enzymatic activity. In case studies of schizophrenia, there was found an altered expression of MAP2 and MAP5 that results in abnormalities in the neuronal cytoarchitecture [6]. In the case of Alzheimer's disease, the primary alteration is the phosphorylation status of axonal MAP–tau and the activity of microtubule bound protein phosphatase 2A (PP2A) regulated via the attachment/detachment to microtubules [98].

We consider the mechanism of the tubulin tail enzymatic action to be generated by vibrationally assisted tunneling — an emerging key concept which was experimentally verified within the last few years (see [103,104] and the discussion later). A locally formed tubulin tail standing breather could promote or suppress conformational tunneling of a molecule attached to the tubulin tail. The effect of vibrations on mixed-tunneling could be either to promote or to suppress the tunneling process, and this depends on certain boundary conditions [105,106]. Formally, the mechanism of the tubulin tail breathing action could be manifestly a form of enzymatic energase action. Motivation comes from the fact that energases do not in themselves provide a source of energy, but rather, induce conformational transitions in a molecule that might have accumulated energy in an intermediate highly energetic conformational state [87]. The accumulated energy is derived from hydrolyzed ATP or GTP in preliminary biochemical steps, so for that reason this energy is usually called "primed energy" and "priming" is the process of energy accumulation in metastable protein states.

The idea that microtubules might be agents of subneuronal processing of information was originally suggested by Hameroff and Watt [58]. In [54], it was conjectured that the energy for computation could be delivered from the tubulin bound GTP molecules. Since it had been already observed that in stable microtubules there is no possibility for tubulin bound nucleotide cycling, we propose that the tubulin tail energase action releases the energy accumulated in metastable conformational states of kinesin, dynein, or phosphorylated MAPs. The metastable states of these proteins are produced via ATP hydrolysis through previous stages of priming. We note that ideas on GTP–hydrolysis, ferroelectric phase, and the CTTs as possible agents of information transfer, have been reported in [43,44,47,92].

17.5   Dynamics of the Water Electric Dipole Field

17.5.1   The Water Electric Dipole Field

As an organizational framework for specific neurobiological processes, the cytoskeleton is the major structure that provides a surface to which water molecules cling, thus facilitating the (water) ordering. Note that the term "water" used here is not quite the same as its usual sense, but instead should be regarded as a protein-like saturated mixture. Ordered (vicinal) water molecules are microscopic dipoles that interact with each other via hydrogen bonds, inducing a relatively high viscosity, surface tension, and dielectric constant. They form the water electric dipole (WEDP)–field occurring on either side of a brain cell. Within the interior of the cell, the water molecules generate a WEDP–field in the vicinity of the cytoskeleton, whereas in the exterior of the cell the molecules form an intercellular flow completing the regions between neighboring cells. Influential are combined electrostatic and van der Waals forces resulting from strong protein surface–water interactions (see e.g., [75]). In a series of papers by Del Giudice *et al.* [30–32], it was proposed that electromagnetic waves arising from the WEDP–field within the body of the cytoskeleton, can create signals compatible in size with the internal diameter of a given microtubule. Later developments were taken up by Jibu *et al.* [63,65–67], which to some extent we follow in this section.

Let $\mathbb{V}$ denote a perimembranous (spatial) region in the vicinity of a cytoskeletal microtubule. The WEDP–field in $\mathbb{V}$ within a cylindrical neighborhood, is represented by a 2–spinor field

$$\psi(\mathbf{x}, t) = \begin{bmatrix} \psi^+(\mathbf{x}, t) \\ \psi^-(\mathbf{x}, t) \end{bmatrix} \tag{17.41}$$

where $\psi^+(\mathbf{x}, t)$ and $\psi^-(\mathbf{x}, t)$ are spinor components. The electric dipole moment is given by

$$\mu = \psi(\mathbf{x}, t)^* \frac{\hbar}{2} \sigma \, \psi(\mathbf{x}, t) \tag{17.42}$$

where $\sigma = [\sigma_1, \sigma_2, \sigma_3]$ is a 3–vector whose components consist of the Pauli spin matrices. The dipole moment μ exhibits the water molecule as having the characteristic of a quantum–mechanical spinning top. In other words, it is due to μ that the water molecules interact dynamically with the quantized electromagnetic field in $\mathbb{V}$. If m_p and e_p denote the proton mass and charge respectively, then the average moment of inertia of a water molecule is estimated as $I = 2m_p d^2$ with $d \approx 0.82\text{Å}$, whereas μ is estimated as $\mu = 2e_p P$, with $P \approx 0.2\text{Å}$. Given $\psi(\mathbf{x}, t) \neq 0$ only holds at each position $\mathbf{x} = \mathbf{x}_k$ of the k–th manifestation of localization, the WEDP–field with N localizations are describable in terms of N spin variables, as given by

$$s^k(t) = \psi(\mathbf{x}_k, t)^* \sigma \, \psi(\mathbf{x}_k, t) \qquad 1 \leq k \leq N \tag{17.43}$$

The Hamiltonian of the WEDP–field for N water molecules with energy difference ϵ, is given by

$$H_{WM} = \epsilon \sum_{k=1}^{N} s_3^k(t) \qquad (17.44)$$

where for a given wave vector k_0, it is convenient to assume that a normal mode has an angular frequency ω_{k_0} resonating to the energy difference between two principal eigenstates for which $\epsilon = \hbar\omega_{k_0}$ ($\epsilon \approx 24.8$ meV), as in accord with the predictions of dominance over other possible energy exchanges [32]. The radiation field of $\mathbb{V}$ is given by a scalar electric field operator $E = E(\mathbf{x}, t)$ whose associated Hamiltonian is given by

$$H_{EM} = \frac{1}{2} \int_{\mathbb{V}} E^2 \, d^3\mathbf{x} \qquad (17.45)$$

A main postulate of [65] is that the dynamics of the WEDP–field and the quantized electromagnetic (EM) field is an energy interchange through creation and annihilation operators of photons. In order to see this, consider a decomposition of the electric field operator $E = E^+ + E^-$ into its positive and negative frequency components. Then the Hamiltonian for the interaction between the WEDP–field and the EM–field is given by

$$H_I = -\mu \sum_{k=1}^{N} \{ E^-(r^k, t)s_-^k + s_+^k E^+(r^k, t) \} \qquad (17.46)$$

where $s_\pm^k = s_1^k \pm \iota s_2^k$. The total Hamiltonian H_{QM}, which governs the quantum mechanical dynamics of the electromagnetic field and the dipolar vibrational field of water molecules along with their interaction, is then expressed by

$$H_{QM} = H_{EM} + H_{WM} + H_I \qquad (17.47)$$

Since parts of the region $\mathbb{V}$ in the vicinity of the cytoskeleton can be considered as a cavity for the electromagnetic wave, we introduce the normal mode expansion of E given by

$$E^\pm(\mathbf{x}, t) = \sum_\lambda E_\lambda^\pm(t) \, \exp[\pm\iota(\lambda \cdot \mathbf{x} - \omega_\lambda t)] \qquad (17.48)$$

17.5.2 Biological Motivation

At this juncture of development, we comment upon several instances from the biological viewpoint, motivating such models that can be compared with mechanisms for propagating optical/acoustical solitons. It was discovered in [75] that the mesoscopic ordering of water near proteins and between surfaces of interacting proteins is favorable for intermolecular electron tunneling. This is an unexpected phenomenon as recently reported in [75] and is referred to as *the water–mediated tunneling effect on electron transfer kinetics*. Secondly, the process of signaling response in synapses is influenced by certain classes of *cellular adhesive molecules* (CAMs) in which the actin cytoskeleton provides a suitable structural mechanism for assimilating the signaling inputs. The formation of functional synapses at an axonal growth cone involves identifying and initiating contacts with suitable companion cells [12]. Recent studies have shown that there are two types of CAMs essential for synaptic formation. These are known as *β-neurexin* and *neuroligin* and it is remarkable that their interaction has the unique capability of acting as a bidirectional trigger for synaptic formation [29]. Once the *β*–neurexin–neuroligin transmembrane protein junction is formed, it triggers synaptic differentiation at the site of contact. *β*–neurexin is located in axons and interacts presynaptically with CASK, a multidomain scaffolding protein that organizes the presynaptic space and emits signals to the actin cytoskeleton via protein 4.1. *β*–neurexin also directly interacts with the synaptic vesicle protein synaptotagmin-1, thus controlling exocytosis and neuromediator release (see Section 17.11). Synaptotagmin-1 *per se* might act as a MAP molecule binding to *β*–tubulin tails and stabilizing microtubules in high Ca^{2+} concentration at the pre-synapse. Neuroligins (neuroligin-1 is specific for excitatory synapses, while neuroligin-2 is specific for inhibitory synapses) are located in dendrites and transmit information to the postsynaptic density protein (PSD-95), which is a multidomain scaffold protein that anchors different ion channels at the active zones of the postsynaptic membrane. PSD-95 is anchored to postsynaptic microtubules via another protein known as CRIPT. Neuroligins on binding with presynaptic β–neurexins, constitute an adhesive system facilitating learning processes manifested as a morphological reorganization of the synapse. We note that in view of these properties the radiation field of (17.48) could thus be considered as falling within this junction as shielded by the ordered water molecules, and consequently it assists the signaling mechanism between neighboring neurons [45,46].

This seemingly over-detailed description of the various proteins forming the cytoskeletal scaffold in the presynaptic and postsynaptic neuronal interior is nevertheless indispensable for the precise understanding of synaptic interneuronal communication. Whereas in the classical computational approach, the synapse is considered as just an electric switch that operates with some delay (on the order of 1 ms), it is now accepted that neuronal communication, *in vivo*, is much richer in depth. For example, the presynaptic release of neurotransmitter molecules via exocytois (see later) occurs across a wide range of probabilities varying from 0.15 to 0.75, and suggests that subneuronal events control the fine–tuning of the release probability of each presynaptic terminal. But this is not the whole story. After the neuromediator is released, it binds to the postsynaptic membrane receptors and triggers the postsynaptic electrical currents. Among the excitatory glutamatergic synapses, of central importance are the NMDA (N-methyl-D-aspartate) receptors that permit Ca^{2+} entry into neurons. The Ca^{2+} ions not only influence the NMDA receptor binding to PSD-95 and other scaffold proteins, but also regulate the NMDA receptor binding to actin via a special protein, known as *drebrin* which is responsible for dendritic formation. The postsynaptic Ca^{2+} flux leads to an actin-based contraction as well as to an alteration in the spine neck diameter as verified by special videomicroscopic techniques [37,79]. Thus, any feasible computational approach should be reflected in a very nonlinear descriptive model of neuronal activity, *in vivo*. The latter is due to the fact that dendritic spines contract (and thus alter their cable length), their spine necks rapidly alter their diameters in millisecond time-scale (thus varying the cable resistance), along with distributed boosting "hot spots" in the dendritic tree that could amplify the postsynaptic potentials (PSPs). One should also keep in mind the aforementioned fine–tuning of the probability for exocytosis. It is clear that a suitable explanation of the very nonlinear stochastic or chaotic behavior of the neuronal activity as observed, may only be realized through the study of the cytoskeletal proteins via their interaction with the ordered vicinal water and the radiant EM field [45,46,64,65].

We suggest a suitable biological motivation for the electric field operator $E = E(x,t)$ leading to the Hamiltonian H_{EM} in (17.45). This concerns a basic property of cytosol as an electrolytic solution, whereby ionic currents are induced by electrogenic currents across the dendritic and axonal membranes. As proposed in [43], such cytosolic currents (and the associated EM field) may generate propagating conformational waves along the CTTs. The ionic cytosolic currents may be projected onto the CTTs of the underlying microtubules, thus inducing a dynamical interplay between the prevailing repulsive effects of the CTTs from the negatively charged surface of the microtubules.

The electric potential φ at a point x in neuronal cytosol varies in time as a result of the opening or closing of the membrane incorporated ligand or voltage-gated ion channels and associated ion fluxes. The spatio–temporal evolution of a single PSP generated by ionic flux via ligand-gated ion channels can be obtained as a solution of the passive *cable equation* as expressed by

$$\varphi(x,t) = \varphi_0 \exp\left[-\frac{x}{\lambda} - \frac{t}{\tau}\right] \tag{17.49}$$

where λ denotes the spatial constant, τ is the time constant of the neuronal projection, $\varphi(x,t)$ is the potential at a point x in time t, and φ_0 is the potential of the cytosol at the site of the PSP generation. In cases where there is a boosting of the electric potential due to a rapid opening of nearby located voltage-gated ion channels, the dynamical equations of the electric potential along the neuronal projections take a nonlinear form, as in the models of *Hodgkin–Huxley* and *Baer–Rinzel* [7,22].

In all events, however, the resulting cytosolic electric field depends upon the gradient of the electric potential in cytosol via the relation $E = -\nabla\varphi$, and is a manifestation of the existing ionic concentration gradients. This is demonstrated by the *Nernst equation* for positive ions of type i with concentration c_i, and valence z_i. The resulting electric field is

$$E = -\frac{\partial\varphi}{\partial x} = -\frac{RT}{z_i F}\frac{1}{c_i}\frac{\partial c_i}{\partial x} \tag{17.50}$$

where $R = 8.314, J\ mol^{-1}K^{-1}$ is the gas constant, T is the temperature, $F = 96500\ C\ mol^{-1}$ is Faraday's constant, and φ is the electric potential.

Even though inside cytosol the electric potential at a given point x depends not only upon the intracellular gradient of diffusible ions, but also upon the screened electrostatic fields contributed by vicinal charged biomolecules, we note that it is the dynamics of the ionic concentration gradients that provide information for external events. Thus, if the subneuronal processing of information occurs, it will surely depend also on internal electric fields, as contributed by charged biomolecules in the cytosol. Yet, in a certain sense the potential difference and its associated electric field between various cytosolic points, due to existing concentration gradients of diffusible ions, will alone determine the component of information that refers to external events. Since neurotransmitters can also trigger G–protein coupled receptors, then part of the external information is delivered to the cytoskeleton in the form of attachment/detachment, and/or activation/inhibition of the various cytoskeletal bound kinases or phosphatases, or other regulatory scaffold proteins activated by G–proteins or subsequent second messengers.

The cytosolic EM field is a direct result of electrogenic transmembrane fluxes that generate a concentration gradient of diffusible ions in cytosol, and thus the potential difference and the electric field intensity can be computed in principle by the Nernst equation in (17.50). Indeed, it is the ionic concentration gradients that motivate, introducing the electric field $E = E(x, t)$ that interacts with the WEDP-field and the CTTs of microtubules.

17.5.3 Collective Dynamical Variables of the Cytosolic Water

Following [65], we introduce collective dynamical variables $S_\lambda^\pm$ for water molecules, given by

$$S_\lambda^\pm(t) = \sum_{k=1}^{N} s_\pm^k(t)\ \exp[\pm\iota(\lambda \cdot \mathbf{x} - \omega_\lambda t)] \tag{17.51}$$

On setting $S \equiv \sum_k s_3^k$, we can express (17.47) in the form

$$H_{QM} = H_{EM} + \epsilon S - \mu \sum_\lambda \{E_\lambda^- S_\lambda^- + S_\lambda^+ E_\lambda^+\} \tag{17.52}$$

Equation (17.52) resembles that of the Hamiltonian for a laser radiation process (cf [33]), so suggesting that the water molecules of $\mathbb{V}$ exhibit a laser-like coherent optical property, *provided the energy is sustained above a certain threshold* (see Equation (17.57) later). The dynamically ordered region of water molecules and quantized EM–field are considered within a coherence length of 50 μm. The explanation is that by increasing the ordering of water on the microtubule surface, spontaneous symmetry breaking occurs (see the following), thus creating Nambu–Goldstone (NG) bosons, the quanta of long-range correlation waves of the aligned electric dipoles referred to as *dipole wave quanta*, denoted DWQ [65,112,113].

The Hamiltonian H_{EM} can also be expressed in terms of canonical operators (observables) $P_\lambda(t)$ and $Q_\lambda(t)$, as defined by

$$P_\lambda(t) = \sqrt{\frac{\hbar\omega_\lambda}{2}}\ \iota(E_\lambda^- - E_\lambda^+)$$

$$Q_\lambda(t) = \sqrt{\frac{\hbar}{2\omega_\lambda}}\ (E_\lambda^- + E_\lambda^+) \tag{17.53}$$

which in [65] are seen to satisfy the well-known canonical commutation relations of the Heisenberg algebra. On making the necessary transformations and substituting into (17.52), this leads to

$$H_{QM} = \frac{1}{2} \sum_{\lambda} \{P_\lambda^*(t) P_\lambda(t) + \omega_\lambda^2 Q_\lambda^*(t) Q_\lambda(t)\} + \epsilon \sum_{k=1}^{N} s_3^k(t)$$

$$- \sqrt{\frac{2}{\hbar}} \mu \sum_{k=1}^{N} \sum_{\lambda} \left\{ \sqrt{\omega_\lambda} Q_\lambda(t) s_1^k - \frac{1}{\sqrt{\omega_\lambda}} P_\lambda(t) s_2^k \right\} \qquad (17.54)$$

17.5.4 Spontaneous Symmetry Breaking

Suppose we consider a system possessing a certain symmetry but where the prevailing vacuum state is not invariant under this symmetry. The vacuum state might possibly be transformed into some other degenerate state, whereas the Lagrangian symmetry is independent of the vacuum solution. In other words, the Hamiltonian may be invariant under the symmetry transformation but the vacuum (or lowest energy) state is not. That is, spontaneous symmetry breaking (SSB) occurs, resulting in massless quanta governed by BE–statistics that are assigned to repair the broken symmetry. The NG bosons are understood to be the quanta of long range coherence induced by the vacuum state, which violated the original dynamical symmetry. Typically, what might otherwise be two massive fields emerge from SSB as one massive and one massless field, the latter in this case is a NG boson. In [65], this is explained when the corresponding Heisenberg equations of (17.54) are considered in studying the dynamically ordered state of the WEDP-field in terms of a long-range alignment of associated spin variables. Under an SO(2)–transformation of the canonical variables, the Hamiltonian H_{QM} is invariant, whereas a time-independent solution is not invariant.

17.6 Langevin Pumping and Energase Action

17.6.1 Coherent Pulse Emission and the Water Laser

The water laser effect of the WEDP had been considered in [32]. We further propose that in order for the coherent emission of photons to have the necessary biological impact, it is necessary to consider time-scales of the order of 10 to 15 picoseconds, which are compatible with that of protein (enzymatic) action. In the presence of a disordered thermodynamic system, thermal fluctuations, noise, and dissipation also have to be taken into account. However, the laser-like emission of coherent photons may still be realized under such circumstances given that the protein molecules can attain an energy sufficient to engage a pumping effect of the WEDP–field. This "slow phenomenon" (dynamic time-scale $10^{-11} s$) involving the water laser is preferred in this situation to the "fast phenomenon" (dynamic time-scale of $10^{-14} s$) of "superradiance" (cf [63]). Jibu and Yasue in [65] consider the relevant system of Heisenberg–Langevin equations governing the collective dynamics of the quantized EM–field in the spatial region $\mathbb{V}$. On assuming a certain coherent state representation, these are seen to reduce to the stochastic Langevin equation

$$\frac{dZ}{dt} = \alpha_1 Z - \alpha_2 \overline{Z} Z^2 + B \qquad (17.55)$$

where $Z = Z(t)$ is a Markov process in $\mathbb{C}$ of the corresponding EM–field operator, $B = B(t)$ is a (complex) Gaussian white noise of thermal fluctuations of quantized EM–field, and α_1, α_2 are particular constants depending on the volume V of the region, thermal fluctuations for the EM and WEDP–field, damping coefficients (denoted γ, γ_0) for the WEDP–field, and a parameter of pumping rate (denoted S_∞) resulting from the interaction of the WEDP–field with the dynamics of the microtubule protein molecules. These parameters in turn are used to define a diffusion constant D, which along with the probability density function $f = f(z, \bar{x}, t)$ of $Z(t)$, transform equation (17.55) equivalently to the associated Fokker–Planck equation

$$\frac{\partial f}{\partial t} = -\frac{\partial}{\partial z}[(\alpha_1 z - \alpha_2 \bar{z} z^2) f] + D \frac{\partial^2 f}{\partial z \, \partial \bar{z}} \qquad (17.56)$$

The required level of excitations of the quantized EM–field, namely the photon emission as induced by the supplied metabolic energy, is attained when the pumping rate S_∞ satisfies the estimate

$$S_\infty > \frac{\hbar^2 V \gamma_0 \gamma}{4\pi \epsilon f^2} \qquad (17.57)$$

Accordingly, it is suggested that the energy for the coherent pulse emission by vicinal water within a proximity of 4 to 5 nm of the microtubule's outer surface could be gained from tubulin electric dipole oscillations and/or from vibrations along the microtubule walls, as well as from C–tubulin tail energase action releasing energy by metastable ADP–kinesin complexes moving on the microtubule surface, and phosphatase action of microtubule anchored phosphatases such as PP2A or other MAPs. The transmission of pulse mode coherent photons is consider to be determined by the Maxwell–Bloch equation as derived from the total Hamiltonian H_{QM}. For $E = E(z, t)$, it is given by the quantum dynamical equation of motion [4,63,65]:

$$\frac{\partial E^\pm}{\partial z} + \frac{1}{c} \frac{\partial E^\pm}{\partial t} = \mp \iota \frac{2\pi \epsilon \mu}{\hbar V} S^\pm \qquad (17.58)$$

In terms of a quantum average, denoted $\langle \rangle_q$, the expression for the electric field is

$$\theta^\pm(z, t) = \frac{2\mu}{\hbar} \int_{-\infty}^{t} \langle E^\pm(z, u) \rangle_q \, du \qquad (17.59)$$

and the latter leads to an SG equation expressed in Lorentzian coordinates by

$$\frac{\partial^2}{\partial t \, \partial \sigma} \theta^\pm = -2A \sin \theta^\pm \qquad (17.60)$$

where $A = \frac{2\pi \epsilon \mu^2 N}{\hbar^2 V}$, $\frac{N}{V}$ is the number of water dipoles per unit of volume, and $\sigma = t + \frac{z}{c}$. The indices $\pm$ indicate the transverse directions of the electric field where it is assumed there is no propagation in the longitudinal direction. Equation (17.60) is an equation characteristic of SIT realized in nonlinear optics, as discussed earlier, and here suggests how the cumulative effects of the WEDP–field might induce a transfer of energy via dissipationless waves.

Time–differentiating (17.59) leads to

$$E = \frac{\hbar}{\mu} \sqrt{A\rho} \ \text{sech} \left[\sqrt{A\rho} \left(t - \frac{z}{c} \right) \right], \quad \text{where } \rho = \frac{v_0}{c - v_0} \qquad (17.61)$$

The preceding equations were considered in [4] relative to a correspondence between information configurations induced by solitonic interactions and the DWQ at levels of excitation. In view of the appearance of the SG equation (17.60), we recognize these cumulative effects, which include the WEDP–field, as propagating energy within a dynamic mechanism conducive to a neurocybernetic process.

17.6.2 Breather Solutions and Energase Action

In [41,42,47] is described how the water dipoles from the tubulin tail hydration shells that form a 4 to 5 nm layer on the outer microtubular surface, strongly interact with the local electromagnetic field and so influence the conformational state of the tiny C–tubulin tails. The model is based on a long-range interaction of the water molecule dipoles and NG bosons, leading to a coherent emission of photon pulses propagating via tunneling. The resulting solitons could be viewed as traveling conformational waves in the tubulin tails that do not dissipate under thermal fluctuations, but could be pumped by the water laser provided the threshold inequality (17.57) is satisfied. This model also considers solutions to the SG equation as providing the necessary dynamics. To facilitate matters, we make a change of parameters from Lorentzian coordinates (as shown earlier) to laboratory coordinates to recover Equation (17.16):

$$\frac{\partial^2 u}{\partial t^2} - \frac{\partial^2 u}{\partial x^2} = \pm \sin u, \quad u \equiv u(x,t) \tag{17.62}$$

Following [34], we recall that a kink soliton involves a twist in a solution, $u = u(x,t)$ say, which moves from one solution $u = 0$ to an adjacent solution $u = 2\pi$. Vacuum states as constant solutions of zero energy, correspond to $u = 0 (mod\ 2\pi)$. In this respect, the traveling solitons of [65,67] can be regarded as tunneling (evanescent) photons coupled with tubulin tail hydration shells. The working hypothesis proposes a biologically feasible coherence time-scale of 10 to 15 ps.

Such a kink (K) solution u_K of (17.62) as given by:

$$u_K = 4 \tan^{-1} \exp[\gamma_K (x - v_K t - x_K)] \tag{17.63}$$

where $0 \le v_K < 1$ is the kink velocity, x_K the kink position at $t = 0$, and

$$\gamma_K^{-1} = (1 - v_K^2)^{\frac{1}{2}} \tag{17.64}$$

is the kink width. The kink energy is given by $E_K = 8\gamma_K$.

On setting $G = \gamma_K (x - v_K t - x_K)$, one also finds the derived equations:

$$\frac{\partial u}{\partial x} = 2\gamma_K \text{ sech } G \quad \text{(magnetic field)}$$

$$\frac{\partial u}{\partial t} = -2\gamma_K v_K \text{ sech } G \quad \text{(electric field)}$$

$$\sin \frac{1}{2} u = \text{sech } G \tag{17.65}$$

(see e.g., [35]). The antikink solutions correspond to reversing the velocity, $v \mapsto -v$, and taking the negative square root in (17.64).

At this stage, we mention the role of *breather* solutions which are manifestly local oscillating waves resulting from how a kink and antikink can merge into a combined state. Breathers admit more structure compared to a usual traveling wave since the former's internal oscillations can evolve without energy activation. In practice, they have been realized as linear phonon modes that are excitable within thermal fluctuations [89]. It was suggested earlier that some class of propagating optical solitons may influence the conformational states of the tubulin tails. Thus, in [41,42,47] several possibilities involving SG kink–antikink–breather soliton collisions are proposed, where for instance, a standing breather soliton could be coupled to the energase action of the tubulin tails through vibrationally assisted tunneling. Further, we recall how the β–tubulin tails may interact with kinesin switches and the role of the α–tubulin tail in activating the kinesin walk [96].

As outlined in [35], the scheme of Bäcklund transformations can be employed to derive three–soliton from two–soliton solutions. In relationship to the kink solution u_K in (17.63), we adopted in [42] the development of [34] in order to describe a three–soliton solution u_{KB} representing the elastic collision (without exchange of energy or momentum) between a kink and a breather, as it is given by the sum

$$u_{KB} = u_K + w_B \tag{17.66}$$

where the term w_B is explained as follows. Firstly, if ω denotes the frequency of the breather, $0 \le \omega < 1$, we set $\eta = (1 - \omega^2)^{\frac{1}{2}}$. Then

$$\tan \frac{w_B}{4} = \left[\frac{2\omega\eta(\sinh D - \cos C \sinh G) + 2\eta\gamma_K \gamma_B(v_K - v_B) \sin C \cosh G}{2\omega\eta(\cos C + \sinh D \sinh G) - 2\omega\gamma_K \gamma_B(1 - v_K v_B) \cosh D \cosh G} \right] \tag{17.67}$$

where we have set $C = -\omega\gamma_B(t - v_B(x - x_B)) + 2\pi m$, m an integer, $D = \eta\gamma_B(x - x_B - v_Bt)$, $\gamma_B^{-1} = (1 - v_B^2)^{\frac{1}{2}}$ is the kink width in which v_B denotes the velocity of the breather $0 \le |v_B| < 1$, and lastly, x_B denotes the position of the breather at time $t = 0$. In the continuum limit, the breather's wavelength λ and period T are related via

$$|v_B| = \frac{\lambda}{T}, \quad \lambda = 2\pi\gamma_B|v_B|\frac{1}{\omega} \tag{17.68}$$

whereas the amplitude A and energy E_B are given by $A = 4\tan^{-1}(\frac{\eta}{\omega})$, $E_B = 16\eta\gamma_B$.

Post-collision of a standing breather ($v_B = 0$) and traveling kink, the velocity and shape are recovered, where the subsequent interaction results in a phase shift of the breather that oscillates at a new position. As given in [42] following [34], the shift Δ_B of the breather is given by the formula

$$\Delta_B = \frac{2\tanh^{-1}\sqrt{(1 - \omega^2)(1 - v_K^2)}}{\sqrt{1 - \omega^2}} \tag{17.69}$$

where v_K is the velocity of the kink. If the original position is denoted x_0, then post-collision the new position will be $x = x_0 + \Delta_B$.

Consequently, the breather through its subsequent phase shift is conjectured to cause a deflection of the tubulin tails so as to sway the kinesin walk across the microtubule surface. Collisions between kinks or antikinks and standing breathers on the microtubule surface can be considered as a kind of computational gate application whereby a breather can serve as a catalytic agent registering the transitions, influencing MAPs besides (as we have noted) the functions of the prevailing motor proteins (kinesin and dynein) through tunneling and energase action. It is possible there are other combinations and permutations of kink/antikink–breather collisions in, say, the pendulum or discrete models [81], or even possibly in a configuration of moving breathers [89].

Remark 17.3   Indeed, for further motivation, we remark that the screened Coulomb interaction between CTTs and the negatively charged microtubule surface restrict the CTT motion within a *thermal cone* where the motion is influenced mainly by thermal fluctuations [43]. Such Coulomb interaction and thermal fluctuations inducing chaotic effects as previously suggested, may indeed cause the necessary perturbation of the SG model to that of the DSG as remarked circa Equation (17.22). It seems significant that such a recourse to a double-well potential model as induced by the WEDP field could have a startling connection to the underlying physics of protein (enzymatic) tunneling.

Let us note that there are the kink etc. counterparts in other solvable or solitary wave systems which, although not solitons in the conventional sense, might also serve as models of regulatory or computational gates that influence cytoskeletal processes involving tubulin conformations, but which involve different dynamics. For instance, in [20] energy releasing effects of GTP–hydrolysis could generate certain kinks and pulsations that propagate along the microtubule via elastic flexing of the dimers. In [92], a liquid crystal property of microtubules is considered relative to kink "shifting" through GTP hydrolysis, whose rate may increase given additional Ca^{2+} and where possible impediments to the kink motion, polymerization, and microtubular caps are taken into account. These models, however, investigate effects in dynamic microtubules that undergo assembly/disassembly while not addressing the contrasting situation for stable microtubules (such as the neuronal class). Another model involving solitonic interactions, as considered in [80], entails possible quantum coherent states of the DWQ on the tubulin dimer walls where the DWQ are paired to electrons in the dimer hydrophobic pockets via Rabi field coupling. This latter model, although different from how we have formulated the breather–energase dynamics in relationship to the CTTs, affords a brief discussion in the section following.

17.6.3   Vibrationally Assisted Tunneling

Enzyme catalyzed H–transfer reactions involving H–tunneling coupled to enzyme dynamics in flavoprotein and quinoprotein were studied by Sutcliffe and Scrutton [103,104]. The (quantum) H-tunneling itself is driven by the thermally induced dynamics of the enzyme. Classical transition state theory as modeled on gas phase reactions falls short of explaining many enzymatic interactions. On the other hand, quantum tunneling by transfer of electrons or protons and wave particle duality may predict such behavior. According to [103], the dynamic fluctuations in the protein molecule (referred to as "thermal breathing") result in a transient compression of the width of the potential energy barrier and equalize the vibrational energy levels on the reactant and product sides of the catalysed biochemical reaction. Compression of the barrier reduces the tunneling distance and increases the probability of transfer, while the equalization of the vibrational energy states is a prerequisite for tunneling to proceed. Once the proton transfer is complete, relaxation from the geometry required for tunneling traps the hydrogen nucleus by preventing quantum leakage to the reactant side of the barrier.

Potential barrier shape analysis in enzymatic hydrogen-tunneling was carried out for tryptamine, dopamine, and benzylamine, and revealed a dynamic movement of enzymes crossing a barrier shape via disallowed classical regions, which would thus only be possible with the assistance of quantum tunneling effects. In other words, the vibrational excitations of molecular units positioned away from an active site may actually increase motion and catalysis (cf [71] for tunneling effects in different enzymatic families).

17.7   Analogies with Quantum Optical Solitons and Entanglement

17.7.1   Microtubules as Cavities

The emission of radiation in ATP hydrolysis together with coherent excitations of single and highly polar modes typical of a soliton, were phenomena hypothesized by Fröhlich in the discussion earlier. When such systems admit holes of some kind, it was conjectured that electron–hole pairing leads to the existence of intra-cellular solitonic structures, and so induces dissipationless information transfer via kink solitons. In [80], microtubules are hypothesized as quantum mechanical isolated quantum electrodynamic (QED) cavities with analogous properties of electro–magnetic cavities since the latter are realized in quantum optics. Thus, it is conjectured that in relationship to the ferroelectric properties of microtubular configurations, an information transfer results under the hypothesis that QED cavities can harbour electromagnetic radiation as relevant in the context of quantum optics. Within the cavity structure there is an interaction of the dipole moments of ordered water molecules with EM–radiation, thus leading to collective coherent modes, namely the dipole quanta. This accompanies an ordering of electric dipoles caused by the interaction of the tubulin dimers with the ordered water environment of the microtubular cavity. Information processing is then considered to be manifest via interactions along the microtubule protofilament chains. Related are alternative mechanisms based on ferroelectric properties employing the descriptive technique of Ising spin chains on a triangular lattice as a statistical–mechanical framework for representing microtubules [13,110]. One might compare this scenario with the previously mentioned correspondence between the QSG and the eight–vertex model which contains the two-dimensional Ising lattice, and further, compare this with the earlier observation that the 'sech' breather solutions of the NLS as quantum (optical) solitons can be derived from the SG by a SVEPA transformation. Observe that the NLS equation (17.14) admits the 'sech' breather solution (17.15)

$$ u = |c|^{-\frac{1}{2}} q \, \mathrm{sech}[q(x-vt)] \exp \iota \left[\left(q^2 - \frac{1}{4}v^2 \right) t + \frac{1}{2}vx \right] \qquad (17.70) $$

a certain class of which are quantum solitions; more specifically they will comprise solutions of the quantum NLS in $(1+1)$–dimensions with the associated Bose commutation relations $[u, u^*] = \hbar \delta(x-x')$ (see [16]).

The dimer/water environment coupling is viewed in [80] as analogous to how atoms interact with coherent modes of the EM–radiation in quantum optical cavities via the Rabi field splitting, and where every QED cavity property may determine a corresponding property for microtubules. In order to see how kink solitons arise, first consider a microtubular chain. Let u_n denote the displacement field of the n th dimer in the chain, and consider the assignment $u_n \mapsto u(x, t)$ where x denotes a spatial coordinate along the microtubular longitudinal axis of symmetry. Vicinal water influences a viscosity $F \sim \frac{\partial u}{\partial t}$, but since $u(x, t)$ is prospectively a soliton, there will be no energy dissipation. Setting $\xi = x - vt$, the relevant equation is

$$\frac{\partial^2 u}{\partial \xi^2} + \rho \frac{\partial u}{\partial \xi} = P(u) \tag{17.71}$$

where ρ is a certain constant and $P(u)$ a polynomial depending on a microtubular interaction potential. Solutions to (17.71) include kink solitons [80] and are of the type

$$u(x, t) \sim c_1(\tanh[c_2(x - vt)] + c_3), \tag{17.72}$$

where c_1, c_2, c_3 are constants. Somewhat in analogy with quantum solitons, a possible consequence is that if two microtubules A and B are entangled with a third C, then there exists a teleportation of coherent states between A and B with dissipationless information as generated by kink solitons along the dimer chains.

Whereas this latter model could in a sense be analogous to our derivation of kink/antikink/breather solitons, let us remark that we have not proposed the coupling of solitons to electrons in the hydrophobic pocket, but instead have described how the optical-like solitons might be coupled to the CTT conformations. The difference is that the electron hopping in the hydrophobic pocket would require a greater energy supply that might never be attained by neuronal currents, whereas the CTTs should be sensitive to neuronal currents as suggested by the calculations of [43].

17.7.2 Analogy with Entanglement

It is plausible that certain quantum–optical–information systems could be modeled on quantum solitons where an NLS soliton may be hypothesized as a quantum soliton representing one qubit of information [16]. This could be supported by the fact that optical solitons as candidates for qubit representation stand a good chance of experimental recognition when compared to others. Given possible grounds for quantization of the (D)SG equation in the biological framework under discussion, it seems plausible to suggest further analogies with this type of information process. A tentative discussion following [16] is as follows.

A typical qubit is the quantum state of a two-level atom/spin-$\frac{1}{2}$ system. Let us denote by $|g\rangle$ and $|e\rangle$ the ground and excited states, respectively. Then, the state $|\psi\rangle$ is given by

$$|\psi\rangle = a|g\rangle + b|e\rangle \tag{17.73}$$

where $a, b \in \mathbb{C}$ and $|a|^2 + |b|^2 = 1$. Now we relate back to the case of self-induced transparency (SIT). Here, a 2π–pulse takes an atom in $|g\rangle$ back to $|g\rangle$ via a passage through $|e\rangle$, and generally, a θ–pulse takes $|g\rangle$ to

$$|\psi\rangle = \cos \frac{1}{2}\theta \, |g\rangle - \iota \sin \frac{1}{2}\theta \, |e\rangle \tag{17.74}$$

This shows that two qubits are quantum states in the Hilbert space spanned by

$$|g_1\rangle |g_2\rangle, \quad |g_1\rangle |e_2\rangle, \quad |e_1\rangle |g_2\rangle, \quad |e_1\rangle |e_2\rangle, \tag{17.75}$$

and so the state $|\psi\rangle$ is given by the linear combination

$$|\psi\rangle = a \, |g_1\rangle |g_2\rangle + b \, |g_1\rangle |e_2\rangle + c \, |e_1\rangle |g_2\rangle + d \, |e_1\rangle |e_2\rangle \tag{17.76}$$

In a successive measurement of this state will be measured any one of those states with a probability proportional to $|a|^2, |b|^2$, etc. We see that $|\psi\rangle$ includes all four states. Now an optical pulse may act on each atom where, for instance, a $\theta = \pi$–pulse yields

$$-\iota a \, |e_1\rangle \, |g_2\rangle - \iota b \, |e_1\rangle \, |e_2\rangle + \iota c \, |g_1\rangle \, |g_2\rangle + \iota d \, |g_1\rangle \, |g_2\rangle \qquad (17.77)$$

Since it is possible to manipulate the four basis vectors (states) of the Hilbert space, then once given N qubits there are 2^N numbers that can be manipulated analogously. In [59] is discussed an engagement of a two-level atom plus a one photon qubit to function as a two-qubits. Here, a two-state basis is used: $|e\rangle|0\rangle$, $|g\rangle|1\rangle$ (note that the right sides are product states). For a θ–pulse, we have

$$|\psi\rangle = \cos\frac{1}{2}\theta \, |g\rangle|1\rangle - \iota \sin\frac{1}{2}\theta \, |e\rangle|0\rangle \qquad (17.78)$$

In particular:

(1) For $\theta = \pi$, we have $|\psi\rangle = |g\rangle|1\rangle \leftrightarrow |e\rangle|0\rangle$
(2) For $\theta = \frac{\pi}{2}$, we have $|\psi\rangle = \frac{1}{\sqrt{2}}[|g\rangle|1\rangle - \iota \, |e\rangle|0\rangle)]$ which amounts to a quantum entanglement of the photon and atom (since $|\psi\rangle$ is not expressed as a single product state).

Referring back to the photon/atom entanglement described by (17.78), we could take for instance the case whereby the atom takes one bit of binary information represented as $|e\rangle \leftrightarrow 1$, $|g\rangle \leftrightarrow 0$, and the photon also takes one qubit, such as $|1\rangle$ or $|0\rangle$.

Remark 17.4   These analogies with quantum optical solitons and entanglement were prompted by our discussion of several soliton (or soliton-like) equations in relationship to specific neurobiological processes, where at the underlying microscopic molecular–atomic level, quantum mechanical effects may occur. Such analogies should be considered, at least for now, as notably apart from the Orchestrated Objective Reduction hypothesis of Hameroff and Penrose [57], whereby qubits are associated to the switching of tubulin conformational states. There [57], the quantum superposition of such states is collapsed by a quantum gravitational effect to classical output states, which in turn somehow influence the axonal hillock potential and the generation of axonal action potentials. However, the validity of this model remains questionable due to biological mis-modeling and is not pursued in this paper since our approach is fundamentally different. In particular, we do not need actin gel-sol isolation cycles during which the microtubules remain coherent for 25 ms, and we do not confine the physiology of mind and memory to dendritic events alone. Our model operates at 10 to 15 ps (the characteristic dynamic time-scale of enzymatic catalysis), and further, we investigate the entire cytoskeletal network as a unitary cybernetic device. The desired output resulting from the subneuronal computation directly affects neuromediator release at the presynaptic site by quantum vibrationally assisted tunneling, a process that indeed agrees with Tegmark's decoherence estimate [107].

17.7.3   Resonances of the Schrödinger Operator

Since a significant part of this essay implicitly involves resonant phenomena, we conclude this section by mentioning a deeper mathèmatical explanation in relationship to Schrödinger's equation. We consider the latter in the form

$$\iota h \frac{\partial \psi}{\partial t} = (-h^2 \Delta + V)\psi \qquad (17.79)$$

where V is regarded as a suitable potential. In particular, we deduce

$$P(h) = (-h^2 \Delta + V) \qquad (17.80)$$

to be the corresponding Schrödinger operator of (17.79). In mathematical terms, the operator (17.80) can be regarded as a closed linear operator on a complex Hilbert space $\mathcal{H}$. The *spectrum* $\sigma(P)$ consists of all complex numbers λ such that $\lambda \mathbf{I} - P$, is non-invertible where each λ is an *eigenvalue* and the complement

$\mathbb{C}\backslash\sigma(P)$ is called the *resolvent set*. Consider a state ψ satisfying

$$P\psi = (-h^2\Delta_V)\psi = \mathbb{E}\psi \tag{17.81}$$

There are two cases to be distinguished. Firstly, on the interval $[0, \pi]$, the set of all $\mathbb{E}$ corresponding to the eigenvalues of P is discrete and the energy levels can be quantized. But on extending to all of the real line $\mathbb{R}$, tunneling can prevent ψ from becoming a genuine wave function (mathematically, the equation might not admit any square integrable solutions)—a particle may wobble about within a potential well as we have mentioned. However, if no damping effects are admitted but instead an escape route to infinity is provided, then resonance can occur. In this respect, the spectral problem is transformed to a scattering problem. The eigenvalues are seen to be the poles of the *resolvent* $\mathcal{R}(z)$ of P, where

$$\mathcal{R}(z) = (P - z)^{-1} \tag{17.82}$$

As an operator on $\mathcal{H}$, the resolvent $\mathcal{R}(z)$ is bounded at each $z \in \mathbb{C}\backslash\mathbb{R}$. Let $\mathcal{H}_c \subset \mathcal{H}$ be the subset of elements that are zero outside of a compact set and let $\mathcal{H}_{\text{loc}}$ denote the space of functions locally in $\mathcal{H}$. Then as an operator

$$\mathcal{R}(z) = (P - z)^{-1} : \mathcal{H}_c \longrightarrow \mathcal{H}_{\text{loc}} \tag{17.83}$$

$\mathcal{R}(z)$ continues meromorphically from $\operatorname{Im} z > 0$ to $\mathbb{C}$. In this strictly mathematical sense, the resonances are manifestly the poles of the meromorphically continued resolvent (for further technical details see [117]).

17.8 Interneuronal Quantum Coherence

We have already discussed how the energy supply in neuronal cytosol might be sufficient to sustain water lasing at a time-scale of 10 to 15 ps, which is characteristic of protein (enzymatic) action. Yet we have not addressed the possibility for a quantum coherent state to extend between neighboring neurons. Although in the case of electric synapses, the cytosol of communicating neurons is connected via gap junctions, the existence of electric synapses *per se* cannot explain the existing long-range correlations between neurons. In a pioneering work by Sperry and Gazzaniga [100], it was reported that human subjects having undergone callosotomy (surgical severing of the axons connecting the two cerebral hemispheres) manifest a prodigious Jekyll–Hyde syndrome—their brain hosts two separate minds each being unaware of the other mind's existence. Each disconnected hemisphere appeared to have its own, largely separate, cognitive domain with its private perceptual, learning and memory experiences, all of which were seemingly oblivious of corresponding events in the other hemisphere [99]. The clinical tests revealed that the disconnected right hemisphere (as a separate mind) was still able to recognize itself among a select array of portrait photos, and in doing so, generated various emotional reactions with an apparent sense of humor in the context of social evaluations [101]. Although such investigations showed that callosotomy is a severe act of splitting the human psyche, the most significant conclusion from those studies appears to be that axons and axo-dendritic communication are critical for the existence of a unitary mind.

17.8.1 The Josephson Junction Effect

In 2002, Georgiev elaborated on the work of Jibu and Yasue [65] and suggested that the macroscopic quantum coherence of the evanescent (tunneling) photons across axo-dendritic synapses might be coupled with protein conformational dynamics on both sides of the synapse as a possible means of explaining the unity of conscious processes [46]. It was shown that CAMs might regulate the synaptic cleft width and might transmit signals between the postsynaptic and presynaptic cytoskeletons [48]. Moreover, it was suggested that there is a reverse transmission of information from the postsynaptic site to the presynaptic active zones, enabling an efficient control of exocytosis and neuromediator release via neuroligin-1/ β–neurexin/ synaptotagmin-1 ligation, and assisted in its function by the activity of high-temperature Josephson junctions (JJs) coupling the perimembranous water lasing in the presynaptic and postsynaptic cytosol with the water collective modes in the synaptic cleft.

In order to model the possible JJ phenomena, we first have to show that not only the inner phospholipid layer but also the outer phospholipid layer of the synaptic membrane is in contact with the evanescent (tunneling) photon water manifesting macroscopic BE–condensation. Since we discussed a laser process pumped by an energy supply, it is clear that although in the extracellular space surrounding the neurons there are also electric currents and an EM field (but in an opposite direction), the extracellular water might not sustain coherence above the time-scale of thermal fluctuations, simply because in the extracellular space there is an absence of energy supply. Therefore, we are not in agreement with [64], which claims the relevant processes are manifested outside of neurons and/or in glial cells.

As presented in [48], the synaptic cleft is an isolated cavity muffled by glial cells and organized by electron dense proteoglycan molecules, such as syndecan, CAT-301, phosphacan, α-dystroglycan, etc. (see Figure 17.2). Inside the synaptic cleft, the ion concentrations may differ from these in the extracellular space. Further, there is an intense metabolic energy supply by ionic pumps, protein transporters for the re-uptake of excess neurotransmitter, and/or enzymes releasing energy from a cleavage of excess neurotransmitters. So it seems feasible that the water ordering and water BE collective dynamics in the synaptic cleft is essential for transmitting information between pre-and postsynaptic proteins. This situation may be favorable for both inner and outer phospholipid bilayers of the synaptic membrane to have contact with evanescent water and assist forming high temperature JJs operating in concert with the CAM molecules.

17.8.2 Synaptic Josephson Oscillation Currents and Plasmon modes

Here we recall some of the basic features of the JJ model as originally presented in [64–66]. Let us focus on a domain of the synaptic membrane smaller than the coherence length ℓ_c in which the sandwich-structured JJ is realized in terms of the macroscopic condensates of evanescent photons in the perimembranous regions inside and outside of the membrane. Note that in principle, the intermediate "sandwich" is considered as sufficiently thin so as to allow the possibility of quantum tunneling [35], an observation which clearly serves our purpose. The electric potential difference $V = V(t)$ between the outer and inner surfaces of the membrane can be considered as the voltage across the JJ. Then, following [65], the standard quantum field theoretical treatment of the electric Josephson effect leads to the Josephson current induced by the voltage $V(t)$, as given by

$$J(t) = J_0 \sin \left(\theta_0 + \frac{q}{\hbar} \int_0^t V(s) \, ds \right) \tag{17.84}$$

where J_0, θ_o, and q are certain constants [114]. The circuit equation for the JJ is then given by

$$C_0 \frac{dV(t)}{dt} = -J(t) \tag{17.85}$$

where C_0 denotes the capacitance parameter of the membrane. Introducing a new variable

$$W(t) \equiv \theta_0 + \frac{q}{\hbar} \int_0^t V(s) \, ds \tag{17.86}$$

we can express Equation (17.85) as follows:

$$\frac{d^2 W(t)}{dt^2} = -\frac{J_0 q}{C_0 \hbar} \sin W(t) \tag{17.87}$$

As pointed out in [66], Equation (17.87) is a nonlinear differential equation resembling the classical equation of motion for a physical pendulum and admits an oscillatory solution $W = W(t)$ represented implicitly

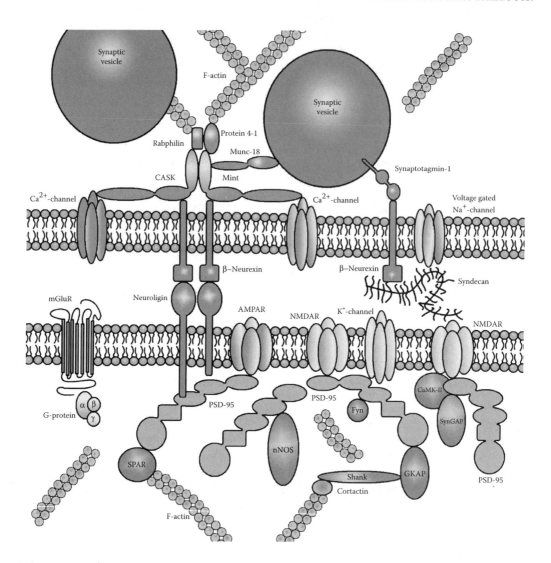

FIGURE 17.2 Schematic representation of a synapse with the synaptic cleft. The β-neurexin-neuroligin adhesion is a central structure for establishing a functional synapse. Presynaptically β-neurexin interacts with scaffold proteins (CASK, Mint, Munc-18), most of which are involved in the synaptic vesicle docking and fusion. The presynaptic scaffold is bound to the actin cytoskeleton via protein anchors (protein 4.1, rabphilin). In the synaptic cleft are present various proteoglycans (such as syndecan). Postsynaptically in the membrane are located ionotropic receptors such as NMDA, AMPA (α-amino-3-hydroxy-5-methyl-4-isoxasole-propionate) receptors or metabotropic ones (such as mGluR) coupled with G-proteins. PSD-95 recruits postsynaptic scaffold proteins (synGAP, GKAP, Shank), various kinases (such as Fyn that modulates NMDAR activity), or neuronal NO-synthase (nNOS). SPAR and cortactin transmit signals to the postsynaptic actin cytoskeleton.

by an elliptic function. Correspondingly, the membrane electric potential difference $V(t)$ manifests a self-excited oscillation characteristic to the JJ:

$$V(t) = \frac{\hbar}{q} \frac{dW(t)}{dt} \qquad (17.88)$$

This is known as the *Josephson oscillation* and a small oscillating current across the JJ is called *the Josephson oscillating current*.

Let us consider a cell membrane as a horizontal plane patched by many JJs smaller than the coherence length across which the Josephson oscillating currents perpendicular to the plane are maintained. Then, the collective modes of the totality of those Josephson oscillating currents emerge as plasmon modes capable of propagating horizontally as solitons along the plane of the presynaptic or postsynaptic membrane without damping. This remarkable fact can be seen from the propagation equation for the plasmon mode obtained by rewriting (17.87) in terms of spatial variables:

$$\frac{\partial^2 W(x,t)}{\partial t^2} - v^2 \frac{\partial^2 W(x,t)}{\partial x^2} = -\frac{J_0 q}{C_0 \hbar} \sin W(x,t) \qquad (17.89)$$

where v denotes the characteristic speed of the plasmon mode, and x a spatial direction along the membrane. In particular, we note that this latter equation is of a SG–type (cf the perturbed SG equation [17.23]).

17.9   Solitons in α–Helix Protein Molecules

17.9.1   Davydov Model

As we had mentioned in the introduction, Davydov [26,27] provides a foundation for applying the theory of certain nonlinear systems to molecular chains, DNA, membraneous flexing, etc. Next, we recall the basic principle of how proteins function in converting chemical into mechanical energy, and when aided by lipids they generate the traffic of ions and molecules in and around cellular membranes. As we have mentioned, the protein chain can coil into a helical form which is manifestly the structure of hydrogen bonded peptide groups within the protein molecule. Protein molecules incorporated into the cytoskeleton create transduction energy and intracellular couplings all of which assist and determine the energy release of hydrolysis of ATP molecules, while at the same time portions of the helix constitute part of the cytoskeleton's protein composition. On the other hand, the excited states of a protein molecule are related to the resonant interaction between peptide groups within distinct chains. According to [26,27,95], a class of solitons evolve at the origin of each chain and so can be created within short intervals of α–helix proteins. The vibrational energy of amide–I oscillators localized in the helix acts via a phonon coupling effect, creating a distortion of the helical structure. In turn, the distortion acts to grasp the amide–I energy and so prevent its dispersion, an effect known as *self–localization* or *self–trapping*. Such effects are relevant to anesthetic actions, which will be mentioned at a later stage.

The propagation of a soliton within a α–helix protein molecule could be either symmetric or asymmetric. Of these, the asymmetric soliton is the more stable and its radiation life span does not depend on velocity and can increase sharply as the angle between the spiral axis and vibrational dipole moment decreases. This explains why the asymmetric solitons are favorable for transferring the energy of ATP hydrolysis without loss of energy along the α–helix protein chain over suitably large distances. In [68] is predicted a helicial structure to the C–terminal domains, and for β C–terminal recombinant peptides, this helicity has been determined with evidence supporting a functional coil to the helix transition at the CTT base. As also seen in [5], each tubulin monomer possesses twelve α–helices (labeled from H1 to H12), so in terms of short-range localization, it is plausible that the preceding asymmetric soliton propagation can be applied.

In order to see how the corresponding solutions arise, consider the Hamiltonian H_{PM} for collective excited states of the protein molecules as given by

$$H_{PM} = \sum_{n,\alpha} \{(\mathcal{E} + D_{n\alpha}) B_{n\alpha}^* B_{n\alpha} + J_{n,\alpha;n+1,\alpha}(B_{n\alpha}^* B_{n+1,\alpha} + B_{n+1,\alpha}^* B_{n\alpha})$$

$$+ J_{n\alpha;n,n+1}(B_{n\alpha}^* B_{n,\alpha+1} + B_{n,\alpha+1}^* B_{n\alpha})\} + H_{ph} \qquad (17.90)$$

(see [26]). In this expression, the $B_{n\alpha}^*$ and $B_{n\alpha}$ are creation/annihilation operators for the excitation $\mathcal{E}$ over the peptide group $n\alpha$; the term $J_{n\alpha;m\beta}$ denotes the energy of the resonant inter-dipolar coupling between the peptide groups $n\alpha$ and $m\beta$; $D_{n\alpha}$ denotes the deformation energy of interaction with neighboring groups arising from excitations of the group $n\alpha$, and H_{ph} is the displacement operator of the groups from their equilibrium position along hydrogen bonds. This last operator is given by

$$H_{ph} = \frac{1}{2} \sum_{n\alpha} \left[\frac{1}{M} P_{n\alpha}^2 + w(U_{n\alpha} - U_{n+1,\alpha})^2 \right] \tag{17.91}$$

where M denotes the effective mass displaced along with the peptide group, w is the elasticity coefficient of the chain along the hydrogen bonds, and $P_{n\alpha}$ is the momentum operator conjugated to the displacement operator $U_{n\alpha}$ of the peptide group along the latter.

Associated to the Hamiltonian H_{PM} is the wave function describing the collective vibrations of the system as given by:

$$|\Psi(t)\rangle = \sum_{n\alpha} a_{na}(t) e^{\sigma(t)} B_{n\alpha}^* |0\rangle \tag{17.92}$$

where $|0\rangle$ denotes a function for which the groups are in the ground-state with vibrationless excitations away from their equilibria, and where

$$\sigma(t) = -\frac{\iota}{\hbar} \sum_{n\alpha} [\beta_{n\alpha}(t) P_{n\alpha} - \pi_{n\alpha}(t) U_{n\alpha}] \tag{17.93}$$

In this last expression, the functions $\beta_{n\alpha}(t)$ and $\pi_{n\alpha}(t)$ depend on the average values for the displacement of the groups $n\alpha$ and their momenta in the preceding state. The coefficient function $a_{n\alpha}(t)$ satisfy $\sum |a_{n\alpha}(t)|^2 = 1$, where the latter corresponds to the distributive probability over the groups $n\alpha$ in their collective excitation states. The complex-valued functions $a_{n\alpha}(t)$ and the real-valued functions $\beta_{n\alpha}(t), \pi_{n\alpha}(t)$ are obtained from minimizing the functional

$$\langle \Psi(t)|H|\Psi(t)\rangle \tag{17.94}$$

and on applying a certain approximation, the following system of equations is deduced. Firstly, since the functions $a_{n\alpha}(t), \beta_{n\alpha}(t)$ are continuous in n, they are replaced by $a_\alpha(\xi, t), \beta_\alpha(\xi, t)$, respectively. The system in question is then:

$$\left\{ i\hbar \frac{\partial}{\partial t} - [\mathcal{E}_0 + W - 2J] - 2\chi \frac{\partial \beta_\alpha}{\partial \xi} \right\} a_\alpha + J \frac{\partial^2 a_\alpha}{\partial \xi^2} - L(a_{\alpha+1} + a_{\alpha-1}) = 0$$

$$\left[\frac{\partial^2}{\partial t^2} - v_\alpha \frac{\partial^2}{\partial \xi^2} \right] \beta_\alpha = \frac{2\chi}{M} \frac{\partial}{\partial \xi} |a_\alpha|^2 \tag{17.95}$$

Here χ is formed from coupling parameters for internal excitations of the peptide groups and their displacements from equilibrium positions; J denotes the resonance energy of inter-dipolar interactions between neighboring groups in the same chain, and L the energy of the same interaction between neighboring groups from different chains ($J \approx 967\ \mu eV$, $L \approx 1537\ \mu eV$). Also, $v_\alpha^2 = w/M$, the term W is the average density for displacement of molecules from the equilibria position, and $\mathcal{E}_0$ is the excitation energy of the peptide group relative to the deformation potential. It is from this system that the symmetric and asymmetric solitons are derived (see [26] for explicit details).

17.9.2 H–N–C=0 Groups and Weakening of Hydrogen Bonds in the α–Helix

ATP–hydrolysis is known to provide an energy source for many types of protein interactions, whereby this energy can be converted via resonant coupling into vibrational excitations within the domain of the protein. As we have noted, [26,27] propose for the case of α–helix proteins (see Figure 17.3) the interaction of amide–I energy within a system of hydrogen bonded peptide groups, leading to a class of solitons (phonons) that characteristically propagate along the chain without changing their form. Molecular vibrations may then be carriers of energy transfer along chains, such as in the spines of the α–helix proteins where "spine" refers to the chain of hydrogen-bonded peptide groups. Therein, amide–I vibrations can be coupled to

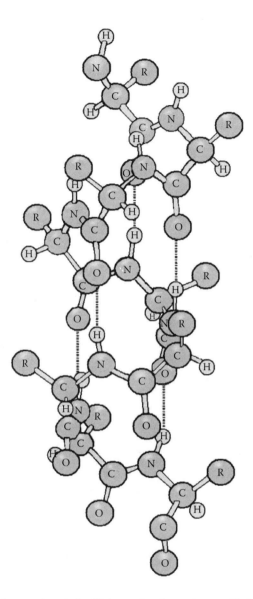

FIGURE 17.3   Structure of the protein α-helix. Hydrogen bonds are shown with dashed lines, C is carbon, N is nitrogen, O is oxygen, H is hydrogen, and R is amino acid radicals.

longitudinal sonic waves in the task of stretching the hydrogen bonds along the helical spine (a more detailed description of the biophysics is outlined in [76]).

Intraveneous general anesthetics containing H–N–C=0 moiety (such as barbiturates, hydantoins, suc-cinimides, and urethanes) are capable of altering localized structures within proteins. For instance, when a barbiturate molecule clings to a protein, there is a severance of hydrogen bonds linking neighboring peptide groups. Simultaneously, new hydrogen bonds are produced by the oxygen and hydrogen atoms of the molecule. The 'perturbed protein hypothesis' of [73,74] suggests that such altering of the normal protein dynamics by anesthetic molecules is the result of actual disturbances of the soliton propagation in the polypeptide chain, since the newly introduced hydrogen bonding to the helix will distort and sever intraprotein bonds which normally maintain the spiral form of the helix.

17.9.3   Modification of the Davydov–NLS Equation

The governing equations require some modification of the original Davydov–NLS equations for solitons relative to amide–I energy. Consider first the following form of the NLS

$$i\hbar\frac{\partial\Psi}{\partial t} + \frac{\partial^2\Psi}{\partial x^2} + G_0|\Psi|^2\Psi = 0 \qquad (17.96)$$

where Ψ is the wave amplitude, $G_0 = 4\chi^2(Jw)^{-1}$ in which χ is the exciton–phonon interaction parameter, J is the longitudinal dipole–dipole coupling energy, and w is the longitudinal spring constant for the hydrogen bond. Since an anesthetic molecule creates distortion within the protein and so disturbs the coupling, the quantity G_0 is likely to vary. Thus, following [74], this latter quantity is replaced by a function $G(x) = G_0 + \epsilon G_1(x)$, so that smooth variations permit a soliton to adjust to the ambient changes only with a slight disturbance. If however, the changes are abrupt, the soliton may not be able to re-adjust and there will be progressive energy loss through radiation.

What has been proposed is the following modification of the original Davydov equation given in [94]:

$$i\hbar\frac{da_{n\alpha}}{dt} = [\mathcal{E}_0 + \chi_1(\beta_{n+1,\alpha} - \beta_{n-1,\alpha})]a_{n\alpha} - J(a_{n+1,\alpha} + a_{n-1,\alpha})$$
$$+ \chi_2[a_{n+1,\alpha}(\beta_{n+1,\alpha} - \beta_{n,\alpha}) + a_{n-1,\alpha}(\beta_{n,\alpha} - \beta_{n-1,\alpha})] + LF_L + NF_N + \cdots + ZF_Z \qquad (17.97)$$

The index n specifies a particular unit cell, in the H–N–C=0 group counted longitudinally along the helix, and $\alpha = 1, 2, 3$ specifies each of the three spines of the latter. The term $a_{n\alpha}$ is the probability amplitude for the vibrational excitation of the amide I group located at spines n, α; $\mathcal{E}_0$ is the quantum energy of the amide–I vibration; $\beta_{n\alpha}$ is the longitudinal displacement of the residue located at n, α; w is the linear spring constant for a hydrogen bond; J is the dipole–dipole coupling energy between a particular amide–I group; χ_1 orders the coupling between exciton and phonon interactions; and χ_2 orders the change in the J term by stretching the helix between cells $n + 1$ and n. The terms $LF_L, NF_N, \ldots, ZF_Z$ are additional dipole coupling terms relative to the helical structure that assist the necessary efficiency of the mechanism. There is also the accompanying equation giving the second variation of the longitudinal displacement

$$M\frac{d^2\beta_{n\alpha}}{dt^2} = w(\beta_{n+1,\alpha} - 2\beta_{n\alpha} + \beta_{n-1,\alpha})$$
$$= \chi_1(|a_{n+1,\alpha}|^2 - |a_{n-1,\alpha}|^2) + \chi_2(a_{n\alpha}^*(a_{n+1,\alpha} - a_{n-1,\alpha}) + (a_{n+1,\alpha}^* - a_{n-1,\alpha}^*)a_{n\alpha}) \qquad (17.98)$$

where M denotes one-third the molecular weight of a unit cell.

In the presence of an anesthetic molecule, Equations (17.97) and (17.98) are modified by adjusting the values of J, w, and χ_1. For instance, the term J, the resonant interaction between adjacent amide–I dipoles, is approximated by

$$J \approx \frac{\mathbf{d}^2}{4\pi\epsilon_0 r^3}(3\cos^2\theta - 1) \qquad (17.99)$$

where $\mathbf{d}$ denotes the dipole moment, r the distance between dipoles, ϵ_0 is the dielectric coefficient, and θ the angle between dipoles. So when an anesthetic molecule severs the hydrogen bond in an α–helix, causing the latter to unwind, the distance r will increase and therefore J will decrease accordingly. The subsequent distortion in the helical coil will cause the extra coupling terms $NF_N, \ldots, ZF_Z$ to be shifted incrementally. The solutions to the system (17.97), (17.98) have been determined by computer-implemented numerical methods and appear as mobile solitons of sech pulse type, as depicted in [74].

One basic observation is that protein conformations are purportedly influenced by weak van der Waals quantum mechanical forces, known as *London dispersion forces* which permit the dissolution of anesthetic gases in the hydrophobic pockets. In short, there is an interplay between the amino acids of the protein pocket and those of the anesthetic molecule which involve a shifting of electron locations to avoid repulsion and maximize the attraction between electrons and positively charged nuclei. This scenario is just a version of the 'perturbed protein hypothesis' and is further discussed in [56] in relationship to how tubulin conformational states might function as qubits, a separate issue which we do not pursue here.

Although the discussion deals with the biophysical aspect of how the anesthetic molecule can perturb protein conformations, it does not by itself completely explain the specific anesthetic actions (that is, how the anesthetic molecules bind selectively only to a few 'target' proteins). In the case of intravenous anesthetics, the main target is either the $GABA_A$ ion channel (GABA abbreviates γ–aminobutyric acid) or the NMDA receptor. In contrast, the class of volatile anesthetics might bind to more than one target, including Na^+ or Ca^{2+} voltage–gated ion channels, two-pore domain K^+ channels, and interestingly enough, the 4-α-helix bundle of the SNARE complex (see Section 17.11.5) that drives synaptic vesicle exocytosis. These and other observations will bear further relevance to the notion of "SNARE zipping" to be discussed in Section 17.11.5.

17.10   Stochastic Resonance in Neural Networks

The underlying theme of this section concerns to an extent how neuron signals are realized as intermittent pulsations through spikes (vis-a-vis action potentials) as propelled along a nerve axon, thus inducing potentials within a configuration of postsynaptic neurons. The postsynaptic potentials (PSPs) undergo spatial and temporal summation in the dendritic tree of each neuron and when the transmembrane voltage at the axonal hillock reaches a threshold value, it triggers an axonal action potential that is transmitted to the presynaptic boutons for subsequent neuromediator release. A physiological principle is that information processing is based mainly on the dynamics of such spiking networks in relationship to the acclaimed model of Hodgkin–Huxley. An action potential on reaching the endpoints of an axon triggers the secretions of neurotransmitters into the synaptic cleft. The neurotransmitters traverse the synaptic cleft and arrive at the membrane of the postsynaptic neuron, activating ligand gated ion channel receptors that allow for electrogenic transmembrane ion fluxes that alter the transmembrane potential and generate postsynaptic electric currents. Such inputs are said to be *depolarizing* or *excitatory* if the spiking rate is increased, and in contrast, they are said to be *hyperpolarizing* or *inhibitory* if the spiking rate is decreased (otherwise said, greater and diminished probabilities for spiking, respectively). Within a spike train may be found differing behavior: regular, or fast spiking, or bursting. The spike count in an essential way influences the rate coding and in some cases can be realized in terms of a stimulus reconstruction (for further discussions, see e.g., [49]).

In the presence of a periodic driving force, particles passing through a double-well potential such as those considered previously (e.g., relative to the ϕ^4–equation), could be prone to bouncing back and forth. When such wobbling between potential minima become synchronized with weak periodic forcing, stochastic resonance (SR) occurs (that is, the prevailing noise within the system amplifies a relatively weak coherent input or signal). Generally, this can be seen as the increased sensitivity to small perturbations at a prevailing (and usually optimal) noise level (see [40]).

As pointed out in [40], the shape and amplitude of the propagating action potential is very stable. In the absence of a neurotransmitter, the so-called *resting potential* of the membrane of the post synaptic neuron is determined by the balancing of resting ionic fluxes. The opening of postsynaptic receptors (such as AMPA, kainate, or NMDA receptors) leads to Na^+ or Ca^{2+} influx that results in an excitatory PSP. The PSP diffuses progressively towards the soma, while at the same time losing about 80% of its amplitude. When the sum of arriving PSPs at the axonal hillock exceeds a certain threshold value (of the order $\approx -55\ mV$), the maximum likelihood of spiking is attained. Phase-locking is seen as a correlation between rates of neuronal discharge and time-dependence of stimuli, as often realized in the encoding of acoustical information on the primary auditory nerve where evidence suggests SR results from the firing of periodically stimulated neurons. Since all of this seems to occur in a noisy environment, it leaves open the possibility that the relevant systems are somehow conditioned to welcoming ambient noise for the purpose of neural processing.

Another example is the visual cortex where quanta of information are encoded not by individual cells, but rather by communities of these. Typically, place cells are representative of encoding information within an environmental frame of reference, whereupon a quorum of cells responds to the demands of a given location. Each constituent putatively breaks down its response in terms of an average, plus a variation in noise (again, neurons can be characteristically noisy, and in turn, can cause noisy synaptic inputs which may impede transmission relay), thus contributing to sequences of spiking, in turn encoding information within the period of stimulus. Eventually, there results an overall cumulative response to the environment in relationship to the direction of motion, color, shape, form, etc., since they are encoded into the appropriate regions of the visual cortex. It is plausible that some type of computational procedure could in part arise from population coding, but be further enhanced by the effects of SR. Whereas maximum likelihood methods can involve substantial data accumulation, a viable alternative suggests establishing an "electoral system" for predicting vectors by regarding the activity of a given cell as a "vote" for taking a preferred direction [86]. Periodic inputs give rise to spike trains, but SR seems an acceptable enough hypothesis for the surpassing of a threshold in order to attain an action potential.

SR appears particularly relevant for NMDA receptors which are necessary for synaptogenesis. The opening of a NMDA channel permits a passage of Ca^{2+} ions, which signal and trigger off complex chain reactions influencing the actual strength of synapses, and in turn the working efficiency of subsequent neural and mental events. More specifically, the Ca^{2+} influx results in fast actin-based dendritic spine plasticity (short-term memory) and is also involved in memory stabilization via activation of Ca^{2+}/calmodulin-dependent kinase II (CaM-K II) and subsequent cytoskeletal reorganization, ion channel modulation and/or gene activation (long-term memory). The NMDA receptors might utilize some form of SR due to the relatively long period during which the receptor remains in an open conformation ($\approx 200 \ ms$) and the gradual deinhibition of the receptor by the Mg^{2+} block. In particular, it has been shown that back-propagating action potentials can deliver sufficient depolarization to postsynaptic NMDA receptors in order to release the Mg^{2+} block, and the capability of the back-propagating action potentials to active NMDA receptors is increased if the timing of the presynaptic glutamate release and postsynaptic firing is further synchronized [70]. This mechanism leaves open the possibility that NMDA receptor–activation may utilize some ambient noise to attain to SR.

17.11   Molecular Bases of Synaptic Transmission

17.11.1   Introduction

Synaptic connections are viewed as forming the principal means for the engagement of memory and environmentally related functions. Operational changes in pre-existing synapses influence synaptic plasticity and cognition as realized by the re-emergence of new synaptic connections, neurotransmitters, and classes of signaling molecules [36].

In the central nervous system (CNS) are two major types of synapses: the chemical and electric type. About 90% of the synapses are chemical and transmit biological information in the form of a neuromediator released in the synaptic cleft that further binds to ionotropic and metabotropic receptors at the postsynaptic neuron. The chemical synapses are evolutionarliy younger, meaning that they possess a greater capability for information transfer than the electric ones. Usually, it is considered that chemical synapses transmit information in a unidirectional sense from a presynaptic site to a postsynaptic site. Yet this is true only when considering the electric potential, and it is generally not true when one considers the protein–protein coupling of presynaptic and postsynaptic neurons. On the other hand, the small percentage of electrical synapses that comprise direct cytosolic linkages between two neurons via protein channels, known as *gap junctions*, allows for a free exchange of small molecules and trophic factors. The passage of electric currents is not impeded, so the voltage spreads into the neighboring neuron without time delay, but is subjected to a spatial decrement as derived from the cable equation (17.49). In the case of axo–axonal gap junctions, the voltage may be beyond the critical threshold and could trigger an action potential in the neighboring axon. This may result in fast 200 Hz axonal coupling. In dendro–dendritic signaling, the electrical synapses may lead to the generation and propagation of Ca^{2+} waves in view of the existence of voltage-gated ion channels in the dendritic spines that are capable of generating action potentials.

17.11.2 The Beck–Eccles Hypothesis

Synaptic vesicle exocytosis is a fundamental process of rapidly releasing rates of high-frequency signaling across chemical synapses whereby efficiency determines the effective synaptic strength. Within each bouton lies a multitude of vesicles, each of which contains 500 to 1000 molecules of synaptic transmitters (such as glutamate, aspartate, GABA, etc.). The neurotransmission property of glutamate causes a momentary decrease in the electric potential across the postsynaptic membrane, creating an excitatory post-synaptic potential (EPSP) of the dendrite. The failure to release such neurotransmitters, or a demise in efficiency, may create so-called "silent" synapses. Consequently, the notion of *synaptic plasticity* is considered the result of a fluctuation between the silencing and awakening of the synapses.

The axonal terminal bouton in the CNS neurons is a bulbous structure containing 30 to 50 synaptic vesicles (SVs) organized by a complex protein scaffold called the *presynaptic vesicular grid* (PVG) or *cytomatrix of the active zone* (CAZ), where the latter term is used more frequently in recent articles [115]. Only those docked vesicles at the active zone of the presynaptic membrane can undergo rapid fusion on neurotransmitter release under Ca^{2+} entry. The number of docked vesicles varies from 3 to 13, and is referred to as a *rapidly releasable vesicle pool*. SVs attached to actin or microtubules via synapsin-1 represent a population of vesicles that can be mobilized only after some activation. In this respect, they comprise the so-called *reserve vesicle pool*. The CAZ, being of a paracrystalline structure, appears suited to the ordering and regulation of exocytosis, and conceivably such ordering, as originally suggested by Beck and Eccles [11], is conducive to long-range interactions and is thus a valid premise for inherent quantum events, whereby intention may take effect on neuronal networks by increasing the probability for exocytosis. The latter process may engender a combination of many probability amplitudes and thus induce a coherent action.

That actual probabilistic events originate from a quantum mechanical process is suggested by a trigger mechanism involving a Ca^{2+} interaction within structures of an extreme microscopic nature. Effectively, when a nerve impulse impinges upon a bouton, depolarization induces an entry of Ca^{2+} ions that combine with synaptotagmin-1 for subsequent triggering of exocytosis. The trigger mechanism is partly modeled by the quantum tunneling of a quasiparticle with one degree of freedom. This quasiparticle is represented by the variable q and denoted by the potential energy of the motion by $V(q)$. The resulting wave function $\psi = \psi(q,t)$ satisfies the NLS

$$i\hbar\,\frac{\partial \psi}{\partial t} + \frac{\hbar}{2M}\frac{\partial^2 \psi}{\partial q^2} - V(q)\cdot\psi = 0 \qquad (17.100)$$

Next, let $p_1(t)$ (respectively, $p_2(t)$) be the probability that exocytosis does (respectively, does not) occur. The wavepacket may reside either "left" or "right" of the potential barrier through which the energy tunnels. Accordingly, we have the probabilities

$$p_1(t) = \int_{left} \psi^*(q,t)\psi(q,t)\,dq$$

$$p_2(t) = \int_{right} \psi(q,t)\psi^*(q,t)\,dq \qquad (17.101)$$

where $p_1(t) + p_2(t) = 1$. Intention procedures could be viewed as assisting in the coupling of these probabilities in order to maximize the likelihood of synaptic exocytosis, and subsequently the formation of the EPSP.

Taking into consideration the energy barrier height as well as the experimentally measured probability for occurence of exocytosis, the mass of the tunneled quasiparticle was calculated to be six times the mass of the proton or less. Exceeding this critical mass would turn the motion of the quasiparticle into classical, thus making it unable to penetrate the potential barrier. Further, in [11], it is suggested that exocytosis might depend upon the tunneling of electrons between the vesicle membrane and the presynaptic membrane lipids. The original suggestion required pure tunneling that is not sensitive to temperature. Since the temperature dependence of exocytosis has been experimentally verified, we consider the pure tunneling idea as unfeasible. Instead, in [47,48], it is suggested that exocytosis undergoes vibrationally assisted tunneling (VAT), hence *the protein thermal breathing is essential to accelerate the process.* The novel VAT model was corroborated by the experimental evidence for VAT in hydrogen tunneling performed by dehydrogenase systems [103,104] and is in close agreement with that calculated for the mass of the tunneled quasiparticle in [11].

Remark 17.5   The Beck–Eccles hypothesis has often been linked to a quantum–theoretic basis for the neurophysiology of mind and intention [11,36,84]. Related to the basic questions is the von Neumann–Wigner theory of Stapp [102], which involves time-scales of $10^{-23}s$ which do not seem feasible for biological processes (we regard a 10- to 15-ps interval as feasible for such processes). In mathematical terms, the possible role of projection–valued probability densities in the context of a Hilbert (state) space, as suggested in [102], does not seem entirely out of line as a possible mechanism for approaching the problem. However, as we have pointed out, since the environment of the cytoplasm appears conducive to the occurrence of SR and chaotic effects, we would thus propose operational probabilities (e.g., in regard to those previously mentioned for exocytosis), as belonging to a wider class representative of a semi-classical/quantum stochastic process exhibiting "fuzzy," rather than "sharp" characteristics (as would be in the case of projection operators). The distinctions between such operational processes are explained in precise mathematical terms in [52].

17.11.3   Active Zones

Active zones (AZ) are presynaptic sites both for vesicle docking and subsequent fusion releasing neurotransmitters into the synaptic cleft. Characteristic properties of the AZ include the following [115]:

(1) Synaptic vesicles cluster, tether, and fuse at the AZ.
(2) The plasma membrane of the active zones appears to be electron dense as viewed by electron microscopy (implying the existence of a perimembraneous protein scaffold).
(3) The AZ is closely aligned with a post synaptic density (PSD) area such that the AZ spans the same width as the PSD and the extracellular space between the two membranes.
(4) The AZ has two gates essential for neurotransmission: one for Ca^{2+} entry (the voltage gated Ca^{2+} channel), and another for the neurotransmitter exit, which is the synaptic vesicle fusion site.

The vesicle fusion sites are so specialized for creating a protein fusion pore (aligned by a circular array of 5 to 8 syntaxin molecules), thus allowing the lipid bilayers of synaptic vesicles and AZ plasma membrane to unite. In this way, the main purpose of the plasma membrane at the AZ during neurotransmitter release is to mediate fusion of synaptic vesicles as they interact with the depolarization-triggered presynaptic Ca^{2+} microdomains. Multidomain scaffold proteins constitute the large protein complexes at the pre- and postsynaptic sites, while pre-assembled dense-core transport vesicles deliver the necessary protein constituents of the CAZ. Although christened with musical names, Piccolo and Bassoon are two types of scaffold proteins underlying the PTV (Piccolo–Bassoon transport vesicle), which transport components of the CAZ. Their vesicular features imply an amount of 3 to 5 PTV's adequate for forming the functional presynapse [72].

17.11.4   Presynaptic Ca^{2+} Microdomains

Ca^{2+} entry through voltage-gated Ca^{2+} channels (for short, VGCCs) results in a highly heterogeneous distribution of the near-membrane Ca^{2+}. Experimental data has shown that micromolar Ca^{2+} concentrations occur within tens of nanometers of sites of Ca^{2+} entry, and that Ca^{2+} concentrations reach basal levels at less than 350 nm from the entry point [10]. Exocytosis is possible when the presynaptic Ca^{2+} level is increased in the range $50 - 150 \ \mu M$. There is a sharp nonlinear dependence of the secretory rate on the level of Ca^{2+} suggesting that at least four Ca^{2+} ions must bind with positive cooperation to a target complex in order to trigger rapid secretion [116]. As a result of the spatial confinement of microdomain Ca^{2+}, only a fraction of the presynaptic vesicles are exposed to a Ca^{2+} trigger during a depolarizing stimulus. In [10], it was determined that the proximity of the Ca^{2+} microdomain to the docked vesicle, and not the absolute Ca^{2+} amplitude, is the major factor for release. Also, the evidence that vesicles and Ca^{2+} microdomains are uncoupled when they are spread more than 300 nm apart, has several implications. Firstly, with the current depolarizing stimulus, only about half of the morphologically docked near-membrane vesicles are within reach of microdomain Ca^{2+}. Secondly, of these vesicles, about 10 % can be instantly released when triggered, suggesting that additional intrinsic factors are responsible for triggering the vesicle fusion [10].

17.11.5   The SNARE Complex

We proceed to describe two important mechanisms corresponding to protein constituents such as synapsin-1 and synaptotagmin-1. In [61], it is proposed that phosphorylation of synapsin-1 by Ca^{2+} dependent kinases, on releasing synaptic vesicles (SVs) from actin filaments, may accelerate the passage of vesicles to the presynaptic membrane. Thus, the Ca^{2+} entry may stimulate the engagement of SVs from the reserve pool making them tether and dock within the active zone where they are available for rapid release. In [62], it is shown that cytoskeletal protein tubulin binds directly to synaptotagmin-1, which promotes tubulin assembly. At the same time, synaptotagmin-1 functions by attaching synaptic vesicles to microtubules in high concentrations of Ca^{2+}.

Hirokawa *et al.* [61] presented quick-freeze deep-etch EM images, where microtubules are occasionally seen to run close to the presynaptic membrane. Thus, presynaptic microtubules may be directly connected to docked SVs, or cross-linked to the SVs by the CAZ scaffold proteins. Although exocytosis *per se*, is driven by the core SNARE protein complex[1] assisted in its action by the C2 domains of synaptotagmin-1 (which bind presynaptic membrane lipids in the presence of Ca^{2+}), we wish to point out that *local solitonic control may be a requirement for tunneling through the energetic barrier for synaptic pore opening and membrane fusion.*

An essential feature of the SNARE complex lies in its assembly, which facilitates fusion of SV and presynaptic membrane lipid bilayers. In the case of neurons, the SNARE complex is composed of the vesicle protein *synaptobrevin* and the presynaptic membrane proteins *syntaxin*-1 and *SNAP-25*. The cytoplasmic domains of these proteins assemble into a four α-helix bundle that unites the vesicle and presynaptic membranes. The Ca^{2+} sensor synaptotagmin-1, together with various accessory proteins, are considered as collaborating with the core SNARE complex in order to create a tightly regulated SV fusion at the synapses. Supplementary agents of the fusion machinery assembly are the *vesicle tethers* (e.g., synaptotagmin-1 and β–neurexin). Synaptobrevin and synaptotagmin on an SV couple with their cognate targets anchored in the plasmalemma (syntaxin, SNAP-25, β-neurexin), thus forming the so-called SNAREpins. The core SNARE complex can be seen to be formed by the zipping of four α–helices, where synaptobrevin and syntaxin contribute one α–helix each, while SNAP-25 contributes two α-helices. Syntaxin and SNAP-25 are localized to the presynaptic plasma membrane, whereby those proteins interact with Ca^{2+} channels. This is the relevant mechanism for close proximity of the fusion machinery to the Ca^{2+} microdomains generated by presynaptic depolarization [19]. Prior to the terminal bouton depolarization, the core SNARE complex is half-zipped. In order for the vesicle fusion to proceed, the full zipping is required and it is at this stage that the described Davydov solitons supported by additional dipolar coupling/moments, should be seen as instrumental. Indeed, according to the estimate of Beck–Eccles, the tunneling quasiparticle could be up to six times the mass of the proton. Considering the calculated mass of the Davydov soliton being approximately 5% of the proton mass and the fact that the fusion pore is formed by 5 to 8 SNARE complexes, one might hypothesize that vibrationally assisted quantum tunneling of a Davydov soliton might trigger zipping of all the SNARE complexes within the fusion pore.

---

[1] Soluble NSF attachment protein receptor, where NSF abbreviates N-ethyl-maleimide-sensitive fusion protein.

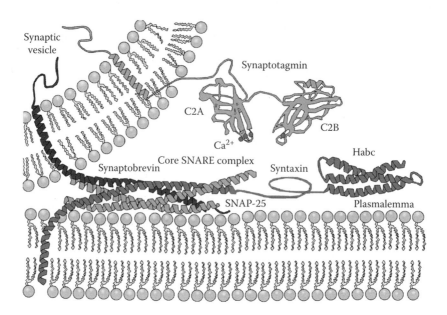

FIGURE 17.4 A nanomolecular mechanism for membrane fusion and exocytosis. The core SNARE complex is formed by four α-helices contributed by syntaxin, SNAP-25, and synaptobrevin. The SNARE zipping could be induced by Davydov soliton dynamics as triggered by a conformational change in the synaptotagmin molecule under a binding of four Ca^{2+} ions within the influence of long-range quantum correlations in the CAZ scaffold. Synaptotagmin C2A and C2B domains also dip into the plasma membrane, thus destabilizing locally the lipid bilayer and exerting an additional traction force upon the docked synaptic vesicle. In the inactive state, the syntaxin molecule is not engaged in the SNARE complex, but instead forms a 4-α-helical bundle with its own Habc domain.

Vesicle tethering complexes containing protein subunits, by their internal structure promote SNARE complex assembly and harbor large multidomain proteins. The α–helix containing polypeptides within several complexes have some influence over the tether length and, hence, in constraining vesicle movement. Syntaxin is bound to n-Sec1 prior to formation of the core complex. The dissociation of n-Sec1 from syntaxin is considered to be aided by Rab proteins, which in turn permit a nucleation binding of the neuronal SNAREs : syntaxin, SNAP-25, and synaptobrevin (see Figure 17.4). Following the fusion process, α–SNAP and NSF are taken from the cytoplasm, and subsequent ATP hydrolysis by NSF causes dissociation of the SNARE complex. Eventually, syntaxin, SNAP-25, and synaptobrevin can be recycled for subsequent neuromediator release.

Volatile anesthetics (VAs) can interfere with the SNARE complex in regulating neuromediator release since both syntaxin-1 and SNAP-25 are VA–hypersensitive (see e.g., [24,25,111]). Of the VAs, halothane has been shown to bind to a hydrophobic pocket with a 4–α–helix bundle [69], suggesting that the hydrophobic core SNARE complex is a suitable target for VAs. Isoflurane binding to the ternary syntaxin/SNAP-25/synaptobrevin core complex was directly measured by 19F-NMR spin–spin relaxation [24], and in addition, it was shown that isoflurane binds to the ternary complex itself and not just to a contaminating monomer. Since most VAs (e.g., halothane, isoflurane, sevoflurane, enflurane, etc.) in anesthetic clinical concentrations dramatically reduce the glutamate release in the cerebral cortex, we suggest that the blocking of exocytosis in the cerebral cortex *per se* may be instrumental in the cause of anesthesia (unconsciousness).

The SNARE complex, while functioning as a fusion mechanism, may be capable of receiving Ca^{2+} signals transmitted by synaptotagmin-1 Ca^{2+} binding, which may result in the fusion of synaptic vesicles with the presynaptic membranes. In relationship to our discussion of kink/antikink solitons, the possibility opens up that a traveling kink, antikink, or breather originating in the CAZ scaffold may function as a necessary supplemental agent for triggering exocytosis, since the Ca^{2+} entry *per se* is insufficient to prevent exocytosis failure. Thus, the various soliton-like systems we have considered may be proposed to model the supply of the additional energy required to ensure the Ca^{2+}–triggered full zipping of the SNARE complex.

The model is consistent to that conjectured in [19], namely that exocytosis results from the cooperative interaction of SNAREs, synaptotagmin-1, and in our case, we include the dynamics of the CAZ scaffold.

17.12 Conclusions

In the context of biophysical nonlinear excitation phenomena we have discussed some mechanisms for microtubule biodynamics. The solitonic-like interactions ambient to microtubular surfaces suggest a number of possibilities for interaction between local EM-fields of electro–neural impulses and the cytoskeletal structure. The broader model as we have proposed includes agencies of protein catalysis, energase action, VAT, stochastic resonance and certain mechanisms afforded by the Davydov solitons to determine structural changes within the α–helices in relationship to the SNARE complex. The cumulative effect of this resonant system is seen to induce an EM-field through the sequence of events

$$EM\text{–}field \Longrightarrow tubulin\text{–}tail\ solitons \Longrightarrow CAZ\ and\ SNARE\ solitons \Longrightarrow exocytosis \Longrightarrow EM\text{–}field$$

By means of this sequence we may envisage a framework supportive of a unitary cybernetic system. Such a progression of ideas seems crucial towards understanding the neurobiological basis for information transmission within neural networks and one that may be relevant to further research into fields of nanomolecular structures involving biosensory processors and protein engineering. Further, we have suggested the possible implementation of related quantum or semi-classical computational schemes such as to be assessed in future work.

References

[1] Ablowitz MJ and Clarkson PA. *Solitons, Nonlinear Evolution Equations and Inverse Scattering.* Cambridge University Press, Cambridge, U.K. 1991.

[2] Ablowitz MJ, Herbst BM, and Schober C. On the numerical solution of the sine–Gordon equation. *J. Comp. Phys.* 126: 299–314, 1996.

[3] Al-Bassam J, Ozer RS, Safer D, Halpain S, and Milligan RA. MAP2 and tau binding longitudinally along the outer ridges of microtubule protofilaments. *J. Cell Biol.* 157: 1187–1196, 2002.

[4] Abdalla E, Marouff B, Melgar BC, and Sedra MB. Information transport by sine–Gordon solitons in microtubules. *Physica A* 301: 169–173, 2001.

[5] Amos LA. Focusing on microtubules. *Curr. Opin. Struct. Biol.* 10: 236–241, 2000.

[6] Arnold SE, Lee VM, Gur RE, and Trojanowski JQ. Abnormal expression of two microtubule-associated proteins (MAP2 and MAP5) in specific subfields of the hippocampal formation in schizophrenia. *Proc. Natl. Acad. Sci. USA* 88: 10850–10854, 1991.

[7] Baer SM and Rinzel J. Propagation of dendritic spikes mediated by excitable synapses: A continuum theory. *J. Neurophysiol.* 65: 874–890, 1991.

[8] Bai J and Chapman ER. The C2 domains of synaptotagmin—partners in exocytosis. *Trends Biochem. Sci.* 29: 143–151, 2004.

[9] Baxter RJ. *Exactly Solved Models in Statistical Mechanics.* Academic Press, New York, 1982.

[10] Becherer U, Moser T, Stuhmer W, and Oheim M. Calcium regulates exocytosis at the level of single vesicles. *Nat. Neurosci.* 6: 846–853, 2003.

[11] Beck F and Eccles JC. Quantum aspects of brain activity and the role of consciousness. *Proc. Natl. Acad. Sci. USA* 89: 11357–11361, 1992.

[12] Brose N. Synaptic cell adhesion proteins and synaptogenesis in the mammalian central nervous system. *Naturwissenschaften* 86: 516–524, 1999.

[13] Brown J and Tuszynski JA. A review of the ferroelectric model of microtubules. *Ferroelectrics* 220: 141–156, 1999.

[14] Bullough RK, Pilling DJ, and Timonen J. Quantum and classical statistical mechanics of the sinh–Gordon equation. *J. Phys. A: Math. Gen.* 19: 955–960, 1986.

[15] Bullough RK, Pilling DJ, and Timonen J. Statistical mechanics of the sine–Gordon field. In Claro F, editor, *Nonlinear Phenomena in Physics*, pages 70–128. Springer, Berlin, 1985.

[16] Bullough RK. Optical solitons: Twenty-seven years of the last millenium and three more years of the new? In Ashour AA and Obada A-SF, editors, *Mathematics and the 21st Century*, pages 69–121. World Scientific, Singapore, 2001.

[17] Calogero F and Degasperis A. *Spectral Transform and Solitons: Tools to Solve and Investigate Nonlinear Evolution Equations*. North Holland, New York, 1982.

[18] Caspi S and Ben-Jacob E. Conformation change and folding of proteins mediated by Davydov's soliton. *Phys. Letts. A* 272: 124–129, 2000.

[19] Chapman ER. Synaptotagmin: A Ca^{2+} sensor that triggers exocytosis? *Nat. Rev. Mol. Cell Biol.* 3: 498–508, 2002.

[20] Chou KC, Zhang CT, and Maggiora GM. Solitary wave dynamics as a mechanism for explaining the internal mechanism during microtubule growth. *Biopolymers* 34: 143–153, 1994.

[21] Coffey WT, Kalmykov YP, and Waldron JT. *The Langevin Equation*. World Scientific, Singapore, 2004.

[22] Coombes S and Bressloff PC. Solitary waves in a model of dendritic cable with active spines. *SIAM J. Appl. Math.* 61: 432–453, 2000.

[23] Coombes S and Timofeeva Y. Sparks and waves in a stochastic fire-diffuse-fire model of calcium release. *Phys. Rev. E.* 68: 021915, 2003.

[24] Crowder CM, Nagele P, and Mendel B. Block of isoflurane binding to syntaxin by addition of N-terminal amino acids. ASA Annual Meeting Abstracts A-885, 2004.

[25] Crowder CM and Berilgen J. Isoflurane binds the rat synaptic protein SNAP-25 at clinical concentration. ASA Annual Meeting Abstracts A-806, 2000.

[26] Davydov AS. *Biology and Quantum Mechanics*. Pergamon Press, Oxford, U.K. 1982.

[27] Davydov AS. *Solitons in Molecular Systems*. Kluwer, Dordrecht, 1991.

[28] Davydov AS. The theory of contraction of proteins under their excitation. *J. Theor. Biol.* 38: 559–569, 1973.

[29] Dean C and Dresbach T. Neuroligins and neurexins: Linking cell adhesion, synapse formation, and cognitive function. *Trends Neurosci.* 29: 21–29, 2006.

[30] Del Giudice E, Doglia S, and Milani M. Self-focusing and ponderomotive forces of coherent electric waves–a mechanism for cytoskeleton formation and dynamics. In Fröhlich H and Kremer F, editors, *Coherent Excitations in Biological Systems*. Springer–Verlag, Berlin, 1983.

[31] Del Giudice E, Doglia S, Milani M, and Vitiello G. Electromagnetic field and spontaneous symmetry breaking in biological matter. *Nucl. Phys. B* 275: 185–199, 1986.

[32] Del Giudice E, Preparata G, and Vitiello, G. Water as a free electric dipole laser. *Phys. Rev. Lett.* 61: 1085–1088, 1988.

[33] Dicke RH. Coherence in spontaneous radiation processes. *Phys. Rev.* 93: 99–110, 1954.

[34] Dmitriev SV, Shigenari T, Vasiliev AA, and Miroshnichenko AE. Effect of discreteness on a sine—Gordon three soliton solution. *Phys. Lett. A* 246: 129–134, 1998.

[35] Dodd RK, Eilbeck JC, Gibbon JD, and Morris HC. *Solitons and Nonlinear Wave Equations*. Academic Press, New York, 1982.

[36] Eccles JC. *How the Self Controls Its Brain*. Springer–Verlag, Berlin, 1994.

[37] Fischer M, Kaech S, Knutti D, and Matus A. Rapid actin-based plasticity in dendritic spines. *Neuron* 20: 847–854, 1998.

[38] Fröhlich H. The extraordinary dielectric properties of biological materials and the action of enzymes. *Proc. Natl. Acad. Sci. USA* 72: 4211–4215, 1975.

[39] Fröhlich H. Long-range coherence and energy storage in biological systems. *Int. J. Quant. Chem.* 2: 641–649, 1968.

[40] Gammaitoni L, Hanggi P, Jung P, and Marchesoni F. Stochastic resonance. *Rev. Mod. Phys.* 70: 223–287, 1988.

[41] Georgiev DD, Papaioanou SN, and Glazebrook JF. Neuronic system inside neurons: Molecular biology and biophysics of neuronal microtubules. *Biomed. Rev.* 15: 67–75, 2004.

[42] Georgiev DD and Glazebrook JF. Dissipationless waves for information transfer in neurobiology – some implications. *Informatica* 30: 221–232, 2006.

[43] Georgiev DD and Glazebrook JF. Conformational dynamics and thermal cones of C–terminal tubulin tails in microtubules. *Neuroquantology* 5(1): 62–84, 2007.

[44] Georgiev DD. Electric and magnetic fields inside neurons and their impact upon the cytoskeletal microtubules. (2003) http://cogprints.org/3190/

[45] Georgiev DD. The β-neurexin/neuroligin-1 interneuronal intrasynaptic adhesion is essential for quantum brain dynamics. (2003) http://arXiv.org/abs/quant-ph/0207093

[46] Georgiev DD. On the dynamic timescale of mind–brain interaction. In *Proceedings of Quantum Mind II : Consciousness, Quantum Physics and the Brain*. The Leo Rich Theater, Tucson, AZ, 2003. http://cogprints.org/4463/

[47] Georgiev DD. Solitonic effects of the local electromagnetic field on neuronal microtubules–tubulin tail sine–Gordon solitons could control MAP attachment sites and microtubule motor protein function. (2004) http://cogprints.org/3894/

[48] Georgiev DD. Solving the binding problem: Cellular adhesive molecules and their control of the cortical quantum entangled network. (2003) http://cogprints.org/2923/

[49] Gerstner W and Kistler W. *Spiking Neuron Models*. Cambridge University Press, Cambridge, U.K., 2002.

[50] Gray HB and Winkler JR. Electron tunneling through proteins. *Q. Rev. Biophys.* 36: 341–372, 2001.

[51] Grundler W and Keilmann F. Sharp resonances in yeast growth proved nonthermal sensitivity to microwaves. *Phys. Rev. Lett.* 51: 1214–1216, 1983.

[52] Gudder S. Noncommutative probability and applications. In Rao MM, editor, *Real and Stochastic Analysis–New Perspectives*, pages 199–238. Birkhauser, Boston, 2004.

[53] Habib S, Khare A, and Saxena A. Statistical mechanics of double sinh–Gordon kinks. *Physica D* 123: 341–356, 1998.

[54] Hagan S, Hameroff SR, and Tuszynski JA. Quantum computation in brain microtubules? Decoherence and biological feasibility. *Phys. Rev. E* 65: 061901, 1–11, 2002.

[55] Hall ZW. *Introduction to Molecular Neurobiology*. Sinauer Assoc., Sunderland, MA 1992.

[56] Hameroff SR. Anesthesia, consciouness, and hydrophobic pockets–a unitary quantum hypothesis of anesthetic action. *Toxicol. Lett.* 101–101: 31–39, 1998.

[57] Hameroff SR and Penrose R. Conscious events as orchestrated space-time selections. *J. Consc. Stud.* 3: 36–53, 1996.

[58] Hameroff SR and Watt RC. Information processing in microtubules. *J. Theor. Biol.* 98: 549–561, 1982.

[59] Haroche S, Nogues G, Rauschenbeutel A, Osnaghi S, Brune M, and Raimond JM. Quantum knitting in cavity QED. In Blatt et al., editors, *Laser Spectroscopy XIV International Conference*, pages 140–149. World Scientific, Singapore, 1999.

[60] Heald R and Nogales E. Microtubule dynamics. *J. Cell Sci.* 115: 3–4, 2002.

[61] Hirokawa N, Sobue K, Kanda K, Harada A, and Yorifuji H. The cytoskeletal architecture of the presynaptic terminal and molecular structure of synapsin-1. *J. Cell Biol.* 108: 111–126, 1989.

[62] Honda A, Yamada M, Saisu H, Takahashi H, Mori KJ, and Abe T. Direct Ca^{2+}–dependent interaction between tubulin and synaptotagmin-1. A possible mechanism for attaching synaptic vesicles to microtubules. *J. Biol. Chem.* 277: 20234–20242, 2002.

[63] Jibu M, Hagan S, Hameroff SR, Pribram KH, and Yasue K. Quantum optical coherence in cytoskeletal microtubules : Implications for brain function. *Biosystems* 32: 195–209, 1994.

[64] Jibu M and Yasue K. *Quantum Brain Dynamics and Consciousness: An Introduction*. Advances in Consciousness Research, vol. 3, John Benjamins, Amsterdam, 1995.

[65] Jibu M and Yasue K. What is mind? – Quantum theory of evanescent photons in brain as quantum theory of consciousness. *Informatica* 21: 471–490, 1997.

[66] Jibu M, Yasue K, and Pribram KH. From conscious experience to memory storage and retrieval: The role of quantum brain dynamics and boson condensation of evanescent photons. *Int. J. Mod. Phys. B* 10: 1735–1754, 1996.

[67] Jibu M, Yasue K, and Hagan S. Evanescent (tunneling) photon and cellular vision *Biosystems* 42: 65–73, 1997.

[68] Jiminez MA, Evangelio JA, Aranda C, Lopez-Braet A, Andreu D, Rico M, Lagos R, Andreu JM, and Monasterio, O. Helicity of α (404–451) and β (394–445) tubulin C–terminal recombinant peptides. *Protein Sci.* 8: 788–799, 1999.

[69] Johansson JS, Scharf D, Davies LA, Reddy KS, and Eckenhoff RG. A designed four-α-helix bundle that binds the volatile general anesthetic halothane with high affinity. *Biophys J.* 78: 982–983, 2000.

[70] Kampa BM and Stuart GJ. NMDA receptor kinetics are tuned for spike–timing dependent synaptic plasticity. *Physiol. News* 58: 29–30, 2005.

[71] Knapp MJ and Klinman JP. Environmentally coupled hydrogen tunneling—Linking catalysis to dynamics. *Eur. J. Biochem.* 269: 3113–3121, 2002.

[72] Kreinenkamp H-J and Dityatev A. The magic of synaptogenesis. *EMBO Reports* 5: 1125–1129, 2004.

[73] Layne SP. The modification of Davydov solitons in the extrinsic H–N–C=0 group. In *Nonlinear Electrodynamics in Biological Systems*, pages 531–548. Plenum Press, New York, 1984.

[74] Layne SP. A possible mechanism for general anesthesia. *Los Alamos Science*, 23–26, Spring 1984.

[75] Lin J, Balabin IA, and Beratan DN. The nature of aqueous pathways between electron–transfer proteins. *Science* 310: 1311–1313, 2005.

[76] Lomdahl PS, Layne SP, and Bigio IJ. Solitons in biology. *Los Alamos Science*, 11–22, Spring 1984.

[77] Luther A. Quantum solitons in statistical physics. In Bullough RK and Caudrey PJ, editors, *Solitons, Topics in Current Physics* 17, pages 355–371. Springer, Berlin, 1980.

[78] Marcus RA and Sutin N. Electron transfer in chemistry and biology. *Biochim. Biophys. Acta* 811: 265–322, 1985.

[79] Matus A. Actin-based plasticity in dendritic spines. *Science* 290: 754–758, 2000.

[80] Mavromatos NE, Mershin A, and Nanopoulos DV. QED–cavity model of microtubules implies dissipationless energy transfer and biological quantum teleportation (2002). *Int. J. Mod. Phys. B* 16: 3623–3642, 2002.

[81] Miroshnichenko AE, Dmitriev SV, Vasiliev AA, and Shigenari T. Inelastic three-soliton collisions in a weakly discrete sine–Gordon system. *Nonlinearity* 13: 837–848, 2000.

[82] Musienko AI and Menevich LI, Classical mechanical analogs of relativistic effects. *Phys. Uspekhi* 47: 797–820, 2004.

[83] Nogales E, Wolf SG, and Downing KH. Structure of the α/β tubulin dimer by electron crystallography. *Nature* 391: 199–203, 1998.

[84] Penrose R. *Shadows of the Mind—A Search for the Missing Science of Consciousness*. Oxford University Press, Oxford, 1994.

[85] Pribram KH. *Brain and Perception*. Lawrence Erlbaum NJ, 1991.

[86] Pouget A, Dayan P, and Zemel R. Information processing within population codes. *Nat. Rev. Neurosci.* 1: 125–132, 2000.

[87] Purich DL. Enzyme catalysis: A new definition accounting for noncovalent substrate- and product-like states. *Trends Biochem. Sci.* 26: 417–421, 2001.

[88] Reitman ET. *Molecular Engineering of Nanosystems*. American Institute of Physics, 2001.

[89] Russell FM, Zolotaryuk Y, and Eilbeck JC. Moving breathers in a chain of magnetic pendulums. *Phys. Rev. B* 55: 6304–6308, 1997.

[90] Sackett DL. Structure and function in the tubulin dimer and the role of the acid carboxyl termini. *Subcellular Biochemistry–Proteins: Structure, function and engineering* 24: 255–302, 1995.

[91] Sataric MV, Zakula RB, and Tuszynski JA. A model of the energy transfer mechanism in microtubules involving a single soliton. *Nanobiology* 1: 445–456, 1992.

[92] Sataric MV and Tuszynski JA. Relationship between the nonlinear ferroelectric and liquid crystal models for microtubules. *Phys. Rev. E* 67: 011901, 1–11, 2003.

[93] Sataric MV, Zakula R, Ivic Z, and Tuszynski JA. Influence of a solitonic mechanism on the process of chemical analysis. *J. Mol. Electron.* 7: 39–46, 1991.

[94] Scott AC. Dynamics of Davydov solitons. *Phy. Rev. A* 26: 578–595, 1982.

[95] Scott AC. Davydov's soliton. *Phys. Rev. Lett.* 217: 1–67, 1992.

[96] Skiniotis G, Cochran JC, Mueller J, Mandelkow E, Gilbert SP, and Hoenger A. Modulation of kinesin binding by the C–termini of tubulin. *EMBO J.* 23: 989–999, 2004.

[97] Söllner TH. Vesicle tethers promoting fusion machinery assembly. *Developmental Cell* 2: 377–378, 2002.

[98] Sontag E, Nunbhakdi-Craig V, Lee G, Brandt R, Kamibayashi C, Kuret J, White CL, Mumby MC, and Bloom GS. Molecular interactions among protein phosphatase 2A, tau, and microtubules: Implications for the regulation of tau phosphorylation and the development of tauopathies. *J. Biol. Chem.* 274: 25490–25498, 1999.

[99] Sperry RW. Some effects of disconnecting the cerebral hemispheres. Nobel Lecture: December 1981.

[100] Sperry RW and Gazzaniga MS. Language following disconnection of the hemispheres. In Millikan CH and Darley FL, editors, *Brain Mechanisms Underlying Speech and Language*, pages 177–184. Grune and Stratton Inc., New York, 1967.

[101] Sperry RW, Zaidel E, and Zaidel D. Self recognition and social awareness in the deconnected minor hemisphere. *Neuropsychologia* 17: 153–166, 1979.

[102] Stapp HP. Attention, intention and will in quantum physics. *J. Consc. Stud.* 6: 143–164, 1999.

[103] Sutcliffe MJ and Scrutton NS. Enzyme catalysis: Over-the-barrier or through-the-barrier? *Trends Biochem. Sci.* 25: 405–408, 2000.

[104] Sutcliffe MJ and Scrutton NS. A new conceptual framework for enzyme catalysis—Hydrogen tunneling coupled to enzyme dynamics in flavoprotein and quinoprotein enzymes. *Eur. J. Biochem.* 269: 3096–3102, 2002.

[105] Takada S and Nakamura H. Wentzel–Kramer–Brillouin theory of multi-dimensional tunneling: General theory for energy splitting. *J. Chem. Phys.* 100: 98–113, 1994.

[106] Takada S and Nakamura H. Effects of vibrational excitation on multi-dimensional tunneling: General study and proton tunneling in tropolone. *J. Chem. Phys.* 102: 3977–3992, 1995.

[107] Tegmark M. Importance of quantum decoherence in brain processes. *Phys. Rev. E* 61: 4194–4206, 2001.

[108] Timonen J, Stirland M, Pilling DJ, Cheng Y, and Bullough RK. Statistical mechanics of the sine–Gordon equation. *Phys. Rev. Lett.* 56: 2233–2236, 1986.

[109] Tripisova B and Brown JA. Ordering of dipoles in different types of microtubule lattice. *Int. J. Mod. Phys. B* 12: 543–578, 1998.

[110] Tuszynski JA, Hameroff SR, Sataric MV, Trpisova B, and Nip MLA. Ferroelectric behavior in micro-tubule dipole lattices: Implications for information processing, signalling and assembly/disassembly. *J. Theor. Biol.* 174: 371–380, 1995.

[111] van Swinderen B, Saifee O, Shebester L, Roberson R, Nonet ML, and Crowder CM. A neomorphic syntaxin mutation blocks volatile-anesthetic action in *Caenorhabditis elegans. Proc. Natl. Acad. Sci. USA* 96: 2479–2484, 1999.

[112] Vitiello G. Quantum dissipation and information: A route to consciousness modeling. *Neuroquantology* 2: 266–279, 2003.

[113] Vitiello G. Dissipation and memory capacity in the quantum brain model. *Int. J. Mod. Phys. B* 9: 973–989, 1995.

[114] Yasue K. Quantum mechanics of non-conservative systems. *Ann. Phys.* 114: 479–496, 1978.

[115] Zhai RG and Bellen HJ. The architecture of the active zone in the presynaptic nerve terminal. *Physiology* 19: 262–270, 2004.

[116] Zucker RS. Exocytosis: A molecular and physiological perspective, *Neuron* 17: 1049–1055, 1996.

[117] Zworski M. Counting scattering poles. In Iwaka M, editor, *Spectral and Scattering Theory*, pages 301–331. Marcel–Dekker, New York, 1994.

18

Electronic and Ionic Conductivities of Microtubules and Actin Filaments, Their Consequences for Cell Signaling and Applications to Bioelectronics

Jack A. Tuszynski

Avner Priel

J.A. Brown

Horacio F. Cantiello

John M. Dixon

18.1   Introduction

The prevailing view of neural information processing is based on passive properties of the membrane derived from the application of linear cable theory to dendrites, e.g., [114]. Recent studies, however, have suggested nonlinear models that accommodate new experimental evidence that appears to be inconsistent with this classical view. In particular, conventional theories do not satisfactorily explain: (a) significant fluctuations of synaptic efficacy over short periods of time in response to a recent burst of activation, e.g., [54], (b) a variety of active ion-channels capable of affecting the local membrane electric properties, and (c) nonlinear responses localized to specific dendritic sites, pointing to highly specialized mechanisms correlated with specific inputs. On the other hand, nonlinearities inherent in the new models, e.g., [74,133], give rise to a wider repertoire of (computational) capabilities, such as multiplication, fast correlation, etc. Moreover, experimental and theoretical support for conductive properties of protein filaments and the ionic clouds around them is becoming available, which indicates a far more complex picture of the signaling and information processing phenomena in neurons and other cells.

In this chapter, we suggest that functional electrodynamic interactions between cytoskeletal structures and ion-channels are central to the neural information processing mechanism. These interactions are supported by long-range ionic wave propagation along microtubule networks and actin filaments, and exhibit subcellular control of ionic channel activity, hence impacting the computational capabilities of the whole neural function. Cytoskeletal biopolymers, including actin filaments (AFs) and microtubules (MTs), constitute the backbone of wave propagation, and in turn interact with membrane components to modulate synaptic connections and membrane channels. Indeed, only recently clear functional interactions between these cytoskeletal structures have become apparent. Association of MTs with AFs in neuronal filopodia appears to guide microtubule growth, and plays a key role in neurite initiation [28]. This is further evidenced by the presence in neurons of proteins that interact with both MTs and F-actin, and proteins that can mediate signaling between both types of filaments. This is likely used to control microtubular invasion. The microtubule associated proteins MAP1B and MAP2, for example, are known to interact with actin in vitro, e.g., [146]. Cross-linking, MAP2 and/or MAP1B is highly probable to associate with both types of filaments contributing to the guidance of MTs along AF bundles. Extensive evidence exists confirming a direct interaction between AFs and ion channels, and a regulatory functional role associated with actin. Thus it is clear that the cytoskeleton has a direct connection to membrane components, in particular channels and synapses.

In the following, we demonstrate how each individual cytoskeletal component is capable of supporting both ionic and electronic wave propagation. In particular, we present a molecular dynamics model of the dendritic microtubule network (MTN) where arrays of MTs are interconnected by MAP2s. In this model, ionic waves propagate along the MTN and interact with C-termini of MTs to generate collective modes of behavior. Also in this chapter, a biophysical model of nonlinear ionic wave propagation along AFs is presented, supported by experimental evidence. Finally, we describe a model for a direct regulation of ion-channels and synaptic strength by AFs and networks of MTs that control the electrical response of the dendrite in particular, and the neuron in general. We begin this chapter by providing a general overview of the structure and function of a nerve cell, emphasizing the role of cytoskeletal filaments, ion channels, and their interactions.

18.2   The Neuron

The neuron is the quintessential communicating cell. A great variety of neural subtypes can be found interconnected to one another in complex electrochemical circuits via multiple synaptic connections. Thus, the amount of information processing by a neural circuit depends on the number of such connections or synapses. Neurons are highly polarized cells, whose level of morphological complexity increases during maturation into distinct subcellular domains. A typical neuron possesses three functional domains. These are the cell body or *soma* containing the nucleus and all major cytoplasmic organelles, a single axon, which extends away from the soma and takes cable like properties, and a variable number of dendrites, different

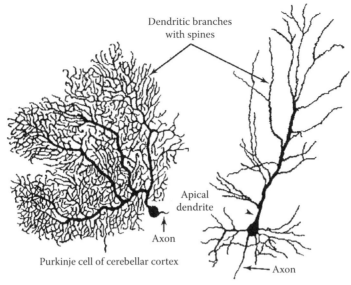

Dendritic branches
with spines

Apical
dendrite

Axon

Purkinje cell of cerebellar cortex

Axon

Pyramidal neuron of neocortex

FIGURE 18.1   Two prominent examples of nerve cells.

in shape and complexity, which emanate from the soma. The axonal terminal region, where the synapse contacts it with other cells, displays a wide range of morphological specialization (see Figure 18.1).

Two major types of synapses can be found: (a) asymmetric, responsible for the transmission of excitatory impulses, and (b) symmetrical, present in inhibitory synapses. Turning morphology into a functional feature, the soma and dendritic tree are the major domains of receptive inputs. Thus, both the dendritic arborization and the axonal ramifications confer a high level of sub-cellular specificity in the localization of particular synaptic contacts on a given neuron. The extent to which a neuron may be interconnected largely depends on the three-dimensional spreading of the dendritic tree. Dendrites contain a large number of dendritic spines, which are small sac-like organelles, projecting from the dendritic trunk. Dendritic spines are more abundant in highly arborized cells such as pyramidal neurons, and scarcer in lowly interconnected smaller-sized dendritic tree inter-neurons.

Spines are the dendritic regions where most excitatory inputs are found, such that each spine contains one asymmetric synapse. The number of excitatory inputs can be correlated with the number of spines present in the dendrite. Dendritic spines are complex saccular enlargements, containing ribosomes, and cytoskeletal structures, including actin, and α- and β-tubulin. Electrically speaking, a single neuronal circuit is closed such that afferent inputs reaching the soma and the dendritic tree are transferred to the axon at the other pole of the cell which, in turn, is responsible for transmitting efferent neural information. Both the morphology of the axon and its course through the nervous system are correlated with the type of information processed by a particular neuron, and by its interconnectivity with other neurons. While the axon leaves the soma with a small swelling known as the axon hillock, some neurons protrude the axonal projection from the main dendrite instead. Cytoskeletal components are highly relevant in the structure/function of the neuron. The axonal hillock contains large "parallel" bundles of MTs. Axons are also located with other cytoskeletal structures, such as neurofilaments, which are more abundant in axons than in dendrites. The dynamics in cytoskeletal structures is central to our contention that information can be processed and "delivered" to the synaptic function by changes in the cytoskeletal structures.

Dendrites are the principal element responsible for synaptic integration and for the changes in synaptic strengths that take place as a function of neuronal activity. While dendrites are the principal sites for excitatory synaptic input, little is still known about their function. The activity patterns inherent to the dendritic tree-like structures, such as in integration of synaptic inputs, are likely based in the wide diversity of shapes and sizes of the dendritic arborizations [116]. Some dendrites are unipolar, while others are multipolar; some have many orders of branching, while others have only one or two; and some branch primarily in two dimensions, while others have complex three-dimensional structures. The size and complexity of dendritic trees increase during development, which has been associated with the ability of the neural system to organize and process information [49]. The majority of synapses, both excitatory and inhibitory, terminate on dendrites, so it has long been assumed that dendrites somehow integrate the numerous inputs to produce single electrical outputs. It is increasingly clear that the morphological functional properties of dendrites are central to its integrative function. Branching in this context is essential to the increasingly complex ability of the neuron to respond to developmental complexity. Nerve branching and complexity are controlled by both external and internal cues. Of current interest to this are substrate and electromagnetic fields, while also recognized are the chemical gradients involved in local adaptation. Highly relevant to adaptation is cytoskeletal dynamics, as the dendritic cytoskeleton plays a central role in the process of ramification, filopodia formation, and more specialized neuronal activity such as long-term potential (LTP) and long-term depression (LTD) circuit dynamics, including memory formation.

The electrotonic or passive properties of dendrites became of great importance to our understanding of dendritic integration of synaptic signals. Early hypotheses were supported by work on motor neurons [34] and sensory receptors [38,67]. In that action, potentials were initiated in the axon hillock region as a result of the graded or algebraic summation of EPSPs and IPSPs occurring in various parts of the neuron. Because of the relatively short space constant calculated for motor neurons [18], it was predicted that excitatory post-synaptic potentials from synapses on the distal portions of dendrites would be ineffective in firing a neuron. Rall [115] revolutionized the field of dendrite physiology by using cable theory for the study of neurons and their dendrites. Action potentials initiated in dendrites by synaptic stimulation were later observed propagating in both directions at velocities of around 0.3 m/s. [7,21,43]. Direct intracellular recordings from dendrites of cerebellar Purkinje cells [84], cortical neurons [57], and hippocampal neurons [153] confirmed that action potentials could be initiated and propagated in dendrites. Very little is known, however, about the active, or voltage-dependent, properties of dendrites, and many questions remain unanswered. Such as the types of ion channels present in dendrites, location, and how active properties alter synaptic integration.

The firing of action potentials elicits characteristic M-shaped Ca^{2+} profiles across the neuron. The Ca^{2+} rise is small in the soma, highest in the proximal apical and basal dendrites, and small again in the distal apical dendrites [62,100,118–120]. The spatial distribution of Ca^{2+} channels explains the distinct Ca^{2+} rise in various cell locations but rather reflects the propagation into the dendrites of trains of action potentials. Imaging with Na^{+}-sensitive dyes has yielded similar patterns, suggesting that Ca^{2+} influx into the dendrites is driven by Na^{+} influx into the dendrites which is driven by Na^{+}-dependent action potentials [62]. Ca^{2+}-dependent action potentials have been demonstrated in hippocampal neurons and dendrites [131]. Most of the action potential induced Ca^{2+} entry is mediated by voltage-gated Ca^{2+} channels [100]. Multiple types of voltage-gated Ca^{2+} channels have been demonstrated in hippocampal neurons [36,40,68,101]. Based on the addition of specific Ca^{2+} channel blockers, the presence of L-, N-, P-, Q-, R-, and T-type Ca^{2+} channels has been reported [16]. This channel distribution may contribute to spike-triggered Ca^{2+} entry into the soma and apical dendrites. Both low-voltage activated (LVA) channels and at least one high-voltage activated (HVA) channel seem to be most predominant in regions of the dendrite at least a 100 mm from the soma. Experimental results strongly suggest a heterogeneous distribution of different types of voltage-gated Ca^{2+} channels within the soma and dendrites of hippocampal neurons [152]. The presence of Na^{+} channels in distal dendrites helps explain the back-propagation of action potentials well into the dendritic tree [88,89,138,139,141]. Action potential propagation in dendrites is rather low, and susceptible to the effects of cable filtering.

A most prominent feature of dendritic ion channel regulation is the potential boosting and enhancement of distal synaptic events. Voltage-gated Na^{+} and Ca^{2+} channels may play an important role in amplifying the magnitude of EPSPs towards the soma. However, channel regulation may also be important for dendritic interactions in the immediate vicinity of the synaptic input. Voltage-gated channels may alter the local input resistance and time constant, which in turn would influence both spatial and temporal summation of EPSPs and IPSPs, thus conveying highly nonlinear interactions; signal amplification [64,137]. Thus, the nonlinear properties of dendrites may serve as electronic devices. In addition to providing an output, the action potentials elicited by synaptic input, whether initiated at the local synaptic sites or at the axon hillock and back-propagated into the dendrites, would essentially reset the dendritic membrane potential. Action potentials might also provide feedback or an associative signal to other synapses active just before or just after the action potential.

Back-propagating action potentials may also provide feedback to dendritic synapses after a neuronal output has occurred. Ca^{2+} signals are implicated in post-synaptically-induced synaptic plasticity such as short-term potentiation (STP) [92], LTP, and LTD [8,15]. Actin has long been postulated to play important roles in neuronal morphogenesis. High concentrations of actin have been found in the leading edge of the growth cones of developing neurons, and in synapses of mature neurons. Actin filament dynamics also plays a role in a number of related phenomena, including axon initiation, growth, guidance, and branching, in the morphogenesis and stability of dendrites and dendritic spines in synapse formation and stability. Neurons conduct two types of cytoplasmic growth processes in axons and dendrites, both of which are functionally and morphologically distinct, but are developmentally led from the growth cone. Growth cones are composed of finger-like filopodia and veil-like lamellipodia that help the growing processes explore their environment. Axons travel long distances, and upon reaching their targets, produce terminal branches where the growth cones are converted into pre-synaptic terminals. Dendrites usually do not extend such long distances, but often branch extensively, giving rise to dendritic trees. Upon proper contact with axons, post-synaptic dendritic specializations create functional synapses by opposition with the pre-synaptic terminals of axons. These are localized to small protrusions on their dendritic trees called dendritic spines, which are the site of most excitatory synapses. In the mammalian brain these multiple synaptic connections are believed to play important roles in learning and memory. Thus, neural morphogenesis requires changes in both the underlying cytoskeleton and specific interactions with the plasma membrane. To generate a membrane protrusive structure such as an axon, coordinated changes in actin and microtubular organization are required. Growth cones of developing neurons, for example, rapidly extend and retract filopodia by concentred changing in cytoskeletal dynamics, namely bundled AFs in filopodia, and composed of a cross-linked actin network in lamellipodia [80,81,155]. In this section, we have outlined the main features of signaling taking place within a neuron. Major emphasis was placed on the role played by the cytoskeleton in the neuron's structure, growth, and internal signaling. The next section discusses the main biophysical properties of the cytoskeleton.

18.3 The Cytoskeleton

The cytoskeleton is a major component of all living cells. It is made up of three different types of filamental structures, including actin-based microfilaments (MFs), intermediate filaments (IFs) (e.g., neurofilaments, keratin), and tubulin-based microtubules (MTs). All of them are organized into networks, which are interconnected through numerous particular proteins, and which have specific roles to play in the functioning of the cell. The cytoskeletal networks are mainly involved in the organization of different directed movements in cell migration, cell division, or in the internal transport of materials. Polymerization of MFs is responsible for cell migration and for the remodeling of the leading edge of cells. Molecular motors are protein complexes associated with the cytoskeleton and drive organelles along MTs and MFs in a "vectorial" transport.

MTs are some of the most basic and important cytoskeletal elements in the morphological scheme of neurons. As for the case of cytochalasins in actin structures, the addition of microtubular modifying agents was first used to assess the role of MTs in axon elongation. Microtubular disrupting agents affect axons first and have no early effects on growth cones and filopodia, suggesting that MTs provide support for the axon. Conversely, microtubular invasion may be a critical element in the formation of neurite projections. Initially, expanded lamellipodia first undergo segmentation at certain spots where MTs accumulate in an ordered array [27,144,156] and gradually "migrate away" from the cell body. These lamellipodia transform into nascent growth cones. Neurite elongation takes place as the MTs in the shaft become compressed into a narrower bundle. In cultured hippocampal and cortical neurons, this process most likely implies the coordinated control of the major cytoskeletal components, including actin, MTs, and associated proteins.

Thus, MTs are essential for neurite development. One particular aspect of microtubular involvement in neurite outgrowth is vesicular transport. Most membrane insertion occurs at the growth cone of axons [22,24]. MAP-stabilized MTs, form parallel or antiparallel arrays, which might act as compression resistant struts inside neurites [61]. Because MTs constantly invade the actin cytoskeleton in lamellipodia of epithelial [125] and neuroblastoma cells [27], such a mechanical role for MTs could also be important during de novo neurite initiation. Neurite-like protrusions, for example, can be induced in non-neuronal cells by the exogenous expression of stabilizing proteins such as MAP2 and tau [72,79,157]. Spontaneous neurite initiation is dependent on both the presence and the dynamic properties of MTs.

A number of proteins associate with MTs, including MT motors, kinases, so-called "structural" MAPs, which alter microtubule structure, and specialized MAPs, which bind to microtubule plus or minus ends. Members of the MAP2/tau family have long been proposed to be important regulators of neurite behavior [10,48,151]. MAP2 and tau stabilize MTs by reducing catastrophe and promoting rescues, leading to prolonged growth periods and thus enhanced net microtubule accumulation [32,45,75]. MAP2 and tau stabilize MTs by binding along the outer ridges of protofilaments [2].

F-actin and MTs cytoskeletal networks are typically thought to fulfill separate, independent cellular roles. Highly dynamic actin networks are known for their role in cell spreading and contraction. The more stable MT cytoskeleton is known for its importance in cell division and organelle trafficking. However, recent studies provide important roles for the actin cytoskeleton in cell division and trafficking and for MTs in the generation and plasticity of cellular morphology. A direct physical association between both cytoskeletons has been suggested, because MTs often preferentially grow along actin bundles and transiently target actin-rich adhesion complexes.

Association of MTs with actin cables in filopodia appears to guide microtubule growth along the most efficient path anti-parallel to retrograde flow [130]. Thus, specific coupling between MTs and actin bundles presumably promotes MT advance. Proteins that are able to interact with both MTs and F-actin, or proteins that can mediate signaling between both cytoskeletons likely control the regulation of MT invasion. This is further strengthened by specific microtubule association of the microtubule-associated proteins MAP1B and MAP2, both known to interact with actin in vitro [20,109,128,134,146]. Furthermore, MAP2 binds to F-actin and efficiently induces actin bundle formation [51,129]. It is likely that by crosslinking, MAP2 and/or MAP1B associated with both cytoskeletons could be involved in the guidance of MTs along actin filament bundles. Alternatively, MAPs could shuttle from MTs to actin and could alter F-actin behavior by actively crosslinking AFs.

The effect of changes in the cytoskeleton on ion channel activity has been the focus of recent attention. Most studies have employed drugs that selectively stabilize or destabilize either AFs or MTs, with resulting effects on specific ion channel activity, either as changes in whole cell conductance or single channel activity by the patch clamping techniques. Although direct binding of specific cytoskeletal proteins to individual channel proteins remains to be demonstrated, recent work in which exogenous purified actin has been added to the cytoplasmic side of channels excised from the cell membrane has in numerous cases directly confirmed an effect of cytoskeletal filaments. A body of earlier evidence suggested that various cytoskeletal components, including actin and actin-associated proteins, anchor, co-localize, and regulate both the spatial stability as well as the function of ion transport proteins. It is possible that actin either binds directly to the channels, or that actin may first interact with actin-binding proteins, which in turn regulate (by binding or other indirect interaction) ion channel function. Independent studies demonstrated that membrane-resident CFTR is functionally regulated by AF organization, which may also involve a direct interaction with actin. Thus, ion channel may be controlled by direct and indirect cytoskeletal interactions.

Before we present an integrated view of the conductive properties of the active and passive structural blocks of the cytoskeleton, we should review the current state of knowledge regarding their conductive properties.

18.4 Overview of Biological Conductivity

Cope [19] has presented a detailed review of the applications of solid-state physics concepts to biological systems. He divided the discussion into seven groups. The first involves semiconduction of electrons across enzyme particles as a rate-limiting process in cytochrome oxidase, evidenced by kinetic patters of enzymes and microwave Hall measurements. Second, pn-junction conduction electrons were suggested by kinetics of photobiological free radicals in the eye and in photosynthesis. Their I-V characteristics conform to the diode equation:

$$I = I_0 exp(V/kT) \tag{18.1}$$

Third, there may be a connection between superconduction and growth in nerves. Fourthly, phonons and polarons are involved in mitochondrial phosphorylation. Next, piezoelectricity and pyroelectricity may be involved in the growth of nerves and bone structures. Then, infrared electromagnetic waves may transmit energy in lipid bilayers of nerves and mitochondria. Finally, complex sodium and potassium ions in structured cell water may be analogous to valence band electrons in semiconductors, free cations being analogous to conduction electrons. The ionic process in cell water resembles electronic conduction in semiconductors. For dried proteins or DNA, Szent-Gyorgi[142] noticed that their conductivities obeyed the relationship:

$$\sigma = \sigma_0 exp(-E_a/kT) \tag{18.2}$$

With activation energies E_a on the order of 1 eV. The presence of water may cause considerable changes in conductivities, (for example, in cytochrome oxidase this may be reduced to 0.3 eV).

Adessi *et al.* [1] have provided a theoretical treatment of transmission through a poly(g)DNA molecule. They found that there was a modification of the rise of a B-DNA form, which can induce a shift of the conduction channel towards a valence one. They found that deformation of the backbone of the molecule has a significance on the hole transport. Furthermore, the presence of ionic species (e.g., Na) in the surrounding of the molecule can create new conduction channels. In photo-excitation experiments, hole transport occurs through coherent transport at short distances and long-range incoherent or hopping transport at large distances. Fink *et al.* [39] studied the metallic behavior in λ DNA using an electron projection technique. Superconducting behavior has been observed by Kasumov *et al.* [66]. Porath *et al.* [111] observed semiconductor behavior with a poly(g)-poly(c)DNA molecule. A scanning force microscope has been used by de Pablo *et al.* [26] to observe insulating behavior. This paper models a single poly(g) DNA lying between electrodes, and transmission was computed with a Landauer–Buettiker formalism based on Green's functions. The electronic properties of O_2-doped DNA have been studied by Mehrez *et al.* [96], who suggested that DNA conduction is affected by O_2-doping, in particular hole doping. However, ab initio atomic structure calculations of O_2-doped DNA indicate negligible charge transfer. The absence of dc conductivity in λ DNA has been investigated by de Pablo *et al.* [26]. First principles electronic structure calculations indicate a minimum DNA resistance of 10^{16} Ω per molecule, and a minimum resistivity of 10^4 Ωm. However, lower energy electron bombardment may dramatically increase conductivity and with electron energies between 50 and 200 eV may lead to metallic conduction with a resistance of 2×10^8 Ω. Fink *et al.* [39] discussing DNA and conducting electrons stated that I-V characteristics of DNA indicate ohmic-like behavior with a resistance of 2 *M*Ω for a 600 nm-long DNA rope corresponding to $\rho = m\Omega$ cm. These authors speculate that DNA may be semiconducting, insulating, or metallic simultaneously depending on the arrangement of molecules in the structure. Endres *et al.* [37] discussed the quest for high-conductance DNA saying that DNA conduction is an unsolved problem first enunciated by Eley and Spivey [35]. Experimental outcomes range from insulating to semiconducting to even superconducting, and the uncertainties depend on: (a) contacts between the electrodes and the DNA and (b) differences in the molecules (sequence, length, arrangement, preparation) within the DNA. It appears that drying DNA can lead to conformations with localized electronic high states, although hole doping of the backbone by counter-ions might also be possible. Wet DNA may support electrical current due to solvent impurity states in the large high-Π^* energy gap. Davies and Inglesfield [25] discussed the embedding method for conductance of DNA calculations. It is believed that the pathway for transport in DNA runs through the bases in the center of the double helix molecule. The dominant transport mechanism for DNA appears to be coherent hopping. Calculations show extreme sensitivity to the choice of contact orbitals.

Marechal [95] discussed the transfer of protons as a third fundamental property of hydrogen bonds. The importance of the cooperative resonance types in cyclic structures is stressed. Proton transfers along H-bonds have been discussed as a mechanism of vital importance for chains, helices, and in cyclic structures such as rings. Consta and Kapral [17] discussed proton transfer in mesoscopic molecular systems. Proton transfer rates and mechanics are studied theoretically with solvent molecules in solution. The proton transfer occurs as a result of orientational fluctuations on the cluster's surface. The environment influences solvent effects and reaction rates. The rate constant was established to be 1/59 ps at 260 K and for the bond length fixed at 2 Å. Morowitz [102] discussed proton semiconductors and energy transduction in biological systems. He presented the possibility of protochemistry paralleling electrochemistry. ATP synthesis was discussed as coupled to proton transport assuming a gated protein semiconductor across a membrane. Protein conduction in biological structures involves the water of hydration and protein molecules.

Ripoll *et al.* [121] investigated ionic condensation and signal transduction. Ion condensation on biopolymers at a critical value of the charge density may lead to intra-cellular signaling due to the diffusion of condensed counter-ions in the near region along cytoskeletal filaments. A feedback mechanism was proposed between condensation/decondensation of calcium and the activation of calcium-dependent enzymes. This is shown to create coherent patters of protein phosphorylation. It was postulated that ion condensation operates in signal transduction. In the case of polymer ionics [117], the conductivity depends on temperature via the formula:

$$\sigma T = A exp[-B/(T - T_o)] \tag{18.3}$$

indicating activated hopping. The concept of the Fermi level in solution has been extended by Bockris and Khan [9] through the introduction of a reversible potential for a redox couple.

Unfortunately, measurements of conductivity of biopolymers (including MTs) are fraught with difficulties due to their structural heterogeneity, strong dependence on the environmental conditions (pH, ionic concentration, temperature, etc.), and the inherent liquid state of the sample. Nonetheless, attempts have been made recently to measure resistance values of MTs, both intrinsically and as ionic conduction cables. Fritzsche *et al.* [41] made electrical contacts to single MTs following dry-etching of a substrate containing gold microelectrodes. Their results indicate intrinsic resistance of a 12 micron-long MT to be in the range of 500 megaohms, giving a value of resistivity of approximately 40 mega-ohms/micron in their dry state. The same group performed measurements on gold-coated MTs [42] which were covered with a 30-nm layer of gold following sputtering. The resistance of the thus metallized MTs was estimated to be below 50 ohms, (i.e., originated entirely due to the metallic coating). Minoura and Muto [99] measured the ionic conductivity of MTs using an electro-orientation method under an alternating electric field. The ionic MT conductivity was calculated as $1.5 + / - 0.5$ mS/m, which is approximately 15 times greater than that of the buffer solution. This was attributed to the counter-ion polarization and is consistent with another recent study of ionic conductivity along MTs. Furthermore, Goddard and Whittier [47] reported their measurements of RF reflectance spectroscopy of samples containing the buffer solution, free tubulin in buffer, MTs in buffer, and finally, MTs with MAPs in buffer. The sample size in each case was 5 ml. The concentration of tubulin was 5 mg/ml and in the last case the concentration of MAP2 and the tau protein was 0.3 mg/ml. The average DC resistance reported by these authors was (a) 0.999 $k\Omega$ (buffer), (b) 0.424 $k\Omega$ (tubulin), (c) 0.883 $k\Omega$ (MTs), and (d) 0.836 $k\Omega$ (MTs + MAPs). It's not straightforward to translate these results into resistivity of MTs without making assumptions about their geometrical arrangement and connectivity. However, assuming all tubulin being polymerized in case (c) and a uniform distribution of MTs forming a combination of parallel and series networks, one can find the resistance of a 10 μm–long MT, forming a basic electrical element in such a circuit, to have approximately an 8 $M\Omega$ resistance. This compares very favorably to an early theoretical estimate of MT conductivity based on the Hubbard model with electron hopping between tubulin monomers. This model predicted the resistance of a 1-μm MT to be in the range of 200 $k\Omega$, hence a 10-μm MT would be expected to have an intrinsic resistance of 2 $M\Omega$, within the same order of magnitude as the result reported by Goddard and Whittier [47]. While direct experimental analysis of intrinsic conductivity of the various cytoskeletal components is still nebulous, theoretical modeling of these effects is very well developed. The next section presents our current understanding of the intrinsic conductivity of MTs.

18.5 Intrinsic Electronic Conductivity of Microtubules

Microtubules (MTs) are ubiquitous in eukaryotic cell biology. Their primary role is to serve as the cellular scaffold and act as a corridor for the transport of vesicles through the cell by motor proteins. The interest in MTs stems from their involvement in a number of crucial cellular processes including mitosis and, more recently, the discovery of their role in the communication between the exterior of the cell and the nucleus [46,94]. The special arrangement of MTs just prior to chromosone separation makes them an ideal candidate for mediating a signal of some sort, which would cause the coordinated breakage of sister chromatids. In the axons of nerve cells, the MTs are arranged in bundles, parallel to the axon [56]. These groups of MTs span the entire length of the axon and might serve to carry signals from the cell body to the nerve terminus or vice versa. No mechanism for this sort of direct feedback is known. We demonstrate here that the MT may be able to conduct a current along its length through a process of electron hopping between binding sites on each dimer.

The MT is made up of a protein called tubulin. Each subunit contains α-tubulin and β-tubulin monomers joined together to form a dimer. The dimers are connected end to end to form a protofilament. These protofilaments, usually 13 in number, are wrapped up to form a hollow tube, the MT. The dimers are themselves 5 nm in diameter and have a length of about 8 nm. A MT has an outer diameter of 25 nm and an inner diameter of 15 nm (see Figure 18.2).

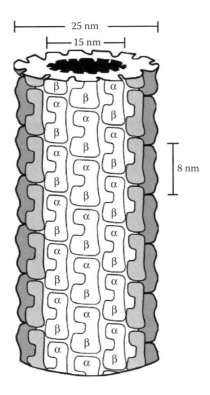

FIGURE 18.2 Structure of a typical microtubule demonstrating how it is formed by a special arrangement of tubulin dimers.

The tubulin dimer of which the MT is composed has at least two different conformational states [58]. That is, while the amino-acid chain of the protein remains fixed, its three-dimensional shape may change. It is believed that these conformational changes are the result of hydrolysis and an associated charge movement from one binding site within the molecule to another. It is precisely this charge movement that may lead to the conduction through the polymer.

One may take a semi-classical view of the movement of electrons along the protofilament. We ignore the fact that the protofilament is nearly one-dimensional and consider its *bulk* properties. Assuming two conduction electrons for each tubulin dimer, the conduction electron density is

$$n_0 = 2/(8nm \times 5nm \times 5nm) \tag{18.4}$$
$$= 1.0 \times 10^{19}/cm^3.$$

In comparison with other materials, this value is slightly higher than typical semi-conductor concentrations, usually $10^{13} - 10^{17}/cm^3$, but is much less than typical metals, which have a conduction electron density of $10^{22} - 10^{23}/cm^3$. However, the electrons are expected to have relatively low mobilities unless they are excited from the ground state configuration. The excitation energy is a function of both the number of electrons within the protofilament and its size. Suppose that the mobility of excited electrons is at least similar to the mobility of electrons within semiconductors, such as Si where the electron mobility, μ_e, is about 1300 cm^2 V^{-1} s^{-1}. The electrical conductivity σ is given in a semi-classical model by the well-known formula

$$\sigma = ne\mu_e \tag{18.5}$$

where e is the electron charge and n is the conduction electron density given by

$$n = n_0 e^{-\Delta/\kappa T} \tag{18.6}$$

where n_0 represents the total number of electrons. If we assume that the gap between valence and conduction bands, Δ, is about 0.40 eV, then at physiological temperature, some of the carriers will be excited and should be available to conduct. Applying the preceding formula (18.6), a conductivity of about 10^{-2} (in SI units) is expected for the protofilament.

$$\sigma \sim (10^{25}e^{-0.40/0/026})(1.6 \times 10^{-19})(0.13) \sim 0.04\Omega^{-1}m^{-1} \tag{18.7}$$

However, if the hybridization of electronic orbitals lowers the gap to a value closer to zero, the expected conductivity would be significantly higher:

$$\sigma \sim (10^{25})(1.6 \times 10^{-19})(0.13) \sim 2 \times 10^5 \Omega^{-1}m^{-1} \tag{18.8}$$

These values should be compared to values in excess of 10^8 for copper and other good conductors, and 10^7 for lesser conductors and semi-metals. Typical semi-conductors such as Si and Ge have intrinsic conductivities of $10^{-3} - 10^{+1}$ at room temperature, but when heavily doped, the conductivity may rise to between 10^4 and 10^5.

In order to model the MT as a semiconductor, the individual tubulin dimer must first be physically analyzed. The $\alpha\beta$-tubulin dimer is comprised of about 890 amino acids which represents about 13000 nuclei. A gross simplification which we introduce is to treat each monomer as having an effective site energy for electronic binding. Fluctuations of the site energy on scales less than the monomer spacing are therefore washed out. It is convenient to model this system as two quantum wells separated by a potential barrier. The two wells represent two binding sites for an electron, which may hop between the monomers, resulting in the change of molecular conformation. We can estimate the relevant parameters of the well depth and the well width from kinetic studies and with a knowledge of the molecule's overall geometry. Once the bound states and their energies are known for this system, this knowledge may be applied to a

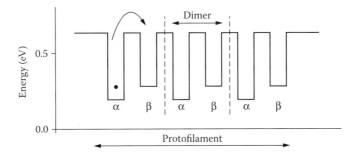

FIGURE 18.3 A one-dimensional chain of quantum wells representing the energy profile along a MT protofilament where the α and β monomers have alternating site energies.

second-quantization scheme. This procedure allows us to make progress without specifically requiring us to solve for the wavefunction of the system.

A 1D chain of quantum double wells is pictured in Figure 18.3. For simplicity, we are initially making each binding site have the same energy. If this is related to GTP hydrolysis, they may differ by about 0.17 eV (4 kcal/mol) [59]. This is roughly the quantity of energy stored in the lattice when GTP is hydrolyzed to GDP [13,60]. The barrier height between dimers has been estimated from kinetic data. The bond strength must be somewhat larger than the hydrolysis energy delivered to the lattice or else hydrolysis of the dimers would cause immediate disassembly. Hopping along the protofilament by electrons presumably has an energy comparable to this binding energy. We have estimated 0.4 eV for intra-dimer hopping and 1.0 eV for the inter-dimer hopping. What may be varied in the model, is the central barrier which must be overcome for each tubulin dimer to change conformation. If it is nearly as large as the barrier to electron movement between dimers, then the protofilament will look like a chain of single quantum wells rather than double quantum wells because the difference between the α and β monomers will not be distinguished. However, if the central barrier is too small, the chain will again look like a series of single quantum wells since the effect of the barrier would be negligible. The depth of the deeper well, f_1, has been selected as 0.4 eV initially. Given the value, we assumed a 25% difference in well depths. If f_1 is much larger, the dimer nature of the protofilament is lost because the wells are effectively the same depth.

The results for the single electron eigenstates of a double quantum-well may be found using a transfer matrix method for the time-independent 1D Schrodinger equation in each interval in which the piece-wise constant potential is defined as

$$\hat{H}\psi = \frac{\hat{p}^2}{2m}\psi + \hat{V}\psi = E\psi \tag{18.9}$$

The oscillatory solution

$$\psi(x) = A\cos(kx) + B\sin(kx) \tag{18.10}$$

applies to $E > V$ where

$$k = a\sqrt{2m(E-V)}/\hbar \tag{18.11}$$

while for $E < V$, the solutions are exponentially damped

$$\psi(x) = Ae^{\kappa x} + Be^{-\kappa x} \tag{18.12}$$

where

$$k = \sqrt{2m(V-E)}/\hbar \tag{18.13}$$

Now, matching conditions for the wave function and its derivative are applied at each interface. This gives a 2 × 2 matrix relating the coefficients in the wave function expansion according to

$$\begin{pmatrix} A_2 \\ B_2 \end{pmatrix} = \begin{pmatrix} ^{2,1}M_{11} & ^{2,1}M_{12} \\ ^{2,1}M_{21} & ^{2,1}M_{22} \end{pmatrix} \begin{pmatrix} A_1 \\ B_1 \end{pmatrix} \tag{18.14}$$

In the preceding equation, the superscript to the left of the M denotes which coefficients are being linked. The subscripts indicate the matrix element. In an iterated fashion, one eventually arrives at the following:

$$\begin{pmatrix} A_n \\ B_n \end{pmatrix} = \begin{pmatrix} ^{n,1}M_{11} & ^{n,1}M_{12} \\ ^{n,1}M_{21} & ^{n,1}M_{22} \end{pmatrix} \begin{pmatrix} A_1 \\ B_1 \end{pmatrix} \tag{18.15}$$

Now, for bound states $A_1 = B_n = 0$ so that the wavefunction decays outside the region of the quantum wells:

$$B_n = 0 \quad \Rightarrow \quad ^{n,1}M_{22} = 0 \tag{18.16}$$

Given the choices for our quantum double well (width, $a = 2$ nm; and height, $V = 1.0$ eV), one finds that there are two bound states for this dimer problem. As the number of dimers is increased, one resorts to numerical analysis. In Figure 18.4, the energy of the two bound states is plotted as the energy of the central barrier is varied. The lowest energy state is the symmetric wavefunction, which increases in energy sharply as the central barrier is increased. The second bound state is the antisymmetric wavefunction it has a node within the barrier, hence its energy increases only marginally.

The energy calculation for a sequence of n such dimers leads to the development of bands as hybrid orbitals develop across the polymer. Roughly speaking, one band is of hybridized symmetric orbitals and the higher lying band of hybridized antisymmetric orbitals. The separation of the two bands and the spread of each individual band determine the values of the parameters which shall be used in the tight-binding Hubbard model used to model the polymer.

The second-quantized Hamiltonian for a many-electron multidimer system becomes

$$\hat{H} = \sum_{i,\sigma} \epsilon_i \hat{c}_{i\sigma}^{\dagger} \hat{c}_{i\sigma} + \sum_{i \neq j,\sigma} t_{ij} \hat{c}_{i\sigma}^{\dagger} \hat{c}_{j\sigma} \tag{18.17}$$

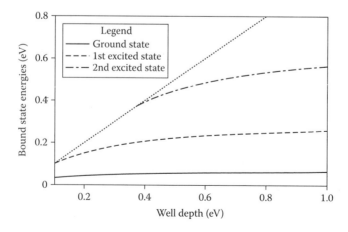

FIGURE 18.4   Variation of the bound state energies with the depth of the double quantum wells, $f_1 = f_2$. The straight line indicates where the bound state energy is equal to the well depth.

where the operators $\hat{c}_{i\sigma}^{\dagger}$ and $\hat{c}_{i\sigma}$ are the usual fermion creation and annihilation operators, and where i runs from 1 to $2n$ representing the two sites on each of the n dimers. The ϵ_i represent the potential energy at each site and the kinetic terms t_{ij} account for hopping from one site to another. There are two different kinds of interactions depending on whether an electron hops within a dimer, or between dimers. Hence, this kinetic parameter has two different values along the chain. Since there are no interactions between the states with $+\frac{1}{2}$ spin and those with $-\frac{1}{2}$ spin, the state space may be divided into two identical subspaces labeled by the electron spin. In each of these subspaces, the matrix corresponding to this chain of dimers is tri-diagonal and may be diagonalized to determine the itinerant eigenstates of the system. The coefficient t_d which represents the movement of an electron within a dimer causes the eigenstates to be split into two bands by an amount, $2t_d$. Within each of these bands, the coefficient t_l of the inter-dimer hopping, causes these two single electron bands to broaden each to about $2t_l$. Now, although the preceding description is qualitative, clearly whenever $t_d \sim t_l$, these bands will overlap and transitions between the states will be important.

Considering the band formation in the quantum well formulation, we can assign values to t_d and t_l. The two bands are separated by about 0.40 eV, so $t_d \sim 0.20$ eV, and using the bottom band to estimate t_l it widens until it is about 0.10 eV thick, so $t_l \sim 0.05$ eV.

The Hubbard model is the simplest second quantized model to capture all of the essential features of a real system of interacting particles [82]. In this model, electrons are the fundamental excitations. All states may be built up from the vacuum using the fermion creation operators. Once the number of sites is selected, n, and the number of electrons is determined, e, there are

$$\binom{2n}{e} = \frac{2n!}{e!(2n-e)!} \tag{18.18}$$

distinct electronic states. These states are built up by applying in normal order each of the combinations of e-electron states.

The second quantized Hamiltonian for the Hubbard model with electron–electron interactions is conventionally written as

$$\hat{H} = \sum_{i,\sigma} \epsilon_i \hat{c}_{i\sigma}^{\dagger} \hat{c}_{i\sigma} + \sum_{i \neq j,\sigma} t_{ij} \hat{c}_{i\sigma}^{\dagger} \hat{c}_{j\sigma} + \sum_i U_i \hat{c}_{i\uparrow}^{\dagger} \hat{c}_{i\uparrow} \hat{c}_{i\downarrow}^{\dagger} \hat{c}_{i\downarrow} \tag{18.19}$$

The last term accounts for electron–electron repulsion now that we have gone to a many electron system. In our simple model, this is limited to an onsite interaction. To this point, we have made each of the site energies equivalent ($\epsilon_i = \epsilon, \forall i$). In what follows, $U = 2.0$ eV which is the energy associated with two bare electrons lying about 0.7 nm apart. This choice seems reasonable within a well of width 2.0 nm. We try to solve the problem by considering the electron–electron term as a perturbation to study how it affects the electronic states of the dimer. The Hamiltonian in the absence of the final U-term may be diagonalized by the following unitary transformation:

$$\alpha_{1'\sigma}^{\dagger} = \frac{1}{\sqrt{2}}(c_{1\sigma}^{\dagger} + c_{2\sigma}^{\dagger}) \tag{18.20}$$

$$\alpha_{1'\sigma}^{\dagger} = \frac{1}{2\sqrt{2}}(c_{1\sigma}^{\dagger} - c_{2\sigma}^{\dagger}) \tag{18.21}$$

The linear combination of operators is simple because we have made the site energies ϵ_1 and ϵ_2 equal. These new operators are easily shown to obey the usual anticommutation relations for fermions and are said to be in the itinerant or distributed electron basis. The Hamiltonian may now be rewritten in terms of these new operators, as

$$\hat{H} = \sum_{\sigma} (\epsilon + t)\alpha_{1\sigma}^{\dagger}\alpha_{1\sigma} + \sum_{\sigma} (\epsilon - t)\alpha_{2\sigma}^{\dagger}\alpha_{2\sigma} + \frac{1}{2} \sum_{\sigma \neq \sigma'} V_{ijlm}\alpha_{i\sigma}^{\dagger}\alpha_{j\sigma'}^{\dagger}\alpha_{l\sigma'}\alpha_{m\sigma} \tag{18.22}$$

the vacuum state is represented by $|\Omega >$. Acting with our Hamiltonian on this basis gives us the matrix form of the Hamiltonian:

$$\hat{H} = \begin{pmatrix} 2(\epsilon + t) + \frac{(U_1+U_2)}{4} & 0 & \frac{(U_1-U_2)}{4} & \frac{(U_2-U_1)}{4} & 0 & \frac{(U_1+U_2)}{4} \\ 0 & 2\epsilon & 0 & 0 & 0 & 0 \\ \frac{(U_1-U_2)}{4} & 0 & 2\epsilon + \frac{(U_1+U_2)}{4} & -\frac{(U_1+U_2)}{4} & 0 & \frac{(U_1-U_2)}{4} \\ \frac{(U_2-U_1)}{4} & 0 & -\frac{(U_1+U_2)}{4} & 2\epsilon + \frac{(U_1+U_2)}{4} & 0 & \frac{(U_2-U_1)}{4} \\ 0 & 0 & 0 & 0 & 2\epsilon & 0 \\ \frac{(U_1+U_2)}{4} & 0 & \frac{(U_1-U_2)}{4} & \frac{(U_2-U_1)}{4} & 0 & 2(\epsilon - t) + \frac{(U_1+U_2)}{4} \end{pmatrix}$$

The Hamiltonian is then diagnolized in order to determine its eigenstates. In the absence of the Coulomb repulsion ($U = 0$), the electrons are localized on the 2' site with energy $2(\epsilon - t)$, the gap to the first excited state being $2t$. However, as U is increased, the effect of the Coulomb interaction raises the energy of the ground state relative to the excited singlet states, which do not feel the Coulomb repulsion. Consequently, the gap between the ground state and the first excited state is actually reduced through the introduction, as shown in Figure 18.5. Thus, when conductivity is being considered, it is important to bear in mind that the gap is reduced by the presence of the electron–electron interactions.

To this point, we have made each of the site energies equivalent, ($\epsilon_i = \epsilon$), and the electron repulsion the same at all sites ($U_i = U$). The lone complication lies within the kinetic term, where the hopping parameter will depend on the two sites between which the electron hops. However, since we first consider the MT protofilament, it shall initially be a constant as well, $t = 0.4$ eV, where the choice comes from the separation of the energy bands in the 1D Schrödinger picture.

Since the Hamiltonian conserves both the particle number and the overall spin of each state, we are free to work in subspaces enumerated by the spin of the states it contains. This reduces the problem slightly since it is possible to consider individually the spin $+1$, spin 0, and spin -1 systems separately in a two-electron system. In addition, the symmetry implies that the spin $+1$ and spin -1 systems have the same conductive properties. We also find that the ground state is always a state consisting of the lowest possible total spin.

When the inter-dimer hopping term is small relative to the intra-dimer hopping term $t_{inter} < t_{intra}$, the form of the ground state is independent of the number of dimers. In each case, the ground state is a linear combination of singlet states on each dimer. The singlet states are those where two electrons are

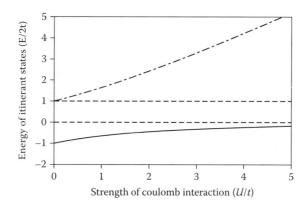

FIGURE 18.5 Energy of the itinerant states as a function of U/t with the energy plotted in units of $2t$.

paired on an individual binding site, or where the state consists of the anti-symmetric combination of an up-electron and a down-electron on neighboring binding sites of the dimer. When the inter-dimer hopping parameter is increased, the density of states becomes more uniform between the ground state and the highest energy eigenstate. With the gap between the ground state and the first excited state of the system becoming diminished. Since the hopping term within a dimer is largest, the system prefers to have two electrons with the same spin within the same dimer in the ground state.

For electronic conductivity, the band formation importantly depends upon the hopping parameter between dimers. When this parameter is small relative to the intra-dimer hopping, $t_{inter}/t_{intra} \sim 0.1$, there remains a rather large gap between the ground state and the first excited state of about $2t_{intra}$ which is reduced as the repulsion term increases. However, when the hopping between dimers is equal to the hopping within dimers, $t_{inter} \sim t_{intra}$, the gap is minimized. In the event that $t_{inter} > t_{intra}$, we would effectively have to relabel the dimers and would then return to the picture where $t_{inter} < t_{intra}$.

Kubo was the first to derive formulae for the electrical conductivity in solids, and his derivation is applied to our case in the present section [77]. In the Coulomb gauge, we write the vector potential for a uniform electric field along the direction of the chain, $A(t)$, as

$$A(t) = \frac{cE(t)}{-i\omega} e^{i\omega t}, \tag{18.23}$$

Adding a field-coupling term to the Hubbard Hamiltonian and using a canonical transformation gives [91]

$$\hat{H}_0' = -t \sum_{i\sigma} [\hat{c}_{i\sigma}^\dagger \hat{c}_{i+1\sigma} e^{-ieA(t)/c} + \hat{c}_{i\sigma}^\dagger \hat{c}_{i-1\sigma} e^{ieA(t)/c}]. \tag{18.24}$$

where a is the lattice spacing and c is the speed of light. Expanding to second order in the vector potential, we find

$$\hat{H}_A = \frac{A}{c} \hat{j} - e\frac{A^2}{c^2} a^2 \hat{H}_0, \tag{18.25}$$

where

$$\hat{j} = i E a t \sum_{i\sigma} (\hat{c}_{i\sigma}^\dagger \hat{c}_{i+1\sigma} - \hat{c}_{i\sigma}^\dagger \hat{c}_{i-1\sigma}). \tag{18.26}$$

So, the energy of the original system is corrected to this order and the interaction term is proportional to the current.

Suppose a time-dependent external electric field is applied to a solid:

$$E_\alpha^{\text{ext}}(r,t) = \Xi_\alpha^{\text{ext}} e^{iq\cdot r - i\omega t}, \tag{18.27}$$

where α represents the Cartesian coordinate directions. In a linear response theory, the induced current is proportional to the applied electric field:

$$J_\alpha(r,t) = \sum_\beta \sigma_{\alpha\beta}'(q,\omega) \Xi_\beta^{\text{ext}} e^{iq\cdot r - \omega t}, \tag{18.28}$$

where $\sigma_{\alpha\beta}'$ is a parameter relating the observed current density to the applied field's direction, its periodicity, and frequency. However, the symbol appearing in (18.28) is not the conductivity we seek. Rather, we want the conductivity which represents the response to the total electric field in the solid, a quantity that can be measured. This conductivity takes into account all of the currents, induced by the external fields, that create their own electric fields. Thus, we seek $\sigma_{\alpha\beta}$ that relates the macroscopic electric field to the currents of the system.

$$J_\alpha(r,t) = \sum_\beta \sigma_{\alpha\beta}(q,\omega) E_\beta e^{iq\cdot r - i\omega t} \tag{18.29}$$

$$E_\alpha(r,t) = \Xi_\alpha e^{iq\cdot r - i\omega t} \tag{18.30}$$

$$\sigma_{\alpha\beta} = Re(\sigma_{\alpha\beta}) + i\,Im(\sigma_{\alpha\beta}) \tag{18.31}$$

We write the Hamiltonian of the system as $\hat{H} + \hat{H}'$ where the latter term contains the electric field, which we shall introduce as a time-dependent perturbation. The evolution of the operators is given as follows [30]:

$$\hat{j}(x,t) = e^{i\hat{H}t}\hat{j}(x)e^{-i\hat{H}t}. \tag{18.32}$$

where $\hat{H}$ is the unperturbed Hamiltonian. Thus, the interaction picture is adopted. The Kubo formula for electrical conductivity gives the result in terms of a current–current correlation function:

$$\sigma_{\alpha\beta}(q,w) = \frac{1}{\omega}\int_0^\infty dt e^{i\omega t} < \Omega|\hat{j}_\alpha^\dagger(q,t)\hat{j}_\beta(q,0) - \hat{j}_\beta(q,0)\hat{j}_\alpha^\dagger(q,t)|\Omega > + i\frac{n_0 e^2}{m\omega}\delta_{\alpha\beta} \tag{18.33}$$

where $|\Omega>$ is the ground state of the system, which has not been perturbed by the electric field. The first term in (18.33) is known as the incoherent contribution to the conductivity. Now, we use (18.32) in the preceding equation and use the fact that the basis we are using is that of the eigenstates. Consequently, with proper normalization the new expression for the real part of the conductivity is [90]

$$\sigma(\omega) = \frac{\pi}{Z}\frac{(1-e^{-\beta\omega})}{\omega}\sum_{n,m}e^{-\beta E_m}|<n|\hat{j}_\alpha|m>|^2\delta[\omega - (E_n - E_m)] \tag{18.34}$$

The DC conductivity is calculated by a procedure [90,91], which employs the sum rule

$$\int_0^\infty \sigma'(\omega)d\omega = -\frac{\pi e^2}{2}a^2 < K > \tag{18.35}$$

where

$$\sigma'(\omega) = D\delta(\omega) + \sigma_{kubo}(\omega) \tag{18.36}$$

and $<K>$ is the expectation value of the kinetic energy term of the Hamiltonian. The strategy of computing $\sigma(\omega)$, and subsequently the Drude contribution, at either zero or finite temperature is reasonably straight forward. The first step is to construct the matrix corresponding to the Hamiltonian [Equation (18.19)] in a subspace which is defined by the geometry of the lattice, the number of electrons, and the total electron spin. This leads to serious computational difficulties since for a moderately sized lattice of say ten dimers and eight electrons of arbitrary spin, the corresponding space is about 77 million states. A matrix of dimension 77 million square has about 5.9×10^{15} elements. Obviously, working on a moderately sized lattice, we are restricted to a low concentration of electrons. Due to electron-hole symmetry, we are also able to study the nearly filled situation. It is useful to keep in mind the relationship

$$\lambda = \frac{1240nm \cdot eV}{\text{Energy}}. \tag{18.37}$$

So that in the plots following, the visible light range is between 1.75 eV and 3.10 eV. Energies larger than this range correspond to ultraviolet, and those smaller to infrared wavelengths. Since cells respond to light in the near-infrared [4], and centrioles comprised of MTs are proposed to be the site of light detection [5], absorption in this energy range is of great interest to cell biology. These energies should also be compared

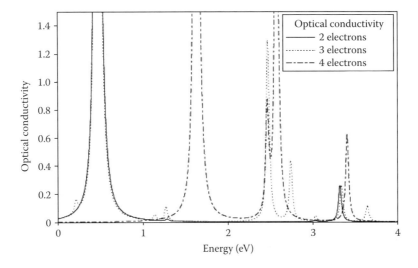

FIGURE 18.6   Optical conductivity of the two dimer system with two, three, and four electrons.

to the energy available from ATP and GTP hydrolysis, which are 0.49 eV and 0.22 eV, respectively. These values provide limits for the interactions of chemical energy in the conduction process.

At zero temperature, the linear polymer results are summarized concisely by saying that there is no DC conductivity that agrees with the previous modeling of Hubbard systems [44].

The following results for the 1D chain of tubulin dimers represent the AC conductivity. There are peaks in $\sigma(\omega)$, also known as the optical conductivity, which correspond to excitations between electronic states of the polymer. Electronic excitation results in electron redistribution and hence the development of a current. We have selected $\epsilon = 0.02$ eV to present our results, which is roughly kT and is smaller than the three energy scales in the Hubbard model, t, U, and $4t^2/U$ [82] so we expect to be able to pick out features corresponding to specific transitions in the predicted optical spectra.

We shall begin with two dimers and $t = 0.4$ eV, $U = 2.0$ eV, and equal site energy on the α and β monomer sites. This system can support up to eight electrons. The results for these systems are presented in Figure 18.6. The predicted spectra for the two and three electron cases are quite similar. The differences include a new but small peak in the three electron spectrum at an energy just below the major peak, and variations in the detail of the absorption band centered about 2.6 eV. However, the qualitative similarity of the two spectra is remarkable in terms of band locations. Contrast these two cases with the four-electron system that corresponds to half-filling. The difference is that the peak in the optical conductivity spectrum increases dramatically from about 0.48 eV to 1.64 eV. This change indicates how the half-filled case is special. The ground state in this system largely consists of electrons spread out to minimize the Coulombic energy. Exciting this system by moving any particular electron will come at the cost of greatly increasing the electron–electron repulsion and, consequently, the location of the required excitation energy corresponds roughly to U.

As the length of the polymer is increased, results remain qualitatively similar. In the three-dimer structure, results are shown for the conductivity in Figure 18.7(a) and for the integrated conductivity in Figure 18.7(b). Once again, the spectra are quite similar for all fillings aside from the half-filled case which consists of six electrons. The shift of the conductivity to higher energies can be clearly seen as the filling fraction increases towards half filling. We now present results where the broadening parameter, ϵ, and the temperature have been varied to see the effect that they have on the results. Figure 18.8 demonstrates that as the width is decreased, the spectrum becomes sharper. Note how the two peaks in this simple spectrum that occur near 2.5 eV are successively blended into a band as the ϵ is increased. Peaks represent transitions from one specific electron configuration to another and are weighted with the associated conductivity.

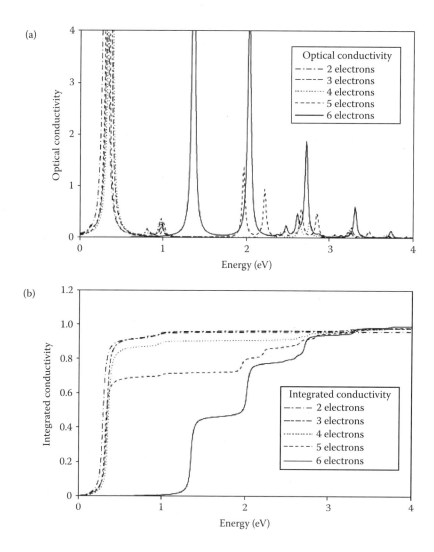

FIGURE 18.7 (a) Optical conductivity of the three-dimer system from two electrons through half filling. (b) Integrated optical conductivity of the three-dimer system shows that indeed the protofilament is an insulator. The gap for excitation of the half-filled polymer is seen easily here.

The effect of changing the temperature is to excite electron configurations other than the ground state. Consequently, transitions from one excited state to another can now contribute to the conductivity. Figure 18.9 shows how this often reduces the gap to conduction: all the major peaks in the 300 K spectrum shift slightly towards lower energies in the 3000 K spectrum, but many more features develop as well as states that previously did not contribute to conductivity may make transitions to excited states. Physiological temperature of 300 K ($kT = 0.026$ eV) is small relative to the system's energy scales, consequently lowering the temperature further causes very little change to the spectrum.

If the α-tubulin and β-tubulin sites have different energies, then the MT A and B lattices can be distinguished. The effect of this change on the optical spectrum seems to be small when individual protofilaments are considered. In addition, the sum rule shows that protofilaments remain insulators. There are a few parameters we can consider individually and which can be compared as the size of the system changes: the threshold to conductivity, and the expectation value of the kinetic energy operator. The threshold to conductivity is simply the location of the first absorption peak in the optical spectrum.

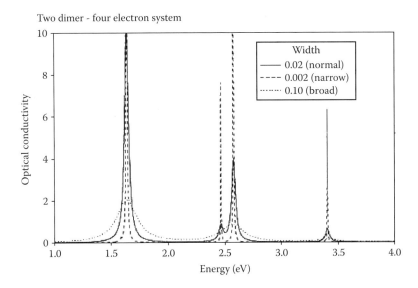

FIGURE 18.8   Optical conductivity spectrum variation with the broadening parameter.

It is a decreasing function of the polymer length. Starting at 0.48 eV for the two-dimer system, it seems to drop a little faster than $1/n$ for systems consisting of n dimers and containing two electrons. The fitted value in Figure 18.10 varies $\alpha 1/n^{1.4}$. Extrapolating to larger values of n, this gap appears certain to fall into the range where these excited states could be thermally excited. The protofilament then develops a finite DC conductivity at finite temperature. It is interesting that although the first peak does move towards zero, the fraction of the conductivity it contains is gradually diminished. Some of the conduction continues to reside in a peak located close to t in the energy spectrum. In the 2D lattice, there is a phase transition in the dimensionality. DC conduction becomes possible even at zero temperature. However, we shall still be concerned about the boundary conditions since calculations are being performed on a small lattice.

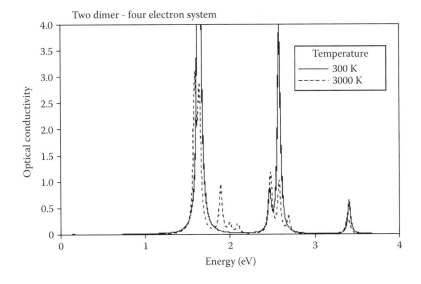

FIGURE 18.9   The variation of the optical conductivity spectrum with temperature is shown.

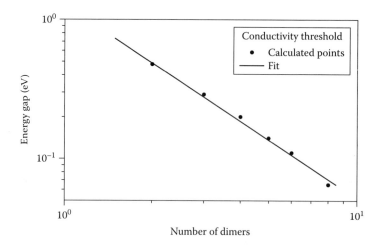

FIGURE 18.10   Threshold to conduction as a function of protofilament length.

Consider the kinetic energy/electron that has been calculated in a Hubbard model with our parameters for t and U and with either one, two or three protofilaments. Figure 18.11 demonstrates how the kinetic energy of each electron approaches the limit of $2t$ or 0.80 eV as the size of the system increases. We can easily see that for more than about 20 electron sites, the average kinetic energy of each electron starts to approach the infinite limit and we expect that the results obtained from our model should converge to the long protofilament limit in a similar manner. In particular, we are considering 28 electron sites in our MT model, which span three protofilaments. While this is not the ideal situation, the boundary conditions are not expected to create huge effects in our results. We are also using open boundary conditions since they minimize the boundary effects when compared with a variety of periodic boundary conditions.

The unit cells depicted in Figure 18.12 are the basis for the calculations. In the A lattice used for the calculations, the α-dimers connect to β-dimers along the $\bar{3}$-start helix, both within the lattice and at the seam when the MT is wrapped up. In the B lattice, the connections are $\alpha - \alpha$ or $\beta - \beta$ along the $\bar{3}$-start helix and never $\alpha - \beta$. For comparison, calculations have also been carried out with a flat tubulin sheet, specifically an A lattice MT lacking periodic boundary conditions in the lateral direction. This simulates the

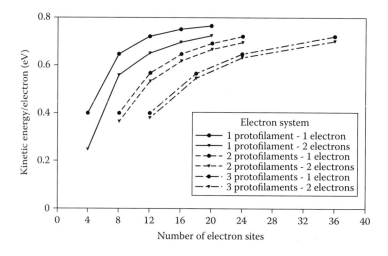

FIGURE 18.11   Kinetic energy per electron as a function of system size.

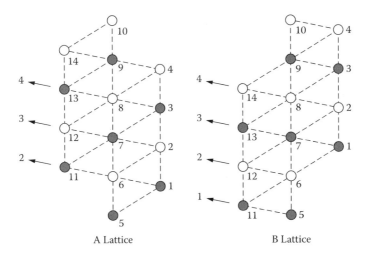

A Lattice   B Lattice

FIGURE 18.12 Lattice unit cells for conductivity calculations. The unit cell on the left corresponds to the MT A lattice, while that on the right corresponds to the MT B lattice.

MT when it has been unwrapped. We recall that the MT lattice is triangular and has three distinct hopping directions. Since protofilaments are strongly bound together along their length but relatively weakly bound to other protofilaments, one also expects the hopping parameter to be much smaller when an electron moves between protofilaments. The calculations have been carried out with the same parameters as in the one-dimensional case, $U = 2.00$ eV, $t = 0.40$ eV along the protofilaments and the site energies have been set equal to zero for all sites. The value of the hopping parameter along the $\bar{3}$-start helix is called t_ℓ (left) and that along the 8-start helix that connects diagonally from a monomer in a direction up and to the right is called t_r (right).

We consider first the results of the MT lattice which has not rolled up to form a tube but is that of a tubulin sheet where $t_r = 0$ and only the value of t_l is varied (see Figure 18.13). Since $t_r = 0$, the nature of the lattice remains rectangular to this point. When $t_l = 0$, this particular system is equivalent to individual protofilaments. The two peaks in the optical spectrum arise from the fact that the protofilaments have different lengths. The absorption peaks of 0.22 eV and 0.40 eV correspond roughly to those of the two-dimer and three-dimer, two electron cases discussed earlier in the chapter. The difference here is that the two electrons are spread over three protofilaments and consequently the effect of the electron repulsion is reduced. However, as t_l is increased from zero, the entire lattice becomes accessible and the protofilament character of the optical spectrum is lost. The two peaks of the t_l spectrum coalesce into a single peak with a large activation energy. A second but smaller absorption peak forms just above 0.50 eV. In the plot of the integrated conductivity, we can see that there is also a smaller peak at 0.16 eV when $t_l = 0.10$ eV. The source of this peak is an absorption for conduction along the direction of t_l. When t_l is increased to 0.40 eV, this peak occurs at 0.61 eV. Thus, this peak seems to occur at roughly $1.5t_l$. The activation energy for conduction along the protofilament axis also increases but is not as sensitive to t_l. Finally, examination of the integrated conductivity demonstrates that the total absorption combines such that the Drude weight remains zero.

The results in Figure 18.14 show a difference between an individual protofilament and a tubulin sheet. The behavior here is inconsistent with that of an insulator since the Drude weight is non-zero. In the optical conductivity, we find that the largest peak occurs at the same location in all directions but with different weightings. The first line traces the results for $t_r = 0.00$ eV and has peaks at 0.38 eV and 0.61 eV. Once the third hopping direction is allowed, the peak absorption is raised slightly to 0.43 eV for $t_r = 0.10$ eV and a smaller peak forms at 0.74 eV. The lower peak consists mainly of hopping along the protofilament while the second peak consists largely of hopping along the t_l direction, which also corresponds to a large

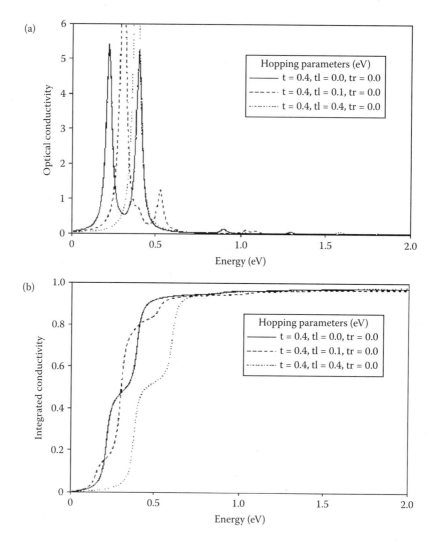

FIGURE 18.13 (a) Optical conductivity along the protofilament axis of the tubulin sheet consisting of a 14-site lattice with two electrons while $t_r = 0$ and t_l is varied. (b) Integrated conductivity of the tubulin sheet with two electrons while $t_r = 0$ and t_l is varied.

hopping parameter. Once t_r is raised to 0.40 eV so that it has the same magnitude as the other hopping parameters, there is only a single peak at 0.70 eV.

The effect of applying the MT A lattice boundary conditions to our system is shown in Figure 18.15, where we see that changing only t_l to a finite quantity while still maintaining t_r as zero is sufficient to result in a non-zero Drude weight. Thus, the periodic boundary conditions in the lateral direction seem to act in a manner similar to the additional hopping directions of the unwrapped lattice. When the kinetic parameters between protofilaments are set to zero, the optical conductivity spectrum shows there are two large absorption peaks at about 0.23 eV and 0.40 eV. As t_l increased, these peaks move higher in energy and the second peak becomes smaller in integrated weight. Eventually, once $t_l = t$, there is a single large peak near 0.47 eV. Introduction of the third hopping direction serves only to raise further the absorption peak, up to 0.70 eV in the case where the hopping parameter is 0.40 eV in all directions.

From Figure 18.15 one can see that more conduction results in the B lattice for a small t_l, such as 0.1 eV, than in the MT A lattice, by comparing it with Figure 18.16. However, even the positioning of absorption

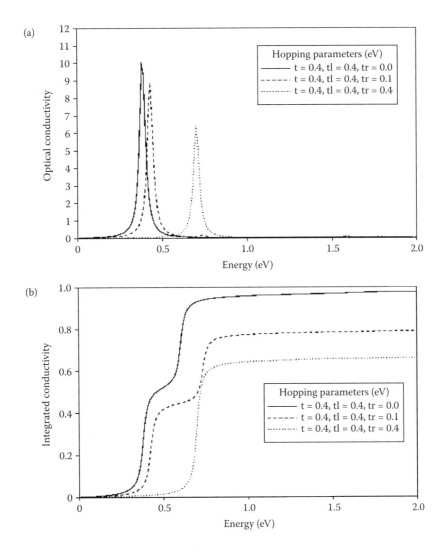

FIGURE 18.14 (a) Optical conductivity along the protofilament axis of the unwrapped 14-site lattice with two electrons while $t_l = 0.4$ and t_r is varied. (b) Integrated conductivity of the unwrapped 14-site lattice with two electrons while $t_l = 0.4$ and t_r is varied.

peaks is similar in the two lattices. Again, it is sufficient for only one of t_l or t_r to be non-zero for the Drude weight to be finite.

Finally, we compare the MT A and B lattices with the unwrapped tubulin sheet as well as with individual protofilaments of tubulin dimers. There is an interesting geometrical interpretation to the results, which shows that the MT A and MT B lattices have identical conduction properties when $t_l = t_r$. This is because in our calculations we have considered all monomers to have the same site energy. Consequently, the B lattice can be viewed as an A lattice with the opposite helicity but the same structure. The sign of the system's helicity does not affect the conduction properties along the major axis. The unwrapped MT lattice is not as good a conductor since some of the hopping freedom has been removed from the lattice but does remain conducting, provided that both t_l and t_r are non-zero.

If we consider the case where one of t_l or t_r is quite small, however, the wrapping of the lattice is much more important. As discussed earlier, the unwrapped tubulin sheet is an insulator while the MT A and B lattices may carry electrons. This observation is especially interesting given that MTs have been observed to zip up and essentially change their structure from a tube to that of a sheet *in vivo*.

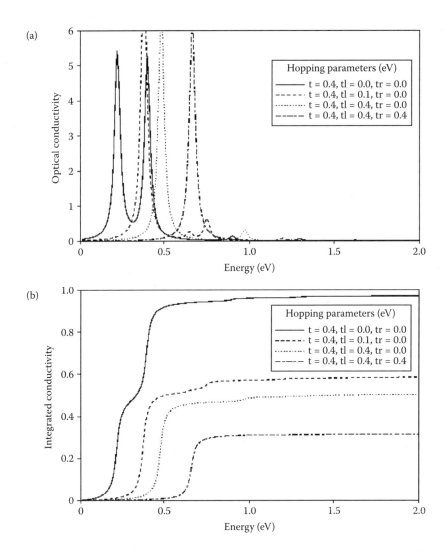

FIGURE 18.15 (a) Optical conductivity along the protofilament direction of the 14-site MT B lattice with two electrons. (b) Integrated conductivity of the 14-site MT B with two electrons shows a non-zero weight for finite t_l.

Our calculations based on the Hubbard model of electron hopping show that the MT is not an insulator, but that instead its conductivity depends on the lattice geometry and whether the MT is wrapped up or not. How well or poorly does one expect the MT to conduct relative to known semi-conductors? Referring back to Equations (18.35) and (18.36), the DC conductivity of the MT is simply given by D as

$$\frac{D}{2} = -\frac{\pi e^2}{2} a^2 < K > - \int_0^\infty \sigma_{kubo}(\omega)d\omega. \tag{18.38}$$

In this equation, the constants can be factored, and what remains is the fraction that can be read from the graphs presented earlier. The Kubo fraction (KF) is the limit of the integral as the frequency tends to infinity.

$$\frac{D}{2} = -\frac{\pi e^2}{2} a^2 < K > (1 - KF). \tag{18.39}$$

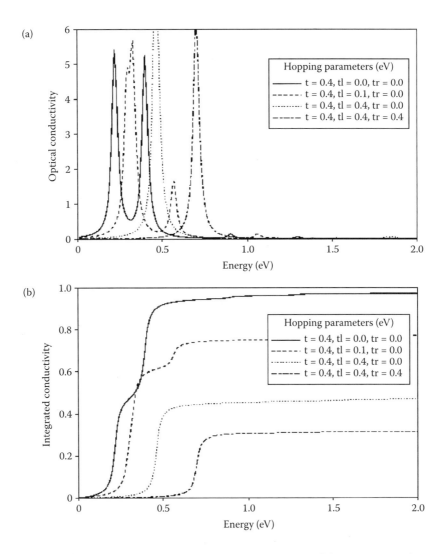

FIGURE 18.16 (a) Optical conductivity along the protofilament direction of the 14-site MT A lattice with two electrons. (b) Integrated conductivity of the 14-site MT A lattice with two electrons shows a non-zero Drude weight for finite t_1.

By returning constants of the system so that the dimensionality of D is correct, we find

$$D = -\frac{\pi n e^2 a^2}{\hbar} \frac{< K >}{t}(1 - KF). \qquad (18.40)$$

Finally, we are in a position to predict the conductivity of individual MTs of varying lattice types. For the following calculations, the electron density has been taken to be of two electrons within the lattice of 14 dimers and corresponds to a density of $1.8 \times 10^{18}/cm^3$. Thus, the results must assume that the conductivity remains constant as the length of the polymer is extended, provided the degree of filling remains the same.

The values in Table 18.1 should be compared with the conductivity of metals, such as copper $(6 \times 10^7 \, \Omega m^{-1})$ and iron $(1 \times 10^7 \, \Omega m^{-1})$. Indeed the values can also be compared to the intrinsic semiconductors germanium $(2.5 \, \Omega m^{-1})$ and silicon $(4 \times 10^{-4} \, \Omega m^{-1})$. Given these comparisons, the MT may indeed be quite a good semi-conductor given our assumptions. What is particularly interesting is the

TABLE 18.1   Calculated Conductivity and Resistance of a 1-μm
Polymer Composed of Tubulin Dimers

Tubulin Structure	Conductivity ($\Omega^{-1}m^{-1}$)	Resistance
Individual Protofilament	Non-Conducting	
Tubulin sheet ($t_1 = t_r = 0.04\ eV$)	9.4×10^3	$418k\ \Omega$
Tubulin sheet ($t_l = t_r = 0.1\ eV$)	2.0×10^4	$195k\ \Omega$
Tubulin sheet ($t_1 = t_r = 0.4\ eV$)	6.3×10^4	$62k\ \Omega$
MT (A lattice, $t_l = t_r = 0.04\ eV$)	1.8×10^4	$219k\ \Omega$
MT (A lattice, $t_l = t_r = 0.1\ eV$)	4.5×10^4	$87k\ \Omega$
MT (A lattice, $t_l = t_r = 0.4\ eV$)	1.6×10^5	$25k\ \Omega$
MT (B lattice, $t_l = t_r = 0.04\ eV$)	1.9×10^4	$211k\ \Omega$
MT (B lattice, $t_l = t_r = 0.1\ eV$)	4.7×10^4	$84k\ \Omega$
MT (B lattice, $t_l = t_r = 0.4\ eV$)	1.6×10^5	$24k\ \Omega$

way the conduction properties depend on the lattice and the particular boundary conditions. The situation can be compared to carbon nanotubes which have a similar size and structure to MTs. In addition, theoretical consideration of nanotube structure predicts that the conduction properties depend on the boundary conditions. In the case of nanotubes, this means the way in which the graphene sheet is wrapped up to form the nanotube. It is quite interesting that particular sets of boundary conditions produce a semi-conducting nanotube while the appropriate choice of wrapping the nanotube gives rise to metallic conduction [31].

The ability to change from an insulator to a conductor by a simple geometrical change could be biologically relevant even without observing a single electron to be conducted along the MT since it is the conductivity property that affects the way the cell views and responds to external electromagnetic fields. Specifically, the reflectivity of MTs to electromagnetic fields is high when they are conducting, and, consequently, they could act to direct infrared signals to the interior of the centrosome. Since the centrioles are maintained at right angles to each other, the cell would be able to determine the location in latitude and longitude of a light source if the MTs are conducting [4,5]. Peaks in the AC conductivity are all at energies significantly above the thermal activation threshold so no conductivity is expected along a darkened MT, but photoconduction in the infrared-visible range is possible. Therefore, it is possible this photo-activated conduction has functional repercussions. In addition to the intrinsic semiconducting behavior of MTs, and possibly AFs, these protein filaments possess very intensely active cable properties when immersed in an ionic buffer solution. In the next section, we discuss the ionic conductivity effects of AFs.

18.6   Actin Filaments Support Nonlinear Ionic Waves

Several experiments [12,83] indicate the possibility of ionic wave generation along AFs. As the condensed cloud of counter-ions separates the filament core from the rest of the ions in the bulk solution, we expect it to act as a dielectric medium between the two. It has both resistive and capacitive components associated with each monomer that makes up the AF. Ion flow is expected to occur at a radial distance from the surface of the filament approximately equal to the Bjerrum length. An inductive component is proposed to emerge due to the actin's double-stranded helical structure that induces the ionic flow in a solenoidal manner. Due to the presence of the sheath of counter-ions around the AF, these polymers act as biological "electrical wires" [83], and have been modeled as nonlinear inhomogeneous transmission lines propagating nonlinear dispersive solitary waves. Recently, Lader *et al.* [78] applied an input voltage pulse with an amplitude of approximately 200 mV and a duration of 800 μs to an AF, and measured electrical signals at the opposite end of the AF, indicating that AFs support ionic waves in the form of axial nonlinear currents. In an earlier experiment [83], the wave patterns observed in electrically stimulated single AFs were remarkably similar to recorded solitary waveforms for electrically stimulated nonlinear transmission lines [85]. Considering the AFs highly nonlinear complex physical structure and the thermal fluctuations of the counter-ionic cloud [105], the observation of soliton-like ionic waves is consistent with the idea of AFs functioning as biological transmission lines.

The electro-conductive medium is a condensed cloud of ions surrounding the polymer and separated from it due to the thermal fluctuations in the solution. The distance beyond which thermal fluctuations are stronger than the electrostatic attractions or repulsions between charges in solution is defined as the so-called Bjerrum length, λ_B. With the dielectric constant of the medium denoted by ε, the Bjerrum length is given by

$$\frac{e^2}{4\pi\varepsilon\varepsilon_0\lambda_B} = k_B T \tag{18.41}$$

for a given temperature T in Kelvin. Here e is the electronic charge, ε the dielectric constant, ε_0 the permittivity of the vacuum, and k_B is Boltzmann's constant. For a temperature of 293 K, we find that $k_B = 7.13 \times 10^{-10} m$. Counter-ion condensation occurs when the mean distance between charges, b, is such that $\lambda_B B/b = S > 1$. Each actin monomer carries an excess of 14 negative charges in vacuum, and accounting for events such as protonation of histidines, and assuming there to be 3 histidines per actin monomer, there exist 11 fundamental charges per actin subunit [143]. Assuming an average of 370 monomers per μm, we find there is approximately $4e/nm$. Thus, we expect a linear charge spacing of $b = 2.5 \times 10^{-10} m$ so $S = 2.85$. As the effective charge, q_{eff}, or renormalized rod charge is the bare value divided by S, we find $q_{eff} = 3.93e/monomer$. Consequently, it can be shown that approximately 99% of the counter-ion population is predominantly constrained within a radius of $8 nm$ [110] around the polymer's radial axis [158]. Significant ionic movements within this "tightly bound" ionic cloud are therefore allowed along the length of the actin, provided that it is shielded from the bulk solution [104,107].

The physical significance of each of the components of the electrical network, and additional details, are described in [148]. For actin in solution, a key feature is that the positively charged end assembles more quickly than the negatively charged end [135]. This results in an asymmetry in the charges at the ends of the filaments and F-actin's electric polarization. Actin monomers arrange themselves head to head to form actin dimers resulting in an alternating distribution of electric dipole moments along the length of the filament [73]. We assume, therefore, that there is a helical distribution of ions winding around the filament at approximately one Bjerrum length. This corresponds to a solenoid in which a fluctuating current flows as a result of voltage gradient between the two ends.

The capacitive element of the electric circuit is obtained following the observation of oppositely charged layers surrounding the filament surface. We envisage the protein surface's negative charge to be distributed homogeneously on a cylinder defining the filament surface. Furthermore, positive counter-ionic charges in the bulk are expected to form another cylinder at a radius greater than the AF itself, approximately one Bjerrum length, λ_B, away from the actin surface, which includes the condensed ions. The permittivity, ε_0, is given by $\varepsilon = \varepsilon_0\varepsilon_r$ where ε_r is the relative permittivity which we take to be that of water, i.e., $\varepsilon_r = 80$. We take the length of an actin monomer typically as $a = 5.4$ nm and the radius of the actin filament, r_{actin}, to be $r_{actin} = 2.5$ nm [14]. The next step is to consider a cylindrical Gaussian surface of length a whose radius is r such that $r_{actin} < r < r_{actin} + \lambda_B$.

Applying of Gauss's law for the total charge enclosed in the cylinder, we have the following expression for the capacitance

$$C_0 = \frac{2\pi\varepsilon a}{ln(\frac{r_{actin}+\lambda_B}{r_{actin}})} \tag{18.42}$$

With the parameters given previously, we estimate the capacitance per monomer is $C_0 = 96 \times 10^{-6}$ pF. The resistive part is obtained from Ohm's law. Taking into account the potential difference and the current I, the magnitude of the resistance, $R = V/I$, for an actin filament is given by

$$R = \frac{\rho ln(\frac{r_{actin}+\lambda_B}{r_{actin}})}{2\pi l} \tag{18.43}$$

where ρ is the resistivity. Typically, for K^+ and Na^+, intracellular ionic concentrations are 0.15 M and 0.02 M, respectively [147]. Kohlrausch's law states that the molar conductance of a salt solution is the sum of the conductivities of the ions comprising the salt solution. Thus

$$\sigma = \Lambda_0^{K^+} c_{K^+} + \Lambda_0^{Na^+} c_{Na^+} = 1.21 (\Omega m)^{-1} \tag{18.44}$$

Using this with $\rho = \sigma^{-1}$, the resistance estimate becomes $R = 6.11$ $M\Omega$, which is much lower than pure water since $R_{water} = 1.8x10^6$ $M\Omega$.

To describe the properties of the whole filament, we simply connect n subcircuits, as described earlier, to obtain an effective resistance, inductance, and capacitance, respectively, such that:

$$R_{eff} = \left(\sum_{i=1}^{n} \frac{1}{R_{2,i}} \right)^{-1} + \sum_{i=1}^{n} R_{1,i} \tag{18.45}$$

$$L_{eff} = \sum_{i=1}^{n} L_i \tag{18.46}$$

and

$$C_{eff} = \sum_{i=1}^{n} C_{0,i} \tag{18.47}$$

where $R_{1,i} = 6.11 \times 10^6$ Ω, $R_{2,i} = 0.9x10^6$ Ω such that $R_{1,i} = 7 R_{2,i}$. Note that we have used $R_{1,i} = R_1$, $R_{2,i} = R_2$, $L_i = L$, and $C_{0,i} = C_0$. For a 1 μm of the AF, we find

$$R_{eff} = 1.2 \times 10^9 \ \Omega \tag{18.48}$$

$$L_{eff} = 340 \times 10^{-12} \ H \tag{18.49}$$

$$C_{eff} = 0.02 \times 10^{-12} \ F \tag{18.50}$$

This solenoidal flow geometry leads to an equivalent electrical element possessing self-inductance. From Faraday's law, we can derive an effective inductance for the actin filament in solution by

$$L = \frac{\mu N^2 A}{l} \tag{18.51}$$

where l is the length of the F-actin and A is the cross-sectional area of the effective coil given by

$$A = \pi (r_{actin} + \lambda_B)^2 \tag{18.52}$$

The number of turns is approximated by simply working out how many ions could be lined up along the length of a monomer. We would then be approximating the helical turns as circular rings lined up along the axis of the F-actin. We also take the hydration shell of the ions into account in our calculation. The hydration shell is then the group of water molecules oriented around an ion. It can be shown that $L = 1.7 pH$ for the length of the monomer.

The electrical model of the AF is an application of Kirchhoff's laws to one section of the effective electrical circuit that is coupled to neighboring monomers. Taking the continuum limit [148] for a large number of monomers along an AF, we derive the following equation which describes the spatio-temporal

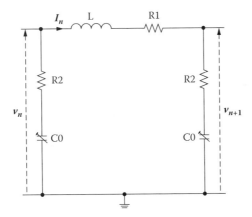

FIGURE 18.17   An equivalent electrical circuit for a segment of an actin filament.

behavior of the potential along the AF:

$$LC_0\frac{\partial^2 V}{\partial t^2} = a^2(\partial_{xx}V) + R_2C_0\frac{\partial[a^2(\partial_{xx}V)]}{\partial t} - R_1C_0\frac{\partial V}{\partial t} + R_1C_02bV\frac{\partial V}{\partial t} \tag{18.53}$$

Motivated by this picture, physical properties of the ionic distribution along a short stretch of the polymer (the average pitch 35 to 40 nm) have been modeled by Tuszynski *et al.*, [148] as an electrical circuit with non linear components (see Figure 18.17). The main elements of the circuit are: (a) a non linear capacitor associated with the spatial charge distribution between the ions located in the outer and inner regions of the polymer; (b) an inductance due to the helical geometry of the filament; and (c) a resistor due to the viscosity imposed by the solution.

From Kirchhoff's laws, one can derive an equation governing the propagation of voltage along the filament (see [148]). One of the key aspects in this model is the nonlinearity of the associated capacitance [86,150] that eventually gives rise to the self-focusing of the ionic waves. The equations developed for the model originate from the application of Kirchhoff's laws to the RLC resonant circuit of a model actin monomer in a filament. Perhaps the most important finding is the existence of the traveling wave, which describes a moving transition region between a high and low ionic concentration due to the corresponding inter-monomeric voltage gradient. The velocity of propagation was estimated to range between 1 and 100 *ms*$^{-1}$ depending on the characteristic properties of the electrical circuit model. It is noteworthy that these values overlap with action potential velocities in excitable tissues [55].

Analysis of the model described earlier reveals the possibility of stationary waves in time that may lead to the establishment of spatial periodic patterns of ionic concentration. As mentioned previously, Lader *et al.* [78] applied an input voltage pulse with an amplitude of approximately 200 *mV* and a duration of 800 *μs* to an AF, and measured electrical signals at the opposite end of the AF, indicating that AFs support ionic waves in the form of axial nonlinear currents. In an earlier experiment [83], the wave patterns observed in electrically stimulated single AFs were similar to recorded solitary waveforms for electrically stimulated nonlinear transmission lines [85]. Considering the AFs highly nonlinear physical structure and thermal fluctuations of the counter-ionic cloud [105], the observation of soliton-like ionic waves is consistent with the idea of AFs functioning as biological transmission lines.

This section only provides an indication as to a realistic model of actin that can support soliton-like ionic traveling waves. Modeling relies on data constrained by experimental conditions, and/or assumptions made, including the charge density, which is calculated based on the net surface charges of actin. It should also be considered that soliton velocity is directly proportional to the magnitude of the stimulus, which in a biological setting has not been formally described. Actin interacts with a number of ion channels of different ionic permeability and conductance. Thus, it is expected that channel opening, single channel currents, and other channel properties, including the resting potential of the cell, may significantly modify the amplitude and velocity of the soliton-supported waves. This should correlate with the velocity of the traveling waves along channel-coupled filaments. Other parameters that may play a role in this type of electrodynamic interaction are the local ionic gradients and the regulatory role of actin binding proteins, which can help "focus" the conductive medium, or otherwise impair wave velocity. The physical significance of each of the components of the electrical network (see Figure 18.17) and additional details, are described in [148]. MTs differ from AFs in several respects. One of the key differences is the presence of peptide chains called C-termini that protrude from their surface. In the following section, we outline their role in ionic wave conductivity along MTs.

18.7 Long-Range Spatio-Temporal Ionic Waves Along Microtubules

MTs are long hollow cylinders made of $\alpha\beta$-tubulin dimers [33], as described in Section 18.5. Recently, it has become apparent that neurons utilize MTs in cognitive processing. Both kinesin and MAP2 that associate with MTs have been implicated in learning and memory [69,154]. Dendritic MTs are implicated in particular, and it is highly probable that the precisely coordinated transport of critical proteins and mRNAs to the post-synaptic density via kinesin along MT tracks in dendrites is necessary for learning, as well as for LTP [70].

The following molecular dynamics (MD) simulation results focus on the C-termini of neighboring tubulins, whose biophysical properties have a significant influence on the transport of material to activated synapses. This affects cytoskeletal signal transduction and processing as well as synaptic functioning related to LTP. Using MD modeling we calculated conformation states of the C-termini protruding from the outer surfaces of MTs and strongly interacting with other proteins, such as MAP2 and kinesin [124]. To elucidate the biophysical properties of C-termini and gain insight into the role they play in the functioning of dendrites, we developed a quantitative computational model based on the currently available biophysical and biochemical data regarding the key macromolecular structures involved, including tubulin, their C-termini, and associated MAP2. In the proposed model of the C-termini microtubular network, the tubulin dimer is considered to be the basic unit. Each dimer is decorated with two C-termini that may either extend outwardly from the surface of the protofilament or bind to it in one of a few possible configurations.

The most dynamic structural elements of the system (i.e., its elastic and electric degrees of freedom) are envisaged as conformational states of the C-termini. Each state of the unbound C-terminus evolves so as to minimize the overall interaction energy of the system. The negatively charged C-termini interact with (a) the dimer's surface, (b) neighboring C-termini and (c) adjacent MAPs. While the surface of the dimer is highly negatively charged overall, it has positive charge regions that attract the C-termini causing them to bend and bind in a "downward" state. The energy difference between the two major metastable states is relatively small, on the order of a few $k_B T$ [112].

To simplify our calculations, we first used a bead-spring model representing the C-terminus as a sequence of beads with flexible connections. We have accounted for the electric field exerted by the dimer, the external field generated by the environment, and various short-range interactions within the simulated C-terminus, including interactions between the beads (i.e., Lennard–Jones potential, angular forces, etc.).

A simplified model of the interaction between MAP2 and its ionic environment via counter-ions was used to investigate the ability of MAP2 to function as a "wave-guide" that transfers the conformational change in a C-terminus state to an adjacent MT (see Figure 18.18). A perturbation applied to the counter-ions at one end of the MAP2 drives them out of equilibrium and initiates a wave that travels along the MAP2 (for details see [112]). Figure 18.18 (right) depicts the main result for a localized perturbation applied for a few picoseconds to the counter-ions near the binding site of MAP2. Wave propagation along the chain of $N = 50$ counter-ions is shown as a counter-ion displacement parallel to MAP2, u_i, where i denotes the ith counter-ion. This perturbation along MAP2 propagates almost as a "kink" whose phase velocity has been found to be $v_{ph} \approx 2nm \cdot ps^{-1}$.

C-terminal conformations (axes in nanometers)

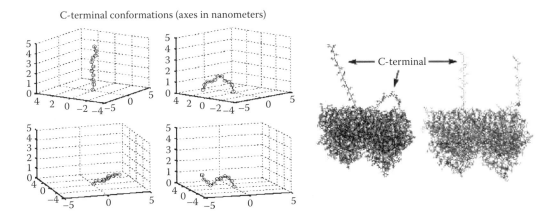

FIGURE 18.18 Examples of the conformational states of the C-termini in a tubulin dimer (right) and results from bead-spring model simulations (left).

Our simulations indicate the ability of an ionic wave to trigger a coupled wave of C-termini state changes from their upright to downward orientations. Four examples of "up" and "down" states of the C-termini obtained through bead-spring model simulations are shown in Figure 18.18 (left) and results from molecular dynamics simulations for an actual tubulin dimer are shown in Figure 18.18 (right). Calculations of the energy minimized positions of the individual beads representing the amino acids of the C-terminus in two equivalent forms reveal that the probability of the down position, which includes all cases of full or partial attachment, is 15%. This means that the system has two major states with a strong bias towards the stretched-up state.

In the remainder of this section, we present an MD simulation study of the interaction between MTs and MAP2, followed by preliminary experimental results of ionic wave conduction by MTs.

A simplified model of the interaction between MAP2 and its ionic environment via counter-ions was used to investigate the ability of MAP2 to function as a "biological wire" that transfers the conformational change in a C-terminus state to an adjacent MT (see Figure 18.19).

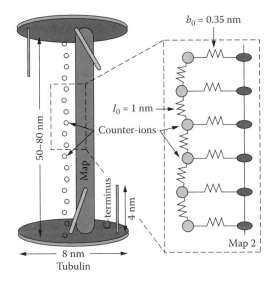

FIGURE 18.19 An illustration of a MT C-terminus conformational change transfer process. Ionic waves travel along an adjacent MAP via local perturbations to the bounded counter-ions.

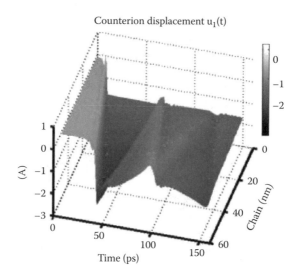

FIGURE 18.20   Propagation of a localized perturbation applied to counter-ions near the binding site of a MAP2.

A perturbation applied to the counter-ions at one end of the MAP2 drives them out of equilibrium and initiates a wave that travels along the MAP2 [112]. Figure 18.20 depicts the main result for a localized perturbation applied for a few picoseconds to the counter-ions near the binding site of a MAP2. Wave propagation along the chain of counter-ions is demonstrated as counter-ions displacement parallel to MAP2.

To validate the proposed model, we have conducted experiments on electro-stimulation of MTs in a solution. A detailed description of the experiment and its results has been published elsewhere [113]. Here we only describe the main aspects of the work on this phenomenon. Figure 18.21 depicts the experimental setup used. Isolated taxol-stabilized MTs have been shown to be able to amplify an electric signal applied to them through a micropipette. The input signal was of 5 to 10 msec duration, with amplitude in the range of ±200 mV. The signal that arrived at the other end of the MT was more than twice as high as the signal recorded in a control experiment where the same two pipettes were immersed in a solution with no MT making contact to them. The calculated conductivity of MTs was found to be on the order of 10 nS, indicating a high level of ionic conductivity along an MT. By comparison, for a typical ion channel, the corresponding value ranges between 5 and 200 pS.

The results of the MD modeling previously described raise the possibility of transmitting electrostatic perturbations collectively among neighboring C-termini, and from C-termini on one MT to those C-termini on another MT via MAP2. The importance of collective conformational states of C-termini and of transmitting perturbations among the C-termini as a novel information processing mechanism operating at a subneuronal level is clear. MAP2 and kinesin bind near to the C-termini on tubulin and the electrostatic properties of C-termini affect this binding [2,145]. Hence the conformational states of the C-termini must

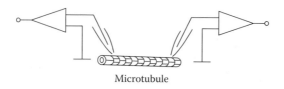

FIGURE 18.21   Schematic representation of the experimental set-up used for the measurements of a MTs ionic conductivity. The MT is attached to two micropipettes that are connected to signal amplifiers.

at some point be taken into account in order to understand neural processing that depends on transport of synaptic proteins inside of neurons. Kim and Lisman [71] have shown that inhibition of MT motor proteins reduces an AMPA receptor-mediated response in hippocampal slice. This means that a labile pool of AMPA-receptors depends on MT dynamics, and MT-bound motors determine the amplitudes of excitatory postsynaptic currents (EPSC). A walking kinesin carries with it a protein or an mRNA molecule. Since kinesin binds to a MT on a C-terminus, as it steps on it, it brings the C-terminus to the MT surface and makes it ineffective in binding for the next kinesin over a period of time that it takes the C-terminus to unbind and protrude outside. From this we can deduce that long stretches of C-termini in the upright position are going to be most efficient at transporting kinesin and kinesin cargo while C-termini that lie flat are expected to be most efficient at detaching kinesin. Thus, in considering the trafficking of many kinesins, collective electrostatic effects of C-termini become crucially important. Collective states that correspond to transport strategies that will send optimal numbers of kinesin molecules to synaptic zones are likely to occur when synapses are activated (e.g., when synapses are generating EPSCs). One type of kinesin cargo associated with learning and memory is the NMDA receptor, which is well known to be associated with LTP. The PSD might be expected to require replenishment of NMDA receptors at a later time than AMPA receptors are replenished (see [71]). Kinesin protein (specifically KIF17) actively transports NMDA receptor 2B subunits (NR2B) to the region of the PSD [52]. Once NR2B is in the vicinity of an active synapse it dissociates from kinesin and then becomes associated with the PSD, presumably using actin transport as an intermediary step. Hence, kinesin-mediated transport of NMDA receptors along MTs has a built-in negative feedback mechanism: whenever too many NMDA receptors are transported to the synaptic site (often located on a spine head), then those NMDA receptors can initiate proteolytic breakdown of MAP2. The latter event would be expected to reduce further transport of NMDA receptors to the synaptic site. The signal transduction molecule CaMKII is also critical for learning and LTP. Similar to the NMDA receptor, CaMKII is transported to active synapses via a kinesin-mediated transport mechanism. However, it appears to be the mRNA for CaMKII to a larger extent than the protein that is transported to spines, since dendrites are enriched with polyribosomes and CaMKII mRNA [140]. Ribosomes are redistributed from dendrites to spines with LTP [106]. Local translation of CaMKIIα in dendrites appears to be necessary for the late phase of LTP: fear conditioning and spatial memory [98]. Moreover, changes in synaptic efficacy are often accompanied by changes in morphology; reorganization of underlying MTs is a fundamental factors for these morphological changes. Having discussed the individual properties and roles played by AFs and MTs in subcellular conduction processes, we now turn to an integrated view of their functioning within a cell. The next several sections discuss this issue in depth.

18.8 Dendritic Cytoskeleton Information Processing Model

Our current hypothesis states that the cytoskeletal biopolymers constitute the backbone for ionic wave propagation that interacts with, and regulates, dendritic membrane components, such as ion channels to effectively control synaptic connections. Figure 18.22 depicts a portion of the dendritic shaft where MTs are decorated by C-termini and interconnected by MAP2 (thick line). Connections between MTs and AFs are shown as well as two types of synaptic bindings. On the upper left side, actin bundles bind to the post-synaptic density (PSD) of a spineless synapse. On the lower right side, a spiny synapse is shown, where actin bundles enter the spine neck and bind to the PSD, which at the other end, is connected to the MTN.

We envision a mechanism in which a direct regulation of ion channels and thus synaptic strength by AFs and associated cytoskeletal structures controls and modifies the electrical response of the neuron. In this picture, MTs arranged in networks of mixed polarity receive signals in the form of electric perturbations, from synapses via AFs connected to MTs by MAP2 [122], or via direct MT connections to postsynaptic density proteins by molecules such as CRIPT [108]. As discussed next, the MTN may be viewed as a high-dimensional dynamic system where the main degrees of freedom are related to the conformational state of the C-termini. The input signals perturb the current state of the system that continues to evolve. Hypothetical integration of the preceding ideas is outlined as follows. Electrical signals arrive at the PSD

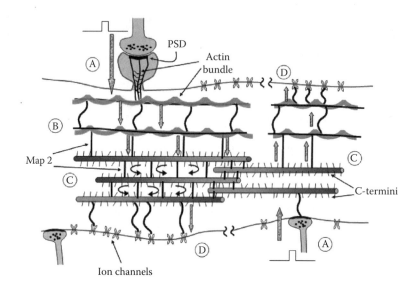

FIGURE 18.22 A scheme of the dendritic shaft with MTs arranged in networks of mixed polarity that receive signals in the form of electric perturbations from synapses via AFs connected to MTs by MAP2, or via direct MT connections to postsynaptic density proteins by molecules such as CRIPT. A spiny synapse is depicted in the bottom right where an actin bundle enters the spine neck.

via synaptic transmission, which in turn elicits ion waves along the associated AFs at the synaptic spine. These dendritic input signals propagate in the form of ionic waves through AFs to the MTN where they serve as input signals. The MTN, operating as a large high-dimensional state machine, evolves these input states, e.g., by dynamically changing C-termini conformation. The output from the MTN is the state of the system that may propagate via AFs to remote ion-channels. These output functions are assumed to regulate the temporal gating state of voltage-sensitive channels. This process subsequently regulates the membrane conductive properties and controls the axon hillock behavior by changing the rate, distribution, and topology of open/close channels. The overall functions of the dendrite and neuron can thus be regulated in this manner.

The attractiveness of the concept that the cytosol, with its cytoskeletal structures, may behave as a large dynamical system is clear since it provides a means for real-time computation without the need for stable attractor states. Moreover, the output is relatively insensitive to small variations in either the MTN (cytoskeletal networks) or the input patterns. It should be noted that the temporal system's state evolves continuously, even without external inputs. Recent perturbations, however, have a long-term effect on the MTN trajectories, i.e., there is a memory effect inherent to this system (not to be mistaken with synaptic LTP, which has a much longer time scale). The output from the MTN may converge at or near ion channels to regulate their temporal behavior. The issue of adaptation requires a feedback mechanism that will, at least locally, enable the change of the output function. In the context of neuronal function, with a focus on processivity, synaptic strengthening, LTP, and memory enhancement, the output function may simply reflect an effect of the MTN on synaptic channel function, such that the desired state of the channel appears to have a higher probability of being open/close upon the presentation of the associated input pattern. One possibility is a Hebbian-based response where more frequent activity of certain subdomains of the MTN output states gives rise to a higher/lower density of AFs connecting to corresponding channels.

We now explain how this integrative view may serve as a regulatory (adaptive) mechanism. The input level denoted by A in Figure 18.22 is associated with external electric perturbations passing through membranes, mainly synaptic inputs arriving from other neurons. These signals arrive at the cytoskeleton and directly affect MTNs, and/or actin filaments (bundles), in spiny synapses. However, the actin cytoskeleton is responsible for the propagation of signals to the MTN (see level B). These propagated signals (level C) are in turn used as inputs to the MTN, viewed as a dynamic system. We further propose that the MTN generates diverse phase space trajectories in response to different input vectors. The requirements from such a system are not too restrictive since the output-state is not an attractor of the system. In other words, information processing at this level is not necessarily based on attractor dynamics but rather on real-time computations. This proposition relies on the observed ability of the MTN to propagate signals and on the specific topological features of MTNs in dendrites, in which shorter MTs of mixed polarity are interconnected by MAP2s. The output from the MTN would be a function of the evolved state vector in certain areas accessed by actin filaments and/or directly linked to ion channels (see both possibilities at level D). These output signals may modify the temporal channel activity, either by directly arriving from the MTN to the channels, or mediated by actin filaments. Hence, the channel-based synaptic membrane conductance is regulated, in particular in the axon hillock region, which is, in most cases, responsible for the generation of action potentials.

The idea that a nonspecific high-dimensional dynamical system may serve as a reservoir of trajectories in the context of liquid state machines (LSMs), has recently been suggested as an explanation for the existence of microcircuits in the brain [87]. The basic structure of an LSM is an excitable medium (hence "liquid") and an output function that maps the current liquid state. The liquid must be sufficiently complex and dynamic to guarantee a universal computational power. This is sufficient to ensure that different input vectors lead to separate trajectories. A network of spiking neurons and a recurrent neural network has been used as a "liquid", whereas the output (readout) function has been implemented by simple perceptrons, threshold functions, or even linear regression functions. Clearly, simpler readout functions restrict the ability of the whole system to capture complex nonlinear dependencies.

18.9   The Inter-Relation Between the Neural Cytoskeleton and the Membrane

Usually, F-actin and microtubular cytoskeletal networks are thought to fulfill separate, independent cellular roles. Highly dynamic actin networks are known for their role in cell motility, in particular the spreading of the leading edge and contraction. The more stable MT cytoskeleton is best known for its importance in cell division and organelle trafficking. However, recent studies provide a more unified role, ascribing important roles to the actin cytoskeleton in cell division and trafficking and important roles for MTs in the generation and plasticity of cellular morphology. Coordination between the actin- and MT- based cytoskeletons has been recently observed during cellular migration and morphogenesis, processes that share some similarities with neurite initiation [28]. A direct physical association between both cytoskeletons has been suggested, because MTs often preferentially grow along actin bundles and transiently target actin-rich adhesion complexes. In neurons, certain plakins and neuron-specific MAPs, like MAP1B and MAP2, may play a role in linking MTs and AFs, helping in the transition from an undifferentiated state to neurite-bearing morphology.

MAP1B and MAP2 are both known to interact with actin in vitro [20,109,128,134,146]. It is likely that by crosslinking, MAP2 and/or MAP1B associated with both cytoskeletons could be involved in guidance of MTs along AF bundles. Alternatively, MAPs could shuttle from MTs to actin and could alter F-actin behavior by actively crosslinking AFs.

Little is known, however, about interactions between neuronal ion channels and the cytoskeleton. Recently, whole-cell and single channel recordings showed that acute disruption of endogenous AFs with cytochalasin D activated voltage-gated K^+ currents in these cells, which was largely prevented by intracellular perfusion with the AF-stabilizer agent, phalloidin. Interestingly, the direct addition of actin to excised, inside-out patches activated, and/or increased single K^+ channels. Thus, acute changes in actin-based cytoskeleton dynamics regulate voltage-gated ion channel activity in bipolar neurons. This may be indicative of a more general and quite appealing mechanism by which cytoskeletal structures control feedback mechanisms in neuronal channels.

Recent theoretical and experimental studies of the electrical properties of AFs and MTs in solution revealed their capability to act as biopolymer wires [83,148]. This means that these protein polymer filaments are capable of conducting nonlinear ionic waves and even amplifying the signal with respect to the conducting solution [112]. We conjecture a mechanism in which a direct regulation of ion channels and thus synaptic strength by AFs and associated cytoskeletal structures controls and modifies the electrical response of the neuron. According to this scenario (see Figure 18.22), MTs arranged in networks of mixed polarity receive signals in the form of electric perturbations from synapses via AFs connected to MTs by MAP2 [122], or via direct MT connections to postsynaptic density proteins by molecules such as CRIPT [108]. These signals propagate in the form of ionic waves. Specific physical properties of the propagating ionic waves will be discussed in sections that follow.

18.10   Relationship to Cognitive Functions

Transport of proteins and receptors in neurons is likely to have an electromagnetic basis to the extent that this function is possibly a result of MT computation. The perspective advanced in this chapter is that a specific fingerprint defined by a particular electromagnetic state of a microtubular array potentially corresponds to a unique unit of cognition (e.g., a basic visual parameter). Recently, it has been shown that visual components can be represented musically [23]; hence, the idea that one common type of energy underlies divergent perceptual and cognitive processes appears likely. Activation of one electromagnetic fingerprint could, in turn, activate another electromagnetic fingerprint, irrespective of sensory input. Moreover, the subjective feels of this widespread pattern of electromagnetic energy can be specified according to those key physical properties of MTs that influence the transport of proteins to synapses. Factors influencing kinesin-mediated transport include the protein conformation of tubulin and the nature of the C-termini (see [112,136]).

Not only is the conformational state of tubulin critical to effective transport, motor proteins appear to alter albeit temporarily, the conformation of tubulin. Kinesin binding and that of MAP, tau, significantly alter the direction of the protruding protofilament ridges along MTs, which in turn influences their further binding abilities [127]. More than mere local adaptation to binding, MTs may alter their conformation ahead of kinesin processivity [76], supporting the notion of long-range cooperative effects between tubulin dimers located along longitudinal protofilaments of MTs. These biochemical relationships have consequences for electromagnetic fields. The dipole moment of tubulin depends on its configuration in the MT [53,97]. Thus, electromagnetic fields among MTs could, in theory, be induced or inhibited by synaptic inputs that affect the protein conformation of tubulin directly or through alterations in kinesin or MAP binding. Synaptic effects upon MTs could be mediated through ionic currents, by propagation via actin filaments [148], or by signal transduction cascades resulting in the phosphorylation of MAPs [126].

Due to lengthwise electric dipoles of tubulin dimers, information in the form of traveling waves propagated along microtubular tracks can, in principle, be transmitted between synapses with high fidelity (see [149]). MAP2 bridges keep microtubular arrays within the dendritic core parallel and antiparallel by aligning portions of polarized MTs. The antiparallel alignment of MTs, which specifically occurs in dendrites, would severely attenuate any electromagnetic field generated by MTs. However, during enhanced kinesin-mediated transport, as is likely to occur with heightened synaptic activity, MAP2 bound to the MT would be perturbed and may even temporarily detach from the MT. A similar phenomenon might also occur due to dynein-mediated transport, which occurs largely in the opposite direction to that of kinesin-mediated transport. Assuming that at least some MAP2 stays attached to the antiparallel MTs, keeping the dendritic array intact, any net unidirectional transport along the MT array should increase the strength of the electromagnetic field associated with the fingerprint and should further result in the spread of that electromagnetic field to adjacent MTs. Once a sufficient number of MTs were engaged in dynamically sending and receiving complementary electromagnetic energies, whole neuronal compartments (e.g., dendrites) might be expected to interact. Due to the parallel/antiparallel arrangement of MTs in cortical dendrites and the ability of electromagnetic fields to pass from one dendrite to adjacent dendrites, information could, in principle, pass between neurons when such electromagnetic fields were sufficiently amplified as a result of changes in the binding of MAPs or kinesin.

While it is true that there are some 100 billion neurons and some 100 trillion synapses in the human cerebral cortex, so too does each sensory field afford the luxury of extremely high resolution. The fovea of the retina, for example, contains roughly 1 million receptor cells that relay information to the cerebral cortex, diverging to eventually drive the synaptic activity of at least 1 billion cortical cells. Although the mammalian cortex has many neurons, and even more synapses, its connectivity can be readily grasped by one recurrent theme: repetitive high-resolution topographic representation. Despite this simple organizational plan, cells in cortical areas, in particular cells in higher sensory or association areas, inexplicably show correlated responses during cognitive tasks. Experience alters the basic structure of dendrites, and as a consequence, it should also alter any electromagnetic fields generated by MTs as a result of transport proteins moving along them. As described earlier, a rearrangement of the cytoskeleton occurs during early development, with learning, and with neurodegeneration underlying dementia.

Although models of changes in synaptic efficacy, e.g., long-term potentiation or depression (LTP or LTD), offer great potential as memory mechanisms (see [93]), one often overlooked problem is that large numbers of synapses are affected in concert. If only a few synapses are changed with each memory, then the entire cortical system would have a near unlimited capacity, but if many synapses participate in each memory, as seems to be the case for LTP, then there could be a serious saturation effect. This relates back to the topographic organization of the cerebral cortex. Strictly synaptic models (electrochemical) suggest that complex neural networks increase synaptic weights (i.e., neurons that fire together wire together). Nonetheless, visual cortical regions, for example, do not have massively random interconnections among all parts of the visuotopic map, thereby making many of the changes in synaptic efficacy necessary for encoding complex perceptual features impossible.

The biophysical properties of MTs are just beginning to be understood at molecular and atomic levels, and recent empirical evidence suggests interesting effects occur between MTs under certain experimentally induced conditions. Two groups, one led by Watt Webb at Cornell and another led by Paul Campagnola and William Mohler at the University of Connecticut, observed that MTs give rise to intense second-harmonic generation: a frequency doubling upon exposure to a sapphire laser in the 880-nm range. (Other frequencies were partially effective.) MTs were one of the few biological materials having electric dipoles that constructively interfered with the dipoles of neighboring MTs [11,29]. This occurred for parallel MTs in axons, but not for antiparallel MTs in dendrites. Could such a phenomenon be expected to occur with natural learning or upon exposure to oscillatory input, an LTP-inducing tetanus or various pharmacological agents? To the extent that these induce electromagnetic energy, it is conceivable. In addition to being sensitive to electromagnetic radiation, MTs may themselves produce this kind of energy. Second-harmonic generation by MTs is consistent with their ferroelectric properties. In second-harmonic generation, the most strongly enhanced wavelength is 880-nm, supporting the frequently overlooked reports of electromagnetic signaling by cells. Guenther Albrecht-Buehler [5] observed electromagnetic energy in the near infrared region overlapping the 880 nm value that was generated by centrioles for the purpose of cell-to-cell communication, leading him to suggest that: "... one of the functions of MTs may be to play the role of cellular 'nerves.'" Infrared light in this range of wavelengths also induces cell aggregation [6]. Since frequency (and hence wavelength) determines the long-range effect of dielectric polarization of a given MT, one would expect this effect to be length-dependent. The distance between MTs was shown to be critical in second-harmonic generation by MTs.

18.11 The Potential for Bioelectronic Applications and Neuromorphic Computing

Traditionally, biologists viewed the cell as a "fluid mosaic bag" containing enzymes, organelles, etc. Recent work has focused on finding the proteins that come together in this cellular "soup" and proceed to effect a change corresponding to a particular biological function. Through the molecular-level analysis of these protein-protein interactions, recent work in various labs has proceeded to build a model of "systems biology," in which these biochemical reactions that have been individually characterized can be brought together to form a cohesive whole, a synergistic dynamics network. From this knowledge, we hope to extract useful information that will affect fields as diverse as nanotechnology, biotechnology, and materials science. Inside each cell, there is an elaborate network that serves a multitude of functions. It is an intracellular "highway" that provides a roadway for motor proteins to carry cargo and travel along in their journey to extend axonal length, for example. It controls synapses, and dendrite outgrowth. It is responsible for pulling the chromosomes apart during mitosis, for cell motility, etc.

As man-made engineered systems become more and more miniaturized, the push to build "nano-machinery" is driving physicists and engineers towards biomimetics, (i.e., the branch of engineering that mimics biological materials). This is highly feasible and much progress has been made in this area. However, due to the limitations inherent in trying to build mechanical devices on such a small scale, there is still much to learn. It is conceivable that through our better understanding of the cell's individual functional and structural components we will be able to advance medicine and technology beyond even our wildest dreams. One area that may undergo a major paradigm shift as a result of it is computer engineering.

As we enter into the 21st century, the dawn of the exciting new world of computational biotechnology is upon us and we are poised for a new generation of computers that do not have to be told what to do, but have the power to learn on their own. The development of a biological computer that is fast, small, and evolvable is no longer considered science fiction. Scientists are combining biological materials with the latest silicon-based technology in an effort to give us not only electronic devices that are smaller, but are also more flexible in terms of structure and function.

Ironically, our brain can serve as proof of concept. Many scientists now believe that individual neurons display wiring patterns and communication powers that resemble a human computer, and therefore may provide the framework for the design of a biological computer. The brain is composed of 10 billion neurons, each of which may be communicating with 1000 others. The neuron transmits electric signals along its arms, or 'axons'. Inside each axon a parallel architecture of microtubules is interconnected with other proteins, not unlike the parallel computer's wiring. In fact, the structure of microtubules may have evolved towards optimal computational efficiency. The piezoelectric properties of these protein filaments allow them to bend as a result of electric fields or currents which makes them ideal candidates as regulators of synaptic plasticity, a mechanism that could explain the popular principle of "use it or lose it." Thus, seldom used synaptic connections would be switched off in favor of more active ones. Biologist Guenter Albrecht-Buehler [3] demonstrated that cells perceive their environment through a tiny organelle called the centriole, composed of microtubules that were shown to interact with electromagnetic radiation. To further display that microtubules themselves can be conductive, the German biotechnology group, led by Eberhard Unger [41,42] succesfully conducted required experiments and is now working on building nanoelectronic components using these and other proteins. Simultaneously, ideas of combining biological and silicon-based materials in hybrid arrangements are gaining more and more support for future bioelectronic applications. Work is underway at various labs around the world to develop a biological chip. The main objective is to design nanoelectronic components using hybrid protein–silicon structures with arrays of proteins as biological oscillators that can be stimulated by electrodes or acoustic couplers. This will provide an analog circuit with rich dynamics.

DNA computers are unlikely to become stand alone competitors for electronic computers. But digital memory in the form of DNA and proteins is a real possibility with exquisitely efficient editing machines that navigate through the cell, cutting and pasting molecular data into the stuff of life. Beyond that, the innate intelligence built into DNA molecules could help fabricate tiny, complex structures: in essence, using computer logic not to crunch numbers but to build things, an idea conceived of by Caltech's Winfree and Rothemund [123]. A single test tube of DNA tiles could perform about 10 trillion additions per second (about a million times faster than an electronic computer). Interfacing DNA or protein-based hybrid structures with living cells will enable a host of medical and technological applications including non-invasive diagnostic and therapeutic medical applications and computational devices.

One major component of this framework is a protein-based subcellular structure called a microtubule, to which we have devoted a large portion of this chapter. Microtubules are composed of 13 protofilaments, arranged in a spiral pattern, and each protofilament is composed of subunits of tubulin. Each tubulin possesses one of two conformational states: either both monomers can align, or one monomer can shift

30° relative to the vertical axis. Based on the conformation an individual tubulin exists in at a given moment, the conformation of its nearest neighbors can be altered as well. This conformational change can be transmitted throughout the microtubule in a similar manner as patterns are transmitted through cellular automata programs. In addition, as we argued earlier in this chapter, tubulin can have several electronic states as well as dipolar states, some of which can be conceivably linked to the conformations of C-termini and the counter-ion clouds around them. The idea that we could input an electric, chemical, or mechanical signal and generate another one at the output end of the system brings computing or information processing to mind. We believe that ordered protein arrays can function as a programmable processor or even a computer either inside cells such as neurons or as nano-scale devices.

Nanotechnology could and should take advantage of biology in a completely novel manner, (i.e., instead of building materials that mimic biological components, we could use the components themselves). In the case of microtubules and tubulin, rather than purify it from traditional sources such as cow or pig brains, tubulin can be readily cloned into an expression vector, and purified from bacteria in mass quantities. Then, purified tubulin of a predetermined sequence (and thus with desirable biophysical and biochemical characteristics) can be routinely polymerized to form microtubules with specific structural and functional properties. Following that, the microtubules produced could be stabilized in a polymer gel capable of undergoing a phase transition at a particular value of the control parameter such as temperature, applied pressure, pH, or ionic concentration. This would then trigger a signal that would initiate computational processes carried out by individual microtubules or their assemblies interconnected by MAPs or even connected to actin filaments in architecures that could mimic subneuronal designs. This technology could also be applied to sensors or artifical intelligence systems in building an evolvable computer. The only limitations in this endeavor are those imposed by our imagination and our ability to dream up new applications.

18.12   Discussion

Considering the abundance of MTNs and AFs in axons and dendritic trees, the findings and theoretical models described earlier may have important consequences for our understanding of the signaling and ionic transport at an intracellular level. Extensive new information (see [63]) indicates that AFs are both directly [14] and indirectly linked to ion channels in both excitable and non-excitable tissues, providing a potentially relevant electrical coupling between these current generators (i.e., channels), and intracellular transmission lines (i.e., AFs and MTs). Furthermore, both filaments are crucially involved in cell motility and, in this context, they are known to be able to rearrange their spatial configuration. In nerve cells, AFs are mainly located in the synaptic bouton region, whereas MTs are located in both dendrites and the axon. Again, it would make sense for electrical signals supported by these filaments to help trigger neurotransmitter release through a voltage-modulated membrane deformation leading to exocytosis [132]. Actin is also prominent in post-synaptic dendritic spines, and its dynamics within dendritic spines has been implicated in the post-synaptic response to synaptic transmission. Kaech *et al.*, [65] have shown that general anesthetics inhibit this actin mediated response. Among the functional roles of actin in neurons, we mention in passing glutamate receptor channels, which are implicated in long-term potentiation. It is therefore reasonable to expect ionic wave propagation along AFs and MTs to lead to a broad range of physiological effects.

In summary, this chapter broadly described the physical conditions that enable cytoskeletal polymers such as AFs and MTs to act as electrical transmission lines for ion flows along their lengths and along MAP-mediated interconnections between adjacent filaments. In the case of AFs, we propose a model in which each protein subunit is equivalent to an electric circuit oscillator. The physical parameters used in the model were evaluated based on the molecular properties of the polymer. Using the general conductivity rules that apply to electrical circuits we analyzed the properties of ionic waves that propagate along AFs and compared these values to those observed in earlier experiments. In the context of the role played by MTs in neurons, we described the dynamics of C-termini states. We discussed both individual MTs and their networks, including the interactions with ions and signal transmission via MAPs. Recent experiments on ionic conductivity along AFs and MTs show the validity of the basic assumptions postulated in our models. In light of these results, we conjectured a new dendritic signaling mechanism that involves ion waves along protein filaments which may travel without significant decay over tens of microns, potentially affecting the function of synapses and ion channels.

We proposed a new model for information processing in dendrites based on electrical signaling involving the cytoskeleton. This model predicts that the dendritic cytoskeleton, including MTs and AFs plays an active role in computations affecting neuronal function. These cytoskeletal filaments are affected by, and regulate, ion channel activity. A molecular dynamics description of the C-termini protruding from the surface of a MT has revealed the existence of several conformational states, which lead to collective dynamical properties of the neuronal cytoskeleton. These collective states of the C-termini on MTs have been shown to have a significant effect on the ionic condensation and ion cloud propagation with physical similarities to those recently found in actin-filaments. We have been able to then provide an integrated view of these phenomena in a bottom-up scheme that demonstrates how ionic wave propagation along cytoskeletal structures may impact channel functions, and thus neuronal computational capabilities.

The possibility of an evolvable, dynamic, and responsive electrical circuitry within the cell provided by actin and MT filaments could be of enormous consequence to our understanding of the way cells operate internally and interact with their environment. In particular, it would cast an entirely new light on cell differentiation, cell division, and cell–cell communication. While an integrated theory of this type of behavior is far from being constructed, its individual elements are gradually taking shape.

Acknowledgments

This research was supported by grants from NSERC (Canada), MITACS and the Leverhulme Trust Fund (UK), and by funding from Technology Innovations, LLC of Rochester, NY.

References

[1] Adessi, Ch., Walch, S., and Anantram, M.P. (2003). Environment and structure influence on DNA conduction. *Phys. Rev. B*, **67**, 081405(R).

[2] Al-Bassam, J., R.S. Ozer, D. Safer, S. Halpain, and R.A. Milligan. (2002). MAP2 and tau bind longitudinally along the outer ridges of microtubule protofilaments. *J. Cell Biol.*, **157**:1187–1196.

[3] Albrecht-Buehler, G. (1994). *Cell Motil. Cytoskel.*, **27**, 262–271.

[4] Albrecht-Buehler, G. (1995). Changes of cell behavior by near-infrared signals. *Cell Motil. Cytoskel.*, **32**, 299–304 .

[5] Albrecht-Buehler, G. (1998). Altered drug resistance of microtubules in cells exposed to infrared light pulses: Are microtubules the "nerves" of cells? *Cell Motil. Cytoskel.*, **40**, 183–192.

[6] Albrecht-Buehler G. (2005). A long-range attraction between aggregating 3T3 cells mediated by near-infrared light scattering. *Proc. Natl. Acad. Sci. USA*, **102**, 5050–5055.

[7] Andersen, P. (1960). Interhippocampal impulses. II. Apical dendritic activation of CAI neurons. *Acta. Physiol. Scand.*, **48**:178–208.

[8] Bliss, T.V.P. and Collingridge, G.L. (1993). A synaptic model of memory: long-term potentiation in the hippocampus. *Nature*, **361**, 31–39.

[9] Bockris, J.O'M. and Khan, S.U.M. (1983). Fermi levels in solution. *Appl. Phys. Lett.*, **42**(1), 124–125.

[10] Caceres A., Mautino J., and Kosik K.S. (1992). Suppression of MAP2 in cultured cerebellar macroneurons inhibits minor neurite formation. *Neuron*, **9**(4):607–18.

[11] Campagnola P.J., Millard A.C., Terasaki M., Hoppe P.E., Malone C.J., and Mohler W.A. (2002). Three-dimensional high-resolution second-harmonic generation imaging of endogenous structural proteins in biological tissues. *Biophys. J.*, **82**, 493–508.

[12] Cantiello, H., Patenande, C., and Zaner, K. (1991). Osmotically induced electrical signals from actin filaments. *Biophys. J.*, **59**:1284–1289.

[13] Caplow, M. Ruhlen, R. and Shanks, J. (1994). *J. Cell. Biol.*, **127**, 779.

[14] Chasan, B., Geisse, N., Pedatella, K., Wooster, D., Teintze, M., Carattino, M., Goldmann, W., and Canteillo, H. (2002). Evidence for direct interaction between actin and the cystic fibrosis transmembrane conductance regulator. *Eur. Biophys. J.*, **30**:617–624.

[15] Christie, B.R. and Abraham, W.C. (1994). Differential regulation of paired-pulse plasticity following LTP in the dentate gyrus. *Neuroreport*, **5(4)**:385–8.

[16] Christie, B.R., Eliot, L.S. , Ito, K. , Miyakawa, H. and Johnston, D. (1995). Different Ca2+ channels in soma and dendrites of hippocampal pyramidal neurons mediate spike-induced Ca2+ influx. *J. Neurophysiol.*, **73(6)**, 2553–2557.

[17] Consta, S. and Kapral, R. (1994). Proton transfer in mesoscopic, molecular clusters. *J. Chem. Phys.*, **101(12)**, 10908–10914.

[18] Coombs, J.S., Eccles, J.C., and Fatt, P. (1955). The specific ionic conductances and the ionic movements across the motoneuronal membrane that produce the inhibitory post-synaptic potential. *J. Physiol.*, **130(2)**: 326–373.

[19] Cope, F.W. (1975). A review of the applications of solid state physics concepts to biological systems. *J. Biologic. Phys.*, **3(1)**, 1–41.

[20] Correas, I., Padilla, R., and Avila, J. (1990). The tubulin-binding sequence of brain microtubule associated proteins... is also involved in actin binding, *Biochem. J.*, **269(1)**: 61–64.

[21] Cragg, B.G. and Hamlyn, L.H. (1955). Action potentials of the pyramidal neurones in the hippocampus of the rabbit. *J. Physiol.*, **129(3)**: 608–627.

[22] Craig, A.M., Wyborski, R.J., and Banker, G. (1995). Preferential addition of newly synthesized membrane protein at axonal growth cones. *Nature*, **375**, 592–594.

[23] Cronly-Dillon J., Persaud K., and Gregory, R.P. (1999). The perception of visual images encoded in musical form: a study in cross-modality information transfer. *Proc. Biol. Sci.*, 1999 Dec 7, **266**(1436): 2427–33.

[24] Dai, J. and Sheetz, M.P. (1995). Mechanical properties of neuronal growth cone membranes studied by tether formation with laser optical tweezers. *Biophys. J.*, **68**, 988–996.

[25] Davies, O.R. and Inglesfield, J.E. (2004). Embedding method for conductance of DNA. *Phys. Rev. B*, **69**, 195110.

[26] de Pablo, P.J., Moreno-Herrero, F., Colchero, J., Gómez Herrero, J., Herrero, P., Baró, A.M., Ordejón, P., Soler, J.M., and Artacho, E. (2000). Absence of dc-conductivity in λ-DNA. *Phys. Rev. Lett.*, **85**, 4992–4995.

[27] Dehmelt, L., Smart, F.M., Ozer, R.S., and Halpain, S. (2003). The role of microtubule-associated protein 2c in the reorganization of microtubules and lamellipodia during neurite initiation. *J. Neurosci.*, **23(29)**:9479–9490.

[28] Dehmelt, L. and Halpain, S. (2004). The MAP2/Tau family of microtubule-associated proteins, *Genome Biol.*, 6.

[29] Dombeck, D.A., Kasischke, K.A., Vishwasrao, H.D., Ingelsson, M., Hyman, B.T., and Webb, W.W. (2003). Uniform polarity microtubule assemblies imaged in native brain tissue by second-harmonic generation microscopy. *Proc. Natl. Acad. Sci. USA*, **100**, 7081–7086.

[30] Doniach, S. and Sondheimer, E. (1974). *Green's Functions for Solid State Physicists* (W.A. Benjamin Inc., Don Mills).

[31] Dresselhaus, M.S., Dreselhaus, G., and Eklund, P.C. (1996). *Science of Fullerenes and Carbon Nanotubes*, (Academic Press: San Diego).

[32] Drewes, G., Ebneth, A., and Mandelkow, E-M. (1998). MAPs, MARKs and microtubule dynamics. *TIBS*, **23**, 307–311.

[33] Dustin, P. (1984). *Microtubules*. (Springer-Verlag, Berlin).

[34] Eccles, J.C. (1964). *The Physiology of Synapses* (Springer, Berlin).

[35] Eley, D.D. (1989). Studies of organic semiconductors for 40 years I. *Mol. Cryst. Liq. Cryst.*, **171**, 1–21.

[36] Eliot, L.S. and Johnston, D. (1994). Multiple components of calcium current in acutely dissociated dentate gyrus granule neurons. *J. Neurophysiol.*, **72(2)**, 762–777.

[37] Endres, R.G., Cox, D.L. and Singh, R. (2004). Colloquium: The quest for high-conductance DNA. *Rev. Mod. Phys.*, **76**, 195.

[38] Eyzaguirre, C. and Kuffler, S.W. (1955). Further Study of Soma, Dendrite, and Axon Excitation in Single Neurons. *J. Gen. Physiol.*, **39**, 121–153.

[39] Fink, H.W. (2001). DNA and conducting electrons. *Cell. Mol. Life Sci.*, **58(1)**.

[40] Fisher, R., Gray, R., and Johnston, D. (1990). Properties and distribution of single voltage-gated calcium channels in adult hippocampal neurons. *J. Neurophysiol.*, **64(1)**, 91–104.

[41] Fritzsche,W., Böhm, K., Unger, E., and Köhler, J.M. (1998). Making electrical contact to single molecules. *Nanotechnology*, **9**, 177–183.

[42] Fritzsche, W., Koehler, J.M., Boehm, K., Unger, E., Wagner, T., Kirsch, R., Mertig M., and Pompe, W. (1999). *Nanotechnology*, **10**, 331–335.

[43] Fujita, Y. and Sakata, H. (1962). Electrophysiological properties of CA1 and CA2 apical dendrites of rabbit hippocampus. *J. Neurophysiol.*, **25**: 209–222.

[44] Fye, R.M., Martins, M.J., and Scalapino, D.J. (1992). Optical-conductivity properties of one-dimensional Hubbard rings: Repulsive and attractive-u cases. *Phys. Rev.*, **B45**, 7311–7314.

[45] Gamblin, T.C., Nachmanoff, K., Halpain, S., and Williams, R.C. Jr. (1996). Recombinant microtubule-associated protein 2c reduces the dynamic instability of individual microtubules. *Biochemistry*, **35 (38)**, 12576–12586.

[46] Glanz, J. (1997). *Science*, **276**, 678.

[47] Goddard, G. and Whittier, J.E. (2006). Biomolecules as nanomaterials: interface characterization for sensor development. *Proc. SPIE*, Varadan, V.K., Ed., Vol. 6172, 617206.

[48] Gonzales-Billault, C., Owen, R., Gordon-Weeks, P.R., Avila, J. (2002). Microtubule-associated protein 1B is involved in the initial stages of axonogenesis in peripheral nervous system cultured neurons. *Brain Res.*, **943**:56–67.

[49] Greenough, W.T. (1975). Experiential modification of the developing brain. *Am. Sci.*, **63(1)**: 37–46.

[50] Greenough, W.T., Black, J.E., Wallace, C.S. (1987). Experience and brain development. *Child Devel.*, **58(3)**, 539–559.

[51] Griffith, L.M. and Pollard, T.D. (1982). The interaction of actin filaments with microtubules and microtubule associated proteins. *J. Biol. Chem.*, **257(15)**, 9143–9151.

[52] Guillaud, L., Setou, M., and Hirokawa, N. (2003). KIF17 dynamics and regulation of NR2B trafficking in hippocampal neurons. *J. Neurosci.*, **23**:131–140.

[53] Hagan, S., Hameroff, S.R., and Tuszynski, J.A. (2002). Quantum computation in brain microtubules: decoherence and biological feasibility. *Phys. Rev. E*, **65**, 061901-1 to -11.

[54] Hempel et al. (2000).

[55] Hille, B. (1992). *Ionic Channels of Excitable Membranes*. (Sinauer Associates, Sunderland, MA.)

[56] Hirokawa, N. (1991). In *The Neuronal Cytoskeleton* (Wiley-Liss, New York), 5–74.

[57] Houchin, J. (1973). Procion yellow electrodes for intracellular recording and staining of neurons in the somatosensory cortex of the rat. *J. Physiol.*, **232**, 67–69.

[58] Howard, W., and Timasheff, S. (1986). *Biochemistry*, **25**, 8292.

[59] Hyman, A. (1992). *J. Cell Biol.*, **127**.

[60] Hyman, A., Chretien, D., Arnal, I., and Wade, R. (1995). *J. Cell. Biol.*, **128**, 117.

[61] Ingber, D.E. (1993). Cellular tensegrity: defining new rules of biological design that govern the cytoskeleton. *J. Cell Sci.*, **104**, 613–627.

[62] Jaffe, D. and Brown, T. (1994). Metabotropic glutamate receptor activation induces calcium waves within hippocampal dendrites. *J. Neurophysiol.*, **72**: 471–474.

[63] Janmey, P. (1998). The cytoskeleton and the cell signaling: component localization and mechanical coupling. *Physio. Rev.*, **78**:763–781.

[64] Jaslove, S.W. (1992). The integrative properties of spiny distal dendrites. *Neuroscience*, **47**:495–519.

[65] Kaech, S., Brinkhaus, H., Matus, A. (1999). Volatile anesthetics block actin-based motility in dendritic spines. *Proc. Natl. Acad. Sci. USA*, **96**, 10433–10437.

[66] Kasumov, A.Y., Kociak, M., Gueron, S., Reulet, B., Volkov, V.T., Klinov, D.V., Bouchiat, H. (2001). Proximity induced superconductivity in DNA. *Science*, **291**, 280–282.

[67] Katz, B. (1950). Depolarization of sensory terminals and the initiation of impulses in the muscle spindle. *J. Physiol.*, **11(3–4)**:261–82.

[68] Kay, A.R. and Wong, R.K. (1987). Calcium current activation kinetics in isolated pyramidal neurones of the Ca1 region of the mature guinea-pig hippocampus. *J. Physiol.*, **392(1)**, 603–616.

[69] Khuchua, Z., Wozniak, D.F., Bardgett, M.E., Yue, Z., McDonald, M., Boero, J., Hartman, R.E., Sims, H., Strauss, A.W. (2003). Deletion of the N-terminus of murine map2 by gene targeting disrupts hippocampal ca1 neuron architecture and alters contextual memory. *Neuroscience*, **119**, 101–111.

[70] Kiebler, M.A., and DesGroseillers, L. (2000). Molecular insights into mRNA transport and local translation in the mammalian nervous system. *Neuron.*, 25:19–28.

[71] Kim, C.H., and Lisman, J.E. (2001). A labile component of AMPA receptor-mediated synaptic transmission is dependent on microtubule motors, actin, and N-ethylmaleimide-sensitive factor. *J. Neurosci.*, 21:4188–4194.

[72] Knops, J., Kosik, K.S., Lee, G., Pardee, J.D., Cohen-Gould , L., and McConlogue , L. (1991). Overexpression of tau in a nonneuronal cell induces long cellular processes. *J. Cell Biol.*, **114**, 725–733.

[73] Kobayasi, S., Asai, H., and Oosawa, F. (1964). Electric birefrigence of actin. *Biochim. Biophys. Acta*, **88**:528–540.

[74] Koch and Segev (2000).

[75] Kowalski, R.J. and Williams, R.C. Jr (1993). Microtubule-associated protein 2 alters the dynamic properties of microtubule assembly and disassembly. *J. Biol. Chem.*, **268(13)**, 9847–9855.

[76] Krebs, A., Goldie, K.N., Hoenger, A. (2004). Complex formation with kinesin motor domains affects the structure of microtubules. *J. Mol. Biol.*, **335 (1)**:139–53.

[77] Kubo, R. (1957). Statistical mechanical theory of irreversible processes, I. General theory and simple applications to magnetic and conduction problems. *J. Phys. Soc. Japan*, **12** 570–586.

[78] Lader, A., Woodward, H., Lin, E., and Cantiello, H. (2000). Modeling of ionic waves along actin filaments by discrete electrical transmission lines. *METMBS'00 International Conference*, 77–82.

[79] LeClerc, N., Kosik, K.S., Cowan, N., Pienkowski, T.P., and Baas, P.W. (1993). Process formation in Sf9 cells induced by the expression of a microtubule associated protein 2C-like construct. *Proc. Nat. Acad. Sci.*, **90**, 6223–6227.

[80] Letourneau, P.C. and Ressler, A.H. (1983). Differences in the organization of actin in the growth cones compared with the neurites of cultured neurons from chick embryos. *J. Cell Biol.*, **97**, 963–973.

[81] Lewis, A.K. and Bridgman, P.C. (1992). Nerve growth cone lamellipodia contain two populations of actin filaments that differ in organization and polarity. *J. Cell Biol.*, **119**, 1219–1243.

[82] Lieb, E.H. (1995). *The Hubbard Model* (Plenum Press, New York).

[83] Lin, E. and Cantiello, H. (1993). A novel method to study the electrodynamic behavior of actin filaments. Evidence of cable-like properties of actin. *Biophys. J.*, 65:1371–1378.

[84] Llinas, R. and Nicholson, C. (1971). Electrophysiological properties of dendrites and somata in alligator Purkinje cells. *Neurophysiol.*, **34**: 532–551.

[85] Lonngren, K. (1978). *Solitons in Action*, (Academic Press, New York).

[86] Ma, Z., Wang, J., and Guo, H. (1999). Weakly nonlinear ac response: Theory and application. *Phys. Rev. B*, **59**:7575.

[87] Maass, W., Natschläger, T., and Markram, H. (2002). Real time computing without stable states: A new framework for neural computation based on perturbations. *Neural. Computation*, 14:2531–2560.

[88] Magee, J.C., and Johnston, D. (1995). Synaptic activation of voltage-gated channels in the dendrites of hippocampal pyramidal neurons. *Science*, **268(5208)**:301–4.

[89] Magee, J.C., and Johnston, D. (1995). Characterization of single voltage-gated Na+ and Ca2+ channels in apical dendrites of rat CA1 pyramidal neurons. *J. Physiol.*, **487(1)**, 67–90.

[90] Mahan, G.D. (1981). *Many-Particle Physics* (Plenum Press, New York).

[91]   Maldague, P.F. (1977). Optical spectrum of a Hubbard chain. *Phys. Rev.*, **B16**, 2437–2446.

[92]   Malenka, R.C. (1991). *Neuron*, **6**, 53–60.

[93]   Malenka, R.C., Bear, M.F. (2004). LTP and LTD: an embarrassment of riches. *Neuron.*, **44**, 5–21.

[94]   Maniotis, A.J., Chen, C.S., and Ingber, D.E. (1997). *Proc. Natl. Sci. USA*, **94**, 849.

[95]   Marechal, Y. (1991). What mechanisms for transfers of protons through H-bonds? *J. Mol. Liq.*, **48(2–4)**, 253–260.

[96]   Mehrez, H., Walch, S., and Anantram, M.P. (2005). Electronic properties of O2-doped DNA. *Phys. Rev. B*, **72**, 035441.

[97]   Mershin, A., Kolomenski, A.A., Schuessler, H.A., Nanopoulos, D.V. (2004a). Tubulin dipole moment, dielectric constant and quantum behavior: computer simulations, experimental results and suggestions. *Biosystems*, **77**, 73–85.

[98]   Miller, S., Yasuda, M., Coats, J.K., Jones, Y., Martone, M.E., and Mayford, M. (2002). Disruption of dendritic translation of CaMKIIalpha impairs stabilization of synaptic plasticity and memory consolidation. *Neuron.*, **36**:507–519.

[99]   Minoura, I., and Muto, E. (2006). Dielectric measurement of individual microtubules using the electroorientation method. *Biophys. J.*, **90**:3739–3748.

[100] Miyakawa, H., Ross, W.N., Jaffe, D., Callaway, J.C., Lasser-Ross, N., Lisman, J.E., Johnston, D. (1992). Synaptically activated increases in Ca2+ concentration in hippocampal CA1 pyramidal cells are primarily due to voltage-gated Ca2+ channels. *Neuron*, **9(6)**:1163–73.

[101] Mogul, D.J., Adams, M.E., Fox, A.P. (1993). Differential activation of adenosine receptors decreases N-type but potentiates P-type Ca2+ current in hippocampal CA3 neurons. *Neuron*, **10(2)**:327–34.

[102] Morowitz, H.J. (1978). Proton semiconductors and energy transduction in biological systems. *AJP - Regulatory, Integrative and Comparative Physiol.*, **235(3)**, 99–114.

[103] Moskowitz, P. and Oblinger, M. (1995). *J. Neurosci.*, **15**, 1545.

[104] Oosawa, F. (1970). Counterion fluctuation and dielectric dispersion in linear polyelectrolytes. *Biopolymers*, **9**:677–688.

[105] Oosawa, F. (1971). *Polyelectrolytes.* (Marcel Dekker, New York.)

[106] Ostroff, L.E., Fiala, J.C., Allwardt, B., and Harris, K.M. (2002). Polyribosomes redistribute from dendritic shafts into spines with enlarged synapses during LTP in developing rat hippocampal slices. *Neuron*, **35**:535–45.

[107] Parodi, M., Bianco, B., and Chiabrera, A. (1985). Toward molecular electronics. Self-screening of molecular wires. *Cell Biophys.*, **7**:215–235.

[108] Passafaro, M., Sala, C. Niethammer, M., and Sheng, M. (1999). Microtubule binding by CRIPT and its potential role in the synaptic clustering of PSD-95. *Nat Neurosci.*, **2**:1063–1069.

[109] Pedrotti, B., and Islam, K. (1996). Dephosphorylated but not phosphorylated microtubule associated protein MAP1B binds to microfilaments. *FEBS Lett.*, **388(2–3)**:131–3.

[110] Pollard, T. and Cooper, J. (1986). Actin and actin-binding proteins. a critical evaluation of mechanisms and functions. *Ann. Rev. Biochem.*, **55**:987–1035.

[111] Porath, D., Bezryadin, A., de Vries, S., and Dekker, C. (2000). Direct measurements of electrical transport through DNA molecules. *AIP Conf.*, **544**, 452–456.

[112] Priel, A., Tuszynski, J.A., and Woolf, N. (2005). Transitions in Microtubule C-Termini Conformations as a Possible Dendritic Signaling Phenomenon, *Eur. Biophys. J.*.

[113] Priel, A., Ramos, A.J., Tuszynski, J.A., and Cantiello, H.F. (2006). A Biopolymer transistor: electrical amplification by microtubules. *Biophys. J.*, **90**:4639–4643.

[114] Rall, W. (1959). Branching dendritic trees and motorneuron sensitivity, *Exp. Neurol.*, **1**, 491–527.

[115] Rall, W. (1957). Membrane time constant of motoneurons. *Science*, **126(3271)**:454.

[116] Ramón-Moliner, E. (1968). The morphology of dendrites. In *Structure and Function of the Nervous Tissue*, Bourne, D.F,, Ed., (Academic Press, New York).

[117] Ratner, M.A., and Shriver, D.F. (1989). Polymer ionics. *MRS Bull.*, **Sept 1989**, 39–52.

[118] Regehr, W.G., Connor, J.A., and Tank, D.W. (1989). Optical imaging of calcium accumulation in hippocampal pyramidal cells during synaptic activation. *Nature*, **341**, 533–536.

[119] Regehr, W.G. and Tank, D.W. (1990). Postsynaptic NMDA receptor-mediated calcium accumulation in hippocampal CAl pyramidal cell dendrites. *Nature*, **345**, 807–810.

[120] Regehr, W.G. and Tank, D.W. (1992). Calcium concentration dynamics produced by synaptic activation of CA1 hippocampal pyramidal cells. *J. Neurosci.*, **12**, 4202–4223.

[121] Ripoll, C., Norris, V., Thellier, M. (2004). Ion condensation and signal transduction. *BioEssays*, **26(5)**, 549–557.

[122] Rodriguez, O.C., Schaefer, A.W., Mandato, C.A., Forscher, P., Bement, W.M., and Waterman-Storer, C.M. (2003). Conserved microtubule-actin interactions in cell movement and morphogenesis. *Nat. Cell Biol.*, **5**:599–609.

[123] Roweis, S., Winfree, E., Burgoyne, R., Chelyapov, N., Goodman, M., Rothemund, P., and Adleman, L.M. (1998). *J. Comput. Biol.*, **5(4)**:615–629.

[124] Sackett, D.L. (1995). Structure and function in the tubulin dimer and the role of the acid carboxyl terminus. In *Subcellular Biochemistry — Proteins: Structure, Function and Engineering*, Biswas, B.B. and Roy, S. Ed. (Kluwer Academic Publishers, Dordrecht). 24:255–302.

[125] Salmon, W.C., Adams, M.C., and Waterman-Storer, C.M. (2002). Dual-wavelength fluorescent speckle microscopy reveals coupling of microtubule and actin movements in migrating cells. *J. Cell Biol.*, **158(1)**, 31–37.

[126] Sanchez, C., Diaz-Nido, J., Avila, J. (2000). Phosphorylation of microtubule- associated protein 2 (MAP2) and its relevance for the regulation of the neuronal cytoskeleton function. *Prog. Neurobiol.*, 2000 June, **61(2)**:133–68.

[127] Santarella, R.A., Skiniotis, G., Goldie, K.N., Tittmann, P., Gross, H., Mandelkow, E.M., Mandelkow, E., and Hoenger, A. (2004). Surface-decoration of microtubules by human tau. *J. Mol. Biol.*, **339**, 539–553.

[128] Sattilaro R.F. (1986).

[129] Sattilaro, R.F. (1986). Interaction of microtubule-associated protein 2 with actin filaments. *Biochemistry*, **25**:2003–2009.

[130] Schaefer, A.W., Kabir, N., and Forscher, P. (2002). Filopodia and actin arcs guide the assembly and transport of two populations of microtubules with unique dynamic parameters in neuronal growth cones. *J. Cell Biol.*, **158(1)**, 139–152.

[131] Schwartzkroin, P.A., and Slawsky, M. (1977). Probable calcium spikes in hippocampal neurons. *Brain Res.*, **135(1)**:157–61.

[132] Segel, L. and Parnas, H. (1991). What controls the exocytosis of neurotransmitters. In Peliti, L., Ed., *Biologically Inspired Physics*, (Plennum Press, New York), 2000.

[133] Segev, I., London, M. (2000). Untangling dendrites with quantitative models. *Science*, **290(5492)**, 744–750.

[134] Selden and Pollard, (1983). Phosphorylation of microtubule-associated proteins regulates their interactions with actin filaments, *J. Biol. Chem.*, 258(11), 7064–7071.

[135] Sept, D., Xu, J., Pollard, T., and McCammon, J. (1999). Annealing accounts for the length of actin filaments formed by spontaneous polymerization. *Biophys. J.*, **77**:2911–2919.

[136] Skiniotis, G., Cochran, J.C., Muller, J., Mandelkow, E., Gilbert, S.P., Hoenger, A. (2004). Modulation of kinesin binding by the C-termini of tubulin. *EMBO J.*, **23**, 989–999.

[137] Softky, W. (1994). Sub-millisecond coincidence detection in active dendritic trees. *Neuroscience*, **58(1)**:13–41.

[138] Spruston, N., Jonas, P., and Sakmann, B. (1995). Dendritic glutamate receptor channels in rat hippocampal CA3 and CA1 pyramidal neurons. *J. Physiol.*, **482(2)** 325–352.

[139] Spruston, N., Schiller, Y., Stuart, G., Sakmann, B. (1995). Activity-dependent action potential invasion and calcium influx into hippocampal CA1 dendrites. *Science*, **268**:297–300.

[140] Steward, O. and Schuman,E.M. (2001). Protein synthesis at synaptic sites on dendrites. *Annu. Rev. Neurosci.*, **24**:299–325.

[141] Stuart, G.J. and Sakmann, B. (1994). Active propagation of somatic action potentials into neocortical pyramidal cell dendrites. *Nature*, **367**, 69–72.

[142] Szent-Gyorgyi, A. (1941). Study of energy levels in biochemistry. *Nature*, **14**, 157–187.

[143] Tang, J. and Janmey, P. (1996). The polyelectrolyte nature of F-actin and the mechanism of actin bundle formation. *J. Biol. Chem.*, **271**:8556–8563.

[144] Tang, D., Goldberg, D.J. (2000). Bundling of microtubules in the growth cone induced by laminin. *Mol. Cell. Neurosci.*, **15(3)**, 303–313.

[145] Thorn, K.S., Ubersax, J.A., and Vale, R.D. (2000). Engineering the processive run length of the kinesin motor. *J. Cell Biol.*, **151**:1093–1100.

[146] Togel, M., Wiche, G., and Propst, F. (1998). Novel features of the light chain of microtubule-associated protein MAP1B: microtubule stabilization, self interaction, actin filament binding, and regulation by the heavy chain. *J. Cell Biol.*, **143**:695–707.

[147] Tuszynski, J.A. and Dixon, J.M. (2001). *Biomedical Applications of Introductory Physics.* (John Wiley & Sons, New York).

[148] Tuszynski, J.A., Portet, S., Dixon, J.M., Luxford, C., and Cantiello, H.F. (2004). Ionic wave propagation along actin filaments, *Biophys. J.*, **86**, 1890–1903.

[149] Tuszynski, J.A., Brown, J.A., and Hawrylak, P. (1998). *Philos. Trans. R. Soc. London Ser. A*, **356**, 1897.

[150] Wang, B., et al. (1999). Nonlinear quantum capacitance. *Appl. Phys. Lett.*, **74**:2887.

[151] Weisshaar, B., Doll, T., and Matus, A. (1992). Reorganisation of the microtubular cytoskeleton by embryonic microtubule-associated protein 2 (MAP2c). *Development*, **116**, 1151–1161.

[152] Westenbroek, R.E., Ahlijanian, M.K., and Catterall, W.A. (1990). Clustering of L-type Ca2+ channels at the base of major dendrites in hippocampal pyramidal neurons. *Nature*, **347**, 281–284.

[153] Wong, R.K.S., Prince, D.A., and Basbaum, A.I. (1979). Intradendritic recordings from hippocampal neurons. *PNAS*, **76(2)**, 986–990.

[154] Woolf, N.J., Zinnerman, M.D., and Johnson, G.V.W. (1999). Hippocampal microtubule-associated protein-2 alterations with contextual memory. *Brain Res.*, **821**:241–249.

[155] Yamada, K.M., Spooner, B.S., and Wessell, N.K. (1971). Ultrastructure and functions of growth cones and axons of cultured nerve cells. *J. Cell Biol.*, **49 (3)**: 614.

[156] Yu, W., Ling, C., and Baas, P.W. (2001). Microtubule reconfiguration during axogenesis. *J. Neurocytol.*, **30(11)**, 861–875.

[157] Yu, X., Shacka, J.J., Eells J.B., Suarez-Quian, C., Przygodzki, R.M., Beleslin-Cokic, B., Lin, C.S., Nikodem, V.M., Hempstead, B., Flanders, K.C., Costantini, F., and Noguchi, C.T. (2002). Erythropoietin receptor signalling is required for normal brain development. *Development*, **129**, 505–516.

[158] Zimm, B. (1986). *Coulombic Interactions in Macromolecular Systems*, 212–215, (American Chemical Society, Washington D.C).

IV

Molecular and Nano Electronics: Device-Level Modeling and Simulation

19

Simulation Tools in Molecular Electronics

Christoph Erlen

Paolo Lugli

Alessandro Pecchia

Aldo Di Carlo

Introduction

Recently, molecular electronics has attracted increasing attention due to appealing features such as very high integration capabilities, low production cost, flexibility in substrate choice, and the possibility of large area deployment. Brief inspection shows two approaches to this field: one can mimic existing devices and substitute active layers with organic ones; and single molecules can be contacted and their transport characteristics exploited to achieve electronic functionalities. While in the former case conventional device simulators can be adapted to the theoretical investigation of electronic and optoelectronic devices, completely new tools must be developed for the latter.

In the following, recent developments in the simulation of organic and molecular systems will be presented based on a series of results obtained from our groups. Examples of the two approaches previously mentioned will be included. The chapter starts by studying quantum transport in single molecules and afterwards moves to the simulation of organic thin film devices.

19.1 Quantum Transport Tools

The idea of using molecular elements and even single molecular units in electronic devices has been proposed for a long time. Great progress has been made in recent years in the field of molecular electronics, as shown in the other contributions to this book. New experimental breakthroughs are achieved every year and the number of publications in the field is growing exponentially. Conduction through single molecules can now be routinely obtained using different experimental configurations.

However, the problems that need to be solved in order to turn a nice principle into real working devices are still quite challenging, due to the difficulty of precisely controlling the molecular building blocks, and due to stability issues related to organic materials.

Despite considerable advances, understanding and controlling transport mechanisms is far from understood. From a theoretical point of view, the description of conduction in molecular systems is quite a challenge, as subtle many-electron correlations may play a significant role, changing transport from the coherent regime to incoherent or Coulomb blockades. Current state-of-the-art atomistic quantum transport tools are unable to describe correlated transport mechanisms and often it is necessary to resort to simplified models. On the other hand, highly accurate quantum chemical methods are by now far too computationally expensive to treat the very large clusters required to simulate realistic systems made up of hundreds of atoms.

Certainly one of the most promising molecular devices is the bi-stable molecular switch [1], which may be used for memory arrays and FPGA logic architectures. The switch is triggered by a threshold voltage inducing a conformational change associated to a change of conductance. Stability of the junction under bias and the switching mechanism are therefore crucial issues to future development of practical devices.

In order to address such issues, we have developed a tool for quantum transport calculations coupled to a thermal dissipation model, which includes first-principle calculations of the electron–vibration coupling, phonon release into the molecules and heat dissipation into the contacts. The code (*gDFTB*) is based on the density functional tight-binding (DFTB) method [2], extended to non-equilibrium Green's functions (NEGF) for the self-consistent computation of charge density and electronic transport [3,4].

The *gDFTB* method allows a nearly first-principle treatment of systems comprising a large number of atoms. The Green's function technique enables the computation of the tunneling current flowing between two contacts in a manner consistent with the open boundary conditions that naturally arise in transport problems. On the other hand, the NEGF formalism allows us to compute the charge density consistently with the non-equilibrium conditions in which a molecular device is driven when biased by an external field.

A brief description of our methodology is given in the first section. Applications to the computation of conductance of various molecular systems are also shown, focusing on IETS of octane-thiols and heat dissipation in a molecular system.

19.1.1 The gDFTB Approach

The density-functional based tight-binding formalism (DFTB) has been described in detail in many articles and reviews (see e.g., [2]). All matrix elements and orbital wavefunctions are derived from density-functional calculations. The advantage of the method relies on the use of a small basis set, and the restriction to two center integrals, allowing extensive use of look-up tables. What distinguishes our approach from empirical methods is the explicit calculation of the basic wave functions, which allows deeper physical insights and better control of the approximations used. The method solves the Kohn–Sham equations consistently, using a Mulliken charge projection [5].

In the traditional DFTB code, a minimal basis set of atomic orbitals is used in order to reduce the matrix dimensions for diagonalization speed-up. This approach has proved to give transferable and accurate interaction potentials, and the numerical efficiency of the method allows molecular dynamic simulations of large super-cells, containing several hundreds of atoms, particularly suitable to study the electronic properties and dynamics of large mesoscopic systems and organic molecules such as CNTs, DNA strands or adsorbates on surfaces, semiconducting heterostructures, etc. (see [6]).

Next, we shall briefly describe the self-consistent DFTB method, developed from an idea first introduced by Foulkes, where the electronic density is expanded as the sum of a reference density, $n^0(r)$ (which can be chosen as the superposition of neutral atomic densities), and a deviation, $\delta n(r)$, such that $n(r) = n^0(r) + \delta n(r)$. The total energy of the system can be described, up to the second order in the local density fluctuations, as:

$$E_{tot}[n] = \sum_k n_k < \Psi_k|H^0|\Psi_k > + E_{rep}[n^0] + E^{(2)}[\delta n], \qquad (19.1)$$

The first term in Equation (19.1) can be written in terms of the TB Hamiltonian, which is given by

$$\begin{cases} H^0_{\mu v} = \varepsilon^{free-atom}_\mu, & \mu = v \\ < \phi_\mu | T + v_{eff}[n^0_i + n^0_j] | \phi_v >, & \mu \in i, v \in j \end{cases} \tag{19.2}$$

where ϕ_μ and ϕ_v are the atomic orbitals localized around the atomic centers i and j; T is the kinetic energy operator, and v_{eff} is the effective one-particle potential that depends on the density of the two atomic centers i and j.

The term $E_{rep}[n^0]$ in Equation (19.1) is the repulsive energy between the ions, screened by the electronic distribution and the exchange energy. The third term in Equation (19.1) is the second order correction, which can be written as

$$E^2[\delta n] = \frac{1}{2} \int\int \left[\frac{1}{|r - r'|} + \frac{\delta^2 E_{xc}}{\delta n(r)\delta n(r')} \right] \delta n(r)\delta n(r')drdr' \tag{19.3}$$

where the Hartree and exchange-correlation (XC) potentials have been separated. This quantity is greatly simplified by retaining only the monopole term in the radial expansion of the atom-centered density fluctuations [7].

Within the LDA approximation, the exchange contribution vanishes for large atomic distances, hence in Equation (19.3) the second order correction to E_{xc} can be neglected with respect to the Coulomb interaction. For short ranges, Coulomb and XC contributions are accounted for with onsite Hubbard parameters, which are calculated for any atom type within LDA-DFT as the second derivative of the total energy of the atom with respect to the occupation number of the highest occupied atomic orbital. These values are therefore neither adjustable nor empirical parameters [5]. Applying the variational principle to the energy functional of Equation (19.1), it is possible to obtain a modified Hamiltonian for the Kohn–Sham equations,

$$H_{\mu v} = H^0_{\mu v} + \frac{1}{2} S_{\mu v} \sum_k (\gamma_{ik} + \gamma_{jk}) \Delta q_k, \quad \forall \mu \in i, \quad v \in j. \tag{19.4}$$

where $S_{\mu v}$ is an overlap matrix between atomic wave functions, γ_{ik} is a shorthand for the interatomic potential [7], computed from Equation (19.3), and Δq_i are the atomic charges. Since the atomic charges depend on the one-particle wave functions Ψ_k, a self-consistent procedure is required. The improvement of the self-consistent over the non-self-consistent procedure is considerable in determining structural and energetic properties of molecular systems [5].

The DFTB approach has been extended to the NEGF formalism (already introduced in this book), by generalizing the calculation of the atomic charges using the non-equilibrium (DFT) density matrix and computing the functional (19.3) using a three-dimensional Poisson solver [3,4].

Despite its mathematical complexity, the non-equilibrium Green's functions (NEGF) method for calculations of quantum transport has gained a great popularity in recent years, mostly because of the versatility and numerical stability of the method, in contrast to wave-function or transfer matrix approaches. The open boundary conditions can be elegantly included by exactly mapping the contacting leads into a finite and small part of the system [8]. Furthermore, the Green's function approach can be generalized for many-body quantum theory, allowing the inclusion of electron-phonon [9] as well as electron-electron interactions [10] within a unified and systematic formalism. Good references in many-body quantum theory can be found in [11,12] and an exhaustive review on NEGF can be found in reference [13].

19.1.2 Incoherent Electron–Phonon Scattering

The usual procedure to treat molecular vibrations is to expand the effective nuclear potential up to the harmonic term, and to decouple the Hamiltonian as a superposition of independent one-dimensional oscillators corresponding to the normal modes of vibrations. Each vibrational mode will be labeled by q.

The harmonic oscillators can be quantized following the usual prescriptions, by making use of the standard relationships between the position operator and the Bose field operator.

The calculation of the electron–phonon scattering requires an explicit evaluation of the electron–vibration couplings matrices, $\gamma_{\mu v}^q$, obtained by expanding the TB Hamiltonian to the first order in the atomic displacements. The couplings are then expressed in terms of derivatives of the Hamiltonian and the overlap matrices, therefore without fitting parameters [14,15], as

$$\gamma_{\mu v}^q = \sqrt{\frac{\hbar}{2\omega_q M_q}} \sum_\alpha \left[\frac{\partial H_{\mu v}}{\partial R_\alpha} - \sum_{\lambda,\sigma} \frac{\partial H_{\mu\lambda}}{\partial R_\alpha} S_{\lambda\sigma}^{-1} H_{\sigma v} - H_{\mu\lambda} S_{\lambda\sigma}^{-1} \frac{\partial H_{\sigma v}}{\partial R_\alpha} \right] \mathbf{e}_\alpha^q \tag{19.5}$$

where R_α are atomic displacements, M_q are the atomic masses, and $\mathbf{e}_\alpha^q$ are the vibrational modes eigenvectors. While crossing the system, electrons interact with the molecular ionic vibrations from which they can be inelastically scattered. The electron–phonon interaction is treated within perturbation theory of the non-equilibrium Green's function formalism, and the current through the junction is computed using the Meir–Wingreen formula [16]:

$$I = \frac{2e}{h} \int Tr[\Sigma^< G^>(E) - \Sigma^> G^<(E)] dE \tag{19.6}$$

where $\Sigma^{<(>)}$ represents the in-scattering of electrons (holes) through the *left* contact of the device, and $G^{<(>)}$ is the electron (hole) correlation function. The current expressed by Equation (19.6) contains both a coherent and an incoherent component, since the *lesser* and *greater* self-energies (SE) are given by the sum of three terms:

$$\Sigma^{<(>)} = \Sigma_L^{<(>)} + \Sigma_R^{<(>)} + \Sigma_{el-ph}^{<(>)} \tag{19.7}$$

The first two contributions come from the contacts, and the third term is related to the scattering processes within the molecule, caused by electron–phonon interactions. Such decoupling is valid within the non-crossing approximation (NCA), which we took care to fulfill by physically separating the molecular subunit—now free to move—from the contacts, where the atoms are assumed fixed.

The electron–phonon self-energy can be evaluated with diagrammatic techniques. The simplest is the first-order Born approximation (BA), which gives

$$\Sigma_{el-ph}^{<,>}(\omega) = i \sum_q 2\pi \int d\omega' \gamma^q G^{<,>}(\omega - \omega') \gamma^q D_{0,q}^{<,>}(\omega') \tag{19.8}$$

where $D_{0,q}^{<,>}(\omega)$ are the correlation functions related to the vibrational modes, assumed Einstein oscillators in thermal equilibrium with a bath:

$$D_{0,q}^{<,>}(\omega) = -2\pi i \left[(N_q + 1)\delta(\omega \pm \omega_q) + N_q \delta(\omega \mp \omega_q) \right]. \tag{19.9}$$

The thermally averaged phonon population, N_q, can be obtained to first approximation by the Bose–Einstein distribution function, using the environmental temperature. A more sophisticated calculation requires a self-consistent computation of a distribution such as the function of applied bias, as discussed in the next section.

19.1.3 Power Dissipation in Molecular Junctions

The electron–phonon interactions can be viewed as an exchange of particles with a *virtual* contact [17] that adsorbs electrons at a given energy and emits them at another energy, accounting for the energy loss in the phonon quanta. In perfect analogy to the real contacts, the virtual contact brakes the phase coherence of the wave function. The amount of power dissipated in the molecule, due to inelastic phonon emission, can be obtained by considering the virtual contact current, as discussed in [14,17]. The power dissipated is given by the net rate of energy transferred to the molecule and can be calculated using the virtual contact current as

$$W = \frac{2}{h} \int_{-\infty}^{+\infty} \omega Tr[\Sigma_{el-ph}^{<}(\omega)G^{>}(\omega) - \Sigma_{el-ph}^{>}(\omega)G^{<}(\omega)]d\omega \qquad (19.10)$$

which simply provides the average energy transfer occurring at the virtual contact.

The power dissipated can be used to compute the rate of phonon emission. This can be done by first observing that Equation (19.10) can be expressed as a sum over the individual vibrational modes, $W = \sum W_q$, allowing us to compute the power dissipated in each mode [14].

In order to compute the non-equilibrium phonon population under bias, we set up a phenomenological rate equation. The rate of phonon emissions can be defined as $R_q(N_q) = W_q/\hbar\omega_q$, representing the ratio between the energy emitted per unit time in the oscillator q divided by its energy. This is generally a function of bias and depends on the phonon population itself via. The rate R_q is actually the net rate of phonon emission, and also includes the absorption rate due to assisted tunneling and electron-hole pair production, something very important in metal contacts. The phonon population of the vibrational modes is then calculated via a rate equation, which includes the rate of emission and dissipation into the leads, as:

$$\frac{dN_q}{dt} = R_q(N_q) - J_q\left(N_q - N_q^0(T)\right), \qquad (19.11)$$

balancing the rate of quanta emitted with the rate of phonon decay to the bath, J_q, discussed in the following. This tends to restore the equilibrium Bose–Einstein distribution, characterized by the contact temperature, T. Under stationary conditions, the phonon distribution on the molecule is given by $N_q = N_q^0(T) + R_q(N_q)/J_q$.

The phonon dissipation rate is by no means a simple quantity to predict from initial calculations since a large number of different mechanisms participate in phonon relaxations. For this reason, J_q can be regarded in many cases as a fitting parameter of the model. Nonetheless, we compute an upper bound of this parameter by considering the elastic coupling of the molecular modes with the contact vibrations.

19.1.4 Power Dissipation in a Si-Styrene-Ag System

Let's now consider the hybrid system shown in Figure 19.1, which comprises a Si(100) substrate, reconstructed 2 × 1, with an adsorbed styrene molecule in a bridge position and an Ag metal contact, which models an STM tip. The position of the tip was chosen in order to obtain a relatively large direct coupling between the tip and the π orbitals of the styrene, which guarantees a sufficiently high conductance. Figure 19.1 shows a unit cell of the system considered, but periodic boundary conditions have been imposed.

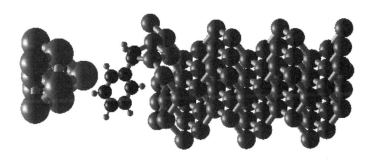

FIGURE 19.1 (see color insert following page 5-6) Representation of the system comprising a styrene molecule adsorbed on a Si(100) 2 × 1 substrate.

The Si substrate is heavily p-doped in order to make it conducting, and the Fermi Level is assumed at the valence band edge. While electrons cross the molecule, they interact with the molecular ionic vibrations, from which they can be inelastically scattered. The electron–phonon scattering within the leads is not considered.

In order to study the electron–vibron coupling, we first accurately relax the structure. (This is done under no applied bias.) Then we compute the vibrations of the molecule constraining the Si and Ag atoms. The electron–vibration couplings are obtained using Equation (19.5).

In order to calculate the rate of phonon dissipation into the contacts, the molecular modes are coupled with the bulk vibrations. Such a calculation is performed using a Green's function approach similar in concept to the electronic calculation. First, the Hessian (dynamical matrix) of the whole contact–molecule–contact device is computed, and then the system is partitioned into molecules and contacts. A Green's function is then built for the generalized vibration eigensystem:

$$\mathbf{G} = \left[\mathbf{M}\omega^2 - \mathbf{H} - \Sigma_L - \Sigma_R \right]^{-1} \tag{19.12}$$

where $\mathbf{H}$ is the Hessian matrix, $\mathbf{M}$ is the diagonal mass matrix, and $\Sigma_{L,R}$ are the self-energies obtained using open boundary conditions on the bulk side of the contact leads. This formulation is possible because the atomic coupling, expressed by the dynamical matrix, is restricted to few neighbors. The imaginary part of the self-energy is used to extract the vibron lifetime. This approach includes first-order one-phonon-to-one-phonon decay processes, but obviously neglects many other mechanisms that may take place when the direct decay is not allowed by energy conservation. This happens, for instance, to the high frequency modes, characteristic of molecular vibrations, generally lying well beyond the vibrational bandwidth of the bulk reservoirs which cannot decay other than via one-to-many phonon channels. Other decay mechanisms may involve coupling with the surrounding molecules and generally depend on the environment or via multistep processes and anharmonic couplings. The result of this calculation is shown in Figure 19.2, reporting the superposition of the phonon density of states of the coupled system (gray area) with the uncoupled molecular frequencies shown as vertical dashed lines. The cut-off frequency of the Si vibrations

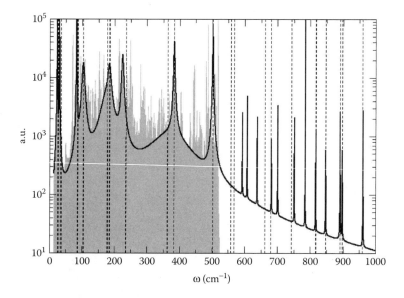

FIGURE 19.2 The vibrational density of states of the system Si-styrene-Ag. The gray area represents the Si d.o.s., vertical lines are the uncoupled molecular modes, and the solid line is the d.o.s. projected on the molecular modes showing their broadening due to coupling with the substrate.

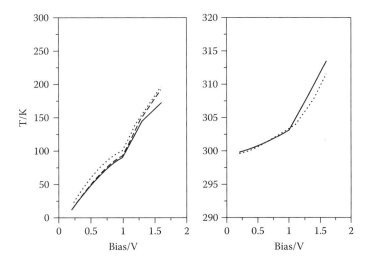

FIGURE 19.3 Molecular temperature as a function of applied bias for different types of approximations in the electron–phonon couplings (see text).

at 500 cm^{-1} clearly appears in this plot. The solid line represents the projected density of states obtained by approximating the self-energies, similar to a broadening and frequency shift. As discussed earlier, the calculation gives realistic results for those molecular modes lying within the Si phonon bands, where the broadening is sizable. The corresponding phonon decay rates are in the 10^{11} to 10^{12} Hz range. For those energy modes lying beyond the contact bandwidth, the broadening sharply decreases, leading to decay rates as slow as 10^2 Hz for the highest modes, which is rather unrealistic. Therefore, the lowest limit for the decay rate has been fixed to $J_{q,min} = 10^6$ Hz in all subsequent calculations.

We find that the out-of-equilibrium phonon population is strongly bias-dependent. Figure 19.3 reports the behavior of the average molecular temperature as a function of applied bias for two different contact temperatures (10 K and 300 K). The molecular temperature, T_{mol}, is computed by setting the energy balance expressed with

$$\sum_q \hbar\omega_q N_q = \sum_q \hbar\omega_q n_q (T_{mol}) \tag{19.13}$$

where $n_q(T_{mol})$ is the Bose–Einstein distribution for the temperature T_{mol}. As the bias increases, the temperature rises since more heat is dissipated in the molecular vibrations. The figure reports three different calculations obtained for different approximations in the treatment of the electron–phonon coupling. The dotted lines correspond to the simplest lowest-order calculation of the electron–phonon self-energy (not including the renormalization of the electron propagator), and the dashed line corresponds to the first order correction, commonly known as first order Born approximation (BA), which includes renormalizations of the propagator to the first order. The solid line corresponds to the self-consistent Born approximation (SCBA) where self-energy and propagators are computed self-consistently. Thus, we can conclude that, because of the small incoherent tunneling current, the inclusion of higher orders in the electron–phonon coupling brings only small corrections to the final result.

In Figure 19.3, it is possible to appreciate the change in slope at the applied bias of 1.0 V. The origin of this effect comes from a molecular resonant state entering the injection window at the applied bias of 0.95 V. This has the effect of strongly increasing the coherent and incoherent currents. On average, all vibrational modes are excited as the bias increases, although some of them are particularly favored (such as the lowest vibrational mode) at the frequency of 10.55 cm^{-1}, corresponding to a rigid oscillation of the whole benzene ring, away and toward the Ag tip. This mode leads to a strong variation of the

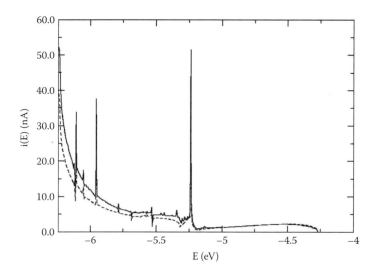

FIGURE 19.4   Coherent and total current density as a function of injection energy.

molecule–metal matrix elements and, therefore, to a large electron–phonon coupling. Once the first resonance is reached, most of the phonon emission occurs in the lowest energy mode, thus leading to a sharp increase in molecular temperature.

The computed current density within the bias window at 2.0 V is shown in Figure 19.4, where it is possible to see the Si valence band edge at −4.26 eV and the molecular resonance peak at −5.25 eV. The figure shows the total current and the coherent component. The incoherent component (not shown) is about two orders of magnitude smaller, except at resonance, where it becomes comparable. The total tunneling current computed at 1.0 V is 1.2 nA.

The molecular temperature sensibly depends on the tunneling current, which can be changed by varying the tip-molecule distance. When this distance becomes approximately 2.0 Å, the flowing current reaches 0.5 μA, leading to a very large temperature increase ($\approx$ 1000 K). Obviously, the molecule is not likely to withstand such temperatures, and probably a desorption is induced.

19.1.5   Applications to Octanethiols

In this application, we show the calculation of power dissipation in an octanedithiol molecule sandwiched between two Au electrodes. The geometry of the molecule within the junction, shown in Figure 19.5, is determined in two steps. First, an optimized geometry is obtained for octanethiol chemisorbed through the terminal sulfur bonded to a single Au(111) surface. Periodic boundary conditions are used; however, the chemisorbed molecules are sufficiently far apart to be considered isolated. The geometry for the full electrode–molecule–electrode system was then generated by symmetrizing about a point of inversion between the C4–C5 bond to have octanedithiol bound to two co-facial Au(111) surfaces. The calculated coherent current across this system agrees well with experimental data [14].

A thorough analysis of the inelastic electron tunneling spectroscopy (IETS) [15], shown in Figure 19.6, reveals that most of the electron–vibron coupling is associated to the longitudinal C-C stretching modes of the molecular backbone (peaks 9–15), in the range of energies between 140 to 180 meV. Significant signals are also found for the CH_2 wagging mode (peak 7) and the C-S stretch mode at 80 meV (peak 6). Relatively large peaks are also found in the low energy segment from 7 meV to 50 meV (peaks 1–5), including the Au-S stretch mode (25 meV). We conclude that these modes are the most involved in absorbing energy from the electrons.

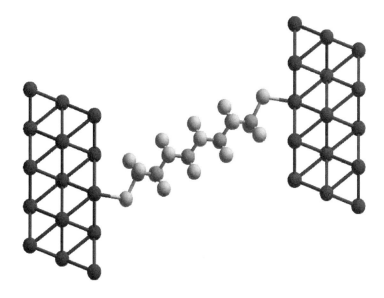

FIGURE 19.5 An octanedithiol molecule between Au contacts.

The phonon decay rates, J_q, are calculated as described in the previous section. The significant difference with respect to Si is that the Au phonon bandwidth extends to only about 30 meV. Furthermore, the modes lying within the contact phonon bands have a decay rate smaller than $\sim 10^{11}$ Hz, indicating a less efficient vibrational coupling with respect to Si. Also, in this case we set the lowest bound for the decay rates beyond the Au bands to an adjustable parameter, $J_{q,min}$. We find that the molecular temperature

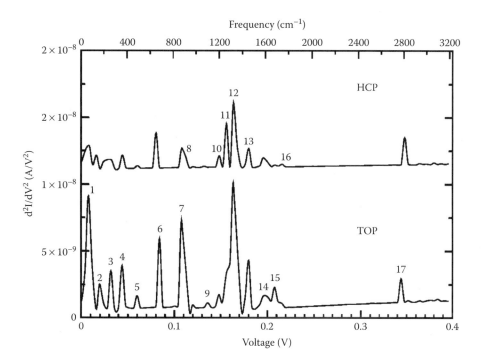

FIGURE 19.6 Simulated IETS spectra of otenedithiol on Au for two adsorption geometries, corresponding to the S atom found on "top" of a Au atom or in the hexagonal closed-packed (hcp) position.

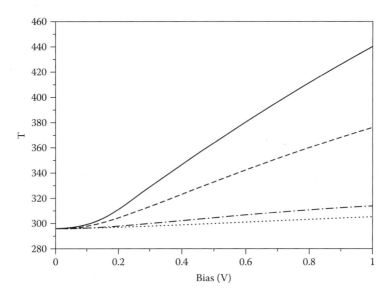

FIGURE 19.7   Molecular temperature as a function of bias for different values of $J_{q,min}$. The dotted line is $J_{q,min} = 1010$ Hz; the dot-dashed line is $J_{q,min} = 5 \cdot 109$ Hz; the dashed line is $J_{q,min} = 109$ Hz; and the solid line is $J_{q,min} = 5 \cdot 108$ Hz.

sensitively depends on this lowest bound. This is shown in Figure 19.7, reporting the calculation of the molecular temperature as a function of bias voltage, and assuming the equilibrium contact temperature of T = 296K. The different lines correspond to different choices of $J_{q,min}$, ranging from $5 \cdot 10^8$ to 10^{10} Hz. These parameters are chosen in order to have results comparable to recent experimental measurements of bias-dependent local temperature in octanedithiol between Au contacts [18].

19.2   Simulation of Organic Thin Film Devices

In recent years, research and development in the field of organic semiconductor materials has advanced tremendously, driven by high expectations to implement new and cheap applications. Ever since the discovery of conductivity and semiconductivity in polymers [19], as well as the discovery of metallic conductivity in poly-acetylene in 1977 [20], the electrical properties of various small molecules and polymers have been investigated. This has enabled the development of electronic devices such as organic light-emitting diodes (OLED), solar cells, and photodiodes based on organic semiconductor thin films. The first organic thin film transistor (OTFT) was reported in 1986 with a device made on an electrochemically grown polythiophene film [21]. The possibility of fabricating OTFTs with small conjugated molecules was shown in 1989 with sexithiophene [22].

In contrast to traditional inorganic materials, organic semiconductors can be deployed on various large area substrates in mass production, reducing clean room costs and other expensive manufacturing factors. Using flexible substrates and being lightweight, possible applications could include integration in clothing or on plastic surfaces. Organic semiconductors introduce a new degree of freedom since modifications can be done on the molecular level. It is possible to design materials with application-specific properties such as a desired absorption and emission spectra. This is a revolutionary approach compared to traditional limitations involving a small set of inorganic semiconductors such as silicon and gallium arsenide.

The field of organic optoelectronics, in particular, has developed rapidly. Here, the advantages of tuning materials via chemical synthesis have become quickly evident. Consequently, it is not surprising that this field is where the first commercial products (like OLEDs) have emerged. Today, OLED-based dot-matrix displays with inorganic backplanes are already on the market. Being thinner, brighter, lighter, cheaper, and consuming less power during manufacturing, all-organic displays, which combine OLEDs and OTFTs, are expected to follow in the near future.

The mobility obtained in organic thin film devices has improved by five orders of magnitude over the past 15 years, and has reached values comparable to amorphous silicon [23]. Additionally, fabrication technology, which turns out to be a key factor, has reached a high level of maturity. However, understanding of the principles governing device behavior is less developed. A general approach is the application of standard device models developed for inorganic semiconductors. Such an approach often falls short due to particular features of organic semiconductors and nontraditional device layouts. Our goal is to therefore set up a simulation framework for modeling and optimizing organic devices and circuits. The framework is based on a commercial drift-diffusion simulation tool, which we modify to account for the special nature of organic semiconductors.

19.2.1 The Drift–Diffusion Method

For our simulations, we have chosen the simulation package ISE TCAD. It is a multidimensional, electrothermal, mixed-mode device and circuit simulator that can handle geometries of up to three dimensions [24]. Generally, the simulator divides the device into a finite element grid and solves the drift–diffusion transport equations self-consistently coupled to the Poisson equation. Main input variables are device geometry, material specifications, and interface properties. Additionally, the tool ISE TCAD offers a versatile software connector that allows the introduction of user-defined physical models, such as mobility or recombination. This is of great importance when accurately describing organic devices.

A brief outline of the simulation method is presented in the following (see also [24]). As mentioned before, the program solves the Poisson equation for the electrical potential ϕ and the continuity equations for the hole and electron densities. The Poisson equation is given as

$$\nabla \varepsilon \cdot \nabla \phi = -q \cdot (p - n + N_{D+} - N_{A-}) \tag{19.14}$$

where ε is the electrical permittivity; q is the elementary charge; n and p are the electron and hole densities; N_{D+} is the number of ionized donors, and N_{A-} is the number of ionized acceptors. The continuity equations for the hole and electron densities are generally written as

$$\nabla \cdot \vec{J}_n = q \frac{\partial n}{\partial t} + q \cdot R_{net}$$

$$-\nabla \cdot \vec{J}_p = q \frac{\partial p}{\partial t} + q \cdot R_{net} \tag{19.15}$$

J_n is the electron current density, J_p is the hole current density and R_{net} is the net electron–hole recombination rate. In the nondegenerate limit, they are connected with the hole and electron quasi-Fermi potentials ϕ_n and ϕ_p by

$$n = n_i \cdot \exp \left\{ \frac{\phi - \phi_n}{U_T} \right\}$$

$$p = n_i \cdot \exp \left\{ \frac{\phi_p - \phi}{U_T} \right\}$$

$$n_i = \sqrt{N_V N_C} \cdot \exp \left\{ -\frac{E_g}{2q U_T} \right\} \tag{19.16}$$

where n_i is the intrinsic density, N_V and N_C are the effective density of the states, E_g is the bandgap, and $U_T = k_B T / q$ is the thermal voltage. In molecular materials, Equation (19.3) represents the Nernst equation describing oxidation and reduction for low concentrations [25]. In the solid state, the gap is the difference between reduction and oxidation potential. So, the molecular or monomer density is used instead of the effective densities of states. For most organic semiconductor materials, it is in the order of $N_C = N_V = 5 \cdot 10^{20}$ cm^{-3}.

To calculate potential, carrier, and current distributions, the simulator solves the equation system:

$$\nabla \varepsilon \varepsilon_0 \left(-\nabla \phi\right) = e\left(p - n + N_{D+} - N_{A-}\right)$$
$$\nabla e n \mu_n \left(-\nabla \phi_n\right) = +e\frac{\partial n}{\partial t} + e \cdot R_{net}$$
$$\nabla e p \mu_p \left(-\nabla \phi_p\right) = -e\frac{\partial p}{\partial t} - e \cdot R_{net} \tag{19.17}$$

The gradients of the two quasi-Fermi potentials are driving the hole and electron currents. They can be expressed as drift and diffusion currents since the mobilities are connected with the diffusion coefficients D_p and D_n by the (nondegenerate) Einstein relations $\mu_p = D_p/U_T$ and $\mu_n = D_n/U_T$, respectively, which are also valid in the case of hopping transport [26].

19.2.2 Organic Semiconductor Materials

Alan J. Heeger, Alan G. MacDiarmid, and Hideki Shirakawa were awarded the Nobel prize in Chemistry "for the discovery and development of conductive polymers [19]." In organic molecules with conjugated systems, the atoms are bonded covalently with alternating single and double bonds. An overlap of the π-orbitals formed by the sp^2-hybridization of the carbon atoms causes delocalization of electrons over a certain molecular region. This delocalization enables the transport of charges when an electrical field is applied. Within the conjugated systems, mobility is considered to be high, whereas the transition between conjugated systems poses a transport barrier. This barrier needs to be overcome by so-called *hopping processes*. The valence band of an inorganic semiconductor corresponds to the highest occupied molecular orbital (HOMO), while the lowest unoccupied molecular orbital (LUMO) represents the conduction band. The difference between the energy levels of the HOMO and LUMO is the bandgap energy. This forms an energetic structure very similar to inorganic semiconductors, but with localized states that are divided by defects and entropy. An energy diagram and typical values for pentacene are given in Figure 19.8.

Organic semiconductors are categorized into *polymers* and *small molecules*. Long molecules with repeating conjugated units form semiconducting polymers (e.g., polythiophene). The class of small molecule semiconductors consists of organic aromatic molecules with a low molecular weight such as anthracene, pentacene, and α–6T. Polymers are highly amorphous, whereas small molecules tend to form crystalline or polycrystalline films. Since there is little order and overlap between the chains, polymer conduction is dominated by interchain hopping transport. Due to a higher order in the crystalline films, a mix of hopping and band-like transport governs small molecule semiconductors.

Monte Carlo simulations (reported in [27]) and considerations in the *Variable Range Hopping* theory predict that the mobility in hopping systems exhibits a Poole–Frenkel field-dependence

$$\mu = \mu_0(T)e^{\sqrt{E/E_0}} \tag{19.18}$$

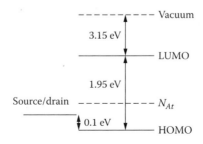

FIGURE 19.8 Energy levels of pentacene with gold source/drain contacts. The acceptor trap level is indicated by a dotted line.

where μ_0 is the low field mobility, and E_0 a critical field parameter equal to $3 \cdot 10^5$ V/cm. We have implemented the field-dependence in an ISE TCAD user-defined model.

The classification of organic semiconductors as *p*-type or *n*-type has a number of peculiar aspects. For inorganic semiconductors, a material is considered *p*-type (*n*-type) if the Fermi energy is shifted towards the valence (conduction) band by the addition of ionized dopants. Organic semiconductors are usually found to have low, unintentional *p*-doping in the order of 10^{15} cm^{-3}. This can be extracted experimentally from the slope of MOS CV-measurements. But, additionally, there are three other important points. First, organic semiconductors have a relatively wide band gap which makes the choice of contact materials crucial for carrier injection [28]. A contact with a work function closer to the LUMO acts as an electron injector, while a contact with a work function closer to the HOMO mainly injects holes. Depending on the barrier conditions, an organic semiconductor can therefore be operated in hole or electron mode. Second, electron and hole mobility in semiconductors can differ by several orders of magnitude. In practice, the term "*p*-type" is therefore also attributed to materials that have negligible electron conductivity, and vice versa. Third, organic semiconductors usually contain traps caused by a relatively large number of impurities and defects. If they mainly immobilize electrons (holes), the material also effectively exhibits *p*-type (*n*-type) conduction. Most organic semiconductor materials, such as P3HT and pentacene, exhibit *p*-type behavior.

The contact barriers at the electrodes are defined as absolute values of the difference between metal work function and the energy of the transport states. For electrons, this is the electron affinity X, and for holes it is the ionization energy $X + E_g$.

19.2.3 Simulation of Organic Thin Film Transistors

Advances in fabrication techniques have resulted in organic thin film transistors (OTFTs), which have become of increasing interest to the electronics industry. For optimization and circuit design, OTFT device behavior needs to be understood and described with appropriate models. In the following, we demonstrate that drift–diffusion is a viable approach to gaining further insight into device behavior.

Figure 19.8 shows a cross-section of a pentacene OTFT in a bottom contact configuration (right panel). The left panel of Figure 19.9 shows the corresponding finite element grid, which is used in the numerical calculation. A typical experimental transfer characteristic of an organic pentacene OTFT on a SiO$_2$ insulator with 20-μm channel length is shown in Figure 19.10. Pronounced in IV characteristics are particular to organic transistors. Being a clear fingerprint of time-dependent processes, they are most frequently encountered as transfer and output curve hysteresis. This effect has three possible causes: polarization, mobile ions, and semiconductor traps. In our structure, the first two are not expected to take effect. The insulating material is a very clean SiO$_2$, which should be very stable. The pentacene is highly purified by gradient sublimation so that no mobile ions should be present in the semiconducting layer. No gas molecules like water can introduce polarization since the measurements of Figure 19.10 are preformed in a vacuum. Therefore, the effect must be assigned to a property built into the semiconductor. The following simulations test an explanation based on trapping effects.

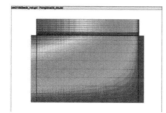 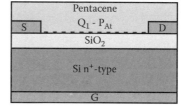

FIGURE 19.9 (see color insert) The finite element grid (left) and a cross-section of the organic thin film transistor in the bottom contact configuration (right). The dashed line indicates fixed charges and traps at the insulator semiconductor interface.

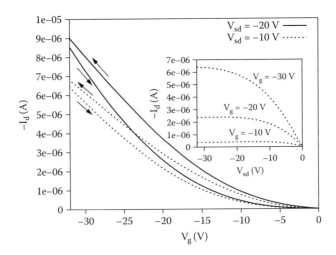

FIGURE 19.10 Transfer characteristic of a fabricated pentacene thin film transistor with reproducible hysteresis. The output characteristic is shown in the inset.

19.2.4   Traps in Organic Thin Films

Literature provides evidence that acceptor-type traps with energies close to the valence band populate the organic/oxide interface [29,30]. In the scope of our simulations, we assume a single trap level with a distinct energy, E_{At}, and an aerial density of N_{At}. As depicted in Figure 19.11, the acceptor type trap is negative if filled with an electron, and is neutral if occupied by a trapped hole [31]. In equilibrium, the trapped hole density p_{At} is related to the occupation probability of holes f_p as follows:

$$p_{At} = N_{At} \cdot f_p \tag{19.19}$$

The total charge density associated with the acceptor traps equals $Q_{At} = -q\,(N_{At} - p_{At})$. If equilibrium conditions are disturbed, trap recharging occurs. Balanced carrier flows to and from the trap level give the following rate equation [31]:

$$\frac{dp_{At}}{dt} = v_{th}^p \sigma_p N_{At} \left[p(1 - f_p) - \frac{p_1}{4} f_p \right]$$
$$p_1 = n_i \cdot \exp(-E_{At}/k_B T) \tag{19.20}$$

where v_{th}^p is the thermal hole velocity, σ_p is the hole capture cross-section, and n_i is the intrinsic electron density in the semiconductor. Equation (19.20) describes only carrier flows between the trap level and the

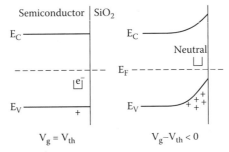

FIGURE 19.11   An acceptor-type interface trap for different gate biases.

valence band. There is an analog term for the transitions between the conduction band and the trap. Due to low electron concentrations in the conduction band, as well as a large energy gap, the contribution of this transition is only minor.

19.2.5 Influence of Interface Traps and Charges on Device Characteristics

The presence of traps and fixed interface charges can strongly affect the electrical characteristics of OTFTs. Performing very slow measurement sweeps of the transfer characteristic, the effect of acceptor traps is mainly visible below threshold. The threshold itself is defined by fixed interface charges Q_f [31,32]. "Very slow" implies that a desired voltage is applied and held until the system is in quasi-equilibrium before the current is measured. Performing a series of such measurements one obtains the quasi-static OTFT characteristics. From experiments, it is known that organic devices can require hold times of several seconds or minutes. Figure 19.11 (left side) shows that below threshold the traps are charged with electrons and attract holes to the interface. These holes lead to an increase of current and a higher inverse sub-threshold slope [29]. For gate voltages above threshold, captured holes neutralize the traps so there is little influence on the quasi-static current in this region (Figure 19.11; right side).

The analysis is more delicate if rapidly varying voltages are considered. Assuming a fast sweep from positive to negative gate voltages, the traps are initially filled with electrons. When the gate voltage crosses the threshold, equilibrium conditions require that they discharge according to rate Equation (19.20). However, if the sweep speed is sufficiently high, the traps do not have time to reach equilibrium before the applied gate voltage is even more negative. In this case, the negative traps add to the total interface charge, which is directly proportional to the change in threshold voltage:

$$\Delta V_T = -\frac{Q_{int}}{C_{ox}} = -\frac{(Q_f + Q_{At})}{C_{ox}} \tag{19.21}$$

The threshold appears to be shifted right for fast up-sweep conditions (see Figure 19.10).

19.2.6 Static and Transient IV Simulations

Figure 19.12 shows experimental transfer characteristics and the corresponding curves which result from a static device simulation. The hysteresis is clearly visible. The simulation strategy is to first fit the threshold voltage by adjusting fixed interface charges Q_f. Since the device has a negative threshold, we obtain

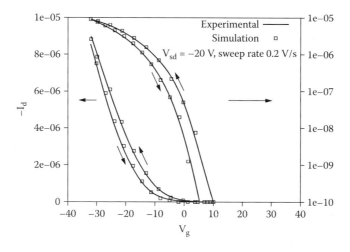

FIGURE 19.12 Separate simulations for up and down sweep with comparison to the measured values. The left pair of curves belongs to the linear axis, while the right pair shows the identical currents on the logarithmic scale.

TABLE 19.1   Parameters Extracted from Two Separate Static
Simulations for Up and Down Sweep

	μ_0 [cm²/Vs]	Q_f [cm^{-2}]	N_{At} [cm^{-2}]	E_{At} [eV]
Up	1.9	$0.5 \cdot 10^{11}$	$9 \cdot 10^{11}$	0.3
Down	2.3	$5.5 \cdot 10^{11}$	$9 \cdot 10^{11}$	0.3

densities of positive charges. The slope above threshold is fitted using the mobility. Subsequently, the sub-threshold characteristic is modeled. The simulation of the sub-threshold region is very sensitive with respect to the trap parameters and ideally suited for their extraction (see [29]). The values used for the simulated curves are presented in Table 19.1.

It is interesting to note that the static simulations lead to identical traps for both up and down sweep. This provides additional confidence that the hysteresis is caused by a single species of traps that is not fully recharged during the sweeps. In the transient simulations, we apply the down sweep parameters $Q_f = 5.5 \cdot 10^{11}$ cm^2 and $\mu_0 = 2.3$ cm²/Vs. The remaining free variable for the fit of the trap properties is according to (19.20) the product $v_{th}\sigma_p$. Figure 19.13 shows the resulting transient curves for different $v_{th}\sigma_p$ values. An optimal fit of the hysteresis can be obtained for a value of $v_{th}\sigma_p = 2.3 \cdot 10^{-19}$ cm³s^{-1}.

By inspecting the simulated curves, it is apparent that the influence of transient acceptor trap behavior has successfully been predicted. The device simulator has reproduced the hysteresis as well as the down bending of the transfer characteristics for the up sweep. For disordered hopping materials with a mobility of ~ 1 cm²/Vs, a thermal velocity v_{th} in the order of 10^3 cm/s is theoretically predicted [29]. A capture cross-section equal to 10^{-22} cm^2 can therefore be estimated for acceptor traps at the pentacene/SiO$_2$ interface. In silicon, acceptor trap cross-sections of $1.5 \cdot 10^{-17}$ cm^2 have been reported [33] at the Si/SiO$_2$ interface with a thermal velocity of $1.56 \cdot 10^7$ cm/s [31]. It can be argued that the combination of a small thermal velocity and small hole capture cross-sections results in the large time constants observed in the transient behavior of pentacene transistors.

The presented organic semiconductor simulations show that the drift–diffusion method is an appropriate way to test physical models. Perfect agreement of experiment and simulations may not always be realized. But this should be the motivation to fine-tune existing models and to add what is missing based on new theoretical or empirical findings.

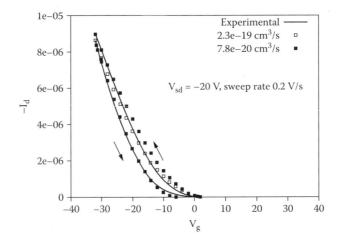

FIGURE 19.13   Transient simulation of a full V_g sweep cycle with two different $v_{th}\sigma_p$ products. The hysteresis is clearly visible. The values for the down sweep are identical in both simulations.

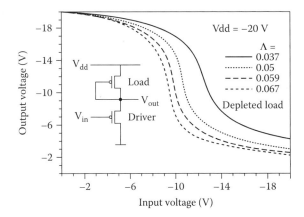

FIGURE 19.14 Influence of the geometry factor Λ on the voltage transfer characteristic of an organic inverter with depleted load.

19.2.7 Analysis and Optimization of Organic Logic Circuits

Drift–diffusion simulations are also a powerful tool in the design, optimization, and verification of organic circuits. Starting from the simulation of single transistor structures, logic circuits can be simulated under both static and transient conditions. The ability to analyze the effects of different transistor geometries on circuit behavior is highly beneficial in setting up robust organic circuits.

19.2.8 Inverter Circuits

The limitation to p-type materials is a constraint in the design of organic electronics. Although research aims at n-type OTFTs [34], materials and fabrication are still immature. Therefore, it is of great interest to design exclusively p-based circuits for digital and analog purposes. The inset in Figure 19.14 shows a possible inverter layout with depleted load. A key advantage of TFT-based designs is the ability to tune the circuit behavior by changing the $\lambda = W/L$ ratios of the transistors. The figure shows the strong influence of the geometry factor

$$\Lambda = \frac{\lambda_{driver}}{\lambda_{load}} = \frac{W_{driver} \cdot L_{load}}{L_{driver} \cdot W_{load}} \tag{19.22}$$

on the voltage transfer characteristic of the inverter. Optimizing the symmetry of the presented transfer voltage curve with respect to the point $(-10 \text{ V}, -10 \text{ V})$ results in a ratio of $L = 0.059$ (Figure 9.14).

To determine the dynamic switching behavior, the inverter circuit is simulated while being connected to a load capacity of $C_L = 20$ fF. The response to a square wave signal with a 40-μs period is plotted in Figure 19.15. The output voltage shows distinct asymmetry for rising and falling edges. The rise time going from 0 to -20 V is approximately 5 μs, while the fall time is close to 0.5 μs. Additionally, the supply current of the circuit is plotted to visualize the high power consumption of the noncomplementary logic circuit. By changing circuit parameters, the simulations can be applied to these properties

At present, research in the field of organic electronics is primarily devoted to the optimization of device behavior, materials, and fabrication processes. Attention gradually shifts to the realization of circuit applications. It is realistic that some of the problems of organic electronics such as strong parameter variations and hysteresis should not solely be addressed on the device level. Future efforts should be directed toward establishing robust circuit concepts, which can tolerate these particularities. The presented TCAD framework for organic circuits will likely be the tool of choice for evaluating new circuit designs.

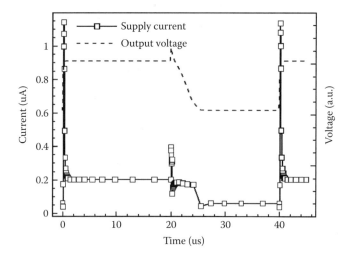

FIGURE 19.15   The simulated transient response of an inverter with depleted load ($\Lambda = 0.059$) and a load capacity CL = 20 fF at the output. A square wave signal with a $T = 40\mu s$ is applied to the inverter input. The dashed line shows the output current, whereas the solid line with symbols shows the inverter supply current.

19.2.9   Application to Organic Ring Oscillators

To show that the ISE TCAD scheme can also handle complex circuits, the simulated signal of a three-stage ring oscillator running at 105 kHz is presented in Figure 19.16.

For logic applications, it is not sufficient to simply build logic gates. It is required that they amplify the input signal so the signal is not lost after a number of stages. Furthermore, the output of one logic gate must be able to drive the input of another. This also requires that input and output voltage ranges match.

Ring oscillators consist of an odd number of inverter stages. The output of one stage is fed into the input of the following. The last inverter connects to the input of the first one closing the ring. The odd number of inverters hinders the system to obtain a stable operation point. The previously mentioned requirements must strictly be obeyed by the inverter stages if oscillating is taking place. Since this is a

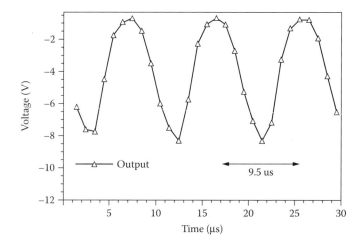

FIGURE 19.16   Simulated output voltage of a three-stage organic oscillator in steady-state operation ($V_{dd} = -20\,\text{V}$). The signal has a period of operates = 9.5 μs, which corresponds to $f = 105$ kHz.

proof-of-concept for the logic capability of organic electronics, much attention has been devoted to the experimental realization of organic ring oscillators. Oscillation frequencies of up to 220 kHz have been reported [35].

19.3 Acknowledgments

We thankfully acknowledge funding for parts of this project by the Deutsche Forschungsgemeinschaft (DFG) in the scope of the Priority Program OFET and quantum transport at the molecular scale.

References

[1] Cai, L. et al., *Nano Lett.*, 5, 2365, 2005.

[2] Frauenheim, T. et al., *J. Phys.: Condensed Matter*, 14, 3015, 2002.

[3] Pecchia, A. and Di Carlo, A., *Rep. Prog. Phys.*, 67, 1497, 2004.

[4] Di Carlo, A. and Pecchia, A., *Introducing Molecular Electronics*, Springer, 2005, Cuniberti, G. et al., Eds.; Pecchia, A. et al., *Molecular Electronics: Analysis, Design, and Simulation*, Elsevier, 2006.

[5] Elstner, M. et al., *Phys. Rev. B*, 58, 7260, 1998.

[6] Maragakis, P. et al., *Phys. Rev. B*, 66, 241 104, 2002.

[7] Porezag, D., *Phys. Rev. B*, 52, 14, 963, 1995.

[8] Caroli, C. et al., *J. Phys. C: Solid State Phys.*, 5, 21, 1972.

[9] Galperin, M. et al., *Nano Lett.*, 4, 1605, 2004.

[10] Faleev, S.V. and Stockman, M.I., *Phys. Rev. B*, 66, 085 318, 2002.

[11] Fetter, A.L. and Walecka, J.D., *Quantum Theory of Many Particle Systems*, Dover Publications, 1971.

[12] Mahan. D.M., *Many Particle Physics*, Plenum Press, 1981.

[13] Haung, H. and Jauho, A.P., *Quantum Kinetics in Transport and Optics of Semiconductors*, 123, Springer Series in Sol. State Sci., 1993.

[14] Pecchia, A. et al., *Nano Lett.*, 4, 2109, 2004.

[15] Solomon, G.C. et al., *J. Chem. Phys.*, 124, 094704, 2006.

[16] Meir, Y. and Wingreen, N.S., *Phys. Rev. Lett.*, 68, 2512, 1992.

[17] Datta, S., *Electronic Transport in Mesoscopic System*, Cambridge University Press, 1995.

[18] Huang, Z. et al., *Nano Lett.*, 6, 1240, 2006.

[19] Shirakawa, H. et al., Synthesis of electrically conducting organic polymers: halogen derivatives of polyacetylene, *Chem. Commun.*, 16:578–580, 1977.

[20] Chiang, C.K. et al., Electrical conductivity in doped polyacetylene, *Phys. Rev. Lett.*, 39, 1098–1011, 1977.

[21] Kurokawa, T. and Ando, Macromolecular electronic device: Field-effect transistor with polythiophene thin film, *Appl. Phys. Lett.*, 49:1210, 1986.

[22] Horowitz, F. et al., A field-effect transistor based on conjugated alpha-sexithienyl, *Solid State Commun.*, 72, 381, 1989.

[23] Shaw, J.M. and Seidler, P.F., Organic electronics: Introduction, *IBM J. Res. & Dev.*, 45, 3, 2001.

[24] ISE-TCAD, *Manual Dessis*, Integrated Systems Engineering, AG, Zurich, Switzerland, 1995–1999.

[25] Paasch, G. et al., Simulation of organic light emitting diodes: influence of charges localized near the electrodes, *Synth. Met.*, 139, 425–432, 2003.

[26] Nesterov, A. et al., Simulation study of the influence of polymer modified anodes on organic led performance, *Synth. Met.*, 130, 165–175, 2002.

[27] Bolognesi, A. et al., Large drift-diffusion and monte carlo modeling of organic semiconductor devices, *Synth. Met.*, 138, 2003.

[28] Schilinsky, P., *Loss Analysis of the Power Conversion Efficiency of Organic Bulk Heterojunction Solar Cells*, Ph.D. thesis, Carl von Ossietzky Universität Oldenburg, 2005.

[29] Scheinert, S. et al., Subthreshold characteristics of field effect transistors based on poly (3-dodecylthiophene) and an organic insulator, *J. Appl. Phys.*, 92, 2002.

[30] Kang, J. et al., Shallow trap states in pentacene thin films from molecular sliding, *Appl. Phys. Lett.*, 86, 2005.

[31] Sze, S., *Physics of Semiconductor Devices*, John Wiley & Sons, New York, 1982.

[32] Bolognesi, A. et al., Effects of grain boundaries, field dependent mobility and interface trap states on the electrical characteristics of pentacene thin film transistors, *IEEE Trans. Electron Dev.*, 51, 2004.

[33] Masson, P. et al., Frequency characterization and modeling of interface traps in HfSixOy/HfO2 gate dielectric stack from a capacitance point-of-view, *Appl. Phys. Lett.*, 81, 2002.

[34] Ahles, M. et al., N-type organic field-effect transistor based on interface-doped pentacene, *Appl. Phys. Lett.*, 85, 19, 2004.

[35] Clemens, W. et al., From polymer transistors toward printed electronics, *J. Mater. Res.*, 19, 7, 2004.

20

Theory of Current Rectification, Switching, and a Role of Defects in Molecular Electronic Devices

A.M. Bratkovsky

Abstract

Devices for nano- and molecular size electronics may allow for an efficient current rectification and switching. A few molecular scale devices are reviewed here on the basis of first-principles and model approaches. Current rectification by molecular quantum dots can produce the rectification ratio $\lesssim 100$. Current switching due to conformational changes in the molecules is slow, on the order of a few kHz. Fast switching (~ 1 THz) may be achieved, at least in principle, in a molecular quantum dot with strong coupling with vibrational excitations. Defects in molecular films result in spurious peaks in conductance, apparent negative differential resistance, and may also lead to unusual temperature and bias dependence of current. The observed switching in most cases is extrinsic, caused by changes in molecule-electrode geometry, molecule reconfiguration, metallic filament formation through the film, etc.

20.1   Introduction

Current interest in molecular electronics is largely driven by expectations that molecules can be used as nanoelectronics components able to complement/replace standard silicon CMOS technology [1, 2] on the way down to $\lesssim$10nm circuit components. The first speculations about molecular electronic devices (diodes, rectifiers) were apparently made in mid-1970s [3]. That original suggestion of a molecular rectifier has generated a large interest in the field and a flurry of suggestions of various molecular electronics components, especially coupled with premature estimates that silicon-based technology cannot scale to below 1μm feature size. The Aviram-Ratner's Donor-insulator-Acceptor construct TTF$-\sigma-$TCNQ ($D^+ - \sigma - A^-$, see details below), where carriers were supposed to tunnel asymmetrically in two directions through insulating saturated molecular $\sigma -$ 'bridge', has never materialized, in spite of extensive experimental effort over a few decades [4]. End result in some cases appears to be a slightly electrically anisotropic *insulator*, rather than a diode, unsuitable as a replacement for silicon devices. This comes about because in order to assemble a reasonable quality monolayer of these molecules in Langmuir-Blodgett trough (avoiding defects that will short the device after electrode deposition) one needs to attach a long 'tail' molecule C18 [$\equiv (CH_2)_{18}$] that can produce enough of a Van-der-Waals force to keep molecules together, but C18 is a wide-band insulator with a bandgap $E_g \approx 9 - 10$eV. The outcome of these studies may have been anticipated, but if one were able to assemble the Aviram-Ratner molecules without the tail, they could not rectify anyway. Indeed, a recent ab-initio study [5] of $D^+\sigma A^-$ prospective molecule showed no appreciable asymmetry of its I-V curve. The molecule was envisaged by Ellenbogen and Love [6] as a 4-phenyl ring Tour wire [1] with dimethylene insulating bridge in the middle directly connected to Au electrodes via thiol groups. Donor-acceptor asymmetry was produced by side NH_2^+ and NO_2^- moieties, which is a frequent motif in molecular devices using the Tour wires [1]. The reason of poor rectification is simple: the "insulating" bridge is too short, it is a transparent piece of one-dimensional insulator, whereas the applied field is three dimensional and it cannot be screened efficiently with an appreciable voltage drop on the insulating group in this geometry. Although there is only 0.7eV energy separation between levels on the D and A groups, one needs about 4eV bias to align them and get a relatively small current because total resonant transparency is practically impossible to achieve. Remember, that the model calculation implied an ideal coupling to electrodes, which is impossible in reality and which is known to dramatically change the current through the molecule (see below). We shall discuss below some possible alternatives to this approach.

It is worth noting that studies of energy and electron transport in molecular crystals [7] started already in early 1960s. It was established in mid-1960s in what circumstances charge transport in biological molecules involves electron tunneling [8]. It was realized in mid-1970s that since the organic molecules are 'soft', energy transport along linear biological molecules, proteins, etc. may proceed by low energy nonlinear collective excitations, like Davydov solitons [9] (see review [10]).

To take over from current silicon CMOS technology, the molecular electronics should provide smaller, more reliable, functional components that can be produced and assembled concurrently and are compatible with CMOS for integration. The small size of units that molecules may hopefully provide is quite obvious. However, meeting other requirements seems to be a very long shot. To beat alternative technologies for e.g. dense (and cheap) memories, one should aim at a few TB/in^2 ($>10^{12} - 10^{13}$ bit/cm^2), which corresponds to linear bit (footprint) sizes of $3 - 10$ nanometers, and an operation lifetime of ~ 10 years. The latter requirement is very difficult to meet with organic molecules that tend to oxidize and decompose, especially under conditions of very high applied electric field (given the operational bias voltage of ~ 1V for molecules integrated with CMOS and their small sizes on the order of a few nanometers). In terms of areal density, one should compare this with rapidly developing technologies like ferroelectric random access (FERAM) [11] or phase-change memories (PCM) [12]. The current smallest commercial nano-ferroelectrics are about 400×400 nm^2 and 20–150 nm thick [13], and the 128×128 arrays of switching ferroelectric pixels bits have been already demonstrated with a bit size $\lesssim$50nm (with density $\sim$TB/in^2) [14]. The phase-change memories based on chalcolgenides GeSbTe (GST) seem to scale even better than the ferroelectrics. As we see, the mainstream technology for random-access memory approaches molecular size very rapidly. For instance, the so-called "nanopore" molecular devices by M.A. Reed et al [15] have comparable sizes and yet to demonstrate a repeatable behavior (see below).

In terms of parallel fabrication of molecular devices, one is looking at *self-assembly* techniques (see, e.g. [16, 17], and references therein). Frequently, the Langmuir-Blodgett technique is used for self-assembly of molecules on water, where molecules are prepared to have hydrophilic "head" and hydrophobic "tail" to make the assembly possible, see e.g. Refs. [18, 19]. The allowances for a corresponding assembly, especially of hybrid structures (molecules integrated on silicon CMOS), are on the order of a *fraction* of an Angstrom, so actually a *picotechnology* is required [2]. Since it is problematic to reach such a precision any time soon, the all-in-one molecule approach was advocated, meaning that a fully functional computing unit should be synthesized as a single supermolecular unit [2]. The hope is that perhaps directed self-assembly will help to accomplish building such a unit, but self-assembly on a large scale is impossible without defects [16, 17], since the entropic factors work against it. Above some small defect concentration ("percolation") threshold the mapping of even a simple algorithm on such a self-assembled network becomes impossible [20].

There is also a big question about electron transport in such a device consisting of large organic molecules. Even in high-quality pentacene crystals, perhaps the best materials for thin film transistors, the mobility is a mere 1-2 cm^2/V·s (see e.g. [21]), as a result of carrier trapping by interaction with a lattice. The situation with carrier transport through long molecules ($>2 - 3nm$) is, of course, substantially different from the transport through short rigid molecules that have been envisaged as possible electronics components. Indeed, in *short* molecules the dominant mode of electron transport would be resonant tunneling through *electrically active* molecular orbital(s)[22], which, depending on the workfunction of the electrode, affinity of the molecule, and symmetry of coupling between molecule and electrode may be one of the lowest unoccupied molecular orbitals (LUMO) or highest occupied molecular orbitals (HOMO) [23, 24]. Indeed, it is well known that in longer wires containing more than about 40 atomic sites, the tunneling time is comparable to or larger than the characteristic phonon times, so that the polaron (and/or bipolaron) can be formed inside the molecular wire [25]. There is a wide range of molecular bulk conductors with (bi)polaronic carriers. The formation of polarons (and charged solitons) in polyacetylene (PA) was discussed a long time ago theoretically in Refs. [26] and formation of bipolarons (bound states of two polarons) in Ref. [27]. Polarons in PA were detected optically in Ref. [28] and since then studied in great detail. There is an exceeding amount of evidence of the polaron and bipolaron formation in conjugated polymers such as polyphenylene, polypyrrole, polythiophene, polyphenylene sulfide [29], Cs-doped biphenyl [30], n-doped bithiophene [31], polyphenylenevinylene(PPV)-based light emitting diodes [32], and other molecular systems. Given the above problems with electron transport through large molecules one should look at the short- to medium-size molecules first.

The latest wave of interest in molecular electronics is mostly related to recent studies of carrier transport in synthesized linear conjugated molecular wires (Tour wires [1]) with apparent non-linear I-V characteristics [negative differential resistance (NDR)] and "memory" effects [15, 33, 34], various molecules with a mobile *microcycle* that is able to move back and forth between metastable conformations in solution (molecular shuttles) [35] and demonstrate some sort of "switching" between relatively stable resistive states when sandwiched between electrodes in a solid state device [36] (see also [37]). There are also various photochromic molecules that may change conformation ("switch") upon absorption of light [38], which may be of interest to some photonics applications but not for the general purpose electronics. One of the most serious problems with using this kind of molecules is *power dissipation*. Indeed, the studied organic molecules are, as a rule, very resistive (in the range of $\sim 1M\Omega - 1G\Omega$, or more). Since usually the switching bias voltage exceeds 0.5V the dissipated power density would be in excess of 10 kW/cm^2, which is orders of magnitude higher than the presently manageable level. One can drop the density of switching devices, but this would undermine a main advantage of using molecular size elements. This is a common problem that CMOS faces too, but organic molecules do not seem to offer a tangible advantage yet. There are other outstanding problems, like understanding an actual switching mechanism, which seems to be rather molecule-independent [37], stability, scaling, etc. It is not likely, therefore, that molecules will displace silicon technology, or become a large part of a hybrid technology in a foreseeable future.

First major moletronic applications would most likely come in the area of chemical and biological sensors. One of the current solutions in this area is to use the functionalized nanowires. When a target analyte molecule attaches from the environment to such a nanowire, it changes the electrostatic potential "seen" by the carriers in the nanowire. Since the conductance of the nanowire device is small, even one chemisorbed molecule could make a detectable change of a conductance [39]. Semiconducting nanowires can be grown from seed metal nanoparticles [16], or it can be carbon nanotubes (CNT), which are studied extensively due to their relatively simple structure and some unique properties like very high conductance [40].

In this paper we shall address various generic problems related to electron transport through molecular devices, and describe some specific molecular systems that may be interesting for applications as rectifiers and switches, and some pertaining physical problems. We shall first consider systems where an elastic tunneling is dominant, and interaction with vibrational excitations on the molecules only renormalizes some parameters describing tunneling. We shall also describe a situation where the coupling of carriers to molecular vibrons is strong. In this case the tunneling is substantially inelastic and, moreover, it may result in current hysteresis when the electron-vibron interaction is so strong that it overcomes Coulomb repulsion of carriers on a central narrow-band/conjugated unit of the molecule separated from electrodes by wide band gap saturated molecular groups like $(CH_2)_n$, which we shall call a molecular quantum dot (molQD). Another very important problem is to understand the nature and the role of imperfections in organic thin films. It is addressed in the last section of the paper.

20.2 Role of Molecule-Electrode Contact: Extrinsic Molecular Switching due to Molecule Tilting

We have predicted some while ago that there should be a strong dependence of the current through conjugated molecules (like the Tour wires [1]) on the geometry of molecule-electrode contact [23,24]. The apparent "telegraph" switching observed in STM single-molecule probes of the three-ring Tour molecules, inserted into a SAM of non-conducting shorter alkanes, has been attributed to this effect [34]. The theory predicts very strong dependence of the current through the molecule on the tilting angle between a backbone of a molecule and a normal to the electrode surface. Other explanations, like rotation of the middle ring, charging of the molecule, or effects of the moieties on the middle ring, do not hold. In particular, switching of the molecules *without* any NO_2 or NH_2 moieties has been practically the same as with them.

The simple argument in favor of the "tilting" mechanism of the conductance lies in a large anisotropy of the molecule-electrode coupling through π-conjugated molecular orbitals (MOs). In general, we expect the overlap and the full conductance to be maximal when the lobes of the p-orbital of the end atom at the molecule are oriented perpendicular to the surface, and smaller otherwise, as dictated by the symmetry. The overlap integrals of a p-orbital with orbitals of other types differ by a factor about 3 to 4 for the two orientations. Since the conductance is proportional to the square of the matrix element, which contains a product of two metal-molecule hopping integrals, the total conductance variation with overall geometry may therefore reach two orders of magnitude, and in special cases be even larger.

In order to illustrate the geometric effect on current we have considered a simple two-site model with p−orbitals on both sites, coupled to electrodes with s−orbitals [24]. For non-zero bias the transmission probability has the resonant form (20.5) with line widths for hopping to the left (right) lead Γ_L (Γ_R). The current has the approximate form (with $\Gamma = \Gamma_L + \Gamma_R$)

$$I \approx \begin{cases} \dfrac{q^2}{h} \dfrac{\Gamma^2}{t_\pi^2} V \propto \sin^4 \theta, & qV \ll E_{\text{LUMO}} - E_{\text{HOMO}}, \\[2ex] \dfrac{8\pi q}{h} \dfrac{\Gamma_L \Gamma_R}{\Gamma_L + \Gamma_R} \propto \sin^2 \theta, & qV > E_{\text{LUMO}} - E_{\text{HOMO}}, \end{cases} \quad (20.1)$$

where θ is the tilting angle [23,24], t_π the hopping integral for π−electrons, q the elementary charge, Figure 20.1. The tilting angle has a large effect on the I-V curves of benzene-dithiolate (BDT) molecules, especially when the molecule is anchored to the Au electrode in the top position, Figure 20.2. By changing

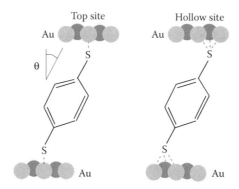

FIGURE 20.1 Schematic representation of the benzene-dithiolate molecule on top and hollow sites. End sulfur atoms are bonded to one and three surface gold atoms, respectively, θ is the tilting angle.

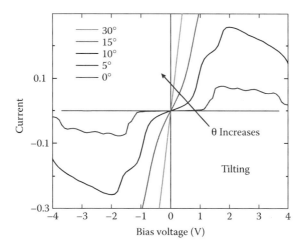

FIGURE 20.2 Effect of tilting on I-V curve of the BDT molecule, Figure 20.1. Current is in units of $I_0 = 77.5\mu A$, θ is the tilting angle.

θ from 5° to just 15°, one drives the I-V characteristic from the one with a gap of about 2V to the ohmic one with a large relative change of conductance. Even changing θ from 10° to 15° changes the conductance by about an order of magnitude. The I-V curve for the hollow site remains ohmic for tilting angles up to 75° with moderate changes of conductance. Therefore, if the molecule in measurements snaps from the top to the hollow position and back, it will lead to an apparent switching [34]. It has recently been realized that the geometry of a contact strongly affects coherent spin transfer between molecularly bridged quantum dots [41]. It is worth noting that another frequently observed *extrinsic* mechanism of "switching" in organic layers is due to electrode material diffusing into the layer and forming metallic "filaments" (see below).

20.3 Molecular Quantum Dot Rectifiers

Aviram and Ratner speculated about a rectifying molecule containing donor (D) and acceptor (A) groups separated by a saturated σ-bridge (insulator) group, where the (inelastic) electron transfer will be more favorable from A to D [3]. The molecular rectifiers actually synthesized, $C_{16}H_{33} - \gamma Q3CNQ$, were of somewhat different $D - \pi - A$ type, i.e. the "bridge" group was conjugated [4]. Although the molecule did show rectification (with considerable hysteresis), it performed rather like an anisotropic insulator with

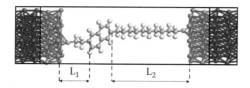

FIGURE 20.3 Stick figure representing the naphthalene conjugated central unit separated from the left(right) electrode by saturated (wide band gap) alkane groups with length $L_{1(2)}$.

tiny currents on the order of 10^{-17} A/molecule, because of the large alkane "tail" needed for LB assembly. It was recently realized that in this molecule the resonance does not come from the alignment of the HOMO and LUMO, since they cannot be decoupled through the conjugated π−bridge, but rather due to an asymmetric voltage drop across the molecule [42]. Rectifying behavior in other classes of molecules is likely due to asymmetric contact with the electrodes [43, 44], or an asymmetry of the molecule itself [45]. To make rectifiers, one should avoid using molecules with long insulating groups, and we have suggested using relatively short molecules with "anchor" end groups for their self-assembly on a metallic electrodes, with a phenyl ring as a central conjugated part [46]. This idea has been tested in Ref. [47] with a phenyl and thiophene rings attached to a $(CH_2)_{15}$ tail by a CO group. The observed rectification ratio was $\lesssim 10$, with some samples showing the ratio of about 37.

We have recently studied a more promising rectifier like $-S-(CH_2)_2$-Naph$-(CH_2)_{10}-S-$ with a theoretical rectification $\lesssim 100$[48], Figure 20.3. This system has been synthesized and studied experimentally [49]. To obtain an accurate description of transport in this case, we employ an ab-initio non-equilibrium Green's function method [50]. The present calculation takes into account only elastic tunneling processes. Inelastic processes may substantially modify the results in the case of strong interaction of the electrons with molecular vibrations, see Ref. [51] and below. There are indications in the literature that the carrier might be trapped in a polaron state in saturated molecules somewhat longer than those we consider in the present paper [52]. One of the barriers in the present rectifiers is short and relatively transparent, so there will be no appreciable Coulomb blockade effects. The structure of the present molecular rectifier is shown in Figure 20.3. The molecule consists of a central conjugated part (naphthalene) isolated from the electrodes by two insulating aliphatic chains $(CH_2)_n$ with lengths $L_1(L_2)$ for the left (right) chain.

The principle of molecular rectification by a molecular quantum dot is illustrated in Figure 20.4, where the electrically "active" molecular orbital, localized on the middle conjugated part, is the LUMO, which lies

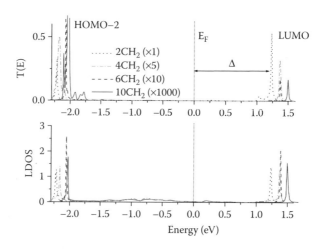

FIGURE 20.4 Transmission coefficient versus energy E for rectifiers $-S-(CH_2)_2-C_{10}H_6-(CH_2)_n-S-, n = 2, 4,$ $6, 10$. Δ indicates the distance of the closest MO to the electrode Fermi energy ($E_F = 0$).

at an energy Δ above the electrode Fermi level at zero bias voltage. The position of the LUMO is determined by the work function of the metal $q\phi$ and the affinity of the molecule $q\chi$, $\Delta = \Delta_{\text{LUMO}} = q(\phi - \chi)$. The position of the HOMO is given by $\Delta_{\text{HOMO}} = \Delta_{\text{LUMO}} - E_g$, where E_g is the HOMO-LUMO gap. If this orbital is considerably closer to the electrode Fermi level E_F, then it will be brought into resonance with E_F prior to other orbitals. It is easy to estimate the forward and reverse bias voltages, assuming that the voltage mainly drops on the insulating parts of the molecule,

$$V_F = \frac{\Delta}{q}(1 + \xi), \qquad V_R = \frac{\Delta}{q}\left(1 + \frac{1}{\xi}\right), \tag{20.2}$$

$$V_F/V_R = \xi \equiv L_1/L_2, \tag{20.3}$$

where q is the elementary charge. A significant difference between forward and reverse currents should be observed in the voltage range $V_F < |V| < V_R$. The current is obtained from the Landauer formula

$$I = \frac{2q^2}{h} \int dE[f(E) - f(E + qV)]g(E, V). \tag{20.4}$$

We can make qualitative estimates in the resonant tunneling model, with the conductance $g(E, V) \equiv T(E, V)/q$, where $T(E, V)$ is the transmission given by the Breit-Wigner formula

$$T(E, V) = \frac{\Gamma_L \Gamma_R}{(E - E_{MO})^2 + (\Gamma_L + \Gamma_R)^2/4}, \tag{20.5}$$

E_{MO} is the energy of the molecular orbital. The width $\Gamma_{L(R)} \sim t^2/D = \Gamma_0 e^{-2\kappa L_{1(2)}}$, where t is the overlap integral between the MO and the electrode, D is the electron band width in the electrodes, κ the inverse decay length of the resonant MO into the barrier. The current above the resonant threshold is

$$I \approx \frac{2q}{\hbar}\Gamma_0 e^{-2\kappa L_2}. \tag{20.6}$$

We see that increasing the spatial asymmetry of the molecule (L_2/L_1) changes the operating voltage range linearly, but it also brings about an *exponential* decrease in current [46]. This severely limits the ability to optimize the rectification ratio while simultaneously keeping the resistance at a reasonable value. To calculate the I-V curves, we use an ab-initio approach that combines the Keldysh non-equilibrium Green's function (NEGF) with pseudopotential-based real space density functional theory (DFT) [50]. The main advantages of our approach are (i) a proper treatment of the open boundary condition; (ii) a fully atomistic treatment of the electrodes and (iii) a self-consistent calculation of the non-equilibrium charge density using NEGF. The transport Green's function is found from the Dyson equation

$$(G^R)^{-1} = \left(G_0^R\right)^{-1} - V, \tag{20.7}$$

where the unperturbed retarded Green's function is defined in operator form as $(G_0^R)^{-1} = (E + i0)\hat{S} - \hat{H}$, H is the Hamiltonian matrix for the scatterer (molecule plus screening part of the electrodes). S is the *overlap* matrix, $S_{i,j} = \langle \chi_i | \chi_j \rangle$ for non-orthogonal basis set orbitals χ_i, and the coupling of the scatterer to the leads is given by the Hamiltonian matrix $V = \text{diag}[\Sigma_{l,l}, 0, \Sigma_{r,r}]$, where l (r) stands for left (right) electrode. The self-energy part $\Sigma^<$, which is used to construct the non-equilibrium electron density in the scattering region, is found from $\Sigma^< = -2i\text{Im}[f(E)\Sigma_{l,l} + f(E + qV)\Sigma_{r,r}]$, where $\Sigma_{l,l(r,r)}$ is the self-energy of the left (right) electrode, calculated for the semi-infinite leads using an iterative technique [50]. $\Sigma^<$ accounts for the steady charge "flowing in" from the electrodes. The transmission probability is given by

$$T(E, V) = 4\text{Tr}[(\text{Im }\Sigma_{l,l})G_{l,r}^R(\text{Im }\Sigma_{r,r})G_{r,l}^A], \tag{20.8}$$

where $G^{R(A)}$ are the retarded (advanced) Green's function, and Σ the self-energy part connecting left (l) and right (r) electrodes [50], and the current is obtained from Equation (20.4). The calculated

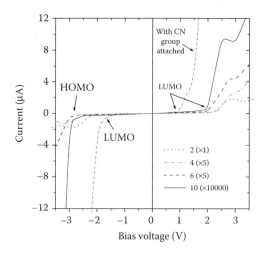

FIGURE 20.5   I-V curves for naphthalene rectifiers $-S-(CH_2)_2-C_{10}H_6-(CH_2)_n-S-$, $n = 2, 4, 6, 10$. The short-dash-dot curve corresponds to a cyano-doped (added group $-C{\equiv}N$) $n = 10$ rectifier.

transmission coefficient $T(E)$ is shown for a series of rectifiers $-S-(CH_2)_m-C_{10}H_6-(CH_2)_n-S-$ for $m = 2$ and $n = 2, 4, 6, 10$ at zero bias voltage in Figure 20.4. We see that the LUMO is the molecular orbital transparent to electron transport, lies above E_F by an amount $\Delta = 1.2 - 1.5$eV. The transmission through the HOMO and HOMO-1 states, localized on the terminating sulfur atoms, is negligible, but the HOMO-2 state conducts very well. The HOMO-2 defines the threshold reverse voltage V_R, thus limiting the operating voltage range. Our assumption, that the voltage drop is proportional to the lengths of the alkane groups on both sides, is quantified by the calculated potential ramp. It is close to a linear slope along the $(CH_2)_n$ chains [48]. The forward voltage corresponds to the crossing of the LUMO(V) and $\mu_R(V)$, which happens at about 2V. Although the LUMO defines the forward threshold voltages in all molecules studied here, the reverse voltage is defined by the HOMO-2 for "right" barriers $(CH_2)_n$ with $n = 6, 10$. The I-V curves are plotted in Figure 20.5. We see that the rectification ratio for current in the operation window I_+/I_- reaches a maximum value of 35 for the "2-10" molecule ($m = 2, n = 10$). Series of molecules with a central *single phenyl* ring [46] do not show any significant rectification. One can manipulate the system in order to increase the energy asymmetry of the conducting orbitals (reduce Δ). To shift the LUMO towards E_F, one can attach an electron withdrawing group, like $-C{\equiv}N$ to the conjugated central group (naphthalene) [48]. The molecular rectification ratio is not great by any means, but one should bear in mind that this is a device necessarily operating in a ballistic quantum-mechanical regime because of the small size. This is very different from present Si devices with carriers diffusing through the system. As silicon devices become smaller, however, the same effects will eventually take over, and tend to diminish the rectification ratio, in addition to effects of finite temperature and disorder in the system.

20.4   Molecular Switches

There are various molecular systems that exhibit some kind of current "switching" behavior [34–37], "negative differential resistance" [33], and "memory" [15]. The switching systems are basically driven between two states with considerably different resistances. This behavior is not really sensitive to a particular molecular structure, since this type of bistability is observed in complex rotaxane-like molecules as well as in very simple alkane chains $(CH_2)_n$ assembled into LB films [37], and is not even exclusive to the organic films. The data strongly indicates that the switching has an extrinsic origin, and is related either to bistability of molecule-electrode orientation [23, 24, 34], or transport assisted by defects in the film [53, 54].

20.4.1 Extrinsic Switching in Organic Molecular Films: Role of Defects and Molecular Reconfigurations

Evidently, large defects can be formed in organic thin films as a result of electromigration in very strong field, as was observed long ago [55]. It was concluded some decades ago that the conduction through absorbed [53] and Langmuir-Blodgett [54] monolayers of fatty acids $(CH_2)_n$, which we denote as Cn, is associated with *defects*. In particular, Polymeropoulos and Sagiv studied a variety of absorbed monolayers from C7 to C23 on Al/Al_2O_3 substrates and found that the exponential dependence on the length of the molecular chains is only observed below the liquid nitrogen temperature of 77K, and no discernible length dependence was observed at higher temperatures [53]. The temperature dependence of current was strong, and was attributed to transport assisted by some defects. The current also varied strongly with the temperature in Ref. [54] for LB films on Al/Al_2O_3 substrates in He atmosphere, which is not compatible with elastic tunneling. Since the He atmosphere was believed to hinder the Al_2O_3 growth, and yet the resistance of the films increased about 100-fold over 45 days, the conclusion was made that the "defects" somehow anneal out with time. Two types of switching have been observed in 3–30 μm thick films of polydimethylsiloxane (PDMS), one as a standard dielectric breakdown with electrode material "jet evaporation" into the film with subsequent Joule melting of metallic filament under bias of about 100V, and a low-voltage (<1 V) "ultraswitching" that has a clear "telegraph" character and resulted in intermittent switching into a much more conductive state [56]. The exact nature of this switching also remains unclear, but there is a strong expectations that the formation of metallic filaments that may even be in a ballistic regime of transport, may be relevant to the phenomenon.

Recently, a direct evidence was obtained of the formation of "hot spots" in the LB films that may be related to the filament growth through the film imaged with the use of AFM current mapping [57]. The system investigated in this work has been Pt/stearic acid (C18)/Ti (Pt/C18/Ti) crossbar molecular structure, consisting of planar Pt and Ti electrodes sandwiching a monolayer of 2.6-nm-long stearic acid (C18H36OH) molecules with typical zero-bias resistance in excess of $10^5\Omega$. The devices has been switched reversibly and repeatedly to higher ("on") or lower ("off") conductance states by applying sufficiently large bias voltage V_b to top Ti electrode with regards to Pt counterelectrode, Figure 20.6.

Interestingly, reversible switching was not observed in symmetric Pt/C18/Pt devices. The local conductance maps of the Pt/C18/Ti structure have been constructed by using an AFM tip and simultaneously measuring the current through the molecular junction biased to $V_b = 0.1V$ (AFM tip was not used as an electrode, only to apply local pressure at the surface). The study revealed that the film showed pronounced switching between electrically very distinct states, with zero-bias conductances 0.17μS ("off" state) and 1.45μS ("on" state), Figure 20.7.

At every switching "on" there appeared a local conductance peak on the map with a typical diameter ~40 nm, which then disappeared upon switching "off", Figure 20.7 (top inset). The switching has been attributed to local conducting filament formation due to electromigration processes. It remains unclear how exactly the filaments dissolve under opposite bias voltage, why they tend to appear in new places after

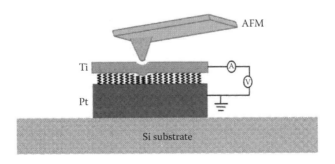

FIGURE 20.6 Experimental setup for mapping local conductance. AFM produces local deformation of top electrode and underlying organic film. The *total* conductance of the device is measured and mapped. (Courtesy C.N. Lau).

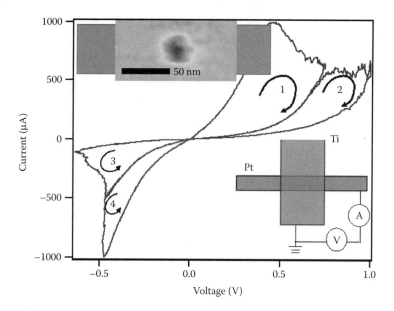

FIGURE 20.7 I-V characteristic showing the reversible switching cycle of the device (bottom inset) with organic film. The arrows indicate sweep direction. A negative bias switches the device to a high conductance state, while the positive one switches it to low conductance state. The mapping according the schematic in Figure 20.6 shows the appearance of "hot spots" after switching (top inset). (Courtesy C.N. Lau).

each switching, and why conductance in some cases strongly depends on temperature. It is clear, however, that switching in such a simple molecule without any redox centers, mobile groups, or charge reception centers should be *extrinsic*. Interestingly, very similar "switching" between two resistive states has also been observed for tunneling through thin *inorganic* perovskite oxide films [58].

There have been plenty of reports on non-linear I-V characteristics like negative differential resistance (NDR) and random switching recently for molecules assembled on metal electrodes (gold) and silicon. Reports on NDR for molecules with metal contacts (Au, Hg) have been made in [33, 44, 59]. It became very clear though that most of these observations are related to molecular reconfigurations and bond breaking and making, rather to any intrinsic mechanism, like redox states, speculated about in the original Ref. [33]. Thus, the NDR in Tour wires was related to molecular reconfiguration with respect to metallic electrode[23, 34], NDR in ferrocene-tethered alkyl monolayers [59] was found to be related to oxygen damage at high voltage [60]. Structural changes and bond breaking have been found to result in NDR in experiments with STM [61–63] and mercury droplet contacts [64].

Several molecules, like styrene, have been studied on degenerate Si surface and showed an NDR behavior [65]. However, those results have been carefully checked later and it was found that the styrene molecules do not exhibit NDR, but rather sporadically switch between states with different current while held at the same bias voltage (the blinking effect) [66].

The STM map of the styrene molecules (indicated by arrows) on the Silicon (100) surface shows that the molecules are blinking, see Figure 20.8. The blinking is absent at clean Si areas, dark (D) and bright (C) defects. This may indicate a dynamic process occurring during the imaging. Comparing the panel (a) and (b) one may see that some molecules are actually decomposing. The height versus voltage spectra over particular points are shown in Figure 20.8c. The featureless curve 1 was taken over a clean silicon dimer. The other spectra were recorded over individual molecules. Each of these spectra have many sudden decreases and increases in current as if the molecules are changing between different states during the measurement causing a change in current and a response of the feedback control, resulting in a change in height so there exists one or more configurations that lead to measurement of a different height. Evidently, these changes

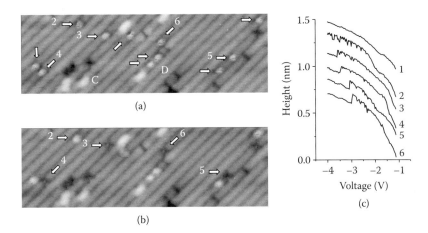

(a)

(b)

(c)

FIGURE 20.8 STM images or styrene molecules on clean Si(100) before and after spectroscopy over the area 75×240 Å. (Top left) Bias -2 V, current 0.7 nA. Only the styrene molecules (indicated by arrows) are blinking during imaging. The clean Si surface, bright defects (marked C), and dark sports (marked D) do not experience blinking. (Bottom left) bias -2 V, current 0.3 nA. STM image of individual styrene molecules indicated with numbers 2–6. Styrene molecules 3-6 have decomposed. Decomposition involves the changing of the styrene molecule from a bright feature to a dark depression and also involves the reaction with an adjacent dimer. Styrene molecule 2 does not decompose and images as usual with no change of position. (Right) Height-voltage spectra taken over clean silicon (1) and styrene molecules (2–6). The spectra taken over molecules show several spikes in height related to blinking in the images. In spectra 3–6, an abrupt and permanent change in height is recorded and is correlated with decomposition, as seen in the bottom left image. Spectrum 2 has no permanent height change, and the molecule does not decompose. (Courtesy J. Pitters and R. Wolkow).

have the same origin as the blinking of molecules in STM images. Figure 20.8b reveals clear structural changes associated with those particular spectroscopic changes. In each case where a dramatic change in spectroscopy occurred, the molecules in the image have changed from a bright feature to a dark spot. This is interpreted as a decomposition of the molecules. A detailed look at each decomposed styrene molecule, at locations 3, 4, 5, and 6, shows that the dark spot is not in precise registry with the original bright feature, indicating that the decomposition product involves reaction with an adjacent dimer [66].

The fact that the structural changes and related NDR behavior are not associated with any resonant tunneling through the molecular levels or redox processes, but are perhaps related to inelastic electron scattering or other extrinsic processes, becomes evident from current versus time records shown in Figure 20.9, Ref. [66]. The records show either no change of the current with time (1), or one or a few random jumps between certain current states (telegraph noise). The observed changes in current at a fixed voltage obviously cannot be explained by shifting and aligning of molecular levels, as was suggested in Ref. [67], they must be related to an adsorbate molecule structural changes with time. Therefore, the explanation by Datta *et al.* that the resonant level alignment is responsible for NDR does not apply [67]. As mentioned above, similar telegraph switching and NDR has been observed in Tour wires [34] and other molecules. Therefore, the observed negative differential resistance apparently has similar origin in disparate molecules adsorbed on different substrates, and has to do with molecular reconformation/reconfiguration on the surface. In fact, to the best of our knowledge, there is no convincing data about intrinsic switching of a resistance state of the molecule subject to external electric field. Various molecules do show conformational switching, but it is limited to either photochromic molecules subject to illumination by photons with certain energy, or induced as a result of a chemical reaction,. like in rotaxanes in solution [38]. What happens to rotaxanes in solid state junctions, is not known, but it is likely to be extrinsic, since rotaxanes behave there in exactly the same manner as benign insulating molecular alkane chains [37].

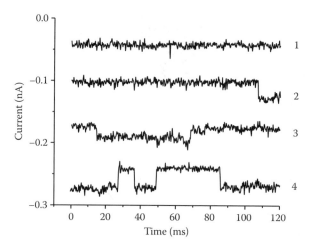

FIGURE 20.9   Variation of current through styrene molecules on Si(100) with time. Tunneling conditions were set at −3 V and 0.05 nA. Abrupt increases and decreases in current relate to changes of the molecule during the spectroscopy. Some experiments show no changes in current (curve 1), others show various kinds of telegraph switching. (Courtesy J. Pitters and R. Wolkow).

20.4.2   Intrinsic Polarization and Extrinsic Conductance Switching in Molecular Ferroelectric PVDF

The only well established, to the best of our knowledge, intrinsic molecular switching (of polarization, not current) under bias voltage was observed in molecular ferroelectric block co-polymers polyvinylidene [68]. Ferroelectric polymer films have been prepared with the 70% vinylidene fluoride copolymer, P(VDF-TrFE 70:30), formed by horizontal LB deposition on aluminum-coated glass substrates with evaporated aluminum top electrodes. The polymer chains contain random sequence of $(CH_2)_n (CF_2)_m$ blocks, fluorine site carries a strong negative charge, and in the ferroelectric phase most of carbon-fluorine bonds point in one direction. The fluorine groups can be rotated and aligned in very strong electric field, ∼5MV/cm. As a result, the whole molecular chain orders, and in this way the macroscopic polarization can be switched between the opposite states. The switching process is extremely slow, however, and takes 1–10 seconds (!) [69, 70]. This is not surprising, given strong Coulomb interaction between charged groups and the metal electrodes, pinning by surface roughness, and steric hindrance to rotation. This behavior should be suggestive of other switching systems based on one of few monolayers of molecules, and other nontrivial behavior involved [70].

The switching of current was also observed in films of PVDF 30 monolayers thick. The conductance of the film was following the observed hysteresis loop for the polarization, ranging from ∼$1 \times 10^{-9} - 2 \times 10^{-6} \Omega^{-1}$ [71]. The phenomenon of conductance switching has these important features: (i) It is connected with the bulk polarization switching; (ii) there is a large ∼1000:1 contrast between the ON and OFF states; (iii) the ON state is obtained only when the bulk polarization is switched in the positive direction; (iv) the conductance switching is much faster than the bulk polarization switching. The conductance switches ON only after the 6s delay, after the bulk polarization switching is nearly complete, presumably when the last layer switches into alignment with the others, while the conductance switches OFF without a noticeable delay after the application of reverse bias as even one layer reverses (this may create a barrier to charge transfer). The slow ∼2s time constant for polarization switching is probably nucleation limited as has been observed in high-quality bulk films with low nucleation site densities [72]. The duration of the conductance switching transition ∼2ms may be limited only by the much faster switching time of individual layers.

The origin of conductance switching by 3 orders of magnitude is not clear. It may indeed be related to a changing amount of disorder for tunneling/hopping electrons. It is conceivable that the carriers are

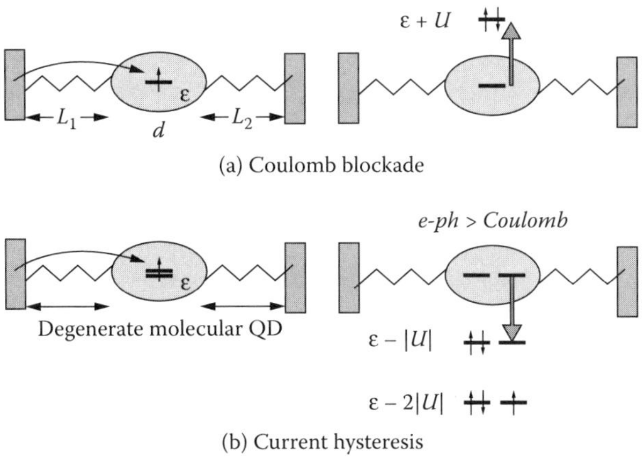

(a) Coulomb blockade

(b) Current hysteresis

FIGURE 20.10 Schematic of the molecular quantum dot with central conjugated unit separated from the electrodes by wide-band insulating molecular groups. First electron tunnels into the dot and occupies an empty (degenerate) state there. If the interaction between the first and second incoming electron is repulsive, $U > 0$, then the dot will be in a Coulomb blockade regime (a). If the electrons on the dot effectively attract each other, $U < 0$, the system will show current hysteresis (b).

strongly trapped in polaron states inside PVDF and find optimal paths for hopping in the material, which is incompletely switched. This is an interesting topic that certainly in need of further experimental and theoretical study.

20.4.3 Electrically Addressable Molecules

For many applications one needs an *intrinsic* molecular "switch", i.e. a bistable voltage-addressable molecular system with very different resistances in the two states that can be accessed very quickly. There is a trade-off between the stability of a molecular state and the ability to switch the molecule between two states with an external perturbation (we discuss an electric field, switching involving absorbed photons is impractical at a nanoscale). Indeed, the applied electric field, on the order of a typical breakdown field $E_b \lesssim 10^7 \text{V/cm}$, is much smaller than a typical atomic field $\sim 10^9 \text{V/cm}$, characteristic of the energy barriers. Small barrier would be a subject for sporadic thermal switching, whereas a larger barrier $\sim 1 - 2\text{eV}$ would be impossible to overcome with the applied field. One may only change the relative energy of the minima by external field and, therefore, redistribute the molecules statistically between the two states. An intrinsic disadvantage of the conformational mechanism, involving motion of ionic group, exceeding the electron mass by many orders of magnitude, is a slow switching speed ($\sim$kHz). In case of supramolecular complexes like rotaxanes and catenanes [35] there are two entangled parts which can change mutual positions as a result of redox reactions (in solution). Thus, for the rotaxane-based memory devices a slow switching speed of $\sim 10^{-2}$ seconds was reported.

We have considered a bistable molecule with $-\text{CONH}_2$ dipole group [73]. The barrier height is $E_b = 0.18\text{eV}$. Interaction with an external electric field changes the energy of the minima, but estimated switching field is huge, $\sim 0.5\text{V/A}$. At non-zero temperatures, temperature fluctuations might result in statistical dipole flipping at lower fields. The I-V curve shows hysteresis in the 3 to 4 Volts window for two possible conformations. One can estimate the thermal stability of the state as 58 ps at room temperature, and 33 ms at 77 K.

20.5 Molecular Quantum Dot Switching

The molecular quantum dot, as we define it, consists of a central *conjugated* unit (containing half-occupied, and, therefore, extended $\pi-$orbitals), Figure 20.10. Frequently, those are formed from the p-states on carbon atoms that are not *saturated* (i.e. they do not share electrons with other atoms forming strong

σ−bonds, with typical bonding-antibonding energy difference about 1Ry). Since the π−orbitals are half-occupied, they form the HOMO-LUMO states. The size of the HOMO-LUMO gap is then directly related to the size of the conjugated region d, Figure 20.10, by a standard estimate $E_{\text{HOMO-LUMO}} \sim \hbar^2/md^2 \sim 2-5$ eV. It is worth noting that in conjugated linear polymers like polyacetylene ($-\overset{|}{C}{=}\overset{|}{C}\,)_n$ the spread of the π−electron would be $d = \infty$ and the expected $E_{\text{HOMO-LUMO}} = 0$. However, such a one-dimensional metal is impossible, Peierls distortion (C=C bond length dimerization) sets in and opens up a gap of about ~1.5eV at the Fermi level [18]. In a molecular quantum dot the central conjugated part is separated from electrodes by insulating groups with saturated σ−bonds, like e.g. the alkane chains, Figure 20.3. Now, there are two main possibilities for carrier transport through the MQD. If the length of at least one of the insulating groups $L_{1(2)}$ is not very large (a conductance $G_{1(2)}$ is not much smaller than the conductance quantum $G_0 = 2e^2/h$), then the transport through the molQD will proceed by resonant tunneling processes. If, on the other hand, both groups are such that the tunnel conductance $G_{1(2)} \ll G_0$, the charge on the dot will be quantized. Then we will have another two possibilities: (i) the interaction of the extra carriers on the dot is *repulsive* $U > 0$, and we have a Coulomb blockade [74], or (ii) the effective interaction is *attractive*, $U < 0$, then we would obtain the current *hysteresis* (see below). Coulomb blockade in molecular quantum dots has been demonstrated in Refs. [75]. In these works, and in Ref. [76], the three-terminal active molecular devices have been fabricated and successfully tested.

20.5.1 Origin of Attractive Correlations in MQD With Strong Electron-Vibron Interactions

Although the correlated electron transport through mesoscopic systems with repulsive electron-electron interactions received considerable attention in the past, and continues to be the focus of current studies, much less has been known about a role of electron-phonon correlations in "molecular quantum dots" (MQD). Some while ago we have proposed a negative−U Hubbard model of a d-fold degenerate quantum dot [77] and a polaron model of resonant tunneling through a molecule with degenerate level [51]. We found that the *attractive* electron correlations caused by any interaction within the molecule could lead to a molecular *switching* effect where I-V characteristics have two branches with high and low current at the same bias voltage. This prediction has been confirmed and extended further in the theory of *correlated* transport through degenerate MQDs with a full account of both the Coulomb repulsion and realistic electron-phonon (e-ph) interactions [51]. We have shown that while the phonon side-bands significantly modify the shape of hysteretic I-V curves in comparison with the negative-U Hubbard model, switching remains robust. It shows up when the effective interaction of polarons is attractive and the state of the dot is d−fold degenerate, $d > 2$. We explicitly calculate I-V curves of the nondegenerate ($d = 1$) and two-fold degenerate ($d = 2$) MQDs to show that there is no switching in those systems (see discussion below)

Let us first consider a simplest model of a single atomic level coupled with a single one-dimensional oscillator using the first quantization representation for its displacement x,

$$H = \varepsilon_0 \hat{n} + fx\hat{n} - \frac{1}{2M}\frac{\partial^2}{\partial x^2} + \frac{kx^2}{2}. \tag{20.9}$$

Here M and k are the oscillator mass and the spring constant, f is the interaction force, and $\hbar = c = k_B = 1$. This Hamiltonian is readily diagonalized with the *exact* displacement transformation of the vibration coordinate x,

$$x = y - \hat{n}f/k, \tag{20.10}$$

to the transformed Hamiltonian without electron-phonon coupling,

$$\tilde{H} = \varepsilon \hat{n} - \frac{1}{2M}\frac{\partial^2}{\partial y^2} + \frac{ky^2}{2}, \tag{20.11}$$

$$\varepsilon = \varepsilon_0 - E_p, \tag{20.12}$$

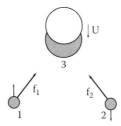

FIGURE 20.11 Two localized electrons at sites **1** and **2** shift the equilibrium position of the ion at site **3**. As a result, the two electrons *attract* each other.

where we used $\hat{n}^2 = \hat{n}$ because of the Fermi-Dirac statistics. It describes a small polaron at the atomic level ε_0 shifted down by the polaron level shift $E_p = f^2/2k$, and entirely decoupled from ion vibrations. The ion vibrates near a new equilibrium position, shifted by f/k, with the "old" frequency $(k/M)^{1/2}$. As a result of the local ion deformation, the total energy of the whole system decreases by E_p since a decrease of the electron energy by $-2E_p$ overruns an increase of the deformation energy E_p.

Lattice deformation also strongly affects the interaction between electrons. When a short-range deformation potential and molecular electron-vibron (phonon) interactions are taken into account together with the long-range Fröhlich interaction, they can overcome the Coulomb repulsion. The resulting interaction becomes attractive at a short distance comparable to a lattice constant. The origin of the attractive force between two small polarons can be readily understood from a similar Holstein-like toy model as above [78], but with two electrons on neighboring sites **1, 2** interacting with an ion **3** between them, Figure 20.11. For generality, we now assume that the ion is a three-dimensional oscillator described by a displacement vector $\mathbf{u}$, rather than by a single-component displacement x as in Eq.(20.9).

The vibration part of the Hamiltonian in the model is

$$H_{ph} = -\frac{1}{2M}\frac{\partial^2}{\partial \mathbf{u}^2} + \frac{k\mathbf{u}^2}{2}, \tag{20.13}$$

Electron potential energies due to the Coulomb interaction with the ion are about

$$V_{1,2} = V_0 - \mathbf{u} \cdot \nabla_{R_{1,2}} V_0(R_{1,2}) = V_0 + \mathbf{u} \cdot \mathbf{f}_{1,2}. \tag{20.14}$$

where $f_{1,2} = Ze^2/a^2$ is the Coulomb force, $\vec{R}_{1(2)}$ is the vector connecting the ion 3 with the electron 1 (2), Figure 20.11. Hence, the Hamiltonian of the model is given by

$$H = E_a(\hat{n}_1 + \hat{n}_2) + \mathbf{u} \cdot (\mathbf{f}_1\hat{n}_1 + \mathbf{f}_2\hat{n}_2) - \frac{1}{2M}\frac{\partial^2}{\partial \mathbf{u}^2} + \frac{k\mathbf{u}^2}{2}, \tag{20.15}$$

where $\hat{n}_{1,2}$ are the site occupation number operators. This Hamiltonian is diagonalized by the same displacement transformation of the vibronic coordinate $\mathbf{u}$ as above,

$$\mathbf{u} = \mathbf{v} - (\mathbf{f}_1\hat{n}_1 + \mathbf{f}_2\hat{n}_2)/k. \tag{20.16}$$

The transformed Hamiltonian has no electron-phonon coupling:

$$\tilde{H} = (\varepsilon_0 - E_p)(\hat{n}_1 + \hat{n}_2) + V_p\hat{n}_1\hat{n}_2 - \frac{1}{2M}\frac{\partial^2}{\partial \mathbf{v}^2} + \frac{k\mathbf{v}^2}{2}, \tag{20.17}$$

and it describes two small polarons at their atomic levels shifted by the polaron level shift $E_p = f_{1,2}^2/2k$, which are entirely decoupled from ion vibrations. As a result, the lattice deformation caused by two electrons leads to an effective interaction between them, V_p, which should be added to their Coulomb repulsion, V_c,

$$V_p = -\mathbf{f}_1 \cdot \mathbf{f}_2 / k. \qquad (20.18)$$

When V_p is negative and larger by magnitude than the positive V_c, the resulting interaction becomes attractive. It is responsible for the hysteretic behavior of MQDs, as discussed below.

20.5.2 Exact Solution for Current Through MQD

Since high-frequency molecular vibrations (vibrons) are involved in mediation of interaction between carriers on a molecule, one might expect almost instantaneous establishment of a correlated two-electron state on a molecule. The attractive energy is the difference of two large interactions, the Coulomb repulsion and the phonon mediated attraction, on the order of 1eV each, hence one can assume that $|U| \sim 0.1$eV. This state is below the one-particle states and it could persist even if the bias voltage is reduced below the initial threshold, leading to a current hysteresis. This is indeed the case when the degeneracy of the level is $d > 2$, since the quantum fluctuations are apparently unable to destroy it, at least below some critical temperature, as we shall see below.

We shall consider a situation pertaining to MQD, so that the coupling to the leads is weak, and the level width $\Gamma \ll |U|$. The current is found from from [79]

$$I(V) = I_0 \int_{-\infty}^{\infty} d\omega [f_1(\omega) - f_2(\omega)] \rho(\omega), \qquad (20.19)$$

$$\rho(\omega) = -\frac{1}{\pi} \sum_\mu \operatorname{Im} \hat{G}_\mu^R(\omega), \qquad (20.20)$$

where $|\mu\rangle$ is a complete set of one-particle molecular states. Here $I_0 = q\Gamma$, $\rho(\omega)$ is the molecular DOS, $\hat{G}_\mu^R(\omega)$ is the Fourier transform of the Green's function $\hat{G}_\mu^R(t) = -i\theta(t)\langle\{c_\mu(t), c_\mu^\dagger\}\rangle$, $\{\cdots,\cdots\}$ is the anticommutator, $c_\mu(t) = e^{iHt}c_\mu e^{-iHt}$, $\theta(t) = 1$ for $t > 0$ and zero otherwise. We calculate $\rho(\omega)$ *exactly* for the Hamiltonian, which includes both the Coulomb U^C and e-ph interactions,

$$H = \sum_\mu \varepsilon_\mu \hat{n}_\mu + \frac{1}{2} \sum_{\mu \neq \mu'} U_{\mu\mu'}^C \hat{n}_\mu \hat{n}_{\mu'} + \sum_{\mu,q} \hat{n}_\mu \omega_q (\gamma_{\mu q} d_q + H.c.) + \sum_q \omega_q (d_q^\dagger d_q + 1/2). \qquad (20.21)$$

Here ε_μ are one-particle molecular energy levels, $\hat{n}_\mu = c_\mu^\dagger c_\mu$ the occupation number operators, c_μ (d_q) annihilates electrons (phonons), ω_q are the phonon (vibron) frequencies, and $\gamma_{\mu q}$ are e-ph coupling constants (q enumerates the vibron modes). We apply the standard Lang-Firsov polaron unitary transformation, integrating phonons out. The electron and phonon operators are transformed as $\tilde{c}_\mu = c_\mu X_\mu$, and $\tilde{d}_q = d_q - \sum_\mu \hat{n}_\mu \gamma_{\mu q}^*$, respectively, $X_\mu = \exp(\sum_q \gamma_{\mu q} d_q - H.c.)$, and the transformed Hamiltonian is

$$\tilde{H} = \sum_i \tilde{\varepsilon}_\mu \hat{n}_\mu + \sum_q \omega_q (d_q^\dagger d_q + 1/2) + \frac{1}{2} \sum_{\mu \neq \mu'} U_{\mu\mu'} \hat{n}_\mu \hat{n}_{\mu'}, \qquad (20.22)$$

where

$$U_{\mu\mu'} \equiv U_{\mu\mu'}^C - 2 \sum_q \gamma_{\mu q}^* \gamma_{\mu' q} \omega_q \qquad (20.23)$$

is the interaction of polarons, which we simplify as $U_{\mu\mu'} = U$. The molecular energy levels are subject to a polaron level shift ,

$$\tilde{\varepsilon}_\mu = \varepsilon_\mu - \sum_q |\gamma_{\mu q}|^2 \omega_q. \qquad (20.24)$$

The retarded GF becomes

$$G_\mu^R(t) = -i\theta(t)[\langle c_\mu(t)c_\mu^\dagger\rangle\langle X_\mu(t)X_\mu^\dagger\rangle + \langle c_\mu^\dagger c_\mu(t)\rangle\langle X_\mu^\dagger X_\mu(t)\rangle]. \tag{20.25}$$

The phonon correlator is simply

$$\langle X_\mu(t)X_\mu^\dagger\rangle = \exp\sum_q \frac{|\gamma_{\mu q}|^2}{\sinh\frac{\beta\omega_q}{2}}\left[\cos\left(\omega t + i\frac{\beta\omega_q}{2}\right) - \cosh\frac{\beta\omega_q}{2}\right], \tag{20.26}$$

where the inverse temperature $\beta = 1/T$, and $\langle X_\mu^\dagger X_\mu(t)\rangle = \langle X_\mu(t)X_\mu^\dagger\rangle^*$. The remaining GFs $\langle c_\mu(t)c_\mu^\dagger\rangle$, are found from the equations of motion *exactly*. Finally, for the simplest case of a coupling to a single mode with the characteristic frequency ω_0 and $\gamma_q \equiv \gamma$ we find [51]

$$G_\mu^R(\omega) = \mathcal{Z}\sum_{r=0}^{d-1} C_r(n)\sum_{l=0}^\infty I_l(\xi)\left[e^{\frac{\beta\omega_0 l}{2}}\left(\frac{1-n}{\omega - rU - l\omega_0 + i\delta} + \frac{n}{\omega - rU + l\omega_0 + i\delta}\right) + (1 - \delta_{l0})e^{-\frac{\beta\omega_0 l}{2}}\right.$$
$$\left.\times\left(\frac{1-n}{\omega - rU + l\omega_0 + i\delta} + \frac{n}{\omega - rU - l\omega_0 + i\delta}\right)\right], \tag{20.27}$$

where

$$\mathcal{Z} = \exp\left(-\sum_q |\gamma_q|^2 \coth\frac{\beta\omega_0}{2}\right) \tag{20.28}$$

is the familiar *polaron narrowing* factor, the degeneracy factor

$$C_r(n) = \frac{(d-1)!}{r!(d-1-r)!}n^r(1-n)^{d-1-r}, \tag{20.29}$$

$\xi = |\gamma|^2/\sinh\frac{\beta\omega_0}{2}$, $I_l(\xi)$ the modified Bessel function, and δ_{lk} is the Kroneker symbol. The important feature of the DOS, Equation (20.19), is its nonlinear dependence on the occupation number n. It contains full information about all possible correlation and inelastic effects in transport, in particular, all the phonon sidebands. We then obtain the exact spectral function for a $d-$fold degenerate MQD (i.e. the density of molecular states, DOS) from Eqs.(20.20, 20.27) [51]:

$$\rho(\omega) = \mathcal{Z}d\sum_{r=0}^{d-1} C_r(n)\sum_{l=0}^\infty I_l(\xi)\left[e^{\beta\omega_0 l/2}[(1-n)\delta(\omega - rU - l\omega_0) + n\delta(\omega - rU + l\omega_0)]\right.$$
$$\left. + (1 - \delta_{l0})e^{-\beta\omega_0 l/2}[n\delta(\omega - rU - l\omega_0) + (1-n)\delta(\omega - rU + l\omega_0)]\right]. \tag{20.30}$$

20.5.3 Nonlinear Rate Equation and Switching

In the present case of molQD weakly coupled with leads one can apply the Fermi-Dirac golden rule to obtain an equation for n. Equating incoming and outgoing numbers of electrons in molQD per unit time we obtain the self-consistent equation for the level occupation n as

$$(1-n)\int_{-\infty}^\infty d\omega\{\Gamma_1 f_1(\omega) + \Gamma_2 f_2(\omega)\}\rho(\omega) \tag{20.31}$$

$$= n\int_{-\infty}^\infty d\omega\{\Gamma_1[1 - f_1(\omega)] + \Gamma_2[1 - f_2(\omega)]\}\rho(\omega), \tag{20.32}$$

where $\Gamma_{1(2)}$ are the transition rates from left (right) leads to molQD, and $\rho(\omega)$ is found from Eqs. (20.27) and (20.19). For $d = 1, 2$ the kinetic equation for n is *linear*, and the switching is *absent*. Switching appears for $d \geq 3$, when the kinetic equation becomes non-linear. For $d = 1, 2$ the kinetic equation for n is linear, and the switching is *absent*. Switching appears for $d \geq 3$, when the kinetic equation becomes non-linear. Taking into account that $\int_{-\infty}^{\infty} \rho(\omega) = d$, Equation (**??**) for the symmetric leads, $\Gamma_1 = \Gamma_2$, reduces to

$$2nd = \int d\omega \rho(\omega)(f_1 + f_2), \tag{20.33}$$

which automatically satisfies $0 \leq n \leq 1$. Explicitly, the self-consistent equation for the occupation number is

$$n = \frac{1}{2} \sum_{r=0}^{d-1} C_r(n) \left[na_r^+ + (1-n)b_r^+ \right], \tag{20.34}$$

where

$$a_r^+ = \mathcal{Z} \sum_{l=0}^{\infty} I_l(\xi)$$
$$\times \left(e^{\frac{\beta\omega_0 l}{2}} [f_1(rU - l\omega_0) + f_2(rU - l\omega_0)] + (1 - \delta_{l0})e^{-\frac{\beta\omega_0 l}{2}} [f_1(rU + l\omega_0) + f_2(rU + l\omega_0)] \right), \tag{20.35}$$

$$b_r^+ = \mathcal{Z} \sum_{l=0}^{\infty} I_l(\xi)$$
$$\times \left(e^{\frac{\beta\omega_0 l}{2}} [f_1(rU + l\omega_0) + f_2(rU + l\omega_0)] + (1 - \delta_{l0})e^{-\frac{\beta\omega_0 l}{2}} [f_1(rU - l\omega_0) + f_2(rU - l\omega_0)] \right). \tag{20.36}$$

The current is expressed as

$$j \equiv \frac{I(V)}{dI_0} = \sum_{r=0}^{d-1} Z_r(n) \left[na_r^- + (1-n)b_r^- \right], \tag{20.37}$$

where

$$a_r^- = \mathcal{Z} \sum_{l=0}^{\infty} I_l(\xi)$$
$$\times \left(e^{\frac{\beta\omega_0 l}{2}} [f_1(rU - l\omega_0) - f_2(rU - l\omega_0)] + (1 - \delta_{l0})e^{-\frac{\beta\omega_0 l}{2}} [f_1(rU + l\omega_0) - f_2(rU + l\omega_0)] \right), \tag{20.38}$$

$$b_r^- = \mathcal{Z} \sum_{l=0}^{\infty} I_l(\xi)$$
$$\times \left(e^{\frac{\beta\omega_0 l}{2}} [f_1(rU + l\omega_0) - f_2(rU + l\omega_0)](1 - \delta_{l0})e^{-\frac{\beta\omega_0 l}{2}} [f_1(rU - l\omega_0) - f_2(rU - l\omega_0)] \right). \tag{20.39}$$

Consider the case $d = 4$, where the rate equation is non-linear, to see if it produces multiple physical solutions. For instance, in the limit $|\gamma| \ll 1$, $T = 0$, where we have $b_r = a_r$, $\mathcal{Z} = 1$, the remaining interaction is $U = U^C < 0$, we recover the negative-U model [77], and the kinetic equation for $d = 4$ is

$$2n = 1 - (1 - n)^3 \tag{20.40}$$

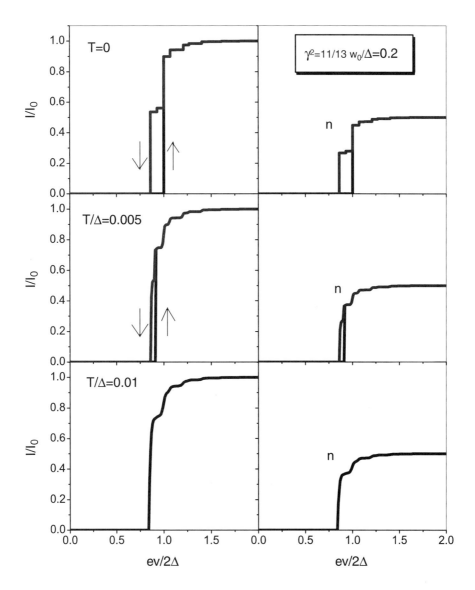

FIGURE 20.12 The I-V curves for tunneling through the molecular quantum dot, Figure 20.10b with the electron-vibron coupling constant $\gamma^2 = 11/13$.

in the voltage range $\Delta - |U| < eV/2 < \Delta$. This equation has an additional nontrivial physical solution $n = (3 - \sqrt{5})/2 = 0.38$. The current is simplified as $I/I_0 = 2n$. The current-voltage characteristics will show a *hysteretic* behavior in this case for $d = 4$. When the voltage increases from zero, 4-fold degenerate MQD remains in a low-current state until the threshold $eV_2/2 = \Delta$ is reached. Remarkably, when the voltage *decreases* from the value above the threshold V_2, the molecule remains in the high-current state down to the voltage $eV_1/2 = \Delta - |U|$ well below the threshold V_2.

The hysteresis of current is there at all values of the electron-phonon constant γ, e.g. $\gamma^2 = 11/13$ (selected in order to have an accidental commensurability of ω_0 and U), Figure 20.12. Indeed, the exact equation for average occupation of the dot reads.

$$2n = (1 - n)^3 \left[na_0^+ + (1 - n)b_0^+ \right] + 3n(1 - n)^2 \left[na_1^+ + (1 - n)b_1^+ \right]$$
$$+ 3n^2(1 - n) \left[na_2^+ + (1 - n)b_2^+ \right] + n^3 \left[na_3^+ + (1 - n)b_3^+ \right]. \qquad (20.41)$$

We solved this nonlinear equation for the case $\omega_0/\Delta = 0.2$, $U^C = 0$, so that the attraction between electrons is $U = -2\gamma^2\omega_0 = -0.4$ (all energies in units of Δ), and obtained a hysteresis curve, Figure 20.12. The bistability region shrinks down with temperature, and the hysteresis loop practically closes at $T/\Delta = 0.01$. As we see from Equations (20.35), (20.36), the electron levels with phonon sidebands $\Delta \pm l\omega_0$, $\Delta + U \pm l\omega_0$, $\Delta + 2U \pm l\omega_0$, $\Delta + 3U \pm l\omega_0$ with $l = 0, 1, \ldots$ contribute to electron transport with different weights, and this creates a complex picture of steps on the I-V curve, Figure 20.12.

Note that switching required a degenerate MQD ($d > 2$) and the weak coupling to the electrodes, $\Gamma \ll \omega_0$. Different from the non-degenerate dot, the rate equation for a multi-degenerate dot, $d > 2$, weakly coupled to the leads has multiple physical roots in a certain voltage range and a hysteretic behavior due to *correlations* between different electronic states of MQD [51].

Summarizing this Section, we have calculated the I-V characteristics of the nondegenerate and two-fold degenerate MQDs showing no hysteretic behavior, and conclude that mean field approximation [82] leads to a non-existent hysteresis in a model that was solved exactly in Ref. [51]. Different from the non-degenerate and two-fold degenerate dots, the rate equation for a multi-degenerate dot, $d > 2$, weakly coupled to the leads, has multiple physical roots in a certain voltage range showing hysteretic behavior due to *correlations* between different electronic states of MQD [51]. Our conclusions are important for searching of the current-controlled polaronic molecular switches. Incidentally, C_{60} molecules have the degeneracy $d = 6$ of the lowest unoccupied level, which makes them one of the most promising candidate systems, if the weak-coupling with leads is secured.

20.5.4 What MQD Can Switch?

Note that switching required a degenerate MQD ($d > 2$) and the weak coupling to the electrodes, $\Gamma \ll \omega_0$. The case of strong coupling of nondegenerate level ($d = 1$) with $\Gamma \gtrsim \omega_0$, where the electron transport is adiabatic, has been considered in [80, 81]. Obviously, there is *no switching* in this case, and this is exactly what these authors have found. This obvious conclusion for molecules strongly coupled to the electrodes can be reached in many ways, see e.g. a rather involved derivation in Ref. [80, 81]. The current hysteresis does not occur in their model, the current remains a single-valued function of bias with superimposed noise. In the case of a double-degenerate MQD, $d = 2$, there are two terms, which contribute to the sum over r, with $C_0(n) = 1 - n$ and $C_1(n) = n$. The rate equation becomes a quadratic one [51]. Nevertheless there is only one physical root for any temperature and voltage, and the current is also single-valued, Figure 20.13.

Later on, however, Galperin *et al* published a paper [82] where they claimed, without discussing earlier results, that a strongly coupled ($\Gamma \gtrsim \omega_0$, $\Gamma \approx 0.1 - 0.3$ eV) molecular bridge, which is nondegenerate ($d = 1$) does exhibits switching. This result is an artefact of their adiabatic approximation, as discussed in detail in Ref. [78]. Indeed, Galperin *et al.* [82] have obtained the current hysteresis in MQD as a result of illegitimate the occupation number operator $\hat{n}$ in the e-ph interaction by the average population n_0 [Equation (2) of Ref. [82], "mean field" approximation for $\hat{n}^2 \approx n_0\hat{n}$ (?) in violation of exact relation $\hat{n}^2 = \hat{n}$] and found the average steady-state vibronic displacement $\langle d + d^\dagger \rangle$ proportional to n_0 (this is an explicit *neglect* of all quantum fluctuations on the dot accounted for in the exact solution). Then, replacing the displacement operator $d + d^\dagger$ in their bare Hamiltonian [cf. Equation (20.21)] by its average, Ref. [82], they obtained a new molecular level, $\tilde{\varepsilon}_0 = \varepsilon_0 - 2\varepsilon_{reorg}n_0$ shifted linearly with the average population of the level. This is in stark disagreement with the conventional constant polaronic level shift, Equations (20.12), (20.24) (ε_{reorg} is $|\gamma|^2\omega_0$ in our notations). The MFA spectral function turned out to be highly nonlinear as a function of the population, e.g. for the weak-coupling with the leads $\rho(\omega) = \delta(\omega - \varepsilon_0 - 2\varepsilon_{reorg}n_0)$, see Equation (20.17) in Ref. [82]. As a result, the authors of Ref. [82] have found multiple ghost solutions for the steady-state population, Equation (20.15) and Figure 20.1, and switching, Figure 20.4 of Ref. [82].

Note that the mean-field solution by Galperin *et al.* [82] applies at any ratio Γ/ω_0, including the limit of interest to us, $\Gamma \ll \omega_0$ where their transition between the states with $n_0 = 0$ and 1 only sharpens, but none of the results change. Therefore, MFA predicts a current bistability in the system where it does not

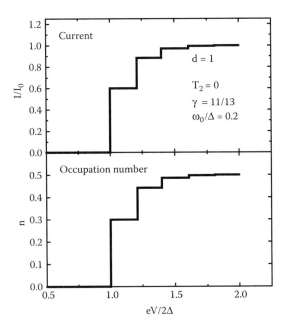

FIGURE 20.13 Current-voltage characteristic of the nondegenerate ($d = 1$) MQD at $T = 0$, $\omega_0/\Delta = 0.2$, and $\gamma^2 = 11/13$. There is the phonon ladder in I-V , but no hysteresis.

exist at $d = 1$. The results in Ref. [82] are plotted for $\Gamma \geq \omega_0$, $\Gamma \approx 0.1 - 0.3$ eV, which corresponds to molecular bridges with a resistance of about a few 100KΩ. Such model "molecules" are rather "metallic" in their conductance and could hardly show any bistability at all because carriers do not have time to interact with vibrons on the molecule. Indeed, taking into account the coupling with the leads beyond the second order and the coupling between the molecular and bath phonons could hardly provide any non-linearity because these couplings do not depend on the electron population. This rather obvious conclusion for molecules strongly coupled to the electrodes can be reached in many ways, see e.g. rather involved derivations in Refs. [80, 81]. While Refs. [80, 81] do talk about telegraph current noise in the model, there is no hysteresis in the adiabatic regime, $\Gamma \gg \omega_0$ either. This result certainly has nothing to do with our mechanism of switching [51] that applies to molecular quantum dots ($\Gamma \ll \omega_0$) with $d > 2$. Such regime has not been studied in Refs. [80, 81, 83] as being a "too challenging problem". This did not prevent Mitra *et al.* [83] from misrepresenting our results [51] claiming that it "lacks of renormalization of the dot-lead coupling" (due to electron-vibron interaction), or "treats it in an average manner". As shown above, the formalism [51] is exact, it fully takes into account the polaronic renormalization, phonon-side bands and polaron-polaron correlations in the exact molecular DOS, Equation (20.30).

As a matter of fact, most of the molecules are very resistive, so the actual molecular quantum dots are in the regime we study, see Ref. [84]. For example, the resistance of fully conjugated three-phenyl ring Tour-Reed molecules chemically bonded to metallic Au electrodes [33] exceeds 1GΩ. Therefore, most of the molecules of interest to us are in the regime that we discussed, not that of Refs. [80, 81].

20.6 Role of Defects in Molecular Transport

Interesting behavior of electron transport in molecular systems, as described above, refers to ideal systems without imperfection in ordering and composition. In reality, one expects that there will be a considerable disorder and defects in organic molecular films. As mentioned above, the conduction through absorbed [53] and Langmuir-Blodgett [54] monolayers of fatty acids $(CH_2)_n$ was associated with *defects*. An absence of tunneling through self-assembled monolayers of C12-C18 (inferred from an absence of thickness dependence at room temperature) has been reported by Boulas et al. [52]. On the other hand, the

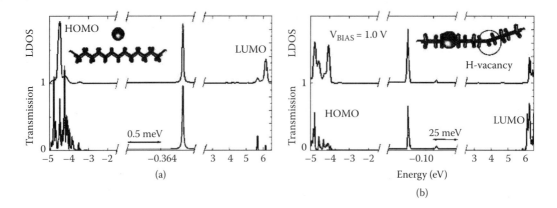

FIGURE 20.14 Local density of states and transmission as a function of energy for (a) C13 with Au impurity and (b) C13 with Au impurity and H vacancy (dangling bond). Middle sections show closeups of the resonant peaks due to deep defect levels with respect to the HOMO and LUMO molecular states. The HOMO-LUMO gap is about 10 eV.

tunneling in alkanethiol SAMs was reported in [85, 86], with an exponential dependence of monolayer resistance on the chain length L, $R_\sigma \propto \exp(\beta_\sigma L)$, and no temperature dependence of the conductance in C8-C16 molecules was observed over the temperatures $T = 80 - 300$K [86].

The electrons in alkane molecules are tightly bound to the C atoms by $\sigma-$bonds, and the band gap (between the highest occupied molecular orbital, HOMO, and lowest unoccupied molecular orbital, LUMO) is large, $\sim 9 - 10$eV [52]. In conjugated systems with $\pi-$electrons the molecular orbitals are extended, and the HOMO-LUMO gap is correspondingly smaller, as in e.g. polythiophenes, where the resistance was also found to scale exponentially with the length of the chain, $R_\pi \propto \exp(\beta_\pi L)$, with $\beta_\pi = 0.35 \text{Å}^{-1}$ instead of $\beta_\sigma = 1.08 \text{Å}^{-1}$ [85]. In stark contrast with the temperature-independent tunneling results for SAMs [86], recent extensive studies of electron transport through 2.8 nm thick eicosanoic acid (C20) LB monolayers at temperatures 2K-300K have established that the current is practically temperature independent below $T < 60$K, but very strongly temperature dependent at higher temperatures $T = 60 - 300$K [87].

A large amount of effort went into characterizing the organic thin films and possible defects there [19, 88, 89]. It has been found that the electrode material, like gold, gets into the body of the film, leading to the possibility of metal ions existing in the film as single impurities and clusters. Electronic states on these impurity ions are available for the resonant tunneling of carriers in very thin films (or hopping in thicker films, a crossover between the regimes depending on the thickness). Depending on the density of the impurity states, with increasing film thickness the tunneling will be assisted by impurity "chains", with an increasing number of equidistant impurities [90]. One-impurity channels produce steps on the I-V curve but no temperature dependence, whereas the inelastic tunneling through pairs of impurities at low temperatures defines the temperature dependence of the film conductance, $G(T) \propto T^{4/3}$, and the voltage dependence of current $I(V) \propto V^{7/3}$ [91]. This behavior has been predicted theoretically and observed experimentally for tunneling through amorphous Si [92] and Al_2O_3 [93]. Due to the inevitable disorder in a "soft" matrix, the resonant states on different impurities within a "channel" will be randomly moving in and out of resonance, creating mesoscopic fluctuations of the I-V curve. The tunneling may be accompanied by interaction with vibrons on the molecule, causing step-like features on the I-V curve [51, 76].

During processing, especially top electrode deposition, small clusters of the electrode material may form in the organic film, causing Coulomb blockade, which also can show up as steps on the I-V curve. It has long been known that a strong applied field can cause localized damage to thin films, presumably due to electromigration and the formation of conducting filaments [55]. The damaged area was about 30nm in diameter in 40-160 monolayer thick LB films [55](a) and 5-10μm in diameter in films 500-5000 Å thick, and showed switching behavior under external bias voltage cycling [55](b). As discussed above, recent spatial mapping of a conductance in LB monolayers of fatic acids with the use of conducting AFM has revealed damage areas 30-100nm in diameter, frequently appearing in samples after a "soft" electrical breakdown, which is sometimes accompanied by a strong temperature dependence of the conductance through the film [57].

A crossover from tunneling at low temperatures to an activation-like dependence at higher temperatures is expected for electron transport through organic molecular films. There are recent reports about such a crossover in individual molecules like the 2nm long Tour wire with a small activation energy $E_a \approx 130$meV [94]. Very small activation energies on the order of 10-100 meV have been observed in polythiophene monolayers [95]. Our present results suggest that this may be a result of interplay between the drastic renormalization of the electronic structure of the molecule in contact with electrodes, and disorder in the film (Figure 20.15, right inset). We report the ab-initio calculations of point-defect assisted tunneling through alkanedithiols $S(CH_2)_nS$ and thiophene T3 (three rings SC_4) self-assembled on gold electrodes. The length of the alkane chain was in the range $n = 9 - 15$.

We have studied single and double defects in the film: (i) single Au impurity, Figures 20.10a, 20.11a, (ii) Au impurity and H vacancy (dangling bond) on the chain, Figures 20.15b, 20.14c, (iii) a pair of Au impurities, Figure 20.14b, (iv) Au and a "kink" on the chain (one C=C bond instead of a C−C bond). Single defect states result in steps on the associated I-V curve, whereas molecules in the presence of two defects generally exhibit a negative differential resistance (NDR). Both types of behavior are generic and may be relevant to some observed unusual transport characteristics of SAMs and LB films [53, 54, 57, 87, 94, 95]. We have used an ab-initio approach that combines the Keldysh non-equilibrium Green's function (NEGF) method with self-consistent pseudopotential-based real space density functional theory (DFT) for systems with *open* boundary conditions provided by semi-infinite electrodes under external bias voltage [48, 50]. All present structures have been relaxed with the Gaussian98 code prior to transport calculations [96]. The conductance of the system at a given energy is found from Equation (20.8) and the current from Equation (20.4).

The equilibrium position of an Au impurity is about 3 Å away from the alkane chain, which is a typical Van-der-Waals distance. As the density maps show (Figure 20.15), there is an appreciable hybridization between the s- and d-states of Au and the sp- states of the carbohydrate chain. Furthermore, the Au^+ ion produces a Coulomb center trapping a 6s electron state at an energy $\epsilon_i = -0.35$eV with respect to the Fermi level, almost in the middle of the HOMO-LUMO ~ 10eV gap in Cn. The tunneling evanescent resonant state is a superposition of the HOMO and LUMO molecular orbitals. Those orbitals have a very complex spatial structure, reflected in an asymmetric line shape for the transmission. Since the impurity levels are very deep, they may be understood within the model of "short-range impurity potential" [97]. Indeed, the impurity wave function *outside* of the narrow well can be fairly approximated as

$$\varphi(r) = \sqrt{\frac{2\pi}{\kappa}} \frac{e^{-\kappa r}}{r}, \tag{20.42}$$

where κ is the inverse radius of the state, $\hbar^2\kappa^2/2m^* = E_i$, where $E_i = \Delta - \epsilon_i$ is the depth of the impurity level with respect to the LUMO, and $\Delta = $ LUMO$-F$ is the distance between the LUMO and the Fermi level F of gold and, consequently, the radius of the impurity state $1/\kappa$ is small. The energy distance $\Delta \approx 4.8$eV in alkane chains $(CH_2)_n$ [52] (≈ 5eV from DFT calculations), and $m^* \sim 0.4$ the effective tunneling mass in alkanes [86]. For one impurity in a rectangular tunnel barrier [97] we obtain the Breit-Wigner form of transmission $T(E, V)$, as before, Equation (20.5). Using the model with the impurity state wave function (20.42), we may estimate for an Au impurity in C13 ($L = 10.9$Å) the width $\Gamma_L = \Gamma_R = 1.2 \times 10^{-6}$, which is within an order of magnitude compared with the calculated value 1.85×10^{-5}eV. The transmission is maximal and equals unity when $E = \epsilon_i$ and $\Gamma_L = \Gamma_R$, which corresponds to a symmetrical position for the impurity with respect to the electrodes.

The electronic structure of the alkane backbone, through which the electron tunnels to an electrode, shows up in the asymmetric lineshape, which is substantially non-Lorentzian, Figure 20.14. The current remains small until the bias has aligned the impurity level with the Fermi level of the electrodes, resulting in a step in the current, $I_1 \approx \frac{2q}{\hbar}\Gamma_0 e^{-\kappa L}$ (Figure 20.14a). This step can be observed only when the impurity

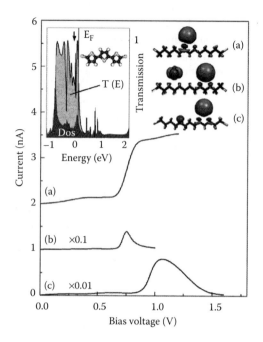

FIGURE 20.15 Current-voltage characteristics of an alkane chain C13 with (a) single Au impurity (6s-state), (b) two Au impurities (5d and 6s-states on left and right ions, respectively), and (c) Au impurity and H vacancy (dangling bond). Double defects produce the negative differential resistance peaks (b) and (c). Inset shows the density of states, transmission, and stick model for polythiophene T3. There is significant transmission at the Fermi level, suggesting an ohmic I-V characteristic for T3 connected to gold electrodes. Disorder in the film may localize states close to the Fermi level (schematically marked by arrow), which may assist in hole hopping transport with an apparently very low activation energy (0.01-0.1eV), as is observed.

level is not very far from the Fermi level F, such that biasing the contact can produce alignment before a breakdown of the device may occur. The most interesting situations that we have found relate to the *pairs* of point defects in the film. If the concentration of defects is $c \ll 1$, the relative number of configurations with pairs of impurities will be very small, $\propto c^2$. However, they give an exponentially larger contribution to the current. Indeed, the optimal position of two impurities is symmetrical, a distance $L/2$ apart, with current $I_2 \propto e^{-\kappa L/2}$. The conductance of a two-impurity chain is [97]

$$g_{12}(E) = \frac{4q^2}{\pi\hbar} \frac{\Gamma_L \Gamma_R t_{12}^2}{|(E - \epsilon_1 + i\Gamma_L)(E - \epsilon_2 + i\Gamma_R) - t_{12}^2|^2}. \tag{20.43}$$

For a pair of impurities with slightly differing energies $t_{12} = 2(E_1 + E_2)e^{-\kappa r_{12}}/\kappa r_{12}$, where r_{12} is the distance between them. The interpretation of the two-impurity channel conductance (20.43) is fairly straightforward: if there were no coupling to the electrodes, i.e. $\Gamma_L = \Gamma_R = 0$, the poles of g_{12} would coincide with the bonding and antibonding levels of the two-impurity "molecule". The coupling to the electrodes gives them a finite width and produces, generally, two peaks in conductance, whose relative positions in energy change with the bias. The same consideration is valid for longer chains too, and gives an intuitive picture of the formation of the impurity "band" of states. The maximal conductivity $g_{12} = q^2/\pi\hbar$ occurs when $\epsilon_1 = \epsilon_2, \Gamma_L \Gamma_R = t_{12}^2 = \Gamma_2^2$, where Γ_2 is the width of the two-impurity resonance, and it corresponds to the symmetrical position of the impurities along the normal to the contacts separated by a distance equal to half of the molecule length, $r_{12} = L/2$. The important property of the two-impurity case is that it produces negative differential resistance (NDR). Indeed, under external bias voltage the impurity levels shift as

$$\epsilon_i = \epsilon_{i0} + qVz_i/L, \tag{20.44}$$

where z_i are the positions of the impurity atoms counted from the center of the molecule. Due to disorder in the film, under bias voltage the levels will be moving in and out of resonance, thus producing NDR peaks on the I-V curve. The most pronounced negative differential resistance is presented by a gold impurity next to a Cn chain with an H-vacancy on one site, Figure 20.14b (the defect corresponds to a dangling bond). The defects result in two resonant peaks in transmission. Surprisingly, the H vacancy (dangling bond) has an energy very close to the electrode Fermi level F, with $\epsilon_i = -0.1$eV (Figure 20.14b, right peak). The relative positions of the resonant peaks move with an external bias and cross at 1.2V, producing a pronounced NDR peak in the I-V curve, Figure 20.15c. No NDR peak is seen in the case of an Au impurity and a kink C=C on the chain because the energy of the kink level is far from that of the Au 6s impurity level. The calculated values of the peak current through the molecules were large: $I_p \approx 90$ nA/molecule for an Au impurity with H vacancy, and ≈ 5 nA/molecule for double Au impurities.

We have observed a new mechanism for the NDR peak in a situation with two Au impurities in the film. Namely, Au ions produce two sets of deep impurity levels in Cn films, one stemming from the 6s orbital, another from the 5d shell, as clearly seen in Figure 20.15b (inset). The 5d-states are separated in energy from 6s, so that now the tunneling through s-d pairs of states is allowed in addition to s-s tunneling. Since the 5d-states are at a lower energy than the s-state, the d- and s-states on different Au ions will be aligned at a certain bias. Due to the different angular character of those orbitals, the tunneling between the s-state on the first impurity and a d-state on another impurity will be described by the hopping integral analogous to the Slater-Koster $sd\sigma$ integral. The peak current in that case is smaller than for the pair Au-H vacancy, where the overlap is of $ss\sigma$ type (cf. Figures 20.15b, c).

Thiophene molecules behave very differently since the π −states there are conjugated and, consequently, the HOMO-LUMO gap is much narrower, just below 2 eV. The tail of the HOMO state in the T3 molecule (with three rings) has a significant presence near the electrode Fermi level, resulting in a practically "metallic" density of states and hence ohmic I-V characteristic. This behavior is quite robust and is in apparent disagreement with experiment, where tunneling has been observed [85]. However, in actual thiophene devices the contact between the molecule and electrodes is obviously very poor, and it may lead to unusual current paths and temperature dependence [95].

We have presented the first parameter-free DFT calculations of a class of organic molecular chains incorporating single or double point defects. The results suggest that the present generic defects produce deep impurity levels in the film and cause a resonant tunneling of electrons through the film, strongly dependent on the type of defects. Thus, a missing hydrogen produces a level (dangling bond) with an energy very close to the Fermi level of the gold electrodes F. In the case of a single impurity, it produces steps on the I-V curve when one electrode's Fermi level aligns with the impurity level under a certain bias voltage. The two-defect case is much richer, since in this case we generally see a formation of the negative-differential resistance peaks. We found that the Au atom together with the hydrogen vacancy (dangling bond) produces the most pronounced NDR peak at a bias of 1.2V in C13. Other pairs of defects do not produce such spectacular NDR peaks. A short range impurity potential model reproduces the data very well, although the actual lineshape is different.

There is a remaining question of what may cause the strong temperature dependence of conductance in "simple" organic films like $[CH_2]_n$. The activation-like conductance $\propto \exp(-E_a/T)$ has been reported with a small activation energy $E_a \sim 100 - 200$meV in alkanes [87, 94] and even smaller, 10-100 meV, in polythiophenes [95]. This is much smaller than the value calculated here for alkanes and expected from electrical and optical measurements on C_n molecules, $E_a \sim \Delta \sim 4$eV [52], which correspond nicely to the present results. In conjugated systems, however, there may be rather natural explanation of small activation energies. Indeed, the HOMO in T3 polythiophene on gold is dramatically broadened, shifted to higher energies and has a considerable weight at the Fermi level. The upward shift of the HOMO is just a consequence of the work function difference between gold and the molecule. In the presence of (inevitable) disorder in the film some of the electronic states on the molecules will be localized in the vicinity of E_F. Those states will assist the thermally activated hopping of holes within a range of small activation energies $\lesssim 0.1$ eV. Similar behavior is expected for Tour wires [94], where $E_F -$ HOMO ~ 1 eV [50](c), if the electrode-molecule contact is poor, as is usually the case.

With regards to carrier hopping in monolayers of saturated molecules, one may reasonably expect that in many studied cases the organic films are riddled with metallic protrusions (filaments), emerging due to electromigration in a very strong electric field, and/or metallic, hydroxyl, etc. inclusions [55, 57]. It may result in a much smaller tunneling distance d for the carriers and the image charge lowering of the barrier. The image charge lowering of the barrier in a gap of width d is $\Delta U = q^2 \ln 4/(\epsilon d)$, meaning that a decrease of about 3.5 eV may only happen in an unrealistically narrow gap $d = 2 - 3\,\text{Å}$ in a film with dielectric constant $\epsilon = 2.5$, but it will add to the barrier lowering. More detailed characterization and theoretical studies along these lines may help to resolve this very unusual behavior. We note that such a mechanism cannot explain the crossover with temperature from tunneling to hopping reported for single molecular measurements, which has to be a property of the device, but not a single molecule [94].

20.7 Conclusions

Studying molecules as possible building blocks for ultradense electronic circuits is a fascinating quest that spans of over 30 years. It was inspired decades ago by the notion that silicon technology is approaching its limiting feature size, estimated at around 1985 to be about $1\,\mu$m [98]. More than thirty years later and with FET gate lengths getting below 10nm [99], the same notion that silicon needs to be replaced at some point by other technologies floats again. We do not know whether alternatives will continue to be steamrolled by silicon technology, which is a leading nanotechnology at the moment, but the mounting resistance to the famed Moore's law requires to look hard at other solutions for power dissipation, leakage current, crosstalk, speed, and other very serious problems. There are very interesting developments in studying electronic transport through molecular films but the mechanisms of some observed conductance "switching" and/or nonlinear electric behavior remain elusive, and this interesting behavior remains intermittent and not very reproducible. Most of the currently observed swithing is extrinsic in nature. For instance, we have discussed the effect of molecule-electrode contact: the tilting of the angle at which the conjugated molecule attaches to the electrode may dramatically change its conductance, and that probably explains extrinsic "telegraph" switching observed in Tour wires [24, 34] and molecule reconfigurations may lead to similar phenomena in other systems [66]. Defects in molecular films have also been discussed and may result in spurious peaks in I-V curves. We have outlined some designs of the molecules that may demonstrate rectifying behavior, which we call molecular "quantum dots". We have shown that at least in some special cases molecular quantum dots may exhibit fast (~THz) intrinsic switching.

The author is grateful to Jeanie Lau for her Figures 20.6, 20.7 and Jason Pitters and Robert Wolkow for Figures 20.8, 20.9. The work has been partly supported by DARPA.

References

[1] J. M. Tour, *Acc. Chem. Res.* **33**, 791 (2000).

[2] C. Joachim, *Nanotechnology* **13**, R1 (2002).

[3] A. Aviram and M. A. Ratner, *Chem. Phys. Lett.* **29**, 277 (1974).

[4] A. S. Martin, J. R. Sambles, and G. J. Ashwell, *Phys. Rev. Lett.* **70**, 218 (1993); R. M. Metzger, B. Chen, U. Höpfner, M. V. Lakshmikantham, D. Vuillaume, T. Kawai, X. Wu, H. Tachibana, T. V. Hughes, H. Sakurai, J. W. Baldwin, C. Hosh, M. P. Cava, L. Brehmer, and C. J. Ashwell, *J. Am. Chem. Soc.* **119**, 10455 (1997).

[5] K. Stokbro, J. Taylor, and M. Brandbyge, *J. Am. Chem. Soc.* **125**, 3674 (2003).

[6] J. C. Ellenbogen and J. Love, *IEEE Proc.* **70**, 218 (1993).

[7] A. S. Davydov, *Theory of Molecular Excitons* (McGraw-Hill, New York, 1962).

[8] F. Gutmann, *Nature* **219**, 1359 (1968).

[9] A. S. Davydov, *J. Theor. Biol.* **66**, 379 (1977); A. S. Davydov and N. I. Kislukha, *Phys. Status Solidi (b)* **75**, 735 (1976).

[10] A. Scott, *Phys. Reports* **217**, 1 (1992).

[11] J. F. Scott, *Ferroelectric Memories* (Springer, Berlin, 2000).

[12] G. Atwood and R. Bez, presentation at ISIF11 (Honolulu, 21-27 April, 2006); W. Y. Cho et al., ISSCC Dig. Tech. papers (2004); G. Wicker, SPIE **3891**, 2 (1999); S.R. Ovshinsky, *Phys. Rev. Lett.* **21**, 1450 (1968).

[13] D. J. Jung, K. Kim, and J. F. Scott, *J. Phys. Condens. Mat.* **17**, 4843 (2005).

[14] Y. Hiranaga and Y. Cho, 11th Intl. Mtg. Ferroel. (IMF 11, Iguassu Falls, Brazil, Sep 5–9, 2005).

[15] C. Li, D. Zhang, X. Liu, S. Han, T. Tang, C. Zhou, W. Fan, J. Koehne, J. Han, M. Meyyappan, A. M. Rawlett, D. W. Price, and J. M. Tour, *Appl. Phys. Lett.* **82**, 645 (2003); M. A. Reed, J. Chen, A. M. Rawlett, D. W. Price, and J. M. Tour, *Appl. Phys. Lett.* **78**, 3735 (2001).

[16] T. I. Kamins, Interface **14**, 46 (2005); Self-assembled semiconductor nanowires, in *The Nano-Micro Interface*, edited by H. J. Fecht and M. Werner (Wiley-VCH, 2004), p. 195.

[17] E. Rabani, D. R. Reichman, P. L. Geissler, and L. E. Brus, *Nature* **426**, 271 (2003).

[18] M. C. Petty, *Langmuir-Blodgett Films* (Cambridge University Press, Cambridge, 1996).

[19] A. Ulman, *Characterization of Organic Thin Films* (Butterworth-Heinemann, Boston, 1995).

[20] G. Snider, P. Kuekes, and R. S. Williams, *Nanotechnology* **15**, 881 (2004).

[21] B. Stadlober, M. Zirkl, M. Beutl, G. Leising, S. Bauer-Gogonea, and S. Bauer, *Appl. Phys. Lett.* **86**, 242902 (2005); J. M. Shaw and P. F. Seidler, IBM *J. Res. Dev.* **45**, 3 (2001).

[22] S. Datta *et al.*, *Phys. Rev. Lett.* **79**, 2530 (1997).

[23] A. M. Bratkovsky and P. E. Kornilovitch, *Phys. Rev. B* **67**, 115307 (2003).

[24] P. E. Kornilovitch and A. M. Bratkovsky, *Phys. Rev. B* **64**, 195413 (2001).

[25] N. Ness, S. A. Shevlin, and A. J. Fisher, *Phys. Rev. B* **63**, 125422 (2001).

[26] W. P. Su and J. R. Schrieffer, *Proc. Natl. Acad. Sci.* **77**, 5626 (1980); S. A. Brazovskii, *Sov. Phys. - JETP* **51**, 342 (1980).

[27] S. A. Brazovskii and N. N. Kirova, Zh. Eksp. Teor. Fiz. Pis'ma Red. **33**, 6 (1981) [JETP Lett. **33**, 4 (1981)].

[28] A. Feldblum *et al.*, *Phys. Rev. B* **26**, 815 (1982).

[29] R. R. Chance, J. L. Bredas, and R. Silbey, *Phys. Rev. B* **29**, 4491 (1984).

[30] M. G. Ramsey *et al.*, *Phys. Rev. B* **42**, 5902 (1990).

[31] D. Steinmuller, M. G. Ramsey, and F. P. *Netzer, Phys. Rev. B* **47**, 13323 (1993).

[32] L. S. Swanson *et al.*, *Synth. Metals* **55**, 241 (1993).

[33] J. Chen, M. A. Reed, A. M. Rawlett, and J. M. Tour, *Science* **286**, 1550 (1999).

[34] Z. J. Donhauser, B. A. Mantooth, K. F. Kelly, L. A. Bumm, J. D. Monnell, J. J. Stapleton, D. W. Price, Jr., A. M. Rawlett, D. L. Allara, J. M. Tour, and P. S. Weiss, *Science* **292**, 2303 (2001); Z. J. Donhauser, B. A. Mantooth, T. P. Pearl, K. F. Kelly, S.U. Nanayakkara, and P. S. Weiss, Jpn. *J. Appl. Phys.* **41**, 4871 (2002).

[35] C. P. Collier, E. W. Wong, M. Belohradský, F. M. Raymo, J. F. Stoddart, P. J. Kuekes, R. S. Williams, and J. R. Heath, *Science* **285**, 391 (1999); C. P. Collier, G. Mattersteig; E. W. Wong, Y. Luo, K. Beverly, J. Sampaio, F. M. Raymo, J. F. Stoddart, J. R. Heath, *Science* **289**, 1172 (2000).

[36] Y. Chen, D. A. A. Ohlberg, X. Li, D. R. Stewart, and R. S. Williams, J. O. Jeppesen, K. A. Nielsen, and J. F. Stoddart, D. L. Olynick, and E. Anderson, *Appl. Phys. Lett.* **82**, 1610 (2003); Y. Chen, G. Y. Jung, D. A. A. Ohlberg, X. Li, D. R. Stewart, J. O. Jeppesen, K. A. Nielsen, J. F. Stoddart, and R. S. Williams, *Nanotechnology* **14** 462 (2003).

[37] D. R. Stewart, D. A. A. Ohlberg, P. A. Beck, Y. Chen, R. S. Williams, J. O. Jeppesen, K. A. Nielsen, and J. F. Stoddart, *Nanoletters* **4**, 133 (2004).

[38] *Molecular Switches*, edited by Ben L. Feringa (Wiley-VCH, 2001).

[39] G. Zheng, F. Patolsky, Yi Cui, W.U. Wang, C.M. Lieber, *Nature Biotechnology* **23**, 1294 (2005).

[40] A. Bachtold, P. Hadley, T. Nakanishi, and C. Dekker, *Science* **294**: 1317 (2001); P. G. Collins, M. S. Arnold, and P. Avouris, *Science* **292**, 706 (2001); T. Rueckes, K. Kim, E. Joselevich, G. Y. Tseng, C.-L. Cheung, and C. M. Lieber, *Science* **289**, 94 (2000).

[41] M. Ouyang and D. D. Awschalom, *Science* **301**, 1074 (2003).

[42] C. Krzeminski, C. Delerue, G. Allan, D. Vuillaume, and R. M. Metzger, *Phys. Rev. B* **64**, 085405 (2001).

[43] C. Zhou, M. R. Deshpande, M. A. Reed, L. Jones II, and J. M. Tour, *Appl. Phys. Lett.* **71**, 611 (1997).

[44] Y. Xue, S. Datta, S. Hong, R. Reifenberger, J. I. Henderson, C. P. Kubiak, *Phys. Rev. B* **59**, 7852(R) (1999).

[45] J. Reichert, R. Ochs, D. Beckmann, H. B. Weber, M. Mayor, and H. v. Löhneysen, *Phys. Rev. Lett.* **88**, 176804 (2002).

[46] P. E. Kornilovitch, A. M. Bratkovsky, and R. S. Williams, *Phys. Rev. B* **66**, 165436 (2002).

[47] S. Lenfant, C. Krzeminski, C. Delerue, G. Allan, D. Vuillaume, *Nanoletters* **3**, 741 (2003).

[48] B. Larade and A. M. Bratkovsky, *Phys. Rev. B* **68**, 235305 (2003).

[49] S. Chang, Z. Li, C. N. Lau, B. Larade, and R. S. Williams, *Appl. Phys. Lett.* **83**, 3198 (2003).

[50] J. Taylor, H. Guo and J. Wang, *Phys. Rev. B* **63**, R121104 (2001); *ibid.* **63**, 245407 (2001); A. P. Jauho, N. S. Wingreen and Y. Meir, *Phys. Rev. B* **50**, 5528 (1994).

[51] A. S. Alexandrov and A. M. Bratkovsky, *Phys. Rev. B* **67**, 235312 (2003).

[52] C. Boulas, J. V. Davidovits, F. Rondelez, and D. Vuillaume, *Phys. Rev. Lett.* **76**, 4797 (1996).

[53] E. E. Polymeropoulos and J. Sagiv, *J. Chem. Phys.* **69**, 1836 (1978).

[54] R. H. Tredgold and C. S. Winter, *J. Phys. D* **14**, L185 (1981).

[55] H. Carchano, R. Lacoste, and Y. Segui, *Appl. Phys. Lett.* **19**, 414 (1971); N. R. Couch, B. Movaghar, and I. R. Girling, *Sol. St. Commun.* **59**, 7 (1986).

[56] I. Shlimak and V. Martchenkov, *Sol. State Commun.* **107**, 443 (1998).

[57] C. N. Lau, D. Stewart, R. S. Williams, and D. Bockrath, *Nano Lett.* **4**, 569 (2004).

[58] J. Rodriguez Contreras, H. Kohlstedt, U. Poppe, R. Waser, C. Buchal, and N. A. Pertsev, *Appl. Phys. Lett.* **83**, 4595 (2003).

[59] R. A. Wassel, G. M. Credo, R. R. Fuierer, D. L. Feldheim, C. B. Gorman, *J. Am. Chem. Soc.* **126**, 295 (2004).

[60] J. He and S. M. Lindsay, *J. Am. Chem. Soc.* **127**, 11932 (2005).

[61] J. Gaudioso, L. J. Lauhon, and W. Ho, *Phys. Rev. Lett.* **85**, 1918 (2000).

[62] S.-W. Hla, G. Meryer, and K.-H. Rieder, *Chem. Phys. Lett.* **370**, 431 (2003).

[63] G. Yang and G. Liu, *J. Phys. Chem. B* **107**, 8746 (2003).

[64] A. Salomon, R. Arad-Yellin, A. Shanzer, A. Karton, and D. J. Cahen, *Am. Chem. Soc.* **126**, 11648 (2004).

[65] N. P. Guisinger, M. E. Greene, R. Basu, A. S. Baluch, M. C. Hersam, *Nano Lett.* **4**, 55 (2004).

[66] J. L. Pitters and R. A. Wolkow, *Nano Lett.* **6**, 390 (2006).

[67] T. Rakshit, G. C. Liang, A. W. Ghosh, and S. Datta, *Nano Lett.* **4**, 1803 (2004).

[68] A. V. Bune, V. M. Fridkin, S. Ducharme, L. M. Blinov, S. P. Palto, A. Sorokin, S. G. Yudin, and A. Zlatkin, *Nature* **391**, 874 (1998).

[69] S. Ducharme, V. M. Fridkin, A. V. Bune, S. P. Palto, L. M. Blinov, N. N. Petukhova, and S. G. Yudin, *Phys. Rev. Lett.* **84**, 175 (2000).

[70] A. M. Bratkovsky and A. P. Levanyuk, *Phys. Rev. Lett.* **87**, 019701 (2001).

[71] A. Bune, S. Ducharme, V. Fridkin, L. Blinov, S. Palto, N. Petukhova, and S. Yudin, *Appl. Phys. Lett.* **67**, 3975 (1995).

[72] T. Furukawa, M. Date, M. Ohuchi, and A. Chiba, *J. Appl. Phys.* **56**, 1481 (1984).

[73] P. E. Kornilovitch, A. M. Bratkovsky, and R. S. Williams, *Phys. Rev. B* **66**, 245413 (2002).

[74] D. V. Averin and K. K. Likharev, in: *Mesocopic Phenomena in Solids*, edited by B.L. Altshuler *et al.* (North-Holland, Amsterdam, 1991).

[75] H. Park, J. Park, A. K. L. Lim, E. H. Anderson, A. P. Alivisatos, and P. L. McEuen, *Nature* **407**, 57 (2000); J. Park, A. N. Pasupathy, J. I. Goldsmith, C. Chang, Y. Yaish, J. R. Petta, M. Rinkoski, J. P. Sethna, H. D. Abruña, P. L. McEuen, and D. C. Ralph, *ibid.* **417**, 722 (2002); W. Liang, M. P. Shores, M. Bockrath, J. R. Long, and H. Park, *ibid.* **417**, 725 (2002).

[76] N. B. Zhitenev, H. Meng, and Z. Bao, *Phys. Rev. Lett.* **88**, 226801 (2002).

[77] A. S. Alexandrov, A. M. Bratkovsky, and R. S. Williams, *Phys. Rev. B* **67**, 075301 (2003).

[78] A. S. Alexandrov and A. M. Bratkovsky, cond-mat/0603467.

[79] Y. Meir and N. S. Wingreen, *Phys. Rev. Lett.* **68**, 2512 (1992).

[80] A. Mitra, I. Aleiner, and A. Millis, *Phys. Rev. Lett.* **94**, 076404.(2006).

[81] D. Mozyrsky, M. B. Hastings, and I. Martin, *Phys. Rev. B* **73**, 035104 (2006).

[82] M. Galperin, M. A. Ratner, and A. Nitzan, *Nano Lett.* **5**, 125.(2005).

[83] A. Mitra, I. Aleiner, and A. Millis, *Phys. Rev. B* **69**, 245302 (2004).

[84] H. Park, J. Park, A. K. L. Lim, E. H. Anderson, A. P. Alivisatos, and P. L. McEuen, *Nature* **407**, 57 (2000); J. Park, A. N. Pasupathy, J. I. Goldsmith, C. Chang, Y. Yaish, J. R. Retta, M. Rinkoski, J. P. Sethna, H. D. Abruna, P. L. McEuen, and D. C. Ralph, *Nature* (London) **417**, 722 (2000);. Liang, W.; *et al. Nature* (London) **417**, 725 (2002).

[85] H. Sakaguchi, A. Hirai, F. Iwata, A. Sasaki, and T. Nagamura, *Appl. Phys. Lett.* **79**, 3708 (2001); X. D. Cui, X. Zarate, J. Tomfohr, O. F. Sankey, A. Primak, A. L. Moore, T. A. Moore, D. Gust, G. Harris, and S. M. Lindsay, *Nanotechnology* **13**, 5 (2002).

[86] W. Wang, T. Lee, and M. A. Reed, *Phys. Rev. B* **68**, 035416 (2003).

[87] D. R. Stewart, D. A. A. Ohlberg, P. A. Beck, C. N. Lau, and R. S. Williams, *Appl. Phys. A* **80**, 1379 (2005).

[88] G. L. Fisher, A. E. Hooper, R. L. Opila, D. L. Allara, and N. Winograd, *J. Phys. Chem. B* **104**, 3267 (2000); K. Seshadri, A. M. Wilson, A. Guiseppi-Elie, D. L. Allara, Langmuir **15**, 742 (1999).

[89] A. V. Walker, T. B. Tighe, O. Cabarcos, M. D. Reinard, S. Uppili, B. C. Haynie, N. Winograd, and D. L. Allara, *J. Am. Chem. Soc.* **126**, 3954 (2004).

[90] M. Pollak and J. J. Hauser, *Phys. Rev. Lett.* **31**, 1304 (1973); I. M. Lifshitz and V. Ya. Kirpichenkov, *Zh. Eksp. Teor. Fiz.* **77**, 989 (1979).

[91] L. I. Glazman and K. A. Matveev, *Sov. Phys. JETP* **67**, 1276 (1988).

[92] Y. Xu, D. Ephron, and M. R. Beasley, *Phys. Rev. B* **52**, 2843 (1995).

[93] C. H. Shang, J. Nowak, R. Jansen, and J. S. Moodera, *Phys. Rev. B* **58**, 2917 (1998).

[94] Y. Selzer, M. A. Cabassi, T. S. Mayer, D. L. Allara, *J. Am. Chem. Soc.* **126**, 4052 (2004).

[95] N. B. Zhitenev, A. Erbe, and Z. Bao, *Phys. Rev. Lett.* **92**, 186805 (2004).

[96] M. J. Frisch, GAUSSIAN98, Revision A.9, Gaussian, Inc., Pittsburgh, PA, 1998.

[97] A. I. Larkin and K. A. Matveev, *Zh. Eksp. Teor. Fiz.* **93**, 1030 (1987).

[98] N. G. Rambidi and V. M. Zamalin, *Molecular Microelectronics: Origins and Outlook* (Znanie, Moscow, 1985).

[99] M. Ieong, B. Doris, J. Kedzierski, K. Rim, M.Yang, *Science* **306**, 2057 (2004).

21

Complexities
of the Molecular
Conductance Problem

Gil Speyer

Richard Akis

David K. Ferry

Abstract

New conductance phenomena observed in isolated individual molecules require theoretical approaches that accurately represent contact geometry and Fermi-level positioning since these factors wield tremendous influence in the behavior of the metal–molecule–metal system. Although this argues that rigorous first-principles calculations must be performed, the various approaches each have significant drawbacks. The linearly independent plane wave basis ensures an accurate potential across the system, but this basis produces an intractably large rank Hamiltonian, and a local orbital basis must then be used. Density functional theory approaches allow the calculation of a Hamiltonian representing a hybrid of a metal and an unsaturated organic compound. However, bandgaps are underestimated, and bond lengths and dissociation energies are incorrectly calculated in this scheme. Using a combination of approaches, including complex bandstructure, density of states calculations, and a highly efficient scattering matrix method, insights about the nature of molecular conduction are drawn in this chapter. In addition, calculations over suites of Hamiltonians—for molecules in both a stretched configuration and at several positions within a vibrational mode—allow the determination of trends despite the known inaccuracies in the calculation. Results investigate the phenomena of conductance changes with stretching and of the unexpected higher conductance of a four-membered oligomer over its three-membered counterpart.

21.1 Introduction

When Mark Reed's experimental group at Yale University published the current–voltage characteristic across a single molecule using their mechanical break junction technique, an entirely new domain of electronic devices became available for research [1]. Shortly after this, other experimental groups were able to perform similar measurements using a variety of different techniques. However, the analysis of the electronic characteristics of isolated individual molecules connected to metallic contacts, although attainable in practice, has proven rather difficult to accomplish in theory. Initial calculations yielded conductance values varying over several orders of magnitude. The details of contact geometry and adsorption chemistry in their myriad variations were discovered to play a pivotal role in theoretical analysis [2]. All the contact specific minutiae could be lumped and parameterized to match experimental measurements, but this ad hoc fitting gave little insight into the nature of molecular conduction. On the other hand, from a first-principles standpoint, the number of contact configurations and variations multiplied like hydra-heads, never seeming to cover all the possibilities, and never satisfactorily producing the quantitative end result.

This chapter examines the problem of molecular conduction with specific attention to the recurrent barriers toward an accurate solution found in the literature. Instead of targeting a numerical range for the conductance, the nature of the problem itself is examined. Foremost stands the electronic potential, the acid test of the effectiveness of the bandstructure calculation. An accurate equilibrium potential calculation establishes a certain confidence in all subsequent non-equilibrium calculations.

Of the multiple approaches that have been employed, each method has its shortcomings. The complex bandstructure gives no insight into the nature of the contact. The inability of localized orbitals to render a metal surface adequately misrepresents the potential. Density functional theory underestimates bandgaps, particularly in unsaturated organic compounds. The aim of using several theoretical methods is to present a constellation of theoretical conclusions by which some insight into experimentally observed phenomena might be understood. This method has been employed by other groups with some success [3].

A companion strategy to this employs the comparison of suites of similar systems. Due to the highly efficient transport kernel used in this work, the conductance can be calculated quickly; therefore, comparisons can be drawn between slightly differing systems to infer insights into experimental phenomena. For this reason, the careful progression of a molecule gradually being stretched and the series of "snapshots" depicting molecular vibrations within a nanojunction are two main examples considered in this work. These specific problems lend themselves well to the analysis used here.

Reviews of the experimental progress in this area initiated by Reed's breakthrough abound. Roughly, methods fall into two camps. While one strives to minimize the size of leads and the gap between them in an attempt to bind individual molecules, the other exploits the regularity of molecules assembled in a monolayer on a gold substrate and employs a conducting atomic force microscope tip in an attempt to contact individual molecules. Figures 21.1 and 21.2 show two pictorial examples of individual molecule binding. The mechanically controllable break (MCB) junction technique marked a critical breakthrough

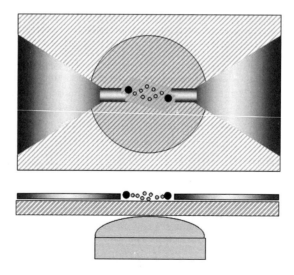

FIGURE 21.1   Rough illustration of MCBJ setup. Top and side views.

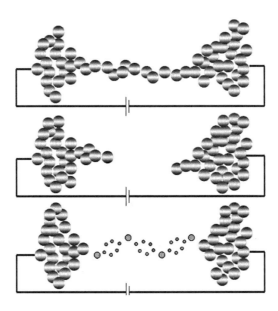

FIGURE 21.2 Electrochemical nanojunction setup. Upon nanowire formation, electromigration bias is briefly reversed, forming a small gap. Molecules can be subsequently deposited.

for molecular conductance investigations. A highly tunable piezotransducer gradually applies pressure to break a narrow gap in a photolithographically defined, vapor-deposited, metallic nanowire attached to a flexible substrate [4]. Within this tiny gap, carefully deposited thiol-terminated molecules in solution could reconnect the contacts, albeit at a much higher resistance. The thiol end groups were key, behaving as molecular "alligator clips," readily relinquishing their hydrogens to bond to the gold surface [5]. A measurement of current across the probes at different biases, allowed for the measurement of conductance as:

$$G(V) = \frac{\partial I(V)}{\partial V}. \tag{21.1}$$

This resulted in reproducible measurements—in the tens of megohm range—for small molecules such as benzenedithiol and xylyldithiol. The second method (shown in Figure 21.2), employs electrochemical junctions, and uses electromigration instead of lithography to fabricate gold nanojunctions [6]. Once again, the technique permits a high level of control over the width of the nanogap, this time via the applied bias. Upon the formation of the atomic bridge, the bias can be slightly reversed, thereby creating a nanogap readily available for molecules. Figure 21.3 shows how molecules arranged on a substrate, typically in a self-assembled monolayer, can be individually probed. In order to facilitate connections to single target molecules, a monolayer of molecules known to be insulating (such as monothiolated alkanes) are sparsely populated with the target molecules. A suspension of gold nanoclusters attach only to the thiolated ends of the target molecules. A gold-plated AFM tip with a nominal applied pressure can then probe the surface. When a connection to a gold nanocluster is made, the connection to individual molecules is measured. Figure 21.3 shows an experiment by Lindsay's group illustrating the aromatic target molecule in the monolayer of dodecanes [7].

FIGURE 21.3 AFM probing of target aromatic molecule embedded in a self assembled monolayer of monothiolated dodecanes.

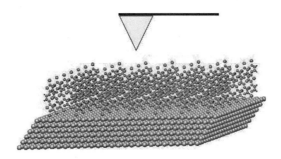

FIGURE 21.4 Terthiophene monolayer prior to lowering and attachment of gold-plated AFM tip.

The experimental setup responsible for the results examined in this chapter employed a gold-plated AFM tip which was repeatedly lowered into a self-assembled monolayer of some chosen molecule on a gold substrate. While monitoring the current from tip to substrate, traces measuring the current would reveal steps corresponding to the quantum conductance through a gold nanowire. On many measurements, even lower conductance steps would be recorded. These unit step multiples indicated individual molecules detaching from the tip until the unit step—indicating one isolated molecule connecting tip to substrate—remained. Figure 21.4 shows an illustration of this setup prior to the gold-plated AFM tip attachment to the thiol of a target molecule in the monolayer—in this case, terthiophene.

Theoretical treatments of the metal–molecule–metal problem must couple the isolated molecule with its electrons bound in discrete molecular orbitals to the metallic contact with its reservoir of electrons filling a continuous band up to the Fermi level. This coupling hybridizes the molecular orbitals with the extended states in the contact defining a combined spectrum for the entire system. An applied bias creates a nonzero transmission across the system which can then be calculated. Hence, two major steps in the calculation, the bandstructure, determined for the equilibrium system, and the transport under bias, can be individually described.

To start with, a discussion of first-principles calculations weighs the merits of various, more rigorous, bandstructure calculations. In our assessment, the advantage of a tractable Hamiltonian offered by the local orbital basis outweighs the inaccurate representation of the potential it brings to the calculation. A review of transport calculation techniques follows, and the rapid and numerically stable scattering matrix method (developed by Usuki and colleagues) is introduced, followed by some discussion of the self-consistent implementation. Complex bandstructure provides further insight into the conductance behavior of the molecules short of describing the dynamics of connecting to a contact.

Conductance calculations over suites of Hamiltonians are used to understand some trends in the behavior of stretched and vibrated molecular systems. Comparisons, which maintain the complexity of the molecular coupling to the metal electrodes, reveal features which can be used to explain phenomena observed in the experiments. For example, variations in molecular orbital coupling to the extended states in the metal help account for the conductance peaks measured with stretching. In a particularly interesting case, a four-membered oligomer shows higher conductance than its three-membered counterpart. The variation of the conductance of these oligomers in vibrational analysis can be used to account for this.

21.2 Theoretical Approaches

21.2.1 Bandstructure

Initial explorations into molecular conductance calculations employed tight-binding models that described the system as atoms with a discrete set of orbitals. Electrons could be characterized as hopping between atomic orbitals as they moved through the system. The Hamiltonian in its matrix representation would have site energies for each orbital j on atom i, as

$$\langle j, i | H | j, i \rangle = \varepsilon_{j,i} \qquad (21.2)$$

for the diagonal terms. Off-diagonal terms, corresponding to the hopping energy for an electron in orbital j of atom i to orbital l on atom k, are

$$\langle j, i | H | l, k \rangle = t_{l,k,j,i}. \qquad (21.3)$$

With this basis chosen, semi-empirical parameters or simple π-orbital terms could then be employed in transmission calculations. Matrices constructed via simple tight-binding, despite their tractability in transport calculations, proved inadequate in the endeavor to understand various molecular conductance phenomena, particularly with respect to the contact. As a result, more rigorous, first-principles approaches were explored.

Electronic structure calculations aim to describe the orbitals (or states) available to carriers, specifically electrons, in the investigated system. After these states are identified, they are populated from lowest to highest, up to a maximum state, the highest occupied molecular orbital (HOMO), which operates as a valence band in conjugated molecules. The very next available state, the lowest unoccupied molecular orbital (LUMO), then serves as a conduction band. If the LUMO is separated by a gap from the filled HOMO (at 0 K), then the Fermi level lies between these two states, emulating an insulator or semiconductor.

The computation of the electronic structure of the combined metal–molecule–metal system couples two disparate systems into one calculation. Trade-offs arise between the advantages and disadvantages of particular theories as they address both systems. More specifically, the density functional theory approach, best suited for the delocalized electron density of the metal, will inevitably be a poor choice to describe the excited states and bond character of the non-uniform density profile of the organic system. In a similar trade-off, while a localized orbital basis can be efficiently employed to give the molecular orbital characteristics, metal surface dipoles and work functions cannot be satisfactorily derived in this framework. The vacuum must be treated, but the plane wave basis, so ideal for a metal surface in isolation, creates a prohibitively high rank Hamiltonian matrix when attempting the fine structure of the combined system.

Density functional theory uses a simple vector, the density functional, to replace the wavefunction with its 10^{23} coordinates. This historic theoretical step, establishing a one-to-one correspondence between a one-electron density and a one-body potential, enabled calculations which had previously been computationally prohibitive. One complexity, however, is that the exact functional is not known, so empirical functions—such as the local density approximation (LDA), generalized gradient approximation (GGA), and hybrid functionals—are employed. The problems don't end there. LDA does not address excited states, and underestimates energy gaps, particularly for unsaturated organic compounds [8–10]. Calculations using GGA overestimate the gaps [11]. Hybrid functional approaches have been shown to treat open-shell systems inadequately and lead to artificial degeneracies in the tunneling states between the electrodes, thereby overestimating the transmission [12]. Even regardless of any approximation, the typical DFT treatment of exchange and correlation in quasi-static nonequilibrium with the standard equilibrium expression has been shown to be unjustified at weak couplings, leading to overestimation of the level broadening and thence the transmission [13].

The Hartree–Fock theory uses a Slater determinant to describe the many-electron wavefunction. This treatment of correlation and exchange differs greatly from the Fermi liquid treatment commonly used in DFT. This latter treatment has been shown to result in improper bond dissociation calculations, such as those needed in this work [14]. Exact exchange has been proposed to ameliorate the low bandgaps that plague DFT, but it introduces considerably more computation. For example, the VASP group estimates two orders of magnitude more computation time for the exact exchange functionals in VASP5 (scheduled for release in mid-2007 [15]). Other groups describe DC conductance corrections to the exchange and correlation functionals using time-dependent current–density–functional theory [16]. Hartree–Fock methods, on the other hand, have their own shortcomings in the calculation of opposite spin correlations,

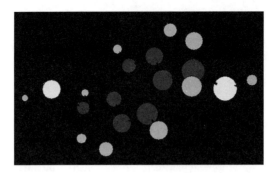

FIGURE 21.5   (see color insert following page 5-6) Isolated xylyldithiol. Carbons are grey, sulfurs yellow, and hydrogens blue. (P)aracarbons and (O)rthocarbons are identified. Nonplanarity of molecule is evident. (From G. Speyer, R. Akis, and D.K. Ferry, *J. Vac. Sci. Tech. B*, 24(4), 1987–1991, 2006. With permission.)

one of which can be corrected using multiple Slater determinants in the computationally demanding configuration–interaction method [17].

It is instructive to examine the differences between the theories. The much investigated xylyl–dithiol molecule, shown in Figure 21.5, will serve as the model molecule. Its basic construction is a benzene ring connected to two methyl groups at the (1,4) positions (or para-carbon sites), which, in turn, connect to two thiol groups. In Figure 21.6, the carbon–carbon distances within the benzene ring are calculated as the ring is stretched apart. All the lines show an abrupt change at the point of detachment. The definition of this point is different between the two theories. For DFT, the molecule was stretched between two incrementally separated gold layers in the calculation. For the Hartree–Fock, the gold slabs were not included, so the outer sulfurs were gradually pulled apart until one detached from its carbon. Since the sulfur–carbon detachment ionizes the molecule differently, the post-detachment behaviors shouldn't be compared. More significantly, the DFT code represents the molecule bonded to gold, and charge transfer occurs in this situation. However, examining the isolated molecule with DFT, the bond lengths within the benzene ring, which is somewhat

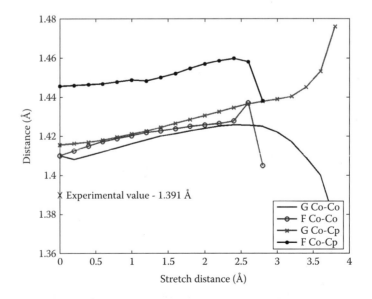

FIGURE 21.6   Interatomic distances for the stretched benzene ring orthocarbons (Co) and paracarbons (Cp) as calculated in DFT (F) and HF (G) codes.

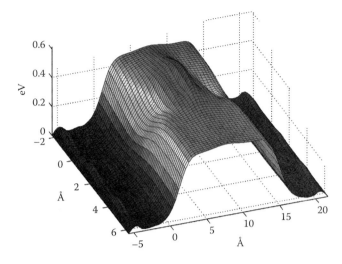

FIGURE 21.7 (see color insert) Inaccurate potential as calculated from local orbitals.

isolated from the contacts, shows the same character. What is of interest is the nature of the conjugation of the π-electrons in the benzene ring. With no stretching, the inter-carbon distances should all be the same as the Hartree–Fock (G) data indicates. Also, this data is close to the well-observed experimental value attained for this distance. Clearly, the DFT code (F) does not show the conjugation well. Moreover, the bond dissociations effected by the stretching would have an expected inaccuracy from DFT [14].

From a theoretical standpoint, plane waves, being an unbiased basis set and having computational simplicity, offer a definite advantage over localized orbitals. In the specific case of metals, which can be characterized as possessing highly delocalized orbitals, the plane wave basis has been proven demonstrably superior [18,19]. This is confirmed in practice, although the potential for the combined system can only be accurately rendered using plane waves. The inability to render a vacuum within the local orbital basis prohibits an accurate surface dipole, as shown in Figure 21.7. This plot has been smoothed to emphasize the work function, given as the potential difference between the contact regions on the left and right and the gap in the middle, which barely reaches 0.5 eV. The potential as calculated in VASP, as shown in Figure 21.8. The plane in which the potential is calculated is shown on the left, chosen in a space between molecules in the monolayer. The top potential plot shows the gold layers from -15 to $2\,\text{Å}$ and beginning again at $13\,\text{Å}$. Between these layers, the vacuum level plateaus around roughly 7 eV. The Fermi level from the calculation is indicated by the blue line at roughly 2 eV. The accuracy of the calculation is therefore demonstrated in the work function of gold, which is known to be ~ 5 eV. Figure 21.9 shows a plane where the molecules are included. The plane of the calculation has been chosen so that the benzene ring of the xylyl is visible. Unfortunately, the large number of plane waves needed to represent the system accurately necessitates a prohibitively large rank Hamiltonian matrix. If we examine projected density of states (PDOS) plots of the molecular atoms, using both the local orbital and the plane wave basis codes, shown in Figure 21.10, we can see that although the bandgaps and general character of the orbitals is similar in the calculations, the Fermi-level position (at the origin) does differ. Since conductance is very sensitive to the location of the Fermi level in the gap, some quantitative error in local orbital calculations, such as the results in this work, must be conceded.

21.2.2 Transport

For the transport calculation, preliminary efforts split into two general directions. The first examined direct diagonalization techniques. In this scheme, an often large rank matrix was diagonalized in order to obtain the eigenvectors for a transmission calculation. With a simple tight-binding approach, such as extended-Hückel-based Hamiltonians, which simplified to the treatment of π-electron systems, these

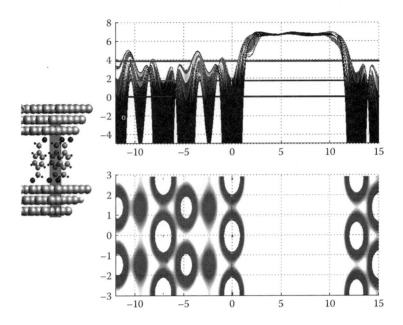

FIGURE 21.8   (see color insert) Potential in plane shown in figure on the left as calculated in VASP. Variation occurs around gold nuclei, increasing up from surface and plateauing at vacuum level. Red (top) and blue (bottom) lines correspond to HOMO and LUMO levels in molecule. Green (middle) line corresponds to metallic Fermi level.

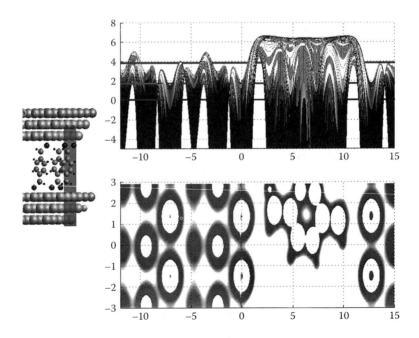

FIGURE 21.9   (see color insert) Potential in plane of molecule as calculated in VASP.

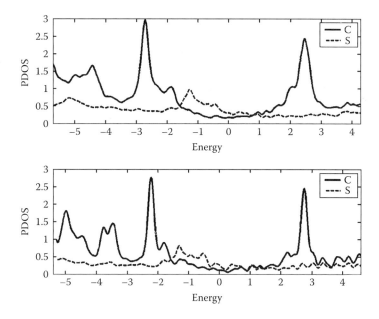

FIGURE 21.10 (see color insert) PDOS of xylyldithiol attached to gold. Local orbitals (top) and VASP (bottom). (From G. Speyer, R. Akis, and D.K. Ferry, *J. Vac. Sci. Tech. B*, 24(4), 1987–1991, 2006. With permission.)

approaches were computationally tractable and yielded promising results [20,21]. The Lippmann–Schwinger formalism could be used to demonstrate the effect of the coupling of the metallic (Bloch) wavefunctions Φ_0 and the isolated molecular orbitals on the wavefunctions for the full interacting system [22]:

$$|\Psi\rangle = |\Phi_0\rangle + G_0(E)V\,|\Psi\rangle,\tag{21.4}$$

where V represents the perturbative interactive coupling, and the propagator G_0 represents the summed separate Green's functions for the isolated left lead, right lead, and molecule. However, in the quest for quantitative accuracy, first-principles approaches, with their atomistic rigorous depth, attracted much attention. As a result, instead of one π-type orbital per atom, multiple atomic orbitals were considered and, for the metal contacts, this implied a full treatment of the atomic d-orbitals over a large enough contact to be considered metallic. Things became complex very quickly.

The second widely used method employed non-equilibrium Green's functions (NEGFs) and partitioned the Hamiltonian between a generalized "extended molecule" H_M, semi-infinite contact regions H_l and H_R, and their interaction terms V[23,24]:

$$H_{tot} = H_L + H_M + H_R + V.\tag{21.5}$$

The "extended molecule" would subsume some amount of the metal atoms into the molecule, thereby garnering all contact and surface-specific information into the finite transport region. In this way, the semi-infinite contacts could be expected to represent bulk material. The rank of the matrices defining the "extended molecule" could be much smaller than the full system used in the direct diagonalization approaches, thereby making the first-principles approaches more tractable.

Typically, a continued fraction transcendental equation determined the left and right semi-infinite contact interaction terms [25]. The trace of the product of interaction terms and the "extended molecule" defined Green's function terms yielded the transmission:

$$T(E) = Tr\left[\Gamma_L(E)G_M^+(E)\Gamma_R(E)G_M(E)\right].\tag{21.6}$$

An integration of the transmission as calculated by either method over some energy window, multiplied by the quantum conductance, would then yield the current through the system:

$$I(V) \approx \frac{2e}{h} \int_{E_F}^{E_F+eV} dE \, T(E). \tag{21.7}$$

At the outset, both of these efforts met some difficulty in including the contacts. As the Fermi level and the contact distance were both difficult to specify, and very influential on the conductance calculation, transmission plots were used to tell much of the story [26,27], and the same two important conclusions were drawn: (1) If the Fermi level were placed at some particular point along the energy axis, the transmission could be expected to vary from metallic to highly insulating behavior; and (2) by increasing or decreasing the coupling constant to the contact, the transmission curve would rise or fall by orders of magnitude. This implied that the conductance, an integration of the transmission over some energy window, would also be dramatically affected. Thus, the most salient point, which emerged from these studies, identified the Fermi level and interface specifics as the key determiners of molecular conductance. These factors were not easily deduced. Conductance calculations varied over several orders of magnitude with arbitrary variation of the Fermi level and contact coupling [28]. One group was even able to demonstrate the possibility that experiments might be measuring conductance through two molecules in interacting monolayers rather than a single metal–molecule–metal connection [29]. The poles of the molecular Green's function could be effectively "tuned" to explain any experimental measurement, or predict myriad phenomena [30]. As a result, the push towards more sophisticated calculations began.

To begin, theorists tackled the conundrum of the interface. Few disputed its significance in molecular conduction, but the manner of its representation in the calculation evoked a wide range of approaches and criticisms [31]. Many approached the problem by treating the interface as a simple Schottky–Mott interface, where the vacuum levels are simply aligned, and the Fermi level determines the electron and hole barriers, φ_e and φ_h. However, this neglects the charge transfer across the interface, which causes a vacuum level shift—and, thus, an interface dipole, Δ, forms. Experimentally, ultraviolet photoemission spectroscopy (UPS) has been used to determine this shift [32]. Results of these types of experiments show that interface dipole barriers form at nearly all interfaces [33]. The nature of the dipole is determined by the extent of the charge transfer between metal and molecule and by the effect of the displacement and rearrangement of the metal surface charge [34].

Several studies noted the significance of the angle at which certain molecules are fixed to the contact, suggesting the π-orbital coupling was maximized when these molecules lay orthogonal to the surface plane [35]. This stems from the directional dependence of the p-orbitals, seen to greater effect in the sp^2 hybrid orbitals characteristic of the unsaturated organic compounds often studied in these systems [36,37]. In order to best represent the statistical distribution of experimental configurations, multiple calculations over a variety of angles have been analyzed to understand the conductance dependence [38].

Thiol endgroups are often employed in experiments because they readily attach to gold. The point of the gold-thiol attachment gives rise to further controversy. Various theoretical calculations have specified the most energetically favorable contact to be the fcc-hollow site [39,40] and the bridge-like position (between fcc-hollow and bridge sites) [41], while an experimental paper favors the on-top site [42]. Another theoretical study suggests that lateral degrees of freedom preclude the preference of one contact over another [43]. Some studies which have examined conduction through all varieties note better conductance in the hollow site and bridge positions, while on-top site is noted for lower conduction, higher sensitivity to tilting angles, and highly nonlinear current-voltage behavior [44–46]. Yet others find better conduction properties in the on-top compared to the hollow site [47–49]. There has even been some suggestion that the hydrogens may not always detach upon sulfur attachment to gold, thereby influencing this contact in yet another unexpected way [50].

One aspect of the interface which may be important involves the surface reconstruction of the Au (111) surface. Only the top layer has been shown to relax, but this relaxation results in a 0.05-Å corrugation, as well as a uniaxial contraction along the [110] direction [51]. A large supercell size, and therefore a large computational load, would be required by a model aiming to include these surface effects adequately.

Many first-principles codes [52,53] cannot render the periodic cell, thereby making the infinite metal slab calculation impossible. Instead, a gold cluster is used to represent the contact [54]. Although the cluster has been known to be a poor approximation for metallic behavior, the idea of using disordered gold contacts, instead of perfect periodic atomic arrangements, could lend some insight into the true behavior of the experimental configurations, even beyond those which explicitly employ gold nanoclusters. Some theoretical investigations into grain-like gold leads have revealed low temperature Coulomb blockade features in the current-voltage trace [55]. Of course, gold contacts need not be the only contact metal, and investigations of graphitic and silver electrodes have revealed interesting trends [56,57].

Next, the potential profile across the device, specifically when a bias has been applied, needs to be addressed. Since initial calculations tend to treat the potential drop as uniform or as some average value when the transmission was calculated, the inaccuracy of calculations seemed to suggest some fault in these approximations [58]. The obvious remedy for this perceived shortcoming involves a self-consistent calculation of the potential in response to the new charge distribution under the applied bias. This new charge distribution comprises not only the static redistribution of charge due to the applied bias, but also the dynamic effects of the current flow through the system. Typically, the first-principles codes, which are used to determine the equilibrium bandstructure of the system, are not equipped to handle dynamic systems. Thus, the self-consistent calculation necessitates a loop within the separate transport kernel. Since this routine tends to exclude the force routines from the bandstructure calculation, the movement of nuclei with bias, for example, cannot be handled.

Certain groups immediately saw that the two routines had to be fully integrated to deal with this level of sophistication. As an initial step, a partial approach looked only at the static redistribution of charge, assuming the tunneling currents made a small contribution to the accurate potential [59]. These revealed a strongly nonlinear potential drop, focused mainly in the contact regions. When the computational complexity of the combined programs was actually confronted, initial findings showed that the current-induced corrections to the conduction and force behavior were only noticeable at markedly large, perhaps even unrealistic, biases [60]. The Trans-Siesta project was one of the first to publish results for the fully developed, combined bandstructure-transport code [61], though many other groups soon followed suit [62–67]. The integration of the lesser Green's function for the density matrix:

$$D = \frac{1}{2\pi i} \int_{-\infty}^{\infty} d\varepsilon\, G^<(\varepsilon), \qquad (21.8)$$

where

$$G_{L,R}^<(\varepsilon) = G_M \Gamma_{L,R} G_M^+, \qquad (21.9)$$

required a complex integration, which means that, for the values along the chosen contour, the Green's function and interaction terms must be calculated. The density matrix is then projected onto the local basis to get the density

$$n(r) = \sum_{\mu,\nu} \phi_\mu(r)\, Re[D_{\mu,\nu}]\phi_\nu(r). \qquad (21.10)$$

This density, in turn, enters Poisson's equation, yielding the Hartree potential used to update the Hamiltonian and the states of the system. Hence, the recursively solved transcendental equation, the complex integration, and all the associated diagonalizations in the self-consistency loop incur a considerable computational burden.

Theoretical studies have revealed that the potential, much like the interface, has its own layers of complexity. Studies into the variation of potential with screening length and molecular wire effective "width" show that the profile cannot be arbitrarily assigned [68]. Molecule charge transfer, polarization by the applied electric field, the molecular screening length, and the various binding geometries all play varying roles [69–71]. Recent investigations into thermoelectric effects have lent another dimension to potential influencing phenomena [72].

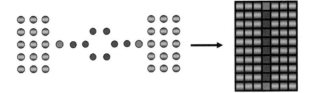

FIGURE 21.11 (see color insert) Pictorial representation of assignment of site terms into scattering matrix slice scheme described in text.

The Landauer formula, which lies at the core of the transport calculation in this work, expresses the conductance of a small bridge between two electron reservoirs [73]. Since the conductance of this bridge is represented as a quantum mechanical transmission probability, the bridge (or, in our case, molecular) dimensions must be on the length scale of the coherence length. For this reason, all of the molecules examined are very small. Coherent transport can be characterized as either resonant with an extended molecular level or exponentially decaying as a tunneling event. Longer wires necessitate the dynamic inclusion of inter-electronic interactions and of vibrational coupling.

Our approach exploits the computational efficiency and numerical stability of a scattering matrix variant of the Lippmann–Schwinger equation [74]. By discretizing the system into sites on a lattice, the transmission is solved through iterative matrix calculations and translations at each slice. Starting with the Hamiltonian matrix elements and interatomic distances for the system, the molecular and gold atomic site energies are mapped onto a discrete lattice, and an incident electron flux is then propagated slice by slice. The transmission coefficients at the last slice enter the Landauer–Büttiker formula to yield the conductance. This scheme is shown pictorially in Figure 21.11.

Because the system is three dimensional, the slices represent a collapsed space where the terms of the Hamiltonian preserve the spatial relationships. In addition, each atom may have multiple orbitals with distinct coupling information to the neighboring atomic orbitals, as well as to other orbitals on the same atom. For the periodic gold slabs, each atomic orbital is assigned a row in the matrix H_l, with nonzero off-diagonal terms representing the hopping terms to other atoms in the slab. Edge atoms have these hopping terms to their neighbors periodically linked across the cell. In fact, the spread of the metallic wavefunction leaves few nonzero matrix elements in the matrix. The molecule is collapsed into a single slice as well. The cell is made large so no nonzero periodic matrix elements enter this slice. In addition the molecular wavefunctions are much more localized, so this matrix is sparser than that for the gold slab slices. The coupling matrices between the slices, $H_{l,l-1}$ and $H_{l,l+1}$, comprise the hopping terms between adjacent atoms. Since this formalism cannot treat next-nearest-slice coupling, this information cannot be included in the calculation. This includes any tunneling information from the left to right contact. Therefore, the conductance calculations represent the transmission exclusively through the molecule.

The vector representing the wavefunction, therefore, has a length N equal to the number of atoms in the slab, usually in some sequential order, multiplied by the number of orbitals per atom. Typically, hydrogens have one orbital; carbons, nitrogens, and sulfurs have four; and golds have nine orbitals. An incoming wavefunction, ψ, through the left contact can be propagated through each slice l as:

$$\begin{bmatrix} \psi^{(l)} \\ \psi^{(l+1)} \end{bmatrix} = T_l \begin{bmatrix} \psi^{(l-1)} \\ \psi^{(l)} \end{bmatrix}, \tag{21.11}$$

where

$$T_l = \begin{bmatrix} 0 & 1 \\ -H_{l,l+1}^{-1} H_{l,l-1} & H_{l,l+1}^{-1}(E_F - H_l) \end{bmatrix}, \tag{21.12}$$

and E_F is the Fermi energy. The slice discretized Schrödinger equation, and the trivial equation, are combined to calculate the progression of the wavefunction through the system. To yield the first slice transfer matrix, solving

$$
\begin{bmatrix} 0 & 1 \\ -H_{0,1}^{-1} H_{0,-1} & H_{0,1}^{-1}(E_F - H_0) \end{bmatrix} \begin{vmatrix} u_m(\pm) \\ \lambda_m(\pm)u_m(\pm) \end{vmatrix} = \lambda_m(\pm) \begin{vmatrix} u_m(\pm) \\ \lambda_m(\pm)u_m(\pm) \end{vmatrix} \tag{21.13}
$$

gives the N eigenvalues, λ_m, and N eigenvectors, u_m, for the forward $(+)$ and backward $(-)$ propagating modes deep in the left lead, which are then used to construct

$$
T_0 = \begin{bmatrix} [u_1(+), \cdots, u_N(+)] & [u_1(-), \cdots, u_N(-)] \\ [u_1(+), \cdots, u_N(+)] \, D[\lambda_1(-), \cdots, \lambda_N(-)] & [u_1(-), \cdots, u_N(-)] \, D[\lambda_1(-), \cdots, \lambda_N(-)] \end{bmatrix},
$$
$$\tag{21.14}$$

with D representing a matrix with the λs on the diagonal. In the calculation performed in this work, the Hamiltonian is orthogonalized. A non-orthogonal overlap could be introduced, but this would require some reformulation of Equations (21.12) and (21.13) to include the overlap matrix.

Traditional transfer matrix methods cascade the T_l matrices. However, this often has led to divergence factors, causing numerical instability. This variant stabilizes the problem by using an iterative scattering matrix technique that shrewdly cancels the divergent factors, thereby making the technique as stable as the Green's function technique. On the other hand, there is no guarantee the inverses in Equation (21.12) are always defined. If, due to large cell size or some uniquely isolated configuration, a column of zeroes was introduced reflecting the non-interaction between slices, the method would be expected to break down.

The advantage over a Green's function method is that the transmission matrix and wavefunction for all points can be derived simultaneously, and this directly yields the density at each point. This saves considerable time in the self-consistency scheme in which the density (i.e., the normalized square of the wavefunction) must be repetitively calculated. As described in the previous section, Green's function approaches require an integration over the density of states to obtain the density, which, due to its fine structure, often implies complex integration [61]. In addition, the interaction terms (Σ terms) are often iteratively solved using a continued fraction or other transcendental equation, creating a further computational burden.

The self-consistency employs a Poisson solver, which calculates the change in electronic potential and adds these corrective terms back into the Hamiltonian matrix.

$$
\nabla^2 \delta V(r) = \delta \rho(r). \tag{21.15}
$$

For this calculation, the density from the transmission calculation must be projected onto the wavefunctions for each atomic orbital centered at the accurate positions in three-dimensional space. The solver employed is the symmetric successive over-relaxation preconditioned bi-conjugate gradient stabilized algorithm (SSOR-BiCGSTAB) originally developed by van der Vorst [75]. To comply with the non-orthogonal mesh given by the unit cell from the energy spectrum code, the solver works over a 15-diagonal matrix. The applied bias is implemented as Dirichlet boundary conditions. It should be added that due to the low applied bias used in the experiments which were analyzed in this work, the self-consistent potential introduced fairly small corrections.

21.2.3 Complex Bandstructure

When the molecular system can be modeled as a series of repeated fundamental units, which characterizes many of the oligomers analyzed in these experiments, the electronic eigenstates can be treated as Bloch states [76]. The treatment of molecular orbitals with periodic boundary conditions gives rise to standing Bloch waves—the real Bloch k-vector oscillating at some rate, but also specifying, in a sense, the dimensions of a molecular "barrier" much like the simple rectangular barrier problem from quantum mechanics. Once this barrier is known, the complex Bloch vectors can be evaluated. These correspond to the electron tunneling coefficient, called β, through finite units, comparable to the spatial decay rate through a barrier. In order to extend the eigenvalue problem to complex k-vectors, the standard paradigm is reversed where now wave vectors are solved in terms of energy, yielding a wider range of wave vector solutions with real and complex parts. Plotting these, the real parts correspond to the traditional bandstructure plot. On the imaginary side, a semi-ellipse connecting the edges of the real HOMO and LUMO gives the β value or tunneling coefficient of the electron probability through the gap. The semi-ellipse gives some indication of the tunneling behavior through a finite number of unit cells, but other aspects of molecular conduction, most significantly, the nature of the contact to gold, are clearly not accounted for. Therefore, complex bandstructure should not be expected to give an exhaustive explanation of molecular conduction behavior.

21.2.4 Vibrational Modes

Diagonalization of the mass-weighted Hessian yields a set of eigenvalues corresponding to the frequency of vibration, and a set of eigenvectors describing the specific nuclear motions associated with the particular mode [77]. The intensities of the modes can be calculated for IR or Raman spectroscopy by calculating the transition rates, or the polarizability, respectively [78]. The vibration of metal–molecule–metal systems typically involves relaxation times around $10^{-11} - 10^{-13}$ s [79]. Since the transport time across the smaller molecules tends to be less than this range, the interaction between relaxation and transport were not considered in this study, thereby making this a purely elastic treatment. Some groups have broadened their scope to examine the effects of bias and current on the vibrational modes. The result of this research tends to show interesting vibronic resonances with applied bias. However, sometimes these biases tend to be much larger than those ever applied in experiment. Indeed, a variety of expected phenomena at high bias, from theoretical calculations, have been reported [60,80–82]. High bias has been predicted to reduce the energy gap [83] as well as rearrange the nuclei of both the molecule and the cluster [84]. Nevertheless, the considerations of high bias cannot be neglected. For example, the electron affinity of a molecule can become comparable to the applied bias at roughly 2 eV, which means that the molecule can no longer be treated as charge neutral [85]. Furthermore, the weakest bond, that of the sulfur–gold connection, has been estimated at somewhere in the 1 to 2eV range [5]. In other words, calculations which apply a bias should consider whether it is too strong to keep the molecule neutral or attached. Recently, important work describing the influence of electron–phonon coupling in molecular conduction has revealed compelling insight into the dynamics of metal–molecule–metal systems [86–88].

21.3 Results

The complex bandstructure makes a good starting point for discussing the effects of stretching on conduction. After isolating two phenyl rings and the coupling nitrogen of the polyaniline cell (as shown in Figure 21.12), the two outermost carbons are each pulled 0.1 Å away from the nitrogen before the whole system is relaxed using the Hartree–Fock theory GAMESS program [52]. This is repeated a few more times, and for each bandstructure calculation, the full cell is reconstructed. Analyzing the nuclear positions in Figure 21.12, the molecule becomes more planar, allowing the π orbitals to interact constructively. As an expected result, the imaginary part of the bandstructure should indicate a diminishing β. Indeed, in Figure 21.13, the red lines indicate the bandstructure of the stretched unit cell. On the real side, we can see the HOMO and LUMO states have broadened, indicating delocalization. In addition, the gap has decreased, extending to the semi-ellipse on the imaginary side, which shows the expected reduction in β. This can explain the general trend seen in stretched nonplanar molecules, although it may not be the whole story.

In previous research, we examined the enhancement of conduction with stretching on the xylyldithiol molecule as it related to observed fluctuations seen in the measured trace from AFM experiments [89].

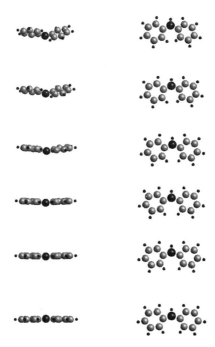

FIGURE 21.12 Polyaniline cell planarization with stretching.

This stretching calculation was performed twice —the first determined the Hellmann–Feynman forces in Fireball 2000, a local atomic orbital-based DFT method in the pseudopotential LDA [90], the other using the Hartree–Fock theory based GAMESS. The main difference between the ways in which the stretching was done relates to the gold–sulfur bonds. Using Fireball, the periodic cell was used which allowed the

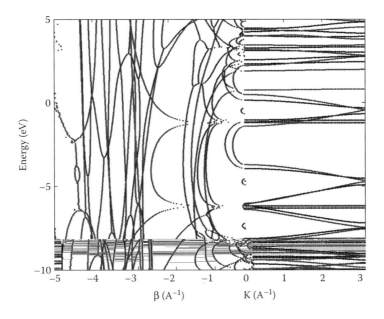

FIGURE 21.13 (see color insert) Bandstructure of polyaniline. Stretched cell is indicated in red.

FIGURE 21.14   (see color insert) Polythiophene. Left: terthiophene (T3); Right: quaterthiophene (T4).

gold layers to be included as an infinite slab. The stretching, therefore, involved moving the slabs a small distance from the sulfurs. In GAMESS, gold could not be represented as an infinite slab, so instead the isolated molecule with hydrogen-terminated thiols was used. In this stretching, the sulfurs themselves were pulled apart and fixed spatially [91]. Top site and hollow site contact geometries differed in magnitude, but showed similar trends, especially a conductance enhanced region when the molecule was most distended [46]. Related work in this area has examined stretching effects on metal nanowires using DFT [92,93]. A conductance calculation on compressed C60 showed a conductance enhancement [94].

The oligothiophenes, the aromatic heterocyclic oligomers shown in Figure 21.14 in the three-membered (T3) and four-membered (T4) forms, present an interesting case. Since the molecule is already planar, as confirmed in experiments via x-ray diffraction, stretching effects would not be expected to improve β [95]. Indeed, most of the steps in the measured traces of an AFM tip setup reveal a noticeable downward slope with stretching [96]. This is confirmed by theory when we examine the complex bandstructure of the infinite polymer, as shown in Figure 21.15. The gap and β, both in fairly good agreement with other theoretical studies [97], noticeably increase with stretching. In addition, the HOMO becomes more localized. Nevertheless, in examining the conductance shown in Figure 21.16, there is a noticeable peak for both T3 and T4 molecules initially upon stretching before the conductance drops off as expected. These peaks, when normalized in stretch distance to the length of the molecules, appear at nearly the same point of distortion. Figure 21.17 shows the distortion of the individual bonds in the thiophene cell. As might be expected, the internal single and double carbon-to-carbon bonds distort less than the single carbon-to-carbon bonds which link the rings. The carbon-to-sulfur single bonds distort the most within the ring. Calculations of the change in the ground state energy with stretching compare well to the measured force from experiment.

Experiment has revealed an interesting phenomenon of oligothiophene conductance: the four-membered T4 oligomer measures higher conductance than the three-membered T3 [98]. This runs counter to the expected length dependence of oligomers [99]. The infinite thiophene polymer unit cell has two isomers, and this makes a good starting point for the analysis of this T3/T4 conductance anomaly. Experimental groups have already noted the quinoidic character of charged oligothiophenes which showed some

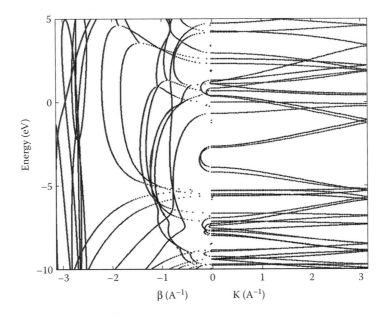

FIGURE 21.15 (see color insert) Complex bandstructure of polythiophene. Stretched cell shown in red.

conductance enhancement over their neutral counterparts [100]. Figures 21.18, 21.19, and 21.20 show the two isomers and their complex bandstructures. Clearly, there is a significant reduction in β in the quinoid variant. But how can this configuration be realized? The sulfur end atoms fix the benzenoid character of at least the two outer rings.

The vibrational spectrum for T3 with IR intensities, as calculated in the Hartree–Fock code, is shown in Figure 21.21. The spectrum calculated for T4 shows similar features. A vibrational analysis, conducted by Nitzan *et al.* [101], agrees well with the results shown here. Note the vibrational spectra analyzed here are not for the isolated T3 and T4, but for these molecules with the end sulfur atoms fixed. Although these

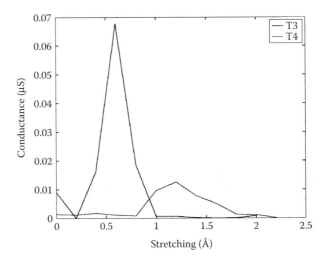

FIGURE 21.16 Left: T3 and T4 conductance.

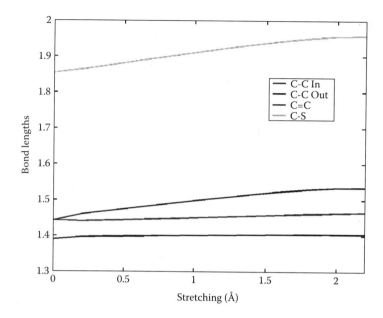

FIGURE 21.17   (see color insert) Individual T3 bond stretching between carbons (C) and sulfurs (S) in and out of the rings.

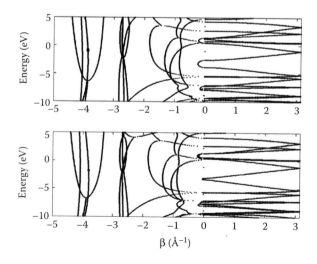

FIGURE 21.18   Unit cells of the two isomers of the infinite thiophene. On the left is the benzenoid; on the right, the quinoid configuration. (From G. Speyer, R. Akis, and D.K. Ferry, *J. Vac. Sci. Tech. B*, 24(4), 1987–1991, 2006. With permission.)

FIGURE 21.19   Complex bandstructure for the two isomers shown in Figure 21.18. Benzenoid is on top; quinoid is below.

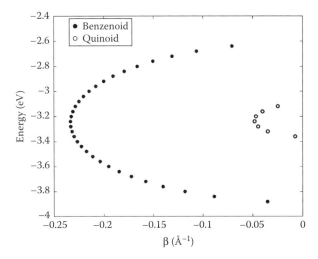

FIGURE 21.20 (see color insert) Comparison of the benzenoid (blue) and quinoid (green) complex bandstructures. (From G. Speyer, R. Akis, and D.K. Ferry, *J. Vac. Sci. Tech. B*, 24(4), 1987–1991, 2006. With permission.)

atoms would be expected to move in the real system, the aim is to approximate the bound oligothiophene. This makes even more sense when we examine the vibrational modes of the stretched system. Regardless, it stands to reason that the nuclei closest to the static gold contacts will move the least. Given that the end rings will be fixed in the benzenoid configuration due to the attached sulfur, we seek the vibrational mode which can give a quinoid character to the two center rings of T4. In the T3 case, this means the one center ring will sway more toward one of the outer benzenoid rings. Nonlinear conductance variations in odd- and even-membered atomic chains have been calculated previously [102,103].

Modes with wave numbers greater than 1600 cm^{-1} typically involve the hydrogen–carbon bonds, so these are ignored. Typically, the *en masse* shifting of monomer nuclei would be expected in the lower part of the spectrum. However, in order to observe the temporary switching between isomers, we seek a shifting in the dimerization of the underlying polyene structure. In agreement with the literature, these vibrations are observed around 1400 cm^{-1} [104]. Upon careful examination of several of the modes shown

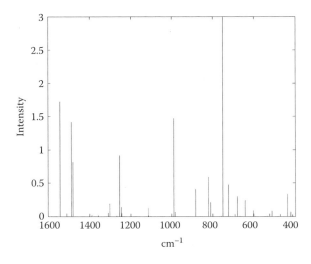

FIGURE 21.21 Vibrational frequencies of T3 as calculated in Hartree-Fock code.

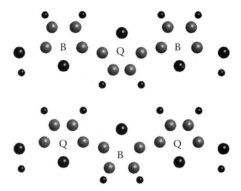

FIGURE 21.22 (see color insert) Nuclear position extrema of T3 molecule for selected mode, the first and last of twenty snapshots used in the calculations.

in Figure 21.21, the mode at 1480 cm^{-1} resembled the type of displacement dynamics that best realizes the phenomenon we seek. For T4, a comparable mode a bit lower in the spectrum was found. The extrema of 20 snapshots in nuclear displacements throughout this vibration are shown for T3 in Figure 21.22, and the corresponding displacements are shown for T4 in Figure 21.23. Because the two inner rings of T4 both shift to the quinoid isomer at the same time, the expectation is that the tunneling coefficient through the center of this molecule would be drastically reduced for some part of the vibration. The key is the unique vibrational configuration for T4 since nearly all the other modes would be expected to scale normally from a three- to four-membered oligomer.

In a similar approach to that used by Olson *et al.* [105], the displacement variation for the selected mode, Δq_λ, was calculated via the equipartition theorem as

$$\Delta q_\lambda = \sqrt{\frac{k_B T}{k_\lambda}}, \tag{21.16}$$

where k_λ represents the force constant of the mode. The individual nuclear displacements can then be scaled as:

$$\Delta r_i = c_{i,\lambda} \Delta q_\lambda. \tag{21.17}$$

At room temperature, this scales the amplitude of the selected vibration by roughly 0.37. By calculating the conductance over several individual displacements of the mode within this variation, the average over the set of calculations yields the overall performance. This approach treats the energy classically. From a quantum standpoint, the energy in the numerator of Equation (21.16) would correspond to the lowest oscillator energy. The zero point motion calculation reveals some discrepancy with the classical amplitude, but it is still in the same order of magnitude.

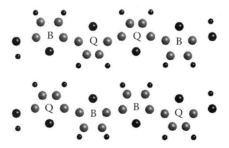

FIGURE 21.23 (see color insert) Nuclear position extrema for T4, the first and last of twenty snapshots used in the calculations. Note inner ring quinoid cycle synchronization.

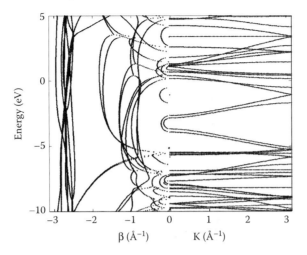

FIGURE 21.24 (see color insert) Complex bandstructure at the extrema (red) and at the unperturbed nuclear positions of the selected vibrational mode of the thiophene cell.

First, however, the unit cell for the vibration can be extracted at the extrema of the vibrational displacement, manifested in quinoid and distorted benzenoid heterocycles, and the bandstructure can be calculated and compared to the unperturbed thiophene unit cell. The result of this calculation is shown in Figure 21.24. An evident reduction in β as well as a reduction of the bandgap, can be observed. The band splitting results from the elimination of the degeneracy of two unperturbed rings when replaced by one quinoid and one benzenoid ring. Of course, this doesn't represent the exact situation in the T4 cell, which doesn't have alternating benzenoid and quinoid rings.

The T3 and T4 conductance plots, shown in Figure 21.25 display a striking difference between the two molecules. The maximum conductance for the T3 occurs when the molecule is only slightly vibronically perturbed (the maximum is around 0.18 μS) toward the configuration with the quinoid rings on the outside, as shown on the bottom in Figure 21.22. The conductance of the completely unperturbed system, shown at the leftmost point in Figure 21.16, is roughly 0.007 μS, which would lie between points 10 and 11 on the plot. For the T4, the configurations with the quinoids in the middle (as seen on the top in Figure 21.23), and decreasing in amplitude from 1 through 10 in Figure 21.25, have a conductance enhancement which nearly tracks the T3 conductance. The configurations with the quinoids on the ends

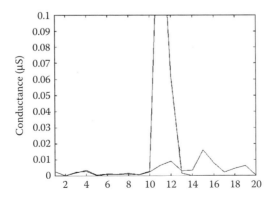

FIGURE 21.25 (see color insert) Conductance over 20 snapshot configurations of the selected mode for T3 (blue) and T4 (green).

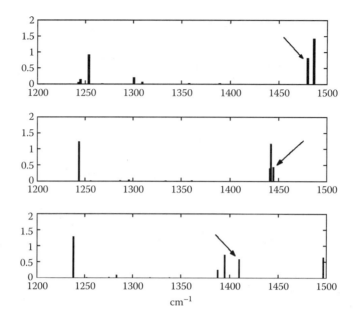

FIGURE 21.26   Change in vibrational spectrum with stretching. Unstretched on top; 0.8 Å in the middle; 1.8 Å on the bottom. Selected mode indicated by arrow.

show even more enhanced conductance. Although the average conductance over all the "snapshots" in the vibrational mode has slightly higher conductance for T3 over T4, it is only by a factor of 3.5, slightly less than the factor of 4 enhancement seen in the unperturbed molecules—not much evidence of the supposition that the vibrations are the source of the experimentally observed conductance enhancement in T4 over T3. Moreover, although the vibrations do enhance the conductance of T4, the quinoid isomer in the middle of the molecule does not give the enhancement that the two distended benzenoid rings in the center give.

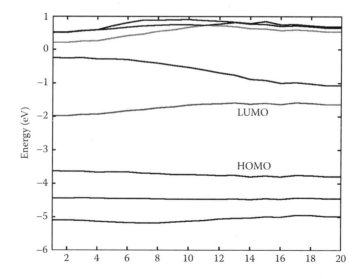

FIGURE 21.27   T3 bands with vibrational movements. Horizontal axis indicates snapshot index.

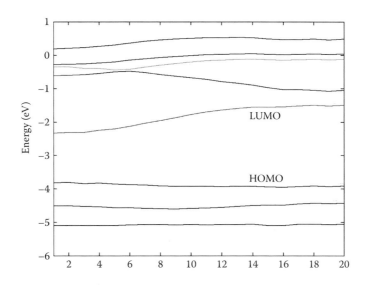

FIGURE 21.28 T3 bands with vibrational movements and stretching. Horizontal axis indicates snapshot index.

This is not the end of the story, however. Since the molecules happened to be in a stretched configuration, it made sense to perform a vibrational analysis of the molecules as they were stretched. Figure 21.26 shows the development of the vibrational spectrum with stretching with the mode under analysis indicated by an arrow. With stretching, a noticeable red shift in this mode indicates that the amplitude of this mode will be higher. This shift would be expected since the bonds weaken and lengthen with stretching.

As can be seen in Figures 21.27 and 21.28, the bands of T3 narrow and widen with vibration, an effect even more noticeable with stretching. Figures 21.29 and 21.30 demonstrate these trends in T4 to even greater effect. Figures 21.31 and 21.32, therefore, follow the expectation of enhanced conductance for T4 over T3 for two stretch distances. Now, the quinoid center rings give the best conductance—at least an order of magnitude enhancement over the T3 counterpart. Due to the size and topology of the systems,

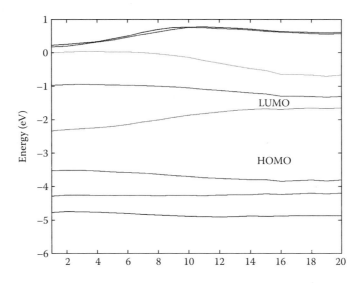

FIGURE 21.29 T4 bands with vibrational movements. Horizontal axis indicates snapshot index.

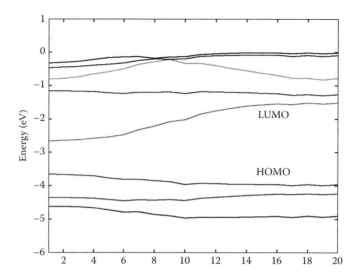

FIGURE 21.30 T4 bands with vibrational movements and stretching. Horizontal axis indicates snapshot index.

VASP PDOS analysis of the Fermi level location within the gap of the attached molecule is not possible as the number of plane waves needed increases the matrix rank beyond the capability of the computer used. However, the PDOS can be examined using Fireball 2000. The PDOS for unstretched and stretched T3 at different points in the vibration are shown in Figures 21.33 and 21.34. Position 1 corresponds to the top of Figure 21.22, while position 20 refers to the bottom configuration. Position 11 would closely resemble the unperturbed molecule. The band narrowing and widening looks similar to that seen in Figure 21.27, with the bandgap minimized and the quinoid in the center. The gap for the stretched molecule is reduced further than that for the unstretched. Interestingly, the Fermi level seems to lock into a fixed distance from the HOMO. The PDOS at the Fermi level, as shown in Figure 21.35, does not change dramatically

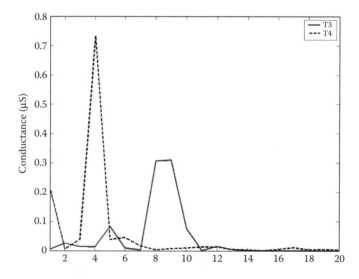

FIGURE 21.31 (see color insert) Conductance over 20 snapshot configurations of the selected mode for T3 (blue) and T4 (green) at 0.8 Å. (From G. Speyer, R. Akis, and D.K. Ferry, *J. Vac. Sci. Tech. B*, 24(4), 1987–1991, 2006. With permission.)

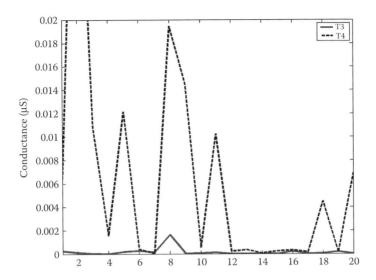

FIGURE 21.32 (see color insert) Conductance over 20 snapshot configurations of the selected mode for T3 (blue) and T4 (green) at 1.8Å.

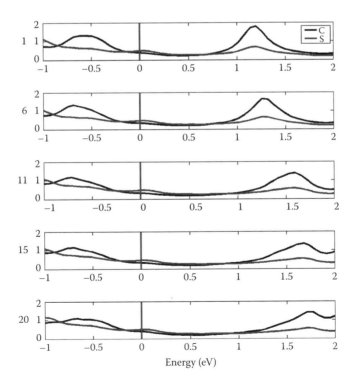

FIGURE 21.33 PDOS of unstretched T3 with selected vibration for carbon (C) and sulfur (S). 1 refers to configuration with benzenoid rings on the ends. 20 refers to configuration with quinoid rings on the ends. (Refer to Figure 21.22.) 11 would closely resemble the unperturbed molecule.

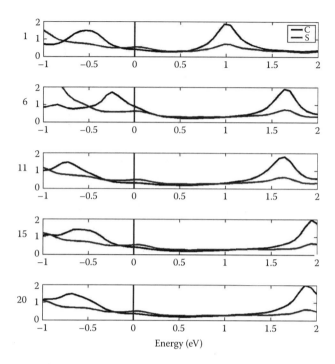

FIGURE 21.34   PDOS of stretched T3 with selected vibration for carbon (C) and sulfur (S). 1 refers to configuration with benzenoid rings on the ends. 20 refers to configuration with quinoid rings on the ends. (Refer to Figure 21.22.) 11 would closely resemble the unperturbed molecule.

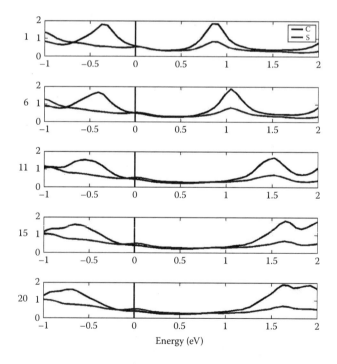

FIGURE 21.35   PDOS of unstretched T4 with selected vibration for carbon (C) and sulfur (S). 1 refers to configuration with benzenoid rings on the ends. 20 refers to configuration with quinoid rings on the ends. (Refer to Figure 21.23.) 11 would closely resemble the unperturbed molecule.

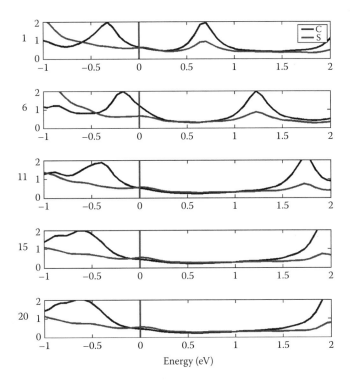

FIGURE 21.36 PDOS of stretched T4 with selected vibration for carbon (C) and sulfur (S). 1 refers to configuration with benzenoid rings on the ends. 20 refers to configuration with quinoid rings on the ends. (Refer to Figure 21.23.) 11 would closely resemble the unperturbed molecule.

across the vibration. Examining the unstretched and stretched T4 PDOS in Figures 21.35 and 21.36, there are similar trends to T3 in the bandgap behavior. The gap is smaller than that of T3 in both cases. Most significantly, Figures 21.37 and 21.38 reveals the quinoid center enhancement of the PDOS we had initially sought for T4 that is absent in T3. The increased carbon PDOS for this unique configuration

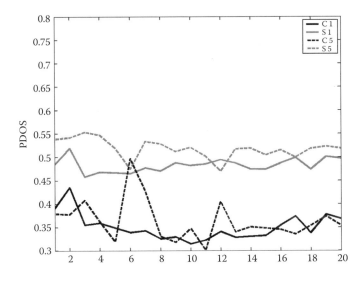

FIGURE 21.37 PDOS of unstretched (1, solid) and stretched (5, dashed) T3 for carbon (C) and sulfur (S) atoms at the Fermi energy for 20 snapshot configurations throughout the vibration.

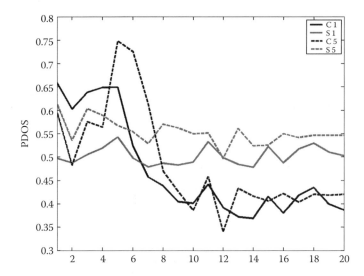

FIGURE 21.38 PDOS of unstretched (1, solid) and stretched (5, dashed) T3 for carbon (C) and sulfur (S) atoms at the Fermi energy for 20 snapshot configurations throughout the vibration.

of T4 indicates conduction channels of higher electron density, thereby providing an explanation for the conductance performance observed in experiment.

21.4   Conclusion

The aim of this work was to present the various approaches and concomitant pitfalls in the pursuit of molecular conductance calculations. The progress of this work edified a number of trade-offs that confront the theory of these systems. The contact specifics and the Fermi level position are critical in the determination of the conductance behavior. Including these accurately requires a large unit cell which can model the metal. This, in turn, creates more computational burdens. The inaccuracy of the surface potential within the local orbital basis prevents quantitative conclusions to be drawn from the calculations [106]. The plane wave basis, which can represent this more accurately, creates an intractably large Hamiltonian. DFT and Hartree–Fock theories each bring their own variety of errors in calculating the combined metal–molecule–metal system Hamiltonian.

As a result, the most compelling conclusions indicated trends over large suites of similar systems. Complex bandstructure analysis confirms that, despite the increased distance between the contacts, certain nonplanar molecules can be expected to improve in conductance with stretching. The effect of vibrational modes on the coupling explains the higher conductance measured on a four-membered oligomer (quaterthiophene) compared to its three-membered counterpart (terthiophene). Stretching combined with the nuclear positions within a specific vibrational mode permit the interior rings of quaterthiophene to form a unique configuration which lowers the tunneling barrier for electrons. This configuration cannot be realized in terthiophene. These specific examples, however, cannot exhaustively describe the conductance properties of molecules. The individual molecular conductance calculation, deceptive in its apparent simplicity, requires the ability to treat the complexities of the contact, the extended states and work function of the metal, and the bandgap and bonds of the organic system.

In this spirit, future efforts might pursue plane wave–based first-principles calculations which would accurately calculate the surface dipoles of the metal. Given the shortcomings of DFT in the LDA, these first-principles theories would have to be able to handle the complexity of the heterogeneous system. This work attempted to strike a balance by stretching the molecules in the Hartree–Fock theory code GAMESS, which handled the conjugated organic systems well, and exploit the metallic slab of DFT for the snapshot calculation. This was due to the fact that GAMESS could only handle gold clusters at best, and Hartree–Fock theory has its own shortcomings in this regard. VASP5 would seem to fit the bill well in this case if not for the prohibitively high rank Hamiltonian it generated. This, however, replaces a theoretical problem with a computational one—and with parallelization and larger faster computers, these calculations could be within reach. A fully integrated bandstructure–transport loop would be able to handle the dynamic effects of applied bias on nuclear displacement, nuclear vibration, and charge rearrangement.

References

[1] M. A. Reed *et al.*, "Conductance of a molecular junction," *Science*, 278, 1705, 1997.

[2] M. Di Ventra, S. T. Pantelides and N. D. Lang, "First-principles calculation of transport properties of a molecular device," *Phys. Rev. Lett.*, 84, 979, 2000.

[3] J. K. Tomfohr and O. F. Sankey, "Complex band structure, decay lengths, and Fermi level alignment in simple molecular electronic systems," *Phys. Rev. B*, 65, 245105, 2002.

[4] J. Reichert *et al.*, "Driving current through single organic molecules," *Phys. Rev. Let.*, 88, 176804, 2002.

[5] A. Mankefors, A. Grigoriev, and G. Wendin, "Molecular alligator clips: A theoretical study of adsorption of S, Se and S-H on Au(111)," *Nanotechnology*, 14, 849, 2003.

[6] H. He and N. J. Tao, "Interactions of molecules with metallic quantum wires," *Adv. Mate.*, 14, 161, 2002.

[7] A. M. Rawlett *et al.*, "Electrical measurements of a dithiolated electronic molecule via conducting atomic force microscopy," *App. Phys. Lett.*, 81, 3043, 2002.

[8] M. Toerker*et al.*, "Electronic transport through occupied and unoccupied states of an organic molecule on Au: Experiment and theory," *Phys. Rev. B*, 65, 245422, 2002.

[9] D. J. Tozer *et al.*, "Does density functional theory contribute to the understanding of excited states of unsaturated organic compounds?" *Mol. Phys.*, 97, 859, 1999.

[10] Z.-L. Cai, K. Sendt, and J. R. Reimers, "Failure of density-functional theory and time-dependent density-functional theory for large extended π systems," *J. Chem. Phys.*, 117, 5543, 2002.

[11] S. Picozzi *et al.*, "Organic/metal interfaces: An ab initio study of their structural and electronic properties," *Surf. Sci.*, 566–568, 628, 2004.

[12] G. C. Solomon, J. R. Reimers, and N. S. Hush, "Single molecule conductivity: The role of junction-orbital degeneracy in the artificially high currents predicted by ab initio approaches," *J. Chem Phys.*, 121, 6615, 2004.

[13] F. Evers, F. Weigend, and M. Koentopp, "Conductance of molecular wires and transport calculations based on density-functional theory," *Phys. Rev. B*, 69, 235411, 2004.

[14] J. R. Reimers *et al.*, "The appropriateness of density-functional theory for the calculation of molecular electronics properties" in *Molecular Electronics III*, J. R. Reimers *et al.* eds., New York Academy of Sciences, 235, 2003.

[15] VASP web site: http://cms.mpi.univie.ac.at/vasp/vasp/node8.html.

[16] N. Sai *et al.*, "Dynamical corrections to the dft-lda electron conductance in nanoscale systems," *Phys. Rev. Lett*, 94, 186810, 2005.

[17] S. P. McGlynn *et al.*, *Introduction to Applied Quantum Chemistry*, Holt, Rinehart and Winston, New York, 1971.

[18] G. Kresse and J. Furthmüller, "Efficient iterative schemes for *ab initio* total-energy calculations using a plane-wave basis set," *Phys. Rev. B*, 54, 11169, 1996.

[19] G. Kresse and J. Furthmüller, "Efficiency of ab-initio total energy calculations for metals and semi-conductor using a plane-wave basis set," *Comp. Mater. Sci.*, 6, 15, 1996.

[20] R. Dahlke and U. Schollwöck, "Electronic transport calculations for self-assembled non-layers of 1,4-phenylene diisocyanide on Au (111) contacts," *Phys. Rev. B*, 69, 85324, 2004.

[21] G. Yoder, B. K. Dickerson, and A.-B. Chen, "Semiempirical method for calculating structure and band gap od semiconducting polymers," *J. Chem. Phys.*, 111, 10347, 1999.

[22] E. G. Emberly and G. Kirczenow, "Antiresonances in molecular wires," *J. Phys.: Condens. Matter*, 11, 6911, 1999.

[23] V. Mujica, M. A. Ratner, and O. Goscinski, "Partitioning technique and transport across molecular interfaces: Many-body effects," *Intl. J. Quantum Chem.*, 90, 14, 2002.

[24] V. Mujica, M. Kemp, and M. A. Ratner, "Electron conduction in molecular wires. I. A Scattering formalism," *J. Chem. Phys.*, 101, 6849, 1994.

[25] J. K. Tomfohr, "Electron Tunneling Transport Theory for Molecules," Ph.D. Dissertation, Arizona State University, 2002.

[26] E. G. Emberly and G. Kirczenow, "Theoretical study of electrical conduction through a molecule connected to metallic nanocontacts," *Phys. Rev. B*, 58, 10911, 1998.

[27] C. Kergueris *et al.*, "Electron transport through a gold-bisthiolterthiophene-gold junction," in *Electronic Properties of Novel Materials: Science and Technology of Molecular Nanostructures*, H. Kuzmany *et al.*, eds., *American Institute of Physics*, 421, 1999.

[28] F. Zahid, M. Paulsson, and S. Datta, "Electrical conduction through molecules," in *Advanced Semiconductors and Organic Nano-Techniques*, H. Morkoc, ed., Academic Press, 2003.

[29] E. G. Emberly and G. Kirczenow, "Models of electron transport through organic molecular monolayers self-assembled on nanoscale metallic contacts," *Phys. Rev. B*, 64, 235412, 2001.

[30] V. Mujica *et al.*, "Molecular wire junctions: tuning the conductance," *J. Phys. Chem. B*, 107, 91, 2003.

[31] A. A. Farajian *et al.*, "Electronic transport through benzene molecule: Effect of gold contacts," *Phys. E*, 18, 253, 2003.

[32] A. Kahn, N. Koch, and W. Gao, "Electronic structure and electrical properties of interfaces between metals and π-conjugates molecular films," *J. Polymer Sci. B*, 41, 2529, 2003.

[33] I. G. Hill, A. Rajagopal, and A. Kahn, "Molecular level alignment at organic semiconductor-metal interfaces," *Appl. Phys. Lett.*, 73, 662, 1998.

[34] G. Witte *et al.*, "Vacuum level alignment at organic/metal junctions: 'Cushion effect' and the interface dipole," *Appl. Phys. Lett.* 87, 263502, 2005.

[35] J. M. Seminario, C. E. De La Cruz, and P. A. Derosa, "A theoretical analysis of metal–molecule contacts," *J. Am. Chem. Soc.*, 123, 5616, 2001.

[36] P. E. Kornilovitch and A. M. Bratkovsky, "Orientational dependence of current through molecular films," *Phys. Rev. B*, 64, 195413, 2001.

[37] A. Onipko, "Analytical model of molecular wire performance: A comparison of π and σ electron systems," *Phys. Rev. B*, 59, 9995, 1999.

[38] Y. Hu *et al.*, "Conductance of an ensemble of molecular wires: A statistical analysis," *Phys. Rev. Lett.*, 95, 156803, 2005.

[39] H. Grönbeck, A. Curioni, and W. Andreoni, "Thiols and disulfides on the Au(111) surface: The headgroup-gold interaction," *JACS*, 122, 3839, 2000.

[40] Y. Yourdshahyan, H. K. Zhang, and A. M. Rappe, "n-alkyl thiol head-group interactions with the Au(111) surface," *Phys, Rev. B*, 63, 81405, 2001.

[41] T. Hayashi, Y. Morikawa, and H. Nozoye, "Adsorption state of dimethyl disulfide on Au(111): Evidence for adsorption as thiolate at the bridge site," *J. Chem. Phys.*, 114, 7615, 2001.

[42] H. Kondoh *et al.*, "Adsorption of thiolates to singly coordinated sites on au(111) evidenced by photoelectron diffraction," *Phys. Rev. Lett*, 90, 66102, 2003.

[43] K. M. Beardmore *et al.*, "Determination of the headgroup-gold(111) potential surface for alkanethiol self-assembled monolayers by ab initio calculation," *Chem. Phys. Lett.*, 286, 40, 1998.

[44] A. Grigoriev *et al.*, "Critical roles of metal-molecule contacts in electron transport through molecular-wire junctions," *Phys. Rev. B*, 74, 045401, 2006.

[45] H. Kondo *et al.*, "Contact-structure dependence of transport properties of a single organic molecule between Au electrodes," *Phys. Rev. B*, 73, 235323, 2006.

[46] G. Speyer *et al.*, "Conductance investigations of stretched molecules," *IEEE Trans. Nano.*, 4, 403, 2005.

[47] J. Tomfohr and O. F. Sankey, "Theoretical analysis of electron transport through organic molecules," *J. Chem. Phys.*, 120, 1542, 2004.

[48] S.-H. Ke, H. U. Baranger, and W. Yang, "Contact atomic structure and electron transport through molecules," *J. Chem. Phys,.* 122, 74704, 2005.

[49] S.-H. Ke, H. U. Baranger and W. Yang, "Molecular conductance: Contact atomic structure and chemical trends of anchoring groups," *JACS*, 126, 15897, 2004.

[50] H. Basch and M. A. Ratner, "Molecular binding at gold transport interfaces. IV. Thiol chemisorption," *J. Chem. Phys.*, 120, 5771, 2004.

[51] N. Takeuchi, C. T. Chan, and K. M. Ho, "Au(111): A theoretical study of the surface reconstruction and the surface electronic structure," *Phys. Rev. B*, 43, 13899, 1991.

[52] M. W. Schmidt *et al.*, "General atomic and molecular electronic-structure system," *J. Comput. Chem.*, 14, 1347, 1993.

[53] Æ. Frisch and M. J. Frisch, *Gaussian 98 Users Reference*, 2nd ed., Gaussian Inc., Pittsburgh, PA, 1998.

[54] C.-K. Wang, Y. Fu, and Y. Luo, "A quantum chemistry approach for current-voltage characterization of molecular junctions," *Phys. Chem. Chem. Phys.*, 3, 5017, 2001.

[55] M. Bowman *et al.*, "Localization and capacitance fluctuations in disordered Au nanojunctions," *Phys. Rev. B*, 69, 205405, 2004.

[56] G. Fagas and A. Kambili, "Conductance properties of carbon-based molecular junctions," *http://arxiv.org/abs/cond-mat/0403694*, 2004.

[57] W. T. Geng, J. Nara, and T. Ohno, "Adsorption of benzene thiolates on the (111) surface of M (M = Pt, Ag, Cu) and the conductance of M/benzene dithiolate/M molecular junctions: A first-principles study," *Thin Solid Films*, 464–465, 379, 2004.

[58] S. K. Pati, "Transport in molecular wire with long-range Coulomb interactions: A mean-field approach," *J. Chem. Phys.*, 118, 6529, 2003.

[59] J. Li, J. K. Tomfohr, and O. F. Sankey, "Electric field effects on the octanedithiol wire," *Phys. Stat. Solid B*, 239, 80, 2003.

[60] M. Di Ventra, S. T. Pantelides, and N. D. Lang, "Current-induced forces in molecular wires," *Phys. Rev. Lett*, 88, 46801, 2002.

[61] M. Brandbryge *et al.*, "Density-functional method for nonequilibrium electron transport," *Phys. Rev. B*, 63, 165401, 2002.

[62] J. Taylor, H. Guo, and J. Wang, "Ab initio modeling of quantum transport properties of molecular electronic devices," *Phys. Rev. B*, 63, 245407, 2001.

[63] M. Galperin and A. Nitzan, "NEGF-HF method in molecular junction property calculations," in *Molecular Electronics III*, J. R. Reimers, C. A. Picconatto, J. C. Ellenbogen and R. Shashidhar, eds., New York Academy of Sciences, 48, 2003.

[64] T. Tada, M. Kondo, and K. Yoshizawa, "Green's function formalism coupled with Gaussian broadening of discrete states for quantum transport: Application to atomic and molecular wires," *J. Chem. Phys.*, 121, 8050, 2004.

[65] O. Berman and S. Mukamel, "Current profiles of molecular nanowires: dft green function representation," *Phys. Rev. B.* 69, 155430, 2004.

[66] P. Derosa and J. M. Seminario, "Electron transport through single molecules: scattering treatment using density functional and green function theories," *J. Phys. Chem.*, 105, 471 (2001).

[67] S.-H. Ke, H. U. Baranger, and W. Yang, "Electron transport through molecules: Self-consistent and non-self-consistent approaches," *Phys. Rev. B*, 70, 85410, 2004.

[68] A. Nitzan *et al.*, "On the electrostatic potential profile in biased molecular wires," *J. Chem. Phys.*, 117, 10837, 2002.

[69] N. D. Lang and P. Avouris, "Understanding the variation of the electrostatic potential along a biased molecular wire," *Nano Lett.*, 3, 737, 2003.

[70] Y. Xue, S. Datta, and M. A. Ratner, "Charge transfer and 'band lineup' in molecular devices: A chemical and numerical interpretation," *J. Chem. Phys.* 115, 4292, 2001.

[71] F. Zahid *et al.*, "Charging-induced asymmetry in molecular conductors," *Phys. Rev. B*, 70, 245317, 2004.

[72] M. Paulsson and S. Datta, "Thermoelectric effect in molecular electronics," *Phys. Rev. B*, 67, 241403, 2003.

[73] D. K. Ferry and S. M. Goodnick, *Transport in Nanostructures*, Cambridge University Press, Cambridge, 1997.

[74] T. Usuki *et al.*, "Numerical analysis of ballistic-electron transport in magnetic fields by using a quantum point contact and a quantum wire," *Phys. Rev. B*, 52, 8244, 1995.

[75] H. A. van der Vorst, "Bi-CGSTAB: A fast and smoothly converging variant of Bi-CG for the solution of non-symmetric linear systems," *J. SIAM J. Sci. Stat. Comp.*, 13, 631, 1992.

[76] J. K. Tomfohr and O. F. Sankey, "Complex band structure, decay lengths, and Fermi level alignment in simple molecular electronic systems," *Phys. Rev. B*, 65, 245105, 2002.

[77] J. B. Foresman and A. Frisch, *Exploring Chemistry with Electronic Structure Methods*, Gaussian, Inc. Pittsburgh, PA, 1993.

[78] J. Neugebauer *et al.*, "Quantum chemical calculation of vibrational spectra of large molecules – raman and ir spectra for Buckminsterfullerene," *J. Comput. Chem.*, 23, 895, 2002.

[79] E. G. Petrov, "Transmission of electrons through a linear molecule: Role of delocalized and localized electronic states in current formation," *Low Temp. Phys.*, 31, 338, 2005.

[80] Y. Chen, M. Zwolak and M. Di Ventra, "Inelastic current–voltage characteristics of atomic and molecular junctions," *Nano Lett.*, 4, 1709, 2004.

[81] E. G. Petrov and P. Hanggi, "Nonlinear electron current through a short molecular wire," *Phys. Rev. Lett.*, 86, 2862, 2001.

[82] E. G. Emberly and G. Kirczenow, "Current-driven conformational changes, charging, and negative differential resistance in molecular wires," *Phys. Rev. B*, 64, 125318, 2001.

[83] Y. Xue and M. A. Ratner, "Microscopic study of electrical transport through individual molecules with metallic contacts. I. Band lineup, voltage drop, and high-field transport," *Phys. Rev. B*, 68, 115406, 2003.

[84] H. Basch and M. A. Ratner, "Molecular binding at gold transport interfaces. III. Field dependence of electronic properties," *J. Chem. Phys.*, 120, 5761, 2004.

[85] Y. Luo, C.-K. Wang, and Y. Fu, "Effects of chemical and physical modifications on the electronic transport properties of molecular junctions," *J. Chem. Phys.*, 117, 10283, 2002.

[86] J. Koch *et al.*, "Current-induced nonequilibrium vibrations in single-molecule devices," *Phys. Rev. B*, 73, 155306, 2006.

[87] M. Galperin *et al.*, "Resonant inelastic tunneling in molecular junctions," *Phys. Rev. B*, 73, 045314, 2006.

[88] A. Pecchia *et al.*, "Role of thermal vibrations in molecular wire conduction," *Phys. Rev. B*, 68, 235321, 2003.

[89] X. Xiao, B. Xu, and N. J. Tao, "Measurement of single molecule conductance: benzenethiol and benzenedimethanethiol," *Nano Lett.*, 4, 267, 2004.

[90] O. F. Sankey and D. J. Niklewski, "*Ab initio* multicenter tight-binding model for molecular-dynamics simulations and other applications in covalent systems," *Phys. Rev. B*, 40, 3979, 1989.

[91] G. Speyer *et al.*, "Self-consistent conductance calculations on molecular calipers using a transfer matrix method," *Superlattices and Microstructures*, 34, 429, 2003.

[92] P. Jelínek *et al.*, "First-principles simulations of the stretching and final breaking of Al nanowires: Mechanical properties and electrical conductance," *Phys. Rev. B*, 68, 85403, 2003.

[93] E. Z. da Silva *et al.*, "Theoretical study of the formation, evolution, and breaking of gold nanowires," *Phys. Rev. B*, 69, 115411, 2004.

[94] M. Paulsson and S. Stafstrom, "Conductance manipulation at the molecular level," *J. Phys.: Condens. Matter*, 11, 3555, 1999.

[95] A. K. Bakhshi, J. Ladik, and M. Seel, "Comparative study of the electronic structure and conduction properties of polypyrrole, polythiophene, and polyfuran and their copolymers," *Phys. Rev. B*, 35, 704, 1987.

[96] Xu, B. Q. *et al.*, "Electromechanical and conductance switching properties of single oligothiophene molecules," *Nano Lett.*, 5, 1491, 2005.

[97] M. Magoga and C. Joachim, "Conductance and transparence of long molecular wires," *Phys. Rev. B*, 56, 4722, 1997.

[98] Ž. Crijen *et al.*, "Nonlinear conductance in molecular devices: Molecular length dependence," *Phys. Rev. B*, 71, 165316, 2005.

[99] N. B. Zhitenev, H. Meng, and Z. Bao, "Conductance of small molecular junctions," *Phys. Rev. Lett.*, 88, 226801, 2002.

[100] J. Cornil, D. Beljonne, and J. L. Bredas, "Nature of optical transitions in conjugated oligomers. II. Theoretical characterization of neutral and doped oligothiophenes," *J. Chem. Phys.*, 103, 842, 1995.

[101] A. Troisi, M.A. Ratner, and A. Nitzan, "Vibronic effects in off-resonant molecular wire conduction," *J. Chem. Phys.*, 118, 6072, 2003.

[102] E. G. Emberly and G. Kirczenow, "Electron standing-wave formation in atomic wires," *Phys. Rev. B*, 60, 6028, 1999.

[103] Ph. Avouris and N. D. Lang, "Oscillatory conductance of carbon-atom wires," *Phys. Rev. Lett.*, 81, 3515, 1998.

[104] E. Yurtsevar and S. Kirmizialtin, "Vibrational spectroscopy of structural defects in oligothiophenes," *Mol. Phys.*, 101, 2725, 2003.

[105] M. Olson *et al.*, "A conformational study of the influence of molecular vibrations on the conductance in molecular wires," *J. Phys. Chem. B*, 102, 941, 1998.

[106] Late in the development of this work, we learned that the Trans Sresta group (61) had attained improved work function calculations using local orbitals.

22

Nanoelectromechanical Oscillator as an Open Quantum System

Lev G. Mourokh

Anatoly Yu. Smirnov

Abstract

We examine electron transport through a nanoelectromechanical oscillator coupled to a tunnel junction, as well as the dynamics of this oscillator in the presence of electron tunneling. We analyze two different models, a nano-cantilever, where electrons tunnel directly from one reservoir to another with the transfer matrix elements being modulated by the mechanical oscillations; and a quantum shuttle, when the mechanical oscillator has its own quantized electron states intermediate for the electrons tunneling between the two reservoirs. In both cases, the nano-oscillator is considered as a dynamic subsystem interacting with a heat bath of conducting electrons, and the previously developed theory of open quantum systems is applied. The nonlinear conductances of the systems, the oscillators' damping rates, and the variances of the oscillator position fluctuations are determined as functions of the applied voltage, temperature, and the electron tunneling length. We also discuss the applications of our analysis to real systems of nano- and molecular electronics.

22.1 Introduction

The rapid development of nanotechnology in recent years has ushered in a new generation of devices, so-called nanoelectromechanical systems (NEMS), where a nanoscale mechanical resonator (oscillator) is coupled to an electronic structure of comparable dimensions [1–42]. Examples of these kinds of devices include small grains embedded in the elastic medium between the leads [9,10,12], single conducting

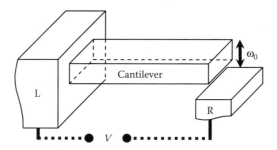

FIGURE 22.1   Schematic of a cantilever coupled to a tunnel junction.

molecules attached to metallic contacts [43] or placed between them [11], carbon nanotubes [26,27], and man-made constructions, such as cantilevers (suspended beams clamped at one end) [14,25], nanobridges (suspended beams clamped at both ends) [31], nanopillars [40], and so on. The frequencies of these mechanical oscillations lie in the range from a few megahertz to about one gigahertz. With many good reviews or books devoted to various aspects of NEMS dynamics [1–8], the main aim of the present paper is to study the mutual interplay of mechanical and electron degrees of freedom within the NEMS with special attention on fluctuations of the oscillator position. Electron transport through the NEMS is usually achieved by electron tunneling to and from the nano-oscillator. Tunnel matrix elements depend exponentially on the objects separation, so the mechanical motion affects the conductance of the system drastically. On the other hand, with the applied bias, conducting electrons serve as a non-Gaussian non-equilibrium heat bath for the nano-oscillator, and fluctuations of the oscillator position no longer can be determined by the Fluctuation–Dissipation Theorem [44]; even the cooling of the oscillator can be accomplished [36,39]. We base our analysis on our previous development of a theory of open quantum systems and its applications to electron dynamics in nanostructures [45–49]. Results obtained in the present paper can be applied both to artificial structures serving as prototypes for nanoelectronic devices and to electronic systems incorporating single molecules as part of an electric circuit. The rest of this chapter is structured as follows: Section 22.2 contains the general formulation of the problem for the nano-cantilever case where electrons tunnel directly from one reservoir to another with the transfer matrix elements being modulated by the mechanical oscillations. An electric circuit including such a cantilever is shown in Figure 22.1 and the model which we use to describe such a system is presented

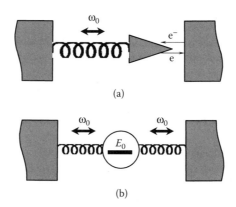

FIGURE 22.2   Models of the devices discussed. (a) Nanomechanical cantilever with one of the leads vibrating and the tunnel distance for electrons oscillating; (b) quantum shuttle containing a single electron energy level. Electron tunneling is a two-step process in this case, from one of the leads to the shuttle level and, subsequently, to the other lead.

in Figure 22.2(a). The equations of motion for the oscillator position operator and the self-consistent set of equations to determine oscillator fluctuations are obtained in terms of the correlation functions of reservoirs. These correlation functions are determined from the analysis of electron gas in reservoirs as a non-equilibrium heat bath for the nano-oscillator in Section 22.3. The expression for the current through such a structure is derived in Section 22.4. Numerical solutions of the equations derived in Section 22.2 are presented in Section 22.5 for various system parameters. The nonlinear conductance of the system, the oscillator damping rate, and the variance of the oscillator position fluctuations are shown as functions of the applied voltage, temperature, and the electron tunneling length. Section 22.6 is devoted to the modification of the system behavior for the shuttle case—i.e., when the mechanical oscillator has its own quantized electron states intermediate for the electrons tunneling between the two reservoirs (Figure 22.2[b]). Both a new formalism and numerical results are presented. And, finally, Section 22.7 represents conclusions of our work.

22.2 General Formalism

The Hamiltonian of the cantilever coupled to the tunnel junction (Figure 22.2[a]) is given by

$$H = H_0 + H_L + H_R + H_{tun} \tag{22.1}$$

where

$$H_0 = \frac{p^2}{2m} + \frac{m\omega_0^2 x^2}{2} \tag{22.2}$$

is the Hamiltonian of the nanomechanical oscillator with the mass m and the resonant frequency ω_0,

$$H_\alpha = \sum_k E_{\alpha k} c_{\alpha k}^+ c_{\alpha k} \tag{22.3}$$

is the Hamiltonian of the leads ($\alpha = L, R$ for the left, right lead, respectively), and

$$H_{tun} = -\sum_{k,q} \left(T_{kq} c_{Lk}^+ c_{Rq} + T_{kq}^* c_{Rq}^+ c_{Lk} \right) e^{-x/\lambda} \tag{22.4}$$

is the tunneling term taking into account the dependence of the transfer matrix elements on the oscillator position with λ as the characteristic tunneling length. One can see that the form of the Hamiltonian, Equation (22.1), is very similar to the Hamiltonian of an open quantum system, with H_0 and H_α being the Hamiltonians of the dynamic subsystem and heat bath, respectively, and $H_{tun} = H_{int}$ playing the role of their interaction. In this case, it makes sense to write H_{tun} as a product of two functions, $H_{tun} = -QF$, where

$$F = e^{-x/\lambda} \tag{22.5}$$

is the function of the dynamic subsystem (nano-oscillator) variable, x, and

$$Q = \sum_{k,q} \left(T_{kq} c_{Lk}^+ c_{Rq} + T_{kq}^* c_{Rq}^+ c_{Lk} \right) \tag{22.6}$$

is the effective heat bath variable. The heat bath thus defined is characterized by a response function $\varphi(t,t_1)$ and a symmetrized correlation function $M(t,t_1)$ of the unperturbed variables $Q^{(0)}(t)$, as

$$\varphi(t,t_1) = \frac{i}{\hbar} \left\langle \left[Q^{(0)}(t), Q^{(0)}(t_1) \right]_- \right\rangle \Theta(t - t_1) \tag{22.7a}$$

and

$$M(t,t_1) = \frac{1}{2} \left\langle \left[Q^{(0)}(t), Q^{(0)}(t_1) \right]_+ \right\rangle \tag{22.7b}$$

where $\Theta(\tau)$ is the unit Heaviside step function; $[\cdots,\cdots]_-$ and $[\cdots,\cdots]_+$ are the commutator and anticommutator, respectively; and the angle brackets refer to averaging over the equilibrium states of both left and right leads. We introduce here one more correlation function which we will use next:

$$R(t, t_1) = \frac{1}{2}\left\langle \left[Q^{(0)}(t), Q^{(0)}(t_1) \right]_- \right\rangle \tag{22.7c}$$

The spectral function $S(\omega)$ and the imaginary part of the susceptibility $\chi''(\omega)$, which are the Fourier time transforms of the response function of the heat bath and the correlator, Equations (7 [a],[b]), respectively,

$$\chi''(\omega) = \int d\tau \sin(\omega\tau)\varphi(\tau) \tag{22.8a}$$

and

$$S(\omega) = \int d\tau e^{i\omega\tau} M(\tau) \tag{22.8b}$$

are connected at equilibrium by the Fluctuation–Dissipation Theorem (FDT) [44]. However, when the heat bath is out of equilibrium (with a voltage bias applied to the junction), FDT is not valid anymore. Nevertheless, the assumption that the tunnel coupling is weak allows us to express the dissipative oscillator rate and its position fluctuations, as well as the electric current through the structure, in terms of the *equilibrium* values of $S(\omega)$ and $\chi''(\omega)$ which are calculated in the following.

If the tunnel coupling is weak, the back action of the oscillator on the heat bath (conducting electrons) can be included in the heat bath variable as [45]

$$Q(t) = Q^{(0)}(t) + \int dt_1 \varphi(t, t_1) e^{-x(t_1)/\lambda} \tag{22.9}$$

This variable is involved in the Heisenberg equation of motion for the position operator of the oscillator as

$$\ddot{x}(t) + \omega_0^2 x(t) = -\frac{1}{m\lambda} Q(t) e^{-x(t)/\lambda} \tag{22.10}$$

The effect of the heat bath on the dynamics of the oscillator (the right-hand side of Equation [22.10]) is twofold. It produces fluctuations in the system and causes oscillator damping. To obtain explicit expressions both for the fluctuation source and for the damping rate, we employ the quantum Furutsu–Novikov theorem [45]:

$$\left\langle Q^{(0)}(t) e^{-x(t)/\lambda} \right\rangle = \int dt_1 \left\langle Q^{(0)}(t) Q^{(0)}(t_1) \right\rangle \left\langle \frac{\delta}{\delta Q^{(0)}(t_1)} e^{-x(t)/\lambda} \right\rangle \tag{22.11}$$

where $\delta/\delta Q^{(0)}(t_1)$ is the functional derivative with respect to the unperturbed heat bath variable $Q^{(0)}(t_1)$. Such a functional derivative of an arbitrary operator $A(t)$ of the dynamical subsystem is proportional to the commutator [45] in the form

$$\frac{\delta A(t)}{\delta Q^{(0)}(t_1)} = \frac{i}{\hbar} [A(t), e^{-x(t_1)/\lambda}]_- \Theta(t - t_1) \tag{22.12}$$

After substitution Equations (22.9), (22.11), and (22.12) into Equation (22.10) and the symmetrization of noncommuting operators, we obtain a non-Markovian stochastic equation for the oscillator position operator, as given by:

$$\ddot{x}(t) + \omega_0^2 x(t) + \frac{1}{m\lambda} \int dt_1 \left(\varphi(t, t_1) \frac{1}{2} [e^{-x(t)/\lambda}, e^{-x(t_1)/\lambda}]_+ \right.$$

$$\left. + M(t, t_1) \frac{i}{\hbar} [e^{-x(t)/\lambda}, e^{-x(t_1)/\lambda}]_- \Theta(t - t_1) \right) = \xi(t) \tag{22.13}$$

The fluctuation source $\xi(t)$, defined as:

$$\xi(t) = -\frac{1}{m\lambda} \left(Q^{(0)}(t) e^{-x(t)/\lambda} - \int dt_1 \left\langle Q^{(0)}(t) Q^{(0)}(t_1) \right\rangle \frac{i}{\hbar} \langle [e^{-x(t)/\lambda}, e^{-x(t_1)/\lambda}]_- \rangle \Theta(t - t_1) \right) \quad (22.14)$$

has a zero mean value according to Equation (22.11). The explicit expression, Equation (22.14), allows us to calculate its correlation functions of any order. In particular, the anticommutator has the following form in the case of weak coupling:

$$\left\langle \frac{1}{2}[\xi(t), \xi(t_1)]_+ \right\rangle = \frac{1}{2m^2\lambda^2} \left(M(t, t_1) \langle [e^{-x(t)/\lambda}, e^{-x(t_1)/\lambda}]_+ \rangle + R(t, t_1) \langle [e^{-x(t)/\lambda}, e^{-x(t_1)/\lambda}]_- \rangle \right) \quad (22.15)$$

It should be emphasized that the non-Markovian stochastic equation, Equation (22.13), takes into account a nonlinearity of the coupling between subsystems, and incorporates the nonlocal character of heat-bath fluctuations. This nonlinearity leads to the non-Gaussian fluctuation source, Equation (22.14), having not only the second-order correlation function, but the high-order correlators as well. In this respect, our treatment goes well beyond the well-known Caldeira–Leggett approach [50]. It should be noted, however, that in the present case, the heat bath is non-Gaussian itself due to the Fermi statistics of electrons.

To simplify Equations (22.13) and (22.14), it is necessary to calculate the (anti)commutators involved. To contribute significantly to the final result, the moments of time t and t_1 have to be separated by less than a correlation time τ_c of the heat bath. We assume that the heat bath, weakly interacting with the oscillator, will have a negligibly small effect on the time evolution of the oscillator during this small correlation time τ_c. Correspondingly, we can evaluate the (anti)commutators in Equations (22.13) and (22.14) with the uncoupled free-evolution equation for the mechanical motion ($t - t_1 = \tau \sim \tau_c$):

$$x(t) = x(t_1) \cos \omega_0 \tau + \frac{p(t_1)}{m\omega_0} \sin \omega_0 \tau \quad (22.16)$$

and with the Baker–Hausdorff theorem as:

$$\frac{1}{2}[e^{-x(t)/\lambda}, e^{-x(t_1)/\lambda}]_+ = \cos \left(\frac{\hbar \sin \omega_0 \tau}{2m\omega_0 \lambda^2} \right) \exp \left(-\frac{x(t) + x(t_1)}{\lambda} \right) \quad (22.17a)$$

and

$$\frac{i}{2}[e^{-x(t)/\lambda}, e^{-x(t_1)/\lambda}]_- = \sin \left(\frac{\hbar \sin \omega_0 \tau}{2m\omega_0 \lambda^2} \right) \exp \left(-\frac{x(t) + x(t_1)}{\lambda} \right). \quad (22.17b)$$

The oscillator position operator can be separated into the mean and fluctuating parts. It should be noted that even if the mean value of the oscillator displacement is much smaller than the tunneling length, λ, position fluctuations can still be important with the root-mean-square of the variance to be of the order of λ. Equation (22.13) can be rewritten for the fluctuating part of the position operator in the form of a non-Markovian Langevin-like equation as:

$$\ddot{\tilde{x}}(t) + \int dt_1 \gamma(t - t_1) \dot{\tilde{x}}(t_1) + \omega_0^2 \tilde{x}(t) = \xi(t) \quad (22.18)$$

Taking the Fourier transformation and using again the weakness of the tunneling, we obtain the following expression for the variance of oscillator position fluctuations:

$$\left\langle \tilde{x}^2 \right\rangle = \frac{K(\omega_0)}{2\omega_0^2 \gamma(\omega_0)} \quad (22.19)$$

where

$$K(\omega) = \int d\tau e^{i\omega\tau} \left\langle [\xi(\tau), \xi(0)]_+ \right\rangle \quad (22.20)$$

Both $K(\omega)$ and $\gamma(\omega)$ have explicit expressions (in terms of the heat bath spectral functions) obtained by the evaluation of Equation (22.15) and the collision term of Equation (22.13), as:

$$
\begin{aligned}
K(\omega) = {}& \frac{\hbar^2}{2m^2\lambda^2} e^{\nu_c} \sum_{l=-\infty}^{l=\infty} I_l(\nu_c) \{J_0(\nu_0) \left[S(\omega - l\omega_0) + S(\omega + l\omega_0)\right] \\
& + \sum_{n=1}^{n=\infty} J_{2n}(\nu_0) \left[S(\omega + 2n\omega_0 - l\omega_0) + S(\omega + 2n\omega_0 + l\omega_0)\right. \\
& \left. + S(\omega - 2n\omega_0 - l\omega_0) + S(\omega - 2n\omega_0 + l\omega_0)\right] \\
& + \sum_{n=0}^{n=\infty} J_{2n+1}(\nu_0) \left[\chi''(\omega - (2n+1)\omega_0 - l\omega_0) + \chi''(\omega - (2n+1)\omega_0 + l\omega_0)\right. \\
& \left. - \chi''(\omega + (2n+1)\omega_0 - l\omega_0) - \chi''(\omega + (2n+1)\omega_0 + l\omega_0)\right]\}
\end{aligned}
\tag{22.21a}
$$

and

$$
\begin{aligned}
\gamma(\omega) = {}& \frac{\hbar}{2m\omega_0\lambda^2} e^{\nu_c} \sum_{l=-\infty}^{l=\infty} I_l(\nu_c)\{J_0(\nu_0)[\chi''(\omega - l\omega_0) + \chi''(\omega + l\omega_0)] \\
& + \sum_{n=1}^{n=\infty} J_{2n}(\nu_0)[\chi''(\omega + 2n\omega_0 - l\omega_0) + \chi''(\omega + 2n\omega_0 + l\omega_0) \\
& + \chi''(\omega - 2n\omega_0 - l\omega_0) + \chi''(\omega - 2n\omega_0 + l\omega_0)] \\
& + \sum_{n=0}^{n=\infty} J_{2n+1}(\nu_0)[S(\omega - (2n+1)\omega_0 - l\omega_0) + S(\omega - (2n+1)\omega_0 + l\omega_0) \\
& - S(\omega + (2n+1)\omega_0 - l\omega_0) - S(\omega + (2n+1)\omega_0 + l\omega_0)]\}
\end{aligned}
\tag{22.21b}
$$

Here, we have introduced parameters

$$
\nu_0 = \frac{\hbar}{2m\omega_0\lambda^2}
\tag{22.22}
$$

characterizing the level of vacuum oscillator fluctuations relative to the tunneling length, and

$$
\nu_c = \frac{\langle \tilde{x}^2 \rangle}{\lambda^2}
\tag{22.23}
$$

characterizing the level of non-equilibrium oscillator fluctuations. J_n and I_l in Equations (22.21a,b) are the ordinary and modified Bessel functions, respectively. One can see that Equations (22.19–22.23) represent a set of equations that have to be solved self-consistently to determine fluctuations of the oscillator position. The solution of this set of equations will be used in calculations of the effective oscillator temperature and the electron current through the structure.

22.3   Heat Bath Correlation Functions

To solve the self-consistent set of equations, Equations (22.19–22.23), we have to determine the unperturbed spectral function of the electronic heat bath and the imaginary part of its susceptibility, $S(\omega)$ and $\chi''(\omega)$, respectively. To achieve this, we assume that the unperturbed creation/annihilation electron operators in the leads obey the Wick theorem; thus, we can express the correlator of heat bath variables in the form

$$
\left\langle Q^{(0)}(t)Q^{(0)}(t_1)\right\rangle = \sum_{k,q} |T_{kq}|^2 \left(\left\langle c_{Lk}^{(0)+}(t)c_{Lk}^{(0)}(t_1)\right\rangle\left\langle c_{Rq}^{(0)+}(t)c_{Rq}^{(0)}(t_1)\right\rangle + \left\langle c_{Lk}^{(0)}(t)c_{Lk}^{(0)+}(t_1)\right\rangle\left\langle c_{Rq}^{(0)}(t)c_{Rq}^{(0)+}(t_1)\right\rangle\right)
$$

$$
\tag{22.24}
$$

We introduce retarded, advanced, and "lesser" Green's functions of electrons in the leads as

$$g_{\alpha k}^{r}(t,t_1) = (-i)\big\langle \big[c_{\alpha k}^{(0)}(t), c_{\alpha k}^{(0)+}(t_1)\big]_{+}\big\rangle \Theta(t-t_1) = -i\exp\left(-\frac{i}{\hbar}E_{\alpha k}(t-t_1)\right)\Theta(t-t_1) \quad (22.25a)$$

$$g_{\alpha k}^{a}(t,t_1) = i\big\langle \big[c_{\alpha k}^{(0)}(t), c_{\alpha k}^{(0)+}(t_1)\big]_{+}\big\rangle \Theta(t_1-t) = i\exp\left(-\frac{i}{\hbar}E_{\alpha k}(t_1-t)\right)\Theta(t_1-t) \quad (22.25b)$$

and

$$g_{\alpha k}^{<}(t,t_1) = i\big\langle c_{\alpha k}^{(0)}(t)c_{\alpha k}^{(0)+}(t_1)\big\rangle = if_{\alpha}(E_{\alpha k})\exp\left(-\frac{i}{\hbar}E_{\alpha k}(t-t_1)\right) \quad (22.25c)$$

where $f_a(E)$ is the Fermi distribution in the a-lead with the chemical potential μ_a. It should be noted that the chemical potentials of the two leads can be different with $\mu_L - \mu_R = eV$, where V is the voltage bias applied to the tunnel junction. The heat bath correlation function and the response function can be expressed in terms of the following Green's functions as

$$M(t,t_1) = \frac{1}{2}\sum_{k,q}|T_{kq}|^2\Big(g_{Rq}^{<}(t,t_1)\big(g_{Lk}^{r}(t_1,t) - g_{Lk}^{a}(t_1,t)\big) + g_{Lk}^{<}(t_1,t)\big(g_{Rq}^{r}(t,t_1) - g_{Rq}^{a}(t,t_1)\big)$$
$$+ g_{Rq}^{<}(t_1,t)\big(g_{Lk}^{r}(t,t_1) - g_{Lk}^{a}(t,t_1)\big) + g_{Lk}^{<}(t,t_1)\big(g_{Rq}^{r}(t_1,t) - g_{Rq}^{a}(t_1,t)\big)$$
$$+ 2g_{Rq}^{<}(t,t_1)g_{Lk}^{<}(t_1,t) + 2g_{Rq}^{<}(t_1,t)g_{Lk}^{<}(t,t_1)\Big) \quad (22.26a)$$

and

$$\varphi(t,t_1) = \frac{i}{\hbar}\Theta(t-t_1)\sum_{k,q}|T_{kq}|^2\Big(g_{Lk}^{<}(t_1,t)\big(g_{Rq}^{r}(t,t_1) - g_{Rq}^{a}(t,t_1)\big) - g_{Rq}^{<}(t,t_1)\big(g_{Lk}^{r}(t_1,t) - g_{Lk}^{a}(t_1,t)\big)$$
$$+ g_{Rq}^{<}(t_1,t)\big(g_{Lk}^{r}(t,t_1) - g_{Lk}^{a}(t,t_1)\big) - g_{Lk}^{<}(t,t_1)\big(g_{Rq}^{r}(t_1,t) - g_{Rq}^{a}(t_1,t)\big)\Big) \quad (22.26b)$$

Correspondingly, their Fourier transforms, the heat bath spectral function and the imaginary part of its susceptibility, are given by:

$$S(\omega) = \pi\sum_{k,q}|T_{kq}|^2\big(\delta(\omega - E_{Lk}/\hbar + E_{Rq}/\hbar) + \delta(\omega + E_{Lk}/\hbar - E_{Rq}/\hbar)\big)$$
$$\times\big(f_R(E_{Rq}) + f_L(E_{Lk}) - 2f_R(E_{Rq})f_L(E_{Lk})\big) \quad (22.27a)$$

and

$$\chi''(\omega) = \pi\sum_{k,q}|T_{kq}|^2\big(\delta(\omega + E_{Lk}/\hbar - E_{Rq}/\hbar) - \delta(\omega - E_{Lk}/\hbar + E_{Rq}/\hbar)\big)\big(f_L(E_{Lk}) - f_R(E_{Rq})\big) \quad (22.27b)$$

We can replace the summation over k and q by the integration over the energy, introducing the densities of states, $D_L(E)$ and $D_R(E)$, for the electrons in the leads as

$$\sum_{k,q}(\cdot) = \int dE_L\int dE_R D_L(E)D_R(E)(\cdot) \quad (22.28)$$

In the wide-band limit, we assume that the tunneling elements $|T_{kq}|^2$ do not depend on energies $|T_{kq}|^2 = |T_0|^2$ and the densities-of-states near the Fermi surface in the leads are also energy independent, $D_{\alpha}(E) = D_{\alpha}(\mu) = D_{\alpha}$. Furthermore, we assume that all energy parameters of our problem $(eV, k_B T, \hbar\omega_0, \cdots)$ are much less than the basic chemical potential μ of the electron gas in the leads.

In this case, Equations (22.27a,b) can be rewritten at low temperatures as

$$S(\omega) = \pi D_L D_R |T_0|^2 \left\{ (\hbar\omega - eV) \coth\left(\frac{\hbar\omega - eV}{2k_B T}\right) + (\hbar\omega + eV) \coth\left(\frac{\hbar\omega + eV}{2k_B T}\right) \right\} \quad (22.29a)$$

and

$$\chi''(\omega) = 2\pi D_L D_R |T_0|^2 \omega \quad (22.29b)$$

One can see that at zero applied voltage, the FDT is satisfied:

$$S(\omega)|_{V=0} = \hbar \chi''(\omega) \coth\left(\frac{\hbar\omega}{2k_B T}\right) \quad (22.30)$$

Concluding this section, we note that the electrons in the lead represent the Ohmic nonwhite heat bath with respect to the mechanical nano-oscillator. We call it Ohmic because the imaginary part of the susceptibility is proportional to ω and we call it nonwhite because the spectral function exhibits the frequency dependence (other than that involved in the hyperbolic cotangent) which makes the difference from the white noise case.

22.4 Electron Transport through the Nanomechanical Oscillator

The electric current through the junction is defined as $I = I_L = -I_R$, where

$$I_\alpha = -e \left\langle \dot{N}_\alpha \right\rangle = -e \frac{d}{dt} \sum_k \left\langle c_{\alpha k}^+(t) c_{\alpha k}(t) \right\rangle \quad (22.31)$$

The equations of motion for the annihilation operators of electrons in the leads can be obtained from the Hamiltonians, Equations (22.3) and (22.4), and have the forms

$$i\hbar \dot{c}_{Lk}(t) = E_{Lk} c_{Lk}(t) - \sum_{q'} T_{kq'} c_{Rq'}(t) e^{-x(t)/\lambda} \quad (22.32a)$$

and

$$i\hbar \dot{c}_{Rq}(t) = E_{Rq} c_{Rq}(t) - \sum_{k'} T_{k'q} c_{Lk'}(t) e^{-x(t)/\lambda} \quad (22.32b)$$

Correspondingly, the expression for the electric current contains the dependence on the oscillator position and is given by

$$I = \frac{ie}{\hbar} \sum_{k,q} \left\langle \left(T_{kq}^* c_{Rq}^+ c_{Lk} - T_{kq} c_{Lk}^+ c_{Rq} \right) e^{-x/\lambda} \right\rangle \quad (22.33)$$

The creation/annihilation operators involved in this expression can be written in terms of the unperturbed operators and the retarded Green's function, Equation (22.25a), as

$$c_{Lk}(t) = c_{Lk}^{(0)}(t) - \sum_{q'} T_{kq'} \int dt_1 g_{Lk}^r(t, t_1) c_{Rq'}^{(0)}(t_1) e^{-x(t_1)/\lambda} \quad (22.34a)$$

and

$$c_{Rq}(t) = c_{Rq}^{(0)}(t) - \sum_{k'} T_{k'q} \int dt_1 g_{Rq}^r(t, t_1) c_{Lk'}^{(0)}(t_1) e^{-x(t_1)/\lambda} \quad (22.34b)$$

Therefore, the electric current is given by

$$I = \frac{e}{\hbar} \sum_{k,q} |T_{kq}|^2 \int dt_1 \left\langle \frac{1}{2} \left[e^{-x(t)/\lambda}, e^{-x(t_1)/\lambda} \right]_+ \right\rangle \left(g^r_{Rq}(t, t_1) g^<_{Lk}(t_1, t) - g^r_{Lk}(t, t_1) g^<_{Rq}(t_1, t) + h.c. \right)$$

(22.35)

Substituting the expressions for the anticommutator, Equation (22.17a), and the Green's functions, Equations (22.25a–c), and performing the integration over t_1, we obtain

$$I = \frac{e}{\hbar} \frac{\pi}{2} e^{v_c} \sum_{l,n=-\infty}^{\infty} I_l(v_c) J_{2n}(v_0) \sum_{kq} |T_{kq}|^2 \left(f_L(E_{Lk}) - f_R(E_{Rq}) \right)$$

$$\times \left(\delta(E_{Lk}/\hbar - E_{Rq}/\hbar + (l - 2n)\omega_0) + \delta(E_{Lk}/\hbar - E_{Rq}/\hbar - (l - 2n)\omega_0) \right.$$

$$\left. + \delta(E_{Lk}/\hbar - E_{Rq}/\hbar + (l + 2n)\omega_0) + \delta(E_{Lk}/\hbar - E_{Rq}/\hbar - (l + 2n)\omega_0) \right)$$

(22.36)

Replacing the summation over k and q by the integration over corresponding energies and employing the wide-band limit, we finally obtain

$$I = G(V)V = 2\pi \frac{e^2}{\hbar} D_L D_R |T_0|^2 e^{2v_c} V$$

(22.37)

where $G(V)$ is the nonlinear conductance of the tunnel junction, which depends on the fluctuation level of the mechanical oscillator. It should be noted that this fluctuation level is also a function of the voltage applied to the junction.

22.5 Results and Discussion of the Cantilever Case

22.5.1 Voltage Dependence

The self-consistent set of equations, Equations (22.19–22.23) is quite lengthy and requires a numerical solution for general conditions. However, it is possible to obtain analytical results of fundamental importance for some limiting situations (such as a low temperature case).

First of all, let us consider the case of weak nonlinearity of the cantilever–junction coupling, $v_0 \ll 1$. Correspondingly, the contribution of vacuum oscillator fluctuations can be neglected and the spectrum of fluctuation forces, Equation (22.21a), has the form

$$K(\omega) = \frac{\hbar^2}{2m^2\lambda^2} e^{v_c} \sum_{l=-\infty}^{\infty} I_l(v_c) \left(S(\omega - l\omega_0) + S(\omega + l\omega_0) \right)$$

(22.38)

It should be noted once again that the fluctuation source involved in the Langevin-like equation, Equation (22.13), is not white noise, because the Fourier transform of its correlation function exhibits the frequency dependence even for the weak heating case, when the root-mean-square amplitude of the cantilever fluctuations is much smaller than the tunneling length (i.e., when $v_c \ll 1$). The reason for such a behavior is the nonwhite character of the electron heat bath producing these oscillator fluctuations. Substituting the expression for the heat bath spectral functions, we obtain for the weak heating case

$$K(\omega) = \frac{\pi D_L D_R |T_0|^2 \hbar^2}{m^2 \lambda^2} e^{2v_c} \left((\hbar\omega - eV) \coth\left(\frac{\hbar\omega - eV}{2k_B T} \right) + (\hbar\omega + eV) \coth\left(\frac{\hbar\omega + eV}{2k_B T} \right) \right)$$

(22.39)

Under the same conditions, the oscillator damping rate has a form

$$\gamma(\omega) = \frac{2\pi\hbar D_L D_R |T_0|^2}{m\lambda^2} e^{2v_c}$$

(22.40)

Using Equations (22.19), (22.39), and (22.40), we obtain the analytic explicit expression for the variance of oscillator position fluctuations as

$$\langle \tilde{x}^2 \rangle = \frac{\hbar}{4m\omega_0} \left(\frac{\hbar\omega_0 - eV}{\hbar\omega_0} \coth\left(\frac{\hbar\omega_0 - eV}{2k_B T} \right) + \frac{\hbar\omega_0 + eV}{\hbar\omega_0} \coth\left(\frac{\hbar\omega_0 + eV}{2k_B T} \right) \right) \qquad (22.41)$$

With no voltage applied to the system, the usual expression for thermal fluctuations of the oscillator position is recovered, and at low temperature these fluctuations are on the vacuum level, $\nu_c = \nu_0$. However, when the applied voltage becomes larger than the frequency of the mechanical oscillations, $eV > \hbar\omega_0$, the variance increases linearly with the voltage, as

$$\langle \tilde{x}^2 \rangle = \frac{\hbar}{2m\omega_0} \frac{eV}{\hbar\omega_0} \qquad (22.42)$$

or

$$\nu_c = \nu_0 \frac{eV}{\hbar\omega_0} \qquad (22.43)$$

This linear increase of the fluctuation level may be interpreted as an increase in *an effective oscillator temperature*, $k_B T_{eff} = eV/2$, from the result previously obtained in the framework of the Caldeira–Legget model [14].

The heating process also affects the bias dependence of the damping rate, Equation (22.40), and the nonlinear conductance, Equation (22.37). For low temperature and low voltage ($eV < \eta\omega_0$), both of these characteristics do not depend on the voltage, as

$$G = 2\pi e D_L D_R |T_0|^2 (1 + 2\nu_0) \qquad (22.44)$$

and

$$\gamma = \frac{2\pi\hbar D_L D_R |T_0|^2}{m\lambda^2} (1 + 2\nu_0) \qquad (22.45)$$

whereas above the threshold of the excitation of the cantilever oscillations ($eV < \omega_0$), the heating process makes a linear contribution to the conductance and to the damping rate, as

$$G(V) = 2\pi e D_L D_R |T_0|^2 \left(1 + 2\nu_0 \frac{eV}{\hbar\omega_0} \right) \qquad (22.46)$$

and

$$\gamma(V) = \frac{2\pi\hbar D_L D_R |T_0|^2}{m\lambda^2} \left(1 + 2\nu_0 \frac{eV}{\hbar\omega_0} \right) \qquad (22.47)$$

For the case of higher temperatures, the self-consistent set of equations, Equations (22.19–22.23), has to be evaluated numerically. The solution of these equations, the parameter ν_c, can be substituted to the expressions for the conductance, Equation (22.37), and the damping rate, Equation (22.19a). These characteristics are presented in Figures 22.3 and 22.4, respectively, as functions of the applied bias voltage, for cantilever mass, $m = 10^{-17}$ g; tunneling length, $\lambda = 10^{-8}$ cm; and various temperatures. For these values of parameters $\nu_0 \ll 1$, the nonlinearity of the cantilever–junction coupling is weak and the quantum effects are small. Both the applied voltage and the temperature are normalized to the fundamental frequency of mechanical oscillations. The conductance is normalized to the value $G_0 = 2\pi e^2 D_L D_R |T_0|^2 / \hbar$. One can see that at small temperatures, both the conductance and the damping rate are voltage-independent until the applied voltage reaches the oscillator frequency consistent with the earlier qualitative analysis. With temperature increasing, the sharp kinks in the voltage dependencies become smoother, and with further increasing, both the conductance and the damping rate have larger absolute values, with the functional shape remaining the same. The temperature dependencies of the system characteristics will be discussed in the next subsection.

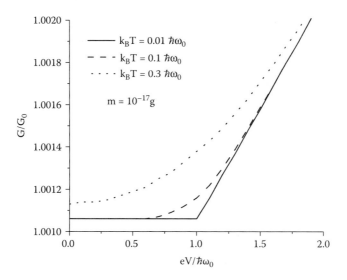

FIGURE 22.3   Voltage dependence of the conductance for mass $m = 10^{-17}g$ and various temperatures.

We have shown that drastic changes in the system behavior appear when the applied voltage becomes comparable to the fundamental frequency of the nano-oscillator. Also, we can expect something unusual when the spatial characteristic parameter of mechanical oscillations, the root-mean-square of the variance of position fluctuations, approaches the tunneling length, λ (i.e., when $\nu_0 \sim 1$). To illustrate that, we plot in Figures 22.5 and 22.6 the same voltage dependencies of the nonlinear conductance and the oscillator damping rate for the cantilever with a smaller mass, $m = 10^{-20}$ g. It is evident from these figures that the quantum effects take place, and that both these parameters increase with increasing voltage even at voltages smaller than the oscillator frequency. Moreover, the absolute values of the conductance and the damping rate are larger than that of the cantilever with larger mass, even at the equilibrium because notwithstanding zero applied voltage, the vacuum events of the electron tunneling between the leads affect the oscillator

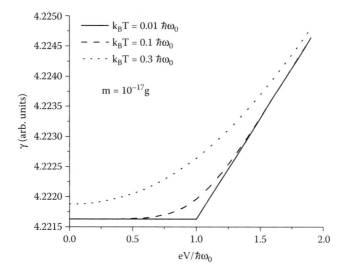

FIGURE 22.4   Voltage dependence of the oscillator damping rate for mass $m = 10^{-17}g$ and various temperatures.

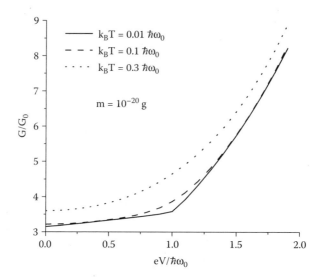

FIGURE 22.5 Voltage dependence of the conductance for mass $m = 10^{-20}$ g and various temperatures.

dynamics and, therefore, change its dynamical characteristics. It should be noted that the small deviation from the classical behavior can also be seen in Figure 22.3, where the equilibrium value of normalized conductance is not exactly one.

To separate quantum, thermal, and non-equilibrium effects, we examine the variance of non-equilibrium position fluctuations normalized to the level of thermal fluctuations of the nano-oscillator decoupled from the leads, as

$$var = \left\langle \tilde{x}^2 \right\rangle \Big/ \frac{\hbar}{2m\omega_0} \coth\left(\frac{\hbar\omega_0}{2k_B T}\right) \tag{22.48}$$

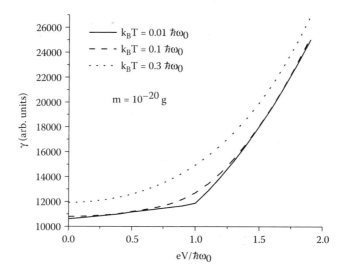

FIGURE 22.6 Voltage dependence of the oscillator damping rate for mass $m = 10^{-20}$ g and various temperatures.

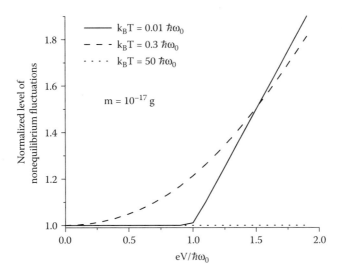

FIGURE 22.7 The level of non-equilibrium fluctuations of the oscillator position as a function of the applied voltage for mass $m = 10^{-17}g$ and various temperatures.

The voltage dependencies of this characteristic are presented in Figures 22.7 and 22.8 for the two values of the oscillator mass, $m = 10^{-17}$ g and $m = 10^{-20}$ g, respectively. One can see that (i) the normalized variance exhibits a sharp kink in the voltage dependence at low temperature when the applied voltage approaches the oscillator frequency, (ii) quantum effects can also be seen for the case of smaller oscillator mass, and (iii) at large temperatures both non-equilibrium and quantum effects produced by the conducting electrons disappear and the nano-oscillator behaves as the decoupled one.

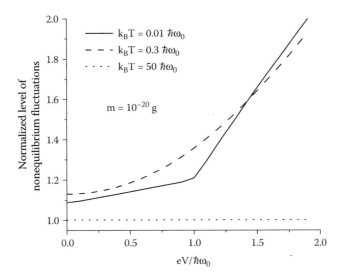

FIGURE 22.8 The level of non-equilibrium fluctuations of the oscillator position as a function of the applied voltage for mass $m = 10^{-20}g$ and various temperatures.

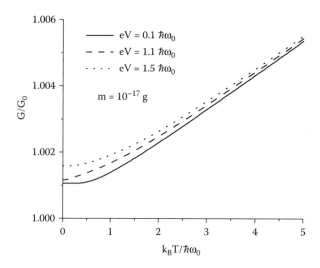

FIGURE 22.9   Temperature dependence of the conductance for mass $m = 10^{-17}g$ and various applied voltages.

22.5.2   Temperature Dependence

The temperature dependencies of the nonlinear conductance are presented in Figures 22.9 and 22.10 for the two values of the oscillator mass, $m = 10^{-17}$ g and $m = 10^{-20}$ g, respectively, and for various voltages applied to the junction. It is evident from these figures that the value of the applied voltage is important only for low temperatures, since at higher temperatures all curves merge. One can see that if the applied voltage is quite different from the oscillator frequency, a plateau occurs at low temperatures, whereas when these two parameters are almost equal, the conductance starts to increase immediately as the temperature increases. This effect can be seen also in Figures 22.3 and 22.5, where the sharp kink (solid lines) in the voltage dependence of the conductance is smoothed for higher temperatures (dashed lines), but the solid and dashed lines coincide below and above the kink.

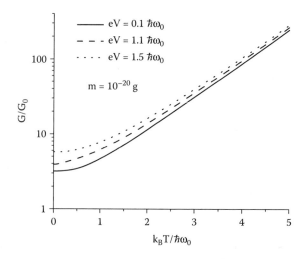

FIGURE 22.10   Temperature dependence of the conductance for mass $m = 10^{-20}g$ and various applied voltages.

The character of the temperature dependence of the conductance is completely different for the two different oscillator masses. For larger masses, when $v_0 \ll 1$, there is just a small almost linear increase of the conductance (Figure 22.9), whereas for smaller masses, when $v_0 \sim 1$, the conductance growth is exponential (the conductance axis is in the logarithmic scale in Figure 22.10). It follows from Equation (22.37) that the conductance is proportional to the exponent of $2v_c$, (i.e., to the exponent of the variance of non-equilibrium position fluctuations). In turn, the variance is proportional to $\coth(\hbar\omega/2k_B T)$, and becomes proportional to T at higher temperatures, so the resulting temperature dependence of the conductance is exponential. This is a very intriguing and important result. Typically, the conductance of quantum systems either exhibits activation dependence (in the case of the over-barrier transport) or is temperature independent (in the case of tunneling) and only incorporation of the mechanical motion of the cantilever brings about such an unusual dependence. The observations of exponential temperature dependencies were reported for the conductance of single molecules placed between metallic electrodes [51] and for transport in manganites [52], but in both cases the full understanding of the underlying process is not yet achieved, so there are no proper publications to date. The physical mechanisms of mechanical oscillations can be the elastic coupling between the molecule and electrodes for the former case, and the oscillations of the oxygen atom between the manganese atoms for the latter case.

To further illustrate the interplay between quantum, non-equilibrium, and thermal effects on the system dynamics, we plot the temperature dependencies of the normalized variance of the position fluctuations, Equation (22.48), in Figure 22.11 for various applied voltages. One can see that at high temperatures all curves approach one, which means that the thermal motion dominates the electron–oscillator interaction. At low temperature and low applied voltage, the electron energy is not enough to excite the mechanical motion of the cantilever, so the nano-oscillator is almost not heated by electrons. The quantum interaction between electron and mechanical degrees of freedom increases with increasing temperature when the thermal electron energy becomes comparable to the frequency of oscillations, resulting in a peak in the temperature dependence. This peak becomes more pronounced for higher voltages and, finally, when the voltage is larger than the oscillator frequency, the peak almost disappears because the oscillator is heated even at low temperature. The behavior of the oscillator with mass $m = 10^{-20}$ g is shown in Figure 22.11 and the oscillator with mass $m = 10^{-17}$ g exhibits the same kind of qualitative behavior, but with smaller heating.

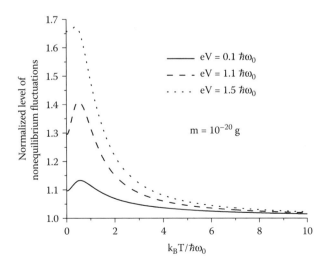

FIGURE 22.11 The level of non-equilibrium fluctuations of the oscillator position as a function of temperature for mass $m = 10^{-20}$g and various applied voltages.

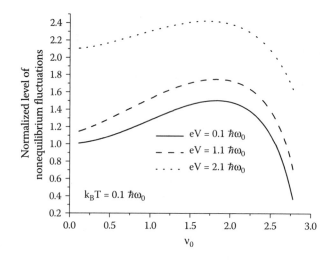

FIGURE 22.12 The level of non-equilibrium fluctuations of the oscillator position as a function of the parameter ν_0 for various applied voltages.

22.5.3 Dependence on the Quantum Parameter ν_0

We have shown in previous subsections that the system's behavior changes drastically when the quantum parameter ν_0 approaches one. The oscillator heating increases, vacuum tunneling events affect the mechanical motion of the cantilever, and the system conductance rises exponentially with temperature. For more careful examination we plot the dependence of the normalized variance of oscillator position fluctuations on the parameter ν_0 in Figure 22.12. It is evident from this figure that after the initial increase of the fluctuation level, the variance drops almost abruptly with the further increasing of ν_0. Such behavior can be easily explained if one notes that at a fixed oscillator mass and frequency, the increase of ν_0 corresponds to the decrease of the tunneling length, λ, which means that the tunneling matrix element involved in Equation (22.4) decreases and, finally, the leads are decoupled. Accordingly, the interaction between electron transport and mechanical motion initially becomes stronger with an increase of ν_0, and, subsequently, it is turned off.

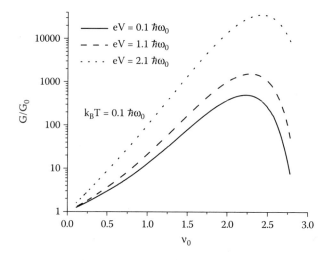

FIGURE 22.13 The conductance as a function of the parameter ν_0 for various applied voltages.

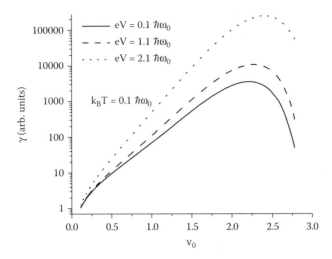

FIGURE 22.14 The oscillator damping rate as a function of the parameter v_0 for various applied voltages.

This effect is very pronounced in the dependencies of the conductance and the damping rate on the quantum parameter v_0, presented in Figures 22.13 and 22.14, where rapid exponential drops follow the initial (also exponential) growth. Dependencies in Figures 22.12 through 22.14 are shown for temperature $k_B T = 0.1\hbar\omega_0$, but the qualitative character is the same for higher temperatures.

In this section, we have used values of the applied voltage and temperature normalized to the frequency of the cantilever. For the frequency of $\omega_0 = 1\,GHz$, $eV = \hbar\omega_0$ corresponds to the voltage bias of approximately $6.55 \cdot 10^{-6}\,V$, and $k_B T = \hbar\omega_0$ corresponds to the temperature of $7.6\,mK$.

22.6 Shuttle Dynamics

22.6.1 Formalism

In previous sections, we have examined the situation when one of the leads in the tunnel junction experiences the mechanical oscillatory motion, but the electron tunneling occurs directly from one lead to another. Here, in Section 22.6, we analyze a different case when the small nano-object (shuttle) having its own quantized electron states oscillates between the two leads (Figure 22.2[b]). These conditions correspond to electron transport through a C_{60}-based transistor [11], a silicon nanopillar [40], or through small grains embedded in the elastic medium between the leads [9,10,12].

The Hamiltonian of the quantum shuttle placed between the leads is given by

$$H = \frac{p^2}{2m} + \frac{m\omega_0^2 x^2}{2} + E_0 a^+ a + \sum_{\alpha,k} E_{\alpha k} c_{\alpha k}^+ c_{\alpha k} + \sum_{\alpha,k} \left(T_{\alpha k} c_{\alpha k}^+ a + T_{\alpha k}^* a^+ c_{\alpha k} \right) e^{x/\lambda_\alpha} \qquad (22.49)$$

Here, a^+/a are the creation/annihilation operators of the quantized electron state of the shuttle with energy E_0 (we consider other states to be energetically inaccessible), $\alpha = L, R$, and $\lambda_L = -\lambda, \lambda_R = \lambda$. The equations of motion derived from this Hamiltonian have forms

$$i\hbar \dot{a}(t) = E_0 a(t) - \sum_{\alpha,k} T_{\alpha k}^* c_{\alpha k}(t) e^{x(t)/\lambda_\alpha} \qquad (22.50a)$$

$$i\hbar \dot{c}_{\alpha k}(t) = E_{\alpha k} c_{\alpha k}(t) - T_{\alpha k} a(t) e^{x(t)/\lambda_\alpha} \qquad (22.50b)$$

and

$$\ddot{x}(t) + \omega_0^2 x(t) = \sum_{\alpha,k} \frac{1}{m\lambda_\alpha} \left(T_{\alpha k} c_{\alpha k}^+(t) a(t) + T_{\alpha k}^* a^+(t) c_{\alpha k}(t) \right) \tag{22.50c}$$

The electric current through the shuttle is defined as $I = I_L = -I_R$, where

$$I_\alpha = -e \frac{d}{dt} \sum_k \left\langle c_{\alpha k}^+ c_{\alpha k} \right\rangle = \frac{ie}{\hbar} \sum_k \left(T_{\alpha k}^* \left\langle a^+ e^{x/\lambda_\alpha} c_{\alpha k} \right\rangle + T_{\alpha k} \left\langle c_{\alpha k}^+ e^{x/\lambda_\alpha} a \right\rangle \right) \tag{22.51}$$

One can see that in contrast to the cantilever case, the operators of the shuttle state are involved in the expression for the current through the lead instead of the operators of the other lead. The expression for the lead electron operator also contains the operators of the shuttle state, as

$$c_{\alpha k}(t) = c_{\alpha k}^{(0)}(t) + i T_{\alpha k} \int dt_1 e^{-i E_{\alpha k}(t-t_1)} a(t_1) e^{x(t_1)/\lambda_\alpha} \Theta(t-t_1) \tag{22.52}$$

where the retarded Green's function, Equation (22.25), is substituted and the Planck constant is omitted. Performing the same operations as we have done in the previous sections for the cantilever case, we obtain the electric current through the shuttle, as given by

$$I_\alpha = \frac{e}{\hbar} \left(\sum_k |T_{\alpha k}|^2 \int dt_1 f_\alpha(E_{\alpha k}) e^{-i E_{\alpha k}(t-t_1)} \Theta(t-t_1) \left\langle \left[a^+(t) e^{x(t)/\lambda_\alpha}, a(t_1) e^{x(t_1)/\lambda_\alpha} \right]_+ \right\rangle \right.$$
$$\left. + \sum_k |T_{\alpha k}|^2 \int dt_1 e^{-i E_{\alpha k}(t-t_1)} \Theta(t-t_1) \left\langle a^+(t) a(t_1) \right\rangle \left\langle e^{x(t)/\lambda_\alpha} e^{x(t_1)/\lambda_\alpha} \right\rangle + h.c. \right) \tag{22.53}$$

To further simplify this expression, we have to evaluate the steady-state electron population of the shuttle

$$N = \left\langle a^+ a \right\rangle \tag{22.54}$$

For the case of the weak shuttle-leads coupling, the electron correlators have forms

$$\left\langle a^+(t) a(t_1) \right\rangle = N e^{\frac{i}{\hbar} E_0 (t-t_1)} \tag{22.55a}$$

and

$$\left\langle a(t_1) a^+(t) \right\rangle = (1-N) e^{\frac{i}{\hbar} E_0 (t-t_1)} \tag{22.55b}$$

The steady-state electron population, N, can be calculated from the condition

$$I_L + I_R = 0 \tag{22.56}$$

Using formulas derived in the previous sections and substituting expressions for the correlators of the shuttle electron operators, Equations (22.55a,b), we obtain

$$I_\alpha = \frac{e\Gamma_\alpha}{\hbar} e^{\nu_c} \sum_{l,n=-\infty}^{+\infty} I_l(\nu_c) J_n(\nu_0) \left(N \left(1 - f_\alpha(E_0 - l\hbar\omega_0 - n\hbar\omega_0) \right) - (1-N) f_\alpha(E_0 + l\hbar\omega_0 + n\hbar\omega_0) \right) \tag{22.57}$$

where ν_0 and ν_c are given by Equations (22.22) and (22.23), respectively, I_l and J_n are the modified and ordinary Bessel functions, respectively, and we have introduced the parameters

$$\Gamma_\alpha = 2\pi \sum_k |T_{\alpha k}|^2 \delta(\omega - E_{\alpha k}/\hbar) \tag{22.58}$$

which are energy-independent in the wide band limit. In the following, we consider the symmetric case, so $\Gamma_L = \Gamma_R = \Gamma$.

The steady-state shuttle electron population can be found from Equation (22.57) as

$$N = \frac{C}{D} \tag{22.59}$$

where

$$C = \sum_{l,n=-\infty}^{+\infty} I_l(\nu_c) J_n(\nu_0) \left(f_L(E_0 + l\hbar\omega_0 + n\hbar\omega_0) + f_R(E_0 + l\hbar\omega_0 + n\hbar\omega_0) \right) \tag{22.60}$$

and

$$D = \sum_{l,n=-\infty}^{+\infty} I_l(\nu_c) J_n(\nu_0) \left(2 - f_L(E_0 - l\hbar\omega_0 - n\hbar\omega_0) \right.$$
$$\left. + f_L(E_0 + l\hbar\omega_0 + n\hbar\omega_0) - f_R(E_0 - l\hbar\omega_0 - n\hbar\omega_0) + f_R(E_0 + l\hbar\omega_0 + n\hbar\omega_0) \right) \tag{22.61}$$

The expressions for the oscillator damping rate and for the correlation function of fluctuation sources involved in the equation for the oscillator position fluctuations can also be obtained on the basis of the previously derived formulas, as

$$\gamma(\omega) = \nu_0 \Gamma e^{\nu_c} B(\omega) \tag{22.62}$$

and

$$K(\omega) = \frac{\hbar\omega_0}{m} \nu_0 \Gamma A(\omega) \tag{22.63}$$

where

$$A(\omega) = \sum_\alpha \sum_{l,n=-\infty}^{+\infty} I_l(\nu_c) J_n(\nu_0)$$
$$\times \left(N \left(2 - f_\alpha(E_0 - l\hbar\omega_0 - n\hbar\omega_0 - \hbar\omega) - f_\alpha(E_0 - l\hbar\omega_0 - n\hbar\omega_0 + \hbar\omega) \right) \right.$$
$$\left. + (1 - N) \left(f_\alpha(E_0 + l\hbar\omega_0 + n\hbar\omega_0 + \hbar\omega) + f_\alpha(E_0 + l\hbar\omega_0 + n\hbar\omega_0 - \hbar\omega) \right) \right) \tag{22.64}$$

and

$$B(\omega) = \sum_\alpha \sum_{l,n=-\infty}^{+\infty} I_l(\nu_c) J_n(\nu_0)$$
$$\times \left(N \left(f_\alpha(E_0 - l\hbar\omega_0 - n\hbar\omega_0 - \hbar\omega) - f_\alpha(E_0 - l\hbar\omega_0 - n\hbar\omega_0 + \hbar\omega) \right) \right.$$
$$\left. + (1 - N) \left(f_\alpha(E_0 + l\hbar\omega_0 + n\hbar\omega_0 - \hbar\omega) - f_\alpha(E_0 + l\hbar\omega_0 + n\hbar\omega_0 + \hbar\omega) \right) \right) \tag{22.65}$$

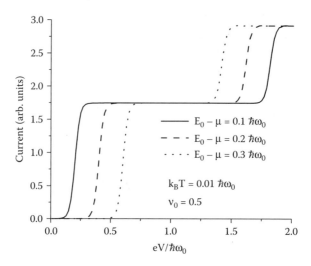

FIGURE 22.15 Current–voltage characteristics of the quantum shuttle for various separations between the equilibrium chemical potential in the leads and the electron energy in the shuttle.

Correspondingly, the variance of oscillator position fluctuations is given by

$$\langle \tilde{x}^2 \rangle = \frac{\hbar}{2m\omega_0} \frac{A(\omega_0)}{B(\omega_0)} \tag{22.66}$$

or

$$\nu_c = \nu_0 \frac{A(\omega_0)}{B(\omega_0)} \tag{22.67}$$

22.6.2 Numerical Results

Equations (22.59–22.61,22.64,22.66,22.67) constitute the self-consistent set of equations to determine the steady-state shuttle electron population and the level of non-equilibrium oscillator position fluctuations. We solve this set of equations numerically and substitute the obtained values into the expression for the electric current, Equation (22.57). The current–voltage characteristics are shown in Figure 22.15 for low temperature ($k_B T = 0.01\hbar\omega_0$) and various separations between the equilibrium chemical potential of the leads, μ, and the energy of the electron state in the shuttle, E_0. It is evident from this figure that there are pronounced steps in the low-temperature current–voltage characteristics at the bias voltages (i) $eV/2 = E_0 - \mu$, and (ii) $eV/2 = \hbar\omega_0 - E_0 + \mu$. The first one occurs when the chemical potential of the left lead ($\mu_L = \mu + eV/2$) passes through the energetic level of the shuttle, while the second one corresponds to the passing of the chemical potential of the right lead ($\mu_R = \mu - eV/2$) through the virtual level with energy $E_0 - \hbar\omega_0$. In the latter case, an electron of the left lead having energy $E_0 - \hbar\omega_0$ tunnels to the shuttle electron state of energy E_0 with absorption of the quantum of the shuttle mechanic motion (phonon). The subsequent tunneling to the state of the right lead having energy $E_0 - \hbar\omega_0$ is accompanied by the phonon emission. Such processes are possible even at equilibrium, but in this case the populations of the left and right leads having energy $E_0 - \hbar\omega_0$ are equal and there is no current flow. Only when the chemical potential of the right lead passes through this energy, the population of the right lead becomes less than that of the left lead, and the second step in the current–voltage characteristics occurs.

The processes of the phonon emission and absorption are more likely with larger coupling between electron and mechanical degrees of freedom of the shuttle. To illustrate this, we plot in Figure 22.16 the

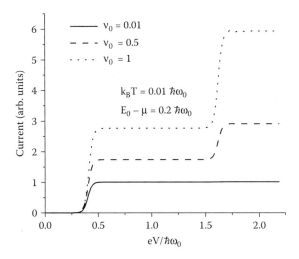

FIGURE 22.16 Current–voltage characteristics of the quantum shuttle for various values of the parameter ν_0.

current–voltage characteristics for various values of the parameter ν_0, characterizing such couplings. One can see that the second step becomes more pronounced with increasing values of ν_0, whereas for the small value, the second step is almost invisible. There is no second step in the voltage dependence of the shuttle population presented in Figure 22.17 for various temperatures, where only the conduction through the shuttle energetic level populates it. (It should be noted that at larger temperatures the shuttle is populated even in the equilibrium.) However, even this negligibly small current flow through the virtual level $E_0 - \hbar\omega_0$ affects the mechanical shuttle characteristics, such as the damping rate (Figure 22.18) and the level of non-equilibrium fluctuations of the oscillator position (Figure 22.19). In both cases, the passing of the applied voltage through the value $eV/2 = \hbar\omega_0 - E_0 + \mu$ is accompanied by the sharp steps at low temperatures, and these steps are smoothed with temperature increasing.

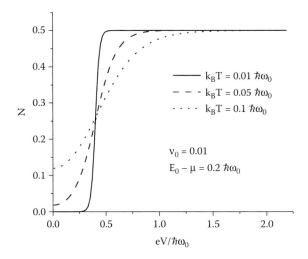

FIGURE 22.17 The population of the shuttle electron state as a function of the applied voltage for various temperatures.

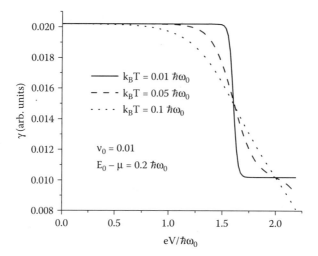

FIGURE 22.18   The shuttle damping rate as a function of the applied voltage for various temperatures.

The current–voltage characteristics at larger values of ν_0 are shown in Figure 22.20 for various temperatures. One can see that the steps are smoothed at higher temperatures and, moreover, there is current growth, in this case, at large voltages. This increasing of the current is a consequence of the shuttle instability that occurs at larger voltages [21,24,35] and starts earlier if the temperature is higher. The features caused by this instability can be also seen in Figures 22.18 and 22.19.

We plot the oscillator damping rate and the level of non-equilibrium fluctuations of the oscillator position in Figures 22.21 and 22.22, respectively, for the larger value of parameter ν_0. Increasing the coupling between mechanical and electron degrees of freedom leads to the occurrence of the steps in these figures not only when the current flow through the shuttle accompanied by the emission/absorption starts,

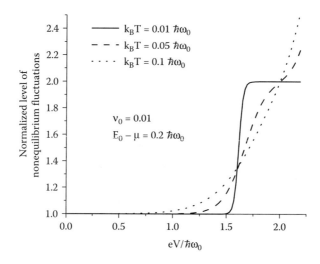

FIGURE 22.19   The level of non-equilibrium fluctuations of the shuttle position as a function of the applied voltage for various temperatures.

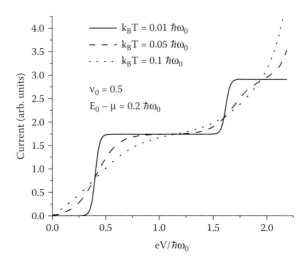

FIGURE 22.20 Current–voltage characteristics of the quantum shuttle for various temperatures.

but also at the voltage corresponding to the first step in the current–voltage characteristics. The steps are also smoothed at larger temperatures and the features associated with the shuttle instability arise.

The temperature dependencies of the electric current through the shuttle are shown in Figures 22.23 through 22.25 for the values of the parameter $\nu_0 = 0.01, 0.1$, and 0.5, respectively. The values of the applied voltage for these figures are taken from the initial, first, and second plateaus. One can see that the qualitative characters of the dependencies are very different for different values of ν_0. At small ν_0, the current decreases with increasing temperature (except for the voltages corresponding to the initial plateau where there is no current at low temperature)—at intermediate ν_0, the current increases, and at larger ν_0, it begins to grow exponentially. The reason for such behavior can be seen from the limiting expression for the current at high temperatures ($k_B T \gg |E - \mu|$). In this case, $\nu_c = \nu_0 (2 k_B T / \hbar \omega_0)$ and the current is described by the

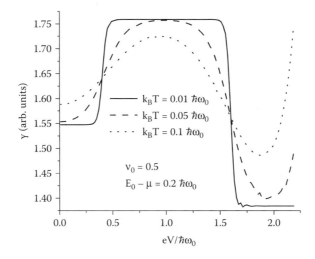

FIGURE 22.21 Voltage dependence of the shuttle damping rate for various temperatures.

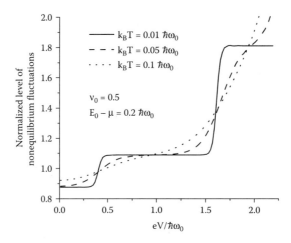

FIGURE 22.22 Voltage dependence of the level of non-equilibrium fluctuations of the shuttle position for various temperatures.

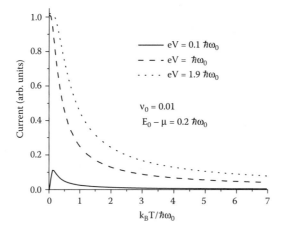

FIGURE 22.23 The electron current through the shuttle as a function of temperature for $\nu_0 = 0.01$ and various applied voltages.

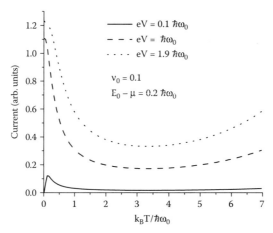

FIGURE 22.24 The electron current through the shuttle as a function of temperature for $\nu_0 = 0.1$ and various applied voltages.

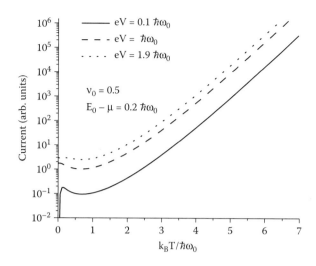

FIGURE 22.25 The electron current through the shuttle as a function of temperature for $v_0 = 0.5$ and various applied voltages.

formula

$$I = \frac{e}{\hbar} \Gamma \exp\left(\frac{4k_B T}{\hbar \omega_0} v_0 \right) \frac{eV}{4k_B T} \tag{22.68}$$

where the last term in the product appears from the Fermi distribution functions. One can see that there is competition between the exponent and $1/T$ term in Equation (22.68). Depending on the value of v_0, one of these two terms dominates, so it is $1/T$ dependence at small v_0, almost no temperature dependence at intermediate v_0, and exponential growth with temperature increasing at larger v_0.

These three regimes are also very pronounced in the temperature dependence of the level of non-equilibrium oscillator position fluctuations presented in Figures 22.26 through 22.28. At small v_0 (Figure 22.26), fluctuations decay to the decoupled value at higher temperatures, and there is no heating for

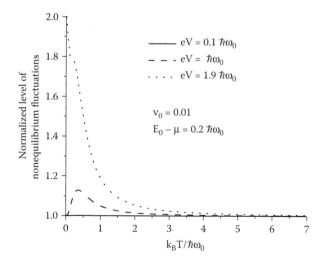

FIGURE 22.26 The level of non-equilibrium fluctuations of the shuttle position as a function of temperature for $v_0 = 0.01$ and various applied voltages.

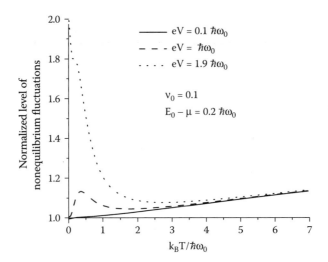

FIGURE 22.27 The level of non-equilibrium fluctuations of the shuttle position as a function of temperature for $v_0 = 0.1$ and various applied voltages.

small applied voltages. The heating decays to the straight line at moderate voltage (Figure 22.27) and to a nonlinear function at a higher voltage (Figure 22.28) with a contrast to the cantilever case where non-equilibrium and quantum effects disappear at high temperature.

The typical shuttle oscillation frequency is about 1 THz. In this case, $eV = \hbar\omega_0$ corresponds to the voltage bias of approximately $6.55 \cdot 10^{-3} V$ and $k_B T = \hbar\omega_0$ corresponds to the temperature of 7.6 K.

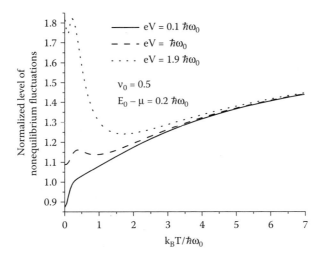

FIGURE 22.28 The level of non-equilibrium fluctuations of the shuttle position as a function of temperature for $v_0 = 0.5$ and various applied voltages.

22.7 Conclusions

We have examined the nonlinear dynamics of a mechanical nano-oscillator coupled to a tunnel junction. We have shown that electron transport through the junction affects the mechanical motion and the oscillator can be considered an open quantum system with conducting electrons representing an Ohmic nonwhite heat bath. The statistical characteristics of such a heat bath have been determined. Based on the previously developed theory, we have derived equations of motion both for the mechanical and electron components of the system.

Two cases have been analyzed: (i) cantilever dynamics when direct tunneling between two electron reservoirs with tunnel matrix element are affected by the mechanical motion; and (ii) shuttle dynamics when the mechanical oscillator has its own quantized electron state, and electrons tunnel from one lead to this state and, subsequently, to the other lead. For both cases the nonlinear conductance, the oscillator damping rate, and the variance of non-equilibrium fluctuations of the oscillator position have been determined as functions of temperature, the voltage applied to the junction, and the ratio of the oscillator parameter $\hbar/2m\omega_0$ and the electron tunneling length squared. This ratio, ν_0, shows how strong the electron–oscillator coupling is and how deep the oscillator is in the quantum regime. In particular, we have obtained that for the cantilever case at low temperature and small ν_0, electron transport is impossible until the applied voltage reaches the fundamental frequency of the mechanical oscillator, and there is a sharp kink in the voltage dependence of the conductance when the voltage passes through this value. At higher ν_0, vacuum electron tunneling events give rise to the oscillator heating even at equilibrium, but the kink can still be observed with temperature increases smoothing the kink. We also show that in the deep quantum regime, when ν_0 approaches one, the conductance grows exponentially with temperature. The effect previously reported for various systems has yet to be understood.

For the shuttle case, we have obtained that the current–voltage characteristics exhibit very pronounced steps at low temperature, when the chemical potential of one of the leads is in resonance with the energy of the shuttle electron state or with virtual states separated from the shuttle state by the energy of the quantum of the mechanical motion, so the electron tunneling is accompanied by the emission and absorption of such quanta. These additional steps are more pronounced for larger values of ν_0. The temperature dependencies of the current through the shuttle exhibit a large variety of behavior, from $1/T$ decreasing (at small ν_0) to almost no dependence (at intermediate ν_0) to exponential growth (at large ν_0).

References

[1] Roukes, M.L., Nanoelectromechanical systems, *Technical Digest of the 2000 Solid-State Sensor and Actuator Workshop*, Transducers Research Foundation, Cleveland, 2000; arXiv:cond-mat/0008187.
[2] ———, Nanophysics: plenty of room, indeed, *Sci. Am.*, 285, 48, 2001.
[3] ———, Nanoelectromechanical systems face the future, *Phys. World*, 14, 25, 2001.
[4] Cleland, A.N., *Foundations of Nanomechanics*, Springer, Heidelberg, 2002.
[5] Lyshevski, S.E., *Mems and Nems: Systems, Devices and Structures*, CRC Press, Boca Raton, FL, 2002.
[6] Blick, R.H. et al., Nanostructured silicon for studying fundamental aspects of nanomechanics, *J. Phys.: Condens. Matter*, 14, R905, 2002.
[7] Shekhter, R.I. et al., Shuttling of electrons and Cooper pairs, *J. Phys.: Condens. Matter*, 15, R441, 2003.
[8] Blencowe, M., Quantum electromechanical systems, *Phys. Rep.*, 395, 159, 2004.
[9] Gorelik, L.Y. et al., Shuttle mechanism for charge transfer in Coulomb blockade nanostructures, *Phys. Rev. Lett.*, 80, 4526, 1998.
[10] Isacsson, A. et al., Shuttle instability in self-assembled Coulomb blockade nanostructures, *Physica B*, 255, 150, 1998.
[11] Park, H., et al., Nanomechanical oscillations in a single-C60 transistor, *Nature*, 407, 58 2000.
[12] Gorelik, L.Y. et al., Coherent transfer of Cooper pairs by a movable grain, *Nature*, 411, 454, 2001.

[13] Erbe, A. et al., Nanomechanical resonator shuttling single electrons at radio frequencies, *Phys. Rev. Lett.*, 87, 096106, 2001.

[14] Mozyrsky, D. and Martin, I., Quantum-classical transition induced by electrical measurement, *Phys. Rev. Lett.*, 89, 018301, 2002.

[15] Armour, A.D. and MacKinnon, A., Transport via a quantum shuttle, *Phys. Rev. B*, 66, 035333, 2002.

[16] Nishiguchi, N., Gate voltage dependence of a single-electron transistor using the shuttle mechanism, *Phys. Rev. B*, 65, 035403, 2002.

[17] Isacsson, A. et al., Mechanical Cooper pair transportation as a source of long-distance superconducting phase coherence, *Phys. Rev. Lett.*, 89, 277002, 2002.

[18] Huang, H.M.H. et al., Nanoelectromechanical systems: Nanodevice motion at microwave frequencies, *Nature*, 421, 496, 2003.

[19] Knobel, R.G. and Cleland, A.N., Nanometer-scale displacement sensing using a single electron transistor, *Nature*, 424, 291, 2003.

[20] Blencowe, M., Quantum physics: Uncertain future, *Nature*, 424, 262, 2003.

[21] Novotny, T. et al., Quantum shuttle in phase space, *Phys. Rev. Lett.*, 90, 256801, 2003.

[22] Smirnov, A.Yu. et al., Nonequilibrium fluctuations and decoherence in nanomechanical devices coupled to the tunnel junction, *Phys. Rev. B*, 67, 115312, 2003.

[23] McCarthy, K.D. et al., Incoherent dynamics of vibrating single-molecule transistors, *Phys. Rev. B*, 67, 245415, 2003.

[24] Fedorets, D., Quantum description of shuttle instability in a nanoelectromechanical single-electron transistor, *Phys. Rev. B*, 68, 033106, 2003.

[25] Rugar D. et al., Single spin detection by magnetic resonance force microscopy, *Nature*, 430, 329, 2004.

[26] Cleland, A.N., Nanophysics: Carbon nanotubes tune up, *Nature*, 431, 251, 2004.

[27] Sazonova V. et al., A tunable carbon nanotube electromechanical oscillator, *Nature*, 431, 284, 2004.

[28] Blencowe, M., Nanomechanical quantum limits, *Science*, 304, 56, 2004.

[29] LaHaye, M.D. et al., Approaching the quantum limit of a nanomechanical resonator, *Science*, 304, 74, 2004.

[30] Mozyrsky, D. et al., Quantum-limited sensitivity of single-electron-transistor-based displacement detectors, *Phys. Rev. Lett.*, 92, 018303, 2004.

[31] Weig, E.M. et al., Single-electron-phonon interaction in a suspended quantum dot phonon cavity, *Phys. Rev. Lett.*, 92, 046904, 2004.

[32] Fedorets, D. et al., Quantum shuttle phenomena in a nanoelectromechanical single-electron transistor, *Phys. Rev. Lett.*, 92, 166801, 2004.

[33] Novotny, T. et al., Shot noise of a quantum shuttle, *Phys. Rev. Lett.*, 92, 248302, 2004.

[34] Armour, A.D. et al., Classical dynamics of a nanomechanical resonator coupled to a single-electron transistor, *Phys. Rev. B*, 69, 125313, 2004.

[35] Smirnov, A.Yu. et al., Temperature dependence of electron transport through a quantum shuttle, *Phys. Rev. B*, 69, 155310, 2004.

[36] Clerk, A.A. and Girvin, S.M., Shot noise of a tunnel junction displacement detector, *Phys. Rev. B*, 70, 121303, 2004.

[37] Armour, A.D., Current noise of a single-electron transistor coupled to a nanomechanical resonator, *Phys. Rev. B*, 70, 165315, 2004.

[38] Flindt C. et al., Current noise in a vibrating quantum dot array, *Phys. Rev. B*, 70, 205334, 2004.

[39] Clerk, A.A., Quantum-limited position detection and amplification: A linear response perspective, *Phys. Rev. B*, 70, 245306, 2004.

[40] Scheible, D.V. and Blick, R.H., Silicon nanopillars for mechanical single-electron transport, *Appl. Phys. Lett.*, 84, 4632, 2004.

[41] Fedorets, D. et al., Spintronics of a Nanoelectromechanical Shuttle, *Phys. Rev. Lett.*, 95, 057203, 2005.

[42] Rodrigues D.A. and Armour, A.D., Noise properties of two single-electron transistors coupled by a nanomechanical resonator, *Phys. Rev. B*, 72, 085324, 2005.

[43] Chen J. et al., Molecular electronic devices, in *Molecular Nanolectronics*, Reed, M. and Lee, T., Eds., American Scientific Publishers, 2003.

[44] Callen, H.B. and Welton, T.A., Irreversibility and generalized noise, *Phys. Rev.*, 83, 34, 1951.

[45] Efremov G.F. and Smirnov, A.Yu., Contribution to the microscopic theory of fluctuations of a quantum system interacting with a Gaussian thermostat, *Sov. Phys. JETP*, 53, 547, 1981.

[46] Efremov G.F. et al., Noise-induced relaxation of quantum oscillator interacting with a thermal bath, *Phys. Lett. A*, 175, 89, 1993.

[47] Mourokh, L.G. and Smirnov, A.Yu., Brownian motion in submicron rings, in *Quantum Dynamics of Submicron Structures*, Cerdeira, H. et al., Eds., NATO ASI Series, 291, Kluwer Academic Publishers, Dordrecht, 1995.

[48] Mourokh, L.G. et al., Electron Transport through a Parallel Double-Dot System in the Presence of Aharonov–Bohm Flux and Phonon Scattering, *Phys. Rev. B*, 66, 085332, 2002.

[49] Puller V.I. et al., Theory of Open Quantum Systems as Applied to Spin Relaxation in Solids, *Problems of Modern Statistical Physics*, 1, 63, 2002, arXiv:cond-mat/0205625.

[50] Caldeira, A.O. and Leggett, A.J., Influence of dissipation on quantum tunneling in macroscopic systems, *Phys. Rev. Lett.*, 46, 211, 1981.

[51] Stewart, D.R. et al., unpublished data, 2003.

[52] Noginova, N., personal communication, 2003.

23

Coherent Electron Transport in Molecular Contacts: A Case of Tractable Modeling

Alexander Onipko

Lyuba Malysheva

We have two goals in this chapter:

1. Our first and most important objective is to provide a researcher with the necessary requisites for independent modeling of coherent electron transport through a single molecule connecting two metal/semiconductor leads. With this idea in mind, we present in Sections 23.1 and 23.2, the methodology and basic physics needed for comparatively easy, sensible calculations of contact conductance and current-voltage characteristics on the basis of model exact formulas. To our knowledge, such a compressed resource does not exist, though there are several comprehensive books, reviews, and summarizing articles [1–12]. The framework of the Lippman–Schwinger formalism is used [13], complemented by a variation of the Löwdin partitioning technique [14]. Within this framework, Datta's trace formula for the contact transmission coefficient [15] is rederived in a more straightforward manner, avoiding a chain of extra definitions. Also, concrete examples of the molecule-lead interaction and molecule Green's function are given in a form ready for usage in "try-and-see" calculations. Our earlier obtained formulas describing tunneling through oligomer molecules are discussed in fine detail.

2. The second goal is to provide a description of the effects of the constant electric field on the molecular spectrum and on electron transmission through a molecule having the field-modified spectrum. The topic is appealing because it has many connections with predictions of Fowler and Nordheim [16], Zenner [17], and Wannier [18]. In Section 23.3, the tunneling exponent, specifically its energy and electric field dependence, receives most of the attention. We describe how the tunneling exponent behaves when the barrier top is biased by external voltage; when further increase of the bias tunes the tunneling energy to a field-induced, triangular-shaped well; and finally, when bias is large enough to create the Wannier–Stark band and to make its bottom lower than the tunneling energy. This examination is carried out employing the tight-binding model of long molecules, where the molecular content of "barrier terminology" is clearly exposed. The analysis shows new features of field effects missing in earlier theories.

23.1 Models of the Wire-to-Lead Connection

We begin with a brief delineation of the system to be discussed, followed by a thorough derivation of the contact transmission coefficient and its analysis. Molecule interaction with leads is described by coupling functions, playing the role of self-energy. They are specified in closed expressions. Links between the present and Green's function-self-energy formulations of contact transmission are displayed.

23.1.1 A Frame of Coherent Transmission

The model in focus, two semi-infinite but otherwise ideal leads connected by a molecule, is sketched in Figure 23.1. It is relevant to many experiments–for example, to STM spectroscopy on metal substrates layered by self-assembled molecules. This is to explain labeling of the leads as *substrate* and *tip*. The choice of coordinate axes for the substrate and tip is shown in the figure. Notations $\mathbf{r} = (r_\perp, r_z)$, $r_\perp = (r_x, r_y)$, will be used for atoms of both leads. The respective sets of coordinates will be identified as $\{\mathbf{r}\}_\alpha$, $\alpha = s, t$. Molecular atoms that belong to coordinate set $\{\mathbf{r}\}_m$ will be specified by numbers.

Fundamental characteristics of coherent (totally elastic) electron transport in conductors with scattering regions (exemplified in Figure 23.1) are provided by the transmission coefficient T. It determines the probability that a stationary monoenergetic electron flux that is incident on an obstacle, passes the obstacle. In terms of electron states of the s and t leads without interruptions (let us denote these states for the

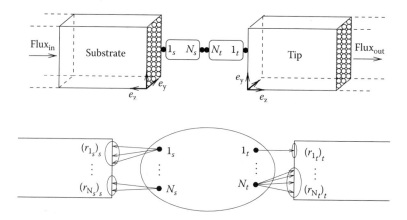

FIGURE 23.1 Schematic representation of lead–molecule–lead contacts. Up: from either sides the molecule is connected via its one atom (black circles 1_s and 1_t) to a domain of lead atoms. Down: molecular atoms $1_s, \ldots, N_s$ are coupled with domains $\{\mathbf{r}_{1_s}\}_s, \ldots, \{\mathbf{r}_{N_s}\}_s$ of atoms of the substrate lead. Connection molecule-tip is shown similarly.

moment as μ and ν, respectively), T is defined as the sum of channel μ to channel ν transmission probabilities [4,15,19–22]:

$$T = \sum_{\nu,\mu} T_{\nu,\mu} \tag{23.1}$$

For our system, index μ represents quantum numbers of an eigenstate which describes electrons moving in the substrate from the left to the right with wave vector k_μ, $\psi^s_{\mu,k_\mu}(\mathbf{r}) \exp(-ik_\mu r_z)$ that is an incident wave. Index ν labels transmitted electron waves $\psi^t_{\nu,k_\nu;\mu,k_\mu}(\mathbf{r}) \exp(ik_\nu r_z)$, appearing in the presence of k_μ waves (i.e., dependent on k_μ and μ). These waves propagate in the same direction but inside the tip. The summation in Equation (23.1), is carried out over all propagating states that correspond to the energy of incident electrons E.[1] Once the wavefunction amplitudes $\psi^s_{\mu,k_\mu}(\mathbf{r})$, $\psi^t_{\nu,k_\nu;\mu,k_\mu}(\mathbf{r})$, and group velocities $v^s_{k_\mu}$, $v^t_{k_\nu}$ in the respective media are known, the channel-to-channel transmission probability can be calculated as the ratio of the ν flux to μ flux:

$$T_{\nu,\mu} = \frac{\displaystyle\sum_{r_\perp \in \{r\}_t} v^t_{k_\nu} \left| \psi^t_{\nu,k_\nu;\mu,k_\mu}(\mathbf{r}) \right|^2}{\displaystyle\sum_{r_\perp \in \{r\}_s} v^s_{k_\mu} \left| \psi^s_{\mu,k_\mu}(\mathbf{r}) \right|^2} \tag{23.2}$$

where summation is performed over any cross-section of the leads t and s.

There are other ways of finding the transmission coefficient T by utilizing, for example, the S-matrix method, Green's function formalism, and other techniques [5,7–12,15,23–26]. However, we shall follow the prescriptions of Equations (23.1) and (23.2), which fit well to our goals.

23.1.2 Hamiltonian and Scattering-Type Eigenstates

Any eigen state Ψ^α of Hamiltonian operator $\hat{H}^\alpha$ of the substrate $\alpha = s$, tip $\alpha = t$, and molecule $\alpha = m$, can be expanded in a series of the respective basis set of atomic orbitals $|\mathbf{r}\rangle$:

$$\Psi^\alpha = \sum_{\mathbf{r} \in \{\mathbf{r}\}_\alpha} \psi^\alpha_\mathbf{r} |\mathbf{r}\rangle \tag{23.3}$$

Let the substrate and tip regions be described by tight-binding Hamiltonians of semi-infinite crystal lattices

$$\hat{H}^\alpha = \sum_{\mathbf{r} \in \{\mathbf{r}\}_\alpha} \varepsilon_\alpha |\mathbf{r}\rangle\langle\mathbf{r}| - \sum_{\substack{\mathbf{r},\mathbf{r}' \in \{\mathbf{r}\}_\alpha \\ |\mathbf{r}-\mathbf{r}'| = a}} L_\alpha |\mathbf{r}\rangle\langle\mathbf{r}'|, \quad \alpha = s,t \tag{23.4}$$

where ε_α is the onsite electron energy and $-L_\alpha$, $L_\alpha > 0$ is the energy of electron–transfer interaction between nearest-neighbor atoms, $|\mathbf{r} - \mathbf{r}'| = a$.

The tip-to–substrate drop of the applied voltage V can be taken into account in Equation (23.4) by a uniform shift of site energies by value of $eU_{s(t)}$, such that $e(U_t - U_s) = eV$. The operator $\hat{H}^m$ needs to be specified only at the stage of an analysis of the electron transmission (determining the ohmic conductance [1,19,22]) for a concrete molecular contact.

---

[1]Here and henceforth, indication of explicit dependence T on E is omitted.

In the tight-binding representation, operator of the lead-molecule interaction is given by:

$$\hat{V}^s + \hat{V}^t = \sum_{i_s=1}^{N_s} \sum_{\mathbf{r}\in\{\mathbf{r}_{i_s}\}_s} V_{i_s,\mathbf{r}}^s \left(|i_s\rangle\langle\mathbf{r}| + |\mathbf{r}\rangle\langle i_s|\right) + \sum_{i_t=1}^{N_t} \sum_{\mathbf{r}\in\{\mathbf{r}_{i_t}\}_t} V_{i_t,\mathbf{r}}^t \left(|i_t\rangle\langle\mathbf{r}| + |\mathbf{r}\rangle\langle i_t|\right) \tag{23.5}$$

where index $i_{s(t)}$, numbers molecular atoms that are facing the substrate (tip) surface and are coupled with a domain $\{\mathbf{r}_{i_{s(t)}}\}_{s(t)}$ of substrate (tip) atoms (see Figure 23.1). Without any loss of generality, substrate–molecule (tip–molecule) coupling constants $V_{i_{s(t)},\mathbf{r}}^{s(t)}$ are supposed to be real. Usually, they have non-zero values for a few atoms lying at the substrate and tip x, y surface/subsurface monatomic layers so that coordinate r_z of the substrate (tip) binding atoms runs over a bounded number of monolayers.

We shall find the expression of the transmission coefficient by using the Lippman–Schwinger equation [13]. To introduce it, we start from the formal solution to the Schrödinger equation with Hamiltonian $\hat{H} = \sum_\alpha \hat{H}^\alpha + \hat{V}^s + \hat{V}^t$

$$\Psi = \Psi^0 + \hat{G}^0(\hat{V}^s + \hat{V}^t)\Psi \tag{23.6}$$

where $\hat{H}^0\Psi^0 = E\Psi^0$, $\hat{H}^0 \equiv \sum_\alpha \hat{H}^\alpha$, $\hat{G}^0 = (E\hat{I} - \hat{H}^0)^{-1}$, $\hat{I}$ is the identity operator, and E is the electron energy.

The wavefunction $\Psi = \sum_\alpha \Psi^\alpha$ [with Ψ^α defined in Equation (23.3)] is supposed to describe a steady-state flux of electrons incident on, transmitted through, and reflected from the molecule connecting substrate and tip. For the incident electron flux propagating in the substrate, as shown in Figure 23.1, we can choose $\Psi^0 = \Psi^{s0} = \sum_{\mathbf{r}\in\{\mathbf{r}\}_s} \psi_\mathbf{r}^{s0}|\mathbf{r}\rangle$, $\hat{H}^s\Psi^{s0} = E\Psi^{s0}$. Then, the expansion coefficients, which determine Ψ^α, obey the following set of equations:

$$\psi_\mathbf{r}^s = \psi_\mathbf{r}^{s\,0} + \sum_{i_s=1}^{N_s} \sum_{\mathbf{r}'\in\{\mathbf{r}_{i_s}\}_s} G_{\mathbf{r},\mathbf{r}'}^s V_{i_s,\mathbf{r}'}^s \psi_{i_s}^m \tag{23.7}$$

$$\psi_\mathbf{r}^t = \sum_{i_t=1}^{N_t} \sum_{\mathbf{r}'\in\{\mathbf{r}_{i_t}\}_t} G_{\mathbf{r},\mathbf{r}'}^t V_{i_t,\mathbf{r}'}^t \psi_{i_t}^m \tag{23.8}$$

for $\alpha = s, t$, and

$$\psi_{i_s}^m = \sum_{i_s'=1}^{N_s} \sum_{\mathbf{r}\in\{\mathbf{r}_{i_s'}\}} G_{i_s,i_s'}^m V_{i_s',\mathbf{r}}^s \psi_\mathbf{r}^s + \sum_{i_t=1}^{N_t} \sum_{\mathbf{r}\in\{\mathbf{r}_{i_t}\}} G_{i_s,i_t}^m V_{i_t,\mathbf{r}}^t \psi_\mathbf{r}^t \tag{23.9}$$

$$\psi_{i_t}^m = \sum_{i_s=1}^{N_s} \sum_{\mathbf{r}\in\{\mathbf{r}_{i_s}\}} G_{i_t,i_s}^m V_{i_s,\mathbf{r}}^s \psi_\mathbf{r}^s + \sum_{i_t'=1}^{N_t} \sum_{\mathbf{r}\in\{\mathbf{r}_{i_t'}\}} G_{i_t,i_t'}^m V_{i_t',\mathbf{r}}^t \psi_\mathbf{r}^t \tag{23.10}$$

for $\alpha = m$.

Notice, $G_{\mathbf{r},\mathbf{r}'}^\alpha = \langle\mathbf{r}|\hat{G}^0|\mathbf{r}'\rangle$, $\mathbf{r}, \mathbf{r}' \in \{\mathbf{r}\}_\alpha$, and matrixes $G^s \equiv [G_{\mathbf{r},\mathbf{r}'}^s]$ and $G^t \equiv [G_{\mathbf{r},\mathbf{r}'}^t]$ should be understood as the retarded Green's functions of the substrate and tip, respectively. (The term Green's function will be used here as an abbreviation of "matrix of the Green's function operator".)

The solution of the scattering problem $\psi_\mathbf{r}^{s\,0}$ for the isolated semi-infinite substrate is supposed to be known. What happens with the incident wave, due to substrate coupling with the tip via molecule is fully determined by the wavefunction amplitudes $\psi_\mathbf{r}^\alpha$, $\alpha = s, t, m$.

23.1.3   Molecule–Lead Connection via a Single Atom

Let us consider the case where $N_s = N_t = 1$, $i_{s(t)} = 1_{s(t)}$, and $\{\mathbf{r}_{1_{s(t)}}\}_{s(t)} = \{\mathbf{r}\}_{s(t)}$ [27]. This is the case, for example, of thiol-terminated molecules connecting Au leads [28]. Then, for the waves transmitted to the tip, solution of Equations (23.7–23.10) gives:

$$\psi_{\mathbf{r}}^t = G_{1_s,1_t} \sum_{\mathbf{r}' \in \{\mathbf{r}\}_s} V_{1_s,\mathbf{r}'}^s \psi_{\mathbf{r}'}^{s0} \sum_{\mathbf{r}'' \in \{\mathbf{r}\}_t} G_{\mathbf{r},\mathbf{r}''}^t V_{\mathbf{r}'',1_t}^t \tag{23.11}$$

where

$$G_{1_s,1_t} = \frac{G_{1_s,1_t}^m}{\left(1 - A^s G_{1_s,1_s}^m\right)\left(1 - A^t G_{1_t,1_t}^m\right) - A^s A^t \left(G_{1_s,1_t}^m\right)^2} \tag{23.12}$$

has the meaning of the Green's function matrix element calculated for molecule m, where the site energies of binding atoms 1_s and 1_t, ε_{1_s} and ε_{1_t}, are changed by complex, E-dependent values A^s and A^t, respectively; $A^{s(t)} \equiv A^{s(t)\mathcal{R}} - iA^{s(t)\mathcal{I}}$:

$$A^{s(t)} = \sum_{\mathbf{r},\mathbf{r}' \in \{\mathbf{r}\}_{s(t)}} V_{1_{s(t)},\mathbf{r}}^{s(t)} G_{\mathbf{r},\mathbf{r}'}^{s(t)} V_{\mathbf{r}',1_{s(t)}}^{s(t)} = \sum_{\mathbf{r},\mathbf{r}' \in \{\mathbf{r}\}_{s(t)}} V_{1_{s(t)},\mathbf{r}}^{s(t)} \left(\mathrm{Re}G_{\mathbf{r},\mathbf{r}'}^{s(t)} + i\mathrm{Im}G_{\mathbf{r},\mathbf{r}'}^{s(t)}\right) V_{\mathbf{r}',1_{s(t)}}^{s(t)} \tag{23.13}$$

In other words, the denominator in the right hand side of Equation (23.12) has exactly the same form as determinant $\det\left(E\hat{I} - \hat{H}^m\right)$ with perturbed site energies $\varepsilon_{1_s} + A^s$ and $\varepsilon_{1_t} + A^t$ [29]. Thus, quantities A^s and A^t determine the shift and broadening of molecular levels due to the metal–molecule interaction.

In the Green's function formulation of the present problem, $A^s + A^t$ has been conditionally termed as the system "self energy" [15]. It is worthwhile noting that a similar conception appeared in the Newns chemisorption theory [30,31] and in other physical contexts, where terms "chemisorption function" and "spectral density" were in use. Without any pretension, we shall refer $A^{s(t)}$ as the coupling function.

To this point, the surface and crystalline structure of the substrate and tip have not been specified. To obtain an explicit form of the solution, we shall make further use of the model assumptions.

Let the contact surfaces be $\mathcal{N} \times \mathcal{N}$ square lattices coinciding with (001) planes of s and t cubic lattices. The atom coordinates can now be represented as $\mathbf{r} = \{r_\perp, r_z\}$, $r_\perp = an_\perp$, $r_z = an_z$, where a is the lattice constant, and $n_\perp \equiv \{n_x, n_y\}$. The leads are of a finite crossection, $n_x, n_y = 1, 2, \ldots, \mathcal{N}$, but they are infinite in length, $n_z = 1, 2, \ldots \infty$. Suppose also that for $\hat{V}^s = 0$, the electron flux in the substrate is described by plane waves with the wave vector $k'_{j'_\perp}/a$,[2] $\exp(-ik'_{j'_\perp} n_z)$ and $\exp(ik'_{j'_\perp} n_z)$, propagating in, respectively, $-z$- and z direction, (see Figure 23.1). In the transverse x and y directions, the electron state is represented by two standing waves. So,

$$\psi_{r_\perp,r_z}^{s0} = -2i\sin(k'_{j'_\perp} n_z)\chi_{j'_\perp}(n_\perp),$$

$$\chi_{j'_\perp}(n_\perp) \equiv \frac{2}{\mathcal{N}+1}\sin\left(\frac{\pi j'_1 n_x}{\mathcal{N}+1}\right)\sin\left(\frac{\pi j'_2 n_y}{\mathcal{N}+1}\right) \tag{23.14}$$

where index $j'_\perp \equiv \{j'_1, j'_2\}$ numbers modes of the transverse electron motion.

For the given energy of incident electrons E, j'_1 and j'_2 can take any integer value from 1 to $\mathcal{N}$, satisfying the energy relation:

$$E_{k'_{j'_\perp}, j'_\perp}^s + eU_s = E, \quad E_{k'_{j'_\perp}, j'_\perp}^s = \varepsilon_s - 2L_s\left(\cos k'_{j'_\perp} + \cos\frac{\pi j'_1}{\mathcal{N}+1} + \cos\frac{\pi j'_2}{\mathcal{N}+1}\right) \tag{23.15}$$

---

[2]In what follows, we use dimensionless wave vectors in units a^{-1}.

The wave vector $k_{j_\perp}^s$ ($k_{j_\perp}^t$) and mode numbers of any reflected (transmitted) wave must obey similar equations:

$$E_{k_{j_\perp, j_\perp}^s}^s + eU_s = E_{k_{j_\perp, j_\perp}^t}^t + eU_t = E \qquad (23.16)$$

where $j_\perp \equiv \{j_1, j_2\}$, $j_1, j_2 = 1, 2, \ldots, \mathcal{N}$, and

$$E_{k_{j_\perp, j_\perp}^\alpha}^\alpha = \varepsilon_a - 2L_\alpha \left(\cos k_{j_\perp}^\alpha + \cos \frac{\pi j_1}{\mathcal{N}+1} + \cos \frac{\pi j_2}{\mathcal{N}+1} \right), \quad \alpha = s, t \qquad (23.17)$$

Equation (23.16) can be satisfied by real and imaginary values of $k_{j_\perp}^s$ and $k_{j_\perp}^t$. Therefore, these wave vectors may acquire real values within the interval $0 \le k_{j_\perp}^\alpha \le \pi$, as well as imaginary values $k_{j_\perp}^\alpha = iq_{j_\perp}^\alpha$ or $k_{j_\perp}^\alpha = iq_{j_\perp}^\alpha + \pi$, $q_{j_\perp}^\alpha > 0$. This means that both kinds of waves, the propagating and evanescent, are generated in the scattering region. The net flux of transmitted and reflected electrons is formed by propagating waves.

To operate with coupling functions, one needs matrixes G^s and G^t. Solutions to equation:

$$\left(E\delta_{\mathbf{r},\mathbf{r}''} - \langle \mathbf{r} | \hat{H}^\alpha | \mathbf{r}'' \rangle \right) G_{\mathbf{r}'',\mathbf{r}'}^\alpha = \delta_{\mathbf{r},\mathbf{r}'}, \quad \alpha = s, t \qquad (23.18)$$

which describe outgoing waves from the molecule-lead interfaces, take the form [27]:

$$G_{r_\perp, r_z; r_\perp', r_z'}^\alpha = -\frac{1}{L_\alpha} \sum_{j_\perp} e^{ik_{j_\perp}^\alpha n_z} \frac{\sin \left(k_{j_\perp}^\alpha n_z' \right)}{\sin k_{j_\perp}^\alpha} \chi_{j_\perp}(n_\perp) \chi_{j_\perp}(n_\perp') \qquad (23.19)$$

It is written for $n_z \ge n_z'$. If $n_z < n_z'$, the replacement $n_z \leftrightarrow n_z'$ has to be made. The summation is carried out over all modes of the transverse electron motion.

These are propagating modes that contribute to the imaginary part of the Green's function $G^{s(t)}$. Hence, according to Equation (23.13), $A^{\alpha\mathcal{I}}$ equals:

$$A^{\alpha\mathcal{I}} = \frac{1}{L_\alpha} {\sum_{j_\perp}}' \sin k_{j_\perp}^\alpha \tilde{\chi}_{j_\perp}^{\alpha\, 2} \qquad (23.20)$$

where the prime indicates that the summation should be performed over the propagating modes. A new notation, introduced in Equation (23.20), stands for:

$$\tilde{\chi}_{j_\perp}^\alpha = \sum_{\mathbf{r} \in \{\mathbf{r}\}_\alpha} \chi_{j_\perp}(n_\perp) \frac{\sin \left(k_{j_\perp}^\alpha n_z \right)}{\sin k_{j_\perp}^\alpha} V_{1_a,\mathbf{r}}^\alpha \qquad (23.21)$$

An explicit expression of $A^{a\mathcal{R}}$ can be obtained from Equations (23.13) and (23.19) in a similar way. However, it is more convenient to use the Kramers–Kronig dispersion relation [32,33]:

$$A^{\alpha\mathcal{R}}(E) = \frac{1}{\pi} P \int_{-\infty}^{+\infty} \frac{A^{\alpha\mathcal{I}}(E')}{E - E'} dE' \qquad (23.22)$$

where P denotes the Cauchy principal value. Some particular expressions of $A^{\alpha\mathcal{I}}$ for on-top and on-hollow positioning of molecule binding atoms with respect to the (001) substrate (tip) surface can be found in [27].

Now, having all the quantities in Equation (23.11) defined, we can complete our calculation of the transmission coefficient. Making use of Equations (23.14) and (23.19) in Equation (23.11), one can obtain the wavefunction of transmitted electrons in the form of a superposition of propagating and evanescent waves:

$$\psi_{\mathbf{r}}^t = \sum_{j_\perp} t_{k_{j_\perp}^t, j_\perp; k_{j_\perp'}', j_\perp'}^t (n_\perp) \exp(ik_{j_\perp}^t n_z) \qquad (23.23)$$

where

$$t_{k^t_{j_\perp},j_\perp;k'_{j'_\perp},j'_\perp}(n_\perp) = \frac{2i \sin k'_{j'_\perp}}{L_t} \chi_{j_\perp}(n_\perp) \tilde{\chi}^s_{j'_\perp} \tilde{\chi}^t_{j_\perp} G_{1_s,1_t} \tag{23.24}$$

According to Equation (23.2), where $v^t_{k_\nu}/v^s_{k_\mu} = v^t_{k'_{j'_\perp}}/v^s_{k'_{j'_\perp}} = L_t \sin k^t_{j_\perp}/(L_s \sin k'_{j'_\perp})$, $|\psi^s_{\mu,k_\mu}(\mathbf{r})|^2 = \chi^2_{j'_\perp}(n_\perp)$, and $|\psi^t_{\nu,k_\nu;\mu,k_\mu}(\mathbf{r})|^2 = |t_{k^t_{j_\perp},j_\perp;k'_{j'_\perp},j'_\perp}(n_\perp)|^2$, the transmission coefficient from channel $j'_\perp$ to channel $j_\perp$ equals

$$T_{k^t_{j_\perp},j_\perp;k'_{j'_\perp},j'_\perp} = \sum_{n_\perp} \frac{v^t_{k_{j_\perp}}}{v^s_{k'_{j'_\perp}}} \left| t_{k^t_{j_\perp},j_\perp;k'_{j'_\perp},j'_\perp}(n_\perp) \right|^2 \tag{23.25}$$

By summing both sides of this equation over all propagating modes (specified by Equations [23.15] and [23.16]) and utilizing relations (23.20) and (23.21), we obtain the total probability of electron transmission at the given energy E. It reads

$$T = \sum_{j'_\perp}' \sum_{j_\perp}' T_{k^t_{j_\perp},j_\perp;k'_{j'_\perp},j'_\perp} = 4A^{s\mathcal{I}} A^{t\mathcal{I}} |G_{1_s,1_t}|^2 \tag{23.26}$$

In this final form, the transmission coefficient was probably obtained for the first time by Mujico *et al.* [34]. However, that model was one-dimensional. An expression of T, not restricted by the system dimensionality and/or the number of molecule binding atoms, was derived by Datta [15] on the basis of the Green's function method and the Lee and Fisher theorem [35]. In Section 23.1.5, we show how Equation (23.26) can be generalized to cover contact structures of a general type (see illustration in Figure 23.1). But first, we proceed with an analysis that brings to light some general regularities of coherent electron transport in molecular contacts.

23.1.3.1 Resonant Transmission

A useful equivalent form of Equation (23.26) can be given by a rearranging denominator $\mathcal{D}_T$ in the expression of $|G_{1_s,1_t}|^2 \equiv G^{m\,2}_{1_s,1_t}/\mathcal{D}_T$ [36]:

$$\mathcal{D}_T(E) = \left[1 - A^{s\mathcal{R}} G^m_{1_s,1_s} - A^{t\mathcal{R}} G^m_{1_t,1_t} + (A^{s\mathcal{R}} A^{t\mathcal{R}} + A^{s\mathcal{I}} A^{t\mathcal{I}}) G^m_\Delta\right]^2$$
$$+ \left[A^{s\mathcal{I}} G^m_{1_s,1_s} - A^{t\mathcal{I}} G^m_{1_t,1_t} - (A^{s\mathcal{I}} A^{t\mathcal{R}} - A^{s\mathcal{R}} A^{t\mathcal{I}}) G^m_\Delta\right]^2 + 4A^{s\mathcal{I}} A^{t\mathcal{I}} G^{m\,2}_{1_s,1_t} \tag{23.27}$$

where

$$G^m_\Delta = G^m_{1_s,1_s} G^m_{1_t,1_t} - G^{m\,2}_{1_s,1_t} \tag{23.28}$$

Such a representation:

$$T = 4A^{s\mathcal{I}} A^{t\mathcal{I}} \frac{G^{m\,2}_{1_s,1_t}}{\mathcal{D}_T} \tag{23.29}$$

exposes the resonance structure of the transmission spectrum. It shows, in particular, that the through molecule transmission without backscattering can be achieved in symmetric contacts, where $A^s = A^t$ and $G^m_{1_s,1_s} = G^m_{1_t,1_t}$. The use of these equalities in Equations (23.27) and (23.29) leads us to the condition of resonant transmission (or resonant tunneling):

$$G^{m\,2}_{1_s,1_s} - G^{m\,2}_{1_s,1_t} = -\frac{1}{|A^s|^2}\left(1 - 2A^{s\mathcal{R}} G^m_{1_s,1_s}\right) \tag{23.30}$$

For electron energies satisfying the above equation, $T = 1$, which means that the per channel conductance reaches its maximal value $g = 2e^2/h$ [19,20]. On the other hand, analysis of Equations (23.27) and (23.29) makes it clear that in asymmetric systems (different leads or/and coupling constants, or/and not central symmetric molecules, presence of external voltage, etc.), full transparency of the molecular contact for electron transmission is hardly possible.[3] Earlier, similar effect of suppression of the resonant tunneling by the system asymmetry was predicted by Ricco and Azbel [37] for conventional double barrier structures.

As an illustration, we take for the role of G^m in Equation (23.30), the Green's function matrix $G^{(C)_N}$ from Equation (23.58), describing a chain of N carbon atoms. By setting $G^m_{1_s,1_s} = G^m_{1_t,1_t} = G^{(C)_N}_{1,1}$, $G^m_{1_s,1_t} = G^{(C)_N}_{1,N}$, Equation (23.30) transforms into

$$\sin[(N+1)\xi] + 2\frac{A^s\mathcal{R}}{\beta}\sin(N\xi) + \frac{|A^s|^2}{\beta^2}\sin[(N-1)\xi] = 0 \qquad (23.31)$$

where $\xi \in \overline{0,\pi}$, ξ is related to electron energy E as $2\beta\cos\xi = \varepsilon - E$, and site energy ε and hopping integral β are introduced in Section 23.2.1. Equation (23.31) exemplifies a general property of molecular contacts: Energies of the resonant transmission at which $T = 1$, can only be found within the bands of molecular levels. (In the case of $(C)_N$ chain, it is only one band, $E \in \varepsilon - 2\beta, \varepsilon + 2\beta$.) For other energies, the transmission probability exhibits a strong dependence on contact quality and the wire length. This dependence is briefly discussed next.

23.1.3.2 Tunneling Mechanism of Transmission

Estimating T for energies which are far from resonances, the term $A^s A^t G^{m\,2}_{1_s,1_t}$ in Equation (23.12) can be neglected. Further assumptions are not principal for final conclusions but greatly facilitate intermediate algebra. Suppose that there exists a dominating molecule–lead interaction, $V^\alpha_{1_\alpha,\mathbf{r}} = V^\alpha\delta_{\mathbf{r},\mathbf{r}_0}$, and let $A^{\alpha\mathcal{R}} \ll A^{\alpha\mathcal{I}} = V^{\alpha\,2}/L_\alpha$. If we choose $(C)_N$ chain as a model of the contacting molecule, $G^m \rightarrow G^{(C)_N} \sim 1/\beta$, the only coupling parameters in Equation (23.26) are $\tilde{V}^s$ and $\tilde{V}^t$, $\tilde{V}^\alpha = \dfrac{V^{\alpha\,2}}{\beta L_\alpha}$. Then, with the help of Equation (23.58), it is straightforward to show that:

$$T = 4\exp(-2N\delta)\frac{\tilde{V}^s\tilde{V}^t[1 - \exp(-2\delta)]^2}{[1 + \tilde{V}^{s\,2}\exp(-2\delta)][1 + \tilde{V}^{t\,2}\exp(-2\delta)]} \qquad (23.32)$$

where $|E - \varepsilon| > 2\beta$, $2\beta\cosh\delta = |E - \varepsilon|$, and $N\delta \gg 1$.

The exponential-like dependence of T on the molecule length $\approx aN$ is a well-known trademark of tunneling mechanism of electron [38] (and excitation energy) through-space transfer. The obtained non-linear dependence of T on coupling parameters, $\tilde{V}^s$ and $\tilde{V}^t$ is probably less familiar. At least, it never appears in perturbation treatments of tunneling, such as Bardeen's tunneling theory [39,40], see also [5].

As it follows from the exact expression of transmission probability, a strong molecule-lead interaction, i.e., large values of either $\tilde{V}^s$ or $\tilde{V}^t$, does not favor the contact transmission, because if $\tilde{V}^\alpha \gg 1$:

$$T \sim \frac{1}{\tilde{V}^\alpha} \ll 1 \qquad (23.33)$$

From this point of view, as molecular contacts of "good quality" should be regarded those, where the value of $\tilde{V}^\alpha$ is close to unity for the both leads $\alpha = s, t$.

For Equation (23.32) to be valid, inequality $N\delta \gg 1$ must hold.[4] This condition cannot be fulfilled at or near the molecular band edges. In particular, for $(C)_N$ chain, we obtain at $E = \varepsilon \pm 2\beta$

$$T = \frac{4}{N^2}\frac{\tilde{V}^s\tilde{V}^t}{(1 + \tilde{V}^{s\,2})(1 + \tilde{V}^{t\,2})} \qquad (23.34)$$

---

[3]It is difficult to prove rigorously the impossibility of unit transmission in asymmetric contacts. However, it is very unlikely that different expressions, which appear in two square brackets of the right-hand side of Equation (23.27), turn to zero at one and the same energy.

[4]Energy dependence of the tunneling exponent is examined in detail in Section 23.2, where, among other results, an explicit expression of δ for $(C)_N$ chain is presented.

showing that at the band edges, the transmission probability has a power-like dependence on the molecule length.

23.1.4 Transmission in STM-Type Contacts

So far, our consideration has been free of any restrictions with regard to the magnitude of coupling constants. Now we assume that the molecule-tip coupling function A^t is small and can be treated as a perturbation. If, in addition, $V^t_{1_t,\mathbf{r}} = \delta_{\mathbf{r},\mathbf{r}_0} V^t$, then, $A^{t\mathcal{I}}$ is proportional to the local density of states (LDOS) $\rho^t_{\mathbf{r}_0}$ at the tip atom $\mathbf{r}_0$: $A^{t\mathcal{I}} = -V^{t\,2}\mathrm{Im}G^{\alpha}_{\mathbf{r}_0,\mathbf{r}_0} = \pi V^{t\,2}\rho^t_{\mathbf{r}_0}$. It is easy to see that, up to terms $\sim V^{t\,2}$, Equation (23.26) reads:

$$T = 4\pi V^{t\,2}\rho^t_{\mathbf{r}_0}\frac{A^{s\mathcal{I}}G^{m\,2}_{1_s,1_t}}{\left|1 - A^s G^m_{1_s,1_s}\right|^2} \tag{23.35}$$

To elucidate the meaning of the fractional cofactor in Equation (23.35), let us consider the system composed of a substrate with molecule m chemisorbed on it. The system is described by Hamiltonian operator $\hat{H} = \hat{H}^0 + \hat{V}^s$ with $\hat{H}^0 = \hat{H}^s + \hat{H}^m$. Its Green's function operator $\hat{G}^{s-m}$ obeys the following equation:

$$\hat{G}^{s-m} = \hat{G}^0 + \hat{G}^0\hat{V}^s\hat{G}^{s-m} \tag{23.36}$$

where matrix elements of $\hat{G}^0$ are $G^0_{\mathbf{r},\mathbf{r}'} = G^{s(m)}_{\mathbf{r},\mathbf{r}'}$, if $\mathbf{r},\mathbf{r}' \in \{\mathbf{r}\}_{s(m)}$; otherwise, $G^0_{\mathbf{r},\mathbf{r}'} = 0$.

For the diagonal matrix element $G^{s-m}_{1_t,1_t}$ that refers to the molecule atom 1_t, Equation (23.36) gives:

$$G^{s-m}_{1_t,1_t} = G^m_{1_t,1_t} + \frac{A^s G^{m\,2}_{1_s,1_t}}{1 - A^s G^m_{1_s,1_s}} \tag{23.37}$$

By definition, $\rho^{s-m}_{1_t} = -\frac{1}{\pi}\mathrm{Im}G^{s-m}_{1_t,1_t}$. Thus, Equation (23.35) takes the form

$$T = 4\pi^2 V^{t\,2}\rho^{s-m}_{1_t}\rho^t_{\mathbf{r}_0} \tag{23.38}$$

where $\rho^t_{\mathbf{r}_0} = -\frac{1}{\pi}\mathrm{Im}G^t_{\mathbf{r}_0,\mathbf{r}_0}$, and

$$\rho^{s-m}_{1_t} = \frac{1}{\pi}\frac{A^{s\mathcal{I}}G^{m\,2}_{1_s,1_t}}{\left(1 - A^{s\mathcal{R}}\right)^2 + \left(A^{s\mathcal{I}}G^m_{1_s,1_s}\right)^2} \tag{23.39}$$

has the meaning of LDOS at the atom of chemisorbed molecule that is probed by STM tip.

Equation (23.38) can be modified to make it applicable to those (experimentally feasible) situations, where the substrate and tip are facing each other by molecules chemisorbed on their surfaces. Let molecule m be composed of two arbitrary parts m_s and m_t, and let them be coupled via atoms $\mathcal{N}_s$ and $\mathcal{N}_t$ (see upper contact in Figure 23.1) by a weak interaction V^{s-t}. In the linear approximation regarding the latter parameter, we have $G^m_{1_s,1_s} = G^{m_s}_{1_s,1_s}$, $G^m_{1_t,1_t} = G^{m_t}_{1_t,1_t}$, and $G^m_{1_s,1_t} = V^{s-t}G^{m_s}_{1_s,\mathcal{N}_s}G^{m_t}_{1_t,\mathcal{N}_t}$. Using these approximate expressions in Equation (23.26), we obtain (up to the lowest order of perturbation V^{s-t}):

$$T = 4\pi^2 (V^{s-t})^2\rho^{s-m_s}_{\mathcal{N}_s}\rho^{t-m_t}_{\mathcal{N}_t} \tag{23.40}$$

where

$$\rho^{\alpha-m_\alpha}_{\mathcal{N}_\alpha} = \frac{1}{\pi}\frac{A^{\alpha\mathcal{I}}G^{m_\alpha\,2}_{1_\alpha,\mathcal{N}_\alpha}}{(1 - A^{\alpha\mathcal{R}})^2 + (A^{\alpha\mathcal{I}}G^{m_\alpha}_{1_\alpha,1_\alpha})^2}, \quad \alpha = s, t \tag{23.41}$$

has the same meaning as $\rho^{s-m}_{1_t}$ in Equation (23.38).

It should be stressed that the appearance of LDOS in the transmission coefficient[5] is bounded to the case, when the tip-sample interaction involves a single atom from either side of the STM contact. Participation of more atoms in the tip-sample interaction should be described by an appropriate coupling function.

Formula (23.40) has been obtained in [27] and utilized in [42] to calculate STM current–voltage characteristics measured for self-assemblies of one- and three phenyl ring oligomers with amine terminus. It was one of the few attempts of a quantitative description of observed peculiarities of electric behavior of molecular contacts.

23.1.5   Contacts with Arbitrary Number of Binding Atoms

The purpose of this section is to obtain a closed form of the transmission probability (an analogue of Equation [23.26]) that makes it possible to address a much broader range of contact structures. Molecules connected via several atoms to metal leads is one example. Molecular self-assemblies imbedded in some way between metal electrodes, e.g., in nanopores [28,43], can also be covered by the model considered next.

Suppose that molecule interacts with substrate and tip via, respectively, N_s and N_t atoms (see lower contact in Figure 23.1). In all other respects, the molecular contact model is the same as described in Section 23.1.1. The basic equations we have to solve now, i.e., Equations (23.7–23.10), contain only those matrix elements of the molecule Green's function that refer to the binding atoms. Therefore, it is convenient to use this (reduced) matrix in the form of four blocks:

$$G^m = \begin{pmatrix} [G^m_{i_s,i'_s}] \ [G^m_{i_s,i_t}] \\ [G^m_{i_t,i_s}] \ [G^m_{i_t,i'_t}] \end{pmatrix} \equiv \begin{pmatrix} G^m_{ss} \ G^m_{st} \\ G^m_{ts} \ G^m_{tt} \end{pmatrix} \tag{23.42}$$

The solution to the set of Equations (23.7–23.10) can be expressed in terms of the Green's function matrix G that satisfies the Dyson equation

$$G = G^m + G^m A G \tag{23.43}$$

with the coupling function matrix A defined as

$$A = \begin{pmatrix} [A^s_{i_s,i'_s}] \ \ \ 0 \\ 0 \ \ \ [A^t_{i_t,i'_t}] \end{pmatrix} \equiv \begin{pmatrix} A^s_{ss} \ \ 0 \\ 0 \ \ A^t_{tt} \end{pmatrix} \tag{23.44}$$

where

$$A^\alpha_{i_\alpha,i'_\alpha} = \sum_{r \in \{r_{i_\alpha}\}_\alpha} \sum_{r' \in \{r_{i'_\alpha}\}_\alpha} V^\alpha_{i_\alpha,r} G^\alpha_{r,r'} V^\alpha_{r',i'_\alpha} \equiv A^{\alpha R}_{i_\alpha,i'_\alpha} - i A^{\alpha I}_{i_\alpha,i'_\alpha} \tag{23.45}$$

Matrix G has the same structure as matrix G^m, i.e.,

$$G = \begin{pmatrix} G_{ss} \ G_{st} \\ G_{ts} \ G_{tt} \end{pmatrix} \tag{23.46}$$

Note also that because matrix elements of A are determined by matrix elements of retarded Green's functions G^s and G^t, matrix G also has the meaning of retarded Green's function.

---

[5]Proportionality of ohmic conductance to the LDOS is the central concept of the Tersoff and Hamman STM theory [40,41].

Solving Equations (23.7–23.10) with the use of Equations (23.42), (23.43), and (23.44) yields

$$\psi_{\mathbf{r}}^t = \sum_{i_s=1}^{N_s} \sum_{i_t=1}^{N_t} G_{i_s,i_t} \sum_{\mathbf{r}' \in \{\mathbf{r}_{i_s}\}_s} V_{i_s,\mathbf{r}'} \psi_{\mathbf{r}'}^{s\,0} \sum_{\mathbf{r}'' \in \{\mathbf{r}_{i_t}\}_t} G_{\mathbf{r},\mathbf{r}''}^t V_{\mathbf{r}'',i_t} \tag{23.47}$$

It is readily verified that for $V_{i_\alpha,\mathbf{r}}^\alpha = \delta_{i_\alpha,1_\alpha} V_{1_\alpha,\mathbf{r}}^\alpha$, the solution of Equation (23.43) for $G_{st} = G_{ts}^* = G_{1_s,1_t}$, has the form of Equation (23.12). Under the same assumption regarding the coupling constants, Equation (23.45) coincides with (23.13). So, as it should be, Equation (23.47) repeats (23.11) in the case of only one atom creating contact with the lead on either molecule terminus.

Algebra similar to what led us to Equation (23.26), gives:

$$T = 4\mathrm{Tr}\left(A_{ss}^{s\mathcal{I}} G_{st} A_{tt}^{t\mathcal{I}} G_{ts}^*\right) = 4\mathrm{Tr}\left(G_{ts}^* A_{ss}^{s\mathcal{I}} G_{st} A_{tt}^{t\mathcal{I}}\right) \tag{23.48}$$

The analogous Datta's formula (3.5.20) from [15], reads:

$$T = \mathrm{Tr}\left(\Gamma_p G^R \Gamma_q G^{R*}\right) \tag{23.49}$$

where

$$G^R = [EI - H_C - \Sigma_p^R - \Sigma_q^R]^{-1} \tag{23.50}$$

$\Gamma_{p(q)} = -2\mathrm{Im}\Sigma_{p(q)}$, H_C (our H^m) represents the scattering region, and indexes p, q correspond to s, t. One can immediately see equivalence of these equations to ours, if matrixes of the self energy $\Sigma_{p(q)}$ and Green's function G^R have the form

$$\Sigma_p^R = \begin{pmatrix} \Sigma_p^R & 0 \\ 0 & 0 \end{pmatrix} \equiv \begin{pmatrix} A_{ss}^s & 0 \\ 0 & 0 \end{pmatrix}, \qquad \Sigma_q^R = \begin{pmatrix} 0 & 0 \\ 0 & \Sigma_q^R \end{pmatrix} \equiv \begin{pmatrix} 0 & 0 \\ 0 & A_{tt}^t \end{pmatrix} \tag{23.51}$$

and

$$G^R = \begin{pmatrix} G_{pp}^R & G_{pq}^R \\ G_{qp}^R & G_{qq}^R \end{pmatrix} \tag{23.52}$$

respectively.

In conclusion, provided the molecule Green's function is known, formulas represented throughout the above discussion allow one to examine independently properties of molecular contacts by using appropriate values for parameters of molecule–lead interaction. To retain the exactness of the results and the possibility to express them analytically (which is the aim of this chapter) molecular Hamiltonians must be sufficiently simple. Fortunately, the π and σ electron subsystems of organic molecules can be reasonably described by one or another modification of the Hückel Hamiltonian [44] that admits analytical treatment. For molecules consisting of few atoms such as small heterocycles and aromatic rings, finding G^m is a simple exercise. Many examples of concrete Green's functions with indication of suitable parameterization can be found in refs. [45,46]. For long molecules built of repeating atomic blocks, a closed form of G^m is also accessible. A concrete explanation of the form will be given in the next section.

23.2 Models of Molecular Wires

This section focuses on the molecule Green's function. Its explicit dependence on energy will be presented in an analytical form, covering a wide class of organic molecules. In our experience, this is useful for developing an understanding of molecular contacts, donor/bridge/acceptor (DBA) systems, and how these can be modified in a desirable way.

In the context of molecular wire contacts, we are interested first of all in organic oligomers $(M)_N$, a linear sequence of N identical chemical entities M connected to its nearest neighbors by a single bond,

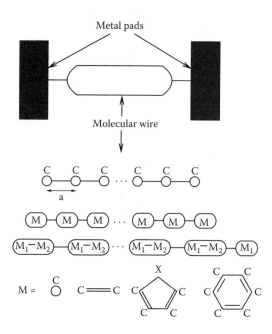

FIGURE 23.2   Molecules which can be used in metal–molecule–metal contacts: hydrocarbon chain (hydrogens are not shown), M– and M_1–M_2-oligomers with examples of monomers; a is the lattice constant.

(see Figure 23.2). Also, we will give a short description of M_1–M_2-oligomers (see lower example in Figure 23.2), which in their formal description are similar to M-oligomers. We shall show that the oligomer Green's function $G^{(M)_N}$ which corresponds to the tight-binding Hamiltonian $\hat{H}^{(M)_N}$ can be explicitly expressed in terms of monomer Green's function G^M which corresponds to the monomer Hamiltonian $\hat{H}^M$. On the one hand, this completes the scheme of analytical modeling described in preceding section. On the other, this gives us an analytically expressed relationship between the molecular electronic structure and the exponential factor which governs ohmic tunnel current in molecular contacts and long-range electron transfer in DBA systems incorporating M- and M_1–M_2-oligomers.

First we discuss briefly the Green's functions for carbon chains with nonalternating, M = C, and alternating, M = C=C, bonds between carbon atoms. Also, eigen energies E_μ and eigen states ψ_n^μ that appear in the standard bilinear expansion of the Green's function

$$G_{n,n_0}^{(M)_N}(E) = \sum_\mu \frac{\psi_n^\mu \psi_{n_0}^{\mu *}}{E - E_\mu} \tag{23.53}$$

M = C, C=C, are given for completeness and as frequently used reference. We continue with the general case of M-oligomers, where using the Dyson equation technique allows one to avoid a lot of tiring calculations in finding a closed form for oligomer Green's function matrix. On these grounds, exact explicit expressions of the dependence of the exponential factor on energy of tunneling electron are obtained for oligomers $(C)_N, (C=C)_N, (M)_N,$ and $(M_1 - M_2)_N$. Basic physics of this dependence is thoroughly examined and methodology is illustrated by several examples.

Model Hamiltonians that form the basis of the forthcoming discussion have a long list of referenced applications. They approach the real electronic structure of organic oligomers in terms of sound parameters that can be evaluated from more advanced semi-empirical schemes, ab initio calculations, or deduced from experiments. In what follows, we summarize our findings on instructive description of electron tunneling through molecular wires, showing how the tunneling rate at the given energy is related to the molecular electronic structure parameters which can be found and checked independently.

23.2.1 Canonical Tight-Binding Model

To begin, we introduce the tight-binding Hamiltonian of monatomic chain $(C)_N$ shown in Figure 23.2. Originally suggested by Hückel to describe aromatic hydrocarbons [44], it can be written in the form:

$$\hat{H}^{(C)_N} = \sum_{n=1}^{N} [\varepsilon |n\rangle\langle n| - \beta(1 - \delta_{n,N})|n\rangle\langle n+1| - \beta(1 - \delta_{n,1})|n\rangle\langle n-1|] \tag{23.54}$$

where n indicates the atom position in the chain. We use the same formal language as in Section 23.1: $|n\rangle \equiv c_n^+|0\rangle$ ($|0\rangle$ is the vacuum wavefunction), c_n^+ and c_n are the Fermi creation and annihilation operators, ε is the electron energy (Coulomb integral) of an isolated carbon atom, modeled by a single level, and $-\beta$ ($\beta > 0$) stands for electron hopping integral between the nearest-neighbor C atoms.

Normalized eigenstates of Hamiltonian (23.54) $\Psi_j^{(C)_N} = \sum_{n=1}^{N} \psi_{n,j}|n\rangle$

$$\psi_{n,j} = \sqrt{\frac{2}{N+1}} \sin\left(\frac{\pi j n}{N+1}\right) \tag{23.55}$$

labeled by index $j = 1, 2, \ldots, N$, correspond to eigenenergies

$$E_j = \varepsilon - 2\beta \cos\left(\frac{\pi j}{N+1}\right) \tag{23.56}$$

so that Green's function in Equation (23.53) takes the form [47]:

$$G_{n,n_0}^{(C)_N}(E) = \langle n|\hat{G}^{(C)_N}|n_0\rangle = \frac{2}{N+1} \sum_{j=1}^{N} \frac{\sin\frac{\pi j n}{N+1} \sin\frac{\pi j n_0}{N+1}}{E - \varepsilon + 2\beta \cos\frac{\pi j}{N+1}} \tag{23.57}$$

An enjoyable feature of this model is that the summation in Equation (23.57) can be performed explicitly [48]

$$G_{n,n_0}^{(C)_N}(E) = -\frac{1}{\beta} \begin{cases} \dfrac{\sin(n_0\xi) \sin[(N - n + 1)\xi]}{\sin(\xi) \sin[(N + 1)\xi]}, & n \geq n_0, \\[2ex] n \leftrightarrow n_0, & n \leq n_0 \end{cases} \tag{23.58}$$

where real values of the wave vector k ($ka \equiv \xi$, $0 \leq \xi \leq \pi$) are related to the electron energy by:

$$2\beta \cos\xi = \varepsilon - E \tag{23.59}$$

within an energy band of the width $E_{bw}^0 = 4\beta$.

For energies outside the band spectrum (23.56), dispersion relation (23.59) is satisfied by complex values of ξ: $\xi = i\delta$, $\delta > 0$, for $E < \varepsilon - 2\beta \equiv E_b^0$ and $\xi = \pi + i\delta$ for $E > \varepsilon + 2\beta \equiv E_t^0$, so that:

$$\pm 2\beta \cosh\delta = \varepsilon - E \tag{23.60}$$

Signs "+" and "−" refer to energies below and above the band, respectively. Expressions of the Green's function matrix elements (23.58) change correspondingly. For example:

$$G_{1,1}^{(C)_N} = G_{N,N}^{(C)_N} = \frac{\text{sign}(E - \varepsilon) \sinh(N\delta)}{\beta \sinh[(N + 1)\delta]} \xrightarrow[{[N\delta \gg 1]}]{} \frac{\text{sign}(E - \varepsilon)}{\beta} e^{-\delta}$$

$$G_{1,N}^{(C)_N} = -\frac{[\text{sign}(\varepsilon - E)]^N \sinh\delta}{\beta \sinh[(N + 1)\delta]} \xrightarrow[{[N\delta \gg 1]}]{} -\frac{[\text{sign}(\varepsilon - E)]^N}{\beta} e^{-N\delta}(1 - e^{-2\delta}) \tag{23.61}$$

The above equations were used in Section 23.1.3 to estimate the ohmic conductance of molecular contacts. Matrix element $G_{1,N}^{(C)_N}$, which refers to the end atoms of a bridging molecule, also determines the electronic factor in the non-adiabatic electron transfer rate k_{et} [49–52]. Both quantities T and k_{et} are proportional to $(G_{1,N}^{(C)_N})^2 \sim \exp(-2N\delta)$. Combining this with Equation (23.60), we get:

$$T \sim k_{et} \sim \exp \left[-2 \ln \left(\frac{|E - \varepsilon|}{2\beta} + \sqrt{\left(\frac{E - \varepsilon}{2\beta} \right)^2 - 1} \right) N \right] \qquad (23.62)$$

for the case where $(C)_N$ chain is acting in the role of contacting (bridging) molecule.

Equation (23.62) states that ohmic tunnel current in molecular contacts and electron transfer rate in DBA systems possess the same dependence on the molecule length as it has the probability of tunneling of a free electron through a rectangular potential barrier on the barrier thickness. However, the energy dependence of tunneling exponent is different. To expose similarities and distinctions between the two models clearly, let us take a closer look at how δ depends on E.

Below the molecular band spectrum, $E < E_b^0$, and close to the band bottom, where $\delta << 1$, we have from Equation (23.60):

$$\delta \approx \sqrt{\left(E_b^0 - E \right)/\beta} \qquad (23.63)$$

Above (and close to) the top of the band, $\delta \approx \sqrt{(E - E_t^0)/\beta}$. Expressing hopping integral β in terms of electron effective mass m^*, $\beta = \hbar^2/(2m^*a^2)$, one can easily recognize in Equation (23.63) the handbook tunneling formula. For convenience, we rewrite it in more common notations: $k^{im} = (\sqrt{2m^*}/\hbar)\sqrt{\Phi - E}$, where $\Phi = E_b^0$ is the height of potential barrier, and k^{im} is modulus of imaginary wave vector of the electron having energy E ($< \Phi$) in the barrier region; hence, $\delta = ak^{im}$.

One to one correspondence between exponential factors, which govern tunneling through a long $(C)_N$ chain and tunneling through a rectangular barrier, holds only within the energy interval which is small in comparison with the band width.[6] When electron energy moves further down from the band edge (that is from the top of the barrier), the tunneling exponent increases slower than the square-root dependence. Far from the band, δ approaches logarithmic asymptotic:

$$\delta \approx \ln \left(\frac{|\varepsilon - E|}{\beta} \right) \qquad (23.64)$$

Since its derivation by McConnel [38] this is a well-known result. It was quoted in numerous papers, although quite often without proper mention of its asymptotic character.

The model discussed above was often exploited for interpretation of electron transfer in DBA systems and in other physical contexts. However, in most cases, it is too simple to be directly compared with the experiment. For example, if the model Hamiltonian pretends to be adequate to organic oligomers in the role of molecular wires, it must take into account the fact that the electronic structure is of the semi-conductor type. Otherwise, the through bandgap tunneling cannot be treated on realistic ground. The next section presents a two band tight-binding Hamiltonian, which describes valence and conduction bands divided by the gap of forbidden states.

---

[6]Formula for the probability of free electron tunneling through a rectangular potential barrier can be retrieved from the tight-binding analogue in the continual limit that implies $a \to 0$, $N \to \infty$, under the condition $a(N - 1) = d$ is constant [53]. In this limit, tight-binding Hamiltonian $H^{(C)_N}$ and free-electron Hamiltonian on segment d give equivalent results.

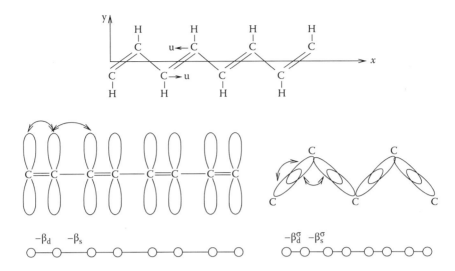

FIGURE 23.3 Upper part. Linear polyene with four double bonds. The hopping integrals $-\beta_s = -\beta \exp(-\eta)$ and $-\beta_d = -\beta \exp \eta$ are related to parameters t_0 (the electron resonance transfer energy between the nearest-neighbor CH groups in an undimerized polyene, $t_0 < 0$), κ (the electron-phonon coupling constant) and u (the C–C-bond alternation along the polyene axis as shown by arrows) [55], by equations $t_0 = -\beta \coth \eta$, $\kappa u = -\beta \sinh \eta$. The parameters C=C = 1.33Å, C–C = 1.45Å, $\angle$C=C–C = 125°, and $u = 0.034$Å, mimic the structure of octatetraene [56]. Parameters $\beta = 3.757$ eV and $\eta = 0.1333$ ($\beta_d = 4.33$ eV, $\beta_s = 3.29$ eV) were deduced by Kohler [57] from high-resolution spectroscopy experiments. Lower part. Schematic representation of the π electronic structure of polyenes [54,55] (to the left) and σ-electronic structure of alkanes according to Sandorfy's C approximation [58,59] (to the right) and their tight-binding equivalents. Hopping integrals for σ-electrons $-\beta_{d(s)}^\sigma = -\beta \exp[+(-)\eta^\sigma]$ with $\eta^\sigma = 0.54$ ($\beta_d^\sigma = 6.447$ eV, $\beta_s^\sigma = 2.189$ eV) are consistent with existing data on saturated linear hydrocarbons [36].

23.2.2   Carbon Chain with Alternating C–C Bonds

To describe a rigid chain of CH groups with alternating length of C–C bonds that was shown to be the case for oligomers of polyene [54] (see Figure 23.3), we use the electronic part of the Su–Schrieffer–Heeger Hamiltonian [55]. For our purposes, it is convenient to write it as a sum of the energy operator of non-interacting monomers and the intermonomer interaction:

$$\hat{H}^{(C=C)_N} = \sum_{n=1}^{N} \hat{H}_n^{C=C} + \hat{V} \tag{23.65}$$

It is assumed that C atoms are coupled by the "double" bond inside monomers

$$\hat{H}_n^{C=C} = \varepsilon(|n_l\rangle\langle n_l| + |n_r\rangle\langle n_r|) - \beta_d(|n_l\rangle\langle n_r| + \text{h.c.}) \tag{23.66}$$

whereas each monomer interacts with its nearest neighbors via "single" C–C bonds:

$$\hat{V} = -\beta_s \sum_{n=1}^{N}(|(n+1)_l\rangle\langle n_r| + |(n-1)_r\rangle\langle n_l|) \tag{23.67}$$

In Equations (23.65–23.67), index n runs over the double bonds (i.e., monomers) the number of which is half of all carbon atoms in the chain; ket vector $|n_{l(r)}\rangle$ has the meaning of the $2p_z$ atomic orbital of C atom, label l (r) refers to the left (right) atom connected by the nth double bond. $|0_r\rangle = |(N+1)_l\rangle = 0$; $-\beta_s = -\beta \exp(-\eta)$ and $-\beta_d = -\beta \exp \eta$ ($\beta > 0$) are the hopping integrals between the nearest neighbors connected by single and double bond, respectively.

Hamiltonian (23.65) was used most frequently in discussions of the π–electronic structure of polyenes and polyacetylene. Under certain model assumptions regarding the Hamiltonian parameters, Equation (23.65) can be applied to σ–electron subsystem in saturated or conjugated hydrocarbons. An outline of the model is as follows (see [36] for more details).

In the tight-binding description, π electrons are delocalized over the entire molecule due to a strong overlapping of π orbitals from the adjacent carbon atoms. The hopping integrals responsible for π-electron delocalization are different because of dimerization of the carbon backbone. Similar picture was introduced by Sandorfy [58] to describe σ–electron delocalization (see Figure 23.3 and ref. [60]). As shown in the figure, $-\beta_d^\sigma$ associates with (strong) electron transfer between sp^3 (or sp^2) hybrids directed along given covalent C–C bond, while $-\beta_s^\sigma$ describes (weak) electron transfer between two sp^3 (or sp^2) hybrids of the same carbon directed along covalent bonds with its neighbor carbons. If different hopping integrals and appropriate values for Coulomb integrals are chosen, Hamiltonian (23.65) can be used to model π–and σ–electron subsystems of linear hydrocarbons with $2N$ and $N-1$ carbon atoms in the chain. This allows one to investigate conducting properties of polyenes and alkanes on equal footings [36], and that can be done in the range unaffordable to more rigorous treatments.

23.2.2.1   Explicit Expressions for Eigenenergies and Eigenstates

Solution of the Schrödinger equation with Hamiltonian (23.65) with cyclic [54,55] and terminus-like [61] boundary conditions gives the π–electron spectrum consisting of valence (v) and conduction (c) bands

$$E_j^{v(c)} = \varepsilon - (+)\beta\sqrt{2\left(\cosh 2\eta + \cos \xi_j^{v(c)}\right)} \tag{23.68}$$

For periodic boundary conditions, $\xi_j^v = 2\pi j/N$, $\xi_j^c = 2\pi(N-j)/N$, $j = 0, 1, 2, \ldots, N-1$. In the following, we focus our interest on the $(C{=}C)_N$ chain with CH_2 termini. In this case, numbers $\xi_j^{v(c)}$ are subject to the solution of the Lennard–Jones equation [62]:

$$D = \exp(-2\eta)\sin(N\xi) + \sin[(N+1)\xi] = 0 \tag{23.69}$$

within the interval $0 \leq \xi \leq \pi$. The roots ξ_j^v and $\xi_j^c = \xi_{N+1-j}^v$, $j = 1, 2, \ldots, N$, are numbered so that ξ_1^v and ξ_1^c correspond to the lowest possible energy within the respective band. This means that the root ξ_1^v is closest to zero, while ξ_1^c stands for the largest (closest to π) root of Equation (23.69).

Note that the spectrum of end-substituted polyenes can be substantially different from that just described. In particular, the perturbation introduced by the substitutions (or not less likely, as a result of molecule interaction with the leads via its end atoms) can give rise to discrete in-gap states, a kind of Tamm/Shockley states [63]. Such states, the presence of which can substantially modify the molecule transmitting properties, have been examined in [29] and papers referenced therein.

In this model of polyene π–electron spectrum, the appearance of the band gap is entirely due to the bond-length alternation (manifestation of Pierls instability) which is described by the alternation parameter η. For $\eta = 0$, $\hat{H}^{(C{=}C)_N} \to \hat{H}^{(C)_{2N}}$. In this case, Equation (23.69) gives $\xi_j^v\big|_{\eta=0} = 2\pi j/(2N+1)$, $\xi_j^c\big|_{\eta=0} = 2\pi(2N+1-j)/(2N+1)$, so that in fact, the two-band spectrum (23.68) represents nothing but the lower and higher parts of the spectrum of $\hat{H}^{(C)_{2N}}$, see Equation (23.56).

To give the full description of polyenes in the Hückel approximation, we present eigenstates of Hamiltonian $\hat{H}^{(C{=}C)_N}$:

$$\Psi_{v(c),j}^{(C{=}C)_N} = \sum_{l=1}^{N}\left(\psi_{2l-1,j}^{v(c)}|2l-1\rangle + \psi_{2l,j}^{v(c)}|2l\rangle\right) \tag{23.70}$$

where [61]

$$\psi_{2l-1,j}^{v(c)} = \pm(-1)^{j+1} B\left(\xi_j^{v(c)}\right) \sin\left[(N+1-l)\xi_j^{v(c)}\right] \underset{\eta=0}{\longrightarrow} \sqrt{\frac{2}{2N+1}} \sin\left(\frac{2l-1}{2}\xi_j^{v(c)}\bigg|_{\eta=0}\right), \quad (23.71)$$

$$\psi_{2l,j}^{v(c)} = B\left(\xi_j^{v(c)}\right) \sin\left(l\xi_j^{v(c)}\right) \underset{\eta=0}{\longrightarrow} \sqrt{\frac{2}{2N+1}} \sin\left(l\xi_j^{v(c)}\bigg|_{\eta=0}\right)$$

and

$$B\left(\xi_j^{v(c)}\right) = \left(\frac{2\sin\xi_j^{v(c)}}{(2N+1)\sin\xi_j^{v(c)} - \sin\left[(2N+1)\xi_j^{v(c)}\right]}\right)^{1/2} \quad (23.72)$$

23.2.2.2 Green's Function

Summation in Equation (23.53), $\sum_\mu \to \sum_{\{\xi_j^c\}} + \sum_{\{\xi_j^v\}}$, utilizing the eigenenergies/eigenstates, which are given in Equations (23.68), (23.71), and (23.72), is rather tricky. To avoid this difficulty, the polyene Green's function was found by using the relationship between the Green's functions for $(C=C)_N$ chains with even and odd number of C atoms [64]. Later, a more general method was developed which gives analytical expression of oligomer Green's function $G^{(M)_N}$ in terms of monomer Green's function G^M [45,46,65]. M-oligomers are discussed next, Section 23.2.3. Here, we illustrate only the applicability of the general relation (23.90) to the case M = C=C. We use this example to highlight essentials of tunneling across molecular wires, when it occurs in the band-gap energy region.

For the matrix elements which refer to C atoms at termini of the chain with alternating length of C–C bonds, we have (for derivation details see Section 23.2.3)

$$G_{1_l,1_l}^{(C=C)_N} = G_{N_r,N_r}^{(C=C)_N} = \frac{(E-\varepsilon)\sin(N\xi)}{\beta^2 D}, \qquad G_{1_l,N_r}^{(C=C)_N} = -\frac{e^\eta \sin\xi}{\beta D} \quad (23.73)$$

where D is defined in Equation (23.69).

In these expressions, the relationship between wave vector ξ and energy E, depends on which interval of electron energy it refers to:

$$\frac{1}{2}\left(\frac{E-\varepsilon}{\beta}\right)^2 - \cosh 2\eta \equiv f\left(E, G^{C=C}\right)$$

$$= \cos\xi \begin{cases} \xi = i\delta, & \delta > 0 & \text{if } E < E_b^v, \\ \xi = \xi^v, & 0 \le \xi^v \le \pi & \text{if } E_b^v \le E \le E_t^v, \\ \xi = \pi + i\delta^{v-c}, & \delta^{v-c} > 0 & \text{if } E_t^v \le E \le E_b^c, \\ \xi = \xi^c = \pi - \xi^v, & & \text{if } E_b^c \le E \le E_t^c, \\ \xi = i\delta, & \delta > 0 & \text{if } E > E_t^c \end{cases} \quad (23.74)$$

The appearance of $f(E, G^{C=C})$, a functional of monomer C=C Green's function matrix elements

$$G_{l,l}^{C=C} = G_{r,r}^{C=C} = \frac{E-\varepsilon}{(E-\varepsilon)^2 - \beta^2 \exp(2\eta)},$$

$$\beta_s G_{l,r}^{C=C} = -\frac{\beta^2}{(E-\varepsilon)^2 - \beta^2 \exp(2\eta)} \quad (23.75)$$

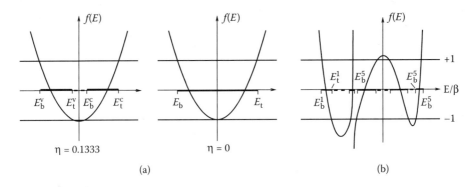

FIGURE 23.4 Examples of polynomial $f(E, G^M)$ which determines dependence of energy on wave vector in the μth energy band $E^\mu(\xi^\mu)$, $\mu = 1, 2, \ldots, N_M$, and dependence of energy on modulus of imaginary part of wave vector inside bandgaps $E(\delta^\mu)$, $\mu = 1, 2, \ldots, N_M - 1$, below, $E(\delta)$, and above, $E(\delta^{\mu=N_M})$, the band spectrum of M-oligomers. a: M is vinylene group C=C, $f(E, G^M)$ – second-order polynomial (see left-hand side of Equation [23.74]) calculated for the value of alternation parameter η found in [57]; $\mu = 1, 2$ correspond to, respectively, valence (v) and conduction (c) bands, the boundaries of which, $E_{b(t)}^{v(c)} = \varepsilon - (+)2\beta \cosh \eta$ and $E_{t(b)}^{v(c)} = \varepsilon - (+)2\beta \sinh \eta$, are given by intersections with horizontal lines ± 1. In Equation (23.74), $\delta^{v-c} \equiv \delta^{\mu=1}$, $\delta|_{E>E_t^c} \equiv \delta^{\mu=2}$. The band and bandgap energy intervals are shown on E/β axis by, respectively, thick solid and dashed segments. Two-band spectrum merges into a single-band spectrum at $\eta = 0$. b: M is furan ring C_4H_2O, $f(E, G^M)$ – fifth-order polynomial (see details of calculations in ref. [65]); five π bands of polyfuran and respective bandgaps are indicated in the same way as π and π^* bands of polyene (notations $E_{b(t)}^\mu$, $\mu = 2, 3, 4$ are omitted); HOMO–LUMO (or v-c) gap in polyfuran corresponds to $E_b^{\mu=4} - E_t^{\mu=3}$.

will be clarified in discussion of the general case of M-oligomers. Figure 23.4a explains notations used in the above equation and shows its link with the dispersion relation for a nonalternating $(C)_N$ chain. Note also that for $\eta = 0$, Equation (23.73) with ξ replaced by 2ξ coincides with Equation (23.58) for the respective matrix elements of the Green's function for $(C)_{2N}$-chain, i.e., $G^{(C=C)_N} \xrightarrow{\eta=0} G^{(C)_{2N}}$.

23.2.2.3 Tunneling Exponent

Let us consider electron tunneling through a polyene/alkane chain assuming that $N\delta \gg 1$ and $N\delta^{v-c} \gg 1$. Then, according to Equations (23.26), (23.73), and (23.74), in the case of through-gap tunneling, we have

$$T \sim \left(G_{1_l, N_r}^{(C=C)_N} \right)^2 \sim \exp\left(-2\delta^{v-c} N \right) \tag{23.76}$$

where

$$\delta^{v-c} = \cosh^{-1}\left[1 + \frac{(E_b^c - E)(E - E_t^v)}{2\beta^2} \right] \tag{23.77}$$

Notice that in terms of the number of carbon atoms N_C, the tunneling exponent equals $2\delta^{v-c}N = \delta^{v-c}N_C$ for polyenes, whereas for the alkane chain with the same N, it is nearly two times larger, $2\delta^{v-c}(N_C - 1)$ [36].

It immediately follows from Equation (23.77) that δ^{v-c} is zero at the gap edges and reaches its maximal value, $\delta_{\max}^{v-c} = 2\eta$, in the gap middle $E = \varepsilon$. For $\delta^{v-c} \ll 1$, i.e., for energies near gap edges or for a narrow gap, $E_g/\beta \ll 1$ (that corresponds to $\eta \ll 1$), one can use instead of Equation (23.77),

$$\delta^{v-c} \approx \frac{1}{\beta}\sqrt{(E_b^c - E)(E - E_t^v)} = \frac{1}{\beta}\sqrt{\Delta E(E_g - \Delta E)} \tag{23.78}$$

where $\Delta E \equiv |E_{b(t)}^{c(v)} - E|$. Equation (23.78) simplifies further, if $\Delta E << E_g$ (an arbitrary width of the bandgap):

$$\delta^{v-c} \approx \frac{1}{\beta}\sqrt{\Delta E E_g} \tag{23.79}$$

In the case of wide gaps $E_g/\beta >> 1$ (equivalently, $\eta >> 1$), and not far from the gap middle $(\delta^{v-c} >> 1)$, the energy dependence of tunneling exponent is well described by:

$$\delta^{v-c} \approx \ln\left[\frac{\left(E_b^c - E\right)\left(E - E_t^v\right)}{\beta^2}\right] \tag{23.80}$$

so that the maximal value is

$$\delta_{max}^{v-c} \approx 2\ln\left(\frac{E_g}{2\beta}\right) \approx 2\eta \tag{23.81}$$

In contrast, for narrow gaps ($\eta << 1$, $E_g \approx 4\beta\eta$, see caption in Figure 23.4), the maximal value of the tunneling exponent is proportional to the band gap

$$\delta_{max}^{v-c} \approx \frac{E_g}{2\beta} \approx 2\eta \tag{23.82}$$

Thus, for the wide-gap and narrow-gap cases, the interrelation between the maximal value of tunneling exponent and band gap differs significantly from each other. In this classification, polyenes ($\eta = 0.1333$ [57]) give an example of a narrow-gap molecular wire, whereas alkanes represent an intermediate case, since $\eta \to \eta^\sigma \approx 0.5$ [36].

Several comments are in order. First, the comparison of the two-band and one-band models shows that in the case of wide gaps, δ_{max}^{v-c} is two times *larger*, than an estimate given by the single-band model for the same energy $E = E_b^0 - E_g/2$. In this case, it follows from Equation (23.64): $\delta \approx \ln[(E_g + E_{bw}^0)/(2\beta)] \approx \ln[E_g/(2\beta)]$.

Second, there is a belief that if the tunneling electron energy is in the bandgap and close to either the valence or conduction band, the effect of the other distant band on the tunneling transport is negligible, and Equation (23.63) is a good approximation. As shown above, this is not true for the two-band model Hamiltonian of polyenes/alkanes.

Third, this consideration, as well as our earlier presented results [45,46,51,52], clearly expose that the effective mass concept (in either framework, the conventional [66,67] or Franz two-band model of tunneling [68]) is of very limited value for the interpretation of electron tunneling through long molecules. We return to this point later.

For energies lying below and above polyene π-electron spectrum, the value of δ increases monotonically from zero value at the band edges. It can be shown that far away from polyene π levels ($|E - E_{b(t)}^{v(c)}| >> \beta$) the increase is governed by the asymptotics which, up to the modulus sign, has the form of Equation (23.80), i.e.,

$$\delta^{v-c} \approx \ln\left|\frac{\left(E_b^c - E\right)\left(E - E_t^v\right)}{\beta^2}\right| \tag{23.83}$$

This observation, in our opinion, is as curious as it is instructive.

Asymptotics of the tunneling exponent for M-oligomers were discussed in detail in ref. [51,52].

23.2.3 Oligomers M–M– · · · –M

For a long time, the tight-binding Hamiltonians of (C)$_N$- and (C=C)$_N$-type oligomers discussed in preceding sections, remained among the few if not the only exactly solvable models of linear organic molecules.

Being probably the most successfully parameterized semi-empirical Hamiltonians that permit one to obtain exact solutions, they significantly contributed to heuristic works in many related branches. As mentioned previously, they were also used for modeling of molecular wires in metal–molecular contacts and bridging pathways in long-distance electron transfer.

From the latter perspective, a generalization to broad families of organic molecules, covered by the structural formulas M–M–...–M (M-oligomers) and M_1–M_2–M_1–M_2–...–M_1–M_2–M_1 [(M_1–M_2)-oligomers], was made with applications to oligomer spectra [65], quantum conductance of molecular wires [45,46], and nonadiabatic electron transfer across bridging molecules in DBA systems [51,52].

23.2.3.1 Hamiltonian and Green's Function of M-Oligomers

The polyene Hamiltonian introduced in Section 23.2.2, has the form that is characteristic for M-oligomers. The π–and σ–electronic structure of such compounds (abbreviated as $(M)_N$, where N is the number of monomers) can be described by:

$$\hat{H}^{(M)_N} = \sum_{n=1}^{N} \hat{H}_n^M + \hat{V} \tag{23.84}$$

where $\hat{H}_n^M$ is the Hamiltonian operator of the nth monomer, and $\hat{V}$ is the energy operator of inter-monomer interaction,

$$\hat{V} = -\beta_{int} \sum_{n=1}^{N} \left[|(n+1)_{\alpha_l}\rangle\langle n_{\alpha_r}| + |(n-1)_{\alpha_r}\rangle\langle n_{\alpha_l}| \right] \tag{23.85}$$

In Equation (23.85), $|0_{\alpha_r}\rangle = |(N+1)_{\alpha_l}\rangle = 0$, the ket $|n_{\alpha_i}\rangle$ has the meaning of π (or σ) orbital of the α_ith atom in the nth monomer, $-\beta_{int}$ is the energy of resonance π (σ) electron transfer between the binding atoms which are denoted as α_l and α_r. Since the further derivation is applicable to any number of atoms within monomers (denoted below as N_M) and any chemical structure of M, operator $\hat{H}_n^M$ can be specified whenever necessary.

The Green's function for the Hamiltonian defined in Equations (23.84) and (23.85) was found in [46] from the Dyson equation

$$\hat{G}^{(M)_N} = \hat{G}^M + \hat{G}^M \hat{V} \hat{G}^{(M)_N} \tag{23.86}$$

In matrix representation, it reads:

$$G_{n_{\alpha_i}, m_{\alpha_j}}^{(M)_N} = \delta_{n,m} G_{\alpha_i,\alpha_j}^M - \beta_{int} \left[G_{\alpha_i,\alpha_l}^M G_{(n-1)_{\alpha_r}, m_{\alpha_j}}^{(M)_N} + G_{\alpha_i,\alpha_r}^M G_{(n+1)_{\alpha_l}, m_{\alpha_j}}^{(M)_N} \right],$$

$$G_{0_{\alpha_r}, m_{\alpha_j}}^{(M)_N} = 0, \qquad G_{(N+1)_{\alpha_l}, m_{\alpha_j}}^{(M)_N} = 0 \tag{23.87}$$

where $G_{\alpha_i,\alpha_j}^M = \langle n_{\alpha_i} | (E\hat{I} - \hat{H}_n^M)^{-1} | n_{\alpha_j}\rangle$, indexes n and m take all possible values from 1 to N, while indexes α_i and α_j identify the location of atoms within the monomer. In particular, setting in Equation (23.87) $\alpha_i = \alpha_l, \alpha_r$, and $\alpha_j = \alpha_l, \alpha_r$, we obtain four equations for oligomer Green's function matrix elements referring to the binding atoms of monomers.

Finding the solutions to the set of four equations that follow from Equation (23.87) and involve only G_{α_l,α_l}^M, G_{α_r,α_r}^M, and G_{α_l,α_r}^M, is substantially simplified, if we know the relation between the energy E and wave vector ξ of electron in M-oligomers. The latter has been shown to have the form [65]:

$$\cos\xi = f(E, G^M), \quad f\left(E, G^M\right) = -\frac{1 - G_\Delta^M}{2\beta_{int} G_{\alpha_l,\alpha_r}^M} \tag{23.88}$$

where notation $G^M_\triangle$ stands for

$$G^M_\triangle \equiv \beta^2_{int} \left[G^M_{\alpha_l,\alpha_l} G^M_{\alpha_r,\alpha_r} - \left(G^M_{\alpha_l,\alpha_r} \right)^2 \right] \tag{23.89}$$

and $f\left(E, G^M \right)$ is a certain polynomial of the order $\leq N_M$, determined by the particular dependence of monomer Green's function matrix elements on energy. Note that to recover the dispersion relation (23.59), that is the case of M = C, one has to set $G^M_{\alpha_{l(r)},\alpha_{l(r)}} = G^M_{\alpha_l,\alpha_r} = (E - \varepsilon)^{-1}$ and $\beta_{int} = \beta$. Substitution of matrix elements (23.75) in Equations (23.88) and (23.89), gives us Equation (23.74).

Representing $G^{(M)_N}_{n_{\alpha_{l(r)}},m_{\alpha_{l(r)}}}$ and $G^{(M)_N}_{n_{\alpha_{l(r)}},m_{\alpha_{r(l)}}}$, as a superposition of incident and reflected waves and exploiting Equation (23.88), after lengthy but straightforward algebra we arrive at the following expressions for these matrix elements

$$G^{(M)_N}_{n_{\alpha_i},m_{\alpha_j}} = \left(G^M_{\alpha_l,\alpha_r} P^{(M)_N}_N \sin \xi \right)^{-1}$$

$$\times \begin{cases} -\beta^{-1}_{int} G^M_{\alpha_l,\alpha_l} \sin[(N+1-m)\xi] P^{(M)_N}_{n-1}, & i = j = l, \ n < m, \\[2mm] n \leftrightarrow m, & i = j = l, \ n \geq m, \\[2mm] G^M_{\alpha_l,\alpha_l} G^M_{\alpha_r,\alpha_r} \sin[(N+1-m)\xi] \ \sin \xi n, & i = l, j = r, \ n < m, \\[2mm] \beta^{-2}_{int} P^{(M)_N}_{m-1} P^{(M)_N}_{N-n}, & i = l, j = r, \ n \geq m, \\[2mm] \beta^{-2}_{int} P^{(M)_N}_{n-1} P^{(M)_N}_{N-m}, & i = r, j = l, \ n < m, \\[2mm] G^M_{\alpha_l,\alpha_l} G^M_{\alpha_r,\alpha_r} \sin[(N+1-n)\xi] \ \sin \xi m, & i = r, j = l, \ n \geq m, \\[2mm] -\beta^{-1}_{int} G^M_{\alpha_r,\alpha_r} \sin(n\xi) \ P^{(M)_N}_{N-m}, & i = j = r, n < m, \\[2mm] n \leftrightarrow m, & i = j = r, n \geq m, \end{cases} \tag{23.90}$$

where

$$P^{(M)_N}_n = \sin(n\xi) + \beta_{int} G^M_{\alpha_l,\alpha_r} \sin[(n-1)\xi] = \sin(n\xi) \ G^M_\triangle - \beta_{int} G^M_{\alpha_l,\alpha_r} \sin[(n+1)\xi] \tag{23.91}$$

It is readily verified that in the case of $(C)_N$ chain, Equation (23.90) transforms into (23.58). The use of the monomer Green's function for M = C=C, see Equation (23.75), and $\beta_{int} = \beta_s$ recovers Equation (23.73) represented in Section 23.2.2.

With the matrix elements listed above, finding the other that refer to indexes α_i, α_j, where $i, j \neq l, r$, requires solution of the set of equations of the $(N_M - 2)$th order. This, in comparison with the $(N \times N_M)$th order of the initial problem, is a substantial simplification. Furthermore, Equations (23.88) and (23.90) give the exact, explicit expression of M-oligomer Green's function in terms of matrix elements of monomer Green's function operator G^M. This suggests a number of advantages in calculations and analysis of electronic properties of oligomers and respective polymers. Some of them are overviewed next.

23.2.3.2 Gross Structure of M-Oligomer One-Electron Spectrum

Equation (23.88) gives full and instructive representation of the electron spectrum gross structure of M-oligomers. As illustrated by examples in Figure 23.4, intersections of curve $f\left(E, G^M \right)$ with horizontal lines ± 1:

$$f\left(E, G^M \right) = \pm 1 \tag{23.92}$$

determine energies of the band boundaries, denoted as $E^\mu_{b(t)}$. Similarly to polyenes, index μ associates with the roots of polynomial $f(E, G^M)$. For each band (i.e., zone of allowed energies), Equation (23.88) plays the role of the dispersion relation, specifying the dependence $E^\mu(\xi^\mu)$. The eigenvalues of ξ^μ are determined by solutions of the generalized Lennard–Jones equation

$$P_N^{(M)_N} = \sin \xi N + \beta_{int} G_{\alpha_l,\alpha_r}^M \sin[(N-1)\xi] = 0 \qquad (23.93)$$

for all the energy intervals $E_b^\mu \le E \le E_t^\mu$, where $|f(E, G^M)| \le 1$ [65]. Usage of the explicit expression of $G_{\alpha_l,\alpha_r}^M = G_{l,r}^{C=C}$ and $\beta_{int} = \beta_s$ converts Equation (23.93) into the Lennard–Jones equation (23.69).

Now, consider the energies lying within the bandgaps, where $|f(E, G^M)| > 1$. In the limit of infinitely long oligomers, Equations (23.88) and (23.93) cannot be satisfied simultaneously, if:

$$\beta_{int} \left| G_{\alpha_l,\alpha_r}^M \right| < 1 \qquad (23.94)$$

This inequality represents the necessary condition for *all* the electron states have their energies within zones of allowed energies.

Alternatively, fulfillment of the condition $\beta_{int}|G_{\alpha_l,\alpha_r}^M| > 1$ means that Equation (23.93) can be satisfied at certain in-gap energies. These energies, let us denote them as E_{loc}, belong to the oligomer spectrum, if

$$G_{\alpha_l,\alpha_l}^M G_{\alpha_r,\alpha_r}^M = 0 \qquad (23.95)$$

In other words, M-oligomer spectrum can contain discrete in-gap states which do not exist under periodic boundary conditions.

A solution of Equation (23.95) acquires another meaning if additionally, equation

$$\beta_{int} \left| G_{\alpha_l,\alpha_r}^M \right| = 1 \qquad (23.96)$$

is fulfilled at the same energy. This is a special case of the bands joining, i.e., closing the corresponding band gap [65].

Let us test the above statements by utilizing the monomer Green's function which corresponds to oligomers of polyene, see Equation (23.75). At $E = \varepsilon$, Equation (23.95) has the root of the second order. At this energy, $\beta_s \left| G_{l,r}^{C=C} \right| > 1$ if $\eta < 0$. So, if an infinitely long dimerized carbon chain is terminated by single C–C bonds, there exists a two-fold degenerate discrete level at the gap middle. Also direct solution of Equations (23.69) and (23.74) with $\eta < 0$ gives two in-gap levels, below and above the gap middle. Energies of these levels move towards each other with the increase of the length of polyene chain.

Furthermore, $\beta_s \left. G_{l,r}^{C=C} \right|_{E=\varepsilon,\eta=0} = -1$. Thus, if $\eta = 0$, both Equations (23.95) and (23.96) are fulfilled at $E = \varepsilon$, which means that the bandgap shrinks down to zero. The bandgap closing in polyenes has been shown already by direct calculations, see Equation (23.68), discussion thereafter, and Figure 23.4.

23.2.3.3 Tunneling Exponent for M-Oligomers

In analogy with previous examples of oligomers $(C)_N$ and $(C=C)_N$, the dispersion relation (23.88) can be extended to the case of imaginary wave vectors by replacing $\xi^\mu \to i\delta^\mu$, $\delta^\mu > 0$, if $f(E, G^M) > 1$, and by $\xi^\mu \to \pi + i\delta^\mu$, if $f(E, G^M) < -1$, i.e.:

$$\begin{aligned} \cosh \delta &= |f(E, G^M)|, \quad E < E_b^{\mu=1}; \\ \cosh \delta^\mu &= |f(E, G^M)|, \quad E_t^\mu < E < E_b^{\mu+1}, \quad \mu = 1, 2, \ldots, N_M - 1 \end{aligned} \qquad (23.97)$$

and $E > E_t^{\mu=N_M}$, if μ refers to the highest energy band. Note that if the order of polynomial $f(E, G^M)$ is less than N_M, the number of electron bands decreases accordingly. Obtained above Equations (23.60) and (23.74) represent just particular cases of Equation (23.97).

This equation together with the definition of $f(E, G^M)$ (23.88) gives a general expression of the tunneling exponent in terms of monomer Green's function matrix elements referring to the monomer-binding atoms. The monomer Green's function plus oligomer length is the only knowledge one needs to calculate the tunneling exponent for any bandgap.

Usually, it is the HOMO(highest occupied molecular orbital)-LUMO(lowest unoccupied molecular orbital) gap, i.e., the gap between the valence and conduction bands, that is of prime interest. Preserving the notation exploited in the case of polyenes, we can write for tunneling through M-oligomers within the gap between the valence and conduction bands

$$T \sim k_{et} \sim \exp(-2\delta^{v-c} N) \tag{23.98}$$

where

$$\delta^{v-c} = \ln\left(|f(E, G^M)| + \sqrt{f^2(E, G^M) - 1}\right) \tag{23.99}$$

To notice, the HOMO (LUMO) level of an oligomer is always lower (higher), than the corresponding conduction (valence) band top (bottom). In the energy intervals $E_t^v - E_{HOMO}$ and $E_{LUMO} - E_b^c$, Equations (23.98) and (23.99) are meaningless, as it follows from the derivation.

For in-gap discrete levels (if any) satisfying Equation (23.95) and inequality $\beta_{int}|G_{\alpha_l,\alpha_r}^M| > 1$, definition of $\delta^{v-c}(E = E_{loc})$ simplifies to:

$$\delta^{v-c}(E = E_{loc}) = \ln\left(\beta_{int}\left|G_{\alpha_l,\alpha_r}^M(E = E_{loc})\right|\right) \tag{23.100}$$

As an example, for a long polyene chain, the above relation gives $\delta^{v-c}(E = \varepsilon) = \delta_{max}^{v-c} = 2|\eta|$ that agrees with Equations (23.82) and (23.82).

It should be stressed at this point that at and near the energies of in-gap discrete levels, approximation:

$$T \sim k_{et} \sim \left|G_{1_{\alpha_l}, N_{\alpha_r}}^{(M)_N}\right|^2 \tag{23.101}$$

is invalid, so that Equation (23.98) should be used with extra precaution. Accurate evaluation of electron transmission in the region of in-gap resonances requires the use of exact formulas represented in Section 23.1.2.

So, only the knowledge of $f(E, G^M)$ (that is, the knowledge of certain matrix elements of the monomer Green's function) is required to establish the dispersion relation for electron waves propagating in the actual band of M-oligomers or to calculate tunneling exponent for the bandgap of interest, at any energy.

Function $f(E, G^M)$ has a specific dependence on energy (i) for the given chemical structure of monomer M and (ii) for the way which monomers are connected to each other to form the respective M-oligomer. Oligomer design reveals itself in oligomer spectral properties via Equations (23.88), (23.93), and their derivatives. In this connection, a few comments are of practical significance.

Bilinear combination of monomer Green's function matrix elements G_Δ^M, which appears in the definition of $f(E, G^M)$ (23.88), has the number of poles bounded by N_M. These poles may coincide with those of the monomer Green's function. Alternatively, the number of poles of G_Δ^M can be larger than the one that has G_{α_l,α_r}^M. In the latter case, $f(E, G^M)$ is a singular function. Also, singular behavior of $f(E, G^M)$ can be caused by zeros of G_{α_l,α_r}^M. An example of this is shown in Figure 23.4. For whatever reason, the tunneling exponent at singular points of $f(E, G^M)$ is infinite, indicating that the given oligomer inherently possesses the property of an electric switch. Transmission spectrum of such molecules contains deeps, i.e., antiresonances, which can appear *only* in the bandgap energy intervals [52]. In the absence of singular points of $f(E, G^M)$, maximal value of the tunneling exponent within the given bandgap corresponds to maximum or minimum of $f(E, G^M)$, see Figure 23.4.

Entirely determined by the monomer electronic structure, the tunneling exponent is *independent* of end effects. By no means should this be understood to imply that the ohmic tunneling conductance is independent of contacts. There are many interesting effects associated with molecule–lead connection. Here is one of them.

Thiol groups which are often used to contact organic molecules to Au substrate, can generate in-gap states. Their presence can give rise to a resonance-like transmission spectrum in the bandgap energy interval. Such resonances (called pseudoresonances to distinguish them from in-band resonances) were investigated theoretically for molecular contacts based on thiol- and dithiolpolyenes, and thiol- and dithiolalkanes [36].

23.2.3.4 (M₁–M₂)-Oligomers

Description of electronic structure and properties of M-oligomers is readily extended to the case of (M_1-M_2)-oligomers which can be considered as M-oligomers, $M = M_1-M_2$, with the end defect: the Nth monomer M is substituted by M_1. In presence of such defect, Equations (23.94) and (23.95) are modified in a certain way [46]. However, the dispersion relation for this family of linear compounds preserves the form of Equation (23.88), where $f(E, G^M)$ should be replaced by $f(E, G^{M_1-M_2})$. Therefore, the only difference in calculations of band gap energies and respective tunneling exponents is that one has to use the monomer Green's function $G^{M_1-M_2}$ instead of G^M. The required matrix elements are

$$G^{M_1-M_2}_{\alpha_l,\alpha_l} = \frac{G^{M_1}_{\alpha_l,\alpha_l} - G^{M_2}_{\alpha_l,\alpha_l} G^{M_1}_\Delta}{1 + \beta^2_{M_1-M_2} G^{M_1}_{\alpha_r,\alpha_r} G^{M_2}_{\alpha_l,\alpha_l}}, \quad G^{M_1-M_2}_{\alpha_r,\alpha_r} = \frac{G^{M_2}_{\alpha_r,\alpha_r} - G^{M_1}_{\alpha_r,\alpha_r} G^{M_2}_\Delta}{1 + \beta^2_{M_1-M_2} G^{M_1}_{\alpha_r,\alpha_r} G^{M_2}_{\alpha_l,\alpha_l}} \tag{23.102}$$

$$G^{M_1-M_2}_{\alpha_l,\alpha_r} = -\beta_{M_1-M_2} \frac{G^{M_2}_{\alpha_l,\alpha_r} G^{M_1}_{\alpha_l,\alpha_r}}{1 + \beta^2_{M_1-M_2} G^{M_1}_{\alpha_r,\alpha_r} G^{M_2}_{\alpha_l,\alpha_l}} \tag{23.103}$$

where $-\beta_{M_1-M_2}$ denotes hopping integral between M_1 and M_2. For the case $M_1-M_2 = $ C=C and $\beta_{M_1-M_2} = \beta \exp(\eta)$, Equation (23.102) reduces to Equation (23.75). From above relations, under replacements $M_1 \rightarrow$ M–M–···–M, where $M = M_1-M_2$; $M_2 \rightarrow M_1$, and $\beta_{M_1-M_2} \rightarrow \beta_{int}$, one obtains the matrix elements of $G^{(M_1-M_2)_N}$ which refer to the oligomer end atoms.

For (M_1-M_2)-oligomers, formula for T and k_{et} (that is an analogue to Equation (23.101)) reads

$$T \sim k_{et} \sim \left| \frac{\beta_{int} G^{M_1}_{\alpha_l,\alpha_r} G^{(M_1-M_2)_N}_{1_{\alpha_l}, N_{\alpha_r}}}{1 + \beta^2_{int} G^{M_1}_{\alpha_l,\alpha_l} G^{(M_1-M_2)_N}_{N_{\alpha_r}, N_{\alpha_r}}} \right|^2 \sim \exp(-2\delta^{v-c} N) \tag{23.104}$$

where

$$\delta^{v-c} = \ln\left(\left| f(E, G^{M_1-M_2}) \right| + \sqrt{f^2(E, G^{M_1-M_2}) - 1} \right) \tag{23.105}$$

and $E^v_t < E < E^c_b$.

23.2.3.5 Other Descriptions of Tunneling through Long Molecules

Similar in spirit but different in technical details description of electron tunneling through long molecules was suggested in ref. [69]. However, instead of functional $f(E, G^M)$ which is explicitly defined and uniquely related to the wave vector, either real $\arccos[f(E, G^M)]$ or imaginary $\cosh^{-1}|f(E, G^M)|$, (see Equation [23.88]), the authors introduced another polynomial [$\Delta(E)$ in their notations] that determines the dependence of the wave vector on energy and has the double higher order, than polynomial $f(E, G^M)$. This leads to unnecessary overcomplications in the calculation of the tunneling exponent.

The formulas obtained in [69] were rearranged so the tunneling exponent was expressed in terms of the energy dependent effective mass [70]. In our opinion, interpretation of experimental data in terms of parameters of the molecular Hamiltonian is much more informative than that interpretation which can be derived from the effective mass description.

23.2.4 Intermediate Synopsis

In this section, an analytically tractable model of organic oligomers — a class of molecular wires, is formulated on the basis of the established Hamiltonians that reproduce essentials of the electronic structure of real conjugated and saturated molecules. Exact explicit expressions of M – and (M_1-M_2)-oligomer Green's functions have been formally elaborated in physically sound contexts, and illustrated by representative examples.

In view of the great complexity of the system to be treated even at the semi-empirical level, to say nothing of huge computational resources needed for accurate ab initio calculations, it is especially attractive that the dependence of heterojunction transmission on electron energy and external voltage is accessible within seconds. Also the role of the main parameters/factors that govern electrical current across molecular contact can be quickly examined in great detail.

It is demonstrated that the exponential factor which governs tunneling through molecular wires is uniquely determined by the molecular length and a certain functional of monomer Green's function. This immediately relates molecular wire/bridging molecule electron transfer efficiency to the monomer electronic structure and the position of binding atoms in it.

23.3 Electric Field Effects

The analytical modeling of molecular contacts outlined in the preceding two sections is sufficient for analysis of ohmic conductance, (see [1,15,19,20] and references therein). Non-linear current-voltage characteristics can also be examined under the restriction that the drop of driving electric potential occurs entirely at the metal–molecule interfaces. In such a case, the applied voltage is easy to take into account by appropriate definition of the site energies in the lead Green's functions, see Section 23.1.

The model which shows a sudden drop of the electric potential has dominated the calculations of non-linear molecular conductance reported during the last decade [5,6,8,10,23–25,42]. The case of linear potential change along the molecular wire, which deserves to be studied as a likely alternative, has received little attention.

In our works [71–74], Hamiltonian of $(C)_N$ chain with field-shifted site energies[7] was used to answer the following questions: How the electron spectrum is modified by the presence of a constant electric field and how these changes manifest themselves in conducting properties of molecular wires. The model is simple enough to expose electric field quantum effects in finite systems in an analytical form.

Following the cited papers, we consider first the quantization of band electron levels in the carbon chain where the electric field is created by the voltage applied to the system termini. The dependence of level spacing on the applied voltage is examined in great detail. With regard to electron quantum transport, classification of electron states details the content of what is called *biased* or *tilted* band (see Figure 23.5), the central concept in the theory of Zenner tunneling, Bloch oscillations, and other related phenomena. This consideration is continued and concluded with the discussion of electric field effects on electron transmission across molecular wires. Emphasis is placed on the tunneling mechanism of electron motion. Compared with previous works in this direction, the most important point is that the presented results concern *finite* band width (in energy) and *finite* band length (in space). Links with earlier studies of electron tunneling in semiconductor heterojunctions and field emission are also brought to light.

23.3.1 Spectrum of $(C)_N$ Chain in a Biased Potential

Suppose that in presence of constant electric field $\mathcal{E}$, the site energies in $(C)_N$ chain form a stairway of equidistant electron levels separated from each other by energy $\varepsilon_{\mathcal{E}} = ae\mathcal{E}$, a is the lattice constant, e is the electron charge. With this change, the tight-binding Hamiltonian operator Equation (23.54) can be rewritten as

$$\hat{H}^{(C)_N} = \sum_{n=1}^{N} \left\{ \left[\varepsilon + \left(n - \frac{N+1}{2} \right) \varepsilon_{\mathcal{E}} \right] |n\rangle\langle n| \right.$$

$$\left. - \beta[(1 - \delta_{n,N})|n\rangle\langle n+1| + (1 - \delta_{n,1})|n\rangle\langle n-1|] \right\} \qquad (23.106)$$

Equivalents of $\hat{H}^{(C)_N}$ repeatedly appeared in studies of electric field effects [75–78].

---

[7] The physical context of these works was directed more towards biased superlattices. However, the same Hamiltonian can reasonably model a variety of systems.

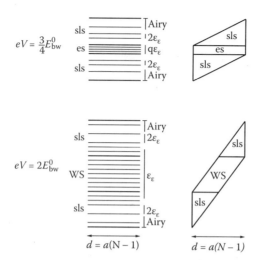

FIGURE 23.5 Signatures of constant electric field in the electron spectrum. Left hand side: subbands with characteristic level spacing for $eV = 3E_{bw}^0/4$ (up) and $eV = 2E_{bw}^0$ (down); thick horizontal lines mark the full-band-, and es-, sls-, and WS-subband edges; thin horizontal lines represent one-electron levels; interruptions in the vertical line indicate energy intervals where the spectrum cannot be described by elementary functions; Airy = Airy-type spectrum. To the right: corresponding tilted bands and their subdivision into es-, sls-, and WS-bands at the same voltages.

In Equation (23.106), the zero field shift of the site energy is placed in the mid of the chain; field parameter $\varepsilon_\mathcal{E}$ has the meaning of Bloch oscillation frequency times the Plank constant; in our notations, the voltage applied to the molecule termini is equal to $V = (N-1)a\mathcal{E} = (N-1)\varepsilon_\mathcal{E}/e$. It is obvious that the spectrum of $\hat{H}^{(C)_N}$ is symmetric with respect to the site energy ε and independent of the sign of $\varepsilon_\mathcal{E}$. Therefore, the field parameter is chosen to be positive and hence, $eV > 0$ hereinafter.

As shown by many authors [72,75–78], the Hamiltonian matrix determinant

$$\mathcal{D}_N(E) = (-1)^N \frac{\epsilon_\epsilon}{\pi\beta} \det[\beta^{-1}(\hat{I}E - \hat{H}^{(c)_N})] \tag{23.107}$$

is represented by a bilinear combination of Bessel functions. For the particular choice of zero electrostatic energy at $n = (N+1)/2$, it has the form [72]

$$\mathcal{D}_N(E) = J_{\mu+(N+1)/2}(z)Y_{\mu-(N+1)/2}(z) - Y_{\mu+(N+1)/2}(z)J_{\mu-(N+1)/2}(z) \tag{23.108}$$

where $\mu \equiv (\varepsilon - E)/\varepsilon_\mathcal{E}$, $z \equiv 2\beta/\varepsilon_\mathcal{E}$, and $J_\mu(z)$ and $Y_\mu(z)$ are the Bessel functions of the first and second kind, respectively.

In certain, dependent on eV energy intervals, solutions of the eigenvalue problem:

$$\mathcal{D}_N(E) = 0 \tag{23.109}$$

can be expressed in elementary functions. The obtained expressions reflect a pronouncedly different nature of electron states, called henceforth extended states (es), Wannier–Stark (WS) states, and surface localized states (sls). Energy levels of these states reside in es-, WS-, and sls-bands, respectively, see Table 23.1. Accordingly, the whole spectrum consists of the central (sub)band of bulk states merging into lower and upper lying (sub)bands of field-induced surface states. The bulk states can be a kind of extended states at lower voltages or WS (localized) states at higher voltages. If the electric potential difference eV is smaller, than the zero-field band width E_{bw}^0, the field dependence of bulk level quantization is not the same as that is observed for $eV > E_{bw}^0$. We term these two cases as, respectively, *low voltages* and *high voltages*, and consider them separately. Boundaries of characteristic subbands appearing at low and high voltages, are defined in Table 23.1.

TABLE 23.1   Dependence of the Band Edges of Characteristic Bands (Listed in the Left Hand Side Column) on the Applied Voltage

Nature of the Band	Bottom of the Band		Top of the Band	
	$eV < E_{bw}^0$	$eV > E_{bw}^0$	$eV < E_{bw}^0$	$eV > E_{bw}^0$
es-band	$E_b^0 + eV/2$	N/E	$E_t^0 - eV/2$	N/E
lower sls-band		$E_b^0 - eV/2$	$E_b^0 + eV/2$	$E_t^0 - eV/2$
upper sls-band	$E_t^0 - eV/2$	$E_b^0 + eV/2$	$E_t^0 + eV/2$	
WS-band	N/E	$E_t^0 - eV/2$	N/E	$E_b^0 + eV/2$

Zero-field band bottom (top) is (see Equation (23.59), $E_{b(t)}^0 = \varepsilon - (+)2\beta$. Abbreviation N/E stands for "non-existent". Notice that the top of lower sls-band goes up with increase of $eV < E_{bw}^0 = 4\beta$ but it goes down at higher voltages. Edges of es- and WS-bands are driven by the field in one direction, up or down.

23.3.1.1   Low Voltages, $eV < E_{bw}^0$: Band of Extended States

At low voltages, the bulk state levels occupy the energy interval $[\varepsilon - 2\beta + eV/2, \varepsilon + 2\beta - eV/2]$. Within, for example, the lower half of es-band, where the ε_ε-order intervals near the half-band edges are excluded $(-2\beta + eV/2 + \varepsilon_\varepsilon < E - \varepsilon < -\varepsilon_\varepsilon)$, determinant (23.108) transforms into

$$\mathcal{D}_N(E) \approx \frac{\varepsilon_\varepsilon}{\pi\beta}\sqrt{\frac{1}{\sin\xi\sin\xi'}}\sin\left[\xi + \xi' - \frac{2\beta}{\varepsilon_\varepsilon}(\Phi_\xi - \Phi_{\xi'})\right] \qquad (23.110)$$

where

$$2\beta\cos\xi = \varepsilon - E + \frac{eV}{2}, \qquad \Phi_\xi = \sin\xi - \xi\cos\xi,$$

$$2\beta\cos\xi' = \varepsilon - E - \frac{eV}{2}, \qquad \Phi_{\xi'} = \Phi_\xi(\xi \to \xi') \qquad (23.111)$$

Thus, electron levels of extended states are quantized according to equation

$$\sin[(N+1)\xi^{es}] = 0 \qquad (23.112)$$

where

$$\xi^{es} = \frac{1}{N+1}\left[\xi + \xi' - \frac{2\beta}{\varepsilon_\varepsilon}(\Phi_\xi - \Phi_{\xi'})\right] \qquad (23.113)$$

Worth noting is that Equation (23.112) is formally identical to that determines the wave vector eigenvalues at zero field, $\sin[(N+1)\xi] = 0$ (see Equation [23.56]). However, ξ^{es} does not have such a simple physical meaning as ξ in Equation (23.59).

It can be shown that in the limit $N \to \infty$, the maximal (and field-independent) value of ξ^{es} is equal to $\pi/2$. The minimal possible value of ξ^{es} is controlled by the applied voltage. It determines half the number of states in the es-band

$$n^{es} = \left[\frac{N+1}{2}\right]_{N/I} + 1 - \frac{N+1}{\pi}\xi^{es}\Big|_{E = \varepsilon - 2\beta + eV/2} \qquad (23.114)$$

where $[\ldots]_{N/I}$ denotes the nearest integer of the argument. The rest of electron states, $N - 2n^{es}$ (N is even) or $N - 1 - 2n^{es}$ (N is odd), are "pooled out" by the field from es-band into the lower and upper sls-bands.

Further analysis of Equations (23.110–23.113) leads us to the conclusion that under the condition $N \gg 1$, the electron levels in the mid part of the es band have the following energies:

$$E_l = \varepsilon \pm \frac{\varepsilon_{\mathcal{E}}^{(m'/m)}}{1 - 2m'/m} \begin{cases} l, & \text{odd } N, \\ (l+1/2), & \text{even } N, \end{cases} \quad l = 0, 1, 2, \dots \quad (23.115)$$

Equation (23.115) holds for specific values of the field parameter $\varepsilon_{\mathcal{E}} = \varepsilon_{\mathcal{E}}^{(m'/m)}$, which are determined by:

$$\beta^{-1} \varepsilon_{\mathcal{E}}^{(m'/m)} = 4 \cos \left(\pi \frac{m'}{m} \right) \begin{cases} 1/(N-1), & \text{odd } N, \\ 1/N, & \text{even } N. \end{cases} \quad (23.116)$$

Here, $m = 3, 4, \dots, m' = 1, 2, \dots < m/2$, and m and m' are relatively prime numbers, i.e., they do not have any common divisor.

The interrelation between the electron spectrum and applied voltage that is represented in Equations (23.115) and (23.116), reveals an interesting regularity: A proper tuning of the applied voltage should give rise to a ladder-like spectrum with the level spacing larger than that predicted by Wannier:

$$\Delta E = q \varepsilon_{\mathcal{E}}, \quad q > 1 \quad (23.117)$$

According to Wannier's theory [18], the bulk states of a crystal subjected to the constant electric field are quantized with the level spacing equal to $\varepsilon_{\mathcal{E}}$. This kind of spectrum is often called Wannier–Stark ladder (WSL), see following. Spectrum (23.115) contrasts with the canonical WSL: it is also equidistant but has ΔE larger than $\varepsilon_{\mathcal{E}}$.

By tuning the applied voltage, one equidistant spectrum can be replaced by the other in a way predicted by Equations (23.115) and (23.116). Consider a particular example. Let N be odd and $eV/\beta = 4\cos(\pi/4)$ which corresponds to $m' = 1, m = 4$, and hence, the double-$\varepsilon_{\mathcal{E}}$ spaced WSL is realized: $E - \varepsilon = 0, \pm 2\varepsilon_{\mathcal{E}}, \pm 4\varepsilon_{\mathcal{E}}, \dots$. If we decrease the voltage to the value V' such that $\arccos(eV'/4\beta)\beta / \arccos(eV/4\beta) = 4/3$, the triple-$\varepsilon_{\mathcal{E}}$ spaced WSL has to emerge instead, i.e., $E - \varepsilon = 0, \pm 3\varepsilon_{\mathcal{E}}', \pm 6\varepsilon_{\mathcal{E}}', \dots$, where $\varepsilon_{\mathcal{E}}' = eV'/(N-1)$.

Summarizing, at lower voltages (preceding the appearance of canonical WSL spectrum at high voltages), electron states of the mid part of the es-band are quantized according (23.117) with V-dependent factor $q > 1$ which, in particular, can acquire integer numbers approaching the limiting value $q = 1$.

23.3.1.2   High Voltages, $eV > E_{bw}^0$: Wannier–Stark Band

At $eV = E_{bw}^0$, the es-band shrinks down to zero which means that $n^{es} = 0$, so that all states belong either to the lower or upper sls-band. At higher voltages, for the energy interval $[\varepsilon + 2\beta - eV/2, \varepsilon - 2\beta + eV/2]$ and for $(N-2n)\varepsilon_{\mathcal{E}}/\beta \gg 1$ the solution to the eigenvalue problem with Hamiltonian (23.106) yields [72]:

$$E_n = \varepsilon \pm n\varepsilon_{\mathcal{E}} \pm \frac{2\varepsilon_{\mathcal{E}}\beta^2}{\pi(eV - 2n\varepsilon_{\mathcal{E}})^2} \left[\frac{2 \cdot 2.7183}{N - 2n} \frac{\beta}{\varepsilon_{\mathcal{E}}} \right]^{(N-2n)} + \begin{cases} 0 & \text{odd } N, \\ \pm \varepsilon_{\mathcal{E}}/2 & \text{even } N, \end{cases} \quad (23.118)$$

$n = 0, 1, 2, \dots$, where $+$ and $-$ refer to the upper and lower half of the spectrum, respectively. The above expression was proved to be accurate except $\varepsilon_{\mathcal{E}}$-order intervals near the band edges. Without the third term in the right hand side, which is exponentially small, spectrum (23.118) retrieves the Wannier classic result. Similar corrections to the Wannier levels were obtained in ref. [76].

23.3.1.3   Bands of Surface Localized States

Both es- and WS-bands represent bulk electron states. From below and above, they border on the lower and upper sls-bands, whose voltage dependent boundaries are specified in Table 23.1. Because of the spectrum symmetry, we can consider either of them; let it be the lower sls-band. In the upper, sls-levels are just the mirror reflection of their counterparts in the lower, with respect to $E = \varepsilon$.

For energies below the top of the sls-band, determinant (23.108) takes the form

$$\mathcal{D}_N(E) \approx \frac{\varepsilon_\varepsilon}{\pi\beta} \sqrt{\frac{1}{\sin\xi' \sinh\delta}} \cos\left(\frac{2\beta}{\varepsilon_\varepsilon}\Phi_{\xi'} + \xi' - \frac{\pi}{4}\right) \exp\left(\frac{2\beta}{\varepsilon_\varepsilon}\Phi_\delta + \delta\right) \tag{23.119}$$

where

$$2\beta\cosh\delta = \varepsilon - E + \frac{eV}{2}, \quad \Phi_\delta = \delta\cosh\delta - \sinh\delta \tag{23.120}$$

As it follows from Equation (23.119), the quantum numbers of sls-band states ξ_n^{sls} are given by solutions to equation:

$$\xi^{\text{sls}} \equiv \frac{2\beta}{\varepsilon_\varepsilon}\Phi_{\xi'} + \xi' + \frac{\pi}{4} = \pi n, \quad n = 1, 2, \ldots, n^{\text{sls}} - 1, n^{\text{sls}} \tag{23.121}$$

where the number of states in the sls-band is defined as

$$\pi n^{\text{sls}} = \xi^{\text{sls}}\big|_{E=\varepsilon - |2\beta - eV/2|} \tag{23.122}$$

Simple analytical expressions of eigenenergies within the sls-band can be obtained near the bottom, mid, and top of the band.

1. **Levels near the bottom of the lower sls-band.**
   According to its definition (23.111), ξ' is small in an energy interval, where $E - \varepsilon + 2\beta + eV/2 \ll \beta$. Then, $\Phi_{\xi'}$ can be expanded in powers of $\xi' \approx \sqrt{2 + (E - \varepsilon)/\beta + eV/(2\beta)} \ll 1$. The use of such an expansion in Equation (23.121) yields:

$$E_n - \varepsilon + 2\beta + eV/2 = \beta^{1/3}\left[\frac{3}{2}\pi\left(n - \frac{1}{4}\right)\varepsilon_\varepsilon\right]^{2/3} - \varepsilon_\varepsilon \tag{23.123}$$

   which is nothing else but the Airy spectrum (see, e.g., ref. [79]). Similar result can be obtained in the effective mass approximation [80]. It comes out from Equation (23.123), if we set $\beta = \hbar^2/(2m^*a^2)$. Under such replacement, Equation (23.123) gives the free electron spectrum for an infinitely deep triangular potential well.

2. **Levels near the mid of ($eV > E_{\text{bw}}^0/2$) sls-band.**
   For energies satisfying $|\varepsilon - E - eV/2| \ll \beta$ and $eV > E_{\text{bw}}^0/2$, electron levels are given by [72]:

$$\varepsilon - E_m \approx \frac{4\beta}{\pi} + 2\varepsilon_\varepsilon\left(\frac{3}{4} - \left[\frac{2\beta}{\pi\varepsilon_\varepsilon} + \frac{3}{4}\right]_{N/I}\right) \pm 2\varepsilon_\varepsilon m + \frac{eV}{2} \tag{23.124}$$

   where $m = 0, 1, 2, \ldots, m \ll \beta/(2\varepsilon_\varepsilon)$.
   This is a noteworthy feature of the spectrum: In the mid of sls-bands, the level spacing is two times larger than it is in the canonical WSL.

3. **Levels near the top of the full-width sls-band.**
   For $\varepsilon + 2\beta - eV/2 - E \ll \beta$ and $eV > E_{\text{bw}}^0$, it follows from Equation (23.121) that

$$\varepsilon + 2\beta - eV/2 - E = 2\beta - \varepsilon_\varepsilon(n - 5/4) \tag{23.125}$$

   where n is of the order of n^{sls}. Thus, in this energy interval (which belong to the sls-band), the level spacing is equal to:

$$E_{n+1} - E_n = \varepsilon_\varepsilon \tag{23.126}$$

   i.e., it coincides with that is within the WS-band.

Combining this result with Equation (23.118) we can conclude that at high voltages, the level spacing with good accuracy equals field parameter $\varepsilon_{\mathcal{E}}$ in the energy interval that extends from below the top of lower sls-band to above the bottom of the upper sls-band. In other words, the WS-type spectrum actually merges into the sls-bands in the sense that near an arbitrarily chosen level, the neighboring levels form a WS-ladder with negligible corrections.

Further away from the WS-band, the level spacing increases. In the middle of sls-bands, the electron levels are quantized with doubled Wannier quantum $2\varepsilon_{\mathcal{E}}$ while closer to the spectrum edges, quantization of sls-levels behaves similarly to zeros of the Airy function.

It is worthwhile mentioning that peculiar features of the sls-band spectrum, which are manifested by the level spacing equal to $\varepsilon_{\mathcal{E}}$ (near edges of WSL-band) and $2\varepsilon_{\mathcal{E}}$ (in the mid of sls-bands), are characteristic for finite systems. A detailed comparison of sls band spectrum in finite and semi-infinite systems was made in [81].

Band electron levels in the finite-length wire (finite-thickness crystal, etc.) subjected to the constant electric fields are sketched in Figure 23.5. The quantum level structure is compared with semi-classical interpretation of the field effect on the spectrum in terms of the tilted band. Commented above explicit expressions that describe different cases of electron level quantization under the influence of electric field relate, on one hand, the level spacing and respective energy intervals and, on the other hand, the external voltage applied to the system termini.

In what follows, we consider electron transmission through the tilted band that can be associated with molecular wires (and also, with thin crystal layers, superlattices, etc.) imbedded between two conducting leads.

23.3.2   Tunneling through a Tilted Band of Finite Length

Manifestation of electric field effects on coherent electron transport is strongly dependent on the relative position of the molecular levels with respect to the Fermi energy in metal leads at zero voltage, E_F^0. To make the discussion concise and related to conventional heterojunctions, we set E_F^0 below the bottom of zero-field tight-binding band.

23.3.2.1   Energy Diagram of the Contact Region

The energy diagram of the metal–molecule–metal contact under consideration is shown in Figure 23.6. Filled levels of conduction bands of the left and right leads are indicated by shaded rectangles. At zero temperature, and under the bias shown, only electrons from the right hand side lead with energies above the upper filled level in the left hand side lead contribute to the net current. This energy interval is termed further on as the "tunneling window". The lead electrons with energies below the sls-band have to pass a trapezium-shaped potential barrier. Tunneling of such electrons is indicated in Figure 23.6 by TB arrow. Those lead electrons, which have their energies within the sls-band, have to tunnel through the triangular potential barrier (FN arrow in the Figure). This situation is similar to the electron field emission described by Fowler and Nordheim [16] on the basis of Sommerfield's electron theory of metals [82]. In the case of electron transport across the wire, tunnelling is assisted by the molecular states in the triangular well. Finally, if the region of the WS-band has become accessible for electron transmission across molecule, tunneling is assisted by electron states localized inside the molecular chain. In barrier language, electrons have to go through two triangular barrier (WS arrow). Before discussing TB-, FN-, and WS-tunneling probabilities, let us briefly summarize fundamental changes in the molecular spectrum under the influence of constant electric field, as described in Section 23.3.1.

In zero field, all molecular electronic states, from the bottom E_b^0 to the top E_t^0 of the band, have extended- or band-like character. The respective band has label es in Figure 23.6. With an increase of the potential energy difference eV, the width of the es-band narrows down and in addition, two bands of field-induced surface states (sls-bands) appear above and below the es-band. The lower (upper) sls-band and es-band occupies the energy intervals $[E_b^0 - eV/2, E_b^0 + eV/2]$ ($[E_t^0 - eV/2, E_t^0 + eV/2]$) and from $E_b^0 + eV/2$ to $E_t^0 - eV/2$, respectively, so that the total width of the spectrum increases proportionally to

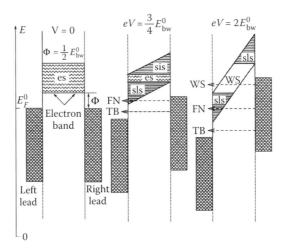

FIGURE 23.6 Band of electron states of $(C)_N$ chain (to the left) and its changes in constant electric field $\mathcal{E}$ when the difference in electrostatic energy between molecule termini eV is less and larger than the band width in zero field E_{bw}^0. The es-, sls-, and WS-bands, and parts of the lead conduction bands filled by electrons are distinguished by shading styles.

the electrostatic energy, $E_{bw}^0 + eV$. Such a gross structure of electron spectrum is valid until eV remains smaller than the zero-field band width E_{bw}^0. At $eV = E_{bw}^0$ the width of es-band is zero. Further increase of the applied voltage results in the appearance of a new band of Wannier–Stark states. The WS-band width grows as $eV - E_{bw}^0$, whereas the sls-band width remains constant and equals E_{bw}^0 (see Table 23.1).

23.3.2.2 Exponential Factor

The preceding qualitative discussion has received a detailed quantitative description that was based on Equation (23.26) and the Green's function G^m specified by the Hamiltonian (23.106). It was shown that under restrictions $N \gg 1$, $\varepsilon_\mathcal{E} = ae\mathcal{E} \ll E_{bw}^0$, and excluding the energy interval of es-band, the transmission coefficient is proportional to an exponential factor. Skipping subtle details concerning resonance structure of the transmission spectrum, which can be found in ref. [73], we concentrate here on analysis of the exponential factor:

$$T \sim \exp(-\kappa N) \tag{23.127}$$

which governs TB-, FN-, and WS-type of tunneling:

$$\kappa = \kappa_{TB}, \quad \text{if} \quad E \leq E_b^0 - eV/2;$$

$$\kappa = \kappa_{FN}, \quad \text{if} \quad \begin{cases} E_b^0 - eV/2 < E \leq E_b^0 + eV/2, & eV \leq E_{bw}^0; \\ E_b^0 - eV/2 < E \leq E_b^0 + E_{bw}^0 - eV/2, & eV > E_{bw}^0; \end{cases} \tag{23.128}$$

$$\kappa = \kappa_{WS}, \quad \text{if} \quad E_b^0 + E_{bw}^0 - eV/2 < E \leq E_b^0 + E_{bw}^0/2.$$

Energy intervals in Equation (23.128) can be easily identified with the help of diagrams in Figure 23.6.

For further discussion, it is instructive to specify the contact model in terms of a potential barrier, introducing the barrier height Φ as it is usually understood in models of semiconductor heterojunctions.[8]

---

[8]Notation Φ has nothing in common with quantities Φ_ξ, $\Phi_{\xi'}$, and Φ_δ used in Section 23.3.1.

Namely, let $E_F^0 + \Phi = E_b^0$, so that Φ has the meaning of the difference between the energy of the es-band bottom and the Fermi level in the absence of external electric field.

In what follows, we restrict ourselves to the case $\Phi < E_{bw}^0$ and consider simplified expressions of κ_{TB}, κ_{FN}, and κ_{WS}, which reasonably reproduce the exact behavior of κ as a function of energy and applied voltage [73]. Electrostatic energy eV is set to be positive. All formulas will be given for $E < \varepsilon$. For $E > \varepsilon$, the dependence of κ on energy is the mirror reflection with respect to the spectrum center $E = \varepsilon$.

1. **Tunneling through trapezoid barrier**.
   For energies $E \leq E_b^0 - eV/2$, the energy and voltage dependence of κ_{TB} can be represented in the form:

   $$\kappa_{TB} = 2\delta \left(1 - \frac{eV}{4\beta\delta^2}\right) \tag{23.129}$$

   where

   $$\beta\delta^2 = \Phi + \frac{eV}{2} - \left(E - E_F^0\right) \tag{23.130}$$

   so that

   $$\kappa_{TB} = \frac{2}{\sqrt{\beta}}\sqrt{\Phi + eV/2 - \left(E - E_F^0\right)} \left[1 - \frac{eV}{4\left[\Phi + eV/2 - \left(E - E_F^0\right)\right]}\right] \tag{23.131}$$

   Equation (23.131) is equivalent to:[9]

   $$\kappa_{TB} = \frac{2}{\sqrt{\beta}}\sqrt{\Phi - \left(E - E_F^0\right)} \tag{23.132}$$

   coinciding with the result (23.63) obtained for the unbiased $(C)_N$ chain.

2. **Tunneling assisted by surface localized states**.
   If $eV \leq \Phi$, one can use $\kappa = \kappa_{TB}$ in Equation (23.127) for all energies within the tunneling window $E_F^0 - eV/2 \leq E \leq E_F^0 + eV/2$. For higher voltages, $\Phi < eV \leq E_{bw}^0 + \Phi$, $\kappa = \kappa_{TB}$ only up to the energy $E = E_b^0 - eV/2$. In the interval $E_b^0 - eV/2 \leq E \leq E_F^0 + eV/2$, we have $\kappa = \kappa_{FN}$, where:

   $$\kappa_{FN} = \frac{4}{3} \frac{\left(E_F^0 + \Phi + eV/2 - E\right)^{3/2}}{\sqrt{\beta}\, eV} \tag{23.133}$$

   Derivation of Equation (23.133) from the tight-binding Hamiltonian is rather complicated [73]. However, it is easy to see, how this result can be obtained from the WKB description of field-emission. In that approximation, the tunneling probability calculated for the field-created triangular barrier is equal to:

   $$T \sim \exp\left(-\frac{4}{3}\frac{\sqrt{2m^*}}{\hbar}\frac{\Phi_E^{3/2}}{e\mathcal{E}}\right) = \exp\left[-\frac{4}{3}\frac{\sqrt{2m^*}}{\hbar}\frac{\left(E_F^0 + \Phi + eV/2 - E\right)^{3/2}}{e\mathcal{E}}\right] \tag{23.134}$$

   where Φ_E is the height of the potential barrier for electron having energy E. In the case of tilted band, this height should be understood as $\Phi_E = E_F^0 + \Phi + eV/2 - E$ (i.e., the top of the lower

---

[9]Slightly better approximation is provided by the WKB exponent: $\kappa_{TB}\kappa_{WKB} = \frac{4}{3eV\sqrt{\beta}}(X_+^{3/2} - X_-^{3/2})$, $X_\pm = \Phi \pm eV/2 - (E - E_F^0)$.

sls-band minus electron energy, see Figures 23.6 and Table 23.1).[10] On the other hand, substitution $\beta = \hbar^2/(2m^*a^2)$ in Equation (23.133) and $\kappa = \kappa_{FN}$ in Equation (23.127) converts the latter into Equation (23.134).

Exponential factor (23.134) is comparable with the Fowler-Nordheim tunneling probability [16]:

$$T_{FN} \sim \exp\left[-\frac{4}{3}\frac{\sqrt{2m^*}}{\hbar}\frac{\left(E_F^0 + \Phi - E\right)^{3/2}}{e\mathcal{E}}\right] \qquad (23.135)$$

only at small voltages $eV \ll \Phi$. As shown in [73], in the through molecular wire transmission factor (23.134) has another meaning: it determines an envelope of the transmission spectrum whose resonance structure is due to the presence of sls-states (see for details the original paper).

3. **Tunneling assisted by Wannier–Stark states.**

In an electric field such that $eV > E_{bw}^0 + \Phi$, the tunneling window overlaps with the WS-band, see Figure 23.6. Hence, through-WS-band tunneling comes into play. In this case, the tunneling exponent in Equation (23.127) has the following dependence on energy: $\kappa = \kappa_{TB}(E)$ for $E_F^0 - eV/2 \leq E \leq E_b^0 - eV/2$; $\kappa = \kappa_{FN}(E)$ for $E_b^0 - eV/2 \leq E \leq E_t^0 - eV/2$; and for energies $E_t^0 - eV/2 < E \leq \varepsilon$, $\kappa = \kappa_{WS}(E)$. Here, the energy intervals are specified for the lower part of the spectrum.

Two kinds of exponential factors can be distinguished in the description of WS tunneling, $T \sim \exp(-\kappa_{WS}^+ N)$ and $T \sim \exp(-\kappa_{WS}^- N)$ [73]. The energy and field dependence of these factors is given by:

$$\kappa_{WS}^{\pm} = \frac{4}{3}\sqrt{\frac{eV}{\beta}}\left[\left(\frac{1}{2} + \frac{E_b^0 - E}{eV}\right)^{3/2} \pm \left(\frac{1}{2} + \frac{E - E_b^0 - E_{bw}^0}{eV}\right)^{3/2}\right] \qquad (23.136)$$

The former factor corresponds to the envelope of the molecular wire transmission spectrum over its minima, whereas the latter envelopes the spectrum maxima. As calculations show, it is the magnitude of κ_{WS}^+ that mostly determines the tunneling current across the wire.

23.3.2.3 Energy/Voltage Dependence of the Tunneling Exponent

For two values of eV (illustrating the cases of low and high voltages) typical behavior of the tunneling exponent as a function of energy is illustrated in Figure 23.7. Bold solid curves represent the accurate calculations [73]; other line styles are used for approximate expressions (23.132), (23.133) and (23.136), see the figure caption for the correspondence. Here are some characteristic features of the dependence of κ on energy and applied voltage which are not apparent from the illustrations.

Function $\kappa_{TB}(V)$ decreases monotonically with an increase of V. In contrast, function $\kappa_{FN}(V)$ exhibits a non-monotonic behavior. For energies $E < E_b^0$, it takes its minimal value:

$$\kappa_{FN}^{min} = \sqrt{\frac{3}{\beta}\left(E_b^0 - E\right)} = \frac{\sqrt{3}}{2}\sqrt{\frac{eV}{\beta}} \qquad (23.137)$$

at $eV = 4(E_b^0 - E)$. For any fixed energy $E > E_b^0$, $\kappa_{FN}(V)$ is monotonically growing function of V, just as it is $\kappa_{WS}^+(V)$. The latter, (see Equation [23.136]), has its minimal value at $E = \varepsilon$

$$\kappa_{WS}^{+min} = \frac{2\sqrt{2}}{3}\sqrt{\frac{eV}{\beta}}\left(1 - \frac{E_{bw}^0}{eV}\right)^{3/2} \qquad (23.138)$$

---

[10]To notice, definition of the top of sls-band at high voltages, see Table 23.1, cannot be used to obtain (23.134). The reason is that the case $eV > E_{bw}^0$ does not have any semi-classical analogy.

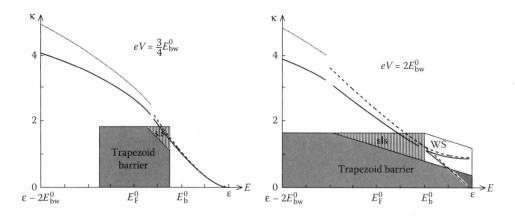

FIGURE 23.7 Dependence of the tunneling exponent κ on energy found in ref. [73] (thick-solid curves). Breaks on the curves correspond to ε_ε-order energy intervals which were excluded to derive accurate analytical expressions of κ. Rectangle to the left and trapezium to the right show which part of the molecular spectrum can be "seen" by electrons having energies within the tunneling window at the indicated voltages (to compare with barrier diagrams in Figure 23.6). Upper and lower curve branches of κ_{WS} correspond to, respectively, κ_{WS}^+ and κ_{WS}^- defined in the text. Approximations are shown for κ_{TB} [Equation (23.132) – thin-solid curve], κ_{FN} [Equation (23.133) – dashed curve], and $\kappa_{WS}^\pm$ [Equation (23.136) – dashed branching curve].

Peculiar field dependence of the tunneling exponent in the region of sls-band has its reflection in the crossing of two curves $\kappa_{FN}(E)$ at different voltages V_1 and V_2: $\kappa_{FN}(E_{cross}, V_1) = \kappa_{FN}(E_{cross}, V_2)$. Let $V_2 > V_1$. Then, for energies $E > E_{cross}$, $\kappa_{FN}(V_2) > \kappa_{FN}(V_1)$ which means that the tunneling probability is *smaller* for *higher* voltages. For energies $E < E_{cross}$, the effect is just the opposite, i.e., tunneling is enhanced by the field. The crossing energy is determined by

$$E_{cross} = E_b^0 - \frac{e}{2}\frac{V_1 V_2}{V_1 + V_2}\left(\sqrt[3]{\frac{V_1}{V_2}} + \sqrt[3]{\frac{V_2}{V_1}} - 1\right) \tag{23.139}$$

indicating that E_{cross} is always smaller than E_b^0.

As already mentioned, exponential factor (23.134) agrees (though does not repeat) with the Fowler–Nordheim result only in the case of very low voltages. For $eV \gg E_F^0 + \Phi - E$, the field dependence changes qualitatively, exhibiting exponentially large suppressing effect on tunneling produced by strong electric fields

$$T \sim \exp(-\kappa_{FN} N) = \exp\left(-\frac{\sqrt{2}}{3}\sqrt{\frac{eV}{\beta}}\,N\right) \tag{23.140}$$

In this limit, tunneling exponent associated with through WS-band tunneling, is two times larger:

$$T \sim \exp(-\kappa_{WS}^+ N) = \exp\left(-\frac{2\sqrt{2}}{3}\sqrt{\frac{eV}{\beta}}\,N\right) \tag{23.141}$$

The above equations contrast both the Fowler–Nordheim (23.135) and the McConnel-type factor (23.62). However, the relevance of these results to real experiments has not been neither confirmed nor denied.

To the best of our knowledge, experimental techniques capable of probing the transmission spectrum have not been applied yet to molecular contacts. To compare predictions of this modeling with I-V measurements, additional effort is required to advance the obtained relations to $I(V)$ curves for concrete molecular contacts. Results of current work in the direction will be reported soon.

23.3.3 Synopsis

We have presented a methodology of analytical modeling of electron transmission and conduction in molecular contacts. The theoretical approach is based on the Green's function technique and tight-binding Hamiltonians describing metal leads and the molecules contacting them. The transmission probability of metal–molecule–metal contact has been presented in a form which is adjusted for easy modeling of major factors (such as characteristics of the contact interfaces, molecule electronic structure, and applied voltage) that influence coherent electron transport in molecular junctions. The principle determiners of the contact properties, the lead and molecule Green's functions are thoroughly defined and exemplified by a number of particular cases that makes the suggested scheme suitable for independent modeling. Some new results have been singled out throughout the discussion.

Presented material is based exclusively on our own findings. By no means do we pretend that this is a complete review of related works in the field. It is one theoretical approach to electron transmission that, despite its limitations, has brought to light some fundamentals of what is current in molecular wires.

Acknowledgments

Authors are indebted to Yuri Klymenko who substantially contributed to the results overviewed in Sections 23.1 and 23.2. Help of Linda Wylie in refining the text is gratefully appreciated.

References

[1] Beenaker CWJ and van Houten H. Quantum transport in semiconductor nanostructures. *Solid State Physics*, 44:1–228, 1991.

[2] Petty MC, Bryce MR, and Bloor D. *Introduction to Molecular Electronics*. Oxford University Press, New York, 1995.

[3] Jortner J and Ratner MA. editors, *Molecular Electronics*. Blackwell, Oxford, 1997.

[4] Ferry DK and Goodnik SM. *Transport in Nanostructure*. Cambridge University Press, Cambridge, 1999.

[5] Nitzan A. Electron transmission through molecules and molecular interfaces. *Annu. Rev. Phys. Chem.*, 52:681–750, 2001.

[6] Reed MA and Lee T. editors, *Molecular Nanoelectronics*. American Scientific Publishers, 2003.

[7] Mujica V and Ratner MA. Molecular conductance junctions: A theory and modeling progress report. *Handbook of Nanoscience, Engineering, and Technology*, 10-1–27, 2003.

[8] Paulsson M, Ferdows Z, and Datta S. Resistance of a Molecule. *Handbook of Nanoscience, Engineering, and Technology*, 10-1–27, 2003.

[9] Joachim C and Ratner MA. Molecular wires: Guiding the super-exchange interactions between two electrodes. *Nanotechnology*, 15:1065–1075, 2004.

[10] Datta S. *Quantum Transport: Atom to Transistor*. Cambridge University Press, Cambridge, 2005.

[11] Cuniberti C, Fagas G, and Richter K. editors, *Introducing Molecular Electronics*. Springer, 2005.

[12] Petrov EG, May V, and Hänggi P. Kinetic rectification of charge transmission through a single molecule. *Phys. Rev. B*, 73:045408-1–12, 2006.

[13] Lippman BA. and Schwinger J. Variational principles for scattering processes. I. *Phys. Rev.*, 79:469–480, 1950;

[14] Löwdin PO. Studies in perturbation theory. IV. Solution of eigenvalue problem by projection operator formalism. *J. Math. Phys.*, 3:969–982, 1962.

[15] Datta S. *Electronic Transport in Mesoscopic Systems*. Cambridge University Press, Cambridge, 1995.

[16] Fowler RH and Nordheim LW. Electron transmission in intense electric fields. *Proc. R. Soc. London*, Ser. A 119:173–181, 1928.

[17] Zener C. A theory of electric breakdown of solid dielectrics. *Proc. R. Soc. London*, Ser. A 145:523–529, 1934.

[18] Wannier GH. Wave functions and effective Hamiltonian for Bloch electrons in an electric field. *Phys. Rev.* 117(2):432–439, 1960.

[19] Landauer R. Spatial variation of currents and fields due to localized scatterers in metallic conduction. *IBM J. Res. Develop.*, 1:223–231, 1957; *Philos. Mag.*, 21:683, 1970.

[20] Büttiker M, Imry Y, Landauer R, and Pinhas S. Generalized many-channel conductance formula with application to small rings. *Phys. Rev. B*, 31(10):6207–6215, 1985;

[21] Büttiker M. Four-terminal phase-coherent conductance. *Phys. Rev. Lett.*, 57(14):1761–1764, 1986.

[22] Imry Y. *Introduction to Mesoscopic Physics*. Oxford University Press, Oxford, 2002.

[23] Datta S. Nanoscale device modeling: the Green's function method. *Superlattices and Microstructures*, 28(4):253–278, 2000.

[24] Tian W, Datta S, Hong S, Reifenberger R, Henderson JI, and Kubiak CP. Conductance spectra of molecular wires. *J. Chem. Phys.*, 109:2874–2882, 1998.

[25] Lang ND and Avouris Ph. Electrical conductance of individual molecules. *Phys. Rev. B*, 64:125323-1–7, 2001.

[26] Emberly EG and Kirczenow G. Multiterminal molecular wire systems: A self-consistent theory and computer simulations of charging and transport. *Phys. Rev. B*, 62:10451–10458, 2000.

[27] Onipko A, Klymenko Yu, and Malysheva L. Conductance of molecular wires: Analytical modeling of connection to leads *Phys. Rev B*, 62:10480–10493, 2000.

[28] Wang W, Lee T and Reed MA. Electron tunneling in self-assembled monolayers. *Rep. Prog. Phys.*, 68:523–544, 2005.

[29] Malysheva LI and Onipko AI. Local state spectrum of end substituted polyenes. *J. Chem. Phys.*, 105:11032–11041, 1996.

[30] Newns DM. Self-consistent model of hydrogen chemisorption. *Phys. Rev.*, 178:1123–1135, 1969.

[31] Desjonquéres MC and Spanjaard D. *Concepts in Surface Physics*. Springer-Verlag, Berlin, 1993.

[32] Kronig RL. On the theory of dispersion of X-rays. *J. Opt. Soc. Am.*, 12:547–555, 1926;

[33] Kramers HA. Some remarks on the theory of absorption and refraction of x-rays. *Nature*, 117:775–778, 1926.

[34] Mujica V, Kemp M, and Ratner M. Electron conduction in molecular wires. I. A scattering formalism. *J. Chem. Phys.*, 101:6849–6855, 1994.

[35] Fisher DS and Lee PA. Relation between conductivity and transmission matrix. *Phys. Rev. B*, 23:6851–6854, 1981.

[36] Onipko A. Analytical model of molecular wire performance: A comparison of π and σ electron systems. *Phys. Rev. B*, 59:9995–10006, 1999.

[37] Rico B and Azbel MY. Physics of resonant tunneling. The one-dimensional double-barrier case. *Phys. Rev. B*, 29:1970–1981, 1984.

[38] McConnell HM. Intramolecular charge transfer in aromatic free radicals. *J. Chem. Phys.*, 35:508–515, 1961.

[39] Bardeen J. Tunnelling from a many-particle point of view. *Phys. Rev. Lett.*, 6:57–59, 1961.

[40] Gottlieb AD and Wesoloski L. Bardeen's tunnelling theory as applied to scanning tunnelling microscopy: a technical guide to the traditional interpretation. *Nanotechnology*, 17:R57–R65, 2006.

[41] Tersoff J and Hamman DR. Theory of the scanning tunneling microscope. *Phys. Rev. B*, 31:805–813, 1985.

[42] Onipko A, Berggren K–F, Klymenko Yu, Malysheva L, Rosink JJWM, Geerligs LJ, van der Drift E, and Radelaar S. Scanning tunneling spectroscopy on π-conjugated phenyl-based oligomers: A simple physical model. *Phys. Rev. B*, 61:11118–111124, 2000.

[43] Wang W, Lee T and Reed MA. Mechanism of electron conduction in self-assembled alkanethiol monolayer devices. *Phys. Rev B*, 68:035416-1–7, 2003.

[44] Hückel E. Quantentheoretische Beitrage zum Benzolproblem. *Z. Phys.*, 70:204–286, 1931.

[45] Onipko A, Klymenko Yu, Malysheva L, and Stafström S. Tunneling across molecular wires: an analytical exactly solvable model. *Solid State Commun.*, 108:555–559, 1998.

[46] Onipko A, Klymenko Yu, and Malysheva L. Green's function of conjugated oligomers: the exact analytical solution with an application to the molecular conductance. *Mater. Sci. Eng. C*, 8-9:281–291, 1999.

[47] Lakatos-Linderberg K, Hemenger RP, and Pearlstein RM. Solutions of master equations and related random walks on quenched linear chains. *J. Chem. Phys.*, 56:4852– 1972.

[48] Karadakov P and Castaño O. A resolvent approach towards the electronic structure of systems constructed of chain fragments. Branched polyenes. *J. Chem. Soc., Faraday Trans. 2*, 78:73–80, 1982.

[49] Ratner MA. Bridge-assisted electron transfer: effective electronic coupling. *J. Phys. Chem.*, 94:4877–4883, 1990.

[50] Evenson JW and Karplus M. Effective coupling in bridged electron transfer molecules: Computational formulation and examples. *J. Chem. Phys.*, 96:5272–5278, 1992.

[51] Onipko A. An extension of the MConnell superexchange formula to the case of real conjugated oligomers. *Chem. Phys. Lett.*, 292:267–272, 1998.

[52] Onipko A and Klymenko Yu. Nonadiabatic electron transfer: exact analytical expression of through-conjugated-bridge effective coupling and its asymptotics and zeros. *J. Phys. Chem. A*, 102:4246–4255, 1998.

[53] Onipko A and Malysheva L, unpublished.

[54] Longuet-Higgins HC and L. Salem L. The alternation of bond lengths in long conjugated chain molecules. *Proc. Roy. Soc. London*, Ser. A 251:172–185, 1959.

[55] Su WP, Schrieffer JR, and Heeger AJ. Soliton excitations in polyacetylene. *Phys. Rev. B*, 22:2099–2111, 1980.

[56] Baughman RH, Kohler BE, Levy IJ, and Spangler C. The crystal structure of trans,trans-1,3,5,7-octatetraene as a model for fully-ordered trans-polyacetylene. *Synth. Met.*, 11:37–52, 1985.

[57] Kohler BE. A simple model for linear polyene electronic structure. *J. Chem. Phys.*, 93:5838–5842, 1990.

[58] Sandorfy C. LCAO MO calculations on saturated hydrocarbons and their substituted derivatives. *Can. J. Chem.*, 33:1337–1351, 1955.

[59] Pople JA and Santry DP. A molecular orbital theory of hydrocarbons. I. Bond delocalization in paraffins. *Mol. Phys.*, 7: 269–295, 1964.

[60] Murrell JL, Kettle SFA, and Clarke TC. *Valence Theory*. Wiley, London, 1965.

[61] Kohler BE, Malysheva LI, and Onipko AI. Molecular orbital coefficients and transition dipoles of real polyenes. *J. Chem. Phys.*, 103:6068–6075, 1995.

[62] Lennard-Jones JE. The electronic structure of some polyenes and aromatic molecules. I—The nature of the links by the method of molecular orbitals. *Proc. Roy. Soc. London*, Ser. A 158:280–296, 1937.

[63] Kouteský J. Quantum Chemistry of crystal surfaces. *Adv. Chem. Phys.*, 9:85–168, 1965.

[64] Malysheva LI and Onipko AI. Exact solution for the Hückel model of heteropolyenes. *Synth. Met.*, 80:11–23, 1996.

[65] Onipko A, Klymenko Yu, and Malysheva L. Analytical one-particle approach to the π-electronic structure of heterocyclic polymers *J. Chem. Phys.*, 107:5032–5050, 1997.

[66] Duke CB. *Tunneling in Solids*. Academic, New York, 1969.

[67] Burstein E and Lundquist S., Eds., *Tunneling Phenomena in Solids*. Plenum Press, New York, 1969.

[68] Franz W. Dielektrischer Durchschlag. *Handbuch der Physik*, Edited by S. Flügge, Springer-Verlag, Berlin, 17:155–263, 1956.

[69] Magoga M and Joachim C. Minimal attenuation for tunneling through a molecular wire. *Phys. Rev. B*, 57(3):1820–1823, 1998.

[70] Joachim C and Magoga M. The effective mass of an electron when tunneling through a molecular wire. *Chem. Phys.*, 281:347–352, 2002.

[71] Onipko A and Malysheva L. Triple-, double-, and fractionally-spaced Wannier-Stark ladders. *Solid State Commun.*, 118:63–67, 2001.

[72] Onipko A and Malysheva L. Noncanonical Wannier-Stark ladders and surface state quantization in finite crystals subjected to a homogeneous electric field. *Phys. Rev. B*, 63:235410-1–10, 2001.

[73] Onipko A and Malysheva L. Signatures of Wannier-Stark and surface states in electron tunneling and related phenomena: Electron transmission through a tilted band. *Phys. Rev. B*, 64:195131–1–14, 2001.

[74] Onipko A and Malysheva L. Tunneling through a tilted tight-binding band. *Molecular Electronics II*, Annals of the New York Academy of Sciences, 960:143-152, 2002.

[75] Saitoh M. Electronic states in a finite linear crystal in an electric field. *J. Phys. C: Solid State Phys.*, 6:3255-3267, 1973.

[76] Stey GC and Gusman G. Wannier-Stark ladders and the energy spectrum of an electron in a finite one dimensional crystal. *J. Phys. C: Solid State Phys.*, 6:650–656, 1973.

[77] Fukuyama H, Bari RA, and Fogedby HC. Tightly bound electrons in a uniform electric field. *Phys. Rev. B*, 8(12):5579–5586, 1973.

[78] Davison SG, English RA, Mišković AL, Goodman FO, Amost AT, and Burrows BL. Recursive green-function study of Wannier-Stark effect in tight-binding systems. *J. Phys.: Condens. Matter*, 9(30):6371–6382, 1997.

[79] Davies JH. *The Physics of Low-Dimensional Semiconductors*. Cambridge University Press, Cambridge, 1998.

[80] Rabinovitch A and Zak J. Electron in crystals in a finite-range electric field. *Phys. Rev. B*, 4(8):2358–2370, 1971.

[81] Malysheva L. Quantization of band states of a crystal layer in a constant electric field. *Ukr. Fiz. Zh.*, 47:1173–1179, 2002.

[82] Sommerfield A and Bethe H. Elektronentheorie der Metalle. *Handbüch der Physik Von Geiger Unf Scheel*, Vol. 24/2. Springer-Verlag, Berlin, 1933.

24

Pride, Prejudice, and Penury of *ab initio* Transport Calculations for Single Molecules

F. Evers

K. Burke

Recent progress in measuring the transport properties of individual molecules has triggered a substantial demand for *ab initio* transport calculations. Even though program packages are commercially available and placed on custom-tailored to address this task, reliable information often is difficult and very time-consuming to attain in the vast majority of cases, namely when the molecular conductance is much smaller than e^2/h. The article recapitulates procedures for molecular transport calculations from the point of view of time-dependent density functional theory. Emphasis is on describing the foundations of the "standard method." Pitfalls will be uncovered and the domain of applicability discussed.

24.1 Introduction

In an impressive sequence of experiments, it has recently been demonstrated that measuring the current voltage (I-V) characteristics of an individual molecule has become feasible [1–6]. Each molecule is an interesting species in itself, exhibiting individual signatures in each IV-curve, such as step positions and heights [7] or inelastic excitation energies [8, 8a]. For this reason, a clear demand for *ab initio* transport calculations of single molecules has emerged in recent years.

Such calculations are a difficult enterprise, because they must simultaneously meet two requirements. Powerful methods exist to deal with each one separately, but the combined problem still is one of the challenging adventures of theoretical physics and quantum chemistry.

Difficulty number one is that a molecule is a genuine many-body system, where the mutual interaction of the particles is important for understanding its properties. These include, in particular, the energy and shape of the (effective) molecular quasiparticle orbitals. Some of the salient aspects, such as the position of the molecule's atoms, the symmetry of molecular orbitals, and their relative energies, are often described accurately by effective single-particle theories, such as density functional theory. For less basic questions, concerning, for instance, excitation energies or details of the electronic charge distribution, polarization, and charging effects, an advanced machinery equipped with methods and codes from quantum chemistry and electronic structure calculations is available.

Difficulty number two is related to the fact that a transport calculation investigates the effect of coupling the molecule to a macroscopic electrode, (i.e., a reservoir with which particles [and energy] can be exchanged). It is the associated broadening of the molecular energy levels that is supposed to be understood quantitatively in transport calculations, and therefore the coupling has to be modeled with great detail and care.

While the first difficulty can be resolved for sufficiently small molecular systems, the second one requires including many electrode atoms, (i.e., a great number of degrees of freedom), in order to properly extrapolate to the macroscopic limit. These conditions are mutually exclusive (almost), and this is the particular challenge in molecular scale transport calculations with *ab initio* methods.

In reality, any *ab initio* transport calculation begins with a compromise accepting strong, often uncontrolled, approximations when dealing with one of the two mentioned difficulties. In the present "standard approach to molecular conductance," [9–11] a drastic simplification on the many-body side is being made. One is accepting the Kohn–Sham energies and orbitals that appear in structure calculations based on ground-state density functional theory (DFT) as the legitimate single particle states for a self-consistent scattering theory of transport. The procedure has the tremendous advantage that including the reservoirs is then a very well-defined process, if sufficient care is taken. For practical purposes, a formulation in terms of non-equilibrium Green's (or "Keldysh") functions is advantageous and therefore often used [12]. For non-interacting particles, the theory is equivalent to a Landauer–Büttiker theory of transport [13,14]. For KS-particles a slight generalization is introduced, in which the electronic charge distribution is calculated self-consistently in the presence of the bias voltage.

Clearly, the use of DFT for the scattering states already includes many non-trivial interaction effects beyond the electrostatic Hartree interaction – Fermi liquid (FL) renormalizations in the language of condensed matter physics – even on the level of local or gradient corrected density approximations (LDA,GGA) [15]. Still, ground and excited states of correlated electron systems are *not* single Slater determinants and therefore the validity of applying scattering theories designed for non-interacting particles to interacting systems is not straightforward to establish. Moreover, including correlation effects beyond FL renormalizations, which can be very important for transport characteristics, like the Coulomb blockade or the Kondo resonance, is not inconceivable in a single determinant theory, but it certainly requires density functionals advanced far beyond LDA. In this respect, it is tempting to use the advanced machinery of quantum chemistry to calculate a better approximation for correlated many-body states including more than one Slater determinant; however, this approach also has a serious drawback because it is limited to relatively small system sizes. The bare molecule appearing in typical transport experiments consists of typically 5 to 10 aromatic rings, which is a size already at the limits of what correlated methods could still reasonably deal with. Including in addition 10 to 200 metal atoms in order to accurately model the coupling to the leads in a controlled way appears to be out of reach at present. For this reason, only very few attempts limited to small molecules have been made in this direction [16].

In this article, we describe three principle approaches to transport calculations based on time-dependent density functional theory (TDDFT). We shall first present a brief account of the basic principle strategies. Then, we explain in more detail one of them, the standard method of *ab initio* transport calculations. In Section 24.2, we discuss an attempt to justify the procedure from the point of view of TDDFT [17–19], list loose ends and apparent conceptual difficulties.

Since the exchange-correlation potential $V_{XC}(\mathbf{x})$ is not known exactly, in any practical calculation approximations like LDA have to be admitted. These are not controlled any more when one deals with realistic molecules. As a consequence, in addition to conceptual problems, appreciable artifacts related to approximate functionals can emerge, which have been well studied for standard DFT applications in quantum chemistry and electronic structure theory, and which carry over to transport calculations as well [20]. A brief list of deficencies most important for transport purposes has been included in Section 24.3.

24.2 TDDFT and Transport

Time-dependent DFT is a well-established generalization of (ground state) density functional theory and has been introduced by Runge and Gross [21] and expanded on by van Leeuwen [22].

RG-Theorem: For any interacting fermion system, there is a unique dual system of *non*-interacting fermionic quasiparticles with the following property: the time-dependent density of original and dual particles is identical for any driving field $V_{ex}(t)$; the time evolution of dual ("Kohn-Schon" or KS) fermions is governed by a Schrödinger-type equation decorated with a Hartree term and an exchange correlation potential $V_{XC}[n]$, which can be expressed as a functional of the time-dependent particle density and its history, $n(\mathbf{x}, t)$. In the general case, $V_{xc}[n]$ depends on the full many-body state at the initial time $t = 0$.

Because the Runge-Gross-Theorem guarantees that the dual system delivers the exact time evolution of the interacting particle density, also longitudinal transport currents can be calculated by exploiting the continuity equation [47]:

$$\dot{n}(\mathbf{x}, t) + \nabla \cdot \mathbf{j}(\mathbf{x}, t) = 0,$$

where a dot denotes a time-derivative. This observation underlies all applications of TDDFT to transport.

Quite generally, transport can be investigated in several different languages, all of which are equivalent in the regime where their validity overlap. Even though we're ultimately interested in the standard method, the TDDFT version of the others will give valuable information, too. Therefore, we shall briefly discuss them as well. We begin, however, by recalling the basic formalism of TDDFT.

24.2.1 TDDFT Formalism

TDDFT is a machinery for propagating a density in time, not a many-body wavefunction. Hence, as a prerequisite for applying the method an initial density ($t = 0$) is required. It needs to be represented as a single Slater determinant $|0\rangle$ constructed from a (complete) set of effective single particle states ϕ_m. This is always possible, if at $t < 0$ the system is in its ground state, then the KS-orbitals of ground state DFT are obvious candidates for ϕ_m. In this case, one has for the density matrix at $t = t' = 0$

$$n(\mathbf{x}, \mathbf{x}') = \sum_{m}^{\text{occ.}} \phi_m^*(\mathbf{x})\phi_m(\mathbf{x}') \tag{24.1}$$

Time evolution of the state $|0\rangle$ together with its density matrix is mediated via H_s:

$$H_s = -\frac{1}{2m} \int d\mathbf{x}\, \psi^\dagger(\mathbf{x})\Delta\psi(\mathbf{x}) + V_H[n(t)] + V_{XC}[n(t)] + V_{ex}(t) \tag{24.2}$$

with

$$V_H[n(t)] = \frac{1}{2} \int d\mathbf{x}\, v_H[n](\mathbf{x}, t)\, \psi^\dagger(\mathbf{x})\psi(\mathbf{x}) \tag{24.3}$$

where $v_H(\mathbf{x}, t) = \int d\mathbf{x}'\, n(\mathbf{x}'; t)/|\mathbf{x} - \mathbf{x}'|$ and

$$V_{XC}[n(t)] = \int d\mathbf{x}\, d\mathbf{x}'\, v_{xc}[n](\mathbf{x}, t)\, \psi^\dagger(\mathbf{x})\psi(\mathbf{x}). \tag{24.4}$$

The nontrivial aspects originate from the fact that the orbitals $\phi_m(t)$ are not eigenstates of H_s at $t > 0$.

H_s can be explicitly time-dependent in the probing potential, $V_{ex}(t)$. An implicit dynamics exists via the Hartree and exchange-correlation terms, that depend on the time-dependent particle density $n(\mathbf{x}, t)$. A few remarks about the potential $v_{xc}[n]$ debuting here are in place.

(a) The exchange correlation potential $v_{xc}[n]$ is not just a density functional. Its precise definition requires specification of the initial many body state at $t = 0$. But, for an initial non-degenerate ground state the dependence on the initial wavefunction is replaced by the initial density, thank to the Hohenberg–Kohn theorem.

(b) TDDFT strictly applies only to *finite* systems, and a generalization that uses the current as the basic variable is needed for infinite systems.

(c) In practice, it might be advantageous for the construction of useful approximations to allow for a more general, offdiagonal structure of $v_{xc}[n](\mathbf{x}, \mathbf{x}'; t, t')$ that could also include (time-dependent) gauge fields. Similarly, one can also consider $v_{xc}[n]$ as a functional of the full density operator $n(\mathbf{x}, \mathbf{x}'; t, t')$, rather than only its diagonal elements, which is the particle density. Thus, additional observables, like the current density, are introduced into the Hamiltonian. In statistical mechanics this is a standard recipe in order to eliminate a history dependence in kinetic equations [23], and here it serves exactly the same purpose [24].

Structure of XC potential if V_{ex} is weak

In this subsection, we analyze the non-equilibrium piece of the $v_{xc}[n]$ potential. The idea is to exploit the fact, that it can be related to known correlation functions if the probing potential V_{ex} is weak.

We begin by recalling some basic facts of the theory of linear response for TDDFT. Since the density evolution of the dual and original system, $n(\mathbf{x}, t)$, coincide, they exhibit in particular the same susceptibility for the density, $\chi(\mathbf{x}, \mathbf{x}', t - t')$, which describes the linear response to $V_{ex}(t)$. What is usually measured is not the response to a probing but rather to the total electric field, which is the sum of external and induced (screening) fields. An important example is the linear conductance. It is the ratio of the current and the measured (total) electrostatic voltage drop at the resistor: I / V_{bias}.

The corresponding response function is the Hartree-irreducible correlator:

$$\chi_{irr}^{-1} \equiv \chi^{-1} + f_H$$

$[f_H(\mathbf{x}, \mathbf{x}') = 1/|\mathbf{x} - \mathbf{x}'|]$. In TDDFT the operator χ_{irr} can be decomposed even further. Namely, the TDDFT Hamiltonian H_s has two pieces that react to density modifications. In addition to V_H incorporating the electrostatic screening, there is also an induced effect on V_{XC}:

$$\chi_{irr}^{-1} = \chi_{KS}^{-1} - f_{xc} \tag{24.5}$$

which can be split off in the same manner as f_H. The truncated correlator χ_{KS} describes the (bare) response of ground state DFT. The promised connection between the correlator χ_{irr} and $v_{xc}[n]$ is mediated via the XC kernel:

$$f_{xc}(\mathbf{x}, \mathbf{x}'; t - t') = \delta v_{xc}[n](\mathbf{x}, t) / \delta n(\mathbf{x}', t'). \tag{24.6}$$

Relations (24.5, 24.6) are very useful, because due to a beautiful series of works by Kohn, Vignale, and collaborators there is a simple approximation to the non-equilibrium contribution to χ_{irr} [24,25]. In fact, these authors reveal the full hydrodynamic structure of χ_{irr}^{-1} by exploiting the relation to the phenomenological theory of quantum liquids.

In their analysis, f_{xc} is the sum of two very different pieces:

$$f_{xc} = f_{xc}^{adia} + f_{xc}^{non-eq} \tag{24.7}$$

where $f_{xc}^{adia} = \delta v_{xc}^{gs}[n]/\delta n$ and a gradient expansion of f_{xc}^{non-eq} is given by f_{VK}. The first term is fully analogous to f_H and it describes the exchange-correlation screening of the ground state DFT Hamiltonian to the probing field. Only in the second piece do many-body effects of a genuinely non-equilibrium nature appear: it incorporates the visco-elastic response of the electronic quantum liquid. Since the emerging term is local and dissipative, the corresponding forces are not conservative. This means, that f_{VK} cannot be incorporated as a pure density coupling – i.e., a potential term – in H_s, but gives rise to a (time-dependent) gauge field instead, and is only described as a current-dependent kernel.

24.2.2   External Driving Field

In this section, we return to the transport problem in TDDFT. To investigate transport currents generated by an external driving field, one needs to supplement H_s with an inhomogenous electric potential, $V_{ext}(q, \omega)$, together with electrodes [51]. *Practical* applications with this approach in TDDFT suffer from the fact, that before one can perform the dc-limit ($\omega \to 0$), one first has to take the thermodynamic limit ($q \to 0$) of infinite system size. In realistic calculations, this is usually a very cumbersome excercise.

From a *conceptual* point of view, thinking about currents as generated by weak external driving fields is rewarding, however, because one is led back to the Kubo formula and the theory of linear response. We thus can directly apply results from the preceding section and in particular investigate the effect of visco-elasticity on the current flow.

24.2.2.1   Kubo Formula

The Kubo formula provides an exact relation between the current density and the driving external and induced (effective) electric fields, that appear in the TDDFT calculation [26]:

$$j(\mathbf{x}, \omega) = \int d\mathbf{x}' \, \sigma_{KS}(\mathbf{x}, \mathbf{x}', \omega) \left[E_{ex} + E_H + E_{xc} \right] (\mathbf{x}', \omega) \tag{24.8}$$

where σ is the nonlocal conductivity tensor and where the electric fields derive from the potential terms given in Equation (24.2). As always, susceptibilities for particle densities χ_{KS} and currents σ_{KS} are related via the continuity equation:

$$\partial_t \chi_{KS}(\mathbf{x}, \mathbf{x}', t) = - \sum_{i,j=1}^{3} \nabla_i \nabla_j \sigma_{KS,ij}(\mathbf{x}, \mathbf{x}', t) \tag{24.9}$$

The structure analysis 24.2.1 suggests to split the full current density into a bare response, j_0, and a remaining piece, j_{VK}:

$$j = j_0 + j_{VK} \tag{24.10}$$

We discuss the second term first. It is driven by a force field, E_{VK}:

$$j_{VK}(\mathbf{x}, \omega) = \int d\mathbf{x}' \sigma_{KS}(\mathbf{x}, \mathbf{x}', \omega) E_{VK}(\mathbf{x}', \omega) \tag{24.11}$$

that is associated with f_{VK}. In the hydrodynamic limit considered by Vignale et al. [25] it has an interpretation as visco-elastic force internal to the electron liquid, that can be described by the stress tensor, ς_{ij} [unperturbed density: $n_0(\mathbf{x})$]:

$$E_{VK,i}(\mathbf{x}, \omega) = n_0(\mathbf{x}')^{-1} \sum_{k=1}^{3} \nabla_k \varsigma_{ik}(\mathbf{x}', \omega). \tag{24.12}$$

The relative magnitude of j_{VK} is small compared to j_0 since the viscosity, η, of the electron liquid is quite small. Based on the homogenous value of η in two and three spatial dimensions a rough analytical consideration can give an estimate [26]. The ratio j_{VK}/j_0 is expected to be typically of the order of 10% or less. In a numerical study, the viscous corrections to the conductance of the benzene-dithiol molecule have been explicitly calculated. The effect is small, roughly 5%, as expected [27].

Note, however, that the previous conclusion is to be taken with a grain of salt. Strictly speaking, the explicit derivation of Equation (24.12) assumes, that the inhomogeneities, which provide the "surface" for the viscous friction to appear, are very smooth: in the period ω^{-1} of the probing field, the electron should travel a distance not larger than the typical spatial scale ℓ on which the inhomogenuous background changes, $v_F/\omega \ll \ell$. Applying this condition to molecules and assuming $k_F \ell \sim 1$, one would get $v_F k_F \ll \ell \omega$, which is satisfied only at optical frequencies of the order of eV, but not in the *dc* limit. Nevertheless, the qualitative finding of the mentioned estimates – namely that viscosity effects tend to be small – should be indicative, because the (transverse) momentum exchange between electrons at low temperatures is a rare process due to phase space constraints. We believe that this basic principle pertains to electrons in a molecule as well [52].

We thus propose that the dominating contribution in Equation (24.10) is given by the response to the reactive forces,

$$j_0(\mathbf{x}) = \int d\mathbf{x}' \, \sigma_{KS}(\mathbf{x}, \mathbf{x}') \left[E_{ex} + E_H - \nabla_{\mathbf{x}'} v_{xc}^{gs} \right] (\mathbf{x}') \tag{24.13}$$

where our notation suppresses the ω-dependence.

24.2.2.2 Bias Voltage and Kohn–Sham Voltage Drop

The expression (24.13) for the current density can be simplified, if we assume that the dependence of the forces on the coordinate, $\mathbf{x}_\perp$, perpendicular to the current path, z, is negligibly weak. This assumption is not necessarily a good one for quantitative questions, but it allows us to discuss more clearly the difference between σ_{KS} and the conductivity, σ_{irr}, measured in typical transport experiments:

$$j(\mathbf{x}) = \int d\mathbf{x}' \, \sigma_{irr}(\mathbf{x}, \mathbf{x}') \left[E_{ex} + E_H \right] (\mathbf{x}'). \tag{24.14}$$

If we make the proposed step and neglect the $\mathbf{x}_\perp$ dependence of the forces, we can integrate both sides of (24.14) over any cross-section, and obtain

$$I = G \, V_{bias} \tag{24.15}$$

with a conductance

$$G = \int d\mathbf{x}_\perp d\mathbf{x}'_\perp \sigma_{irr}(\mathbf{x}_\perp, \mathbf{x}'_\perp; z, z') \tag{24.16}$$

Due to particle number conservation, the cross-sectional integrals render the sum independent of z, z' in the *dc*-limit, $\omega \to 0$. As usual, the bias voltage is given by

$$V_{bias} = \int_{\mathbf{x}_l}^{\mathbf{x}_r} d\mathbf{s} \, [E_{ex} + E_H[n]] (\mathbf{s}). \tag{24.17}$$

where $\mathbf{s}(t)$ is any path connecting the left with the right-hand side of the molecule. V_{bias} should be picked up between points $\mathbf{x}_l$ and $\mathbf{x}_r$ sufficiently far away from the scattering region, in the near asymptotics, where the electrostatic potential energy surface has turned constant.

The same procedure can be repeated also for Equation (24.13)

$$I = G_{KS} \, V_{KS} \tag{24.18}$$

with an expression for G_{KS} completely analogous to (24.16). We have introduced the KS-voltage drop, V_{KS}, given by the sum of *all* KS-forces, Equation (24.13), along $\mathbf{s}(t)$:

$$V_{KS} = V_{bias} + V_{xc}; \qquad V_{xc} = -\int_{\mathbf{x}_l}^{\mathbf{x}_r} d\mathbf{s} \, E_{xc}(\mathbf{s}) \tag{24.19}$$

In the spirit of Equation (24.10), we can decompose the deviation, V_{xc}, of the measured bias and V_{KS} into two pieces

$$V_{xc} = \int_{\mathbf{x}_l}^{\mathbf{x}_r} d\mathbf{s}\, \nabla_s v_{xc}^{gs}[n](\mathbf{s}) - \int_{\mathbf{x}_l}^{\mathbf{x}_r} d\mathbf{s}\, E_{VK}(\mathbf{s}). \tag{24.20}$$

The first term can be integrated trivially: $v_{xc}^{gs}[n](\mathbf{x}_l) - v_{xc}^{gs}[n](\mathbf{x}_r)$. If the left- and right-hand side electrodes consist of the same material, then this difference can be non-vanishing only due to long-range terms in v_{KS}^{gs}. In particular, local approximations like LDA or GGA cannot give a finite contribution in the XC part. The second term in Equation (24.20) describes the genuine non-equilbrium forces, that result from the viscosity of the elctron liquid discussed earlier above.

Equations (24.18) and (24.19) demonstrate, that a KS-particle behaves under bias very differently from a physical quasiparticle. The only long-range forces, that the physical particle realizes upon applying V_{ex} are of the pure Coulomb type. For this reason, the bias voltage must be exactly equal to the difference in electro-chemical potentials:

$$V_{bias} = \mu_{\mathcal{L}} - \mu_{\mathcal{R}}$$

Interaction terms (beyond the Hartree-level) do not occur. By contrast, the KS-particle experiences an effective voltage V_{KS}, that can be quite different from V_{bias}. This way of including interaction effects by adding the corrective term V_{xc} to the voltage drop is not very physical. A difficulty appears, that comes back on us in the next two subsections.

24.2.3   Initial Value Problem

The second access to transport investigates an initial value (so called "relaxation") problem without any reference to a driving external field V_{ex}. One considers a molecule and two reservoirs, left and right ($\mathcal{L}, \mathcal{R}$). At $t < 0$, the molecule is coupled to, and in equilibrium with, $\mathcal{L}$. At $t = 0$, a coupling to $\mathcal{R}$ is being switched on and the time evolution with H_s begins, as described in Section 24.2.1. A current, I, starts to flow at $t > 0$, if $\mathcal{L}$ and $\mathcal{R}$ are not in equilibrium with one another. I is related to the time derivative of the number of particles in the electrodes, $N_{\mathcal{L},\mathcal{R}}$:

$$I(t) = -\dot{N}_{\mathcal{L}} = \dot{N}_{\mathcal{R}} \tag{24.21}$$

As in the previous case, the current may be obtained via the continuity equation from an explicit TDDFT propagation of the electronic density. Since relaxation involves processes on all time scales, in principle the response function $\chi(\omega)$ can be extracted at all frequencies larger than the inverse observation time.

Also, the drawbacks of this approach are similar to the previous case: one has to include big reservoirs and to propagate many time steps if long time, low frequency properties, such as steady-state currents, are to be addressed. Due to this difficulty, the relaxation method with TDDFT has been applied mainly to obtain the high frequency response. In an incarnation where a step potential is switched on at $t = 0$, it has also been used for transport studies in non-interacting model systems, but not for a full-fledged realistic TDDFT calculation, yet [28]. However, we mention that an application of the method for model studies of strongly correlated transport in interacting Hubbard chains has recently been very successful using the density renormalization group method [29].

For our purpose, the formulation of transport in terms of an initial value problem is conceptually important, because it allows us to link time propagation of the density (TDDFT) with DFT-scattering theory.

Indeed, let us perform the following "Gedanken experiment" in which we allow ourselves to work with perfect reservoirs, and where we can do the time propagation of the density up to any time we wish. At the initial stages, $0 < t \ll t_{trans}$, we shall encounter transient phenomena, which render the particle density near the contact region time-dependent. Only at a much later stage, $t \gg t_{trans}$ do we arrive at the asymptotic non-equilibrium situation.

In order for the usual scattering formalism to be applicable, the asymptotic current carrying state of TDDFT, $|QS\rangle$, should meet the following conditions:

c1: At zero temperature, $|QS\rangle$ is a single Slater determinant of left and right moving scattering states, $\psi_{l,r}$, which are eigenstates of the asymptotic TDDFT Hamiltonian H_{qs}. The associated quasi-static KS-density matrix (definition see Equation [24.22]) is invariant under time translations: $n_{qs}(\mathbf{x}, \mathbf{x}'; t - t')$.

c2: The potential $v_{xc}[n]$ takes an asymptotic form which is independent of the history.

c3: The KS-scattering states are occupied according to Fermi-Dirac distributions, $f_{\mathcal{L},\mathcal{R}}$, carrying the temperature and chemical potential of the reservoirs that they emanate from.

Discussion: Very little is known about the true nature of the non-equilibrium state of the interacting electron liquid. For this reason, our discussion can be no other but very qualitative.

Condition **c1** puts a strong requirement on the physical relaxation process. Because the equal time density matrix $n_{qs}(\mathbf{x}, \mathbf{x}'; 0)$ is time-independent at $t \gg t_{diss}$, particle and current densities must have become stationary:

R: After transient phenomena have died out, at $t \gg t_{trans}$ a quasi-stationary non-equilibrium state is reached. "Quasi-stationarity" in this context is meant in the strong sense, in which the time evolution of the particle and current densities have come to a standstill.

It is plausible, that this requirement is always fullfilled in the linear regime of small voltages $\mu_{\mathcal{L}} - \mu_{\mathcal{R}}$ [53]. For non-interacting particles, this is certainly true also in the non-linear case [54]. The situation is much less clear for interacting particles in the non-linear voltage range. In fact, due to the non-linear nature of the kinetic equations one suspects that phenomena like turbulence should occur [30]. This would imply densities and currents fluctuating in time even at $t \gg t_{trans}$, so that **R** is not strictly satisfied [55].

If indeed the quantum liquid goes turbulent, then part of the memory of the intial conditions is never lost. This is, because even small microscopic details in the initial values of the relevant kinetic fields (particle densities, currents, etc.) will in general invoke different dynamics at later times. For this reason, a validity of **c2** is not guaranteed. However, usually there is no interest in a precise set of initial conditions. Possibly, a suitable averaging procedure (either over initial conditions or over a small time interval) could reestablish **c2** in an effective sense.

Returning to the case of small biases, let us emphasize that even if $|QS\rangle$ is a single Slater determinant, validity of **c3** is not automatically guaranteed. In order to see this, we imagine a surplus of particles in one reservoir ($T = 0$), so that an electro-chemical potential difference maintains a particle flow to the other reservoir. If it is correct, that all current is carried by the scattering states in an energy window situated between $\mu_{\mathcal{L}}$ and $\mu_{\mathcal{R}}$, then we find that the current linear in the voltage is necessarily given by $G_{KS}(\mu_{\mathcal{L}} - \mu_{\mathcal{R}})$. The correction term V_{xc} is missing, so we arrive at a statement contradicting Equation (24.18). This is the difficulty, already alluded to at the end of the last subsection. We will come back to it again in the following subsection.

24.2.4 Scattering Approach

It is *the* advantage of scattering theory that all information is encoded in scattering states and no reference to time propagation is being made. The idea is to replace the initial value problem by an equivalent boundary problem. This is the philosophy adopted by the standard method of molecular transport calculations. Its persuasive, charming aspects are: (i) the method is stationary and (ii) the reservoirs are relatively easy to include.

The most important problematic aspect is that up to now a rigorous justification of the approach has not been given, and it is not obvious it should exist. The validity of the scattering formalism has been rigorously established only for non-interacting particles. Whether the conditions **c1**-**c3** formulated in the preceding section are really met, so that the treatment is legitimate also for KS-particles, is not fully clear at present. If they are taken for granted, the following self-consistency procedure can be justified, which in essence is the standard method.

One starts with a guess for the equal time density operator $n_{qs}(\mathbf{x}, \mathbf{x}')$. Condition **c2** ensures that nothing more is needed in order to construct a first approximation for the Hamiltonian H_{qs} – provided the functional $v_{xc}[n]$ is given, of course.

In order to start the next iteration, one should know how to construct a better guess for the density operator from H_{qs}. This is where **c3** and again **c1** kick in: according to **c1**, scattering states can be found as the eigenstates of H_{qs}, which then can be filled up successively in order to obtain $|QS\rangle$ and the improved density matrix. The procedure is completely analogous to the case of non-interacting particles:

$$n_{qs}(\mathbf{x}, \mathbf{x}') = \sum_l f_{\mathcal{R}}(\epsilon_l) \, \psi_l^*(\mathbf{x}) \psi_l(\mathbf{x}')$$
$$+ \sum_r f_{\mathcal{L}}(\epsilon_r) \, \psi_r^*(\mathbf{x}) \psi_r(\mathbf{x}') \tag{24.22}$$

At the end of the iteration cycle, self-consistency is reached. This means that scattering states have been found in an effective potential that incorporates already the shifts in the charge distribution characteristic of the non-equilibrium situation that also generates the current.

The bottom line is that under the assumptions **c1-c3**, there are only two minor differences between a ground state calculation and a standard *dc*-transport calculation:

1. A non-equilibrium density operator calculated from Equation (24.22) replaces the ground state expression Equation (24.1) (to be obtained with $f_{\mathcal{L}} = f_{\mathcal{R}}$).
2. In principle, a quasi-stationary functional should replace the ground state functional. In actuality, due to lack of any better choice, common ground state functionals are employed leading to additional artifacts (see Section 24.3).

24.2.5 Standard Method and Kubo Formula: Discrepancies

The calculation of the *dc*-current can proceed directly from Equation (24.22) which leads to a Landauer–Buttiker type description:

$$I = \int dE \, T(E, \mu_{\mathcal{L}}, \mu_{\mathcal{R}}) [f_{\mathcal{L}} - f_{\mathcal{R}}] \tag{24.23}$$

where the transmission function, T, has been introduced. The kernel $T(E, \mu_{\mathcal{L}}, \mu_{\mathcal{R}})$ can be expressed by the resolvent operator

$$G(E) = (E - H_{qs} - \Sigma_{\mathcal{L}} - \Sigma_{\mathcal{R}})^{-1}$$

and self-energies $\Sigma_{\mathcal{L},\mathcal{R}}$, which represent the boundary conditions at the surface of the electrodes [31]. One has

$$T(E, \mu_{\mathcal{L}}, \mu_{\mathcal{R}}) = \text{tr} \, \Gamma_{\mathcal{L}} G \Gamma_{\mathcal{R}} G^{\dagger} \tag{24.24}$$

where $\Gamma_{\mathcal{L},\mathcal{R}} = i(\Sigma - \Sigma^{\dagger})_{\mathcal{L},\mathcal{R}}$ [17]. For non-interacting particles, Equations (24.23, 24.24) are equivalent to the Landauer formula and give the exact current [9]. The Landauer conductance, $G_{qs} = T(\epsilon_F)e^2/h$, is the response of the current linear in the applied electro-chemical potential difference

$$I = G_{qs}(\mu_{\mathcal{L}} - \mu_{\mathcal{R}}) \tag{24.25}$$

(We choose the nomenclature in the spirit of Section 24.2.2.2.) For non-interacting particles, rigorous results exist, which show that the Landauer conductance and the conductance obtained from the

Kubo-conductivity, σ_{irr} coincide [32]. This is true, because in the absence of interactions

$$G = G_{\text{KS}} = G_{\text{qs}}. \tag{24.26}$$

However, the KS-particles are *not* truely non-interacting. This expresses itself, in the fact that $G \neq G_{\text{KS}}$, in general, because $V_{\text{XC}} \neq 0$. In the present standard approach, one has $G_{\text{qs}} = G_{\text{KS}}$. Therefore, a term in the current proportional to V_{XC} is ignored. In order to include this term, one would have to manipulate the XC-functional used in H_{qs} such that the *bare dc*-current response becomes χ_{irr} rather than χ_{KS}. Note, that it is known that even within LDA the terms ignored can be important. Including them on the level of "adiabatic" LDA can shift resonances and therefore have a substantial impact on transmission characteristics [33].

24.3   Ground State DFT: Artifacts of Local Density Functionals

In this section, we discuss some well-known limitations of common density functional approximations, and their implications for transport calculations [46].

By common density functional approximations, we will mean the original local density approximation of Kohn and Sham [37], the generalized gradient approximation, employing both the density and its gradient, e.g., PBE [38], and hybrids of GGA with exact exchange, such as PBE0 [39] and B3LYP [40–42].

These approximations have a variety of related deficiencies. The first is that they fail for one-electron systems, in which the exchange energy should exactly cancel the Hartree, while the correlation energy should vanish. The preceding density functionals generally fail this requirement, and are said to have a *self-interaction* error, meaning that the electron is incorrectly interacting with itself [45].

A related difficulty is that the ground-state KS potential in such approximations is poorly behaved. For a neutral system, the exact KS potential decays as $-1/r$ for large distances. But with these approximations, the potential decays too rapidly, in fact exponentially with distance, due to the local dependence on the density. (Hybrid functionals do have a fraction of $-1/r$, but not the right amount.) This leads to potentials that are far too shallow overall, and HOMO's (highest occupied molecular orbitals) that are insufficiently deep.

The exact KS HOMO can be proven to be equal to the negative of the ionization potential, but this is not even roughly true for approximate KS potentials, for reasons given earlier. Thus, the charge density in tail regions, i.e., where the density is low, is inaccurate. Furthermore, whenever a localized system is in weak contact with a reservoir, so that the average particle number on the system can be continuous, the exact KS potential jumps by a (spatially) constant amount whenever the particle number passes through an integer [43,44]. This behavior is entirely missed by the preceding approximations, which smoothly interpolate between either side of this discontinuity.

These difficulties are either largely or totally overcome by the use of orbital-dependent functionals. The first popular one of these was SIC-LDA, the self-interaction corrected local density approximation, as introduced by Perdew and Zunger [45]. These days, many codes have been developed to handle orbital-dependent functionals and to find the corresponding KS potential, via the optimized potential method (OPM), a.k.a., optimized effective potential (OEP). Exact exchange potentials have the correct decay, their HOMO's are close to the negative of the ionization potential, and they jump discontinuously at integer particle numbers [50].

How does all this affect transport calculations? There are two principal effects, one obvious, the other less so.

In the first, since transport is often a weak tunneling process, the position of the molecular levels relative to the leads greatly affects the calculated current. If a molecule is weakly coupled to the leads, there is every reason to think that standard functional approximations will make huge errors in the calculation of currents. The levels will be misaligned not only in the equilibrium situation, but will respond completely wrong to the transfer of charge into a localized molecular orbital [48].

These deficiencies in common ground state functionals may be the reason why present DFT-calculations fail to correctly reproduce elementary ground state properties that manifest themselves in the transport characteristics. The most prominent example is the Coulomb blockade phenomenon, which usually is not reproduced quantitatively [34]. It is less well appreciated, that also the Kondo-effect belongs to this category of phenomena. Its manifestation is an extra resonance in the spectral function of the molecule at ϵ_F, the *Abrikosov–Suhl resonance*. In principle, this resonance is a ground state property. It affects the total charge on the molecule and for this reason it should be detectable with DFT, provided an appropriate (so far unknown) functional is used.

Furthermore, density functional approximations will artificially smear out a sharp resonance into a much weaker peak, spread out sometimes over several eV, between the LUMO of the uncharged molecule and the HOMO of the charged molecule [26]. This effect has already been demonstrated in calculations on simple models [48], and has recently been seen in a full OEP calculation [49]. Of course, for molecules that are chemically bonded to the leads, there is no region of very low density, and the potential should be reasonably accurate within the common approximations. So another question is: How big an effect is this in real experiments?

The second effect is more subtle, but could be more important in the chemically bonded situation, perhaps. The standard approximations, being local in nature, yield no XC correction to the potential drop across the molecule. In the language of the previous section, V_{KS} is identical to the real electrostatic potential drop, V_{bias}. But there is no reason this should be true in reality, or in a more accurate calculation. Thus an exchange calculation should produce a finite effect. As mentioned earlier, a current-dependent approximation (VK) indeed produces a small, but finite, effect. Such calculations, performed self-consistently and at finite bias, need to account for this drop, and correct the Landauer formula to account for it. This means that the conductance is *not* just proportional to the transmission through the self-consistent KS potential. In one spatial dimension this means, that the current must be calculated using the total potential drop, including the XC contribution, and this must be divided by the electrostatic potential drop. In this case, it is hard to see how a simple single-particle effective potential could produce the exact conductance [26].

24.4   Conclusion

The various issues that we have discussed – validity of a scattering approach, neglect of V_{XC}, deficiencies of common density functional approximations – raise serious doubts about the accuracy of the present standard method for transport calculations. How quantitatively significant these errors are is only poorly understood at the moment. For such an estimate, time and orbital-dependent calculations are an important tool [28]. To gain further insight, it is important to also go beyond (TD)DFT. Work in this direction is in progress. Proposals include approaches based on a configuration interaction [16], the GW-method [35], and the LDA+U formalism [36]. All of the proposed directions have their virtues and drawbacks. It remains to be seen which one of them turns out to be most suitable to deliver conductances of real, large molecules that require a controlled handling of electrode effects.

24.5   Acknowledgments

FE expresses his gratitude to P. Schmitteckert, P. Wölfle, and E. K. U. Gross for valuable discussions. Also, we thank V. Meded for useful comments on the manuscript, as well as the Center for Functional Nanostructures at Karlsruhe University and the DOE under grant DE-FG02-01ER45928 for their financial support (FE).

References

[1] D. Djukic, K. S. Thygesen, C. Untiedt, R. H. M. Smit, K. W. Jacobsen, and J. M. van Ruitenbeek, *Phys. Rev. B* **71**, 161402 (2005).

[2] S. Kubatkin, A. Danilov, M. Hjort, J. Cornil, J.-L. Brédas, N. Stuhr-Hansen, P. Hedegård, and T. Bjørnholm, *Nature* **425**, 698 (2003).

[3] Z. Li, B. Han, G. Meszaros, I. Pobelov, and T. Wandlowsky, Faraday discussions (2005).

[3a] L. Venkataraman, J. E. Klare, C. Nuckolls, M. S. Hybertsen, and M. L. Steigerwald, *Nature Lett.* 442, 904 (2006).

[4] E. Loertscher, J. W. Ciszek, J. Tour, and H. Riel, *Small* **2**, 973 (2006).

[4a] Z. K. Keane and D. Natelson, *Nano Lett.* 6, 1518 (2006).

[5] M.-H. Jo, J. E. Grose, K. Baheti, M. M. Deshmukh, J. J. Sokol, E. M. Rubmberger, D. N. Hendrickson, J. R. Long, H. Park, and D. C. Ralph, *Nano Lett.* **6**, 2014 (2006).

[6] J. Reichert, R. Ochs, D. Beckmann, H. Weber, M. Mayor, and H. von Löhneysen, *Phys. Rev. Lett.* **88**, 176804 (2002).

[7] M. Elbing, R. Ochs, M. Koentopp, M. Fischer, C. von Hänisch, F. Weigend, F. Evers, H. B. Weber, and M. Mayor, *Proc. Natl. Acad. Sci. U.S.* **102**, 8815 (2005).

[8] H. B. Heersche, Z. de Groot, J. A. Folk, H. S. J. van der Zant, C. Romeike, M. R. Wegewijs, L. Zobbi, D. Barreca, E. Tondello, and A. Cornia, *Phys. Rev. Lett.* **96**, 206801 (2006).

[9] M. Brandbyge, J.-L. Mozos, P. Ordejon, J. Taylor, and K. Stokbro, PRB **65**, 165401 (2002).

[10] Y. Xue, S. Datta, and M. Ratner, *J. Chem. Phys.* **115** (2001).

[11] M. Paulsson, F. Zahid, and S. Datta, *Handbook of Nanoscience, Engineering and Technology* (CRC Press, 2002), cond-mat/0208183.

[12] L. P. Kadanoff and G. Baym, *Quantum Statistical Mechanics* (Addison-Wesley, 1962).

[13] R. Landauer, *IBM J. Res. Dev.* **1**, 233 (1957).

[14] M. Büttiker, PRL **57**, 1761 (1986).

[15] R. Dreizler and E. Gross, *Density Functional Theory* (Springer–Verlag, 1990).

[16] P. Delaney and J. C. Greer, *Phys. Rev. B* **73**, 241314 (2006).

[17] F. Evers, F. Weigend, and M. Koentopp, *Phys. Rev. B* **69**, 235411 (2004).

[18] G. Stefanucci and C.-O. Almbladh, *Phys. Rev. B* **69**, 195318 (2004).

[19] G. Stefanucci and C.-O. Almbladh, *Europhys. Lett.* **67**, 14 (2004).

[20] J. R. Reimers, Z.-L. Cai, A. Bilič, and N. S. Hush, *Ann. N.Y. Acad. Sci.* **1006**, 235 (2003).

[21] E. Runge and E. Gross, *Phys. Rev. Lett.* **52**, 997 (1984).

[22] R. van Leeuwen, *Phys. Rev. Lett.* **82**, 3863 (1999).

[23] W. Brenig, *Statistical Theory of Heat: Nonequilibrium Phenomena*, vol. II (Springer Verlag, 1989).

[24] G. Vignale and W. Kohn, *Phys. Rev. Lett.* **77**, 2037 (1996).

[25] G. Vignale, C. A. Ullrich, and S. Conti, *Phys. Rev. Lett.* **79**, 4878 (1997).

[26] M. Koentopp, K. Burke, and F. Evers, *Phys. Rev. B* **73**, 121403 (2006).

[27] N. Sai, M. Zwolak, G. Vignale, and M. D. Ventra, *Phys. Rev. Lett.* **94**, 186810 (2005).

[28] S. Kurth, G. Stefanucci, C.-O. Almbladh, A. Rubio, and E. K. U. Gross, *Phys. Rev. B* **72**, 035308 (2005).

[29] D. Bohr, P. Schmitteckert, and P. Wölfle, *Europhys. Lett.* **73**, 246 (2005).

[30] R. D'Agosta and M. D. Ventra, cond-mat/0512326 (2005).

[31] F. Evers and A. Arnold, *Proceedings of the Summerschool on Nano-Electronics*, Bad Herrenalb Sept. 1–4, 2005, Springer, 2006.

[32] H. U. Baranger and A. D. Stone, *Phys. Rev. B* **40**, 8169 (1989).

[33] X. Gonze and M. Scheffler, *Phys. Rev. Lett.* **82**, 4416 (1998).

[34] A. Arnold and F. Evers, unpublished (2006).

[35] K. S. Thygesen and A. Rubio, condmat/0609223v1 (2006).

[36] C. D. Pemmaraju, T. Archer, D. Sànchez-Portal, and S. Sanvito, condmat/0609325v1 (2006).

[37] W. Kohn and L.J. Sham, *Phys. Rev.* **140**, A 1133 (1965).

[38] J.P. Perdew, K. Burke, and M. Ernzerhof, *Phys. Rev. Lett.* **77**, 3865 (1996); **78**, 1396 (1997) (E).

[39] J. P. Perdew, K. Burke, and M. Ernzerhof, Local and gradient-corrected density functionals, in *Chemical Applications of Density-Functional Theory*, B.B. Laird, R.B. Ross, and T. Ziegler, Eds., ACS Symposium Series 629 (ACS Books, Washington DC, 1996).

[40] A.D. Becke, *Phys. Rev. A* **38**, 3098 (1988).

[41] C. Lee, W. Yang, and R.G. Parr, *Phys. Rev. B* **37**, 785 (1988).

[42] A.D. Becke, *J. Chem. Phys.* **98**, 5648 (1993).

[43] J.P. Perdew, R.G. Parr, M. Levy, and J.L. Balduz, Jr., *Phys. Rev. Lett.* **49**, 1691 (1982).

[44] J. P. Perdew and M. Levy, *Phys. Rev. Lett.* **51**, 1884 (1983).

[45] J.P. Perdew and A. Zunger, *Phys. Rev. B* **23**, 5048 (1981).

[46] C. Chang, M. Koentopp, K. Burke, and R. Car, in prep.

[47] M. Di Ventra, T.N. Todorov, *J. Phys. Cond. Matt.* **16**, 8025 (2004).

[48] C. Toher, A. Filippetti, S. Sanvito, and K. Burke, *Phys. Rev. Lett.* **95**, 146402 (2005).

[49] S.-H. Ke, H. U. Baranger, and W. Yang, cond-mat/0609637v1.

[50] E. Engel, Orbital-dependent functionals for the exchange-correlation energy: A third generation of density functionals, in *A Primer in Density Functional Theory*.

[51] Equivalently, a time-dependent vector potential may also be introduced.

[52] An additional caveat should be mentioned, also: the viscosity η not only has a real (dissipative) but also an imaginary (reactive) piece. The effect of the latter has not been investigated thus far and we have ignored it here.

[53] For a thorough discussion of a related problem, see Chapter 3 of [23].

[54] G. Stefanucci, S. Kurth, A. Rubio, and E.K.U. Gross, condmat/0701279; condmat/0607333.

[55] G. Stefanucci, condmat/0608401.

25

Molecular Electronics Devices

Anton Grigoriev

Rajeev Ahuja

List of Acronyms

CNT	carbon nanotube
GF	Green's function
DF	density functional
DFT	density functional theory
DOS	density of states
HOMO	highest occupied molecular orbital
IVC or IV	current-voltage characteristic
LCAO	linear combination of numerical atomic orbitals
LDA	local density approximation
LUMO	lowest unoccupied molecular orbital
ME	molecular electronics
MM	metal–molecular
MMM	metal–molecule–metal
MO	molecular orbital
MOSFET	metal–oxide–semiconductor field-effect transistor
NEGF	non-equilibrium Green's function

SAM	self assembled monolayer
SIESTA	Spanish Initiative for Effective Simulations with Thousands of Atoms, computer program
STM	scanning tunneling microscope
TranSIESTA	computer program, initially designed as Transport package for SIESTA, currently a self-contained program

25.1   Introduction

The subject of this chapter is exceedingly broad. Thus, it is not possible to review all the research in this area in a single chapter. Instead, the motivation for studying molecules as conductors will be given when the basic principles of the theoretical description and modeling are considered. Though these are both fundamental topics, they are usually left out by the researchers involved, who think them obvious, thus making it difficult for non-experts to wade into the field.

Molecular electronics (ME) devices can be defined as interconnected electronic devices with critical dimensions below 5 nm. This definition reflects a conceptually important view on the subject. According to the encyclopedia definition, a molecule is the smallest bit of a substance that retains its chemical properties. The ME device is typically the size of a molecule, and we will see there is good reason to use molecules for building electronics devices. However, it was recently recognized that significant progress in the description and design of nanoelectronic devices could be made if the nanodevice (silicon transistor, nanotube, etc.) is seen as a big molecule. From this viewpoint, ME becomes an important playground for nanoelectronics, and even traditional silicon-based microelectronics become naturally linked to ME on the way to miniaturizing such devices.

The idea behind building ME is, at first, to beat or at least extend Moore's Law [1,2]. Gordon Moore made his famous observation in 1965, just four years after the first planar integrated circuit was fabricated. The press called it *Moore's Law* and the name has stuck ever since. In his original paper, Moore observed an exponential growth in the number of transistors per integrated circuit and predicted that this trend would continue. So far, Moore's Law (that the transistor density of integrated circuits will double every two years) has been essentially maintained, and is expected to hold true until at least the end of this decade. There is a principle limit for the expansion of the law since at some point the size of the transistor will shrink to that of just a few atoms. Yet another limit is economical considerations since the production cost grows rapidly with desired miniaturization. ME aims to beat both limitations of Moore's Law by introducing cheaper technological production processes as well as new possibilities for processor logic and architecture with the use of quantum effects on electron confinement.

ME is based on these two cornerstone ideas. To make things smaller means to use the smallest possible building blocks, which are molecules by definition. To make things cheap, one has to use the cheapest material, which is usually the most common one. The chemical methods of production of the artificial molecules frequently used in ME are typically aimed at production of a fraction of a mole ($\sim 6.022 \times 10^{23}$ molecules) of the material, which is on the order of a few grams. This, however, yields an enormous amount of identical molecules, diminishing the cost of a single molecule.

While the conventional electronics design traditionally assumes driving the current through metallic leads or semiconductors, one instead defines ME design as driving current either through the molecules and metal–molecular (MM) systems or along the conducting surfaces and controlling their properties by means of attached molecules [3]. Building the ME devices implies the connection of molecules to surfaces, and molecules to molecules. This can also be achieved by chemical means, additionally minimizing the cost of production. Many molecules can self-assemble in solution or on the surface and form predefined regular patterns.

In contrast with conventional electronics, which utilizes the so-called "top-down" approaches when technology is used to miniaturize existing devices, ME makes use of a "bottom-up" approach. The molecules are, in a sense, the smallest possible building blocks that can be maintained at normal conditions (room temperature and pressure), therefore ME devices are already the smallest possible ones and the problem is to build up a full system and connect all the elements together [2]. The bottom-up approach is based on self-assembling and other chemical means to build predefined structures.

Will ME devices be useful? They may allow implementation of many functions in the same molecule due to pure quantum laws and effects governing their properties. Hence, a rich "zoo" of ME devices can be anticipated. The second reason to believe in a rich functionality for future ME devices is the actual existence of molecular "devices" in nature. A common example is an ant's brain which consumes milliwatts of energy and works at a frequency of about a few Hz, but is still capable of solving image recognition tasks. Given this, highly specialized ME-based devices may be the only alternative in, for example, medicine.

The analysis predicts, that, if realized, the ME devices will at least complement the limiting conventional silicon devices. There is much room for cheap yet functional electronics (e.g., one can easily imagine developing ultra-portable ready-to-use inkjet printers that can be used anywhere with a special ink or generating instantaneous readouts from medical tests using molecular sensors). Hence, the future of ME depends on the progress in production technology—conventional methods alongside those more advanced, like self-assembling, etc.

Recently significant breakthroughs have occurred in producing molecules for the purpose of ME [3,4], as well as in measuring their transport properties in various experimental setups. Nevertheless, the field is still at the pioneering stage, and theoretical modeling is essential to give proper insight into the experimental results and to design new functional ME devices and components.

This chapter is organized as follows. First, the scope of the devices considered is defined. A molecule or some definite atomic arrangement connected to metallic electrode(s) will be considered. Then the most common experimental setups and achievements in the engineering of ME devices are reviewed. Electron transport processes in molecular devices will be described, and their potential applications in molecular electronics explored. The most definitive problem in ME, the interface between a molecule and an electrode is considered separately. Since the aim is to analyze real devices, theory must follow explanation, and thus possibly predict the results of current and planned experiments. A short review of theoretical approaches to the calculation of transport and structural properties of ME devices will be given. After that, simulation tools available will be discussed. To conclude, possible applications from biosensors to alternative computational logics will be discussed.

25.2 Scope

In this part, the scope of the ME subject will be discussed and the important terminology introduced. This chapter starts with the definition of ME as interconnected electronic devices with critical dimensions below 5 nm. It seems that any definite atomic arrangement connected to metallic electrode(s) fits this definition; however, what exactly does the word "molecular" mean?

Devices entirely made of molecules and functioning as very small computers capable of performing very complex tasks, such as image recognition, exist in nature—for example, the brain of an insect. However, the manufacture of such a device is still a dream. What can presently be done is to interconnect (macroscopic) electrodes by molecular wires [5,6]. We will see later, that such a wire can bear substantial functionality, be it a rectifier (diode) or an electrical switch.

The phenomenon of molecular conductivity and intramolecular donor–acceptor charge transport is well known from chemistry, where the charge transfer in reactions is studied, treating the molecule as a bridge between donor and acceptor groups.

The reason to use molecules in devices is due to the word "smallest" in the definition of molecule. The smallest possible chip has to be made of the smallest possible compounds, which are the smallest possible bits of substances: molecules. Miniaturization of electronics in turn may provide faster, less power-consuming, and cheaper devices [2].

Molecules, however, are not the only objects considered in ME research. Carbon nanotubes (CNT) [7], gold [8–10], and silicon [11] nanowires are discussed in the context of the electrical interconnection of nanodevices. Metal atomic chains can be fabricated by embedding metal atoms in a matrix formed by

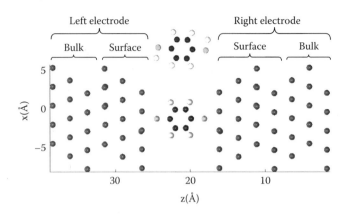

FIGURE 25.1 (See color insert following page 5-6) The 1,4-benzenedithiol molecule and the corresponding metal–molecule–metal sandwich.

organic molecules (especially DNA [12] and stacks of porphyrins [13]) or by assembling metal atoms on the surface of other materials. These structures are not precisely molecules, which has prompted the suggested name of *molecular cousin* [14]. Although they are stable and retain their properties at ambient conditions, their size is determined upon fabrication. According to the encyclopedia definition, a molecule is the smallest bit of substance retaining its chemical properties.[1] From this perspective, CNTs and other wires can be cut in shorter parts, retaining the properties of the longer wire. This implies substantial differences in the description and behavior of such systems.

In general, though with a number of valuable exceptions, molecular wires differ from the molecules (in solution or crystal phase) of a substance, because of the strong interaction (often chemisorption) with electrodes. The molecule loses its protective groups; instead the bond is created between the molecule and the electrode (see Figure 25.1). The electronic structure of the molecule becomes substantially altered by the interaction with the electron states in and on the surface of the electrode.

From this perspective, there is a good reason to call CNTs, polymer chains, etc. sandwiched or in other ways connected to electrodes molecular wires. In practice, only a short segment of the wire is used in experiments as a connector between electrodes. This tiny piece of material essentially behaves like a big molecule, and can be described and studied with ME methods.

Let us now discuss the metal–molecule–metal (MMM) structure in Figure 25.1 in detail. Electric current is flowing from the left to right electrode, along the direction of the z-axis. The z-axis is traditionally reserved to be aligned with current since in most experiments in the past the scanning tunneling microscope (STM) setup was used, where the STM tip was scanning along the horizontal $x - y$ plane, conducting substrate with a molecular film adsorbed, and the vertical position of the tip was naturally measured along the z-axis of the coordinate system.

It has already been mentioned that the protection group is removed either prior to, or upon, adsorption on the electrode surface. In the example in Figure 25.1, the H atom is removed from both SH groups, and the metal:S bond is established upon the surface of the electrode. Deep inside the electrode, the material is undisturbed and has a bulk-like structure and properties. Even if an electrode with a flat surface is considered, the surface region would differ from bulk due to the broken translational symmetry, and additionally due to reactions with adsorbed molecules. The surface region of the electrode can also contain surface defects, which promote molecular adsorption or represent STM tip.

---

[1]To be precise, a molecule is "the smallest identifiable unit into which a pure substance can be divided and still retain the composition and chemical properties of that substance." (From the *Encyclopædia Britannica*.)

If the contact between two flat electrodes via a molecular monolayer is considered, relaxation and reconstruction of the surface atomic layers should be contemplated. For dense films, the variation of density between the metal and vacuum is smoothed by the presence of the molecular film, and the reconstruction is often lifted. However, for the dense films, the charge transport between the surface and adsorbate becomes important, and the corresponding change of atomic and electronic structure should be taken into account.

The central region of the MMM junction consisting of the electrode surfaces and the adsorbed molecule is referred to as the *extended molecule*. In this region, which is considered between two periodic bulky metallic left and right half-spaces, the entire scattering processes introduced by the junction to the propagating electrons are localized. The surface and the molecule-related phenomena are also included in extended molecules, together with the MM bond in the case of a chemisorbed molecule. Ideally, the electronic and structural changes in the extended molecule are negligible when it is connected to (or disconnected from) the bulk electrodes in the equilibrium and no bias voltage is applied.

Within the framework of ME, the analysis of a MMM junction usually proceeds from the analysis of the corresponding free molecule or infinite molecular wire (the latter in case of CNT, etc.) towards the extended molecule at equilibrium. The electronic states of the free molecule, surface states, and their interaction constitute the combined structure of the extended molecule. Connecting the extended molecule to the electrodes, the zero bias electron transmission properties of the system can be calculated. At low bias voltage the current is linear and the zero bias transmission determines the conductance. At higher bias voltage, the system is driven completely out of equilibrium and the population of the states and structure of the extended molecule can substantially change.

In this perspective, there is no difference in considering a molecule or a short piece of wire constituting the extended molecule in MMM junction. Both molecular states and the states induced from the electrodes can provide the basis for transmission resonances. However, in the case of longer junctions, the wire itself becomes the bottle neck for electron transport, and the upper bound of the transmission can be assessed via the band structure of the infinite wire. In this case, the Bloch states in the wire become the main current carriers, and the poor MM contact can only decrease transmission.

Although the ME field is not limited by molecular wires, even the molecule in the MMM junction can offer substantial functionality. As it is shown in Ref. [2] for CNTs, a molecular wire can operate in the Coulomb blockade regime due to charging effects. Molecular oxidation (reduction) is another charging effect intrinsically changing the molecular structure and properties, making it possible to switch on and off molecular conductance. The other switching mechanism is conformational change between different isomers of the same molecule. Such a change can be induced by electric or light pulse or done mechanically (especially for CNTs). Obviously, when the molecule is switched by the change in the environment it becomes an electric sensor. However, there is another use for switchable molecules: molecular memory and (re)configurable networks.

While the current–voltage characteristics (IVCs) from some of the devices are symmetric, others can be significantly asymmetric with respect to the swap of the bias voltage [5], making it possible to build a molecular rectifier or a diode. Significant interest has arisen regarding molecular transistors consisting of a molecular wire junction and a third gate electrode (usually in the form of an oxide-covered metal template). Gate voltage changes effective potential on the molecule altering transmission properties.

Technologically, the "magical" process that can help in building more sophisticated devices than MMM junctions is self-assembly [2]. Many molecules, small metallic particles and even CNTs tend to self-organize in highly ordered films or other structures. Chemically functionalized molecules can recognize and attach to specific sites to create sophisticated arrangements [15]. Indeed, this is the mechanism for creation of a double helix of a DNA molecule from two adjacent strands.

All the aforementioned devices (and many others) constitute the field of ME research. The advantage of the ready molecular device is its size. In the following section, the fabrication of such a small device is considered in detail.

25.3   Experimental Techniques

Before dealing with the electron transport properties of the system, it is essential to understand its physical structure. The first challenge here is to create contacts with a gap that is the size of the molecule. Alternatively, the molecule could be grown to the size of the gap. The latter approach is successful for CNTs, but for the long molecules, conductance decreases exponentially with length [16] and it is quite difficult to manipulate the exponent.

The design of the electrode shape is chosen to permit molecules to adsorb on them and establish electrical contact. Another requirement could be to enable control over the structure. One can distinguish contacts aimed to contact single (or few) non-interacting molecules, or single (or few) molecules out of a film or many molecules (in a film) at once. Currently, the precise control over the configuration could be established with scanning probe microscopy (see Figure 25.2), which means that at least one of the electrodes is flat and the probe can scan over it. On such a flat surface, individual molecules tend to adsorb parallel to the surface. When the concentration of the molecules increases, a film of the molecules standing almost perpendicular to the surface is formed and could be contacted by evaporation of metal from the top (crossbar lithography [19] or nonporous technique [20]). An individual molecule from the film can be (in principle) contacted by a scanning probe microscope tip, especially if it is a single long molecule protruding from a film of short ones, or a conducting molecule embedded into a film consisting of insulating molecules [21]. The recently recognized technique is to use small (nano) metal particles placed on a film and acting as solder between the film and the STM tip. The well-controlled shape of the particles and their small size guarantee contact with very few molecules in the film, while the nature of contact of the scanning probe microscope (STM) to the metal particle is well defined. For the films made of a mixture of protected and de-protected (see the following) molecules, the actual contact might be established to only one molecule. The drawback of the technique is that the film screens the metal–molecular interface, and thus interaction of the neighboring species in the film should be taken into account.

To contact an individual molecule, contacts with very small contact area are created. The most common techniques are break junction [5], the angle evaporation technique with hanging mask [22], and electromigration [6] (see Figure 25.3).

The first setup (break junction) involves the mechanical breaking of metal wire and creates a very stable gap. The process of calibrating the gap, filling in some solution with molecules and measuring, is time-consuming (one can not repeat the measurement to, say, produce IVs for a different number of molecules or a different type of molecules using the same junction) and the number of successfully contacted molecules and the configuration itself is not known.

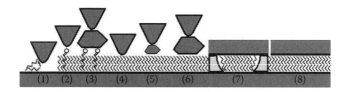

FIGURE 25.2 Establishing contact to the molecules adsorbed on a flat electrode, the role of inter-molecular interactions is increasing from left to right. (1) STM of a single molecule by spontaneous attachment and detachment monitored in the time domain [18]; (2) STM of a single molecule embedded in a less-conductive molecular film [21]; (4) STM of a SAM; (3), (5), and (6) a gold cluster used to define the contact area makes it possible to measure current from different numbers of molecules, distinguishing the contribution of a single molecule into total conductance [24–28]; (7) the micropore technique [25]; (8) a macroscopic, yet very small, "pad" contact on top of molecular film [27]. The two latter techniques often result in a shortcut between the electrodes, and are primarily used for thicker multilayer films.

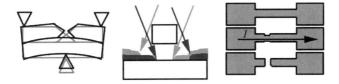

FIGURE 25.3 Break junction (left), angle evaporation with hanging mask (center) and electro migration (right) methods for making electrodes with nanometer separation. The latter two methods are often combined with metal cluster deposition (electrostatic trapping) in the gap to reduce the electrode separation. The separation is often measured by recording the electrode through-vacuum tunneling rate between the electrodes.

For the angle evaporation method with hanging mask, electrodes are produced lithographically by evaporation of gold at some finite angle so the shadow area under the mask shrinks. The electrodes configuration here is under better control, but the number of adsorbed molecules is again not known and experiments are very time- and resource-consuming.

For the electromigration method, thin metallic wires connect the electrodes and electric current is driven through the wire until it breaks. When the contact is lost, a gap between the broken pieces of the wire is created. An array of such electrodes is created and covered with molecules in a diluted solution. When the solution is dried, some molecules or molecular agglomerates are adsorbed on approximately 20% of the electrode pairs. This means that the probability to adsorb more than one molecular structure is small. Few measurements can be carried at the same time. The results, however, are in striking contrast to the break junction method since the device here is always in the Coulomb blockade regime. Moving the electrodes in break junction makes it possible to establish good contact and to bridge the electrodes with the molecule. For the electromigration method, the molecule is adsorbed somewhere, and it is not clear whether it bridges the electrodes or just attaches close to the gap. For example, it was shown [23] that the fine structure in the two-dimensional differential conductance ($\partial I/\partial V$) plots for the C_{60} molecule between nanometer scale electrodes is due to oscillations of the molecule on the electrode close to the gap.

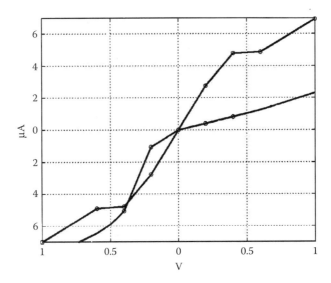

FIGURE 25.4 (See color insert) Calculated IVs for 1,3- 1,4-di thiol bicyclo[2.2.2]octa-2,5,7-triene molecule with asymmetric (red) and symmetric (blue) coupling to a gold (111) surface with a vacancy defect. Open circles mark the points for which the calculated transmission spectrum is presented in Figure 25.5.

The use of a more concentrated solution will not increase the probability of bridging the electrodes since the molecules tend to create islands of self-assembled film rather than spread uniformly across the wafer. Thus, the probability of bridging the gap will increase, but the bridging will be most probably done by an island of self-assembled monolayer, not by isolated molecules.

It is striking that none of the methods aimed at establishing contact to a single, or at least a few molecules, produced the actual IVs for 1, 2, 3, ... etc. molecules contacted within the same setup, as was done for the conducting molecules embedded into the insulating film and contacted through a gold cluster [24–28], as illustrated in Figure 25.2 (panel 3). One would expect that as the number of actual molecules contacted in sequential experiments changes, the comparison of the data would show a change in conductance, associated with some integer number of molecules. However, the complexity of the experiments makes it difficult to successfully repeat sequential measurements done in the same setup.

The practical method to measure molecular conductance based on break junction was proposed only recently [17]. The retracting electrodes produce a characteristic conduction steps pattern typical of metallic constriction. When the direct contact is lost, conductance peaks due to through-molecule conduction appear on a different, and thus a more distinguishable, scale. However, it seems that due to a complicated structure of the interface, there can be many different processes recorded contributing to conduction at intermediate scales between the pure metallic and pure MMM junction [18,29].

It should also be noticed that, owing to their well-controlled size and shape, the metal clusters are also proposed and used [30] as (a part of) electrodes. It is thus often argued that, in this case, establishing a good contact to the cluster becomes a problem and the measured molecular-related phenomena becomes superimposed with Coulomb blockade effects due to cluster charging [26].

The second challenge is to contact the molecule to the metal surface. As we have seen, it is possible to position the molecule between the electrodes; however, that does not necessarily lead to the establishing of a strong bond between the molecule and surfaces of the electrodes. In the experimental setup for the electromigration method of contacts fabrication, the strong bond between a molecule and a metal is expected to minimize molecular mobility on the surface, thus preventing the molecule from moving and eventually bridging the gap. For this setup, the more common situation is when the molecule does not bond to the second electrode.

Although it is possible to contact the molecule to the metal surface, many properties of molecular wires are still studied indirectly, without interfacing to the electrodes. For example, charge transport is studied chemically, looking at charge transfer in reactions and treating the molecule as a bridge between donor and acceptor groups.

The molecular adsorption on the electrode could take place either as chemi- or physisorption. The molecules with active "alligator clips" groups (i.e., sulphur-terminated) can chemisorb on gold. On the other hand, such an active termination as an alligator clip will lead to the lack of stability for the free molecule since it will tend to establish a chemically stable bond. The active molecular head groups are usually protected with some protecting group. Both de-protected prior to adsorption and protected molecules can form some contact with electrodes. Protected molecules, as well as molecules without alligator clips lie over the surface of the wafer, and, possibly, across the gap between electrodes. De-protected molecules with clips adsorb on the metal surface. The structure of the interface is often hidden by the molecular tail and is difficult to determine in experiment. For low coverage, it is known that the molecules form a "striped phase" monolayer [31], where their head groups lie in rows and their tails lie flat on the surface.

For higher coverage, the molecules tend to form a standing monolayer with sulphur bonded to the metal. The sulphur adsorption site depends on the substrate (facet, roughness [vacancies] thickness [if just about a few atomic layers]), and moreover, sulphur atoms might well form dimers and other arrangements [31] so that not all molecules are bonded to gold, but only one or two. When molecules are adsorbed from solution, they are likely to be terminated with a hydrogen atom. It is not clear what happens to it upon adsorption [32]. If there is a possibility for sulphur to be detached from the molecule, and how exactly the configuration of the film depends on the coverage, is not yet fully understood, especially for longer molecules where the attractive interaction between the molecular tails in the film plays an important role [33].

25.4 Molecular Conductance

This section will concentrate on the conductance of molecular wires. For a broader description of the quantum conductance phenomena, please see Ref. [5,34,35(page 6)].

As a first approximation, one can view molecular wires as a constriction between two macroscopic (i.e., comparatively very big) electrodes. Quantum-world phenomena such as inelastic scattering by phonons, electron–electron collisions, spin-flip processes, etc. lead to the loss of coherence between electrons traveling along the junction. The relation between the characteristic length of the collisions with impurities or phonons and the characteristic width of the constriction define the diffusive and ballistic regimes of transport [5].

Let us consider a molecular wire whose length can be increased by repeating some similar building block (for example, see Ref. [22]). When the length of such a wire increases, will it still conduct current? The answer is "yes," but not in the same way as a short wire. When more realistic corrections are introduced in the theory (scattering by phonons, etc.) the coherent transport is not possible over long distances, and conductance decays exponentially with the length of the wire. But incoherent (diffusive) transport does not decay, rather saturate, and starting from some overall length (coherent and incoherent) conductance is constant, independent of the length. However, this saturation is reached only when the molecule is sufficiently long; therefore, the conductance of longer molecules is still smaller than that of the smaller ones, which is in line with intuition.

Returning to the view of the wire as a constriction formed between two electrodes, and assuming this constriction is ideal (no phonon scattering, etc.) one can, analogous to a usual wave guide, relate the conductance to the number of transverse modes available to the electrons. As long as no bias voltage is applied, the electron flow from the left is the same as the electron flow from the right, and no net current flows. When a voltage V is applied, the Fermi levels of metal electrodes are shifted relative to each other by e^*V and the electron flow from the left (higher potential) is not compensated by the flow from the right (lower potential), thus a net current can flow from left to right. Since electrons are waves, they will carry charge (and the current will flow) only if there exists a solution for the Schr dinger equation with a nonvanishing k vector component pointing along the constriction. For the long ideal constriction, one can separate the variables so that each mode is represented by some standing wave pattern across the constriction. As soon as the energy of this standing wave enters the range of energies from the open energy window (between the left and right electrode potentials relatively shifted by eV), electrons from the left can use this channel to flow to the right electrode and the current will start flowing through the channel. Since the conductance of such a channel is constant, it will be added to the overall conductance of the constriction. This implies that the conductance of the ideal constriction is finite and increases stepwise with voltage at 0 K temperature. Due to the thermal broadening of the Fermi distribution, the steps are smoothed at finite temperatures. The same picture holds for a curvilinear constriction, but the steps are smoothed then also due to geometry [36].

This picture allows a simple interpretation for molecular conductance: since molecules are not by any means ideal constriction and since their electron structure are modified by the applied bias, the conductance channels can not only enter the voltage window but also exit from it, and the steps are significantly smoothed.

Figure 25.5 demonstrates the typical voltage dependent on molecular transmission [37]. Transmission is symmetric for the negative bias, as expected for the symmetrical system (right). However, if the molecule, electrodes, or metal–molecule interfaces are not symmetrical (left), the transmission will be different for different signs of the bias, leading to rectification in current–voltage dependence [38].

Transport properties of the molecule depend on its structure. Surprisingly, the same molecule can be conductive or not, depending on which isomer (which shape) it takes. The change here can be either spontaneous or induced. For the ordered film structure, the embedded conducting molecules on the high-resolution STM scan will appear as light stable spots. For the less ordered or sparser film, molecule switches much faster, and the light spots start blinking [39,40], suggesting that the benzene rings can rotate easier, and "on" or "off" states correspond to the rotation of the central benzene ring.

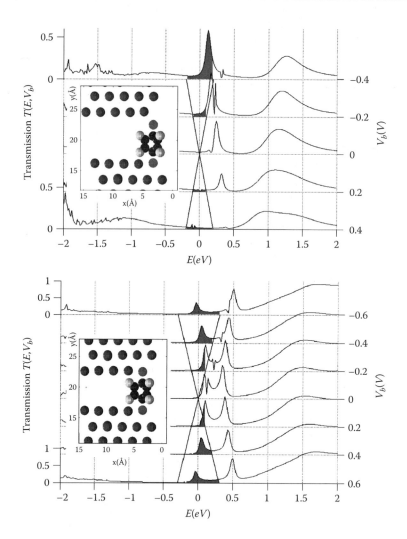

FIGURE 25.5 (See color insert) 1,3-1,4-dithiolbicyclo[2.2.2]octa-2,5,7-triene transmission spectrum with asymmetric (top) and symmetric (bottom) coupling to a gold (111) surface with a vacancy defect. The voltage window and corresponding spectrum parts are marked with grey.

In the same way as the density of the film, other environment factors influence the molecular conduction. In many technologically important applications, molecular electronics devices are expected to work at room temperatures, and it is desirable to produce such devices not in a high vacuum, but in some gas or atmosphere or even in air. Although it is possible to create such a device [41], its properties and structure will be sensitive to the environment [42,43]. Similar problems exist for the molecules that assemble in self-assembled monolayer (SAM) from a solution.

For the nanotubes, adsorption of hydrogen or oxygen on the surface of the tube and defects in the tube structure create randomly located scattering centers [43]. The same is expected for longer molecules or molecular complexes between electrodes. For molecules adsorbed on the surface, even the coadsorption of other species close to the metal–molecular interface significantly changes the transmittance [42].

A molecular wire is, by itself, a very small conductive object. When (or if) it is poorly connected to the electrodes, the potential barriers at the contacts are sufficient to keep electrons inside the wire, thus charging it. Then an attempt to put another electron on the wire will cost not only the work of penetrating the contact barrier, but also the work of overcoming the Coulomb repulsion between two electrons (i.e., effectively the work of charging a capacitor). Since we have already assumed that the contact barriers are high, the two allowed mechanisms of charge transfer are resonant tunneling through the double barrier system and inelastic processes, when electrons sequentially tunnel through the barriers that are exponentially small at low temperature. If the effective capacitance C is small, the potential barrier introduced by charging a molecule with n electrons is high (ne/C) and the resonance tunneling is blocked. Additionally, if the energy needed to overcome this barrier is comparable with thermal energy, $k_B T < ne^2/C$, then inelastic processes are inefficient and all electron transport is blocked (Coulomb blockade). Such a system can be opened again by external potential applied to the wire (gate voltage), like in Ref. [22] where the barriers between electrodes and the molecule were created by depositing protected molecules across the gap. The charging of the molecule can be interpreted as a chemical process associated with oxidation [22,39].

This is about as far as one can go without analyzing the MM interface in detail. It was previously discussed that the full description of the MMM junction is done via an extended molecule, including the MM interface. The MM interface still presents technological and theoretical challenges though: it is difficult to contact molecular wires to metal electrodes, and upon contact the metal–molecular interface gets hidden between the metal and the molecular film. The properties of the interface are crucial for understanding both structural (self-assembling) and transport (i.e., Coulomb blockade regime) properties of the molecular wires.

25.5 Molecular Adsorption on Metal Surfaces and Role of the Electrodes

As has already been pointed out, the molecule should be connected both physically and electronically to the metal surface. Although this is possible, there are still many challenges present both theoretically and technologically in the creating of the desired contact.

We have no intention of covering all of the surface physics and chemistry in this chapter. The field has matured regarding the study of surface catalysis, and an enormous amount of material is available concerning molecular adsorption on metal surfaces. Instead, we will continue the discussion from the previous chapter and concentrate on the adsorption of molecular wires on a metal electrode.

The contact between a molecule and a metal is especially interesting, and not simply because of its technological importance. This is a contact between two objects of a very different nature: metal is often viewed as a gas of nearly free electrons confined in a potential well made by positively charged nuclei, while in an isolated molecule electrons are localized and their orbitals adopt a particular, strictly defined shape. If the atomic orbitals are taken as a starting point, one often refers to this shape as hybridized atomic orbitals. These orbitals have some particular defined energy while in metal electrons fill the band and all the energies in the band are populated with a certain probability.

Upon adsorption, the metal and molecule mutually interact, and their properties change in the contact region. A molecule can in no way modify the properties of the bulk metal away from the surface. Neither is the molecule's tail (which is away from the metal) significantly affected; however, the changes at the interface are often striking. The electrons in molecular orbitals in contact with the metal band get the possibility to tunnel to the metal and back. Thus, the energy of the orbital becomes uncertain (broadened) as the electrons do not stay forever, but the state decays.[2] In other words, it is occupied only for a certain *lifetime*. The metal surface is affected by charge transfer to (from) the molecule, and the strong local field (*near field* [44]) is created by such a dipole.

Due to this metal–molecular interaction, both the properties of the metal surface and the molecule affect the configuration and properties of the interface complex. Individual molecules connect often to the surface defects (steps, vacancies), and the structure of adsorbed molecular fields often reflects the state of the surface (crystallographic orientation, reconstruction, stress, or strain). On the other hand, dense molecular films affect the surface (charge depletion, creation of defects [45]).

---

[2]This connection is discussed in detail in Section 25.5.2.

The nature of molecular adsorption on the electrode could be either chemi- or physisorption. The molecules with active "alligator clips" groups (i.e., sulphur-terminated on gold) can chemisorb. On the other hand, such an active termination as an alligator clip will lead to a lack of stability for the free molecules since sulphur tends to establish a stable chemical bond, and the molecules are usually protected with some protecting group. Both protected molecules and those de-protected prior to adsorption, can form some contact with electrodes. Protected molecules, as well as molecules without alligator clips lie over the surface of the wafer, and, possibly, across the gap between electrodes. If the protection group is not removed, no chemical bond is created. However, as the interaction between the surface and the molecule leads to an energetically favorable configuration, the molecule is referred to as *physisorbed*.

25.5.1   Role of Surface Defects

What is the relation between physi- and chemisorption? Since the origin of the attractive forces is very general (any molecule or atom will experience van der Waals and, possibly, electrostatic attraction) the final configuration depends on the details of the interplay of the two processes.

The origin of the van der Waals force is in dipole–dipole interaction between the charged particles in the adsorbent, and their images in the substrate. This is a long-range interaction, however. When the atom or molecule is adsorbed on the surface, the charge density of the adsorbate comes in contact with the density of the substrate, and the electrons "meet" their images. Hence the interaction becomes local and we can employ the DFT kind of description [46].

DFT within LDA is successful in part because it describes the repulsion between electrons caused by the exchange-correlation interaction (and usually expressed as the Pauli exclusion principle). This electron–electron repulsion can be seen as an electron–hole attraction, where the exchange-correlation hole around the electron describes the absence of other electrons due to repulsion [46]. This is exactly the attraction between the electron and its image charge (hole). Although DFT within LDA cannot handle this interaction nonlocally, it works when the charge and the images approach each other; hence DFT is suitable for the local description of the physisorbed species. All our DFT-calculated results will thus indirectly include contributions from physisorption. Let us estimate this contribution to the energy by the integral

$$- \int d\vec{r}\, d\vec{r}'\, \frac{\rho_{xc}(\vec{r})\, \rho_{el}(\vec{r}')}{|\vec{r} - \vec{r}'|}$$

where ρ_{xc} stands for the density of exchange-correlation holes, and ρ_{el} represents the electron's charge density (positive, hence the minus sign). To minimize this energy and to make the bond stronger, the distance between the image charge and the electron should be minimized and the overlap between the two densities maximized. Clearly, this is achieved in the presence of the surface defects, when the adsorbent can be physisorbed in a vacancy or by a surface step so it is effectively surrounded by the surface. Additionally, as the electrons' charge density tends to smooth sharp surface defects, the charge attracted to the surface upon physisorption will further minimize the total energy. In the same way, filling the vacancy with the adsorbed atom is favorable.

25.5.2   Chemisorption

Chemisorption is characterized by the creation of a chemical bond between the adsorbent and the substrate. The complex can be considered as a single big molecule. For example, de-protected molecules with "alligator clips" (i.e., terminated by a sulphur atom), are known to adsorb on gold. If we consider the changes in the charge distribution in this system upon adsorption, we find two reasons for charge redistribution. One reason is structural changes (changes in atomic positions) caused by the bond formation. Also the surface becomes screened with the adsorbent, and this partly removes the surface relaxation of atoms. Another reason is that charge redistribution strictly represents the formation of a chemical bond. If we consider only this contribution (i.e., looking at the charge redistribution only, and keeping the atoms fixed), we can calculate the charge density difference between the whole system and the molecule/substrate alone. In

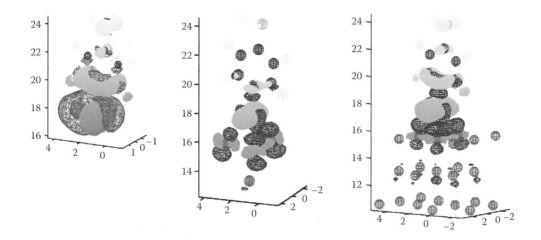

FIGURE 25.6 (See color insert) Charge redistribution for the bond formation. Left to right: DTB S-H bond; DTB bond to 4 Au atoms cluster; DTB on flat surface at hcp site. Isosurfaces enclose the regions with maximum charge redistribution and contain 40% of the displaced (solid) or removed (facet) charge. The distances are in Angstroms.

the same way, one may consider "chemisorption" of the protection group (we use hydrogen atom) on the same molecule—i.e., the usual chemical bond formation.

As one can see in Figure 25.6, the surface bond and the bond with hydrogen are different. The reason is both the chemistry (hydrogen is very different from gold) and geometry (the surface bond occurs to 3 Au atoms). But qualitatively, the mechanism is the same—charge is moved to π-type orbitals at the molecule (and p-type on the sulphur), which is very similar in shape to the region around the bonding atom (H or Au). At the same time, charge is moved from the bonding atom and from the "figure eight" lobes around sulphur. Since hydrogen has only an s-type orbital, charge is moved from the spherical region around it. For the gold, d-type orbitals [47] are involved and the charge is displaced from the "figure eight"–shaped lobes.

The mechanism of the chemical bond in the molecule can be illustrated with a standard two-level example, reminiscent of the hydrogen molecule bond. When two electron levels are brought in contact, their wave functions overlap and orthogonalize (hybridize), hence two levels appear in the molecule at different energies, usually one higher (antibonding) and one lower in energy (bonding). If only two electrons occupy these levels initially, they will end up at the bonding level and the total energy will be lowered. The other reason the energy is being lowered is because of the shell structure. If there is a way to fill the shell, energy is gained by that process. If the gain is more than the loss (filling antibonding levels, polarization, etc.) a bond will be formed.

Two important implications should be mentioned here. The orthogonalization depends on symmetry, which depends on (and determines) the structure. In particular, that implies the molecular bonding will differ, at least quantitatively, from the bonding of the head group alone because the MOs will generally have different symmetry. The other implication is that for many electron systems, the same orbital will be antibonding for some atoms and yet be filled since it is bonding for another electron pair. For example, for two *sp* systems, the *s—s* antibonding state will be filled by the former *p* electrons since it is lower in energy than the hybridized *p* states.

In the case of the surface bond to a metal, the so-called *resonant level model* is applicable. The difference comes from the fact that there are no isolated levels from the surface side. Instead, a continuous density of states (DOS) exists within some energy band in the substrate. Qualitatively, the situation can be understood on a simple one-dimensional example. At sufficient separation, the metal and adsorbent do not interact. One can see these two systems as two potential wells: one for the metal and another for the adsorbent (left and right in Figure 25.7). Assume for simplicity's sake there is only one energy level in the adsorbent well.

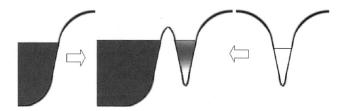

FIGURE 25.7   One-dimensional example of the resonant level model, see text for details. Filled states are gray-shaded.

As the adsorbent approaches the surface, interaction between the systems will slightly shift the energy of this level. But the main effect is that it becomes connected to the metal. The probability for the electron to escape to the metal thus changes from zero for the isolated system to some finite value where the level broadens, acquiring a Lorentzian shape.

This model explains the charge transfer happening upon adsorption. Indeed, if the Lorentzian of the induced DOS peaks below the Fermi level, the metal electrons will gain in energy by filling these states, and the charge will be transferred to the adsorbent. If it peaks above the Fermi level, the electrons from the adsorbent will try to minimize their energy by joining the metal's electrons and the charge will be transferred to the surface. In both cases, the bond will be of the ionic type. If the induced states are at the Fermi level, the bond will be covalent and the metal and the adsorbent will effectively share the electrons of each other.

There is an important modification to this simple model for the transition metals: When the adsorbent moves closer to the surface, its electrons interact with the d-levels of the metals in the way they would interact in the molecule (as illustrated in Figure 25.8). Instead of the d-level in the metal (and, say, the s-level in the adsorbent), two hybridized levels are formed, one bonding and one antibonding with respect to the electrons in the d and s-levels. Now these hybridized levels interact with the continuum spectrum of electron states in the metal.

In our example in Figure 25.6, the charge is moved from the substrate to the molecule, but the bond is not completely ionic since the charge is not only moved towards sulphur but also is redistributed around the sulphur and between the sulphur and gold, indicating the covalent character of the bond. The charge build up around the surface gold atoms indicates the interaction with the d-levels of gold.

The situation, when both the bonding and antibonding states have their energy below the Fermi level and are filled, is easily imagined. The filling of the antibonding states causes repulsion of the adsorbent. If such repulsion overcomes the bonding attraction, the surface stays clean. If this is the case for most adsorbents, the metal is regarded as noble. In the classical study of metal nobleness [48], Hammer and Norskov gave two examples of the metals where the antibonding states are filled upon hydrogen chemisorption: copper

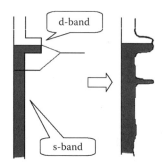

FIGURE 25.8   Schematic DOS for adsorption on the transition metal from the resonant level model.

and gold. The difference between the two is that Au is filling its 5d levels, while Cu fills only to 3d. The intuitive picture that 5d orbitals are bigger than 3d indeed holds, and the 5d orbitals have bigger overlap with hydrogen's 1s than 3d has. Strictly speaking, the overlap is bigger because of the stronger interaction (the s-d coupling element in the Hamiltonian) between 5d on Au and S on H. The bigger overlap leads to a bigger Pauli repulsion, or, equivalently, to a bigger orthogonalization energy spent on orthogonalization of the overlapping orbitals and stored by the electrons filling these orbitals in the form of kinetic energy. In the case of hydrogen on gold chemisorption, the energy required to fill the antibonding states becomes larger than the energy gained on filling the new bonding states, and the process thus becomes energetically unfavorable.

In the presented model for chemisorption, the position of the adsorbent electron levels is considered relative to the Fermi level of the metal. The original unperturbed energy of this level in the isolated adsorbent is changed by interaction with substrate due to two competing effects: interaction of the electrons with their images in the substrate, and the interaction with the effective potential that constitutes the solid surface barrier.

The interaction with the image is attractive and raises the energy by $e^2/4z$, where z is the average distance between the electron and the surface.

As electrons spill out of the effective potential well created by positive nuclei, some of the positive charge is left uncompensated inside the well, creating a dipole layer at the surface. The interaction with this dipole layer is repulsive, and it lowers the energy of the level.

Since the object of this work is the study of electronic devices for charge transport, the charge redistribution and level alignment upon adsorption becomes especially important. The accumulation of charge creates a local electric field that provides a barrier for charge transport.

The absence of the electric current in the nonbiased systems is due to the equal electron transfer rate between the electrodes. This means that as the number of electrons transferred in the unit time from, say, right to left is equal to the one from left to right, no net current flows. For the biased system at low bias voltage, the process of electron transfer is unchanged, but now the chemical potentials of the metals are different by eV, where V stands for voltage, and in the energy range just around the Fermi level, the electron flow from, say, the left is not compensated by the one from the right. Naturally, for the electrodes connected with a molecule, the appearance of a molecular level close to the Fermi level leads to enhanced conductivity (see Figure 25.9).

25.5.3 Alligator Clips for Molecular Electronics

From now on, we will leave the basic theory of adsorption and turn to the molecular electronics and related research. The small and chemically well-defined contact should be clean, conductive, and easy to fabricate. The choice of the most common noble metal, gold, as the material is straightforward. Sulfur- (or selenium-) terminated molecules are known to chemisorb on gold. This is good news since the chemisorbed state is a fairly stable one. Indeed, the successful experiments on electron transport in such molecules were done in room atmosphere.

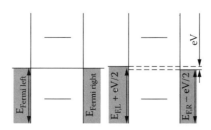

FIGURE 25.9 The Fermi level aligned with the molecular state leads to enhanced conductivity at low bias voltage.

The pioneering reports of the successful SAM deposition on gold were published more than 20 years ago [49], although organic SAMs on metals were used to control wetting properties of the surface since the 1930s [50]. The mechanism of adsorption through gold–sulphur head group interaction was finally established in the 1990s. Although it is commonly accepted now that the H atom is dissociated upon adsorption, the discussion of thiol versus thiolate bonding interaction continues [51].

The overall scheme of the adsorption mechanism is as follows: The de-protected molecules adsorb at low coverage with the S atom attached to gold, with the H atom cleaved of and the tail of the molecule lying flat on the surface [47]. The so-called "striped" phase of the SAM is formed with the tails aligned parallel to each other, forming visible stripes in the STM image. SAMs in this phase are usually ordered so the sulphur head groups lie in rows.

As noted earlier, the molecules tend to form a standing monolayer at higher coverage with sulphur bonded to the metal. In the standing phase, the structure of the interface is hidden from direct observation by the molecular tail, and is difficult to determine in experiment. The arrangement of the molecular tails are highly ordered and are often referred to as crystallized. Usually the tails form a parquet structure, with the adjacent molecules rotated so that molecular plains intersect. However, it was shown that the details of this structure depend on the length of the chain. As the interchain interaction becomes more important than the S-Au or S-S one, the structure of the molecular tails stacking changes [33].

However, not only the conformation, but even the chemical structure is usually unknown since it is not clear what exactly happens with the molecular terminating group upon adsorption. When molecules are adsorbed from solution, they are likely to be terminated with a hydrogen atom, but it is not clear what happens to them upon adsorption [47].

For example, does the hydrogen atom of a thiol group terminating the molecule stay intact upon adsorption [47], dissociate from the group and attach to the gold electrode [52], or dissociate and then desorb from the gold surface [33]? For the pure hydrogen adsorption, the H_2 molecule dissociates on the surface, but hydrogenation of the gold surface does not happen because it is energetically unfavorable [48]. For the thiol head group, the dissociation of the hydrogen atom happens upon adsorption [52], but now the environment is different and it is not clear whether hydrogen will desorb from the gold covered with the molecular film or not.

If there is a possibility for sulphur to be detached from the molecule, and how exactly the configuration of the film depends on the coverage, is not yet clear. Our results reported in Ref. [53] suggest such a possibility.

25.6   The Theory of Electron Transport in Molecules

The theoretical description of electron transport through ME systems is a complex many-particle problem, where the full solution is not possible in practice. The available approaches use certain approximations; however, the basis for transport calculations is the description of the ME device itself, which implies that approaches to the simulation of transport and structure cannot be chosen independently.

A diverse spectrum of approaches to the transport calculations was developed to study the whole spectrum of problems ranging from exact many-particle problem treatment within some simplified models (e.g., the two-barrier potential) for the system, to semi-empirical models for transport through quite realistic structures consisting of few hundred atoms. Although some of these methods were developed for purposes apparently different from the ME design, like intramolecular donor–acceptor charge transfer or STM image calculation, they are often applicable to ME systems.

If the goal is to study realistic systems, then one wants to base the investigation on the atomistic descriptions of both the molecule and the metal electrode, and to be able to use methods of computational chemistry. Two classes of methods are available: semi-empirical and ab initio. Within the first class, adjustable parameters are used in the calculation. To determine these parameters, results are compared to the experiment. Predictions based on semi-empirical calculations are made with an educated hope that the chosen set of parameters is quite general and does not change (at least not significantly) for some range of experimental situations. Since we are not concerned with the chemistry itself, but rather with the connection to the electronics (i.e., we want to extend our research to open systems), it is preferable to have no parameters and use ab initio methods for the basic structure calculations.

Ab initio methods are often called first principles calculations. The distinguishing feature from the other, semi-empirical, type is that the setup and algorithms do not use any fitting parameters, but formally start from the geometry of the system and universal constants like electron mass, charge, etc. Although the quality of the results depends on the degree of approximations used, the obtained result is free from any empirical estimations and expectations. The advantage is that the approximations used in the numerical methods and theoretical model can—at least in principle (and often in practice)—to a very large degree be improved on, and thus the convergence of the result checked. This provides a rigorous assessment procedure for the estimation of the quality of the results. The price paid in practice for the ab initio background of the theory is the complexity of the calculations that consequently demand significant computational resources.

Making a full classification of the available approaches for transport calculations is a thankless task though, not because of their diversity but more due to the lack of any established classification. In addition, most approaches could be rewritten in a variety of equivalent ways.

According to Nitzan[54], two distinct approaches to the transport calculations are suitable for proper accounting of the full chemical structure of the system: the *Green's function* (GF) method and the *density functional* (DF) method. The conceptual difference is that in the GF approach the interactions within the system are treated in the phenomenological way, in the sense that transport calculations and calculations of system properties are separated, thus enabling any level of input to be used. The range of examples varies from semi-empirical models [55] to discrete tight-binding [56] or continuous DF [57] or HF level [58] descriptions. Within the DFT framework, the charge carrier density is the unique parameter, hence the current is naturally calculated simultaneously with other system properties that depend on or constitute the density. Consequently, at least in theory, the system could be seamlessly restructured under the bias, but neither the tight-binding nor semi-empirical level of the description is naturally allowed within the DF method. Hence, GF is often used in a semi-empirical way to fit the experimental data. The fitting parameter is usually the shape of the voltage drop over the system [59].

Another important approach is often referred to as the *Laue expansion (representation)* method [60,61]. In the simplest model, where the system is represented by a hard-wall potential with a circular cross section of varying radius, it coincides with the transfer-matrix method. For a general form of the potential, the Laue expansion is used for wave functions (and for the potential). In the Laue representation, the Schrödinger equation appears as a system of linear one-dimensional differential equations, which could be discretized on a grid. This leads to a linear matrix equation [61]. Numerical routines could be different (e.g., recursive-matrix method could be used [60]), discretization on the grid could be replaced by step-by-step integration, etc. Alternatively, the Schrödinger equation could be discretized first and the discretized version could be used with the Laue expansion [62].

Since the essence of the method is a direct solution of the Schrödinger equation, it is worthwhile to notice other possibilities to solve it directly, especially for simplified empirical Hamiltonians [63,64].

While for a small model system the aforementioned methods might seem identical, the conceptual difference becomes apparent when the application to a realistic device is considered. Such a device consists of millions of atoms, and the task is to make it tractable for a modern computer. Hence, only an "open" part is considered at one time, where "open" means it is considered in connection with the rest of the world, taking into account, say, incoming and outgoing current, and so on. The description of "openness" of the system is handled differently in different methods.

In the following, the Löwdin partitioning method is considered, which is the basis for the NEGF method. For the ab initio atomistic description of the molecular part, metal electrode, and the interface, DFT was used. In the rest of this chapter, we will concentrate on this framework, in which the electron quantum transport calculation will be discussed in detail. Many texts exist on the theory of transport in open systems and the NEGF method [34,63], so there is no reason to repeat them here. Instead, the main results will be outlined only for completeness and to establish systematic notation. The relevant scales will be established and the main concepts will be introduced, each followed by a list of the most important results.

25.6.1 Quantum Current

The current in the molecular wire originates from the electron state in the electrode and propagates through the transverse channel or mode. The latter leads to two important conclusions: first, if the current is quantized in the sense that the cross-section of the wire is varied, the current changes only when the new channel appears or disappears in the wire. Second, in the long wire $k \sim 1/L_z$, the spectrum in the z direction is almost continuous. For the short wire, or the wire with varying cross-sections, the variable cannot be separated in such a simple way, and transport will occur not through the channels but through the eigenstates of the whole Hamiltonian of the wire. Such a transport is definitively ballistic and the electrons are considered to be in a coherent state throughout the wire. If one picks an electron's wavefunction at one side of a wire, one can tell precisely its phase and magnitude everywhere. This is not possible in the presence of incoherent, phase-breaking scattering.

The limited availability of incident electron states in the electrodes also leads to quantization of the current. The electron states in the electrodes are not necessarily available for transport, unless these states enter the wire. For example, the geometry of the interface between the bulk metal and the wire can block certain states from propagating, even if they are available in the metal and can propagate through the cannel in the wire. This difficulty does appear artificially in our model and it can be solved in two equivalent ways. We can assume that the electronic reservoirs are connected to the wire via reflectionless leads. Alternatively, we can define the reservoir-wire interface to be located inside the bulk, so that the interface part belongs to the scattering region. Note that the latter way does not permit us to separate the variables in the wire Hamiltonian. We have seen in the previous section that the current is determined by integrating the current density over any surface separating the electrodes. Hence, the attribution of the part of the physical electrode to the wire does not change our result for the current.

This concept is referred to as the *Landauer approach*. Assume that there are N channels open—that is, $\int dS \psi^*_{xy} \psi_{xy} = 2N$—because every level can support two electron states (xy stays for cross-section). Assume further that the temperature is low, and that the Fermi distribution is well approximated by the step function. We thus obtain the *Landauer formula*

$$I = \frac{2e^2}{h} VN \tag{25.1}$$

The Landauer formula defines the *conductance quantum*—the change in conductance when one channel is opened or closed.

$$G_0 = \frac{2e^2}{h} \approx 1/12.9k\Omega \tag{25.2}$$

The most important conclusion to draw from the Landauer picture of the quantum transport is that even for the ideal wire (such that electrons are not scattered back along it), for reflectionless leads, when connecting the wire and the electron reservoirs one still obtains a finite conductance. The inverse of this is called *contact resistance*.

The same result can be obtained for special shapes of the wire that permit the separation of the variables in the Schrödinger equation, and simultaneously allow the partitioning of the system into the electrodes and the wire (constriction), thus placing the borders in the bulk of the electrodes. The examples are parabolic [65] and hyperbolic [66] constrictions.

25.6.2 Relevant Length Scales

In the Landauer approach, the current is calculated as an integral over the open energy window, $f_L(\varepsilon) - f_R(\varepsilon)$, which is approximately equal to 1 at low temperatures, and is spread roughly over the range of energies from the Fermi level of the left electrode to the Fermi level of the right electrode, thus spanning the interval eV. Therefore, electrons that participate in transport have the energy close to the Fermi level, and this permits an estimation of the bounds for the geometrical size of the wire, allowing observation of quantum effects in transport.

The quantum effects become important when the characteristic size of the system (length, in the case of the wire) becomes compatible with the de Broglie wavelength of the particles. Otherwise, the wave nature of the current carriers can be ignored. As we can see from the Landauer picture of the electron transport applied to the electrons in the metal, the Fermi wavelength is the important length scale. Typically, it is about 1 nm. For the semiconductors, the electron concentration is lower, and so is the Fermi energy. Hence, the Fermi wavelength in semiconductors can be up to 10 nm. These numbers give an estimate of the limiting length of a quasi one-dimensional quantum wire.

The transport is coherent, when for the characteristic length of the wire L holds $L < L_\varphi$, where L_φ is the *phase relaxation* length—the length over which the electron's wave function retains information about its initial phase. Then the interference phenomena can be observed. Inelastic processes, like interaction with phonons, destroy memory about the phase. In the presence of weak inelastic scattering, the nearly degenerate electron states can become degenerate as one of the states loses energy. The length over which the two electrons travel coherently is called the *thermal dephasing* length.

25.6.3 Scattering

The Landauer formula can be extended to situations when scattering of the electrons from some initial electron state in the electrode (labeled with index i) to the final state in the electrode (labeled with j) takes place. The state could be localized in the same electrode (back scattering), or a system connected simultaneously to many electrodes could be considered.

For identical electrodes, the transmission is usually symmetric, $t_{i \to j} = t_{j \to i}$ yielding

$$I = \frac{2e}{h} \int d\varepsilon T(\varepsilon)(f_L(\varepsilon) - f_R(\varepsilon)) \xrightarrow[T \to 0]{} \frac{2e^2}{h} V \int_{E_{F,L}}^{E_{F,R} = E_{F,L} + eV} d\varepsilon T(\varepsilon) \qquad (25.3)$$

where $T(\varepsilon) = \sum_{i,j \in L} t_{i \to j}(\varepsilon) \equiv \sum_{l,r \in R} t_{l \to r}(\varepsilon)$.

For the scattering region connected to more than two electrodes (a *multiterminal* case), the appropriate summation over the terminal index should be added to calculate current in a particular junction.

Surprisingly, one can "guess" the formula for the current in its form (25.3). Indeed, if, at a certain energy, the transmission probability is T, then the probability count of a single transmission event from left to right of one electron per unit of time is proportional to the probability of having one electron at the left electrode f_L times T times the probability of having an empty state at the right electrode, $(1 - f_R)$. Since in our case no reflection happens at the electrode, the transport from right to left is independent from the transport from left to right, so for the symmetric transmission we have at a given energy

$$I \sim eT(f_L(1 - f_R) - f_R(1 - f_L) = T(f_L - f_R)$$

The rest is just to integrate over the energies.

The transmission can be calculated from the scattering matrix. Elements of the scattering matrix, S, by definition connect the incoming and outgoing states. In other words, the absolute value of each element squared gives the probability of transmission from a particular incoming state to the corresponding outgoing. If the elements are arranged in blocks representing transmission $\mathbf{t}$ and reflection $\mathbf{r}$, then (25.3) is recovered:

$$S = \begin{pmatrix} \mathbf{r} & \mathbf{t} \\ \mathbf{t}^* & \mathbf{r}^* \end{pmatrix} \to T(\varepsilon) = tr(\mathbf{t}^*\mathbf{t}) \qquad (25.4)$$

Indeed, the total probability is the sum over channel-to-channel probabilities:

$$T(\varepsilon) = \sum_{\substack{i,j=1 \\ i \in L, j \in R}} |t_{i \to j}|^2 = \sum_{\substack{i,j=1 \\ i \in L, j \in R}} t_{i \to j}^* t_{i \to j} = tr(\mathbf{t}^*\mathbf{t}). \qquad (25.5)$$

The description of the transport with scattering formalism enables us to conveniently discuss the symmetry of transmission from left to right versus transmission from right to left. The declared form of the scattering matrix is not the most general one. Since the matrix we have written is unitary, $S^*S = 1$, and both transmission and reflection are normalized to $\mathbf{r}^*\mathbf{r} + \mathbf{t}^*\mathbf{t} = 1$, the action of the matrix on the set of input states does not change the norm of the input vector, but only introduces a rotation, redistributing the incoming current over the outgoing scattered states. Thus, the form of the current (25.3) is valid and automatically ensures current conservation. The current will flow only if the occupation of the connected states is different due to the difference in chemical potentials of the electrodes (generally caused by applied bias voltage) and/or temperature differences.

25.6.4   Sequential Transport

So far, we have considered only the coherent tunneling processes. However, the example of incoherent processes is easily imagined from the coherent theory. What if three electrode reservoirs are connected sequentially, and the electrons stay in the central reservoir for long enough that the information about the initial phase of the wave function is lost? This has an important application to the transport through quantum dots, tiny metallic islands weakly connected to the electrodes. In fact, one of the first successful models for the electron transport was that involving the sequential transport model [5].

To model such a system, the central region representing a quantum dot or a molecule is considered as an open system itself, connected to the outer world through the transport junctions. Then, the density of states in the "electrodes" becomes $f_L(\varepsilon) \sum_{i,l \in L} t_{l \to i}(\varepsilon)$, since it is no more the states in the metal, but transport states induced on the central island through the junction. If the system is in a stationary state, the probability distribution of the states of the central island does not change. For the metal, it would be the Fermi distribution. For the metallic cluster, the electronic states of the valence electrodes would resemble the shell structure of the atom, and for the molecule that would be some set of states $\{i\}$ occupied with the probability P_i. As the electrons transported to the island occupy some state on it, the population of the states changes. The probabilities to change the occupation of the state i from the state j in the left electrode are

$$T_{i \to j}^L(\varepsilon) = f_L(\varepsilon)t_{j \to i}(\varepsilon); \quad T_{j \to i}^L(\varepsilon) = (1 - f_L(\varepsilon))t_{i \to j}(\varepsilon) \tag{25.6}$$

However, the occupation probability does not change. That gives a stationary condition

$$\frac{d}{dt}P_i = \sum_j T_{j \to i}P_j - T_{i \to j}P_i = 0 \tag{25.7}$$

which simply reads that the probability to populate a state is equal to the probability to depopulate it: $T_{j \to i} = T_{j \to i}^L + T_{j \to i}^R$. The stationary condition is a matrix equation for the vector P. When solved, it gives the density of states on the island. Then the current can be calculated in any of the junctions as the rate of the change of the occupation of states on the island:

$$I = e \sum_{i,j} P_i \left(T_{j \to i}^L - T_{i \to j}^L \right) = e \sum_{i,j} P_i \left(T_{j \to i}^R - T_{i \to j}^R \right) \tag{25.8}$$

Note, that this form is more convenient here than the form (25.3) since the DOS on the electrode is represented by a continuous Fermi distribution, and the DOS on the island is discrete. To calculate their difference, the broadening of the levels should be introduced.

25.6.5   The Non-Equilibrium Green's Function Method

25.6.5.1   Green's Functions and Self-Energies

Turning back to the picture established by the Landauer formulation of quantum currents, a DFT description of the metal–molecule–metal system is introduced. Based on Löwdin's partitioning, the indices L, C, and R for the left electrode, central region, and right electrode are defined. The MMM system is described by the self-adjoint Hamiltonian H

$$H = \begin{pmatrix} H_L & \tau_L & 0 \\ \tau_L^+ & H_C & \tau_R^+ \\ 0 & \tau_R & H_R \end{pmatrix} \tag{25.9}$$

The interaction τ between the left and right electrode is assumed to be zero, and $\tau_{L(R)}$ stands for the interaction between the left (right) electrode and the central region.

The Green's functions are defined as $(E - H)\,G\,(E) = 1$, where

$$G = \begin{pmatrix} G_L & G_{CL} & G_{RL} \\ G_{LC} & G_C & G_{RC} \\ G_{LR} & G_{CR} & G_R \end{pmatrix} \tag{25.10}$$

$$(E - H_{L,C,R})g_{L,C,R} = 1 \tag{25.11}$$

The Green's functions give the impulse response of the system—in this case, the response of the system described by the Schr dinger equation to the delta function perturbation.

It is easier to partition the whole system into different regions and calculate the Green's functions because we are not interested in the processes inside the electrodes, and the central region is connected only to the surface of the electrode, hence one expects τ_i, $i = L, R$ to be much smaller in size than H_i. From Equations (25.9) through (25.11), it follows that

$$\left(-\tau_L^+ g_L \tau_L + (E - H_C) - \tau_R^+ g_R \tau_R\right) G_C = 1$$

That means G_C is also a Green's function for the effective Hamiltonian $H_{eff} = H_C + \Sigma_L + \Sigma_R$, where

$$\Sigma_{L,R} = \tau_{L,R}^+ g_{L,R} \tau_{L,R} \tag{25.12}$$

are the so-called *self-energies*.

It is clear from (25.9) that H_{eff} is much smaller than total Hamiltonian H in the sense that it acts in the subspace of the central region only, which has much smaller dimensions. However, it does not yet simplify the problem, since to partition the equation and separate H_{eff}, one has to obtain $g_{L,R}$, which is a formidable task. The problem is simplified if the coupling between the central region and the electrodes described by $\tau_{L,R}$ is limited to the surface region of the electrode [63]. In that case, it is sufficient to calculate $g_{L,R}$ near the surfaces since the $\tau_{L,R}$ that enter (25.12) are zero otherwise.

Note that, formally, $G_C = (1 - g_C(\tau_L^+ g_L + \tau_R^+ g_R))^{-1} g_C$. This formula reveals, in part, certain illustrational complexities in the formalism. For the simple example of a one-dimensional box connected to a semi-infinite electrode, the Green's function is built from a specially chosen solution of the homogeneous equation.[3] However, as the positions of the boundaries enter the solution, the formulas become quite big when compared to, for example, the transfer matrix method.

However, the Green's function method is extremely powerful, as we will see shortly, and for real-life problems it is sufficiently compact. For example, the Green's function for the jellium surface [67] is

$$G(\vec{r}, \vec{r}') = \frac{1}{2\pi} \int d^2 k_{\|} e^{i\vec{k}_{\|}(\vec{R} - \vec{R}')} \frac{u_{\min(z,z')}^L u_{\max(z,z')}^R}{W(u_z^L, u_z'^R)} \tag{25.13}$$

---

[3] For the free electron, the potential does not depend on x. In the representation $HG = \sum_n a_n \frac{\partial^n}{\partial x^n} G(x) = \delta(x)$, the solution is $G(x) = \theta(x)\psi(x)$, where θ is the Heaviside function, and ψ is a solution of the homogeneous equation, satisfying $\frac{\partial^n}{\partial x^n}\psi(0) = \delta_{1,n}$, $n = 0, 1, 2$.

where W is the Wronskian, and $u^{L,R}$ are solutions of the homogeneous Schrödinger equation corresponding to the outgoing plane wave in the left (right) electrode.

To appreciate the significance of self-energy (25.12), consider a single level connected to the electrodes (or electrode, in the case of adsorption) as described by the effective Hamiltonian $H_{eff} = H_C + \Sigma_L + \Sigma_R$. Since the self-energies are energy-dependent and non-Hermitian, the energy becomes complex. The imaginary part of the energy takes the place of the δ, however, and is a finite number, describing exactly the broad resonance shape the level takes upon hybridizing with the continuum of states in the electrode. This broadening, since it follows from the standard theory of resonances, is inversely proportional to the lifetime of the state, the time the electron (on average) spends in the state localized at the central region or adsorbate.

In our description, the parameter δ, identical in its role to the imaginary part of the self-energy, appears initially to prevent unphysical divergences. One can say it is responsible for the "embedding" of the discrete spectrum associated with the central region or adsorbate into the continuous spectrum in the electrodes. Indeed, it is possible to imagine the situation when the electrode's Hamiltonian also possesses only a discrete spectrum, a set of tightly spaced eigenvalues. When such an electrode is connected to the central region, the wave functions associated with the electrode's energetic structure spill into the central region and hybridization happens—this time between the discrete levels. The local density of states then becomes a set of levels, occupied according to the envelope $\frac{\delta}{(E-\varepsilon_k)^2+\delta^2}$, where δ is the imaginary part of the self-energy. As the limit to the continuum spectra is made, it is the small parameter, denoted symbolically as $i0$, that broadens these spikes into a continuous DOS.

The difference between continuous and discrete spectra can be seen in the time domain, where the population of the resonance decays with time, while only the average population decay with time for the discrete spectrum, but every single level under the envelope is populated and emptied now and then.

Due to the importance of the imaginary part of the self-energy, a special notation is introduced

$$\Gamma = i(\Sigma - \Sigma^+) \tag{25.14}$$

Note, that the real part of the self-energy is being added to the self-adjoint Hamiltonian and shifts the energy of the levels localized at the central region, as compared to the free (disconnected) central Hamiltonian. Interestingly, if the coupling of the central region is asymmetric, as discussed in [68], the energy spectrum will be most influenced by the stronger coupled contact, and the spectrum of the effective Hamiltonian will dominantly follow the change in the potential of this electrode.

25.6.5.2 Transport Problem

Now we can come back to the transport (for definiteness, we will talk about transport from left to right). First, we look at the isolated left contact

$$H_L |v\rangle = E' |v\rangle \tag{25.15}$$

where $|v\rangle$ is a solution, obtained from the Green's function g_L. One can solve the full Schrödinger equation to obtain the retarded solution $|\psi\rangle$:

$$|\psi\rangle = \begin{vmatrix} \psi_L \\ \psi_C \\ \psi_R \end{vmatrix} = \begin{pmatrix} (g_L \tau_L G_C \tau_L^+ + 1) \\ G_C \tau_L^+ \\ g_R \tau_R G_C \tau_L^+ \end{pmatrix} |v\rangle \tag{25.16}$$

This has a clear physical sense, since knowing the incoming solution for the electrode we can build the total solution, and knowing the electron population in the contacts (by generalizing), we can fill different transport states of the system.

In particular, we can calculate the charge density

$$\rho = e \sum_k f(E_k, \mu) |\psi_k\rangle \langle \psi_k| \tag{25.17}$$

where f denotes the occupation of the state with energy E_k, and μ is chemical potential of the electrode. For the metal, it coincides with the Fermi distribution. The charge density induced from the left electrode is

$$\rho_L = e \sum_k f(E_k, \mu_L) G_C \tau_L^+ |\nu_k\rangle \langle \nu_k| \tau_L G_C^+ \tag{25.18}$$

where the summation runs over all states, including degeneracy. This can be rewritten in the compact form, most often reproduced in the literature:

$$\rho_L = \frac{e}{2\pi} \int dE f(E, \mu_L) G_C \Gamma_L G_C^+ \tag{25.19}$$

The full charge density is a sum over contacts:

$$\rho = \frac{e}{2\pi} \int dE \sum_{i=L,R} f(E, \mu_i) G_C \Gamma_i G_C^+ \tag{25.20}$$

25.6.5.3 Transmission

As long as the current flows in the system from left to right, the left reservoir with the electrons is being emptied, while the right is being filled. This is, of course, a computational problem, but we assume (and it is true in the real experimental setup) that the process of refilling the reservoirs takes place far away from the central region of the system, and the current flow in the center has already been switched on for a long time, so it is stationary and all the transitional oscillations associated with switching on have died out due to some weak inelastic processes that again take place far away in the reservoirs. That gives

$$0 = \frac{\partial \langle \psi_C | \psi_C \rangle}{\partial t} \tag{25.21}$$

which means that the density of states is constant in the scattering region. Physically, this is equivalent to the situation where the electrodes are not being refilled, but are instead enormously big (the number of electrons in the electrode reservoirs compared to the number of electrons transferred from left to right in the unit time), something that still (25.21) holds during our experiment. But then

$$\frac{\partial |\psi\rangle}{\partial t} = \frac{i}{\hbar} H |\psi\rangle \tag{25.22}$$

Combining (25.21) and (25.22) and expanding, we get a probability current from the jth contact at the energy E, where $j = L, R$. Since the electric current (at the energy E) is the charge carried by the electrons crossing any of the surfaces separating the left electrode from the right one, and denoting the charge of the electron by $-e$, we get the electric current into the system from the contact j (at the energy E):

$$i_j = -\frac{ie}{\hbar} \left(\langle \psi_C | \tau_j^+ | \psi_j \rangle - \langle \psi_j | \tau_j | \psi_C \rangle \right) \tag{25.23}$$

If we have more than two contacts, we only can say that $\sum_j i_j = 0$ (i.e., there is no charge accumulating in the system). Instead, we need to calculate the current from the contact j (at the energy E) out of the system due to the incoming current (at the energy E) in the contact l. Indeed, it is convenient to use even more detailed information, namely the current due to the incoming state $|\nu\rangle$ in the contact l, using (25.16):

$$i_{l, \nu \to j} = -i_j \big|_{|\psi_c\rangle = G_C \tau_l^+ |\nu\rangle} = \frac{ie}{\hbar} \left(\langle \nu | \tau_l G_C^+ \tau_j^+ | \psi_j \rangle - \langle \psi_j | \tau_j G_C \tau_l^+ | \nu \rangle \right) \tag{25.24}$$

If the electrodes are connected to the reservoirs with the electron gas at the chemical potential μ_j, then the states in the electrodes are filled in turn according to the Fermi distribution $f(\mu_j, E)$, and the total current between two electrodes is obtained by integrating over the energy:

$$I_{l \to j} = \frac{e}{\hbar} \int_{-\infty}^{\infty} \sum_{v'} \langle v | \tau_l G_C^+ \Gamma_j G_C \tau_l^+ | v \rangle f(\mu_l, E) dE \tag{25.25}$$

where the sum over the v' takes into account the actual population of the states $|v\rangle$ at the energy E, including the factor 2 for spin. With the use of Equations (25.12) and (25.14) we get:

$$I_{l \to j} = \frac{e}{\pi \hbar} \int_{-\infty}^{\infty} Tr \left(G_C^+ \Gamma_j G_C \Gamma_l \right) f(\mu_l, E) dE \tag{25.26}$$

Now we are in the position to calculate the current in the device between two reservoirs at chemical potentials μ_L and μ_R:

$$I = I_{L \to R} - I_{R \to L} = \frac{e}{\pi \hbar} \int_{-\infty}^{\infty} Tr \left(G_C^+ \Gamma_R G_C \Gamma_L \right) (f(\mu_L, E) - f(\mu_R, E)) dE \tag{25.27}$$

where we use the property of gamma $\Gamma_j^+ = (i \tau_j^+ (g_j - g_j^+) \tau_j)^+ = -i \tau_j (g_j^+ - g_j) \tau_j^+ = \Gamma_j$ and the trace $Tr(G_C^+ \Gamma_R G_C \Gamma_L) = Tr(G_C \Gamma_L G_C^+ \Gamma_R) = Tr(G_C^+ \Gamma_L G_C \Gamma_R)^* = Tr(G_C^+ \Gamma_L G_C \Gamma_R)$.

Equation (25.27) is the famous Landauer formula for transport. Transmission is $T(E) = Tr(G_C^+ \Gamma_R G_C \Gamma_L)$.

Let us discuss the properties of the obtained current (25.27). First, in the equilibrium, when both contacts have the same chemical potential and temperature, the total current is absent, because the current from left to right exactly cancels the current in the opposite direction.

In the case where the system is close to equilibrium, or if the response to the external voltage is linear, one can expand the difference $f(\mu_L, E) - f(\mu_R, E)$ as

$$f(\mu_L, E) - f(\mu_L + \delta \mu, E) \approx \frac{\delta f}{\delta \mu} \delta \mu = -\frac{\partial f}{\partial E} e V \tag{25.28}$$

obtaining for the current

$$I = \frac{2e^2}{h} V \int_{-\infty}^{\infty} T(E) \left(-\frac{\partial f}{\partial E} \right) dE \xrightarrow[T \to 0]{} \frac{2e^2}{h} V T(E_F) \tag{25.29}$$

This result emphasizes the importance of the zero bias transmission, studied in [37].

25.7 Computational Tools and Algorithms

In the following, we will discuss the computational methods used for simulation of atomic structure and transport properties of the molecular systems.

25.7.1 DFT and NEGF for Transport Calculations

As of 2006, there was a limited choice of ab initio electron transport simulation codes [69–73]. Let us consider the TranSIESTA program [73–75]. It is an ab initio electronic structure simulation program capable of modeling electrical properties of nanostructured systems coupled to semi-infinite electrodes. It allows for atomic scale modeling of systems where an external bias is driving a current through the system. The two electrodes could, for instance, be (semi-infinite) metal crystals and the nanostructure could consist of molecules between the metal surfaces. Other typical examples are nanotubes (as electrodes or a nanostructure) or an interface region between two materials.

The described setup models many systems of scientific and industrial interest like molecular devices, metal wires, or nanotube junctions. In the ab initio calculation, electron transport is assumed to be coherent. The ab initio NEGF method is used, treating the electrons as quantum particles propagating coherently through a scattering region. The interactions with other electrons are described through an effective potential obtained from the DFT. The effective potential is determined self-consistently and gives information about the voltage drop through the device. Electron current or the current induced forces can be also obtained. Additionally, from the program output itself or after post-processing the data transmission curves, current induced, and total forces, the voltage drops across the junction, 3D scattering states, and total energies can be extracted. Current–voltage characteristics can be obtained after successful calculation for a range of voltages. Equilibrium geometry or molecular dynamics can be simulated trough the DF LCAO calculation.

The main distinct feature of the realized DFT approach is its ability to treat open systems. One can say that special boundary conditions are used in the selected direction denoted traditionally by the z coordinate label as reminiscent of an STM setup, where current flows from the substrate to the tip so that the bottom electrode is "left" and the top electrode is "right." Three regions are chosen: left electrode, scattering region, and right electrode. The electrode regions play the role of the boundary regions, where the "boundary" condition consists of that the effective potential retains its bulk value. This value (and, if needed, the structure of the electrodes) is obtained from a separate self-consistent DFT bulk calculation. By bulk calculation we mean that the usual periodic boundary conditions are used.

The effective potential in the scattering region is calculated self-consistently using a non-equilibrium Green's function (NEGF) technique. The open system calculation is based on NEGF techniques and is capable of treating situations in which the two electrodes have different electrochemical potentials, which means an external bias voltage is applied across the structure.

The NEGF techniques are combined with the DFT for the electronic structure calculations. The method is capable of modeling the electronic structure of geometries where two semi-infinite atomic systems (viewed as periodic crystals) are coupled via a scattering region.

Periodic boundary conditions are employed in the x and y direction, while in the z direction the division of the system into three regions is obeyed, as discussed earlier. Left and right electrode regions are assumed to be completely described by bulk parameters obtained from the separate bulk calculation, and typically this condition is satisfied by extending the scattering region into the first few layers of a metal surface.

Before the open system calculation can start, a separate calculation of the bulk phase of the electrodes must be performed in order to obtain the bulk Hamiltonian parameters. This Hamiltonian is used for calculating the surface Green's function of the electrode. Next, a computational supercell is defined that includes all atoms in the left electrode, scattering region, and right electrode. The interaction range of the localized basis set determines the number of atoms needed to describe the electrode region. In the open system calculation, all Hamiltonian parameters within the supercell are calculated within a DFT self-consistent approach and passed to the NEGF subroutine. The Hamiltonian parameters of the electrode atoms are substituted with bulk Hamiltonian parameters and determines the NE density matrix of the system by combining the supercell Hamiltonian with the remainder of the electrodes using the surface Green's functions. New Hamiltonian parameters are determined from the NE density, and the steps are repeated self-consistently.

To start the calculation, a starting guess for the potential in the scattering region is used, typically from a separate bulk calculation on the supercell region. When a bias voltage is applied, the semi-infinite electrode systems have different electrochemical potentials, and the better starting guess across the scattering region is to start from the unbiased system potential. The Hamiltonian matrix is calculated from this starting guess using a localized basis set. This matrix is used to set up the NEGF of the system. In the next step, the non-equilibrium density matrix is calculated from the NEGF. The density matrix defines the effective potential in the scattering region, and thereby the new Hamiltonian parameters. The steps are repeated until a self-consistent solution is found.

Once the self-consistent Hamiltonian is obtained, post-processing tools can be used to extract additional information. It is possible to calculate transmission coefficients within an energy window and separates transmission into eigenchannels. It also calculates the total current from transmission. Additionally, left or right propagating scattering states for a particular energy can be obtained and separated into eigenchannels. Eigenstates of projected self-consistent Hamiltonians are also accessible.

Historically, the initial attempt on DFT for open systems [76] developed into the McDCal and Tran-SIESTA programs. Currently, as the next cycle of the development, TranSiesta-C (a TranSIESTA-based program written in the C programming language) was developed based on the new SIESTA, McDCal, and TranSIESTA packages.

25.7.2 General Algorithms

In this section, we will present the very general technical approach of our research. Such considerations are usually left out as obvious by the researchers involved. This part of the work is nevertheless added to preserve generality and make the work readable and possibly useful for nonexperts.

The first step in describing the adsorbed or sandwiched molecular system is to separately study the metal and the molecule. Indeed, the DFT program based on the plane wave basis set is ideal for structure calculation of the infinite metal crystal. Making use of the periodic boundary conditions, one can describe the fcc structure of gold with just 4 atoms in the cubic cell. For the molecule, a rectangular cell of sufficient size is required to minimize the effect of the periodically located neighbors along the surface normal.

The goals for the preliminary studies of the metal bulk and the molecules are, however, different. The bulk crystal structure is usually known, and the main interest is to establish the cell parameters to be used for the surface calculation so the surface is placed on the relaxed bulk and no artificial stress or strain is introduced. Additionally, the symmetry of the cell chosen for the bulk should correspond to the symmetry of the surface cell.

For the molecules, the primary interest is in the structure. Many molecules have different isomers, not to mention that the exact position of the hydrogen atoms is not known a priori. To model a particular reaction at the surface, one should use the adsorbent in the corresponding state (i.e., the molecule can be considered relaxed or have some groups removed if dissociation happens at the surface). In the latter case, the relaxation might be unnecessary since the bond is reestablished to the surface, but usually the energy of the relaxed de-protected molecule will enter the energy balance equations.

In the case of the molecular film where the interactions between molecular tails play a significant role in determining the structure, the relevant question for the preliminary study is the size of the cell for the crystalline molecules stacked as they would stack to form such a film. It is known that the distance between the molecules and the film structure cannot be obtained from such a calculation since the molecules (even if the interaction between their tails dominate the adsorption energy and the interaction of the head group with the substrate is weak) tend to spread over the surface to fit the surface structure of the adsorbate. However, such a study gives the limiting coverage of the adsorbate and will not only assist in getting the correct result, but can also considerably reduce simulation time since for the surface + molecule system the self-consistent iterations are costly with respect to computer resources.

The next natural step is to study the surface. As the surface breaks the symmetry of the crystal and, in the first place, translational symmetry, the periodic boundary conditions in the direction along the surface normal are now excessive. However, there is no way to change the setup without the use of some special mathematical methods like Green's functions. Instead of changing the program, one can use the supercell approach. The suitable bulk cell is extended along the surface normal vector to include the vacuum region. In other words, we simulate the slab (with two surfaces) periodic along the supercell vectors parallel to the surface and surrounded by vacuum "layers." Naturally, such a layer should be sufficiently thick to exclude interactions with the periodic copy of the slab and, possibly, to host the adsorbent.

Two things are expected to happen to the cleaved slab (i.e., the slab whose atoms were allowed to relax from bulk positions). Each surface of the slab is expected to relax, meaning that atomic layers will shift slightly inwards or outwards along the surface normal, as compared to the bulk position. The surface can also reconstruct, meaning that atoms on and close to the surface will change their lateral positions in the surface plane. If the thickness of the slab is sufficient, the central inner atomic layers will be undisturbed and form the bulk of the material.

If the slab is thick enough to relax to the surface–bulk–surface atomic configuration, it is not necessarily thick enough to allow the inner bulky part to stay unperturbed when the adsorbent is placed on one of its surfaces. Charge transfer between the slab and the adsorbent and the formation of chemical bonds to the surface can significantly increase the depth of the surface relaxation. At the same time, the degree of the reconstruction should be reduced, as the surface becomes covered with adsorbent.

Different measures could be taken to preserve the bulky "basement" for the surface. The most common reason for additional relaxation is significant polarization of the surface due to the charge transfer to, or from, the adsorbent. Placing the same adsorbent symmetrically on the other surface of the slab, so that the whole system retains up-down symmetry and has no net dipole moment caused by the surface polarization, one can minimize the influence of the polarization.

The crucial limit for the thickness and size of the slab is the computer time and resources needed to store the data. To minimize the demands, it is desirable to consider, if possible, only one surface. For the metal, the nearly free electrons screen effectively all the defects, including surfaces, so they do not affect the structure a few layers away from the defect. The common practice is to use this effect and to leave one surface cleaved, putting appropriate constraints on the forces to extend the bulky inner region of the slab and to save time on reconstruction/relaxation of two surfaces, thus allowing for one surface only to relax/reconstruct. Such an approach is somewhat artificial, and one should always consider testing the setup for the appropriate number of fixed unconstrained layers.

The reconstruction, adsorption, or study of certain surface defects brings up the problem of choosing the proper cell for the calculation. As we have already discussed, the cell size in the lateral direction should be consistent with the relaxed bulk periodic crystal structure. However, if there is only one surface atom in the supercell, like in a $1 \times g$ cell for the Au(111) surface, no reconstruction is possible since the neighbors of this atom at the surface are its periodically repeated copies. There is often a mismatch between the bulk and the surface primitive cell (see Figure 25.10). This often leads to the requirement to use some other supercell for the slab, which does not have the full symmetry of the bulk, but permits the periodic repetition of the adsorbate layer structure while keeping the overall size of the system reasonable.

As the size and complexity of the system grows, the role of having some good initial guess increases dramatically. Inaccurate initial guesses make the simulation slower and often impossible, because the structure "explodes" in smaller simple pieces due to the original structure being too strained. Different approaches apply in this situation. The goal is to start with as simple (and small) of a system as possible, and to reuse the obtained structure as an initial guess for the bigger calculation. A good rule to follow is to divide the structure into relatively independent modules, still bearing some functional identity of the whole structure. A trivial example from the previous discussion is a consecutive study of the bulk, slab, molecule, and the molecule adsorbed on the slab. Less trivial is the study of the crystalline layer of the molecules as an adsorbate layer without the substrate. Another example is the study of the metal–molecule interface, when only the adsorption of the head group of the molecule, without a molecular tail, is considered. This method

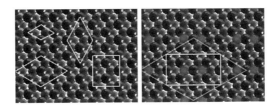

FIGURE 25.10 (See color insert) Left: 111 surface; height of the atoms is proportional to the shading intensity. Possible supercells are marked. Right: the same surface with an adsorbate. Note that the minimal surface cell has low symmetry, while the cell with the full symmetry is twice as big.

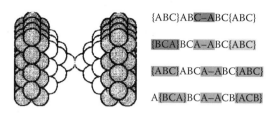

{ABC}AB**C–A**BC{ABC}

{BCA}BCA–ABC{ABC}

{ABC}ABCA–ABC{ABC}

A{BCA}BCA–ACB{ACB}

FIGURE 25.11 (See color insert) Left: imaginable symmetric setup for the transport calculation through a point contact. Right: Different possible stackings for the 111 surface. ABC letters denote the different distinct atomic layers— the infinite, periodically repeated to the left and right lattice is added as . . . and the symmetric (unsymmetric) setup is shaded green (red and yellow).

clearly separates the function of the substrate–adsorbate head group interaction. It cannot by any means predict the whole structure since the amount of charge on the actual molecular tail could determine transfer and intermolecular interactions. However, the size of the system becomes much smaller, thus permitting an accurate investigation of the local structure of the interface. It is generally expected that the knowledge of the preferred/unfavored sites for head group adsorption in connection with the known interaction strength between molecular tails will lead to the possibility of optimizing the structure of adsorbate, thus making a compromise between the head-to-substrate and tail-to-tail preferences. However, this method should be used with caution since for dense adsorbed layers the properties can significantly change. Thus, an initial guess proposed by studying the interface region will not be adequate and will not permit recovery of the ground state configuration of the system.

When the metal–molecule and metal–molecule–metal structure is determined, such a structure can be used in transport calculations. However, as the transport calculation is also based on DFT, the same problem as for the surface calculation (when the translational symmetry is broken) appears. The system in the transport supercell, containing the electrodes and the central scattering region, should be periodic. As one can see, periodicity in the transport direction is a serious constraint. For the "perfect" model system, the left-right symmetry is essential to exclude unwanted effects not related to the studied structure but rather to electrode stacking mismatch. However, such a setup leads to the mismatch at the border of the supercell (as indicated in Figure 25.11). For the "real world" system, when different electrodes are connected from the left and right side, the artificial interface appears on the supercell border. The suggested solution is to separate the electrodes with a buffer region. The atoms in this region serve as screening protection of the bulky structure of the electrodes from any sort of interface at the supercell border. Though the buffer space could be filled with a cleaved surface-vacuum-cleaved surface arrangement, the most computationally efficient is to use a stacking fault, screened with a few layers of bulk because this minimizes the number of atoms in the buffer region and the cell size.

For the metal–molecule–metal sandwich setup, the electrode separation plays a crucial role. The separation is defined as the distance measured between the left bulk electrode and right bulk electrode metal layers. This is an external parameter and can be controlled in experiment. For the theoretical calculation, when the variation of this parameter is not studied, different initial conditions could be used, corresponding to the different experimental situations modeled, as discussed next.

For the molecular monolayer formed on the surface, it is natural to start from the molecular adsorption calculation and proceed by adding the second electrode symmetrically (i.e., starting the sandwich relaxation from the relaxed metal–molecular interfaces).

For the molecule deposited in the gap between electrodes, the natural choice would be the free molecule between two free surfaces, giving (for the same gold surface: the sulphur atom distance) different distances between gold bulk regions, and, after the relaxation of the structure, slightly different interface geometries.

The molecule itself has to be positioned to bridge the electrodes. One can argue, that if the adsorption of both the head and end group on the electrodes is energetically favorable, and the molecules do not cling to each other, the upright standing molecules in the film will bridge the narrowing gap as soon as the sulphur atom in the tail group is able to reach the second electrode (i.e., when the molecule is stretched on its full length perpendicular to the surface). The current measurement can be conducted either just after the contact is established or, possibly, at the moment just before the contact is known to be lost, so that the molecules are perpendicular to the electrode surface and the adsorbed film has the maximum possible thickness.

However, there are different experimental setups, and not all allow the gap between the electrodes to be varied, as in the break junction method. It is easy to imagine the setup with the measurement done on the compressed or tilted film. So far, no generally accepted algorithms are available for handling the sandwiched molecules.

We would like to conclude this description of methodology by looking back at the initial guess given to the DF code as a starting point. The molecular or the crystal structure, and in some cases the structure of the adsorbate, could be checked in experiment. However, the structure of the molecule inside the metal–molecule–metal sandwich is hidden and is very difficult to reveal. Given that the calculated configuration of the atoms and molecules depends on the initial guess, how can one trust the results of such calculations? This might be seen as a rather philosophical question, but it indeed brings the whole idea of using ab initio self-consistent calculations for the problem into question.

The ground state (GS) of the system is selected according to the minimal energy principle. To find the GS of a system we have very limited or no information about (e.g., the metal–molecule–metal sandwich), three steps should be followed: find the set of initial guesses that satisfy the known structural properties of the system; find the corresponding relaxed configurations; find the minimal energy configuration. This "simple" procedure is formidably difficult for systems with many degrees of freedom. Only in rare situations can one follow this scheme, really checking that none of the initial and relaxed states are missing [77].

However, we can still trust the self-consistent ab initio calculations to yield physically reasonable results. When the ab initio self-consistent calculation is used with an unreasonable initial guess, the relaxation will not converge (no attraction) or will depend significantly on input parameters or disturbances of the proposed input structure (instability). This gives a great advantage to such a calculation, because it can quite easily rule out an unrealistic setup, even if little or no information is available *a priori*.

25.7.3 Modeling of the Electrodes

For the MMM sandwich setup, the metal electrode separation is an external parameter that plays a crucial role. Depending on the experimental conditions, the distance between the metal electrodes can be determined by fixed (or variable) geometry metal contacts, or by the molecular layer itself, with added metal contacts. For the theoretical calculations to model these different experimental situations, different initial conditions may be appropriate.

For a molecular monolayer formed on the surface, it is natural to start from a molecular adsorption calculation and proceed by adding the second electrode symmetrically—that is, starting the sandwich relaxation from the relaxed monolayer interfaces (our choice). For molecules deposited in the gap between electrodes, the natural choice could be the free molecule between two free surfaces, giving (for the same gold–sulfur distance) different distances between gold bulk regions, and, after the relaxation of the structure, different interface geometries.

Making a mechanical contact (like in break junctions, or in the method of the gold cluster deposition) to a molecular film when the film is formed on one electrode and another electrode is initially clean, leaves less freedom in controlling the electrode separation: the formation of the chemical bond to the second electrode takes place at some particular separation and can be sensed through the I-V regime of the device, since before the chemical bond is established the expected I-V is asymmetric. However, when the molecule is chemically bonded to both contacts, the force on the molecule is controlled by the break junction or by AFM contact to the gold cluster [24–28].

Let's now consider the shape of the electrodes. Electron transport through a molecule at low bias voltage critically depends on the nature and position of the molecular energy levels. Different effects are associated with the level structure within the scattering region viewed as an extended molecule, which includes the

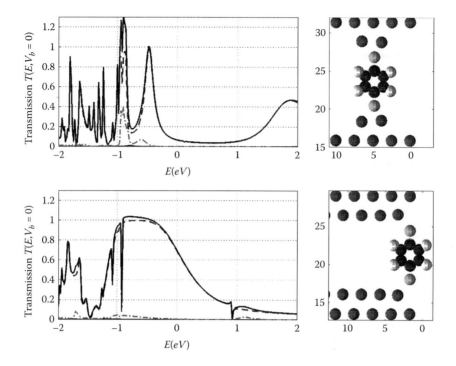

FIGURE 25.12   (See color insert) Zero-bias transmission (total and eigenchannel, $T = \sum_i T_i$) of the dithiol-benzene molecule on a "tip" of three adatoms (top) and on the flat surface (bottom). Two geometric effects are identified: the narrowing of the transmission cone, and the increase in distance between the electrodes. Two peaks become clearly distinguished when the molecule is adsorbed on the tip, although it was verified [37] that the bonding mechanism is essentially unchanged.

electrode surfaces. The conductance is additionally limited by the material of the leads. For a ME device like a sensor, the functionality that depends on the position of the levels in the extended molecule should not be smeared by the band broadening due to transverse motion (x–y band-structure) in the electrode (as illustrated in Figure 25.12). On the other hand, such a smearing will result in the improved reproducibility and stability of the device properties. For the same molecule and the same lead material, the range of the devices can be designed varying the cone of transmission angles up to the extreme case when the transmission is governed by the band structure of the wire-like lead. The discussed problem is especially important when an electrode is being designed as a part of the existing device.

25.8   Conclusions

We have discussed the unique properties of ME devices. Still, one can hardly imagine the STM or break junction setup as a part of the future electronic device. On the other hand, the reproducible self-assembly of the complicated devices from molecular parts has not been demonstrated. So what are the most probable applications for ME devices in the near future?

   We would like to point out two possible areas. First, the highly nonlinear IVCs and switchable properties of the molecules can be utilized within the devices using unconventional logic. As an example, consider devices made entirely from wires. Any number between 0 and 1 can be represented by the probability to measure finite current in the wire where a random sequence of current pulses is propagating during a sufficient interval of time. By simply connecting wires together, one can then perform the addition of two uncorrelated signals corresponding to two numbers between 0 and 0.5. Such a device is clearly simple, though its speed is inversely proportional to its accuracy. The relation can be improved if a bunch of N wires is used to transmit a number during a $1/N$ shorter time interval.

Another area is connected to the traditional lithography. Miniaturization is a problem for conventional electronics only when one considers useful (and actually working) devices. However, an equally spaced array of metal bars of very small size and separation can be fabricated [78], especially if we allow some of the bars to be randomly broken. A pair of such grids can be a base for a crossbar array. Two problems regarding this are currently being solved: a way of actually contacting every bar in such a grid; and the algorithms to configure the switchable molecules between the bars (thus performing useful operations like addressing or memory function and excluding the broken bars) must be developed.

Additionally, the specific role of input/output devices as sensors (components of optical devices like LCD screens, etc.) should not be forgotten. Both the high sensitivity of the molecular wire transport properties to the environment and the possibility to probe the structure with transverse current along the molecule [79] has recently become an active field for the design of new ME setups useful in medicine, chemistry, and biology.

Acknowledgments

Authors gratefully acknowledge the financial support from the Carl Trygger Foundation and the Royal Swedish Academy of Sciences (KVA) to carry out this work.

References

[1] www.intel.com/research/silicon/mooreslaw.htm; G.E. Moore Cramming More Components onto Integrated Circuits, *Electronics*, Vol. 38, No. 8, 1965, www.intel.com/research/silicon/moorespaper.pdf.

[2] M.A. Reed and J.M. Tour, Computing with Molecules, *Scientific American*, June 20, 2000, www.sciam.com.

[3] A. Vilan and D. Cahen, How Organic Molecules Can Control Electronic Devices, *Trends in Biotechnology*, Vol. 20, No.1, January 2002, http://tibtech.trends.com

[4] J.M. Tour, A.M. Rawlett, M. Kozaki, Y. Yao, R.C. Jagessar, S.M. Dirk, D.W. Price, M.A. Reed, C.-W. Zhou, J. Chen, W. Wang, and I. Campbell, Synthesis and Preliminary Testing of Molecular Wires and Devices, *Chemical European Journal*, Vol. 7, No. 23, 2001.

[5] C. Kergueris, J.-P. Bourgoin, S. Palacin, D. Esteve, C. Urbina, M. Magoga, and C. Joachim, Electron Transport through a Metal–Molecule–Metal Junction, *Physical Review B*, Vol. 59, No. 19, 1999.

[6] J. Park and A.N. Pasupathy, Wiring up single molecules, *Thin Solid Films*, 439, 457–461, 2003.

[7] C. Dekker, Carbon Nanotubes as Molecular Wires, *Physics Today*, May 1999.

[8] M. Brandbyge, J. Schiotz, M.R. Sorensen, P. Stoltze, K.W. Jacobsen, J.K. Norskov, L. Olesen, E. Laegsgaard, L. Stensgaard, and F. Besenbacher, Quantized Conductance in Atom-sized Wires between Two Metals, *Physical Review B*, Vol. 52, No. 11, p. 8500, 1995.

[9] Y.M. Galperin *Introduction to Modern Solid State Physics*, 2001, http://edu.ioffe.ru/lib/galperin/.

[10] A. Grigoriev, N.V. Skorodumova, S.I. Simak, G. Wendin, B. Johansson, and R. Ahuja, *Electron Transport in Stretched Monoatomic Gold Wires*, submitted, 2006.

[11] F. Patolsky and C.M. Lieber, Nanowire Nanosensors, *Materials Today*, Vol. 8, pp. 20–28, 2005.

[12] A. Calzolari, R. Di Felice, E. Molinari, and A. Garbesi, Electron Channels in Biomolecular Nanowires, *Journal of Physical Chemistry B*, Vol. 108, p. 2509, 2004; R. Di Felice, A. Calzolari, and H. Zhang, Towards Metalated DNA-Based Structures, Nanotechnology, Vol. 15, p. 1256, 2004.

[13] X. Crispin, J. Cornil, R. Friedlein, K.K. Okudaira, V. Lemaur, A. Crispin, G. Kestemont, M. Lehmann, M. Fahlman, R. Lazzaroni, Y. Geerts, G. Wendin, N. Ueno, J.-L. Brédas, and W.R. Salaneck, Electronic Delocalization in Discotic Liquid Crystals: A Joint Experimental and Theoretical Study, *Journal of American Chemical Society*, Vol.126, No.38, pp. 11889–11899, 2004.

[14] J.G. Cramer, *The Carbon Nanotube–Miracle Material*, Alternate View Column AV-109, http://www.npl.washington.edu/AV/altvw109.html.

[15] E. Braun, Y. Eichen, U. Sivan, and G. Ben-Yoseph, DNA-Templated Assembly and Electrode Attachment of a Conducting Silver Wire, *Nature*, 391, 1998, pp. 775–778.

[16] V. Mujica, A. Nitzan, Y. Mao, W. Davis, M. Kemp, A. Roitberg, and M.A. Ratner, Electron Transfer in Molecules and Molecular Wires: Geometry Dependence, Coherent Transfer, and Control, *Advances in Chemical Physics*, Vol. 107 (Electron Transfer: From Isolated Molecules to Biomolecules, Pt. 2), pp. 403–429, 1999.

[17] B. Xu and N.J. Tao, Formation of Molecular Junctions, *Science*, Vol. 301, p. 1221, 2003.

[18] W. Haiss, R.J. Nichols, H. van Zalinge, S.J. Higgins, D. Bethell, and D.J. Schiffrin, Measurement of Single Molecule Conductivity Using the Spontaneous Formation of Molecular Wires, *Physical Chemistry Chemical Physics*, Vol. 6, pp. 4330 – 4337, 2004.

[19] Y. Chen, G.-Y. Jung, D.A.A. Ohlberg, X. Li, D.R. Stewart, J.O. Jeppesen, K.A. Nielsen, J.F. Stoddart, and R.S. Williams, Nanoscale Molecular-Switch Crossbar Circuits, *Nanotechnology*, Vol. 14, pp. 462–468, 2003.

[20] C. Zhou, M.R. Deshpande, M.A. Reed, K. II Jones, and J.M. Tour, Nanoscale Metal/Self-Assembled Monolayer/Metal Heterostructures, *Applied Physics Letters*, Vol. 71 No. 5, 611–613, 1997.

[21] M.T. Cygan, T.D. Dunbar, J.J. Arnold, L.A. Bumm, N.F. Shedlock, T.P. Burgin, L. II Jones, D.L. Allara, J.M. Tour, and P.S. Weiss, Insertion, Conductivity, and Structures of Conjugated Organic Oligomers in Self-Assembled Alkanethiol Monolayers on Au{111}, *Journal of the American Chemical Society*, Vol. 120, No.12, pp. 2721–2732, 1998.

[22] S. Kubatkin, A. Danilov, M. Hjort, J. Cornil, J.-L. Brédas, N. Stuhr-Hansen, P. Hedegård, and T. Bjørnholm, Single-Electron Transistor of a Single Organic Molecule with Access to Several Redox States, *Nature*, Vol. 425, pp. 698–701, 2003; www.phantomsnet.com/nidconference5/abstracts/Bjornholm_abstract.pdf, for the description of the technique, see, i.e., S.E. Kubatkin, A.V. Danilov, H. Olin, and T. Claeson, Tunneling through a Single Quench-Condensed Cluster, *Journal of Low Temperature Physics*, 2000, Vol. 118, No.5/6, pp. 307–316.

[23] H. Park, J. Park, A.K.L. Lim, E.H. Anderson, A.P. Alivisatos, and P.L. McEuen, Nanomechanical Oscillations in a Single-C60 Transistor, *Nature*, Vol. 407, No. 7, 2000.

[24] K.W. Hipps, It's All About Contacts, *Science*, Vol. 294, 2001.

[25] J. Chen, M.A. Reed, A.M. Rawlett, and J.M. Tour, Large On-Off Ratios and Negative Differential Resistance in a Molecular Electronic Device, *Science*, Vol. 286, No. 5444, pp. 1550–1552, 1999.

[26] G.K. Ramachandran, J.K. Tomfohr, J. Li, O.F. Sankey, X. Zarate, A. Primak, Y. Terazono, T.A. Moore, A.L. Moore, D. Gust, L.A. Nagahara, and S.M. Lindsay, Electron Transport Properties of a Carotene Molecule in a Metal–(Single Molecule)–Metal Junction, *Journal of Physical Chemistry B*, Vol. 107, pp. 6162–6169, 2003.

[27] R.M. Metzger, B. Chen, U. Hopfner, M.V. Lakshmikantham, D. Vuillaume, T. Kawai, X. Wu, H. Tachibana, T.V. Hughes, H. Sakurai, J.W. Baldwin, C. Hosch, M.P. Cava, L. Brehmer, and G.J. Ashwell, Unimolecular Electrical Rectification in Hexadecylquinolinium Tricyanoquin-odimethanide, *Journal of American Chemical Society*, Vol. 119, pp. 10455–10466, 1997.

[28] A. Nitzan and M.A. Ratner, Electron Transport in Molecular Wire Junctions, *Science*, Vol. 300, 2003.

[29] K.-H. Müller, Effect of the Atomic Configuration of Gold Electrodes on the Electrical Conduction of Alkanedithiol Molecules, *Physical Review B*, Vol. 73, p. 045403, 2006.

[30] T. Dadosh, Y. Gordin, R. Krahne, I. Khivrich, D. Mahalu, V. Frydman, J. Sperling, A. Yacoby, and I. Bar-Joseph, Measurement of the Conductance of Single Conjugated Molecules, *Nature*, Vol. 436, pp. 677–680, 2005.

[31] P. Fenter, F. Schreiber, L. Berman, P. Eisenberger, G. Scoles, and M. Bedzyk, On the Structure and Evolution of the Buried S/Au Interface in Self-Assembled Monolayers: X-Ray Standing Wave Results, *Surface Science*, Vol. 412, pp. 213–235, 1998.

[32] K. Stokbro, J. Taylor, M. Brandbyge, J.-L. Mozos, and P. Ordejon, Theoretical Study of the Nonlinear Conductance of Di-Thiol Benzene Coupled to Au(111) Surfaces via Thiol and Thiolate Bonds, *Computational Materials Science*, Vol. 27, pp. 151–160, 2003.

[33] J.J. Gerdy and W.A. Goodard, III, Atomistic Structure for Self-Assembled Monolayers of Alkanethiols on Au(111) Surfaces, *Journal of American Chemical Society*, Vol. 118, pp. 3233–3236, 1996.

[34] S. Datta, *Electronic Transport in Mesoscopic Systems*, Cambridge University Press, Cambridge, 1995.

[35] B. Nikolic, *Quantum Transport in Finite Disordered Electron Systems*, Ph.D. thesis, State University of New York at Stony Brook, 2000, http://www.physics.udel.edu/~bnikolic/PDF/teza_v2.0.pdf.

[36] J. A. Torres, J. I. Pascual, and J. J. Sáenz, Theory of Conduction through Narrow Constrictions in a Three-Dimensional Electron Gas, *Physical Review B*, Vol. 49, No. 23, pp. 16581–16584, 1994.

[37] A. Grigoriev, J. Skoldberg, G. Wendin, and Z. Crljen, Critical Roles of Metal–Molecule Contacts in Electron Transport through Molecular-Wire Junctions, *Physical Review B*, Vol. 74, No. 4, p. 045401, 2006 (Preprint: cond-mat/0603518); included in the Virtual *Journal of Nanoscale Science*, 14 (3) July 2006.

[38] J. Taylor, M. Brandbyge, and K. Stokbro, Theory of Rectification in Tour Wires: The Role of Electrode Coupling, *Physical Review Letters*, Vol. 89, No. 23, 2002.

[39] Dan Feldheim, Flipping a Molecular Switch, *Nature*, Vol. 408, p. 45, 2000.

[40] Z.J. Donhauser, B.A. Mantooth, K.F. Kelly, L.A. Bumm, J.D. Monnell, J.J. Stapleton, D.W. Price Jr., A.M. Rawlett, D.L. Allara, J.M. Tour, and P.S. Weiss, Conductance Switching in Single Molecules Through Conformational Changes, *Science*, Vol. 292, p. 230, 2001.

[41] J. Reichert, R. Ochs, D. Beckmann, H.B. Weber, M. Mayor, and H. von Lohneysen, Driving Current through Single Organic Molecules, *Physical Review Letters*, Vol. 88, No. 17, pp. 176804/1-176804/4, 2002.

[42] N. D. Lang and P. Avouris, Effects of Coadsorption on the Conductance of Molecular Wires, *Nano Letters*, Vol. 2, No.10, pp. 1047–1050, 2002.

[43] P. G. Collins, K. Bradley, M. Ishigami, and A. Zettl, Extreme Oxygen Sensitivity of Electronic Properties of Carbon Nanotubes, *Science*, Vol. 287, p. 1801, 2000.

[44] C. Girard, C. Joachim, and S. Gauthier, *The Physics of the Near-Field*, Reports on Progress in Physics, Vol. 63, No. 6, pp. 893–938, 2000.

[45] F. Schreiber, Structure and Growth of Self-Assembling Monolayers, *Progress in Surface Science*, Vol. 65, pp. 151–256, 2000.

[46] A. Zangwill, *Physics at Surfaces*, Cambridge University Press, Cambridge, 1988.

[47] R. Di Felice, A. Selloni, and E. Molinari, DFT Study of Cysteine Adsorption on Au(111), *Journal of Physical Chemistry B*, 107, pp. 1151–1156, 2003.

[48] B. Hammer and J.K. Norskov, Why Gold Is the Noblest of All the Metals, *Nature*, Vol. 376, pp. 238–240, 1995.

[49] R.G. Nuzzo and D.L. Allara, Adsorption of Bifunctional Organic Disulfides on Gold Surfaces, *Journal of the American Chemical Society*, Vol. 105, No. 13, p. 4481, 1983.

[50] G.E. Poirier, Characterization of Organosulfur Molecular Monolayers on Au(111) Using Scanning Tunneling Microscopy, *Chemical Review*, Vol. 97, pp. 1117–1127, 1997.

[51] S. Letardi and F. Cleri, Interaction of Benzene Thiol and Thiolate with Small Gold Clusters, *Journal of Chemical Physics*, Vol. 120, No. 21, pp. 10062–10068, 2004.

[52] H. Grolnbeck, A. Curioni, and W. Andreoni, Thiols and Disulfides on the Au(111) Surface: The Headgroup-Gold Interaction, *Journal of American Chemical Society*, Vol. 122, No. 16, pp. 3839–3842, 2000.

[53] S. Mankefors, A. Grigoriev, and G. Wendin, Molecular Alligator Clips: A Theoretical Study of Adsorption of S, Se And S–H on Au(111), *Nanotechnology*, Vol.14, No. 8, pp. 849–858, 2003.

[54] A. Nitzan, Electron Transmission through Molecules and Molecular Interfaces, *Annual Review of Physical Chemistry*, Vol. 52, pp. 681–750, 2001.

[55] Y. Luo, C.K. Wang, and Y. Fu, Effects of Chemical and Physical Modifications on the Electronic Transport Properties of Molecular Junctions, *Journal of Chemical Physics*, Vol. 117, No. 22, pp. 10283–10290, 2002.

[56] T.N. Todorov and G.A.D. Briggs, Effects of Compositional Impurities and Width Variation on the Conductance of a Quantum-Wire, *Journal of Physics-Condensed Matter*, Vol. 6 No. 13, pp. 2559–2572, 1994.

[57] N. Kobayashi, M. Aono, and M. Tsukada, Theoretical Approach for Electron Transport through Nanostructures, *RIKEN Review*, No. 37, 2001.

[58] P. Damle, *Nanoscale Device Modeling: From MOSFETs to Molecules*, Ph.D. thesis, Purdue University, West Lafayette, IN, 2003.

[59] W. Tian, S. Datta, S. Hong, R. Reifenberger, J.I. Henderson, and C.P. Kubiak, Conductance Spectra of Molecular Wires, *Journal of Chemical Physics*, Vol. 109, No. 7, p. 2874, 1998.

[60] N. Kobayashi, M. Brandbyge, and M. Tsukada, First-Principles Study of Electron Transport through Monatomic Al and Na Wires, *Physical Review B*, Vol. 62, No.12, pp. 8430–8437, 2000.

[61] S. Lorenz, C. Solterbeck, W. Schattke, J. Burmeister, and W. Hackbusch, Electron-Scattering States at Solid Surfaces Calculated with Realistic Potentials, *Physical Review*, Vol. 55, No. 20, p. R13432,1997.

[62] M. Brandbyge, K.W. Jacobsen, and J.K. Norskov, Scattering and Conductance Quantization in Three-Dimensional Metal Nanocontacts, *Physical Review B*, Vol. 55, No. 4, pp. 2637–2650, 1997.

[63] S. Datta, Nanoscale Device Modeling: The Green's Function Method, *Superlattices and Microstructures*, Vol. 28, No. 4, p. 253, 2000.

[64] E.G. Emberly and G. Kirczenow, Theoretical Study of Electrical Conduction through a Molecule Connected to Metallic Nanocontacts, *Physical Review B*, Vol. 58, pp. 10911–10920, 1998.

[65] M. Buttiker, Quantized Transmission of a Saddle-Point Constriction, *Physical Review B*, Vol. 41, No. 11, p. 7906, 1990.

[66] V. Yosefin and M. Kaveh, Conduction in Curvilinear Constrictions: Generalization of the Landauer Formula, *Physical Review Letters*, Vol. 64, No. 23, pp. 2819–2822, 1990.

[67] N.D. Lang, Resistance of Atomic Wires, *Physical Review B*, Vol. 52 No. 7, p. 5335, 1995.

[68] A. W. Ghosh, F. Zahid, P. S. Damle, and S. Datta, *Insights into Molecular Conduction from I-V Asymmetry*, arXiv:cond-mat/0202519 v1, 2002.

[69] J. J. Palacios, A. J. Pérez-Jiménez, E. Louis, E. SanFabián, J. A. Vergés, and Y. García, Molecular Electronics with GAUSSIAN98/03, *Current Trends in Computational Chemistry*, Vol. 9, p. 1, 2005.

[70] http://www.wannier-transport.org; see also A. Calzolari, N. Marzari, I. Souza, and M. Buongiorno Nardelli, Ab initio Transport Properties of Nanostructures from Maximally Localized Wannier Functions, *Physical Review B*, Vol. 69, p. 035108, 2004.

[71] http://www.smeagol.tcd.ie/; V. Garcia-Suarez, A.R. Rocha, S.W. Bailey, C.J. Lambert, S. Sanvito, and J. Ferrer, Conductance Oscillations in Zigzag Platinum Chains – Suppression of Parity Effects, *Physical Review Letters*, Vol. 95, p. 256804, 2005.

[72] http://www.physics.byu.edu/research/lewis/fireball/; see also P. Jelínek, H. Wang, J.P. Lewis, O.F. Sankey, and J. Ortega, Multicenter Approach to the Exchange-Correlation Interactions in ab initio Tight-Binding Methods, *Physical Review B*, Vol. 71, p. 235101, 2005.

[73] www.atomistix.com; see also J. Taylor, H. Guo, and J. Wang, Ab initio Modeling of Quantum Transport Properties of Molecular Electronic Devices, *Physical Review B*, Vol. 63, p. 245407, 2001; M. Brandbyge, J-L. Mozos, P. Ordejon, J. Taylor, and K. Stokbro, Density Functional Method for Nonequilibrium Electron Transport, *Physical Review B*, Vol. 65, p. 165401, 2002.

[74] J. Taylor, H. Guo, and J. Wang, Ab initio Modeling of Quantum Transport Properties of Molecular Electronic Devices, *Physical Review B*, Vol. 63, p. 245407, 2001.

[75] M. Brandbyge, J.-L. Mozos, P. Ordejon, J. Taylor, and K. Stokbro, Density Functional Method for Nonequilibrium Electron Transport, *Physical Review B*, Vol. 65, p. 165401, 2002.

[76] J. Taylor, Ab-initio Modelling of Transport in Atomic Scale Devices, Ph.D. thesis, McGill University, Montreal, 2000.

[77] Y. Yourdshahyan, C. Ruberto, M. Halvarsson, L. Bengtsson, V. Langer, B.I. Lundqvist, S. Ruppi, and U. Rolander, Theoretical Structure Determination of a Complex Material: Kappa-Al_2O_3, *Journal of the American Ceramic Society*, Vol. 82, No. 6, pp. 1365–1380, 1999.

[78] K. Likharev, Simplifying Hybrid Semiconductor-Nanodevice Circuits, *The International Society for Optical Engineering Newsroom: Nanotechnology*, 2006, http://newsroom.spie.org/x3865.xml.

[79] J. Li, M. Gershow, D. Stein, E. Brandin, and J. A. Golovchenko, DNA Molecules and Configurations in a Solid-State Nanopore Microscope, *Nature Materials*, Vol. 2, pp. 611–615, 2003.

26

An Electronic Cotunneling Model of STM-Induced Unimolecular Surface Reactions

Vladimiro Mujica

Thorsten Hansen

Mark A. Ratner

Abstract

The STM-induced unimolecular decomposition of a molecule adsorbed at a surface is analyzed within the framework of a theory of electron transport involving three different electronic states. Transitions between the electronic states are considered to be provoked by an energy transfer associated with electron cotunneling. The subsequent chemical reaction is then produced by the nuclear dynamics in the excited electronic state, a similar phenomenon to what occurs in gas phase organic photochemistry. We apply our model to account for the main features of some recent experiments. The idea of cotunneling provides a simple physical picture of the mechanism for energy transfer involved in the initial stage of the chemical reaction. It also explains the most important features observed in the experiment: the existence of a voltage threshold; the weak current dependence; and the relatively long-time-scale of variation in the measured current.

26.1 Introduction

The introduction of local probes (e.g., STM and AFM) has made possible the study of a whole host of single-molecule processes on surfaces. Considerable theoretical effort has been devoted to understanding the current through nanojunctions where the STM tip plays the role of one of the electrodes. Although a quantitative agreement with the experiments is still far from complete, there is a rather consistent understanding of the basic aspects of conduction at the molecular scale [1–4].

Recent experiments have shown that relatively weak tunneling currents passing through a molecular junction may induce different processes ranging from heating to desorption and even chemical reactions [5–9]. We are especially interested in the kinds of processes described in reference [5], a remarkable experiment that shows the successive breaking of the two C-H bonds in an acetylene molecule adsorbed on the (111) surface of copper, under the influence of an STM tip. The results of these experiments have prompted us to examine theoretically the mechanisms that could be responsible for the observed experimental behavior.

Generally speaking, we can consider two different mechanisms for surface processes associated with tunneling currents:

(a) Current-induced processes that do not exhibit a voltage threshold. Here we include desorption, junc-
        tion heating, and mechanical vibrations of the type studied by Seideman and co-workers [10–14]
(b) Current-induced processes with voltage control. These processes, like STM-induced chemical reac-
        tions, do show a voltage threshold.

One can think of (*a*) as a sort of molecular Joule effect whereby the absorption of phonons associated with heating of the junction occurs gradually. In contrast, (*b*) would be associated with a different energy transfer mechanism between the tunneling electron and the molecular bridge. Such a mechanism would involve at least two electronic states of the substrate-adsorbate system, with the coupling between them mediated by the contacts.

We will consider as our prototype experiment the one reported in reference [5]. The key result of this work is that the STM current achieves a stationary value that remains constant for voltages below a critical value. For biases larger than the threshold, the current exhibits a step-like structure. The different values of the current are understood as being associated to the different stages of the sequential unimolecular decomposition of the adsorbate induced by the STM current.

Several important questions arise in the context of these experiments:

1. What is the energy transfer mechanism involved in the process?
2. What is the physical origin of the observed voltage threshold?
3. Once the threshold voltage is achieved, what is the dynamics that determines the time-scale of the
        transition between the two stationary values of the current?

We have examined these questions by invoking cotunneling [15–17] as the process responsible for the energy transfer that, having occurred, unfolds a sequence of physical events similar to the ones observed in gas phase photochemistry [18]. Other theoretical approaches have appeared in the literature that consider different explanations and levels of description for related phenomena [14,19–23]. Although questions about the actual physical pathway remain a challenge requiring additional confrontation with experiment, our model presents a description of the process that leads to a chemical transformation under the STM within the same theoretical and conceptual framework is used to describe the current itself.

Section 26.2 of this article presents the cotunneling model for the STM-induced unimolecular decomposition and the description of the observable current. In Section 26.3, we present a simple example of the formalism that applies to the interpretation of the experiment reported in [5]. Finally, Section 26.4 presents a summarized discussion of our results and their implication in the broader context of the description of electronic transport in nanojunctions involving several electronic states.

26.2 A Cotunneling Model

In what follows, we will disregard inelastic processes associated with electron-phonon and electron-electron interaction that are of importance in describing IETS [24–26] and Coulomb blockade and charging in molecular junctions [27]. Instead, we concentrate on the component of the current arising from the coherent elastic and inelastic one-electron tunneling. The inelastic component of tunneling is also known as cotunneling [15–17] and corresponds to a process where the tunneling electron transfers energy to the molecular bridge, which is then left in an electronic excited state. Elastic tunneling corresponds essentially to

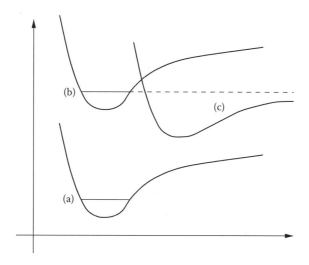

FIGURE 26.1   Potential energy surfaces.

the super exchange mechanism commonly considered in non-adiabatic intramolecular electron transfer. Both of these processes (electronic tunneling and cotunneling) are of importance in the regime where Coulomb blockade can be disregarded (i.e. when the coupling between the molecule and the lead is comparable to the energy separation between molecular electronic charging levels). Electron transport in the weak molecule–electrode coupling regime is dominated by sequential tunneling, which is an incoherent process, and the system may exhibit Coulomb blockade [15–17].

To be specific, we will consider three electronic states involved in the STM-induced reaction: The ground state $|a\rangle$; an excited state, $|b\rangle$, which we assume to have essentially the same geometry as $|a\rangle$, and a third state $|c\rangle$ that corresponds to a different chemical species which is the result of the chemical reaction. In the experiment described in [5], $|a\rangle$ and $|c\rangle$ would be associated with the ground states of $C_2 H_2$ and the radical $C_2 H \cdot$. The state $|b\rangle$ would play the role of what is known as a doorway state in the literature on photochemical processes [18]. One-dimensional sketches of the electronic potential energy surfaces corresponding to the three electronic states are represented in Figure 26.1. The overlap between the excited state and a dissociation channel of state $|c\rangle$ is physically responsible for the chemical reaction involved in the uni-molecular bond-breaking reported in the experiment. Not shown in the picture is the coupling between the electronic states and the states in the electrodes, which is described by the spectral density [2].

One-electron tunneling and cotunneling are coherent second order processes in the molecule–metal coupling. Disregarding other contributions, the stationary state-to-state current through the molecule is then given by the electron charge times the state-to-state rate [28]. The expression for the state-to-state rate is then (the specification of the initial and final states will be given next)

$$I = -e w_{fi} \tag{26.1}$$

The current can be rewritten as a sum of an elastic term and an inelastic term:

$$I = I^{elastic} + I^{inelastic} \tag{26.2}$$

The elastic term usually dominates the actual current. The main effect of the inelastic term is to introduce a small, but finite, probability for the molecule to be excited to a dissociating state. The different contributions to the current, arising from transport through the electronic states $|a\rangle$ and $|c\rangle$, as a function of time are sketched in Figure 26.2.

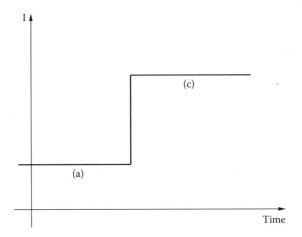

FIGURE 26.2   Current versus time.

26.2.1   Excitation Rate

We consider the excitation of a molecule from the ground state, $|a\rangle$, to the excited state, $|b\rangle$, that is coupled to a state, $|c\rangle$, with a dissociative scattering channel corresponding to the bond-breaking process. For a photodissociation process, where the excitation is induced by an optical field, we can think of treating the process using Fermi's golden rule:

$$w_{fi} = \frac{2\pi}{\hbar}|V|^2\delta(E_i - E_f) \tag{26.3}$$

If instead of an optical excitation, we consider excitations induced by the voltage applied across the metal substrate–molecule–STM probe junction, we must consider processes to a higher order in V, thus replacing the coupling V with the transition matrix T.

26.2.2   Self-Energy

The transition matrix T for a scattering process can be expressed in terms of the retarded Green's function, which is defined in terms of the system Hamiltonian:

$$G^{ret}(E) = \frac{1}{E - H + i\eta} \tag{26.4}$$

where $\eta = 0+$ is an infinitesimally small positive number. When the Hamiltonian, $H = H_0 + V$, can be written as a relatively simple part, H_0, plus a perturbation, V, Green's function can be expanded as

$$G = G_0 + G_0 V G_0 + G_0 V G_0 V G_0 + \cdots \tag{26.5}$$

In this case, however, we consider tunneling through a molecule in a situation where the charging energy is large. Thus, charged states appear only virtually and only terms that are even order in V will appear in the expansion of the Green's function:

$$G = G_0 + G_0 V G_0 V G_0 + G_0 V G_0 V G_0 V G_0 + \cdots \tag{26.6}$$

Defining the self-energy, $\Sigma = V G_0 V$, we can write a Dyson's equation for the Green's function (26.6):

$$G = G_0 + G_0 \Sigma G \tag{26.7}$$

Keeping only terms that are even order in V, the transition matrix expands to

$$T = VG_0^{ret}V + VG_0^{ret}VG_0^{ret}VG_0^{ret}V + \cdots$$
$$= VG^{ret}V \tag{26.8}$$

When we consider cotunneling processes, meaning including only the first term of the expansion (26.8), we have $T = \Sigma^{ret}$. And the generalized Fermi's golden rule now reads

$$w_{fi} = \frac{2\pi}{\hbar}|\Sigma^{ret}|^2\delta(E_i - E_f) \tag{26.9}$$

Equation (26.9) is an important result because it represents the link between the electron transport process and the chemical reaction. A word of caution is needed here because the self-energy Σ^{ret} is a non-hermitian operator describing an open system. Further implications of this kind of approach will be discussed in a forthcoming publication [29].

26.2.3 Cotunneling Rate

In order to apply Equation (26.9) to describe the cotunneling rate, we must specify the initial and final state in consideration. We describe the molecule with states $|a\rangle$ and $|b\rangle$. However, since the states in the generalized Fermi's golden rule, Equation (26.9), must be states of the entire system including metal substrate and STM tip, we construct the following initial and final states:

$$|i\rangle = |L\rangle|a\rangle|R\rangle$$
$$|f\rangle = c_l|L\rangle|b\rangle c_r^{\dagger}|R\rangle$$

This choice of final state corresponds to a situation where an electron is transferred from the left electrode to the right. For reverse bias and higher temperatures, more final states must be taken into account.

The Hamiltonian for the entire system can be divided into three parts:

$$H = H^{mol} + H^{lead} + V \tag{26.10}$$

The molecule is described in terms of many-particle states, $H^{mol} = \sum_j E_j|j\rangle\langle j|$, where the sum is taken over not only the two molecular states of interest, $|a\rangle$ and $|b\rangle$, but also over two charged states, a cationic state $|n-1\rangle$ and an anionic state $|n+1\rangle$, where n is the number of electrons in states $|a\rangle$ and $|b\rangle$. The two charged states will appear as virtual intermediate states during a cotunneling process.

The electrons in the leads, the metal substrate and the tip, are described in the usual way

$$H^{lead} = \sum_{k\in L}(\varepsilon_k - \mu_L)c_k^{\dagger}c_k + \sum_{k\in R}(\varepsilon_k - \mu_R)c_k^{\dagger}c_k \tag{26.11}$$

where μ_L and μ_R denote the chemical potentials in the left and right electrodes, respectively. Finally, $V = V^{n-1} + V^{n+1}$ describes the coupling between the molecule and the leads. We have

$$V^{n-1} = \sum_{j=a,b}\sum_{k\in L,R}\{V_{kn-1,j}c_k^{\dagger}|n-1\rangle\langle j| + V_{kn-1,j}^*|j\rangle\langle n-1|c_k\} \tag{26.12}$$

$$V^{n+1} = \sum_{j=a,b}\sum_{k\in L,R}\{V_{kj,n+1}c_k^{\dagger}|j\rangle\langle n+1| + V_{kj,n+1}^*|n+1\rangle\langle j|c_k\} \tag{26.13}$$

Physically, the operators V^{n-1} and V^{n+1} describe all processes involving electron transfer between the one-electron manifolds associated to the electrodes and the many-electron states corresponding to the molecule [30]. These operators are consistent with Landauer's view that electron–electron interaction needs to be considered for the computation of the current only in the scattering region. Far from it, in the leads, electrons can be considered as non-interacting.

Calculating the inelastic cotunneling rate using the generalized Fermi's golden rule, Equation (26.9), now leads to the following expression for the total transition rate:

$$
w_{ba} = \frac{\Gamma^L \Gamma^R}{2\pi\hbar} \int d\varepsilon_l \int d\varepsilon_r \left| \frac{1}{E_a - (E_{n-1} + \varepsilon_r)} + \frac{1}{E_a - (E_{n+1} - \varepsilon_l)} \right|^2
$$
$$
\times f_L(\varepsilon_l)[1 - f_R(\varepsilon_r)]\delta(E_a - E_b - \varepsilon_r + \varepsilon_l) \tag{26.14}
$$

The elastic tunneling rate, w_{aa}, corresponding to electron transport involving only the ground state $|a\rangle$, can now be obtained simply by choosing $E_b = E_a$,

$$
w_{aa} = \frac{\Gamma^L \Gamma^R}{2\pi\hbar} \int d\varepsilon_l \int d\varepsilon_r \left| \frac{1}{E_a - (E_{n-1} + \varepsilon_r)} + \frac{1}{E_a - (E_{n+1} - \varepsilon_l)} \right|^2
$$
$$
\times f_L(\varepsilon_l)[1 - f_R(\varepsilon_r)]\delta(\varepsilon_l - \varepsilon_r) \tag{26.15}
$$

One sees in these two equations that, in general, the elastic contribution will be larger than its inelastic counterpart because the energy constraint implied by the energy conservation built in the argument of the delta functions in the integrand of Equation (26.14) is harder to satisfy in terms of availability of states than the corresponding one in Equation (26.15), which involves only states in the leads. This difference will eventually translate into the appearance of a voltage offset of the inelastic contribution, a fact that is intimately related to the experimental observation of a voltage threshold for the chemical reaction to occur.

26.3   A Simple Example

Consider the inelastic cotunneling process. Assume, for simplicity, that one intermediate state, say the virtual cation, denoted as $|n-1\rangle$, is dominant. In this case Equation (26.14) reduces to

$$
w_{ba} = \frac{\Gamma^L \Gamma^R}{2\pi\hbar} \int d\varepsilon_l \int d\varepsilon_r \frac{f_L(\varepsilon_l)[1 - f_R(\varepsilon_r)]}{[E_a - (E_{n-1} + \varepsilon_r)]^2}\delta(E_a - E_b - \varepsilon_r + \varepsilon_l)
$$
$$
= \frac{\Gamma^L \Gamma^R}{2\pi\hbar} \int d\varepsilon_l \frac{f_L(\varepsilon_l)[1 - f_R(E_a - E_b + \varepsilon_l)]}{(E_b - E_{n-1} - \varepsilon_l)^2} \tag{26.16}
$$

In the zero temperature limit, the rate can be expressed analytically. Define the positive energies $A = E_{n-1} - E_a$ and $B = E_{n-1} - E_b$. Substitute the variable, $u = B + \varepsilon_l$, and rewrite the rate as (with ϑ the Heaviside step function)

$$
w_{ba} = \frac{\Gamma^L \Gamma^R}{2\pi\hbar} \int du \frac{1}{u^2}[1 - \vartheta(u - B - \mu_L)]\vartheta(u - A - \mu_R)
$$
$$
= \frac{\Gamma^L \Gamma^R}{2\pi\hbar} \int_{A+\mu_R}^{B+\mu_L} du \frac{1}{u^2}
$$
$$
= \frac{\Gamma^L \Gamma^R}{2\pi\hbar} \left[\frac{1}{A + \mu_R} - \frac{1}{B + \mu_L} \right] \tag{26.17}
$$

The corresponding elastic tunneling rate is

$$
w_{aa} = \frac{\Gamma^L \Gamma^R}{2\pi\hbar} \left[\frac{1}{A + \mu_R} - \frac{1}{A + \mu_L} \right]
$$
$$
= \frac{\Gamma^L \Gamma^R}{2\pi\hbar} \frac{\mu_L - \mu_R}{(A + \mu_R)(A + \mu_L)} \tag{26.18}
$$

For symmetric electrode coupling, the chemical potentials are given by $\mu_L = -W + \frac{eV}{2}$ and $\mu_R = -W - \frac{eV}{2}$, where W is the work function of the metal, and the expression for the elastic tunneling becomes

$$w_{aa} = \frac{\Gamma^L \Gamma^R}{2\pi\hbar} \frac{eV}{(A - W)^2 + \frac{e^2 V^2}{4}}$$

For the inelastic contribution, we obtain the result

$$w_{ba} = \frac{\Gamma^L \Gamma^R}{2\pi\hbar} \frac{eV - \Delta E}{(A - W)(B - W) + \frac{eV}{2}\Delta E + \frac{eV^2}{4}} \qquad (26.19)$$

where $\Delta E = E_b - E_a$.

The positiveness condition for the transition probability, requires that both, w_{aa} and w_{ba}, be larger than zero. We see immediately that this condition is always satisfied for w_{aa}. For the inelastic contribution, we observe that the denominator in (26.19) is positive for the parameters of the problem, therefore the sign of w_{ba} depends on the value of the numerator. The state availability alluded to in the previous sections translates into a voltage offset in w_{ba}. That is, the inelastic contribution vanishes for voltages such that $eV < \Delta E$, which is the physical origin of the voltage threshold observed in the experiment.

26.3.1 Time-Scales

We mentioned in the Introduction section that there were three relevant questions raised by the experimental results. We have already addressed the two referring to the energy transfer mechanism and the origin of the voltage threshold. The process dominating the relatively long time-scale in which the transition from the current associated with the electronic state $|a\rangle$ to that corresponding to the state $|c\rangle$ occurs, is the nuclear dynamics that takes place in the excited state $|b\rangle$ as the the bond dissociation occurs. These kinds of processes have been extensively discussed in the literature in photochemical processes [18] and also in reference [12]. The main conclusion of these works is that a dissociation time τ dominates the bond-breaking process and hence the time it takes for the turn-over between the two values of current corresponding to the two different chemical species on the surface.

26.4 Discussion

Deciphering the actual physical mechanism responsible for chemical reactions and other processes induced by tunneling currents remains a challenge. Part of the explanation for this situation is that the inclusion of electron–electron correlation and electron-phonon interaction, the most important interactions involved in inelastic scattering, is a complex problem. Cotunneling is a second-order process that can be essentially described in the stationary regime. We argue that such a process can be of importance particularly in the context of kinetic–electronic energy conversion.

Some of the most conspicuous features of STM-induced chemical reactions on surfaces can be rationalized if electronic cotunneling is indeed the mechanism for energy transfer. Our model takes into account in a natural way the role of the self-energy in providing the actual coupling between the electronic states.

We are currently working on extending our model to consider applications to more realistic systems and to assess the range of molecular and coupling parameters that could make cotunneling a viable mechanism in contexts where consecutive phonon absorption is not the dominant mechanism.

26.5 Acknowledgments

We thank DARPA, the Mol Apps program, the NSF-MRSEC program, the NASA URETI program, and CNM at Argonne National Laboratory of support of the research reported in this work.

References

[1] Ferry, D. and Goodnick, S.M., *Transport in Nanostructures*, Cambridge University Press, New York, 1997.

[2] Datta, S., *Electronic Transport in Mesoscopic Systems*, Cambridge University Press, New York, 1995.

[3] Datta, S., *Quantum Transport: Atom to Transistor*, Cambridge University Press, New York, 2005.

[4] Nitzan, A., *Chemical Dynamics in Condensed Phases: Relaxation, Transfer, and Reactions in Condensed Molecular Systems*, Oxford University Press, New York, 2006.

[5] Lauhon, L.J., and Ho, W., *Phys. Rev. Lett.*, 84, 1527, 2000.

[6] Ho, W., *J. Chem. Phys.*, 117, 11033, 2002.

[7] Nakamura, Y. et al., *Appl. Phys. Lett.*, 85, 5242, 2004.

[8] Tsukamoto, K. and Nozoye, H., *Appl. Surf. Sci.*, 241, 189, 2005.

[9] Hla, S. and Rieder, K., *Annu. Rev. Phys. Chem.*, 54, 307, 2003.

[10] Seideman, T., *J. Phy. Cond. Mat.*, 15, R521, 2003.

[11] Seideman, T., *J Mod. Opt.*, 50, 2393, 2003.

[12] Seideman, T. and Guo, H., *J. Theo. Comp. Chem.*, 3, 439, 2003.

[13] Kaun, C.C. and Seideman, T., *Phys. Rev. Lett.*, 94, 226801, 2005.

[14] Jorn R. and Seideman, T., *J. Chem. Phys.*, 124, 084703, 2006.

[15] Bruus, H. and Flensberg, K., *Many-Body Quantum Theory in Condensed Matter Physics. An Introduction*, Oxford University Press, New York, 2004.

[16] Averin, D.V. and Nazarov, Yu.V., *Phys. Rev. Lett.*, 65, 2446, 1990.

[17] Averin, D.V. and Nazarov, Yu.V., in *Single Charge Tunneling, NATO Proc.*, Plenum Press, 1992.

[18] Schinke, R., *Photodissociation Dynamics*, Cambridge University Press, New York, 1993.

[19] Ueba, H. and Persson, B.N.J., *Surf. Sci.*, 566–568, 1, 2004.

[20] Kuznetsov, A.M. and Medvedev, I.G., *Elektrokhimiya*, 41, 273, 2005.

[21] Hasegawa, K. et al., *J. Phys. Soc. Jpn.*, 71, 569, 2002.

[22] Yuan, L. et al., *J. Chem. Phys.*, 116, 3104, 2202.

[23] Stipe, B.C. et al., *Phys. Rev. Lett.*, 78, 4410, 1997.

[24] Galperin, M. et al., *J. Phys. Chem. B*, 109, 8519, 2005.

[25] Galperin, M. et al., *Nano Lett.*, 5, 125, 2005.

[26] Galperin, M. et al., *J. Chem Phys.*, 121, 11965, 2004.

[27] Bihary, Z. and Ratner, M.A., *Phys. Rev. B*, 72, 115439, 2005.

[28] Mujica, V. et al., *J. Chem. Phys.*, 101, 6849, 1994.

[29] Hansen, T. et al., manuscript in preparation.

[30] Hubbard, J.C., *Proc. Roy. Soc. A*, 285, 542, 1965.

Index

G

P

Q